ALGEBRA AND TRIGO

Quadratic equation: $ax^2 + bx + c = 0$ $\qquad x = \dfrac{-b + \sqrt{b^2 - 4ac}}{2a}$

Right triangle:

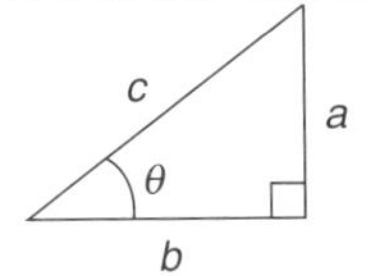

Pythagorean theorem: $c^2 = a^2 + b^2$

$$\sin\theta = \frac{a}{c}$$

$$\cos\theta = \frac{b}{c}$$

$$\tan\theta = \frac{a}{b}$$

$$\sin(\alpha \pm \beta) = \sin\alpha\cos\beta \pm \cos\alpha\sin\beta$$

$$\cos(\alpha \pm \beta) = \cos\alpha\cos\beta \mp \sin\alpha\sin\beta$$

$$\cos\alpha + \cos\beta = 2\cos\left(\frac{\alpha+\beta}{2}\right)\cos\left(\frac{\alpha-\beta}{2}\right)$$

COMMONLY USED PREFIXES FOR POWERS OF 10

Power	*Prefix*	*Abbreviation*	*Example*
10^{-18}	atto-	a	
10^{-15}	femto-	f	
10^{-12}	pico-	p	picofarad (pF)
10^{-9}	nano-	n	nanometer (nm)
10^{-6}	micro-	μ	micrometer (μm)
10^{-3}	milli-	m	millimeter (mm)
10^{-2}	centi-	c	centimeter (cm)
10^{-1}	deci-	d	decibel (dB)

Power	*Prefix*	*Abbreviation*	*Example*
10^{1}	deka-	da	(rarely used in physics)
10^{2}	hecto-	h	(rarely used in physics)
10^{3}	kilo-	k	kilogram (kg)
10^{6}	mega-	M	megawatt (MW)
10^{9}	giga-	G	gigajoule (GJ)
10^{12}	tera-	T	
10^{15}	peta-	P	
10^{18}	exa-	E	

CONVERSION FACTORS OCCASIONALLY NEEDED

Length

1 inch = 2.54 cm

1 mile = 1.653 km

1 light-year = 9.461×10^{15} m

1 AU = 1.496×10^{11} m

Time

1 d = 8.6400×10^{4} s

1 y = 3.156×10^{7} s

Speed

1 km/h = 0.2778 m/s

Volume

1 m^3 = 1000 liters

Energy

1 eV = 1.602×10^{-19} J

Pressure

1 atm = 1.013×10^{5} Pa

UNIVERSITY PHYSICS

UNIVERSITY PHYSICS

RONALD LANE REESE
Washington and Lee University

Brooks/Cole Publishing Company
I(T)P® An International Thomson Publishing Company

Pacific Grove • Albany • Belmont • Boston • Cincinnati
Johannesburg • London • Madrid • Melbourne • Mexico City
New York • Scottsdale • Singapore • Tokyo • Toronto

Sponsoring Editors: *Keith Dodson and Beth Wilbur*
Editorial Assistant: *Nancy Conti*
Production Coordinator: *Tessa McGlasson Avila*
Production Management: *Electronic Publishing Services Inc., NYC*
Marketing: *Steve Catalano*
Interior Design: *Electronic Publishing Services Inc., NYC*
Cover Design: *Roy R. Neuhaus*
Cover Photo: *Ronald Lane Reese*
Interior Illustration: *Electronic Publishing Services Inc., NYC*
Photo Researcher: *Electronic Publishing Services Inc., NYC*
Typesetting: *Electronic Publishing Services Inc., NYC*
Printing and Binding: *Von Hoffman Printing Company*

For more information, contact:

BROOKS/COLE PUBLISHING COMPANY
511 Forest Lodge Road
Pacific Grove, CA 93950
USA

International Thomson Publishing Europe
Berkshire House 168-173
High Holborn
London WC1V 7AA
England

Thomas Nelson Australia
102 Dodds Street
South Melbourne, 3205
Victoria, Australia

Nelson Canada
1120 Birchmount Road
Scarborough, Ontario
Canada M1K 5G4

International Thomson Editores
Seneca 53
Col. Polanco
11560 México, D. F., México

International Thomson Publishing GmbH
Königswinterer Strasse 418
53227 Bonn
Germany

International Thomson Publishing Asia
60 Albert Street
#15-01 Albert Complex
Singapore 189969

International Thomson Publishing Japan
Hirakawacho Kyowa Building, 3F
2-2-1 Hirakawacho
Chiyoda-ku, Tokyo 102
Japan

Printed in the United States of America.

10 9 8 7 6 5 4 3 2

Library of Congress Cataloging-in-Publication Data

The Library of Congress has cataloged the combined volume as follows:

Reese, Ronald Lane
University physics / Ronald Lane Reese.
p. cm.
Includes index.
ISBN 0-534-24655-9
1. Physics. I. Title.
QC21.5.R435 1998
530—dc21 98-41666
CIP

Magna opera Domini:
exquisita in omnes voluntates ejus

[*Psalmi* CXI, v 2]*

The past:
In loving memory of my mother
Edith Lemberg Reese
(1906–1984)

and mother-in-law
Bertha Marie Carlson
(1907–1981)

and in honor of my father
Harold Augustus Reese Sr.
(1906–)

The present:
With grateful thanks for the love
and devotion of my wife
Edith Joanne Carlson Reese

The future:
For the priceless blessings
of a wonderful son and daughter
Daniel Austin Reese
Anna-Loren Reese

* The Hexaplar Psalter, Samuel Bagster and Sons, London, 1843.

PREFACE

GOALS

In recent years much active discussion and debate has revolved around just what body of knowledge and skills science and engineering students should take from a university physics course. An obvious related issue is how best to achieve the desired learning goals. As the primary instructional resource for the student outside the classroom and the professor's office, the textbook naturally has been a focus of these discussions. Over the last 30 years or so a "standard model" of a university physics textbook has evolved to the point of extensive refinement. Several generations of future scientists and engineers have been introduced to the powerful ideas of physics through these texts, and certainly we should acknowledge the many strengths of these books while considering *how we can improve on them as learning tools for today's physics students.*

When writing this text, I decided from the onset that the text should follow the twin educational commandments of Alfred North Whitehead: "Do not teach too many subjects, and what you teach, teach thoroughly." I believe that Whitehead's statement, at a basic level, reflects two of the most common themes emerging from what has come to be known as the physics reform movement. Thus, while this text was not written as a reform textbook, it nonetheless embraces the spirit of many reform goals, such as better integration of modern physics topics, a stronger emphasis on conceptual understanding, and an attention to different learning styles. Most importantly, however, this book is written for students, to allow them not only to learn the tools that physics provides but also to see why they work and the beauty of the ideas that underlie them.

TEXT OVERVIEW

A Focused Perspective

One of the great triumphs of physics is the amount of understanding that comes from a relatively small investment of fundamental ideas and principles. Students, however, often see the course as a random assortment of 25 to 30 topics deemed worthy of chapter status. Unifying concepts, such as conservation laws and field theory, can be lost amidst the mountainous amount of material. Students frequently fail to see just how little must be known to describe as much of nature as possible. Thus, a central goal of this text is to help students develop a thorough *understanding* of the *principles* of the basic areas of physics: kinematics, dynamics, waves, thermodynamics, electromagnetism, optics, relativity, and modern physics. It is better to build technical knowledge upon a firm foundation of fundamental principles than on a large collection of mere formulas.

Since most of us do not innately discern simplifying patterns and connections when faced with the seemingly complex, we become good and experienced students of physics through steady practice. This is a fundamental pedagogical issue, and one that this text addresses clearly through focusing on many of the difficulties encountered by students when studying physics, problems mentioned by Arnold Arons in *A Guide to Introductory Physics Teaching*, and by others in the educational literature of physics. Thus, the book

- continually integrates the most significant material from previous chapters into new material, in keeping with Arons's admonition to "spiral back" frequently, for greater insight and retention.
- provides an accurate conceptual understanding of fundamental physical principles by placing great emphasis on these principles and how they arose.
- recognizes and points out the limits of applicability of the theories and equations of physics. It can be just as important, after all, to know what doesn't apply as what does.
- stresses connections between topics by incorporating many aspects of contemporary physics into a mix of traditional topics. This goal is carried through in all aspects of the text—exposition, examples, questions, problems, and investigative projects.

A Thorough Development

Some recent texts have jockeyed to outdo each other by reducing the number of overall pages. While brevity is often a laudatory goal, it can sometimes also work to defeat other, more important purposes. For true conceptual understanding to take place, a "fewer pages is more" approach can make the physics learning experience similar to trying to extrapolate the beauty and subtleties of a Shakespearean drama by reading a summary of the plot line. This text, while no longer than many other university physics texts, has been written with the primary philosophy that students need a text that lays a careful, detailed groundwork for strong conceptual understanding and the development of mature problem-solving skills. For example, much research has recently been done on the different learning styles that students apply when first studying new material, but for a text to try to implement pedagogical structure to these different learning styles and goals (such as multiple problem-solving approaches or collaborative learning techniques), it is inevitable that the lesser goal of brevity must be sacrificed. In a similar vein, students often complain that the examples in the text do not prepare them well for the more challenging homework problems, where more than one idea may be addressed. Page length can be kept down by focusing on just the most straightforward examples, but students also need to see how the principles can be applied to more involved scenarios. I have placed special emphasis on thoroughly preparing students for the homework sets through strong emphasis (and reemphasis) on problem-solving techniques, by frequent references to and explanations of common misunderstandings, and by providing a set of examples that address both single-concept problem solving and the application of fundamental principles to longer, multiconcept problems. The ability to question whether results are reasonable has been fostered throughout these examples.

The text contains an ample selection of sections from which individual instructors can design a course compatible with their academic institution and student audience. Numerous sections, typically at the end of chapters, are listed as optional (designated by a *) and may be omitted by instructors preferring a leaner course. Others may want to choose their own path, including some of the optional sections while omitting others.

Features

STYLE

Physics is a great story, and in this text I have attempted to tell that story in as lively, clear, and precise a manner as possible. Students sometimes fail to see how the topics connect to each other or to the world outside the classroom. Thus, I have placed great emphasis on introducing each new topic by describing how it relates to experiences and phenomena with which the student is already familiar or to topics previously discussed. There is also no reason that reading a physics text shouldn't be fun (or at least not a chore). My philosophy is that occasional lapses into whimsy are a small price to pay if the result is that students stay more engaged with the reading. Finally, by filling in the details that are sometimes left unstated, this text should help students better bridge the gaps where misconceptions can arise.

STRATEGIC EXAMPLES AND OTHER EXAMPLES

A strong emphasis has been placed on beginning almost all Examples with a few, fundamental principles and equations, rather than specialized equations of secondary importance. Strategic Examples address the application of fundamental principles to longer problems; they are discussed in great detail, which students find particularly helpful in developing their own problem-solving abilities. Moreover, many of the end-of-chapter problems mirror the methodological details of the Strategic Examples.

A unique feature of this text is that many of the Examples are solved in more than one way. All too often students suffer from the perception that they must be doing a problem incorrectly because a fellow student or even the professor has set it up differently. By working selected problems using different choices of signs, coordinate axes, or even overall approach, these Examples

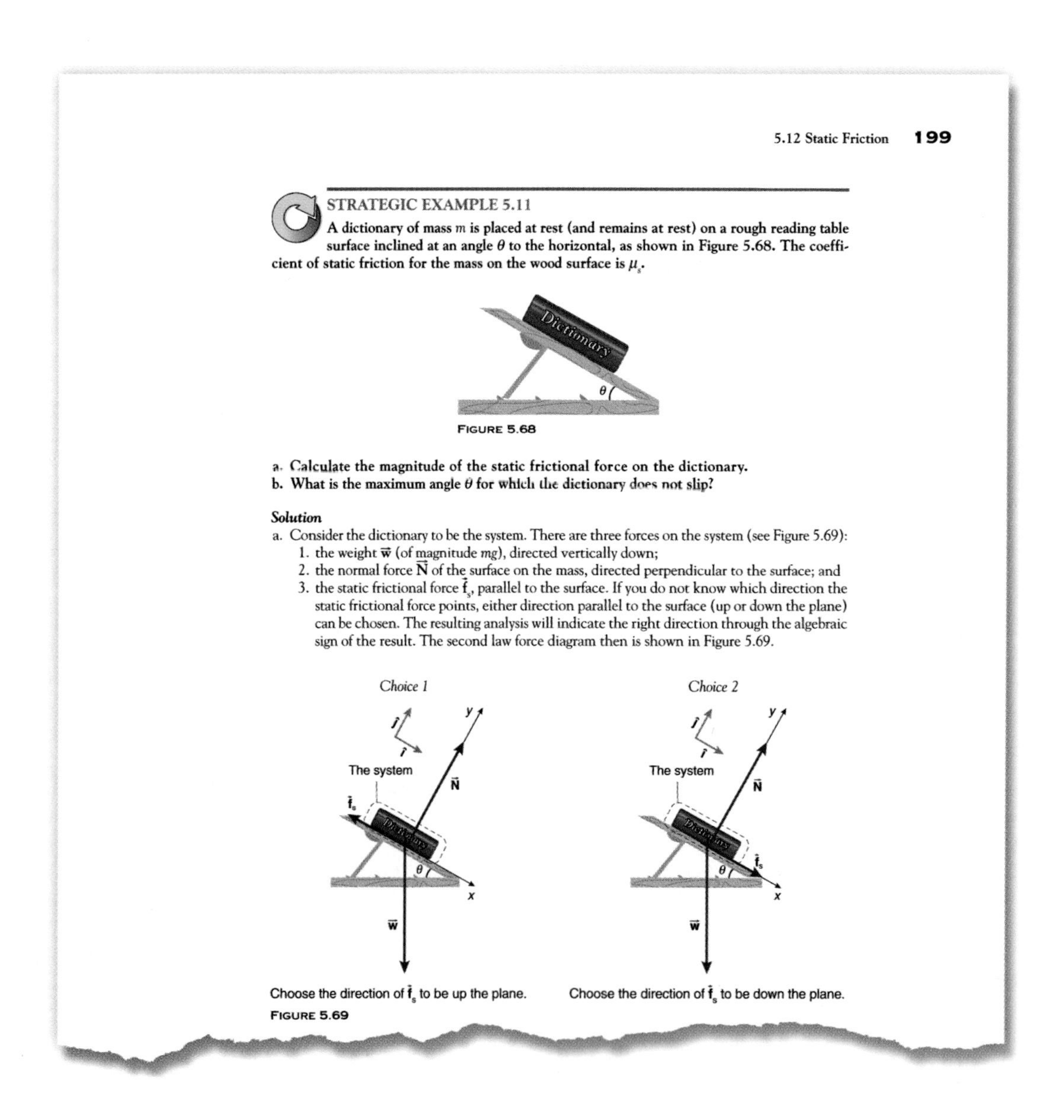

5.12 Static Friction 199

STRATEGIC EXAMPLE 5.11

A dictionary of mass m is placed at rest (and remains at rest) on a rough reading table surface inclined at an angle θ to the horizontal, as shown in Figure 5.68. The coefficient of static friction for the mass on the wood surface is μ_s.

FIGURE 5.68

a. Calculate the magnitude of the static frictional force on the dictionary.
b. What is the maximum angle θ for which the dictionary does not slip?

Solution

a. Consider the dictionary to be the system. There are three forces on the system (see Figure 5.69):
1. the weight $\vec{w}$ (of magnitude mg), directed vertically down;
2. the normal force $\vec{N}$ of the surface on the mass, directed perpendicular to the surface; and
3. the static frictional force $\vec{f}_s$, parallel to the surface. If you do not know which direction the static frictional force points, either direction parallel to the surface (up or down the plane) can be chosen. The resulting analysis will indicate the right direction through the algebraic sign of the result. The second law force diagram then is shown in Figure 5.69.

Choose the direction of $\vec{f}_s$ to be up the plane. Choose the direction of $\vec{f}_s$ to be down the plane.

FIGURE 5.69

also help students develop intuition on how certain choices can simplify later calculations compared with other alternatives.

CONCEPTUAL NOTES

Throughout the text, key points of each section have been highlighted with shading. Importantly, these Conceptual Notes are not just the most important equations—they focus on the principal ideas and concepts (and sometimes equations) that a student should take from each section. My students have found the Conceptual Notes very useful as a reviewing tool for tests and quizzes.

PROBLEM-SOLVING TACTICS

In addition to useful problem-solving hints, the Problem-Solving Tactics also provide warnings to students about common errors and how to avoid them. Often these important tips of the trade are also integrated into the text discussion. For example, specific Problem-Solving Tactics are often cross-referenced in some of the Examples. At the end of each chapter a summary of that chapter's Problem-Solving Tactics is included, with a page reference to the related text discussion for each tactic.

QUESTIONS

A common student lament is that "I understand the material; I just can't do the problems." The questions within and at the end of each chapter test a student's understanding of concepts before the student is asked to apply these concepts to more complex or quantitative situations in the problems. Some of these questions entail a short qualitative explanation, whereas others may require a short back-of-the-envelope calculation or even a quick and dirty experiment to determine the approximate magnitude of a quantity.

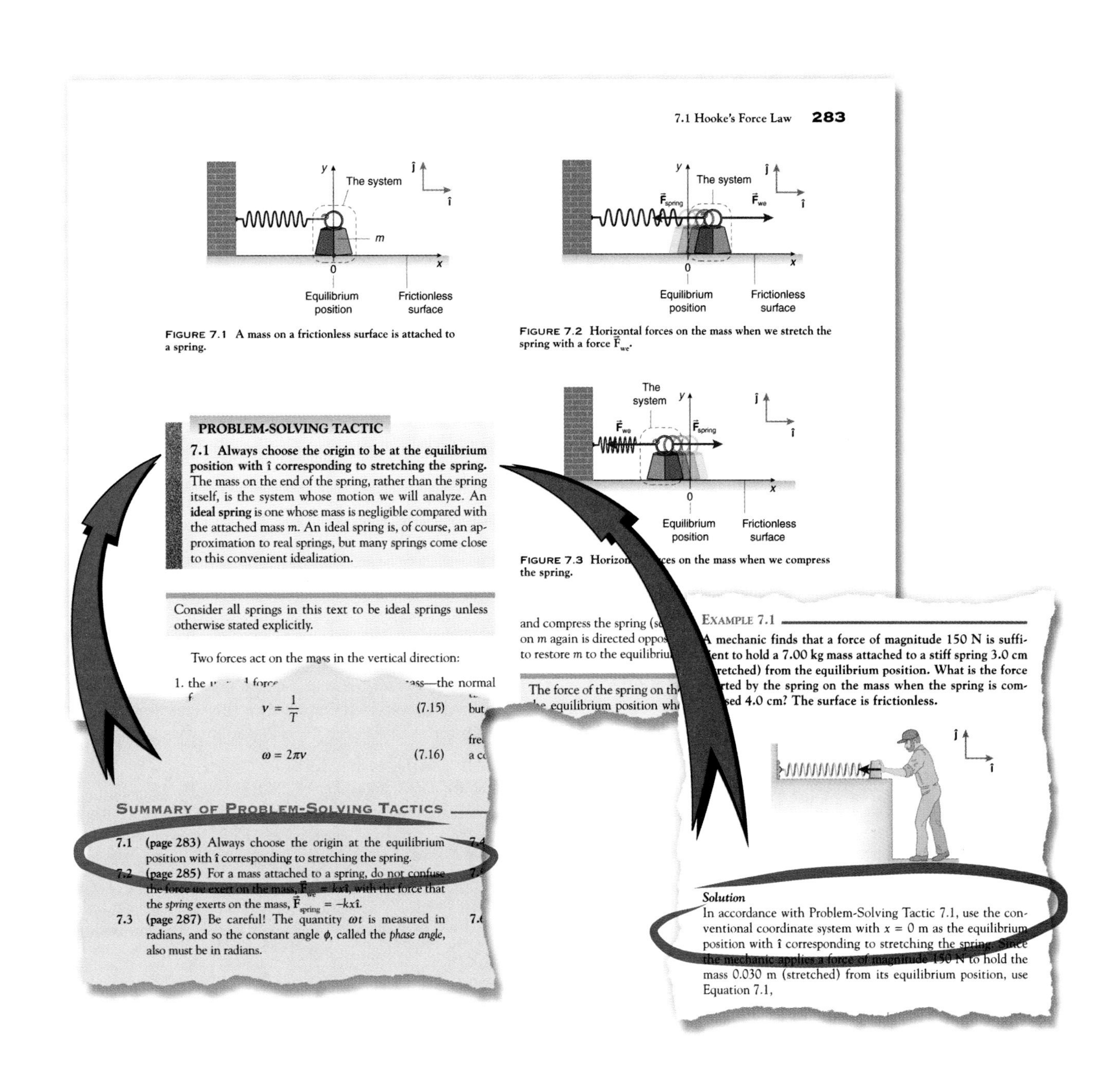

7.1 Hooke's Force Law 283

FIGURE 7.1 A mass on a frictionless surface is attached to a spring.

FIGURE 7.2 Horizontal forces on the mass when we stretch the spring with a force $\vec{F}_{we}$.

FIGURE 7.3 Horizon... ces on the mass when we compress the spring.

PROBLEM-SOLVING TACTIC

7.1 Always choose the origin to be at the equilibrium position with $\hat{\imath}$ corresponding to stretching the spring. The mass on the end of the spring, rather than the spring itself, is the system whose motion we will analyze. An **ideal spring** is one whose mass is negligible compared with the attached mass m. An ideal spring is, of course, an approximation to real springs, but many springs come close to this convenient idealization.

Consider all springs in this text to be ideal springs unless otherwise stated explicitly.

Two forces act on the mass in the vertical direction:

1. the ... force ... ass—the normal

$$\nu = \frac{1}{T} \quad (7.15)$$

$$\omega = 2\pi\nu \quad (7.16)$$

and compress the spring (s... on m again is directed oppo... to restore m to the equilibriu...

The force of the spring on th... equilibrium position wh...

SUMMARY OF PROBLEM-SOLVING TACTICS

7.1 (page 283) Always choose the origin at the equilibrium position with $\hat{\imath}$ corresponding to stretching the spring.

7.2 (page 285) For a mass attached to a spring, do not confuse the force *we* exert on the mass, $\vec{F}_{we} = kx\hat{\imath}$, with the force that the *spring* exerts on the mass, $\vec{F}_{spring} = -kx\hat{\imath}$.

7.3 (page 287) Be careful! The quantity ωt is measured in radians, and so the constant angle ϕ, called the *phase angle*, also must be in radians.

EXAMPLE 7.1

A mechanic finds that a force of magnitude 150 N is sufficient to hold a 7.00 kg mass attached to a stiff spring 3.0 cm (stretched) from the equilibrium position. What is the force exerted by the spring on the mass when the spring is compressed 4.0 cm? The surface is frictionless.

Solution

In accordance with Problem-Solving Tactic 7.1, use the conventional coordinate system with $x = 0$ m as the equilibrium position with $\hat{\imath}$ corresponding to stretching the spring. Since the mechanic applies a force of magnitude 150 N to hold the mass 0.030 m (stretched) from its equilibrium position, use Equation 7.1,

PROBLEMS

The development of successful problem-solving techniques is an essential goal for all introductory physics students. To help students hone these abilities, many of the problems involve a multistep approach in which students are guided through the problem with specific questions that enable them first to find all the pieces and then to put them together. Additionally, many of the problems mimic the approach of the Strategic Examples so that there is a strong correlation between the presentation of the material and the problems that a student is expected to be able to solve.

Since the real world is awash with information, the problems occasionally include irrelevant or superfluous information. This teaches a student to discriminate between what is needed and what is not. It also may be necessary to consult an appropriate table in the chapter or on the front and back inside covers to find numerical values of standard constants or parameters. In both the examples and problems, attention is paid to the consistent use of significant figures.

Three levels of difficulty are provided, with unbulleted, bulleted (•), and double-bulleted (:) problems representing straightforward, moderate, and more difficult problems, respectively. Those problems with red numbers include answers at the back of the book and are solved in the Student's Solutions Manual. I have personally solved all of the problems, and the problem sets have been additionally fine-tuned by actual classroom usage over the course of several years.

INVESTIGATIVE PROJECTS

These projects are highly amenable to collaborative group work and are of the following types:

- *Expanded Horizons*—These projects are well suited to journal club research, discussion, and supplementary reading.
- *Lab and Field Work*—Doing physics is an important part of studying physics. In these projects, students are asked to design and carry out experiments either with other students or the professor.
- *Communicating Physics*—A key developmental goal for any student is the ability to write about and discuss technical topics. Practice on these written and oral communication skills is provided by these project topics, which are ideal for writing-intensive assignments, public speaking, and community service opportunities.

The projects are interesting to read, even if never performed or assigned, since they indicate the breadth and depth of applications of chapter material. As such, they can help stimulate inquiry, class discussion, and faculty–student interaction. Most of the projects are provided with references that serve as a guide (and entryway) to the appropriate literature.

SUMMARIES

Each chapter concludes with an extensive summary that, when combined with the Conceptual Notes and the Summary of Problem-Solving Tactics, provides an ideal in-text study guide for the student.

QUOTATIONS

I have used these frequently throughout the text to cast the subject matter in a different light, be it serious or whimsical. Great writing (communication!) from the past is central to a real understanding of any discipline, even physics.

MATHEMATICAL LEVEL

This text assumes a familiarity with calculus comparable to what a student would obtain from a high school calculus course (with or without advanced placement credit). Of course, additional calculus is useful when taken concurrently with this course. No prior knowledge of physics is presumed.

PRECISION

Effective, unambiguous communication in physics requires clear and consistent use of the technical vocabulary and a solid understanding of the meaning of the technical notation. This tenet has informed the presentation throughout the book.

Chapter Contents

CHAPTER 1 PRELUDES

An overview of physics is presented along with an introduction to measurement standards of the SI unit system, distinguishing them from common units of convenience. The various meanings of the equal sign are discussed as well as estimation and order of magnitude calculations. The distinction is made between precision and accuracy. The notion of significant figures is discussed in the context of common mathematical operations such as multiplication (and division) and addition (and subtraction). Having made these points, the text does not ignore their use and makes consistent use of significant figures throughout its examples and problems so that students realize their importance even outside a laboratory context.

CHAPTER 2 A MATHEMATICAL TOOLBOX: AN INTRODUCTION TO VECTOR ANALYSIS

The proper and consistent use of vectors is very important to success in physics. This chapter and the rest of the book distinguish clearly among a vector, its magnitude, and its components with respect to a chosen coordinate system. Vector addition and subtraction are designated by boldface **+** and **−** to distinguish the operations clearly from their scalar counterparts, a source of much student confusion in problem solving.

CHAPTER 3 KINEMATICS I: RECTILINEAR MOTION

The notion of a particle is addressed. A one-dimensional vector approach is used so its extension to two- and three-dimensional motion in Chapter 4 is seamless and painless. The choices that a student must make in establishing a coordinate system for a problem and the consequences of that choice in tailoring the (few) fundamental kinematic equations to a problem are stressed throughout. Consistent use of vectors and vector terminology takes the mystery out of the choice for the signs associated with the various terms in the equations of kinematics.

CHAPTER 4 KINEMATICS II: MOTION IN TWO AND THREE DIMENSIONS

The vector approach of Chapter 2 easily allows extension to motion in two and three dimensions. Relative velocity addition is examined. Uniform and nonuniform circular motion are approached by introducing both the angular velocity and angular acceleration vectors so students are ready for more advanced work with these vectors in upper-division mechanics or dynamics courses in physics and engineering.

CHAPTER 5 NEWTON'S LAWS OF MOTION

An overview of fundamental particles and forces is presented. The concept of force and its measurement are introduced stressing how important it is to define clearly the system under

INVESTIGATIVE PROJECTS

A. Expanded Horizons

1. Investigate the dynamics associated with throwing a bola, a traditional South American hunting weapon.
D. L. Mathieson, "Wrap up rotational motion by throwing a bola," *The Physics Teacher*, 30, #3, pages 180–181 (March 1992).

2. Yo-yos are fascinating examples of spin and orbital motion. Investigate the physics of the yo-yo.
William Boudreau, "Cheap and simple yo-yos," *The Physics Teacher*, 28, #2, page 92 (February 1990).
Wolfgang Bürger, "The yo-yo: a toy flywheel," *American Scientist*, 72, #2, pages 137–142 (March–April 1984).
Edward Zuckerman, "Quest for the perfect yo-yo," *Science Digest*, 93, #7, pages 54–55, 60 (July 1985).

3. Investigate the use of an ultracentrifuge to determine molecular masses of complex molecules.
I. W. Richardson and Ejler B. Neergaard, *Physics for Biology and Medicine* (Wiley-Interscience, New York, 1972), pages 158–160.

4. Investigate the dynamics associated with boomerangs.
Vernon Barger and Martin Olsson, *Classical Mechanics: A Modern Perspective* (2nd edition, McGraw-Hill, New York, 1995), pages 195–202.
Allen L. King, "Project boomerang," *American Journal of Physics*, 43, #9, pages 770–773 (September 1975).
Henk Vos, "Straight boomerang of balsa wood and its physics," *American Journal of Physics*, 53, #6, pages 524–527 (June 1985).
Michael Hanson, "The flight of the boomerang," *The Physics Teacher*, 28, #3, pages 142–147 (March 1990).
Jacques Thomas, "Why boomerangs boomerang," *New Scientist*, 99, #1376, pages 838–843

...tigate the phys...
...H. D. N...

9. Measurements of the moment of inertia are common in physics laboratories. On the other hand, it is quite another story to *measure* the moment of inertia of the Moon or a planet such as Venus or Mars; they are not easily tinkered with in the lab! Measurements of such moments of inertia enable astronomers and geologists to model the interiors of the moons and planets. How might the moments of inertia of moons and planets be measured?
William B. Hubbard, *Planetary Interiors* (Van Nostrand Reinhold, New York, 1984).
Ralph Snyder, "Two-density model of the Earth," *American Journal of Physics*, 54, #6, pages 511–513 (June 1986).

10. Investigate the physics of the rotating, orbiting space colony stations proposed by Gerald K. O'Neill.
Gerald K. O'Neill, "The colonization of space," *Physics Today*, 27, #9, pages 32–40 (September 1974).

11. High angular speed flywheels are used as energy storage devices in hybrid race cars. Investigate the state of this technology.
William B. Scott, "Satellite control concepts bolster civil, defense systems," *Aviation Week and Space Technology*, 142, pages 43, 46 (6 March 1995).

B. Lab and Field Work

12. Many science museums have a dramatic pendulum in their entrance foyers called a Foucault pendulum. The slow rotation of the plane of oscillation of the pendulum is experimental proof that the Earth rotates. Investigate the dynamics of a Foucault pendulum to explain why this is the case. Imagine such a

C. Communicating Physics

16. Take two quarters (or any two identical coins). Orient the two coins so they are both heads up with both profiles of George Washington upright as indicated in Figure I.16a. Roll one of the coins, say the one on the left, around *half* the circumference of the other coin as indicated. It is amazing to discover that the profile of the rolling coin is *not* upside down but right side up after this maneuver. Explain why. This phenomenon is related to another curious aspect of rolling motion. Suppose a wheel (see Figure I.16b) rolls through one complete turn, carrying it from point A to point B. The distance between A and B is, therefore, equal to the circumference of the wheel. Notice, however, that point C, along the radius from the center of the wheel *also* executes one revolution as the wheel moves from A to B. Since the distance AB is equal to the distance CD, "we are confronted with the evident absurdity that the circumference of the small circle is equal to the circumference of the large circle." Resolve this paradox.
See James R. Newman, *The World of Mathematics* (Simon & Schuster, New York, 1956), volume 3, pages 1937–1939; the quotation is from this work, page 1938.

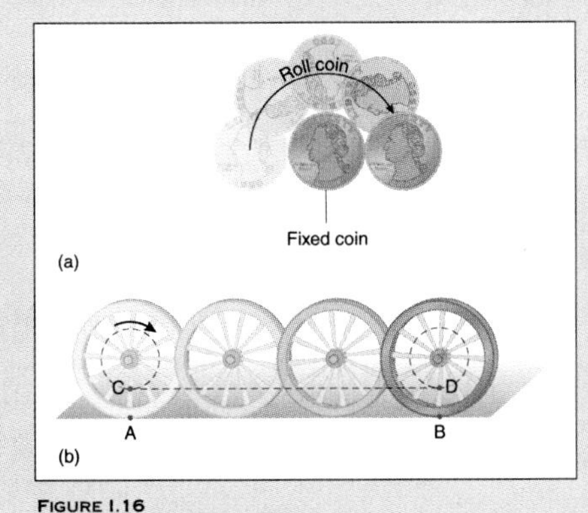

FIGURE I.16

17. Call or visit a tire retailer and determine the manufacturer of a dynamic tire balancing machine. Communicate with engineers at the manufacturer to learn the physics associated with dynamic balancing of tires. Write a summary of your findings appropriate for an audience of your peers.
Richard C. Smith, "Static vs spin balancing of automobile wheels," *American Journal of Physics*, 40, #1, pages 199–201 (January 1972).

18. The tippe top is a toy that has fascinated even great physicists like Niels Bohr and Wolfgang Pauli, perhaps even you. Demonstrate and explain its peculiar dynamics.
Richard J. Cohen, "The tippe top revisited," *American Journal of Physics*, 45, #1, pages 12–17 (January 1977).
Ivars Peterson, "Topsy-turvy top," *Science News*, 146, page 108 (13 August 1994).

consideration. The significance and importance of all three of Newton's laws of motion are addressed. Both second law and third law force diagrams are discussed. The intricacies of special forces such as weight, tension, and static and kinetic friction are explored. Inertial and noninertial reference frames are contrasted.

CHAPTER 6 THE GRAVITATIONAL FORCE AND THE GRAVITATIONAL FIELD

Newton's law of universal gravitation is *not* presented as a *fait accompli* as if inscribed in stone; rather the process by which Newton deduced the law is explored. The gravitational shell theorems are discussed and applied (and proved in an appendix). Kepler's laws of planetary motion are discussed, along with a convenient simplification to the third law commonly used in astronomy (using customized units: years and astronomical units). Newton's form for Kepler's third law is derived. The concept of the gravitational field is introduced so that parallels with it may be exploited later when studying electricity and magnetism. Gauss's law for the gravitational field is proved, so that further parallels with electromagnetism can be made. Many of the problems consider contemporary astronomical applications.

CHAPTER 7 HOOKE'S FORCE LAW AND SIMPLE HARMONIC OSCILLATION

Hooke's law for springlike forces is introduced. A horizontal and a vertical spring (with the additional gravitational force) are compared. Simple harmonic oscillation and its relationship to uniform circular motion are discussed. The simple pendulum is introduced as well as the oscillatory gravitational motion through a uniform sphere. Damped simple harmonic oscillation and forced oscillation with resonance are explored.

CHAPTER 8 WORK, ENERGY, AND THE CWE THEOREM

Students typically think of work and energy as the same thing. The similarities and distinctions between work and energy are explored as well as the concept of power. The classical work–energy theorem (called the CWE theorem) and its limitations caused by the neglect of thermal effects are thoroughly examined to set the stage for a more general and encompassing conservation of energy theorem when we consider thermodynamics in Chapter 13. The importance of the choice made for the zero of a potential energy function is emphasized. The connection between the local form for the gravitational potential energy (mgy, with $\hat{\jmath}$ up) and the more general form ($-GMm/r^2$) is discussed. Applications to astrophysical problems such as the escape speed and black holes are explored. The concept of energy diagrams also is introduced at this early juncture to set the stage for their later use in modern physics.

CHAPTER 9 IMPULSE, MOMENTUM, AND COLLISIONS

The general principles are stressed rather than a plethora of specialized equations for collisions. The contemporary idea of force transmission by particle exchange is explored by means of a classical example for repulsive forces. The center of mass is introduced and the dynamics of a system of particles is explored.

CHAPTER 10 SPIN AND ORBITAL MOTION

The similarities and distinctions between spin and orbital motion are explored. The rotational dynamics of rigid bodies with at least one symmetry axis through the center of mass is examined, emphasizing the parallels to analogous equations in linear dynamics. The shape of the spinning Earth and the precession of tops and of the spinning Earth also are explored. Rolling motion and a model of a wheel also are examined to explain the difficult but common observation that less force is needed to roll rather than to drag a massive system.

CHAPTER 11 SOLIDS AND FLUIDS

The mechanical properties of solids and fluids are investigated. The variation of pressure with depth or height in a liquid is examined, leading to an equation giving students great freedom to approach a problem with many different coordinate choices. Archimedes' principle and the stability of floating systems are explored (why things such as submarines, ships, or poles will float in one orientation but are unstable in another orientation). Bernoulli's principle is derived from the CWE theorem. Capillary action, nonideal fluids, and viscous flow also are discussed.

CHAPTER 12 WAVES

General waves and their wavefunction, waveform, and oscillatory behavior are discussed. Nonsinusoidal periodic waves are discussed before sinusoidal periodic waves so the distinction between the oscillatory behavior of the wave at a fixed place can be clearly distinguished from its waveform at a fixed time. The classical wave equation is derived so that it can be contrasted with the Schrödinger equation in Chapter 27. Waves on strings are introduced. A unique section explores the nature of a sound wave and the relationship between the particle position and the pressure or density wave. The measurement of sound intensity and sound level is discussed so that students become aware of common sound levels that can damage their hearing. The acoustic Doppler effect treats motion of the source and/or observer as well as the effect of a wind. Superpositions of waves to form standing waves are applied to strings and both open and closed pipes. The superposition leading to wave beats is explored as well as the distinction between phase and group speeds. A simplified introduction to Fourier analysis leads to wave uncertainty relations that appear later in Chapter 27 as the Heisenberg uncertainty principle.

CHAPTER 13 TEMPERATURE, HEAT TRANSFER, AND THE FIRST LAW OF THERMODYNAMICS

The definition of a simple thermodynamic system is presented. The intuitive yet difficult concepts of temperature and heat transfer are introduced. Thermal effects in solids, liquids, and gases as well as mechanisms of heat transfer are examined. A general statement of energy conservation is developed that specializes to the CWE theorem of mechanics and to the first law of thermodynamics. Various thermodynamic processes for gases are explored.

CHAPTER 14 KINETIC THEORY

The kinetic theory of an ideal gas is presented as well as its limitations. The notion of degrees of freedom and the effect of quantum mechanics on the effective number of degrees of freedom are discussed. Adiabatic processes for ideal gases also are presented.

CHAPTER 15 THE SECOND LAW OF THERMODYNAMICS

The need for this great unifying principle is discussed in the context of why some things happen and others do not. Thermodynamic models for engines and refrigerators are presented and related to the second law. The nonintuitive concept of entropy is carefully developed as well as a classical model that explores the Boltzmann statistical interpretation of the meaning of entropy.

CHAPTER 16 ELECTRIC CHARGES, ELECTRIC FORCES, AND THE ELECTRIC FIELD

The chapter begins with an exploration of how electrical effects were distinguished from magnetic effects. The question of just what is meant by the term electric charge is confronted by a careful exploration of the experiments that led to the discovery of the two types of charge property and why Franklin's subsequent naming of them (as positive and negative charges) was particularly useful and convenient. Charge quantification is distinguished from charge quantization. The concept of the electric field is developed by exploring the similarities and differences between electricity and gravitation. Gauss's law for the electric field is developed from the parallel law for the gravitational field.

CHAPTER 17 ELECTRIC POTENTIAL ENERGY AND THE ELECTRIC POTENTIAL

The often confused and subtle distinction between these two concepts is thoroughly explored. The electron-volt energy unit is introduced and its convenience illustrated. Lightning rods also are discussed

CHAPTER 18 CIRCUIT ELEMENTS, INDEPENDENT VOLTAGE SOURCES, AND CAPACITORS

CHAPTER 19 ELECTRIC CURRENT, RESISTANCE, AND DC CIRCUIT ANALYSIS

The terminology and methodology used for circuit analysis conforms to the standard conventions used in electrical engineering so the transition between physics and electronics can be made easily. The text clearly explains why positive charge flowing one way is equivalent to negative charge flowing in the opposite direction, a point of much mystification to students.

CHAPTER 20 MAGNETIC FORCES AND THE MAGNETIC FIELD

The need for a magnetic field is introduced by contrasting it with the electric field and its effects on electric charge. North and south magnetic poles are defined clearly rather than assumed to be obvious or innate. Applications include a velocity selector, mass spectrometer, and the Hall effect for determining the sign of the charge carriers of a current. Magnetic forces on currents lead to the torque on a current loop and the electric motor. The source of a magnetic field is introduced by exploring the parallels with both gravitational and electric fields. Gauss's law for the magnetic field, Ampere's law, the concept of a displacement current, and the Ampere–Maxwell law are discussed. The magnetic field of the Earth and how its reversals were discovered (via sea-floor sediments) also are explored to connect with another exciting discipline of the sciences that students see as remote from physics. *Nothing* is remote from physics!

CHAPTER 21 FARADAY'S LAW OF ELECTROMAGNETIC INDUCTION

The technological importance of Faraday's law is presented, leading to the development of an ac generator. The Maxwell equations are celebrated. The Maxwell equations in a vacuum are examined, leading to self-sustaining electromagnetic waves and the identification of such waves with light. Inductors and ideal transformers as standard circuit elements are explored using the standard engineering conventions.

CHAPTER 22 SINUSOIDAL AC CIRCUIT ANALYSIS

The typical approach to ac circuits in physics makes them seem impossibly complicated to students. In contrast, a brief introduction to complex variables permits the treatment of sinusoidal ac circuits via an extension of dc circuit analysis techniques, as is standard practice in electrical engineering. The use of current, potential difference, and voltage source phasors and the concept of impedance mean that sinusoidal ac circuit analysis then is reduced to the algebra of complex numbers.

CHAPTER 23 GEOMETRIC OPTICS

The simple Cartesian sign convention is used for mirrors, single surface refraction, and lenses, rather than a host of different, complex, and difficult to memorize mirror and lens conventions. Applications include the vertebrate eye, cameras, microscopes, and telescopes.

CHAPTER 24 PHYSICAL OPTICS

Interference via wavefront division (single, double, and multiple slit experiments as well as diffraction gratings) and amplitude division (thin-film interference) all are explored. Polarization and optical activity are discussed.

CHAPTER 25 THE SPECIAL THEORY OF RELATIVITY

Classical Galilean relativity is reviewed as well as the need for change. With the two postulates of special relativity, time dilation and length contraction are explored and used to derive the Lorentz transformation equations. The apparent relativistic paradox that *each* reference frame measures clocks in the *other* reference frame to run slow and lengths parallel to the motion to be shorter is confronted directly and resolved with a specific example. The existence of superluminal jets in astrophysics is found to be an optical illusion. The relativistic Doppler effect is explored, leading to the startling realization that for a source approaching with a nonzero impact parameter, the transition from a blue to a red shift occurs *before* the source is transverse to the line of sight. Questions of energy, momentum, the CWE theorem, and the relationship among mass, energy, and particles all are explored. The reason that the speed of light is an unreachable speed limit for material particles is discussed. The so-called mass–energy equivalence is clearly and properly addressed. Space–time diagrams are introduced and used to show why travel into the past (an idea with much student interest in view of contemporary culture) is forbidden in special relativity. The electromagnetic implications of relativity also are examined. The general theory of relativity and its classical tests are discussed using a qualitative approach.

CHAPTER 26 AN APERITIF: MODERN PHYSICS

The fortuitous discoveries of the electron, x-rays, and radioactivity are explored. The nuclear model of the atom is developed from the viewpoint that it was quite a radical proposal by Rutherford, rather than being simply obvious. The photoelectric effect and Compton scattering are used to justify the existence of the photon. The Bohr model and its limitations are explored. The biological effects of radiation and dosage units are discussed. The de Broglie hypothesis is introduced and questions raised about the meaning of a particle-wave.

CHAPTER 27 AN INTRODUCTION TO QUANTUM MECHANICS

The Heisenberg uncertainty principle is explored as well as the famous double slit experiment. The meaning of the wavefunction is assessed. Heuristic arguments lead to the Schrödinger equation.

A COMPLETE ANCILLARY PACKAGE

The following comprehensive teaching and learning package accompanies this book.

For the Student

MEDIA RESOURCES

Brooks/Cole Physics Resource Center is Brooks/Cole's website for physics, which contains a homepage for *University Physics*. All information is arranged according to the text's table of contents. Students can access flash cards for all glossary terms, supplementary practice and conceptual problems, practice quizzes for every chapter, and hyperlinks that relate to each chapter's contents.

InfoTrac® College Edition is an online library available FREE with each copy of each volume of *University Physics*. (Due to license restrictions, *InfoTrac College Edition* is only available to college students in North America upon the purchase of a new book.) It gives students access to full-length articles—not simply abstracts—from more than 700 scholarly and popular periodicals, updated daily and dating back as much as four years. Student subscribers receive a personalized account ID that gives them four months of unlimited Internet access—at any hour of the day—to readings from *Discover, Science World,* and *American Health* magazines.

Other Study Aids

Student Solutions Manual in two volumes by Ronald Lane Reese, Mark D. Semon, and Robin B. S. Brooks includes answers and solutions to every other odd-numbered end-of-chapter problem.

For the Instructor

Complete Solutions Manual in two volumes, by Ronald Lane Reese, Mark D. Semon, and Robin B. S. Brooks, contains answers and solutions to all end-of-chapter problems in the text.

Assessment Tools and Materials

Test Items for University Physics, by Frank Steckel, includes a copy of the test questions provided electronically in *Thomson World Class Learning™ Testing Tools,* Review Exercise Worksheets, and answers to the test item questions. The notation of the test items carefully follows that of the main text.

Thomson World Class Learning™ Testing Tools is a fully integrated suite of test creation, delivery, and classroom management tools. This invaluable set of tools includes World Class Test, Test Online, and World Class Manager software. World Class Test allows instructors to create dynamic questions that regenerate the values of variables and calculations between multiple versions of the same test. Tests, practice tests, and quizzes created in World Class Test can be delivered via paper, diskette or local hard drive, LAN (Local Area Network), or the Internet. All testing results can then be integrated into a complete classroom management tool with scoring, gradebook, and reporting capabilities.

With World Class Test, instructors can create a test from an existing bank of objective questions including multiple-choice, true/false, and matching questions or instructors can also easily edit existing questions and add their own questions and graphics. The online system can automatically score *objective* questions. *Subjective* essay and fill-in-the-blank questions that the instructor evaluates can also be added. Results can be scored, merged with final test results, and entered automatically into the gradebook.

Using *World Class Course,* you can quickly and easily create and update a web page specifically for a course or class. Post your own course information, office hours, lesson information, assignments, sample tests, and links to rich web content.

Presentation Tools and Online Resources

Transparencies in full color include more than 200 illustrations from the text, enlarged for use in the classroom and lecture halls.

CNN Physics Video, produced by Turner Learning, can stimulate and engage your students by launching a lecture, sparking a discussion, or demonstrating an application. Each physics-related segment from recent CNN broadcasts clearly demonstrates the relevancy of physics to everyday life.

With *Brooks/Cole's PhysicsLink,* a cross-platform CD-ROM, creating lectures has never been easier. Using multi-tiered indexing, search capabilities, and a comprehensive resource bank that includes glossary, graphs, tables, illustrations, photographs, and animations, instructors can conduct a quick search to incorporate these materials into presentations and tests. And, any *PhysicsLink* file can be posted to the web for easy student reference.

WebAssignOnline homework, a versatile, web-based homework delivery system, saves time grading and recording homework assignments and provides students with individual practice and instant feedback on their work. It delivers, collects, grades, and records customized homework assignments over the Internet. Assignments can be customized so each student can receive a unique question to solve. Access to *WebAssign* is secured by passwords and each student has access only to his or her record. *WebAssign* ©1998–99 by North Carolina State University.

Acknowledgments

A wise man will hear, and will increase learning;
and a man of understanding shall attain unto wise counsels.

Proverbs 1:5

My interest in physics was sparked long ago by two gifted mentors at Middlebury College: Benjamin F. Wissler, whose wonderfully friendly Cheshire-cat-like grin I still can see today, and Chung-Ying Chih, whose appreciation of elegance was always apparent in his approach to physics. Their excitement for the subject was contagious and their rigor and demands upon their students legendary. Subsequent mentors included Herman Z. Cummins (then at The Johns Hopkins University, now at CUNY) and my colleagues at Washington and Lee University and elsewhere, gentlemen and gentlewomen whose gifts for teaching and research continue to be admirable role models, worthy of emulation.

The unrequited help of many persons involved with this project gives me great faith in the benevolence of humanity. My colleagues in the sciences, mathematics, and the libraries at Washington and Lee University withstood incessant questions about all manner of subjects. The conviviality, camaraderie, and good humor of the Department of Physics and Engineering are especially appreciated. All have fostered an academic environment where teaching and scholarship are emphasized in an atmosphere of mutual respect, dignity, and honor.

The reviewers of this manuscript are listed alphabetically below. Their thorough and insightful reading and frank and honest critiques were invaluable to the creation of this book.

Royal G. Albridge, Vanderbilt University
C. David Andereck, Ohio State University
Gordon Aubrecht, Ohio State University
Rene Bellwied, Wayne State University
Van Bluemel, Worcester Polytechnic Institute
Neal M. Cason, University of Notre Dame
Kenneth C. Clark, University of Washington
Richard M. Heinz, University of Indiana
Daniel G. Montague, Willamette University
Richard Muirhead, University of Washington
Richard Ditteon, Rose-Hulman Institute of Technology
Charles Scheer, University of Texas, Austin
Mark Semon, Bates College
William S. Smith, Boise State University
Karl Trappe, University of Texas, Austin
Ronald E. Zammit, California Polytechnic State University

Two reviewers deserve special accolades. Their dedication went well beyond the call.

Professor Kenneth C. Clark meticulously read and reread, critiqued and recritiqued *every* draft of the manuscript from its humblest beginnings, making innumerable suggestions for clean and clever ways to elucidate many phenomena. Special thanks also go to Professor Mark Semon, whose attention to detail in his reviews was similar: insightful, thoughtful, and brimming with constructive suggestions.

At Brooks/Cole, the dedication and diligence of the entire staff and the pre-production and production teams warrant praise. Their expertise exemplify the work of true professionals. Among them are Physics Editor Beth Wilbur, Senior Developmental Editor Keith Dodson, Production Editor Tessa McGlasson Avila, Assistant Editor Melissa Henderson, Editorial Assistants Georgia Jurickovich and Nancy Conti, and Cover Designer Roy Neuhaus. Also my sincere thanks to past editors Harvey Pantzis and Lisa Moller, Developmental Editors Maxine Effenson Chuck and Casey FitzSimons, Marketing Manager Steve Catalano, and Cartoonist Tom Wentzel.

The entire staff at Electronic Publishing Services deserves kudos for a wonderful job creating the book. They were a real pleasure to work with: Senior Production Editor Rob Anglin, Photo Researcher Francis Hogan, Copyeditor and Accuracy Checker Andrew Schwartz, Art Editor Michael Gutch, Creative Supervisor Stephanie McWilliams, Illustrator Matthew McAdams, their fine coterie of artists, Page Layout Specialists Linda Harms and Brent Burket, Operations Manager Patty O'Connell, Proofreader Cheryl Smith Robbins, and Indexer Lee Gable.

To my students, both recent and more venerable, I owe a *huge* and special debt. The many students here at Washington and Lee (as well as some at Bates College) who endured a photocopied manuscript and critiqued the early drafts deserve kudos and heartfelt thanks. Students Jennifer Strawbridge and William Kanner greatly assisted with the page proofs. I am humbly and profoundly grateful to all of you for your good humor, patience, enthusiasm, and encouragement. God bless you all.

During the decade this manuscript has been in preparation, the personal friendship of many people was a constant blessing. I particularly want to thank Charlotte and Chuck Gilmore, Susan and Doug Blevins, Karen Swan, Ned Wisnefski, Fran and Dick Hodges, Pastor Mark Graham and all other ministers at St. John Lutheran Church (Roanoke, Virginia), Bethany Arnold, Ed Reed, Lynn and Fred Genheimer, and the Chanthavongsa family. Finally, I treasure the warmth and love of my extended family: foremost, my dear and venerable father, as well as Harold Augustus Reese Jr., Christine Reese McCulloch, Elaine Hildebrand, Tom McCulloch, Robert Richards, Betty Fake, Liz Tisdale, Jackie and Mel Lockwood, Kristen Lockwood, Kim and Guy Boros, and Tory and George Bolten.

Clearly, any errors remaining within the text are my responsibility. I would be very grateful to readers who bring errors of any kind to my attention [reeser@wlu.edu]. I truly welcome all your comments, critiques, and suggestions.

Shalom aleichem

PREFACE TO STUDENTS

The supreme task of the physicist is to arrive at those universal elementary laws from which the cosmos can be built up by pure deduction. There is no logical path to these laws; only intuition, resting upon sympathetic understanding of experience, can reach them.

*Albert Einstein (1879–1955)**

What a wonderful and exciting privilege it is to study and to teach physics! Physics is the bedrock of all the sciences and technology. Whether it be chemistry, geology, biology, medicine, engineering, or astronomy, our descriptions of nature involve understanding how particles move and interact individually and collectively. This understanding is the fundamental domain of physics. So, if you want to become a scientist, engineer, doctor, or a natural philosopher, or simply to understand nature at its most fundamental level as an intelligent citizen-scientist, you need to begin with a voyage through the foundations of physics.

The mission of this text is

- to present the subject in a logical, clear, and comprehensible style;
- to stress how *little* needs to be known in order to understand as much as possible;
- to recognize that many aspects of physics, while quantifiable, remain fundamentally abstract and mysterious—we do not have the answers to many profound questions; and
- to lighten the stress over technical gobbledygook with occasional humor.

I hope this book conveys the excitement of a fascinating search for a fundamental understanding of nature. The discipline of physics is, after all, the observation, explanation, and integration into a conceptual whole of as much as we can see, perceive, and infer during our all-too-short intellectual rendezvous with this amazing universe.

The beauty and coherence of physics may be obvious to professors, but may be less clear to you, our students, since physics likely seems less romantic than the beauty of the starlit sky. Understanding the celestial dances of the firmament, though, really is physics. So is understanding why the night sky is dark, why clouds are white, why the sky is blue, why bubbles appear in a bottle of beer or a glass of champagne, and what drives the wondrous biochemical processes that make life itself possible and gives us the opportunity to wonder at the beauty, not only of each other, but also of the natural world surrounding us.

We have an exciting time of study ahead. From a practical viewpoint, careful and thoughtful multiple readings of the text material are encouraged. Your first reading might omit the example problems in order to gain a conceptual overview of the material sufficient to address the Questions; a second reading then can include sufficient examples (particularly the Strategic Examples) for you to gain the necessary familiarity to approach the Problems.

In the conceptual development, examples, and problems, great emphasis is placed on the choices you have to make and the consequences of those choices. This is designed to gradually build confidence in your ability to tackle new and different situations. It will also teach you to be alert for the unexpected, or unanticipated result. Most significant discoveries in science begin with simply noticing something unexpected or peculiar. "What's that?" has led to many a significant "Aha!" in the history of science and technology.

*"Principles of Research," *Ideas and Opinions by Albert Einstein*, edited by Carl Seelig (Crown Publishers, New York, 1954), page 226.

ABOUT THE AUTHOR

Professor Reese teaches physics and astronomy at Washington and Lee University in Lexington, Virginia. He received his undergraduate degree in physics from Middlebury College and his Ph.D. in physics from The Johns Hopkins University. He has been teaching introductory physics, astronomy, and various advanced physics courses for almost 30 years. He also has performed consulting research on the interaction of visible and microwave electromagnetic radiation with matter at the Naval Research Laboratory in Washington, D.C., and has been a Visiting Fellow at University College, Oxford, during several sabbaticals.

BRIEF CONTENTS

CONTENTS

VOLUME I

CHAPTER 1
Preludes 1

CHAPTER 2
A Mathematical Toolbox 35

AN INTRODUCTION TO VECTOR ANALYSIS

CHAPTER 3
Kinematics I 73

RECTILINEAR MOTION

CHAPTER 4
Kinematics II 117

MOTION IN TWO AND THREE DIMENSIONS

PHILOSOPHIÆ
NATURALIS
PRINCIPIA
MATHEMATICA.

Autore J S. NEWTON, Trin. Coll. Cantab. Soc. Matheseos Professore Lucasiano, & Societatis Regalis Sodali.

IMPRIMATUR.
S. PEPYS, Reg. Soc. PRÆSES.
Julii 5. 1686.

LONDINI,
Jussu Societatis Regiæ ac Typis Josephi Streater. Prostat apud plures Bibliopolas. Anno MDCLXXXVII.

CHAPTER 5
Newton's Laws of Motion 169

CHAPTER 6
The Gravitational Force and the Gravitational Field 231

CHAPTER 7

CHAPTER 8

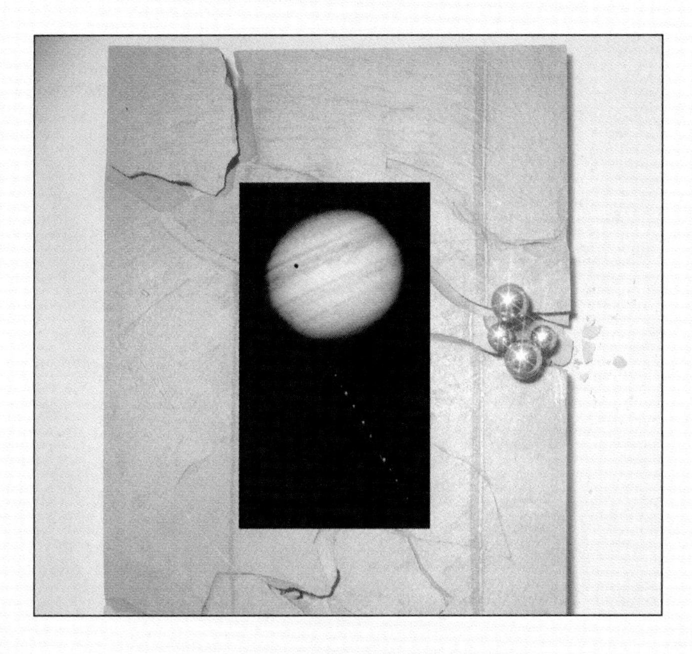

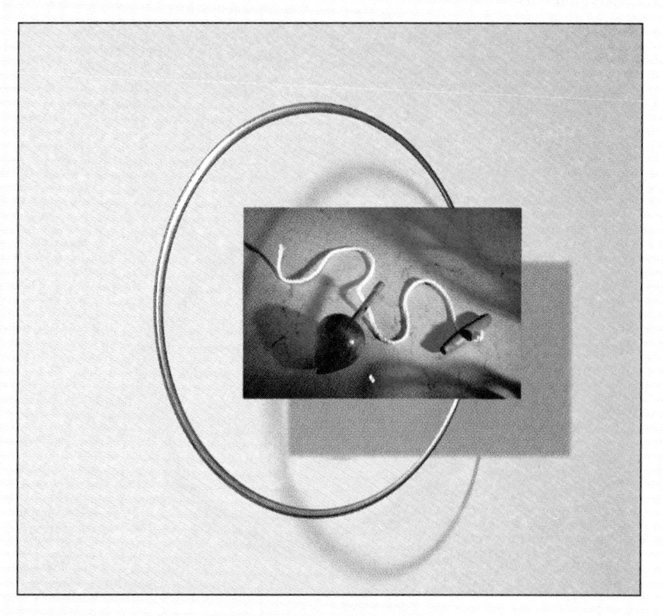

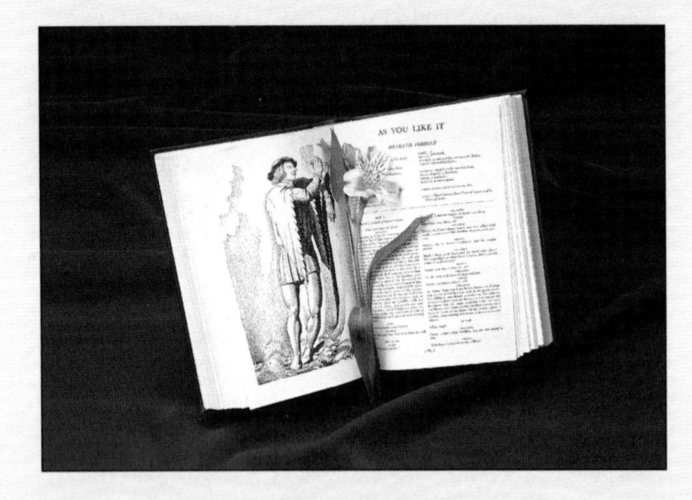

VOLUME II

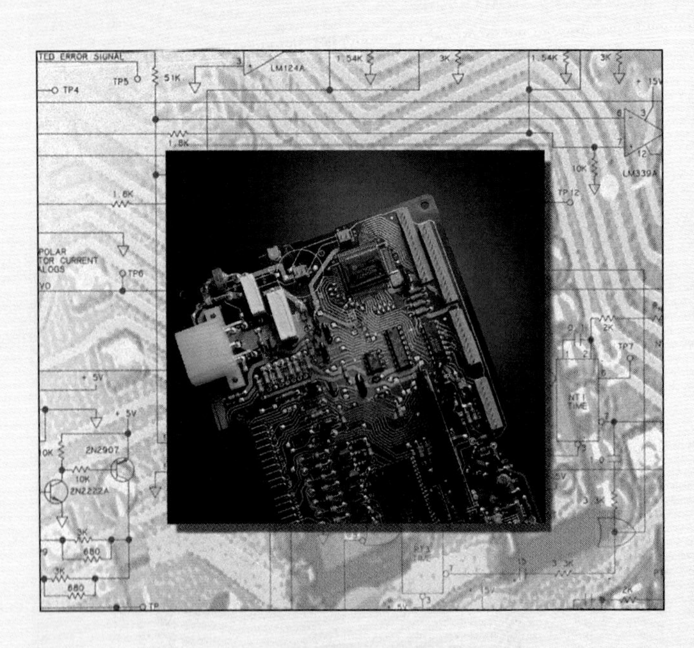

CHAPTER 23
Geometric Optics 1041

CHAPTER 24
Physical Optics 1103

CHAPTER 25
The Special Theory of Relativity 1149

CHAPTER 26
An Aperitif: Modern Physics 1205

TER

Electric Charges, Electrical Forces, and the Electric Field

Electricity is [the] Soul of the Universe

*John Wesley (1703–1791)**

It is hard to imagine life without electricity. Try it sometime. A blackout will do. Backpacking in the wilderness is another way (no flashlights permitted!). No lights, washing machines, dishwashers, microwave ovens, central heat, air conditioners, or snooze alarms either. No stereos, radios, TV, Xerox machines, ink jet printers, no convenient plastic kitchen wrap—essentially no advanced technology. There is one small benefit: no annoying static cling in our clothes. It is a primitive existence without electricity. But it is much worse than that: the electrical force is central to our very being. Without the electrical force, atoms would not exist, no molecules, no chemistry (no chemistry tests!), no life. Worst of all, there would be no physicists.[†] The harnessing of electricity for productive use is a marvel of our collective technological prowess and keeps many of us gainfully employed and challenged.

What is electricity? Simple, you might say. *Everyone knows what a dragon looks like*[‡]; everyone knows what electricity is. It is positive and negative charge, protons and electrons, and "that kind of stuff." So, *what is charge?* Come to think of it, maybe it is not so simple to explain what electricity is . . . at least if you have to explain it under the watchful eye of your professor! The importance of language, not equations, comes back to haunt us again. The concept of electricity, or what we casually call electric charge, is not quite as tangible or as easy to fathom as mass; and as we have seen, even the concept of mass is quite subtle and mysterious.

How do we know electricity even exists? Unlike gravitation, it is not readily apparent in our natural environment[§]; you have to look for it. For years, you likely have been told that electric charge comes in two varieties or flavors: positive charge and negative charge. Protons have positive charge, electrons have negative charge. It is hard to imagine two more different particles than the electron and the proton (they differ in mass by a factor of 1836), yet they have exactly the same absolute value (magnitude) of electric charge—why is that? We address that question in Section 16.4. As a consequence of the equality, normal atoms (with equal numbers of protons and electrons) are electrically neutral (zero total charge), which is why electricity is not as apparent as gravitation.

In this chapter, we first address what we mean by electric charge. We then explore how charges interact quantitatively, and introduce a useful construct for imagining how the electrical interaction is transmitted between charges: the *electric field*. Finally, we discover some of the characteristic properties of the electric field.

16.1 The Discovery of Electrification

Hence have arisen some new terms among us: we say B (and bodies like circumstanced) is electrized positively; A, negatively. Or rather, B is electrized plus; A minus. And we daily in our experiments electrize plus or minus, as we think proper.

Benjamin Franklin (1706–1790)[#]

*(Chapter Opener) *The Desideratum: or Electricity Made Plain and Useful* (London, 1759); from the Contents of the 2nd edition (Bailliere, Tindall, and Cox, London, 1871). Wesley was the founder of the Methodist Protestant Christian religious denomination.

[†] This is a line made famous by George Wald in a taped lecture, "Design in the Universe," part of a lecture series *Cosmic Evolution: Are We Alone in the Universe* (American Association for the Advancement of Science, 1974).

[‡] The title of a children's book by Jay Williams (Four Winds, New York, 1976).

[§] It was only during the 18th century that Benjamin Franklin first demonstrated that lightning had electrical properties.

The wit and wisdom of Benjamin Franklin still are widely admired. He was the first American scientist to gain an international reputation (for his research into electricity).

The discovery of electrical phenomena begins with an unlikely material: amber, the beautiful, yellowish-brown, fossilized resin, valued for ornamental jewelry. A sample of amber is shown in Figure 16.1 encasing a venerable entomological intruder.

The ancient Greeks discovered, likely while polishing the material, that when rubbed vigorously, amber has the ability to attract small bits of matter such as straw and grain hulls.[¶] From such chaff was born a revolution in science and technology, though several millennia later. The interaction between the amber and chaff only exists after the amber is rubbed (the chaff is not rubbed) and, even more perplexing, the force gradually dissipates, leaks away, or diminishes as time passes, particularly on humid days. This is quite unlike gravitation; the gravitational interaction

[#] Letter to Peter Collinson, 1 September 1747; Benjamin Franklin, *Experiments and Observations on Electricity* (E. Cave, London, 1751), page 15. More recently in Duane Roller and Duane H. D. Roller, "The development of the concept of electric charge," *Harvard Case Studies in Experimental Science* (Harvard University Press, Cambridge, Massachusetts, 1954), volume 2, pages 541–693; the quotation is on page 598.

[¶] In modern parlance, the chaff is initially *not* charged but is electrically neutral. We address in Section 16.2 how the attraction between electrified material and neutral material arises.

FIGURE 16.1 **A sample of amber with an ancient insect trapped inside.**

The 16th-century age of exploration precipitated much interest in magnetism and it attracted William Gilbert to explore the relationship between it and amber-like effects.

between masses always is present and certainly does not depend on the weather. There is evidently quite a different interaction going on between the amber and chaff.

The concept of electricity and the word itself arose during the 17th century from careful observation of amber-like effects in other materials. In 1600, a century before Newton, William Gilbert (1540–1603), a scientist of Renaissance interests and physician to Queen Elizabeth I, discovered that glass and many other substances, when rubbed with silk, attracted small bits of matter, just as amber did. He described the observations by saying the materials had become *electrified*, meaning they had "become like amber" or "amberized"; *electron* [ἤλεκτρον] is the Greek word for amber.

Could this attraction be what we know as magnetism? Another naturally occurring material known as lodestone [composed of the mineral magnetite (Fe_3O_4)] was known to the ancient Chinese and Greeks (though not as Fe_3O_4!), and has peculiar attractive properties for a restricted list of materials, typically containing iron. Lodestones gave rise to *magnetism* [the word comes from Magnesia, the region of Asia Minor (Anatolia) in present-day Turkey where the mineral was found]. We study magnetic effects later (Chapter 20). For now, we simply note that *every* magnet *attracts* iron that is not itself a magnet (resembling the attraction of rubbed amber for bits of straw).

So why is electricity different from magnetism? If you play with any *two* magnets as you likely did as a child,* or *two* electrified materials, there is a distinct difference. Any two magnets can attract *or* repel each other depending on the orientation of the ends of the magnets (what we now call the poles of the magnet) that are brought closest together or presented to each other, as shown in Figure 16.2.

This phenomenon is very different from what is observed with electrification forces. By playing with two different pieces of rubbed amber, you find they *always* repel each other; it makes no difference which end of each piece is presented to the other, as indicated in Figure 16.3.

Electric and magnetic effects also were seen to be different phenomena because rubbing seemed intimately connected with electrical effects in amber and other materials, whereas magnetic effects are essentially permanent in lodestones. During the 19th century, electrical and magnetic effects were shown to be closely related. Rubbing is not necessary for some electrical effects, nor are all magnets permanent. We now consider both electrical and magnetic effects to be aspects of **electromagnetism**, to reflect their intimate connection. We shall see how they are related in Chapter 21, after first exploring them separately.

It was the Frenchman Charles-François Cisternay Dufay (1698–1739) who discovered in 1733 that while two glass rods electrified by rubbing with silk *repelled* each other (see Figure 16.4), and two rubber or amber rods electrified by rubbing with fur *also* repelled each other (see Figure 16.5), an electrified glass rod *attracted* an electrified rubber rod or amber (see Figure 16.6). (The latter two figures are on page 709.)

In other words, identical materials electrified by the same procedure always repel each other. Different materials, electrified by different means (rubbing with different materials such as silk or fur), may repel or attract each other depending on the specific materials and means of electrification. If the materials are not electrified by appropriate rubbing, they apparently do not interact (except weakly by gravitation, of course). Thus electrification

*No need to stop such playing now. A fascination with magnets was one of the things that led Albert Einstein into physics.

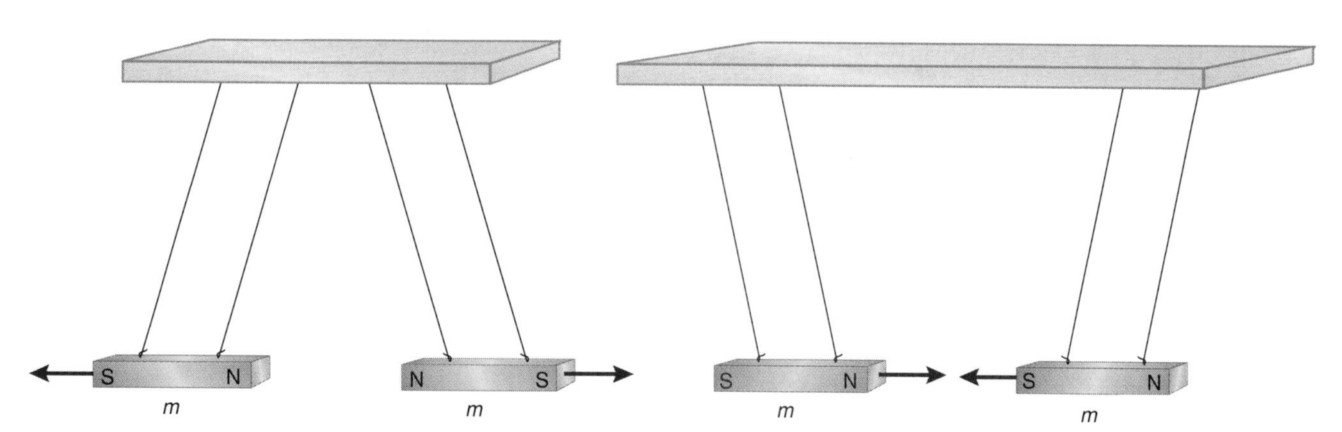

FIGURE 16.2 Two magnets will attract or repel each other depending on which way they face each other.

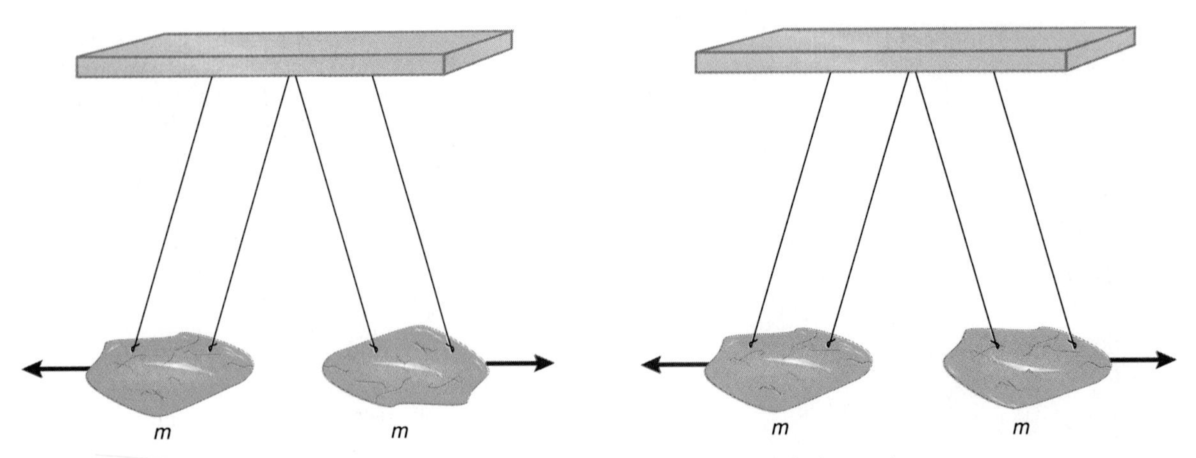

FIGURE 16.3 Two pieces of rubbed amber always repel each other, regardless of which way they face each other.

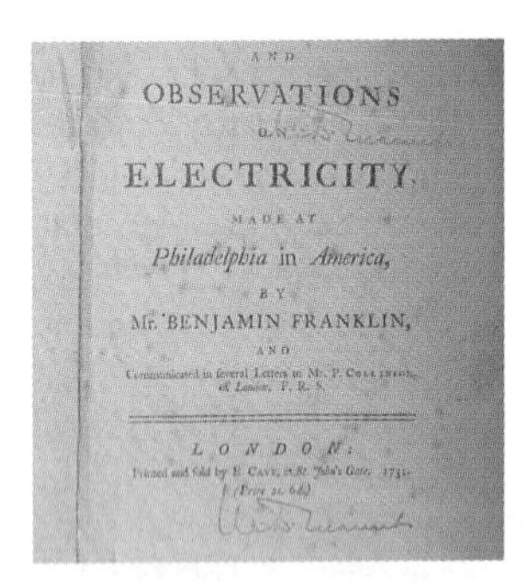
AND

OBSERVATIONS

ON

ELECTRICITY,

MADE AT

Philadelphia in *America*,

BY

Mr. BENJAMIN FRANKLIN,

AND

LONDON:

Benjamin Franklin published his researches into electrical phenomena in 1751, well before the American Revolution.

forces are attractive, repulsive or zero.* Confusing? Yes. But Dufay was the first person to realize the following:

> *Two* different kinds of electrification properties are needed to explain *all* of these observations.

It is fortunate for us that all electrical phenomena can be described by assuming only two (and not more) different kinds of electrification properties. Dufay called the two types of electrification properties *vitreous electricity* (that similar to glass) and *resinous electricity* (that similar to amber). One never observes an electrified material that repels (or attracts) both an electrified glass rod and an electrified rubber rod; such an observation would imply a third type of electrified state or electrification property. Since this never has been observed, we can state with confidence that only two types of electrified state or property exist.

Furthermore, electrical forces are quite distinct from gravitational forces in several respects:

1. Electrical forces between electrified materials are quite apparent even with small pieces of electrified matter, whereas the gravitational force between such small masses is almost negligible and detectable only with the most sensitive types of equipment. Thus electrical forces evidently are intrinsically much stronger than gravitational forces (which are appreciable only when one or both of the masses is huge by laboratory standards).
2. Gravitation is always and only observed as an attractive force, and so we have need for only one kind of mass, positive mass.† Electrical forces are observed to be *either* attractive *or* repulsive, hence the need for two types of electrification property, the vitreous and resinous electrifications of Dufay.

m m

FIGURE 16.4 Two glass rods rubbed with silk repel each other.

*Of course, you know that magnetic forces have this dichotomy also, and so it really is a tribute to these experimentalists that they recognized differences between electrical and magnetic effects.

† Even antiparticles such as the positron (a positively charged electron) and the antiproton (a negatively charged proton) have positive mass.

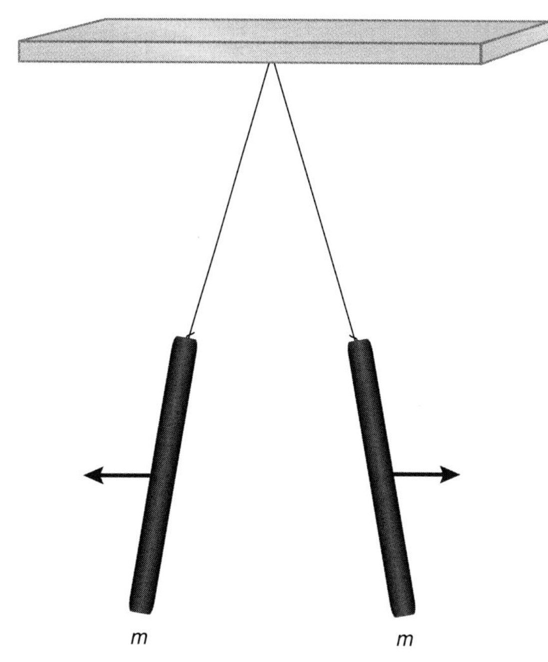

FIGURE 16.5 Two rubber or amber rods rubbed with fur repel each other.

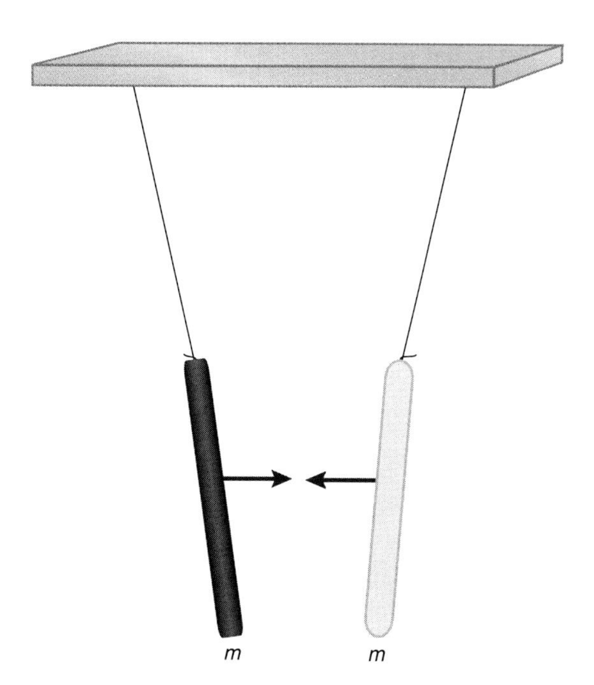

FIGURE 16.6 An electrified glass rod attracts an electrified rubber or amber rod.

The electric and gravitational forces are *similar* in two ways:

1. Both are observed to be *central forces*. They act along the line connecting pointlike materials causing the force.
2. Both are *conservative forces*. The work done by the force around a closed path is zero (equivalently, the work done by the force along a path connecting any two points in space is independent of the path between the two points). We explore the conservative nature of the electrical force in more detail in the next chapter.

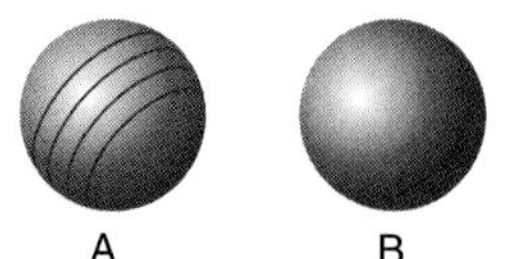

(a) An electrified conducting and an unelectrified conducting sphere.

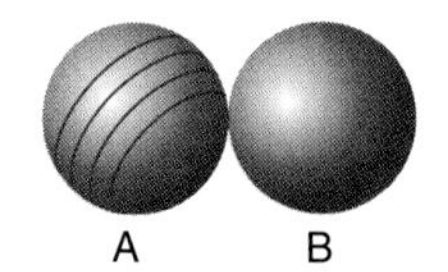

(b) Touch them together.

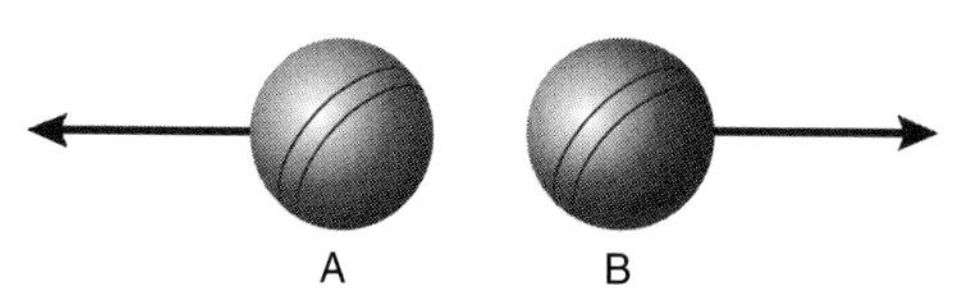

(c) Separate them.

FIGURE 16.7 Transferring charge.

It was Benjamin Franklin who introduced the now common definitions associated with the two properties of electrification:

> A particle or mass is said to be **positively electrified** if it is repelled by a glass rod that has been freshly rubbed with silk.

All glass rods thus rubbed also have positive electrification, since they individually repel each other.

> A particle or mass is said to be **negatively electrified** if it is repelled by rubber or amber that has been freshly rubbed with fur.

Hence the rubber or amber itself has a negative electrification property.

The names for the electrification properties are arbitrary. Dufay called them vitreous and resinous electricity; Franklin called them positive and negative electricity. One could have called the two electrification properties *fat* and *lean* electricity, *hairy* and *bald* electricity, or even *male* and *female* electricity.

The two electrification properties are easily transferred through and shared among materials. Some materials easily let the electrification property move from one place to another; these materials are called **conductors** (from the verb conduct, to transport). With other materials the electrical property lacks mobility (at least over short time intervals); these are called **insulators** (from the verb insulate, to isolate). Insulators also are called **dielectrics**.

If an electrified conducting sphere and an identical but unelectrified conducting sphere (see Figure 16.7a) are brought into contact (Figure 16.7b), and then separated (Figure 16.7c), we find that *both spheres now are electrified* and repel each other. We can measure the repulsive force that exists between the two spheres at a fixed separation. If two more identical conducting but unelectrified spheres are each now brought into contact with one of the identically electrified spheres, as in Figure 16.8, we find that the repulsive force between *any two* of the four electrified spheres, when separated by the same distance, is one-fourth what it was between the original two electrified spheres (see Figure 16.9).

The condition of electrification thus is quantifiable as measured by the forces.

Combining materials having equal amounts of opposite electrification properties exactly cancels their total effectiveness.

The two electrification states or properties thus are quantifiable and behave algebraically and arithmetically as scalars.

This makes Franklin's positive and negative terminology much more convenient than the vitreous and resinous characterizations of Dufay and explains why the latter terms quickly were abandoned after Franklin's time. We examine the quantification of the electrification property in Section 16.3. Electrical forces of attraction and repulsion on materials are, of course, vector quantities, as are all forces.

Analogously to gravitation, we could call the electrification properties of matter the positive or negative *electrical masses*,* but to avoid confusion with gravitation, the name used for the two electrification properties is **electrical charge**.

The word **charge** means to endow with electricity (or the electrification property).†

When something *has* charge, it is endowed with one or the other electrification property. We designate the electric charge property of matter symbolically by q (or Q), which may be either a positive or negative *scalar* according to Franklin's convention.

The concept of electric charge is no less nor more abstract than the concept of mass. Mass itself quantifies the property we called inertia or resistance to a change in motion. Both concepts thus are defined operationally by experiment. We just are more familiar with the concept of mass because we think of it as substance, and so regard it as more real and tangible than charge. Mass and charge both are abstractions used to describe the way things in nature behave in certain experiments. We use such technical terms to disguise our fundamental ignorance of what the property really is. The terms mass and charge (and those naming other properties‡) are our ways of describing the response of a simple or complex system to certain types or classes of experiments.

Thus, when we say an object has a **positive charge**, we mean the object has the electrification property that makes it repelled by a glass rod that has been freshly rubbed with silk, as shown in Figure 16.10. When we say an object has a **negative charge**, we mean the object has the electrification property that makes it repelled by a rubber rod that has been freshly rubbed by fur, as shown in Figure 16.11. This is what we mean by the terms positive or negative charge; and, like Humpty Dumpty, *nothing more*. (See quote on page 3.)

From our perspective on fundamental particles at the dawn of the third millennium of the common era, we find that those with nonzero mass§ are the only kinds of fundamental particles that exhibit electric charge (i.e., can have one or the other electrification property), but that not all particles with mass have nonzero total charge (e.g., the neutron has zero total charge).#

*In fact this was the term used by Charles Coulomb in his experiments during the late 18th century (see Section 16.3).

† What we call mass also could be called gravitational charge (endowed with gravitation) or inertial charge (endowed with inertia).

‡ Another example is the *spin* of fundamental particles.

§ The particles of light (photons) have zero mass.

Another distinction between magnetic and electrical phenomena is worth mentioning here: so-called magnetic poles (we define them in Chapter 20) *always* occur in pairs, whereas electric charge is capable of being isolated as either positive or negative.

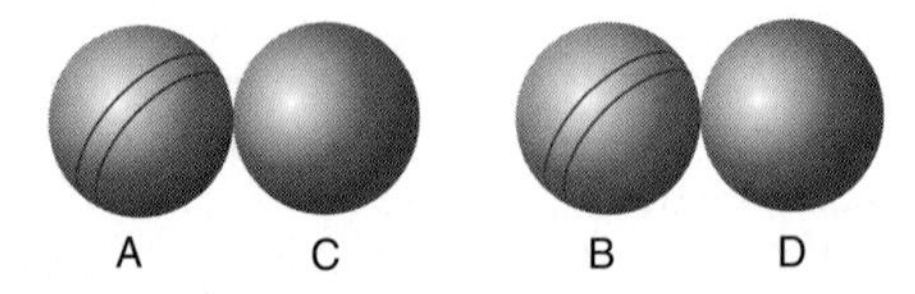

FIGURE 16.8 Touch two more identical conducting spheres to those in Figure 16.7c.

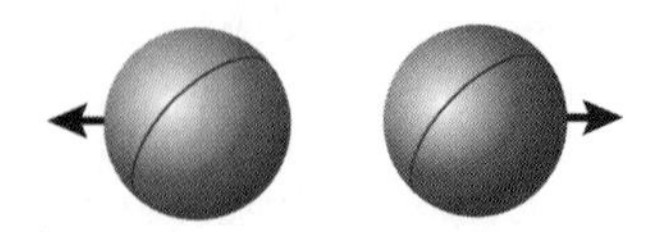

FIGURE 16.9 Any two of the four spheres now repel each other with a force one-fourth as great as that in Figure 16.7c.

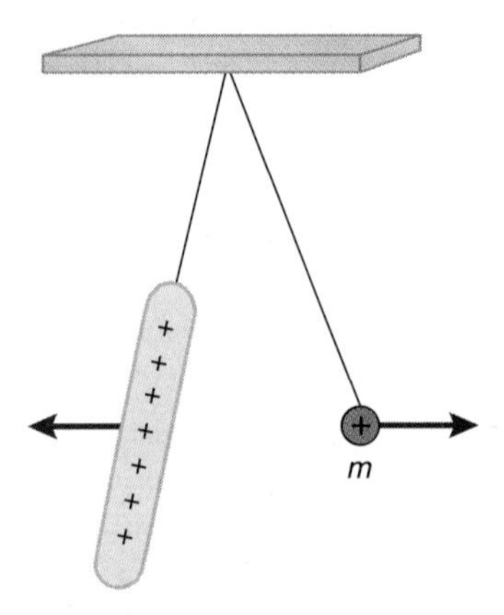

FIGURE 16.10 A positive charge is repelled by a glass rod freshly rubbed with silk.

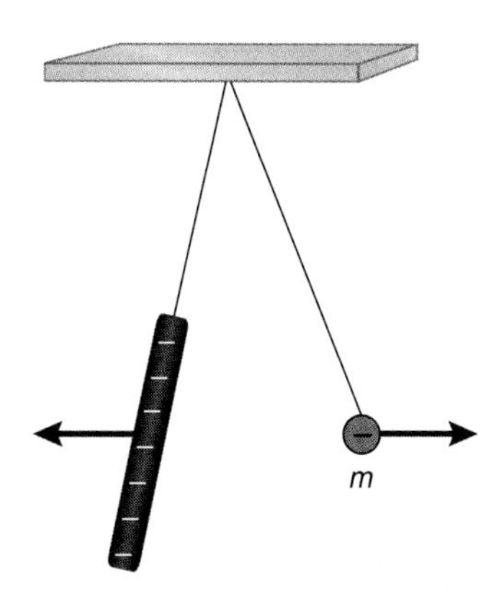

FIGURE 16.11 A negative charge is repelled by a rubber rod freshly rubbed with fur.

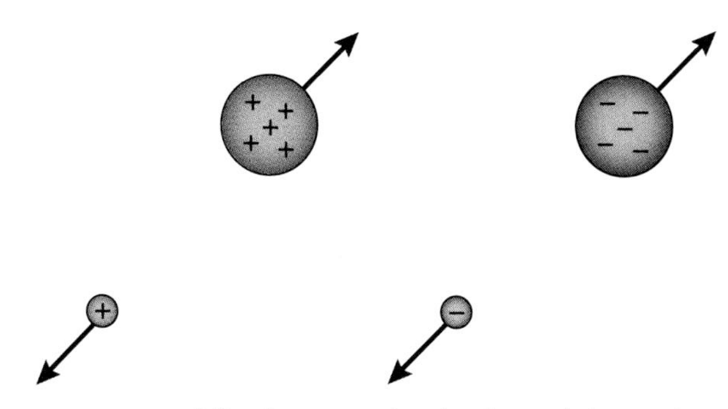

FIGURE 16.12 Like charges repel each other with forces of equal magnitude.

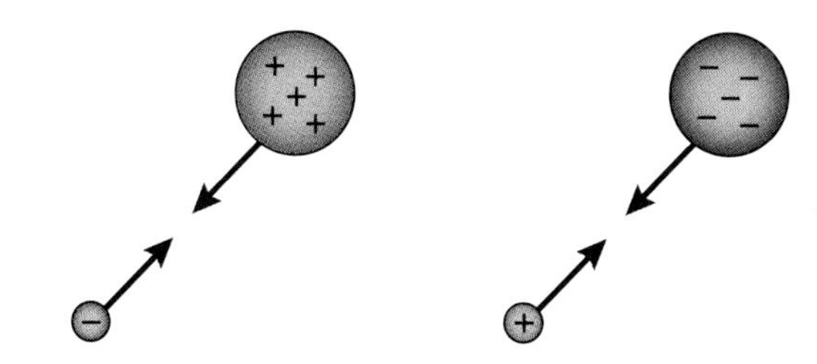

FIGURE 16.13 Unlike charges attract each other with forces of equal magnitude.

With this terminology, the defining experiments indicate that two particles with the same electrification property (flavor), either both positive or both negative, will feel repulsive electrical forces of equal magnitude on each. If two particles have the same kind of electrification property (flavor) (i.e., both positive or both negative), we say the charges are **like charges**; this does *not* mean that charges are of equal magnitude, only that they have the same type of electrification property. Any like charges produce *repulsive electrical forces* of the same magnitude on each other (see Figure 16.12).

If the electrical charges are of *opposite* flavors (one charge positive, the other negative), the electrical force of the charged particles on each other is attractive (see Figure 16.13). If the two particles have opposite electrification flavors (one positive, one negative), we say the charges are **unlike charges**; these produce attractive electrical forces of equal magnitude on each other.

The electrical forces that two charged masses exert on each other are of equal magnitude and opposite in direction, *regardless* of the quantity of charge each has, in accordance with Newton's third law of motion.

Founding father Franklin also discovered that when glass is rubbed with silk, the glass and silk have opposite charges: the charge of the glass rod is positive (by definition) and the silk negative. Likewise when rubber is rubbed with fur, the rubber and fur have opposite charges: the rubber is negative (by definition) and the fur is positive. The implication of this observation is that the process of electrification by rubbing involves charge transfer, say, from the glass rod to the silk or from the fur to the rubber. We now know that the negatively charged electrons are the charges transferred in the rubbing electrification process. Electrons are transferred from the glass to the silk. The silk acquires a negative charge (because of a surplus of electrons), and the glass acquires an equal magnitude of positive charge (because the positive nuclear charges in the glass slightly outnumber the remaining negative electron charges).

Franklin's many ingenious experiments were the first hint of a profound aspect of the electric charge property, now well demonstrated by experiment:

> The total amount of electric charge is *conserved* in any process involving an isolated system; we call this observation **conservation of charge.**

Charge can be shuffled around among the constituents of the isolated system, but the total amount of charge at the beginning and end of a process involving an isolated system is the same. These processes may be chemical reactions, particle decays, nuclear transformations, particle creations, collisions, or what not. For example, an isolated, naked neutron (n) is unstable and breaks up or decays into a proton (p), an electron (here designated by β^-), and another exotic particle known as an antineutrino ($\overline{\nu}$). The reaction is written symbolically as

$$n \rightarrow p + \beta^- + \overline{\nu}$$

The total charge initially is zero (the neutron has zero charge); the total charge resulting from the reaction also is zero because the antineutrino has no charge and the charges on the proton and electron are of equal magnitude (absolute value) but opposite sign. We began with zero total charge and ended with zero total charge in this process. No experiment ever has violated the principle of conservation of charge, and so we take it as no more nor less fundamental than energy conservation. Nature apparently just behaves this way.

QUESTION 1

What is the charge of your cat after petting it with rubber gloves? What experiment can you do to see if the cat has the same type of charge when you pet it with your bare hands?

16.2 Polarization and Induction

Two objects with the same state of electrification, that is, both positive or both negative, repel each other. Two objects with opposite states of electrification (one positive, the other negative) attract each other. But the original experiments involving electricity (the ones with amber and chaff) indicate that any charged object (whether positive or negative) attracts uncharged, electrically neutral matter. This may seem quite surprising; we have to explain how this comes about.

We must be careful to distinguish (via experiments exploring the interaction of the materials with charged glass and rubber rods) whether *attractive* electrical forces (1) arise because the objects have opposite states of electrification, or (2) occur because any charged object will attract an electrically neutral object. An electrically neutral object will be attracted *either* to a positively charged glass rod or to a negatively charged rubber rod. But a charged object will be attracted to one and repelled by the other, in which case the sign of the charge on the unknown object can be determined. If it is repelled by the glass rod, the unknown charge is positive; if it is repelled by the rubber rod, the unknown charge is negative.

> Hence it is *repulsion*, not attraction, that enables us to determine the sign of an unknown charge.

How is it that electrically neutral matter is attracted to a charged object? If, for example, an electrically neutral object is brought close to a positively charged probe, some type of *charge separation* must occur in the neutral object so that negative charges on the neutral object are closer to the positive probe. The total electrical charge on the neutral object is zero; charge separation is the only phenomenon that occurs. The result of the charge separation is (1) an attraction between the positive probe and the closer negative charges on the neutral object, and (2) a repulsion between the more distant positive charges on the neutral object and the positive probe.

The first force is greater in magnitude than the second, and so the total force is attractive; this implies that the electrical force depends inversely on some power of the distance between the charges. We explore the distance dependence in the next section. If a negative probe is used, charge separation occurs in the neutral object so that positive charge is closer to the negative probe, thus causing the attraction.

> Charge separation in electrically neutral materials, caused by the presence of another nearby charged object, is called electrical **polarization.**

It is *not possible* to tell from these experiments which electric charges (positive, negative, or both) in the neutral material are mobile and thus responsible for the charge separation. In fact, one can explain these experiments assuming that *either* positive or negative charges or both are mobile; you might try the explanations as an exercise.

We model electrical polarization phenomena in the following way, assuming the modern atomic and molecular view of matter. In *conducting materials*, we find (using other experiments we discuss in Chapter 20) that the mobile charge carriers are electrons, producing charge separation because of electron migration. When such an uncharged conductor is brought close to a positively charged probe, electrons in the conductor migrate to the region of the conductor closest to the positive probe, leaving behind electrically unbalanced positively charged atoms at the remote end of the conductor. The attractive force on the (closer) negative electrons is greater than the repulsive force on the (more distant) positive charges, resulting in a total force on each that is attractive, as shown in Figure 16.14.

Similarly, if a negative probe is used (see Figure 16.15), some electrons in the conductor are repelled to the remote regions of the conductor. The attractive force on the closer positive charges is stronger than the repulsive force on the more distant negative charges, and the resultant total force is attractive.

So in conducting materials, charge separation occurs over the *macroscopic* distances on the order of the physical size of the conductor itself. Since the conductor as a whole is uncharged, the macroscopic charge separation creates what we call a macroscopic **electric dipole:** charges of equal magnitude, but opposite sign, separated by some distance.

Insulators, however, are by definition materials on which the electrification property (i.e., charge) is not mobile. It is enormously more difficult to transfer charge in them. So how are electrically neutral insulators attracted to charged probes in the vicinity? In this case charge separation—that is, electrical polarization—must occur only on a *local* atomic or molecular scale. But we have great numbers of atoms or molecules to distort slightly. Throughout the interior, this small distortion of charge position for each atom or molecule is balanced out by the same effect on all its nearest neighbors (see Figure 16.16). But at the surface the cancellation is not complete. The excess of positive or negative charge over the near and far surfaces is what gives a resultant total force of attraction toward the probe. Thus an electrically neutral insulator, just as an electrically neutral conductor, is attracted to the charged probe whether the probe is positive or negative.

Let's try yet another experiment.

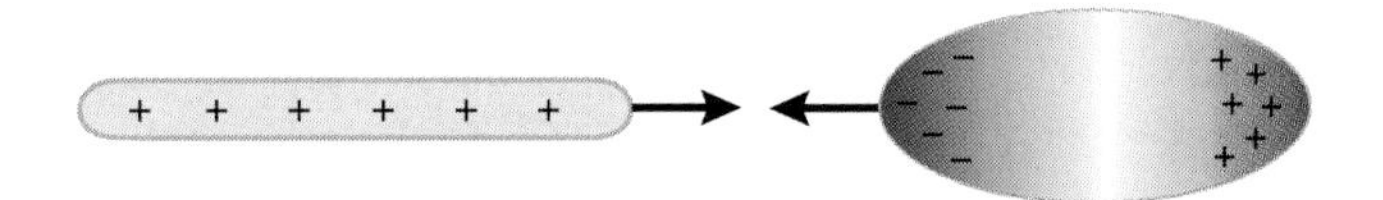

FIGURE 16.14 Attraction of an uncharged conductor to a positively charged object.

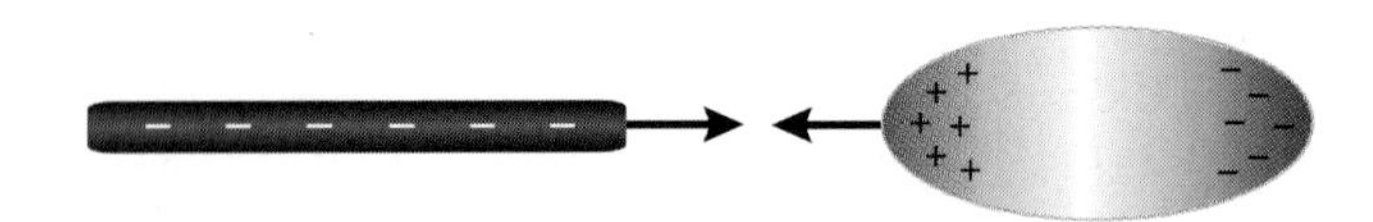

FIGURE 16.15 Attraction of an uncharged conductor to a negatively charged object.

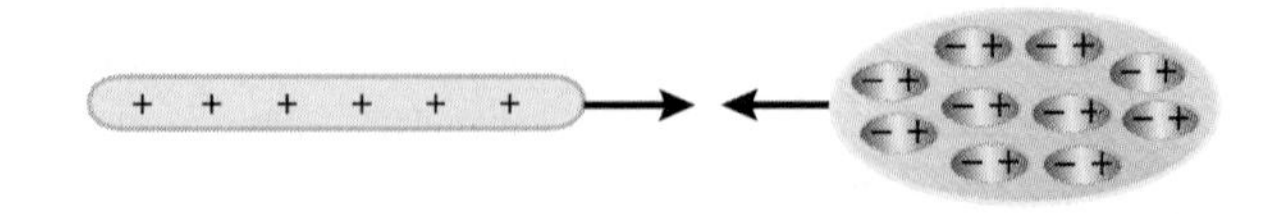

FIGURE 16.16 Attraction of an uncharged insulator to a charged object.

1. We take a negatively charged object and hold it fixed near an uncharged conductor. The uncharged conductor becomes polarized as we explained previously. Electrons migrate in the conductor to the more distant end of the conductor, since they are repelled by the negative probe, as shown in Figure 16.17a.
2. Now we attach another conductor (a wire) to the conductor with the other end of the wire connected to the Earth (a reasonably good conductor itself), as shown in Figure 16.17b. The electrons repelled by the negative probe now can escape to the Earth itself, which is part of the total conducting body. Connecting a conductor to the giant Earth by means of another conductor (typically a wire) is called **grounding** the conductor.
3. If the wire now is removed (see Figure 16.17c), the electrons have no way to get back to the conductor from which they fled, and so the conductor now has a positive charge and is no longer electrically neutral. The negative probe finally can be removed (Figure 16.17d), and the initially neutral conductor remains positively charged and entirely separated. This process is called charging by **induction**.

You might be able to see that the process of electrification by induction operates equally well if begun with a positive probe and a neutral conductor; the result in this case is a negatively charged conductor. Try to explain the corresponding process in your own words.

QUESTION 2

A small object is attracted to an electrified object. Does this imply that the small object is charged opposite to the electrified object? How can you determine whether the small object is charged or not and, if so, whether it is positively or negatively charged?

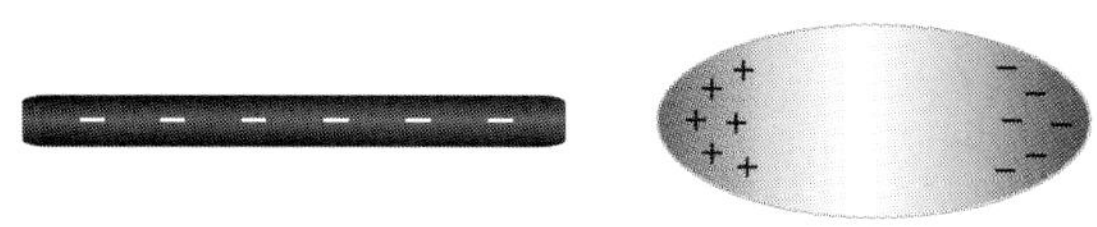

(a) An uncharged conductor placed near a charged object.

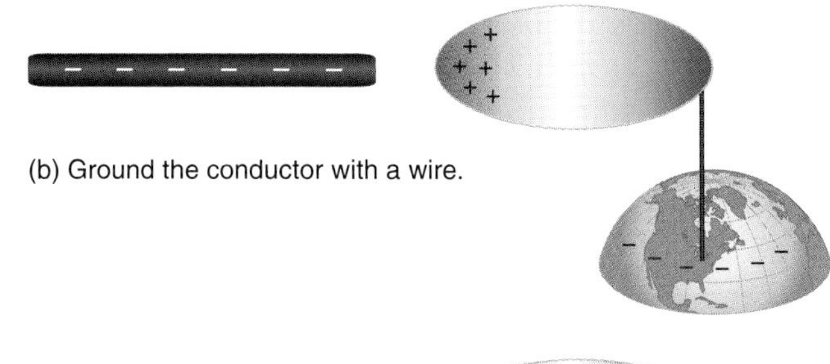

(b) Ground the conductor with a wire.

(c) Remove the grounding wire.

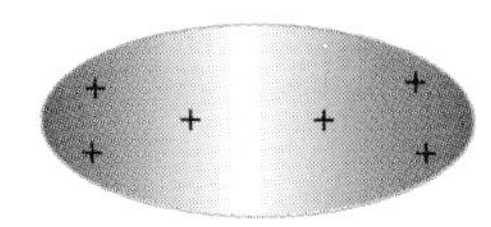

(d) Remove the charged probe and the isolated conductor remains charged.

FIGURE 16.17 Charging by induction.

16.3 COULOMB'S FORCE LAW FOR POINTLIKE CHARGES: THE QUANTIFICATION OF CHARGE

Between 1785 and 1787, Charles Augustin Coulomb (1736–1806) performed a critical and difficult series of experiments on electrified materials using a sensitive torsion balance that he invented for measuring small forces.* He discovered that, as was true with gravitation, the magnitude of the mutual electrical force of attraction or repulsion on each of two small, pointlike, electrified objects varied inversely as the square of the distance of their separation. The magnitude of the electrical force on each pointlike charge is an *inverse square law*, just like gravitation:

$$F_{\text{elec}} \propto \frac{1}{r^2}$$

Modern experiments have verified the inverse square nature of the electrical force to better than 16 significant figures for the exponent of the distance.† By means of experiments similar to those we described earlier with identical conducting spheres, Coulomb discovered that the electrification property (charge) is quantifiable in terms of the forces that electrified objects exert on each other. Coulomb also found that the magnitude of the electrical force that two pointlike charges q and Q exert on each other is proportional to the product of the magnitudes‡ of the charges:

$$F_{\text{elec}} \propto \frac{|q||Q|}{r^2} \qquad (16.1)$$

*A similar balance was used by Henry Cavendish during the 18th century to accurately measure gravitational forces between small masses and thus determine the numerical value of the universal gravitational constant G.

†See Investigative Projects 1 and 2 at the end of this chapter.

‡The word magnitude, when referring to electric charge, means *absolute value*, rather than the magnitude of a vector, though both quantities are nonnegative scalars. Electric charge is a scalar, not a vector.

Aside from his notable researches into electrical forces, Coulomb also did significant work on the mechanical properties of materials and engineering consulting work with water systems, canals, and harbors.

The direction of the force on each charged mass is determined using the observation that like charges repel, unlike charges attract each other.

The forces that *any* two pointlike charges exert on each other are of equal magnitude but in opposite directions, in accordance with Newton's third law. The force on each charge has the same magnitude even if q and Q are of different magnitude, as shown in Figure 16.18.

The units for quantifying the electric charge property of matter depend on the units used for measuring force and distance. In the SI system of units, the unit of charge is defined in terms of *charge flow* or **electric current**. The SI unit for electrical current is the *ampere* (A) and will be defined later more precisely (in Chapter 20). A current of one ampere means that one **coulomb** (C) of charge flows past a specified region during one second. The coulomb turns out to be a very big charge unit; smaller divisions, such as millicoulombs (mC) (= 10^{-3} C), microcoulombs (μC) (= 10^{-6} C), and nanocoulombs (nC) (= 10^{-9} C) also are employed.*

If we measure electric charge using the SI unit, the coulomb, the value of the proportionality constant in Coulomb's force law on each of two pointlike charges, Equation 16.1, is found to be

$$8.987\,552\,425 \times 10^{9}\ \text{N}\cdot\text{m}^2/\text{C}^2$$

For reasons of convenience that are not apparent at this juncture,† we write this proportionality constant in SI units in a peculiar way as

$$8.987\,552\,425 \times 10^{9}\ \text{N}\cdot\text{m}^2/\text{C}^2 \equiv \frac{1}{4\pi\varepsilon_0} \tag{16.2}$$

whose value usually is taken to be approximately

$$9.00 \times 10^{9}\ \text{N}\cdot\text{m}^2/\text{C}^2$$

unless very precise calculations must be made. The constant ε_0 thus has the value (this value is exact)‡

$$\begin{aligned}\varepsilon_0 &= \frac{1}{4\pi(8.987\,552\,425 \times 10^{9}\ \text{N}\cdot\text{m}^2/\text{C}^2)} \\ &\equiv 8.854\,187\,817\ldots \times 10^{-12}\ \text{C}^2/(\text{N}\cdot\text{m}^2)\end{aligned} \tag{16.3}$$

For abstruse reasons we will not pursue, ε_0 is called the **permittivity of free space**.

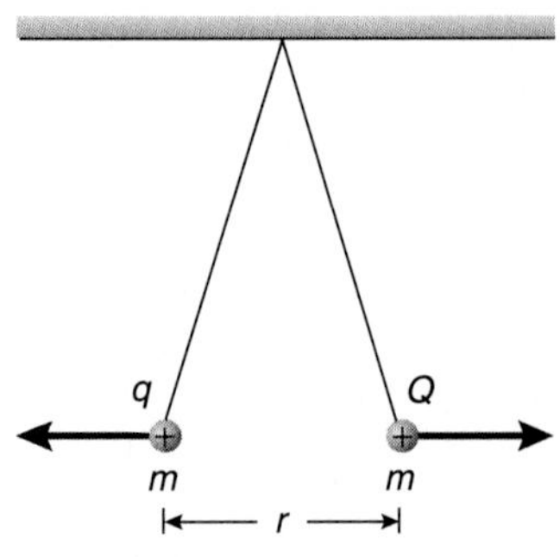

FIGURE 16.18 The electrical force on each of two charges satisfies Newton's third law.

Therefore, in the SI unit system, Coulomb's law for the *magnitude* of the force that two pointlike charges exert on each other is

$$F_{\text{elec}} = \frac{1}{4\pi\varepsilon_0}\frac{|q||Q|}{r^2} \tag{16.4}$$

Coulomb's law is very similar in form to the gravitational force law that gives the magnitude of the force that two pointlike masses m and M exert on each other:

$$F_{\text{grav}} = G\frac{Mm}{r^2}$$

If more than two pointlike charges are present, as in Figure 16.19, experiments indicate that the electrical force on any specific charge is found from the vector sum of the forces that *each* of the *other* charges exerts on the specific charge as if *only* that pair of charges were present; see Figure 16.20. The *total electrical force* on the given charge then is the vector sum of the electrical forces caused by the other charges, calculated as if each acted alone, as indicated in Figure 16.21.

This result is known as the **principle of superposition**. The principle is applicable as long as the charges $Q_1, Q_2, \ldots$ exerting forces on q are fixed in their positions—that is, *static*.

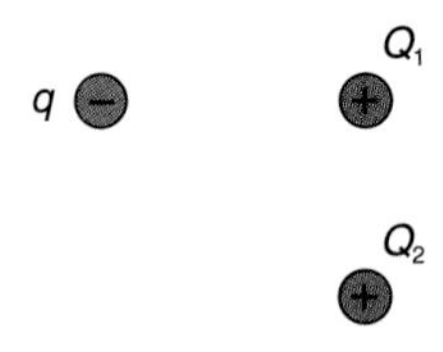

FIGURE 16.19 Three pointlike charges.

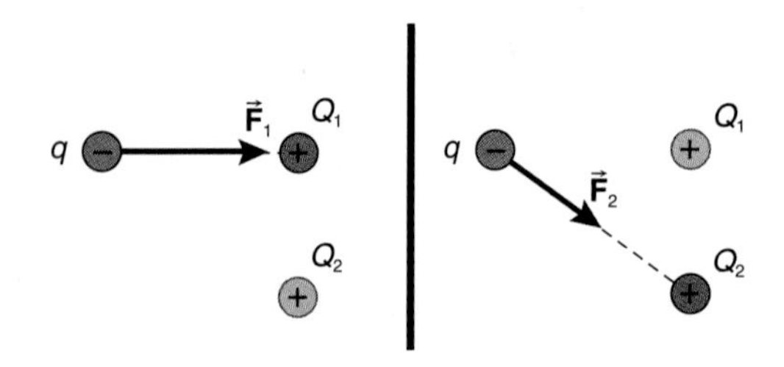

FIGURE 16.20 To find the force on q, find the force that each of the other charges individually exerts on q.

*The coulomb is such a large unit of charge that it is almost as awkward as assessing your tuition in units of the Gross Domestic Product (GDP), about a nanoGDP = 10^{-9} GDP.

† The reason has to do with making one of the fundamental equations of electromagnetism, Gauss's law for the electric field, look prettier. We examine Gauss's law in Section 16.11.

‡ The effect of defining the meter to make the speed of light exactly $2.997\,924\,58 \times 10^{8}$ m/s ripples into electromagnetism and makes the value of ε_0 exact as well; this connection may not be apparent to you at this juncture, so do not fret about it needlessly.

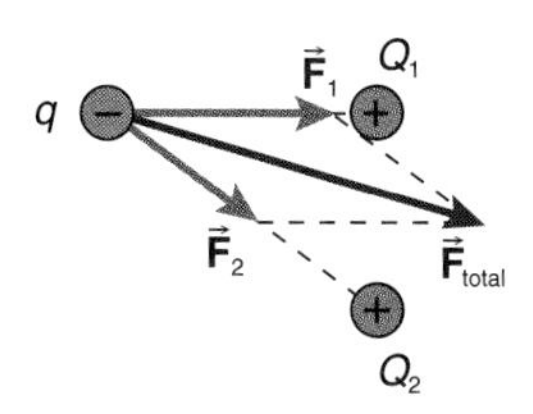

FIGURE 16.21 The total force on q is the vector sum of the individual forces on q.

PROBLEM-SOLVING TACTICS

16.1 When using Coulomb's law for the force between pointlike charges, be sure to express the distance between the charges in meters, not centimeters, and the charge in coulombs. Equation 16.4 is appropriate for SI units, which means the distance must be in the SI unit, meters, and the charge in coulombs.

16.2 To avoid many confusing problems with signs, employ the following procedure when using Coulomb's law. Calculate the *magnitude* of the force (a *positive scalar* quantity) using Coulomb's force law, Equation 16.4. Sketch the situation. Then determine the direction of the force on each charge after examining the types of the charges involved:

- if the charges are *both* positive *or both* negative (so-called like charges), the force on each is repulsive; or
- if the charges are of *opposite* flavors (unlike charges), the force on each is attractive.

Indicate on your sketch the direction of the forces on the charges. Finally, if appropriate, introduce a Cartesian coordinate system and write the forces in standard Cartesian vector form.

QUESTION 3

Two cats named Skimbleshanks and Mr. Mistoffolees are petted vigorously.* Can the electrical force of the cats on each other be calculated using Coulomb's law? Explain why or why not. What assumption must be made to calculate the approximate magnitude of the electrical force of the cats on each other? Why is this an approximation to the actual force?

EXAMPLE 16.1

A pointlike charge, $-2.00\ \mu C$, is located 15.0 cm from another pointlike charge, $+3.50\ \mu C$.

a. What is the magnitude of the electrical force that each exerts on the other?
b. Sketch the situation and indicate the direction of the electrical force on each charge.

Solution

a. Since the charges are pointlike, use Coulomb's law (Equation 16.4) for the magnitude of the force that each exerts on the other. Keep in mind Problem-Solving Tactic 16.1—use meters for the distance and coulombs for the charge:

$$F_{elec} = \frac{1}{4\pi\varepsilon_0}\frac{|q||Q|}{r^2}$$
$$= (9.00 \times 10^9\ N\cdot m^2/C^2) \times \frac{|-2.00 \times 10^{-6}\ C||3.50 \times 10^{-6}\ C|}{(0.150\ m)^2}$$
$$= 2.80\ N$$

b. Since the charges are unlike charges, the force that each exerts on the other is an attractive force, as indicated in Figure 16.22. Notice that the force each exerts on the other satisfies Newton's third law.

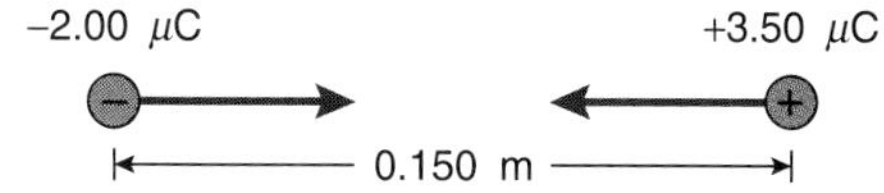

FIGURE 16.22

EXAMPLE 16.2

Two pointlike charges are located as shown in Figure 16.23.

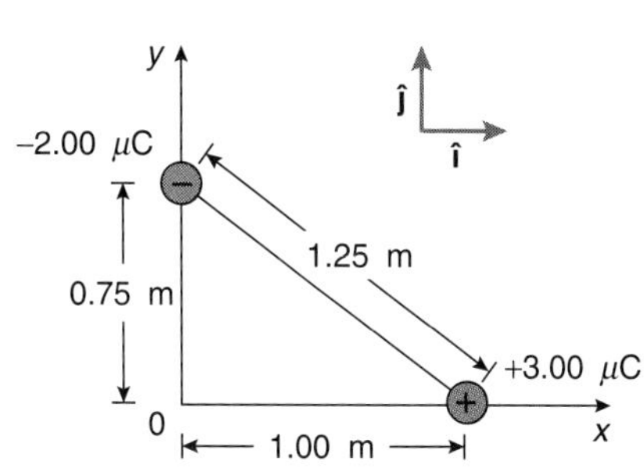

FIGURE 16.23

a. Find the electrical force that each charge exerts on the other and express these forces in terms of the indicated coordinate system.
b. Do these forces represent a Newton's third law force pair?

Solution

a. The magnitude of the electrical force that each charge exerts on the other is found by using the absolute values of the charges and substituting into Coulomb's law, Equation 16.4. Distances must be expressed in meters and the charges in coulombs (Problem-Solving Tactic 16.1):

$$F = \frac{1}{4\pi\varepsilon_0}\frac{|q||Q|}{r^2}$$
$$= (9.00 \times 10^9\ N\cdot m^2/C^2) \times \frac{|-2.00 \times 10^{-6}\ C||3.00 \times 10^{-6}\ C|}{(1.25\ m)^2}$$
$$= 3.46 \times 10^{-2}\ N$$

*The names are from T. S. Eliot, *Old Possum's Book of Practical Cats* (Faber and Faber, London, 1939).

Following Problem-Solving Tactic 16.2, the pair of charges are unlike charges (one positive, one negative), and so the forces are attractive and are directed as indicated in Figure 16.24.

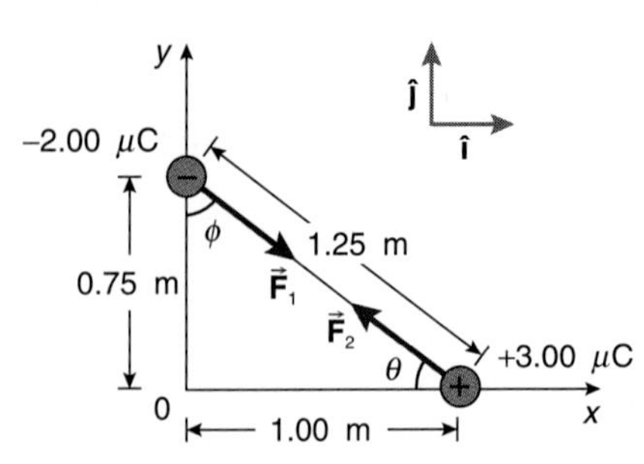

FIGURE 16.24

The forces are expressed in terms of the indicated coordinate axes by writing the force vectors in terms of the Cartesian unit vectors:

$$\begin{aligned}\vec{F}_1 &= (F \sin \phi)\hat{\imath} - (F \cos \phi)\hat{\jmath} \\ &= (3.46 \times 10^{-2}\ \text{N})(0.800)\hat{\imath} - (3.46 \times 10^{-2}\ \text{N})(0.600)\hat{\jmath} \\ &= (2.77 \times 10^{-2}\ \text{N})\hat{\imath} - (2.08 \times 10^{-2}\ \text{N})\hat{\jmath}\end{aligned}$$

Likewise

$$\begin{aligned}\vec{F}_2 &= (-F \cos \theta)\hat{\imath} + (F \sin \theta)\hat{\jmath} \\ &= (-3.46 \times 10^{-2}\ \text{N})(0.800)\hat{\imath} + (3.46 \times 10^{-2}\ \text{N})(0.600)\hat{\jmath} \\ &= (-2.77 \times 10^{-2}\ \text{N})\hat{\imath} + (2.08 \times 10^{-2}\ \text{N})\hat{\jmath}\end{aligned}$$

b. Notice that $\vec{F}_1 = -\vec{F}_2$. The forces form a Newton's third law force pair because they represent the forces that each charge exerts on the other. The forces are of equal magnitude and opposite direction and act on different systems.

STRATEGIC EXAMPLE 16.3

Three charges are located at fixed positions as indicated in Figure 16.25.

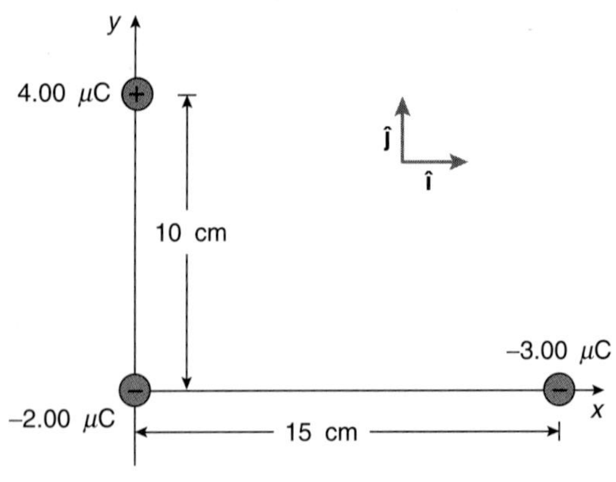

FIGURE 16.25

a. Find the total electrical force on the charge at the origin.
b. Find the magnitude of the total force on this charge.
c. Find the angle that the total force makes with the positive *x*-axis.

Solution

a. Use Coulomb's law and the principle of superposition. You *must* use SI units to use Coulomb's law as given by Equation 16.4. Hence the *distances must be expressed in meters and the charges in coulombs* (Problem-Solving Tactic 16.1).

1. First, find the magnitude of the force $\vec{F}_1$ that the charge out along the *x*-axis exerts on the charge at the origin, pretending that the other charge (the one out along the *y*-axis) is not present; see Figure 16.26 The magnitude of the force is found by using Coulomb's law (Equation 16.4):

$$F_1 = \frac{1}{4\pi\varepsilon_0}\frac{|q||Q|}{r^2}$$

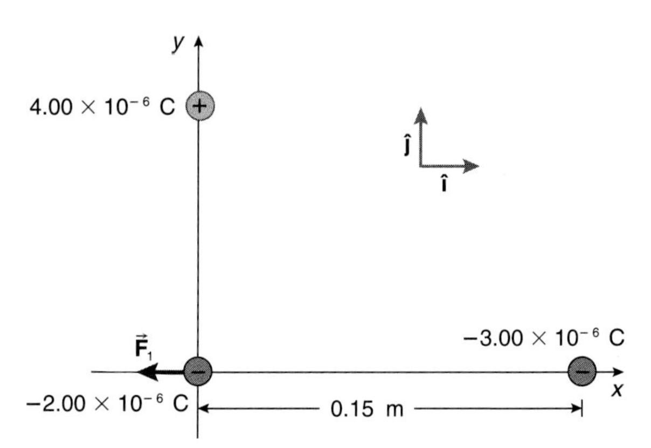

FIGURE 16.26

We take q to be the charge at the origin and Q to be the charge out along the *x*-axis. Making the substitutions, you get

$$\begin{aligned}F_1 &= (9.00 \times 10^9\ \text{N}\cdot\text{m}^2/\text{C}^2) \\ &\quad \times \frac{|-2.00 \times 10^{-6}\ \text{C}||-3.00 \times 10^{-6}\ \text{C}|}{(0.15\ \text{m})^2} \\ &= 2.4\ \text{N}\end{aligned}$$

Since the charge out along the *x*-axis and the charge at the origin are like charges (both are negative), the force on each is repulsive; hence, the force on the charge at the origin is directed toward $-\hat{\imath}$. Thus the force that the charge out along the *x*-axis exerts on the charge at the origin is

$$\vec{F}_1 = (-2.4\ \text{N})\hat{\imath}$$

2. Next, find the magnitude of the force $\vec{F}_2$ exerted by the charge out along the *y*-axis on the charge at the origin, pretending that the other charge (the charge out along the *x*-axis) is not present, as shown in Figure 16.27. The magnitude of the force is found using Coulomb's law (Equation 16.4):

$$F_2 = \frac{1}{4\pi\varepsilon_0}\frac{|q||Q|}{r^2}$$

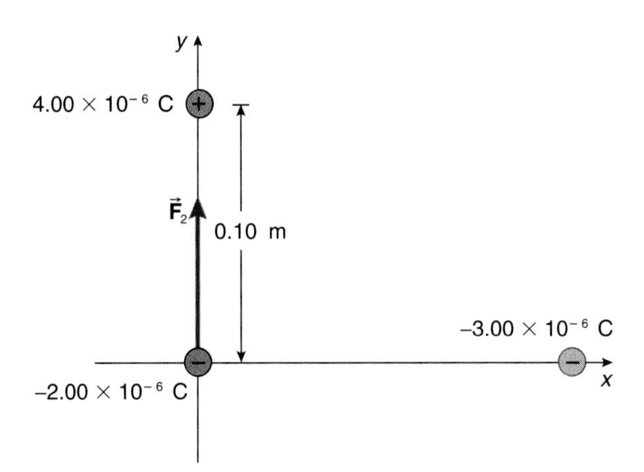

FIGURE 16.27

where we again let q be the charge at the origin and now let Q be the charge out along the y-axis. Making the substitutions,

$$F_2 = (9.00 \times 10^9 \text{ N}\cdot\text{m}^2/\text{C}^2) \times \frac{|-2.00 \times 10^{-6} \text{ C}||4.00 \times 10^{-6} \text{ C}|}{(0.10 \text{ m})^2} = 7.2 \text{ N}$$

Since the charge out along the y-axis and the charge at the origin are unlike charges (one is positive, the other negative), the force of these two charges on each other is attractive. Thus the force on the charge at the origin is directed toward $+\hat{\jmath}$. Hence

$$\vec{F}_2 = (7.2 \text{ N})\hat{\jmath}$$

3. The total force on the charge at the origin is the *vector sum* of the forces $\vec{F}_1$ and $\vec{F}_2$ (see Figure 16.28):

$$\vec{F}_{\text{total}} = \vec{F}_1 + \vec{F}_2 = (-2.4 \text{ N})\hat{\imath} + (7.2 \text{ N})\hat{\jmath}$$

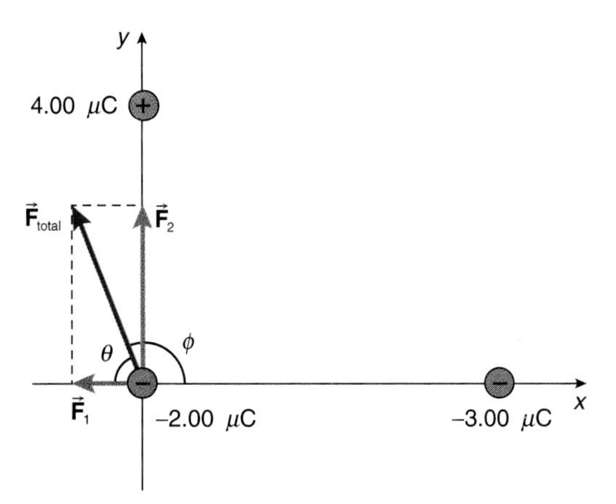

FIGURE 16.28

b. The total force is of magnitude

$$F_{\text{total}} = [(-2.4 \text{ N})^2 + (7.2 \text{ N})^2]^{1/2} = 7.6 \text{ N}$$

c. The total force makes an angle θ with the *negative* x-axis that is found from (see Figure 16.28)

$$\tan\theta = \frac{|F_{y\,\text{total}}|}{|F_{x\,\text{total}}|} = \frac{7.2 \text{ N}}{2.4 \text{ N}} = 3.0$$

or

$$\theta = 72°$$

The angle ϕ that the total force makes with the *positive* x-axis is

$$\phi = 180° - \theta$$

or

$$\phi = 108°$$

16.4 CHARGE QUANTIZATION

I believe there are 15,747,724,136,275,002,577,605,653,961,181,555,468, 044,717,914,527,116,709,366,231,425,076,185,631,031,296 protons in the universe, and the same number of electrons.

Arthur Eddington (1882–1944)*

The development of chemistry during the 19th century, particularly the concept of valence associated with chemical bonding, the rise of the atomic view of matter with developments in the kinetic theory of gases (Chapter 14), as well as the discovery of charged particle radiation and radioactivity in 1896, led to a search for an elementary, smallest quantity of charge. The search finally bore fruit with the work of Robert A. Millikan (1868–1953) in an epic series of experiments begun in 1909.

By means of tedious, eye-straining microscopic observations of individual tiny oil droplets, charged by squirting (rubbing) them through a perfume bottle atomizer,† Millikan discovered that every tiny droplet carried only a positive or negative *integral* multiple of a small, elemental amount of charge. Millikan was awarded the 1923 Nobel prize in physics for this work. We now call the magnitude of this elemental quantity of the electrification property (i.e., charge) the **fundamental unit of electric charge**, designated‡ e; the unit of elemental charge e is considered intrinsically positive. The value of the fundamental unit of charge, in SI units, is approximately

$$e = 1.602\,177 \times 10^{-19} \text{ C} \tag{16.5}$$

The Philosophy of Physical Science (Macmillan, New York, 1939), page 170.

†He could change the charge on each drop by using radiation from radium.

‡Do not confuse the fundamental unit of charge e with the base of natural logarithms, $e = 2.718\,282\ldots$

Coming from a classic midwestern farm background, Millikan wrote a widely used physics text and from there took on the difficult challenge of determining if an elemental quantity of charge existed.

This elemental amount of charge e is the magnitude of the charge on both the electron and proton.

All free electric charges in nature are *integral multiples* of the fundamental unit of electric charge e. We say charge is **quantized,** meaning that only certain values of charge are permitted or allowed to exist.

Charge comes in integral multiples of the discrete and elemental bit of charge e. Any isolated particle in nature has a charge q that is

$$q = ne \qquad (16.6)$$

where n is a positive or negative integer, known as the **charge quantum number.** The electron has a charge $q = -e$, and the proton has a charge $q = +e$, so their charge quantum numbers are -1 and $+1$ respectively. Particles or masses with zero charge (such as the neutron) have a charge quantum number $n = 0$. Masses with zero total electric charge are said to be electrically **neutral.** The proton and electron therefore can be viewed, respectively, as having one charge unit e added to or subtracted from a state of zero charge. The equality of the magnitudes of the charges on the proton and electron thus comes about naturally from this quantum viewpoint.

Contemporary physics views the proton and neutron (and other particles called hadrons) as *composite* particles, made of more elementary particles called quarks. Quarks have *fractional* units of the fundamental unit of charge: $\pm e/3$ and $\pm 2e/3$, thus seeming to violate charge quantization. In this model the proton consists of two quarks with charge $+2e/3$, whimsically called *up* quarks, and one quark with charge $-e/3$, called a *down* quark, for a total charge of

$$\frac{2}{3}e + \frac{2}{3}e - \frac{1}{3}e = e$$

The neutron is composed of one up quark with charge $+2e/3$ and two down quarks (each with charge $-e/3$), for a total charge of

$$\frac{2}{3}e - \frac{1}{3}e - \frac{1}{3}e = 0 \text{ C}$$

Isolated quarks have *never* been observed and there are theoretical reasons to think they never will be observed. Thus the statement that *isolated* particles have integral multiples of the fundamental unit of charge still holds; the quark model does not violate the charge quantization hypothesis.

The amount of charge e is very small. For macroscopic quantities of charge, the charge quantum number may be so huge that the discrete nature of charge is undetectable and the quantity of charge is considered essentially continuous in the mathematical sense; mass has similar properties on the microscopic and macroscopic scales.

PROBLEM-SOLVING TACTIC

16.3 Remember that we treat the fundamental unit of charge $e = 1.602 \times 10^{-19}$ C as positive, and so the charge on the proton is $+e$, while that on the electron is $-e$.

QUESTION 4

Explain the difference between the terms charge quantification and charge quantization.

EXAMPLE 16.4

What is the charge quantum number of a particle with a charge of $+1.00\ \mu$C?

Solution

Since charge is quantized, use Equation 16.6:

$$q = ne$$

The charge quantum number n is

$$n = \frac{q}{e} = \frac{1.00 \times 10^{-6}\ \text{C}}{1.602 \times 10^{-19}\ \text{C}} = 6.24 \times 10^{12}$$

This is a huge charge quantum number and demonstrates why charge quantization typically is not detectable with charges produced by macroscopic methods such as rubbing or induction.

EXAMPLE 16.5

It is interesting to compare the magnitude of the repulsive electrical force that two electrons exert on each other when they are a distance r apart with the magnitude of the attractive gravitational force of the two for each other at the same separation. Calculate the numerical value of this ratio.

Solution

The magnitude of the electrical force is found using Coulomb's law, Equation 16.4:

$$F_{\text{elec}} = \frac{1}{4\pi\varepsilon_0}\frac{|-e||-e|}{r^2}$$

The magnitude of the gravitational force between the two electrons is

$$F_{\text{grav}} = G\frac{m_{\text{electron}}m_{\text{electron}}}{r^2}$$

Thus the ratio of the magnitudes of the forces is

$$\frac{F_{\text{elec}}}{F_{\text{grav}}} = \frac{1}{4\pi\varepsilon_0}\frac{e^2}{Gm_{\text{electron}}^2}$$

Notice that the ratio is *independent* of the separation of the two electrons. The ratio thus is a measure of the relative intrinsic strength of the electric and the gravitational forces.

Substituting numerical values for the various quantities, you find the ratio is

$$\frac{F_{\text{elec}}}{F_{\text{grav}}} = 4.17 \times 10^{42}$$

This incredibly large number means that the electrical force repelling the electrons is about 10^{42} times stronger than their attractive gravitational force. If we represent the gravitational force vector by an arrow a mere 1.00 cm long, as in Figure 16.29, the arrow representing the electrical force is 10^{42} cm long; 10^{42} cm = 10^{40} m, *far, far greater* than the size of the observable universe ($\sim 10^{26}$ m)! Needless to say, we cannot draw this arrow to the appropriate scale in Figure 16.29.

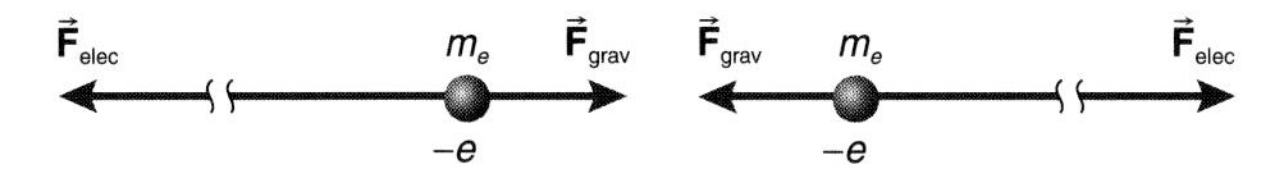

FIGURE 16.29

In the hydrogen atom, if you compute the ratio of the magnitude of the electrical force of attraction that a proton and an electron exert on each other to the magnitude of their mutual gravitational force of attraction, the result is 2.28×10^{39}, still so large that we can comfortably neglect the gravitational force. As a result, *the gravitational force is irrelevant in most atomic and molecular physics*. It is the electrical force that dominates the interactions.

EXAMPLE 16.6

The Bohr model of the electrically neutral hydrogen atom consists of an electron orbiting a single proton. The proton is the only charged constituent of the nucleus of this atom. Actually *both* the electron *and* the nucleus orbit the center of mass of the two-particle system. However, since the mass of the proton is 1836 times that of the electron, the center of mass of the system is essentially coincident with the proton. Thus we say the electron orbits the proton (thinking of the latter as relatively fixed in position). The electron has a number of different possible (or allowed) orbital paths around the central nucleus, but let us consider the electron to be in the orbit closest to the nucleus. This orbit has an average radius of 5.29×10^{-11} m.

a. Find the magnitude and direction of the electrical force on the electron in the hydrogen atom when it is in this orbit.

b. Calculate the magnitude of the acceleration of the electron in this orbit, and compare this magnitude with that of the local acceleration due to gravity ($g = 9.81$ m/s^2).

Solution

a. The *magnitude* of the force on the electron is found using Coulomb's law, Equation 16.4:

$$F = \frac{1}{4\pi\varepsilon_0}\frac{|q||Q|}{r^2}$$

Since the electron and the proton each have the same *magnitude* of charge (the fundamental unit of charge e), the magnitude of the force is

$$F = \frac{1}{4\pi\varepsilon_0}\frac{e^2}{r^2}$$

Substituting numerical values for the various quantities, you obtain

$$F = \frac{(9.00 \times 10^9 \text{ N}\cdot\text{m}^2/\text{C}^2)(1.602 \times 10^{-19} \text{ C})^2}{(5.29 \times 10^{-11} \text{ m})^2} = 8.25 \times 10^{-8} \text{ N}$$

This is the magnitude of the force *on* the electron *by* the proton (as well as the force of the electron on the proton, thanks to Newton's third law). The force on the electron is directed radially toward the proton because the two are unlike charges.

b. The electrical force is essentially the *only* relevant force acting on the electron since the gravitational force between them is so small; see Example 16.5. Newton's second law enables you to find the magnitude of the acceleration of the electron. In terms of the magnitudes of the total force and the acceleration,

$$F_{\text{elec}} = m_e a$$

The acceleration is parallel to the total force and so is directed toward the nucleus. The magnitude of the acceleration thus is

$$a = \frac{F_{\text{elec}}}{m_e}$$

Use the result of part (a) for the electrical force on the electron and substitute for the mass of the electron; you obtain

$$a = \frac{8.25 \times 10^{-8} \text{ N}}{9.11 \times 10^{-31} \text{ kg}} = 9.06 \times 10^{22} \text{ m/s}^2$$

Dividing by the magnitude of the local acceleration due to gravity ($g = 9.81$ m/s^2), you find the ratio of the accelerations is

$$\frac{a}{g} = \frac{9.06 \times 10^{22} \text{ m/s}^2}{9.81 \text{ m/s}^2} = 9.24 \times 10^{21}$$

The magnitude of the acceleration of the electron in the hydrogen atom thus is on the order of 10^{22} so-called "gs." That is *some* acceleration!

16.5 THE ELECTRIC FIELD OF STATIC CHARGES

The Coulomb force law between charges is similar in form to the gravitational force law. We exploit the similarity by recalling several features of the gravitational force that we studied in Chapter 6.

When we discussed gravitational forces in Chapter 6, we saw that a mass m experiences a gravitational force because of the presence of other masses. The effect of these other masses on m is conveyed by a gravitational field $\vec{g}$ created in space by the other masses. The gravitational field $\vec{g}$ at a point in space was defined as the gravitational force per unit mass at the point in question:

$$\vec{F}_{\text{grav on } m} \equiv m\vec{g} \tag{16.7}$$

The gravitational field is the free-fall acceleration due to gravity at the point in question. In Chapter 6 we also found that the specific expression for the gravitational field depends on the shape and distribution of the masses creating the field, *not* the mass m placed *in* the field. The gravitational fields of various mass distributions were summarized in Table 6.1. The only condition subtly attached to the definition of the gravitational field, Equation 16.7, is the assumption that the presence of m *does not alter the shape or distribution of the masses creating the field.* This is typically the case in most gravitational situations,* because in many instances the mass m that we place in the gravitational field is much smaller than the mass that creates the field into which m is placed. For example, the masses m that we place in the gravitational field of the Earth, such as rocks and rockets, humans and heros, beams and beans, even the Moon, have masses that are much less than that of the Earth. These small masses do not affect the distribution of mass on or within the Earth.† Likewise, planetary masses m in the gravitational field of the Sun are very small compared with the mass of the Sun, and so the presence of a planetary mass does not affect the mass distribution within the Sun.

Analogously to gravitation, we introduce an **electric field $\vec{E}$** to convey the electrical force.

> The electric field $\vec{E}$ at a point in space is a vector defined via
>
> $$\boxed{\vec{F}_{\text{elec on } q} \equiv q_{\text{test}}\vec{E}} \tag{16.8}$$
>
> where $\vec{F}_{\text{elec on } q}$ is the electrical force on q_{test} at the point where it is located.

The electric field at the point in space where q_{test} is placed is created by other charges, not q_{test} itself, and so the value of the field at a point in space does not depend on the value of the charge placed in the field. We consider the charges that create the field to be fixed in space, and so we call them **static charges** and the field $\vec{E}$ an **electrostatic field**. The field created by these static charges depends on their locations and distributions in space (just as the specific form for the gravitational field depends on the shape and distribution of the mass creating the field, as indicated in Table 6.1).

*Binary stars with small spatial separations are exceptions.

†The Moon does cause very small tidal bulges on the fluid oceans of the Earth. The heights of these familiar tidal bulges are approximately 2 m.

The reason for introducing the field concept is the same as with the gravitational field: we can talk about the electrical effects of the static charges creating the field, without the presence of another charge (q_{test}) complicating the picture. We think of the field as conveying or transmitting the force to the mass or charge placed in the gravitational or electric field respectively. In the gravitational case, we could talk about the gravitational field of the Earth (the gravitational force per kilogram, or equivalently, the local acceleration due to gravity) without necessarily having to place a mass m in the field.

What is the reason for the subscript "test" on q_{test} in the defining Equation 16.8? Just as in gravitation, we need to ensure that q_{test} is small enough so that its presence does not alter the locations or distribution of the other charges that are producing the field. Since charges on conductors readily move, *this condition is not as easily ensured with electrical phenomena* as it is with gravitation. But we can imagine q_{test} to be a very small magnitude of charge that innocently tests or determines the value of the field at a point in space through the force on it at that location.‡

> From the defining Equation 16.8, we see that the electric field at a point in space is the force per coulomb (force on one coulomb) at that point. Thus the SI units of the electric field $\vec{E}$ are N/C.

However, the coulomb is such a large quantity of charge that, in practice, q_{test} almost never can be considered on the order of a coulomb.

> For our treatment of electrical phenomena, we will always assume that the presence of a test charge q_{test} in an electric field $\vec{E}$ does not alter the distribution or motion of the charges creating the field.

Thus all electric fields we consider in this chapter are electrostatic fields. This assumption ensures that the force on a charge q_{test} placed in an electric field $\vec{E}$ (created in space by other charges, not q_{test} itself) can be found using Equation 16.8. So we will eliminate the subscript on the test charge placed in the electric field and simply call it $q \equiv q_{\text{test}}$.

> It is important to realize that the charge q placed in the field need not be fixed or static but may accelerate in response to the electrical force it experiences. It is all the many *other* charges that create the field that are static, not necessarily the charge placed in the field of these static charges.

> Just as the gravitational field depends on the shape and distribution of the masses producing the field (e.g., the shape and distribution of mass within the Earth), the electric field $\vec{E}$ depends on the specific arrangement of the charges that create the field.

‡It is the same with gravitation. We *measure* the gravitational field (the force per kilogram or the local acceleration due to gravity) only by placing a test mass m in the gravitational field and measuring the force on or the acceleration of m.

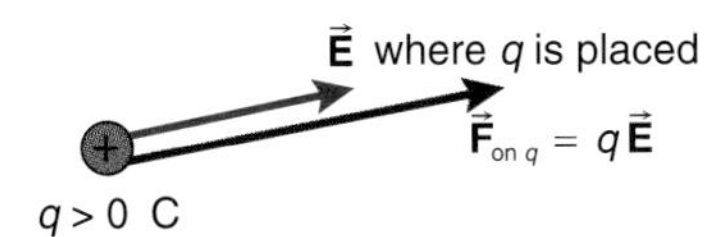

FIGURE 16.30 A positive charge placed in an electric field experiences an electrical force parallel to the field.

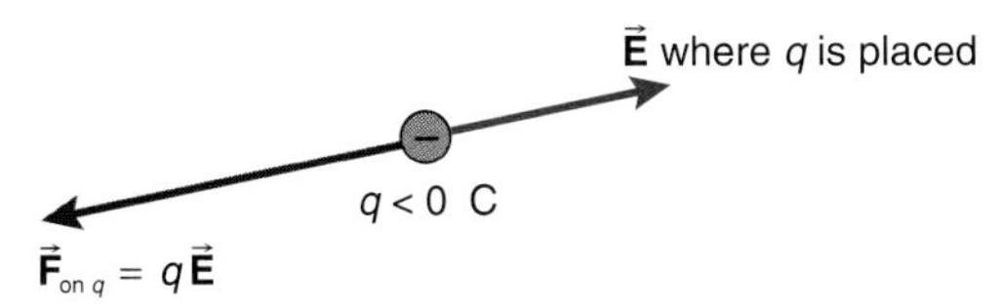

FIGURE 16.31 A negative charge placed in an electric field experiences a force antiparallel to the field.

We will later see how to calculate the electric field for certain distributions of charges. Whatever the field is, the force on *another* charge q, placed *in* the field created by these other charges, is found using Equation 16.8 (as long as the presence of q does not affect the distribution of charge creating the field in which q is placed*).

In the gravitational case (Equation 16.7), the gravitational force on a mass m is *always parallel* to the gravitational field because there is only one type of mass: positive mass. For the electrical case, the direction of the force relative to the field depends on the sign of the charge q placed in the field. If the charge q is *positive*, the electrical force on q is *parallel* to the field $\vec{E}$; as shown in Figure 16.30, the right-hand side of Equation 16.8 then is just a positive scalar times the vector $\vec{E}$. On the other hand, if the charge q is *negative*, the right-hand side of Equation 16.8 is a negative scalar times the vector $\vec{E}$; thus the electrical force on a negative charge q is directed *antiparallel* to $\vec{E}$, as shown in Figure 16.31.

PROBLEM-SOLVING TACTIC

16.4 The algebraic sign of the charge q is very important in Equation 16.8. The sign of the charge q determines whether the force is directed parallel (q positive) or antiparallel (q negative) to the direction of the electric field.

QUESTION 5

The gravitational field has two interpretations: (1) the force per kilogram at a point in space, and (2) the local acceleration of gravity at the point in space. The electric field is defined as the electrical force per coulomb at a point in space. Can the electric field also be interpreted as an acceleration of the charged mass caused by electrical forces? Explain your answer.

EXAMPLE 16.7

a. A proton finds itself in a uniform electric field of magnitude 125 N/C directed as indicated in Figure 16.32. Find the force on the proton.

*If the presence of q *does* alter the distribution of charges creating the field, then q experiences the field produced by the *new arrangement* of the original charges. For our purposes, we will never consider such effects.

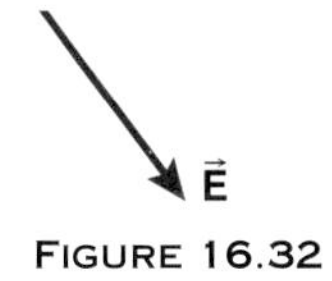

FIGURE 16.32

b. Repeat the problem if the charge placed in the field is an electron.

Solution

Introduce the coordinate system shown in Figure 16.33. With this coordinate system, the electric field vector is

$$\vec{E} = (-125 \text{ N/C})\hat{j}$$

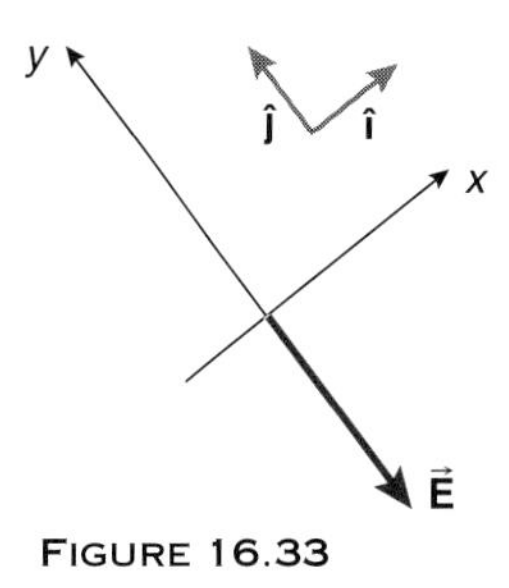

FIGURE 16.33

a. The force on the proton is found using Equation 16.8:

$$\vec{F} = q\vec{E}$$

where $q = +1.602 \times 10^{-19}$ C. Make the appropriate substitutions:

$$\begin{aligned}\vec{F} &= (+1.602 \times 10^{-19} \text{ C})[(-125 \text{ N/C})\hat{j}] \\ &= (-2.00 \times 10^{-17} \text{ N})\hat{j}\end{aligned}$$

Notice that the force on the positively charged proton is *parallel* to the direction of the electric field in which it is placed (see Figure 16.34), in keeping with Equation 16.8 and Problem-Solving Tactic 16.4.

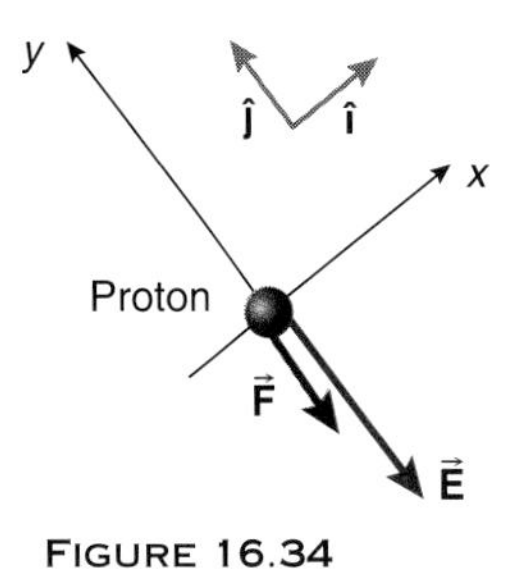

FIGURE 16.34

b. The force on an electron also is found using Equation 16.8:

$$\vec{F} = q\vec{E}$$

But now $q = -1.602 \times 10^{-19}$ C and $\vec{\mathbf{E}}$ is unchanged: $(-125 \text{ N/C})\hat{\jmath}$. With these substitutions, you find

$$\begin{aligned}\vec{\mathbf{F}} &= (-1.602 \times 10^{-19} \text{ C})[(-125 \text{ N/C})\hat{\jmath}] \\ &= (2.00 \times 10^{-17} \text{ N})\hat{\jmath}\end{aligned}$$

Notice that the force on the negatively charged electron is *antiparallel* to the direction of the electric field in which it is placed (see Figure 16.35), in keeping with Equation 16.8 and Problem-Solving Tactic 16.4.

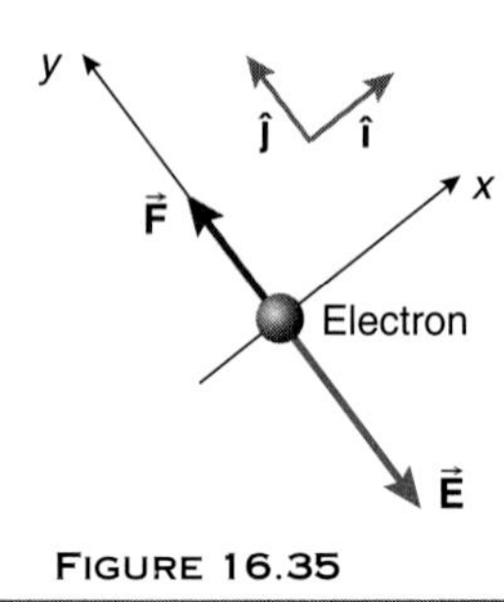

FIGURE 16.35

16.6 The Electric Field of Pointlike Charge Distributions

Protons and electrons are essentially pointlike charges, and so it is of some import in physics and chemistry to know how to find the electric field of pointlike charges. In this section we determine the electric field of a static pointlike charge Q.

We first assume Q is positive. Another positive pointlike charge q finds itself in the pleasant company of Q, as shown in Figure 16.36. Since the charges are like charges (here both positive), q experiences a repulsive electrical force caused by Q.

The magnitude of the force on q is found using Coulomb's law:

$$F_{\text{elec on } q} = \frac{1}{4\pi\varepsilon_0}\frac{|q||Q|}{r^2} \qquad (16.9)$$

The absolute value signs here are redundant since both charges are positive. The charge Q also experiences a force of the same magnitude (but in the opposite direction, according to Newton's third law of motion). Here we are *not* interested in the force on Q but, rather, the force *of Q on q*.

We want to look at the force on q differently. The fixed charge Q establishes an electric field $\vec{\mathbf{E}}$ in the surrounding space. The field then acts on the charge q that is placed in the field of Q. According to this view, the force on q when it is placed in the field $\vec{\mathbf{E}}$ is found from the defining Equation 16.8:

$$\vec{\mathbf{F}}_{\text{elec on } q} = q\vec{\mathbf{E}}_{\text{of } Q}$$

The magnitude of the force is

$$F_{\text{elec on } q} = |q| E_{\text{of } Q} \qquad (16.10)$$

We equate Equations 16.9 and 16.10 for the magnitude of the force on q since they represent the same force:

$$|q| E_{\text{of } Q} = \frac{1}{4\pi\varepsilon_0}\frac{|q||Q|}{r^2}$$

Hence we find that the magnitude of the electric field created by the positive point charge Q is

$$E_{\text{of } Q} = \frac{1}{4\pi\varepsilon_0}\frac{|Q|}{r^2}$$

Notice that the magnitude of the electric field of Q depends only on Q, and not on the charge q that was placed in the field.

What about the direction of the electric field created by Q at the point where q was placed? Since the charge q placed in the field $\vec{\mathbf{E}}$ is positive, the force on q is *parallel* to the direction of the field $\vec{\mathbf{E}}$ according to the defining equation for the field (Equation 16.8): $\vec{\mathbf{F}}_{\text{elec on } q} = q\vec{\mathbf{E}}_{\text{of } Q}$. Since the force on q is repulsive, we see that the electric field created by the positive point charge Q must point *radially away* from positive charge Q, as shown in Figure 16.37.

Thus the electric field of the positive charge Q at a distance r from Q can be written in vector form as

$$\vec{\mathbf{E}} = \frac{1}{4\pi\varepsilon_0}\frac{Q}{r^2}\hat{\mathbf{r}} \qquad (16.11)$$

where $\hat{\mathbf{r}}$ is a unit vector pointing in a radial direction away from Q.

If the positive charge q is placed in the vicinity of a *negative* charge Q (i.e., $Q < 0$ C), the electrical force on q now is an *attractive* force since the two are unlike charges (one is positive, one is negative), as indicated in Figure 16.38. In this case the electrical force on q when it is placed in the field of Q still is given by Equation 16.8: $\vec{\mathbf{F}}_{\text{elec on } q} = q\vec{\mathbf{E}}_{\text{of } Q}$. Since q is positive, the force on q is *parallel* to the field of Q. The force is attractive. Thus the electric field of a negative charge Q must point *toward* Q, as indicated in Figure 16.39.

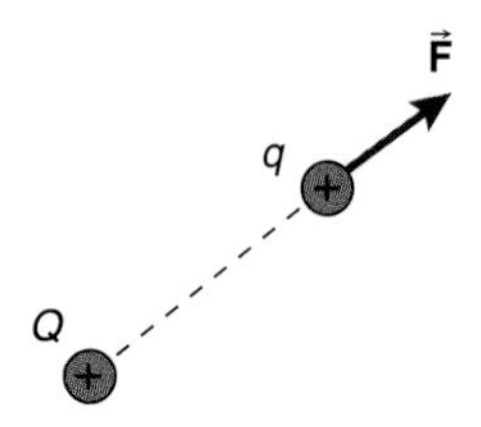

FIGURE 16.36 Positive charge q finds itself near a static, positive charge Q.

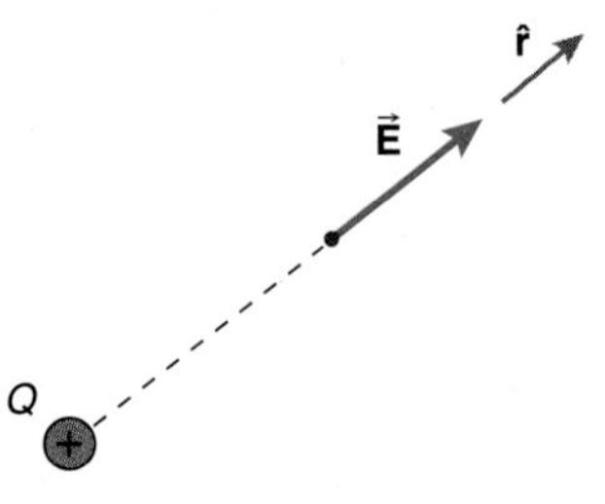

FIGURE 16.37 The electric field of a positive, pointlike charge Q points radially away from Q.

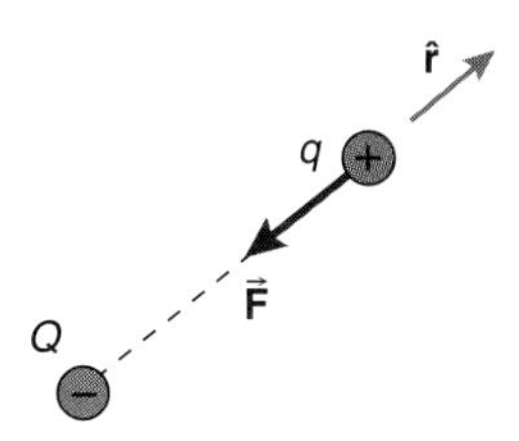

FIGURE 16.38 Positive charge q is attracted to negative charge Q.

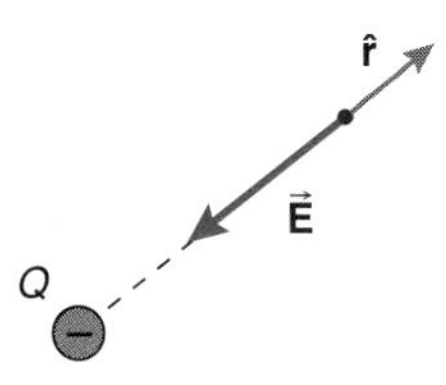

FIGURE 16.39 The electric field produced by a negative, pointlike charge Q is directed toward Q.

Notice that in Equation 16.11, if Q is negative, we have

$$\vec{\mathbf{E}} = -\frac{1}{4\pi\varepsilon_0}\frac{|Q|}{r^2}\hat{\mathbf{r}}$$

and so the field of Q is directed radially *toward* Q, in the direction of $-\hat{\mathbf{r}}$.

Therefore Equation 16.11 gives the electric field of a point charge Q for both positive and negative charges Q.

This result is very nice! Notice also that since we have already defined $\vec{\mathbf{E}}$ in terms of the force on a positive test charge, we can come back with any new test charge q of either sign and $q\vec{\mathbf{E}}$ will give the correct direction (and magnitude) of the force on q by the field $\vec{\mathbf{E}}$.

PROBLEM-SOLVING TACTIC

16.5 Use the following three-step procedure to find the electric field a distance r from a pointlike charge.

1. First find the *magnitude* of the field at the point in question:

$$E = \frac{1}{4\pi\varepsilon_0}\frac{|Q|}{r^2}$$

This always is a positive quantity, because it is the magnitude of a vector.

2. Determine the direction of the field at the desired point and indicate it in a sketch:
   - radially away from Q if Q is positive
   - radially toward Q if Q is negative
3. Finally, introduce a coordinate system and write the electric field vector $\vec{\mathbf{E}}$ in terms of your chosen coordinate system.

If several pointlike charges Q_1, Q_2, . . . are distributed in space (see Figure 16.40) and we need the field of the collection at a point P in space, we use the principle of superposition. The

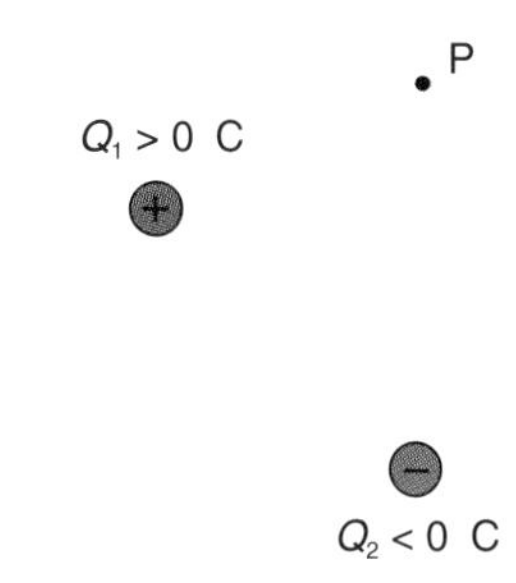

FIGURE 16.40 Several pointlike charges.

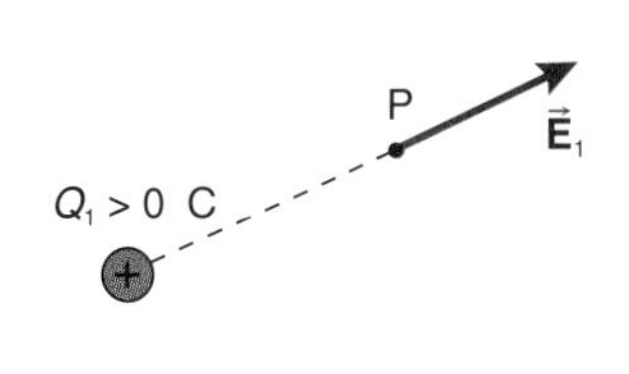

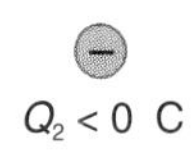

FIGURE 16.41 Charge Q_1 produces electric field $\vec{\mathbf{E}}_1$ at the given point P.

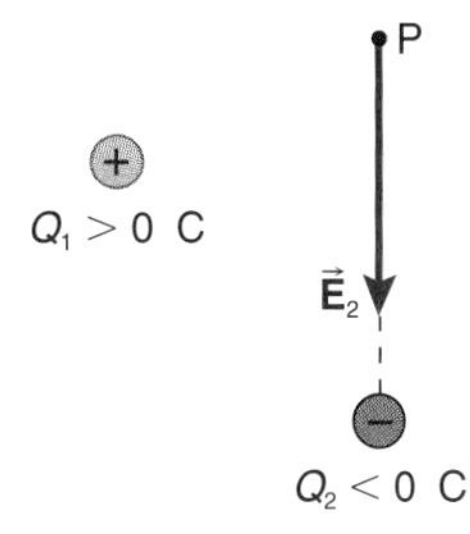

FIGURE 16.42 Charge Q_2 produces an electric field $\vec{\mathbf{E}}_2$ at point P.

charge Q_1 *acting alone* produces an electric field $\vec{\mathbf{E}}_1$ at point P, as indicated in Figure 16.41. Charge Q_2 *acting alone* produces an electric field $\vec{\mathbf{E}}_2$ at the *same* point P; see Figure 16.42.

The total electric field $\vec{\mathbf{E}}_{\text{total}}$ at the point when *all* the pointlike charges are present simultaneously at their fixed positions is the *vector sum* of the individual fields (see Figure 16.43):

$$\vec{\mathbf{E}}_{\text{total}} = \vec{\mathbf{E}}_1 + \vec{\mathbf{E}}_2 + \cdots \tag{16.12}$$

The superposition of fields stems from the underlying principle of superposition associated with the forces: the total force is the vector sum of the individual forces acting on the system.

As always, a charge q, whether positive or negative, placed in this field experiences a force found from Equation 16.8:

$$\vec{\mathbf{F}}_{\text{on } q} = q\vec{\mathbf{E}}_{\text{produced by other charges, not } q}$$

EXAMPLE 16.8

a. Determine the magnitude of the electric field of a proton (the nucleus of a hydrogen atom) a distance 5.29×10^{-11} m away from the proton. Write the field in vector form.

b. Sketch the array of electric field arrows at this distance from the proton.

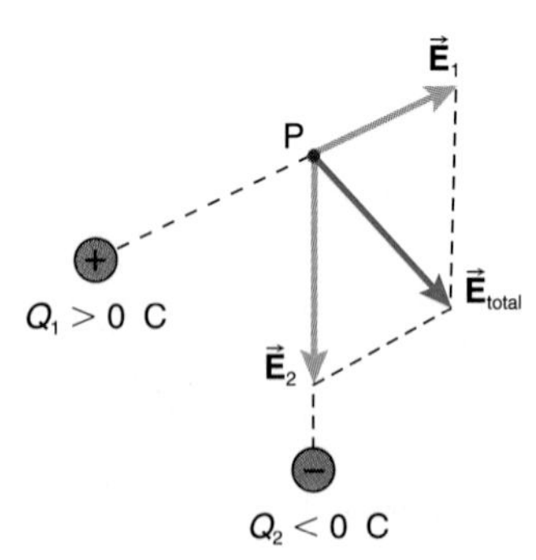

FIGURE 16.43 The total electric field is the vector sum of the individual fields.

c. Place an electron at a point this distance from the proton. Find the force on the electron placed in the field at this location. Sketch the force on a diagram.

Solution

a. The proton is a pointlike charge, and so you can use Equation 16.11. The magnitude of the field is

$$E = \frac{1}{4\pi\varepsilon_0}\frac{|Q|}{r^2}$$

where $Q = +e = 1.602 \times 10^{-19}$ C and $r = 5.29 \times 10^{-11}$ m. Making these substitutions, you find

$$E = (9.00 \times 10^9 \text{ N}\cdot\text{m}^2/\text{C}^2)\frac{1.602 \times 10^{-19} \text{ C}}{(5.29 \times 10^{-11} \text{ m})^2}$$

$$= 5.15 \times 10^{11} \text{ N/C}$$

This is an electric field with a very large magnitude!

The electric field of the proton is directed radially away from the proton, since it is a positive charge. Therefore

$$\vec{E} = (5.15 \times 10^{11} \text{ N/C})\hat{r}$$

b. The field arrows are shown in Figure 16.44.

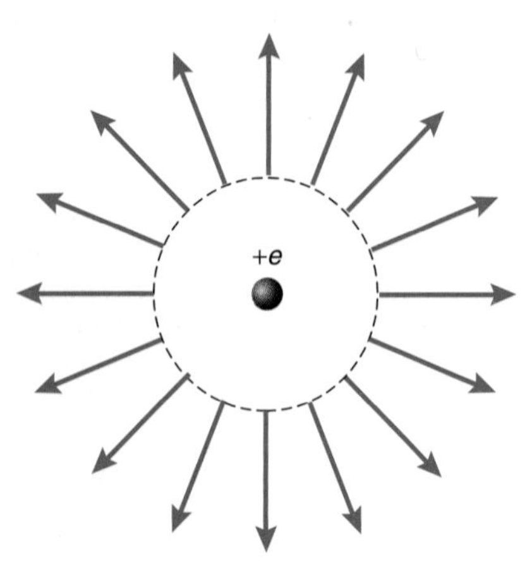

FIGURE 16.44

c. An electron placed in this field experiences a force given by

$$\vec{F} = q\vec{E}$$

$$= (-1.602 \times 10^{-19} \text{ C})[(5.15 \times 10^{11} \text{ N/C})\hat{r}]$$

$$= (-8.25 \times 10^{-8} \text{ N})\hat{r}$$

The force on the electron is directed along $-\hat{r}$, antiparallel to the electric field in which it is placed (which is along $\hat{r}$); the force is toward the proton, as shown in Figure 16.45.

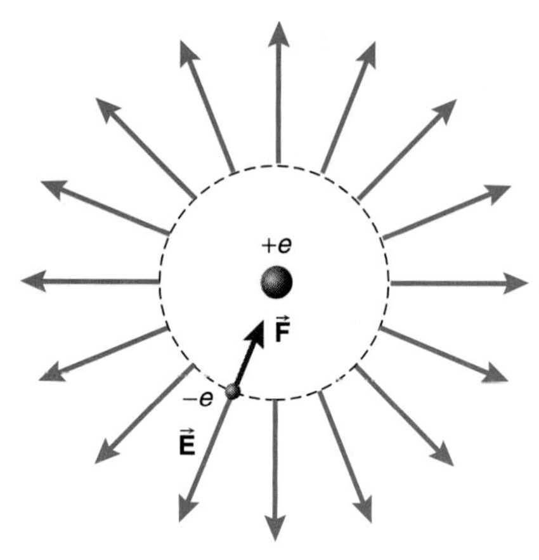

FIGURE 16.45

STRATEGIC EXAMPLE 16.9

a. Find the total electric field at the origin caused by the two pointlike charges shown in Figure 16.46. Determine the magnitude of the total field.

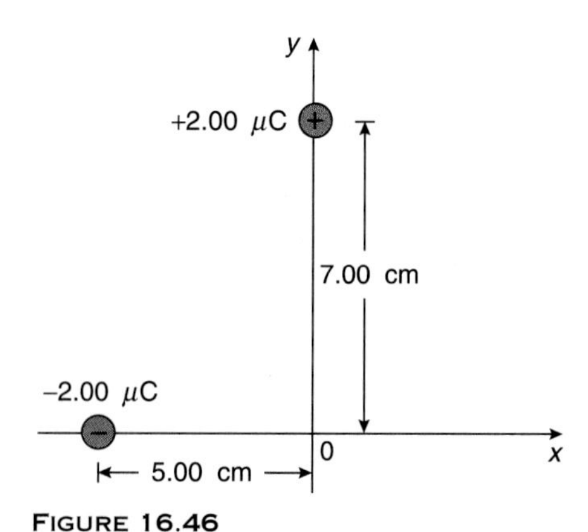

FIGURE 16.46

b. Determine the force on a charge $q = -3.00$ μC placed at the origin. What is the magnitude of the force on this charge?

Solution

a. First find the electric field of each charge at the origin (as if each charge were acting alone) with no charge at the origin. The total electric field is then the vector sum of these fields.

1. **The Electric Field of the −2.00 μC Charge**
   Pretend that the other charge (the +2.00 μC charge) does not exist; see Figure 16.47. The −2.00 μC charge

is a pointlike charge. The field of the charge has a magnitude

$$E_1 = \frac{1}{4\pi\varepsilon_0}\frac{|Q|}{r^2}$$

where $Q = -2.00 \times 10^{-6}$ C and the distance r between the charge and the origin is 5.00 cm or 5.00×10^{-2} m. *You must use coulombs for all charges and meters for all distances.* Making these substitutions,

$$E_1 = \frac{(9.00 \times 10^9\ \text{N}\cdot\text{m}^2/\text{C}^2)\,|-2.00 \times 10^{-6}\ \text{C}|}{(5.00 \times 10^{-2}\ \text{m})^2}$$
$$= 7.20 \times 10^6\ \text{N/C}$$

This field is directed toward the negative charge, as shown in Figure 16.48. Using the given coordinate system, write $\vec{\mathbf{E}}_1$ as a vector:

$$\vec{\mathbf{E}}_1 = (-7.20 \times 10^6\ \text{N/C})\hat{\imath}$$

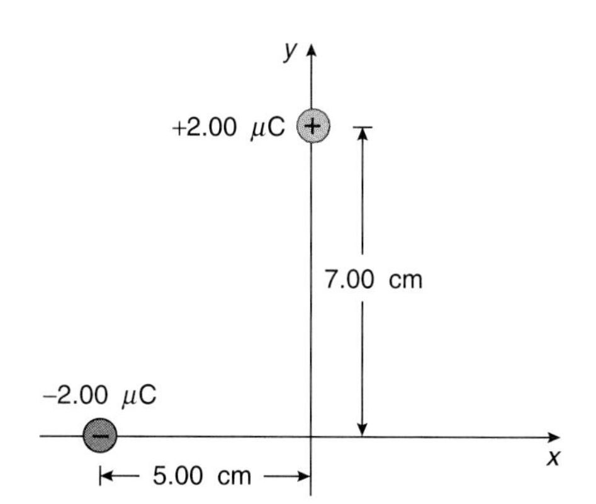

FIGURE 16.47

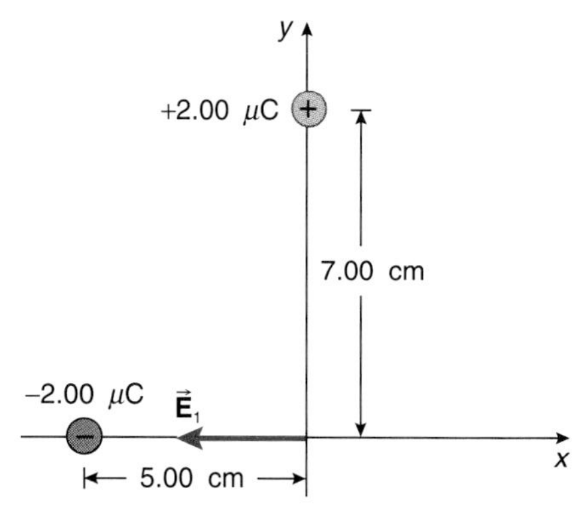

FIGURE 16.48

2. **The Electric Field of the +2.00 μC Charge**
Now pretend that the −2.00 μC charge does not exist, as in Figure 16.49. The +2.00 μC charge is a pointlike charge. The field of the charge has magnitude

$$E_2 = \frac{1}{4\pi\varepsilon_0}\frac{|Q|}{r^2}$$

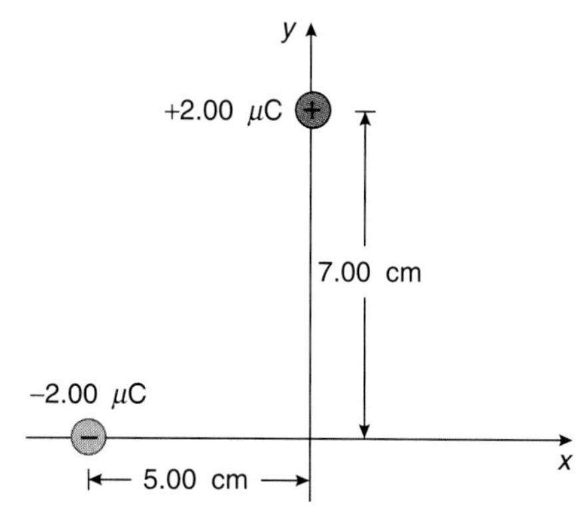

FIGURE 16.49

where $Q = +2.00 \times 10^{-6}$ C and the distance r between the charge and the origin is 7.00×10^{-2} m. You must use coulombs for the charge and meters for all distances. Making these substitutions,

$$E_2 = \frac{(9.00 \times 10^9\ \text{N}\cdot\text{m}^2/\text{C}^2)\,|+2.00 \times 10^{-6}\ \text{C}|}{(7.00 \times 10^{-2}\ \text{m})^2}$$
$$= 3.67 \times 10^6\ \text{N/C}$$

This field is directed away from the positive charge, as indicated in Figure 16.50. Using the given coordinate system, write $\vec{\mathbf{E}}_2$ as a vector:

$$\vec{\mathbf{E}}_2 = (-3.67 \times 10^6\ \text{N/C})\hat{\jmath}$$

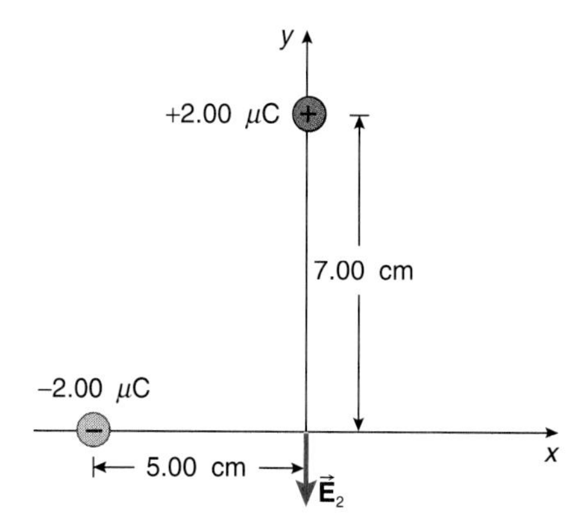

FIGURE 16.50

The total electric field at the origin when both charges are present simultaneously is the *vector sum* of the fields $\vec{\mathbf{E}}_1$ and $\vec{\mathbf{E}}_2$, as indicated in Figure 16.51:

$$\vec{\mathbf{E}}_{\text{total}} = \vec{\mathbf{E}}_1 + \vec{\mathbf{E}}_2$$
$$= (-7.20 \times 10^6\ \text{N/C})\hat{\imath} - (3.67 \times 10^6\ \text{N/C})\hat{\jmath}$$

The magnitude of this field is

$$E_{\text{total}} = [(-7.20 \times 10^6\ \text{N/C})^2 + (-3.67 \times 10^6\ \text{N/C})^2]^{1/2}$$
$$= 8.08 \times 10^6\ \text{N/C}$$

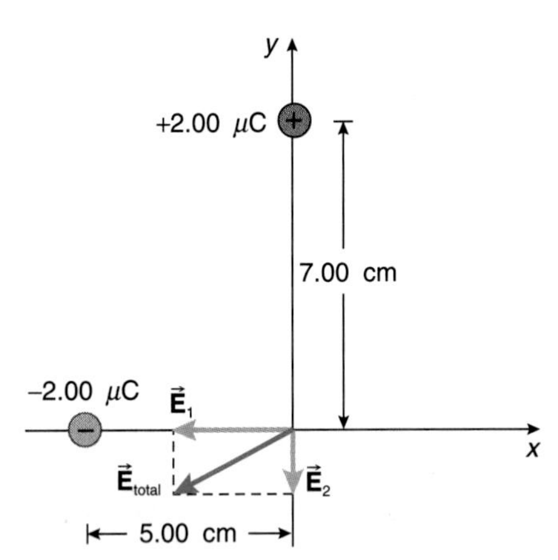

FIGURE 16.51

b. Since you know the electric field at the origin, the force on charge q placed at the origin is found using Equation 16.8:

$$\begin{aligned}\vec{F} &= q\vec{E}\\ &= (-3.00 \times 10^{-6}\ \text{C})[(-7.20 \times 10^{6}\ \text{N/C})\hat{\imath}\\ &\quad - (3.67 \times 10^{6}\ \text{N/C})\hat{\jmath}]\\ &= (21.6\ \text{N})\hat{\imath} + (11.0\ \text{N})\hat{\jmath}\end{aligned}$$

This force is shown in Figure 16.52.

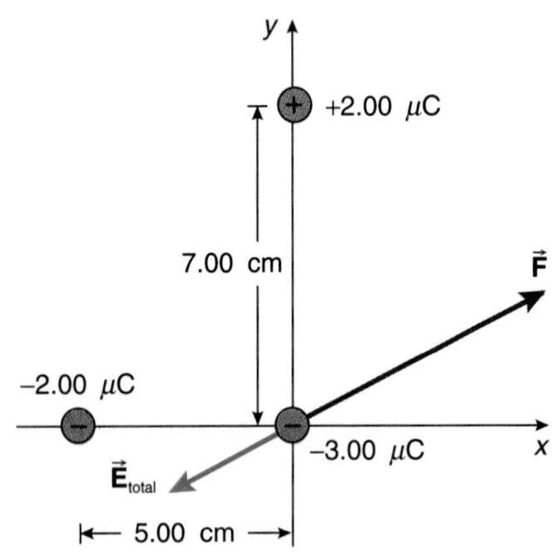

FIGURE 16.52

The magnitude of the force can be found two ways.

Method 1

Find the magnitude of the vector $\vec{F}$:

$$\begin{aligned}F &= [(21.6\ \text{N})^2 + (11.0\ \text{N})^2]^{1/2}\\ &= 24.2\ \text{N}\end{aligned}$$

Method 2

Use the magnitude of the total field and take the magnitude of $\vec{F} = q\vec{E}$:

$$\begin{aligned}F &= |q|E\\ &= (3.00 \times 10^{-6}\ \text{C})(8.08 \times 10^{6}\ \text{N/C})\\ &= 24.2\ \text{N}\end{aligned}$$

16.7 A Way to Visualize the Electric Field: Electric Field Lines

The electric field is a vector. We represent vectors geometrically by arrows with the arrow pointing in the direction of the vector and with the length of the arrow drawn proportional to the magnitude of the vector quantity. The field vectors at several different locations surrounding a positive and negative pointlike charge are indicated in Figures 16.53 and 16.54. Nothing new about this; we did the same for the gravitational fields in Chapter 6.

There is a useful alternative geometric representation of the electric field. In this scheme, we draw continuing straight or curved *lines* that at any point are *tangent* to the direction of the field vector at that point; we then place an arrowhead somewhere along the line to indicate the direction to associate with the field vectors at all points along the line. These lines are called **electric field lines.*** For a positive pointlike charge, the electric field lines are radially symmetric, as shown in Figure 16.55, with arrows pointing away from the positive charge since the electric field of a positive pointlike charge is directed away from the charge. Electric field lines therefore are said to *begin* on positive charges.

A similar pattern of lines is associated with a negative pointlike charge, except that the arrows now are directed *toward* the negative charge (see Figure 16.56), since the electric field of such a charge is directed toward the charge. Therefore we say that electric field lines *end* on negative charges.

The *number* of electric field lines drawn we make proportional to (not necessarily equal to) the magnitude (absolute value) of the charge.

*The scheme is frequently used for gravitational fields as well, giving rise to *gravitational field lines*.

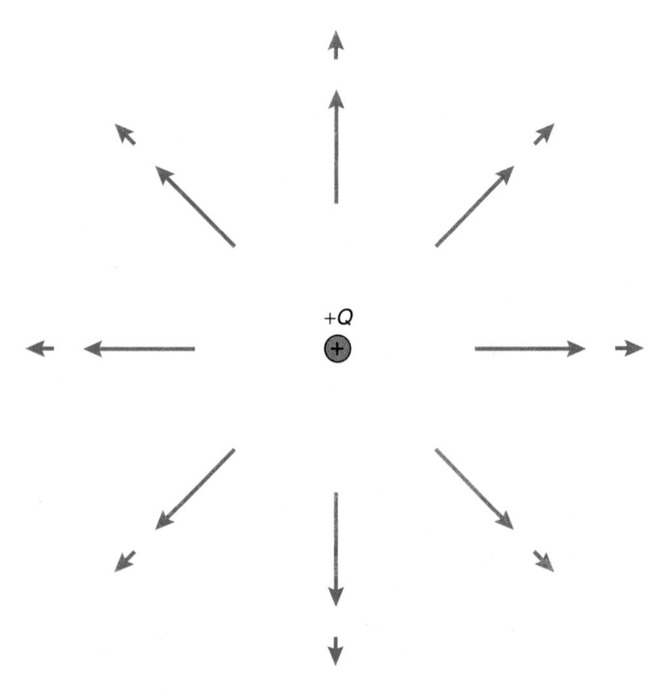

FIGURE 16.53 Electric field vectors at various distances from a positive pointlike charge. (For any arrow, take r as the distance to its tail.)

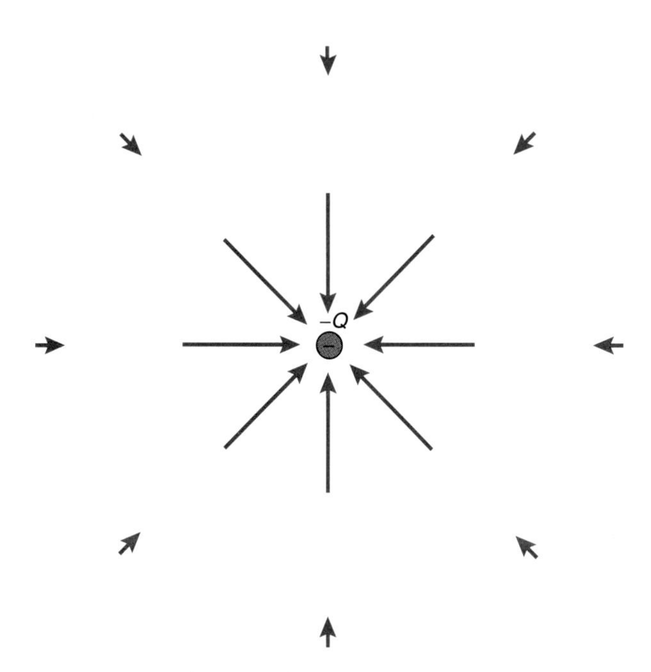

FIGURE 16.54 Electric field vectors at various distances from a negative pointlike charge.

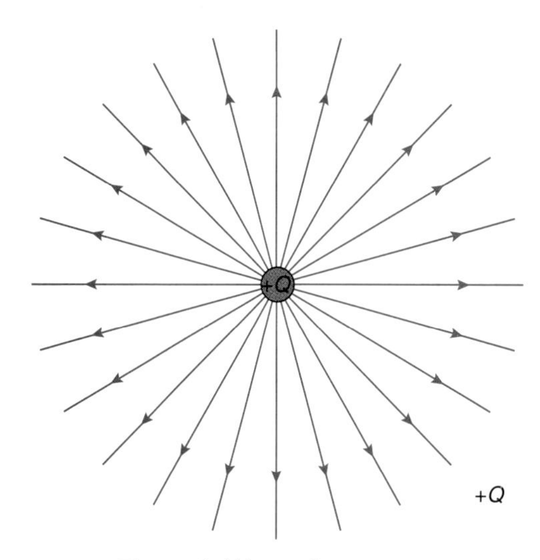

FIGURE 16.55 Electric field lines of a positive pointlike charge.

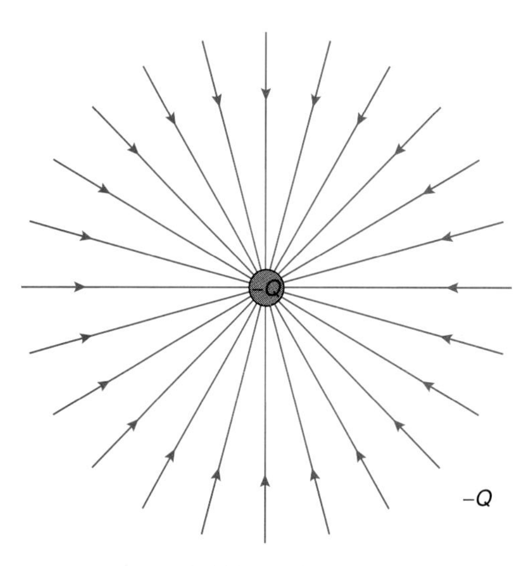

FIGURE 16.56 Electric field lines of a negative pointlike charge.

The area of a sphere of radius r surrounding such pointlike charges is $4\pi r^2$, and so the area *increases* with the square of the radial distance. The electric field of a pointlike charge *decreases* in magnitude as $1/r^2$. Hence the *number of lines* per square meter on the sphere, measured perpendicular to the lines, is a measure of the relative magnitude of the electric field at that location.

There is no reason to restrict the approach to single pointlike charge distributions. With the following rules, you can draw electric field lines for *any* distribution of charges:

1. The electric field lines begin on positive charges and end on negative charges.
2. Very close to pointlike charges, the lines are radially symmetric and their number is a measure of the magnitude of the charge.
3. The number of lines passing through a square meter oriented perpendicular to the lines is proportional to the magnitude of the electric field in that region.

Since the electric field has a *unique* direction at every point in space, no two field lines can intersect. If they were imagined to intersect, there would be two tangents to the field lines at that point, thus implying two different directions for the electric field vector at that point and two different force directions on a charge placed there, which is never the case.*

*Electric field line patterns can be shown in the laboratory by using small bits of insulating or conducting threads suspended in oil to prevent their translational motion. The bits of thread become polarized by the field and align themselves along the direction of the electric field at the points where they are suspended in the oil.

Using the foregoing conventions, the electric field lines surrounding several pointlike charge distributions can be determined in a qualitative way.

For a positive charge Q and negative charge $-Q$ of equal magnitude, the number of lines leaving the positive charge is equal to the number of lines approaching the negative charge, since the charges have equal magnitude. Very close to each pointlike charge, the lines are radially symmetric. As the distance from both charges increases, we can see that *every* electric field line leaving the positive charge eventually ends on the negative charge, as shown in Figure 16.57.

For two positive charges Q of equal magnitude, the number of lines leaving each charge is the same and the patterns are radially symmetric very close to either charge; see Figure 16.58. Very far from the charges, the distribution is similar to that of a single pointlike charge of magnitude $2Q$, for the pattern becomes essentially radially symmetric, with a total number of lines equal to twice the number emerging from each charge.

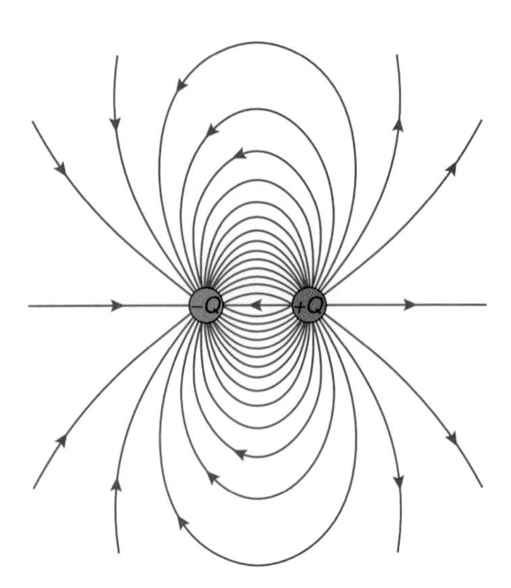

FIGURE 16.57 Electric field lines in the vicinity of a positive and negative charge of equal magnitude.

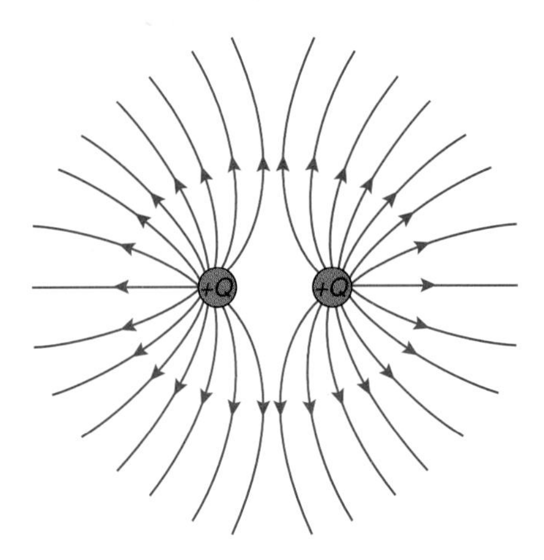

FIGURE 16.58 Electric field lines in the vicinity of two positive charges of equal magnitude.

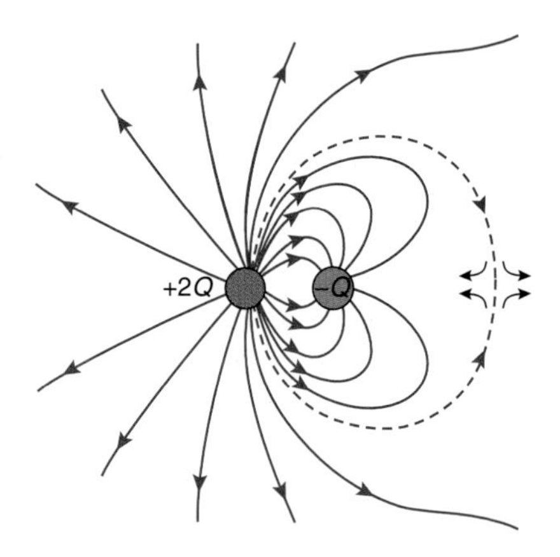

FIGURE 16.59 Electric field lines in the vicinity of a positive and negative charge of unequal magnitude.

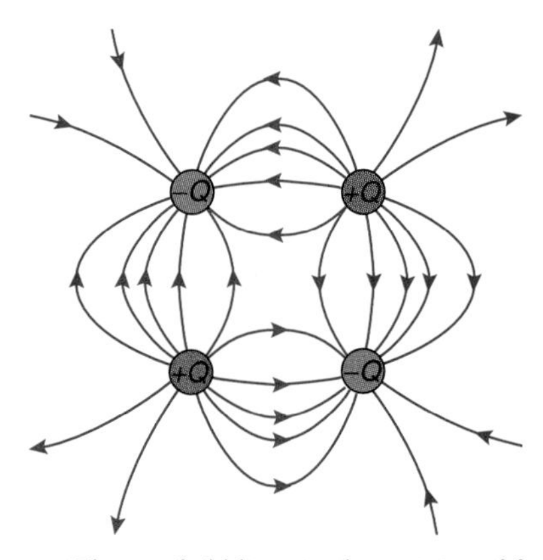

FIGURE 16.60 Electric field lines in the vicinity of four charges of equal magnitude, two of which are positive and two negative.

Figure 16.59 shows the electric field line pattern for a positive charge of magnitude 2Q and a negative charge of magnitude –Q. In this case the number of lines leaving the positive charge is twice that entering the negative charge, reflecting their difference in magnitude. The remaining half of the lines emerging from charge 2Q are essentially radial at great distances from the charge distribution, indicative of the fact that at such distances the charge distribution looks like a single pointlike charge of total charge $2Q - Q = Q$.

The electric field lines of four point charges of equal magnitude, two of which are positive and two negative, arranged on the corners of a square, are shown in Figure 16.60.

QUESTION 6

Explain why electric field lines of static charges (a) never intersect each other and (b) never form closed loops.

16.8 A COMMON MOLECULAR CHARGE DISTRIBUTION: THE ELECTRIC DIPOLE

Many simple molecules like water and more complex molecular structures like DNA, while electrically neutral as a whole, have their positive and negative charges distributed so that they can be modeled as one or more electric dipoles. The prevalence of such molecular structures makes a study of electric dipoles of some importance in chemistry, biology, and physics.

An electric dipole consists of two pointlike charges of equal magnitude $|Q|$, but opposite sign, separated by a distance d, as shown in Figure 16.61. The dipole as a whole is electrically neutral. This implies that the electric field approaches zero at

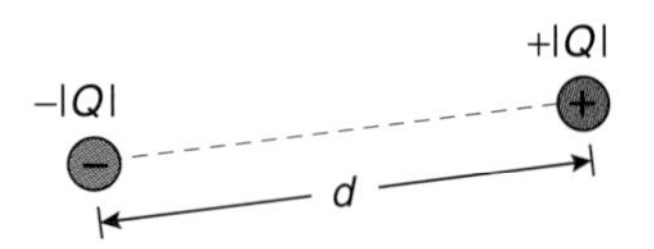

FIGURE 16.61 An electric dipole.

points far removed from the dipole (at distances much greater than the separation d of the charges). However, since the charges are not coincident with each other, the dipole produces a nonzero and distinct electric field in the space immediately surrounding the dipole (at distances comparable to the separation distance d).

In this section we first examine the electric field produced by an electric dipole. Then we examine what happens when an electric dipole is placed in an electric field caused by *other* distributions of charge, not that of the particular dipole itself.

It will be convenient to introduce a new vector called the electric **dipole moment** $\vec{p}$, defined as the product of the magnitude $|Q|$ of either charge times the position vector $\vec{d}$ of the positive charge with respect to the negative charge, as indicated in Figure 16.62:

$$\vec{p} \equiv |Q|\vec{d} \qquad (16.13)$$

Therefore the direction of the dipole moment vector is *from* the negative charge *to* the positive charge. A mnemonic for remembering this direction is to think of the plus sign + as composed of two short intersecting line segments; the head of the vector arrow also is composed of two short line segments: →.

As an example, Figure 16.63 shows how the water molecule is modeled as an electric dipole. The fractional units of the

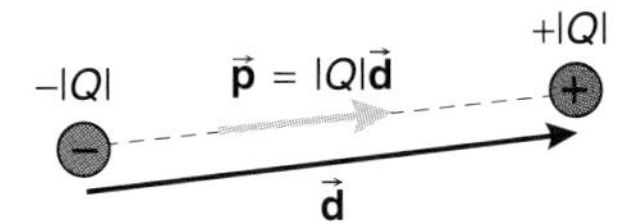

FIGURE 16.62 The electric dipole moment $\vec{p}$.

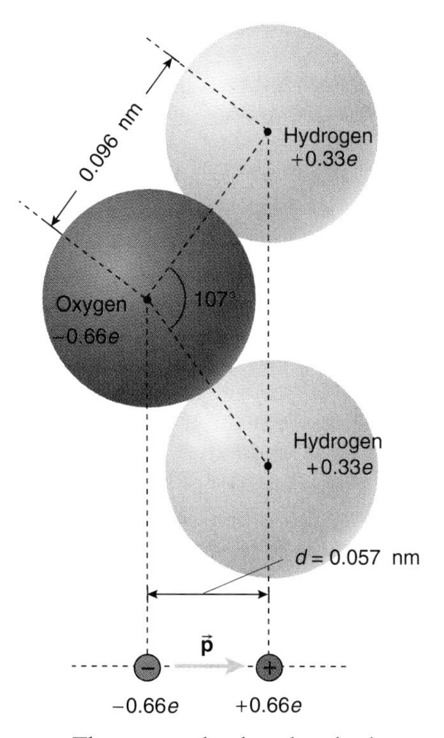

FIGURE 16.63 The water molecule and its dipole moment.

fundamental unit of charge e do not violate charge quantization. Rather, the fractions arise because the surrounding electrons spend more time near the oxygen atom than the hydrogen atoms, giving rise to these effective values of the charge distribution. The magnitude of the electric dipole moment of the water molecule is

$$\begin{aligned} p_{\text{water}} &= 0.66ed \\ &= 0.66(1.602 \times 10^{-19}\ \text{C})(0.057 \times 10^{-9}\ \text{m}) \\ &= 6.0 \times 10^{-30}\ \text{C}\cdot\text{m} \end{aligned}$$

The magnitudes of the electric dipole moments of several common molecules and molecular bonds are indicated in Table 16.1. In chemistry and biology, dipole moments frequently are quoted not in the SI unit of C·m, but in another unit: the *debye* (D). We will not use the debye unit but, if needed, the conversion factor is

$$1\ \text{D} = 3.33 \times 10^{-30}\ \text{C}\cdot\text{m}$$

Electric Field of a Dipole

An expression for the electric field along the axis of the dipole is derived in Example 16.10. The electric field in the plane that is the perpendicular bisector of the line connecting the two charges is derived in Example 16.11. The results of these examples show that the magnitude of the electric field of a dipole along these directions is proportional to the magnitude of the dipole moment and decreases with the *inverse cube* of the distance from the center of the dipole. While the examples demonstrate this distance dependence of the field for directions along the axis or in the bisecting plane, the statement is valid for other directions as well; but in these other directions the field also depends on the angle between the axis of the dipole and the direction from the center of the dipole to the point where the field is to be calculated. The electric field lines associated with an electric dipole have the appearance shown in Figure 16.57.

TABLE 16.1 The Magnitude of the Electric Dipole Moment for Selected Molecules and Molecular Bonds

Molecule	p
Acetone (C_3H_6O)	9.7×10^{-30} C·m
Ammonia (NH_3)	5×10^{-30} C·m
Carbon dioxide (CO_2)	None
Carbon monoxide (CO)	0.4×10^{-30} C·m
Ethyl alcohol (C_2H_5OH)	5.7×10^{-30} C·m
Hydrochloric acid (HCl)	3.6×10^{-30} C·m
Oxygen (O_2)	None
Sulfur dioxide (SO_2)	5.5×10^{-30} C·m
Water (H_2O)	6.0×10^{-30} C·m
Molecular bonds	p
H—O	5×10^{-30} C·m
C—Cl	5×10^{-30} C·m
C—O	2.5×10^{-30} C·m
C=O	7.7×10^{-30} C·m
C—Cl	4.9×10^{-30} C·m
C—N	0.73×10^{-30} C·m
H—N	3.0×10^{-30} C·m

A Dipole in a Uniform Electric Field

If an electric dipole is placed in a *uniform* electric field (produced by other charges), the total force on the dipole is *zero*; see Figure 16.64. The forces on each of the charges are of the same magnitude but are oppositely directed and thus vector sum to zero.

> While the total force on the dipole is zero in a uniform field, the total torque on the dipole depends on the orientation of the dipole moment vector with respect to the direction of the electric field.

Example 16.13 shows that the total torque $\vec{\tau}$ on the dipole in a uniform electric field $\vec{E}$ is

$$\vec{\tau} = \vec{p} \times \vec{E} \tag{16.14}$$

Problem 42 asks you to show that Equation 16.14 is the torque on the dipole *independent of the origin* about which the torque is taken.

Equation 16.14 implies that if the dipole moment $\vec{p}$ is not parallel to $\vec{E}$, and the dipole is free to rotate, it will spin to orient itself until $\vec{p}$ is parallel to $\vec{E}$. When $\vec{p}$ is parallel to $\vec{E}$, the torque is zero.* Equation 16.14 yields the torque on the molecular dipoles in any electric field that is approximately uniform over the small physical size of such dipoles.

The electric field is quite large in the immediate vicinity of molecular electric dipoles (see Example 16.12). The strong nearby field means that *other* electric dipoles or charges in the vicinity experience large electrical forces. Molecular electric dipole–dipole interactions are essential in the study of inter- and intramolecular bonding forces in chemistry and biology.

The unusually high dipole moment of water (for a nonacid and a nonbase) makes it an excellent polar solvent. The relatively strong electric field in the vicinity of a water molecule is able to exert electrical forces on other molecules (particularly ions and molecules with large dipole moments) that disrupt their molecular or intermolecular bonding, causing them to go easily into solution. Hydrocarbons (gasoline and motor oils) have small dipole moments and are not easily dissolved in water. Indeed the ability of water to act as a polar solvent for many materials accounts for its importance in biology for the transport of nutrients and wastes in living organisms. Water is the only solvent from which blood can be made.

*If the dipole moment is antiparallel to $\vec{E}$, the total torque also is zero—but this is an unstable equilibrium position. The slightest jitter, and the dipole will flip around so that $\vec{p}$ is parallel to $\vec{E}$. We show this from an energy viewpoint in the next chapter.

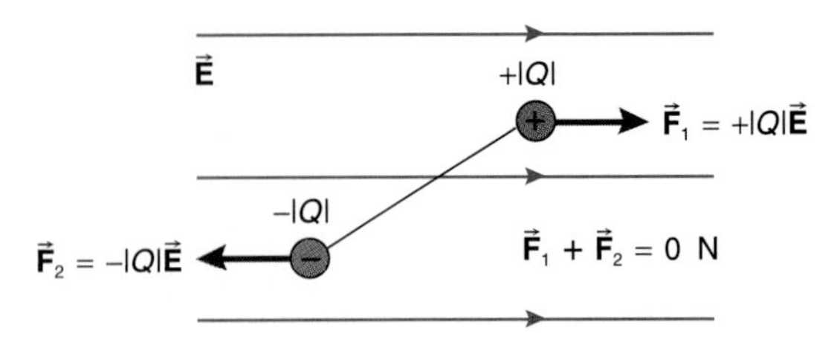

FIGURE 16.64 The total force on a dipole in a uniform electric field is zero.

EXAMPLE 16.10

Calculate the electric field of a dipole at a point P located a distance z from the center of the dipole along its axis, as shown in Figure 16.65. The axis of a dipole customarily is chosen to be the z-axis of a Cartesian coordinate system.

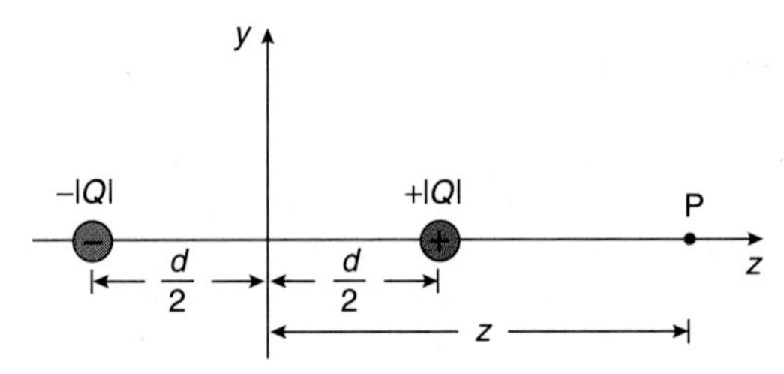

FIGURE 16.65

Solution

The positive charge is a distance

$$z - \frac{d}{2}$$

from point P. The positive charge $+|Q|$ produces a field $\vec{E}_1$ at P that is directed away from $+|Q|$, as shown in Figure 16.66.

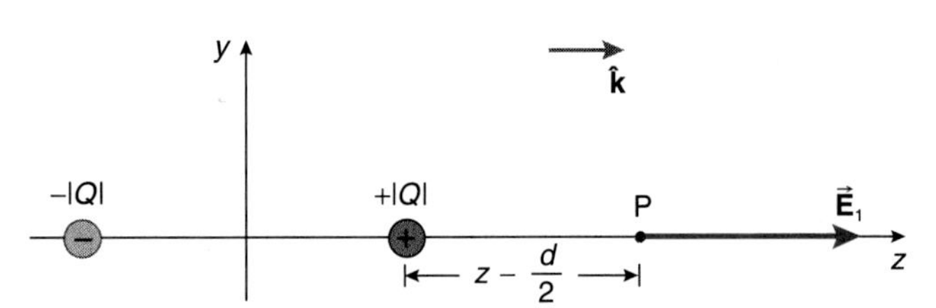

FIGURE 16.66

Expressing Equation 16.11 for the electric field of a pointlike charge in terms of the Cartesian coordinate system here, you get

$$\vec{E}_1 = \frac{1}{4\pi\varepsilon_0}\frac{|Q|}{\left(z - \frac{d}{2}\right)^2}\hat{k}$$

The negative charge $-|Q|$ is a distance

$$z + \frac{d}{2}$$

from the point P. The negative charge produces a field $\vec{E}_2$ at P that is directed toward $-|Q|$, as shown in Figure 16.67.

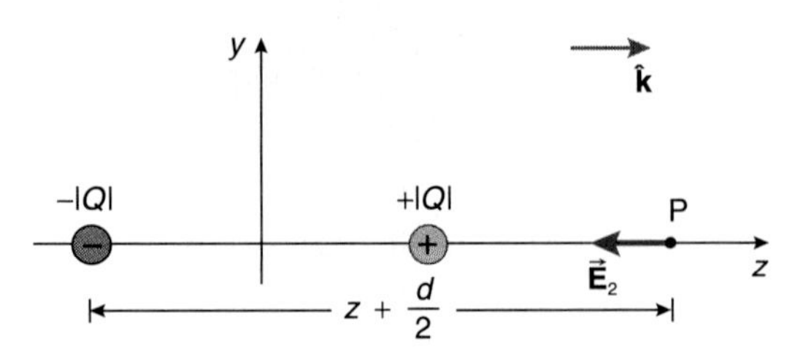

FIGURE 16.67

Using Equation 16.11, again adapted to the coordinate system and direction of the field of the negative charge, you find

$$\vec{E}_2 = \frac{1}{4\pi\varepsilon_0}\frac{|Q|}{\left(z + \frac{d}{2}\right)^2}(-\hat{k})$$

The total field at P is the vector sum of these fields:

$$\vec{E} = \vec{E}_1 + \vec{E}_2$$

$$= \frac{1}{4\pi\varepsilon_0}\left[\frac{|Q|}{\left(z - \frac{d}{2}\right)^2} - \frac{|Q|}{\left(z + \frac{d}{2}\right)^2}\right]\hat{k}$$

After some factoring and rearranging, you have

$$\vec{E} = \frac{1}{4\pi\varepsilon_0}\frac{|Q|}{z^2}\left[\left(1 - \frac{d}{2z}\right)^{-2} - \left(1 + \frac{d}{2z}\right)^{-2}\right]\hat{k}$$

If $z >> d$, you can use the binomial expansion to simplify this unwieldy expression. Specifically

$$(1 + \alpha)^n \approx 1 + n\alpha \qquad \text{for } \alpha << 1$$

With this approximation, the field is approximately

$$\vec{E} \approx \frac{1}{4\pi\varepsilon_0}\frac{|Q|}{z^2}\left\{\left[1 + (-2)\left(\frac{-d}{2z}\right)\right] - \left[1 + (-2)\left(\frac{d}{2z}\right)\right]\right\}\hat{k}$$

$$\approx \frac{1}{4\pi\varepsilon_0}\frac{|Q|}{z^2}\frac{2d}{z}\hat{k}$$

Expressed in terms of the magnitude of the dipole moment ($p = |Q|d$), the field is

$$\vec{E} \approx \frac{1}{4\pi\varepsilon_0}\frac{2p}{z^3}\hat{k} \qquad (1)$$

For $z >> d$, the electric field of the dipole decreases as the inverse cube of the distance from the dipole.

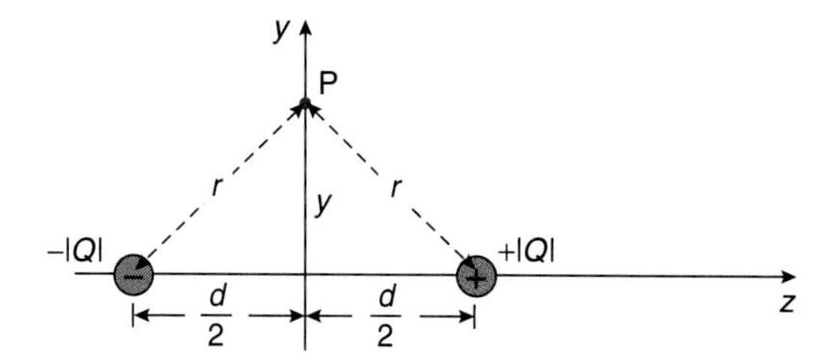

FIGURE 16.68

EXAMPLE 16.11

Calculate the total electric field caused by the two charges of an electric dipole at a point P in the plane that is the perpendicular bisector of the line connecting the two charges; see Figure 16.68. This plane customarily is chosen to be the x–y plane of a Cartesian coordinate system.

Solution

Because the charges are of the same magnitude and the point P in the bisector plane is equidistant from both charges, the electric field of *each* of the two charges (acting alone) at the point P has the same magnitude. From Equation 16.11, the magnitude of the field of each pointlike charge is

$$E = \frac{1}{4\pi\varepsilon_0}\frac{|Q|}{r^2}$$

From the geometry of Figure 16.68 and the Pythagorean theorem,

$$r^2 = y^2 + \frac{d^2}{4}$$

Hence the magnitude of the field each charge produces at P is

$$E = \frac{1}{4\pi\varepsilon_0}\frac{|Q|}{\left(y^2 + \frac{d^2}{4}\right)}$$

The field at P of the positive charge points away from the charge, while the field at P of the negative charge points toward the charge, as indicated in Figure 16.69.

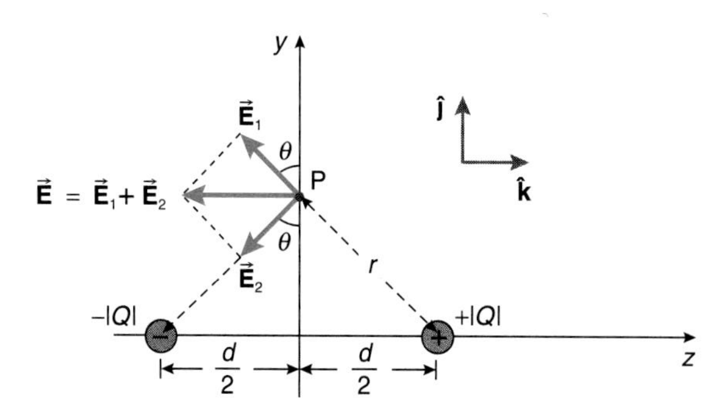

FIGURE 16.69

Use the coordinate system in Figure 16.69. The individual fields in vector form are

$$\vec{E}_1 = \frac{1}{4\pi\varepsilon_0}\frac{|Q|}{\left(y^2 + \frac{d^2}{4}\right)}\left[(\cos\theta)\,\hat{j} - (\sin\theta)\,\hat{k}\right]$$

and

$$\vec{E}_2 = \frac{1}{4\pi\varepsilon_0}\frac{|Q|}{\left(y^2 + \frac{d^2}{4}\right)}\left[(-\cos\theta)\,\hat{j} - (\sin\theta)\,\hat{k}\right]$$

According to the principle of superposition, the total electric field at the point P when *both charges* are present is the *vector sum* of the fields of the individual charges:

$$\vec{E}_{total} = \vec{E}_1 + \vec{E}_2$$

$$= \left[\frac{-2}{4\pi\varepsilon_0} \frac{|Q|}{\left(y^2 + \frac{d^2}{4}\right)} \sin\theta \right] \hat{k} \qquad (1)$$

But from the geometry of Figure 16.69,

$$\sin\theta = \frac{d/2}{\left(y^2 + \frac{d^2}{4}\right)^{1/2}}$$

Making this substitution for $\sin\theta$ in equation (1), you find the total field at the point P is

$$\vec{E}_{total} = \frac{-1}{4\pi\varepsilon_0} \frac{|Q|d}{\left(y^2 + \frac{d^2}{4}\right)^{3/2}} \hat{k}$$

Since the magnitude of the dipole moment $\vec{p}$ is equal to $|Q|d$, the field at P is

$$\vec{E}_{total} = \frac{-1}{4\pi\varepsilon_0} \frac{p}{\left(y^2 + \frac{d^2}{4}\right)^{3/2}} \hat{k} \qquad (2)$$

For distances y from the dipole that are much greater than the separation of the charges in the dipole—that is, for $y >> d$—the d^2 term in the denominator of the field can be neglected; the field then becomes

$$\vec{E}_{total} \approx \frac{-1}{4\pi\varepsilon_0} \frac{p}{y^3} \hat{k} \qquad (y >> d) \qquad (3)$$

For such great distances y, the electric field of the dipole decreases as the inverse *cube* of the distance.

By comparing equation (1) of Example 16.10 and equation (3) here, you can see that at equally great distances from the dipole, the magnitude of the field in the bisector plane is only half that along the axis of the dipole.

Example 16.12

The water molecule has a dipole moment of magnitude 6.0×10^{-30} C·m, as indicated in Table 16.1. Find the magnitude of the electric field at a distance of 1.00×10^{-9} m = 1.00 nm from the axis of the dipole in the plane of the perpendicular bisector to the line connecting the two charges representing the dipole.

Solution
The electric field in the plane of the perpendicular bisector was calculated in Example 16.11. Use the result of that example here [equation (2) of Example 16.11] as well as the data indicated in Figure 16.63 for the water molecule. Specifically for the magnitude of the field, you have

$$E_{total} = \frac{1}{4\pi\varepsilon_0} \frac{p}{\left(y^2 + \frac{d^2}{4}\right)^{3/2}}$$

where $p = 6.0 \times 10^{-30}$ C·m, $y = 1.00 \times 10^{-9}$ m, and d = 0.057 nm = 0.057×10^{-9} m. Making these substitutions, you find

$$E_{total} = 5.4 \times 10^7 \text{ N/C}$$

Using the approximation of equation (3) of Example 16.11 yields the same result since $y >> d$. This is an electric field of substantial magnitude!

Example 16.13

Show that an electric dipole in a uniform electric field $\vec{E}$ experiences a torque $\vec{\tau}$ about the midpoint of the dipole that is given by

$$\vec{\tau} = \vec{p} \times \vec{E}$$

where $\vec{p}$ is the dipole moment vector. See Figure 16.70.

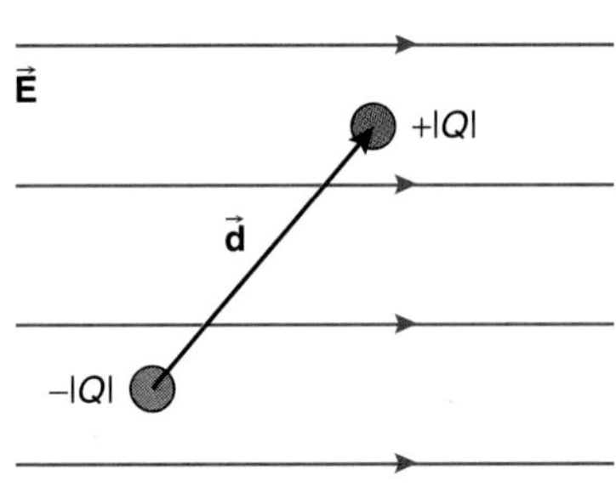

FIGURE 16.70

Solution
Each charge of the dipole experiences an electrical force:

the force on the positive charge is $\vec{F}_1 = +|Q|\vec{E}$; and
the force on the negative charge is $\vec{F}_2 = -|Q|\vec{E}$.

The torque of a force is $\vec{\tau} = \vec{r} \times \vec{F}$, where $\vec{r}$ is the position vector locating the point of application of the force with respect to the point about which the torque is taken. The example asks you to take torques about the midpoint of the dipole; see Figure 16.71.

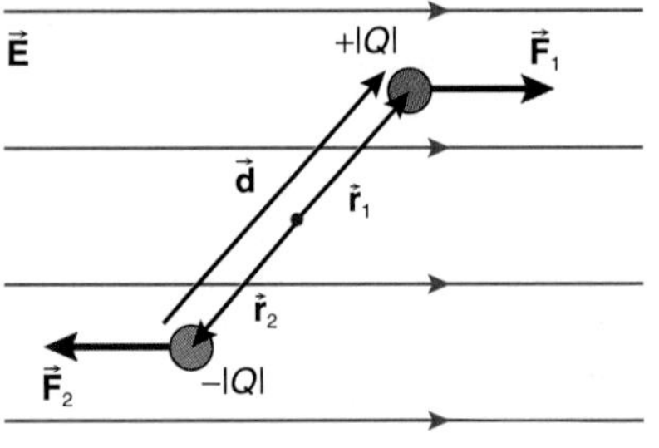

FIGURE 16.71

The position vectors of the points of application of the forces are the locations of the charges themselves:

$$\vec{\mathbf{r}}_1 = \frac{\vec{\mathbf{d}}}{2}$$

$$\vec{\mathbf{r}}_2 = -\frac{\vec{\mathbf{d}}}{2}$$

Hence the torque $\vec{\boldsymbol{\tau}}_1$ of $\vec{\mathbf{F}}_1$ is

$$\begin{aligned}\vec{\boldsymbol{\tau}}_1 &= \frac{\vec{\mathbf{d}}}{2} \times \vec{\mathbf{F}}_1 \\ &= \frac{\vec{\mathbf{d}}}{2} \times |Q|\vec{\mathbf{E}}\end{aligned}$$

The torque $\vec{\boldsymbol{\tau}}_2$ of $\vec{\mathbf{F}}_2$ is

$$\begin{aligned}\vec{\boldsymbol{\tau}}_2 &= \vec{\mathbf{r}}_2 \times \vec{\mathbf{F}}_2 \\ &= -\frac{\vec{\mathbf{d}}}{2} \times \left(-|Q|\vec{\mathbf{E}}\right) \\ &= \frac{\vec{\mathbf{d}}}{2} \times |Q|\vec{\mathbf{E}}\end{aligned}$$

The total torque $\vec{\boldsymbol{\tau}}$ is the vector sum of the two torques:

$$\begin{aligned}\vec{\boldsymbol{\tau}} &= \vec{\boldsymbol{\tau}}_1 + \vec{\boldsymbol{\tau}}_2 \\ &= \frac{\vec{\mathbf{d}}}{2} \times |Q|\vec{\mathbf{E}} + \frac{\vec{\mathbf{d}}}{2} \times |Q|\vec{\mathbf{E}} \\ &= |Q|\vec{\mathbf{d}} \times \vec{\mathbf{E}}\end{aligned}$$

Since $|Q|\vec{\mathbf{d}}$ is the dipole moment $\vec{\mathbf{p}}$, you find

$$\vec{\boldsymbol{\tau}} = \vec{\mathbf{p}} \times \vec{\mathbf{E}}$$

16.9 THE ELECTRIC FIELD OF CONTINUOUS DISTRIBUTIONS OF CHARGE

For charge distributions that are not pointlike in character, we need to invoke the principle of superposition to determine the total electric field at a point in space. The charge distribution is imagined to be composed of a continuous sea of pointlike charges, each of which contributes a bit to the total electric field at the point in space where the field is to be calculated; see Figure 16.72. Each small, differential bit of charge dq produces a differential bit of electric field $d\vec{\mathbf{E}}$ at the point P given by a variation of Equation 16.11 for a pointlike charge:

$$d\vec{\mathbf{E}} = \frac{1}{4\pi\varepsilon_0}\frac{dq}{r^2}\hat{\mathbf{r}} \qquad (16.15)$$

where $\hat{\mathbf{r}}$ is a unit vector pointing *from* the bit of charge dq *to* the point P where the field is to be found.

The contribution $d\vec{\mathbf{E}}$ points away from dq if it is positive and toward dq if it is negative, just as with all pointlike charges.

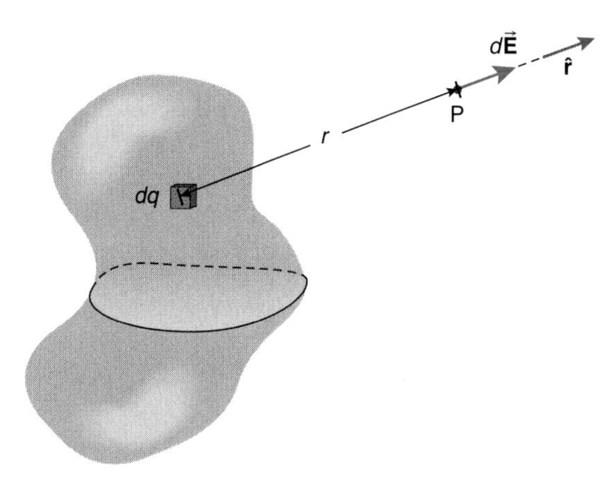

FIGURE 16.72 Each differential bit of charge produces a differential bit of electric field at point P. The situation pictured is for $dq > 0$ C

According to the principle of superposition, the total electric field is found by a *vector sum* of all the contributions from all the bits of charge that make up the charge distribution. Here the vector summation is a *vector integration* (a continuous sum) over the physical extent of the charge distribution:

$$\vec{\mathbf{E}} = \frac{1}{4\pi\varepsilon_0}\int_{\substack{\text{charge}\\\text{distribution}}}\frac{dq}{r^2}\hat{\mathbf{r}} \qquad (16.16)$$

You have to be careful when performing this integration because the direction of the unit vector $\hat{\mathbf{r}}$ typically changes as dq sweeps over the charge distribution. The distance r between dq and P also changes in many cases. Such vector integrations are trickier to carry out than ordinary scalar integrations, as we saw in our study of the gravitational field in Chapter 6. The reason for the trickiness is that the contributions $d\vec{\mathbf{E}}$ typically point in different directions and we have to account for this fact while doing the vector summation implicit in the integration.

In actually doing such integrations, a certain amount of foresight is involved, foresight that only can be gleaned by carefully studying and profiting from a few examples of the technique. Consequently, in the following examples, we calculate the electric field of several different charge distributions to help you gain the needed insight to do similar calculations.

The examples illustrate that the electric field at a point in space depends on several factors:

- the charge creating the field and its sign;
- the geometric shape of the charge distribution creating the field;
- the distance of the point from the charge distribution; and
- the placement of the point with respect to the charge distribution (in particular, the location of the point with respect to any symmetry axes associated with the charge distribution).

Table 16.2 summarizes the electric fields created by various charge distributions.

TABLE 16.2 The Electric Field of Various Charge Distributions

Pointlike charge:

$$\vec{\mathbf{E}} = \frac{1}{4\pi\varepsilon_0}\frac{Q}{r^2}\hat{\mathbf{r}}$$

Dipole along the axis ($z >> d$):

$$\vec{\mathbf{E}} \approx \frac{1}{4\pi\varepsilon_0}\frac{2p}{z^3}\hat{\mathbf{k}}$$

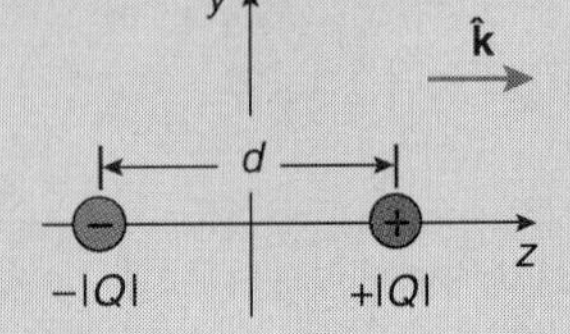

in the perpendicular bisector plane:

$$\vec{\mathbf{E}}_{\text{total}} = \frac{-1}{4\pi\varepsilon_0}\frac{p}{\left(y^2 + \frac{d^2}{4}\right)^{3/2}}\hat{\mathbf{k}}$$

$$\vec{\mathbf{E}}_{\text{total}} \approx \frac{-1}{4\pi\varepsilon_0}\frac{p}{y^3}\hat{\mathbf{k}} \quad (y >> d)$$

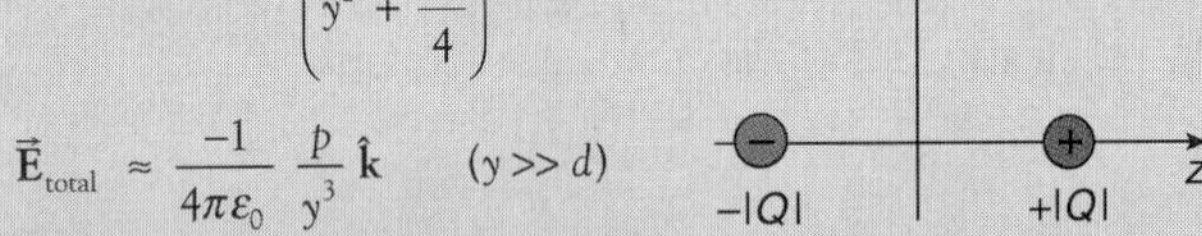

Uniformly charged ring (on the axis):

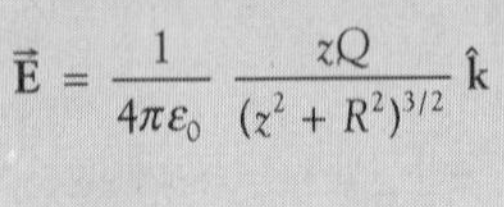

$$\vec{\mathbf{E}} = \frac{1}{4\pi\varepsilon_0}\frac{zQ}{(z^2 + R^2)^{3/2}}\hat{\mathbf{k}}$$

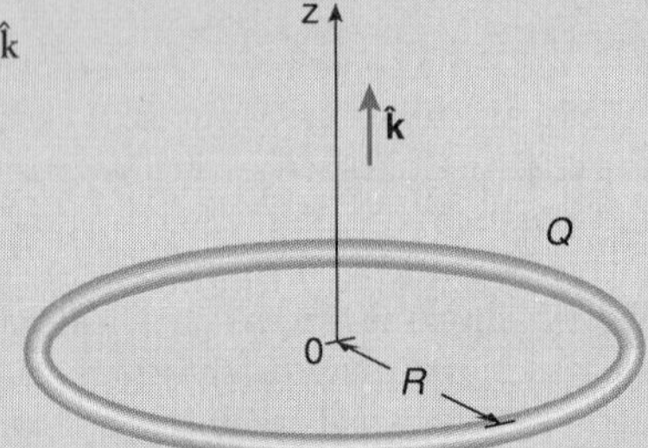

Uniformly charged disk (on the axis):

$$\vec{\mathbf{E}} = \frac{1}{4\pi\varepsilon_0}\frac{2Q}{R^2}\left[1 - \frac{z}{(z^2 + R^2)^{1/2}}\right]\hat{\mathbf{k}}$$

Uniformly charged infinite sheet:

$$E = \frac{\sigma}{2\varepsilon_0}$$

Two infinite uniformly charged sheets with opposite charge:

$$E = \frac{\sigma}{\varepsilon_0}$$

Uniformly charged spherical shell:

$\vec{\mathbf{E}} = 0$ N/C (inside, $r < R$)

$\vec{\mathbf{E}} = \dfrac{1}{4\pi\varepsilon_0}\dfrac{Q}{r^2}\hat{\mathbf{r}}$ (outside, $r > R$)

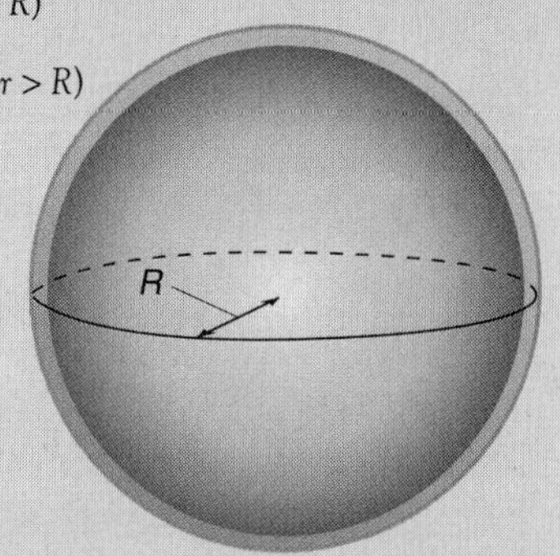

Uniformly charged sphere:

$\vec{\mathbf{E}} = \dfrac{1}{4\pi\varepsilon_0}\dfrac{Q}{R^3}r\,\hat{\mathbf{r}}$ ($r < R$)

$\vec{\mathbf{E}} = \dfrac{1}{4\pi\varepsilon_0}\dfrac{Q}{r^2}\hat{\mathbf{r}}$ ($r > R$)

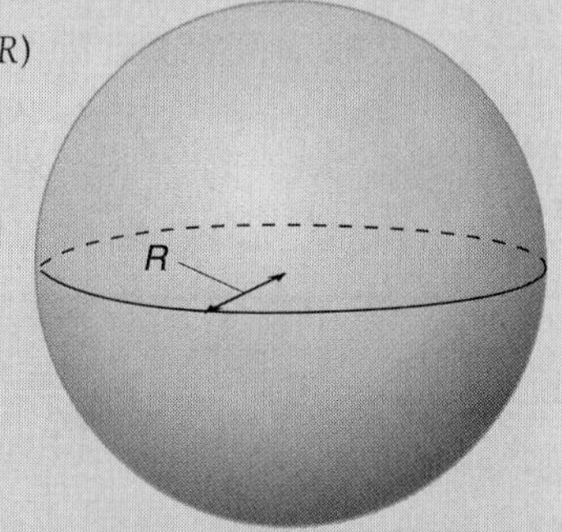

An infinitely long straight-line charge:

$\vec{\mathbf{E}} = \dfrac{1}{4\pi\varepsilon_0}\dfrac{2\lambda}{r}\hat{\mathbf{r}}$ (where $\hat{\mathbf{r}}$ is a unit vector pointing radially away from the line charge)

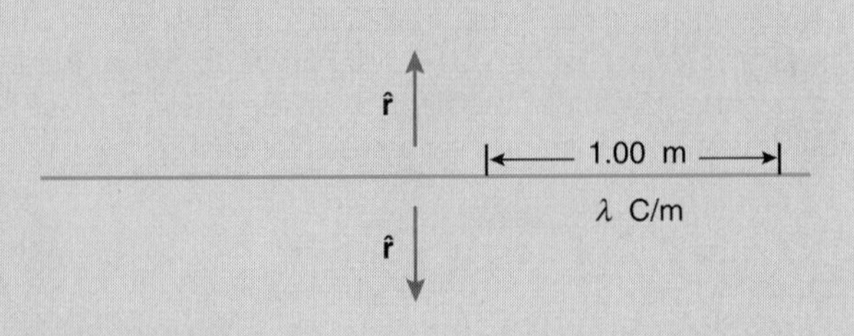

STRATEGIC EXAMPLE 16.14

A ring of radius R has a total charge Q smeared out uniformly along its circumference, as shown in Figure 16.73. Calculate the electric field of the ring at a point P along the axis of the ring, a distance z from the plane of the ring.

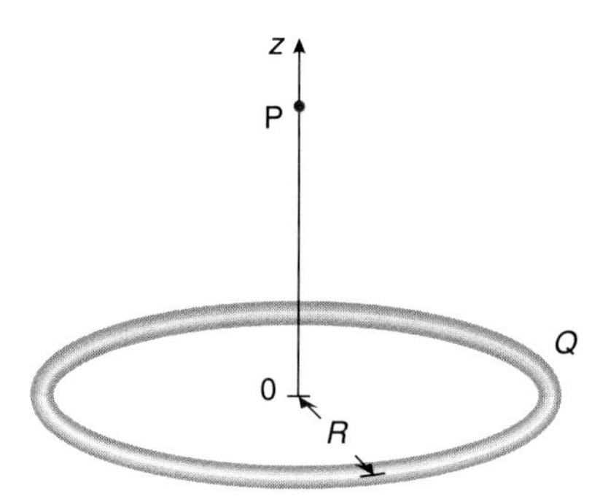

FIGURE 16.73

Solution

The charge Q is smeared out uniformly along the circumference of the ring. If the total charge Q is divided by the circumference of the ring ($2\pi R$), you obtain the number of coulombs of charge along each meter of the circumference of the ring; this quantity is known as a *linear charge density* λ:

$$\lambda = \frac{Q}{2\pi R}$$

Look at the contribution to the field at P from a small differential length ds of the circumference, as shown in Figure 16.74. The length ds has a charge dq on it equal to the product of the linear charge density λ and ds:

$$dq = \lambda\, ds$$

Assume the charge Q on the ring is positive; then this pointlike charge dq produces a bit of electric field $d\vec{\mathbf{E}}$ at the axial point P that is directed away from dq. If dq is negative, the field points in the opposite direction.

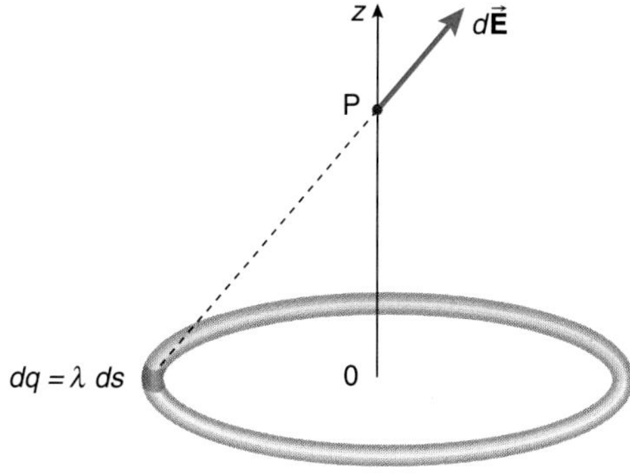

FIGURE 16.74

The magnitude of $d\vec{\mathbf{E}}$ is found using Equation 16.15:

$$\begin{aligned} dE &= \frac{1}{4\pi\varepsilon_0}\frac{dq}{r^2} \\ &= \frac{1}{4\pi\varepsilon_0}\frac{\lambda\, ds}{r^2} \end{aligned}$$

Notice that a corresponding piece of the ring on the opposite side *also* produces a bit of field of the same magnitude at the point P, as shown in Figure 16.75. The components of the two differential fields perpendicular to the axis will vector sum to zero (see Figure 16.76), but the components of the fields along the axis add. Hence, of the piece of field $d\vec{\mathbf{E}}$ produced at P by the charge dq on ds, *only* the component of the field along the axis will survive the vector summation as you integrate around the ring. All the differential components of the field perpendicular to the axis pairwise vector sum to zero.

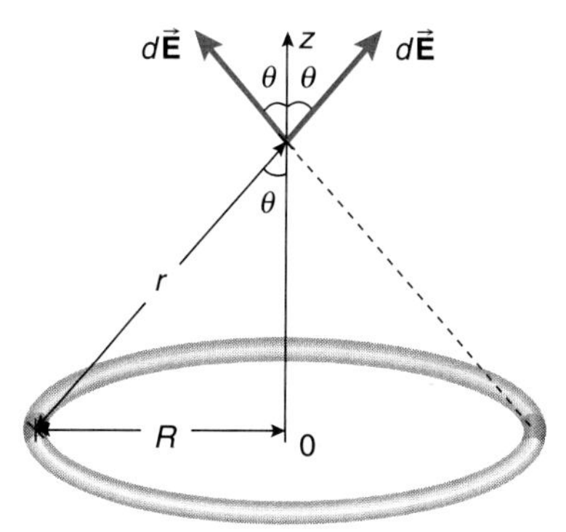

FIGURE 16.75

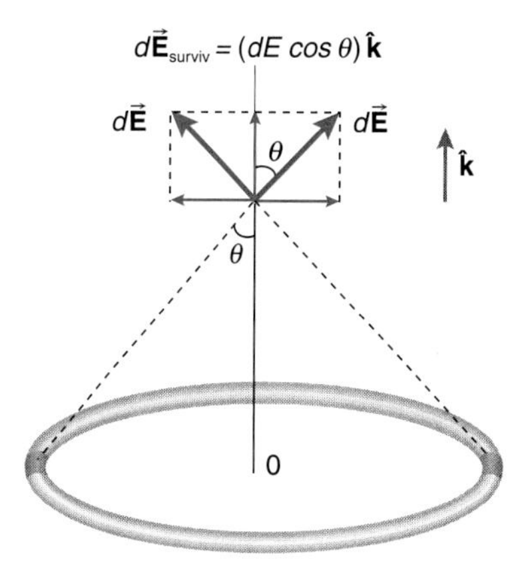

FIGURE 16.76

Hence the *surviving* part of the field caused by the charge on ds is directed along $\hat{\mathbf{k}}$ and is

$$d\vec{\mathbf{E}}_{\text{surviv}} = \left(\frac{1}{4\pi\varepsilon_0}\frac{\lambda\, ds}{r^2}\cos\theta\right)\hat{\mathbf{k}}$$

Since all the surviving pieces of the field are parallel to each other, the original vector integral over the ring is now a scalar integral for the surviving part of the field:

$$\vec{E}_{surviv} = \frac{1}{4\pi\varepsilon_0}\int_{ring}\left(\lambda\frac{ds}{r^2}\cos\theta\right)\hat{k}$$

As you integrate around the ring, r is a constant because all the differential charges on the ring are equidistant from the field point P. Likewise, $\cos\theta$ also is constant as you integrate around the ring. Thus the integral for the surviving part of the field reduces to

$$\vec{E}_{surviv} = \frac{1}{4\pi\varepsilon_0}\lambda\frac{\cos\theta}{r^2}\int_{ring} ds\,\hat{k}$$

The integral of ds (the arc length) around the ring is the circumference of the ring: $2\pi R$. Thus the surviving field is

$$\vec{E}_{surviv} = \left(\frac{1}{4\pi\varepsilon_0}\lambda\frac{\cos\theta}{r^2}2\pi R\right)\hat{k}$$

But $2\pi R\lambda$ is the total charge Q on the ring. The Pythagorean theorem gives us r^2:

$$r^2 = z^2 + R^2$$

The geometry in Figure 16.75 implies that

$$\cos\theta = \frac{z}{\left(z^2+R^2\right)^{1/2}}$$

Making these substitutions into the expression for $\vec{E}_{surviv}$, you get

$$\vec{E}_{surviv} = \frac{1}{4\pi\varepsilon_0}\frac{Qz}{(z^2+R^2)^{3/2}}\hat{k} \qquad (1)$$

For positive Q, and for $z > 0$ m, the field points along $+\hat{k}$; for $z < 0$ m, the field is parallel to $-\hat{k}$. In both cases the field is directed along the axis and away from the positively charged ring. You should convince yourself that equation (1) gives the appropriate direction for $\vec{E}$ for a negatively charged ring ($Q < 0$ C) as well: toward the ring.

Notice that when $z = 0$ m in equation (1), the point is at the center of the ring and the surviving field there is zero. Symmetry considerations yield the same conclusion. As one moves out along the axis of the ring, the field first increases and then eventually decreases.

Let's check to see that equation (1) for $\vec{E}$ makes sense if $z >> R$. When we are very far from the ring, the field of the ring is essentially that of a pointlike charge, because the ring then looks so small. If $z >> R$, you can neglect R^2 compared with z^2 in the denominator of equation (1). In this case $\vec{E}_{surviv}$ reduces to

$$\vec{E}_{surviv} \approx \frac{1}{4\pi\varepsilon_0}\frac{Qz}{(z^2)^{3/2}}\hat{k}$$

$$\approx \frac{1}{4\pi\varepsilon_0}\frac{Q}{z^2}\hat{k}$$

which is the field of a pointlike charge at a distance z along the z-axis.

To derive an equation for the field of the ring at a point P that is not on the axis is quite difficult and beyond the scope of a course in introductory physics. You may see such an analysis in more advanced courses or use appropriate software to approximate such a field.

EXAMPLE 16.15

Find the electric field at a distance z along the axis of a uniformly charged circular disk of radius R and charge Q (see Figure 16.77).

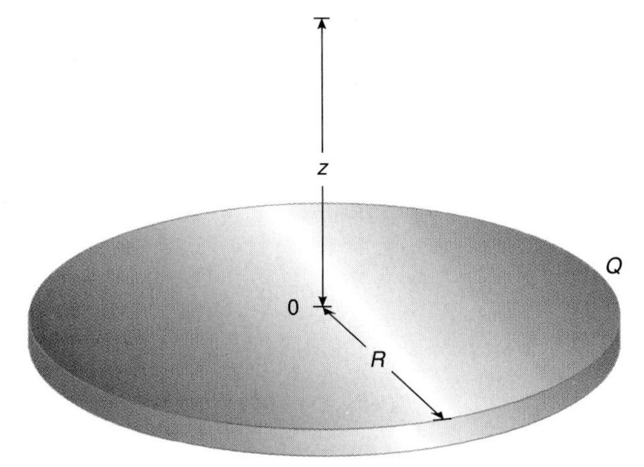

FIGURE 16.77

Solution

The charge is smeared out uniformly over the disk so that the *surface charge density* σ (the number of coulombs on each square meter of the disk) is constant and equal to

$$\sigma = \frac{Q}{\pi R^2}$$

Use the result for the electric field along the axis of a uniformly charged circular ring (Example 16.14). Look at a circular ring of radius r and of differential width dr on the disk; see Figure 16.78.

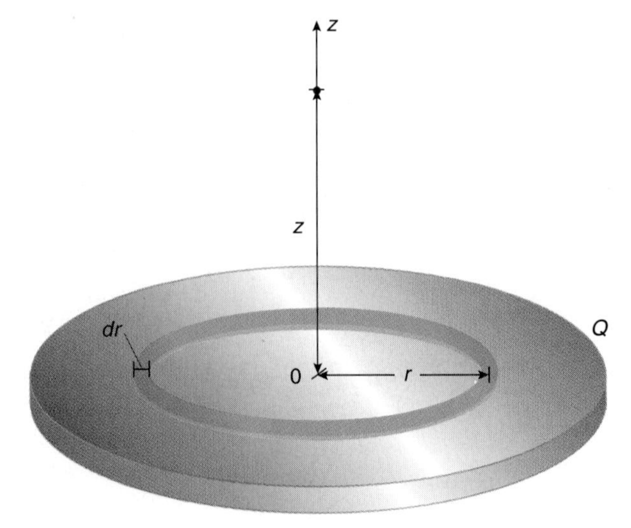

FIGURE 16.78

Imagine unwrapping the ring into a straight-line strip, as in Figure 16.79. The strip has length $2\pi r$ and width dr, and so the differential ring has area

$$2\pi r\ dr$$

FIGURE 16.79

The charge dq on the differential ring is equal to its area times the surface charge density:

$$dq = (2\pi r\ dr)\sigma$$

The charge on the differential ring produces an electric field directed along the axis of the disk. Adapt equation (1) of Example 16.14 for the field of a ring of charge, but let r be the radius of the ring; also, the charge Q on the ring is now $dq = (2\pi r\ dr)\sigma$. Hence the differential ring of charge produces a differential electric field along the axis that is given by

$$d\vec{E} = \frac{1}{4\pi\varepsilon_0}\frac{z\,(2\pi r\ dr\ \sigma)}{(z^2 + r^2)^{3/2}}\hat{k}$$

Now integrate over the disk, since each ring produces an electric field along the axis as well:

$$\vec{E} = \frac{1}{4\pi\varepsilon_0}\int_{0\text{ m}}^{R}\frac{z\,(2\pi r\ dr\ \sigma)}{(z^2 + r^2)^{3/2}}\hat{k}$$

After removing constant terms from the integral, you obtain

$$\vec{E} = \frac{1}{4\pi\varepsilon_0}z\,(2\pi\sigma)\int_{0\text{ m}}^{R}\frac{r\ dr}{(z^2 + r^2)^{3/2}}\hat{k}$$

The integral is a standard form of the type

$$\int u^n\ du = \frac{u^{n+1}}{n+1}$$

Performing the integration, you find

$$\vec{E} = \left[\frac{1}{4\pi\varepsilon_0}z(2\pi\sigma)\frac{-1}{(z^2 + r^2)^{1/2}}\right]\Bigg|_{0\text{ m}}^{R}\hat{k}$$

After evaluating the limits and simplifying slightly, you have

$$\vec{E} = \frac{1}{4\pi\varepsilon_0}2\pi\sigma\left[1 - \frac{z}{(z^2 + R^2)^{1/2}}\right]\hat{k} \qquad (1)$$

Expressing this in terms of the total charge Q on the disk,

$$\vec{E} = \frac{1}{4\pi\varepsilon_0}\frac{2Q}{R^2}\left[1 - \frac{z}{(z^2 + R^2)^{1/2}}\right]\hat{k} \qquad (2)$$

The expression for the field is a bit complicated, but there is nothing that guarantees that all results or equations will be either simple or pretty!

Note that for $Q > 0$ C, the field along the axis is directed away from the disk; for $Q < 0$ C, it is directed along the axis toward the disk.

EXAMPLE 16.16

Find the electric field of a uniformly charged, infinite sheet of charge a distance z from the plane of the sheet (see Figure 16.80).

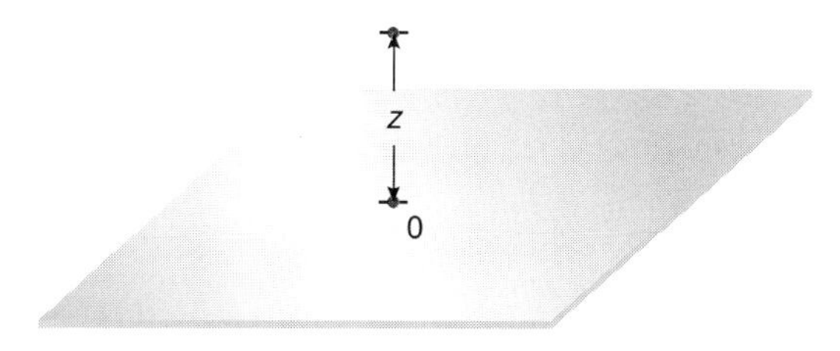

FIGURE 16.80

Solution

Equation (1) of Example 16.15 for the electric field of a uniformly charged circular disk can be used to find the electric field of the uniformly charged infinite sheet by letting the radius R of the disk approach infinity:

$$\vec{E} = \lim_{R\to\infty\text{ m}}\frac{1}{4\pi\varepsilon_0}2\pi\sigma\left[1 - \frac{z}{(z^2 + R^2)^{1/2}}\right]\hat{k}$$

As the radius R increases without bound with z fixed at any chosen value, the second term in the brackets becomes vanishingly small and you are left with a field of magnitude

$$E = \frac{\sigma}{2\varepsilon_0} \qquad (1)$$

If the charge on the sheet is positive, the field is directed perpendicularly *away* from the sheet, as shown in Figure 16.81. If the sheet has negative charge, the field is directed toward the sheet, as indicated in Figure 16.82.

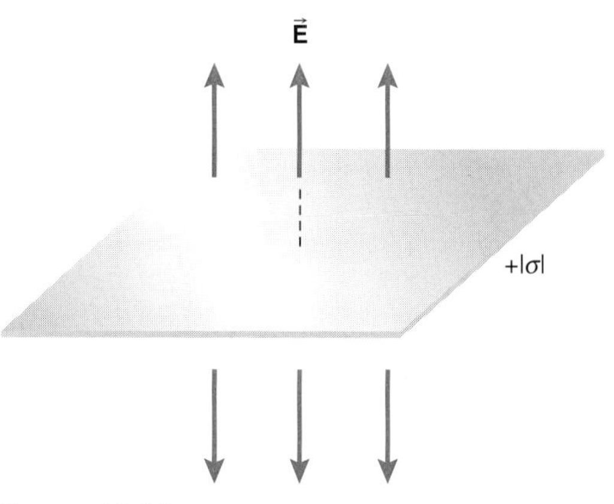

FIGURE 16.81

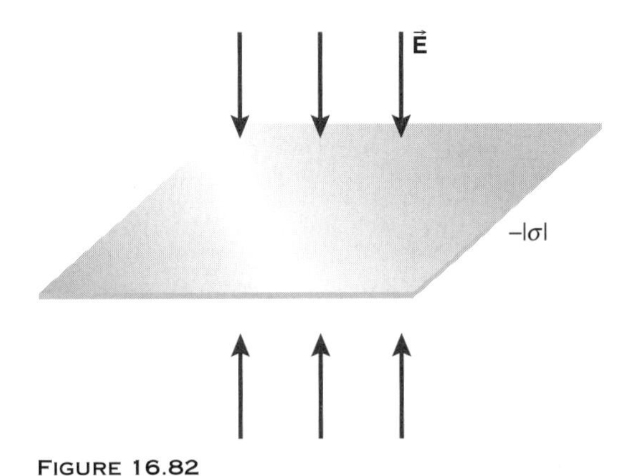

FIGURE 16.82

Such an infinite charged sheet produces a constant electric field. Since the sheet is of infinite extent and no finite distance is significant when compared with infinity, the field is independent of the x–y location *and* the distance z from the sheet.

EXAMPLE 16.17

Two oppositely charged infinite sheets are separated by a distance d; see Figure 16.83. Find the electric field both in the region between the plates and the regions on either side of the plates.

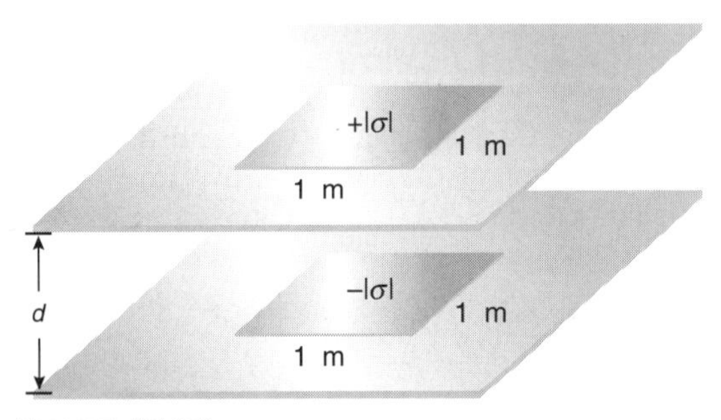

FIGURE 16.83

Solution
We find the electric field by using the principle of superposition. The fields produced by each sheet are indicated separately in Figure 16.84 for three regions.

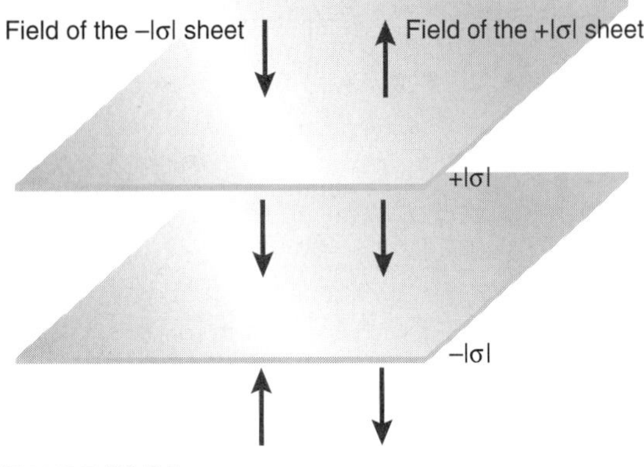

FIGURE 16.84

Notice that below the negative sheet as well as above the positive sheet, the fields of the individual sheets vector sum to zero: the two fields of equal magnitude are antiparallel. In the region between the two sheets, the two fields are parallel. The resulting field between the sheets is then of magnitude

$$E = \frac{\sigma}{2\varepsilon_0} + \frac{\sigma}{2\varepsilon_0}$$

$$= \frac{\sigma}{\varepsilon_0} \qquad (1)$$

where σ is the magnitude of the surface charge density on either sheet.

Such an arrangement of twin, parallel, oppositely charged, infinite sheets is known as an infinite, *parallel plate capacitor* and is the way in which constant electric fields typically are produced in the laboratory. Of course, truly infinite sheets of charge are a bit unwieldy (even if they could be constructed!), and so in practice two finite sheets are placed close together so that the sheets effectively look infinite as long as the location of interest between them is not too near the edges. In the central region between such finite sheets, the electric field is essentially uniform and of magnitude σ/ε_0.

EXAMPLE 16.18

Find the electric field at a point that is a distance r from the center of a uniformly charged, hollow spherical shell of radius R with charge Q; see Figure 16.85. Consider both of the situations $r < R$ and $r > R$.

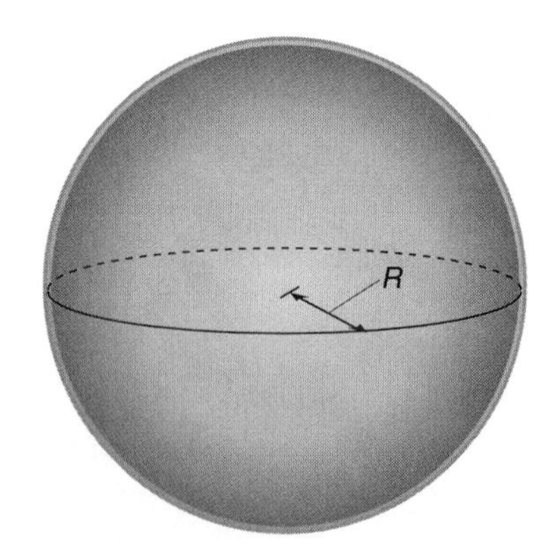

FIGURE 16.85

Solution
Both the gravitational force between pointlike masses and the electrical force between pointlike charges are inverse-square-law forces. The formal similarity between the mathematical expressions for the gravitational force and the electrical force means that the calculations we did in Chapter 6 in deriving the form for the gravitational fields of certain distributions of mass can be applied equally well to the electric fields of the same shape charge distribution.

For example, we discovered (Sections 6.3 and 6.4 and Appendix A) that the gravitational force on a mass m located inside a uniform spherical mass shell was zero:

$$\vec{\mathbf{F}}_{m \text{ inside uniform shell}} = 0 \text{ N}$$

Since the gravitational force on m is related to the gravitational field $\vec{g}$ where m is placed by $\vec{F}_{\text{grav on } m} = m\vec{g}$, the gravitational field inside the shell must be zero:

$$0 \text{ N} = m\vec{g}$$

so

$$\vec{g} = 0 \text{ N/kg}$$

Ultimately, this striking result came about because of the inverse-square-law nature of the gravitational interaction.

Since the electrical interaction also is inverse-square-law, we can say without doing the entire proof again that the electrical field inside a uniform spherical shell of charge also is zero:

$$\vec{E}_{\text{inside uniform shell}} = 0 \text{ N/C} \qquad (\text{any point } r < R)$$

Hence a charge q placed inside the shell (that does not disturb the uniform distribution of charge on the shell) experiences zero total electrical force:

$$\begin{aligned}\vec{F}_{\text{elec on } q} &= q\vec{E}\\ &= q(0 \text{ N/C})\\ &= 0 \text{ N}\end{aligned}$$

Correspondingly the other shell theorem in Chapter 6 means that here the electric field of a uniformly charged spherical shell is the same as that of an equal point charge located at the center of the sphere:

$$\vec{E} = \frac{1}{4\pi\varepsilon_0}\frac{Q}{r^2}\hat{r} \qquad (\text{any point } r > R)$$

Example 16.19

Find the electric field at a distance r from the center of a sphere with charge Q distributed uniformly throughout the volume of the sphere of radius R; see Figure 16.86. Consider both cases: $r < R$ and $r > R$.

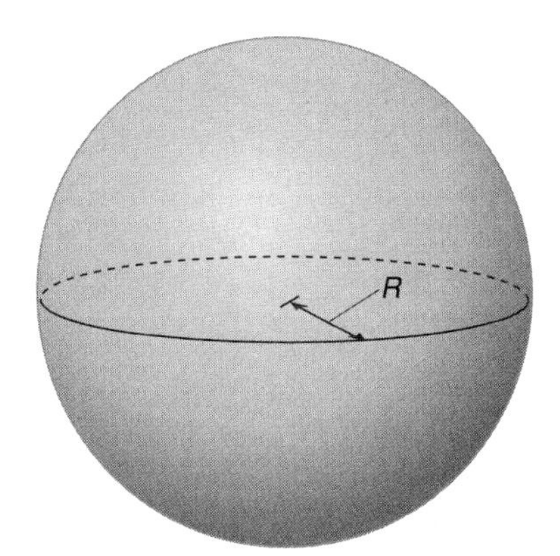

FIGURE 16.86

Solution

Analogously to the gravitational situation, the electric field at a point r *within* a uniformly charged sphere is caused only by the charge located closer to the center of the sphere than the point itself. We modify the case from Table 6.1:

$$\vec{E} = \frac{1}{4\pi\varepsilon_0}\frac{Q}{R^3}r\hat{r} \qquad (r < R)$$

Likewise, the electric field at a point *outside* the spherical charge distribution, a distance r from its center, is the same as the field produced by an equal point charge located at the center of the sphere:

$$\vec{E} = \frac{1}{4\pi\varepsilon_0}\frac{Q}{r^2}\hat{r} \qquad (r > R)$$

16.10 Motion of a Charged Particle in a Uniform Electric Field: An Electrical Projectile

The motion of a charged particle in a uniform electric field has important technological applications in low-energy particle accelerators, television tubes, and CRT displays, among other things. As we have seen (Example 16.17), a uniform electric field exists in the region between two large, oppositely charged parallel plates. There is a strong similarity between the motion of a charged particle in a uniform electric field and the more familiar motion of a projectile in a uniform gravitational field. Since the electric field is uniform, the force is constant and the acceleration of the charge is constant. You can use the kinematic equations for motion under a constant acceleration (Chapters 3 and 4).

Question 7

A charge q is moving with velocity $\vec{v}$ perpendicular to an electric field. Is the electrical force on the charge zero? Describe the analogous gravitational situation.

Strategic Example 16.20

A particle of mass m and charge q is released at rest close to one of two parallel plates in a region where there is a uniform electric field of magnitude E_0. The particle accelerates toward the other plate, a distance d away. Determine the speed at which it makes impact with the opposite plate.

Solution

Use the coordinate system indicated in Figure 16.87. The electric field then is

$$\vec{E} = -E_0\hat{j}$$

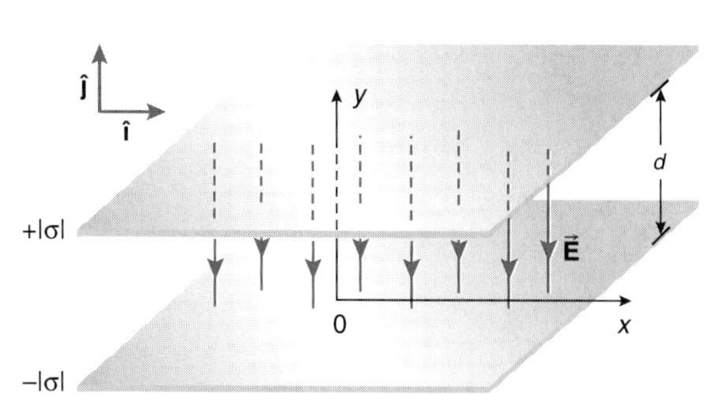

FIGURE 16.87

The charge q experiences an electrical force

$$\begin{aligned}\vec{\mathbf{F}} &= q\vec{\mathbf{E}}\\ &= q(-E_0\hat{\mathbf{j}})\\ &= -qE_0\hat{\mathbf{j}}\end{aligned}$$

We want to determine the speed at which it makes impact with the other plate. The particle is a one-dimensional electrical projectile.

Method 1

If q is positive, the force on it is parallel to the field direction, and so the particle should be released near the upper plate in Figure 16.87. If negative, the force is antiparallel to the field, and the particle should be released near the lower plate in Figure 16.87. The acceleration $\vec{\mathbf{a}}$ of the charge is found using Newton's second law:

$$\vec{\mathbf{F}} = m\vec{\mathbf{a}}$$

The force is constant, and so the acceleration is constant as well:

$$\begin{aligned}\vec{\mathbf{a}} &= \frac{\vec{\mathbf{F}}}{m}\\ &= -\frac{qE_0}{m}\hat{\mathbf{j}}\end{aligned}$$

The acceleration is purely along the y-direction, and the y-component of the acceleration (a_y) is constant and equal to $-qE_0/m$. Since the acceleration is constant, you can apply the one-dimensional kinematic equations of Chapter 3 to determine the impact speed of the particle as it hits the other plate. The problem is analogous to releasing a mass at rest in a uniform gravitational field. Specifically, for motion along the y-direction,

$$v_y = v_{y0} + a_y t \qquad (1)$$

and

$$y = y_0 + v_{y0}t + \frac{1}{2}a_y t^2 \qquad (2)$$

Since the particle was released at rest, $v_{y0} = 0$ m/s. If you assume q is positive, it is released at the coordinate $y_0 = d$. Thus equation (1) simplifies to

$$\begin{aligned}v_y &= a_y t\\ &= -\frac{qE_0}{m}t \qquad (3)\end{aligned}$$

and equation (2) becomes

$$\begin{aligned}y &= d + \frac{1}{2}a_y t^2\\ &= d - \frac{1}{2}\frac{qE_0}{m}t^2 \qquad (4)\end{aligned}$$

Impact occurs at coordinate $y = 0$ m. Thus equation (4) becomes

$$0 \text{ m} = d - \frac{1}{2}\frac{qE_0}{m}t^2$$

or

$$d = \frac{1}{2}\frac{qE_0}{m}t^2$$

Solving for t,

$$t = \left(\frac{2dm}{qE_0}\right)^{1/2}$$

The velocity component on impact, using equation (3), is

$$\begin{aligned}v_y &= -\frac{qE_0}{m}\left(\frac{2dm}{qE_0}\right)^{1/2}\\ &= -\left(2d\frac{qE_0}{m}\right)^{1/2}\end{aligned}$$

Method 2

You also can use the CWE theorem to determine the speed. The work done by all the forces acting on the particle is equal to the change in the kinetic energy of the particle:

$$W_{\text{all}} = \Delta\text{KE}$$

The only force acting is the electrical force

$$\begin{aligned}\vec{\mathbf{F}} &= q\vec{\mathbf{E}}\\ &= -qE_0\hat{\mathbf{j}}\end{aligned}$$

which is constant. The work W done by this constant force as the particle zips between the plates, changing its position vector by $\Delta\vec{\mathbf{r}}$, is

$$W = \vec{\mathbf{F}} \cdot \Delta\vec{\mathbf{r}} \qquad (\text{constant force } \vec{\mathbf{F}})$$

If the particle has a positive charge, it is released near the upper plate, and so the change in the position vector of the charge is

$$\begin{aligned}\Delta\vec{\mathbf{r}} &= (0 \text{ m})\hat{\mathbf{j}} - d\hat{\mathbf{j}}\\ &= -d\hat{\mathbf{j}}\end{aligned}$$

The work is then

$$\begin{aligned}W &= (-qE_0\hat{\mathbf{j}}) \cdot (-d\hat{\mathbf{j}})\\ &= qE_0 d\end{aligned}$$

The particle has no initial kinetic energy, and its kinetic energy just before impact is

$$\frac{mv_y^2}{2}$$

Thus the CWE theorem

$$W_{\text{all}} = \Delta\text{KE}$$

becomes

$$qE_0 d = \frac{mv_y^2}{2} - 0 \text{ J}$$

Solving for v_y, you find

$$v_y = -\left(2d\frac{qE_0}{m}\right)^{1/2}$$

The negative root is chosen because the positive charge is moving toward decreasing values of y.

If a small hole is bored through the plate toward which the particle is traveling, the particle emerges and traverses the region outside the plates at constant velocity, since there is no electric field in that region. We have created a charged particle peashooter, an electrical rifle, known more elegantly as a charged particle accelerator. As long as the particles are released from rest, they will emerge from the hole at the constant speed v given by

$$v = \left(2d \frac{qE_0}{m}\right)^{1/2} \quad (5)$$

EXAMPLE 16.21

A particle of mass m with charge q, initially moving at constant velocity, encounters a region with a uniform electric field at right angles to its initial velocity; see Figure 16.88. Determine the deflection of the particle when it emerges from the region of uniform field. Assume it does not strike either charged sheet.

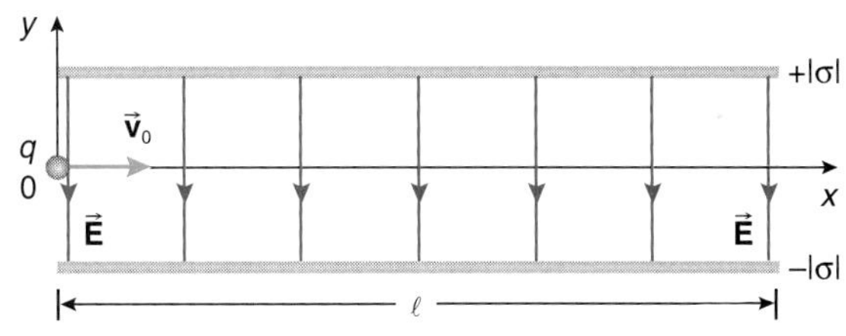

FIGURE 16.88

Solution

Use the given coordinate system. The initial velocity is

$$\vec{v}_0 = v_{x0}\hat{\imath}$$

The uniform electric field is

$$\vec{E} = -E_0\hat{\jmath}$$

The electrical force on the particle is found from the field:

$$\begin{aligned}\vec{F} &= q\vec{E}\\ &= q(-E_0\hat{\jmath})\end{aligned}$$

The force is perpendicular to the initial velocity of the particle. Therefore the particle continues with a constant x-velocity component v_{x0}, but the force in the y-direction causes it to accelerate in the y-direction. You have a two-dimensional electrical projectile. If q is positive, the particle is deflected in the direction of the field; if q is negative, the particle is deflected antiparallel to the field.

The particle is within the region of field during a time interval t found by dividing the length ℓ by the constant x-velocity component v_{x0}:

$$t = \frac{\ell}{v_{x0}}$$

During this interval, the particle is deflected along the y-direction. The deflection of the particle as it leaves the field is found using the one-dimensional kinematic equation for constant acceleration:

$$y = y_0 + v_{y0}t + \frac{1}{2}a_y t^2$$

With the coordinates in Figure 16.88, you have $y_0 = 0$ m and $v_{y0} = 0$ m/s. Hence the equation for y reduces to

$$y = \frac{1}{2}a_y t^2$$

The acceleration component a_y is the force component F_y divided by the mass:

$$a_y = -\frac{qE_0}{m}$$

The time of flight you have already found to be ℓ/v_{x0}. Thus the deflection is

$$y = \frac{1}{2}\left(-\frac{qE_0}{m}\right)\left(\frac{\ell}{v_{x0}}\right)^2 \quad (1)$$

The direction of the deflection is different for positive and for negative charges.

By controlling the magnitude of the field, the deflection can be varied at will. This is the essential principle governing the operation of all CRT (cathode ray tube) displays. TV, some computer screens, oscilloscopes, and the like are based on the motion of charged particles in electric fields. In such devices the particles used are negative charges: electrons. In equation (1), note that the deflection depends inversely on the mass of the charged particle. The small mass of electrons means that they respond to the action of electrical forces more dramatically than would more massive protons in the same electric field.

16.11 GAUSS'S LAW FOR ELECTRIC FIELDS*

Here we extend the idea of Gauss's law to electric fields. We want to know what result is obtained for the flux of the electric field through a closed surface. Fortunately, most of the calculations have been done already. We can build on what we did in formulating Gauss's law for the gravitational field, though we have to account for the fact that there are two different kinds of electric charge but only one kind of mass.

We first encountered the idea of the flux of a vector in our study of the gravitational field in Chapter 6. The differential flux $d\Phi$ of any vector $\vec{A}$ through a differential area $d\vec{S}$ is a measure of the extent to which the vector $\vec{A}$ passes through the area $d\vec{S}$. Specifically we defined the differential flux of a vector $\vec{A}$ via the scalar product:

$$d\Phi = \vec{A} \cdot d\vec{S}$$

The flux of the vector through a finite surface S then is found by integrating $d\Phi$ over the area in question:

$$\Phi = \int_{\text{area } S} \vec{A} \cdot d\vec{S}$$

Mathematically, a flux can be associated with any vector; but only certain vectors have a flux that is significant from a physical viewpoint. One of those vectors is the gravitational field $\vec{g}$; another is the electric field $\vec{E}$.

In our study of the gravitational field in Chapter 6, we discovered (in Section 6.15) a curious general relationship between the flux of the total gravitational field through *any closed surface* and the amount of mass trapped within the surface. Specifically this relationship is *Gauss's law for the gravitational field* (Equation 6.55):

$$\int_{\substack{\text{clsd} \\ \text{surface } S}} \vec{g} \cdot d\vec{S} = -4\pi G M_{\text{within } S}$$

Recall that for a closed surface, the direction of the differential area vector $d\vec{S}$ is perpendicular to the surface and directed outward. In the derivation of Gauss's law in Chapter 6, the minus sign arose from the scalar product of $\vec{g}$ with $d\vec{S}$: with $\vec{g}$ directed into the surface (toward the mass M) and $d\vec{S}$ directed perpendicularly outward from the surface, the angle between the two vectors is greater than 90° (see Figure 16.89) and thus the scalar product is negative (the cosine of angles greater than 90° is negative).

The negative right-hand side of Gauss's law for the gravitational field means that the gravitational field lines have a net influx through the surface S; they thread in but not out. If the surface S has a total mass M within it, then the flux of the gravitational field through the closed surface S is $-4\pi GM$.

If the closed surface S has no mass within it, the flux of the gravitational field through S is zero. Every line of $\vec{g}$ enters the surface but also leaves it (see Figure 16.90); there is no net threading of the surface.

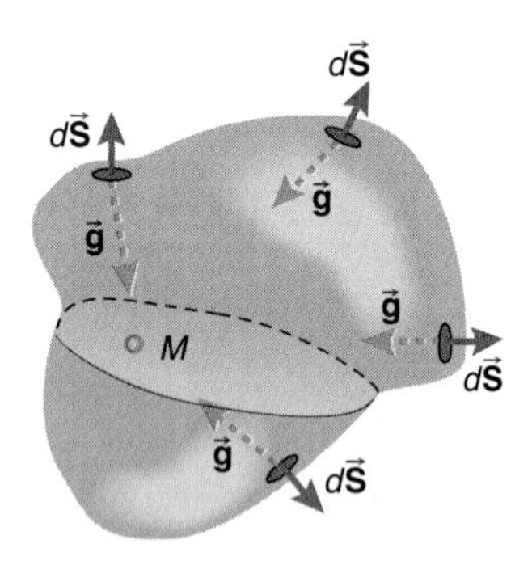

FIGURE 16.89 **For the gravitational case with M inside S, the field $\vec{g}$ is inward, while $d\vec{S}$ is outward.**

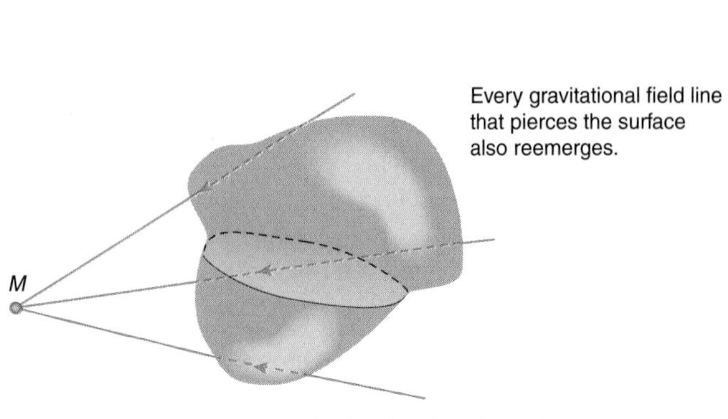

The shape of the closed surface S is arbitrary; what matters is whether there is mass trapped inside it or not. For a weirdly shaped surface (such as in Figures 16.89 and 16.90), the surface integral on the left-hand side of Gauss's law is difficult to evaluate. But the *result* of the complicated integration is just $-4\pi GM_{\text{within } S}$. That is, the right-hand side of Gauss's law is easy to evaluate, because it is simple to account for the mass within the surface S.

The parallels between gravitational and electrical phenomena arise because of the similarity of the underlying force laws between masses and between charges. There are differences between the interactions, of course, notably (1) in the intrinsic strength of the interactions (cf. Example 16.5) and (2) in the direction of the forces: gravitation *always* is an attractive force (there is only one kind of mass, positive mass); whereas electrical forces can be *either* attractive or repulsive, as there are two types of charges.

When a (positive) mass M is *anywhere* inside the closed surface, the flux of the gravitational field through the surface is

$$\int_{\substack{\text{clsd} \\ \text{surface } S}} \vec{g} \cdot d\vec{S} = -4\pi G M$$

The gravitational field $\vec{g}$ is directed toward the mass M as shown in Figure 16.89. The electric field of a *negative charge* Q also points toward the charge, as shown in Figure 16.91.

So to convert Equation 6.55 to its electrical analog, replace

1. the field $\vec{g}$ with the field $\vec{E}$;
2. the mass M with the charge $|Q|$; and
3. the universal gravitational constant G with the corresponding electrical constant $1/4\pi\varepsilon_0$.

Thus we have

$$\int_{\substack{\text{clsd surface} \\ \text{enclosing } Q < 0\,\text{C}}} \vec{E} \cdot d\vec{S} = -4\pi \frac{1}{4\pi\varepsilon_0} |Q| = -\frac{|Q|}{\varepsilon_0}$$

Every gravitational field line that pierces the surface also reemerges.

M

FIGURE 16.90 **If M is outside the closed surface, there is no net threading of the gravitational field through the surface.**

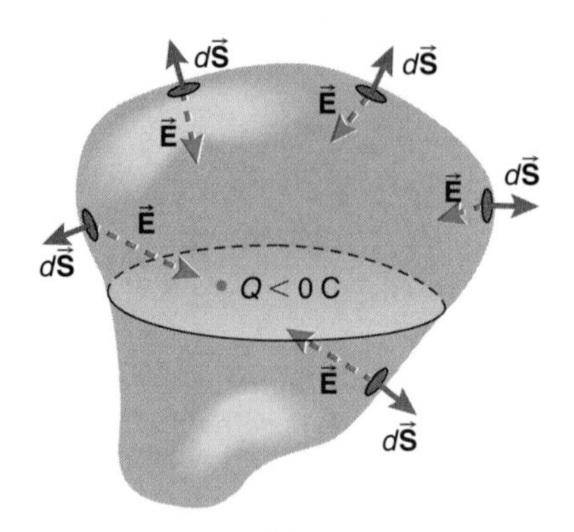

FIGURE 16.91 **The electric field of a negative charge points toward the charge.**

However, since Q itself is negative, we eliminate the absolute value signs by simply writing this as

$$\int_{\substack{\text{clsd surface}\\ \text{enclosing } Q<0\text{ C}}} \vec{E} \cdot d\vec{S} = \frac{Q}{\varepsilon_0}$$

What happens if a *positive charge* is inside the closed surface? Now the electric field of the charge points radially away from the charge (see Figure 16.92), opposite to the direction of the gravitational field $\vec{g}$.

Thus, for a positive charge within closed surface S, we can convert Equation 6.55 to the electrical case by

1. changing the field from $\vec{g}$ to $-\vec{E}$;
2. changing $|M|$ to $|Q|$; and
3. changing the universal gravitational constant G to the electrical constant $1/4\pi\varepsilon_0$.

Thus the flux of the electric field through a closed surface surrounding a positive charge is

$$-\int_{\substack{\text{clsd surface}\\ \text{enclosing } Q>0\text{ C}}} \vec{E} \cdot d\vec{S} = -4\pi\frac{1}{4\pi\varepsilon_0}|Q|$$

or

$$\int_{\substack{\text{clsd surface}\\ \text{enclosing } Q>0\text{ C}}} \vec{E} \cdot d\vec{S} = 4\pi\frac{1}{4\pi\varepsilon_0}|Q|$$

Since Q is itself positive, we can remove the absolute value signs and write

$$\int_{\substack{\text{clsd surface}\\ \text{enclosing } Q>0\text{ C}}} \vec{E} \cdot d\vec{S} = \frac{Q}{\varepsilon_0}$$

In the gravitational case, if there is zero mass within the closed surface, there is no net flux of the gravitational field through the closed surface. The same is true for the electrical case: if there is zero total charge within the closed surface S, there is no net flux of the electric field vector through S.

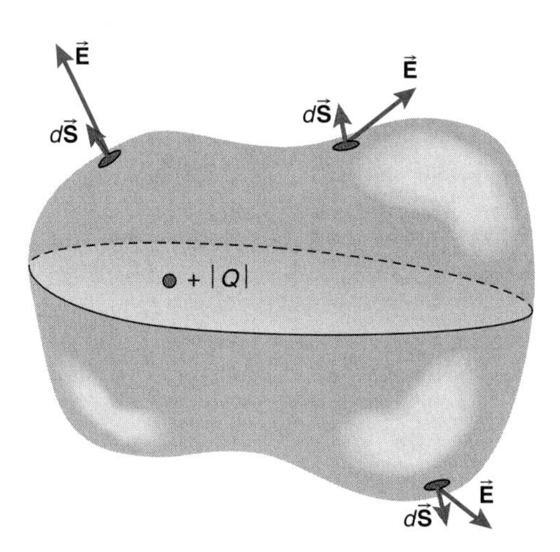

FIGURE 16.92 The electric field of a positive charge points away from the charge.

The results for both a positive or negative charge or zero charge within the closed surface can be summarized with the same equation:

$$\int_{\substack{\text{clsd}\\ \text{surface } S}} \vec{E} \cdot d\vec{S} = \frac{Q_{\text{within } S}}{\varepsilon_0} \tag{16.17}$$

where the appropriate sign is used for Q.

What happens if several different charges are enclosed by the surface S, as in Figure 16.93? Since the electric fields of the charges follow a linear principle of superposition, the fluxes of each field through the surface add algebraically. That is, the flux of the total field $\vec{E}$ through the closed surface is

$$\begin{aligned}\int_{\substack{\text{clsd}\\ \text{surface}}} \vec{E} \cdot d\vec{S} &= \int_{\substack{\text{clsd}\\ \text{surface}}} \vec{E}_1 \cdot d\vec{S} + \int_{\substack{\text{clsd}\\ \text{surface}}} \vec{E}_2 \cdot d\vec{S} \\ &\quad + \int_{\substack{\text{clsd}\\ \text{surface}}} \vec{E}_3 \cdot d\vec{S} + \cdots \\ &= \frac{Q_1}{\varepsilon_0} + \frac{Q_1}{\varepsilon_0} + \frac{Q_1}{\varepsilon_0} + \cdots\end{aligned}$$

where it is important to use the appropriate sign for the individual charges.

Any charge *outside* the closed surface S (such as charge Q' in Figure 16.93) contributes zero flux through the surface; we already showed this for mass in Chapter 6.

Thus Gauss's law for the electric field can be summarized in the following way:

$$\int_{\substack{\text{clsd}\\ \text{surface } S}} \vec{E} \cdot d\vec{S} = \frac{\text{total (net) charge enclosed by S}}{\varepsilon_0} \tag{16.18}$$

Gauss's law is a very general relationship between the flux of the *total* electric field through *any closed surface* and the *total (net) charge enclosed* by the surface.

PROBLEM-SOLVING TACTIC

16.6 To evaluate the flux of the electric field through a closed surface, you need only tally the total (i.e., net) charge enclosed within the surface and divide by ε_0. You need not perform the typically complicated integration on the left-hand side of Gauss's law.

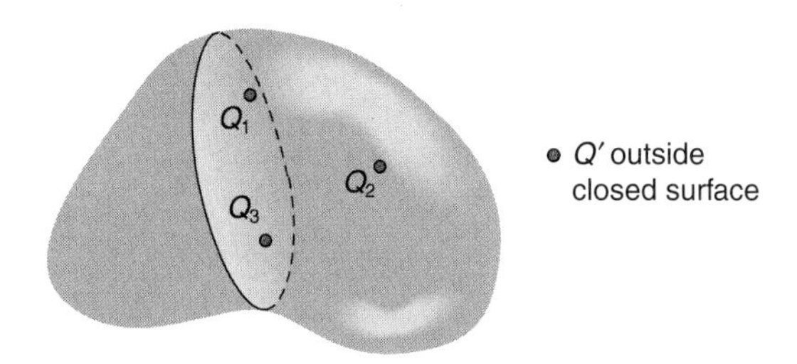

FIGURE 16.93 Several charges within a closed surface.

It is important to realize that Φ is the flux of the field caused by all charges anywhere, even those not enclosed by the surface. The imaginary closed surface involved in Gauss's law is known as a **Gaussian surface**, and we can choose its shape any way that seems useful.

Gauss's law (Equation 16.18) is one of four fundamental equations of electromagnetism called the **Maxwell equations**, named after James Clerk Maxwell (1831–1879), the great 19th-century Scottish physicist who first merged electric and magnetic phenomena into a single encompassing theory. Gauss's law stems from the Coulomb force law between charges; but physicists regard Gauss's law (rather than Coulomb's law) as the fundamental equation, since it expresses the relationship between electric fields and electric charge.

We also finally can see why the peculiar numerical factor of 4π was introduced into Coulomb's law (via $1/4\pi\varepsilon_0$). With the 4π in Coulomb's law (Equation 16.4), there is *no* factor of 4π in Gauss's law for the electric field (Equation 16.18). On the other hand, in the gravitational context, the gravitational force law *lacks* the factor of 4π because the force law has the constant written simply as G, and so a factor of 4π appears in the gravitational version of Gauss's law. The reason a similar juggling of factors of 4π was not done with the gravitational force law was simply because of the difficulty of changing notation. By the time Gauss's law was formulated for electromagnetism during the 19th century, the gravitational theory of Newton had been around for two centuries in the form we still know it. It was just too difficult to change notation at that point in time. This gives you a feeling for the collective inertia associated with the notation of any established theory or practice!

QUESTION 8

If Benjamin Franklin had defined electric charge the other way around, so that protons had negative charge and electrons had positive charge, how would this have affected the form of Gauss's law?

EXAMPLE 16.22

What is the flux of the total electric field through each of the closed Gaussian surfaces in Figures 16.94a–d?

Solution

According to Gauss's law (Equation 16.18), in each case you tally the total (net) charge enclosed within each closed Gaussian surface and divide by ε_0.

a. The total charge enclosed by the surface is zero, and so the flux of the electric field through the closed surface is zero.
b. The total charge enclosed by the surface is +2.00 μC. Thus the total flux is

$$\Phi = \frac{+2.00 \times 10^{-6}\ \text{C}}{\varepsilon_0}$$

$$= \frac{2.00 \times 10^{-6}\ \text{C}}{8.85 \times 10^{-12}\ \text{C}^2/(\text{N}\cdot\text{m}^2)}$$

$$= 2.26 \times 10^{5}\ \text{N}\cdot\text{m}^2/\text{C}$$

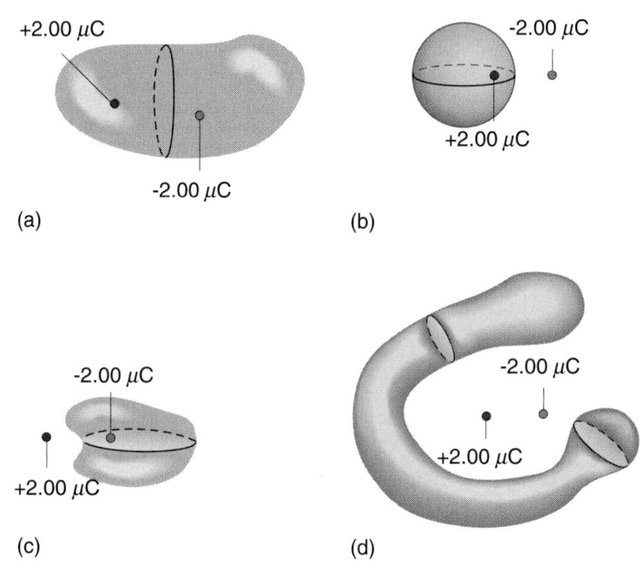

FIGURE 16.94

c. The total charge enclosed by the surface is −2.00 μC, and so the flux of the electric field through the surface is

$$\Phi = \frac{-2.00 \times 10^{-6}\ \text{C}}{8.85 \times 10^{-12}\ \text{C}^2/(\text{N}\cdot\text{m}^2)}$$

$$= -2.26 \times 10^{5}\ \text{N}\cdot\text{m}^2/\text{C}$$

d. The total charge enclosed by the surface is zero, and so the flux of the electric field through the surface is zero.

16.12 Calculating the Magnitude of the Electric Field Using Gauss's Law*

Now we discover how Gauss's law gives us amazingly rapid answers for the magnitude of the electric field for several highly symmetric distributions of charge. This method avoids the use of the complicated vector integration involved with Equation 16.16.

The idea of using Gauss's law to calculate the magnitude of an electric field stems from the very generality of the law: the shape of the closed surface involved is arbitrary. This means that in some situations we may be able to judiciously *choose* a surface that will assist us in finding the magnitude of the field.

The idea behind the method is to concoct an imaginary closed Gaussian surface that enables us to explicitly calculate the flux integral on the left-hand side of Gauss's law. We want the integral for the flux to be as easy as possible. Thus we devise a closed surface of such a shape that a constant-magnitude electric field exists over all or at least part of the surface. The flux integral on the left-hand side of the law then is calculated and set equal to the net charge enclosed by the surface, divided by ε_0.

The symmetry of the charge distribution is a clue to choosing the shape of the Gaussian surface over which to evaluate the flux integral. But it is only for situations involving high symmetry that we can use Gauss's law to calculate the magnitude of the

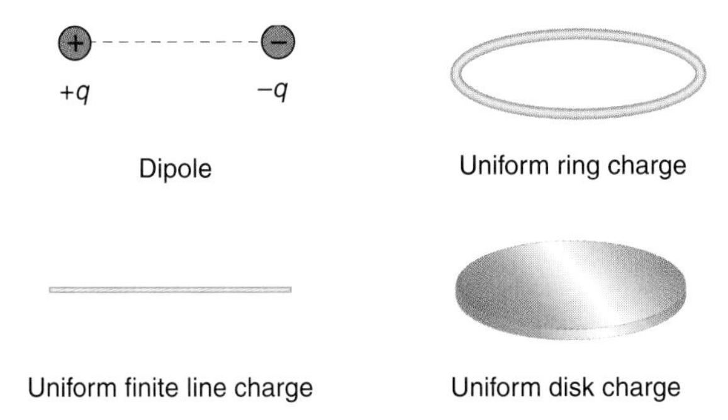

FIGURE 16.95 Charge distributions that *lack* the necessary high symmetry needed to use Gauss's law to find their electric fields.

field. What we mean by the term *high symmetry* is best seen by studying a number of specific examples. There is only a small collection of charge distributions with sufficient symmetry to permit Gauss's law to be usefully employed to find the magnitude of the electric field. In most situations, the charge distribution does not possess enough symmetry to enable us to find a convenient Gaussian surface over which the field has a constant magnitude. Several common charge distributions that *lack* sufficient symmetry are shown in Figure 16.95.

Gauss's law *still applies* to any closed surface; we just cannot easily use the law to find the field because the flux integral on the left-hand side of the law is too complicated to evaluate with ease or in closed form. However, we still can use Gauss's law to find the value of the total electric flux. For these geometries, we must resort to the vector integration techniques of Equation 16.16 to calculate the electric field at any point. This was just what we did in Section 16.9.

The Magnitude of the Electric Field of a Single, Pointlike Charge

We already know what this field is, but the point here is to see how the field can be found using Gauss's law in a neat way. Consider Q to be a positive charge. The field will point radially away from Q, as shown in Figure 16.96.

The symmetry of the situation is the key. Symmetry implies that as long as we remain at a fixed distance r away from the charge, the field $\vec{\mathbf{E}}$ has a constant magnitude. Thus the magnitude of the field is constant over the entire surface of a sphere of radius r centered on the charge. So the spherical symmetry indicates that we should choose a sphere of radius r, centered on Q, to be the mathematical Gaussian surface over which to calculate explicitly the flux of the electric field—that is, over which to evaluate the integral on the left-hand side of Gauss's law.

The field is constant in magnitude over the entire surface and directed radially outward. The various differential surface area vectors $d\vec{\mathbf{S}}$ also are directed radially outward and so are parallel everywhere to the local $\vec{\mathbf{E}}$ at their locations. Thus the scalar product on the left-hand side of Gauss's law becomes

$$\int_{\text{sphere}} \vec{\mathbf{E}} \cdot d\vec{\mathbf{S}} = \frac{Q_{\text{net enclosed}}}{\varepsilon_0}$$

$$\int_{\text{sphere}} E\, dS \cos 0° = \frac{Q_{\text{net enclosed}}}{\varepsilon_0}$$

$$\int_{\text{sphere}} E\, dS = \frac{Q_{\text{net enclosed}}}{\varepsilon_0}$$

Since the magnitude of the electric field is constant over the entire surface, E can be brought out from the integral and we can integrate the *left*-hand side of Gauss's law, obtaining

$$E\left(4\pi r^2\right) = \frac{Q_{\text{net enclosed}}}{\varepsilon_0}$$

because $4\pi r^2$ is the surface area of the spherical Gaussian surface.

The charge enclosed by the Gaussian surface is just Q. Thus Gauss's law becomes

$$E\left(4\pi r^2\right) = \frac{Q}{\varepsilon_0}$$

When we solve for the magnitude of the electric field E, a familiar result emerges:

$$E = \frac{1}{4\pi\varepsilon_0}\frac{Q}{r^2}$$

If the pointlike charge is *negative*, finding the field using Gauss's law proceeds along similar lines but with a few variations. The field of a *negative* charge points radially toward the charge, as shown in Figure 16.97.

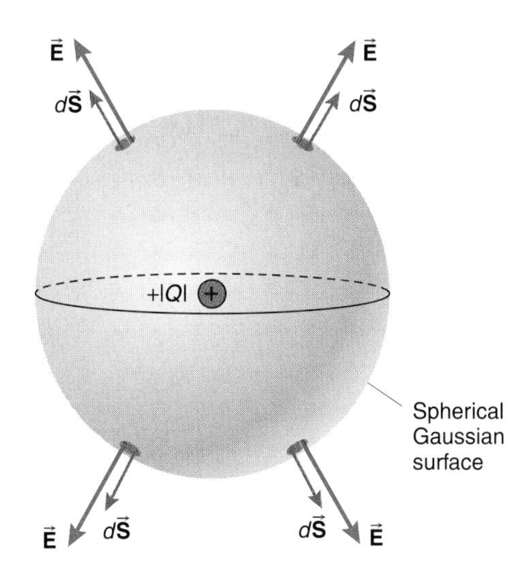

FIGURE 16.96 The electric field a distance r from a pointlike charge.

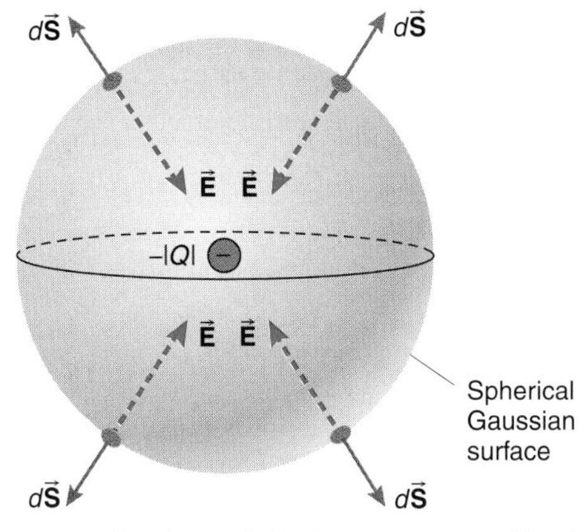

FIGURE 16.97 The electric field of a negative pointlike charge points radially toward the charge.

The symmetry indicates that at a fixed distance r away from the charge, the magnitude of the field is constant. Thus the magnitude of the field is constant over the surface area of a sphere centered on the charge. As before, we choose this sphere as the Gaussian surface over which to evaluate the flux of the electric field. Let E be the magnitude of the field. The differential surface area vector $d\vec{S}$ is *antiparallel* to $\vec{E}$ over the entire Gaussian surface. Thus the left-hand side of Gauss's law becomes

$$\int_{\text{sphere}} \vec{E} \cdot d\vec{S} = \frac{Q_{\text{net enclosed}}}{\varepsilon_0}$$

$$\int_{\text{sphere}} E\, dS \cos 180° = \frac{Q_{\text{net enclosed}}}{\varepsilon_0}$$

$$-\int_{\text{sphere}} E\, dS = \frac{Q_{\text{net enclosed}}}{\varepsilon_0}$$

Since E is constant over the entire surface, the left-hand side of the law becomes

$$-E \int_{\text{sphere}} dS = \frac{Q_{\text{net enclosed}}}{\varepsilon_0}$$

$$-E\left(4\pi r^2\right) = \frac{Q_{\text{net enclosed}}}{\varepsilon_0}$$

The charge enclosed by the surface is $-|Q|$. Gauss's law becomes

$$-E\left(4\pi r^2\right) = -\frac{|Q|}{\varepsilon_0}$$

and so the magnitude of the field (a positive number!) is

$$E = \frac{1}{4\pi\varepsilon_0}\frac{|Q|}{r^2}$$

Once again, a familiar result.

The Magnitude of the Electric Field of an Infinite, Uniform Line Charge Distribution

Let there be $+|\lambda|$ coulombs of charge on each meter of the line, so that the linear charge density is $+|\lambda|$, as shown in Figure 16.98.

The symmetry of the charge distribution indicates that the field will everywhere point radially away from the line and that if we maintain a fixed distance r from the line, the magnitude of the field will be constant. The cylindrical symmetry indicates that an appropriate Gaussian surface for this problem is a cylinder of radius r, concentric with the line charge, as shown in Figure 16.99.

The magnitude of the electric field is constant over the lateral area of the Gaussian surface but is not constant over the ends. This latter fact may seem like a problem, but we will see, in fact, it is not.

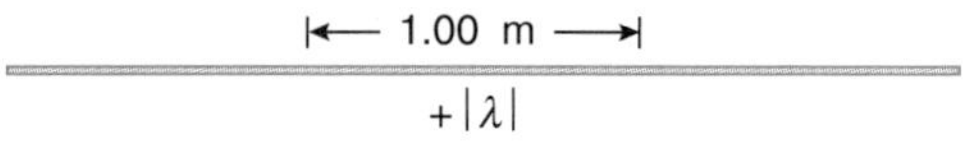

1.00 m

$+|\lambda|$

FIGURE 16.98 An infinite line charge with λ coulombs of charge per meter of length.

The flux of the electric field through this closed, cylindrical Gaussian surface consists of the flux through the cylindrical lateral area plus the flux through the two ends:

$$\int_{\text{lateral}} \vec{E} \cdot d\vec{S} + \int_{\text{end 1}} \vec{E} \cdot d\vec{S} + \int_{\text{end 2}} \vec{E} \cdot d\vec{S}$$

Over the lateral area, the differential surface area vectors $d\vec{S}$ are everywhere parallel to $\vec{E}$ at their locations. The field $\vec{E}$ also is constant in magnitude over the lateral area, and so the flux through the lateral area is

$$\int_{\text{lateral}} \vec{E} \cdot d\vec{S} = \int_{\text{lateral}} E\, dS = E \int_{\text{lateral}} dS = E\left(2\pi r\ell\right)$$

where ℓ is the length of the Gaussian cylinder. Over each end surface, the differential area vectors $d\vec{S}$ always are perpendicular to the direction of $\vec{E}$ at every location. Thus over the ends, the scalar product everywhere is zero:

$$\vec{E} \cdot d\vec{S} = 0 \text{ N}\cdot\text{m}^2/\text{C}$$

Hence there is zero flux through the end surfaces. (Notice that the lines of $\vec{E}$ do not thread the end surfaces of the Gaussian cylinder.) Hence the flux of the electric field vector through the entire closed surface is just that through the lateral area. The charge enclosed by the Gaussian surface is the charge on the length ℓ of the line within the Gaussian cylinder: $\lambda\ell$. Thus Gauss's law becomes

$$E\left(2\pi r\ell\right) = \frac{\lambda\ell}{\varepsilon_0}$$

Solving for the magnitude of the field, we find

$$E = \frac{\lambda}{2\pi r\varepsilon_0} \tag{16.19}$$

$$E = \frac{1}{4\pi\varepsilon_0}\frac{2\lambda}{r}$$

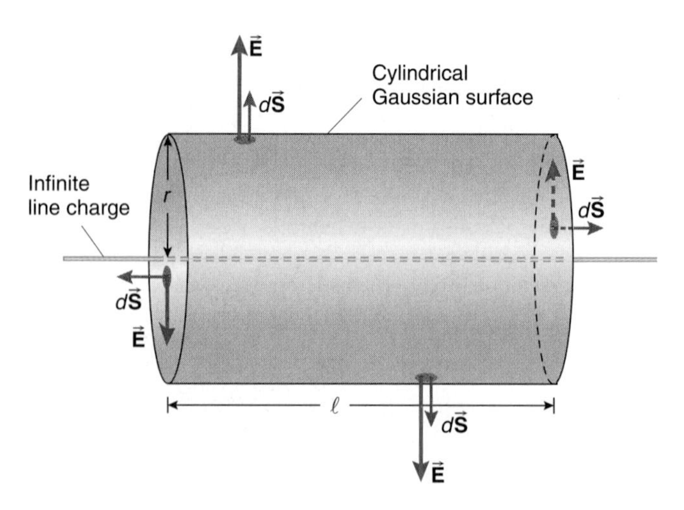

FIGURE 16.99 A cylindrical Gaussian surface concentric with the line charge.

The Magnitude of the Electric Field Inside a Uniformly Charged Solid Sphere

Let charge $+|Q|$ be distributed uniformly throughout a spherical volume of radius R. We want to find the magnitude of the electric field inside the charge distribution at a distance $r < R$ from the center, as shown in Figure 16.100.

The spherical symmetry indicates that the electric field will be in the radial direction. The magnitude of the field will be same for a constant value of r. Hence we choose a spherical Gaussian surface of radius r concentric with the spherical charge distribution. The field has a constant magnitude over this surface and is in the same direction as the differential surface area vectors $d\vec{S}$ at all points on the surface. Thus the left-hand side of Gauss's law is

$$\int \vec{E} \cdot d\vec{S} = \frac{Q_{\text{net enclosed}}}{\varepsilon_0}$$

$$E \int dS = \frac{Q_{\text{net enclosed}}}{\varepsilon_0}$$

$$E\left(4\pi r^2\right) = \frac{Q_{\text{net enclosed}}}{\varepsilon_0}$$

The amount of charge within the Gaussian surface is just that fraction of the total charge that lies closer to the center than r. Let ρ be the volume charge density of the charge distribution:

$$\rho \equiv \frac{|Q|}{\frac{4}{3}\pi R^3}$$

Then the charge within the Gaussian sphere of radius r is

$$\frac{4}{3}\pi r^3 \rho$$

or

$$\frac{r^3}{R^3}|Q|$$

Hence Gauss's law becomes

$$E\left(4\pi r^2\right) = \frac{\frac{r^3}{R^3}|Q|}{\varepsilon_0}$$

Solving for the magnitude of the field, we find

$$E = \frac{1}{4\pi\varepsilon_0}\frac{|Q|}{R^3}r \qquad (16.20)$$

Notice that the field increases in magnitude linearly with r until $r = R$. This is analogous to the gravitational field inside a uniform sphere of mass, which also increases linearly with r inside the sphere, as indicated in Table 6.1. If the charged sphere is positive, the field is directed radially outward; if negative, the field is directed radially inward.

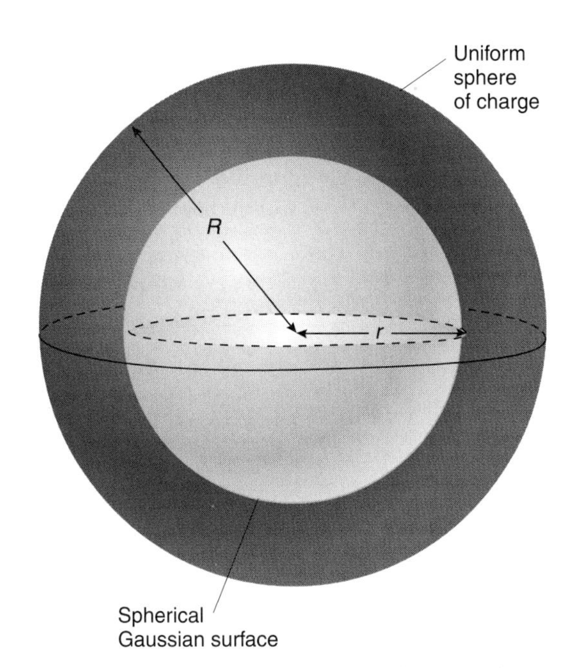

FIGURE 16.100 A uniform spherical distribution of charge.

The Magnitude of the Electric Field of a Uniformly Charged Infinite Sheet

Let the uniform surface charge density be $+|\sigma|$. The symmetry of the problem indicates that the electric field will point perpendicularly away from the sheet. As long as we are the same distance z from the sheet, the field should be constant. For this symmetry we can choose a Gaussian surface in the shape of a cylindrical hockey puck, as shown in Figure 16.101.*

The flux of the electric field vector through this surface is the sum of the flux through each end of the puck plus the flux through the lateral area. The flux through the lateral area, however, is zero because the electric field everywhere is

*This shape sometimes is called a pillbox, since it looks like the small boxes persons occasionally use to store pills.

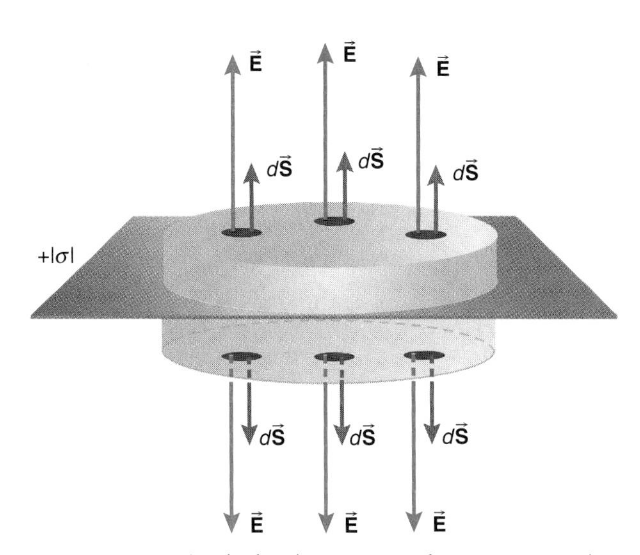

FIGURE 16.101 A cylindrical Gaussian surface intersecting the sheet of charge.

perpendicular to the differential surface area vectors $d\vec{S}$ over the lateral area. Thus

$$\int_{\text{lateral area}} \vec{E} \cdot d\vec{S} = 0 \text{ N}\cdot\text{m}^2/\text{C}$$

Over the entire top of the puck, the electric field vector is constant and parallel to every $d\vec{S}$. Thus the flux of $\vec{E}$ through the top of the puck is

$$\int_{\text{top}} \vec{E} \cdot d\vec{S} = \int_{\text{top}} E\, dS = E \int_{\text{top}} dS = EA$$

where A is the area of the top of the puck. The electric field vector is constant over the entire bottom of the puck too; it also is parallel to $d\vec{S}$ over the entire area of the bottom of the puck. Thus the flux through the bottom of the puck is

$$\int_{\text{bottom}} \vec{E} \cdot d\vec{S} = \int_{\text{bottom}} E\, dS = E \int_{\text{bottom}} dS = EA$$

The total flux through the hockey puck surface thus is

$$EA + EA = 2EA$$

and this is the left-hand side of Gauss's law.

The charge enclosed by the puck is the area A times the surface charge density $+|\sigma|$. Thus Gauss's law becomes

$$2EA = \frac{|\sigma| A}{\varepsilon_0}$$

The magnitude of the electric field is, therefore,

$$E = \frac{|\sigma|}{2\varepsilon_0}$$

as we found before (Example 16.16) for an infinite charged sheet.

QUESTION 9

To calculate the magnitude of the electric field of a uniformly charged infinite sheet using Gauss's law, is it permissible to use a cubical Gaussian surface with the charged sheet intersecting the cube parallel to two of its faces? Is it necessary to have the sheet bisect the cube?

16.13 CONDUCTORS*

A conductor is a material in which the electrons at the outer periphery of an atom have no great affinity for any particular individual atom; they are not bound or tied to individual atoms. These so-called **conduction electrons** are essentially free to move readily and quickly in response to electric fields.

The Electrostatic Field Inside a Conductor

Let a conducting plate be placed in an electric field, as shown in Figure 16.102. The conduction electrons in the plate respond to the field and quickly move to the left, opposite to the field direction; the electrons leave behind positive ions. The positive ions in turn steal conduction electrons from atoms further to the right in the material. When charges are in motion we have an **electrodynamic** situation. In short order, what results is a negative charge accumulation on the left surface of the conductor in Figure 16.102 and a positive charge accumulation on the right surface. The effect of this charge separation within the conductor is to create another electric field within the conductor, directed opposite to the external field.

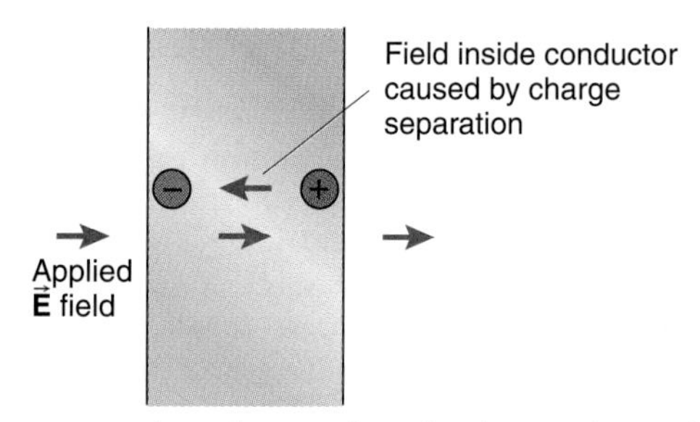

FIGURE 16.102 A conducting plate placed in an electric field.

Charge separation continues until such time that the total electric field within the conductor is zero.

When the total electric field inside the conductor is zero, we have an **electrostatic** situation with no further charge flow.

If electric charge is placed on a conductor (so that the net charge on the conductor is not zero), the charge distributes itself on the conductor to ensure that the electric field within the conductor is zero. Any such so-called free charge on a conductor must reside on the surface of the conductor. To see why this is the case, we consider a Gaussian surface located entirely *just inside* the conductor but infinitesimally close to the surface, as shown in Figure 16.103.

Since the electric field is zero *inside* the conductor in electrostatics, $\vec{E} = 0$ N/C at *every* $d\vec{S}$ over the entire Gaussian surface. The left-hand side of Gauss's law vanishes, because the field itself is zero over the whole Gaussian surface. Therefore the right-hand side of Gauss's law also must be zero. In other words, there is no net electric charge within this Gaussian surface. Other smaller Gaussian surfaces can be considered any and everywhere within the conductor, as in Figure 16.104, and the same argument shows that no net charge exists within them either.

Hence no free charge can exist anywhere within the conductor.

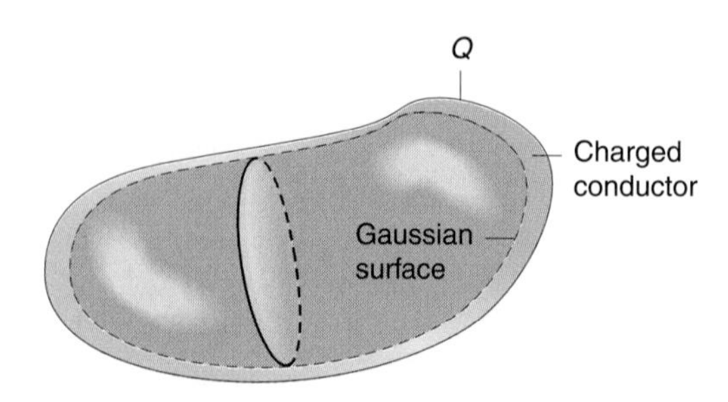

FIGURE 16.103 A Gaussian surface located just inside the surface of a conductor.

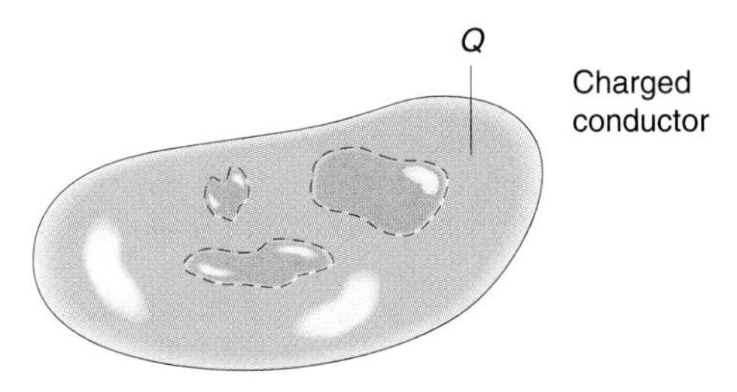

FIGURE 16.104 Additional Gaussian surfaces within the conductor.

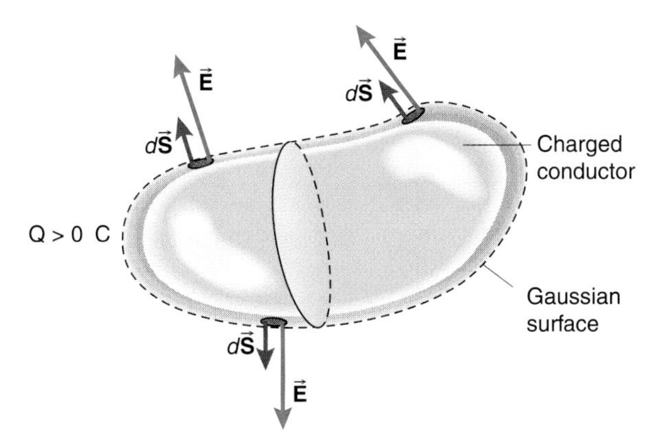

FIGURE 16.105 A Gaussian surface just outside the surface of a conductor.

A Gaussian surface placed differentially just *outside* the surface of the charged conductor, as in Figure 16.105, certainly will have a flux through it, since there is an electric field *outside* the charged conductor (due to the free charge Q on the conductor). The flux through this Gaussian surface must be Q/ε_0 according to Gauss's law.

Thus the free charge Q on a conductor does not lie within it but must reside on its surface.

The Orientation of the Electric Field at the Surface of a Conductor

We also can show that the angle that an electrostatic electric field makes with the actual differential surface area vector at any point on the surface of a conductor is 0°. If the angle were anything other than 0°, as in Figure 16.106, the electric field at the actual surface of the conductor would have a component parallel to the surface. This tangential field then would cause the conduction electrons in the conductor to move. The situation then is not electrostatic, since charges are in motion.

So in electrostatics, the electric field vector at the surface of a conductor must be perpendicular to the surface.

The Magnitude of the Electric Field at the Surface of a Conductor

Examine a small section of the surface of a charged conductor, as in Figure 16.107. Let σ be the value of the surface charge density

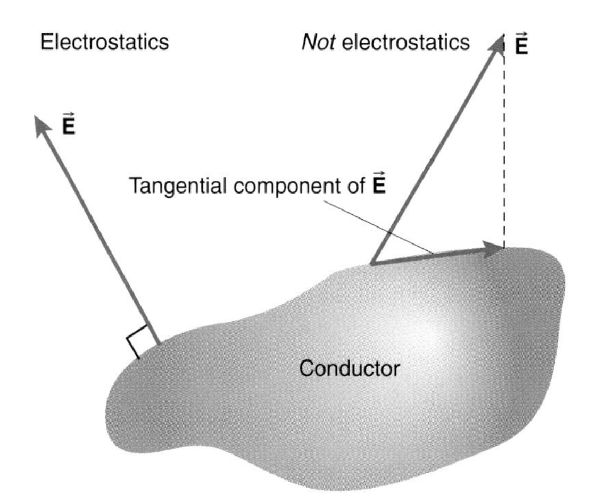

FIGURE 16.106 In electrostatics, the electric field must be perpendicular to the surface of a conductor.

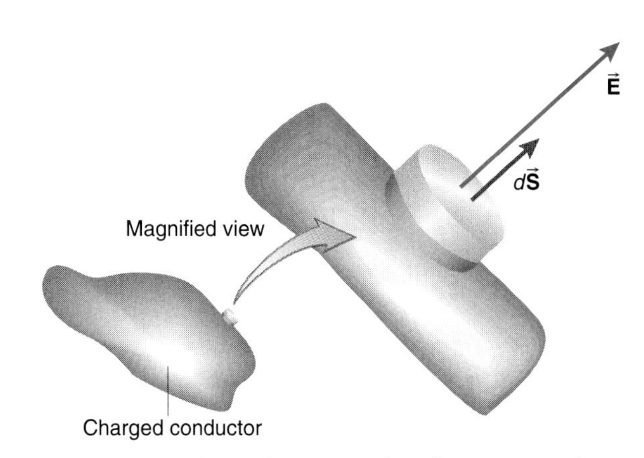

FIGURE 16.107 A tiny Gaussian surface that intercepts the surface of the conductor.

at this location on the surface of the conductor. Assume for simplicity that $\sigma > 0$ C/m^2. Imagine a very small Gaussian hockey puck surface, one end of which is *inside* the conductor and the other end just *outside* the conductor. Let the differential area $d\vec{S}$ of the ends of the puck be small enough so that the field is essentially constant across the end of the closed Gaussian surface.

The flux through the end of the puck lying just outside the conductor then is just $E\,dS$, since $\vec{E}$ is parallel to the differential area vector on this part of the Gaussian surface (since we assumed $\sigma > 0$ C/m^2). Since the field is zero inside the conductor, there is no flux through the end of the puck inside the conductor.

There also is no flux through the tiny portion of the lateral area just outside the conductor, because the electric field vector is perpendicular to the differential area elements over this lateral area. The flux through the portion of the lateral area within the conductor is zero since the field is zero inside this region.

Thus the total flux through the Gaussian surface is $E\,dS$. The area dS is small enough so that the surface charge density is constant over the differential area of the end of the puck. The total charge enclosed by the puck is $\sigma\,dS$. Hence Gauss's law becomes

$$E\,dS = \frac{\sigma\,dS}{\varepsilon_0}$$

The magnitude of the field is

$$E = \frac{\sigma}{\varepsilon_0}$$

Therefore the magnitude of the field at any location on the surface of any conductor is equal to the magnitude of the local surface charge density σ divided by ε_0.

You might try to reach the same conclusion if the surface charge density is assumed to be negative.

The surface charge density on a conductor is independent of position (i.e., is constant or uniform) *only* for isolated spherically shaped conductors, infinite cylindrical conductors, or infinite conducting planes. For differently shaped conductors, or for all shapes of conductors in the presence of other conductors or point charges, the surface charge density varies with position on the conductor.

16.14 Other Electrical Materials*

In contrast to conductors, in insulators (dielectrics) electrical charges remain fixed in position (except for polarization effects).

Semiconductors (such as germanium and silicon) are neither good conductors nor good insulators, thus making their name appropriate. They could equally well be called semi-insulators, but this terminology is not used. The exploitation of semiconducting materials through the judicious addition of small amounts of other elements (a process called **doping**) led to the invention of the transistor (in 1949) and the electronic revolution of the last half of the 20th century.

Superconductors are quite unusual. These materials lose all resistance† to the flow of electric charges below characteristic temperatures that vary with the material. These materials are literally perfect conductors. Superconductivity was discovered in 1911 by Kamerlingh Onnes (1853–1926) when he found that mercury lost all electrical resistance below the rather frigid temperature of 4.2 K. Such extremely cold temperatures are difficult and expensive to achieve (and are typically achieved through the use of liquid helium that, coincidentally, has a boiling point also at about 4.2 K). This low temperature has restricted the use of such materials to applications that can justify the expense involved. The discovery in 1987 of materials that exhibit superconductivity at warmer temperatures [above 100 K, well above the boiling point of liquid nitrogen (77 K), a much cheaper refrigerant than liquid helium] has led to a resurgence of interest in superconductivity and its potential technological applications in a wide variety of fields, including electrical power transmission, medical imaging technologies, and high-speed computing. Explanations of superconductivity involve aspects of quantum mechanics.

†We introduce this concept formally in Chapter 19.

Chapter Summary

An *electric charge* is said to be *positive* if it is repelled by a glass rod that has been freshly rubbed with silk; the charge is *negative* if it is repelled by a rubber rod (or amber) that has been freshly rubbed with fur. *Like charges* have the same sign (but may have different amounts of charge); *unlike charges* have opposite signs (and also may have different amounts of charge).

Conductors are materials in which charges are free to move; silver, copper, gold, and iron are examples of conducting materials. *Insulators (dielectrics)* are materials in which charges are not free to move; examples of insulators include glass, rubber, and wood.

The total electric charge is conserved in any process in an isolated system; this is called *conservation of charge*.

The magnitude of the electrical forces that two pointlike charges q and Q exert on each other is found using *Coulomb's law:*

$$F_{\text{elec}} = \frac{1}{4\pi\varepsilon_0}\frac{|q||Q|}{r^2} \tag{16.4}$$

where, in SI units,

$$\frac{1}{4\pi\varepsilon_0} = 8.987\,552 \times 10^9\ \text{N}\cdot\text{m}^2/\text{C}^2$$

$$\approx 9.00 \times 10^9\ \text{N}\cdot\text{m}^2/\text{C}^2$$

The SI unit for the quantity of charge is the *coulomb* (C). The electrical force is repulsive for like charges and attractive for unlike charges.

Charge is *quantized*. All free charges q in nature are integral multiples of the *fundamental unit of charge* $e = 1.602 \times 10^{-19}$ C:

$$q = ne \tag{16.6}$$

where n is an integer known as the *charge quantum number*. The proton has a charge quantum number of +1, while the electron has a charge quantum number of −1; the neutron has a charge quantum number of zero. Macroscopic amounts of charge have huge charge quantum numbers.

The electrical force is conveyed by an *electric field* $\vec{\mathbf{E}}$ created in space by a *static* (i.e., fixed) arrangement of charges. The electric field is the force per unit charge on a small test charge placed at the point where the field exists. The electric field thus is measured in newtons per coulomb (N/C) when using SI units. We make the assumption that the test charge is small enough so that its presence does not affect the distribution of charges creating the field at the point in question. In this way, a charge q, placed at a point in space where the electric field is $\vec{\mathbf{E}}$, experiences a force that is

$$\vec{\mathbf{F}}_{\text{elec on } q} = q\vec{\mathbf{E}} \tag{16.8}$$

The electric field in space depends on the geometric arrangement of the charges producing the field. In particular, the electric field at a point in space a distance r from a pointlike charge Q is

$$\vec{\mathbf{E}} = \frac{1}{4\pi\varepsilon_0}\frac{Q}{r^2}\hat{\mathbf{r}} \tag{16.11}$$

The field points radially away from an isolated positive charge ($Q > 0$ C), and radially toward a negative charge ($Q < 0$ C).

The electric field vector of a static arrangement of charges obeys a principle of linear superposition: the total field at any point is the vector sum of the individual fields produced by each charge acting alone:

$$\vec{E}_{total} = \vec{E}_1 + \vec{E}_2 + \cdots \qquad (16.12)$$

The electric field can be pictured by drawing *electric field lines*. Such field lines begin on positive charges and end on negative charges. The electric field at any point is tangent to the electric field line at the location in question and has a magnitude proportional to the number of field lines per square meter oriented perpendicular to the lines.

A special arrangement of charges called an *electric dipole* consists of two charges $\pm|Q|$ of equal magnitude, separated by a distance d. The electric *dipole moment* $\vec{p}$ is defined to be

$$\vec{p} \equiv |Q|\vec{d} \qquad (16.13)$$

where the vectors $\vec{d}$ and $\vec{p}$ are directed from the negative to the positive charge. A dipole placed in an electric field $\vec{E}$ experiences a torque $\vec{\tau}$ given by

$$\vec{\tau} = \vec{p} \times \vec{E} \qquad (16.14)$$

The principle of superposition implies that the electric field created by a continuous distribution of charge is found by a vector integration (summation) of the pointlike contributions to the field, where the integration is taken over the extent of the charge distribution:

$$\vec{E} = \frac{1}{4\pi\varepsilon_0} \int_{\substack{\text{charge}\\\text{distribution}}} \frac{dq}{r^2}\,\hat{\mathbf{r}} \qquad (16.16)$$

The electric fields created in space by various continuous distributions of charge are summarized in Table 16.2 on page 734.

Gauss's law for the electric field of a static arrangement of charges states that the flux of the total electric field vector through a closed Gaussian surface is equal to the total (net) charge enclosed by the surface, divided by ε_0:

$$\int_{\substack{\text{clsd Gaussian}\\\text{surface}}} \vec{E} \cdot d\vec{S} = \frac{Q_{\text{net enclosed}}}{\varepsilon_0} \qquad (16.18)$$

For charge distributions with sufficiently high symmetry, Gauss's law can be used to find the magnitude of the electric field of the charge distribution.

Static charges on conductors reside on their surface. The static electric field inside a conductor is zero. The static electric field at any point on the surface of a conductor is perpendicular to the surface and of magnitude $|\sigma|/\varepsilon_0$, where $|\sigma|$ is the magnitude of the local surface charge density at the location in question.

Summary of Problem-Solving Tactics

16.1 **(page 715)** When using Coulomb's law for the force between pointlike charges, be sure to express the distance between the charges in meters, not centimeters, and the charge in coulombs.

16.2 **(page 715)** To avoid many confusing problems with signs, employ the following procedure when using Coulomb's law. Calculate the *magnitude* of the force (a positive scalar quantity) using Coulomb's force law, Equation 16.4. Sketch the situation. Then determine the direction of the force on each charge after examining the types of the charges involved:
- if the charges are *both* positive *or both* negative (so-called like charges), the force on each is repulsive; or
- if the charges are of *opposite* flavors (unlike charges), the force on each is attractive.

Indicate on your sketch the direction of the forces on the charges. Finally, if appropriate, introduce a Cartesian coordinate system and write the forces in standard Cartesian vector form.

16.3 **(page 718)** Remember that we treat the fundamental unit of charge $e = 1.602 \times 10^{-19}$ C as positive, and so the charge on the proton is $+e$, while that on the electron is $-e$.

16.4 **(page 721)** The algebraic sign of the charge q is very important in Equation 16.8.

16.5 **(page 723)** Use the following three-step procedure to find the electric field a distance r from a pointlike charge.

1. First find the *magnitude* of the field at the point in question:

$$E = \frac{1}{4\pi\varepsilon_0}\frac{|Q|}{r^2}$$

This always is a positive quantity, because it is the magnitude of a vector.

2. Determine the direction of the field at the desired point and indicate it in a sketch:
   - radially away from Q if Q is positive
   - radially toward Q if Q is negative
3. Finally, introduce a coordinate system and write the electric field vector $\vec{E}$ in terms of your chosen coordinate system.

16.6 **(page 743)** To evaluate the flux of the electric field through a closed surface, you need only tally the total (i.e., net) charge enclosed within the surface and divide by ε_0.

Questions

1. **(page 711)**; **2.** **(page 713)**; **3.** **(page 715)**; **4.** **(page 718)**; **5.** **(page 721)**; **6.** **(page 728)**; **7.** **(page 739)**; **8.** **(page 744)**; **9.** **(page 748)**

10. What experimental evidence exists to suggest that there are two, and only two, types of charge?

11. Is the vitreous electricity of Dufay positive or negative charge? What is his resinous electricity?

12. If a bit of material is attracted to a glass rod freshly rubbed with silk, can you be certain that the material is negatively charged? Explain. If the material is repelled by the glass rod, can you be certain it is positively charged? Explain.

13. The gravitational force is intrinsically much weaker than the electrical force. Electricity was discovered well before gravitation, yet we are much more aware of gravitational forces than electrical forces in our common experience. Explain why.

14. Two balloons are of identical size. One is now charged by rubbing. Are the balloons still exactly the same size?

15. Explain why it can be said correctly that like charges dislike each other, and that unlike charges like each other.

16. How is the process of electrification similar to heat transfer? How is it different? What is transferred in each process?

17. A thin stream of water falling from a pipette is attracted to a negatively charged rubber rod as indicated in Figure Q.17. Explain why. If a positively charged glass rod is used instead, is the stream repelled? Explain.

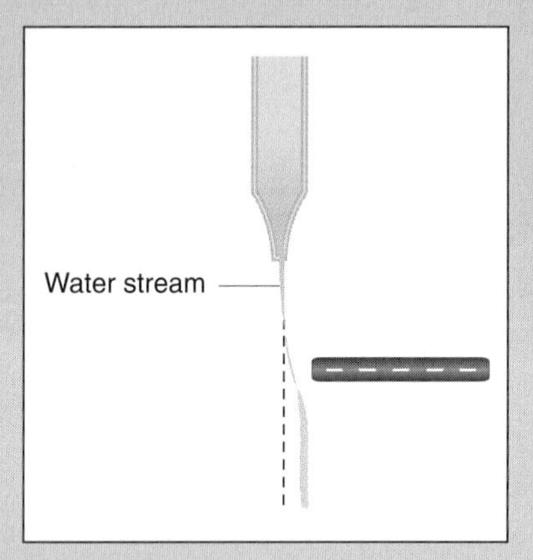

FIGURE Q.17

18. An electroscope (see Figure Q.18a) is a device consisting of twin thin gold foils connected to a metal sphere via a conducting wire. The system typically is enclosed in a glass case to avoid drafts and to prevent us from touching the delicate gold foils with our grubby paws. The gold foils normally hang vertically. (a) A charged rubber rod is brought near the metal sphere. The two gold foils are found to separate and remain so as long as the rod is near. If the rod is removed, the foils return to their initial vertical position (see Figure Q.18b). Explain why. (b) The charged rubber rod now is brought in contact with the metal sphere and the foils separate once again. While maintaining contact with the metal sphere, a grounding wire is connected between the sphere and a nearby water pipe. The foils return to the vertical position as soon as this is done. Explain why. The grounding wire now is removed, and then the charged rubber rod is removed. What happens to the two gold foils? Is the electroscope charged? If so, with what type of charge? Explain.

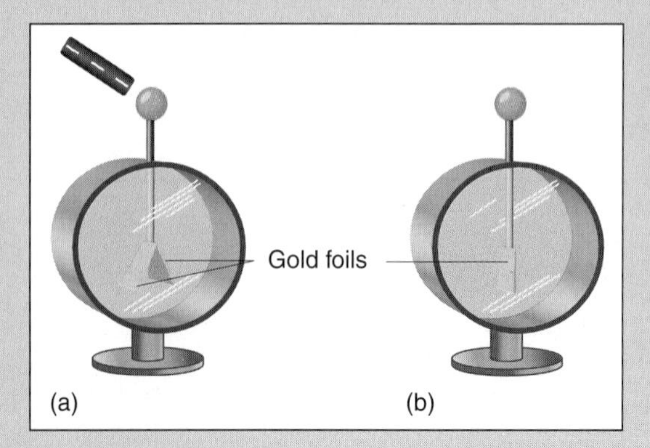

FIGURE Q.18

19. A piano factory polishes each piano when its assembly is complete. Why will the piano be likely to attract dust after the buffing? Can you prevent the same phenomenon from occurring after you polish your car? If so, how?

20. Does the term *total charge* on a conductor mean the same thing as the *free charge* on the conductor? Explain.

21. A pencil lies near this text. Explain why we do not normally notice the gravitational and electrical forces that each exerts on the other.

22. The fundamental unit of charge e occasionally is called the *quantum of charge*, meaning it is the smallest bit of isolated charge that exists in nature. Such quantum ideas are useful in other fields as well. (a) What is the quantum of currency in your country? (b) Could a biologist talk about a quantum of life? (c) Could you define a quantum of the written word or are many quanta necessary? (d) Could a computer scientist define a quantum of information or is more than one necessary?

23. Carefully distinguish between the terms *no charges present* and *zero charge present*.

24. In the electric field line diagram of a positive pointlike charge (Figure 16.55 on page 727), where are the negative charges on which the field lines end?

25. If a pointlike mass m with charge q is released at rest in an electric field, will the path followed by the particle always be *necessarily* coincident with the entire electric field line through the point of release? If so, explain why. If not, specify a situation where the particle does follow the field line and another situation where it does not.

26. Describe an experiment which can demonstrate that the electrical force between two charges is independent of their mass.

27. It is more difficult to electrify (charge) an object on humid days than on days with low humidity. What does this imply about the type of electrical material air is on these days?

28. Notice in Equation 16.11 (on page 722) that the magnitude of the electric field of a pointlike charge increases dramatically and approaches an infinite magnitude as we approach the charge in Figure 16.37; that is, as $r \to 0$ m, $|E| \to \infty$ N/C. Why does the charge not experience an infinite force? Explain carefully.

29. An electron is placed at a point in an electric field, and the electrical force on it and its acceleration are measured. The electron is removed and a proton is placed at the same location, and the electrical force on it and its acceleration also are measured. How are the forces on the electron and proton related to each other? How are the accelerations related to each other?

30. An electron is moving in the electric field of another pointlike charge Q. Is the orbital angular momentum of the electron about the charge Q conserved? Explain why or why not. Does your answer depend on the sign of the charge Q?

31. You likely have noticed that when clothes are removed from a dryer, they tend to cling together and to you. Explain why. Is the effect changed if a fabric softener is used in the drying process? How might you explain this phenomenon?

32. In Gauss's law, is the electric field in the flux integral the field of only the charges inside the closed Gaussian surface, or is it the total field of all the charges present whether inside or outside the Gaussian surface? Explain your answer.

33. Examine Table 16.2 on page 734 summarizing the electric fields of several charge distributions. A pointlike charge q placed in any of these fields experiences a force $\vec{F} = q\vec{E}$. Many of these forces are *not* inverse-square-law forces. Is this a violation of Coulomb's law? Explain.

34. The model of the water molecule shown in Figure 16.63 (on page 729) uses *fractional* values of the fundamental unit of charges e. The fractional units in the model are about the same magnitude as the charges on the up and down quarks, but this is purely fortuitous (coincidental). Do the fractional values of the fundamental unit of charge in the water molecule violate charge quantization?

35. A pseudo-scientist claims that the force keeping us on the Earth is an electrical force between unlike charges rather than gravity. What experiments can you cite or perform to show that this hypothesis is incorrect?

36. The electric field at a point P located a distance r from the center of a uniform, spherically symmetric charge distribution of total charge Q is given in Table 16.2. If the charge distribution shrinks symmetrically to one-third its former diameter, what happens to the value of the electric field at points $r > R$?

37. At one point along the line connecting two pointlike charges, the total electric field of both charges is found to be zero. Are the charges like or unlike charges? With this information alone, is it possible to tell if the charges are of equal magnitude or what the signs of the individual charges are?

38. The protons in an atom are all in the nucleus. The surrounding electrons are bound to the atom by the electrical force. The protons, of course, repel each other with a substantial electrical force, yet they do not fly out of the nucleus. Explain why these observations indicate the need for another fundamental force that is intrinsically even stronger than the electrical force, yet which must have a very limited range of effectiveness. This short-range, but very strong force, is (uninspiringly) called the *strong force*.

39. A rectangular strip of paper of length ℓ and width d is placed in a uniform electric field of magnitude E_0 as indicated in Figure Q.39. What is the flux of the electric field through the surface of the paper? Does this violate Gauss's law? Why or why not?

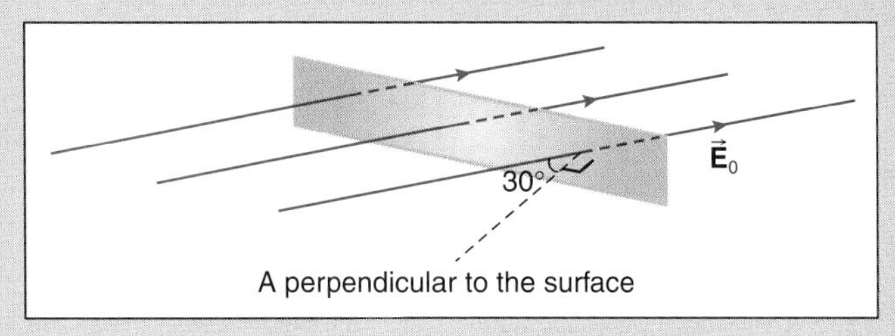

FIGURE Q.39

40. Take a rectangular strip of paper of length ℓ and width d, twist one end 180° about the long axis of the paper, and tape it to the other end. Such a surface is called a *Möbius strip* (see Figure Q.40). If you run a magic marker along the strip, you will find that the marker comes back to where you began; no side of the paper lacks a marker line, and so the strip can be considered one-sided. If such a Möbius strip surface is imagined to be in a uniform electric field of magnitude E_0, what is the total flux of the electric field through the Möbius strip surface?

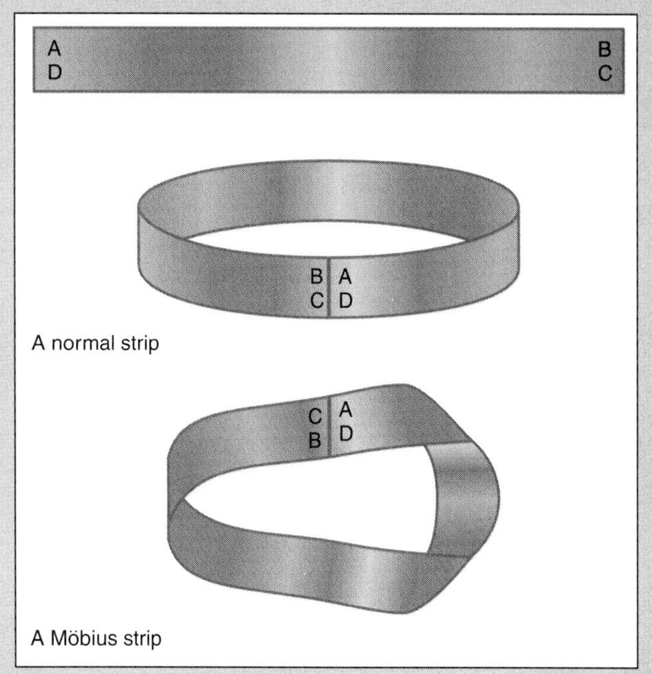

FIGURE Q.40

41. A *Klein bottle* is an extension of the Möbius strip to three dimensions; it is a surface with no inside or outside. A point charge Q is located near a Klein bottle as indicated in Figure Q.41. Can a Klein bottle be used as a Gaussian surface in Gauss's law?

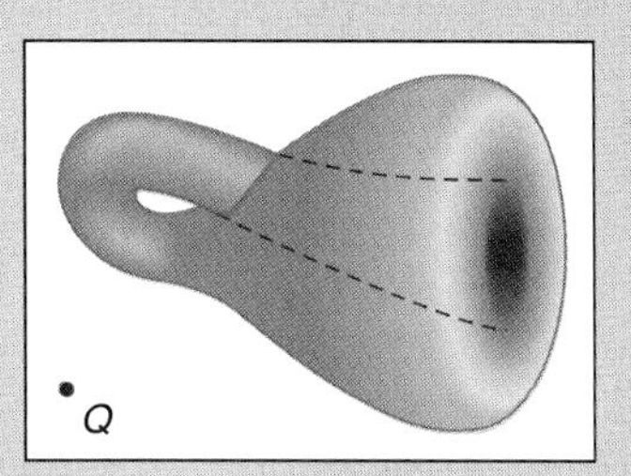

FIGURE Q.41

42. A point charge lies at the center of a spherical Gaussian surface. If the charge is moved to another location inside the Gaussian surface, is the flux of the electric field through the Gaussian surface changed when the charge is in its new location? Explain your answer.

43. If the flux of the electric field through a closed surface is zero, does this imply that the electric field is zero on the surface? Could it be zero on parts of the surface? Explain your answer.

44. A cube with a charge of $+Q$ has a uniform volume charge density. To calculate the magnitude of the electric field of the cubical charge distribution, a student suggests using Gauss's law with a cubical Gaussian surface with surfaces equidistant from each face of the charged cube. Will this enable the student to calculate the magnitude of the electric field of the cube? Why or why not? What is the flux of the electric field through such a Gaussian surface?

45. Two charges of opposite sign and unequal magnitude are the only two charges inside a closed and cozy Gaussian surface. (a) Is the flux of the electric field through the Gaussian surface changed if the positions of the two charges are interchanged? (b) Is the value of the electric field at various points on the Gaussian surface changed by such a procedure? Explain why or why not.

46. The gravitational force, the electrical force, and the Hooke's law spring force all are conservative forces. When studying gravitation, we introduced a gravitational field as a mechanism for the transmission of the force between masses; in this chapter, we introduced an electric field for similar purposes. Discuss whether it is appropriate and/or useful to invent a "spring field" for the study of the Hooke's law force.

47. If you accept the Eddington quote at the beginning of Section 16.4 on page 717, what is the largest charge quantum number?

PROBLEMS

Sections 16.1 The Discovery of Electrification
16.2 Polarization and Induction
16.3 Coulomb's Force Law for Pointlike Charges: The Quantification of Charge
16.4 Charge Quantization

1. A mole of electrons has what total charge? The magnitude of this charge is known as a *faraday* of charge.

2. Find the charge quantum number associated with a macroscopic charge of 1.000 nC.

3. What is the total positive charge contained in 1.00 kg of hydrogen gas (H_2)?

4. What is the total positive charge contained in 1.00 kg of water (H_2O)?

5. How much positive charge (in coulombs) exists in Avogadro's number of hydrogen atoms?

6. How many protons are in 1.00 C of positive charge? How many electrons are in a negative charge of −1.00 C?

•7. Electrical liftoff. A charge q is glued on top of your head and another like charge of equal magnitude is glued to the floor beneath your shoes. If you are 1.90 m tall and have a mass of 70.0 kg, what magnitude charge q will make the normal force of the floor on you equal to zero?

•8. Two M&M candies (yummy!), each of mass m, have equal charges $q > 0$ C bestowed on them and are placed on a frictionless surface a distance d apart as indicated in Figure P.8. (a) What is the magnitude of the electrical force on each M&M? What are the directions of the forces on each? (b) What is the magnitude of the acceleration of each M&M? (c) If the mass of each M&M is 8.6×10^{-4} kg and d is 0.100 m, how large must q be so that the acceleration of each M&M is 0.50 m/s^2?

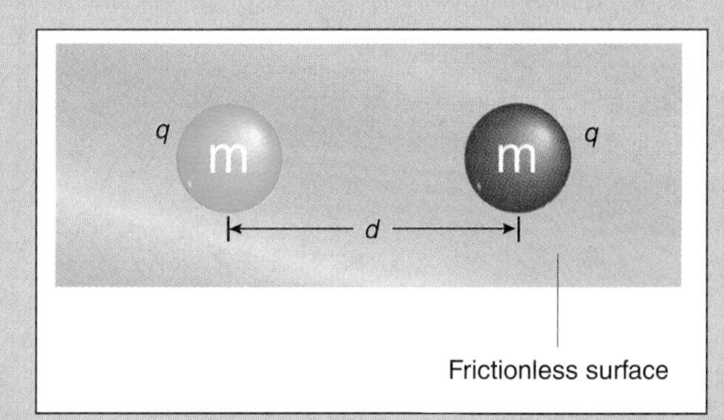

FIGURE P.8

•9. The nucleus of a gold atom has a charge quantum number of +79. An α-particle (a helium nucleus) has a charge quantum number of +2. For this problem *neglect* the electrons surrounding the nucleus of each atom. (a) What is the magnitude of the electrical force of the gold nucleus on the α-particle when it is 1.00×10^{-14} m from the nucleus? (b) What is the magnitude of the electrical force of the α-particle on the gold nucleus when the α-particle is at the same position as in part (a)?

•10. After tedious measurements, you find that five oil droplets in a Millikan-type experiment have the following charges on them:

Drop number	*Charge*
1	1.282×10^{-18} C
2	1.762×10^{-18} C
3	2.083×10^{-18} C
4	3.364×10^{-18} C
5	4.006×10^{-18} C

Without knowing or using the value of the fundamental unit of charge e, describe and use a procedure for determining the numerical value of e in coulombs *from these data alone*, making only the assumption that each charge is an integral multiple of the (unknown) fundamental unit of charge.

•11. Two small, identical, pointlike conducting spheres have charges $q_1 = -2.00$ nC and $q_2 = +6.00$ nC. (a) Calculate the magnitude of the electrical force on each when they are separated by 3.00 cm. Are the forces attractive or repulsive? (b) The two spheres are brought momentarily into contact and then separated by a distance of 3.00 cm. Now what is the magnitude of the electrical force on each? Are the forces attractive or repulsive?

•**12.** A pointlike charge $q_1 = 2.00 \times 10^{-10}$ C is 0.300 m from a second pointlike charge $q_2 = -3.00 \times 10^{-10}$ C. The attractive force on q_1 is found to be of magnitude 6.00×10^{-9} N. (a) What is the ratio of the magnitude of the electrical force on q_1 to the magnitude of the electrical force on q_2? (b) Calculate the ratio

$$\frac{Fr^2}{|q_1||q_2|}$$

(c) Is the ratio in part (b) the same in another experiment with different values for the charges and distances? Explain.

•**13.** Pretend gravity did not exist. Consider the Earth and Sun as pointlike masses and assume the Sun is fixed in position with the Earth in a circular orbit. If the Sun and the Earth had opposite charges of the same magnitude $|q|$, the Earth could orbit the Sun in much the same manner that it does under the influence of the gravitational force, as shown in Figure P.13. What magnitude of charge on each is necessary to accomplish this?

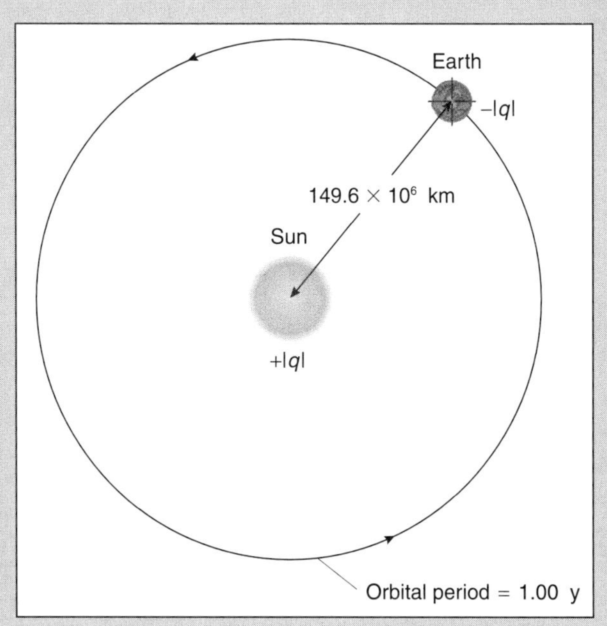

FIGURE P.13

•**14.** You and a friend each have mass 60 kg and each are holding 1.00 C of charge. Watch out! At what distance from each other is the magnitude of the electrical force on the charges equal to the magnitude of your weight on the surface of the Earth?

•**15.** Two spheres of identical mass m and radius have charges $+2.00 \times 10^{-9}$ C and -8.00×10^{-9} C. The spheres are attached to insulating, uncharged pucks of mass M that are free to glide on a frictionless airtable in a physics laboratory as indicated in Figure P.15. (a) Draw four second-law force diagrams and indicate all the forces acting on each sphere and each puck. If forces are of equal magnitude, draw the arrows of equal length. Use words such as "this is the force of _______ on _______" to describe each force. (b) In the four second-law force diagrams in (a), indicate which forces form Newton's third law force pairs. (c) For the horizontal direction, write Newton's second law for one sphere and for the puck on which it rides, letting a be the magnitude of their common horizontal acceleration.

This problem is modeled after and is an extension of a problem suggested by Arnold Arons, *A Guide to Introductory Physics Teaching* (Wiley, New York, 1990), pages 156–157; the figure is from page 157.

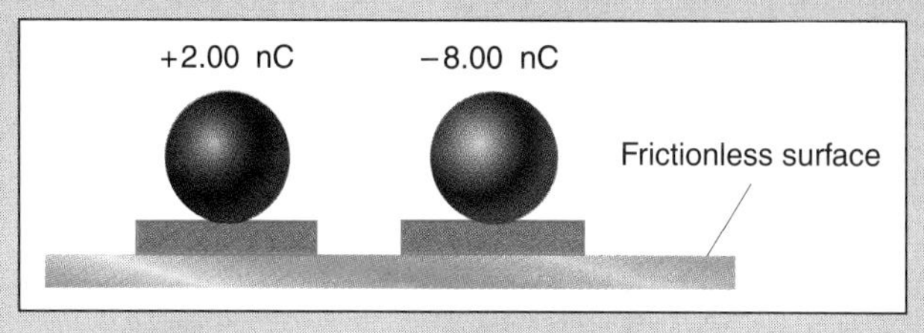

FIGURE P.15

•**16.** Two positive pointlike charges are a fixed distance r apart. The *total* charge of the pair is Q. (a) What is the charge on each such that the electrical force they exert on each other is of *maximum* magnitude? (b) What charge is on each such that the electrical force they exert on each other is of *minimum* magnitude?

•**17.** Two equal pointlike masses m, each with positive charge q, are suspended from a common point by threads of equal length ℓ and negligible mass. Because of their mutual repulsion, the threads make an angle θ with the vertical direction as indicated in Figure P.17. Do *not* neglect the weight of the masses. (a) Draw a second law force diagram indicating the forces on each mass. (b) Show that the charge on each mass can be expressed as

$$q = (2\ell \sin \theta)(4\pi\varepsilon_0 mg \tan \theta)^{1/2}$$

(c) Evaluate the charge if $m = 10$ g, $\ell = 0.500$ m, and $\theta = 15.0°$. (d) With the data in part (c), what is the approximate charge quantum number of each charge?

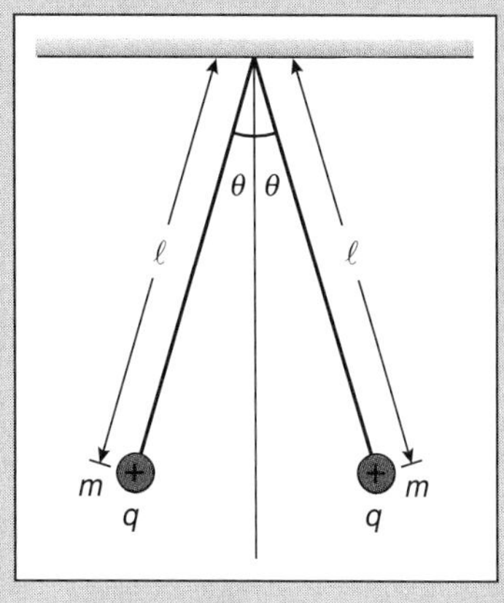

FIGURE P.17

•**18.** Three pointlike charges Q are located on three successive vertices of a regular hexagon with sides ℓ as indicated in Figure P.18. Find the electrical force on another charge q located at the center of the hexagon. Assume all the charges are like charges and express your result in terms of the coordinate system indicated.

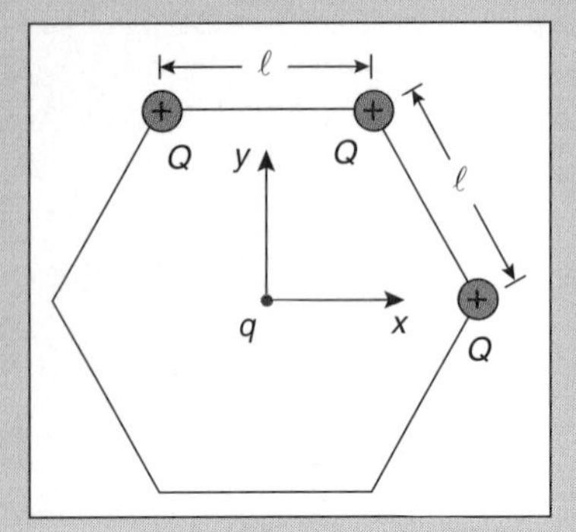

FIGURE P.18

•**19.** The +3.00 μC charge of the light brigade is situated relative to two other charges as shown in Figure P.19. (a) Find the total force acting on the charge of the light brigade. (b) Determine the magnitude of the total force. (c) Determine the acute angle that the total force makes with the x-axis.

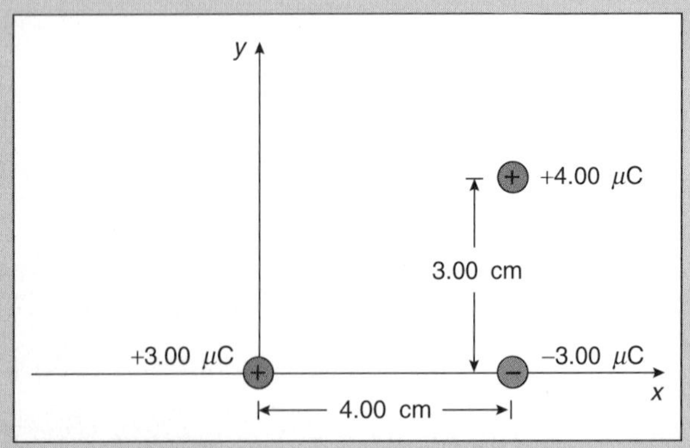

FIGURE P.19

20. A positron is a particle with the same mass as an electron but with a *positive* charge. Positrons are created in many radioactive particle decays and are present in cosmic rays (where they were first discovered). A positron and an electron can briefly form an unusual atom known as *positronium*. Imagine a situation where the two particles are in a circular orbit about their center of mass as indicated in Figure P.20. Since the particles have equal mass, the center of mass is midway between them. Let r be the *separation* of the particles (so the orbits are each of radius $r/2$). (a) Show that the orbital period T is related to the separation distance r by

$$T^2 = \frac{16\pi^3\varepsilon_0}{e^2}\frac{m_e m_p}{m_e + m_p} r^3$$

This is a consequence of Kepler's third law for electrical orbits. This expression simplifies slightly because the mass of the electron m_e is equal to the mass of the positron m_p:

$$T^2 = \frac{8\pi^3\varepsilon_0 m_e}{e^2} r^3$$

(b) Show that if an electron and a *proton* are in circular orbits about *their* center of mass (which is *not* at the midway point between them but much closer to the proton), then the same expression results:

$$T^2 = \frac{16\pi^3\varepsilon_0}{e^2}\frac{m_e m_p}{m_e + m_p} r^3$$

where m_p is now the mass of the proton: In this case $m_p >> m_e$, so that the expression simplifies to

$$T^2 = \frac{16\pi^3\varepsilon_0 m_e}{e^2} r^3$$

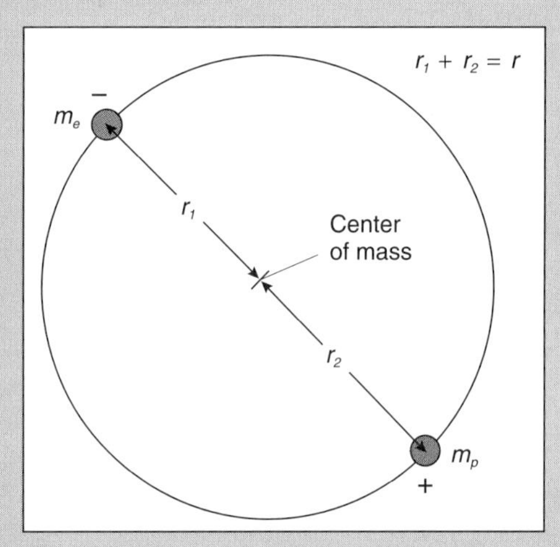

FIGURE P.20

21. When considering collections of masses and extended distributions of mass in Chapter 9, we introduced the concept of the center of mass. Since mass and charge play similar roles in their respective force laws, the idea of a *center of charge* may be useful. Discuss how you could extend the definition of the location of the center of mass to define an analogous position vector for the center of charge of a charge distribution. Under what circumstances might the concept of the center of charge be undefined?

Sections 16.5 The Electric Field of Static Charges 16.6 The Electric Field of Pointlike Charge Distributions

22. A uniform electric field of magnitude 500 N/C exists in a region of space, directed as indicated in Figure P.22. (a) An electron is placed in the field. Indicate the direction of the force on the electron relative to the field direction. Find the magnitude of the force and determine the magnitude of the acceleration of the electron. (b) A proton is placed in the field. Indicate the direction of the force on the proton relative to the field direction. Calculate the magnitude of the force on the proton as well as the magnitude of its acceleration. (c) Calculate the ratio of the magnitude of the acceleration of the electron to the magnitude of the acceleration of the proton when they are placed in the same field. Neglect any electrical interaction between the proton and electron. Compare your result with the ratio of the proton to the electron masses. Explain.

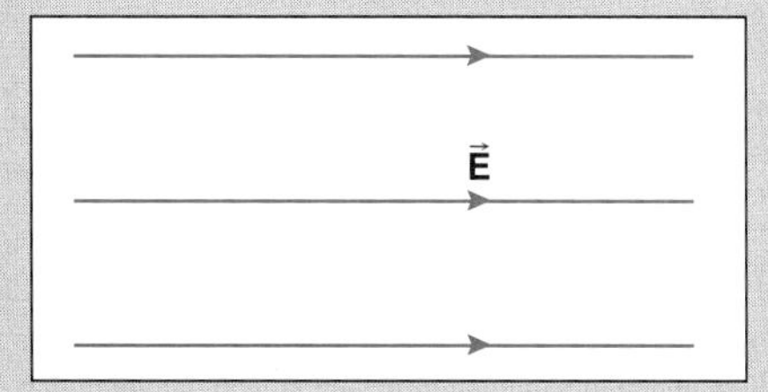

FIGURE P.22

23. At what distance from a proton is the magnitude of its electric field equal to 3.0×10^6 N/C?

24. (a) What is the magnitude of the weight of a proton on the surface of the Earth? (b) What magnitude electric field is sufficient to create an electrical force on the proton of magnitude equal to the magnitude of its weight? (c) How far away should another proton be placed to create an electric field of this magnitude?

25. A botany professor is in an electric field. The uniform electric field has a magnitude of 500 N/C. An electrical force of magnitude 6.0 N is found to act on the professor of mass 75 kg in the direction opposite to that of the field. What is the charge of the professor?

•26. Assume the electron in the ground state of the hydrogen atom is located 5.29×10^{-11} m from the nuclear proton. (a) What is the magnitude of the electric field of the proton at the position of the electron? Make a sketch to indicate the direction of this electric field at the position of the electron. (b) What is the magnitude of the electrical force on the electron? Make a sketch to indicate the direction of this force on the electron. (c) What is the magnitude of the electric field of the electron at the position of the proton? Make a sketch to indicate the direction of this electric field. (d) What is the magnitude of the electrical force on the proton? Make a sketch to indicate the direction of this force on the proton.

•27. A +2.00 μC charge is located at the origin and a −3.00 μC charge is located as indicated in Figure P.27. (a) Find the total electric field at the point P. (b) Suppose Pickett's charge of −4.00 μC now is placed at point P. What is the total force on this charge? (c) What is the magnitude of the force on this charge?

•28. Imagine the two electrons in the helium atom are in a circular orbit at the same radius ($r = 2.64 \times 10^{-11}$ m) about the central nuclear pair of protons but on opposite sides of the nucleus, as shown in Figure P.28. The nucleus is *quite* small compared with the size of the electron orbit; thus the nuclear charge can be considered as a point charge at the center of the orbit. (a) What is the magnitude of the total electric field at the position of one of the electrons? Make a sketch to indicate the direction of this field. (b) What is the magnitude of the electrical force on the electron at this location? Make a sketch to indicate the direction of the force on the electron. (c) What is the magnitude and direction of the electric field caused by each of the electrons at the position of the nuclear protons? What is the total electric field of the two electrons at the position of the nucleus? (d) What is the magnitude of the total electrical force that the two electrons exert on the nuclear protons?

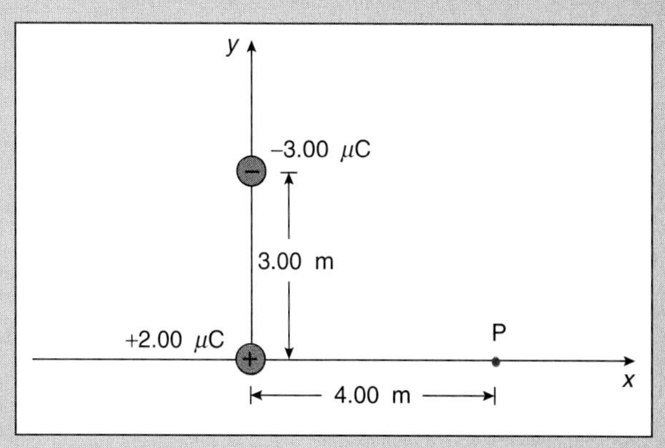

FIGURE P.27

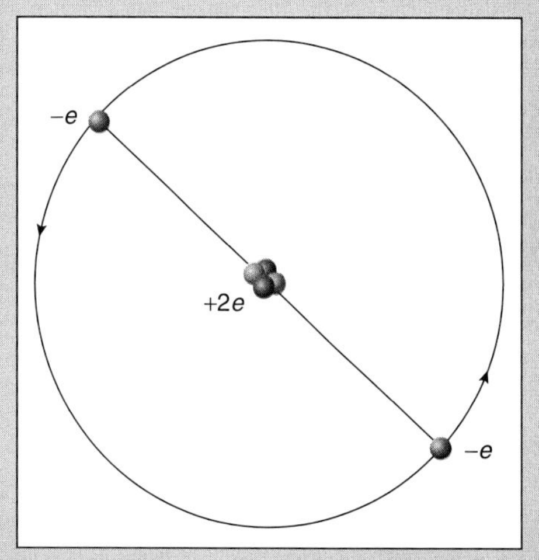

FIGURE P.28

•29. A small charged sphere of mass 20.0 g is suspended at rest by means of a thin (massless) thread of length 50.0 cm in a uniform electric field of magnitude 100 N/C directed as indicated in Figure P.29. (a) What is the sign of the charge on the sphere? (b) Draw a second law force diagram indicating all the forces acting on the little sphere. (c) Find the charge on the sphere.

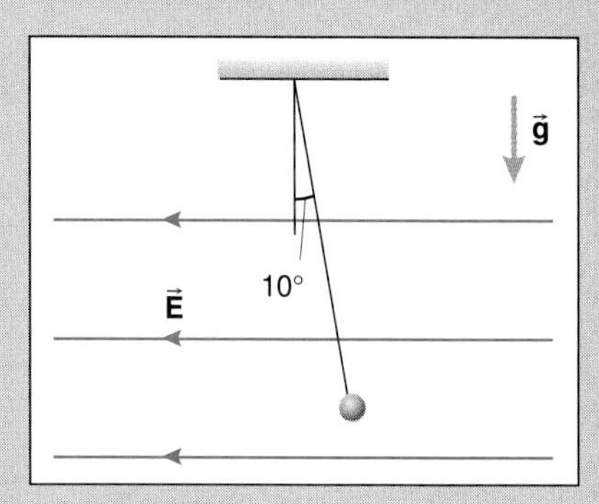

FIGURE P.29

•**30.** To produce an electric field, you arrange four charges of equal magnitude at the corners of a square corral of side ℓ as shown in Figure P.30. (a) Introduce an appropriate Cartesian coordinate system to analyze the problem, clearly indicating the origin and the direction of the positive axes. (b) Find the total electric field at the position of the charge in the upper right corner of the corral caused by the other three charges, expressing your result in Cartesian vector form. Schematically indicate the direction of the electric field on a sketch. (c) Calculate the force on the charge in the upper right corner of the corral and sketch the direction of the force in relation to the electric field calculated in part (b).

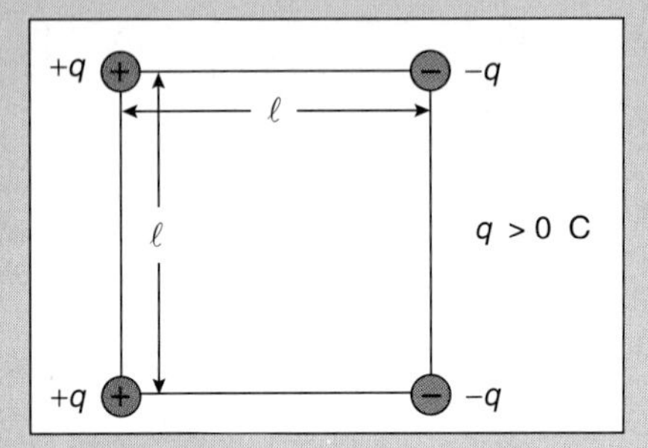

FIGURE P.30

•**31.** You have a pointlike charge of *magnitude* 3.00×10^{-3} C that is attracted to Milly Coulomb yet repelled by Billy Coulomb as shown in Figure P.31. (a) What is the sign of your charge? (b) Find the electric field of Milly and Billy at the position of your charge. Express your result as a Cartesian vector. (c) Find the total electrical force on your charge. Express your result as a Cartesian vector. (d) What is the magnitude of the electrical force on your charge?

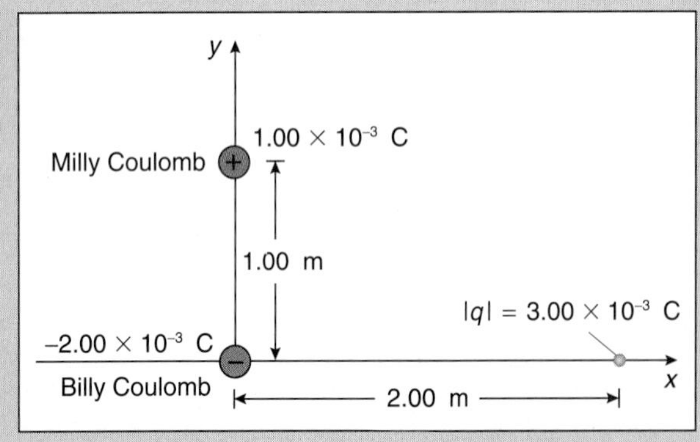

FIGURE P.31

•**32.** You are out cultivating a few charges and find two of them located as shown in Figure P.32. (a) Calculate the total electric field at the origin, expressing it in appropriate Cartesian vector form. (b) What is the magnitude of the electric field at the origin? (c) If a proton now is placed at the origin, find the force on the proton, expressing it in appropriate Cartesian vector form. (d) Find the *magnitude* of the acceleration of the proton.

Section 16.7 A Way to Visualize the Electric Field: Electric Field Lines

•**33.** Figure P.33 depicts the electric field lines associated with a distribution of three pointlike charges. Determine the signs of the various charges and their relative magnitudes.

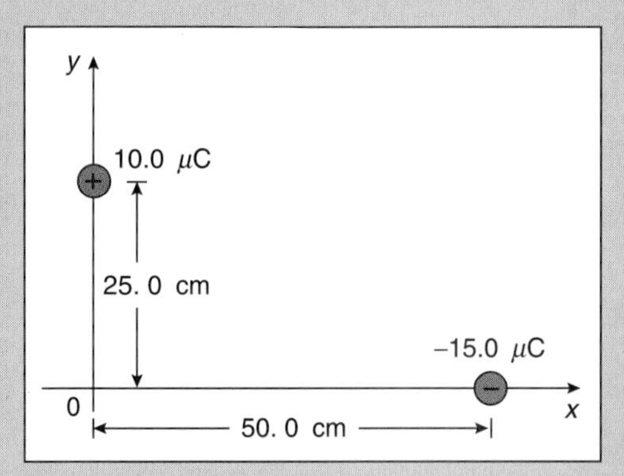

FIGURE P.32

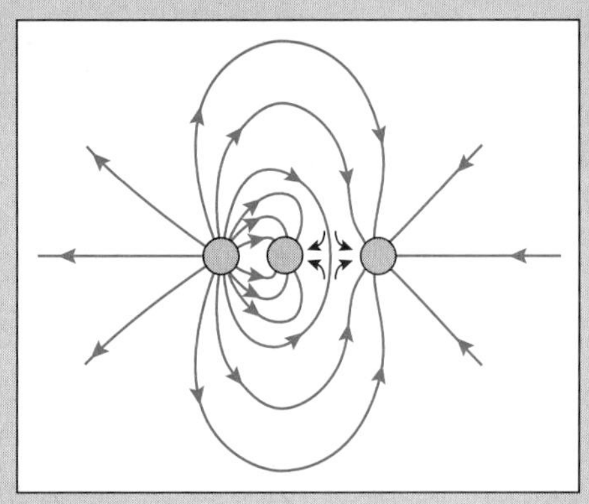

FIGURE P.33

Diagram from William Taussig Scott, *The Physics of Electricity and Magnetism* (Wiley, New York, 1959), page 27.

•**34.** Using the appropriate rules for drawing electric field lines, sketch the electric field lines in the vicinity of two negative charges of equal magnitude placed a distance *d* apart.

•**35.** Three charges of equal magnitude lie on the corners of an equilateral triangle. Two of the charges are positive and one is negative. Make a schematic diagram of the electric field lines of this charge distribution.

Section 16.8 A Common Molecular Charge Distribution: The Electric Dipole

•**36.** Show that the total electrical force on an electric dipole in a *uniform* electric field is zero, regardless of the orientation of the dipole.

•**37.** Notice in Table 16.1 on page 729 that the carbon dioxide molecule (CO_2) has no dipole moment but each carbon–oxygen bond does. What does this information imply about the structure of the molecule?

•**38.** During a physics lab, you and your paramour construct the beautiful electric dipole shown in Figure P.38. (a) Calculate the electric field at the point P, expressing it in Cartesian vector form. (b) A −5.00 μC charge is now placed at the point P. What is the electrical force on this new charge? Express your result in Cartesian vector form. (c) What is the magnitude of the force on the −5.00 μC charge?

•**39.** Determine the torque on the electric dipole shown in Figure P.39.

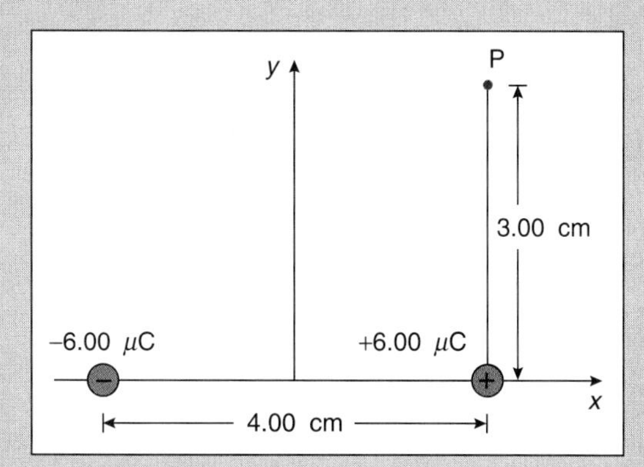

FIGURE P.38

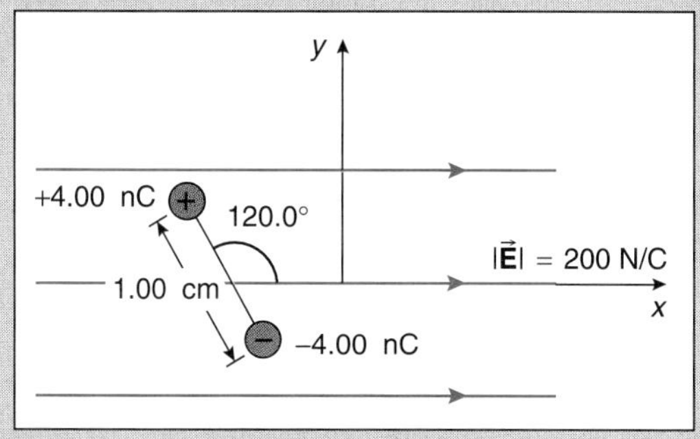

FIGURE P.39

•**40.** The water molecule has an electric dipole moment with a magnitude of about 6.0×10^{-30} C·m. The molecule is placed in a uniform electric field of magnitude 250 N/C with the orientation shown in Figure P.40. What is the torque on the water molecule in its present orientation?

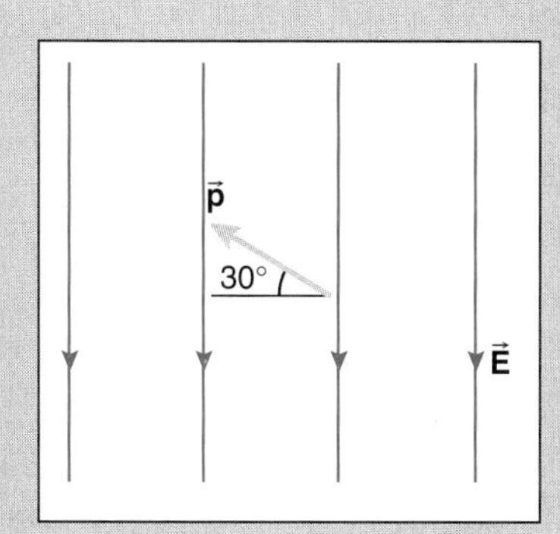

FIGURE P.40

•**41.** A pointlike charge $Q = -2e$ exists at the origin. A water molecule with a dipole moment of magnitude 6.0×10^{-30} C·m is oriented with its dipole moment vector directed toward Q as indicated in Figure P.41. The charge Q is 10.0 nm from the effective positive charge of the dipole. (Figure 16.63 on page 729 may provide pertinent information for this problem.) (a) What is the approximate magnitude of the total force on the water molecule? (b) What is the magnitude of the torque on the water molecule when it is in this orientation?

•**42.** Show that the torque on an electrical dipole placed in a uniform electric field is

$$\vec{\tau} = \vec{p} \times \vec{E}$$

independent of the origin about which the torque is calculated.

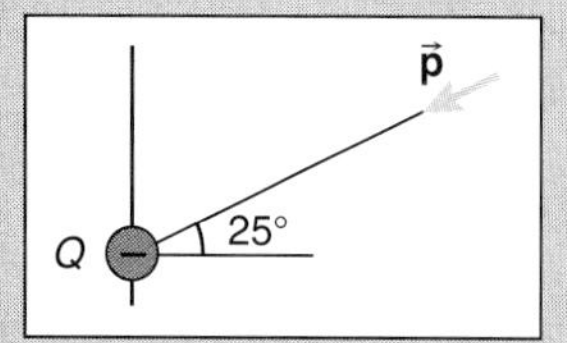

FIGURE P.41

•**43.** The general definition of the dipole moment of two charges Q_1 and Q_2 is

$$\vec{p} \equiv Q_1\vec{r}_1 + Q_2\vec{r}_2$$

where $\vec{r}_1$ and $\vec{r}_2$ are the position vectors of the two charges with respect to some origin. Show that the dipole moment of two charges $+|Q|$ and $-|Q|$, separated by a distance d, is independent of the choice of the origin of the coordinate system and is equal to $\vec{p} = |Q|\vec{d}$, as in Equation 16.13 where $\vec{d}$ is the position vector of the positive charge with respect to the negative charge.

•**44.** One model of the hydrogen atom imagines the electron in a circular orbit about a fixed proton. The arrangement consists of two charges of equal magnitudes but opposite sign, separated by a fixed distance. Explain why the arrangement has a time-averaged dipole moment of zero.

45. Use $\vec{\tau} = I\vec{\alpha}$ to show that if an electric dipole with dipole moment of magnitude p and moment of inertia I is oriented with its dipole moment making a small angle θ with the direction of an external electric field of magnitude E, the dipole will execute simple harmonic oscillations about the field direction with a frequency ν given by

$$\nu = \frac{1}{2\pi}\left(\frac{pE}{I}\right)^{1/2}$$

Section 16.9 The Electric Field of Continuous Distributions of Charge

•**46.** At what points along the symmetry axis perpendicular to the plane of a uniformly charged circular ring is the electric field of maximum magnitude?

•**47.** At what points along the symmetry axis perpendicular to the plane of a uniformly charged circular disk of radius R is the magnitude of the electric field a maximum?

•**48.** Show that the electric field along the symmetry axis perpendicular to the plane of a uniformly charged circular disk approaches the field of an infinite sheet as the distance z approaches zero.

•**49.** A 2.00 g cork with a charge +3.00 μC floats motionless 1.50 cm above a large, uniformly charged, horizontal pane of glass near the surface of the Earth. What is the surface charge density of the glass pane, assuming it to be an infinite sheet?

•**50.** Air becomes a good conductor if the magnitude of the electric field in air exceeds about 3.0×10^6 N/C; the precise value depends on many factors such as the relative humidity and pressure. (a) For oppositely charged parallel plates, what surface charge density is sufficient to create a field of this magnitude? Thunderstorms can be modeled crudely in this way. (b) How many fundamental units of charge e per square millimeter does this surface charge density represent?

•**51.** Imagine a +1.00 nC charge spread uniformly over the surface of a conducting sphere of radius R. If R is sufficiently large, the magnitude of the electric field on the surface of the conductor is less than 3.0×10^6 N/C, at which value air becomes a good conductor. For what value of R is the magnitude of the electric field on the surface of the conductor equal to 3.0×10^6 N/C? This problem indicates that the electric field at the surface of severely curved (sharply pointed) conductors is greater than in more rounded conductors with the same charge. Explain how this principle is used to advantage in lightning rods.

•**52.** A uniformly charged semicircular hoop of radius R has a total charge Q. Show that the magnitude of the electric field at the center of the semicircle is

$$E = \frac{1}{4\pi\varepsilon_0}\frac{2Q}{\pi R^2}$$

•**53.** A uniformly charged semicircular ring of length 50.0 cm has a total charge of −8.00 nC. Find the magnitude of the electric field at the center of the semicircle. Indicate its direction in a sketch.

•**54.** A rod with λ coulombs of charge per meter of its length has the shape of a circular arc of radius R. The rod subtends an angle θ as indicated in Figure P.54. Show that the magnitude of the electric field at the center of the circular arc is

$$E = \frac{1}{4\pi\varepsilon_0}\frac{2\lambda}{R}\sin\left(\frac{\theta}{2}\right)$$

If $\lambda > 0$ C/m, indicate the direction of the field in a sketch.

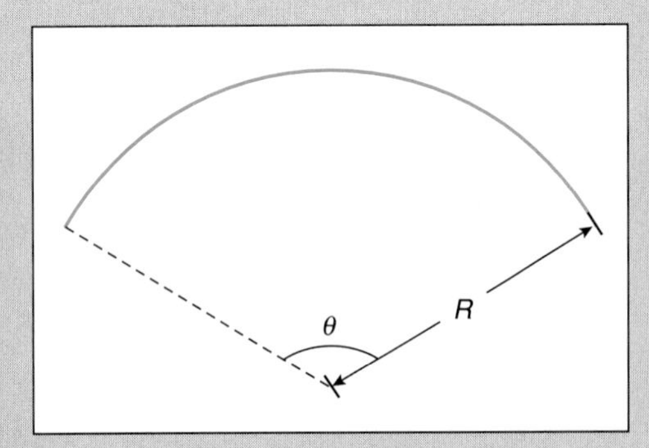

FIGURE P.54

•**55.** A long straight line of uniform positive linear charge density λ coulombs per meter of length begins at the origin and extends out to infinity along the $+x$-axis as shown in Figure P.55. Find the electric field at the point $x = -a$.

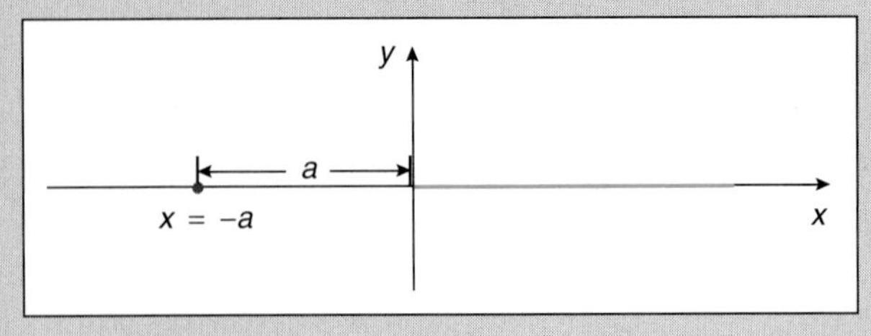

FIGURE P.55

•**56.** Show by direct integration of Equation 16.15 on page 733 that the magnitude of the electric field a distance a from an infinitely long, uniformly charged line charge (with λ coulombs per meter) is given by

$$E = \frac{\lambda}{2\pi a\varepsilon_0}$$

‡**57.** A tiny mass m with charge q is attached to an infinite sheet of charge with surface charge density σ by means of an insulating, massless cord of length ℓ. Neglect gravitational effects. (a) Specify the signs of the charge q and σ such that the cord is taut. (b) Show that if the mass is pulled slightly in a direction parallel to the plane and then released, the mass executes simple harmonic oscillation with frequency ν, where

$$\nu = \frac{1}{2\pi}\left(\frac{q\sigma}{2\varepsilon_0 m\ell}\right)^{1/2}$$

Section 16.10 Motion of a Charged Particle in a Uniform Electric Field: An Electrical Projectile

•**58.** A proton is initially at rest in a uniform electric field of magnitude 300 N/C. (a) How long will it take the proton to travel 1.00 m? (b) How long will it take an electron to do the same thing? (c) What is the ratio of the proton time to the electron time? How is this ratio related to the mass ratio of the proton to the electron?

•**59.** When the magnitude of the electric field in air exceeds about 3×10^6 N/C, it becomes a good conductor. (a) What magnitude acceleration would an electron experience in a field of this magnitude? (b) How long will it take such an electron, initially at rest, to reach a speed equal to 1.00% of the speed of light? Through what distance will such an electron travel to reach this speed?

•**60.** A medical x-ray tube accelerates electrons from rest to a speed of 8.00×10^6 m/s using a uniform electric field of magnitude 5.0×10^3 N/C. Through what distance do the electrons move to acquire this speed?

•**61.** An electron traveling at a speed of 5.00×10^6 m/s is shot into a uniform electric field of magnitude 100 N/C directed along the initial path of the electron to slow the electron from its blistering pace. (a) Is the electric field directed parallel to or antiparallel to the velocity of the electron? (b) Introduce a Cartesian coordinate system to analyze the motion of the electron after it enters the electric field. (c) Calculate the acceleration component of the electron while it is in the field. (d) How much time does it take for the electron to come to rest? (e) What distance will the electron travel into the field region to reach zero speed? (f) Will the electron remain at rest once stopped? If not, what happens to it?

•**62.** An electron is projected at speed 2.00×10^7 m/s into a region with a constant electric field directed perpendicular to the initial velocity of the electron as indicated in Figure P.62. (a) What is the direction of the electrical force on the electron? (b) What is the maximum magnitude for the electric field so that the electron does not strike the sides of the apparatus generating the electric field?

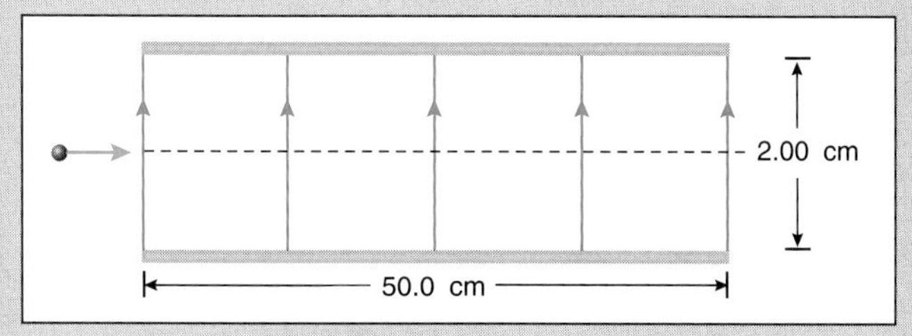

FIGURE P.62

•63. An electron finds itself about to begin an exciting trip in a uniform electric field of magnitude 500 N/C, directed as shown in Figure P.63. (a) What is the magnitude of the electrical force on the electron? (b) What is the magnitude of the acceleration of the electron? (c) If the electron begins its journey at rest, how long will it take the electron to travel 1.00 m in the field? (d) Indicate at which end of the field the electron began its journey.

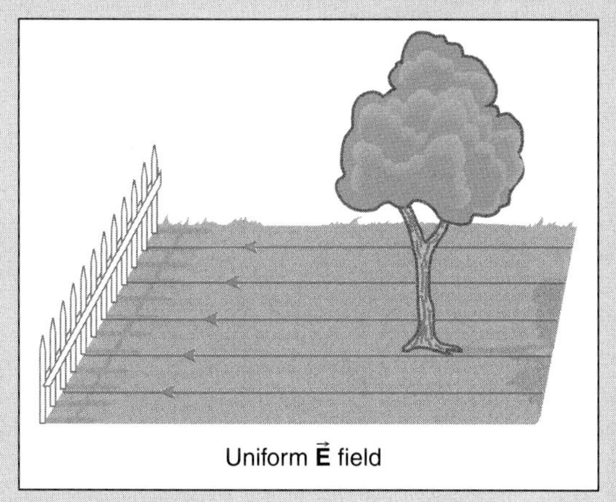

FIGURE P.63

•64. A *positron* has the same mass as an electron (9.11×10^{-31} kg) but a *positive* charge ($+1.602 \times 10^{-19}$ C). (a) In the arrangement shown in Figure P.64, near which plate should a positron be released at rest so it will accelerate toward the other plate? (b) What is the speed of the positron just before impact? (c) If the positron is released from rest, what is its kinetic energy the instant before striking the other plate? Express your result in electron-volts (1 eV $\equiv 1.602 \times 10^{-19}$ J).

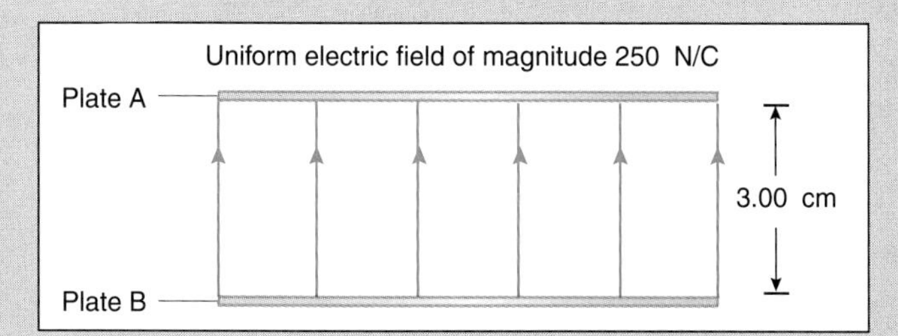

FIGURE P.64

•65. An electron is projected with an initial speed of 5.00×10^6 m/s into a uniform electric field of magnitude 500 N/C as shown in Figure P.65. How far into the field will the electron travel before reversing direction?

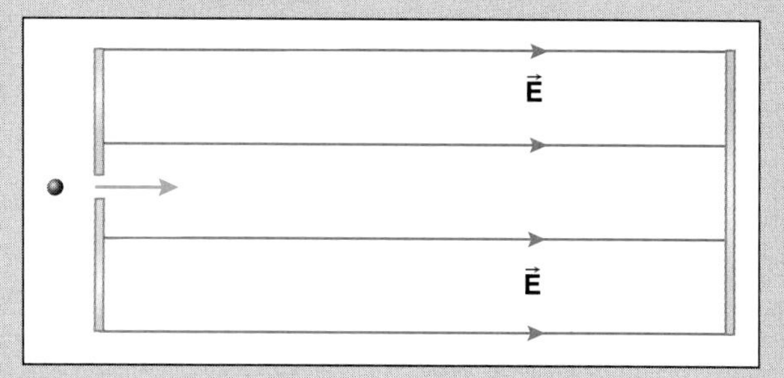

FIGURE P.65

•66. An electron is projected with an initial speed 5.00×10^6 m/s into a uniform electric field of magnitude 250 N/C. The angle between the initial velocity vector and the electric field is 30.0° as indicated in Figure P.66. Assume the electron does not strike the upper plate. What horizontal distance does the electron travel before striking the lower conducting plate?

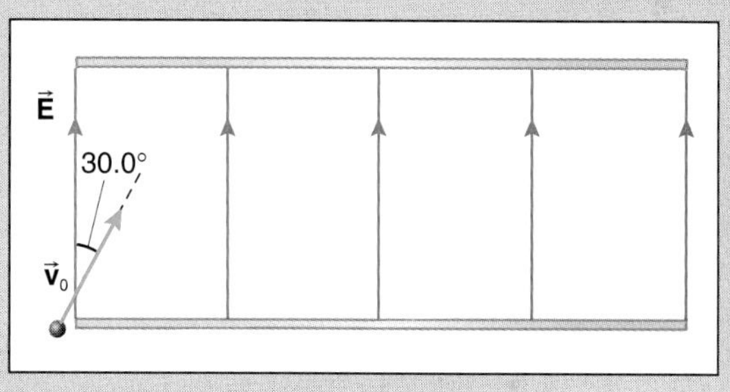

FIGURE P.66

Sections 16.11 Gauss's Law for Electric Fields*
16.12 Calculating the Magnitude of the Electric Field Using Gauss's Law*
16.13 Conductors*
16.14 Other Electrical Materials*

67. A uniform electric field of magnitude 100 N/C threads a map of the state of Colorado of dimensions 0.030 m by 0.040 m. See Figure P.67. What is the flux of the vector through the area? This is a state of flux (groan!).

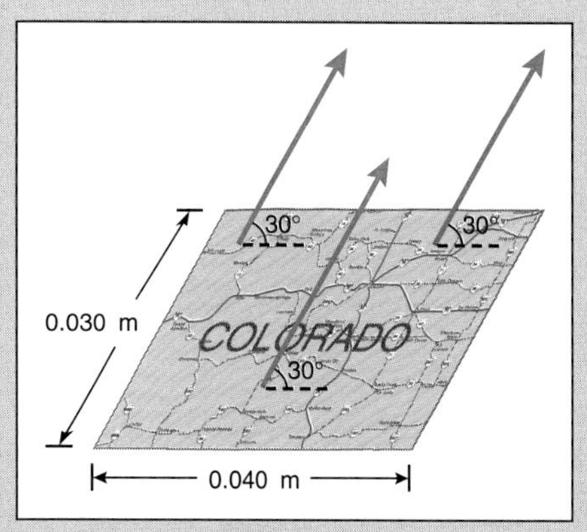

FIGURE P.67

68. For the sheer pleasure of racking up brownie points, calculate the flux of the indicated electric field vector through the surface indicated in Figure P.68.

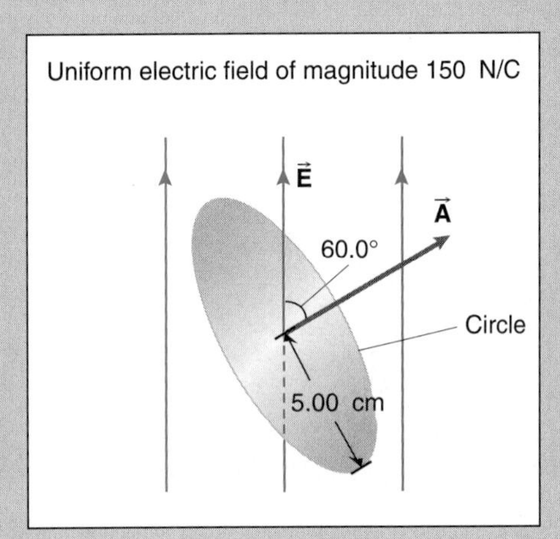

FIGURE P.68

69. What is the total flux of the electric field through a surface that completely encloses an electric dipole?

70. What is the total flux of the electric field vector through the closed surface S in Figure P.70?

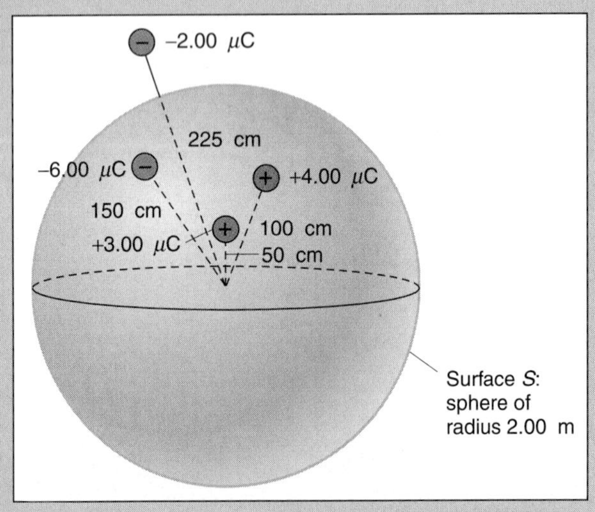

FIGURE P.70

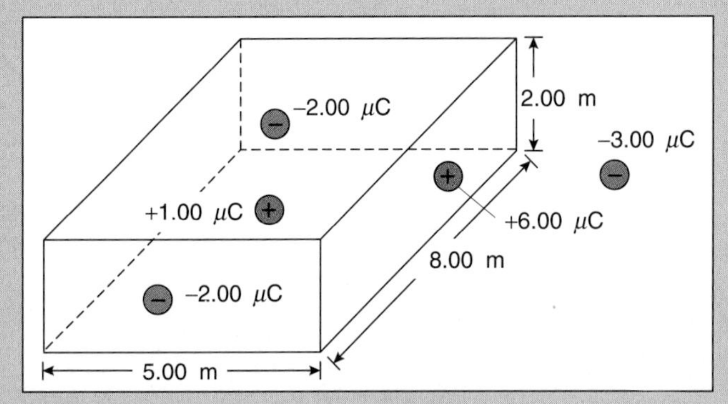

FIGURE P.71

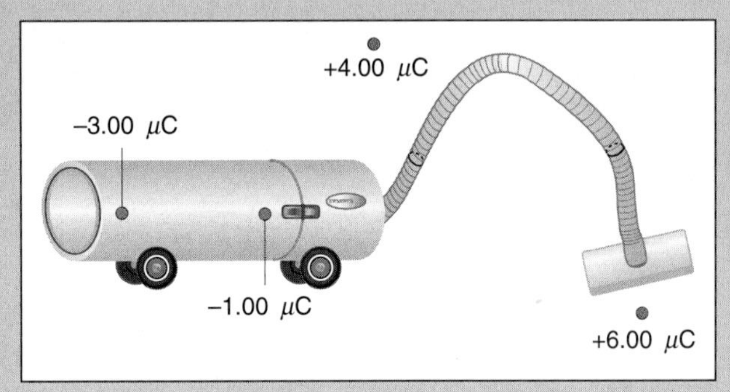

FIGURE P.72

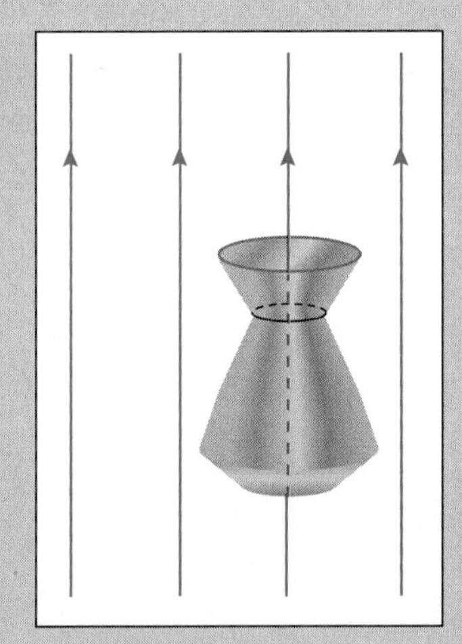

FIGURE P.73

•**71.** The great physicist Dr. E. Fields has discovered that one of her four charges has escaped from the lecture room as indicated in Figure P.71. Free at last! Calculate the total flux of the electric field from all the charges through the walls, floor, and ceiling of the classroom.

•**72.** Calculate the total electrolux . . ., rather, the electric flux through the closed surface indicated in Figure P.72.

•73. Consider a surface shaped like a spittoon. The circular open top has a radius of 6.0 cm and a narrower radius of 4.0 cm along its neck. The symmetry axis of the surface is aligned with a uniform electric field of magnitude 200 N/C as shown in Figure P.73. What is the flux of the electric field through the surface?

•**74.** A small copper spherical BB of radius a is located at the center of a larger hollow copper spherical shell of inner radius b and outer radius R as shown in Figure P.74. A charge of $+q$ is on the small BB. The hollow copper shell has zero charge on it. (a) What is the electric field within the BB (for radii $r < a$)? (b) What is the electric field inside the copper shell (that is, for radii r that satisfy $b < r < R$)? (c) Draw a closed Gaussian surface within the copper of the shell as indicated in Figure P.74. What is the total flux of the electric vector through this Gaussian surface? This result implies that charge must lie on the inside surface of the spherical shell. What charge must reside on the inside surface of the copper shell? Since the copper shell has a total charge of zero, what charge must reside on the outer surface of the copper shell?

•**75.** Two concentric, fixed, hollow, thin uniform spherical shells have charges and radii as indicated in Figure P.75. (a) What is the total force on a pointlike charge q = 10.0 nC if it is located

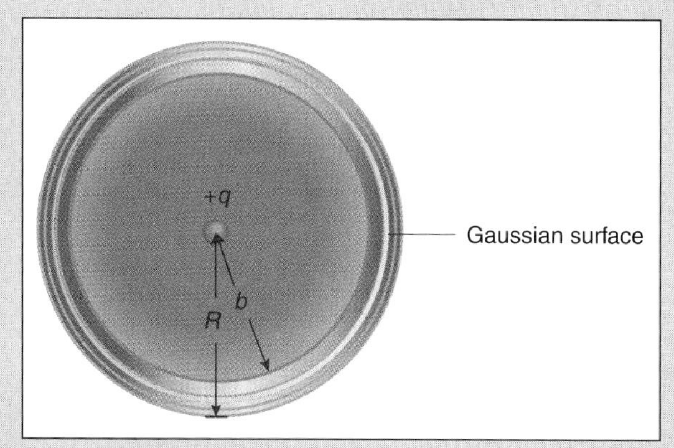

FIGURE P.74

at $r = 0.100$ m from the center? (b) What is the total force on the same charge q if it is located at $r = 0.800$ m from the center? (c) What is the total force on the same charge q if located $r = 2.00$ m from the center? (d) At what value of r does q have the maximum magnitude force on it? Calculate this maximum force.

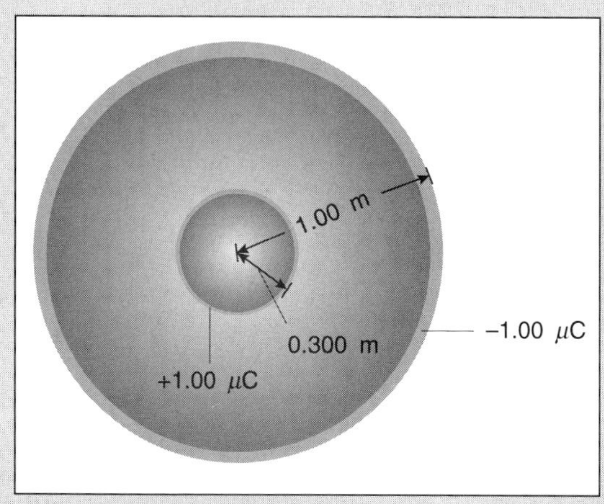

FIGURE P.75

•**76.** Early in the 20th century, there were experimental indications that atoms had internal structure. A crude model of a nuclear atom consists of a central pointlike nucleus of charge $+Ze$ surrounded by a thin uniform spherical shell of radius R and charge $-Ze$, where Z is what we call the atomic number (the number of protons in the nucleus). (a) Sketch the model. (b) Use Gauss's law to determine the magnitude of the electric field caused by this charge distribution for distances $r > R$. (c) Use Gauss's law to determine the magnitude of the electric field caused by this charge distribution for distances $r < R$.

•**77.** Instead of the model atom proposed in Problem 76, imagine a model consisting of a central pointlike nucleus of charge $+Ze$ surrounded by a sphere of radius R with charge $-Ze$ uniformly distributed throughout its volume. (a) Sketch the model. (b) Use Gauss's law to determine the magnitude of the electric field caused by this charge distribution for distances $r > R$. (c) Use Gauss's law to determine the magnitude of the electric field caused by this charge distribution for distances $r < R$.

78. An insulating, thin, hollow sphere has a uniform surface charge density σ. (a) Show that the magnitude of the electric field at the surface of the sphere is $|\sigma|/\varepsilon_0$ (b) A tiny hole is drilled through the shell, thus removing a negligible bit of the charge. Show that the magnitude of the electric field in the hole is $|\sigma|/2\varepsilon_0$.

79. An early model of the hydrogen atom, known picturesquely as the plum pudding model, imagined a sphere of radius R with positive charge $+e$ distributed uniformly throughout its volume with an electron embedded, say, at the center of the sphere. (a) Using Gauss's law, calculate the magnitude of the electric field of the uniform spherical distribution of positive charge at a distance r from the center of the sphere, where $r < R$. (b) Using Newton's second law, show that an electron of mass m at a distance r from the center of the sphere (with $r < R$) executes simple harmonic motion about the center. That is, show that the motion of the electron of mass m is described by an equation of the form

$$\frac{d^2r}{dt^2} + \omega^2 r = 0 \text{ m/s}^2$$

where the angular frequency ω of the motion is

$$\omega = \left(\frac{1}{4\pi\varepsilon_0}\frac{e^2}{R^3 m}\right)^{1/2}$$

(c) Calculate the frequency of the simple harmonic oscillation of the electron using $R \approx 1.0 \times 10^{-10}$ m as an estimate of the size of the atom. It was originally thought that the oscillating electron in this model might account for light emission from the hydrogen atom. The model predicts a single oscillation frequency for the electron and for the emitted light. The model had to be abandoned because it was unable to account for the observation that hydrogen emits light at many different specific frequencies.

80. Some time ago, R. A. Lyttleton and H. Bondi investigated the consequences of assuming the charges on the electron and proton were of slightly different magnitude.* Such an asymmetry could account for the observed expansion of the universe using purely Newtonian physics, as we demonstrate in this problem. The idea is intriguing, but the *actual* expansion of the universe we now know is *not* caused by this mechanism. Nonetheless, an exploration of Lyttleton and Bondi's hypothesis is an interesting application of electricity and dynamics. For simplicity, assume that the entire universe is composed of hydrogen.† Assume also that the charges on the electron and proton are of slightly different magnitude. A hydrogen atom then has a slight charge of magnitude δe, where δ is a purely numerical factor that likely is very small, since matter appears to be quite closely electrically neutral. Imagine such a slightly charged hydrogen atom of mass m on the surface of a large, uniform spherical mass M of radius r of similar hydrogen atoms.

*R. A. Lyttleton and H. Bondi, "On the physical consequences of a general excess of charge," *Proceedings of the Royal Society of London, A252*, pages 313–333, (1959).

†The universe is composed of about 70% hydrogen and 30% helium with just a smattering of heavier elements. For Lyttleton and Bondi's hypothesis, it is not necessary to assume the entire universe is composed of hydrogen but it is easier to do so.

(a) Sketch the situation. (b) Show the giant spherical mass has a charge of magnitude

$$\frac{M}{m}\delta e$$

(c) Show that the hydrogen atom (with charge δe) on the surface of M experiences an electrical *repulsive* force of magnitude

$$\frac{1}{4\pi\varepsilon_0}\frac{M}{m}\frac{\delta^2 e^2}{r^2}$$

Charged matter lying *outside* the large mass M, if also spherical in shape, has no effect on the atom in question because of the shell theorems associated with inverse-square-law forces that we examined in our study of gravitation in Chapter 6. (d) Show that the *ratio* of the magnitude of the repulsive electrical force to the magnitude of the attractive gravitational force on m is

$$\begin{aligned}\frac{\text{repulsion}}{\text{attraction}} &= \delta^2\frac{1}{4\pi\varepsilon_0}\frac{e^2}{Gm^2}\\ &\approx 1.23\times 10^{36}\,\delta^2\\ &\equiv \mu\delta^2\end{aligned}$$

where μ is unitless and has the numerical value of about 1.23×10^{36}. (e) The forces of repulsion and attraction are of *equal* magnitude if the ratio in part (d) is equal to 1. Show that the ratio is equal to 1 if δ has the value

$$\begin{aligned}\delta &\approx 9.02\times 10^{-19}\\ &\approx 10^{-18}\end{aligned}$$

In other words, if the magnitudes of the charges on the electron and proton differ by about one part in 10^{18}, then the repulsive electrical force on the atom on the surface of the sphere is equal in magnitude to the attractive force of gravitation. (f) Assume that δ is bigger than the value calculated in part (e). Let ρ be the mass density of M. Show that the magnitude of the total repulsive force on the atom by the sphere is

$$F_{\text{repulsive}} = \frac{4}{3}\pi\rho Gm(\mu - 1)r$$

The important point to note here is that the magnitude of the net force of repulsion on the atom is proportional to the radius of the sphere; the magnitude of the force is proportional to the distance.* Write this relation as

$$F_{\text{repulsive}} = Kr$$

(g) Apply Newton's second law to m to show that

$$Kr = m\frac{d^2r}{dt^2}$$

The second derivative of r with respect to t is proportional to r itself. Show that one solution to this equation is†

$$r(t) = Ae^{Ht}$$

where

$$H \equiv \frac{K}{m}$$

(h) Show that the expansion (repulsive) speed v of the atom in its motion is

$$\begin{aligned}v &= \frac{dr}{dt}\\ &= Hr\end{aligned}$$

Thus the speed of recession of the atom is proportional to its distance. It has been known since the late 1920s that the universe is expanding and that the recessional speed of distant galaxies is proportional to their distance from us; the proportionality constant H is known as the *Hubble constant*. Although the mechanism of charge asymmetry leads to an expanding universe, the actual expansion of the universe is not caused by this mechanism, but by effects associated with the creation of the universe in the so-called Big Bang event.

*This is not the same as the Hooke's law spring force. The spring force is attractive; the force here is repulsive.

† In this solution e represents the base of natural logarithms, *not* the fundamental unit of charge.

Investigative Projects

A. Expanded Horizons

1. The procedures used by Charles Coulomb in the latter part of the 18th century to determine the force law between static charges are well worth a look on your part. Write a précis about them.
   Peter Heering, "On Coulomb's inverse square law," *American Journal of Physics*, 60, #11, pages 988–994 (November 1992).
   Steven Dickman, "Could Coulomb's experiment result in Coulomb's law?" *Science*, 262, #5133, pages 500–501 (22 October 1993).

2. Experiments with increasing precision test the accuracy of the inverse-square nature of Coulomb's law. Trace the development of these experiments and report on the current precision.
   Peter Heering, "On Coulomb's inverse square law," *American Journal of Physics*, 60, #11, pages 988–994 (November 1992).
   John David Jackson, *Classical Electrodynamics* (2nd edition, John Wiley and Sons, New York, 1975), pages 5–9.
   E. R. Williams, J. E. Faller, and H. A. Hill, "New experimental test of Coulomb's law: a laboratory limit on the photon rest mass," *Physical Review Letters*, 26, #12, pages 721–724 (22 March 1971).
   R. E. Crandall, "Photon Mass Experiment," *American Journal of Physics*, 51, #8, pages 698–702 (August 1983).
   Jibayo Akinrimisi, "Note on the experimental determination of Coulomb's law," *American Journal of Physics*, 50, #5, pages 459–460 (May 1982).
   P. H. Wiley and W. L. Stutzman, "A simple experiment to demonstrate Coulomb's law," *American Journal of Physics*, 46, #11, pages 1131–1132 (November 1978).

3. Investigate how electrostatics is applied to xerography. Describe the principles in a cogent report.
H. Richard Crane, "Physics in the copy machine," *The Physics Teacher*, *22*, #7, pages 454–461 (October 1984).
Charles D. Hendricks, "Electrostatic imaging," *Electrostatics and Its Applications*, edited by A. D. Moore (Wiley, New York, 1973), pages 281–306.
John H. Dessauer and Harold E. Clark (editors), *Xerography and Related Processes* (Focal Press, New York, 1965).
J. Mort, *The Anatomy of Xerography: Its Invention and Evolution* (McFarland, Jefferson, North Carolina, 1989).
Edgar M. Williams, *The Physics and Technology of Xerographic Processes* (Krieger, Malabar, Florida, 1992).

4. Investigate and report on the physics underlying the use of electrostatic precipitators in reducing smokestack emissions.
Myron Robinson, "Electrostatic precipitation," *Electrostatics and Its Applications*, edited by A. D. Moore (Wiley, New York, 1973), pages 180–220.
Harry J. White, *Industrial Electrostatic Precipitation* (Addison-Wesley, Reading, Massachusetts, 1963).
David A. Lloyd, *Electrostatic Precipitator Handbook* (Adam Hilger, Philadelphia, 1988).

5. Investigate the electrostatics associated with lightning as well as the classic kite experiment of Benjamin Franklin that demonstrated that lightning was electricity. Only by luck did Franklin escape electrocution by lightning. Describe the experiment.
Richard E. Orville, "The lightning discharge," *The Physics Teacher*, *14*, #1, pages 7–13 (January 1976).
Martin A. Uman, *All About Lightning* (Dover, New York, 1986).
Martin A. Uman, *The Lightning Discharge* (Academic Press, Orlando, 1987).
Martin A. Uman, *Lightning* (Academic Press, Orlando, 1987).
Leon E. Salanave, *Lightning and Its Spectrum* (University of Arizona Press, Tucson, 1980).
Rudolph Heinrich Golde, *The Physics of Lightning* (Academic Press, New York, 1977).
Peter E. Viemeister, *The Lightning Book* (MIT Press, Cambridge, Massachusetts, 1972).

6. Electrostatics is used to sort seeds by size, to remove dirt, hulls, and rodent excrement from cereal grains, to concentrate various minerals in mining, and in recycling reusable wastes. Investigate some of these applications in greater detail.
A. D. Moore, "Electrostatics," *Scientific American*, *226*, #3, pages 46–58 (March 1972); additional references on page 126.
James E. Lawver and W. P. Dyrenforth, "Electrostatic separation," *Electrostatics and Its Applications*, edited by A. D. Moore (Wiley, New York, 1973), pages 221–249.

7. Tiny charged ink droplets are used in common computer ink jet printers. Investigate the elements involved in the design of such printers.
Larry Kuhn and Robert A. Myers, "Ink-jet printing," *Scientific American*, *240*, #4, pages 162–178 (April 1979); additional references on page 190.
Hewlett-Packard Journal, *45*, #1 (February 1994); the entire issue is devoted to ink jet technology.
A. J. Rogers, "Ink jet takes off," *Byte*, *16*, #10, pages 163–168 (October 1991).

8. Some fish, among them sharks, eels, catfish, and torpedo fish, use electric fields to detect and/or stun and kill prey. Investigate the electric activity of such fish.
Joseph Bastien, "Electrosensory organisms," *Physics Today*, *47*, #2, pages 30–37 (February 1994).
Chau H. Wu, "Electric fish and the discovery of animal electricity," *American Scientist*, *72*, #6, pages 598–607 (November–December 1984).
Harry Grundfest, "Electric fishes," *Scientific American*, *203*, #4, pages 115–125 (October 1960); additional references on page 220.
Louis Roule, *Fishes and Their Ways of Life* (Norton, New York, 1935) pages 156–171.
William N. McFarland, F. Harvey Pough, Tom J. Cade, and John B. Heiser, *Vertebrate Life* (Macmillan, New York, 1979), pages 240–248.
David Hafemeister, "Resource Letter BELFEF-1: Biological effects of low-frequency electromagnetic fields," *American Journal of Physics*, *64*, #8, pages 974–981 (August 1996); this contains many references.
Peter Moller, *Electric Fishes: History and Behavior* (Chapman & Hall, London, 1995).
Victor Percy Wittaker, *The Cholinergic Neuron and Its Target: The Electromotor Innervation of the Electric Ray "Torpedo" as a Model* (Birkhäuser, Boston, 1992).

B. Lab and Field Work

9. Next time you dry your clothes in a dryer and find your socks sticking together, perform some simple electrical experiments to determine the charge on the socks. Explain your procedure and your results. Investigate the dependence of the effect on the type of fabric.

10. Investigate the physics underlying the Millikan oil drop experiment, the method first used to determine the magnitude of the fundamental unit of charge e. Most physics departments have an apparatus to perform an experiment similar to that of Millikan. Design and perform an experiment to determine e with such an apparatus.
Robert Andrews Millikan, *The Electron* (University of Chicago Press, Chicago, 1917).
R. A. Millikan, "On the elementary electrical charge and Avogadro's constant," *Physical Review*, *II*, #2, pages 109–143 (August 1913).
Ray C. Jones, "The Millikan oil-drop experiment: making it worthwhile," *American Journal of Physics*, *63*, #11, pages 970–977 (November 1995).
William M. Fairbank Jr. and Allan Franklin, "Did Millikan observe fractional charges on oil drops?," *American Journal of Physics*, *50*, #5, pages 394–397 (May 1982).
Mark A. Heald, "Millikan oil-drop experiment in the introductory laboratory, *American Journal of Physics*, *42*, #3, pages 244–246 (March 1974).

11. The electric analog of a magnet is called an *electret* or *ferroelectric* material. Ferroelectrics are not as readily available as common permanent magnets (ferromagnets) because only certain materials (among them barium titanate [$BaTiO_3$]) can sustain a permanent electric polarization near room temperature. Other materials have ferroelectric properties at cryogenic (very low) temperatures (122 K for potassium dihydrogen phosphate [KH_2PO_4]). Secure a sample crystal of barium titanate and explore the electric field in its vicinity.
Oleg D. Jefimenko and David K. Walker, "Electrets," *The Physics Teacher*, *18*, #9, pages 651–659 (December 1980).
R. N. Varrey and H. T. Hahn, "Electrets and electrostatic measurement," *American Journal of Physics*, *43*, #6, pages 509–513 (June 1975).
Deborah Schurr and Tim Usher, "Demonstrating hysteresis in ferroelectric materials," *The Physics Teacher*, *33*, #1, pages 30–31 (January 1995).
Thomas Kallard, *Electret Devices for Air Pollution Control* (Optosonic Press, New York, 1972).

C. Communicating Physics

12. Electrostatic motors and generators date back to Benjamin Franklin in the 18th century. The most well-known such device was invented about 1878 by James Wimshurst (1832–1903) and is called a Wimshurst machine. Investigate how this historic device operates. Your physics department may have such a Wimshurst electrostatic generator in its attic or museum. As an exercise in technical writing, write an extensive explanation describing how the machine separates and accumulates charge. Design a poster explaining its operation for use in a hallway display of a Wimshurst machine in your science building or a local science museum.
    Encyclopedia Britannica (11th edition, 1910), volume 9, pages 178–179.
    A. D. Moore, *Electrostatics* (Anchor Doubleday, Garden City, New York, 1968).
    Oleg D. Jefimenko, "Electrostatic motors," *Electrostatics and Its Applications*, edited by A. D. Moore, editor (Wiley, New York, 1973), pages 131–147.
    M. J. Mulcahy and W. R. Bell, "Electrostatic generators," *Electrostatics and Its Applications*, edited by A. D. Moore, editor (Wiley, New York, 1973), pages 148–179.
13. Write a paragraph comparing and contrasting the gravitational and electrical *forces* and another paragraph comparing and contrasting the gravitational and electrical *fields*.

SOLVING YOUR HOMEWORK PROBLEMS BY INDUCTION

Chapter 17

Electric Potential Energy and the Electric Potential

Vive la différence!

In this chapter we bring the concepts of work, potential energy, the CWE theorem, and also a new and convenient idea (that of the electric potential) to bear on electrical phenomena. Such notions are used in the study of electricity and its technological applications as much as or more than they are in gravitation and mechanics. Electrical phenomena are ubiquitous in physics, engineering, and even biology, chemistry, and medicine, particularly at the cellular and molecular level. It is therefore absolutely essential to grasp the ideas of work, energy, and the electric potential, because they manifest themselves in all the sciences—there is no avoiding them.

Fortunately, as we saw in Chapter 16, there are strong parallels between the electrical force on charges and the gravitational force on masses (though, as we have seen, the forces also have distinctly different features). In this chapter we continue to exploit these parallels in the realms of work and energy.

We found in Chapter 8 that the concepts of work and energy were useful and convenient ways of reformulating mechanics that permitted us to solve many problems without a detailed, point-by-point knowledge of the forces on a particle as it moved along a path. In some cases, such knowledge of the forces is unavailable, difficult to determine, or analytically complex. Neglecting thermal effects, the CWE theorem often is a useful tool for the study of the mechanics of particles. The energy approach is particularly useful for the cases where only conservative and/or zero-work forces act on the system, for then the total mechanical energy of the system, $E \equiv \text{KE} + \text{PE}$, is conserved.

17.1 Electrical Potential Energy and the Electric Potential

In Chapter 8 we defined a conservative force in the following equivalent ways:

1. It is a force that does work whose quantity is independent of the path followed by a particle as it moves from an initial to a final position.
2. It is a force that does zero work around any closed path.

We here can make quick use of what we already learned in Chapter 8, where we found that the gravitational force (an inverse-square-law force) is a conservative force. The gravitational and static electrical forces have the same mathematical form; it is merely the letter symbols that formally distinguish them. Since the gravitational force is a conservative force, identical reasoning shows the following:

> The electrical force caused by static charges is a conservative force.

Therefore (1) the work done by the static electrical force is independent of the path that a charge follows from an initial to a final position; equivalently, (2) the work done by the static electrical force around any closed path is zero.

The work done by a conservative force on a particle is the negative of the change in the potential energy associated with the conservative force:

$$\begin{aligned} W_{\text{consrv}} &\equiv -\Delta\text{PE} \\ &= -(\text{PE}_f - \text{PE}_i) \end{aligned} \tag{17.1}$$

The motivation behind the introduction of a *scalar* potential energy associated with a conservative force was so that the work done by the force could be accounted for via the negative of the change in the associated potential energy. The change in the potential energy then can join the change in the kinetic energy on the right-hand side of the by-now-famous CWE theorem. Only the work done by nonconservative forces then remains on the left-hand side of the theorem. That is, the left-hand side of the CWE theorem is divided into the work done by nonconservative forces and the work done by conservative forces:

$$W_{\text{noncon}} + W_{\text{consrv}} = \Delta\text{KE}$$

Invoking Equation 17.1, relating the work done by conservative forces to changes in their respective potential energy, implies that

$$W_{\text{noncon}} + (-\Delta\text{PE}) = \Delta\text{KE}$$

and so

$$W_{\text{noncon}} = \Delta\text{KE} + \Delta\text{PE} \tag{17.2}$$

Following Equation 17.1, we define the work done by the static electrical force in moving a charge q from an initial position to a final position as the negative of the change in the electrical potential energy of the charge q:

$$W_{\text{elec}} \equiv -\Delta\text{PE} \tag{17.3}$$

Once again, notice that it is only a *change* in the potential energy of the charge that is physically significant. The place where the electrical potential energy is set equal to zero is arbitrary. We can choose it to be zero anywhere we like; however, in some situations, the choice is made for us by convention.* The electrical potential energy of a charge (just like the gravitational potential energy of a mass) is a function of the location or position of the charge in space relative to *other* charges.

Using the general definition of the work done by a force, Equation 8.2, we see that the work done by the electrical force on a charge q as it moves from an initial to a final location is the integral over the path of the scalar product of the electrical force with the differential change in the position vector $d\vec{\mathbf{r}}$ of the charge. Equation 17.3 thus becomes

$$\begin{aligned} \int_i^f \vec{\mathbf{F}}_{\text{elec}} \cdot d\vec{\mathbf{r}} &= -\Delta\text{PE} \\ &= -\left(\text{PE}_f - \text{PE}_i\right) \end{aligned} \tag{17.4}$$

The integral on the left-hand side of Equation 17.4 is performed over the path connecting the initial and final locations of the charge. Since the electrical force is conservative, the result of the integration is independent of the path used to get from the initial to final locations.

The electrical force on a charge q, when it finds itself in an electric field $\vec{\mathbf{E}}$ created by other static (fixed) charges, is given by Equation 16.8†:

*We encountered these choices and conventions in the gravitational context as well.

† It is good to recall, as we mentioned in Chapter 16, that we have made the assumption that the presence of the charge q does not affect the distributions of the charges that create the electric field $\vec{\mathbf{E}}$ in which q is placed. The charge q, therefore, occasionally is referred to as a small "test" charge.

$$\vec{\mathbf{F}}_{\text{elec}} = q\vec{\mathbf{E}}$$

Substituting this into Equation 17.4, we find

$$q\int_{i}^{f} \vec{\mathbf{E}} \cdot d\vec{\mathbf{r}} = -(\text{PE}_f - \text{PE}_i) \tag{17.5}$$

Equation 17.5 involves both the charge q and the electric field $\vec{\mathbf{E}}$ it experiences, caused by other fixed (static) charges. We would like to eliminate the reference to q and focus exclusively on the electric field created by the static charges. We do this by dividing Equation 17.5 by the charge q:

$$\int_{i}^{f} \vec{\mathbf{E}} \cdot d\vec{\mathbf{r}} = -\left(\frac{\text{PE}_f}{q} - \frac{\text{PE}_i}{q}\right) \tag{17.6}$$

The right-hand side of Equation 17.6 represents the difference in the *potential energy per unit charge*, or the potential energy per coulomb. Now we introduce new terminology.

Define the **electric potential** V at a point in space to be the potential energy per unit charge at that point*:

$$V \equiv \frac{\text{PE}_{\text{of } q}}{q} \tag{17.7}$$

The SI unit for the electric potential thus is joules/coulomb (J/C). The unit is used so frequently that it is given its own special name: the **volt** (V) = J/C.

Perhaps the volt is a familiar unit, but now you know what it really means and where it comes from in physics. The unit honors the Italian physicist Alessandro Volta (1745–1827) who performed important experiments on electrical phenomena, particularly concerned with the development of a crude, early battery called the voltaic pile.

To find the electric potential energy of a charge q, when q is placed where the electric potential has the value V, we rearrange Equation 17.7 to yield

$$\boxed{\text{PE}_{\text{of } q} = qV} \tag{17.8}$$

Equation 17.8 is a very important definition of the relationship between the potential energy of the charge q and the potential V at the location where the charge is placed. Equation 17.6 then can be rewritten in terms of the difference in electrical potential between the two points:

$$\int_{i}^{f} \vec{\mathbf{E}} \cdot d\vec{\mathbf{r}} = -\left(V_f - V_i\right)$$

The change in the electric potential between two points thus is

$$\boxed{V_f - V_i = -\int_{i}^{f} \vec{\mathbf{E}} \cdot d\vec{\mathbf{r}}} \tag{17.9}$$

*We also could have defined a gravitational potential energy per unit mass, a gravitational potential, but such a potential is used infrequently (except in celestial mechanics) and we omitted mentioning it in Chapters 6 and 8.

The change in the value of the electric potential between two points also is called the **potential difference** between the two points. Since the static electrical force is conservative, the line integral of the static electric field on the right-hand side of Equation 17.9 is independent of the path used to get from the initial to final locations in space.

Notice that Equation 17.9 implies that the SI unit for quantifying the electric field, newtons per coulomb (N/C), also can be expressed equivalently as volts per meter (V/m): N/C = V/m.

In the gravitational situation near the surface of the Earth, we often choose the zero for the gravitational potential energy to be at ground level. We did not always do this, nor did we have to.

In the electrical case, any place where the electric potential energy and, from Equation 17.7, the electric potential are zero is defined to be an **electrical ground.**†

We typically have a choice about what point to take as the zero of the electric potential energy and electric potential, but once the choice is made, that place is an electrical ground. In other words, an electrical ground means that $V = 0$ V at that point, and if a charged particle is placed at that point, the electric potential energy of that charge is 0 J there as well.

There are a number of important points that need to be stressed about the relationship between the electrical *potential* and the electric *potential energy*:

PROBLEM-SOLVING TACTICS

17.1 The electric potential is not the same thing as the electric potential energy. Do not confuse them. The electric potential is the potential energy per unit charge and is expressed in joules per coulomb (J/C), or volts (V) in the SI unit system. Electric potential is a property of a *point in space*, whether or not a charge is placed at that point in space. The numerical value of the potential can be positive, negative, or zero. Electrical potential energy is expressed in joules (J) and is something that a charged particle has by virtue of its location in space relative to other charges.‡

17.2 It is important to use the appropriate sign for the charge q in Equation 17.8:

$$\text{PE} = qV \tag{17.8}$$

The numerical value of the potential energy of the charge q depends on *both* the sign of q and the sign of V at the point where q is placed.

17.3 The electric potential and the electric potential energy both are scalar quantities, not vectors. It is the electric field and electrical force that are vectors.

†Some countries, such as Great Britain, use the word *earth* for an electrical ground.

‡The same is true of gravitational potential energy. The gravitational potential energy of a mass m depends on its location in space relative to other masses (say, the Earth).

Alessandro Volta resisted familial pressure to become an attorney and followed his interests and talents into electrical research. Although active in politics, it is unlikely he wore a toga during the 18th and 19th centuries!

The bottom of a charged thundercloud and the surface of the Earth can be modeled as a pair of large, parallel, charged sheets.

PROBLEM-SOLVING TACTIC

17.4 The direction of the electric field always is from regions with higher values of the electric potential to regions with lower values of the electric potential. This fact is demonstrated in Example 17.1.

QUESTION 1

Your professor asks the class why the electrical force of static charges is conservative. Three students respond as follows:

- **because the force is proportional to the product of the charges;**
- **because the force is a central force;**
- **because the force is an inverse-square-law force.**

Which student(s), if any, provided a correct response?

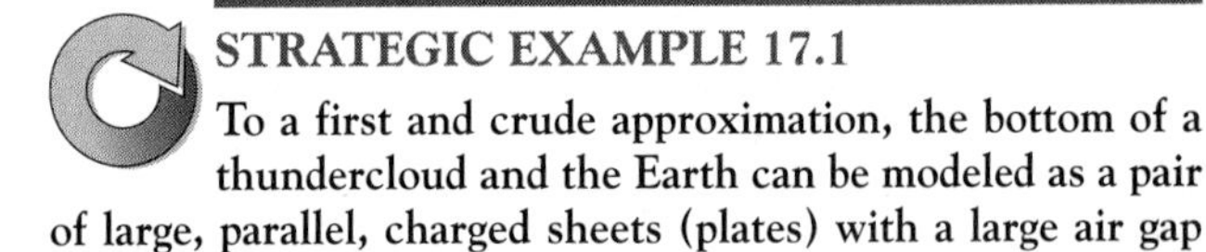

STRATEGIC EXAMPLE 17.1

To a first and crude approximation, the bottom of a thundercloud and the Earth can be modeled as a pair of large, parallel, charged sheets (plates) with a large air gap between them.

a. **Calculate the electric potential at a point x between two infinite, uniformly charged plates, separated by a distance d, that create a uniform electric field of magnitude E between them. For convenience, orient the charged sheets vertically with the x-axis horizontal, as shown in Figure 17.1.**
b. **What is the potential difference between the two charged plates?**
c. **Choose the place for the electrical ground to be at one or the other plate, and graph $V(x)$ versus x.**

Solution

The electric field is directed from the positively charged plate to the negatively charged plate. With the coordinate system

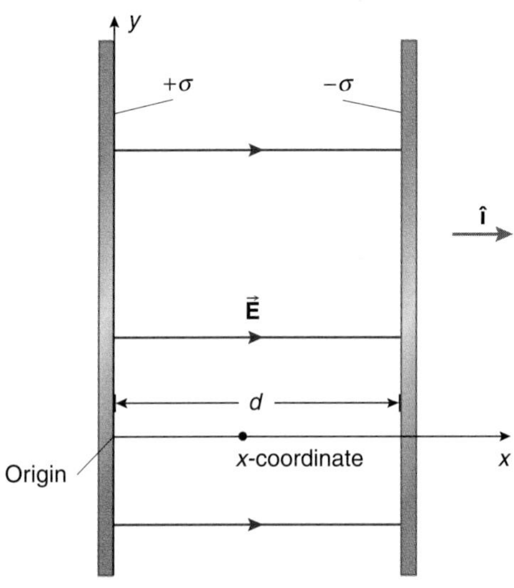

FIGURE 17.1

indicated in Figure 17.1, the electric field between the two plates is, in vector form,

$$\vec{E} = E\hat{\imath}$$

where E is positive

a. To calculate the electric potential at point x, begin with Equation 17.9. Take the origin as the initial position and coordinate x as the final position:

$$V(x) - V(0 \text{ m}) = -\int_{0 \text{ m}}^{x} \vec{E} \cdot d\vec{r} \qquad (1)$$

As always, take the one-dimensional differential change in the position vector $d\vec{r}$ to be

$$d\vec{r} = dx\,\hat{\imath}$$

and let the limits of the integration indicate which way you move along the x-axis from the initial to final positions of the charge. Equation (1) becomes

$$V(x) - V(0 \text{ m}) = -\int_{0 \text{ m}}^{x} E\hat{\imath} \cdot dx\,\hat{\imath}$$

Since E is constant, you can bring it outside the integral:

$$V(x) - V(0\text{ m}) = -E\int_{0\text{ m}}^{x} dx$$

The integral is the kind we all like to see: quite easy! Thus

$$V(x) - V(0\text{ m}) = -Ex \qquad (2)$$

The quantity $-Ex$ is negative, since E is the magnitude of the field and the coordinate x is positive according to Figure 17.1. You can conclude that location x has a lower value of the electric potential than that at the origin where $x = 0$ m.

Equation (2) indicates that the electric potential decreases linearly with x in the same direction as the uniform field. Therefore the direction of the electric field is *from* regions with higher values of the electric potential *to* regions of lower values of the electric potential. This is *always* the case, no matter what the charge configuration.

b. The position $x = d$ is that of the negatively charged plate and, from equation (2), you have

$$V(d) - V(0\text{ m}) = -Ed \qquad (3)$$

which is the potential difference between the locations $x = d$ and $x = 0$ m.

c. Let's consider the ways you might choose the ground.

Choice 1

Since the position $x = d$ is at a lower electric potential than the origin, you might choose the point $x = d$ to be the electrical ground. This means $V(d) = 0$ V. With this choice, equation (3) becomes

$$V(0\text{ m}) = Ed$$

and equation (2) becomes

$$V(x) - Ed = -Ex$$

or

$$V(x) = Ed - Ex \qquad (4)$$

The potential $V(x)$ versus x is graphed in Figure 17.2.

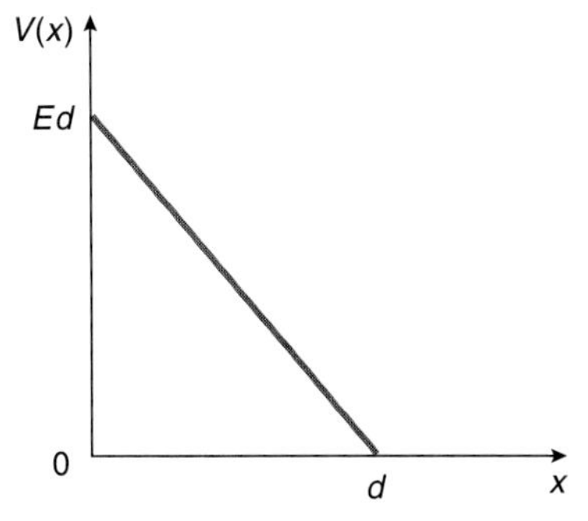

FIGURE 17.2

Choice 2

You can choose the electric ground to be anywhere that is handy! Suppose you choose $x = 0$ m to be the electrical ground. Then $V(0\text{ m}) = 0$ V, and equation (3) becomes

$$V(d) - 0\text{ V} = -Ed$$

or

$$V(d) = -Ed$$

With this choice for the ground, equation (2) becomes

$$V(x) - 0\text{ V} = -Ex$$

or

$$V(x) = -Ex \qquad (5)$$

This function is graphed in Figure 17.3. Notice that location $x = d$ still is at a lower electric potential than is $x = 0$ m. The difference in the electric potential, $V(d) - V(0\text{ m}) = -Ed$, is the same regardless of where you choose the ground. It is only the *difference* in the electric potential between two points that is physically significant; the change in electric potential between the points $x = d$ and $x = 0$ m is the same in both Figure 17.2 and Figure 17.3.

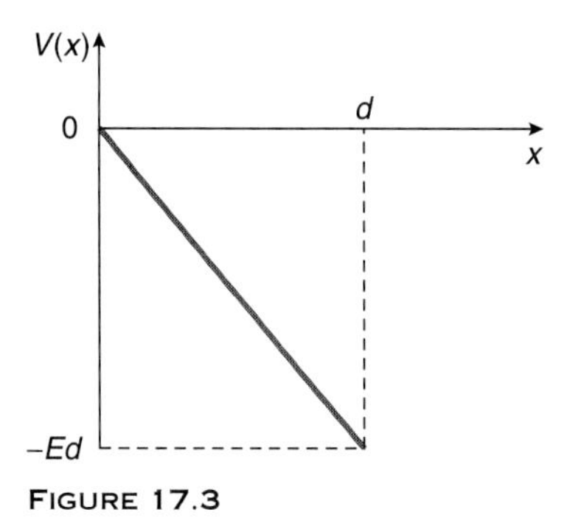

FIGURE 17.3

17.2 The Electric Potential of a Pointlike Charge

Electrons and protons are essentially pointlike charges, and so it is of some import to know the electric potential in the vicinity of a pointlike charge. The electric field a distance r away from a point charge Q is given by Equation 16.11:

$$\vec{E} = \frac{1}{4\pi\varepsilon_0}\frac{Q}{r^2}\hat{r}$$

The field is directed radially away from the charge if $Q > 0$ C and toward the charge if $Q < 0$ C. We want to find the potential difference between the points at $\vec{r}_f$ and $\vec{r}_i$, shown schematically in Figure 17.4.

We begin with the defining equation for the electrical potential difference between two points, Equation 17.9:

$$V(r_f) - V(r_i) = -\int_i^f \vec{E} \cdot d\vec{r}$$

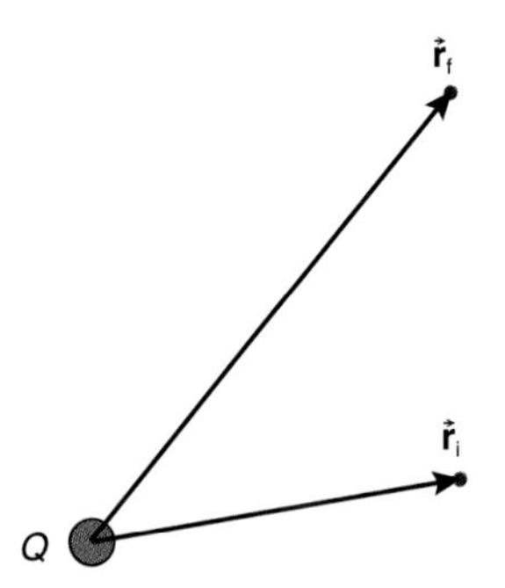

FIGURE 17.4 The vicinity of a pointlike charge Q.

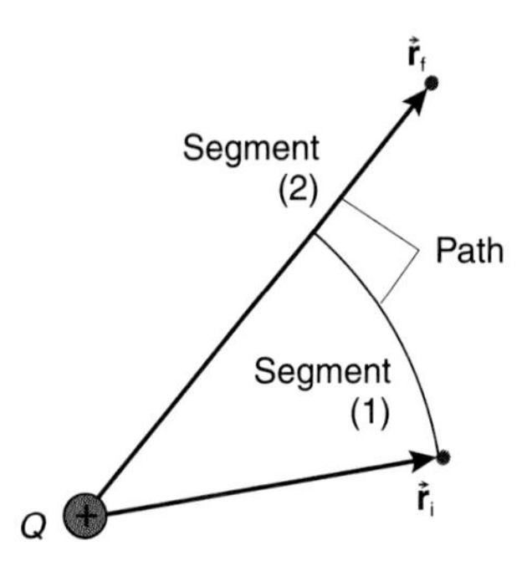

FIGURE 17.5 A path between $\vec{r}_i$ and $\vec{r}_f$ consisting of a circular arc and a radial segment.

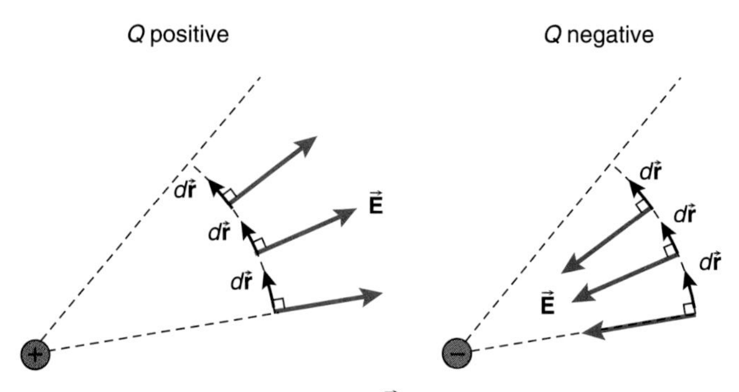

FIGURE 17.6 The scalar product $\vec{E} \cdot d\vec{r}$ is zero at each point along the circular arc.

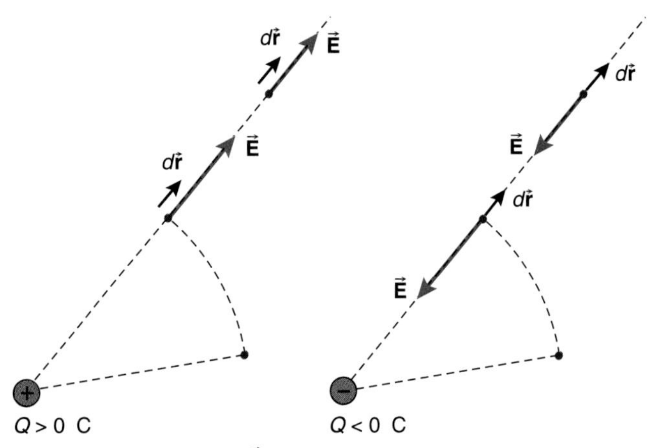

FIGURE 17.7 The field $\vec{E}$ is parallel or antiparallel to $d\vec{r}$ along the radial segment of the path, as Q is positive or negative respectively.

The potential difference between the two points is *independent of the particular path used* to get from $\vec{r}_i$ to $\vec{r}_f$, so we can choose a path to ease the evaluation of the integral. We use a path consisting of an arc of a circle of radius r_i centered on the point charge and subsequently a radial segment out to the point at the tip of $\vec{r}_f$, as indicated in Figure 17.5.

At every point along the circular segment of the path, the scalar product $\vec{E} \cdot d\vec{r}$ is zero because the radially directed electric field $\vec{E}$ is perpendicular to the change in the position vector $d\vec{r}$ along this segment (see Figure 17.6). Thus we need only consider the second segment of the path. The change in the position vector along this segment of the path is in the radial direction, and so we write $d\vec{r}$ as

$$d\vec{r} = dr\,\hat{r}$$

where as always the limits of integration take care of the direction we are moving in the radial direction. Along this second segment, the electric field is along the path, parallel to $d\vec{r}$ if $Q > 0$ C and antiparallel to $d\vec{r}$ if $Q < 0$ C, as shown in Figure 17.7. Equation 17.9 thus becomes

$$\begin{aligned} V(r_f) - V(r_i) &= -\frac{1}{4\pi\varepsilon_0} Q \int_i^f \frac{\hat{r}}{r^2} \cdot dr\,\hat{r} \\ &= -\frac{1}{4\pi\varepsilon_0} Q \int_{r_i}^{r_f} \frac{dr}{r^2} \\ &= -\frac{1}{4\pi\varepsilon_0} Q \left(-\frac{1}{r} \right) \Bigg|_{r_i}^{r_f} \\ &= \frac{1}{4\pi\varepsilon_0} Q \left(\frac{1}{r_f} - \frac{1}{r_i} \right) \end{aligned} \qquad (17.10)$$

For the electric potential of a point charge, it is convenient to choose the location of the electrical ground to be at infinity. In other words, for a point charge, the location of the zero of the electric potential is chosen for us by convention.

That is, when $r_f = \infty$ m, $V(r_f) = 0$ V. This is the *only* choice made for the ground of a point charge. With this choice for the location of the zero of the potential, Equation 17.10 becomes

$$0\text{ V} - V(r_i) = 0\text{ V} - \frac{1}{4\pi\varepsilon_0}\frac{Q}{r_i}$$

or

$$V(r_i) = \frac{1}{4\pi\varepsilon_0}\frac{Q}{r_i}$$

Since the point r_i could be anywhere, we can drop the subscripts and say the following:

The electric potential of a point charge Q a distance r from the charge is

$$V(r) = \frac{1}{4\pi\varepsilon_0}\frac{Q}{r} \qquad (17.11)$$

Notice that as $r \to \infty$ m, the electric potential of the point charge Q approaches 0 V, as we chose.

The electric potential at all points surrounding a positive point charge ($Q > 0$ C) is positive. If the point charge Q is negative, the potential is negative at all points.

In the x–y plane, where $r = (x^2 + y^2)^{1/2}$, the electric potentials of positive and negative pointlike charges have rather spectacular graphs (see Figures 17.8 and 17.9). These potentials approach zero for large r and diverge as $r \to 0$ m.

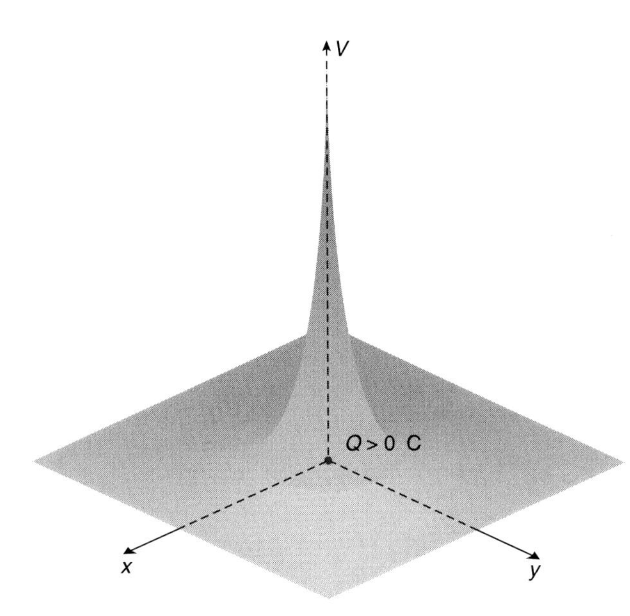

FIGURE 17.8 A graph of the electric potential V of a positive pointlike charge at the origin in the x–y plane.

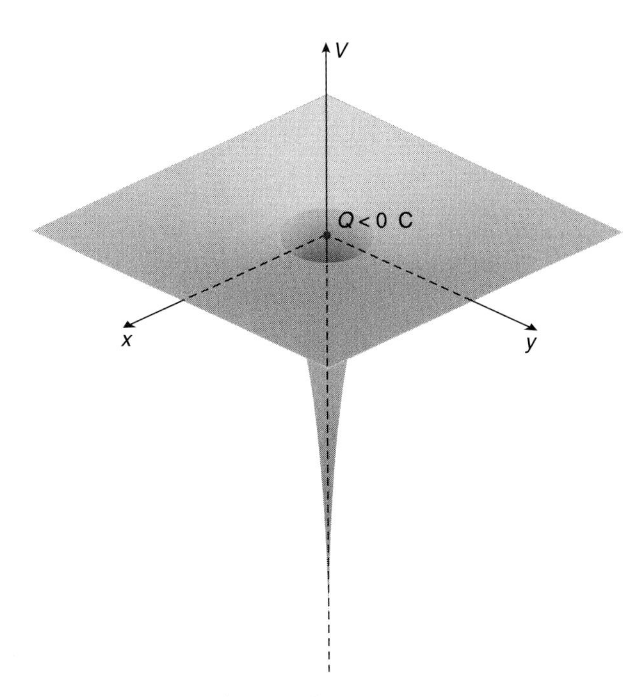

FIGURE 17.9 A graph of the electric potential V of a negative pointlike charge at the origin in the x–y plane.

PROBLEM-SOLVING TACTICS

17.5 Note that in Equation 17.11 for the electric potential of a point charge, the distance r in the denominator is to the first power. The expression for the electric *field* of a point charge (Equation 16.11) has r to the *second* power in the denominator. Remember also that the electric potential is a scalar whereas the electric field is a vector. Keep these distinctions straight and you will avoid many pitfalls in problem solving.

17.6 In Equation 17.11 for the electric potential of a point charge, as well as for other charge distributions in Table 17.1 on page 779, remember to use the appropriate sign for Q.

EXAMPLE 17.2

a. Calculate the electric potential caused by a proton (the nucleus of the hydrogen atom) at a point located 5.29×10^{-11} m away.

b. What is the electric potential energy of an electron placed at this distance from the proton? The electron in hydrogen "orbits" the nucleus at approximately this radius when closest to the nucleus.

Solution

a. The electric potential of a point charge is given by Equation 17.11:

$$V(r) = \frac{1}{4\pi\varepsilon_0}\frac{Q}{r}$$

Since the charge on the proton is $+e$, the electric potential is

$$\begin{aligned} V &= \left(9.00 \times 10^9 \text{ N}\cdot\text{m}^2/\text{C}^2\right)\frac{1.602 \times 10^{-19} \text{ C}}{5.29 \times 10^{-11} \text{ m}} \\ &= 27.3 \text{ J/C} \\ &= 27.3 \text{ V} \end{aligned}$$

b. The potential energy of any charge placed at this location is found from Equation 17.8. Remember that an electron is a negative charge, $q = -e$:

$$\begin{aligned} \text{PE} &= qV \\ &= (-1.602 \times 10^{-19} \text{ C})(27.3 \text{ V}) \\ &= -4.37 \times 10^{-18} \text{ J} \end{aligned}$$

EXAMPLE 17.3

A pointlike small rubber ball is rubbed vigorously with fur and secures a -3.00 nC charge. Find the electric potential at a point 5.00 cm away from the ball.

Solution

All units *must* be SI units, so the distance r is 5.00×10^{-2} m and the charge is -3.00×10^{-9} C. Use Equation 17.11 for the potential of a pointlike charge:

$$V(r) = \frac{1}{4\pi\varepsilon_0}\frac{Q}{r}$$

and make the appropriate substitutions, remembering that the charge Q is negative:

$$V = \left(9.00 \times 10^9 \text{ N}\cdot\text{m}^2/\text{C}^2\right)\frac{(-3.00 \times 10^{-9}\text{ C})}{5.00 \times 10^{-2}\text{ m}}$$
$$= -540 \text{ V}$$

17.3 The Electric Potential of a Collection of Pointlike Charges

For a collection of point charges, such as that in Figure 17.10, the total electric potential at a point P is the algebraic scalar sum of the potentials of each charge, taken individually as if each were the only charge present. That is,

$$V = V_1 + V_2 + V_3 + \cdots \tag{17.12}$$
$$V = \frac{1}{4\pi\varepsilon_0}\frac{Q_1}{r_1} + \frac{1}{4\pi\varepsilon_0}\frac{Q_2}{r_2} + \frac{1}{4\pi\varepsilon_0}\frac{Q_3}{r_3} + \cdots$$
$$= \frac{1}{4\pi\varepsilon_0}\left(\frac{Q_1}{r_1} + \frac{Q_2}{r_2} + \frac{Q_3}{r_3} + \cdots\right)$$

Some terms in the sum may be positive (those for positive charges) and some negative (those for negative charges). Equation 17.12 is a scalar sum because the electric potential is a scalar quantity. Electric fields and forces are vectors, but electric potential and potential energy are scalars.

17.4 The Electric Potential of Continuous Charge Distributions of Finite Size

The beauty and utility of the electric potential is that it is a scalar quantity. You do not have the complication of worrying about associating any direction with it. Huzza! We saw that the electric potential at a distance r from a point charge Q was given by Equation 17.11:

$$V(r) = \frac{1}{4\pi\varepsilon_0}\frac{Q}{r}$$

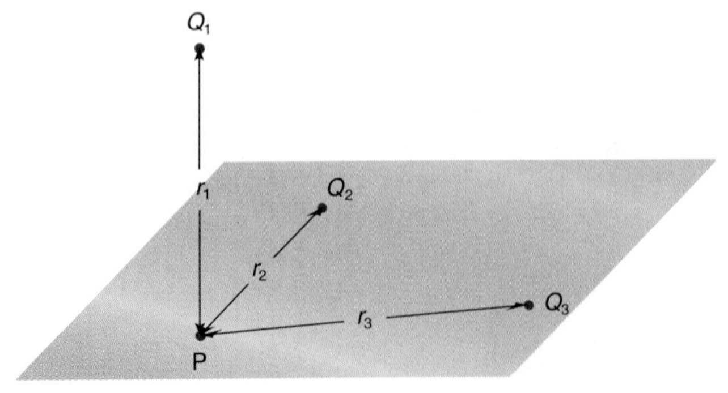

FIGURE 17.10 The electric potential at point P produced by a collection of pointlike charges is the scalar sum of the individual electric potentials at that location.

Electrons and protons are well approximated as pointlike charges. However, macroscopic charges are hardly pointlike: certainly glass and rubber rods are not points, let alone your cat, unless you are very far away from them. If the charge Q is smeared out over some region or object of finite size, as in Figure 17.11, we break up the extended charge distribution into a large collection of differential pointlike charge elements dQ. Each differential charge element dQ produces a differential amount of electric potential dV at point P:

$$dV = \frac{1}{4\pi\varepsilon_0}\frac{dQ}{r}$$

To find the total electric potential at the point P, we then sum these *scalar* contributions from all the differential charge elements. The sum is a continuous one, however: an integration over the charge distribution, be it a line, surface, or volume:

$$\boxed{V = \frac{1}{4\pi\varepsilon_0}\int_{\substack{\text{finite charge}\\ \text{distribution}}}\frac{dQ}{r}} \tag{17.13}$$

This is a scalar integration because V is a scalar quantity. It is unlike the situation for calculating the electric field of an extended charge distribution in Chapter 16, which involved a trickier vector integration. We need not worry about any directions for V: there are none, a cause for celebration.

There is only one string attached to Equation 17.13, but it is an important one: *the charge distribution must be of finite size.*

If the charge distribution extends off to infinity, you cannot use Equation 17.13. Why this restriction? Equation 17.13 was formulated by considering a superposition of pointlike differential charge elements. For pointlike charges, the electric potential was chosen conveniently to be zero at infinity. If there is charge at infinity (as there would be for a charge distribution of infinite extent), this choice does not make sense.

If the charge distribution does extend to infinity [e.g., infinite sheets of charge, infinite linear (line) charge distributions], we must revert to the defining Equation 17.9 to calculate the electric potential difference between two points. Notice that in Example 17.1 for the twin infinite sheets of charge (producing a uniform electric field between them) you did this: you calculated the potential at any point x between the infinite plates using Equation 17.9.

FIGURE 17.11 Each differential bit of charge dQ produces a differential amount of potential dV at point P.

Table 17.1 (on page 779) summarizes the electric potential in the vicinity of variously shaped charge distributions.

EXAMPLE 17.4

a. Find the electric potential at a point P located a distance z along the axis of a uniformly charged circular ring of radius R with total charge Q (see Figure 17.12).

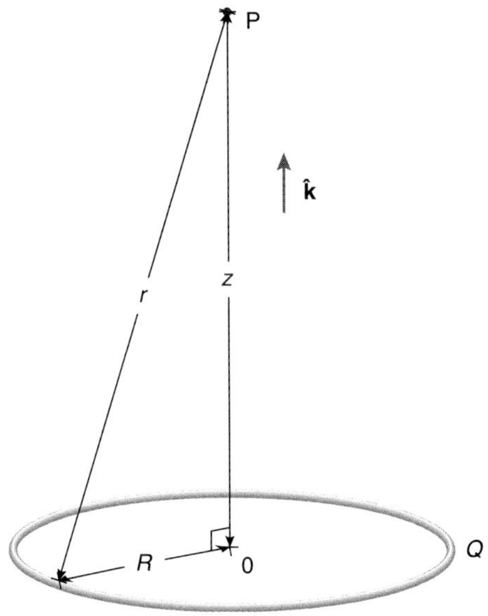

FIGURE 17.12

b. Make two schematic graphs of V as a function of z, one for positive Q and one for negative Q.

Solution

a. All of the charge Q is at the same distance r from the point at coordinate z along the axis. Since the potential is a scalar, you can sum it over the ring of charge. From the geometry, you have

$$r = (z^2 + R^2)^{1/2}$$

The potential at the point z thus is

$$\begin{aligned} V &= \frac{1}{4\pi\varepsilon_0}\int \frac{dQ}{r} \\ &= \frac{1}{4\pi\varepsilon_0}\frac{Q}{r} \\ &= \frac{1}{4\pi\varepsilon_0}\frac{Q}{\left(z^2 + R^2\right)^{1/2}} \end{aligned}$$

Notice that if $Q > 0$ C, $V > 0$ V for all z; if $Q < 0$ C, then $V < 0$ V for all z.

b. Graphs of V as a function of z for positive and negative Q are shown in Figure 17.13.

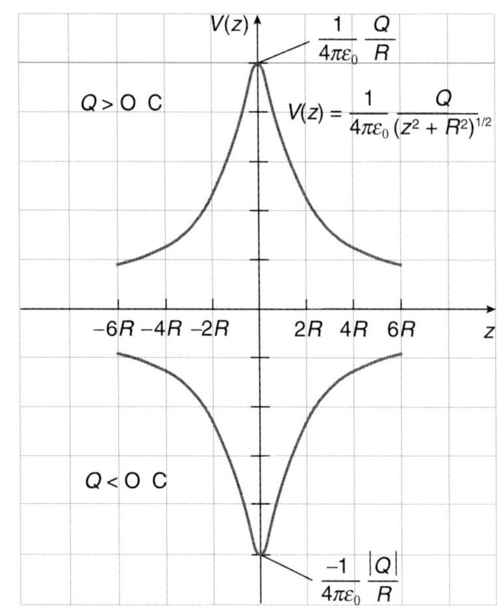

FIGURE 17.13

STRATEGIC EXAMPLE 17.5

a. Find the electric potential at a point P located on the axis of a uniformly charged circular disk of radius R and charge Q a distance z from the plane of the disk; see Figure 17.14.

b. Make two schematic graphs of V as a function of z, one for positive Q and one for negative Q.

Solution

a. The disk of radius R with total charge Q has a surface charge density of

$$\sigma = \frac{Q}{\pi R^2}$$

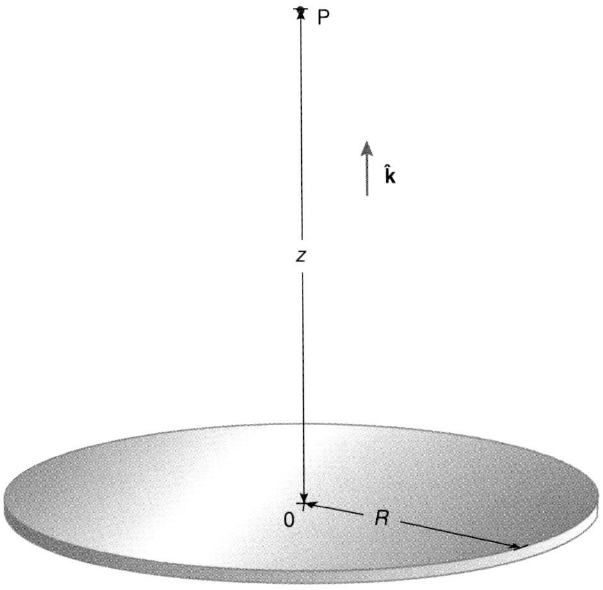

FIGURE 17.14

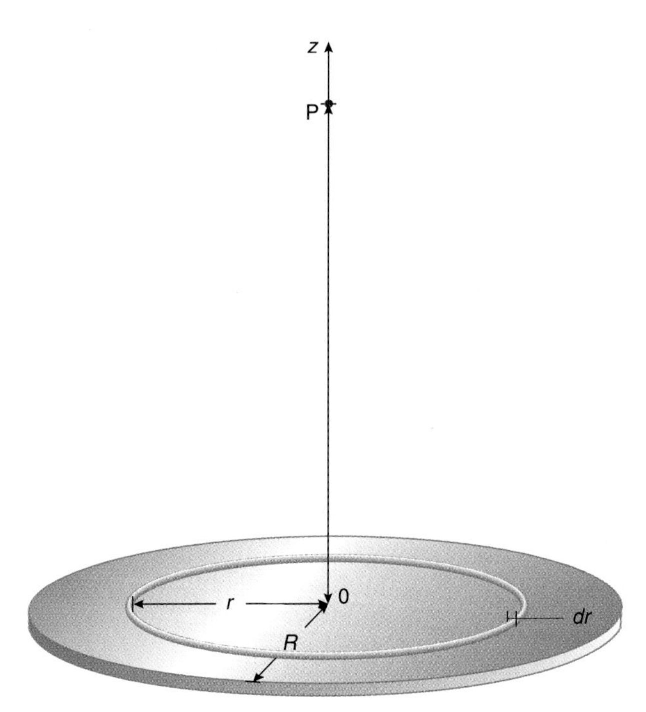

FIGURE 17.15

A ring of radius r and differential width dr (see Figure 17.15) has an area of

$$(\text{circumference})(\text{width}) = 2\pi r\,dr$$

The charge dQ on this differential ring is given by the product of the differential area and the surface charge density:

$$dQ = \sigma(2\pi r)\,dr \qquad (1)$$

Adapt the result of Example 17.4 to find the contribution to the electric potential at point P from this ring of charge:

$$dV = \frac{1}{4\pi\varepsilon_0}\frac{dQ}{(z^2 + r^2)^{1/2}}$$

Now substitute equation (1) for dQ and then integrate over the disk:

$$V = \frac{1}{4\pi\varepsilon_0}\int_{0\text{ m}}^{R}\frac{\sigma(2\pi r)\,dr}{(z^2 + r^2)^{1/2}}$$

The integral is of the form $u^n\,du$, where $u = (z^2 + r^2)$ and $n = -1/2$. Performing the integration, you obtain

$$V = \frac{1}{4\pi\varepsilon_0}2\pi\sigma\left[(z^2 + R^2)^{1/2} - z\right] \qquad (2)$$

To express this result in terms of the total charge Q on the disk, substitute for the surface charge density $\sigma = Q/\pi R^2$:

$$V = \frac{1}{4\pi\varepsilon_0}\frac{2Q}{R^2}\left[(z^2 + R^2)^{1/2} - z\right] \qquad (3)$$

How V on the axis of a charged disk changes with the radius R of the disk may be seen from the previous two equations. Two different scenarios of V variation with R are possible: (1) either hold the surface charge density constant and include more total charge as R increases; or (2) hold the total charge constant and spread it across a greater area as R increases. The first choice of constant σ is more informative: if z is constant, V will increase slowly for small R, but eventually grows linearly with R. On the other hand, if Q is constant, it spreads more thinly over a disk as the radius grows, and then V will eventually decrease as $1/R$ for large values of R.

b. Schematic graphs of equation (3) for V as a function of z for fixed Q and R are shown in Figure 17.16 for both positive and negative Q.

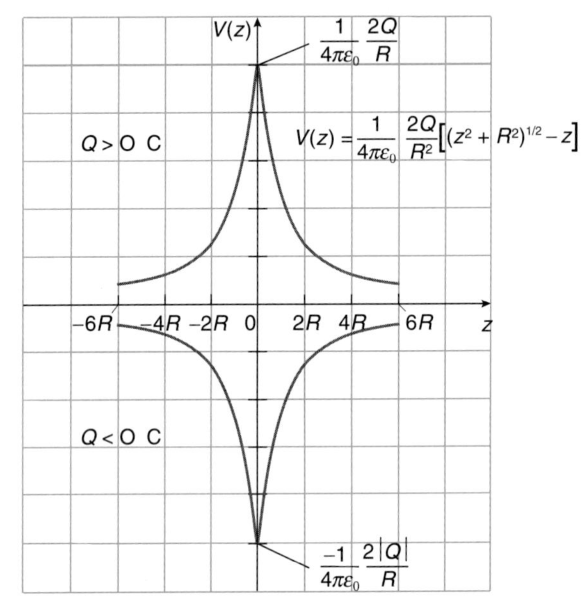

FIGURE 17.16

EXAMPLE 17.6

Find the electric potential at a point P located a distance r from a sphere of radius R that has a total charge Q distributed uniformly throughout its volume, where $r > R$; see Figure 17.17. Find the value of the electric potential on the surface where $r = R$.

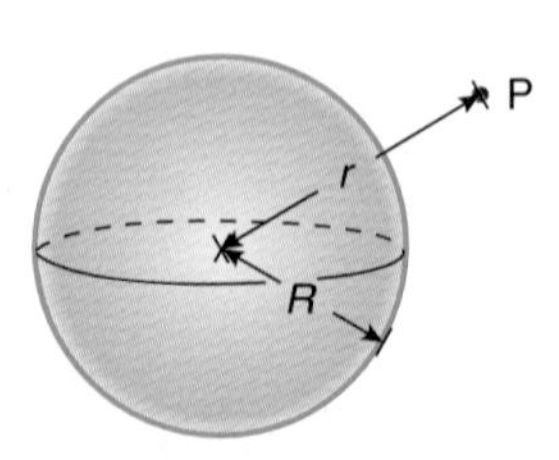

FIGURE 17.17

Solution

You can use Equation 17.13 to find the electric potential since the charge distribution is of finite extent. However, the integration over the sphere is complicated. To avoid this complexity, it is easier to use Equation 17.9 to find the electric potential, since you know the electric field of such a charge distribution from Chapter 16 (Example 16.19):

$$\vec{E} = \frac{1}{4\pi\varepsilon_0}\frac{Q}{r^2}\hat{r}$$

Use Equation 17.9 to find the electric potential at r by integrating the field from an infinite distance away to the radial distance r:

$$V(r_f) - V(r_i) = -\int_i^f \vec{E} \cdot d\vec{r}$$

where the initial position is at infinity and the final position is a point a distance r from the center of the spherical charge distribution. As usual, let the differential change in the position vector $d\vec{r}$ be $dr\,\hat{r}$ and let the limits of integration account for the direction you are moving (here, toward decreasing values of r). Remember that the result is independent of the path (since the electrical force is conservative), so you can come in along a radial direction to the final position. Thus

$$\begin{aligned} V(r) - V(\infty \text{ m}) &= -\frac{1}{4\pi\varepsilon_0} Q \int_{\infty \text{ m}}^{r} \frac{\hat{r}}{r^2} \cdot dr\,\hat{r} \\ &= -\frac{1}{4\pi\varepsilon_0} Q \int_{\infty \text{ m}}^{r} \frac{dr}{r^2} \\ &= -\frac{1}{4\pi\varepsilon_0} Q \left(-\frac{1}{r}\right)\Bigg|_{\infty \text{ m}}^{r} \\ &= \frac{1}{4\pi\varepsilon_0}\frac{Q}{r} \end{aligned}$$

By convention, zero potential is chosen to be where $r = \infty$ m, so $V(\infty \text{ m}) = 0$ V and you have

$$V(r) = \frac{1}{4\pi\varepsilon_0}\frac{Q}{r} \qquad (1)$$

Notice that this expression is the same as the potential of a pointlike charge Q located at the center of the sphere. So for distances $r > R$, the potential of a spherically symmetric charge distribution of radius R is the same as that of a point charge Q located at the center of the sphere. The sphere can be either a uniformly charged conducting sphere (with the charge on the outer surface of the conductor) or an insulating sphere with a uniform charge distribution either on its surface or distributed throughout its volume, as long as the charge distribution is spherically symmetric.

The potential on the surface of the charge distribution is found by setting $r = R$:

$$V(R) = \frac{1}{4\pi\varepsilon_0}\frac{Q}{R} \qquad (2)$$

EXAMPLE 17.7

a. If the spherical charge distribution of Example 17.6 is on a *conducting* sphere, what is the potential within the sphere at distances $r < R$?

b. Make a graph of V as a function of r, including regions where $r < R$ and $r > R$.

Solution

a. The static charge on a conductor resides on its surface; there is no charge within the conductor for $r < R$, even if the conductor is hollow (see Section 16.13). The interior of a conductor has the same potential everywhere. The potential is constant within the sphere and equal to the value of the potential on its surface:

$$V(r) = \frac{1}{4\pi\varepsilon_0}\frac{Q}{R} \qquad (r < R) \qquad \text{(conductor)}$$

b. The electric potential of the uniformly charged conducting sphere is graphed as a function of r in Figure 17.18 for both the positively and negatively charged cases.

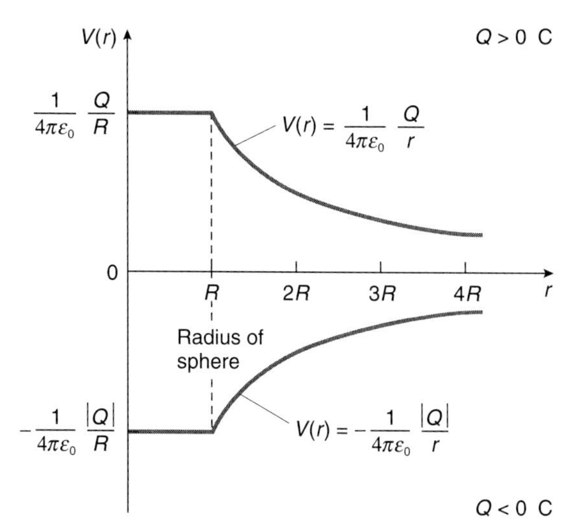

FIGURE 17.18

EXAMPLE 17.8

a. Find the electric potential at a distance r from the center of a *nonconducting* (i.e., insulating) sphere of radius R where the charge Q is distributed uniformly throughout the volume of the sphere and $r < R$; see Figure 17.19.

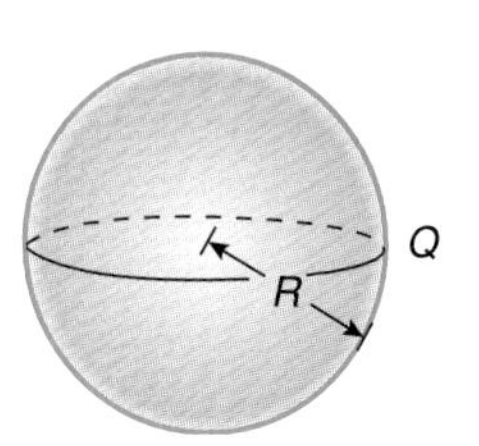

FIGURE 17.19

b. Make a graph of V as function of r, including the regions where $r < R$ and $r > R$.

Solution

a. To avoid a complicated integral over the sphere, begin with Equation 17.9 for the definition of the electric potential difference between two points:

$$V(r_f) - V(r_i) = -\int_i^f \vec{E} \cdot d\vec{r}$$

We found in Section 16.12 that the electric field within such a charge distribution a distance r from the center is given by

$$\vec{E} = \frac{1}{4\pi\varepsilon_0}\frac{Q}{R^3} r\,\hat{r}$$

Integrate from the surface of the sphere R to the interior point r. Once again, let $d\vec{r} = dr\,\hat{r}$ and the limits of integration will account for the radial direction you are moving between the initial and final positions. Hence

$$\begin{aligned} V(r) - V(R) &= -\frac{1}{4\pi\varepsilon_0}\frac{Q}{R^3}\int_R^r r\,\hat{r} \cdot dr\,\hat{r} \\ &= -\frac{1}{4\pi\varepsilon_0}\frac{Q}{R^3}\int_R^r r\,dr \\ &= -\frac{1}{4\pi\varepsilon_0}\frac{Q}{R^3}\frac{r^2}{2}\bigg|_R^r \\ &= \frac{1}{4\pi\varepsilon_0}\frac{Q}{R^3}\left(\frac{R^2}{2} - \frac{r^2}{2}\right) \end{aligned}$$

The potential $V(R)$ at the surface of such a spherical charge distribution is given by equation (2) of Example 17.6; substituting this expression, you get

$$V(r) - \frac{1}{4\pi\varepsilon_0}\frac{Q}{R} = \frac{1}{4\pi\varepsilon_0}\frac{Q}{R^3}\left(\frac{R^2}{2} - \frac{r^2}{2}\right)$$

After some algebraic rearrangement, you find

$$V(r) = \frac{1}{4\pi\varepsilon_0}\frac{Q}{2R}\left(3 - \frac{r^2}{R^2}\right) \quad (r < R) \quad \text{(uniformly charged insulating sphere)}$$

Notice that when $r = R$, you get the potential on the surface of such a spherical charge distribution. The potential at the center of the spherical insulator distribution, where $r = 0$ m, is

$$V(0\text{ m}) = \frac{1}{4\pi\varepsilon_0}\frac{3Q}{2R}$$

b. Graphs of the potential as a function of r for $r < R$ and for $r > R$ are shown in Figure 17.20.

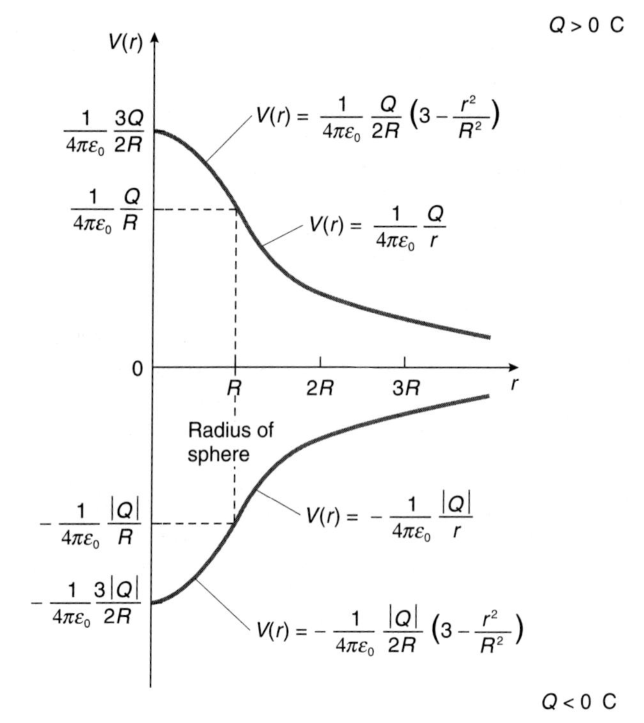

FIGURE 17.20

17.5 EQUIPOTENTIAL VOLUMES AND SURFACES

The electric potential difference between two points is defined by Equation 17.9:

$$V(r_f) - V(r_i) = -\int_i^f \vec{E} \cdot d\vec{r}$$

Let the initial and final positions be very closely (differentially) spaced; indeed, if the points are separated by the differential change in the position vector $d\vec{r}$ (see Figure 17.21), then the differential change in the electric potential is

$$dV = -\vec{E} \cdot d\vec{r} \tag{17.14}$$

Equation 17.9 is the integral of Equation 17.14.

The locus of points or regions for which the electric potential has a constant value are called **equipotential regions** or, more simply, **equipotentials**. Such equipotentials can be volumes, surfaces, and even lines.

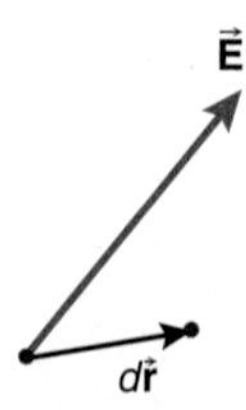

FIGURE 17.21 **Two points separated by differential distance $d\vec{r}$ have a potential difference $dV = -\vec{E} \cdot d\vec{r}$.**

TABLE 17.1 Summary of the Electric Potential of Various Charge Distributions

Two infinite sheets, oppositely charged (uniform electric field):

$$V(x) - V(0) = -Ex$$

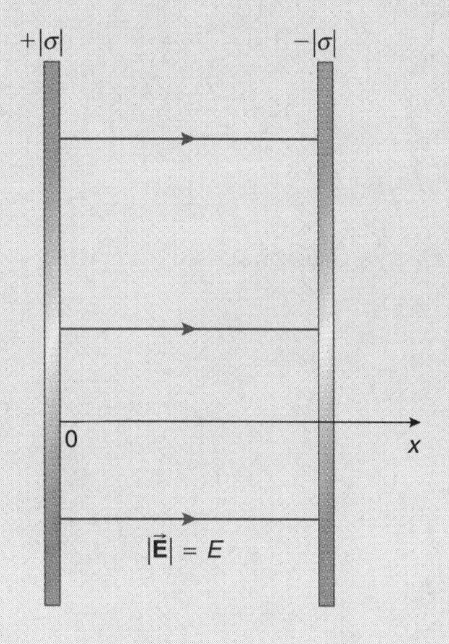

Pointlike charge:

$$V(r) = \frac{1}{4\pi\varepsilon_0}\frac{Q}{r}$$

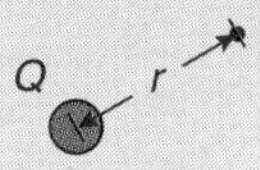

Uniformly charged ring, a distance z along the axis:

$$V = \frac{1}{4\pi\varepsilon_0}\frac{Q}{\left(z^2 + R^2\right)^{1/2}}$$

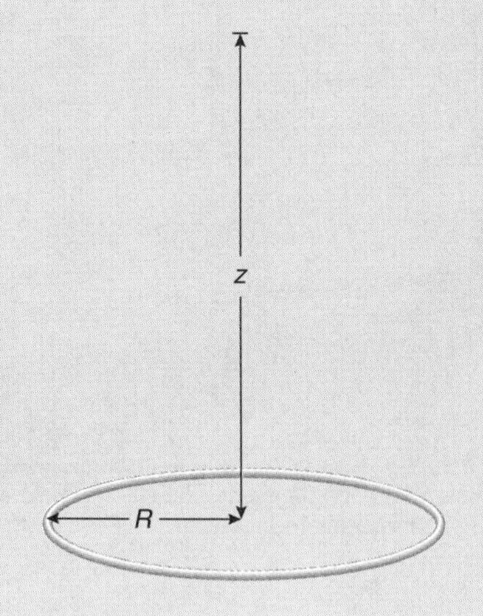

Uniformly charged disk, a distance z along the axis:

$$V = \frac{1}{4\pi\varepsilon_0}\frac{2Q}{R^2}\left[\left(z^2 + R^2\right)^{1/2} - z\right]$$

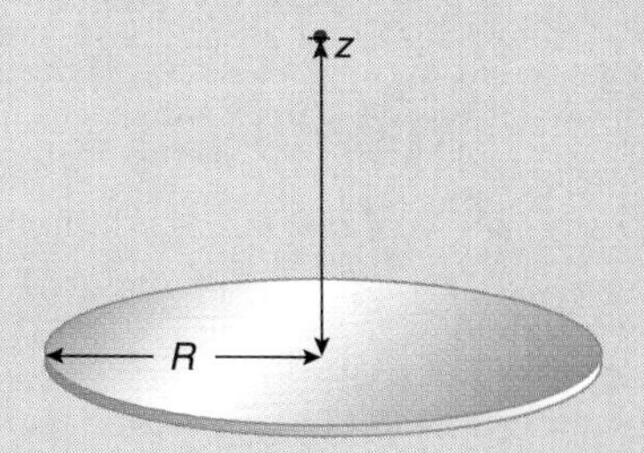

Uniformly charged sphere, $r > R$:

$$V(r) = \frac{1}{4\pi\varepsilon_0}\frac{Q}{r}$$

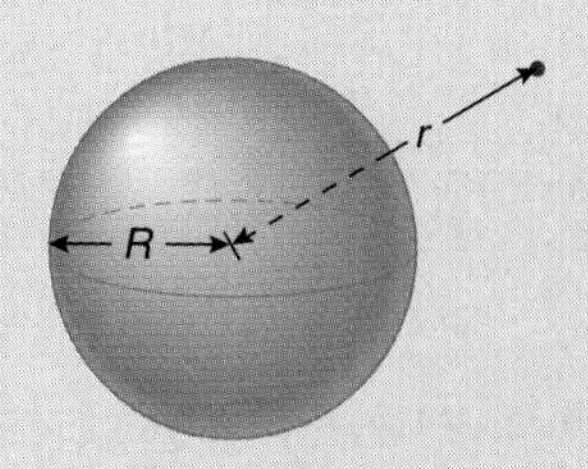

Uniformly charged conducting sphere, $r < R$:

$$V(r) = \frac{1}{4\pi\varepsilon_0}\frac{Q}{R} \quad \text{(constant)}$$

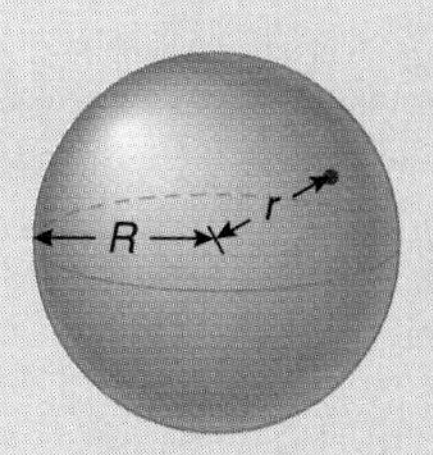

Uniformly charged insulating sphere, $r < R$:

$$V(r) = \frac{1}{4\pi\varepsilon_0}\frac{Q}{2R}\left(3 - \frac{r^2}{R^2}\right)$$

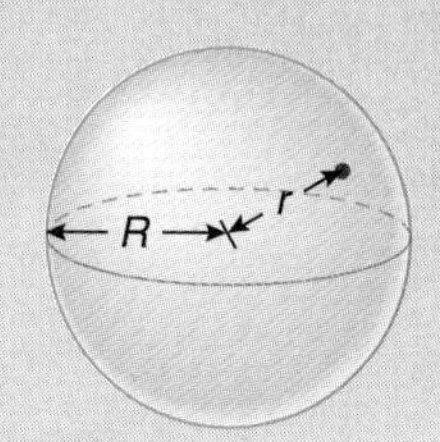

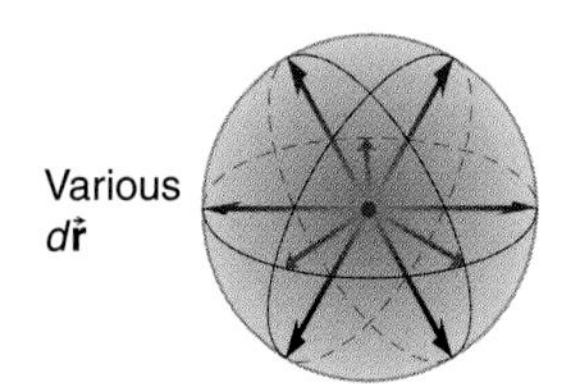

FIGURE 17.22 Within any equipotential volume, the change in the electric potential in any direction is zero.

Within an equipotential volume, the value of V is the same everywhere. If you move in any direction away from any point within the volume (see Figure 17.22), the change in the potential is zero. Hence

$$dV = -\vec{E} \cdot d\vec{r}$$
$$0 \text{ V} = -\vec{E} \cdot d\vec{r}$$

regardless of the direction of the differential change in the position vector $d\vec{r}$.

The only way $\vec{E} \cdot d\vec{r}$ can be zero for every $d\vec{r}$ radiating from a given point is for $\vec{E}$ itself to be zero at the point. This argument is applicable to any point within an equipotential volume.

Hence the electric field must be zero within an equipotential volume.

We saw in Chapter 16 that in an electrostatic situation, the electric field inside a conductor is zero. This means that the interior of a conductor is an example of an equipotential volume. You discovered this fact in Example 17.7.

The same is true if the conductor is *hollow* and there are no charges inside the hollow.* To show this, imagine a Gaussian surface within the conductor surrounding the cavity, as shown in Figure 17.23.

Since the static electric field is zero within the conductor, the left-hand side of Gauss's law is zero, implying there is *zero total charge* within the Gaussian surface. We imagine some positive charge to be on one side of the inside surface of the hollow conductor and some negative charge on the other, as shown in Figure 17.24, so the total charge is zero, consistent with the prediction of Gauss's law. If such charge separation exists on the inside surface of the cavity, there will be electric field lines within the hollow from the positive to the negative charges.

If we use Equation 17.9 and integrate the electric field along one of these field lines from a positive charge to a negative charge, the result will *not be zero* since $d\vec{r}$ is parallel to $\vec{E}$ over the whole path. This implies that there is a potential difference

*No charges are present at all within the hollow; this may be distinct from a situation of zero charge, where the total charge sums to zero.

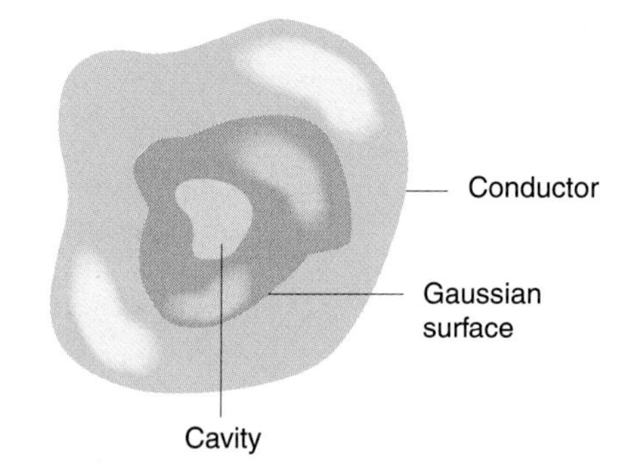

FIGURE 17.23 A Gaussian surface surrounding a hollow in a conductor.

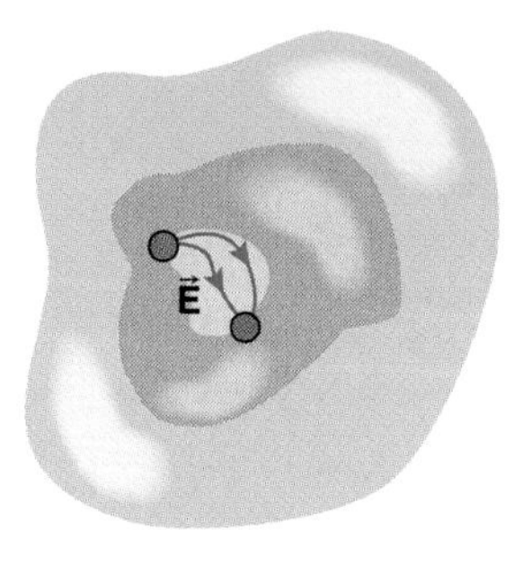

FIGURE 17.24 Electric field lines from a positive to a negative charge.

This stereo amplifier is encased in a metal box to shield its circuitry from outside electrical influences.

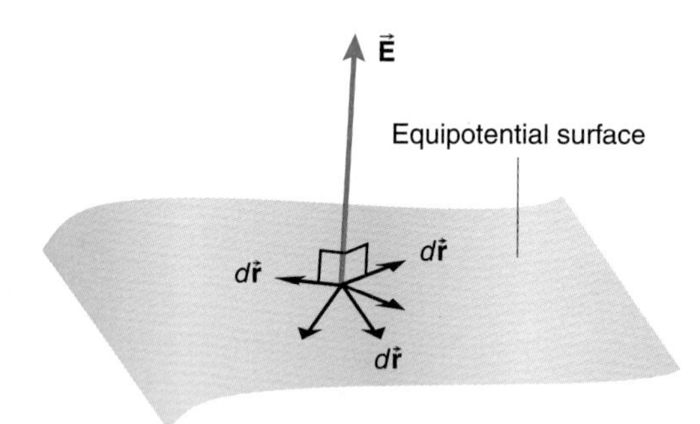

FIGURE 17.25 If you move in any direction on an equipotential surface, the potential is unchanged. The electric field must be perpendicular to an equipotential surface.

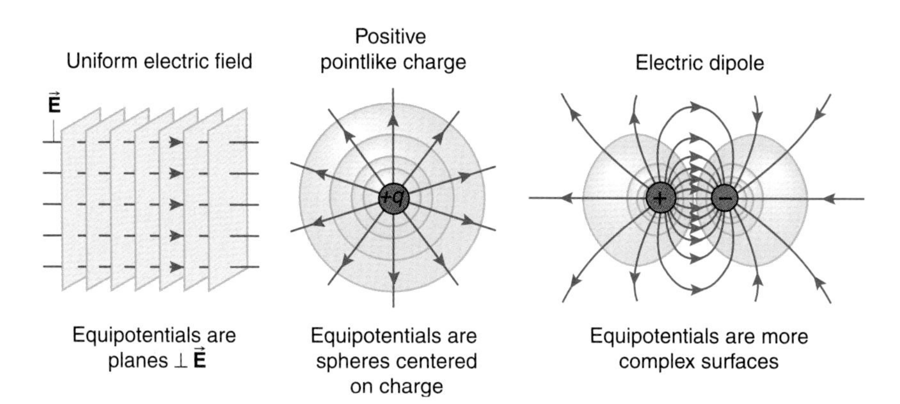

FIGURE 17.26 Equipotential surfaces.

between the two points at opposite ends of the field line on the inside surface of the cavity. But the conductor is an equipotential, so there can be no such potential difference. Hence there can be no electric field lines anywhere within the hollow: the electric field inside the cavity is zero.

Therefore, electric field–free regions of space can be created by surrounding the region with a conductor, as long as there are *no charges* within the region. Nowhere inside such metal cavities or boxes is there any electric field. This procedure is called **electrostatic shielding**. Sensitive electronic instruments are manufactured with their circuitry inside metal boxes specifically to shield various components from electrical influences (interference) from charges outside the box.

On an equipotential surface, if we move in any direction on the surface (see Figure 17.25), the value of the potential is unchanged. That is,

$$dV = -\vec{E} \cdot d\vec{r}$$
$$0 \text{ V} = -\vec{E} \cdot d\vec{r}$$

for any $d\vec{r}$ confined to the surface. Therefore any electric field $\vec{E}$ must have a perpendicular equipotential surface associated with it at every point. The surface of a conductor in electrostatics (i.e., with electric charges at fixed locations on its surface) exemplifies an equipotential surface.

Equipotential surfaces need not be physical surfaces (such as that of a conductor). Any imaginary surface over which the electric field is everywhere perpendicular to it is called an equipotential surface. Various equipotential surfaces along with the associated electric fields are shown in Figure 17.26 for a number of different charge distributions. The surfaces are chosen and drawn so that the potential difference between successive surfaces is a constant value. Two-dimensional diagrams of equipotential surfaces are called lines of equipotentials; they also are typically drawn with a fixed value for the potential difference between the lines, much like the contour interval between contour lines on topographic maps.

Any equipotential volume is enclosed within a surface that itself must be an equipotential surface. The converse, however, is *not* true: an equipotential surface need not surround an equipotential volume.

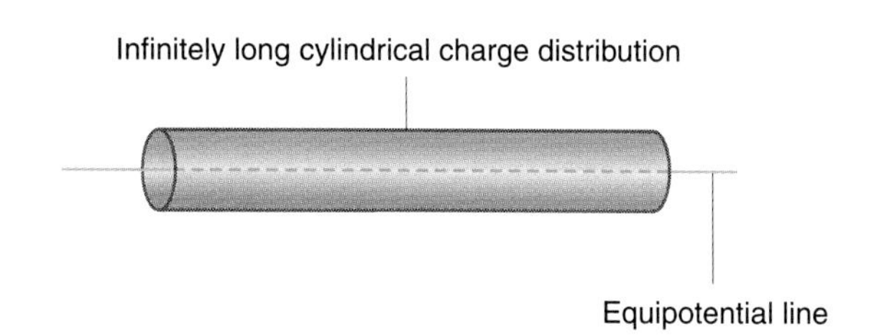

FIGURE 17.27 An equipotential line exists along the axis of a cylindrical charge distribution.

Since a conductor is an equipotential volume, the surface of a conductor is an equipotential surface and, therefore, any electric field at the surface of the conductor must be perpendicular to the surface for any static distributions of charge.

The only charge distribution that can produce a true equipotential line in space is a uniformly charged infinite cylindrical shell or cylindrical volume, as shown in Figure 17.27. An equipotential line exists along the axis of the cylinder; in this rather exotic situation, the electric field is zero along the axis as well.

QUESTION 2

Explain why two equipotential surfaces cannot intersect.

17.6 The Relationship Between the Electric Potential and the Electric Field

In the previous section, we saw that the differential change dV in the electrical potential between two points that are separated by the differential change $d\vec{r}$ in the position vector is given by Equation 17.14:

$$dV = -\vec{E} \cdot d\vec{r}$$

It frequently happens that the electric field is a function of only one coordinate, say s, and has a single vector s-component; this can be seen by examining Table 16.2, which lists the electric field for charge distributions with various geometries. Because of the scalar product in Equation 17.14, the greatest decrease in the value of the electric potential is in the direction parallel to the field direction. If we choose $d\vec{\mathbf{r}}$ to be in this s-direction, then

$$dV = -E_s\, ds$$

where E_s is the component of the field in the direction of the coordinate s. The electric field has *only* this s-component, since we took the direction of the coordinate s to be parallel to the field. Hence we have

$$E_s = -\frac{dV}{ds} \qquad (17.15)$$

Therefore the single s-component of the electric field is the negative of the derivative of the electric potential with respect to the single s-coordinate.

Geometrically, Equation 17.15 implies that the component of the field at a given point is the negative of the slope of a graph of V versus s at that point.

For example, we saw that with the geometry of Figure 17.1, the electric potential associated with the uniform field between two plates is, from Example 17.1,

$$V(x) - V(0\text{ m}) = -Ex$$

where $V(0\text{ m})$ is the electric potential where $x = 0$ m. Hence

$$V(x) = V(0\text{ m}) - Ex$$

Here the coordinate x plays the role of the coordinate s just discussed.

If we use Equation 17.15, adapted to the coordinate x, we find the (single) component of the electric field is

$$\begin{aligned} E_x &= -\frac{dV}{dx} \\ &= -(-E) \\ &= E \end{aligned}$$

The electric field, therefore, is $\vec{\mathbf{E}} = E_x\hat{\mathbf{\imath}} = E\hat{\mathbf{\imath}}$, as in Figure 17.1. Note also in Figures 17.2 and 17.3, that the field component E is the negative of the slope of the graph of V versus x.

If the potential function is a function of the single radial coordinate r—that is, $V(r)$—then the radial component of the electric field is

$$E_r = -\frac{dV}{dr} \qquad (17.16)$$

For the case of a pointlike charge Q, we found the electric potential is a function of the radial coordinate r only (Equation 17.11):

$$V = \frac{1}{4\pi\varepsilon_0}\frac{Q}{r}$$

We find the (single) electric field component E_r using Equation 17.16:

$$\begin{aligned} E_r &= -\frac{dV}{dr} \\ &= -\frac{1}{4\pi\varepsilon_0}\left(-\frac{Q}{r^2}\right) \\ &= \frac{1}{4\pi\varepsilon_0}\frac{Q}{r^2} \end{aligned}$$

So the electric field vector is

$$\begin{aligned} \vec{\mathbf{E}} &= E_r\hat{\mathbf{r}} \\ &= \frac{1}{4\pi\varepsilon_0}\frac{Q}{r^2}\hat{\mathbf{r}} \end{aligned}$$

as we know from Equation 16.11. The component of the field, E_r, is the negative of the slope of the graph of $V(r)$ versus r.

We can just as easily generalize to three dimensions. If the electric potential is a function of the three Cartesian coordinates x, y, and z, the various components of the electric field are

$$E_x = -\frac{\partial V}{\partial x}$$

$$E_y = -\frac{\partial V}{\partial y} \qquad (17.17)$$

$$E_z = -\frac{\partial V}{\partial z}$$

where the derivatives are partial derivatives. Each Cartesian component of the field is the negative of the slope of the graph of V versus that coordinate.

A special differentiation operator, known as the **del operator** ∇ is defined as

$$\nabla \equiv \frac{\partial}{\partial x}\hat{\mathbf{\imath}} + \frac{\partial}{\partial y}\hat{\mathbf{\jmath}} + \frac{\partial}{\partial z}\hat{\mathbf{k}} \qquad (17.18)$$

Then Equations 17.17 can be summarized by writing

$$\vec{\mathbf{E}} = -\nabla V \qquad (17.19)$$

The operation ∇V produces what is called the **gradient** of the scalar potential function V. The gradient is a vector.

Thus we find there is an easier way to find the electric field than via the typically complicated vector integration techniques of Chapter 16. First find the electric potential V, which typically is easier than finding $\vec{E}$ by vector integration, because V is a *scalar* function. Then get $\vec{E}$ from V by taking the negative gradient of V using the ∇ operator.

But remember, we can only find V using Equation 17.13 for charge distributions that are of finite extent:

$$V = \frac{1}{4\pi\varepsilon_0} \int_{\substack{\text{finite charge}\\ \text{distribution}}} \frac{dQ}{r}$$

If the charge distribution extends to infinity, then we can only find V from the defining Equation 17.9:

$$V(r_f) - V(r_i) = -\int_i^f \vec{E} \cdot d\vec{r}$$

which means we have to know the field to begin with; the vector integration techniques to obtain the field then cannot be avoided.

Example 17.9

For a uniformly charged circular ring at a point z along the axis of the ring, you found the potential to be (see Example 17.4)

$$V = \frac{1}{4\pi\varepsilon_0} \frac{Q}{\left(z^2 + R^2\right)^{1/2}}$$

Use the potential to find the electric field along the z-axis of the ring.

Solution

The potential is a function of only the coordinate z; hence the partial derivative of V with respect to z is the same as the ordinary derivative of V with respect to z. The electric field along the axis thus has only a z-component, and it is found using Equation 17.15 where the general coordinate s is here z

$$\begin{aligned} E_z &= -\frac{\partial V}{\partial z} \\ &= -\frac{dV}{dz} \\ &= -\frac{1}{4\pi\varepsilon_0} Q\left(-\frac{1}{2}\right)\frac{2z}{\left(z^2 + R^2\right)^{3/2}} \\ &= \frac{1}{4\pi\varepsilon_0} \frac{Qz}{\left(z^2 + R^2\right)^{3/2}} \end{aligned}$$

The electric field is $\vec{E} = E_z\hat{k}$. This result just what you found in Example 16.14.

17.7 Acceleration of Charged Particles Under the Influence of Electrical Forces

The motion of charged particles under the influence of the electrical force has important technological applications in devices as diverse as x-ray tubes, particle accelerators, and TV tubes. The CWE theorem provides a simple way to analyze the dynamics of such motions.

If the charged particle moves *only* under the influence of the conservative electrical force, the work done by nonconservative forces is zero. Thus the left-hand side of the CWE theorem

$$W_{\text{noncon}} = \Delta\text{KE} + \Delta\text{PE} \qquad (17.20)$$

is zero, and the theorem simplifies to

$$\begin{aligned} 0\ \text{J} &= \Delta\text{KE} + \Delta\text{PE} \\ 0\ \text{J} &= \Delta(\text{KE} + \text{PE}) \qquad (17.21) \end{aligned}$$

In other words, the total mechanical energy of the charged particle is conserved throughout its motion under the influence of the static electrical force.

Consider a charged particle q in space, initially at a point where it has kinetic energy KE_i and electric potential energy $\text{PE}_i = qV_i$. When the particle has moved under the influence of the electrical force to another position where the electric potential is V_f, the potential energy of the charge is $\text{PE}_f = qV_f$ and its kinetic energy is KE_f. Conservation of mechanical energy, Equation 17.21, implies that the change in the kinetic energy ΔKE is

$$\begin{aligned} 0\ \text{J} &= \Delta\text{KE} + (qV_f - qV_i) \\ 0\ \text{J} &= \Delta\text{KE} + q\,\Delta V \end{aligned}$$

or

$$\Delta\text{KE} = -q\,\Delta V \qquad (17.22)$$

Equation 17.22 indicates that a *positive* charge in an electric field, moving under the influence of only the electrical force, will *increase* its kinetic energy (while decreasing its electric potential energy) if it moves to regions of *lower* electric potential (since then $V_f < V_i$ and $\Delta V < 0$ V, and so $\Delta\text{KE} > 0$ J). The total mechanical energy is constant according to the CWE theorem. Correspondingly, a *negative* charge moves to regions of *higher* electric potential to increase its kinetic energy (and decrease its electric potential energy) (since then $V_f > V_i$ and $\Delta V > 0$ V but $q < 0$ C, giving $\Delta\text{KE} > 0$ J).

Question 3

A charge, initially at rest, moves under the action of only the electrical force in a direction toward increasing values of the electric potential. What can be said, if anything, about the sign of the charge? What happens to the potential energy of the charge?

STRATEGIC EXAMPLE 17.10

An electron enters a region with a uniform electric field created by two parallel, uniformly charged plates at potentials +100 V and −150 V, as shown in Figure 17.28. The initial speed of the electron is 5.00×10^6 m/s. Determine the speed of the electron when it emerges from the field. This type of apparatus is used to increase the kinetic energy of electrons in TVs, CRT (cathode ray tube) computer displays, and pieces of laboratory equipment you may encounter even in this course, such as an apparatus to measure the charge to mass ratio (e/m) of the electron.

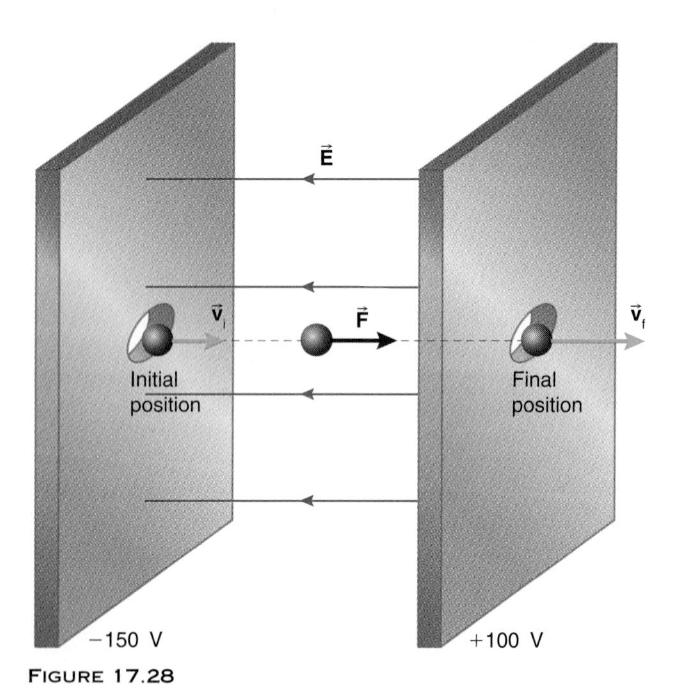

FIGURE 17.28

Solution

The electric field between the charged plates is directed from regions of higher electric potential to regions of lower electric potential. The electric field thus is antiparallel to the initial velocity of the electron. The electrical force on a charged particle in an electric field $\vec{E}$ is given by Equation 16.8:

$$\vec{F} = q\vec{E}$$

For the negatively charged electron ($q = -e$), the electrical force is antiparallel to the field. Hence in this problem the electrical force on the negatively charged electron is parallel to its initial velocity; you can anticipate that the electron will emerge from the device with a higher speed. You find the new speed by applying the CWE theorem to the initial and final positions of the electron.

Method 1

Use the CWE theorem in the guise of Equation 17.21, slightly rearranged:

$$\text{KE}_f + \text{PE}_f = \text{KE}_i + \text{PE}_i$$

$$\frac{mv_f^2}{2} + qV_f = \frac{mv_i^2}{2} + qV_i$$

$$\begin{aligned}\frac{mv_f^2}{2} &= \frac{mv_i^2}{2} + q(V_i - V_f)\\ &= \frac{mv_i^2}{2} + (-e)(V_i - V_f)\\ &= \frac{mv_i^2}{2} + e(V_f - V_i)\end{aligned}$$

Method 2

You also could use Equation 17.22 directly:

$$\Delta\text{KE} = -q\,\Delta V$$

$$\frac{mv_f^2}{2} - \frac{mv_i^2}{2} = -(-e)(V_f - V_i)$$

$$\frac{mv_f^2}{2} = \frac{mv_i^2}{2} + e(V_f - V_i)$$

Substituting numerical values, you find

$$\begin{aligned}\frac{mv_f^2}{2} &= \frac{(9.11 \times 10^{-31}\ \text{kg})(5.00 \times 10^6\ \text{m/s})^2}{2}\\ &\quad + (1.602 \times 10^{-19}\ \text{C})\left[100\ \text{V} - (-150\ \text{V})\right]\\ &= 5.15 \times 10^{-17}\ \text{J}\end{aligned}$$

Solving for the square of the final speed of the electron,

$$\begin{aligned}v_f^2 &= \frac{2 \times 5.15 \times 10^{-17}\ \text{J}}{9.11 \times 10^{-31}\ \text{kg}}\\ &= 1.13 \times 10^{14}\ \text{m}^2/\text{s}^2\end{aligned}$$

The final speed then is

$$v_f = 1.06 \times 10^7\ \text{m/s}$$

17.8 A NEW ENERGY UNIT: THE ELECTRON-VOLT

When specifying the energy of electrons, protons, and other fundamental particles with small charge quantum numbers, it is convenient to use another energy unit called the **electron-volt** (eV). An electron-volt is defined to be the *change* in the kinetic energy of a negative fundamental unit of charge $-e$ when it accelerates through a potential difference of exactly one volt. From Equation 17.22, we see that if an electron accelerates through a potential difference of exactly one volt, the change in the kinetic energy in joules is

$$\begin{aligned}\Delta KE &= -q\,\Delta V \\ &= -(-1.602 \times 10^{-19}\ \text{C})(1.000\ \text{V}) \\ &= 1.602 \times 10^{-19}\ \text{J}\end{aligned}$$

By definition this change in the kinetic energy is one electron-volt. Therefore the conversion factor relating the SI energy unit J to the *convenient* energy unit called the electron-volt (eV) is approximately

$$1\ \text{eV} = 1.602 \times 10^{-19}\ \text{J} \qquad (17.23)$$

Note that the conversion factor is, by definition, *numerically* equal to the number representing the fundamental unit of charge in coulombs.

Although the electron-volt energy unit is defined using the negative fundamental unit of charge, the energy unit can and is equally well applied to protons, helium nuclei, or other charged particles.

The utility of the new energy unit manifests itself by examining Equation 17.22,

$$\Delta KE = -q\,\Delta V$$

If the particle has a charge $q = ne$, where n is the charge quantum number, then

$$\Delta KE = -ne\,\Delta V \qquad (17.24)$$

To convert the kinetic energy change in joules to the new energy unit of electron-volts, use the conversion given by Equation 17.23. The effect of this conversion is to divide e out of Equation 17.24, leaving us with

$$\Delta KE_{\text{in eV}} = -n\,\Delta V \qquad (17.25)$$

where n is the charge quantum number of the particle.

It is interesting to note the following special case of Equation 17.25.

> Protons and electrons have charge quantum numbers ±1 respectively. If they are accelerated by electrical forces alone between two points, the *absolute value* of the change in their kinetic energy between the two points, expressed in electron-volts, is numerically equal to the *absolute value* of the difference in electric potential between the two points.

The same is true for any particle with charge quantum number ±1. That is,

$$|\Delta KE_{\text{in eV}}| = |\Delta V| \quad \text{(for any particle with charge quantum number } n = \pm 1)$$

As noted previously, positive particles such as protons increase their kinetic energy (and decrease their electric potential energy) by moving to regions of lower electric potential; negative charges increase their kinetic energy (and decrease their electric potential energy) by moving to regions of higher electric potential.

PROBLEM-SOLVING TACTICS

17.7 Kinetic energies in electron-volts (eV) must be converted to joules (J) to calculate speeds in meters per second (m/s). Not to do so results in inconsistent units, which should become apparent as you substitute numerical values.

QUESTION 4

A proton and an electron are accelerated (in opposite directions) through a potential difference of the same absolute magnitude. What is the ratio of the change in their kinetic energies?

EXAMPLE 17.11

Calculate the change in the kinetic energy of the electron in Example 17.10 in both electron-volts and joules.

Solution

Since an electron has a charge of $q = ne = -e$, the charge quantum number of an electron is $n = -1$. The change in the electric potential is

$$\begin{aligned}\Delta V &= V_f - V_i \\ &= 100\ \text{V} - (-150\ \text{V}) \\ &= 250\ \text{V}\end{aligned}$$

Use Equation 17.25 for the change in kinetic energy in electron-volts, being careful with the signs:

$$\begin{aligned}\Delta KE_{\text{in eV}} &= -n\,\Delta V \\ &= -(-1)(250\ \text{V}) \\ &= 250\ \text{eV} \\ &= (250\ \text{eV})(1.602 \times 10^{-19}\ \text{J/eV}) \\ &= 4.01 \times 10^{-17}\ \text{J}\end{aligned}$$

EXAMPLE 17.12

A proton is released at rest in the apparatus shown in Figure 17.29, which consists of two flat, conducting plates at the indicated potentials. The proton accelerates under the action of the electrical force to the other plate.

a. Near which plate should it be released to gain the most speed?
b. Calculate the kinetic energy of the proton as it strikes the other plate, expressing your result in electron-volts.
c. Determine the impact speed in meters per second.
d. Did we forget to give the plate separation?

Solution

a. The electric field is directed from higher to lower electric potential regions, so that it is directed as shown in Figure 17.30. The proton has a positive charge, and so the electrical force on a proton is parallel to the field according to $\vec{F} = q\vec{E}$. Hence the proton should be released near the plate with potential +250 V.

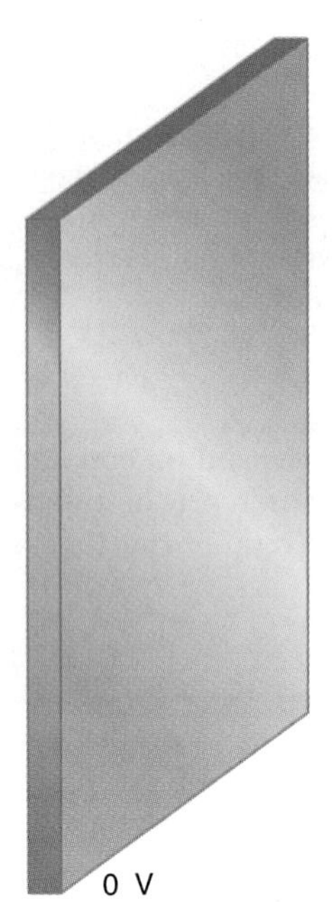

FIGURE 17.29

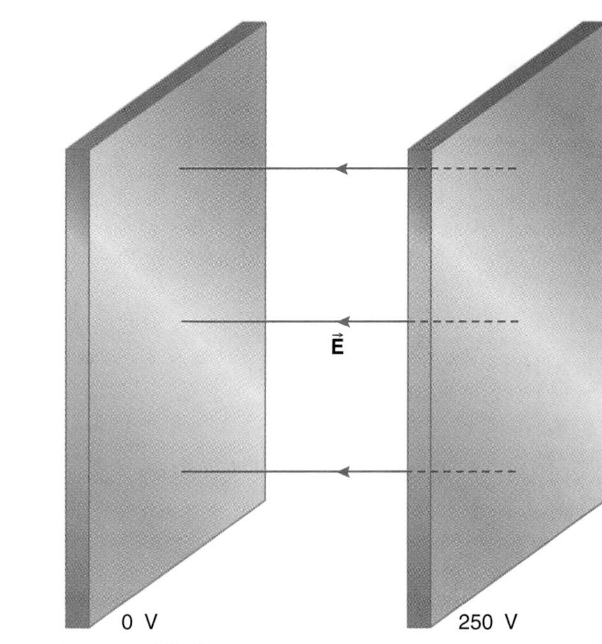

FIGURE 17.30

b. Since the charge on the proton is $q = +e$, the charge quantum number of the proton is $n = +1$. Use Equation 17.25 to find the change in the kinetic energy in electron-volts:

$$\begin{aligned}\Delta KE_{\text{in eV}} &= -n\,\Delta V \\ &= -n(V_f - V_i) \\ &= -1(0\ \text{V} - 250\ \text{V}) \\ &= 250\ \text{eV}\end{aligned}$$

The proton is released at rest and so has zero initial kinetic energy. The final kinetic energy thus is equal to the change in the kinetic energy:

$$\begin{aligned}\Delta KE &= KE_f - KE_i \\ &= KE_f - 0\ \text{J}\end{aligned}$$

The final kinetic energy in electron-volts thus is

$$KE_{f\ \text{in eV}} = 250\ \text{eV}$$

c. To find the speed, the kinetic energy must be expressed in joules as mentioned in Problem-Solving Tactic 17.7. Using the conversion indicated in Equation 17.23, you obtain

$$KE = \frac{mv^2}{2}$$

$$(250\ \text{eV})(1.602 \times 10^{-19}\ \text{J/eV}) = \frac{mv^2}{2}$$

$$4.01 \times 10^{-17}\ \text{J} = \frac{mv^2}{2}$$

Substitute for the mass of the proton, and solve for the speed:

$$\begin{aligned}v &= \left(\frac{2 \times 4.01 \times 10^{-17}\ \text{J}}{1.67 \times 10^{-27}\ \text{kg}}\right)^{1/2} \\ &= 2.19 \times 10^5\ \text{m/s}\end{aligned}$$

d. There is no need to know the separation of the plates!

EXAMPLE 17.13

An α-particle (with charge $+2e$) is accelerated from rest through a potential difference of -1.00 kV. What is the final kinetic energy of the α-particle?

Solution

The charge quantum number n of the α-particle is +2 since its charge is $q = +2e$. The change in the kinetic energy of the α-particle is found from Equation 17.25:

$$\begin{aligned}\Delta KE_{\text{in eV}} &= -n\,\Delta V \\ &= -(+2)(-1.00 \times 10^3\ \text{V}) \\ &= 2.00\ \text{keV}\end{aligned}$$

Since the initial kinetic energy of the α-particle is zero, the final kinetic energy is 2.00 keV, or

$$(2.00 \times 10^3\ \text{eV})(1.602 \times 10^{-19}\ \text{J/eV}) = 3.20 \times 10^{-16}\ \text{J}$$

17.9 An Electric Dipole in an External Electric Field Revisited

Recall from Section 16.8 that an electric dipole consists of two charges of equal magnitude but opposite sign separated by a distance d, as pictured in Figure 17.31. The electric dipole moment $\vec{p}$ was defined by Equation 16.13:

$$\vec{p} = |Q|\vec{d}$$

where $\vec{d}$ is the vector that locates the positive charge with respect to the negative charge (taken as an origin).

Many molecules such as water and DNA have electric dipole moments; hence this charge distribution is important in physics, chemistry, and molecular biology.

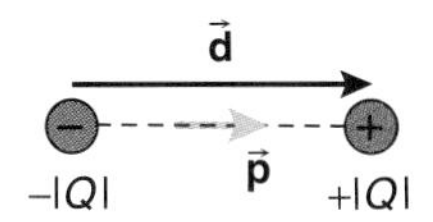

FIGURE 17.31 An electric dipole.

We saw (Example 16.13) that if an electric dipole is in an electric field $\vec{E}$, the field exerts forces on the charges (see Figure 16.71) in opposite directions, creating a torque $\vec{\tau}$ on the dipole given by

$$\vec{\tau} = \vec{p} \times \vec{E}$$

The torque is zero when $\vec{p}$ is parallel to $\vec{E}$ and when $\vec{p}$ is antiparallel to $\vec{E}$, as shown in Figure 17.32.

However, if the dipole moment $\vec{p}$ is antiparallel to $\vec{E}$, any slight change in the orientation allows the forces on the charges to flip the dipole into the parallel position. The antiparallel position thus is a position of unstable equilibrium. We now show this is the case from energy considerations.

Consider a dipole in a uniform electric field, as shown in Figure 17.33. (Uniformity is not really a restriction, because molecular dipoles are so small that most laboratory fields can be considered uniform over the physical size of the dipole.) For a uniform field of magnitude E, the electric potential $V(x)$ at any position x (with the coordinate choice of Figure 17.33) is given (from Example 17.1) by

$$V(x) - V(0\text{ m}) = -Ex$$

or

$$V(x) = V(0\text{ m}) - Ex$$

The potential energy of the charge $-|Q|$ located at the origin is

$$\text{PE}_{-|Q|} = -|Q|V(0\text{ m})$$

The potential energy of the positive charge $+|Q|$ at position x is

$$\begin{aligned}\text{PE}_{+|Q|} &= +|Q|V(x)\\ &= |Q|[V(0\text{ m}) - Ex]\end{aligned}$$

The total potential energy of the dipole in the field is the sum of the individual potential energies of the two charges:

$$\begin{aligned}\text{PE}_{\text{dipole}} &= \text{PE}_{-|Q|} + \text{PE}_{+|Q|}\\ &= -|Q|V(0\text{ m}) + |Q|V(0\text{ m}) - |Q|Ex\\ &= -|Q|Ex\end{aligned}$$

But geometrically (see Figure 17.33), $x = d\cos\theta$, and so

$$\text{PE}_{\text{dipole}} = -|Q|Ed\cos\theta$$

Note that $|Q|d$ is the magnitude of the dipole moment p, and so

$$\text{PE}_{\text{dipole}} = -pE\cos\theta$$

We can write this relationship conveniently using the scalar product because θ is the angle between the vectors $\vec{p}$ and $\vec{E}$.

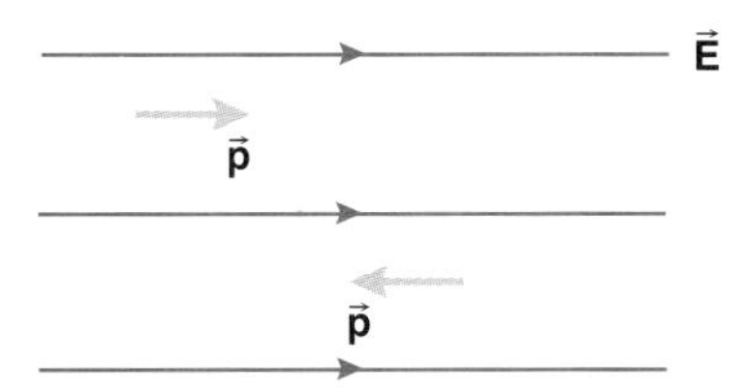

FIGURE 17.32 The torque on a dipole is zero when $\vec{p}$ is either parallel or antiparallel to $\vec{E}$.

Thus the potential energy of the dipole in an electric field is written succinctly as

$$\text{PE}_{\text{dipole}} = -\vec{p}\cdot\vec{E} \tag{17.26}$$

Notice that when $\vec{p}$ is parallel to $\vec{E}$, the potential energy is

$$\begin{aligned}\text{PE}_{\text{dipole parallel}} &= -pE\cos 0°\\ &= -pE\end{aligned}$$

When $\vec{p}$ is antiparallel to $\vec{E}$, the potential energy is

$$\begin{aligned}\text{PE}_{\text{dipole antiparallel}} &= -pE\cos 180°\\ &= +pE\end{aligned}$$

So the potential energy is a maximum when $\vec{p}$ is antiparallel to $\vec{E}$ and is a minimum when $\vec{p}$ is parallel to $\vec{E}$. This confirms what we stated about the antiparallel orientation: if the dipole is free to rotate, the dipole will lower its electrical potential energy by rotating to the parallel orientation. The antiparallel orientation is unstable because it has the maximum potential energy.

Thus when a dipole is placed in an electric field, it will tend to orient itself so that $\vec{p}$ is parallel to $\vec{E}$. Molecular dipoles behave in a similar fashion.

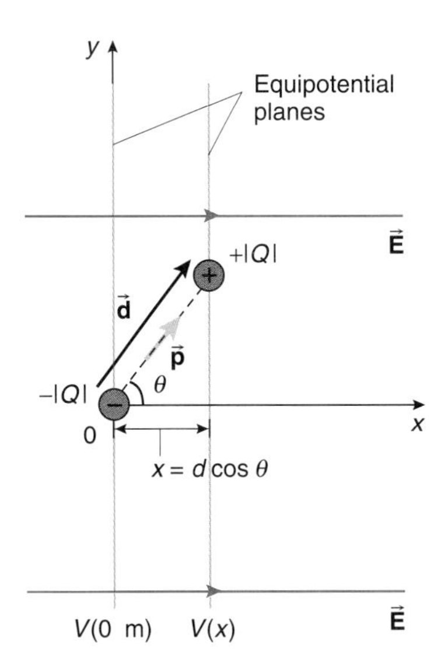

FIGURE 17.33 A dipole in a uniform electric field.

17.10 The Electric Potential and Electric Field of a Dipole*

In Chapter 16, you calculated the electric field caused by a dipole for two special geometries: (1) along the axis of the dipole (see Example 16.10), and (2) in the equatorial perpendicular bisecting plane of the dipole (see Example 16.11). Using the concept of the potential, it is possible to obtain a concise expression for the electric field of a dipole at *any* location r as long as $r >> d$, the separation of the charges in the dipole. Here we see how that is accomplished by taking advantage of the fact that the potential is a scalar.

The dipole produces an electric potential at a point in space that is the algebraic sum of the electric potentials at the point in question caused by each of the two charges (see Figure 17.34):

$$V_{\text{dipole}} = \frac{1}{4\pi\varepsilon_0}\frac{|Q|}{r_+} + \frac{1}{4\pi\varepsilon_0}\frac{(-|Q|)}{r_-} \tag{17.27}$$

where r_+ is the distance of the point P from the positive charge and r_- is the distance of P from the negative charge of the dipole.

Equation 17.27 can be rearranged slightly into the form

$$V_{\text{dipole}} = \frac{|Q|}{4\pi\varepsilon_0}\frac{r_- - r_+}{r_+ r_-} \tag{17.28}$$

If the distance of point P from the dipole is much greater than the separation of the charges d, the dipole is called a pointlike dipole; see Figure 17.35.

Far from a pointlike dipole, we can make the good approximation that

$$r_+ \approx r_- \approx r$$

where r is the distance of the point P from the center of the pointlike dipole. The geometry of Figure 17.35 indicates that when $r >> d$, the difference in the distances of the charges to the point P is approximately

$$r_- - r_+ \approx d \cos\theta$$

where the angle θ is measured from the axis of the dipole, parallel to the dipole moment $\vec{\mathbf{p}}$. Molecular dipoles are very small, so that these approximations are excellent for such dipoles.

We use these approximations in Equation 17.28. The potential of the dipole at P then is

$$V_{\text{dipole}} = \frac{|Q|}{4\pi\varepsilon_0}\frac{d\cos\theta}{r^2}$$

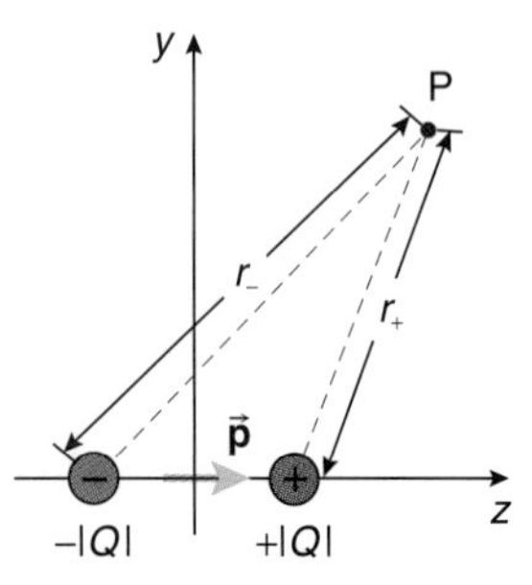

FIGURE 17.34 The electric potential at P is the sum of the potentials caused by $+|Q|$ and $-|Q|$.

But $|Q|d$ is the magnitude of the dipole moment vector $\vec{\mathbf{p}}$. Therefore we have

$$V_{\text{dipole}} = \frac{1}{4\pi\varepsilon_0}\frac{p\cos\theta}{r^2} \tag{17.29}$$

The potential is zero in the equatorial plane of the dipole, where $\theta = 90°$. At a fixed distance r, the potential has a maximum positive value when $\theta = 0°$, that is, along the axis of the dipole closer to the positive charge; the potential has its maximum negative value when $\theta = 180°$, along the axis of the dipole closer to the negative charge. The potential is symmetric about the axis of the dipole.

To calculate the electric field of the dipole at a point that is a great distance from the dipole itself, that is, for values of $r >> d$, we use Equation 17.29 for the potential. To use the del operator in Cartesian coordinates, the potential also must be expressed in Cartesian coordinates. We introduce the Cartesian coordinate system indicated in Figure 17.35, with the dipole moment $\vec{\mathbf{p}}$ oriented parallel to $\hat{\mathbf{k}}$ along the z-axis and with the x–y plane the equatorial plane of the dipole. Then, from the geometry,

$$r = (x^2 + y^2 + z^2)^{1/2}$$

and

$$\cos\theta = \frac{z}{r} = \frac{z}{(x^2 + y^2 + z^2)^{1/2}}$$

Substituting for r and $\cos\theta$ in Equation 17.29, we have

$$V = \frac{p}{4\pi\varepsilon_0}\frac{z}{(x^2 + y^2 + z^2)^{3/2}} \tag{17.30}$$

Using Equations 17.30 and 17.17 to find the Cartesian components of the electric field (after appropriate differentiations), we obtain

$$E_x = -\frac{\partial V}{\partial x} = \frac{p}{4\pi\varepsilon_0}\frac{3zx}{(x^2 + y^2 + z^2)^{5/2}} \tag{17.31}$$

$$E_y = -\frac{\partial V}{\partial y} = \frac{p}{4\pi\varepsilon_0}\frac{3zy}{(x^2 + y^2 + z^2)^{5/2}} \tag{17.32}$$

$$E_z = -\frac{\partial V}{\partial z} = -\frac{p}{4\pi\varepsilon_0}\left[\frac{1}{(x^2 + y^2 + z^2)^{3/2}} - \frac{3z^2}{(x^2 + y^2 + z^2)^{5/2}}\right] \tag{17.33}$$

Along the axis of the dipole, where $x = y = 0$ m, the field has purely a z-component; Equations 17.31–17.33 reduce to the result in Example 16.10, which was calculated using vector methods in Chapter 16. In the equatorial plane, with $z = 0$ m, the field reduces to the result in Example 16.11 (with $x = 0$ m for the y-axis).

One also could find the field of an electric dipole using the potential in polar form as in Equation 17.29.* However, there is

*In physics, when working in three dimensions, the polar angle θ is measured with respect to the z-axis. This differs from the convention typically used in mathematics for three dimensions. In only two dimensions (say x and y), the polar angle is measured with respect to the x-axis both in physics and mathematics.

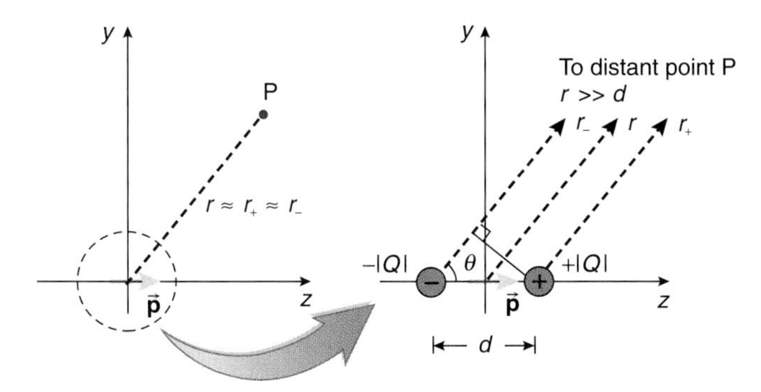

FIGURE 17.35 A pointlike dipole.

one complication. Equation 17.29 for the potential is expressed in polar coordinates r and θ. To use Equation 17.19 to calculate the electric field,

$$\vec{E} = -\nabla V$$

the del operator ∇ also must be expressed in polar coordinates. The expression for ∇ in polar coordinates is more complicated than in Cartesian coordinates (Equation 17.18), since the polar unit vectors $\hat{\mathbf{r}}$ and $\hat{\boldsymbol{\theta}}$ change their orientation in space depending on the location of the point in question, as indicated in Figure 17.36.

The del operator in polar coordinates has the form†

$$\nabla = \frac{\partial}{\partial r}\hat{\mathbf{r}} + \frac{1}{r}\frac{\partial}{\partial \theta}\hat{\boldsymbol{\theta}}$$

where at any point $\hat{\mathbf{r}}$ is the unit vector in the direction of increasing r and $\hat{\boldsymbol{\theta}}$ is a unit vector in the direction of increasing θ.

We use the expression for the del operator in polar coordinates; the dipole electric field then is

$$\begin{aligned}\vec{E} &= -\nabla V \\ &= -\left(\frac{\partial}{\partial r}\hat{\mathbf{r}} + \frac{1}{r}\frac{\partial}{\partial \theta}\hat{\boldsymbol{\theta}}\right)\frac{1}{4\pi\varepsilon_0}\frac{p\cos\theta}{r^2}\end{aligned}$$

Factoring out the constant terms, we have

$$\vec{E} = -\frac{p}{4\pi\varepsilon_0}\left(\frac{\partial}{\partial r}\hat{\mathbf{r}} + \frac{1}{r}\frac{\partial}{\partial \theta}\hat{\boldsymbol{\theta}}\right)\frac{\cos\theta}{r^2}$$

† This identity is proved in many texts on vector analysis. See, for example, Mary Boas, *Mathematical Methods in the Physical Sciences* (2nd edition, Wiley, New York, 1983), page 252.

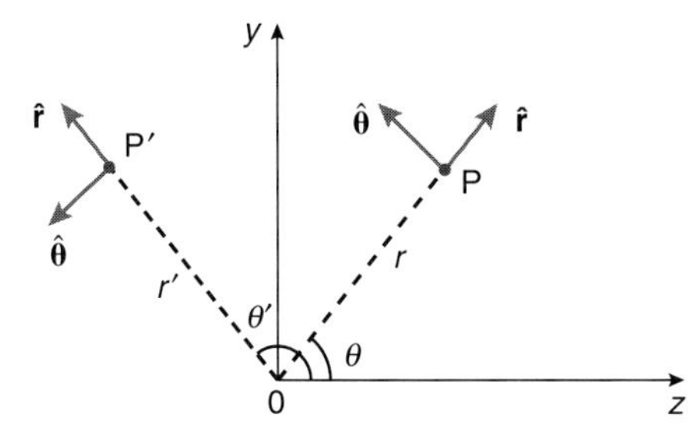

FIGURE 17.36 The orientation of the polar coordinate unit vectors $\hat{\mathbf{r}}$ and $\hat{\boldsymbol{\theta}}$ depend on where the point P is located.

Now we perform the indicated partial differentiations, remembering that when differentiating with respect to r, we treat θ as a constant, and when differentiating with respect to θ, we treat r as a constant. The field is

$$\begin{aligned}\vec{E} &= -\frac{p}{4\pi\varepsilon_0}\left[(\cos\theta)\left(-\frac{2}{r^3}\right)\hat{\mathbf{r}} - \frac{\sin\theta}{r^3}\hat{\boldsymbol{\theta}}\right] \\ &= \frac{p}{4\pi\varepsilon_0}\left(\frac{2\cos\theta}{r^3}\hat{\mathbf{r}} + \frac{\sin\theta}{r^3}\hat{\boldsymbol{\theta}}\right)\end{aligned}$$

Notice that when $\theta = 0°$, the point is along the axis of the dipole (see Figure 17.36), and the field is purely along $\hat{\mathbf{r}}$ since $\sin 0° = 0$:

$$\vec{E}_{\text{axis}} = \frac{1}{4\pi\varepsilon_0}\frac{2p}{r^3}\hat{\mathbf{r}}$$

This corresponds to the result calculated using vector methods in Example 16.10.

Likewise, in the equatorial plane of the dipole, $\theta = 90°$, and the field has no radial component but is purely in the direction of $\hat{\boldsymbol{\theta}}$ (see Figure 17.36):

$$\vec{E}_{\text{equatorial plane}} = \frac{1}{4\pi\varepsilon_0}\frac{p}{r^3}\hat{\boldsymbol{\theta}}$$

This corresponds to a result derived by different vector methods in Example 16.11.

17.11 The Potential Energy of a Distribution of Pointlike Charges

Work is required to assemble a collection of point charges distributed in space, as in Figure 17.37.

Initially all the charges are infinitely far away from each other and so do not interact. We take the potential energy of this initial (widely dispersed) distribution to be zero. Since there is no electric field present initially, no work is done by any electrical force to bring the first charge Q_1 to its new but isolated location in the distribution of Figure 17.38.

The same cannot be said for the second charge. The electric potential created in space by Q_1 at the closer location where Q_2 is next to be placed is

$$V = \frac{1}{4\pi\varepsilon_0}\frac{Q_1}{r_{12}}$$

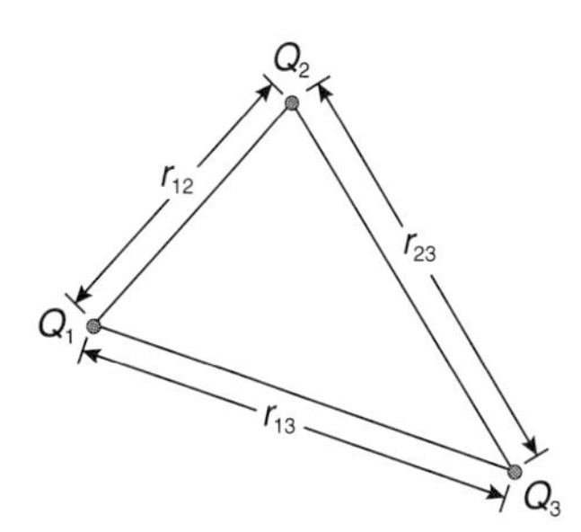

FIGURE 17.37 A collection of point charges.

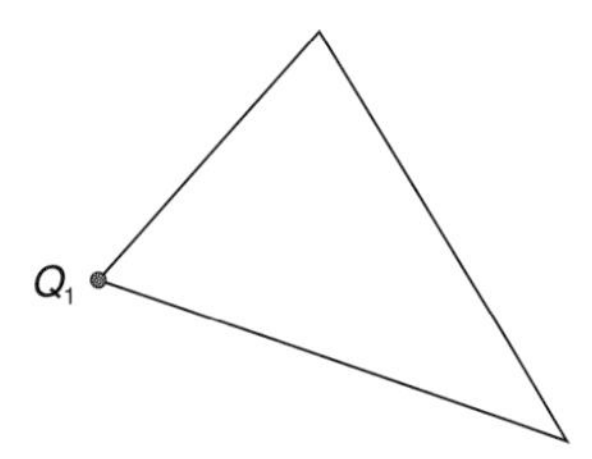

FIGURE 17.38 No work is required to bring the first charge to its location.

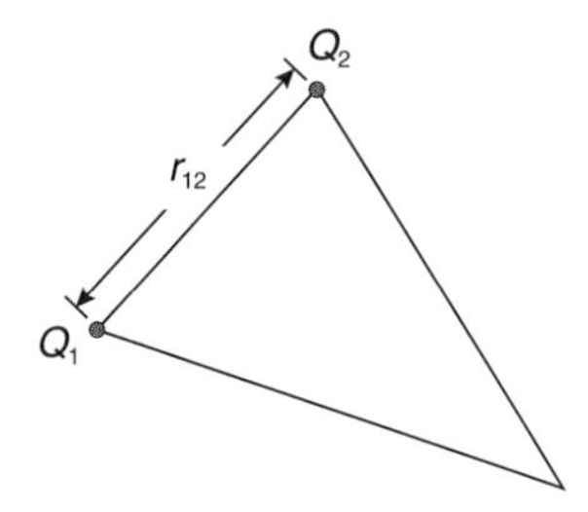

FIGURE 17.39 The second charge is placed at its location.

where r_{12} is the distance between Q_1 and the location where Q_2 is to be placed; see Figure 17.39. When Q_2 is placed at this location, Q_2 has the potential energy Q_2V or

$$\text{PE} = \frac{1}{4\pi\varepsilon_0}\frac{Q_1Q_2}{r_{12}} \quad (17.34)$$

The change in the potential energy of the second charge is

$$\Delta\text{PE} = \text{PE}_f - \text{PE}_i$$

The initial potential energy of Q_2 is zero (since it was infinitely far away from the other charge), and its final potential energy is given by Equation 17.34. The change in the potential energy is the negative of the work done by electrical forces as Q_2 is brought from infinity to its location at a distance r_{12} from Q_1. This work is independent of the path taken by Q_2 because the static electrical force is a conservative force. The potential energy change also can be imagined as the work *we* would need to do to drag Q_2 back from its location near Q_1 to infinity with no change in its kinetic energy; you should try to convince yourself of this statement from the CWE theorem. Equation 17.34 also results if we imagine bringing in Q_2 first (no work done by electrical forces to do this), followed by Q_1.*

The electric potential caused by Q_1 and Q_2 at the location where the third charge Q_3 is to be placed is

$$V' = \frac{1}{4\pi\varepsilon_0}\frac{Q_1}{r_{13}} + \frac{1}{4\pi\varepsilon_0}\frac{Q_2}{r_{23}}$$

where r_{13} is the distance between Q_1 and the place where Q_3 is to be placed, and r_{23} is the distance between Q_2 and the place where Q_3 is to be located; see Figure 17.40. If Q_3 now is brought in from infinity in the presence of the other two charges, it has a potential energy of Q_3V'.

The change in the potential energy of Q_3 is

$$\text{PE}_f - \text{PE}_i = \left(\frac{1}{4\pi\varepsilon_0}\frac{Q_1Q_3}{r_{13}} + \frac{1}{4\pi\varepsilon_0}\frac{Q_2Q_3}{r_{23}}\right) - 0\text{ J}$$

and so the *total* potential energy of the three charges in place is†

$$\begin{aligned}\text{PE}_{\text{total}} &= \frac{1}{4\pi\varepsilon_0}\frac{Q_1Q_2}{r_{12}} + \frac{1}{4\pi\varepsilon_0}\frac{Q_1Q_3}{r_{13}} + \frac{1}{4\pi\varepsilon_0}\frac{Q_2Q_3}{r_{23}} \\ &= \frac{1}{4\pi\varepsilon_0}\left(\frac{Q_1Q_2}{r_{12}} + \frac{Q_1Q_3}{r_{13}} + \frac{Q_2Q_3}{r_{23}}\right) \quad (17.35)\end{aligned}$$

*This result is reflected in the symmetry of Q_1 and Q_2 in Equation 17.34.

† Equation 17.35 also does not depend on the order the particles are moved to their positions, which is manifested by the symmetry of Q_1, Q_2, and Q_3 in the equation.

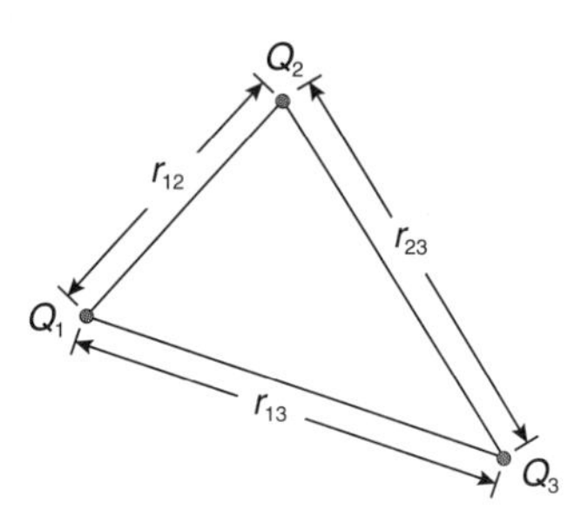

FIGURE 17.40 The third charge is placed at its location.

The addition of a fourth charge leads to an expression similar to Equation 17.35 but with six terms, one for each distinct pair of charges present. Appropriate signs for the respective charges must be used in Equation 17.35 and similar expressions.

The work done by the electrical force is, by definition, the negative of the change in the potential energy:

$$\begin{aligned}W_{\text{elec}} &= -\Delta\text{PE} \\ &= -(\text{PE}_{\text{total f}} - \text{PE}_{\text{total i}})\end{aligned}$$

Since the initial potential energy was zero with the charges infinitely separated, $\text{PE}_{\text{total i}} = 0$ J and

$$W_{\text{elec}} = -\text{PE}_{\text{total f}}$$

If the total final potential energy of such a collection of charges is negative, electrical forces did positive work in assembling the charges and, according to the CWE theorem, *we* will need to do positive work to separate the charges back to infinite distances apart with zero change in their kinetic energies.

QUESTION 5

Explain the distinction between the electric potential of two pointlike charges and the electric potential energy of two pointlike charges.

EXAMPLE 17.14

The charges that model the water molecule have a magnitude of 0.66*e* and an effective separation of 0.057 nm as was shown in Figure 16.63.

a. What is the electric potential energy of this dipolar charge configuration? Express your result in both J and eV.

b. Did electrical forces do positive, negative, or zero work when the dipole formed from charges initially separated by infinite distances?

Solution

a. When the charges are part of the dipole, their potential energy is found from Equation 17.34:

$$\text{PE} = \frac{1}{4\pi\varepsilon_0}\frac{Q_1Q_2}{r_{12}}$$

The two charges are $+0.66e$ and $-0.66e$, with a separation of 0.057 nm. Making these substitutions, you find

$$\begin{aligned}\text{PE} &= \left(9.00 \times 10^9 \text{ N}\cdot\text{m}^2/\text{C}^2\right)\frac{(-0.66e)(+0.66e)}{0.057 \times 10^{-9} \text{ m}} \\ &= -1.8 \times 10^{-18} \text{ J} \\ &= \frac{-1.8 \times 10^{-18} \text{ J}}{1.602 \times 10^{-19} \text{ J/eV}} \\ &= -11 \text{ eV}\end{aligned}$$

b. When the charges initially are infinitely far apart, they have zero potential energy. Since the final potential energy of the dipolar molecule is negative and its initial potential energy was zero, the change in the potential energy is negative. This means that electrical forces did positive work when the dipole formed (since $W_{\text{elec}} = -\Delta\text{PE}$). *You* would need to do positive work to move the charges back to an infinite separation. Energy must be supplied to separate the charges to infinite distances. This is why the water molecule does not fall apart of its own accord (fortunately for us!).

EXAMPLE 17.15

The helically shaped strands of the important genetic macromolecule DNA (*deoxyribonucleic acid*) are held together by electrical forces. Along the double helix are many pairs of adenine and thymine molecules. Each adenine molecule in one DNA strand has a dipole moment of magnitude 3.0×10^{-30} C·m, caused by a hydrogen–nitrogen bond, while each thymine molecule in the other DNA strand has a dipole moment of 7.7×10^{-30} C·m, caused by a double carbon–oxygen bond. The pair of dipoles is arranged with their dipole moments parallel, as shown in Figure 17.41.

The absolute magnitude of the effective charge on the adenine dipole is $0.19e$, while that on the thymine dipole is $0.40e$. Note that the magnitude of the two dipole moments can be computed with the information in Figure 17.41. Calculate the electric potential energy of one dipole in the presence of the other when they are configured as in Figure 17.41. Consider each dipole to be already formed and that the potential energy is that of one dipole in the presence of the other.*

Solution

You might anticipate that the potential energy of this charge formation is negative because the two dipoles, in the orientations shown, attract each other. The positive charge of the adenine dipole is closer to the negative rather than to the positive charge of the thymine dipole. Electrical forces therefore did positive work to bring them together; the total potential energy (change) of the system is negative since $W_{\text{elec}} = -\Delta\text{PE}$.

The dipoles are quite close to each other in space, and so you *cannot* use Equation 17.29 for the electric potential of a dipole. That equation was derived assuming the distance r is much greater than the separation of the charges in the dipole.

Although four charges are present, since you consider each dipole to be already formed, you are interested in the potential energy of each charge in one dipole in the presence of the two charges in the other dipole.

In the adenine (H—N) dipole, let Q_{H} and Q_{N} represent the effective charges on the hydrogen and nitrogen atoms respectively, so that

$$Q_{\text{H}} = 0.19e$$
$$Q_{\text{N}} = -0.19e$$

In the thymine (C=O) dipole, let Q_{C} and Q_{O} represent the effective charges of the carbon and oxygen atoms respectively, so that

$$Q_{\text{C}} = 0.40e$$
$$Q_{\text{O}} = -0.40e$$

The potential energy of one dipole in the presence of the other is found using a variation of Equation 17.35:

$$\text{PE} = \frac{1}{4\pi\varepsilon_0}\left(\frac{Q_{\text{H}}Q_{\text{C}}}{r_{\text{HC}}} + \frac{Q_{\text{H}}Q_{\text{O}}}{r_{\text{HO}}} + \frac{Q_{\text{N}}Q_{\text{C}}}{r_{\text{NC}}} + \frac{Q_{\text{N}}Q_{\text{O}}}{r_{\text{NO}}}\right) \quad (1)$$

where the distances between the various atoms are indicated with subscripts. In Figure 17.41, these distances are

$$r_{\text{HC}} = 0.30 \text{ nm}$$
$$r_{\text{HO}} = 0.18 \text{ nm}$$
$$r_{\text{NC}} = 0.40 \text{ nm}$$
$$r_{\text{NO}} = 0.28 \text{ nm}$$

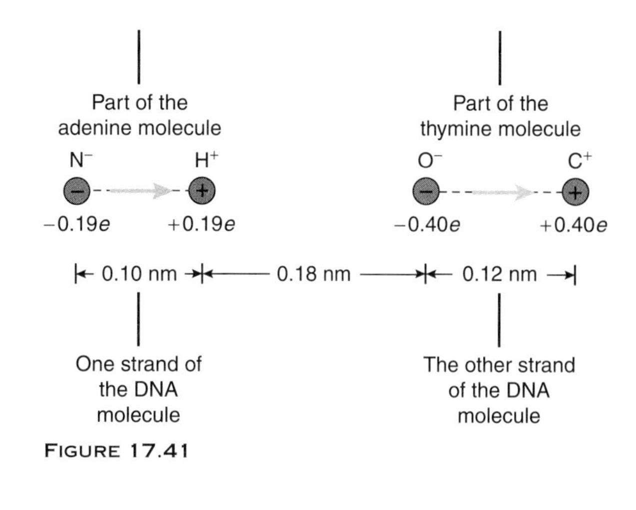

FIGURE 17.41

*This problem is modeled after one in Douglas C. Giancoli, *General Physics* (Prentice-Hall, Englewood Cliffs, New Jersey, 1984), page 474.

Making these substitutions using proper SI units into equation (1) for the potential energy, you have

$$\begin{aligned}
\text{PE} &= \left(9.00 \times 10^{9}\ \text{N}\cdot\text{m}^2/\text{C}^2\right)\left[\frac{(0.19e)(0.40e)}{0.30 \times 10^{-9}\ \text{m}}\right. \\
&\quad \left. + \frac{(0.19e)(-0.40e)}{0.18 \times 10^{-9}\ \text{m}} + \frac{(-0.19e)(0.40e)}{0.40 \times 10^{-9}\ \text{m}} + \frac{(-0.19e)(-0.40e)}{0.28 \times 10^{-9}\ \text{m}}\right] \\
&= \left(9.00 \times 10^{9}\ \text{N}\cdot\text{m}^2/\text{C}^2\right)\frac{(0.19e)(0.40e)}{10^{-9}\ \text{m}} \\
&\quad \times \left(\frac{1}{0.30} - \frac{1}{0.18} - \frac{1}{0.40} + \frac{1}{0.28}\right) \\
&= -2.0 \times 10^{-20}\ \text{J} \\
&= -0.13\ \text{eV}
\end{aligned}$$

Since the potential energy is negative, electrical forces did positive work as the dipoles were brought together; equivalently, *you* need to do positive work to separate the two dipoles. Therefore, the dipole–dipole interaction between such pairs of adenine and thymine molecules along the double helix helps keep the strands of DNA together.

17.12 Lightning Rods

One of the first practical inventions involving electrical phenomena was the lightning rod by Benjamin Franklin in the 18th century; see Figure 17.42. Not wishing to profiteer from such a practical device, Franklin altruistically placed his patent in the public domain to expedite its immediate use by humanity.

Such sharp-pointed conducting rods are used atop structures and even large trees, such as at George Washington's Mount Vernon estate on the banks of the Potomac River in Virginia, to protect them from the destructive effects of lightning. The conducting lightning rod is connected to the Earth by means of a grounding wire. Without lightning rods, portions of trees or buildings literally can explode and/or catch fire when struck by lightning, because of the vaporization and rapid expansion of water in the wood. The purpose of the rod is to discharge nearby charged clouds harmlessly by providing an alternative conducting path to the Earth that is not through the tree or structure itself. Though it may sound like an oxymoron, such rods "attract" lightning (by literally becoming "the lightning rod") and make the effects of lightning quite harmless (other than the noise!). The lightning discharge in air rapidly heats and expands the air along its path, creating dramatically loud acoustic wave pulses: a complicated way to say thunder!

To see how lightning rods remarkably and safely initiate an electrical discharge, first consider two spherical conductors of different radii r and R connected by a long conducting wire, as shown in Figure 17.43. Thanks to the connecting wire, the system represents a single conductor. Since the two spheres and the connecting wire are electrically a single conductor, the two spheres must be at the same electric potential V because the surface of a conductor is an equipotential surface. Let Q be the charge on the conductor of radius R, and q the charge on the conductor of radius r. Since the two spheres are well separated in space, the charge distribution on them is essentially uniform and spherically symmetric, and we can treat them as spherically symmetric charges for purposes of calculating their common electric potential. Therefore for the small sphere we have

$$V = \frac{1}{4\pi\varepsilon_0}\frac{q}{r}$$

while for the larger sphere we have

$$V = \frac{1}{4\pi\varepsilon_0}\frac{Q}{R}$$

FIGURE 17.42 A fancy lightning rod with a nearby elaborate weather vane.

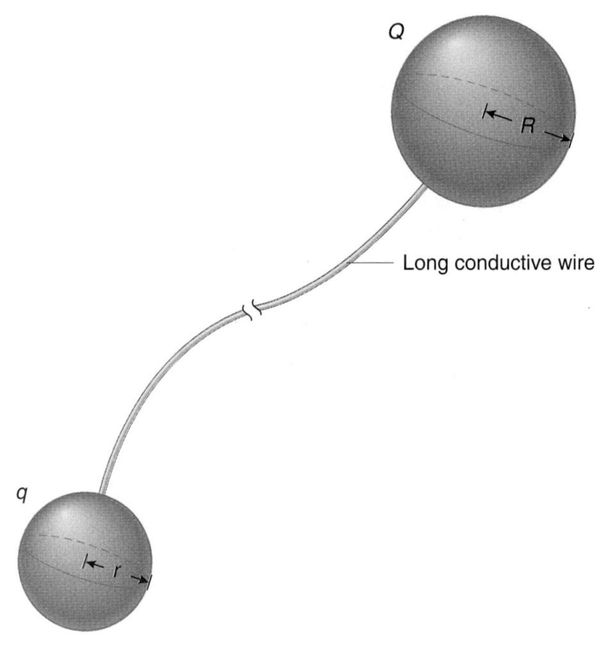

FIGURE 17.43 Two conducting spheres of different radii connected by a long conducting wire.

The potentials V are the same; hence, by dividing the preceding two equations, we find that the ratio of the charges on the two spherical conductors is

$$\frac{q}{Q} = \frac{r}{R} \tag{17.36}$$

The electric field magnitudes at the surface of each spherical conductor also can be approximated by that of spherically symmetric charges. Call them $E_{\text{at surface of } R}$ and $E_{\text{at surface of } r}$. Then

$$E_{\text{at surface of } r} = \frac{1}{4\pi\varepsilon_0}\frac{q}{r^2}$$

and

$$E_{\text{at surface of } R} = \frac{1}{4\pi\varepsilon_0}\frac{Q}{R^2}$$

Dividing these equations, we find the ratio of the field magnitudes:

$$\frac{E_{\text{at surface of } r}}{E_{\text{at surface of } R}} = \frac{q/r^2}{Q/R^2}$$

But we know $q/Q = r/R$ from Equation 17.36. Therefore we finally get

$$\frac{E_{\text{at surface of } r}}{E_{\text{at surface of } R}} = \frac{R}{r} \tag{17.37}$$

This means that the magnitude of the electric field is greater at the surface of the sphere of smaller radius.

This result is quite significant. In a qualitative way, it shows that for a conductor of arbitrary shape, the electric field at its surface has the greatest magnitude near those portions with the smallest radius of curvature; see Figure 17.44.* In other words, if a conductor has a sharp point, the electric field near the point is of much greater magnitude than near other regions of the conductor, as reflected in the number of field lines. This is the purpose of the sharp points on lightning rods. Charged clouds in the vicinity attract or repel electrons in the conducting rod. Since the rod has a sharp tip, the magnitude of the electric field in the vicinity of the tip easily will reach the critical magnitude of about 3×10^6 N/C at which air ceases to be an insulator and becomes an conductor. Thus the electrical discharge is initiated near the tip of the lightning rod rather than near other objects such as a building structure or tall tree.

QUESTION 6

Explain why a lightning rod is connected to the Earth with a grounding wire for the rod to safely do its thing.

*There are exotic exceptions to this rule; see the article by Richard H. Price and Ronald J. Crowley, "The lightning-rod fallacy," *American Journal of Physics*, 53, #9, pages 843–848 (September 1985).

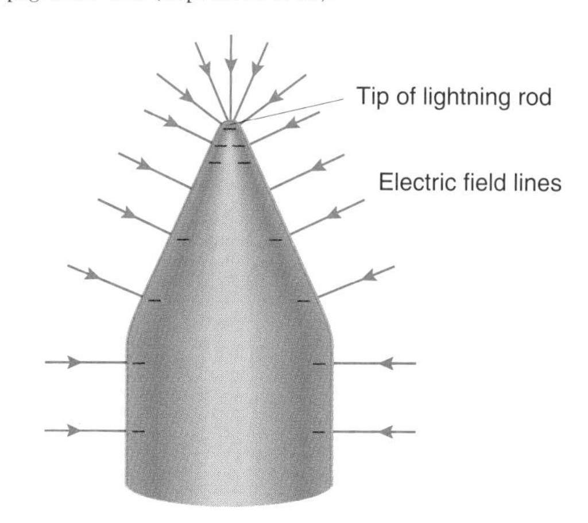

FIGURE 17.44 **The magnitude of the electric field is greater near parts of a conductor with a small radius of curvature (sharper points).**

CHAPTER SUMMARY

The electrical force caused by static charges is a conservative force. The work done by the electrical force on a charge q is the negative of the change in the electric potential energy of q:

$$W_{\text{elec}} = -\Delta\text{PE} \tag{17.3}$$

The electrical force on q is $\vec{\mathbf{F}} = q\vec{\mathbf{E}}$, where $\vec{\mathbf{E}}$ is the electric field at the point where q is located, and so

$$q\int_{\text{i}}^{\text{f}} \vec{\mathbf{E}} \cdot d\vec{\mathbf{r}} = -(\text{PE}_{\text{f}} - \text{PE}_{\text{i}}) \tag{17.5}$$

The potential energy per unit charge at a point in space is defined to be the *electric potential* V at that point:

$$V \equiv \frac{\text{PE}_{\text{of } q}}{q} \tag{17.7}$$

or

$$\text{PE}_{\text{of } q} = qV \tag{17.8}$$

Hence the difference in the electric potential between two points in space is

$$V_{\text{f}} - V_{\text{i}} = -\int_{\text{i}}^{\text{f}} \vec{\mathbf{E}} \cdot d\vec{\mathbf{r}} \tag{17.9}$$

The electric potential of a pointlike charge Q is

$$V(r) = \frac{1}{4\pi\varepsilon_0}\frac{Q}{r} \tag{17.11}$$

The electric potential at a point caused by a collection of pointlike charges Q_i is found from the principle of superposition:

$$V = V_1 + V_2 + V_3 + \cdots \tag{17.12}$$

where V_i is the potential caused at the point by charge Q_i as if it were the only charge present.

The electric potential of a continuous distribution of charge of finite extent is found by extending the principle of superposition from a sum to an integral over the charge distribution:

$$V = \frac{1}{4\pi\varepsilon_0}\int_{\substack{\text{finite charge}\\ \text{distribution}}} \frac{dQ}{r} \tag{17.13}$$

The electric potential at a point in space arising from various charge distributions is summarized in Table 17.1 on page 779.

A volume of space in which the electric potential is constant is called an *equipotential volume*. The electric field is zero within an equipotential volume. The interior of a conductor is an equipotential volume. A hollow conductor, with no charges within the cavity, also is an equipotential volume, and thus is shielded from the electric fields of charges outside or on the outer surface of the conductor.

A surface over which the electric potential has a constant value is called an *equipotential surface*. Electric field lines always are perpendicular to an equipotential surface. The surface of a conductor is an example of an equipotential surface. Imaginary equipotential surfaces can be constructed around any charge distribution.

If the electric potential depends on a single coordinate s, the electric field has a single component along s that is

$$E_s = -\frac{dV}{ds} \qquad (17.15)$$

More generally, the electric field is the negative gradient of the potential:

$$\vec{E} = -\nabla V \qquad (17.19)$$

The potential energy associated with a pair of two pointlike charges Q_1 and Q_2 is

$$PE = \frac{1}{4\pi\varepsilon_0}\frac{Q_1Q_2}{r_{12}} \qquad (17.34)$$

where r_{12} is the separation of the two charges. This potential energy represents the negative of the work done by the electrical force on Q_2 in bringing Q_2 from an infinite distance to its location a distance r_{12} from Q_1. The equation can be extended to larger collections of pointlike charges by writing similar terms for each distinct pair of charges in the charge distribution.

A new energy unit of convenience, used particularly for atomic and molecular physics, is the *electron-volt* (eV). One electron-volt is the change in the kinetic energy of an electron when it freely accelerates through a potential difference of exactly one volt. One electron-volt is approximately 1.602×10^{-19} J:

$$1 \text{ eV} = 1.602 \times 10^{-19} \text{ J} \qquad (17.23)$$

Summary of Problem-Solving Tactics

17.1 (page 769) The electric potential is not the same thing as the electric potential energy.

17.2 (page 769) It is important to use the appropriate sign for the charge q in Equation 17.8:

$$PE = qV \qquad (17.8)$$

17.3 (page 769) The electric potential and the electric potential energy both are scalar quantities, not vectors.

17.4 (page 770) The direction of the electric field always is from regions with higher values of the electric potential to regions with lower values of the electric potential.

17.5 (page 773) Note that in Equation 17.11 for the electric potential of a point charge, the distance r in the denominator is to the first power.

17.6 (page 773) In Equation 17.11 for the electric potential of a point charge, as well as for other charge distributions in Table 17.1, remember to use the appropriate sign for Q.

17.7 (page 785) Kinetic energies in electron-volts (eV) must be converted to joules (J) to calculate speeds in meters per second (m/s).

Questions

1. (page 770); 2. (page 781); 3. (page 783); 4. (page 785); 5. (page 790); 6. (page 793)

7. Show that the SI units for the electric field N/C and V/m are equivalent.

8. In what direction, relative to the direction of an electric field line, is it possible to move a charge so that the electrical force does zero work on it?

9. Given an electric field line of a charge distribution, in what directions (relative to the field line) is it possible for another charge to move without changing its potential energy?

10. The electric potential varies linearly with a coordinate z through a region. What is the nature of the electric field in this region?

11. Shuffle your feet across a carpet with a deep pile. Now bring your head slowly very near a water pipe or faucet and describe what happens to your hair. Move your head away and then bring your finger close to the faucet. A spark may now jump between your finger and the faucet. Explain why.

12. Two positive charges are separated by a distance ℓ. Is there a point on the line connecting the charges where the electric field is zero? Is there a point on the same line where the electric potential is zero? If your answer to both questions is yes, are the two points at the same location?

13. A particular charged particle lowers its electric potential energy while moving to regions of higher electric potential. Is this possible under the action of electrical forces alone? If so, what is the sign of the charge?

14. Consider the following technical terms: electric potential; electric potential difference; electric potential energy; change in electric potential energy. Carefully explain what is meant by each term.

15. How would you define a *proton*-volt unit of energy so that it has the same value as the electron-volt?

16. A hollow, closed conductor has no charges within the hollow. Is it possible to create a static electric field within the hollow using charges exterior to the hollow conductor? If so, explain how. If not, explain why not.

17. A hollow metal conductor can shield the interior from electrical forces caused by charges exterior to the conductor. A corresponding hypothetical gravitational shield would insulate masses from gravitational forces of exterior masses. Since a mass anywhere inside a hollow spherical shell experiences zero gravitational force due to the shell, is such a shell a gravitational shield in the same sense as the electrical shield? Explain.

18. For the electric field line distributions of Figures 16.55–16.60 in Chapter 16, sketch some equipotential lines representing end-on views of equipotential surfaces.

19. Let the bottom of a 10 m high aluminum flagpole be an electrical ground. What is the potential at the top of the pole?

20. What is the sign of the potential energy of a charge distribution of: (a) Two like charges? (b) Two unlike charges? (c) What is the significance of the signs of the two potential energies in (a) and (b)?

21. (a) Is the sign of the electric potential caused by two negative charges the same at every point in space? If so, what is it? If not, explain why the sign may vary depending on the point in space. (b) Is the sign of the electric potential caused by two positive charges the same at every point in space? If so, what is it? If not, explain why the sign may vary depending on the point in space. (c) If two charges are unlike charges, is the sign of the electric potential caused by the two charges the same at every point in space? If so, what is it? If not, explain why the sign may vary depending on the location of the point in space.

22. The electric field is zero at a particular point. Is the electric potential necessarily zero at the same point? If so, explain why. If not, cite an example that illustrates the contrary situation.

23. The electric potential is zero at a particular point. Is the electric field necessarily zero at the same point? If so, explain why. If not, cite an example that illustrates the contrary situation.

24. From the graphs of the potentials in Figures 17.13 and 17.16, explain how you can tell that the magnitude of the electric field at the center of the circular ring is zero where $z = 0$ m while that of the charged disk is not zero where $z = 0$ m.

25. The static electric field is always zero within a conductor. Does it necessarily follow that the potential within the conductor is zero? If so, explain why. If not, explain why not.

26. The electric potential energy of two like charges is positive whereas that of two unlike charges is negative. What does this imply about the work done by electrical forces in forming the respective charge distributions in the two situations?

27. Is it possible to have two pointlike charges, separated by a finite distance, with zero potential energy? If so, explain how. If not, explain why not.

28. Is it possible to have three pointlike charges, separated by finite distances, with zero potential energy? If so, explain how. If not, explain why not.

29. Few people survive being struck by lightning (zap!). What steps or precautions might you take while hiking if it is not possible to reach shelter before a thunderstorm overtakes you? Such precautions might save your life and are well worth discussing. See "Don't get shocked," *Safety & Health*, *149*, #4, page 89 (April 1994), for some tips.

30. If two points in space are at the same electric potential, what is the work done by electrical forces in moving a charge from one point to the other? Does it necessarily follow that the two points are on the same equipotential surface?

31. Let's make an analogy between the diagrams indicating electric field lines and equipotentials and the topographic maps used frequently by hikers, campers, and geologists. On a topographic map, what corresponds to equipotentials? What corresponds to the electric field? In view of this analogy, explain why the word *gradient* is appropriate when expressing the relationship between the electric field and the electric potential, as in $\vec{E} = -\nabla V$.

32. The proper operation of a sensitive instrument necessitates an environment free from stray electric fields. How can you ensure that the instrument is located in such a region of space?

33. Why do you suppose many electrical instruments are housed in metal boxes?

34. Printed circuit boards typically are stored and shipped in metal foil packages. Why?

35. A word used in early times (and occasionally in some countries today) for the electric potential is the *electrical tension*. You likely have heard high-voltage power lines called high-tension lines. Discuss the ways in which electric potential is similar to and different from the mechanical concept of tension.

36. Is it proper to say that a conductor in electrostatics has a given fixed value for the potential everywhere on its surface? Is it meaningful to say that an insulator has a given fixed value of the potential everywhere on its surface? Explain.

PROBLEMS

Sections 17.1 Electric Potential Energy and the Electric Potential
17.2 The Electric Potential of a Pointlike Charge
17.3 The Electric Potential of a Collection of Pointlike Charges

1. What work is done by the electrical force in moving a $-5.00\ \mu C$ charge from an electrical ground to a place where the electric potential is 150 V?

2. Two large parallel plates are separated by 1.00 cm. The potential difference between the plates is 120 V. What is the magnitude of the electric field between the plates?

3. In the Bohr model of the hydrogen atom, the circular electron orbit closest to the nucleus has a radius of 5.29×10^{-11} m. (a) Find the electric potential of the proton at the position of this orbit of the electron. (b) Calculate the electric potential energy of the electron at this location.

•4. A potential difference of 3.00 kV exists between two parallel conducting plates. What plate separation produces an electric field of magnitude 3.0×10^6 N/C in the region between the plates? This magnitude of field causes air to conduct electricity, and a lightning discharge between the plates will neutralize them.

•5. A thunderstorm cloud is 1.00 km overhead. Air will conduct electricity if the field exceeds a magnitude of about 3.0×10^6 V/m. Model the cloud–ground system as a pair of parallel plates. (a) What is the potential difference between the cloud and ground when the field has this magnitude? (b) If the bottom of the cloud has an area of 2.0 km^2, what static charge resides on the cloud?

•**6.** Consider an atom with a single electron in a circular orbit of radius r about a nuclear charge of $+Ze$, where Z is the atomic number (identical to the charge quantum number of the nucleus). Such single-electron atoms are known as hydrogenic atoms. (a) What is the electric potential of the nucleus at the position of the electron? (b) What is the potential energy of the electron at this location? (c) Use the Coulomb force law for the interaction between the orbiting electron and the nucleus and write Newton's second law of motion for the orbiting electron, remembering that the acceleration in circular motion is a centripetal acceleration of magnitude v^2/r, where v is the speed of the orbiting electron. Show that the kinetic energy of the electron and its electric potential energy are related by

$$\text{KE} = -\frac{1}{2}\text{PE}$$

This result is a consequence of a general theorem in theoretical mechanics, known as the virial theorem, applied to inverse-square-law forces. (cf. Chapter 8, Problem 41.)

•**7.** (a) Find the electric potential at the origin in Figure P.7. (b) Find the electric field at the origin.

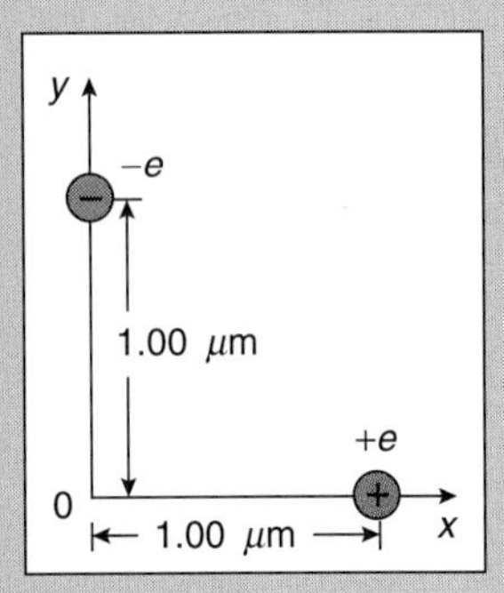

FIGURE P.7

•**8.** Two charges are nailed by a carpenter to the coordinate grid in Figure P.8 at the indicated locations. (a) Find the total electric potential at the origin. (b) Find the total electric potential an infinite distance away from the origin. (c) What is the potential energy of a proton placed at the origin? (d) How much work is done by electrical forces if the proton of part (c) is moved from the origin to infinity? (e) Does your answer to part (d) depend on the particular path used to take the charge to infinity? (f) If released at the origin, will the proton wander off to infinity of its own accord (under only the influence of the electrical force) or will *we* have to drag it there kicking and screaming all the way?

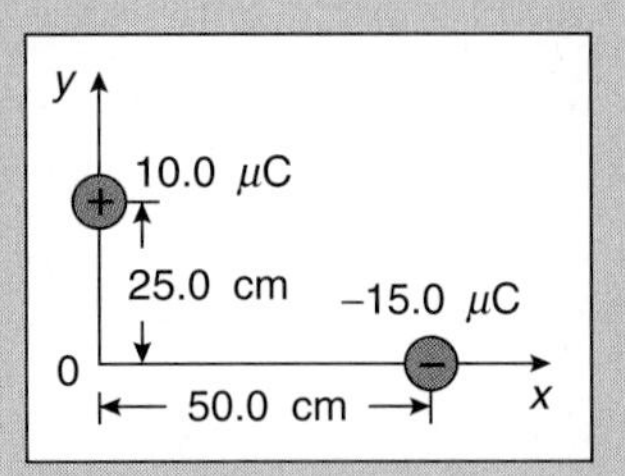

FIGURE P.8

•9. (a) Use the definition of the electric potential as a guide to define carefully the gravitational potential at a point in space. (b) An initial point is located 3.00 m above the ground. A final point is located 8.00 m above the ground and 2.50 m horizontally east with respect to the initial point. What is the gravitational potential difference between the two points?

•**10.** A constant electric field $\vec{\mathbf{E}} = (2.0 \times 10^3 \text{ N/C})\hat{\imath}$ exists in a region of space. What is the potential difference between an initial point at $x = 0.00$ cm and a final point $x = 3.00$ cm in the field?

•**11.** To pacify your professor, consider the electric dipole in Figure P.11. (a) Calculate the total electric potential at the point P. (b) What is the potential energy of a proton placed at the point P? (c) How much work is done by the electrical force on the proton if it is dragged from point P to infinity?

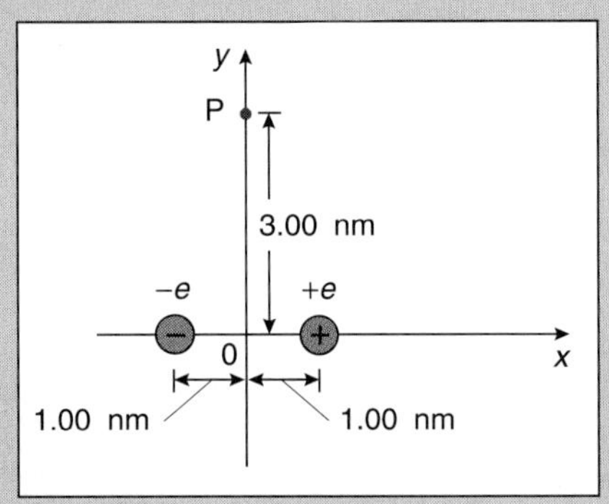

FIGURE P.11

•**12.** A charge $+2q$ is at the origin while another charge $-q$ is 2.00 m away, out along the x-axis as indicated in Figure P.12. Assume $q > 0$ C. (a) At what point along the x-axis will the electric field vanish? Is this point unique or are there other such points along the x-axis? (b) At what point along the x-axis will the electric potential vanish? Is this point unique or are there other such points along the x-axis?

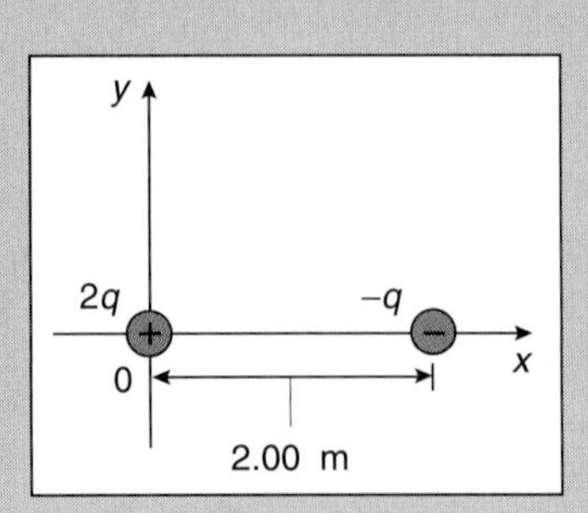

FIGURE P.12

•13. Two large (effectively infinite) horizontal metal plates are separated by 10.0 cm and the electric field between them has a magnitude 1.00×10^4 N/C, directed as shown in Figure P.13. (a) What charge is on each square meter of the upper surface? (b) What charge is on each square meter of the lower surface? (c) What is the electric potential of the upper plate if the lower plate is grounded? (d) A small particle of mass 1.00×10^{-5} kg, placed in the vacuum between the plates, has a charge q and is in equilibrium under the influence of gravitational and electrical forces. What is the charge q?

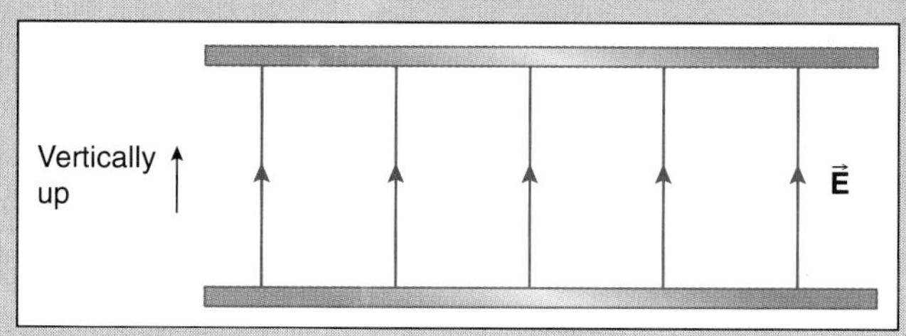

FIGURE P.13

•14. Two charges are located as indicated in Figure P.14. (a) Find the total electric field at point P. (b) Find the magnitude of the electric field at P. (c) Find the angle that the total field at P makes with $\hat{\jmath}$. (d) Find the total electric potential at P. (e) Another charge, $-3.00\ \mu$C, now is placed at P. Find the force on this charge and its potential energy.

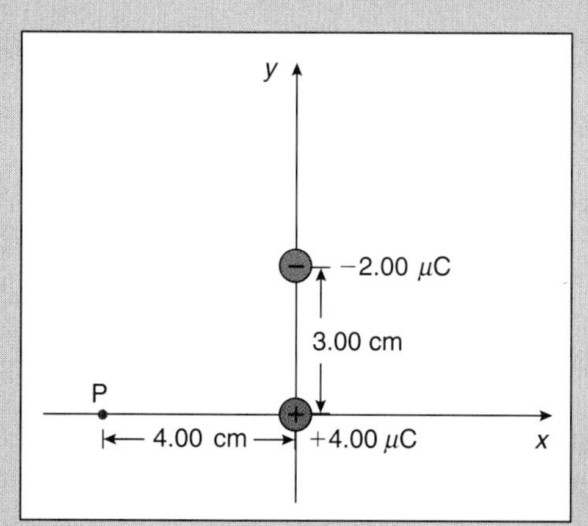

FIGURE P.14

•15. Charge separation (polarization) typically exists across cell membranes such as nerve cells. The inside of the membrane has a low concentration of potassium ions (K^+) while the outside has a high concentration of sodium ions (Na^+). Thus the inside of the membrane is at a lower electric potential than the outside; the potential difference typically is about 90 mV across an insulating membrane roughly of thickness 5 nm. (a) What is the magnitude of the electric field in the membrane if you model the system as a pair of parallel plates? (b) Is the answer to part (a) greater or less than the magnitude of electric field ($\sim 3 \times 10^6$ V/m) that causes air to break down and become a good conductor? When the nerve is stimulated by a mechanical, thermal, or electrical stimulus, the potential difference across the membrane changes (biologists say *depolarizes*) over a time interval of milliseconds in the way schematically illustrated in Figure P.15, and this depolarization propagates along the nerve cell. Sodium ions migrate (biologists say are *pumped*) to restore the original potential difference, so that the cell is ready for further stimulation.

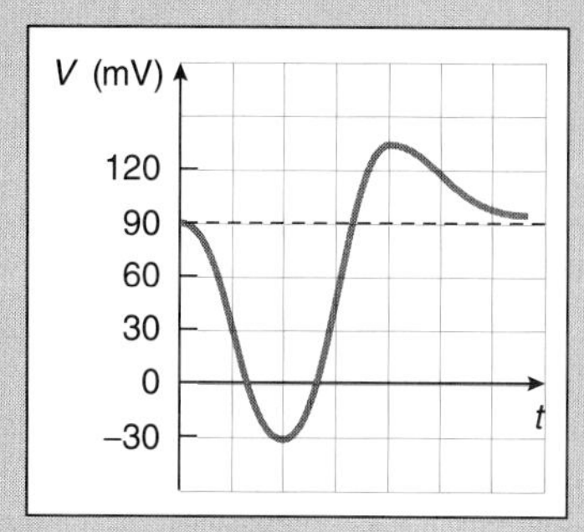

FIGURE P.15

•16. The fundamental unit of charge ($e = 1.602 \times 10^{-19}$ C) was discovered in a classic series of experiments by Robert Andrews Millikan early in the 20th century. In the experiments, tiny oil droplets are sprayed into a region in which there is a constant electric field as shown in Figure P.16. The individual oil droplets are large enough to be seen with a microscope using strong illumination, but small enough to have little weight and charge. The droplets obtain a static electrical charge from frictional effects (rubbing) as they emerge from the tiny orifice into the chamber. Surface tension causes the droplets to assume a spherical shape. To a first approximation, the droplets are subjected to two forces in the region between the plates: (1) the electrical force and (2) the gravitational force. Neglect the buoyant force of the air on the droplet. Let a droplet with mass m and charge $+q$ be in equilibrium within the chamber under the action of the two forces. (a) Show that the charge q is

$$q = \frac{mgd}{V_0}$$

(b) Let ρ be the density of the oil. Express the mass m of the drop in terms of its density and radius. (c) If (i) the oil has a density of 800 kg/m^3; (ii) the potential difference between the plates is 2000 V when the drop is in equilibrium; (iii) the separation of the plates creating the field is 2.00 cm; and (iv) the radius of the oil drop is measured (with a microscope) to be 3.629 μm; how many fundamental units of charge are on the drop?

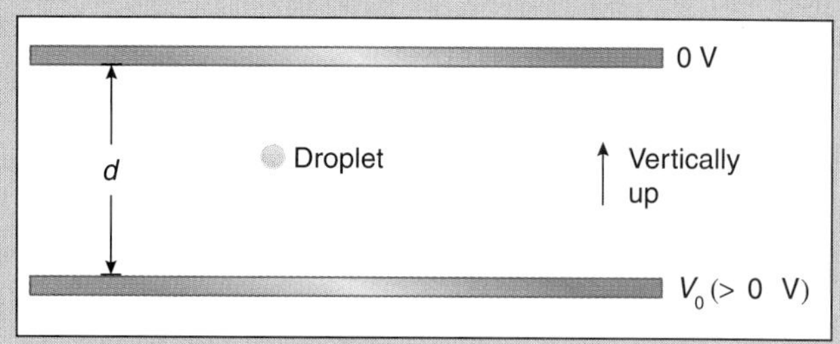

FIGURE P.16

•17. The electric potential is the electric potential energy per coulomb at a point in space. Analogously, the gravitational potential is the gravitational potential energy per kilogram at a point in space. Follow Example 17.1 to formulate the gravitational potential for a uniform gravitational field, such as that near the surface of the Earth.

Sections 17.4 The Electric Potential of Continuous Charge Distributions of Finite Size
17.5 Equipotential Volumes and Surfaces

18. An isolated conducting sphere of radius 3.00 cm is found to have a potential of 300 V on its surface. What charge is on the sphere?

19. Two parallel plates a distance 5.00 cm apart have potentials of −100 V and +100 V. Equipotential surfaces are drawn corresponding to differences in potential of 10 V. How far apart are the equipotential surfaces?

•20. An isolated conducting sphere of radius 3.00 cm has a potential of 300 V on its surface. What are the radii of the equipotential surfaces corresponding to 200 V, 100 V, and 1.00 V respectively?

•21. A spherical conductor is to have an electric potential of 25.0 kV without exceeding the maximum electric field magnitude of 3.0×10^6 N/C that causes air to become a conductor. What is the minimum radius for the sphere? Explain why this is a minimum radius and not a maximum radius.

•22. An insulating rod of length ℓ is bent into a circular arc of radius R that subtends an angle θ from the center of the circle; see Figure P.22. The rod has a charge Q distributed uniformly along its length. (a) Find the electric potential at the center of the circular arc. (b) The rod now is stretched along its length so that it forms a complete circle of the same radius. The total charge on the rod remains unchanged. What is the potential at the center of the circle?

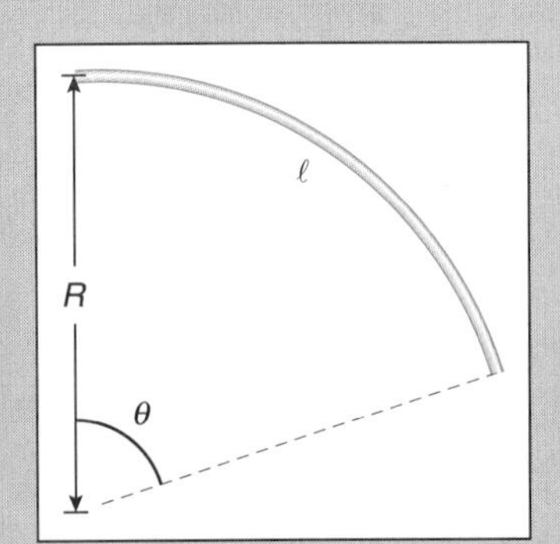

FIGURE P.22

•23. A metal sphere of radius R with a charge q is surrounded by a concentric metal spherical shell as shown in Figure P.23. The outer surface of the spherical shell has a charge of $+30.0 \times 10^{-6}$ C and the inner surface of the shell has a charge of $+25.0 \times 10^{-6}$ C. (a) Find q. (b) Sketch qualitative graphs of (i) the radial electric field component E_r and (ii) the electric potential V as functions of r.

•24. An early crude nuclear model of the atom imagined a very small central nucleus of charge $+Ze$ surrounded by a thin uniform spherical shell of radius R with charge $-Ze$. Show that the electric potential of this charge distribution is

(i) $V = 0$ V for distances $r > R$

and

(ii) $V = \dfrac{Ze}{4\pi\varepsilon_0}\left(\dfrac{1}{r} - \dfrac{1}{R}\right)$ for distances $r < R$

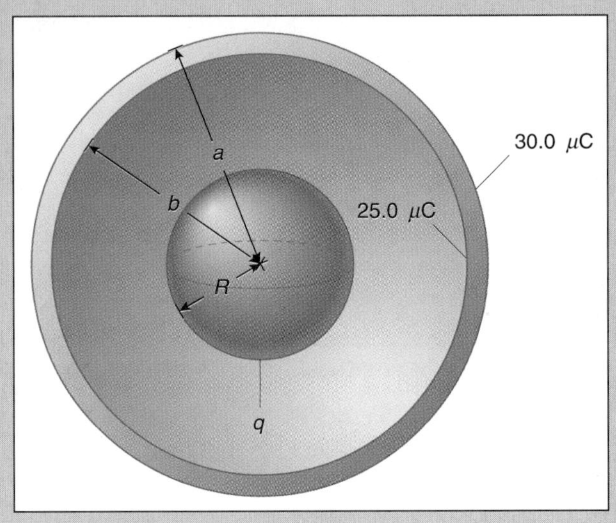

FIGURE P.23

•25. Instead of the model atom proposed in Problem 24, imagine a model consisting of a very small central nucleus of charge $+Ze$ surrounded by a sphere of radius R with charge $-Ze$ uniformly distributed throughout its volume. Find the electric potential of this charge distribution at positions: (a) $r > R$; (b) $r < R$.

•26. At a distance 10.0 cm from the center of a uniformly charged sphere, the electric potential is found to be −100 V. On the surface of the sphere, the potential is −250 V. (a) What is the charge on the sphere? (b) What is the radius of the sphere?

•27. What surface charge density σ on a conductor is needed to create an electric field with a magnitude sufficient to cause air to become a conductor (3.0×10^6 N/C)?

•28. A hollow spherical conductor of radius 5.00 cm has 6.00 nC of charge distributed uniformly over its surface. (a) What is the electric potential at the center of the sphere? (b) If the conductor is solid rather than hollow, what is the potential at the center of the sphere?

•29. A solid, insulating sphere of radius 5.00 cm has a charge of 6.00 nC distributed uniformly throughout its volume. (a) What is the electric potential at the surface of the sphere? (b) What is the electric potential at the center of the sphere?

•30. A hollow, spherical conductor of radius R has a charge of Q. Inside the cavity and concentric with the shell is another conductor of radius $r_0 < R$ with a charge of q as shown in Figure P.30. (a) What is the electric field between the inner conductor and the outer shell? (b) Use the result of (a) and Equation 17.9 to show that the potential difference between the two spheres is

$$V(r_0) - V(R) = \frac{1}{4\pi\varepsilon_0} q \left(\frac{1}{r_0} - \frac{1}{R}\right)$$

Explain why this potential difference is *independent* of the charge Q on the outer spherical shell.

•31. (a) At what distance from a +3.00 nC charge is the equipotential surface corresponding to 100 V located? (b) Where is the equipotential surface corresponding to 200 V? (c) Is the distance between the 100 V and 200 V equipotential surfaces the same as the distance between the 200 V and 300 V equipotential surfaces? Explain why it is or is not.

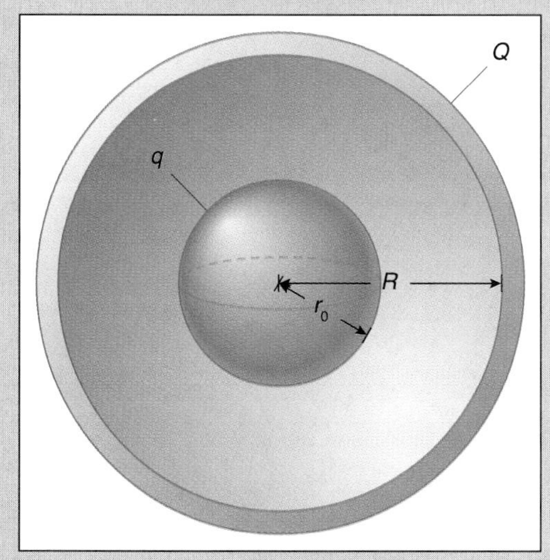

FIGURE P.30

•**32.** A long thin conducting wire connects two conducting spheres of radii 1.00 m and 0.10 m; see Figure P.32. The total charge on the connected pair is -2.00×10^{-7} C. (a) What is the electric field *inside* the wire? (b) What is the difference in potential between the ends of the wire? (c) What is the ratio of the charge on the larger sphere to the charge on the smaller sphere?

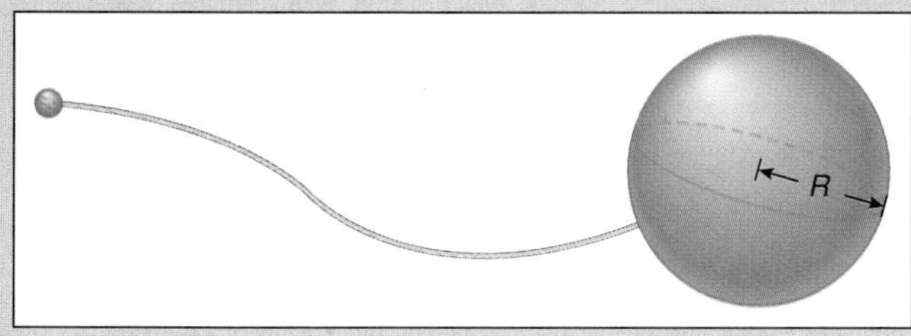

FIGURE P.32

•**33.** Two conducting spheres of radii 2.00 cm and 4.00 cm are far apart and each has charge +3.00 nC. (a) Calculate the approximate value of the electric potential on the surface of each conductor. (b) The two spheres now are connected by a conducting wire. What is the approximate potential on each surface now? What quantity of charge was exchanged by the two spheres?

•**34.** The gap between the electrodes in a spark plug is approximately 1 mm. Estimate the potential difference needed to produce an electric field of magnitude 3×10^6 N/C between the electrodes, sufficient to cause the air to become a conductor (and create a spark).

•**35.** A charge Q is uniformly distributed along an insulating straight wire of length ℓ as shown in Figure P.35. Find an expression for the electric potential at a point located a distance d from the distribution along its perpendicular bisector.

Section 17.6 The Relationship Between the Electric Potential and the Electric Field

•**36.** A graph of the electric potential as a function of x is given in Figure P.36. Make an accurate plot of the electric field component E_x as a function of x over the same domain for x. Ignore the points on the graph of V where the slope of the graph changes discontinuously.

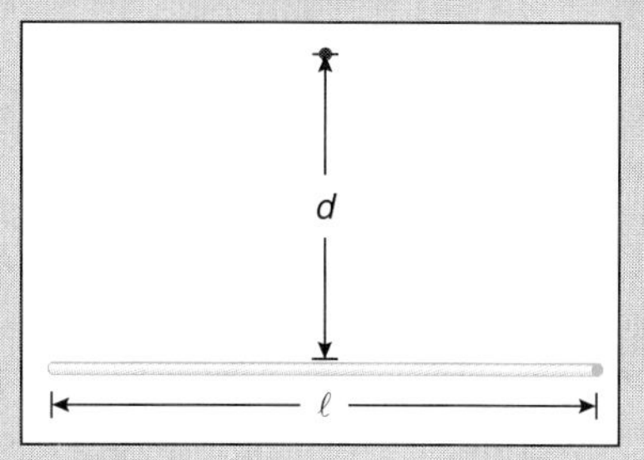

FIGURE P.35

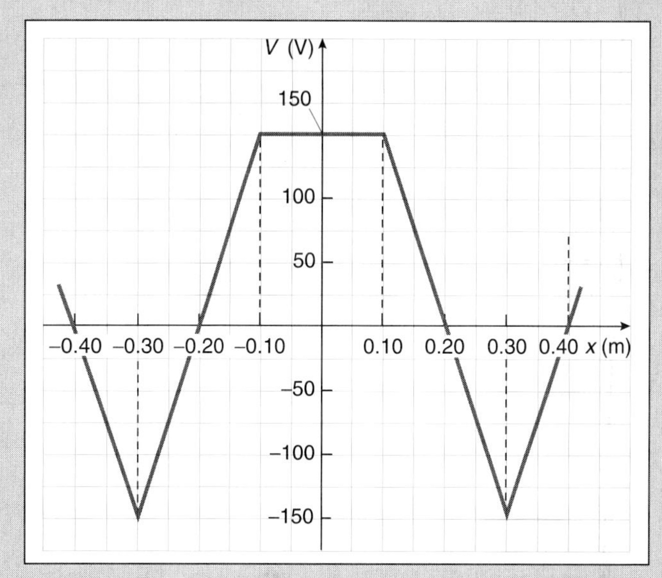

FIGURE P.36

•**37.** (a) Using the expression for the electric field of a uniformly charged, infinite line charge (see Table 16.2), and choosing the electrical ground to be at a distance a from the line, find the electric potential at a distance r from the line (see Figure P.37). (b) For what values of r is $V(r) > 0$ V? (c) For what values of r is $V(r) < 0$ V? (d) Explain why we cannot choose the electrical ground to be at $r = 0$ m or at $r = \infty$ m. (e) Verify that

$$E_r = -\frac{dV}{dr}$$

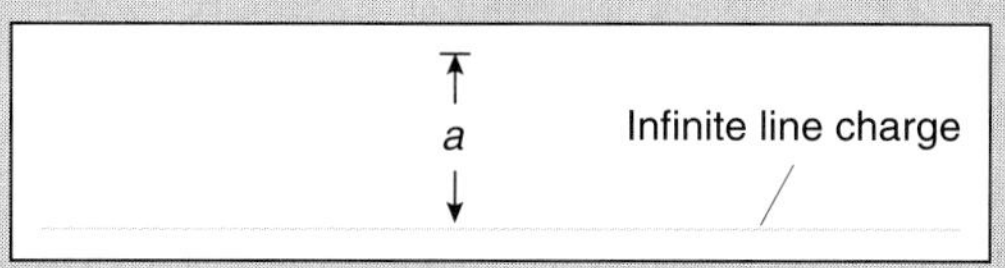

FIGURE P.37

•**38.** Explain why the graph of the electric potential as a function of x in Figure P.38 is not possible.

•**39.** The electric potential within a specific spherical charge distribution varies with the radial coordinate r as

$$V(r) = V_0 \frac{r^2}{2R^2}$$

where V_0 and R are constants and $r \leq R$. (a) Find $E_r(r)$. (b) Is the charge on the sphere positive or negative? Explain how you make this determination. (c) Make a schematic graph of E_r versus r for $0 \text{ m} \leq r \leq R$.

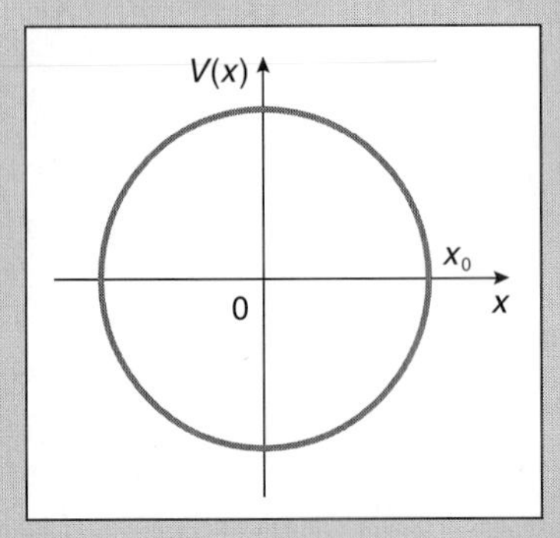

FIGURE P.38

40. An insulating washer (see Figure P.40) with an inner radius a and an outer radius b has a charge Q uniformly distributed on its surface. (a) Calculate the electric potential at a point P located a distance z from the washer along the symmetry axis perpendicular to the washer. (b) What is electric field component E_z as a function of z? (c) What is the electric field at $z = 0$ m?

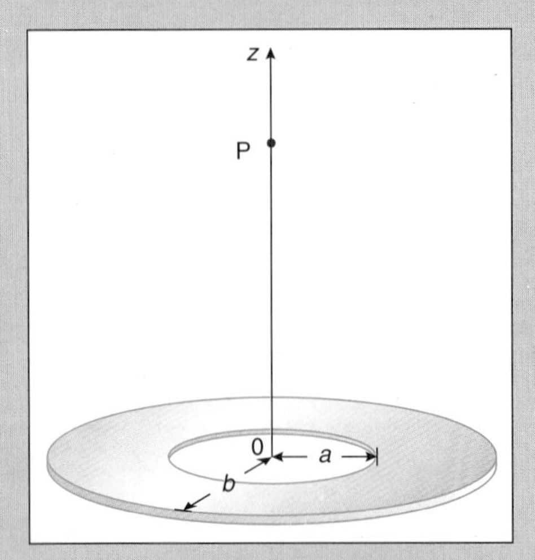

FIGURE P.40

Sections 17.7 Acceleration of Charged Particles Under the Influence of Electrical Forces
17.8 A New Energy Unit: The Electron-Volt

41. An electron is accelerated through a potential difference of 1.00 kV. (a) What is the increase in the kinetic energy of the electron in eV? (b) If the electron began the process at rest, what is its final speed?

•**42.** Express the average kinetic energy of a hydrogen molecule (H_2) at temperature 300 K in electron-volts. Will oxygen molecules (O_2) have the same average kinetic energy per molecule at the same temperature?

•**43.** An electron at rest is surprised to find itself suddenly released very close to one of the conducting infinite sheets shown in Figure P.43 (which are at the indicated potentials). (a) What is the direction of the electric field between the sheets? (b) What is the potential energy of the electron near the top sheet? (c) What is the potential energy of the electron near the bottom sheet? (d) Near which sheet should the electron be released so that it accelerates toward the other sheet? (e) What is the kinetic energy of the electron the instant before it strikes the sheet opposite the one from which it was released? Express your answer in both joules and electron-volts. (f) Calculate the speed of the electron just before impact.

FIGURE P.43

•**44.** An electron is projected at speed 4.00×10^6 m/s into a region with a uniform electric field of magnitude 300 N/C directed parallel to the initial velocity of the electron; see Figure P.44. How far into the region will the electron travel before reversing its direction?

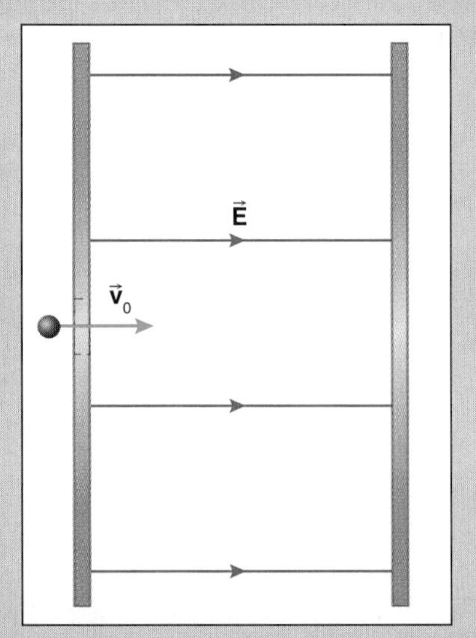

FIGURE P.44

•**45.** A positron has the same mass as an electron but a positive charge of $+e$. (a) In the arrangement shown in Figure P.45, near which conducting sheet should the positron be released so that it accelerates toward the other sheet? (b) If the positron is released at rest, what is its kinetic energy the instant before striking the other charged sheet? Express your result in both electron-volts and joules. (c) What is the speed of the positron at the instant before impact?

FIGURE P.45

•**46.** What kinetic energy in eV must an α-particle (a helium nucleus with charge quantum number +2) have so that it is capable of coming to within a distance of 4.7×10^{-15} m of a gold nucleus (with charge quantum number +79) before coming momentarily to rest? This small distance, much smaller than the size of the atom ($\sim 10^{-10}$ m), is about the radius of the gold nucleus. Experiments with such α-particles (produced by radioactive decay) led Ernest Rutherford (1871–1937) to propose the nuclear model of the atom in 1910.

•**47.** The two infinite conducting sheets shown in Figure P.47 have the indicated potentials. (a) Indicate on a sketch the direction of the electric field between the two sheets. (b) Calculate the magnitude of the electric field between the sheets. (c) What is the electric potential midway between the sheets? (d) At what distance from the lower sheet is the electric potential equal to zero? (e) If a proton is placed at rest midway between the sheets, what is the potential energy of the proton in eV? (f) If the proton in part (e) is released, toward which of the sheets does it accelerate? (g) What is the kinetic energy of the proton just before it strikes the appropriate sheet? (h) What is the speed of the proton just before impact?

FIGURE P.47

•**48.** The Stanford Linear Accelerator (SLAC), located near the San Andreas Fault near Stanford University in California, can accelerate electrons from rest to kinetic energies of about 2.0×10^{10} eV over a straight-line distance of approximately 1.6 km. Assume the acceleration is caused by a uniform electric field directed along the length of the accelerator (in fact, not so). Determine the magnitude of this electric field.

•**49.** An electron is propelled at a speed of 5.0×10^7 m/s into a region where there is a uniform electric field as indicated in Figure P.49. (a) Sketch the direction of the electric field in the region between the two plates. (b) Indicate the direction of the electrical force on the electron in the region of the field. (c) What is the total mechanical energy of the electron at point A? (d) What is the total mechanical energy of the electron at point B? (e) Determine the speed at which the electron exits the device.

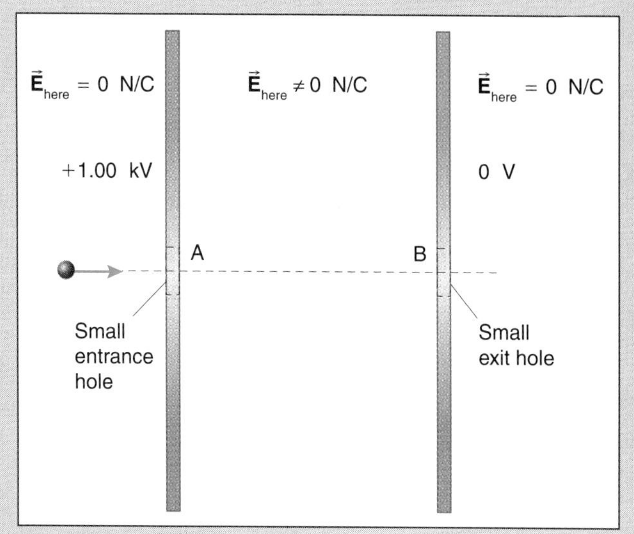

FIGURE P.49

•**50.** A tiny snowflake in air has a charge quantum number of -1.00×10^6 and floats at rest in a vertical electric field of magnitude 1.00×10^5 N/C. (a) Is the electric field directed up or down? (b) What is the mass of the snowflake? (c) In a like manner, is it possible to suspend a charged droplet of oil mist at rest in the same electric field if the mass of the droplet is 2.50×10^{-15} kg? Give quantitative reasoning to justify your answer. (d) Describe how a field like this could be produced with only a potential difference of 2.00 kV available. Provide a labeled diagram, appropriate dimensions, and numerical information.

•**51.** Two positive charges Q are held a distance d apart. (a) One charge now is released at rest and escapes to an infinite distance away. What is the kinetic energy of the escaped charge when it is a great distance away from the fixed charge? (b) The second charge now is released at rest. What happens to it?

•**52.** Three equal charges Q are fixed initially at the corners of an equilateral triangle with sides ℓ. (a) One charge is released at rest from its position and is allowed to escape to an infinite distance away. What is the final kinetic energy of this charge? (b) The second charge now is released at rest from its position. What is the final kinetic energy of this charge after it has escaped to an infinite distance? (c) The third charge now is released at rest from its position. What happens to it?

•**53.** (a) An α-particle (a helium nucleus with charge quantum number +2) moves through a potential difference of +20 V. What is the change in the kinetic energy of the α-particle in electron-volts? (b) If the potential difference was −30 V, what is the change in the kinetic energy of the α-particle, expressed in electron-volts?

•**54.** (a) What potential difference is needed to change the kinetic energy of an α-particle by +500 eV? (b) Is the α-particle moving to regions of higher or lower electric potential as its kinetic energy changes? Explain your reasoning.

•**55.** The nuclear structure of the atom was deduced by sending high-energy α-particles (helium nuclei each with a charge quantum number of +2) into a very thin gold foil. Gold was chosen not because such a foil looks pretty and is expensive, but rather because gold is very malleable and thus can be made into exceedingly thin foils. Most of the α-particles impinging on the foil pass through it with only small deflections. An occasional α-particle, however, makes a head-on approach to a gold nucleus (with a charge quantum number of +79). In such a head-on encounter, the α-particle is brought momentarily to rest before retreating along the path it came; such occurrences are called *back scattering*, to use the lingo of the trade. It is with these privileged α-particles that we can estimate the size of the gold nucleus. (a) Let an incident α-particle initially have a kinetic energy of 5.0×10^6 eV. What is its kinetic energy in joules? The initial electrical potential energy of the α-particle is zero, since it is essentially infinitely far away from the nucleus (and in any case the distant α-particle sees an atom that is electrically neutral, so that the electrical potential is essentially zero outside the atom itself). Thus the initial total mechanical energy of the incident α-particle is purely kinetic energy. (b) What is the electric potential of the gold nucleus at a distance r from the nucleus? (Consider the nucleus to be a spherical distribution of charge of radius $R < r$. *Neglect* any effects due to surrounding electrons.) (c) Use the CWE theorem to find the distance of closest approach for an incident head-on α-particle. This distance is a measure of the *upper limit* on the size of the gold nucleus since $R < r$. When such α-particle scattering experiments were first performed by Geiger and Marsden between 1909 and 1913, the results yielded a distance of closest approach that was so much smaller than the known size of an atom (about 10^{-10} m) that Rutherford proposed the nuclear model of the atom.

•**56.** A collection of electrons has a speed 6.00×10^7 m/s; this is too fast for an experiment you wish to perform. Using a uniform electric field provided by two parallel charged plates, you wish to slow them to 4.00×10^7 m/s. (a) Sketch the arrangement, indicating the direction of the electric field and the direction of the velocity of the electrons. (b) Indicate which charged plate is at the higher electric potential. (c) What potential difference $|\Delta V|$ is necessary to slow the electrons to the desired speed?

57. A charge $+Q$ is distributed uniformly throughout a sphere of radius R. (a) What is the potential on the surface of the sphere? (b) A small negative charge $-|q|$ of mass m is placed on the surface of the sphere. What is the potential energy of the charge $-|q|$ at this location? (c) Now give the negative charge a speed v_{escape} in the radial direction sufficient to (barely) escape to infinity from the sphere of charge $+Q$. Use the CWE theorem to derive an expression for the escape speed v_{escape} of the small charge in terms of $|q|$, R, Q, m, and appropriate electrical constants (such as ε_0). (d) In Chapter 8 we derived an expression for the analogous escape speed for the gravitational situation. The *gravitational* escape speed is *independent* of the mass m of the escaping object. The electrical escape speed [in part (c)] depends on the mass of the escaping particle. Why is there a difference between the two situations? (e) If the small negative charge is an electron and the positively charged sphere has the dimensions of a large nucleus such as uranium, where $R \approx 10^{-14}$ m, what charge Q makes the escape speed equal to the speed of light? About how many protons does this charge represent? This is the electrical analog of a gravitational black hole. (f) The answer to part (e) is a surprisingly small number of protons. Why, then, are most atoms not black holes? To approach this question, recalculate part (e) if the electron is placed initially at a distance from the nucleus equal to a typical *atomic dimension* ($\approx 10^{-10}$ m).

Sections 17.9 An Electric Dipole in an External Electric Field Revisited
17.10 The Electric Potential and Electric Field of a Dipole*

•**58.** Calculate the work done by the electrical force to bring a charge $+2e$ from an infinite distance to the midpoint of the line connecting the charges of an electric dipole composed of charges $\pm e$ separated by 2.00 nm.

•**59.** How far from a dipole along its axis must a point be located so that the approximate expression for the electric potential along the axis, Equation 17.29 (with $\theta = 0°$), is within 1.0% of the exact expression for the potential, Equation 17.27 (applied to a point on the axis). Express your result in terms of d, the separation of the charges in the dipole.

•**60.** The water molecule has an electric dipole moment of about 6.0×10^{-30} C·m. The molecule is placed with the orientation shown in Figure P.60 between two infinite conducting sheets with the indicated potentials. (a) Indicate the direction of the electric field between the sheets. (b) Find the magnitude of the electric field between the sheets. (c) What is the torque on the water molecule in its present orientation? (d) What is the potential energy of the water molecule in its present orientation? Express your result in electron-volts. (e) What is the minimum value for the potential energy of the dipole in the field?

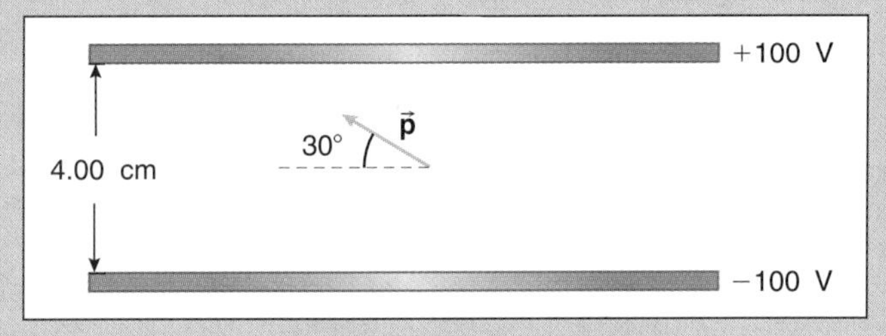

FIGURE P.60

•**61.** (a) Find the total electric field at the point P in Figure P.61. (b) Find the electric potential at the same point. (c) A dipole with dipole moment $\vec{p} = (6.0 \times 10^{-30}\ \text{C·m})\hat{j}$ is placed at this point. Find the potential energy of the dipole in its initial orientation. (d) Calculate the torque on the dipole.

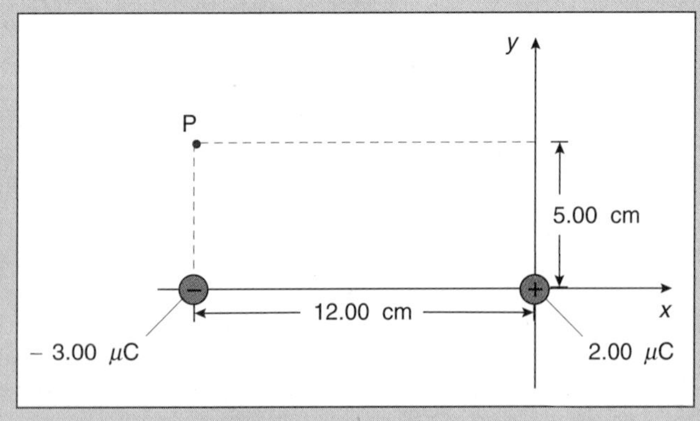

FIGURE P.61

62. A *linear quadrupole* is an arrangement of two oppositely directed dipole moments as shown in Figure P.62. (a) At a distance $|z| > d$ along the axis of the quadrupole, calculate the electric potential. (b) Show that if $|z| >> d$, the expression for V becomes

$$V(z) \approx \frac{|Q|}{4\pi\varepsilon_0} \frac{2d^2}{z^3}$$

The expression $2|Q|d^2$ is known as the quadrupole moment. (c) Use the potential found in (b) to find the electric field component E_z when $z >> d$.

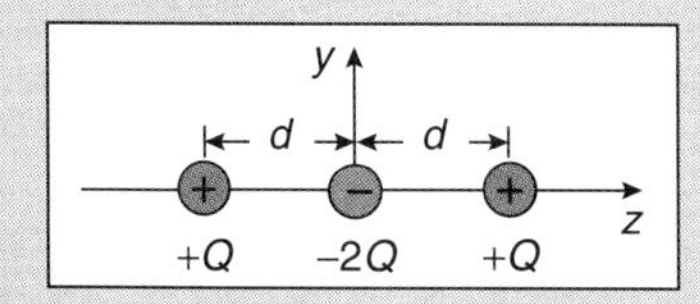

FIGURE P.62

63. Show that the potential of the linear quadrupole at point P in Figure P.63, when $r >> d$, is approximately

$$V \approx \frac{1}{4\pi\varepsilon_0} Qd^2 \frac{3\cos^2\theta - 1}{r^3}$$

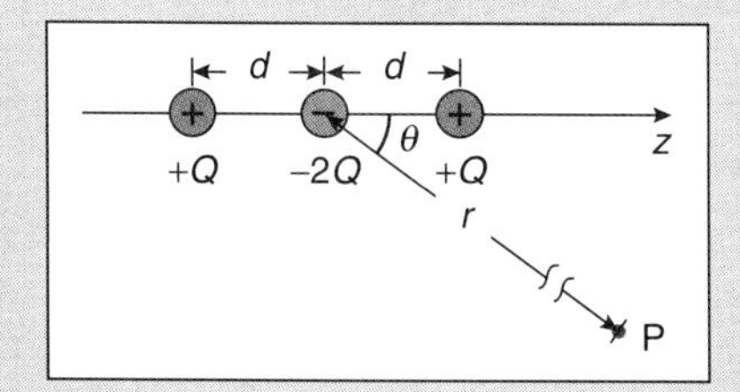

FIGURE P.63

Section 17.11 The Potential Energy of a Distribution of Pointlike Charges

64. What is the electric potential energy of the proton and electron charge distribution in the hydrogen atom when they are normally separated by 5.29×10^{-11} m? Express your result in both joules and electron-volts.

•**65.** A proton is thought to be a composite particle composed of two so-called up quarks with charges $2e/3$ and one down quark with charge $-e/3$. Imagine the three quarks to be at the vertices of a small equilateral triangle with sides 1.30×10^{-15} m. Determine the electric potential energy of the charge distribution.

•**66.** Singly ionized helium has a single electron orbiting the nucleus at a distance of about 2.65×10^{-11} m. The nuclear charge is $+2e$. What is the electric potential energy of this charge configuration? Express your result in electron-volts.

•**67.** When the nucleus of the uranium isotope ${}^{235}_{92}U$ captures a slow neutron, the nucleus splits (we say *fissions*), into two so-called fission fragments. Assume that just after fission, the fragments are of equal mass and charge $(+46e)$ and find themselves initially at rest separated by a distance of about 7.0×10^{-15} m. (a) Calculate the initial electric potential energy of the fragments. Express this energy in both joules and electron-volts. (b) What will the electrical force do to the fragments? (c) What is the kinetic energy of each fragment when the two are effectively at an infinite distance apart? (d) The kinetic energy of the fission fragments is used for destructive purposes in bombs or for constructive purposes in nuclear power plants. Assume the fission fragments have most of the energy released in the fission process (this assumption is in fact reasonable). How many fissions per second are needed to provide 10 MW of power in a reactor? How many fissions are needed to create a bomb that releases 9.8×10^{13} J of energy?* What mass of ${}^{235}_{92}U$ corresponds to this number of fission reactions?

•**68.** Two identical 1.00 kg masses have opposite charges q and $-q$ separated by distance r in space, well away from other masses (such as the Earth or Sun). What magnitude of charge q makes their electric potential energy equal to their gravitational potential energy?

69. Two pointlike dipoles are oriented with their dipole moments p_1 and p_2 parallel to each other and separated by a distance r that is much greater than the separation of the two charges in each dipole (see Figure P.69). Show that the potential energy of one dipole in the presence of the other is approximately

$$\text{PE} \approx -\frac{1}{4\pi\varepsilon_0} \frac{2p_1p_2}{r^3}$$

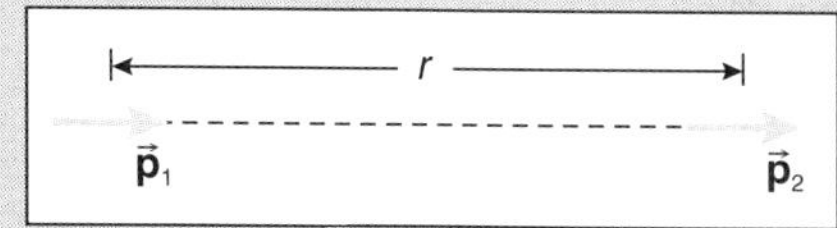

FIGURE P.69

70. Nuclear fusion is difficult to achieve because the nuclei of atoms are all positively charged and repel each other with the electrical force. In order to fuse, nuclei must be brought to within about 10^{-15} m of each other, at which distance the strong nuclear force becomes effective and fuses them. A deuteron is the nucleus of a hydrogen atom that consists of a proton and a neutron (so-called heavy hydrogen). (a) If two deuterons are 1.0×10^{-15} m apart, what is the potential energy of the pair? (b) If the deuterons are momentarily at rest at this location, what will be the kinetic energy of each deuteron when they are infinitely far away from each other? This is the kinetic energy each deuteron needs in order to approach to 1.0×10^{-15} m from one another so that the strong nuclear force can fuse the nuclei. (c) If the deuterons are part of a hot gas and have this average kinetic energy, what is the temperature of the gas? The actual fusion process is more complicated than this calculation, but your result indicates why fusion reactions of a controlled nature are difficult to produce on the Earth. Uncontrolled fusion produces a so-called hydrogen bomb.

*This corresponds to about 20 000 tons of the explosive TNT and is about the size of the bomb dropped on the unfortunate city of Hiroshima, Japan, in August 1945. The energy released in a nuclear bomb typically is expressed in kilotons or megatons of TNT. There are approximately 4.9×10^{12} joules per kiloton of TNT.

Investigative Projects

A. Expanded Horizons

1. Large electric potentials can be produced with a Van de Graaff generator, invented by Robert Jemison Van de Graaff (1901–1967). Investigate and report on how such generators are able to produce high potentials.
   Richard E. Berg, "Van de Graaff generators: theory, maintenance, and belt fabrication," *The Physics Teacher*, 28, #5, pages 281–285 (May 1990).
   Peter H. Rose and Andrew B. Wittkower, "Tandem Van de Graaff accelerators," *Scientific American*, 223, #2, pages 24–33 (August 1970); additional references are on page 128.
   A. D. Moore, *Electrostatics* (Anchor Doubleday, Garden City, New York, 1968).
   Oleg D. Jefimenko, "Electrostatic motors," *Electrostatics and Its Applications*, edited by A. D. Moore (Wiley, New York, 1973), pages 131–147.
   M. J. Mulcahy and W. R. Bell, "Electrostatic generators," *Electrostatics and Its Applications*, edited by A. D. Moore (Wiley, New York, 1973), pages 148–179.
   A. W. Simon, "Theory of the frictional Van de Graaff electrostatic generator," *American Journal of Physics*, 43, #12, pages 1108–1110 (December 1975).

2. The electrical analog of a permanent magnet (a ferromagnet) is known as a ferroelectric (also called an electret); it is a material in which the molecular electric dipole moments all are aligned permanently in the same direction. Investigate the nature of ferroelectricity and its technological importance.
   Charles Kittel, *Introduction to Solid State Physics* (7th edition, Wiley, New York, 1996).
   Werner Känzig, *Ferroelectrics and Antiferroelectrics* (Academic Press, New York, 1957).

3. Stresses applied to certain materials (such as quartz and tourmaline) produce an electric polarization in the material, an effect known as *piezoelectricity*. Such materials have significant technological importance in, among other things, strain gauges. Investigate the piezoelectric effect and some of its applications.
   Walter Guyton Cady, *Piezoelectricity* (Dover, New York, 1964), Volumes I and II.
   Charles Kittel, *Introduction to Solid State Physics* (7th edition, Wiley, New York, 1996).

4. Static electricity is a major hazard in many industrial and chemical settings, causing fires and explosions of chemicals, aerosols, and airborne dust particles such as in grain elevators. Explore the ways such electrically induced fires and explosions can be prevented.
   Paul Cartwright, "Electrical hazards!" *Chemtech*, 21, #1, pages 682–685 (November 1991); other references are in this article.

5. The charge distributions in the heart muscle that give rise to an electrocardiogram can be modeled electrostatically. Those of you interested in medical careers might enjoy investigating and reporting on the model developed in the first reference below.
   Russell K. Hobbie, "The electrocardiogram as an example of electrostatics," *American Journal of Physics*, 41, #6, pages 824–831 (June 1973).
   D. W. Kammer and J. A. Williams, "Some experiments with biological applications for the elementary laboratory," *American Journal of Physics*, 43, #6, pages 544–547 (June 1975).
   Robert Paine, *Generation and Interpretation of the Electrocardiogram* (Lea & Felsriger, Philadelphia, 1988).

B. Lab and Field Work

6. One way of separating proteins, amino acids, and DNA fragments is via a process called *gel electrophoresis*, which uses electric fields. It is a common technique in the biological sciences. Investigate the technique and its application of electrostatic principles. Visit a member of the biology department who uses the technique. Discuss, design, and perform an experiment using the technique, paying particular attention to the physics of the process in your report of the experiment.
   Shyamsunder Erramilli, Fredrik Österberg, and Bruce Vogelaar, "Undergraduate laboratory: principles of gel electrophoresis," *American Journal of Physics*, 63, #7, pages 639–643 (July 1995).
   John M. Clark Jr. and Robert L. Switzer, *Experimental Biochemistry* (2nd edition, Freeman, New York, 1977), pages 43–55.
   Frederick A. Ausubel, *Current Protocols in Molecular Biology* (Current Protocols, New York, 1994), pages 2.5.1–2.5.15.
   Henry M. Zeidan and William V. Dashek, *Experimental Approaches in Biochemistry and Molecular Biology* (W. C. Brown, Dubuque, Iowa, 1996).
   Rodney F. Boyer, *Modern Experimental Biochemistry* (2nd edition, Benjamin Cummings, Redwood City, California, 1993).

C. Communicating Physics

7. Most nonscientists know of the unit of electric potential, the volt, but have no idea what it represents. Ask a few nonscientist friends or relatives to explain what they think the term means. Take notes or tapes of their explanations, without passing judgment on them. Use your survey as a guide and, by making appropriate analogies to gravitation, explain the meaning of the unit of electric potential in language appropriate to a nonscientist. Have your nonscientist friends critique your explanation to improve its clarity, and then make any necessary revisions.

CHAPTER 18

CIRCUIT ELEMENTS, INDEPENDENT VOLTAGE SOURCES, AND CAPACITORS

I sing the body electric

Walt Whitman (1819–1892)*

Storing energy for future use is an important aspect of wise energy management, even survival. Rural residents in northern areas who heat their homes with wood spend considerable time gathering and storing fuel for use during the winter, lest they freeze. Primitive cultures even in tropical climates need to accumulate and store wood for cooking. Squirrels and other animals store acornzs and other foods (sources of energy) for future use. Our bodies store surplus food as fat.

Mechanical energy also can be stored. A rapidly spinning flywheel stores kinetic energy. Lift and hold a set of barbells over your head and you increase and store the gravitational potential energy of the barbells, energy that can be suddenly tapped by simply dropping them. If you either compress or stretch a spring, you increase and store the potential energy of a mass attached to it. In this chapter we examine devices that increase and store the electrical potential energy of charges.

Devices that increase the electrical potential energy of charges are known by several names: **sources of emf**† or, equivalently, **independent voltage sources**. The terms mean the same thing. The common battery is an example of such a device. Calculators, automobiles, flashlights, portable CD players, notebook PCs—all use batteries in abundance. Batteries provide a convenient and portable source of electrical energy that originates from electrochemical reactions within the battery. In the next chapter, we will see how we can model real batteries that, much to our occasional consternation, age and die and need to be replaced (usually when we most need them, according to Murphy's law). Other examples of independent voltage sources (or sources of emf) include, among other things:

- electric generators used to provide the potential differences at the electrical outlets common in our homes, classroom, labs, and offices;
- solar cells used on spacecraft and in remote geographic locations; and
- fuel cells used on board the space shuttle and other spacecraft.

Electrical potential energy is accumulated and stored in devices called **capacitors**. Though they may be less familiar to you than batteries, capacitors are used widely in electronic applications. Capacitors are used in flash units on cameras to accumulate electrical potential energy to be released suddenly in the flash lamp when the shutter is pressed (analogous to dropping a set of barbells). In laser power supplies, as well as in medical defibrillation units used by medical trauma and emergency personnel, capacitors also are used to accumulate large amounts of electrical potential energy for sudden use.

Capacitors have other electronic uses too. When you change a radio station or TV channel by turning the dial (or using the remote), you are changing the value of a capacitor in a special electronic circuit known as an electrical oscillator or tuning circuit. Electric utility companies use capacitors to maintain the desired relationship between various electrical parameters in the electric grid around town. Many PC keyboards use capacitors to sense which key is depressed, so that capacitors are literally at your fingertips every time you sit at your PC. Capacitors also are used to count traffic and in smart traffic light systems to sense when a car approaches an intersection. Capacitors are indeed very useful electrical devices!

Solar panels on spacecraft are independent voltage sources. This is the famous orbiting Hubble Space Telescope.

In this chapter we also study how independent voltage sources (sources of emf) and capacitors, as examples of two-terminal circuit elements, are connected together in various useful ways. This serves as a precursor to a systematic approach to the study of electric circuits in the next chapter and Chapter 22.

18.1 Terminology, Notation, and Conventions

We have seen many times now that language is an important component of physics because it explains what we mean when we discuss the laws of physics; mathematics is employed as quantitative shorthand. It is important to be sure we have a good common understanding of the meaning of various technical terms and phrases.

Imagine two points A and B that are at electric potentials V_A and V_B respectively, as shown in Figure 18.1. The *difference* in the electric potential between the two points is called the **potential difference**‡ between the two points. Since the unit of electric potential in the SI system of units is the volt, the potential difference between two points also is expressed in volts. Of course, differences can be taken two ways. We need

V_A

V_B

Figure 18.1 The potential at two points in space.

*(Chapter Opener) Title of a poem in his collection *Leaves of Grass* (David McKay, Philadelphia, 1900) page 98.

†The term emf arises from the antiquated phrase *electromotive force*, which arose early in the history of explorations into electromagnetism. Emfs are not forces and the term electromotive force is misleading. The term independent voltage source, while also not ideal, is used in more advanced courses in circuits and electronics, but this term is more indicative of what the device actually does: raise the electric potential energy of charges passing through it.

‡In engineering, the term *voltage* is used frequently to mean potential difference. We prefer to use the term potential difference to emphasize that the values of the electric potential at *two different points in space* are involved.

some convention to indicate which way the difference is to be taken: $V_A - V_B$ or $V_B - V_A$.

When you want the potential difference between A and B, take the potential at A and subtract from it the potential at B:

$$\text{potential difference between A and B} \equiv V_A - V_B$$

When you want the potential difference between B and A, take the potential at B minus the potential at A:

$$\begin{aligned}\text{potential difference between B and A} &\equiv V_B - V_A \\ &= -(V_A - V_B)\end{aligned}$$

The potential difference between two points can have a value that is positive, negative, or zero. For example, if the electric potential at the point A is 100 V and that at B is 50 V, the potential difference between A and B is

$$100\ \text{V} - 50\ \text{V} = 50\ \text{V}$$

On the other hand, the potential difference between B and A is

$$50\ \text{V} - 100\ \text{V} = -50\ \text{V}$$

Thus you must be careful in the way you express the potential difference between two points. Two points at the same electric potential have a potential difference of 0 V.

A pair of **polarity markings** (+) and (−) is used to designate which of two given points is at the higher electric potential (+) and which is at the lower potential (−).

You likely have noticed such paired polarity markings on common batteries.

Since the location of the zero of electric potential is arbitrary, the (+) symbol at a point does *not* necessarily mean that the potential at the point has a positive value; the symbol means that the point is at a *higher* electric potential than the point marked with the (−). Likewise the (−) symbol does *not* necessarily mean that the value of the electric potential at that point is negative; it means that the point is at a lower electric potential than the point marked (+). This convention regarding polarity markings is illustrated in Example 18.1.

Polarity markings are typically indicated on most batteries.

In prior chapters, we used the letter V to designate the value of the electric potential at a *single point* in space. Thus the potential at point A is designated V_A. Alas, we need to shift notational gears, not because it is particularly desirable to do so, but because the tide of convention dictates we have no choice but to go with the flow.

In this chapter and the next, the symbol V typically represents the *potential difference* between two points.

We really should call the potential difference ΔV, but the idea of the potential difference between two points occurs so frequently in the study of practical electric devices that it becomes a chore to continually write the delta, meaning difference or change; hence the delta is dropped for notational simplicity. Thus we have to now think of V as typically the *difference* in the electric potential between *two* points. Keep in mind, however, that we occasionally still will need to designate the potential at a *single* point in space and we continue to use, for example, V_A to designate the potential at a point labeled A. The context typically will indicate whether V is the potential at a *single* point in space, or the potential difference between *two* points in space. Of course, in cases where we choose to call $V_B = 0$ V, then $\Delta V \equiv V = V_A - V_B = V_A - 0$ V all have the same value.

QUESTION 1

Point A has an electric potential of −50 V. Point B has a potential 10 V lower. (a) What is the potential at point B? (b) Indicate which of the two points has the positive (+) and the negative (−) polarity markings. (c) What is the potential difference between A and B? (d) What is the potential difference between B and A?

EXAMPLE 18.1

Two points A and B have the potentials indicated in Table 18.1.

TABLE 18.1

B	A
• 50 V	• 100 V
• 0 V	• 50 V
• −25 V	• 25 V
• −50 V	• 0 V
• −75 V	• −25 V

a. **For each case, indicate which point is at the higher electric potential with a (+) polarity marking and the one with the lower electric potential with a (−) polarity marking.**
b. **For each case, determine the potential difference V between A and B.**

Solution

Point B (polarity markings in parentheses)	Point A	Potential difference between points A and B
(−) • 50 V	(+) • 100 V	100 V − 50 V = 50 V
(−) • 0 V	(+) • 50 V	50 V − 0 V = 50 V
(−) • −25 V	(+) • 25 V	25 V − (−25 V) = 50 V
(−) • −50 V	(+) • 0 V	0 V − (−50 V) = 50 V
(−) • −75 V	(+) • −25 V	−25 V − (−75 V) = 50 V

In this particular example, note that the potential difference V between the given points A and B is the same regardless of the various specific values of the potentials at A or B. This is *not* always the case.

18.2 Circuit Elements

With Lego building blocks, you can build simple or amazingly complex structures. Circuit elements are the electronic building blocks used in the many simple and complex applications of electricity that surround us. You likely know of several circuit elements: batteries and light bulbs* are circuit elements. There are others as well. As is often the case in physics, we need to distinguish between ideal and real circuit elements: the difference is analogous to the difference between the ideal frictionless surfaces and real surfaces in mechanics. Real circuit elements are modeled by using combinations of ideal circuit elements.

An **ideal circuit element** has several characteristics:

1. Think of a generic ideal circuit element schematically as a box with two **ideal wires,** made from perfectly conducting material,† coming out of it; see Figure 18.2. The two lines coming out of the schematic symbol represent the two ideal wires and are called the **terminals** of the circuit element or, equivalently, the **leads**‡ of the circuit element.
2. An ideal circuit element cannot be subdivided into other ideal circuit elements.
3. Specific symbols are used to represent each distinct kind of ideal circuit element. The symbol used in Figure 18.2 simply means *any* ideal circuit element.

By convention, we consider every symbol for a circuit element to be an ideal circuit element.

The two leads of a circuit element may be at (or have) different electric potentials; when this situation arises, the circuit element has a nonzero potential difference between its two terminals or leads. When this is the case, we say there exists a potential difference *across* the circuit element, conventionally taken as the higher minus the lower value of the electric potential.

18.3 An Independent Voltage Source: A Source of EMF

An independent voltage source (equivalently, a source of emf) is a circuit element that has the following electrical characteristic: come hell or high water, it always maintains a constant potential difference V_0, also known as its emf, between its two terminals.§

Think of an independent voltage source as an ideal battery, one like the Energizer battery; as the pink bunny says, "It just keeps going and going." It never "dies."

The circuit symbol for an independent voltage source is shown in Figure 18.3. The terminal of the independent voltage source at the higher electrical potential is marked with a (+) polarity sign while that at the lower electrical potential is marked with a (−) polarity sign. The emf V_0 of the independent voltage source is the potential difference between the terminal marked (+) and that marked (−); the value V_0, expressed in volts, is positive.

We use the word independent voltage source in a general way. The term encompasses not only the common batteries (assuming they are ideal) you are most familiar with for your personal electronic gadgets, but also any device that maintains a constant potential difference between its two terminals. Electric generators, solar cells, fuel cells, and other devices

*Light bulbs are a special class of resistors; we study general resistor circuit elements in the next chapter.

† We shall see in the next chapter that a perfectly conducting material has zero electrical resistance. Materials with zero resistance are called superconductors. Such materials in fact exist but are rarely used for the actual leads of real circuit elements.

‡ Leads is pronounced "leeds" like the verb, not like the chemical element Pb.

§ Some physics books use the symbol $\mathcal{E}$ for the emf or, equivalently, the potential difference between the terminals of an ideal independent voltage source. Since the emf is measured in volts, for notational simplicity we prefer to use the symbol V_0 for the potential difference between the terminals of an independent voltage source. This is the notational convention used in engineering; there is no reason for us to use a different notation in physics.

FIGURE 18.2 A generic ideal circuit element.

FIGURE 18.3 The circuit symbol for an independent voltage source.

An independent voltage source never "dies."

can be thought of as independent voltage sources in an ideal or first approximation.

An independent voltage source is an unusual device: it raises the electrical potential energy of positive charges moving from the lower-potential terminal, marked (−), to the higher-potential terminal, marked (+).* The potential difference V_0 is the work done by the independent voltage source on one coulomb of positive charge as it moves from the negative to the positive terminal inside or through the source. The potential difference in volts thus is measured in joules per coulomb.

In gravitational physics, when you lift one kilogram of mass through a vertical distance d to increase its gravitational potential energy by an amount $(1\ \text{kg})gd$, you are acting as a gravitational battery. The increase in gravitational potential energy per kilogram is gd, and so gd in that circumstance is analogous to what we mean by V_0 for an electrical independent voltage source. For common batteries, the increase in electrical potential energy of a positive charge moving from the minus to plus terminal through the battery arises from chemical reactions within the battery itself; in other independent voltage sources, the energy originates from other sources such as sunlight (for solar cells) or mechanical energy (for electric generators).

An independent voltage source maintains the specified potential difference V_0 between the (+) terminal and the (−) terminal, no matter what else is connected to its terminals, or for how long. An independent voltage source thus has an infinite lifetime. This surely is unrealistic in actual situations, but is just as useful an idealization as frictionless surfaces are in mechanics.

*Equivalently, it raises the electrical potential energy of negative charges moving from its (+) terminal to its (−) terminal.

Real voltage sources such as real batteries do not maintain a constant potential difference V_0 between their terminals as they are used over very long periods of time, sometimes to our frustration. Real batteries age and die and must be replaced. Real batteries also have a potential difference between their terminals that is smaller than V_0 under most circumstances. Later, in the next chapter, we see how you can model the behavior of *real* batteries with an ideal independent voltage source and another ideal circuit element (a resistor). We use ideal circuit elements to *model* real circuit elements.

QUESTION 2

Examine the packages of batteries for sale in your campus bookstore, supermarket, or discount outlet. What is the potential difference between the terminals of a common D-cell battery? Is it the same for C-cell, AA, AAA batteries, and common calculator and camera batteries? Make a list of the types of batteries and the labeled potential difference between the terminals of each type.

18.4 Connections of Circuit Elements

Circuit elements are not of much use in isolation; they need to be connected together for various useful purposes. In this section we will learn how to recognize whether circuit elements are connected in several common ways. It will be important to be able to recognize these connections when we study electric circuits.

The place where the leads (terminals) of two or more circuit elements are connected together is called a **node**; see Figure 18.4. A node always is at a specific electric potential whose value depends on, among other things, the reference location where the electric potential is chosen to be zero (the electrical *ground*).

For our purposes, circuit elements may be connected together in several different ways.†

Series Connection

If the collection of circuit elements is connected together as shown in Figure 18.5, the circuit elements are said to be connected in **series**. As we march along the chain of circuit

† More advanced courses in electric circuits consider other types of specialized connections known as Y and Δ connections.

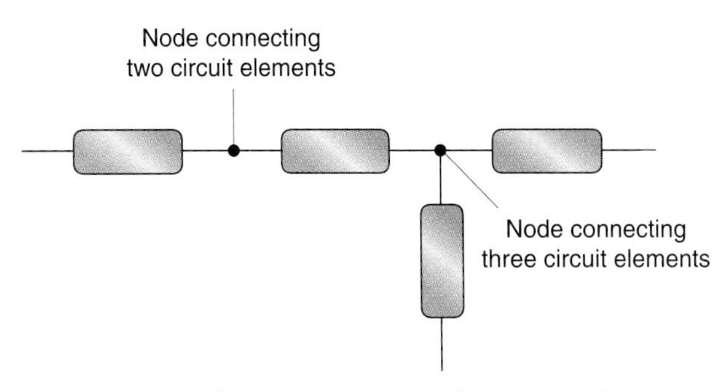

FIGURE 18.4 **Nodes connecting two or three circuit elements.**

FIGURE 18.5 Circuit elements connected in series.

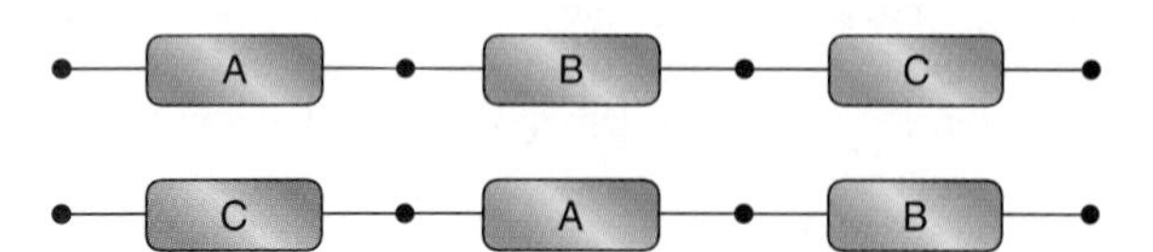

FIGURE 18.6 The order of circuit elements in series is not important.

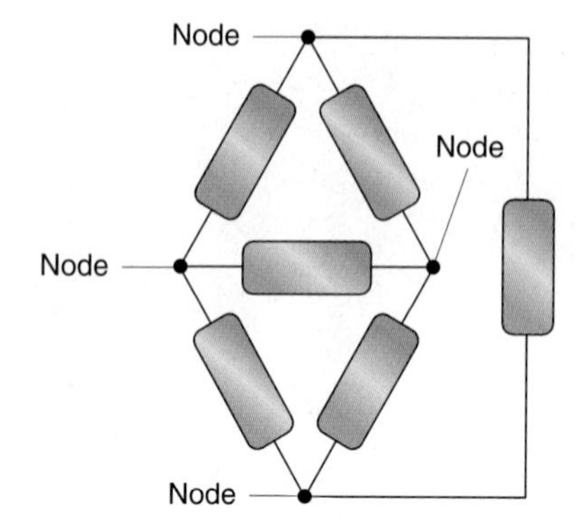

FIGURE 18.8 Six circuit elements connected neither in series nor in parallel.

elements, we encounter each circuit element in turn, or in a serial fashion, one after the other.

The specific order of circuit elements in series is not important (see Figure 18.6), since it has no effect on the potential difference across each one.*

Parallel Connection

If the terminals of various circuit elements are connected to the *same* two distinct nodes (see Figure 18.7), the circuit elements are said to be connected in **parallel**. The parallel connection shown in Figure 18.7a is equivalent to that shown in Figure 18.7b. We prefer schematic drawings like Figure 18.7b because of an aesthetic preference for straight lines and sharp corners in our diagrams.

Since all the leads from one side of each circuit element are connected together, they are all at the same electric potential. The same is true for the leads from the other side of each circuit element in parallel.

Thus the potential difference V across each circuit element in a parallel connection is the same.

Connections Neither in Series nor in Parallel

The circuit elements in Figure 18.8 are connected neither in series nor in parallel. In other words, do not get the impression that everything must be in series or parallel!

We also use another notational convention when we picture on a flat page any three-dimensional arrangement of wires. If two wires or leads are connected together, a dot symbolizes the connection, as indicated to the left of Figure 18.9. If two wires are not connected but merely cross over each other (like a freeway overpass), no dot is drawn, as indicated to the right of Figure 18.9. In a series connection of circuit elements, the dots typically are omitted since it is apparent that the leads are unambiguously connected together; see Figure 18.10.

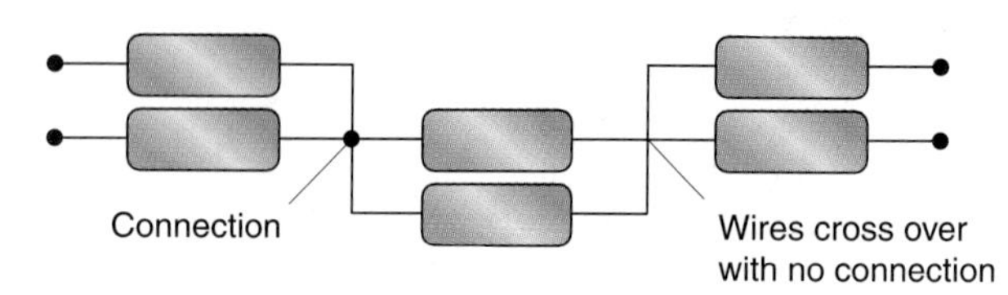

FIGURE 18.9 A dot symbolizes a connection of two or more wires. Two wires that cross over each other but are not connected have no dot drawn where they meet on the diagram. The dots on the far left and right indicate places where other circuit elements may be connected to the given collection.

FIGURE 18.10 Dots are omitted for a series connection.

The number of dots is *not* an indication of the number of nodes. For example, the two dots on each side of Figure 18.11a, symbolizing electrical connections, really are single nodes, as shown in Figure 18.11b. Both dots on the left (and right) side of Figure 18.11a are at the same electric potential. The extra ideal wire segment connecting the two dots ensures that they are at the same electric potential. Thus, be careful in counting the number of nodes; it is *not* necessarily the same as the number of dots in the diagram.

*The electric current through each, a concept we introduce in the next chapter, also is unaffected by permuting the order of circuit elements in series.

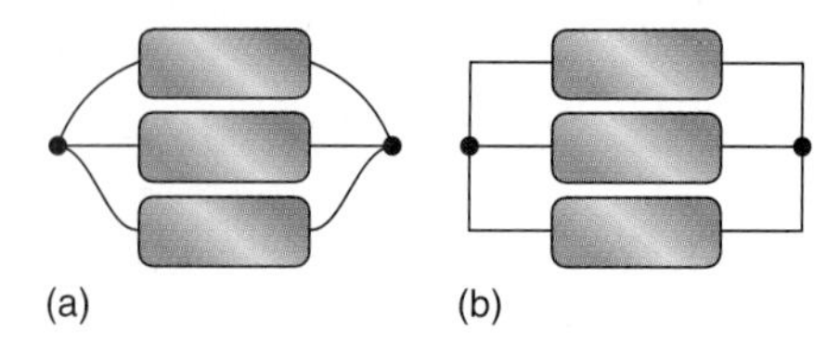

FIGURE 18.7 Two equivalent ways of drawing three circuit elements in parallel; that shown in (b) is preferred.

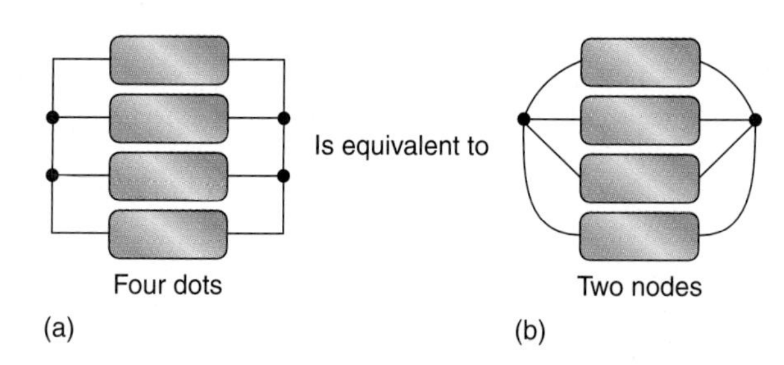

FIGURE 18.11 The number of dots may not be the same as the number of nodes.

Shorting Circuit Elements

Imagine a connection of circuit elements like that shown in Figure 18.12. Later, we shall see that if a wire is connected between two nodes that are at different electric potentials, such as in Figure 18.13, the wire very quickly makes the potential difference between the two nodes equal to zero. The act of doing this is called **shorting** the nodes, or shorting the circuit elements connected to those nodes.

When nodes are shorted, the previously distinct nodes reach the same electric potential and become a single node. Any circuit elements connected to two shorted nodes then must have zero potential difference between their terminals; such circuit elements are called **shorted out**. The two leads of a shorted circuit element now are connected to the *same* node.

The effect of shorting is to make any circuit element connected to the shorted nodes irrelevant because the leads of such a circuit element now are connected to the same node. There is 0 V potential difference between the terminals of a shorted circuit element. Shorted circuit elements can be removed with no effect on the remaining circuit elements, as indicated in Figure 18.14.

In Figure 18.14 the dots at (1) and (2) are obviously equivalent. Thus the extra wire connecting the dots at (1) and (2) in Figure 18.13 effectively obliterates any circuit elements connected to nodes (1) and (2) and renders them the same node. The connection in Figure 18.13 between what were distinct nodes (1) and (2) in Figure 18.12 is what is known as a **short circuit**; it means the new arrangement of circuit elements is simpler (shorter).

Notice that an ideal independent voltage source (a source of emf), which always maintains the potential difference V_0 between its terminals, should *never* be shorted (see Figure 18.15) since it leads to a logical contradiction.*

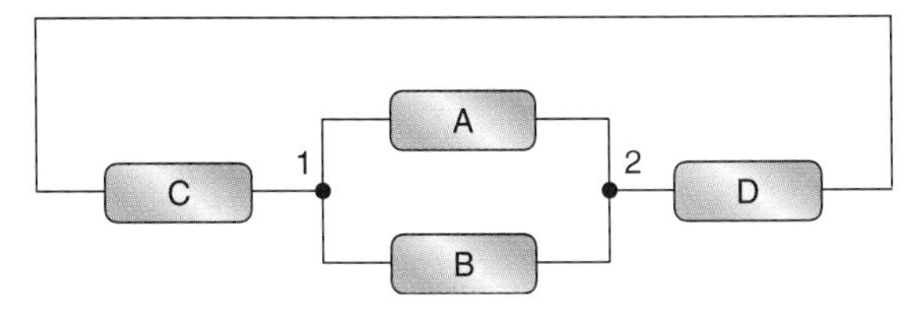

FIGURE 18.12 A connection of circuit elements.

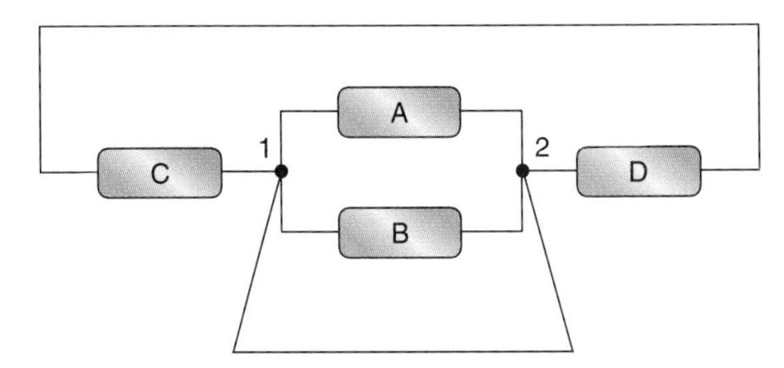

FIGURE 18.13 Shorting circuit elements A and B.

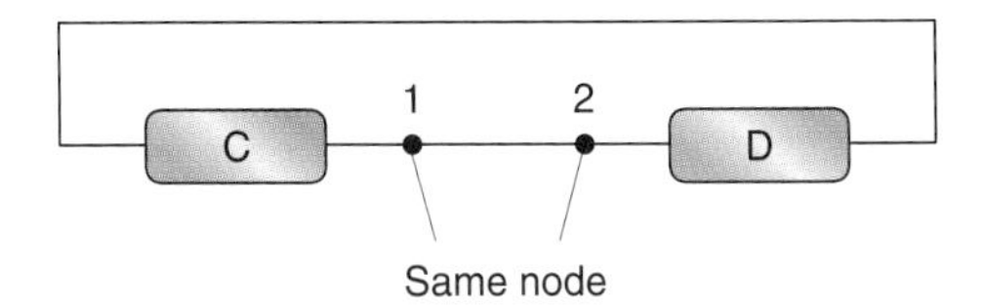

FIGURE 18.14 The circuit of Figure 18.13 with the shorted circuit elements removed.

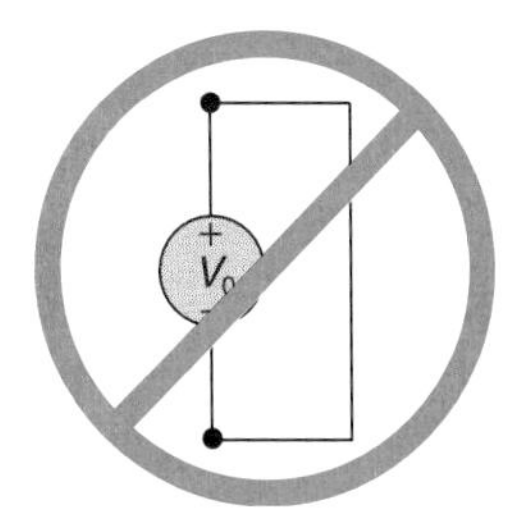

FIGURE 18.15 Never short-circuit an independent voltage source.

QUESTION 3

In a carefully worded sentence or two, explain what is meant by the term node. How does the use of the word in this context differ from its use when associated with standing waves?

QUESTION 4

Carefully explain what is meant by the term shorting out a circuit element.

18.5 INDEPENDENT VOLTAGE SOURCES IN SERIES AND PARALLEL

Many times you likely have replaced the multiple batteries in a flashlight, radio, or portable CD player without thinking about it much, except for the bother. The multiple batteries in these electronic gadgets are connected in series. On the other hand, if you ever have had the misfortune to discover that your car battery is dead, and used jumper cables with the battery in a friend's car or service truck to start your car, the two batteries were connected

*For *real* independent voltage sources such as real batteries, shorting the terminals results in very large electric currents that may destroy the battery, even explosively. We shall see why when we examine real batteries in Chapter 19.

The batteries in this device are connected in series.

To jump start a car, the batteries are connected in parallel.

in parallel. Here we discover what happens when you connect independent voltage sources (sources of emf) in series or parallel.

If several independent voltage sources are placed in series, as shown in Figure 18.16, they can be replaced with a single equivalent independent voltage source (see Figure 18.17). The potential difference across the terminals of the equivalent independent voltage source is equal to the algebraic sum of the potential differences of the individual independent voltage sources.

For example, consider the series of independent voltage sources in Figure 18.16. The potential difference of the equivalent independent voltage source is found by starting at terminal A and going to terminal B: if the positive terminal of an independent voltage source is encountered first, the emf of that source is accounted for in the equation for the emf of the equivalent source with a plus sign; if the negative terminal is encountered first, that source of emf is entered into the equation for the emf of the equivalent source with a minus sign. So, for the series combination of Figure 18.16, the potential difference (or emf) of the equivalent independent voltage source is

$$V_{eq} = V_1 + V_2 + V_3$$

For the series combination shown in Figure 18.18, the equivalent independent voltage source has a potential difference (or emf) of

$$V_{eq} = V_1 + V_2 + (-V_3)$$

From a practical viewpoint, independent voltage sources connected in series are arranged as in Figure 18.16, with their polarity markings all oriented the *same* way.

You probably already know this. When multiple batteries are placed in flashlights, radios, or other electronic equipment, the batteries are inserted in series with the positive terminal of each one connected to the negative terminal of the next one in the series. This arrangement results in the greatest combined potential difference.

Independent voltage sources also occasionally are connected in parallel, as shown in Figure 18.19.

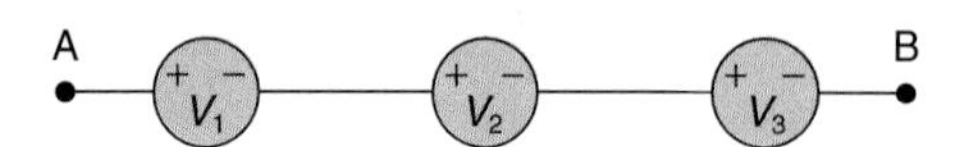

FIGURE 18.16 Independent voltage sources connected in series.

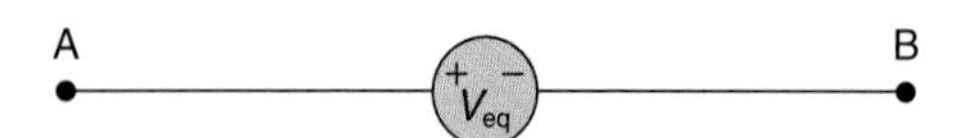

FIGURE 18.17 The equivalent independent voltage source of a series connection of such sources.

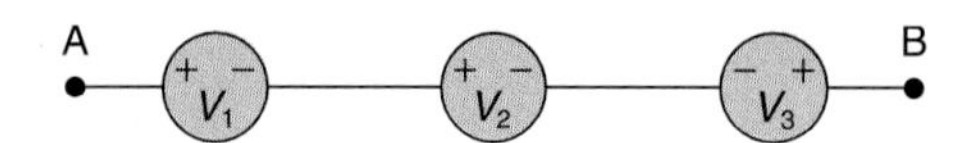

FIGURE 18.18 A possible (but impractical) series connection of independent voltage sources.

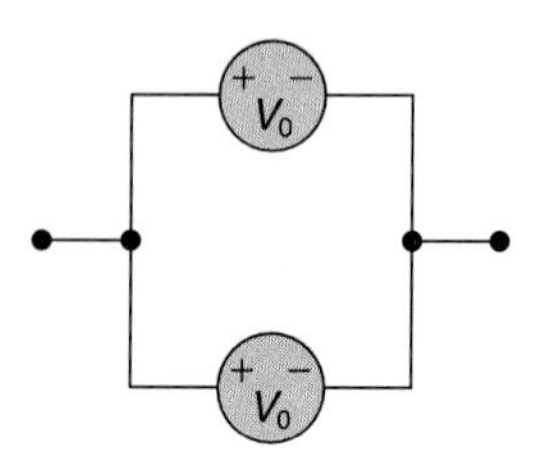

FIGURE 18.19 Independent voltage sources in parallel must have the same emf and must be oriented with similar polarities connected to the same node.

When in parallel, the individual sources *must* have the same emf (or potential difference) between their terminals and must be connected with negative terminals together at one node and positive terminals together at the other node.

The equivalent single independent voltage source has the *same* potential difference (or emf) as either of the sources in parallel.

Never connect two independent voltage sources the way shown in Figure 18.20, since this produces a logical impasse, with each source implying the other node is at the higher potential.

If ideal independent voltage sources with *different* potential differences (or emfs) are connected in parallel, we encounter another logical contradiction (see Figure 18.21). Each source implies a different potential difference between the same two terminals, and yet the potential difference across any two elements in parallel must be the same.

From a practical viewpoint, real batteries with identical emfs are connected in parallel to increase the charge flow (electric current) that the equivalent battery can provide.

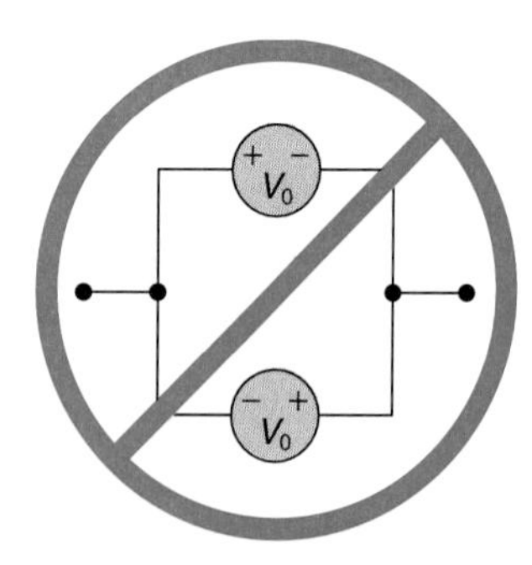

FIGURE 18.20 Never connect two independent voltage sources this way.

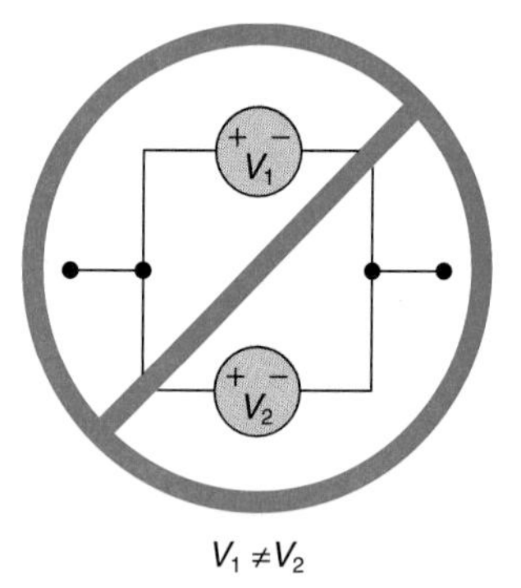

FIGURE 18.21 Never connect in parallel two independent voltage sources with different emfs.

Thus, when your car is jump started, a good real battery is connected in parallel with the dead real battery. The good battery effectively compensates for, or replaces, the dead battery by providing the charge flow needed by the starter motor of the car. Both the good and dead batteries should have the same emf; if this advice is not heeded with real batteries, they may explode. We will see why when we model a real battery in the next chapter.

Real independent voltage sources (such as these batteries) come in a wide variety of sizes and emfs.

EXAMPLE 18.2

Four batteries, modeled as independent voltage sources, each with a potential difference (or emf) of 1.50 V between their terminals, are connected in series for a flashlight, as shown in Figure 18.22.

FIGURE 18.22

a. What is the potential difference (or emf) of the equivalent battery?
b. If one of the batteries is accidentally connected backward (as in Figure 18.23), what is the potential difference (or emf) of the equivalent battery?

FIGURE 18.23

Solution

a. In Figure 18.22, the batteries are connected in series with the positive terminals connected to the corresponding negative terminals of the next battery in the series. Moving from terminal A to terminal B, the potential difference between A and B is the potential difference (or emf) of the equivalent battery V_{eq}:

$$\begin{aligned} V_{eq} &= 1.50\ \text{V} + 1.50\ \text{V} + 1.50\ \text{V} + 1.50\ \text{V} \\ &= 6.00\ \text{V} \end{aligned}$$

b. In Figure 18.23, the potential difference between terminal A and B is the equivalent potential difference (or emf) V_{eq}:

$$\begin{aligned} V_{eq} &= 1.50\ \text{V} + 1.50\ \text{V} + (-1.50\ \text{V}) + 1.50\ \text{V} \\ &= 3.00\ \text{V} \end{aligned}$$

This potential difference could be obtained with simply two of the batteries properly connected in series with their polarities correspondingly oriented, as shown in Figure 18.24.

FIGURE 18.24

18.6 CAPACITORS

A **capacitor** is a circuit element that stores charge and electrical potential energy. It (typically) consists of two isolated (separated) conductors with equal and opposite charges, as shown in Figure 18.25. The two conductors of a capacitor are known as **plates** (thin, flat, circular disks) even though they often do not possess that geometry.

The circuit symbol for an ideal capacitor is shown in Figure 18.26. The straight line of the capacitor symbol represents the conductor that is, or will be, at a higher electrical potential than the second conductor, represented by the slightly curved line.

When the two conductors have equal and opposite charges $\pm|Q|$, the capacitor is said to be **charged**, in which case polarity markings are used to indicate which conductor is at the higher electric potential. If both conductors of the capacitor have zero charge, the capacitor is said to be **uncharged**, in which case the polarity markings are superfluous.

Some capacitors, typically electrolytic capacitors, have their polarity markings inscribed on them. Others, typically ceramic capacitors, do not have polarity markings and can be used with either terminal at the higher potential.

Notice that, whether charged or uncharged, the total electric charge on a capacitor as a whole is zero. A charged capacitor keeps opposite charges of equal magnitudes separated in space. For an isolated, ideal, charged capacitor, the charge separation can last indefinitely.

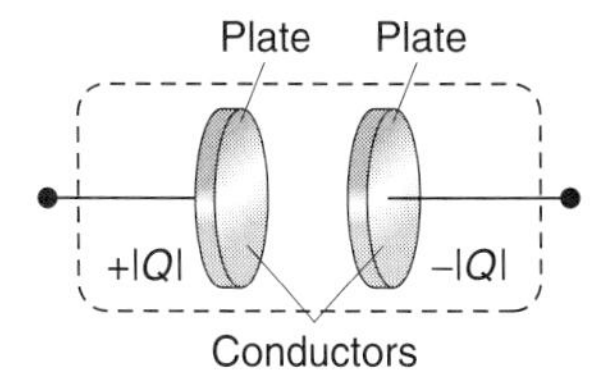

FIGURE 18.25 **A capacitor consists of two separated conductors.**

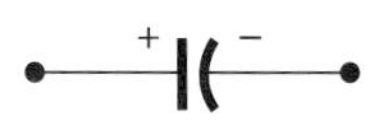

FIGURE 18.26 **The circuit symbol for a capacitor.**

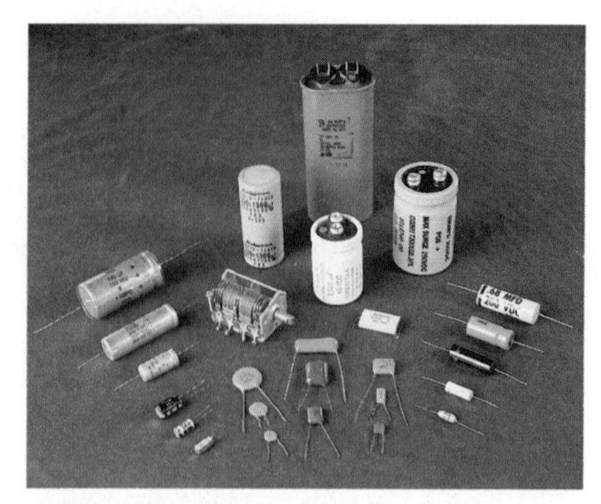

Some of these capacitors have polarity markings on them. When such capacitors are used, the (+) polarity should be at the higher electric potential.

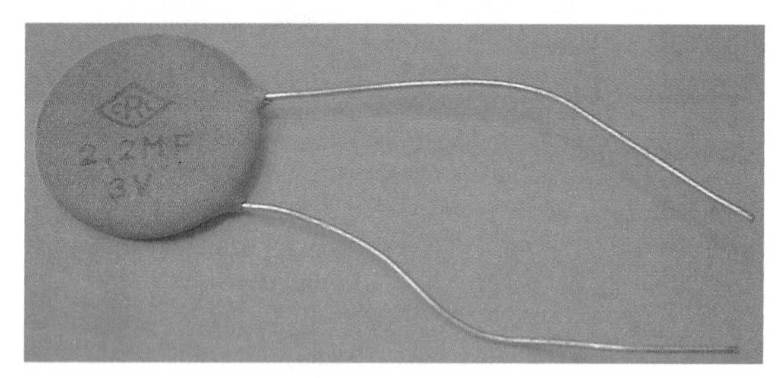

Some capacitors lack polarity markings, and either side may be at the higher electric potential when used.

There is a difference in electric potential between the conductors of a charged capacitor. Let the potential difference between the plate at the higher potential and the plate at the lower potential be V (≥ 0 V).

The **capacitance** C of a capacitor is defined to be the ratio of the absolute value of the charge on either conductor to the absolute value of the potential difference between them:

$$C \equiv \frac{|Q|}{|V|} \tag{18.1}$$

The capacitance always has a positive value.* The SI unit of capacitance is the coulomb/volt (C/V) and is renamed a **farad** (F).

The farad unit is named after Michael Faraday (1791–1867), an early 19th-century experimental genius who studied electrical and magnetic phenomena extensively. The farad (F) is a very large unit; most common capacitances are measured in microfarads (μF = 10^{-6} F) or picofarads (pF = 10^{-12} F), although it now is technologically possible to fabricate small-sized capacitors with capacitances exceeding a farad.

The capacitance of a capacitor is a measure of its capacity for holding (storing) charge. A large capacitance means the capacitor has the capacity to hold a significant amount of charge.

*The absolute value sign on the potential difference V is superfluous since we said V was the potential difference between the plate at the higher potential and the plate at the lower potential. We include the absolute value signs on V simply to emphasize that a positive value of the potential difference between the two plates is to be used to find the capacitance.

Ironically, despite the fact that Equation 18.1 defining the capacitance involves both Q and V, the value of the ratio is *independent* of either one! The potential difference between the conductors always is proportional to the charge on them (a manifestation of the principle of superposition), and so neither Q nor V independently affects C. The capacitance depends on the geometric arrangement of the conductors, the physical size and shape of the conductors, and the material medium separating them.

This is seen in the examples that follow.

QUESTION 5

Your roommate says that a charged capacitor carries a charge $|Q|$. You say the charged capacitor has zero charge. You both are right. Explain.

EXAMPLE 18.3

a. Find the capacitance of a large parallel plate capacitor whose plate areas each are A, separated by a distance d; see Figure 18.27. Similar capacitors frequently are used in electronic devices such as oscilloscopes, TVs, and other CRT displays.

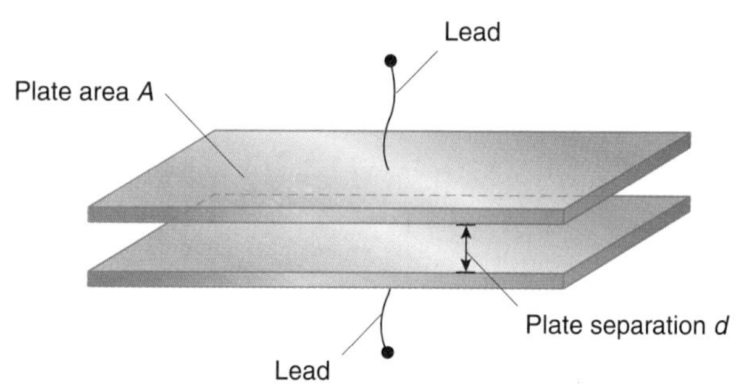

FIGURE 18.27

b. If the capacitor consists of two square plates, each 10.0 cm on an edge, separated by 0.500 cm, find the capacitance.

Solution

a. Imagine charge $+|Q|$ to be on one plate of the capacitor and charge $-|Q|$ on the other. Ignoring the small edge effects, the surface charge density σ on each plate has a magnitude $|Q|/A$. If the spacing d is much less than $\sqrt{A}$, the arrangement is tantamount to a set of parallel infinite plates. The electric field between two such plates is uniform and of magnitude $E = \sigma/\varepsilon_0$ (see Table 16.2). The potential difference between such an arrangement of plates has a magnitude $|V| = |Ed|$, from Example 17.1. The capacitance of the arrangement is, by definition (from Equation 18.1),

$$C = \frac{|Q|}{|V|}$$

Substituting $V = Ed$, you get

$$C = \frac{|Q|}{Ed}$$

Then, using $E = \sigma/\varepsilon_0$,

$$C = \frac{|Q|}{\frac{\sigma d}{\varepsilon_0}}$$

Finally, substituting for σ using $\sigma = |Q|/A$, you find

$$C = \frac{|Q|}{\frac{|Q|d}{A\varepsilon_0}}$$

$$= \varepsilon_0 \frac{A}{d} \qquad (1)$$

Equation (1) shows that the capacitance is a function of geometric specifications, such as the area of the plates and their separation. The dependence on the medium between the plates is represented through ε_0, the permittivity of free space, since the medium between the capacitor plates was assumed here to be a vacuum.

b. Use the given dimensions and substitute into equation (1), remembering to use SI units (m) rather than units of convenience (cm):

$$C = \frac{[8.85 \times 10^{-12}\ \text{C}^2/(\text{N}\cdot\text{m}^2)](0.100\ \text{m})(0.100\ \text{m})}{5.00 \times 10^{-3}\ \text{m}}$$

$$= 1.77 \times 10^{-11}\ \text{C/V}$$

$$= 17.7\ \text{pF}$$

The SI units $\text{C}^2/(\text{N}\cdot\text{m}) = \text{C}^2/\text{J} = \text{C/V}$ are equivalent to farads (F).

Example 18.4

A conducting sphere of radius R with a charge Q is a charged capacitor even though only one conductor is evident; see Figure 18.28. The second conductor is considered to be at infinity (with charge $-Q$). To a first approximation, the Earth itself can be considered to be such a capacitor (see Problem 18).

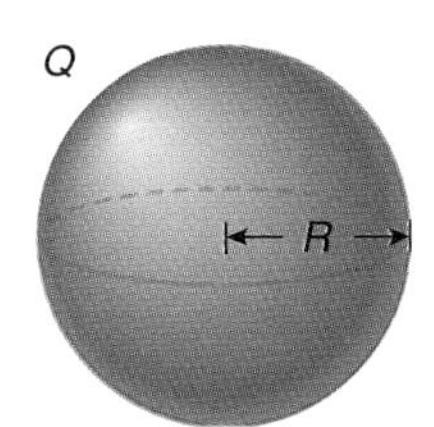

FIGURE 18.28

a. Find the capacitance of the sphere.
b. Evaluate the capacitance for a sphere of radius 10.0 cm.

Solution

a. The electric potential at infinity is 0 V. That on the surface of the conducting sphere is, from Table 17.1,

$$V = \frac{Q}{4\pi\varepsilon_0 R}$$

Thus the potential difference between the conductors has absolute value

$$|V| = \frac{|Q|}{4\pi\varepsilon_0 R}$$

Using Equation 18.1, you find the capacitance of the system is

$$C = \frac{|Q|}{|V|}$$

$$= \frac{|Q|}{\frac{|Q|}{4\pi\varepsilon_0 R}}$$

$$= 4\pi\varepsilon_0 R \qquad (1)$$

Once again you see that the capacitance depends on the geometric specifications of the capacitor (here the radius of the sphere) and the medium (a vacuum), the latter through the permittivity of free space ε_0.

b. For a sphere of radius 10.0 cm, $R = 0.100$ m, since you must use SI units. Making this substitution in equation (1) for the capacitance, you have

$$C = 4\pi\varepsilon_0 R$$

$$= [4\pi \times 8.85 \times 10^{-12}\ \text{C}^2/(\text{N}\cdot\text{m}^2)](0.100\ \text{m})$$

$$= 1.11 \times 10^{-11}\ \text{F}$$

$$= 11.1\ \text{pF}$$

From equation (1), note that the permittivity of free space, ε_0, has the equivalent SI units of farads per meter (F/m).

Example 18.5

A cylindrical capacitor consists of two concentric conducting cylinders of radii a and b, as shown in Figure 18.29. A Geiger counter, used frequently in nuclear physics to detect subatomic particles, typically has this geometry. Find the capacitance assuming the length ℓ of the cylinders is much greater than either radius.

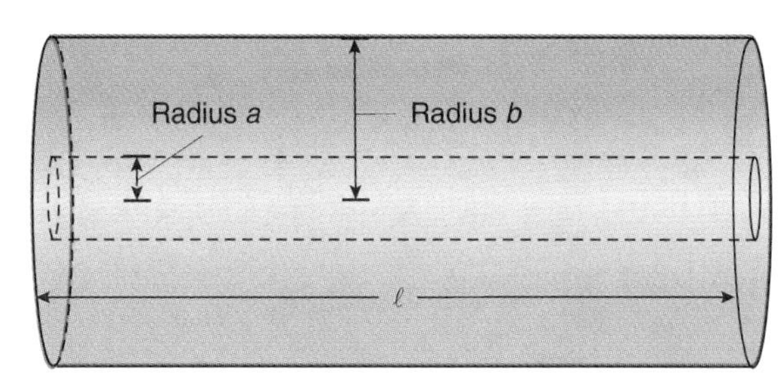

FIGURE 18.29

Solution

Model the system as a pair of infinite concentric cylinders. Imagine a charge of $+|Q|$ on the inner conductor and of $-|Q|$ on the outer conductor. The field between the two is approximately that of an infinitely long wire, and so you can use Equation 16.19 in vector form:

$$\vec{E} = \frac{1}{4\pi\varepsilon_0}\frac{2\lambda}{r}\hat{r}$$

where λ is the charge per unit length $\lambda = |Q|/\ell$.

Since the model charge distribution is not of finite extent, you must use Equation 17.9 to find the potential difference between the two cylinders:

$$\begin{aligned} V_{\text{outer}} - V_{\text{inner}} &= -\int_a^b \vec{\mathbf{E}} \cdot d\vec{\mathbf{r}} \\ &= -\frac{1}{4\pi\varepsilon_0}\int_a^b \frac{2\lambda}{r}\hat{\mathbf{r}} \cdot dr\,\hat{\mathbf{r}} \\ &= -\frac{2}{4\pi\varepsilon_0}\lambda \ln\left(\frac{b}{a}\right) \end{aligned}$$

The magnitude of the potential difference between the conductors is

$$\begin{aligned} |V| &= \frac{2}{4\pi\varepsilon_0}\lambda \ln\left(\frac{b}{a}\right) \\ &= \frac{2}{4\pi\varepsilon_0}\frac{|Q|}{\ell}\ln\left(\frac{b}{a}\right) \end{aligned}$$

The capacitance is found using the definition, Equation 18.1:

$$\begin{aligned} C &= \frac{|Q|}{|V|} \\ &= \frac{4\pi\varepsilon_0\ell}{2\ln\left(\dfrac{b}{a}\right)} \end{aligned}$$

18.7 Series and Parallel Combinations of Capacitors

It is occasionally useful to connect capacitors in parallel or series to secure larger or smaller capacitances.

Parallel Combinations of Capacitors

Consider a collection of, say, four capacitors in parallel, as shown in Figure 18.30. Since the leads of circuit elements in parallel are connected to the same two nodes, the potential difference is the same across each capacitor. Thus, from the definition of the capacitance (Equation 18.1), we can write the following for each capacitor:

$$C_1 = \frac{|Q_1|}{|V|} \qquad C_2 = \frac{|Q_2|}{|V|} \qquad C_3 = \frac{|Q_3|}{|V|} \qquad C_4 = \frac{|Q_4|}{|V|} \tag{18.2}$$

To simplify the arrangement, we want to find a single equivalent capacitor of capacitance C_{eq} that can be placed between the same two terminals (A and B) and maintain the same potential difference between them as the four given capacitors. That is, we want to replace the four capacitors with a single capacitor C_{eq} such that

$$C_{\text{eq}} = \frac{|Q|}{|V|}$$

where $|V|$ is the same potential difference that is across each of the individual capacitors; see Figure 18.31.

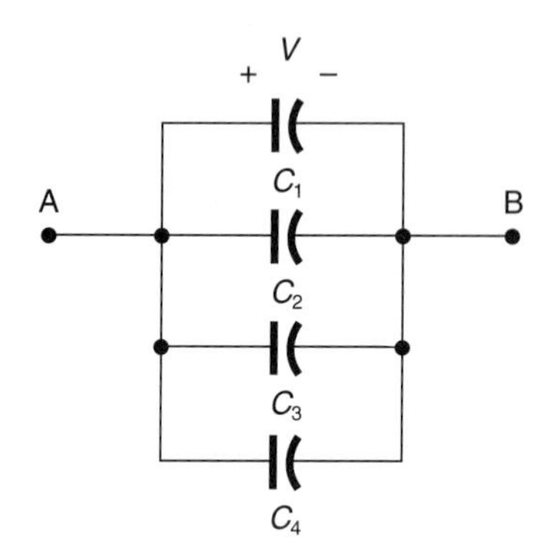

FIGURE 18.30 Four capacitors connected in parallel.

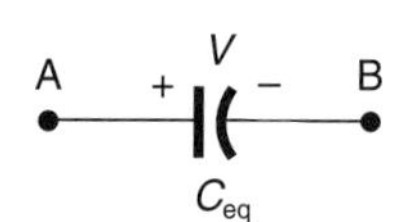

FIGURE 18.31 The equivalent capacitance.

Since all the conducting plates on one side of the capacitors in Figure 18.30 are connected by conducting wires, the total charge on one side of the equivalent capacitor must be the sum of the charges on the individual capacitor plates; the same is true for the other side of the capacitors. That is, the magnitude of the charge $|Q|$ on each plate of the equivalent capacitor must be

$$|Q| = |Q_1| + |Q_2| + |Q_3| + |Q_4|$$

Substituting for the charges using Equation 18.2, we obtain

$$C_{\text{eq}}|V| = C_1|V| + C_2|V| + C_3|V| + C_4|V|$$

and so

$$C_{\text{eq}} = C_1 + C_2 + C_3 + C_4 \qquad \text{(capacitors in parallel)} \tag{18.3}$$

The equivalent capacitance is the simple sum of the individual capacitances.

Since the capacitance always is positive, we have no need to worry about signs when using Equation 18.3. The form of Equation 18.3 applies to any number of capacitors connected in parallel; we simply add the capacitances:

$$C_{\text{eq}} = C_1 + C_2 + C_3 + \cdots \qquad \text{(capacitors in parallel)} \tag{18.4}$$

Series Connection of Capacitors

Four capacitors in series are shown in Figure 18.32. Let their capacitances be C_1, C_2, C_3, and C_4. Let V_1 be the potential difference between the points A and D in Figure 18.32; this is the potential

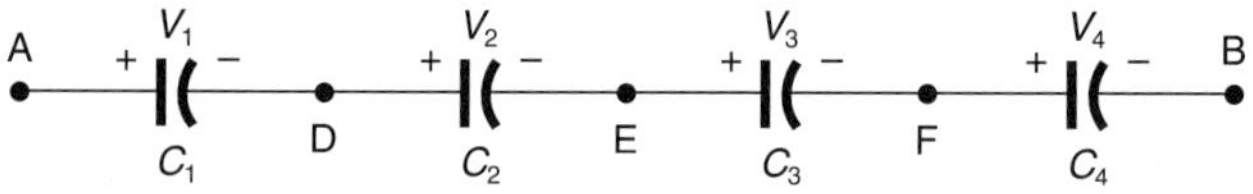

FIGURE 18.32 Four capacitors in series.

difference across capacitor C_1. Likewise, let V_2 be the potential difference between points D and E in Figure 18.32; and so on for the other two capacitors. The total potential difference V between the points A and B is the sum of the potential differences across each capacitor:

$$V = V_1 + V_2 + V_3 + V_4 \qquad (18.5)$$

Once again we want to find a single equivalent capacitor C_{eq} that can be placed between points A and B to maintain the same potential difference V between the two points; see Figure 18.33.

A charge $+|Q|$ placed on the left plate of capacitor C_1 in Figure 18.32 causes charge separation to take place on the isolated and neutral extended conductor that comprises the right-side plate of C_1 and the left-side plate of capacitor C_2. The charge $+|Q|$ on the left side of C_1 attracts a charge $-|Q|$ to the right side of C_1 and repels $+|Q|$ to the left side of C_2. This charge separation ripples through the series of capacitors. In this way we see that each capacitor in the series collection has the *same* magnitude charge on each plate.

The left and right plates of the equivalent capacitor have the same charge as the left plate of C_1 and the right plate of C_4 because they are connected to the points A and B, respectively, in Figures 18.32 and 18.33. Thus the equivalent capacitor and the four individual capacitors have the same magnitude of charge $|Q|$ on each of their plates.

From the definition of the capacitance, Equation 18.1, we have

$$C_{eq} = \frac{|Q|}{|V|}$$

and

$$C_1 = \frac{|Q|}{|V_1|} \qquad C_2 = \frac{|Q|}{|V_2|} \qquad C_3 = \frac{|Q|}{|V_3|} \qquad C_4 = \frac{|Q|}{|V_4|}$$

Using these equations, substitute into Equation 18.5; we find

$$\frac{|Q|}{|C_{eq}|} = \frac{|Q|}{C_1} + \frac{|Q|}{C_2} + \frac{|Q|}{C_3} + \frac{|Q|}{C_4}$$

Dividing by $|Q|$, we have

$$\frac{1}{C_{eq}} = \frac{1}{C_1} + \frac{1}{C_2} + \frac{1}{C_3} + \frac{1}{C_4} \qquad \text{(capacitors in series)} \qquad (18.6)$$

The reciprocal of the equivalent capacitance is the sum of the reciprocals of the individual capacitances.

This rule applies to any number of capacitors connected in series:

$$\frac{1}{C_{eq}} = \frac{1}{C_1} + \frac{1}{C_2} + \frac{1}{C_3} + \cdots \qquad \text{(capacitors in series)} \qquad (18.7)$$

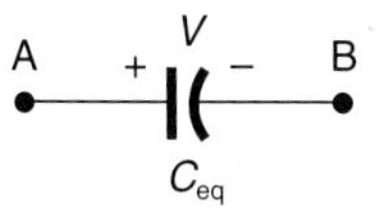

FIGURE 18.33 The equivalent capacitance.

One implication of Equation 18.7 for capacitors in series is that the equivalent capacitance always is less than the smallest capacitor in the series combination.

PROBLEM-SOLVING TACTICS

18.1 When using Equation 18.7, be sure you remember how to add fractions. In particular, for two capacitors in series, note that

$$\frac{1}{C_1} + \frac{1}{C_2} \neq \frac{1}{C_1 + C_2}$$

In other words, don't do crazy things like saying

$$\frac{1}{2} + \frac{1}{3} = \frac{1}{5} \qquad \text{(no!)}$$

This may seem like a trivial point and an insult to your intelligence, but you would be surprised at what is done under the pressure of an examination!

18.2 Be sure to realize that after summing the reciprocals of the individual capacitances in Equation 18.7 that the equivalent capacitance is the reciprocal of the sum. You might correctly add the reciprocals of the individual capacitances but then forget to use the $1/x$ key on your calculator to find the final result for C_{eq}.

18.3 The equivalent capacitance of two capacitors in series is the product of their capacitances divided by the sum. This convenient rule follows from Equation 18.7 applied to two capacitors. The equivalent capacitance is (using Equation 18.7)

$$\frac{1}{C_{eq}} = \frac{1}{C_1} + \frac{1}{C_2}$$

Put the right-hand side over a common denominator and simplify:

$$\frac{1}{C_{eq}} = \frac{C_2 + C_1}{C_1 C_2}$$

or

$$C_{eq} = \frac{C_1 C_2}{C_1 + C_2} \qquad \text{(only two capacitors in series)} \qquad (18.8)$$

This is an easier way to calculate the equivalent capacitance because the product divided by the sum usually can be performed mentally. But be aware that this rule (Equation 18.8) is appropriate only for finding the equivalent of *two* capacitors at a time in series. For three capacitors in series, you can either revert to Equation 18.7 or combine them pairwise (see Example 18.6).

QUESTION 6

Three capacitors are to be connected together to produce various equivalent capacitances for a medical defibrillation device. (a) How should they be connected together to produce as large an equivalent capacitance as possible? (b) How should the three be connected to produce the minimum equivalent capacitance?

STRATEGIC EXAMPLE 18.6

You find three capacitors in your electronics shop inventory, 3.0 μF, 6.0 μF, and 8.0 μF, but they all are too large for your purposes. You combine them in series to make a smaller capacitance. Find the equivalent capacitance of the three capacitors in series (see Figure 18.34).

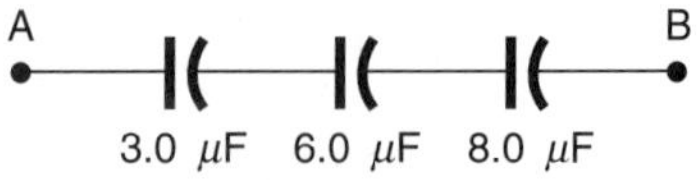

A B

3.0 μF 6.0 μF 8.0 μF

FIGURE 18.34

Solution

Method 1

You can use the shortcut mentioned in Problem-Solving Tactic 18.3 to combine the two capacitors in series on the left side of Figure 18.34. They have an equivalent capacitance given by the product of their capacitances divided by their sum:

$$\frac{(3.0\ \mu\text{F})(6.0\ \mu\text{F})}{3.0\ \mu\text{F} + 6.0\ \mu\text{F}} = 2.0\ \mu\text{F}$$

Note that the equivalent capacitance of the two in series is less than the smallest one that was combined. You now have a collection of two capacitors in series, as shown in Figure 18.35.

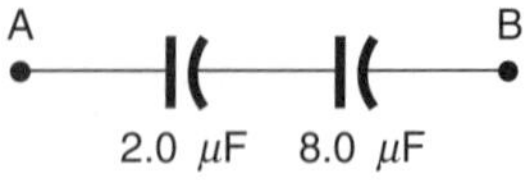

A B

2.0 μF 8.0 μF

FIGURE 18.35

This pair can then be combined using the same rule (product divided by the sum):

$$\frac{(2.0\ \mu\text{F})(8.0\ \mu\text{F})}{2.0\ \mu\text{F} + 8.0\ \mu\text{F}} = 1.6\ \mu\text{F}$$

You also first could combine the 6.0 μF and 8.0 μF capacitors as a pair, to get 3.4 μF. Then combining the resulting 3.4 μF capacitor in series with the remaining 3.0 μF capacitor, you get 1.6 μF. In other words, for capacitors in series, you can combine them by pairs beginning with *any two* of them; they need not be taken in the order in which they appear in the series connection.

Method 2

Use Equation 18.7 for the original arrangement of three capacitors in series (Figure 18.34):

$$\begin{aligned}\frac{1}{C_{eq}} &= \frac{1}{C_1} + \frac{1}{C_2} + \frac{1}{C_3} \\ &= \frac{1}{3.0\ \mu\text{F}} + \frac{1}{6.0\ \mu\text{F}} + \frac{1}{8.0\ \mu\text{F}} \\ &= \frac{15}{24\ \mu\text{F}}\end{aligned}$$

Don't forget to take the reciprocal (Problem-Solving Tactic 18.2):

$$\begin{aligned}C_{eq} &= \frac{24}{15}\ \mu\text{F} \\ &= 1.6\ \mu\text{F}\end{aligned}$$

This is the same result as that obtained by Method 1.

EXAMPLE 18.7

You need a 250 μF capacitor for manufacturing each of a collection of laser power supplies, but no 250 μF capacitors are in stock or available except by special order at significant cost. An electronics supply firm is having a special on 50 μF capacitors and you realize it will be cheaper to use several of them rather than a single 250 μF capacitor even after accounting for the additional time for assembly. How many will you need for each power supply and how should the capacitors be connected to secure the desired 250 μF capacitance? Your supervisor is well pleased with your cost-consciousness.

Solution

Connect the capacitors in parallel because then the equivalent capacitance is the simple sum of the individual capacitances. You will need five of the 50 μF capacitors:

$$\begin{aligned}C_{eq} &= 50\ \mu\text{F} + 50\ \mu\text{F} + 50\ \mu\text{F} + 50\ \mu\text{F} + 50\ \mu\text{F} \\ &= 250\ \mu\text{F}\end{aligned}$$

EXAMPLE 18.8

One part of a complicated circuit diagram has the combination of capacitors indicated in Figure 18.36. Find the equivalent capacitance of the three capacitors.

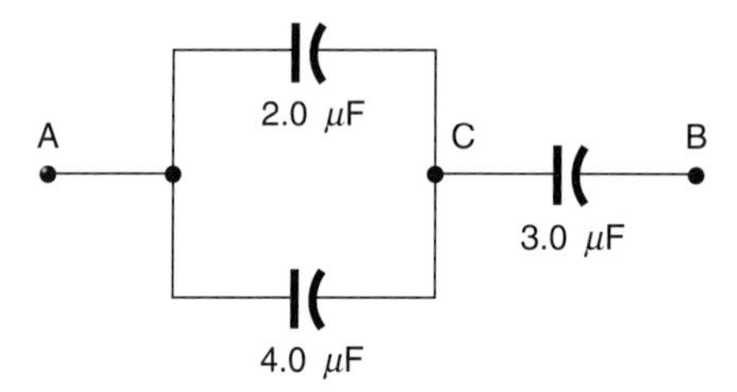

FIGURE 18.36

Solution

The 2.0 μF and 4.0 μF capacitors are in parallel and thus have an equivalent capacitance equal to their sum:

$$2.0\ \mu\text{F} + 4.0\ \mu\text{F} = 6.0\ \mu\text{F}$$

The 6.0 μF equivalent is in series with the 3.0 μF capacitor, as shown in Figure 18.37. The final equivalent capacitance of this pair in series is their product divided by their sum:

$$\begin{aligned}C_{eq} &= \frac{(6.0\ \mu\text{F})(3.0\ \mu\text{F})}{6.0\ \mu\text{F} + 3.0\ \mu\text{F}} \\ &= 2.0\ \mu\text{F}\end{aligned}$$

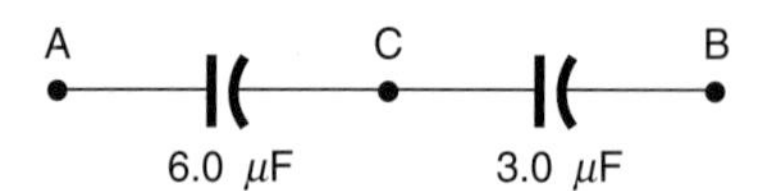

FIGURE 18.37

18.8 Energy Stored in a Capacitor

One widespread use of capacitors is to store electrical potential energy for sudden use in devices such as flash attachments to cameras, laser power supplies, or defibrillators. Here we see how the energy storage arises.

Capacitors do not charge themselves. To take a capacitor from an uncharged to a charged state, work must be done to accomplish the charge separation. We want to figure out how much work must be done to charge a capacitor.

Represent a capacitor schematically as shown in Figure 18.38. Since the location of the zero of electric potential (and potential energy) is arbitrary (it is only changes that are physically significant), we choose plate B to be at zero potential and let plate A have a potential V. In this way V represents the potential difference between plates A and B. If we assume $V > 0$ V, the electric field is directed from plate A to plate B because the electric field always points toward regions of lower electric potential.

Plate A has a positive charge and plate B has a negative charge. Imagine slowly carrying a small, differential positive charge dq from plate B to plate A. The work done by the electrical forces on dq will be independent of the path, since the force is conservative. Obviously, though, we are going to have to do work on the charge to get it over to plate A, or something else is going to have to do the work for us. The charge dq will not go of its own accord since the electrical force on it is directed toward plate B, not toward plate A.

Imagine pulling the charge over to plate A so the kinetic energy of the charge is unchanged. The force we have to exert on the charge then is equal in magnitude to the electrical force on dq but, of course, must point in the opposite direction.* The familiar CWE theorem states that

$$W_{\text{total}} = \Delta\text{KE}$$

*You might ask yourself if the equality of the force magnitudes stems from Newton's third law or from the second law.

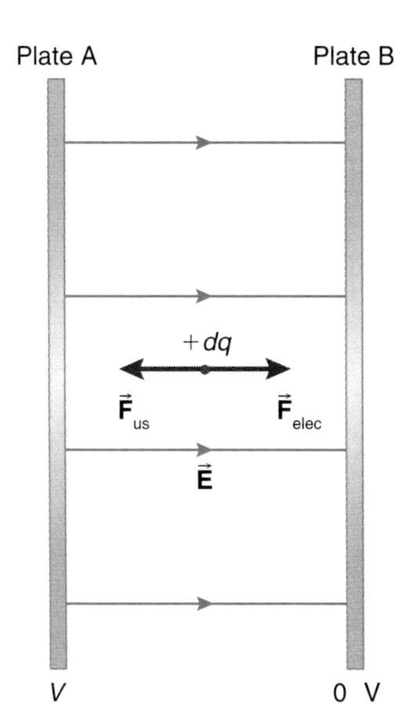

FIGURE 18.38 A schematic representation of a capacitor.

or for differential changes,

$$dW_{\text{total}} = d(\text{KE})$$

There is no change in the kinetic energy because we move the charge slowly at constant speed. We do differential work dW_{us} and the electrical force on dq does differential work dW_{elec} on the charge. Thus the CWE theorem has the form

$$dW_{\text{us}} + dW_{\text{elec}} = 0 \text{ J} \qquad (18.9)$$

But the work done by the electrical force is the negative of the change in the electrical potential energy of the charge. We take differentials of Equation 17.3, obtaining

$$dW_{\text{elec}} = -d(\text{PE})$$

Thus Equation 18.9 becomes

$$dW_{\text{us}} - d(\text{PE}) = 0 \text{ J}$$

or

$$dW_{\text{us}} = d(\text{PE})$$

The differential work we do is equal to the differential change in the potential energy of the charge. But the potential energy of any charge is the product of the charge times the electric potential at the point in question (Equation 17.8). The initial potential energy of dq at plate B is

$$dq(0 \text{ V}) = 0 \text{ J}$$

and the final potential energy of dq at plate A is

$$dq\, V$$

So the change in the potential energy of the charge (final minus initial values, as usual) is

$$\begin{aligned} d(\text{PE}) &= V\,dq - 0 \text{ J} \\ &= V\,dq \qquad (18.10) \end{aligned}$$

From the definition of the capacitance, Equation 18.1, the potential difference across the capacitor is related to the charge on the plates by

$$C = \frac{|q|}{|V|}$$

Since both q and V are positive quantities here, we can dispense with the absolute value signs. Solving for V, we get

$$V = \frac{q}{C}$$

Thus Equation 18.10 becomes

$$d(\text{PE}) = \frac{q}{C}\,dq$$

Now we integrate this expression from an uncharged state to a final charge Q on the capacitor. The capacitance is a constant for a given capacitor and depends neither on q nor on V, so that we have

$$\int_{0\text{ J}}^{\text{PE}} d(\text{PE}) = \int_{0\text{ C}}^{Q} \frac{q}{C}\,dq$$

Take the initial electrical potential energy of the uncharged capacitor to be 0 J. After integration, the accumulated potential energy of the charged capacitor is

$$\text{PE} = \frac{1}{2}\frac{Q^2}{C} \qquad (18.11)$$

Hence in charging a capacitor the electrical potential energy of the capacitor system increases.

With the definition of the capacitance, Equation 18.11 can be written in a number of equivalent ways:

$$\text{PE} = \frac{1}{2}\frac{Q^2}{C} = \frac{1}{2}|Q||V| = \frac{1}{2}CV^2 \qquad (18.12)$$

Only one version of Equation 18.12 need be remembered, since the others can be found from it using the definition of the capacitance.

Consider the special case of a parallel plate capacitor. The capacitance is (from Example 18.3)

$$C = \varepsilon_0 \frac{A}{d}$$

where A is the area of the plates and d is their separation. The magnitude of the potential difference between the plates is related to the magnitude of the electric field between the plates via

$$V = Ed$$

Substituting into the expression for the electrical potential energy stored in the charged capacitor, Equation 18.12, we obtain

$$\begin{aligned}\text{PE} &= \frac{1}{2}CV^2 \\ &= \frac{1}{2}\varepsilon_0 \frac{A}{d}(Ed)^2 \\ &= \frac{1}{2}\varepsilon_0 E^2 Ad\end{aligned}$$

Where is the potential energy stored? We consider it to be distributed throughout the volume of the space wherever there is some electrical field. The product Ad is the volume of the capacitor. If we divide the potential energy by the volume of the capacitor, we have a potential energy per unit volume or an **energy density**. The energy density is then

$$\frac{1}{2}\varepsilon_0 E^2$$

The electrical potential energy density stored in the capacitor is considered to be stored in the electric field.

This result turns out to be quite general:

Any electric field $\vec{\mathbf{E}}$ is said to have associated with it a potential energy density given by

$$\frac{1}{2}\varepsilon_0 E^2 \qquad (18.13)$$

The energy stored in the field of a charged capacitor is quite apparent if the capacitor is discharged suddenly by connecting a wire from one plate (terminal) to the other, thus shorting the capacitor.* The wire provides a path for the charges to decrease their electrical potential energy. Indeed, large capacitors used in common electronic instruments and devices are quite dangerous; if your fingers accidentally become the "wire" that shorts the terminals of a charged capacitor, you can receive quite an electrical shock, even a lethal one if sufficient energy is stored in the charged capacitor. A good capacitor can hold its charge for hours after a piece of equipment is shut off, and so the danger of electrocution is not diminished just because the device is turned off. In brief, *extreme caution* is needed to ensure that all large capacitors are first discharged by connecting a wire to bridge the two terminals (shorting the terminals momentarily) before servicing any piece of electronic equipment.

EXAMPLE 18.9

One of several 100 μF capacitors in a defibrillator has a potential difference of 220 V between its plates.

a. What is the magnitude of the charge on each plate of the capacitor?
b. What is the energy stored in the capacitor?
c. If this energy were fully utilized to lift a 100 g doughnut with no change in its kinetic energy, how high could the doughnut be raised?

Solution

a. From the definition of the capacitance (Equation 18.1), you have

$$C = \frac{|Q|}{|V|}$$

and so

$$\begin{aligned}|Q| &= C|V| \\ &= (100 \times 10^{-6}\ \text{F})(220\ \text{V}) \\ &= 2.20 \times 10^{-2}\ \text{C}\end{aligned}$$

b. The energy stored is found from Equation 18.12:

$$\begin{aligned}\text{PE} &= \frac{1}{2}CV^2 \\ &= \frac{1}{2}(100 \times 10^{-6}\ \text{F})(220\ \text{V})^2 \\ &= 2.42\ \text{J}\end{aligned}$$

c. To raise a mass through a vertical distance Δy takes work $mg\,\Delta y$, where $\hat{\jmath}$ is vertically up. Hence

$$mg\,\Delta y = 2.42\ \text{J}$$

Substituting values, you get

$$(0.100\ \text{kg})(9.81\ \text{m/s}^2)\,\Delta y = 2.42\ \text{J}$$

which gives

$$\Delta y = 2.47\ \text{m}$$

EXAMPLE 18.10

An isolated, charged, parallel plate capacitor of plate area A and plate separation d has opposite charges of magnitude

*We investigate this discharging phenomenon in Section 19.16.

$|Q|$ on its plates. Now increase the separation of the plates to $2d$.

a. By what factor does the electrical potential energy stored in the capacitor change?
b. Where does the increase in potential energy come from, or the decrease go to?

Solution

a. The capacitance of a parallel plate capacitor is, from Example 18.3,

$$C_{\text{parallel plate}} = \varepsilon_0 \frac{A}{d}$$

When the separation between the plates is increased to $2d$, the capacitance falls to half its previous value.

The energy stored in the capacitor is found from Equation 18.12:

$$PE = \frac{1}{2}\frac{Q^2}{C}$$

When the separation between the plates of the isolated capacitor is increased, the charge on the plates is unaffected since the capacitor is isolated. Thus, with $|Q|$ fixed and the capacitance decreasing to half its former value, the potential energy stored increases by a factor of two.

b. The plates of the capacitor have equal and opposite charges on them. They therefore attract each other under the influence of the electrical force. Thus, in order to increase the plate separation, *you* are going to have to pull the plates apart. The work you do to separate the plates is the source of the increase in the electrical potential energy of the system.

The problem is analogous to increasing the gravitational potential energy of a mass. To increase the gravitational potential energy of your physics book, you have to lift the book to a higher elevation. The work you do is equal to the increase in the gravitational potential energy of the text, provided the work is done so that the kinetic energy of the text does not change.

18.9 Electrostatics in Insulating Material Media*

In our study of electrostatics we have assumed that the charges were located in a vacuum. Indeed, the constant ε_0 that appears in so many of our equations is known as the permittivity of free space; the term free space is the electromagnetic buzzword for a vacuum. The equations for Coulomb's law for point charges, the electric field and electric potential of various charge distributions, Gauss's law, and the explicit expressions we calculated for the capacitance early in this chapter—all involve the constant ε_0, the permittivity of free space.

What happens to all these equations if the electrostatic charges are located not in a vacuum but in an insulating material medium?

Empirically, in many insulating materials we find that all the changes can be accounted for by replacing the permittivity of free space ε_0 with a multiple of it†:

$$\varepsilon = \kappa\varepsilon_0 \tag{18.14}$$

where ε is the **permittivity of the material**.

So all the equations look identical except that ε appears instead of ε_0. The pure number κ (not necessarily an integer) is called the **dielectric constant** of the material. Recall from Chapter 16 that a dielectric‡ is the technical term for a material that is commonly called an insulator. The dielectric constant of a vacuum is exactly 1. Table 18.2 gives the value of the dielectric constant for various materials. The dielectric constant for air (under standard conditions) is very close to 1, indicating that the fundamental electrical equations in air and in a vacuum do not differ appreciably.

The specific value of the dielectric constant κ depends not only on the material but also on a host of environmental factors such as temperature and pressure, and in electrodynamic situations other factors such as the frequency of the oscillating fields. In this respect the dielectric constant is analogous to the specific heat in thermodynamics: changes in the dielectric constant are macroscopic manifestations of submicroscopic (atomic or molecular) changes in the material. Thus studies of the dielectric constants of materials and their variation with temperature, pressure, and other parameters are subjects of ongoing interest to physicists, engineers, chemists, and biologists.

† In anisotropic materials such as certain crystals, the relationship is more complicated; we will not consider such materials here.

‡ The term dialectic is a philosophical term, unrelated to the term dielectric.

TABLE 18.2 Approximate Values of the Dielectric Constant and Dielectric Strength of Various Materials

Material	*Dielectric constant κ (dimensionless) at ≈20 °C*	*Dielectric strength (N/C = V/m)*
Air	1.0006	3×10^6
Aluminum oxide	8.5	670×10^6
Bakelite	4.9	24×10^6
Barium titanate ($BaTiO_3$)	~ 10^4	
Epoxy resin	~ 4	
Ethyl alcohol	26	
Lucite	~ 3	
Mica	~ 5	~ $10–50 \times 10^6$
Paper (kraft)	3.7	16×10^6
Paraffin	~ 2.5	
Polyethylene	~ 2.3	50×10^6
Pyrex glass	5.6	14×10^6
Quartz	~ 3.8	8×10^6
Rubber	~ 3	
Silicone oil	2.5	15×10^6
Tantalum oxide	26	500×10^6
Teflon	2.1	60×10^6
Transformer oil	~ 5	~ 10×10^6
Vacuum	1 (exact)	∞
Water	80	

18.10 CAPACITORS AND DIELECTRICS*

Most capacitors in electronic applications have dielectric materials between the two conductors. In this section we see that the effect of completely filling the space between the plates of a capacitor with an insulating material of dielectric constant κ is to increase the capacitance by the factor κ over that with a vacuum between the plates. We saw previously that the capacitance depends on the geometry of the capacitor (in Examples 18.3–18.5) and the material between the conductors; we explore the latter topic in this section. From a practical viewpoint, the insulating material also conveniently provides some structural rigidity to the capacitor.

How does the increase in the capacitance come about with a dielectric between the plates? Insulating materials are classified into two broad families depending on the nature of the atoms or molecules of the material: polar and nonpolar materials.

Polar Materials

Polar materials have molecules with a permanent electric dipole moment. Water is such a molecule. The center of charge (analogous to the center of mass) of the positive charges in the molecule is at a different location than the center of charge for the negative charges in the molecule. This separation of the centers of the two types of charges means that the molecule acts like a small electric dipole (see Figure 16.63).

Recall that the potential energy of a dipole in an electric field is $PE = -\vec{p} \cdot \vec{E}$. When such molecules are placed in an electric field, they tend to orient themselves with the dipole moment parallel to the field in order to lower their electrical potential energy. The alignment is not complete or perfect because of competing thermal motions of the molecules, but increasing the electric field increases the degree of alignment of the molecules with the field.

Nonpolar Atoms or Molecules

Nonpolar materials have no permanent electric dipole moment; the centers of charge for the positive and negative charges in the molecule normally coincide. However, if the material is placed in an electric field, the (much less massive) negatively charged electrons and their center of charge are displaced slightly with respect to the center of the positive charge distribution (formed from the much more massive, less mobile, positively charged nuclei). The result is an induced electric dipole moment in the molecules of the material.

Now we perform two different experiments.

Experiment 1: Adding a Dielectric to a Capacitor

We begin with an isolated and charged parallel plate capacitor with air or a vacuum between its plates, as shown in Figure 18.39.

Let $+Q_0$ be the charge on one plate, $-Q_0$ the charge on the other plate (we consider Q_0 itself to be positive). The surface charge density on the plates then is of magnitude $\sigma = Q_0/A$. The surface charge density on the plates is the **free surface charge density** because the charges on the conducting plates are free to move.

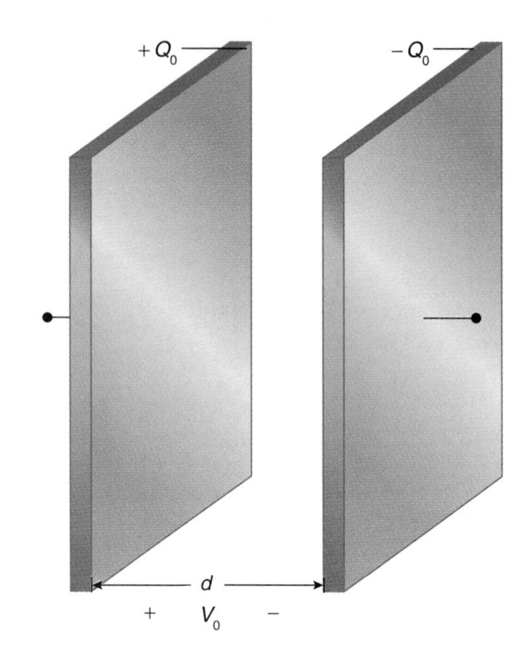

FIGURE 18.39 An isolated, charged parallel plate capacitor with air or a vacuum between its plates.

Let V_0 be the potential difference between the plates, considered so $V_0 > 0$ V. Since the capacitor is isolated, the charge on each plate is fixed. The capacitance of the system is

$$C_0 = \varepsilon_0 \frac{A}{d}$$

as we found before (in Example 18.3). The magnitude E_0 of the electric field between the plates is related to the potential difference by

$$V_0 = E_0 d$$

as we have used on a number of occasions.

Now insert a dielectric slab (with dielectric constant κ), completely filling the space between the plates of the capacitor, as shown in Figure 18.40.

The electric field orients the dipoles of a polar dielectric or induces dipoles in the nonpolar dielectric, resulting in the situation depicted schematically in Figure 18.41. On the surfaces of the dielectric facing the plates of the capacitor lie some unbalanced charges: negative charges on the dielectric surface facing the capacitor plate with charge $+Q_0$, and positive charges on the dielectric surface facing the plate with charge $-Q_0$. These unbalanced charges induced on the surfaces of the dielectric are not free to move around, since they are tied to the molecules (if the charges were free to move, we would have a conductor, not an insulator). These charges on the surface of the dielectric are called **bound charges**; there are **bound surface charge densities** $\pm\sigma_{\text{bound}}$ on the surfaces of the dielectric facing the capacitor plates.

The bound surface charges produce an electric field of their own in the direction opposite to the original field of the charges on the conducting plates of the capacitor. The electric field of the bound surface charges is of magnitude $\sigma_{\text{bound}}/\varepsilon_0$ while that due to the free charges on the capacitor plates is of magnitude σ/ε_0.

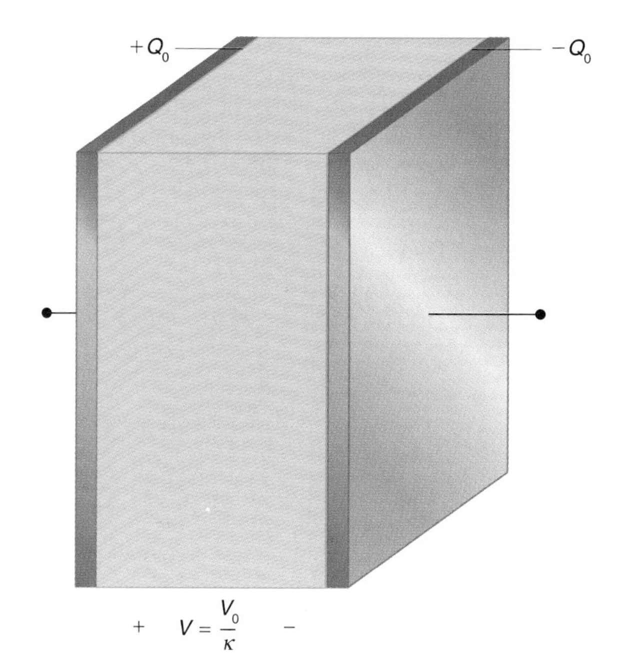

FIGURE 18.40 A slab of dielectric constant κ completely fills the space between the plates of the capacitor.

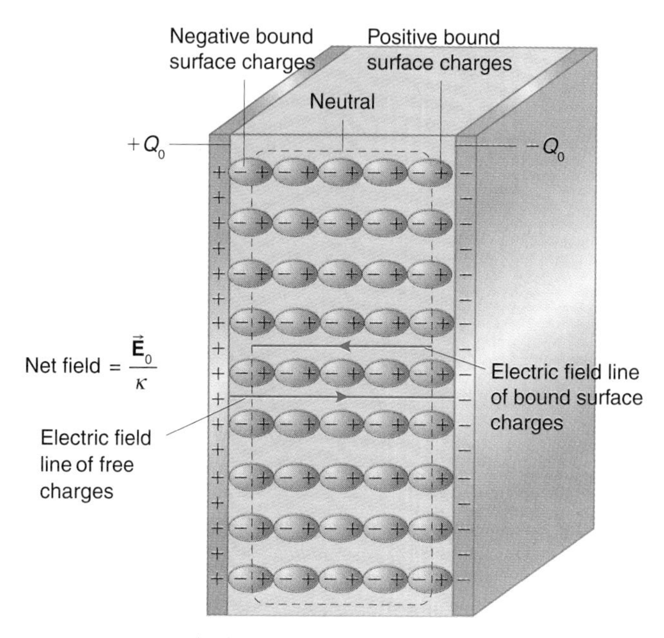

FIGURE 18.41 The free and bound charges each produce an electric field within the dielectric.

Experimentally we find that the potential difference between the capacitor plates is decreased by the factor κ by the presence of the dielectric material:

$$V = \frac{V_0}{\kappa}$$

Since the plate separation is unchanged, the field between the plates must decrease by the same factor κ since $V_0 = E_0 d$:

$$V = \frac{E_0}{\kappa} d \qquad (18.15)$$

The total electric field between the plates of the capacitor is evidently reduced by the factor of the dielectric constant:

$$E = \frac{E_0}{\kappa}$$

The charge Q_0 on the conducting plates of the capacitor is unaffected by all this because the capacitor was isolated. The capacitance of the arrangement is

$$C = \frac{Q_0}{V}$$

Substituting for V using Equation 18.15, we find

$$\begin{aligned} C &= \frac{Q_0}{\frac{E_0 d}{\kappa}} \\ &= \kappa \frac{Q_0}{E_0 d} \\ &= \kappa \frac{Q_0}{V_0} \\ &= \kappa C_0 \end{aligned}$$

This is a remarkable result: the capacitance increases by the factor of the dielectric constant κ.

We also can find the relationship between the free and bound surface charge densities to see which one is greater. The total electric field between the plates is of magnitude E_0/κ, and this is the difference between the magnitude of the field of the free charges and that of the bound charges because these fields are in opposite directions:

$$E_0 - E_{\text{bound}} = \frac{E_0}{\kappa}$$

Writing the field magnitudes in terms of the surface charge densities, we have

$$\frac{\sigma}{\varepsilon_0} - \frac{\sigma_{\text{bound}}}{\varepsilon_0} = \frac{\sigma/\varepsilon_0}{\kappa}$$

Solving for σ_{bound}, we find

$$\sigma_{\text{bound}} = \frac{\kappa - 1}{\kappa} \sigma \qquad (18.16)$$

The bound surface charge density always is less than the free surface charge density on the plates of the capacitor. If $\kappa = 1$, a vacuum, we have no dielectric and zero bound surface charge.

Experiment 2: Adding an Independent Voltage Source

Now we take a parallel plate capacitor with a vacuum or air between its plates and connect it to an independent voltage source (a source of emf), as shown in Figure 18.42.

The independent voltage source maintains a constant potential difference V_0 between its terminals. Since the capacitor is connected to the same two terminals, the two are in parallel

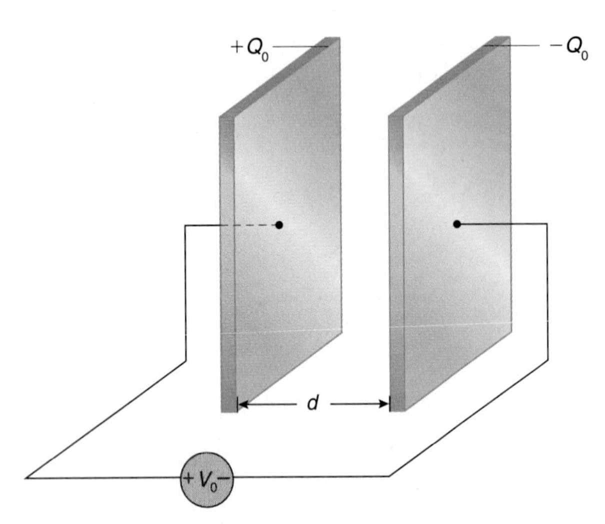

FIGURE 18.42 A parallel plate capacitor with a vacuum or air between its plates connected to an independent voltage source.

and the potential difference across the capacitor also must be V_0. Let $\pm Q_0$ be the charges on the capacitor plates. Thus the capacitance is

$$C_0 = \frac{|Q_0|}{|V_0|}$$

If a dielectric now is inserted into the space between the plates of the capacitor (see Figure 18.43), completely filling them as in Experiment 1, the independent voltage source maintains the same potential difference V_0 between the plates as before. Thus the magnitude of total electric field between the plates must remain equal to E_0, since $V_0 = E_0 d$.

Once again, the dielectric becomes polarized in the field and produces bound surface charge on the surfaces facing the capacitor plates. The bound surface charge produces an electric field in the direction opposite to the original field between the plates. To maintain the same total field as before, additional charge must be supplied to the conducting plates of the capacitor by the battery. The superposition of the field from the charges on

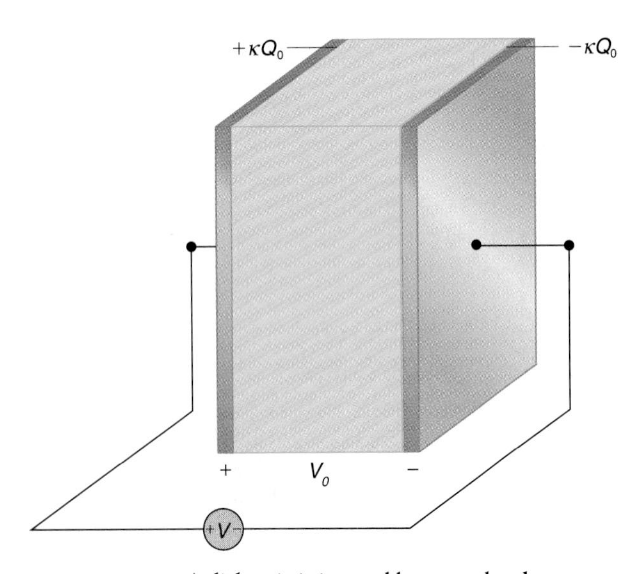

FIGURE 18.43 A dielectric is inserted between the plates.

the conducting plates and the field caused by the bound charges on the surfaces of the dielectric produces a total field equal to the field that existed between the plates when no dielectric was between the plates. The magnitude of charge on the capacitor plates with the dielectric in place is found to be $|Q| = \kappa |Q_0|$. The potential difference between the plates still is V_0. So the new capacitance of the capacitor is

$$\begin{aligned} C &= \frac{|Q|}{V_0} \\ &= \frac{|\kappa Q_0|}{V_0} \\ &= \kappa C_0 \end{aligned}$$

Again, the capacitance increases by the factor κ with the insertion of the dielectric.

QUESTION 7

On the conducting plates of a capacitor with a dielectric filling the space between the plates, what dielectric constant is needed in order to make the bound surface charge density of the dielectric surface half the free surface charge density?

EXAMPLE 18.11

Coaxial cables consist of a central cylindrical wire surrounded by an insulating layer, another cylindrical conductor, and finally another layer of insulating material, as shown in Figure 18.44. Such cables are common in electronics and in communication systems, such as the cable bringing TV signals to your set. A coaxial cable can be approximated as a long cylindrical capacitor. The capacitance per unit length of the cable is important in determining the electronic properties of the cable and the character of the TV signal received.

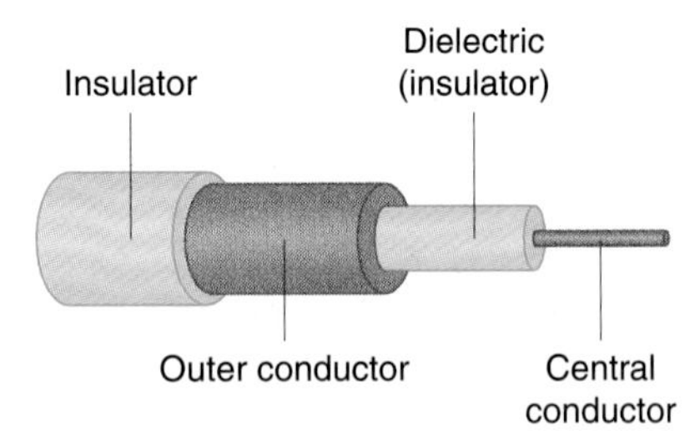

FIGURE 18.44

Use the result of Example 18.5 to determine the capacitance per meter of length of RG59 coaxial cable (used for common household cable TV) that has an inner conductor of diameter 0.812 mm and an outer conductor of diameter 3.66 mm. The insulating material between the two conductors has a dielectric constant of 1.35.

Solution

By the result of Example 18.5, the capacitance of a cylindrical capacitor with a vacuum (or air) between the plates is

$$C = \frac{4\pi\varepsilon_0 \ell}{2 \ln\left(\frac{b}{a}\right)}$$

In a coaxial cable with a material of dielectric constant κ between the conductors, the capacitance increases by the factor κ. Hence the capacitance of the cable is

$$C = \kappa \frac{4\pi\varepsilon_0 \ell}{2 \ln\left(\dfrac{b}{a}\right)}$$

The capacitance per unit length thus is

$$\frac{C}{\ell} = \kappa \frac{4\pi\varepsilon_0}{2 \ln\left(\dfrac{b}{a}\right)}$$

Substituting the numerical values given, you have

$$\begin{aligned} \frac{C}{\ell} &= (1.35) \frac{4\pi[8.85 \times 10^{-12} \ \mathrm{C^2/(N \cdot m^2)}]}{2 \ln\left[\dfrac{(3.66 \times 10^{-3} \ \mathrm{m})/2}{(0.812 \times 10^{-3} \ \mathrm{m})/2}\right]} \\ &= 4.99 \times 10^{-11} \ \mathrm{F/m} \\ &= 49.9 \ \mathrm{pF/m} \end{aligned}$$

or about 50 pF/m.

Lightning is caused by the dielectric breakdown of air in an extremely strong electric field.

18.11 DIELECTRIC BREAKDOWN*

If a dielectric material is subjected to an electric field of sufficient magnitude, the outer electrons of the atoms and molecules in the field are ripped free by the electrical force and then travel through the material (in the direction opposite to the electric field, of course). These liberated electrons in turn collide with other atoms, freeing additional electrons. This multiplication, or avalanche effect, is **dielectric breakdown**. The dielectric then no longer is an insulator but a conductor, because the charges freely move through the material. The maximum electric field that a dielectric can sustain before dielectric breakdown occurs is known as the **dielectric strength** of the material. Table 18.2 indicates the dielectric strength of various dielectric materials. The most spectacular natural example of dielectric breakdown is lightning.

Dielectric breakdown also creates the spark in a spark plug that explodes the gasoline in the pistons of the engine in your car, boat, or lawn mower.†

† Diesel engines do not use spark plugs; they ignite the fuel by compressing it to the point where it spontaneously explodes from the increase in temperature.

The spark in a spark plug creates radio wave static that can be detected by nearby radio telescopes. Hence, engines with spark plugs are prohibited near such telescopes.

QUESTION 8

In a cylindrical capacitor such as a coaxial cable, if the potential difference between the conductors is too great, dielectric breakdown will occur. Will such breakdown begin near the inner or the outer conductor? Explain your reasoning.

CHAPTER SUMMARY

Polarity markings indicate which of two points is at the higher (+) or lower (−) electric potential. A polarity marking (+) does *not* necessarily imply the potential at the indicated point is positive; the marking merely indicates that the point is at a higher electric potential than the corresponding point with the (−) polarity marking.

An *ideal circuit element* is imagined as a device with two *ideal wires*, called *leads* or *terminals*, emerging from it. Real circuit elements are modeled with ideal circuit elements.

An *independent voltage source*, also known as a *source of emf*, is an ideal circuit element that supplies and maintains a potential difference, called its *emf*, between its two terminals. An ideal battery is an example of an independent voltage source.

The places where the leads or terminals of two or more circuit elements are connected together are called *nodes*. Circuit elements connected in *parallel* are connected to the same two distinct nodes and have the same potential difference across them. Circuit elements connected in *series* are strung along sequentially like beads on a string.

When independent voltage sources are connected in series, the effective equivalent independent voltage source has an emf equal to the algebraic sum of the emfs of the collection in series:

$$V_{eq} = V_1 + V_2 + V_3 + \cdots$$

Independent voltage sources are most commonly connected with their polarities all directed in the same sense (plus to minus, plus to minus, etc.).

When two or more ideal independent voltage sources are connected in parallel, they all *must* have the same emf and be connected with their positive polarity terminals at the same node.

A *capacitor* is a circuit element that consists of two conductors, called *plates*, separated in space. If the two plates have equal and opposite charges on them, the capacitor is said to be *charged*; if each plate has zero charge, the capacitor is said to be *uncharged*.

The *capacitance* of a capacitor is the ratio of the absolute magnitude of the charge $|Q|$ on either plate to the absolute magnitude of the potential difference $|V|$ between the plates:

$$C \equiv \frac{|Q|}{|V|} \qquad (18.1)$$

The SI unit of capacitance is the farad (F), which is equivalent to a coulomb per volt (C/V). The capacitance of a capacitor depends on the physical size and shape of the conducting plates, their geometric arrangement in space, and the medium separating the plates.

The equivalent capacitance of a parallel connection of capacitors is the sum of the individual capacitances:

$$C_{eq} = C_1 + C_2 + C_3 + \cdots \quad \text{(capacitors in parallel)} \qquad (18.4)$$

A series connection of capacitors has an equivalent capacitance of

$$\frac{1}{C_{eq}} = \frac{1}{C_1} + \frac{1}{C_2} + \frac{1}{C_3} + \cdots \quad \text{(capacitors in series)} \qquad (18.7)$$

For *two* capacitors in series, the equivalent capacitance is the product of their capacitances divided by the sum.

Work is required to charge a capacitor. The electrical potential energy stored in a charged capacitor is

$$\text{PE} = \frac{1}{2}\frac{Q^2}{C} = \frac{1}{2}|Q||V| = \frac{1}{2}CV^2 \qquad (18.12)$$

If an insulating material, called a dielectric, is placed between the plates of a capacitor, the capacitance increases by a factor κ, known as the *dielectric constant* of the material.

When *dielectric breakdown* occurs, a dielectric material ceases to be an insulator and becomes a conductor. The *dielectric strength* is the maximum magnitude of electric field a dielectric material can sustain before breakdown occurs.

Summary of Problem-Solving Tactics

18.1 **(page 817)** When using Equation 18.7, be sure you remember how to add fractions.

18.2 **(page 817)** Be sure to realize that after summing the reciprocals of the individual capacitances in Equation 18.7 that the equivalent capacitance is the *reciprocal* of the sum.

18.3 **(page 817)** The equivalent capacitance of two capacitors in series is the product of their capacitances divided by the sum.

Questions

1. **(page 807)**; 2. **(page 809)**; 3. **(page 811)**; 4. **(page 811)**; 5. **(page 814)**; 6. **(page 817)**; 7. **(page 824)**; 8. **(page 825)**

9. For each of the two points A and B in Figure Q.9, label the appropriate polarity signs (+) and (−) taking into account the indicated potentials at the two points. Also indicate the potential difference V between A and B, with its correct sign.

10. In a carefully worded sentence or two, explain what is meant by the terms series connection and parallel connection.

11. Draw an arrangement of two circuit elements that are simultaneously in series and parallel with each other.

12. What does it mean when we say a circuit element is shorted out? What is the potential difference between the leads of the circuit element when it is shorted?

13. Why is it meaningless to short out an independent voltage source? Shorting a voltage source also can be dangerous, as we will see in Chapter 19.

14. Two independent voltage sources have potential differences $V_1 \neq V_2$. They can be connected in series but should never be connected in parallel. Why one way and not the other?

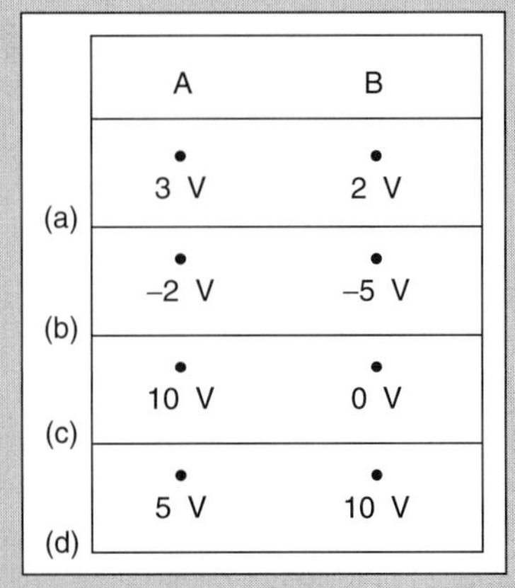

FIGURE Q.9

15. Is Figure 18.8 the only way six circuit elements can be connected together so that they all are neither in series nor in parallel?

16. How many nodes are in the arrangement of circuit elements shown in Figure Q.16?

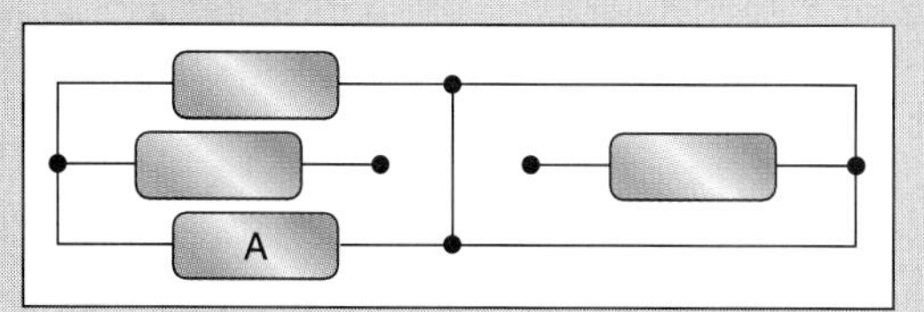

FIGURE Q.16

17. What must be done in Figure Q.16 to short out the circuit element labeled A?

18. Several independent voltage sources are connected as in Figure Q.18. What must be the potential differences V_1, V_2, and V_3?

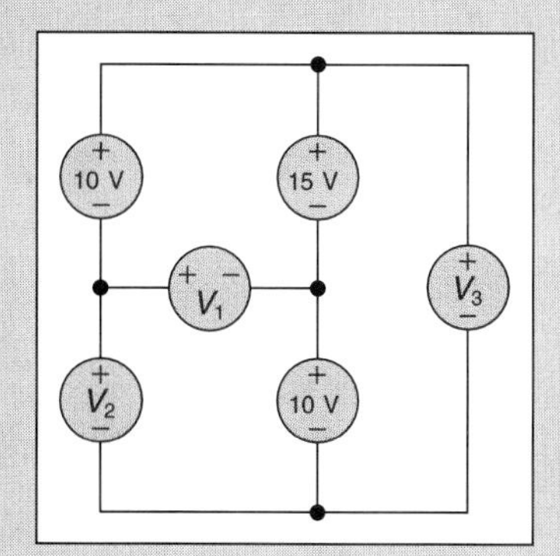

FIGURE Q.18

19. Explain why the capacitance of a capacitor is independent of both the magnitude of the charge on its plates and the potential difference between the plates even though the capacitance is defined as their ratio (Equation 18.1).

20. (a) Sketch a graph of the magnitude of the charge on either plate of a capacitor versus the magnitude of the potential difference between the plates. (b) What is the slope of the curve,

$$\frac{d|Q|}{d|V|}$$

and what are alternative units for the slope besides C/V?

21. Sketch a graph of the magnitude of the potential difference between the plates of a capacitor and the magnitude of the charge on each plate. This is the reverse of the plot in the previous question. Suppose a capacitor has charges $\pm Q_0$ on its plates and a potential difference of magnitude $|V_0|$ between them. Schematically indicate this point on the graph. What is the physical significance of the area under the curve between the origin and the point $(|Q_0|, |V_0|)$?

22. The electrical potential energy of a charge q placed at a point where the electrical potential is V is qV (see Equation 17.8). Why, then, is the potential energy stored on a capacitor not $|Q||V|$, rather than the correct expression

$$\frac{1}{2}|Q||V|$$

(Equation 18.12)?

23. If two capacitors, with capacitances $C_1 = 2C_2$, store equal amounts of potential energy, what is the relationship between the potential differences across them? Can you exclude the possibility that these capacitors are connected in series? In parallel? Explain.

24. A thin conducting sheet is carefully inserted between but not touching the plates of a parallel plate capacitor as shown in Figure Q.24. Will the capacitance increase, decrease, or be unaffected? Justify your answer.

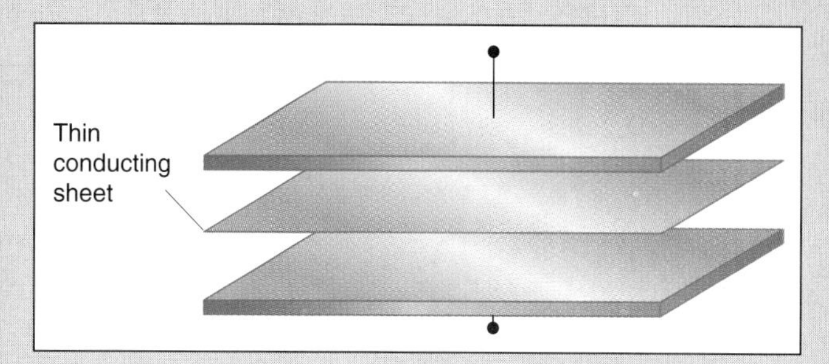

FIGURE Q.24

25. Your instructor likely can provide you with a variable capacitor used for tuning certain circuits in radios. An example of such a capacitor is shown in Figure Q.25. Examine the device carefully. Are the various capacitors that form the variable capacitor connected in series or parallel?

FIGURE Q.25

26. Let C be the capacitance of a parallel plate capacitor of plate area A and plate separation d. If four plates of area A are arranged as in Figure Q.26, separated by distance d, what is the equivalent capacitance?

27. Let C be the capacitance of a parallel plate capacitor of plate area A and plate separation d. If four plates of area A are arranged as in Figure Q.27, separated by distance d, what is the equivalent capacitance?

28. For a polar material, would you expect the dielectric constant to increase or decrease if the temperature is raised? Give arguments to support your answer.

29. Water has a dielectric constant that is quite large (see Table 18.2) but almost never is used as a dielectric between the plates of a capacitor. Why?

30. A charged capacitor has a large potential difference across its plates and a large capacitance. Explain why it is dangerous to touch the terminals of such a capacitor, even well after the battery used to charge it has been disconnected.

31. An independent voltage source maintains a constant potential difference across the plates of a parallel plate capacitor. Does the potential energy increase, decrease, or remain the same if

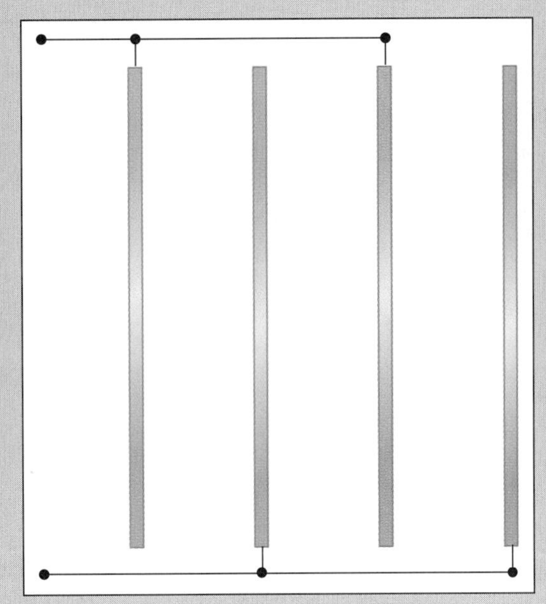
FIGURE Q.26

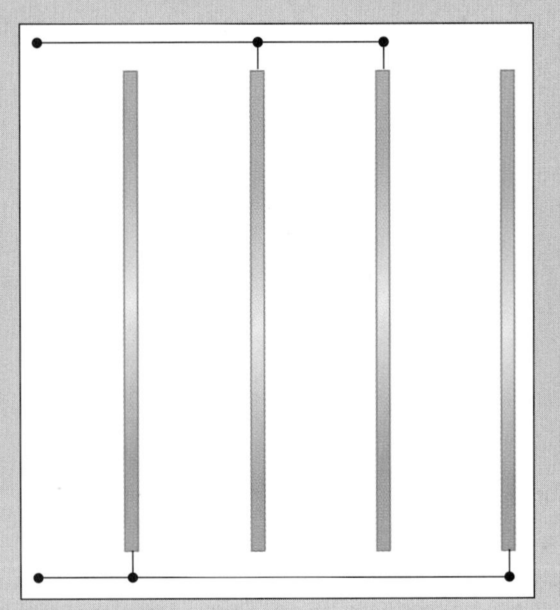
FIGURE Q.27

a sheet of glass is inserted between the plates? If the energy increases, where does the increase in potential energy come from? If it decreases, where does the potential energy go?

32. How is a spark in a spark plug similar to lightning?

33. Notice that, from Example 18.3, the capacitance of a parallel plate capacitor can be increased by decreasing the plate separation d. For a given potential difference V between its plates, what happens to the magnitude of the electric field between the plates as d decreases? If the distance between the plates becomes too small for an air-filled capacitor (or for any capacitor filled with a dielectric), what will happen?

34. Shuffle you feet across a fluffy rug on a nice dry day and then slowly bring your finger near a metal doorknob. Estimate the distance between your finger and the doorknob when the spark jumps between them. From this distance, *estimate* the charge you accumulated. State what assumptions you make to arrive at your estimate.

PROBLEMS

Sections 18.1 Terminology, Notation, and Conventions
18.2 Circuit Elements
18.3 An Independent Voltage Source: A Source of Emf
18.4 Connections of Circuit Elements

1. How many nodes are in the connection of circuit elements shown in Figure P.1?

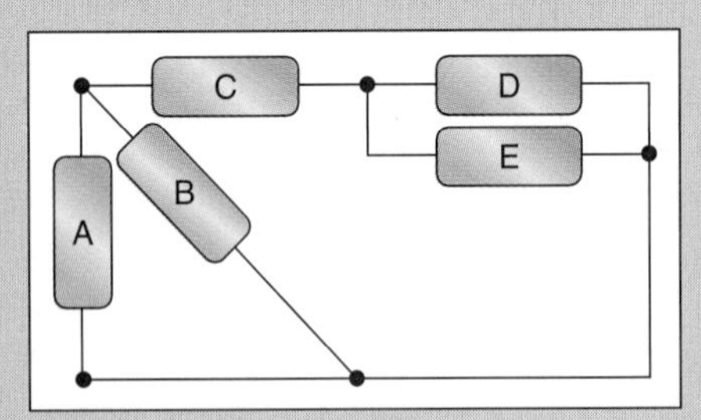

FIGURE P.1

2. How many nodes are in the connection of circuit elements shown in Figure P.2?

3. How many nodes are in the connection of circuit elements shown in Figure P.3?

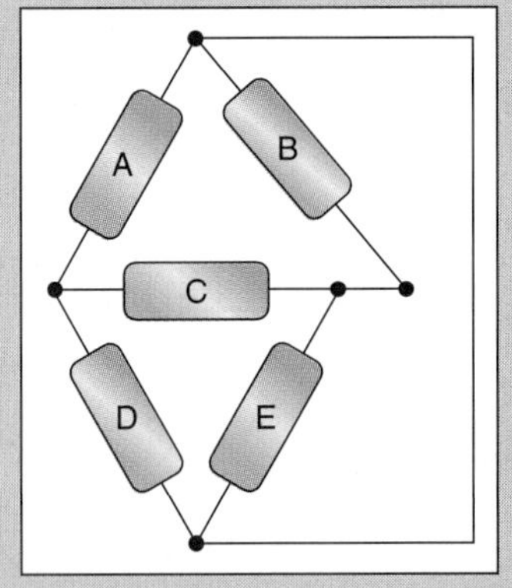

FIGURE P.2

4. Which circuit elements (if any) in Figure P.1 are in series with each other? Which (if any) are in parallel?

5. Which circuit elements (if any) in Figure P.2 are in series with each other? Which (if any) are in parallel?

6. Which circuit elements (if any) in Figure P.3 are in series with each other? Which (if any) are in parallel?

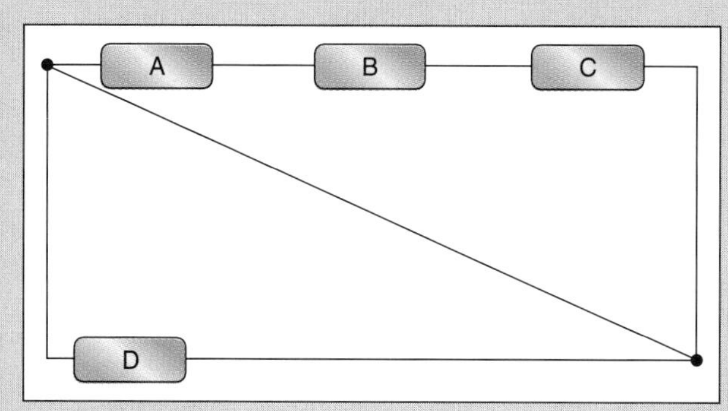

FIGURE P.3

•**7.** In Figures P.1, P.2, and P.3, are any circuit elements shorted out? If so, specify which one(s) and redraw the circuit with extraneous circuit elements removed.

•**8.** (a) In the collection of circuit elements shown in Figure P.8, what is the potential difference between points A and B? (b) What is the potential difference between points B and C? (c) What is the potential difference between points C and A? (d) What is the potential difference between points A and C?

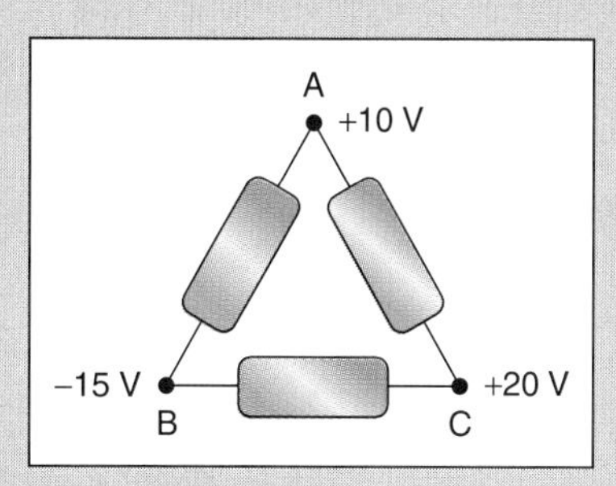

FIGURE P.8

Section 18.5 Independent Voltage Sources in Series and Parallel

9. The lead–acid electrochemical cells used to create the emf in automotive batteries (an independent voltage source, to a first approximation) have an emf of about 2.0 V. How many such cells are needed to create an independent voltage source with an emf of 12.0 V? Sketch how the cells should be connected together to produce the 12.0 V emf.

10. Four D-cell batteries, each with an emf of 1.5 V, are connected in series in a flashlight, with their positive terminals all facing the same way. What is the equivalent emf of the four batteries?

•**11.** The instructions with a set of battery jumper cables say to connect the red wire to the positive polarity terminals and the black wire to the negative polarity terminals of each battery. (a) Are the batteries connected in series or parallel? (b) Will there be any problem if you use the black wire in place of the red wire and the red wire in place of the black wire?

•**12.** A truck driver asks you to help jump start his engine with the help of the 12.0 V battery in your car. You notice, however, that the truck engine has several 12.0 V batteries connected in series as indicated in Figure P.12. (a) Indicate on a sketch a permissible way to connect your battery to the collection. (b) Indicate on another sketch several ways you should *not* connect your battery to the collection and explain why each connection is impermissible.

Section 18.6 Capacitors

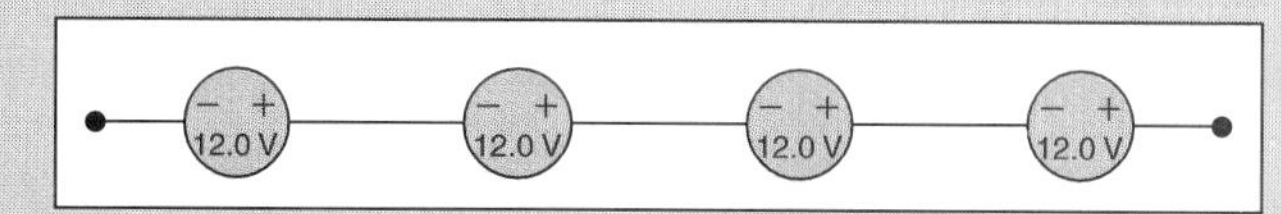

FIGURE P.12

13. A 15 pF capacitor has a potential difference of 1.50 V between its plates. What is the absolute magnitude of the charge on each plate?

14. The plates of a capacitor have charges of ±8.0 nC. A potential difference of −120 V exists between the plates. Find the capacitance.

15. What is the charge on each plate of a 10 μF capacitor with a potential difference of 120 V between its plates?

•**16.** To show you how difficult it is to make capacitors with large capacitances, consider a hypothetical parallel plate capacitor with a capacitance of 1.0 F with plates separated by 1.0 cm. (a) If the plates are square, what is the length of the sides of the square? (b) If the plates are circular, what is the radius of the circles?

•17. A parallel plate capacitor consists of two circular disks of radius R and separation $R/1000$. By what factor does the capacitance change if R is doubled, affecting both the area and separation?

•**18.** Calculate the capacitance of a sphere with a radius equal to the average radius of the Earth.

•**19.** What is the radius of a spherical capacitor with a capacitance of 1.00 F? What is the ratio of this radius to the average separation of the Earth and the Moon?

•**20.** The temperature of a brass spherical capacitor of radius 5.00 cm increases by 100 °C. By how much does the capacitance increase? (Consult Table 13.1 on page 596.)

•21. A conducting sphere of radius a is surrounded by a concentric, thin conducting spherical shell of radius b. Show that the capacitance of the system is

$$C = 4\pi\varepsilon_0 \frac{ab}{b - a}$$

•**22.** Show that the result of Problem 21 reduces to the capacitance of a parallel plate capacitor if the separation distance $d \equiv b - a$ is much smaller than the radius a of the inner conductor.

•**23.** Use the result of Problem 21 to show that as $b \to \infty$ m, the capacitance approaches that of a sphere of radius a.

Section 18.7 Series and Parallel Combinations of Capacitors

For Problems 24–32, find the single equivalent capacitance that can be placed between the terminals A and B.

24.

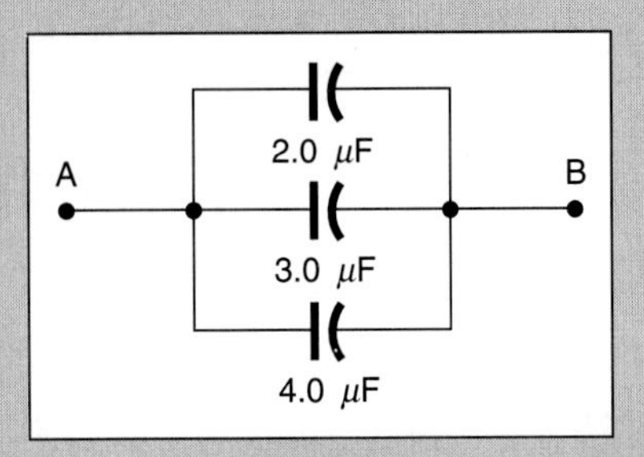

FIGURE P.24

25.

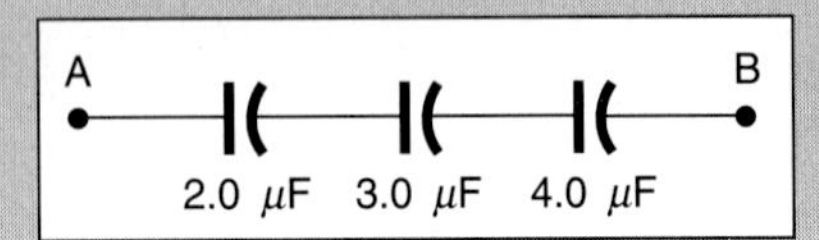

FIGURE P.25

•26.

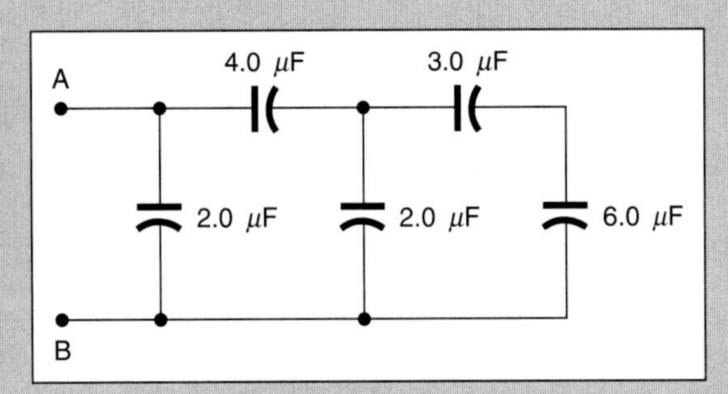

FIGURE P.26

•27.

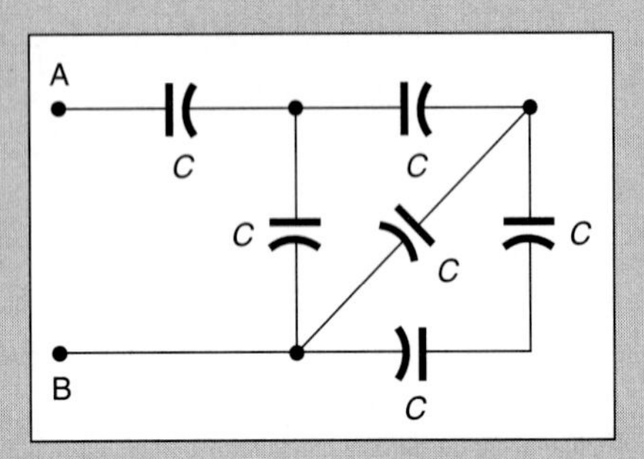

FIGURE P.27

•28.

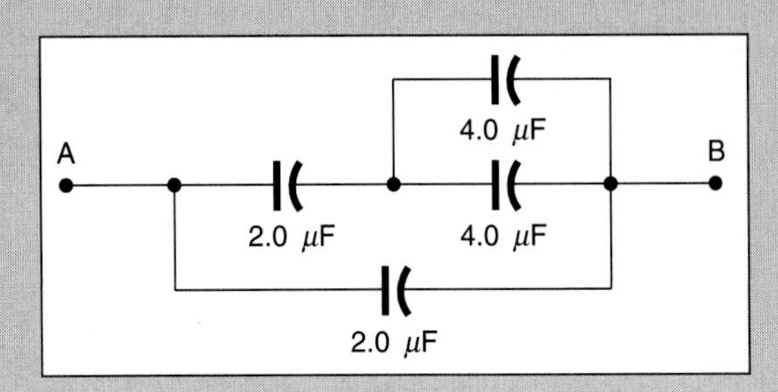

FIGURE P.28

•29.

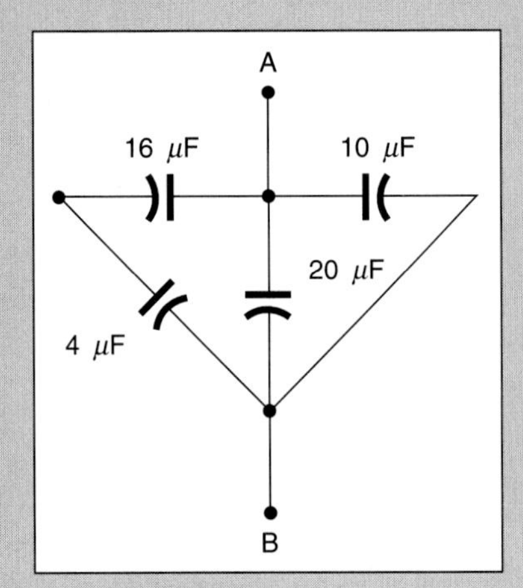

FIGURE P.29

•30.

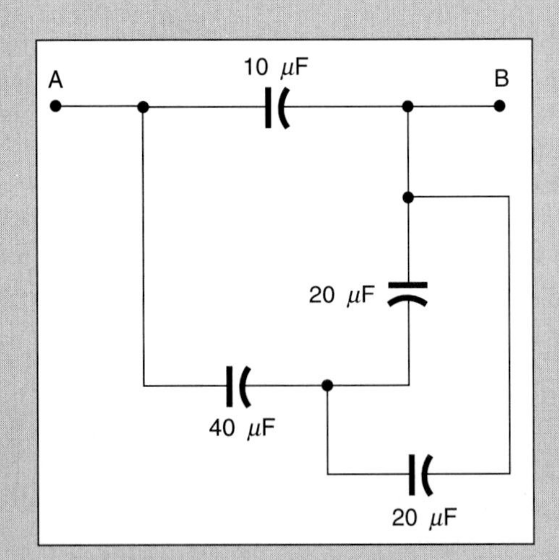

FIGURE P.30

•31.

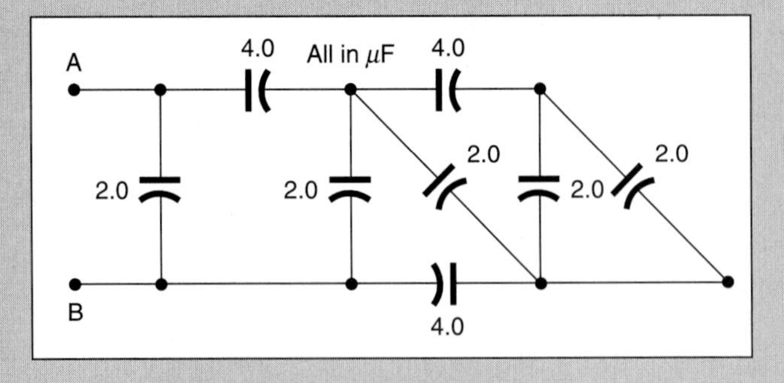

FIGURE P.31

•32.

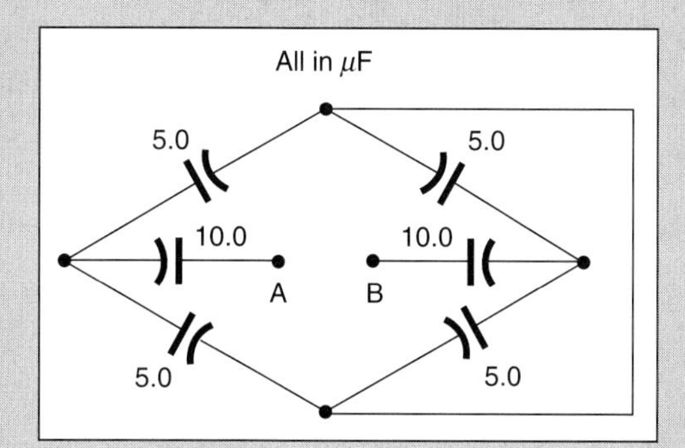

FIGURE P.32

•33. (a) If a single 10 μF capacitor has a potential difference between its plates of 120 V, what magnitude of charge is on its plates? (b) How can you connect other 10 μF capacitors to this one to store additional charge? (c) How many 10 μF capacitors connected to a 120 V independent voltage source are needed to store a total charge of magnitude 1.0 C?

•34. (a) In Figure P.34, find the magnitude of the potential difference across each capacitor. Indicate the polarities of the potential differences. (b) Find the magnitude of the charge on the plates of each capacitor.

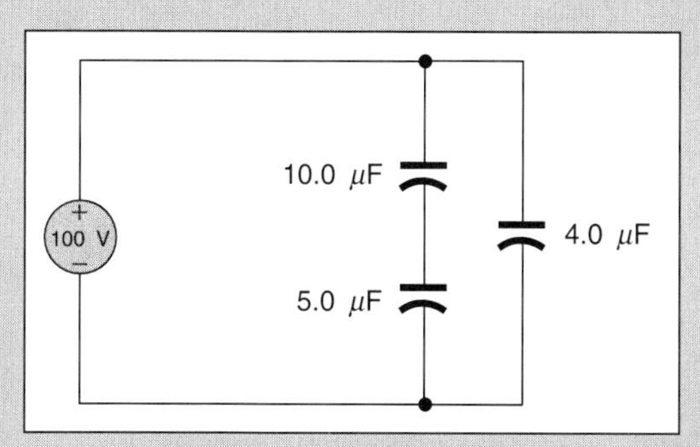

FIGURE P.34

•35. (a) In Figure P.35, what is the potential difference across each capacitor? (b) Determine the magnitude of the charge on each of the plates of the capacitors. (c) What is the equivalent capacitance of the two capacitors?

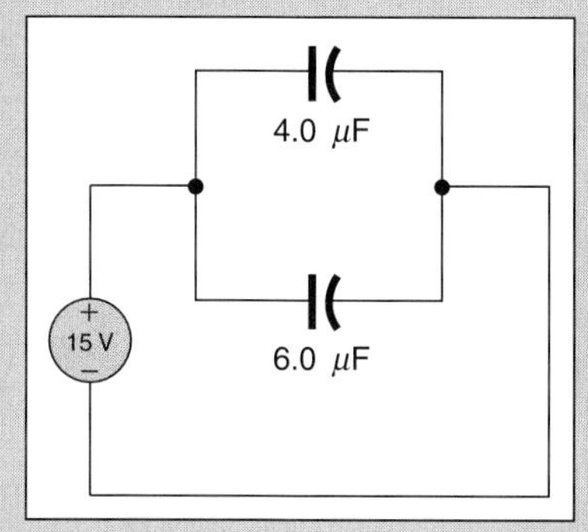

FIGURE P.35

•36. Two capacitors connected in parallel have an equivalent capacitance of 12 μF. When connected in series, the equivalent capacitance is one-third the capacitance of one of the pair. What are the values of the two capacitances?

•37. Nuts. You need a 10 μF capacitor but all you have is a bin of 15 μF capacitors. How can you use these 15 μF capacitors to make one with an equivalent capacitance of 10 μF? Sketch the way the capacitors should be connected together.

•38. Two capacitors are connected in series with an independent voltage source as indicated in the Figure P.38. Show that the potential differences across the capacitors are

$$V_1 = \frac{C_2}{C_1 + C_2} V_0$$

and

$$V_2 = \frac{C_1}{C_1 + C_2} V_0$$

In *real* capacitors (as opposed to the ideal capacitors we have treated), this result is rarely the case because real capacitors also have leakage resistance.* For a discussion of the situation with real capacitors, see L. Kowalski, "A myth about capacitors in series," *The Physics Teacher*, 26, #5, pages 286–287 (May 1988).

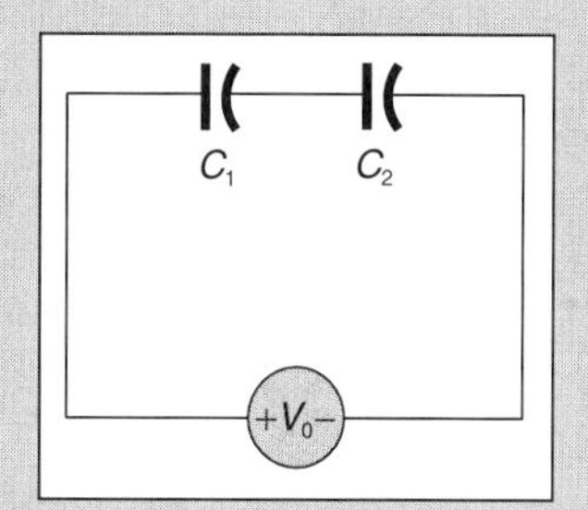

FIGURE P.38

39. Show that the equivalent capacitance for a collection of capacitors connected in series is smaller than the smallest capacitance in the series.

Section 18.8 Energy Stored in a Capacitor

40. A capacitor has a potential difference V between its plates. To double the potential energy stored by the capacitor, by what factor must the potential difference be changed?

41. For a laser power supply, you want to store 1.0 J of energy in a capacitor with a potential difference of 120 V between its plates. What capacitance is needed?

42. A 12 V automotive battery stores 2.0 MJ of energy. If a large parallel plate capacitor, with a plate separation of 0.50 cm and potential difference between its plates of 120 V, stores the same amount of energy, what is the length ℓ of a side of its plates if they are square? The size of this parallel plate capacitor is quite unrealistic.

•43. To store about 10 J of electrical potential energy for a cardiac defibrillator, a capacitor bank is used. The bank consists of a parallel connection of 100 μF capacitors with a potential

*We consider this in more detail in the next chapter.

difference of 100 V across each of them. How many capacitors are in the bank?

•**44.** Professor Milly Coulomb finds it takes 2.0 J of work to drag, at constant speed, a −2.0 mC charge between the plates of a parallel plate capacitor. (a) What is the work done by the electrical force in taking the charge between the plates? (b) What is the potential difference between the plates of the capacitor?

•**45.** You are given an independent voltage source and two identical capacitors. Sketch how you would arrange these circuit elements so that the capacitors store the most total electrical potential energy. Prove that this arrangement is the best choice.

•**46.** Near the surface of the Earth there is an electric field with a magnitude of about 100 N/C directed vertically down. Assume this field is caused by a net free charge on the Earth. (a) What are the magnitude and sign of the charge on the total Earth that produces this field? (b) What electrostatic energy density is stored in the field of the Earth near its surface?

•**47.** Three capacitors in series, with capacitances 3.0 μF, 5.0 μF, and 6.0 μF, are connected in series to a 21.0 V independent voltage source. (a) Sketch the arrangement. (b) What is the equivalent capacitance? (c) What charges are on the plates of each capacitor? Indicate which charge is on what plate of your sketch. (d) To the nearest volt, what is the potential difference across each capacitor? (d) Which capacitor is storing the greatest amount of electrical potential energy? What is this energy?

•**48.** Let n identical capacitors be connected (i) in series and (ii) in parallel. (a) If the equivalent capacitor has a potential difference V between its plates, what is the total electrical potential energy stored by each collection? (b) If the equivalent capacitor has a charge of magnitude $|Q|$ on its plates, what is the total electrical potential energy stored by each collection?

Sections 18.9 Electrostatics in Insulating Material Media*
18.10 Capacitors and Dielectrics*
18.11 Dielectric Breakdown*

49. An air-filled capacitor is connected to the terminals of an independent voltage source. If an insulator of dielectric constant κ is next inserted to completely fill the air gap between the plates, by what factor does the potential energy stored in the capacitor increase?

50. A small ceramic capacitor has plates 1.20 cm in diameter. A dielectric 0.25 mm thick, with a dielectric constant $\kappa = 5.0$, completely fills the space between the plates. Find the capacitance.

•**51.** Real capacitors have a maximum potential difference V_{max} that can be placed across their plates before dielectric breakdown ensues. You have a collection of identical capacitors each with a V_{max} of 50 V and a capacitance of 10 μF. You need an equivalent capacitor with a capacitance of 10 μF, but with a potential difference of 100 V across it. Devise an appropriate connection of the 50 V, 10 μF capacitors to do the job.

•**52.** You are faced with a problem: a number of 0.50 μF capacitors are available but the potential difference across each cannot exceed 200 V without causing dielectric breakdown. A capacitor of 0.50 μF capacitance is needed but must be connected across a potential difference of 400 V for who knows what purpose. Show via a diagram how an equivalent capacitor having the desired capacitance can be obtained using the given collection of capacitors, so that the maximum potential difference across any one of them is 200 V.

•**53.** A parallel plate capacitor with plate area A and plate separation d is filled with two dielectrics as shown in Figure P.53. Show that the capacitance of the arrangement is

$$C = \frac{\varepsilon_0 A}{d}\frac{\kappa_1 + \kappa_2}{2}$$

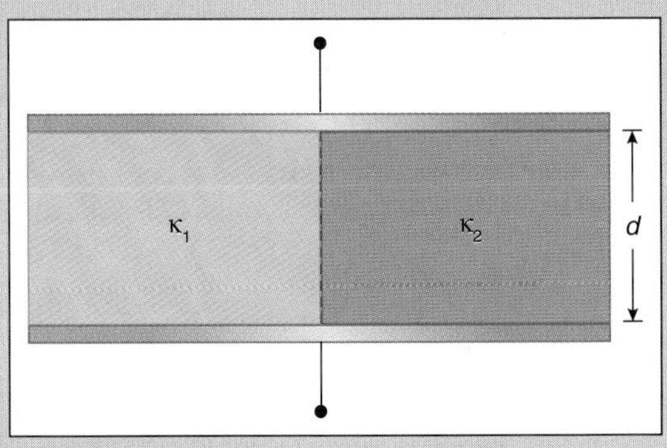

FIGURE P.53

•**54.** A parallel plate capacitor with plate area A and plate separation d is filled with two dielectrics as shown in Figure P.54. Show that the capacitance of the arrangement is

$$C = \frac{2\varepsilon_0 A}{d}\frac{\kappa_1 \kappa_2}{\kappa_1 + \kappa_2}$$

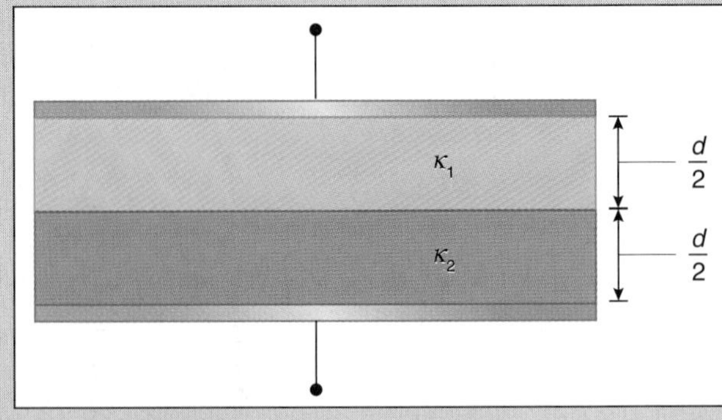

FIGURE P.54

•**55.** A block of paraffin is placed in an electric field of magnitude 1.20×10^4 N/C. What is the magnitude of the electric field within the paraffin? Refer to Table 18.2 on page 821.

•**56.** A charged 50 μF capacitor with a potential difference between its plates of 100 V is about to be connected to an initially uncharged 50 μF capacitor as shown in Figure P.56. (a) Determine the energy stored in the initially charged capacitor. The switch at S now is closed. (b) Are the two capacitors connected in parallel? (c) Determine the magnitude of the potential difference V across each capacitor. (d) Determine the total potential energy stored in the new arrangement. The two energies in (a) and (d) are not equal to each other. This might seem to contradict conservation of energy. With idealized capacitors and wires, it is not possible to explain where the missing energy went. For *real* capacitors and wires, the difference in the two energies appears as an increase in the thermodynamic internal energy of the wires connecting them (because of their finite electrical resistance*), with a small fraction also radiated away as electromagnetic radiation. See R. A. Powell, "Two-capacitor problem: a more realistic view," *American Journal of Physics*, 47, #5, pages 460–462 (May 1979).

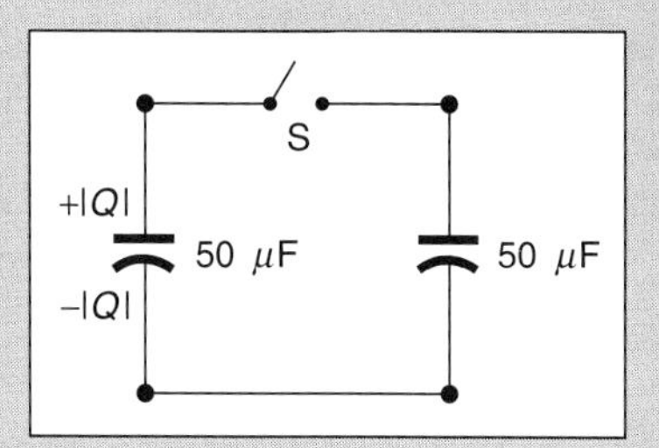

FIGURE P.56

•57. Thunderstorms are quite spectacular. Given the dielectric strength of air in Table 18.2, and that the flat bottom of a thundercloud of area 25 km^2 is on the order of 2 km from the ground, estimate the capacitance of the cloud–Earth system and the electrical potential energy stored in the electric field between the cloud and the Earth as breakdown ensues.

•58. A 100 μF capacitor has tantalum oxide (dielectric constant $\kappa = 26$) completely filling the space between its plates. The capacitor has a potential difference of 50.0 V between its plates. If the tantalum oxide is removed from the capacitor, what is its capacitance?

*We consider the concept of electric resistance in the next chapter.

INVESTIGATIVE PROJECTS

A. Expanded Horizons

1. Investigate and write a report on the uses of capacitors in modern integrated circuit chip technology.
   Donald M. Trotter Jr., "Capacitors," *Scientific American*, 259, #1, pages 86–90B (July 1988).

2. A *Leyden jar* was the first type of capacitor. Investigate and prepare a report on the history of its discovery.
   Donald M. Trotter Jr., "Capacitors," *Scientific American*, 259, #1, pages 86–90B (July 1988).
   Encyclopaedia Britannica (11th edition, 1910), volume 16, pages 528–529.

3. Investigate the ways capacitors are fabricated and the progress toward manufacturing capacitors of small dimensions with increased capacitance.
   Ray Marston, "Capacitors," *Electronics Now*, 64, #3, pages 47–64 (March 1993).
   Josef Bernard, "All about capacitors," *Radio-Electronics*, 60, #5, pages 49–53 (May 1989); and #8, pages 56–59 (August 1989).
   William Hyland, "Tantalum capacitors keep getting better," *Electronics*, 61, #11, pages 93–95 (May 1988).

4. Lightning is and example of dielectric breakdown. Investigate and prepare a report on the nature of this spectacular natural phenomenon, including lightning sprites.
   Earle Williams, "The electrification of thunderstorms," *Scientific American*, 259, #5, pages 88–99 (November 1988).
   Martin A. Uman, *The Lightning Discharge* (Academic Press, Orlando, 1987), International Geophysics Series, volume 39.
   Martin A. Uman, *All About Lightning* (Dover, New York, 1986).
   Tom Koppel, "Lightning lure," *Scientific American*, 268, #2, page 105 (February 1993).
   William R. Newcott, "Lightning: nature's high-voltage spectacle," *National Geographic*, 184, #1, pages 80–103 (July 1993).
   Jeff Rosenfeld, "Lightning: among the sprites and jets," *Weatherwise*, 48, #6, pages 9–10 (December 1995).
   Dennis J. Boccippio, "Sprites, ELF transients, and positive ground strokes," *Science*, 269, #5227, pages 1088–1091 (25 August 1995).
   Richard A. Kerr, "Lofty flashes come down to Earth," *Science*, 270, #5234, page 235 (13 October 1995).

B. Lab and Field Work

5. Design an experiment to measure the dielectric constant of an insulator such as glass or plastic. Specify the equipment needed for the experiment. Make a few back-of-the-envelope calculations to determine the approximate magnitudes of the quantities to be measured to determine the dielectric constant. Secure the equipment from your professor, perform the experiment, and write a report of your results, detailing the design, procedure, and results. The report should be complete enough to guide a peer in reproducing the experiment.

6. Your local electric company uses physically large capacitors in the electric grid around your town. The function of these capacitors is not energy storage, but rather is associated with keeping the oscillatory potential differences and currents in step with each other. Call or visit the engineering department of the utility to determine the capacitances of their capacitors and the potential difference at which each typically is used. Discover what is used as the dielectric material and the value of its dielectric constant. If connected to an independent voltage source with this emf, what maximum energy is stored by the capacitors?

7. Your physics lab may have a meter that measures capacitance. If so, measure the capacitance of several capacitors provided by your instructor to see how close their measured value is to the nominal value typically imprinted somewhere on the capacitor itself.

8. Design an experiment to test the premise and results of Problem 38.

C. Communicating Physics

9. Ask a few nonscience student compatriots what is meant by the term *shorting* in an electrical context. Given this information (or misinformation), write a short paragraph, appropriate for a lay audience, explaining what is meant by the term, specifically addressing any misconceptions you elicit from your fellow students.

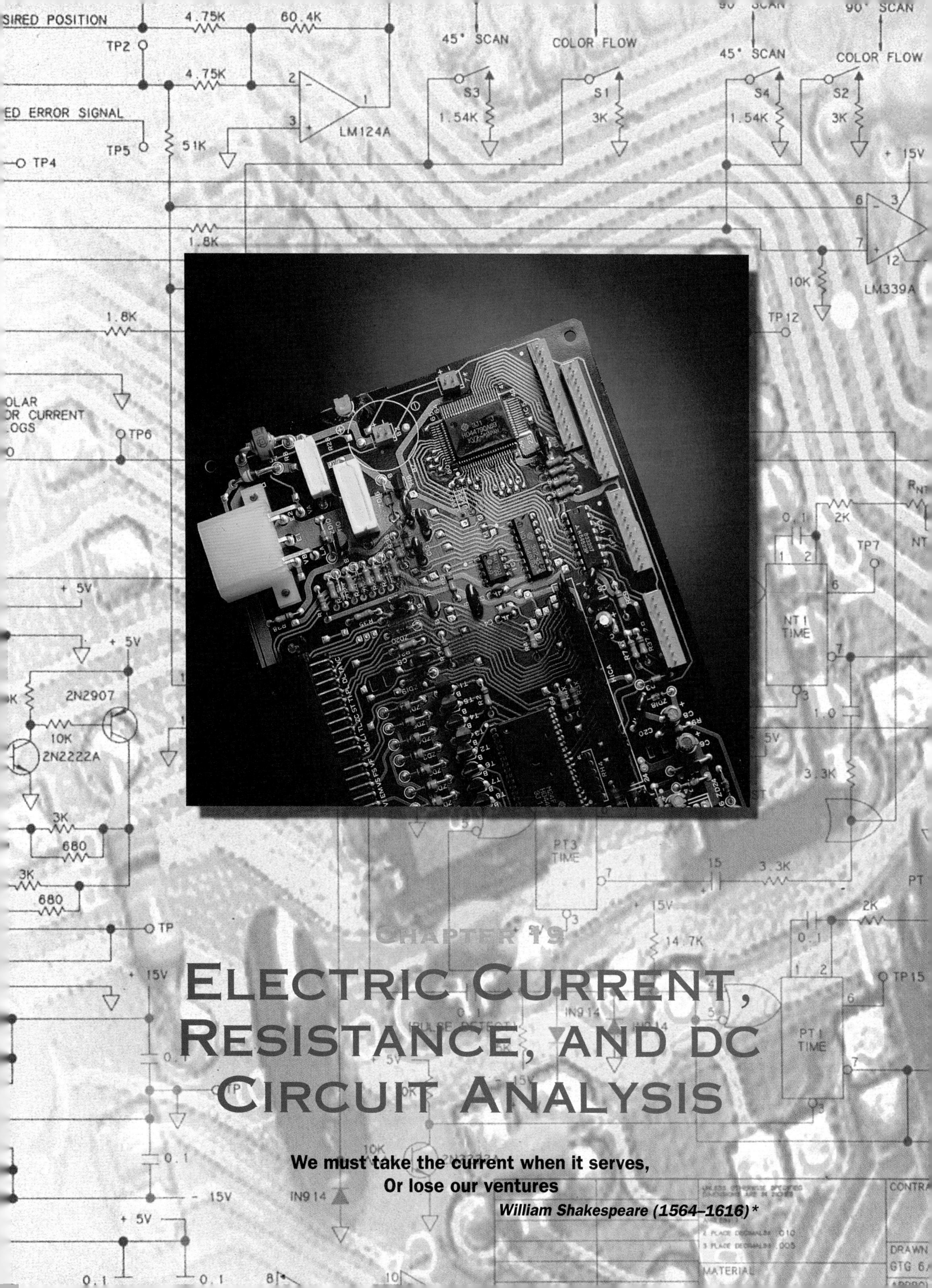

Chapter 19

Electric Current, Resistance, and DC Circuit Analysis

We must take the current when it serves,
Or lose our ventures

William Shakespeare (1564–1616)*

In the previous three chapters, we were concerned principally with electric charges at rest: electrostatic situations. Here we begin to see what happens as electrical charge moves or flows from one place to another, creating an electric current. Electric currents are double-edged swords: beneficial, such as when they are used to revive persons suffering from cardiac arrest, or deadly when they cause electrocution.

Electric currents are associated with many physical phenomena, from lightning to our central nervous system, from toasters to steel furnaces, from TV tubes to the Internet, from the motion of electrons in atoms to the vast whorls of swirling, charged matter in stars and galactic nebulae. We shall see in the next chapter that electric currents also are responsible for magnetism.

Charged particles may flow unimpeded through a vacuum, for example, as particle beams of electrons inside a TV or computer monitor. When charged particles flow in materials such as common wires, friction-like effects, called resistance, decrease the electrical potential energy of the charges. In superconducting materials at very low temperatures, charge flows without such resistance.

The concepts of current and resistance lead us to a study of electric circuits and a panoply of electronic devices. In this chapter we study steady "direct" currents of electrical charge, used widely in many devices such as computer circuits. The study and use of such *direct* (steady) currents form what is called *dc electronics*. Electric currents that reverse direction regularly with time, so-called *alternating* currents (*ac electronics*), we reserve for Chapter 22.

The motion of charged particles is an electric current even if there is no wire, as in this nebula in the constellation Orion.

19.1 THE CONCEPT OF ELECTRIC CURRENT

The concept of electric current, or the flow of the electric charge property of matter, has been with our technological culture so long that it is difficult to appreciate how the concept arose before the advent of the atomic view of matter and the discovery of the electron and proton. A few experiments can show us the origins of the idea of an electric current.[†] We first examine a charged capacitor and see how it can be discharged or neutralized by transferring charge from one plate to the other. We then examine a few experiments with a battery and some light bulbs that also elucidate aspects of a current of electric charge.

Experiments with a Charged Capacitor

Experiment 1: A Capacitor and a Pendulum

We begin with an uncharged, isolated, parallel plate capacitor having its plates maintained at a fixed distance apart and with an isolated independent voltage source.[‡] Experimentally, by connecting the two plates of the uncharged capacitor momentarily to the independent voltage source and then disconnecting the source, we are left with a charged and isolated capacitor. The two plates have equal magnitudes of charge but with opposite signs: $\pm|Q|$. Now imagine one plate fixed with the other plate connected to a spring that measures the attractive force that each plate exerts on the other due to their opposite charges (see Figure 19.1). The stretch of the spring needed to keep the plates a fixed distance apart can tell us if the capacitor is charged (nonzero force) or uncharged (zero force). A small ball of cork, covered with a conducting foil, is suspended by an insulating thread between the two plates of the capacitor as a simple pendulum.

If the ball initially is closer to the positive plate, it will be slightly attracted to that plate because of induction. On contact with the positive plate, some of the plate's positive charge is transferred to the ball by charge sharing, since they then briefly form a common conductor. The positively charged ball then is repelled by the positive plate and attracted to the negative plate. Upon reaching the negative plate, the kinetic energy of the ball is mostly converted into thermodynamic internal energy of the negative plate. The positive charge on the ball neutralizes some of the negative charge on the negative plate. The ball also then becomes negatively charged by charge sharing and subsequently is repelled by the negative plate and attracted back to the positive plate.

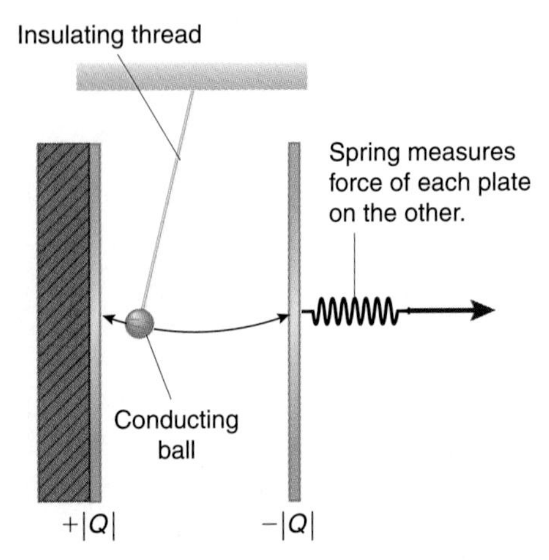

FIGURE 19.1 A spring measures the force needed to keep the plates of a charged capacitor a fixed distance apart.

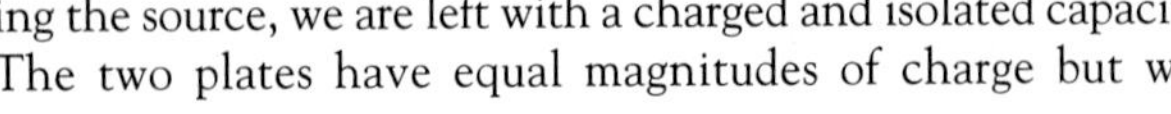

*(Chapter Opener) *Julius Caesar*, Act IV, Scene 3, lines 223–224.

†The author is indebted to Arnold Arons, *A Guide to Introductory Physics Teaching* (John Wiley, New York, 1990) for these ideas.

‡A common battery will do, since, to a first approximation, it is an independent voltage source.

The process continues with the electric pendulum swinging back and forth between the two plates until essentially all of the charge on the capacitor is neutralized and the capacitor is discharged. We imagine positive charge transferred one way, negative charge the other way until the two plates are discharged. The spring enables us to monitor the force that each plate exerts on the other as a function of time. We observe that the force decreases with each swing of the pendulum, confirming our account of the neutralization or discharge of the two plates. Once discharged, the plates of the capacitor are at the same electric potential, the field between them is zero, they exert zero electric force on each other, and the ball hangs motionless.

If we repeat the experiment with an initially charged capacitor but, instead of using the pendulum to transfer charge, connect a conducting wire between the two charged plates and short the plates, the force of each plate on the other decreases to zero almost instantaneously. The charge transfer occurs much faster via the wire than with the oscillating pendulum. We also might wonder about the fate of the potential energy that used to be stored on the charged capacitor; more about this problem toward the end of this chapter.*

Experiment 2: A Capacitor, a Pendulum, and an Independent Voltage Source

Now we repeat the experiment with the independent voltage source continuously connected to the charged capacitor; see Figure 19.2.

The pendulum once again is set swinging, and we discover that the spring now indicates there is *no* decrease in the force of each capacitor plate on the other. The pendulum swings indefinitely. Even though the pendulum still transfers positive charge to the negative plate and negative charge to the positive plate, just as before, there is no decrease in the total charge on each plate, because the force of each plate on the other remains constant, as confirmed by the spring. Hence we must conclude that the independent voltage source replenishes the charges on the plates as fast as they are transferred. We have a primitive electric **circuit**, a closed conducting pathway that permits the continuous transfer or flow of charge. In these experiments, the pendulum provides a way for charge to move or flow across the insulating air gap between the capacitor plates.

*To give the punch line away, the energy appears as an increase in the thermodynamic internal energy of the wire because of its resistance: the wire becomes warm.

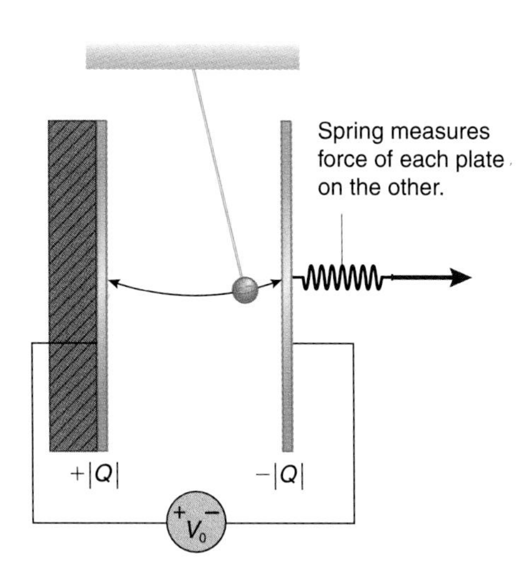

FIGURE 19.2 An independent voltage source connected to the capacitor.

Some Experiments with Light Bulbs

Now we perform a few other experiments, this time using an independent voltage source, some thick wire (imagined to be ideal wire†), and several small, identical light bulbs. A light bulb consists of a thin filament of wire, one end of which is connected to the outer metal sheath and the other connected to the metal contact at the bottom of the bulb, as shown in Figure 19.3.

The contact at the bottom of the bulb is electrically insulated from the outer metal sheath. A light bulb, therefore, is a circuit element with two leads: one is the outer metal sheath and the other is the contact at the bottom of the bulb. We represent the light bulb schematically with the symbol shown in Figure 19.4.

Experiment 1: A Light Bulb and an Independent Voltage Source

Construct the arrangement shown in Figure 19.5. If a string or wood pencil connects points A and B, we find that the light bulb does not light. Materials such as string or wood are insulators.

On the other hand, if a metal wire connects points A and B, as shown in Figure 19.6, the bulb lights. We say an electric circuit exists: a continuous closed conducting path or loop connecting

†An ideal wire has no resistance; we introduce the concept of resistance later in the chapter.

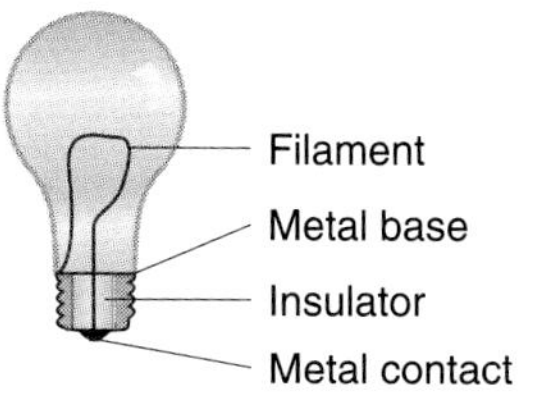

FIGURE 19.3 A light bulb.

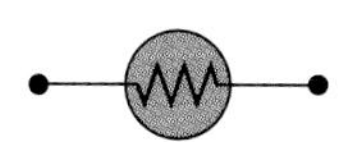

FIGURE 19.4 Schematic representation of a light bulb.

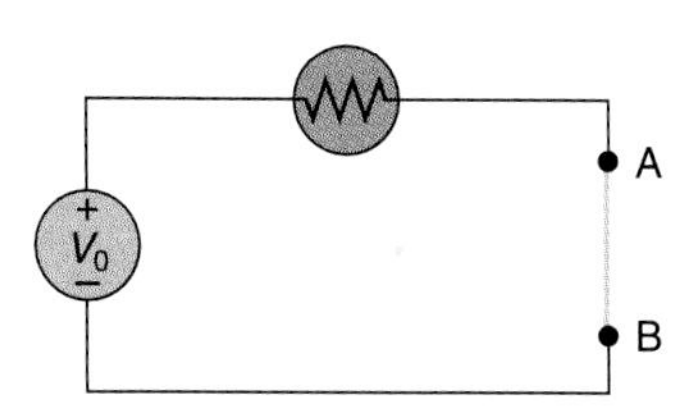

FIGURE 19.5 If an insulator connects points A and B, the bulb does not light.

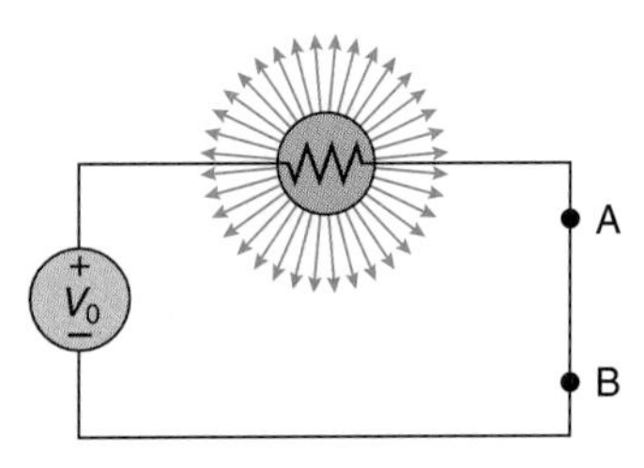

FIGURE 19.6 If a conductor connects points A and B, the bulb lights.

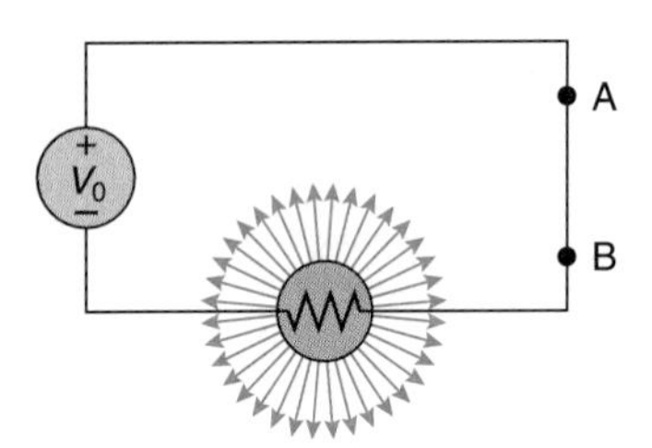

FIGURE 19.7 The bulb lights up with the same brightness, no matter where the bulb is placed.

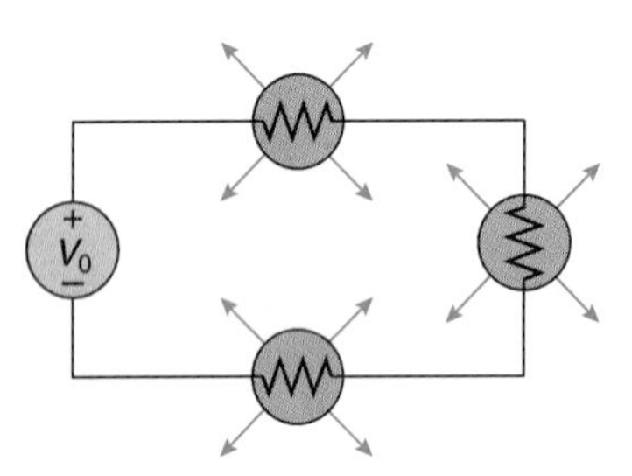

FIGURE 19.8 Several bulbs all light up equally bright, but not as bright as with just one bulb in the conducting loop.

the circuit elements. Note also that the light bulb becomes quite hot, indicating that energy is transferred continuously to the bulb and then goes to the environment or to our fingers via heat transfer. These observations indicate that whatever is happening is not a static electrical effect but is fundamentally dynamic, since energy transfer is occurring continuously. It makes no difference where the bulb is placed along the loop; the effect is the same, as shown in Figure 19.7. The bulb lights up equally bright regardless of where it is placed along the loop.

Experiment 2: Several Light Bulbs

Place several identical bulbs along the closed loop path, as indicated in Figure 19.8. All the bulbs light up equally bright, but not as bright as with only a single bulb.

This observation indicates that whatever is taking place is happening at all points equally along the closed conducting path, not just at one point. The more bulbs used, the dimmer each is, but all the dim bulbs in a series are equally bright.

Experiment 3: Real Wire and an Independent Voltage Source

Similar effects can be observed with no bulb and some real copper or aluminum wires of various lengths connected to a small battery. In Figure 19.9, a very long such real wire is

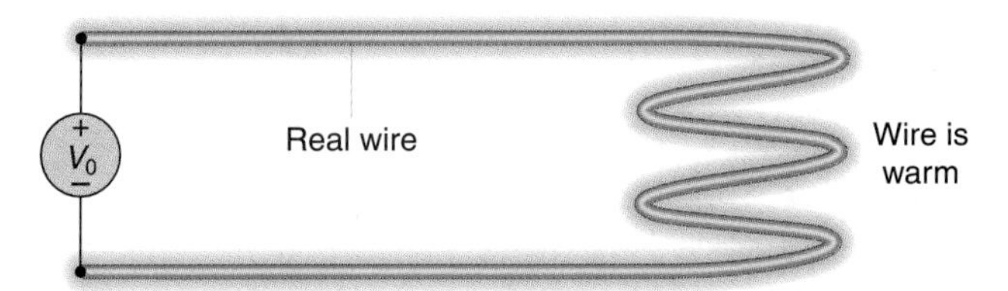

FIGURE 19.9 A long piece of real wire connected to an independent voltage source becomes equally warm along its entire length.

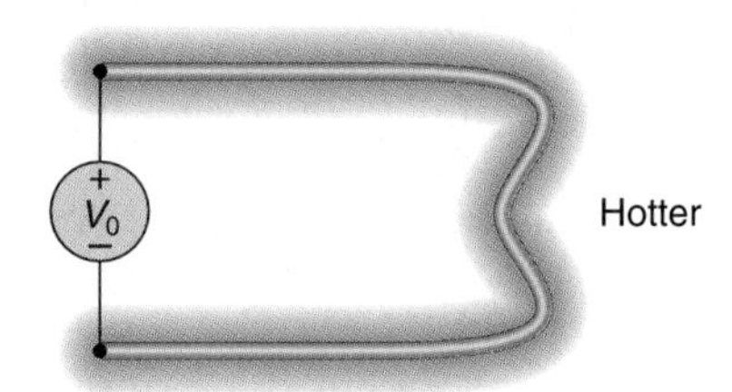

FIGURE 19.10 A shorter wire gets even hotter than a longer wire.

connected to the two terminals of the battery. The real wire is found to become equally warm at all points along its length.

If a shorter length of real wire made of the same material and with the same cross section is used, as in Figure 19.10, the wire becomes even hotter than the very long wire. Hence the length of the real wire somehow affects the amount of energy transfer per unit length to the real wire.

Such experiments and many others indicated to early researchers that whatever was happening was rather like the flow of a fluid through a closed loop of pipe. The battery acts like a pump, increasing the potential energy of the charges transferred through it. The ideal wires act as pipes, and the light bulb or the real wire lengths act like frictional impediments or resistances in the pipe that decrease the potential energy of the charges passing through them. The loss of electrical potential energy increases the thermodynamic internal energy of the material (it becomes warmer), and the energy eventually is passed along to the environment via heat transfer.

From such experiments arose the concepts of an electric **current** and electric **resistance**. From other experiments, such as those we performed with a charged capacitor, we know that the flow involved is a flow of electric charge. Typically it is of negative electrons, though this is not apparent from the experiments mentioned here.* We now explore the concept of electric current, or the flow of electric charge, in quantitative detail.

19.2 Electric Current

The flow of air is called a wind; we could just as well call the flow of the charge property an electric wind, but such terminology is not used. Moving charges constitute an electric current, just as

*In Chapter 20 we will investigate an experiment (the Hall effect) that is used to determine the sign of the charge carriers involved in electric currents. In conducting materials, the moving charge carriers are electrons. In semiconducting materials, positive and/or negative charge carriers are present.

the flow of water is called a current. A quantitative measure of the current is the amount of the charge property that passes a given location per unit time. You can imagine a current meter as something like a traffic counter that tallies the coulombs of charge that move by some point during one second.

> The time rate of charge flow is defined to be the electric current I:
>
> $$I \equiv \frac{dQ}{dt} \tag{19.1}$$
>
> The SI unit for current is the **ampere** (A), one coulomb per second (C/s).

The unit is named for the Frenchman André-Marie Ampère (1775–1836), an early researcher into electrical phenomena.

The particles that carry the charge property are called *charge carriers*. Charge carriers in motion constitute an electric current. The carriers may have positive charge or negative charge. In some materials the charge carriers are exclusively negative charge carriers (ordinary conductors); in other materials the charge carriers are predominantly positive charge carriers, predominantly negative charge carriers, or both positive and negative charge carriers. A current exists anytime there is a charge in motion. The charges need not be in conducting wires. For example:

1. The electron orbiting the proton in the hydrogen atom represents a charge in motion and therefore constitutes an electric current.
2. The moving collection of charged particles in electron beams, proton beams, positron beams, antiproton beams, or ion beams—each constitutes an electric current. Charged particles moving in a vacuum or a material are a current whether the velocity of the particles is constant or not.

Ampere was fascinated by numbers even as a toddler. The deeply religious Ampere made contributions to mathematics, philosophy, botany, taxonomy, chemistry, and the physics of electricity and magnetism.

3. Electrons not attached to any particular atoms are abundantly available in metals, and currents result from their motion in response to an applied electric field.
4. In a pure semiconductor (called an intrinsic semiconductor), currents result from the motion of equal numbers of negative and positive charge carriers in response to an applied electric field; the negative charges are electrons, the propagating vacancies left by electron migration (which act like they have positive charge) are called holes. Small amounts of impurity elements are added to pure semiconducting materials, a process called **doping**, to make two other types of semiconducting materials: (a) n-type semiconducting materials, in which most of the moving charges are electrons (the n stands for *n*egative charge carrier); (b) p-type semiconducting materials, in which most of the moving charges are the positive holes (p stands for *p*ositive charge carrier).

When considering the atomic view of matter, we must be careful with what we mean by the motion or flow of charge. If we take an isolated uncharged material, whether a conductor or not, and set it in motion, there are charges in motion (the positive nuclei and the surrounding electrons), but the current is zero because there is no motion associated with the net or total charge. On the other hand, a charged particle in motion does represent a current.

Likewise, for an uncharged conductor at rest, we find that some (but not all) electrons are continually moving randomly through the conductor, because they are very loosely attached to the nuclei. The thermodynamic internal energy of the material is sufficient to liberate the outer electrons from individual atoms, enabling the electrons to travel about in the material. But just as many electrons are moving one way as in the opposite direction, and the net flow of charge past any location is zero; there is zero current. This is the situation in electrostatics where the total electric field inside the conductor (whether charged or not) is zero. On the other hand, if an independent voltage source maintains even a slight potential difference between two sides (or ends) of a conductor, the electrons preferentially move in the direction opposite to that of the electric field established in the conductor by the independent voltage source; we have an electrodynamic situation where there is a net motion of charge, and so a current exists.

An electric field causes charges to accelerate. An electric field inside a material can be maintained by connecting the material to the terminals of an independent voltage source (say, an ideal battery), as in Figure 19.11, where we pretend the charges in motion are positive for reasons that will become apparent shortly.

The independent voltage source maintains a fixed potential difference between the ends of the material. The mobile charges

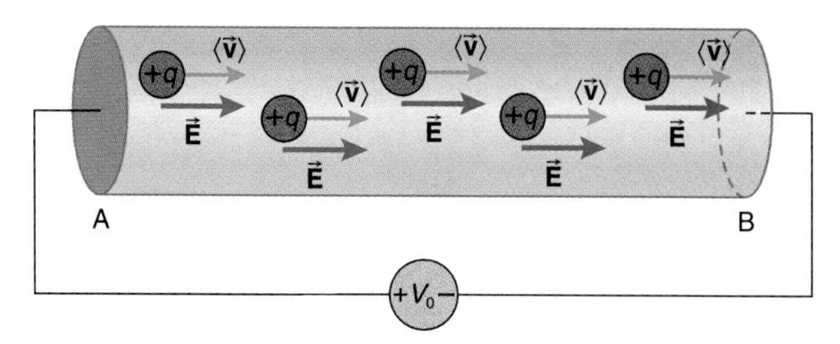

FIGURE 19.11 An independent voltage source maintains a current through a material.

in the material accelerate under the influence of the field (parallel to the field for positive charge carriers, antiparallel to the field for negative charge carriers). However, the accelerating charges collide very frequently with other particles inside the material. Such collisions mean that the velocities of the mobile charge carriers continually change their magnitude and direction almost randomly, though the overall motion or drift is more or less parallel or antiparallel to the field (depending on the sign of the charge carriers).

This rather messy situation is made tractable by introducing an **average drift speed** $\langle v \rangle$ for the mobile charges. As its name implies, the average drift speed represents the average speed at which the mobile charge carriers migrate through the material, parallel or antiparallel to the field. Although $\langle v \rangle$ is called the drift speed, $\langle v \rangle$ is more properly thought of as the (single) *component* of the average velocity vector of the charge carriers.

We wish to relate the current to this velocity component. Let the density of mobile charges (the charge carriers) be n; this represents the number of charge carriers per unit volume of the material. Let q be the charge on each charge carrier. Imagine the material to be cylindrically shaped, as in Figure 19.12. The electric current is the time rate of charge flow:

$$I = \frac{dQ}{dt}$$

where dQ is the number of coulombs of charge that pass a reference location of cross-sectional area A during a time interval dt. Charge conservation implies that the amount of charge per second arriving at and leaving any point along a wire is the same, thus preventing an otherwise embarrassing growing accumulation of charge (a pileup!) at any location within the wire. All the charge carriers within a distance $\langle v \rangle\, dt$ to the left of the reference location in Figure 19.12 will pass the location during the time interval dt. The number of charge carriers within this volume is

$$\text{(number of charge carriers per unit volume)(volume)} = n(A\langle v \rangle\, dt)$$

Each charge carrier has a charge q, and so the charge dQ within this volume is

$$dQ = qnA\langle v \rangle\, dt \qquad (19.2)$$

The current I is the time rate of charge flow past the location, or

$$I = \frac{dQ}{dt}$$

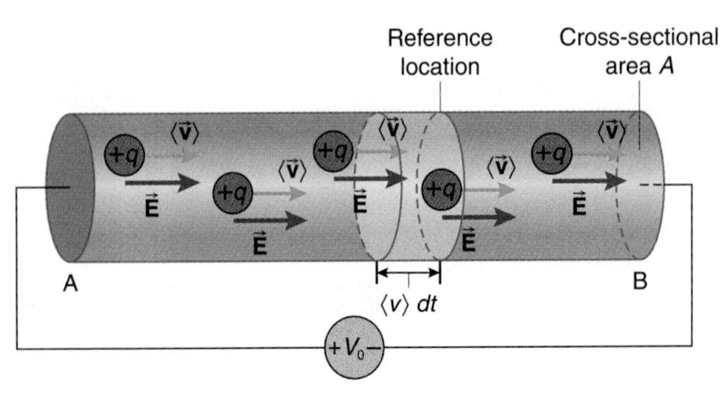

FIGURE 19.12 An electric current.

So, from Equation 19.2, the current is

$$I = qnA\langle v \rangle \qquad (19.3)$$

Note that in Equation 19.3, if we reverse the sign of *both* q and its velocity component $\langle v \rangle$, the *same* current I results. This is an important point; it means that positive charge flowing in one direction is *equivalent* to negative charge flowing in the opposite direction: the same current I results.

But if the flow of negative charge in one direction is equivalent to the flow of positive charge in the opposite direction, why, then, is there any current at all? The reason is because equivalent is not the same thing as actual. The two viewpoints can substitute for each other but are not in addition to each other.

Experiments now indicate the charge carriers in metal conductors are electrons, though this is not at all obvious. There is no way we can tell by examining, say, the experiments we performed in the last section whether the electric current in the wires is due to positive charge flowing one way, negative charge flowing the other way, or both.

By a convention dating back to Benjamin Franklin, the direction of a current is taken to be the direction of positive charge flow, whatever the actual sign of the charge carriers.

It is only through other experiments that we know the charge flow or current in conductors is caused by the motion of the negative electrons. In other materials, such as semiconductors, plasmas, or ionic solutions, the charge carriers may be positive, negative, or both.

In Example 19.1 you will calculate the average drift speed of the electrons in a common copper wire. The drift speed is surprisingly small; most snails easily outrun the electrons, and a terrapin is a fleet-footed demon by comparison. It takes on the order of *hours* for the electrons to drift through a distance of several meters in a common wire conductor! This should be surprising to you. Why does a distant light bulb come on almost instantaneously when you turn on the switch, if it takes the electrons so long to travel even one meter? Why does it not take hours for the bulb to light once the switch is turned on?

That is a good question and it warrants more than a lousy answer. An analogy may help. Hook up a long hose to a water faucet and turn the faucet on; it takes some time for the hose to fill with water and for a stream to emerge from the nozzle at the other end; the longer the hose, the longer the wait. On the other hand, if the hose is already full of water and the faucet is turned on, the water emerges from the other end as soon as the pressure increase propagates through the water in the hose to the other end. Since water is quite incompressible, the pressure pulse travels at great speed down the length of the water column in the hose and the water emerges from the nozzle almost immediately after the faucet is turned on.

Within a wire, the charge carriers (electrons) are distributed all along the wire and, like the water in a filled hose, this means that the electrical pipe (the wire) is already filled and ready for the faucet (the switch) to be turned on. The electric field that instigates the charge drift motion, analogous to the pressure pulse in the hose, propagates down the wire at a speed nearly equal to the speed of light. Thus the electrons all along the wire feel the field almost instantaneously and pick up their drift speed very fast. The current appears all along the wire almost instantaneously.

It is occasionally convenient to eliminate the dependence on the cross-sectional area in Equation 19.3, by defining a **current density** J across it to be

$$J = \frac{I}{A} = qn\langle v\rangle \quad (19.4)$$

The current density J has the SI units of amperes per square meter (A/m^2). We make the current density a full-fledged vector by giving it the direction of the average drift velocity vector $\langle\vec{v}\rangle$:

$$\vec{J} = qn\langle\vec{v}\rangle \quad (19.5)$$

In general, the current I is the *flux* of the current density through an area:

$$I = \int_{\text{area}} \vec{J}\cdot d\vec{S} \quad (19.6)$$

The flux of $\vec{J}$ through the cross-sectional area A, shown in Figure 19.13, is easy to calculate, since $\vec{J}$ is parallel to the differential area vector everywhere across the cross-sectional area.

We also assumed the current density was constant over the entire area A, and so we have

$$I = \int_{\text{area}} \vec{J}\cdot d\vec{S} = J\int_{\text{area}} dS = JA$$

Thanks to the scalar product, we obtain the same current, or flux of the current density $\vec{J}$, regardless of the orientation of the area slice through the wire through which the flux is taken (see Figure 19.14). We have to, because it is the same total current everywhere.

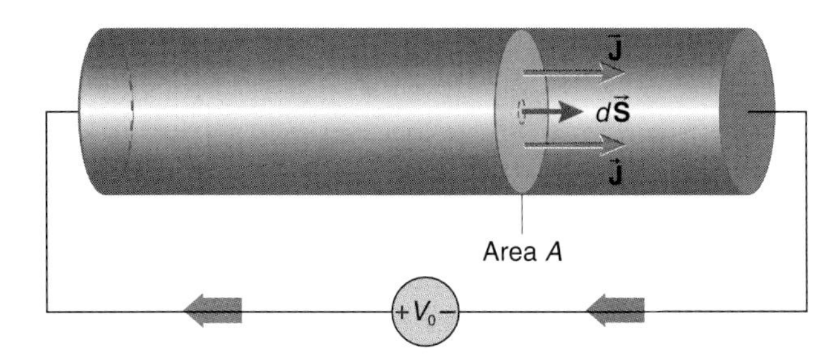

FIGURE 19.13 The flux of the current density through a cross-sectional area of the wire is the current.

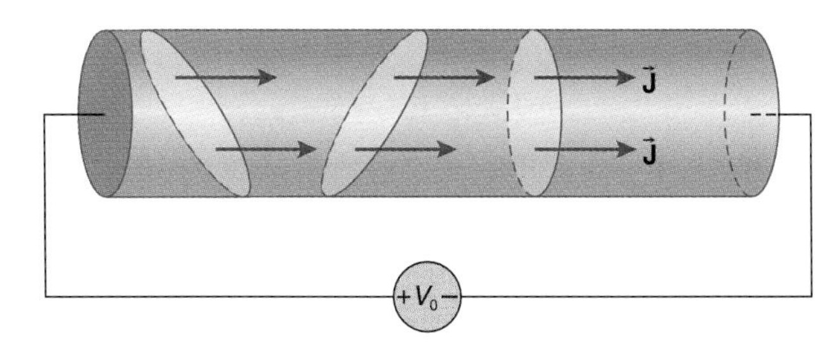

FIGURE 19.14 The current is the same through any cross section of the wire.

QUESTION 1

Geometrically, what represents the electric current at an instant *t* on a graph of charge *Q* versus time?

EXAMPLE 19.1

A 12 gauge copper wire, used for some household wiring, has a diameter of 2.05 mm and carries a current of 15.0 A.

a. Assume one electron per copper atom is responsible for the current. Find the drift speed of the electrons in the wire. The density of copper is 8.93×10^3 kg/m^3 and its molar mass is 63.5 g/mol.

b. How long does it take each charge carrier in this wire to travel 1.00 m?

Solution

a. The number of charge carriers per unit volume, n, is the same as the number of copper atoms per unit volume, since you are assuming one charge carrier per atom. One cubic meter of copper has a mass of 8.93×10^3 kg and one mole of copper has a mass of 63.5 g. Hence one cubic meter of copper represents

$$\frac{8.93\times10^3\ \text{kg/m}^3}{63.5\times10^{-3}\ \text{kg/mol}} = 1.41\times10^5\ \text{mol/m}^3$$

Each mole has Avogadro's number of copper atoms. Hence the number of copper atoms per cubic meter is

$$(1.41\times10^5\ \text{mol/m}^3)(6.02\times10^{23}\ \text{atoms/mol}) = 8.49\times10^{28}\ \text{atoms/m}^3$$

With one charge carrier per atom, the number of charge carriers per cubic meter is this same number:

$$n = 8.49\times10^{28}\ \text{electrons/m}^3$$

The circular cross-sectional area of the wire is

$$\begin{aligned} A &= \pi r^2 \\ &= \pi[(2.05\times10^{-3}\ \text{m})/2]^2 \\ &= 3.30\times10^{-6}\ \text{m}^2 \end{aligned}$$

The charge carriers are electrons, which have a charge of magnitude $e = 1.602\times10^{-19}$ C. From Equation 19.3, the *average drift speed* is

$$\begin{aligned} \langle v\rangle &= \frac{I}{qnA} \\ &= \frac{15.0\ \text{A}}{(1.602\times10^{-19}\ \text{C})(8.49\times10^{28}\ \text{m}^{-3})(3.30\times10^{-6}\ \text{m}^2)} \\ &= 3.34\times10^{-4}\ \text{m/s} \end{aligned}$$

This speed is only a few tenths of a millimeter per second! This is a very slow speed. You can see that the current in the wire results not from the rapid flow of a huge charge, but rather from the slow flow of a humongous number of very small charges.

b. The time it takes the electrons to travel one meter in this wire is the distance divided by the average drift speed:

$$\begin{aligned} t &= \frac{1.00 \text{ m}}{3.34 \times 10^{-4} \text{ m/s}} \\ &= 2.99 \times 10^{3} \text{ s} \\ &= 49.8 \text{ min} \end{aligned}$$

or slightly less than the leisurely duration of a typical physics lecture.

19.3 The Pièce de Résistance: Resistance and Ohm's Law

Now we come to grips with the resistive effects to charge flow in conducting materials, phenomena analogous to the resistive effect of obstructions or friction experienced by water when flowing in pipes. This introduces a very important new circuit element: a **resistor**.

The potential difference V between the points A and B along a length of material (see Figure 19.15) is found from the defining equation for the potential difference between two points, Equation 17.9:

$$V_B - V_A = -\int_A^B \vec{E} \cdot d\vec{r}$$

If the electric field $\vec{E}$ is constant over the entire length ℓ of the material, Equation 17.9 becomes

$$V_B - V_A = -E\ell$$

where V_A is greater than V_B (note the direction of the electric field in Figure 19.15). By convention, we take the potential difference V between the ends of the material to be the higher value minus the lower value; we say the potential difference is *across* the material:

$$V \equiv V_A - V_B = E\ell \tag{19.7}$$

For many materials, it is found experimentally that the magnitude of the current density in the material is proportional to the magnitude of the applied electric field:

$$J = \sigma E \tag{19.8}$$

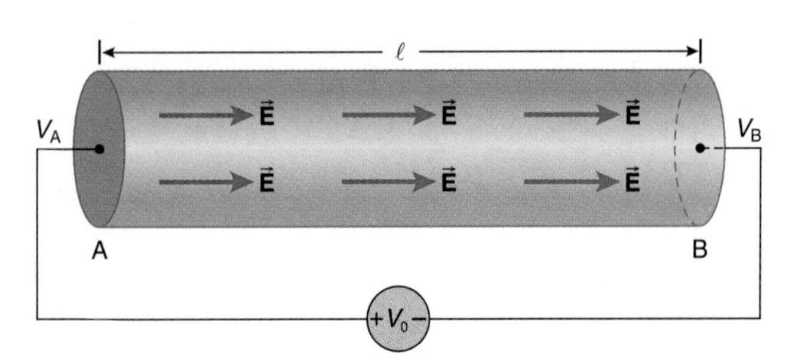

FIGURE 19.15 A material connected to an independent voltage source.

where σ is a proportionality constant called the **conductivity** of the material.* Such materials are called **ohmic materials**; Equation 19.8 is one version of **Ohm's law**, though we will soon recast the law into a more useful form. It is important to realize that Ohm's law is an empirical law and does *not* characterize all materials. Using the ohmic approximation in Equation 19.7 for the potential difference, we find

$$\begin{aligned} V &= E\ell \\ &= \frac{J}{\sigma}\ell \end{aligned} \tag{19.9}$$

We assume the current density is uniform across the cross-sectional area of the material.† Using Equation 19.4 to substitute for J in Equation 19.9, we find

$$V = I\frac{\ell}{\sigma A} \tag{19.10}$$

The reciprocal of the conductivity is called the **resistivity** (ρ) of the material:

$$\rho \equiv \frac{1}{\sigma} \tag{19.11}$$

The resistivity of various materials is listed in Table 19.1.

In terms of the resistivity, Equation 19.10 becomes

$$V = I\frac{\rho\ell}{A} \tag{19.12}$$

We define the **resistance** R of the material to be

$$R \equiv \frac{\rho\ell}{A} \tag{19.13}$$

The SI units of resistance are V/A or **ohms** (Ω), after Georg Simon Ohm (1787–1854), a German physicist.

Equation 19.13 indicates that the resistance is proportional to the length ℓ of an ohmic material of constant cross-sectional area A. Empirically, the resistance is found to be inversely proportional to the cross-sectional *area* A of the

*Do not confuse the symbol for conductivity with same symbol for the surface charge density in electrostatics.

†This assumption is confirmed for dc currents by experiments we mention shortly; the assumption is not necessarily true for ac currents.

Ohm delved into electrical research, discovering his now famous law, to escape a pedestrian life as a tutor and to secure a university position. You better do your homework for him!

TABLE 19.1 The Resistivity and Temperature Coefficient of Resistivity for Various Materials at 20 °C

Material	*Resistivity* ρ ($\Omega\cdot$m)	*Temperature coefficient of resistivity* α (K^{-1})
Aluminum	2.82×10^{-8}	3.9×10^{-3}
Carbon	3500×10^{-8}	-0.5×10^{-3}
Copper*	1.77×10^{-8}	3.8×10^{-3}
Germanium	0.46	-48×10^{-3}
Gold	2.44×10^{-8}	3.4×10^{-3}
Glass	$\sim 10^{12}$	
Iron	10×10^{-8}	5.0×10^{-3}
Lead	22×10^{-8}	4.3×10^{-3}
Mercury	96×10^{-8}	0.9×10^{-3}
Nichrome†	100×10^{-8}	0.4×10^{-3}
Platinum	10×10^{-8}	3.92×10^{-3}
Silicon	640	-75×10^{-3}
Silver	1.59×10^{-8}	3.8×10^{-3}
Tungsten‡	5.6×10^{-8}	4.5×10^{-3}

*Most common electrical wire is made of copper.

†This is a nickel–chromium–iron alloy used widely in electrical heaters such as toasters.

‡Tungsten is used widely for light bulb filaments.

material. If the resistance were found to be inversely proportional to the *radius* of the conductor, one could infer that the current was confined to a certain outer layer of atoms at the perimeter of the circular cross section; instead, it is inversely proportional to the *square* of the radius. Therefore, we can infer that conduction in ohmic materials takes places in the bulk material and not on only its surface, which partially justifies the assumption we made that the current density in the material is uniform across the area A.

For convenience, Table 19.2 lists a few common wire sizes by their gauge numbers, their corresponding diameters in millimeters, and cross-sectional areas in square meters. You may find this table useful in computing the resistance of various real wires. Household wiring typically is 14 gauge, but 12 gauge is used for kitchens and baths, which utilize large currents; 8 gauge wire is used for range ovens and clothes dryers. Common extension cords range from 18 gauge to 10 gauge, depending on the amount of current they are expected to carry.

TABLE 19.2 Wire Gauge Number, Wire Diameter, and Cross-Sectional Area at 20 °C for Common Wire Sizes

*Gauge number**	*Diameter*	*Area*
0	8.25 mm = 8.25×10^{-3} m	53.5×10^{-6} m^2
2	6.54 mm = 6.54×10^{-3} m	33.6×10^{-6} m^2
4	5.19 mm = 5.19×10^{-3} m	21.1×10^{-6} m^2
6	4.12 mm = 4.12×10^{-3} m	13.3×10^{-6} m^2
8	3.26 mm = 3.26×10^{-3} m	8.37×10^{-6} m^2
10	2.59 mm = 2.59×10^{-3} m	5.26×10^{-6} m^2
12	2.05 mm = 2.05×10^{-3} m	3.31×10^{-6} m^2
14	1.63 mm = 1.63×10^{-3} m	2.08×10^{-6} m^2
16	1.29 mm = 1.29×10^{-3} m	1.31×10^{-6} m^2
18	1.02 mm = 1.02×10^{-3} m	0.823×10^{-6} m^2
20	0.812 mm = 0.812×10^{-3} m	0.517×10^{-6} m^2
22	0.644 mm = 0.644×10^{-3} m	0.326×10^{-6} m^2
24	0.511 mm = 0.511×10^{-3} m	0.205×10^{-6} m^2
26	0.405 mm = 0.405×10^{-3} m	0.129×10^{-6} m^2
28	0.321 mm = 0.321×10^{-3} m	0.081×10^{-6} m^2

*For wire diameters of odd number gauge sizes, consult a handbook on electrical engineering.

We now combine Equations 19.12 and 19.13.

> We find the potential difference across an ohmic material (high value − low value) is proportional to the current through it:
>
> $$V = IR \qquad (19.14)$$
>
> This is the most useful form for Ohm's law.

The standard circuit symbol for an **ideal resistor** is shown in Figure 19.16.* As with every ideal circuit element, we consider the leads to be *ideal wires* with zero resistance—that is, infinitesimal compared with real values for circuit resistors. Charges flow through an ideal wire with no loss of potential energy. We use ideal wires in circuit diagrams to connect discrete circuit elements. We have used the notion of ideal wires already as a way to connect other idealized circuit elements: independent voltage sources (ideal batteries) and ideal capacitors.

The direction of current is taken, by convention, to be that of positive charge flow. If an electric current exists in a resistor, the current enters it at the high-potential end and leaves at the low-potential end.

> Thus we make polarity markings on a resistor consistent with the direction of the current through it, as shown in Figure 19.17. The current enters the resistor circuit element at the (+) polarity lead and exits at the (−) polarity lead of the circuit symbol.

*You can think of the resistor symbol as representing the filament of a light bulb, but keep in mind that not all resistors are light bulbs!

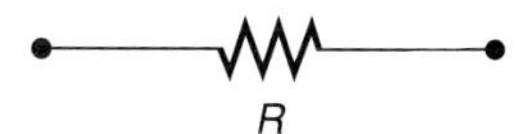

FIGURE 19.16 Circuit symbol for a resistor.

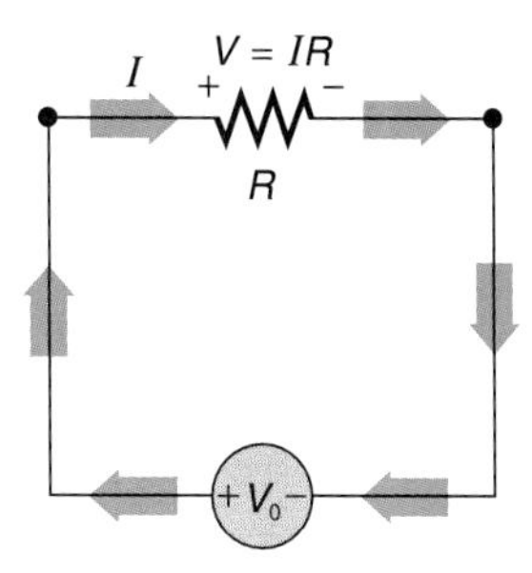

FIGURE 19.17 The polarity marking of a resistor is written so that the current enters it at the (+) side.

The potential difference V between the resistor lead marked (+) and that marked (−) is given by Ohm's law, Equation 19.14.

According to Ohm's law, an ideal resistor has a linear relationship between the potential difference V across it and the current through it. If you plot the current versus the potential difference across an ideal resistor, a straight line is obtained whose slope is the inverse of the resistance, as shown in Figure 19.18.

This linear relationship is an identifying characteristic of all ohmic materials. A negative current means it is directed opposite to that indicated in Figure 19.17; the potential difference $V = IR$ also then is negative, which means that the terminal marked (+) actually is at a lower potential than the one marked (−). This does *not* mean a mistake was made; the significance of these negative values will be noted later in the chapter (Problem-Solving Tactic 19.4 and Example 19.8). Materials or devices with *nonlinear* current–voltage relationships are called **nonohmic** materials or devices.

Off-the-shelf resistors, such as those available at electronic suppliers, have pretty little colored bands encircling them to indicate the approximate numerical value of the resistance, as shown in Figure 19.19. The colored bands are a code to determine the *nominal* value of the resistance in ohms. The *actual* resistance value is within a certain percentage of the nominal value; the percentage is specified by an additional color band indicating the tolerance.

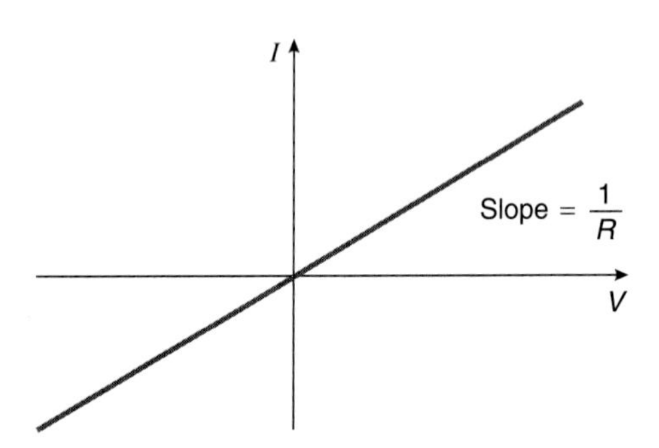

FIGURE 19.18 The current through a resistor plotted versus the potential difference across it; the inverse of the slope is the resistance.

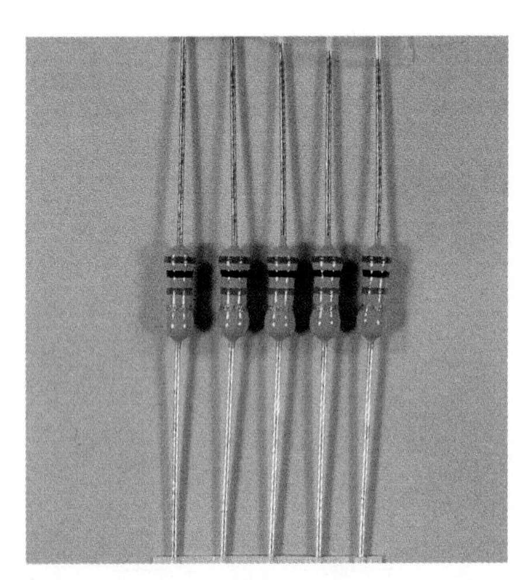

FIGURE 19.19 Resistors with a nominal value of 1 kΩ have the colored bands brown, black, and red with the fourth color band (here, gold) indicating the tolerance.

The resistor color code is as follows. Begin at the lead closest to the color bands. The first two colored bands signify the first two numbers of the value of the resistance according to the scheme in Table 19.3. The third colored band, called the multiplier band, indicates the exponent of the factor 10 by which to multiply the first two numbers to find the nominal value of the resistance in ohms.

The value of the resistance is determined according to the scheme

$$[\text{number number}] \times 10^{(\text{multiplier})}$$

The accuracy associated with the nominal value of the resistance is indicated by a fourth colored band according to the following scheme:

Color	*Tolerance*
Gold	5%
Silver	10%
(No fourth band)	20%

For examples, the resistors in Figure 19.19 have the colored bands brown–black–red–gold. The first two numbers of the value of the resistance are

$$\text{brown} = 1 \quad \text{and} \quad \text{black} = 0$$

or

$$10$$

and the multiplier band is red (2) so the multiplying factor is 10^2. The nominal resistance thus is

$$\begin{aligned}[\text{number number}] \times 10^{(\text{multiplier})} &= [10] \times 10^2\ \Omega \\ &= 1.0\ \text{k}\Omega\end{aligned}$$

The gold band indicates that the actual numerical value of the resistance is within 5% of 1000 Ω. Since 5% of 1000 is 50 Ω, the actual resistance could be anywhere from 950 Ω to 1050 Ω. As an another example, a resistor with bands colored yellow (4), violet (7), and orange (3) with *no* fourth band has a nominal value of its resistance of

$$47 \times 10^3\ \Omega = 47\ \text{k}\Omega$$

with a tolerance of 20% (±9.4 kΩ).

TABLE 19.3 Resistor Color Code

Color	*Number*	*or Multiplying factor*
Black	0	$10^0 = 1$
Brown	1	10^1
Red	2	10^2
Orange	3	10^3
Yellow	4	10^4
Green	5	10^5
Blue	6	10^6
Violet	7	10^7
Gray	8	10^8
White	9	10^9

A resistance measuring device known as an **ohmmeter** is used to measure the actual resistance of individual resistors; we will see how to use an ohmmeter later in the chapter (Section 19.15).

Your physics class might run a contest to create a tasteful and unprofane mnemonic to remember the order of the colors in the resistor color code. For example (and you likely can do better):

"Black Bears Roar, Orangutans Yell, Goats Bleat Violently, Go Weep!"

QUESTION 2

A resistor has the colored bands orange, black, yellow, and gold, in that order. Between what two values is the actual resistance of this resistor supposed to be according to the color code?

EXAMPLE 19.2

Calculate the resistance of 10.0 m of 12 gauge copper wire. Such wire is used frequently in home wiring.

Solution

The resistance is given by Equation 19.13:

$$R = \frac{\rho \ell}{A}$$

The resistivity of copper is found in Table 19.1 on page 843 and the cross-sectional area of a 12 gauge wire is found in Table 19.2. Making the appropriate substitutions, you have

$$\begin{aligned} R &= \frac{(1.77 \times 10^{-8}\ \Omega\cdot\text{m})(10.0\ \text{m})}{3.31 \times 10^{-6}\ \text{m}^2} \\ &= 5.35 \times 10^{-2}\ \Omega \end{aligned}$$

This indicates that the resistance of even long pieces of common copper wire is quite small, so such wires can be treated as ideal wires to an excellent approximation.

EXAMPLE 19.3

Calculate the magnitude of the electric field inside a 10.0 m length of 12 gauge copper wire that carries a current of 15.0 A. Assume the field is uniform along the entire length of the wire.

Solution

In Example 19.2, you found the resistance of such a piece of wire was $5.35 \times 10^{-2}\ \Omega$. The potential difference between the ends of the wire is found using Ohm's law, Equation 19.14:

$$\begin{aligned} V &= IR \\ &= (15.0\ \text{A})(5.35 \times 10^{-2}\ \Omega) \\ &= 0.803\ \text{V} \end{aligned}$$

If you assume the field is uniform along its length, the magnitude of the potential difference is related to the magnitude of the electric field by Equation 19.7:

$$V = E\ell$$

Solving for E, you have

$$\begin{aligned} E &= \frac{V}{\ell} \\ &= \frac{0.803\ \text{V}}{10.0\ \text{m}} \\ &= 8.03 \times 10^{-2}\ \text{V/m} \end{aligned}$$

Recall that volts/meter (V/m) are equivalent to newtons/coulomb (N/C). This example shows that the magnitude of the electric field which causes the electric current inside a conductor is quite small. Notice that since both R and V are proportional to the length ℓ, the value of E is the same for any length of wire of a fixed gauge carrying a given current.

19.4 RESISTANCE THERMOMETERS

For many ohmic materials, the resistivity ρ and the resistance R are linear functions of temperature within certain temperature extremes. That is, we have the empirical relations of the following form:

$$\rho = \rho_0[1 + \alpha(T - T_0)] \tag{19.15}$$

and

$$R = R_0[1 + \alpha(T - T_0)] \tag{19.16}$$

where α is a constant known as the **temperature coefficient of resistivity**. Values of α for various materials are tabulated in Table 19.1 on page 843. The quantity R_0 is the resistance of the material at a reference temperature T_0 while ρ_0 is the corresponding resistivity. The temperature dependence of the resistance is exploited to make resistance thermometers, commonly used for medical applications and in physical laboratories.

In a resistance thermometer, a measurement of the resistance is used to find the temperature of the resistor. If the resistor is in thermal equilibrium with its environment, the temperature found via Equation 19.16 also gives the temperature of the environment. Resistance thermometers have several advantages over traditional mercury bulb thermometers:

1. Resistance thermometers can be fabricated with small physical size (dimensions) and mass, so that it is possible to measure the temperature of small systems easily and reliably. Recall that a thermometer typically must be small in mass and size compared with the system whose temperature is to be measured, unless the system is in thermal contact with a thermodynamic reservoir.
2. Resistance is an electrical quantity, so it is easy to use the electrical measurement of temperature as the first stage in the electronic manipulation of data.

If the resistance is not a linear function of temperature, then of course Equation 19.16 cannot be used. Manufacturers of resistance thermometers supply calibration curves that plot empirically determined temperature versus resistance data.

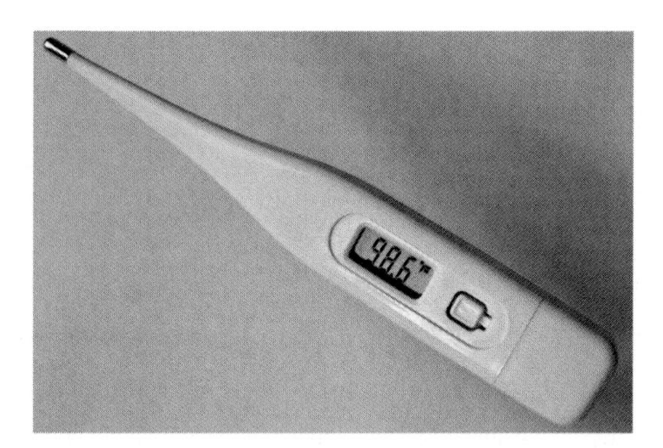

An electric thermometer with a digital readout is quite useful and convenient for both research and medical applications.

A resistance thermometer is an example of a **transducer**, a device for converting a physical quantity into (usually) an electrical quantity or vice versa.

QUESTION 3

Notice in Table 19.1 on page 843 that carbon, germanium, and silicon all have negative temperature coefficients of resistance; the resistance decreases as the temperature increases. Does this imply that their resistance goes to zero at sufficiently high temperatures? What might prevent this?

EXAMPLE 19.4

What temperature change produces a 10% increase in the resistance of a copper wire?

Solution

The resistance of the wire depends on temperature via Equation 19.16:

$$R = R_0[1 + \alpha(T - T_0)]$$

where R is the resistance at temperature T and R_0 is the resistance at temperature T_0. For a 10% increase, you want R to be equal to $1.10R_0$; hence you have

$$1.10R_0 = R_0(1 + \alpha\Delta T)$$

Solve for ΔT:

$$\Delta T = \frac{0.10}{\alpha}$$

The temperature coefficient of resistance α of copper is found from Table 19.1 on page 843, and so you obtain

$$\Delta T = \frac{0.10}{3.8 \times 10^{-3}\ \text{K}^{-1}} = 26\ \text{K}$$

Hence a temperature increase of 26 K increases the resistance of a copper wire by 10%.

EXAMPLE 19.5

A tungsten filament of a light bulb has a resistance of 18 Ω at a room temperature of 20 °C. The bulb is connected to an independent voltage source, as shown in Figure 19.20, and when the potential difference across the bulb has a magnitude of 30.0 V, the current through it is 0.185 A. What is the temperature of the filament of the bulb?

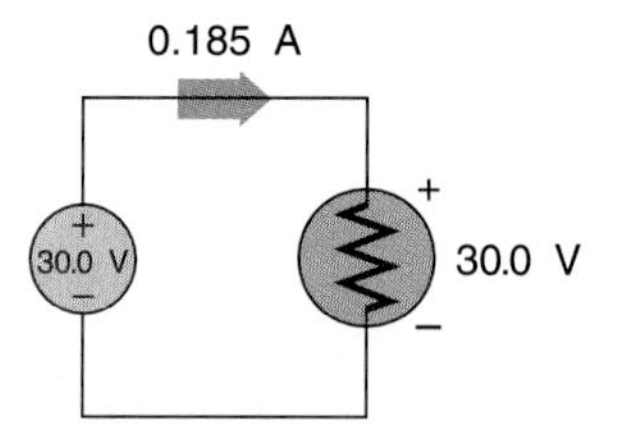

FIGURE 19.20

Solution

The temperature coefficient of resistivity for tungsten is found in Table 19.1 (on page 843) to be

$$\alpha = 4.5 \times 10^{-3}\ \text{K}^{-1}$$

Use Equation 19.16 and let R_0 be the resistance of the bulb at room temperature T_0; that is, $R_0 = 18\ \Omega$ when $T_0 = 293$ K. The resistance of the bulb when lit is found using the given data and Ohm's law, Equation 19.14:

$$V = IR$$

Solve for R:

$$R = \frac{V}{I} = \frac{30.0\ \text{V}}{0.185\ \text{A}} = 162\ \Omega$$

Using this information in Equation 19.16, you get

$$162\ \Omega = (18\ \Omega)[1 + (4.5 \times 10^{-3}\ \text{K}^{-1})(T - 293\ \text{K})]$$

Remember the sizes of a kelvin and a celsius degree are the same. Solving for T, you find

$$T = 2.1 \times 10^3\ \text{K}$$

Hot stuff! For comparison, this temperature is about 30% of the surface temperature of the Sun (about 6×10^3 K).

19.5 CHARACTERISTIC CURVES

No, this section is not about the grading curve of your professor.

A graph of the current through a circuit element versus the potential difference across the element is called the **characteristic curve** of the circuit element.

The potential difference is usually the most easily controlled electrical quantity, and so the potential difference always is graphed as the abscissa (the horizontal axis) with the current as the ordinate (the vertical axis).

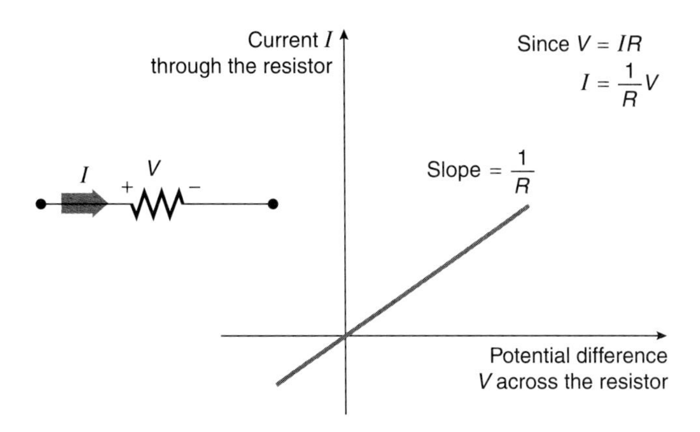

FIGURE 19.21 The characteristic curve of a resistor.

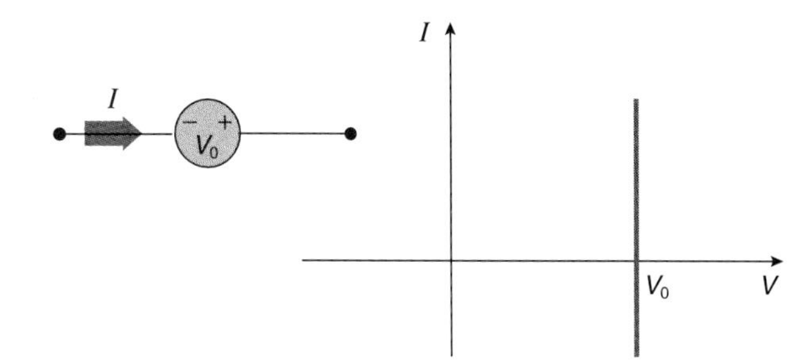

FIGURE 19.22 The characteristic curve of an independent voltage source.

From Ohm's law, Equation 19.14, the characteristic curve of an ideal resistor, a plot of I versus V,

$$I = \frac{1}{R}V$$

is a straight line inclined to the axes, as shown in Figure 19.21. An ideal resistor is called a linear device because its characteristic curve is linear with a positive slope. The slope of the characteristic curve of a resistor is the *reciprocal* of the resistance. For negative values of the potential difference, the current is negative and indicates that the current is in the direction opposite to that implied by the polarity markings.

The characteristic curve of an independent voltage source (such as an ideal battery) is a straight vertical line at the specified emf of the battery, as shown in Figure 19.22. The resistance of an ideal battery is zero, since the slope of its characteristic curve is infinite.

A distinctly nonlinear device is shown in Figure 19.23. It is the characteristic curve of a semiconducting device known as a pn junction. Such pn junctions form one of the fundamental building blocks of semiconducting technology. We define the (varying) resistance of the device as the inverse of the slope of the characteristic curve at any point. That is, more generally, the resistance at any point along *any* characteristic curve is defined as the inverse of the slope at that point:

$$R \equiv \frac{1}{\frac{dI}{dV}} = \frac{dV}{dI} \tag{19.17}$$

For a pn junction, the resistance depends on the value of the potential difference because the (inverse) slope of the curve changes. An ideal resistor has a resistance that is independent of the potential difference across it, since the relationship between the current and the potential difference is linear.

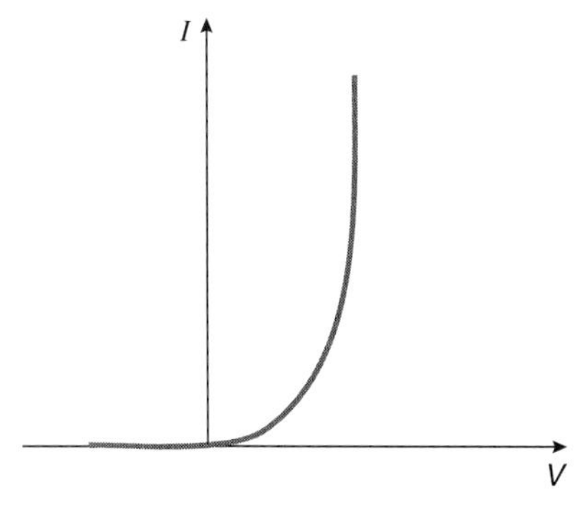

FIGURE 19.23 The characteristic curve of a semiconducting pn junction, a nonlinear device.

QUESTION 4

For the characteristic curve of the pn junction shown in Figure 19.23, what is the resistance of the device for negative values of V? How much current can pass through a resistance of such magnitude? What does the resistance approach for increasingly positive values of V? A resistance of zero is equivalent to a short circuit. A pn junction is called a *diode*, because it conducts current easily for positive values of V and has essentially zero current for negative values of V. A diode has the circuit symbol shown in Figure Q.4, where the arrow indicates the direction that charge flows easily through the device (the direction that significant current can pass through the device).

FIGURE Q.4

19.6 SERIES AND PARALLEL CONNECTIONS REVISITED

We now examine a series connection from the standpoint of electric current. Since charge is conserved, the electric current through every element in a series connection must be the same. What goes in one terminal comes out the other, just like water in a single pipe.

> The current is the same in all circuit elements in series with each other.

> Recall that the potential difference is the same across all circuit elements in parallel with each other.

19.7 Resistors in Series and in Parallel

The light bulb resistors in your home are connected in parallel; cheap strings of Christmas tree lights are connected in series. In this section we see how to combine resistors in series and parallel into equivalent single resistors.

Resistors in Series

Figure 19.24 shows a series connection of three resistors between terminals A and B, connected to an independent voltage source. Current I passes through each of the resistors, instigated by the independent voltage source. Because of the direction of the current, we must mark the resistors with the indicated polarities: the current enters at the high-potential end of each resistor [polarity marking (+)] and leaves at the low-potential end [polarity marking (–)].

The potential difference across each resistor is found from Ohm's law:

$$V_1 = IR_1 \qquad V_2 = IR_2 \qquad V_3 = IR_3 \tag{19.18}$$

We want to replace the series of resistors with a single equivalent resistor R_{eq} connected between the same two terminals A and B, as shown in Figure 19.25.

The current in the equivalent resistor also is I because that is the current between terminals A and B. The total potential difference between A and B is

$$V = V_1 + V_2 + V_3 \tag{19.19}$$

Since the potential difference between A and B across R_{eq} also must be V, we have $V = IR_{eq}$. Using Ohm's law for each potential difference in Equation 19.19, we obtain

$$IR_{eq} = IR_1 + IR_2 + IR_3$$
$$R_{eq} = R_1 + R_2 + R_3$$

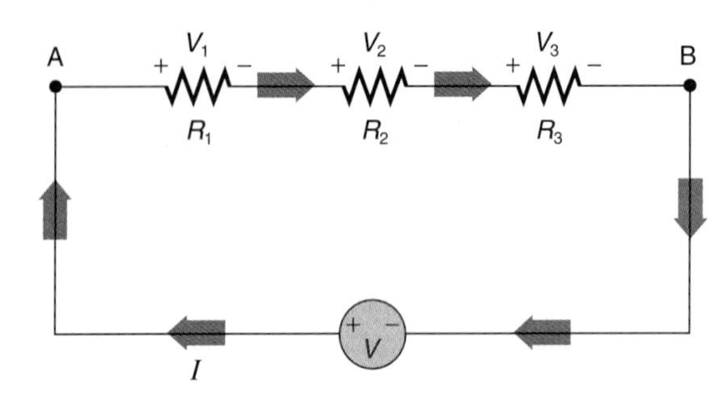

FIGURE 19.24 A series connection of three resistors.

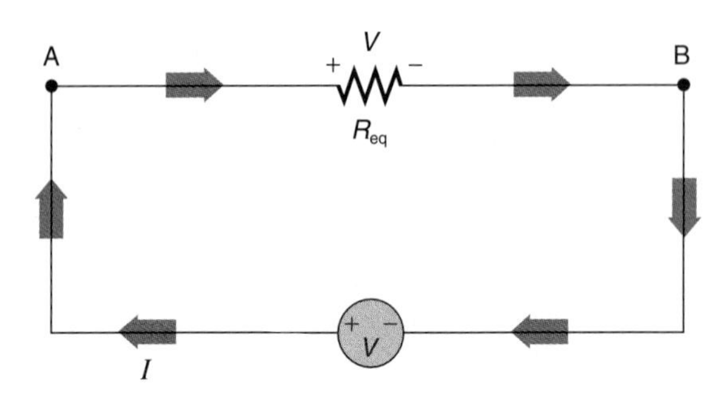

FIGURE 19.25 The equivalent resistor between the same two terminals A and B.

This argument can be extended to any number of resistors in series:

$$R_{eq} = R_1 + R_2 + R_3 + \cdots \qquad \text{(resistors in series)} \tag{19.20}$$

The equivalent resistance is the sum of the individual resistances. Resistors in series combine like capacitors in parallel.

Resistors in Parallel

If several resistors are connected in parallel, they all are connected to the same two distinct nodes, as shown in Figure 19.26. Because the collection is connected to the same two nodes, the potential difference V across each resistor is the same. We want to replace the set with an equivalent resistor with resistance R_{eq} (see Figure 19.27) so that the potential difference across R_{eq} is the same as across any of the resistors in parallel.

The current I leaving terminal A in Figure 19.26 divides (much like water dividing up among multiple pipes), with only a fraction going through each of the individual resistors. But conservation of charge implies that the currents through the individual resistors must sum to the current entering terminal A:

$$I = I_1 + I_2 + I_3 \tag{19.21}$$

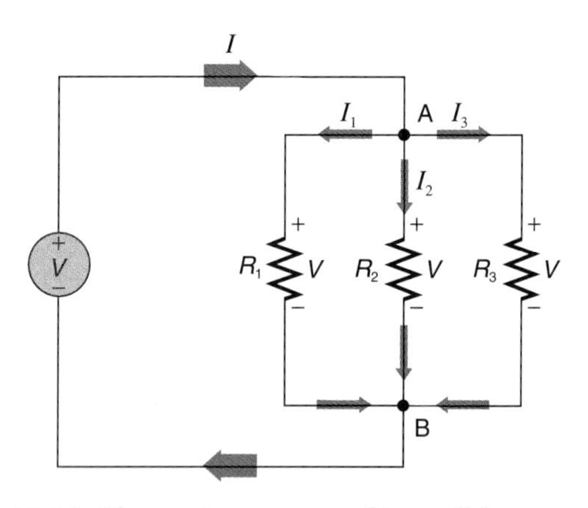

FIGURE 19.26 Three resistors connected in parallel.

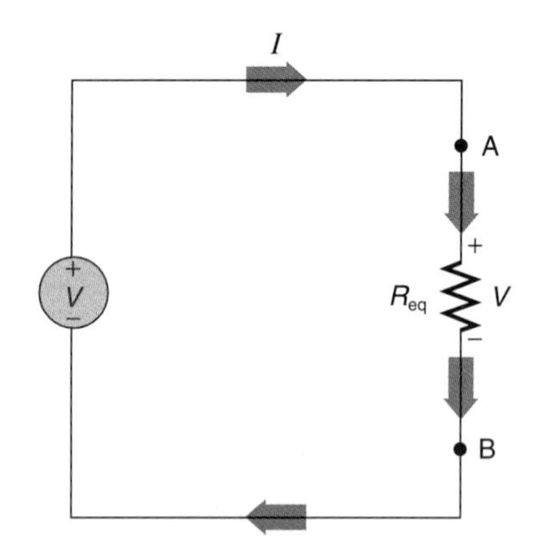

FIGURE 19.27 The equivalent resistor.

For each resistor we write Ohm's law, remembering that the potential difference across each is the same (since they are in parallel):

$$V = I_1R_1 \qquad V = I_2R_2 \qquad V = I_3R_3 \qquad \text{and} \qquad V = IR_{eq} \tag{19.22}$$

We solve these Ohm's law equations for the individual currents, and substitute into Equation 19.21. We obtain

$$\frac{V}{R_{eq}} = \frac{V}{R_1} + \frac{V}{R_2} + \frac{V}{R_3}$$

$$\frac{1}{R_{eq}} = \frac{1}{R_1} + \frac{1}{R_2} + \frac{1}{R_3} \tag{19.23}$$

Equation 19.23 can be generalized for any number of resistors in parallel:

$$\frac{1}{R_{eq}} = \frac{1}{R_1} + \frac{1}{R_2} + \frac{1}{R_3} + \cdots \quad \text{(resistors in parallel)} \tag{19.24}$$

So resistors in parallel combine like capacitors in series.

As with the latter, the equivalent resistance R_{eq} always is less than the smallest resistance that was combined.

PROBLEM-SOLVING TACTICS

19.1 As with the capacitors in series, be careful using Equation 19.24 in (1) adding the fractions correctly and (2) remembering that the sum of the reciprocals is the *reciprocal* of the equivalent resistance.

19.2 The equivalent resistance of two resistors in parallel is the product of their resistances divided by the sum. For *two* resistors in parallel we can use the same shortcut we developed for capacitors in series. That is, for two resistors in parallel the equivalent resistance is

$$\frac{1}{R_{eq}} = \frac{1}{R_1} + \frac{1}{R_2} = \frac{R_2 + R_1}{R_1R_2}$$

$$R_{eq} = \frac{R_1R_2}{R_1 + R_2} \tag{19.25}$$

This shortcut only is appropriate for two resistors at a time; parallel combinations of more than two resistors must be tackled pairwise to use this shortcut.

QUESTION 5

Time for the holidays. You and a friend each have bought a set of tree lights. When one bulb on the set you bought blows out, the entire string of lights goes out; just your luck! With your friend's set, if one light blows out, the others stay lit. What is the difference between the way the two sets of lights are wired? Are the bulbs in the two sets likely to be interchangeable?

EXAMPLE 19.6

What is the equivalent resistance between the terminals A and B of the resistors shown in Figure 19.28?

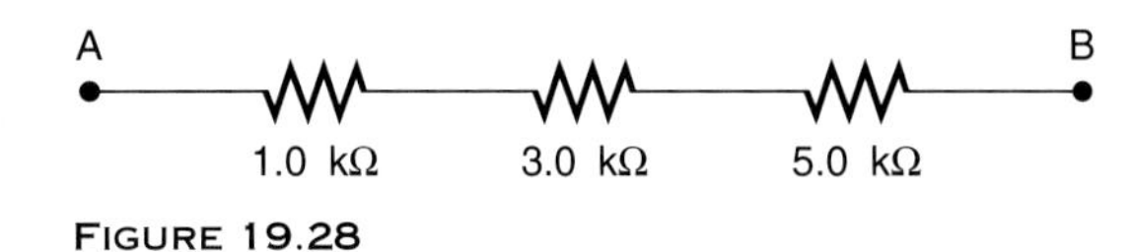

FIGURE 19.28

Solution

The resistors are in series, and so the equivalent resistance between the terminals A and B is the sum of the resistances (Equation 19.20), as shown in Figure 19.29:

$$R_{eq} = 1.0\ \text{k}\Omega + 3.0\ \text{k}\Omega + 5.0\ \text{k}\Omega$$
$$= 9.0\ \text{k}\Omega$$

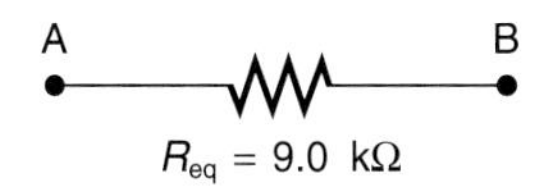

FIGURE 19.29

EXAMPLE 19.7

Find the equivalent resistance of the resistors shown in Figure 19.30.

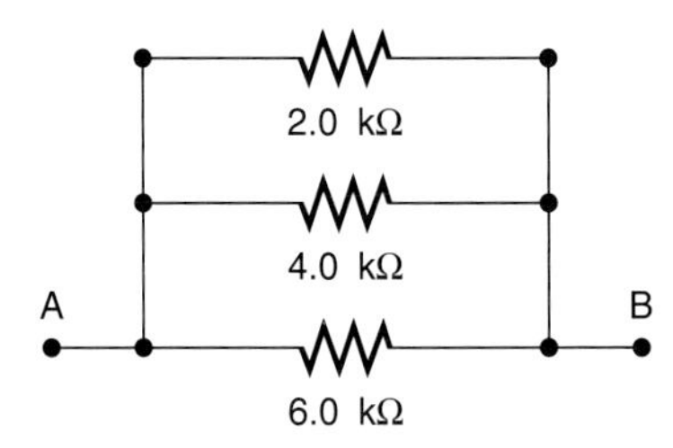

FIGURE 19.30

Solution

All three resistors are connected to the same two distinct nodes and so the three are connected in parallel.

Method 1

Use Equation 19.24:

$$\frac{1}{R_{eq}} = \frac{1}{2.0\ \text{k}\Omega} + \frac{1}{4.0\ \text{k}\Omega} + \frac{1}{6.0\ \text{k}\Omega} = \frac{11}{12.0\ \text{k}\Omega}$$

so

$$R_{eq} = \frac{12.0}{11}\ \text{k}\Omega = 1.1\ \text{k}\Omega$$

Note that the equivalent resistance is smaller than the smallest of the resistances that were combined.

Method 2

Combine the resistances pairwise using the product over the sum rule, Equation 19.25 in Problem-Solving Tactic 19.2. Combining the 2.0 kΩ and 4.0 kΩ resistances, you get an equivalent resistance of

$$\frac{(2.0\ \text{k}\Omega)(4.0\ \text{k}\Omega)}{2.0\ \text{k}\Omega + 4.0\ \text{k}\Omega} = \frac{8.0}{6.0}\ \text{k}\Omega = 1.3\ \text{k}\Omega$$

Then combine this pairwise with the 6.0 kΩ resistance using the same product divided by sum rule:

$$\frac{(1.3\ \text{k}\Omega)(6.0\ \text{k}\Omega)}{1.3\ \text{k}\Omega + 6.0\ \text{k}\Omega} = 1.1\ \text{k}\Omega$$

The resistors also could be combined pairwise in another order, say, by combining the 2.0 kΩ resistance with the 6.0 kΩ resistance, yielding 1.5 kΩ; then this result with the 4.0 kΩ resistance, again yielding 1.1 kΩ.

19.8 Electric Power

The light bulb in your study lamp burning brightly on your desk, long into the night while pouring over this and other texts, likely is a 100 W or 200 W bulb. The wattage rating of a light bulb resistor is an indication of the electric power which is absorbed by (transferred to) the bulb when it is on and then released to the environment (transferred to it) as light and heat transfer. Stove burners and wires that glow inside your toaster also are resistors. Here we see how to calculate the electrical power absorbed by (transferred to) *any* circuit element.

To calculate the power absorbed, we need to account for the changes in the electrical potential energy of charges flowing through a circuit element as a function of time. To do this, we consider a circuit element with a potential difference $V = V_A - V_B$ across it, as shown in Figure 19.31. Point A is at the higher electric potential, so the polarities are marked as indicated.

A small, differential charge dQ at position A has a potential energy $dQ\,V_A$, and at position B it has potential energy $dQ\,V_B$. The differential change in the potential energy of the charge is (final value − initial value):

$$\begin{aligned} d(\text{PE}) &= dQ\,V_B - dQ\,V_A \\ &= dQ\,(V_B - V_A) \\ &= dQ\,(-V) \end{aligned}$$

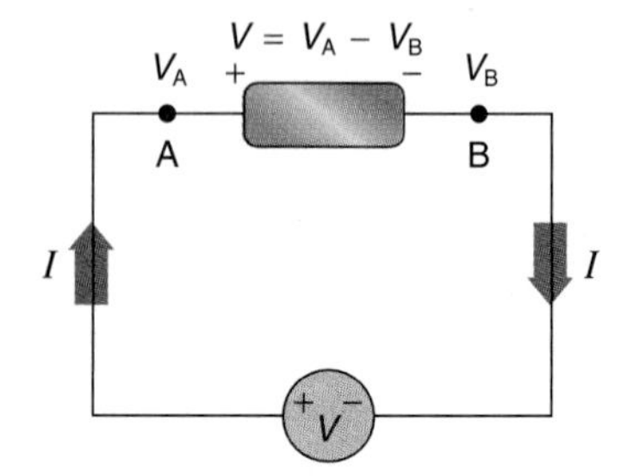

FIGURE 19.31 The potential difference V across a circuit element.

For positive dQ, the change in the electrical potential energy of the charge is negative since $V_B < V_A$. The potential energy of the positive charge decreases. You should convince yourself that the potential energy of a negative charge going the other way (from B to A) *also* decreases. What happens to this loss of electrical potential energy? We say the circuit element *absorbs* the energy; the energy is transferred to it. The circuit element continually converts the electrical potential energy into other forms of energy (typically thermodynamic internal energy and perhaps light) and shares it with the world (via heat transfer).

For a resistor circuit element, the energy increase appears as an increase in the thermodynamic internal energy of the device: it gets warmer. In other words, the energy lost by the charge is gained by the circuit element. The energy gained (absorbed) by the circuit element (transferred to it) is therefore

$$+dQ\,V$$

so that the total energy of the system (charge + circuit element) is conserved. Let the process occur during a differential time dt. The time rate at which the circuit element absorbs energy from the change in the electrical potential energy of the charges flowing through it is the **electric power absorbed** by (transferred to) the circuit element. Hence the electric power absorbed by a circuit element is

$$P = \frac{dQ}{dt}V$$

But

$$\frac{dQ}{dt} = I$$

the current through the circuit element.

Thus the electrical power absorbed by (transferred to) a circuit element is the product of the current through the element and the potential difference V across the element:

$$P = IV \qquad (19.26)$$

Notice the direction of the current indicated in Figure 19.31 and the polarity of the potential difference across the element.

To use Equation 19.26 the current should be directed *into* the positive polarity terminal of the circuit element, as shown in Figure 19.32.

If the current is directed *out* of the positive polarity terminal, the current $-I$ is directed into it, as indicated in Figure 19.33. This result follows since a positive charge moving in one direction is equivalent to a negative charge moving in the opposite direction, as we showed in Section 19.2.

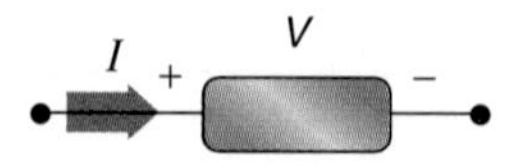

FIGURE 19.32 For Equation 19.26, the current I should be directed into the positive polarity terminal.

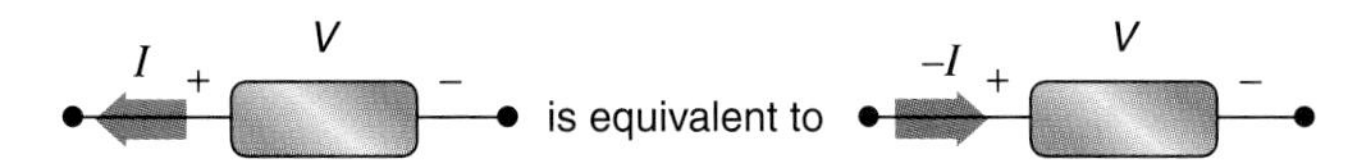

FIGURE 19.33 **Equivalent representations of the current.**

If the result of using Equation 19.26 for the power absorbed is positive, charges passing through the circuit element are losing potential energy (per unit time), the circuit element gains (absorbs) this energy (per unit time), and the energy manifests itself in various other forms. For a resistor, the energy absorbed appears as an increase in its thermodynamic internal energy: the resistor becomes warmer. If an independent voltage source absorbs positive power, it is storing the energy in other forms (such as chemical energy). An independent voltage source absorbing positive electric power is said to be **charging**. When a capacitor absorbs positive power, it stores the energy in the electric field between its plates.

On the other hand, if the result of using Equation 19.26 for the power is negative, we have negative power absorbed: energy (per unit time) is generated (or provided) by the circuit element, transferred from it to electrical charges flowing through it. Negative power absorbed does *not* mean the circuit element gets cooler! It simply means that the circuit element increases the potential energy of positive charges that we imagine constituting the current (the direction of positive charge flow); the energy for this may arise from electrochemical reactions or more exotic sources. Since independent voltage sources typically increase the electrical potential energy of positive charges flowing through them, independent voltage sources absorb negative power under these circumstances.* It is for this reason that independent voltage sources also are called *power supplies*, that is, suppliers of electrical power (electric energy per unit time). When a capacitor absorbs negative power, it is discharging; the capacitor is decreasing the energy stored in the electric field between its plates and increasing the electrical potential energy of the positive charges that we imagine constituting the current.

So electric power absorbed can be either positive or negative. Positive power absorbed is real power absorption by the circuit element (transferred to it) from the positive charges that we imagine constitute the current. Negative power absorbed by a circuit element is power generated by the circuit element (transferred from it) to increase the potential energy of the positive charges that create the current.

Conservation of energy (per unit time) means that the total power absorbed by all the circuit elements in a circuit must be zero:

$$\text{total power absorbed by a circuit} = 0 \text{ W} \qquad (19.27)$$

That is, the sum of the individual powers absorbed by all the circuit elements must be zero.

*If a battery is charging, it absorbs positive electrical power.

The common unit known as the **kilowatt-hour** (kW·h) appears inexorably on the bill that you or your parents receive from the electric company every month. A kilowatt-hour is equivalent to a power of exactly 1000 W ≡ 1 kW used for exactly one hour.

Thus a kilowatt-hour is an *energy* unit, *not* a power unit.

The amount of energy in one kilowatt-hour is exactly

$$\begin{aligned} 1 \text{ kW·h} &= (1000 \text{ W})(1 \text{ h}) \\ &= (1000 \text{ J/s})(3600 \text{ s}) \\ &= 3.600 \times 10^6 \text{ J} \qquad \text{(exact conversion)} \end{aligned}$$

A bill from the electric company, therefore, is not a bill for power; it is a bill for energy. Your electric company is an *energy* company.

PROBLEM-SOLVING TACTICS

19.3 Since resistors always become warm or hot when they have a nonzero current, resistors always absorb positive power (or zero power if the current through the resistor is zero). This follows from Equation 19.26. Since Ohm's law states that $V = IR$, with the current I going into the positive polarity terminal of the resistor, the power absorbed by a resistor is

$$\begin{aligned} P_{\text{res}} &= IV \\ &= I(IR) \\ &= I^2R \qquad \text{(for a resistor } only\text{)} \end{aligned} \qquad (19.28)$$

which is never negative. When solving a problem, if you ever find that the *power* absorbed by a resistor comes out negative, you can be certain that an error was made somewhere in the calculation, not that you have a new patentable invention that will make you a millionaire overnight!

19.4 If you find the current through a resistor is negative, $-|I|$, Ohm's law leads to a negative potential difference, $-|V|$. The power absorbed by the resistor still is positive. This result follows since

$$\begin{aligned} P &= IV \\ &= (-|I|)(-|V|) \\ &= |I|\,|V| \end{aligned}$$

See Example 19.8 for an illustration of this.

19.5 Independent voltage sources absorb either positive or negative power depending on the direction of the current through them. Positive power is absorbed when the current *enters* the positive polarity terminal; negative power when the current *emerges from* the positive polarity terminal.

19.6 If a current I is emerging from the positive polarity terminal of a circuit element, current $-I$ is entering the positive polarity terminal. See Figure 19.33.

QUESTION 6

Equation 19.28 states that the power absorbed by a resistor is proportional to R: $P = I^2R$. Using Equation 19.26 for the power, $P = IV$, and using I from Ohm's law,

$$I = \frac{V}{R}$$

the power absorbed by a resistor is

$$\frac{V^2}{R}$$

which shows that the power absorbed by the resistor is inversely proportional to R. How do you explain the apparent contradiction?

STRATEGIC EXAMPLE 19.8

A 12.0 V independent voltage source is connected to a 10.0 Ω resistor, as shown in Figure 19.34.

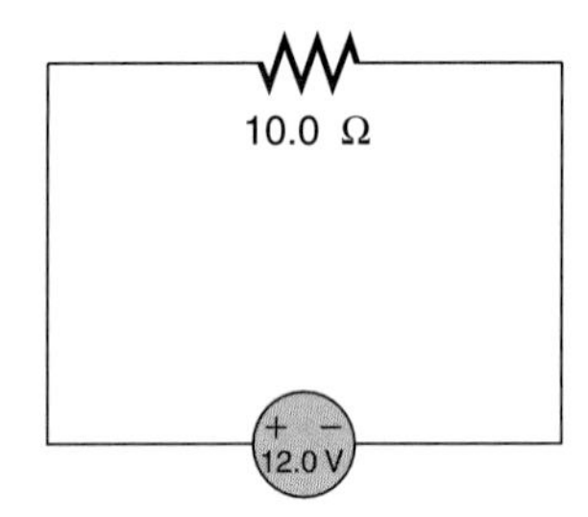

FIGURE 19.34

a. Determine the current through the resistor.
b. Calculate the power absorbed by the resistor.
c. Calculate the power absorbed by the independent voltage source.
d. Verify that the total power absorbed by all the circuit elements is 0 W.

Solution

a. Since the resistor is connected to the same two terminals as the independent voltage source, the potential difference across the resistor is 12.0 V. The two circuit elements are simultaneously in parallel and in series! Use Ohm's law to find the current through the resistor:

$$V = IR$$
$$12.0 \text{ V} = I(10.0 \ \Omega)$$

Solving for I, you get

$$I = 1.20 \text{ A}$$

The current through a resistor is from the high-potential end to the low-potential end, so the direction of the current and the polarity marking of the resistor are as indicated in Figure 19.35.

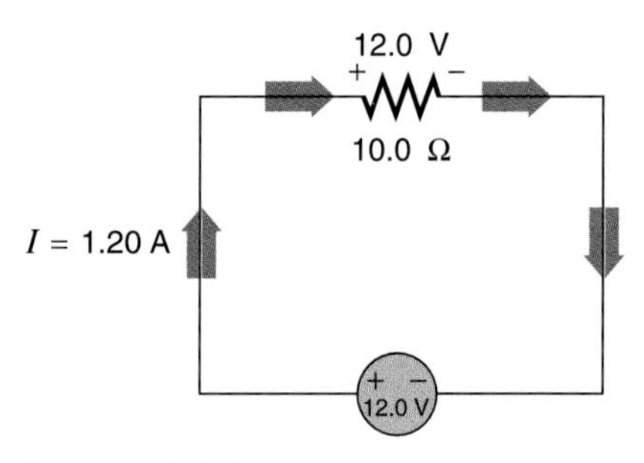

FIGURE 19.35

b. The power absorbed by the resistor can be found in two ways.

Method 1

The power absorbed is the product of the current into the positive polarity terminal of the resistor times the potential difference across the resistor:

$$\begin{aligned} P &= IV \\ &= (1.20 \text{ A})(12.0 \text{ V}) \\ &= 14.4 \text{ W} \end{aligned}$$

Method 2

The power absorbed by a resistor is the product of the square of the current and the resistance (Equation 19.28):

$$\begin{aligned} P &= I^2R \\ &= (1.20 \text{ A})^2(10.0 \ \Omega) \\ &= 14.4 \text{ W} \end{aligned}$$

c. The power absorbed by the independent voltage source is the product of the current into its positive polarity terminal times the potential difference across the circuit element. The current 1.20 A is coming out of the positive terminal, hence −1.20 A is going into the positive terminal (Problem-Solving Tactic 19.6), as shown in Figure 19.36. Hence the power absorbed by the independent voltage source is

$$\begin{aligned} P &= (-1.20 \text{ A})(12.0 \text{ V}) \\ &= -14.4 \text{ W} \end{aligned}$$

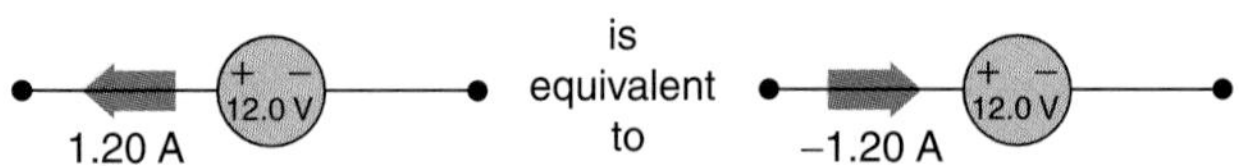

FIGURE 19.36

The power absorbed is negative, indicating that the circuit element actually is generating power for (transferring power to) the rest of the circuit.

d. The total power absorbed by all the circuit elements is

$$\begin{aligned} P_{\text{total}} &= P_{\text{res}} + P_{\text{independent voltage source}} \\ &= 14.4 \text{ W} + (-14.4 \text{ W}) \\ &= 0 \text{ W} \end{aligned}$$

19.9 Electrical Networks and Circuits

Any collection of circuit elements that are connected together is an electrical **network**, as in Figure 19.37.

An electrical **circuit** is a network with at least one closed conducting path or loop, as shown in Figure 19.38.

All circuits are networks, but not all networks are circuits. It is possible for a part of a network to be a circuit; in Figure 19.38, part of the network is a circuit. Circuit elements that are not part of a circuit can be disregarded (see Figure 19.39), since there will be no currents through these circuit elements; they are extraneous circuit elements. So the significant part of a network is just the part that is a circuit. We are specifically interested in electrical circuits since it is around such closed paths that electrical currents can exist.

The simplest circuit has one elementary loop of two circuit elements, as indicated in Figure 19.40. The two circuit elements are connected to the same two distinct nodes, are in parallel, and have the same potential difference between their terminals. But the current in the loop also has no choice but to flow through one element and then the other. The current thus is the same in both circuit elements. The two circuit elements are *simultaneously* connected in series and parallel.

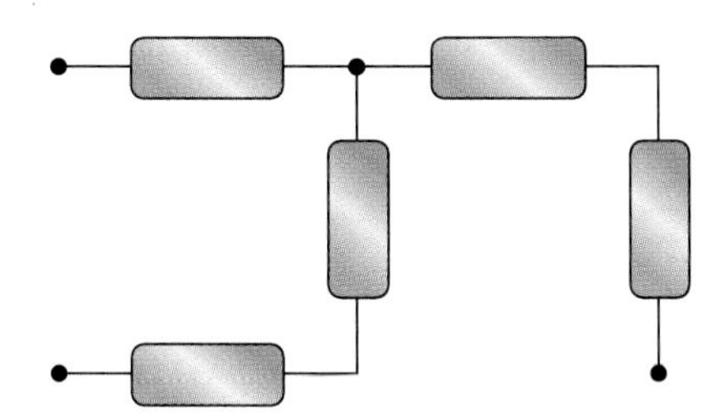

FIGURE 19.37 An electrical network.

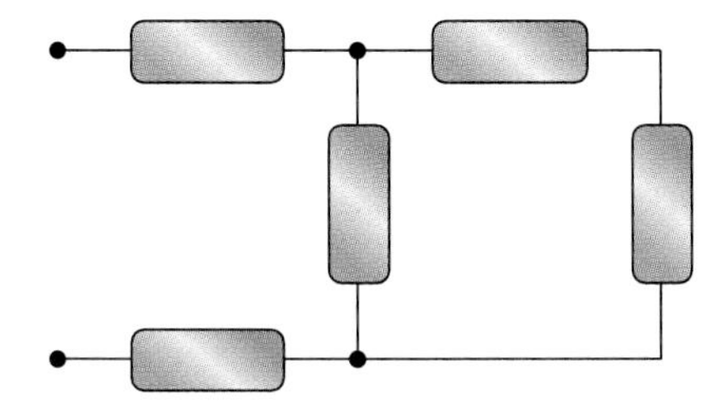

FIGURE 19.38 An electrical circuit.

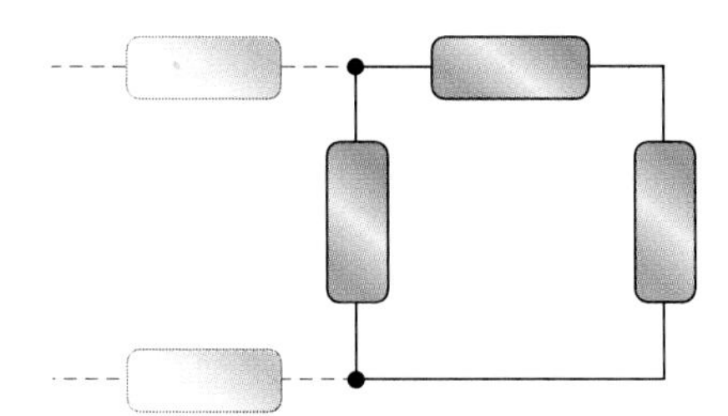

FIGURE 19.39 Circuit elements that are not part of the circuit can be disregarded.

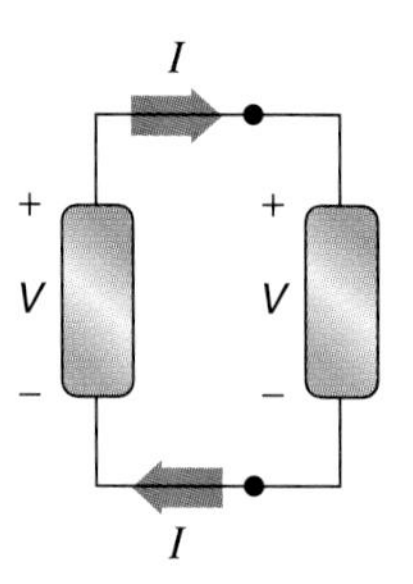

FIGURE 19.40 The simplest circuit: two circuit elements simultaneously in series and parallel.

19.10 Electronics

Electronics is the application of physics to electric charge, its motion, and circuits. The subject is divided into several broad areas to indicate the principal focus of inquiry.

1. In **analog electronics**, the specific values of the currents through and potential differences across the various circuit elements are important to know and are the focus of concern. Common household circuits, flashlights, and most household appliances are examples of applications of analog electronics.
2. In **digital electronics**, such intimate knowledge of specific potential differences and current values is, by and large, of secondary importance. In digital electronics the objective typically is to determine whether various circuit elements are *on* (meaning either conducting current or with some standard potential difference across them) or *off* (not conducting current or with another standard potential difference across them). Much of the electronics associated with computers and other digital equipment are applications of digital electronics.

We will be concerned in this text with analog electronics because the ideas of digital electronics rest on analog principles. It is important to grasp the fundamental analog ideas first; digital electronics then comes quite easily.

Analog electronics is further subdivided into three broad areas:

1. **dc circuit analysis**: In this area, the various currents through the circuit elements and potential differences across the circuit elements are independent of time, as shown in Figure 19.41. In other words, the various potential differences and currents associated with the various circuit elements have steady, constant values. The term dc comes from the words *d*irect *c*urrent, meaning currents that are not functions of time.
2. **ac circuit analysis**: In this area, the various currents and potential differences are periodic functions of time; see Figure 19.42. That is, they fluctuate in time but periodically repeat the pattern of the fluctuations. The fluctuations need not be sinusoidal; they are merely periodic in time. We study sinusoidally varying ac circuits in Chapter 22.

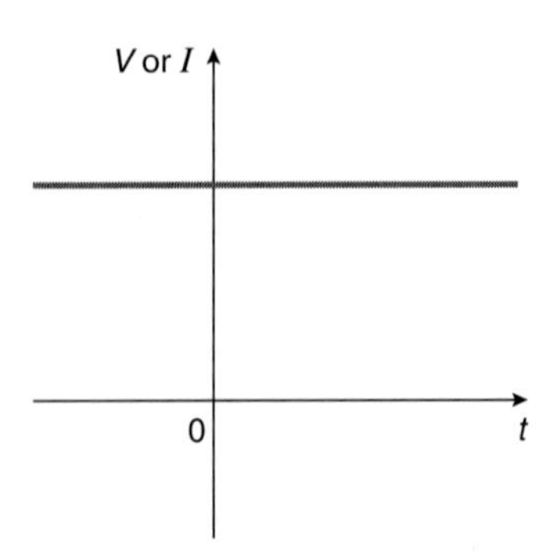

FIGURE 19.41 Currents and potential differences are independent of time in dc circuit analysis.

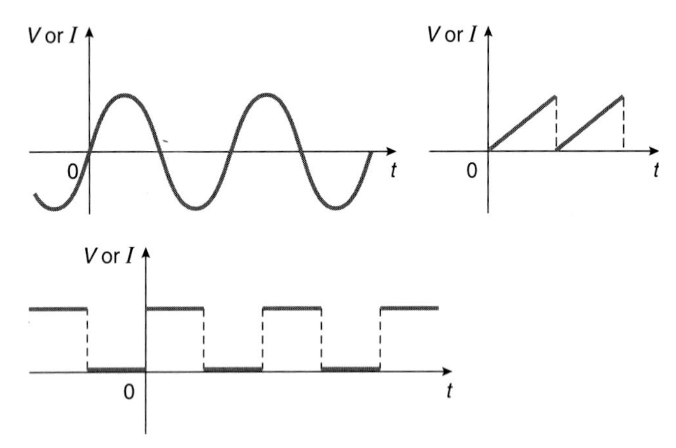

FIGURE 19.42 Periodic currents and potential differences characterize ac circuit analysis.

3. **Transient circuit analysis**: This area investigates what happens when, for example, circuits are turned on or off. Transient analysis investigates the changes that happen when a circuit goes from one steady state (in a dc or ac situation) to another steady state. We examined one aspect of transient circuit analysis in Chapter 18, the charging and discharging of a capacitor.

We begin our study of electronics with an investigation of dc circuits. The fundamental ideas of dc circuit analysis also are used to study ac circuits. We touch only lightly on transient analysis, since the study of transients is perhaps the most difficult of the three fields.

19.11 Kirchhoff's Laws for Circuit Analysis

O, two silver currents, when they join

William Shakespeare*

The analysis of both dc and ac circuits relies on just two pillars: the **Kirchhoff laws** of circuit analysis, named after Gustav Robert Kirchhoff (1824–1887), a German pioneer in the study of electrical phenomena. Kirchhoff's laws are restatements of physical ideas that already are quite familiar to you.

*King John, Act 2, Scene 1, line 441.

Kirchhoff was a gifted German researcher in thermodynamics, spectra, and electricity. He was known for his cheerful disposition. An accident left him disabled and he needed crutches or a wheelchair for mobility.

The Kirchhoff Current Law

The Kirchhoff current law (KCL) states that the algebraic sum of the currents leaving any node of a circuit is zero.

One could state, equivalently, that the algebraic sum of the currents entering any node of a circuit is zero. But in courses in engineering, the statement conventionally is expressed in terms of currents leaving a node, and there is no reason to depart from that convention here.

What does the KCL mean physically?

From the standpoint of physics, the KCL is a statement of *charge conservation.*

The node neither can accumulate more and more charge nor go into complete charge bankruptcy and beyond.

The application of the KCL can be seen with a specific example. Figure 19.43 depicts an arbitrary node, extracted from a circuit. A "bird's nest" of wire leads from various circuit elements are tied together at the node. Notice that some currents are leaving the node and some currents are entering the node. The KCL states that the algebraic sum of all the currents leaving a node is zero. Currents actually leaving the node are placed into the sum of the currents with a plus sign. If a current actually is entering a node (such as current I_2), the current $-I_2$ is leaving the node; so the negative current is placed in the sum of the currents leaving the node. The KCL applied to the node depicted in Figure 19.43 results in the following equation:

$$I_1 + (-I_2) + I_3 + (-I_4) + I_5 = 0 \text{ A}$$
$$I_1 - I_2 + I_3 - I_4 + I_5 = 0 \text{ A}$$

The KCL can be applied to *any* node of a circuit. In practice, one uses the KCL only at nodes connecting three or more circuit elements. Why? Consider a node connecting just two circuit

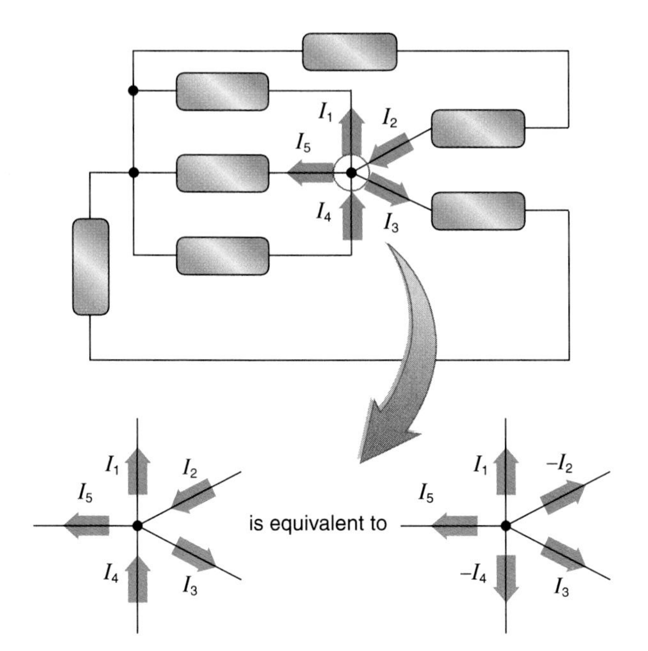

FIGURE 19.43 One node of a circuit.

elements in series, as in Figure 19.44. Let I_1 and I_2 be the currents in the two wires at the node. The KCL applied to this node yields

$$-I_1 + I_2 = 0 \text{ A}$$

or $I_1 = I_2$. That's nice but not too interesting. The two elements are connected in series and we know the current is the same for each element connected in series. So the application of the KCL to such a node did not yield any information that we did not already know.

PROBLEM-SOLVING TACTIC

19.7 The KCL really becomes useful only at nodes connecting three or more circuit elements.

The Kirchhoff Voltage Law

The Kirchhoff voltage law (KVL) states that the algebraic sum of the potential differences around any closed loop of an electrical circuit is zero.

The word voltage in the name of the law really means potential difference. We use the word potential difference consistently in our discussions, but will call the KVL the Kirchhoff voltage law in conformity with common practice.

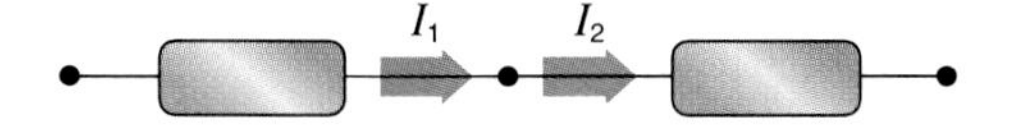

FIGURE 19.44 A node connecting two circuit elements in series.

What does the KVL mean physically? The law is a reflection of the conservative nature of the electrical force: the work done by a conservative force on a charge taken around a closed path is zero.

The KVL is a statement of *conservation of energy*.

The KVL can be applied to *any* loop of a circuit. An elementary loop contains no other closed loops within itself. While a KVL equation can be written for any closed loop in the circuit, in practice we fortunately do not have to write KVL equations for every possible loop in the circuit. The reason for this will be explained shortly. Consider the elementary circuit loop in Figure 19.45. The various circuit elements around the loop have potential differences across them, say, with the indicated values and polarity markings.

The KVL involves walking around the loop and algebraically summing the various potential differences encountered. Three things to note about the stroll around the loop:

1. We can go around the loop in either direction: clockwise (CW) or counterclockwise (CCW); it makes no difference because the overall sum of the potential differences is zero.
2. It also makes no difference where we begin the journey around the loop; we merely have to go all the way around the loop, returning to wherever we began.
3. As we go around the loop, if we encounter the (+) polarity marking of a circuit element first, we place the potential difference across that circuit element into the sum with a plus sign; if we encounter the (−) polarity marking first, the potential difference for that circuit element is placed into the equation for the loop with a minus sign. So in Figure 19.45, if we go around the loop clockwise and begin in the lower left corner, the KVL takes the following form:

$$V_1 + (-V_2) + (-V_3) + V_4 = 0 \text{ V}$$
$$V_1 - V_2 - V_3 + V_4 = 0 \text{ V} \quad (19.29)$$

If we begin in the lower left corner and go around the loop counterclockwise, the KVL takes the form

$$(-V_4) + V_3 + V_2 + (-V_1) = 0 \text{ V}$$
$$-V_4 + V_3 + V_2 - V_1 = 0 \text{ V} \quad (19.30)$$

which differs from Equation 19.29, obtained by executing the loop clockwise, by only a minus sign. Equations 19.29 and 19.30 are not independent equations. So it makes no difference which way we go around the loop. Starting the walk around the loop at a different place in the loop merely

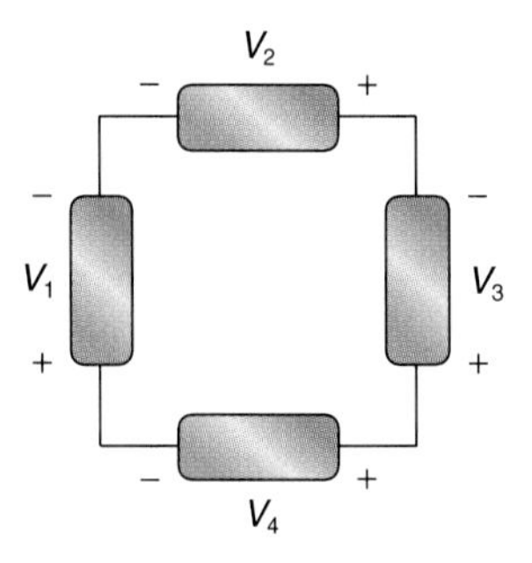

FIGURE 19.45 An elementary loop of a circuit.

rearranges the order of the terms in the sum; we do not get a new equation.

PROBLEM-SOLVING TACTIC

19.8 Write KVL equations only for each elementary loop in a circuit.

The parts of an elementary loop that contain either one circuit element or two or more in series with each other are known as *branches*.

The branches of the circuit in Figure 19.46 are indicated. There are three elementary loops to the circuit in Figure 19.46, each highlighted in Figure 19.47. There are four more complex loops, shown in Figure 19.48.

The KVL applies to all the loops. But we write the KVL *only* for the elementary loops. Why? From a mathematical viewpoint, of the seven possible loop equations for this circuit, only three are independent equations; the others are surplus because they are linear combinations of the three independent equations. In other words, the number of *independent* loop equations is equal to the number of elementary loops in the circuit. It is simply easier to write the KVL equation just for the elementary loops; then we know we have all the independent equations that can be secured using the KVL.

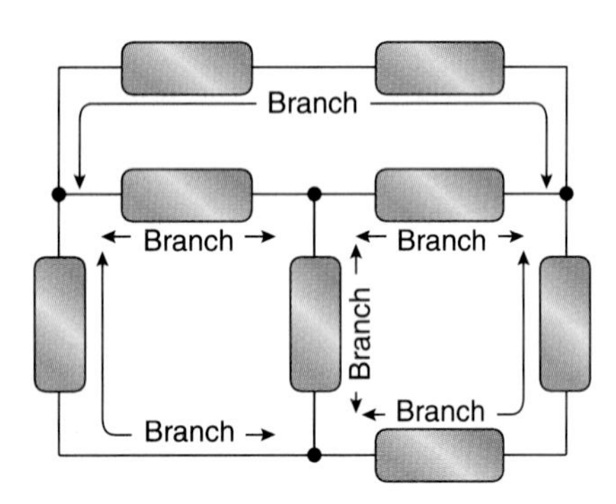

FIGURE 19.46 The branches of a circuit.

The specific steps to follow for solving any dc circuit problem are now outlined. After reading the procedure, we will apply it to several specific examples to see what is really involved.

1. See if the circuit can be simplified by combining resistors in series or parallel, or independent voltage sources in series or parallel. Not all circuits can be simplified. Sketch the circuit in its simplest form.
2. Locate and identify the significant nodes of the simplified circuit, those nodes connecting three or more circuit elements. Nodes connecting only two circuit elements are not really significant (see our earlier discussion of the KCL).
3. Choose current directions for each distinct branch of the simplified circuit. Introduce appropriate current variables I_1, I_2, . . . for each branch; we are trying to find these currents. Recognize that circuit elements in series have the same current in every element: thus the current through all circuit elements in a given branch is the same. The directions chosen for the currents in a branch are arbitrary. In practice, for a branch with an independent voltage source, we usually choose the direction of the current to *emerge* from the (+) terminal of the independent voltage source. This choice for current direction is not required, but it usually makes things easier, as we will see. Any circuit element in series with an independent voltage source will have the same current and current direction through it, as it is part of the same branch.
4. Taking account of the current directions, label the polarity for each resistor consistent with the directions chosen for the current through it. That is, for every resistor, the lead along which the current is going into a resistor indicates the (+) terminal of that resistor and the lead along which

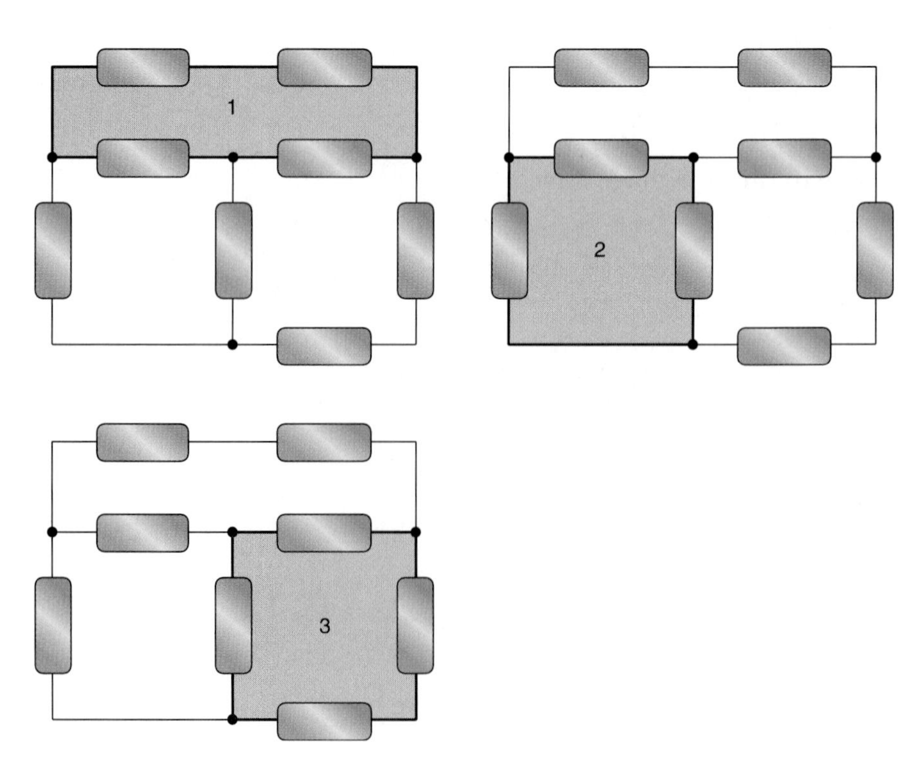

FIGURE 19.47 Elementary loops of the circuit in Figure 19.46.

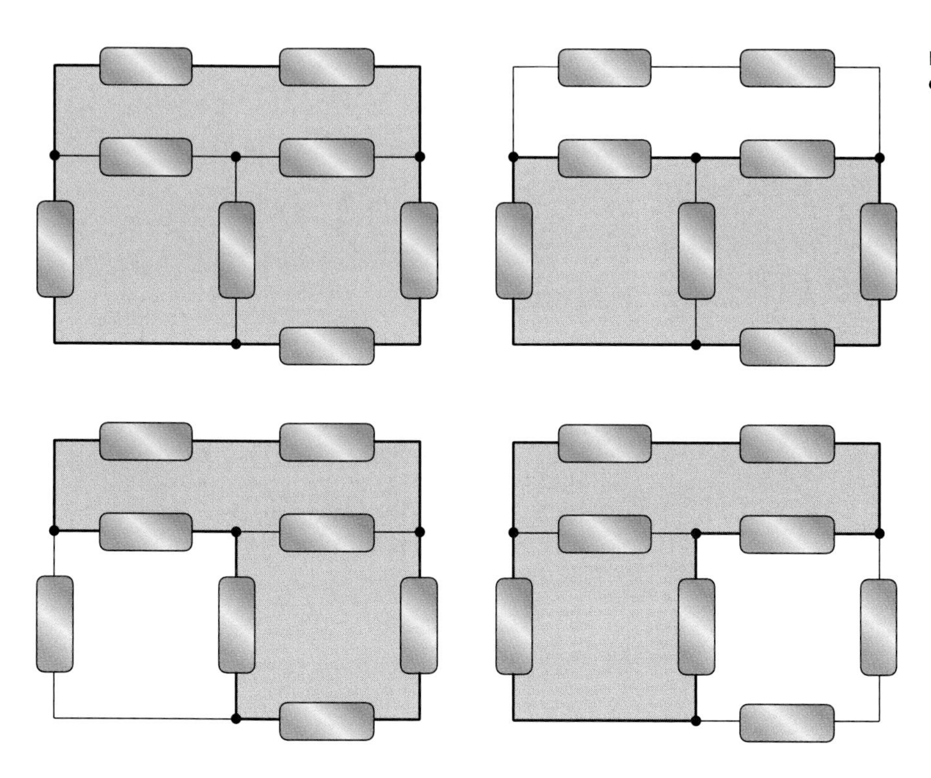

FIGURE 19.48 Four loops that are not elementary loops.

the current is exiting the resistor indicates the (−) polarity terminal of the resistor.

5. Apply the KVL to each elementary loop in the circuit.
6. Assess whether there are sufficient equations to solve for the number of unknown currents. For n unknown currents, n equations are needed. Determine the number of additional equations needed, if any. Then apply the KCL to the significant nodes (connecting three or more circuit elements) to secure any additional equations needed.
7. Solve the equations for the unknown currents. Once the currents are known, everything about the circuit can be determined from them. If one or more of your currents emerge as negative, there is *no need* to go back and redo anything; a negative current simply means the current actually is directed opposite to the way you chose in your diagram. You can use the negative current to determine any needed potential difference and power absorbed.

STRATEGIC EXAMPLE 19.9

Consider the circuit shown in Figure 19.49, consisting of an independent voltage source (an ideal battery) and a single resistor.

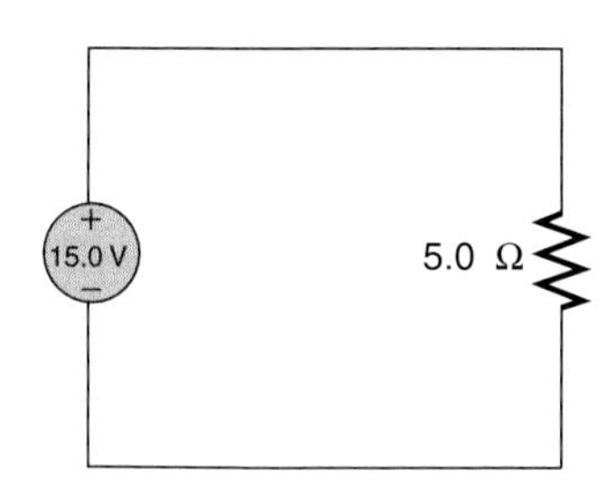

FIGURE 19.49

a. Find the current in the circuit.
b. Find the power absorbed by (transferred to) each circuit element.

Solution

The circuit cannot be simplified. Notice that the resistor is in series with the independent voltage source: the current is the same in both elements since there is no choice of path for the current as it goes around the loop. But this is more interesting than it seems: the resistor also is in parallel with the ideal battery, because the two elements are connected to the same two distinct nodes or terminals. Hence the potential difference across each circuit element is the same. This is an example of the simplest circuit (see Section 19.9), one with only two circuit elements.

Method 1

If an independent voltage source (an ideal battery) is present in a branch, it is convenient to choose the direction of the current to emerge from its positive polarity terminal. Thus choose the current direction indicated in Figure 19.50. This choice for the current direction forces you to label the resistor polarity as indicated in

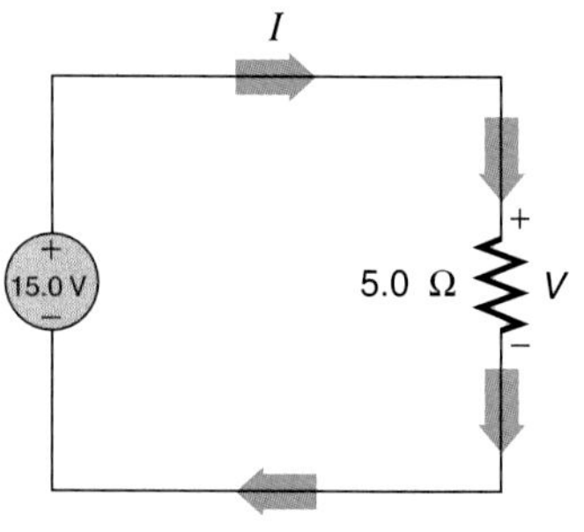

FIGURE 19.50

Figure 19.50, since the current through a resistor always goes from the high- to the low-potential terminal—that is, into the (+) polarity terminal of the resistor and out of its (–) polarity terminal.

a. The potential difference across the ideal battery (an independent voltage source) is 15.0 V; the resistor is in parallel with the ideal battery, and so the potential difference across the resistor also is 15.0 V. Ohm's law, Equation 19.14, states that the potential difference across a resistor is the product of the current through it times the resistance:

$$V = IR$$
$$15.0 \text{ V} = I(5.0 \ \Omega)$$

Thus the current through the resistor and the current in the single loop is

$$I = \frac{15.0 \text{ V}}{5.0 \ \Omega} = 3.0 \text{ A}$$

b. The power absorbed by a circuit element is the product of the current into its positive terminal and potential difference across the circuit element (Equation 19.26). For the resistor, you have

$$P = IV = (3.0 \text{ A})(15.0 \text{ V}) = 45 \text{ W}$$

You also could calculate the power absorbed by the resistor using Equation 19.28:

$$P_{res} = I^2R = (3.0 \text{ A})^2(5.0 \ \Omega) = 45 \text{ W}$$

The power absorbed by the ideal battery also is the product of the current into the positive terminal and the potential difference across it (its emf), Equation 19.26. The current *out* of the positive polarity terminal of the ideal battery is +3.0 A; therefore, the current *into* the positive polarity terminal is –3.0 A. Hence the power absorbed by the ideal battery is

$$P_{bat} = IV = (-3.0 \text{ A})(15.0 \text{ V}) = -45 \text{ W}$$

The power absorbed by the ideal battery is negative, indicating that the battery actually is generating (or providing) power to the circuit.

Notice that the total power absorbed by the circuit, the sum of the power absorbed by the battery and by the resistor, is zero:

$$P_{total} = -45 \text{ W} + 45 \text{ W} = 0 \text{ W}$$

Method 2

a. Choose the current direction indicated in Figure 19.51. Then the polarity of the potential difference across the resistor must be labeled as shown in Figure 19.51, because the current goes into the (+) polarity lead of the resistor and out of the (–) polarity lead. Using the KVL around the elementary loop, going clockwise and beginning in the lower left corner, you obtain

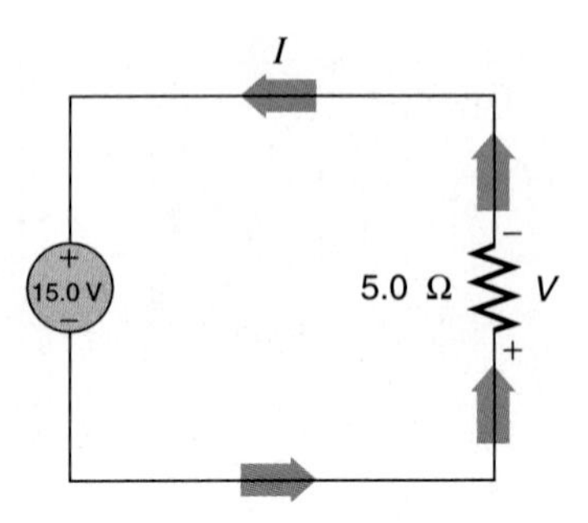

FIGURE 19.51

$$-15.0 \text{ V} - I(5.0 \ \Omega) = 0 \text{ V}$$

Solving for I,

$$I = -3.0 \text{ A}$$

The minus sign indicates that the current actually is in the direction opposite to the one chosen. No matter. You do not have to go back and redo everything! The potential difference across the resistor is, according to Ohm's law,

$$V = IR = (-3.0 \text{ A})(5.0 \ \Omega) = -15 \text{ V}$$

using the polarity indicated in Figure 19.51. The minus sign here indicates that the terminal marked (+) in Figure 19.51 actually is at a lower potential than the terminal marked (–).

b. To calculate the power absorbed by the resistor, use the power Equation 19.26:

$$P = IV = (-3.0 \text{ A})(-15 \text{ V}) = 45 \text{ W}$$

You also could use Equation 19.28, specifically for the power absorbed by a resistor:

$$P_{res} = I^2R = (-3.0 \text{ A})^2(5.0 \ \Omega) = 45 \text{ W}$$

Either way, the power absorbed by the resistor is positive.

The power absorbed by the ideal battery also is found using Equation 19.26 for any circuit element:

$$P_{bat} = IV$$

where the current I is directed into the (+) polarity terminal of the ideal battery. Notice that the current direction chosen in Figure 19.51 is going into the positive polarity terminal of the ideal battery; you do not have to reverse its direction to use the power equation, and so you have

$$P_{\text{bat}} = (-3.0\ \text{A})(15.0\ \text{V})$$
$$= -45\ \text{W}$$

The total power absorbed by all the circuit elements is zero.

STRATEGIC EXAMPLE 19.10

a. Simplify the circuit in Figure 19.52 as much as possible.

b. Find the potential difference and the current through each circuit element.

c. Find the power absorbed by each circuit element and verify that the total power absorbed by all the circuit elements is zero.

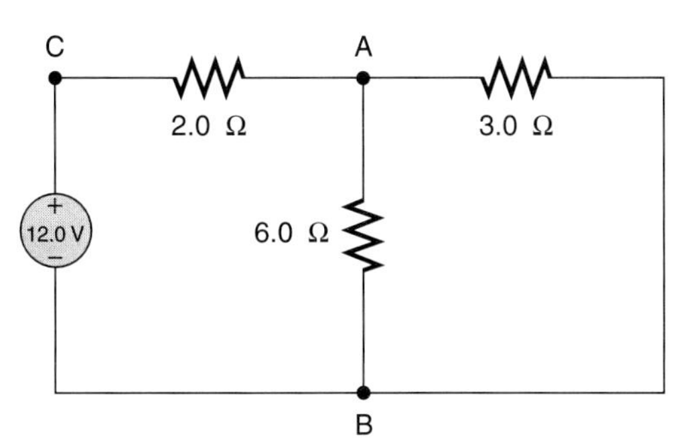

FIGURE 19.52

Solution

a. The 3.0 Ω and 6.0 Ω resistors are in parallel since they are connected to the same two distinct nodes of the circuit (labeled A and B in Figure 19.52). The equivalent resistance of this parallel pair of resistors is (using the product/sum rule for two resistors in parallel, Equation 19.25)

$$\frac{(3.0\ \Omega)(6.0\ \Omega)}{3.0\ \Omega + 6.0\ \Omega} = 2.0\ \Omega$$

The circuit now has the appearance shown in Figure 19.53. The nodes labeled A and B are the same nodes as in Figure 19.52. The pair of 2.0 Ω resistors in Figure 19.53 are in series and so have an equivalent resistance equal to their sum:

$$R_{\text{eq}} = R_1 + R_2$$
$$= 2.0\ \Omega + 2.0\ \Omega$$
$$= 4.0\ \Omega$$

The circuit now has the appearance shown in Figure 19.54, where the nodes labeled C and B are the same as in the original circuit (Figure 19.52). This is the simplest form for the given circuit.

b. The circuit of Figure 19.54 is an example of the simplest circuit, a single loop with two circuit elements. The 4.0 Ω resistor is connected to the same two nodes (C and B) as the independent voltage source, so that the 4.0 Ω resistor has a potential difference of 12.0 V across it with the polarity indicated in Figure 19.55. The current in the single loop circuit is assigned the direction indicated in Figure 19.55, consistent with the resistor polarity. The value of the current is found in one of two ways.

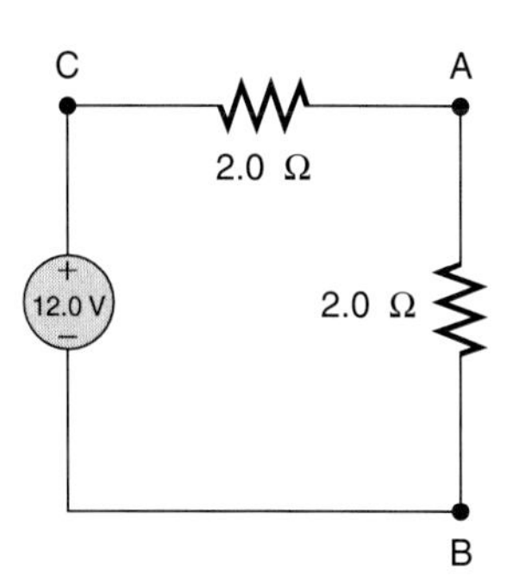

FIGURE 19.53

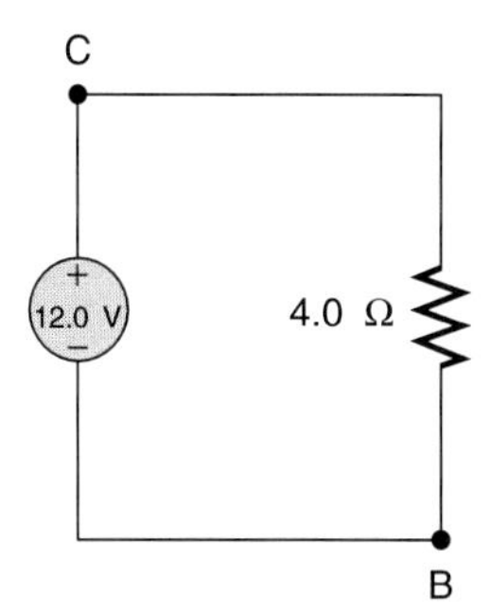

FIGURE 19.54

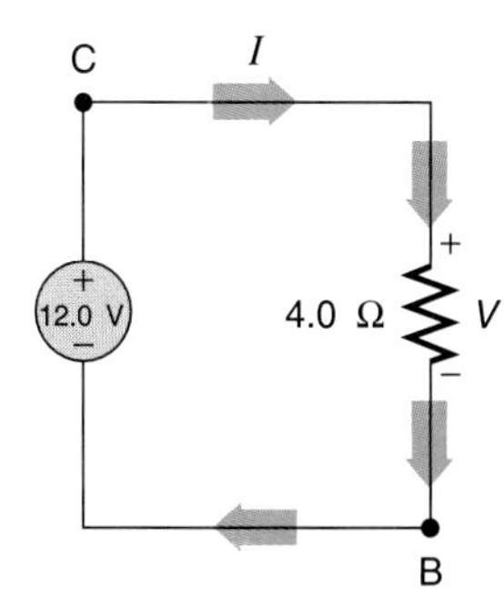

FIGURE 19.55

Method 1

Use Ohm's law applied to the 4.0 Ω resistor:

$$V = IR$$
$$12.0\ \text{V} = I(4.0\ \Omega)$$

Solving for I, you find

$$I = 3.0\ \text{A}$$

Method 2

You also could first choose the current direction as indicated in Figure 19.55, then label the resistor polarity consistent with this choice, and apply the KVL to the elementary loop.

Going around the loop clockwise beginning in the lower left corner, you get

$$-12.0 \text{ V} + I(4.0\ \Omega) = 0 \text{ V}$$

which yields

$$I = 3.0 \text{ A}$$

with the direction indicated in Figure 19.55.

Having found the current, now trace your way back to the original circuit (Figure 19.52) by considering each equivalent simplified circuit. Using the simplified circuit of Figure 19.53, each of the 2.0 Ω resistors has the current 3.0 A since they are both in series with the independent voltage source. The potential difference across each 2.0 Ω resistor is the same (since they have the same resistance and current) and is found using Ohm's law:

$$\begin{aligned} V &= IR \\ &= (3.0 \text{ A})(2.0\ \Omega) \\ &= 6.0 \text{ V} \end{aligned}$$

The lower 2.0 Ω resistor was the parallel equivalent of the 3.0 Ω and 6.0 Ω resistors in the original circuit. Thus the potential difference across both the 3.0 Ω and the 6.0 Ω resistors of the original circuit is 6.0 V. The potential difference across these resistors is the same since they are connected in parallel with each other. Apply Ohm's law to each of these resistors to find the current through them:

3.0 Ω resistor	*6.0 Ω resistor*
$6.0 \text{ V} = I_3\,(3.0\ \Omega)$	$6.0 \text{ V} = I_6\,(6.0\ \Omega)$
$I_3 = 2.0 \text{ A}$	$I_6 = 1.0 \text{ A}$

You thus have found the current through each circuit element and the potential difference across each element. These results are summarized in Figure 19.56.

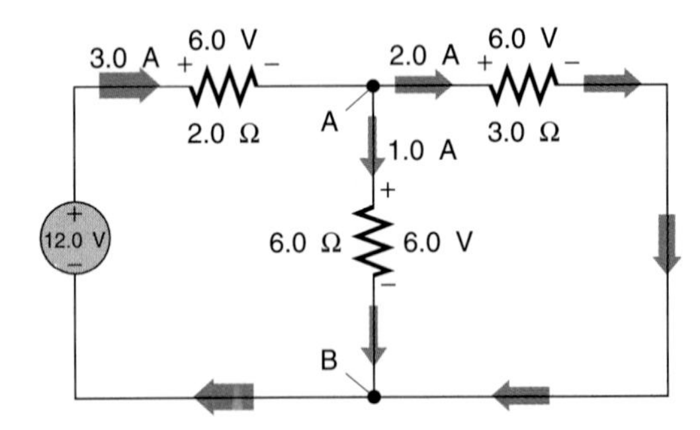

FIGURE 19.56

Note, parenthetically, that the KCL is satisfied at the node at A connecting the three resistors; since currents leaving the node are considered positive, the KCL implies

$$-3.0 \text{ A} + 1.0 \text{ A} + 2.0 \text{ A} = 0 \text{ A}$$

You also can see that the KVL is satisfied around each elementary loop. Going clockwise around each loop beginning in the lower left corner of each, you find the following:

$$\begin{aligned} \text{Left loop:} &\quad -12.0 \text{ V} + 6.0 \text{ V} + 6.0 \text{ V} = 0 \text{ V} \\ \text{Right loop:} &\quad -6.0 \text{ V} + 6.0 \text{ V} = 0 \text{ V} \end{aligned}$$

c. To find the power absorbed by each circuit element, apply Equation 19.26:

$$P = IV$$

where V is the potential difference across the circuit element and I is the current into its positive polarity terminal. Use the directions of the currents to assign polarities to the resistors as indicated in Figure 19.56. Applying the power equation (Equation 19.26) to each resistor, you obtain

$$\begin{aligned} 2.0\ \Omega \text{ resistor:} &\quad P_{2\ \Omega} = (3.0 \text{ A})(6.0 \text{ V}) = 18 \text{ W} \\ 3.0\ \Omega \text{ resistor:} &\quad P_{3\ \Omega} = (2.0 \text{ A})(6.0 \text{ V}) = 12 \text{ W} \\ 6.0\ \Omega \text{ resistor:} &\quad P_{6\ \Omega} = (1.0 \text{ A})(6.0 \text{ V}) = 6.0 \text{ W} \end{aligned}$$

(Note that each resistor absorbs positive power, as every resistor must.) Equivalently, you can find the power absorbed by each resistor by using Equation 19.28:

$$\begin{aligned} 2.0\ \Omega \text{ resistor:} &\quad P_{2\ \Omega} = (3.0 \text{ A})^2(2.0\ \Omega) = 18 \text{ W} \\ 3.0\ \Omega \text{ resistor:} &\quad P_{3\ \Omega} = (2.0 \text{ A})^2(3.0\ \Omega) = 12 \text{ W} \\ 6.0\ \Omega \text{ resistor:} &\quad P_{6\ \Omega} = (1.0 \text{ A})^2(6.0\ \Omega) = 6.0 \text{ W} \end{aligned}$$

The power absorbed by the independent voltage source is found from the general power equation $P = IV$, remembering that I must be the current into the positive polarity terminal. The current into the positive terminal of the ideal battery is –3.0 A. Thus the power absorbed by the independent voltage source is

$$\begin{aligned} P_{\text{bat}} &= (-3.0 \text{ A})(12 \text{ V}) \\ &= -36 \text{ W} \end{aligned}$$

The total power absorbed by the circuit elements must be zero (cross your fingers!):

$$\begin{aligned} P_{\text{bat}} &+ P_{2\ \Omega} + P_{3\ \Omega} + P_{6\ \Omega} \\ &= -36 \text{ W} + 18 \text{ W} + 12 \text{ W} + 6.0 \text{ W} \\ &= 0 \text{ W} \end{aligned}$$

Ta da! If the sum of the powers were not zero, you could be certain that a mistake was made somewhere in the calculations. On the other hand, if the sum of the powers is zero you can be quite confident (but not certain*) that all is well and you are headed for an A+ on the next test.

*The powers might accidentally sum to zero, so while it is necessary they sum to zero, that fact is not sufficient for you to conclude your solution is correct with certainty. Drat.

STRATEGIC EXAMPLE 19.11

Consider the circuit shown in Figure 19.57. Find

- **the current through each circuit element;**
- **the potential difference across each resistor; and**
- **the power absorbed by each circuit element.**

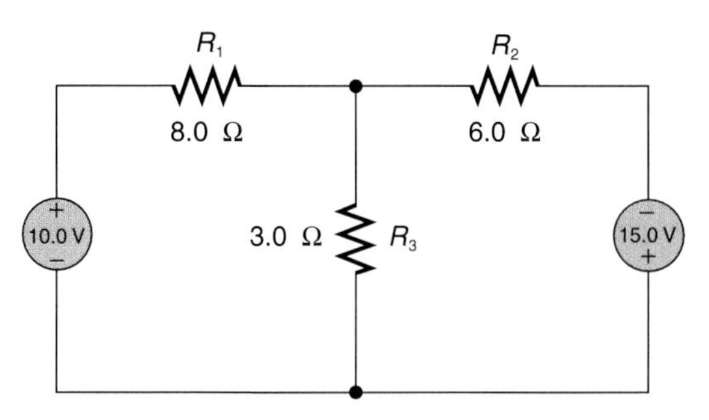

FIGURE 19.57

Solution

Step 1: See if the circuit can be simplified by combining resistors in series or parallel. The resistors in this circuit are neither in series nor parallel with any of the other resistors. The independent voltage sources also are not in series or parallel with each other. This circuit cannot be simplified. Note that the 10.0 V source is in series with the 8.0 Ω resistor, and so they have the same current; the 15.0 V source is in series with the 6.0 Ω resistor, and so they also have a common current.

Step 2: Locate and identify the nodes of the circuit. The nodes of this circuit are indicated in Figure 19.58. You are only interested in the significant nodes, those with three or more leads leaving them. There are two such significant nodes indicated in Figure 19.58.

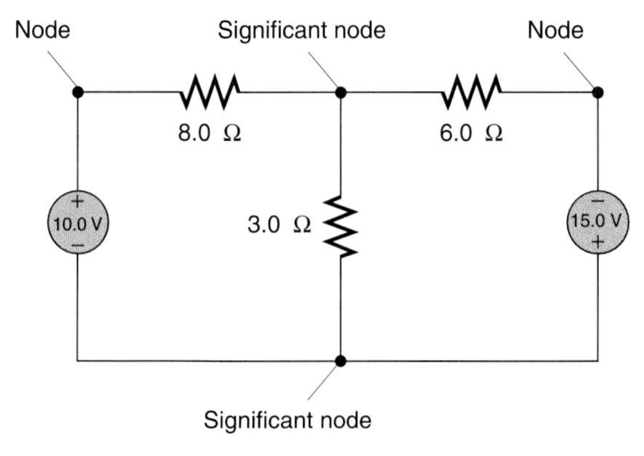

FIGURE 19.58

Step 3: Assign current directions to the individual branches of the circuit. The directions chosen for the currents are arbitrary but, if an independent voltage source is in a particular branch, you should usually choose the current to emerge from its positive polarity terminal. Thus take the directions of I_1 and I_2 to be as indicated in Figure 19.59. The current through the 3.0 Ω resistor can be taken to be in either direction; choose the current I_3 to be in the direction indicated in Figure 19.59.

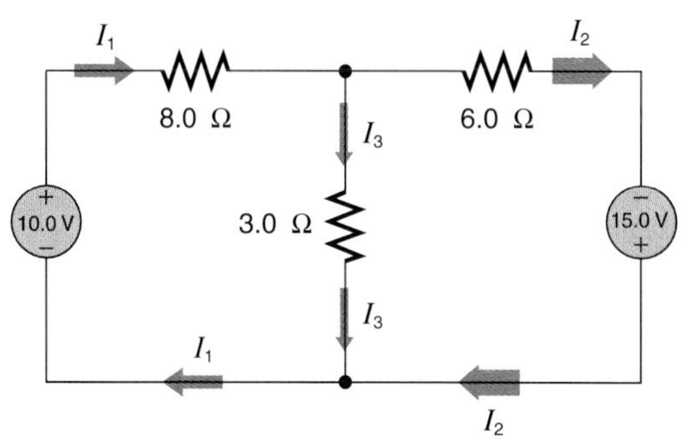

FIGURE 19.59

Step 4: Assign polarities to the resistors consistent with the current direction choices: current enters a resistor at the (+) polarity terminal of the resistor and exits at its (−) polarity terminal. Thus you emerge with the resistor polarities indicated in Figure 19.60.

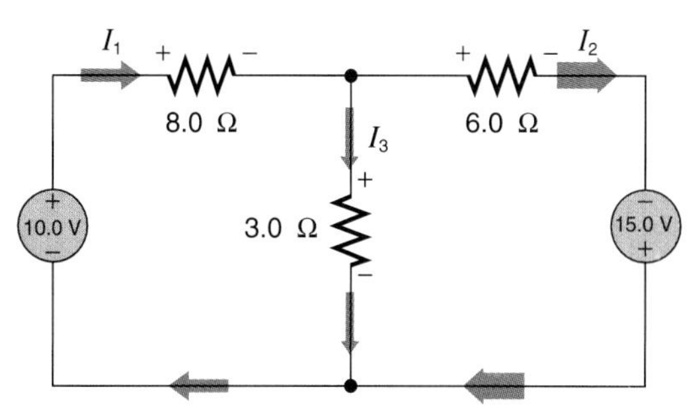

FIGURE 19.60

Step 5: Apply the KVL around each elementary loop. This circuit has two elementary loops; you write a KVL equation for each. The KVL states that the sum of the potential differences around any closed loop is zero. Remember that the direction you travel around the loop is arbitrary and you can begin anywhere along the path of the loop; you simply have to traverse the whole loop. Go around each loop in a clockwise direction beginning in the lower left corner of each loop, as shown in Figure 19.61.

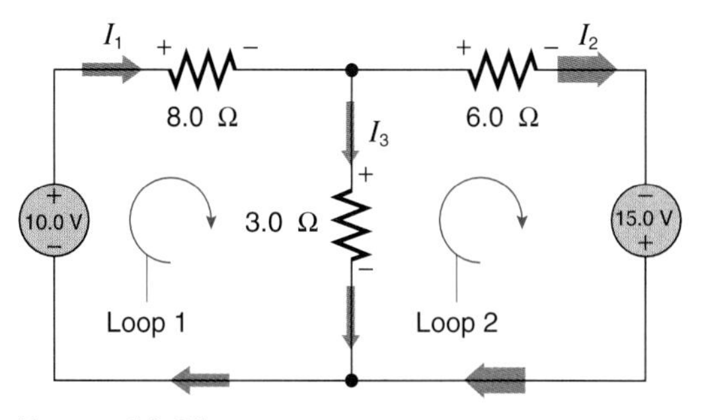

FIGURE 19.61

If a (−) terminal of a circuit element is the terminal first encountered as you go around the loop, the potential difference across that circuit element is put into the KVL equation with a minus sign; if a (+) terminal is the first encountered, that potential difference is put into the KVL equation with a plus sign. The potential difference across each resistor is given by Ohm's law, Equation 19.14. Thus for each loop the following equations emerge:

Loop 1

$$-10.0\ \text{V} + I_1(8.0\ \Omega) + I_3(3.0\ \Omega) = 0\ \text{V}$$
$$(8.0\ \Omega)I_1 + (3.0\ \Omega)I_3 = 10.0\ \text{V}$$

Loop 2

$$-I_3(3.0\ \Omega) + I_2(6.0\ \Omega) - 15.0\ \text{V} = 0\ \text{V}$$
$$(6.0\ \Omega)I_2 - (3.0\ \Omega)I_3 = 15.0\ \text{V}$$

Step 6: Assess whether there is sufficient information to solve for the unknown currents. The KVL yielded two equations but you have three unknown currents. Another equation is needed. So apply the KCL to the significant nodes to secure the additional equations needed. Here you need only one additional equation, and so you can apply the KCL to either of the significant nodes (here the same equation will result, except for a minus sign, regardless of which of the two significant nodes is used). Use the upper node, as shown in Figure 19.62.

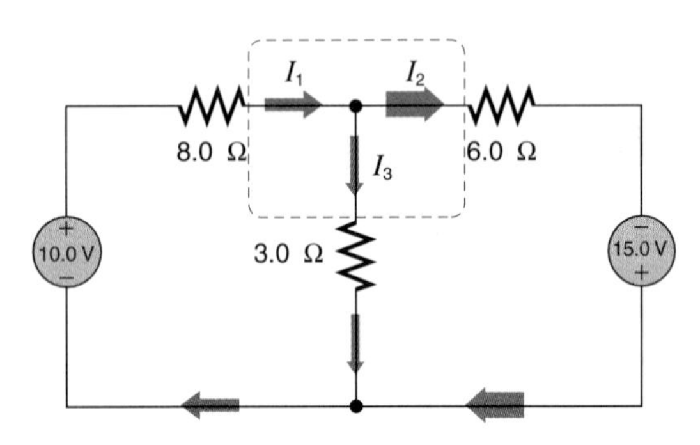

FIGURE 19.62

The KCL states that the sum of the currents leaving the node is zero. Thus the KCL applied to the upper node yields the equation

$$-I_1 + I_2 + I_3 = 0\ \text{A}$$

You now have the three equations needed for the three unknown currents.

Step 7: Solve for the unknown currents; this is a mathematical problem. There are several different ways this can be handled:

a. hire a mathematician;
b. use algebraic brute force;
c. use Kramer's rule methods involving suitable determinants; or
d. use appropriate software to do the grunt work.

Whichever method is chosen, the results for the three currents are

$$I_1 = 1.5\ \text{A}$$
$$I_2 = 2.2\ \text{A}$$
$$I_3 = -0.67\ \text{A}$$

The significance of the minus sign for the current I_3 means that the current through the 3.0 Ω resistor actually is directed opposite to the way indicated in Figure 19.59. No matter; you do *not* need to go back and change this current direction.

With the currents in the circuit known, you can find other quantities of interest.

The potential difference across each resistor is found by using Ohm's law, Equation 19.14:

$$V_1 = I_1R_1 = (1.5\ \text{A})(8.0\ \Omega) = 12\ \text{V}$$
$$V_2 = I_2R_2 = (2.2\ \text{A})(6.0\ \Omega) = 13\ \text{V}$$
$$V_3 = I_3R_3 = (-0.67\ \text{A})(3.0\ \Omega) = -2.0\ \text{V}$$

If a voltmeter* is connected across the various resistors with the (+) lead of the voltmeter on the (+) side of each resistor indicated in Figure 19.60, and the (−) lead on the corresponding (−) side of the resistor, the voltmeter will indicate the values we calculated for V_1, V_2, and V_3—*including the signs*! Here a voltmeter so placed across R_3 will indicate a negative value for V_3. The negative result for V_3 means that the higher-potential end of the 3.0 Ω resistor actually is not the one marked (+); it is the other end. Notice how your results for the currents and potential differences tell you whether the choices made for the current directions are the actual directions or not. A negative current in one direction is identical to a positive current in the opposite direction.

Now calculate the power absorbed by each circuit element. The electrical power absorbed by any circuit element is the product of the current into the (+) polarity terminal of that circuit element and the potential difference across the element (Equation 19.25).

For the 8.0 Ω resistor:

$$P_{8\ \Omega} = I_1V_1 = (1.5\ \text{A})(12\ \text{V}) = 18\ \text{W}$$

For the 6.0 Ω resistor:

$$P_{6\ \Omega} = I_2V_2 = (2.2\ \text{A})(13\ \text{V}) = 29\ \text{W}$$

For the 3.0 Ω resistor (note the signs carefully!):

$$P_{3\ \Omega} = I_3V_3 = (-0.67\ \text{A})(-2.0\ \text{V}) = 1.3\ \text{W}$$

Every resistor absorbs positive power. They get hot because the decrease in the electrical potential energy of the charges passing through them increases the thermodynamic internal energy of the resistors. The power absorbed by the resistors also

*We examine the characteristics of voltmeters in Section 19.15. A voltmeter is used to measure the potential difference between two points.

can be calculated using Equation 19.28, $P = I^2R$, with the same results.

Now for the independent voltage sources. Notice that the currents I_1 and I_2 originally were chosen to emerge from the positive terminals of the two independent voltage sources; the power equation needs to have the current *into* the positive terminal. So you must substitute $-I_1$ and $-I_2$ into the power equation for these circuit elements:

For the 15.0 V independent voltage source:

$$\begin{aligned} P_{15\text{ V}} &= (-2.2\text{ A})(15.0\text{ V}) \\ &= -33\text{ W} \end{aligned}$$

For the 10.0 V independent voltage source:

$$\begin{aligned} P_{10\text{ V}} &= (-1.5\text{ A})(10.0\text{ V}) \\ &= -15\text{ W} \end{aligned}$$

Both absorb negative power; this fact indicates that they are actually generating or providing (transferring) power to the rest of the circuit, derived from chemical or mechanical energy within the sources.

Now for the crucial test: the sum of the electrical power absorbed by all the circuit elements must be zero to conserve energy (per unit time). You can see if this is the case:

$$\begin{aligned} &P_{8\,\Omega} + P_{6\,\Omega} + P_{3\,\Omega} + P_{15\text{ V}} + P_{10\text{ V}} \\ &= 18\text{ W} + 29\text{ W} + 1.3\text{ W} - 33\text{ W} - 15\text{ W} \\ &= 48\text{ W} - 48\text{ W} \\ &= 0\text{ W} \end{aligned}$$

Voilà! Things check.

EXAMPLE 19.12

You have an independent voltage source (an ideal battery) with emf V_0 and two resistors with resistances R_1 and R_2. To have the power absorbed by the resistors be greatest, should they be connected in series to the battery, as shown in Figure 19.63, or in parallel, as in Figure 19.64?

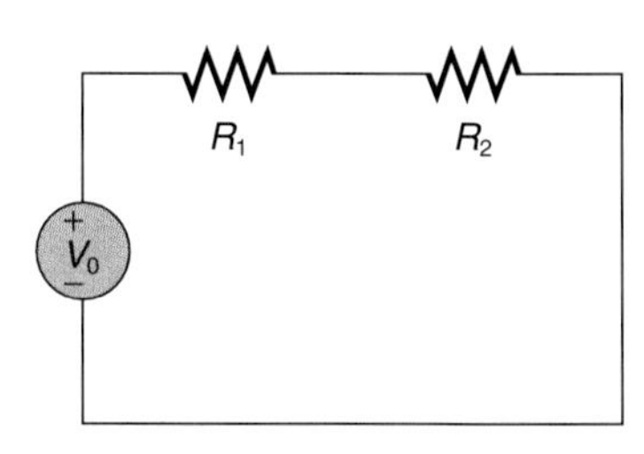

(a) Series connection

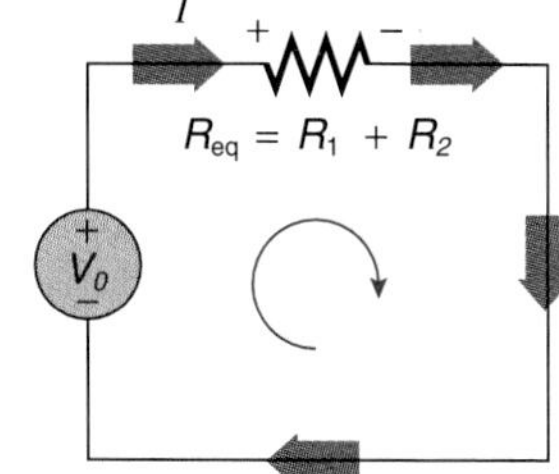

(b) Equivalent

FIGURE 19.63

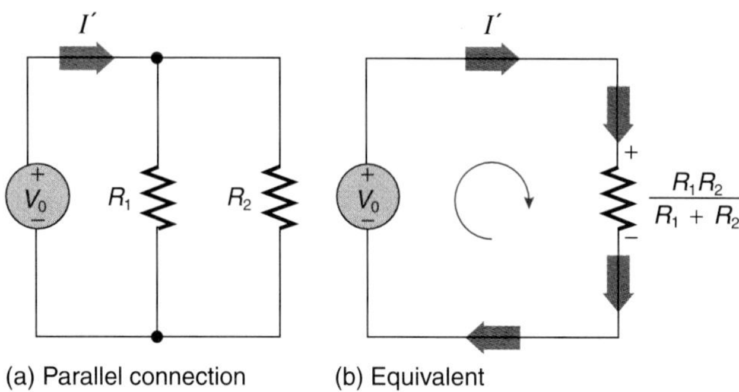

(a) Parallel connection (b) Equivalent

FIGURE 19.64

Solution

You solve this problem in three steps: (1) compute the power absorbed when the resistances are connected in series; (2) compute the power absorbed for the parallel connection; and (3) compare the computed values of power absorbed.

1. For the resistors in series, the equivalent resistor is the sum of the pair: $R_1 + R_2$. Apply the KVL to the equivalent circuit, with the current direction as chosen in Figure 19.63b; go around the circuit clockwise:

$$-V_0 + I(R_1 + R_2) = 0\text{ V}$$

yielding

$$I = \frac{V_0}{R_1 + R_2}$$

Note that the equivalent resistor is in parallel (and series) with the voltage source. The power absorbed by the equivalent resistor is, from Equation 19.26,

$$\begin{aligned} P_{\text{res series}} &= IV \\ &= \left(\frac{V_0}{R_1 + R_2}\right)V_0 \\ &= \frac{V_0^2}{R_1 + R_2} \end{aligned}$$

2. For a parallel connection of two resistors, the equivalent resistor is their product divided by their sum:

$$\frac{R_1R_2}{R_1 + R_2}$$

Apply the KVL around the single equivalent loop shown in Figure 19.64b; go clockwise around the equivalent loop:

$$-V_0 + I'\frac{R_1R_2}{R_1 + R_2} = 0\text{ V}$$

yielding

$$I' = \frac{V_0(R_1 + R_2)}{R_1R_2}$$

Note again that the equivalent resistor is in parallel (and series) with the voltage source. The power absorbed by the equivalent resistor then is found from Equation 19.28:

$$P_{\text{res parallel}} = I'V$$

Substituting for I' and the $V = V_0$, you get

$$\begin{aligned} P_{\text{res parallel}} &= \left[\frac{V_0(R_1 + R_2)}{R_1R_2}\right] V_0 \\ &= V_0^2\,\frac{R_1 + R_2}{R_1R_2} \end{aligned}$$

3. The ratio of the power absorbed by the resistors when they are in parallel to that when they are in series is

$$\begin{aligned} \frac{P_{\text{res parallel}}}{P_{\text{res series}}} &= \frac{V_0^2\,\dfrac{R_1 + R_2}{R_1R_2}}{\dfrac{V_0^2}{R_1 + R_2}} \\ &= \frac{(R_1 + R_2)^2}{R_1R_2} \\ &= \frac{R_1^2 + R_2^2 + 2R_1R_2}{R_1R_2} \end{aligned}$$

The numerator is larger than the denominator, and so the power ratio is greater than unity. Thus a parallel connection of the two resistors absorbs more power from a given voltage source than a series connection of the same resistors.

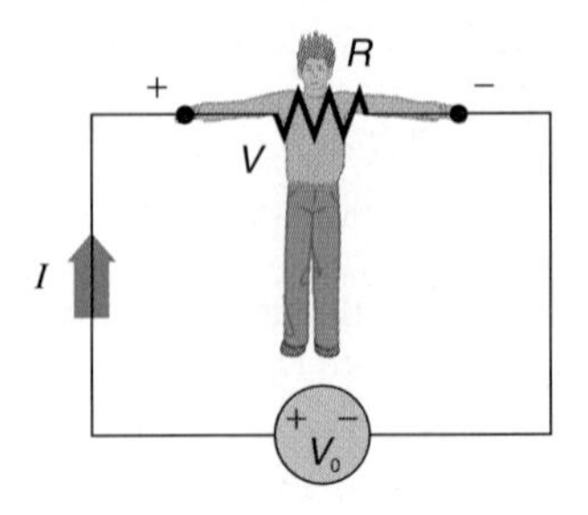

FIGURE 19.65 You can complete a circuit, but it may be dangerous to your health.

19.12 ELECTRIC SHOCK HAZARDS*

Awaiting the sensation of a short, sharp shock

William S. Gilbert (1836–1911) and Arthur Sullivan (1842–1900)†

If you ever have been shocked by an automotive spark plug wire, you have been jolted with a potential difference of over a kilovolt. Potential difference is not what is inherently dangerous to you; it is electric *current* that is dangerous.

You may not think of yourself in these terms, but you are a resistor. If you connect yourself to an independent voltage source (accidentally or intentionally), the current in the circuit depends on both the resistance of your body and the potential difference across it (see Figure 19.65).

Your resistance depends on the nature of the electrical contact the skin makes with the wires that form the rest of the circuit. Dry skin contact with the wires has a much higher resistance than wet skin contact. The condition of your skin and the area of contact determine your effective resistance. Your body resistance can be measured easily with a common laboratory ohmmeter (see Section 19.15) simply by holding each ohmmeter lead with the fingers of each hand. The resistance of the body from one hand to the other is on the order of hundreds of thousands of ohms to several million ohms if the fingers are dry. If the fingers are wet, the resistance is reduced drastically; the smaller body resistance allows a correspondingly larger current from a fixed voltage source. This means that wet contact with an independent voltage source is much more dangerous than dry contact.

Depending on how you are connected to the independent voltage source, the current through your body may flow across your chest and can trigger dangerous consequences if the current is sufficiently large. A current of only about 1 mA is painful; 10–25 mA is sufficient to prevent you from controlling your muscles in order to release the wires; 50 mA to 3 A can be fatal, since such currents interfere with biological electrical signals regulating your heartbeat. **Ventricular fibrillation** results, a condition in which the heart loses its normal rhythmic pumping action. The ventricle wall ripples and the heart is unable to pump any blood; death ensues within just a few minutes.

Ironically, currents above about 3 A are less dangerous. With such large currents, breathing can stop (which is not too good for the health!); but if breathing can be induced once again quickly via artificial respiration, the odds of surviving are better than with the smaller currents that induce ventricular fibrillation. It takes longer to suffocate than to die from a stopped heart. Indeed, large current pulses are used by hospital trauma physicians and rescue squads to *stop* ventricular fibrillation. A large current pulse causes the heart muscles to contract violently. With the cessation of the current pulse, the heart muscles relax and often resume normal activity. During the procedure, bellows are used to pump air into the lungs of such patients until normal breathing functions can be restored. Most people hit by lightning die because their heart or breathing stops rather than from burns.

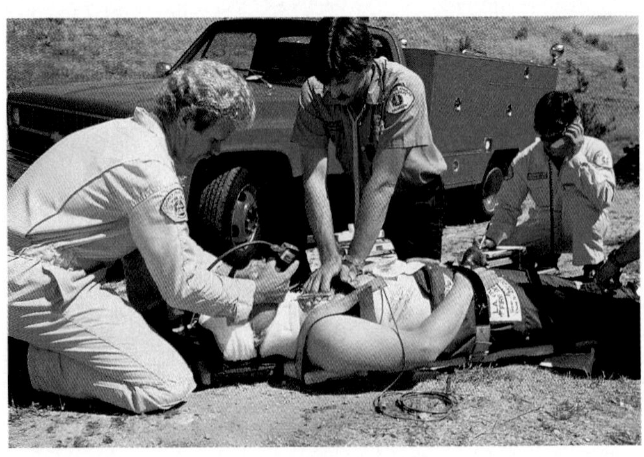

Large currents are used by emergency personnel to stop ventricular fibrillation.

†*The Mikado* (1885), Act I, #10, "Trio."

QUESTION 7

Why are birds not electrocuted when they perch on bare wires at a high electrical potential?

FIGURE Q.7

EXAMPLE 19.13

If your body resistance is 5.0 × 10⁴ Ω with wet hands, what potential difference will cause a painful and dangerous current of 1.0 mA through your body?

Solution

Use Ohm's law, Equation 19.14. A current of 1.0 mA through a resistance of 5.0×10^4 Ω represents a potential difference across the resistance of

$$\begin{aligned} V &= IR \\ &= (1.0 \times 10^{-3}\ \text{A})(5.0 \times 10^4\ \Omega) \\ &= 50\ \text{V} \end{aligned}$$

Common electrical outlets have effective potential differences of about 120 V and so are quite dangerous under these circumstances.

19.13 A MODEL FOR A REAL BATTERY

We now can formulate a model of a real battery to explain why they "die."

An independent voltage source such as an ideal battery has a vertical line for its current–voltage characteristic curve, as shown in Figure 19.66. An independent voltage source maintains the same potential difference V_0 between its terminals for all time. The potential difference across its terminals is independent of the current.

Real batteries do not have these ideal characteristics. Real batteries in circuits get warm, indicating they absorb some of the potential energy of the charges that constitute the current.

> To a good approximation, a real battery is modeled by means of an independent voltage source in series with a resistance r, as shown in Figure 19.67.

The series resistor r, known as the **internal resistance** of the real battery, always absorbs power when there is a current through it, thus accounting for the observation that real batteries get warm when used.

The points A and B in Figure 19.67 represent the terminals of the real battery. The potential difference between A and B is called the **terminal potential difference** of the real battery. Notice that if the real battery is not connected to a circuit, the terminal potential difference is the same as the emf of the independent voltage source in the model. No current flows through r, so there is no potential difference across it.

For a fresh real battery, the internal resistance r is very small. When $I > 0$ A, the finite internal resistance makes the terminal potential difference $V_{\text{term}} < V_0$, so that the characteristic curve of a fresh real battery is nearly a vertical line with negative slope, as shown in Figure 19.68.

Connect a real battery to an external resistor R (such as a light bulb in a flashlight), making a single loop circuit, as in

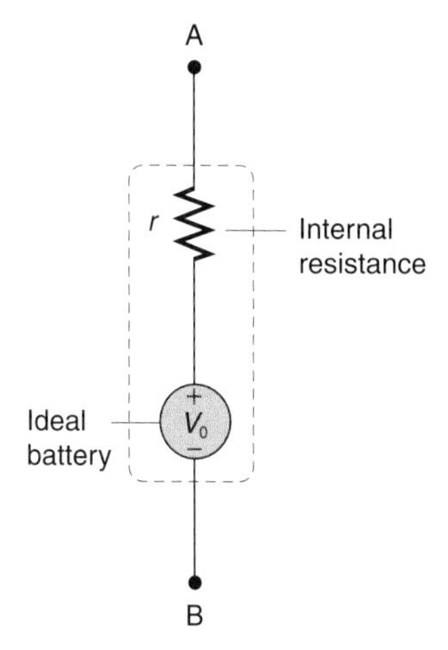

FIGURE 19.67 A model of a real battery.

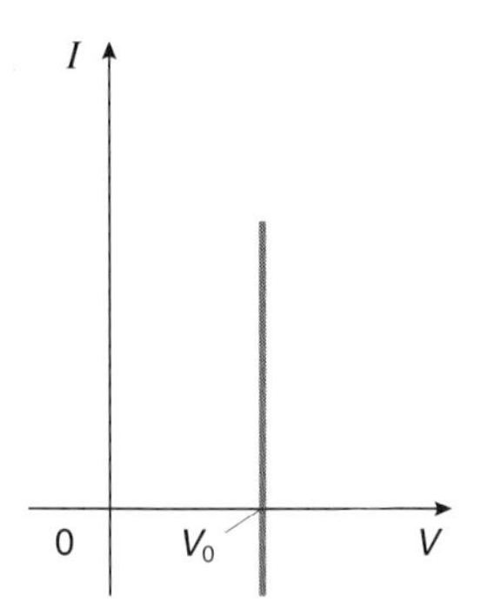

FIGURE 19.66 The characteristic curve of an ideal battery.

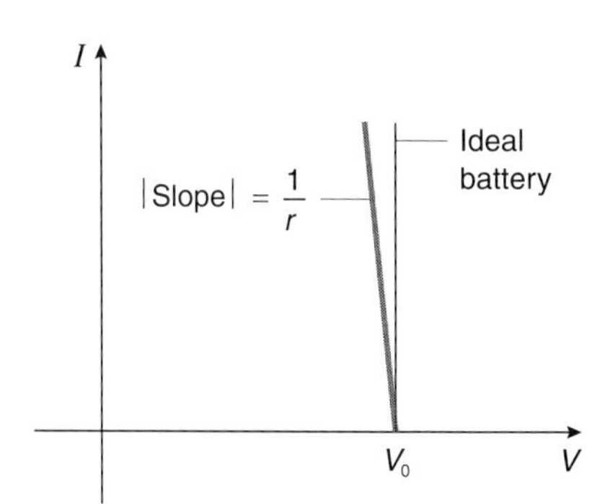

FIGURE 19.68 The characteristic curve of a real battery.

Figure 19.69. The external resistor R sometimes is called a **load resistor**. Since the circuit consists of a single loop, we let I be the current in the loop with the direction chosen as indicated in Figure 19.69. As a result of the current direction choice, the resistors have the polarities indicated in Figure 19.69. Applying the KVL around the single loop in a clockwise direction beginning at point B, we get

$$-V_0 + Ir + IR = 0 \text{ V}$$

and so

$$I = \frac{V_0}{R + r} \tag{19.31}$$

The terminal potential difference of the real battery is the potential difference between A and B. Going from A to B in the circuit and algebraically summing the potential differences, we find

$$-Ir + V_0$$

So the terminal potential difference is

$$V_{\text{term}} = V_0 - Ir \tag{19.32}$$

Using Equation 19.31 for I, the terminal potential difference becomes

$$\begin{aligned} V_{\text{term}} &= V_0 - \frac{V_0}{R + r} r \\ &= V_0 \frac{R}{R + r} \end{aligned} \tag{19.33}$$

For a fresh real battery, the internal resistance r is very small and the terminal potential difference approaches the value of the emf V_0 of the ideal battery. But as a real battery is used, the internal resistance gradually increases with time because of irreversible chemical reactions within the battery.†

†In a rechargeable battery the chemical reactions can be reversed by charging the battery; this involves forcing a current into the positive terminal of the battery with another independent voltage source, so that the battery absorbs positive power.

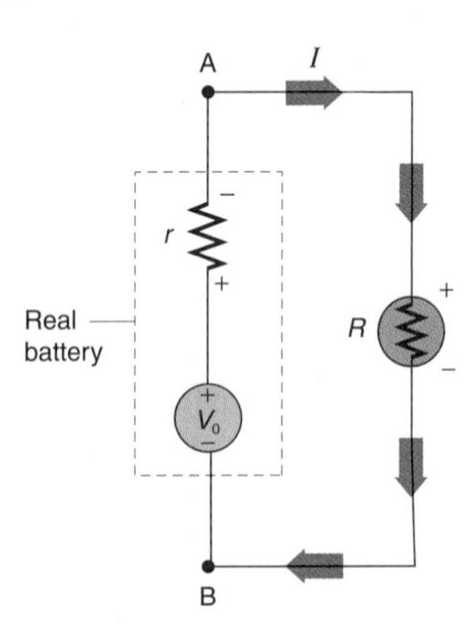

FIGURE 19.69 A real battery connected to an external resistor, such as a light bulb.

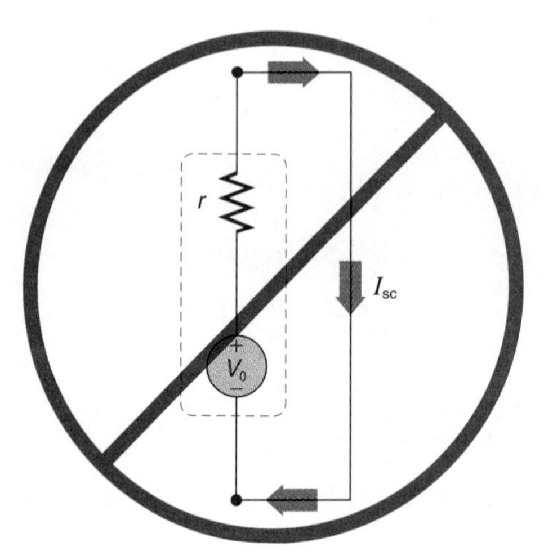

FIGURE 19.70 Shorting a real battery. *Do not do this!*

As a real battery is used, r eventually becomes quite large. As $r \to \infty\ \Omega$, both the current I (see Equation 19.31) and the terminal potential difference (see Equation 19.33) approach zero. The real battery then is called a **dead battery**: there is no current in the circuit to the external load resistor and R absorbs zero power. If R represents the light bulb of your flashlight, the bulb will not light because the current in the circuit is zero.

It is dangerous to short-circuit a real battery. If a fresh real battery with a small internal resistance r is shorted, as shown in Figure 19.70, the external load resistance R is zero and the short circuit current becomes, from Equation 19.31,

$$I_{\text{sc}} = \frac{V_0}{r}$$

Using Equation 19.28, the power absorbed by the internal resistance is

$$\begin{aligned} P_r &= I_{\text{sc}}^2 r \\ &= \frac{V_0^2}{r} \end{aligned}$$

Since r is small, the power absorbed can be quite large. The power absorbed rapidly increases the thermodynamic internal energy of the real battery, which results in a rapid temperature increase, sufficient to explode the battery as chemicals are vaporized within it.

QUESTION 8

Under what circumstances (if any) can the terminal potential difference of a real battery exceed the potential difference of the ideal battery used to model it?

EXAMPLE 19.14

A real battery is connected in series to a 10.0 Ω resistor, as shown in Figure 19.71.

a. Determine the current in the circuit.
b. Find the terminal potential difference of the real battery.
c. Determine the potential difference across the load resistor.
d. What power is absorbed by the internal resistance of the battery?

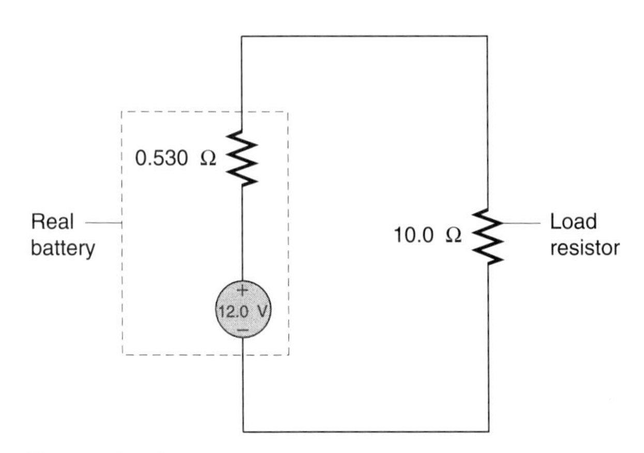

FIGURE 19.71

Solution

a. With the current direction chosen as in Figure 19.72, the resistors have the polarity shown for the potential difference. Use the KVL clockwise around the elementary loop:

$$-12.0 \text{ V} + I(0.530 \ \Omega) + I(10.0 \ \Omega) = 0 \text{ V}$$

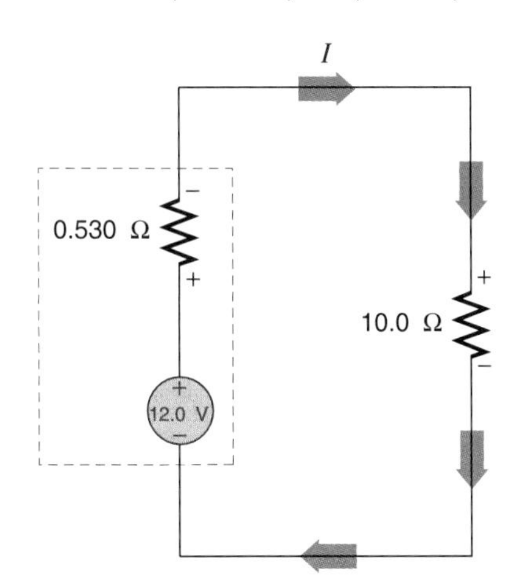

FIGURE 19.72

Solving for I, you get

$$I = 1.14 \text{ A}$$

b. The potential difference across the internal resistance of the real battery is

$$\begin{aligned} V_r &= Ir \\ &= (1.14 \text{ A})(0.530 \ \Omega) \\ &= 0.604 \text{ V} \end{aligned}$$

From Equation 19.32, the potential difference between the terminals of the real battery then is

$$\begin{aligned} V_{term} &= 12.0 \text{ V} - 0.604 \text{ V} \\ &= 11.4 \text{ V} \end{aligned}$$

c. The potential difference across the load resistor is found using Ohm's law:

$$\begin{aligned} V &= IR \\ &= (1.14 \text{ A})(10.0 \ \Omega) \\ &= 11.4 \text{ V} \end{aligned}$$

d. The power absorbed by the internal resistance of the real battery is found using Equation 19.28 for the power absorbed by any resistor:

$$\begin{aligned} P &= I^2 r \\ &= (1.14 \text{ A})^2(0.530 \ \Omega) \\ &= 0.689 \text{ W} \end{aligned}$$

The power absorbed by the internal resistance of the real battery causes it to become warm.

Example 19.15

Find current and the power absorbed by the internal resistance of the real battery shown if its terminals are accidentally short-circuited, as in Figure 19.73.

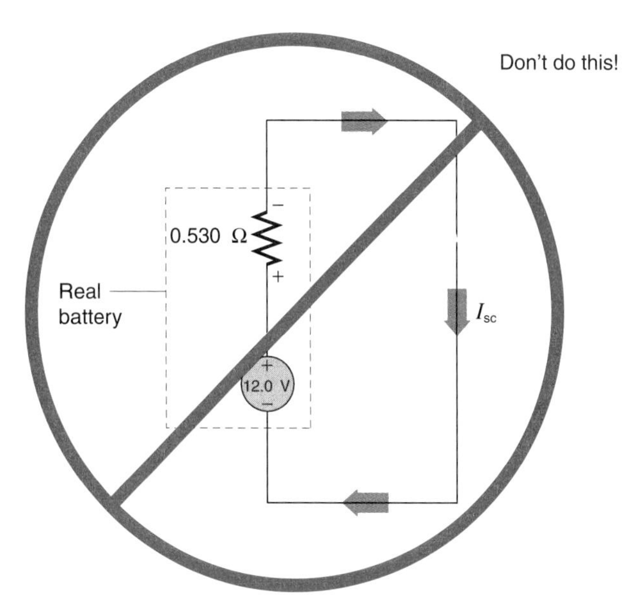

FIGURE 19.73

Solution

The short circuit current I_{sc} in the circuit is found by applying the KVL to the loop:

$$-12.0 \text{ V} + I_{sc}(0.530 \ \Omega) = 0 \text{ V}$$

Solving for I_{sc}, you find

$$I_{sc} = 22.6 \text{ A}$$

The power absorbed by the internal resistance then is found from Equation 19.28 to be

$$\begin{aligned} P_r &= (I_{sc})^2 r \\ &= (22.6 \text{ A})^2(0.530 \ \Omega) \\ &= 271 \text{ W} \end{aligned}$$

This is a considerable amount of power and may cause the battery to explode.

19.14 MAXIMUM POWER TRANSFER THEOREM

If you are an audiophile, you know that your stereo speakers must be matched to the power supply (the real voltage source) of your audio system for optimal performance. The criterion used is an application of the **maximum power transfer theorem.**

It is important to know what load resistor R maximizes the power absorbed by it from a real battery (see Figure 19.74). The power absorbed by any resistor is found from Equation 19.28:

$$P = I^2R$$

The current through the load resistor we found (Equation 19.31) to be

$$I = \frac{V_0}{R + r}$$

and so the power absorbed by the load resistor R is

$$P(R) = \left(\frac{V_0}{R + r}\right)^2 R \qquad (19.34)$$

Equation 19.34 indicates that $P = 0$ W when $R = 0\ \Omega$ and when $R = \infty\ \Omega$, and so the power must have a maximum for a finite value of R, as shown schematically in Figure 19.75. Keeping V_0 and r fixed, we want to discover what load resistance R maximizes the power P delivered by the real battery to the load. This is an maximum problem in calculus. So we take the derivative of P with respect to R, set it equal to zero, and solve the resulting equation for R:

$$\frac{dP}{dR} = 0\ \text{W}/\Omega$$

We use Equation 19.34 for P; its derivative with respect to R is

$$\frac{dP}{dR} = V_0^2\,\frac{(R + r)^2 - R\,[2(R + r)]}{(R + r)^4} = 0\ \text{W}/\Omega$$

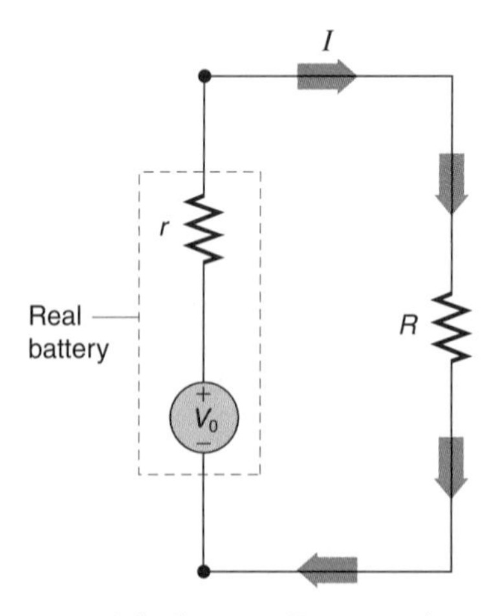

FIGURE 19.74 A load resistor R connected to a real battery.

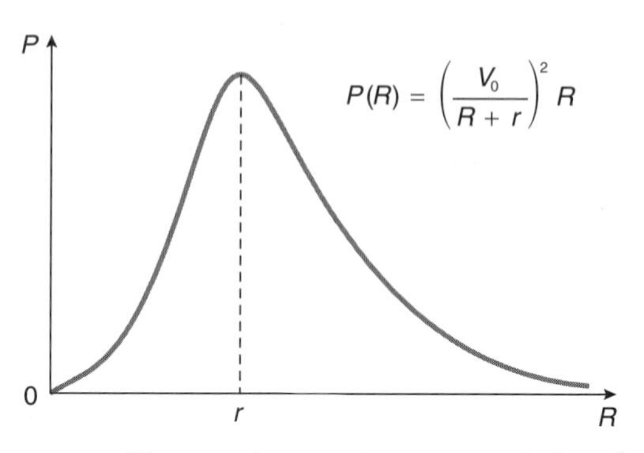

FIGURE 19.75 The power has a maximum at a particular value of R.

Solving for R, we have

$$(R + r)^2 - R(2)(R + r) = 0\ \Omega^2$$
$$(R + r) - 2R = 0\ \Omega$$
$$R = r \qquad (19.35)$$

In other words, for a given real battery, the load resistance R that maximizes the power absorbed by R has a value equal to the internal resistance r of the real battery. This result is the maximum power transfer theorem. If R is chosen to match (be equal to) r, the resulting condition is called an **impedance match.** It should be understood that the load resistance R may be a single resistor (as it is in Figure 19.74) or R may be the single *equivalent* resistance of a collection of resistors connected to the terminals of the real battery (see Example 19.16).

The maximum power transfer theorem is quite general and holds for any real voltage source connected to a load, not just the typical real battery. When you match stereo speakers to a power supply you are impedance matching for maximum power transfer to the speakers.

The model for a real battery consisting of an ideal battery (or ideal voltage source) in series with a resistance r is a model used for many complex electronic devices. Indeed, if you pursue electronics by taking courses in circuit analysis, you will learn both analytical and experimental techniques to reduce many complex circuits, such as the "bird's nest" in Figure 19.76, to just an ideal independent voltage source in series with an internal resistance, with the model connected to a load resistor R, as shown in Figure 19.77. The simple model for the complex circuit is known as the **Thévenin equivalent circuit** of the complex circuit.

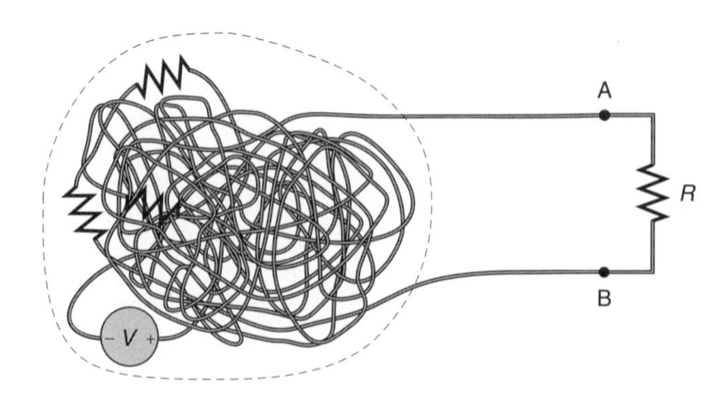

FIGURE 19.76 A complicated circuit.

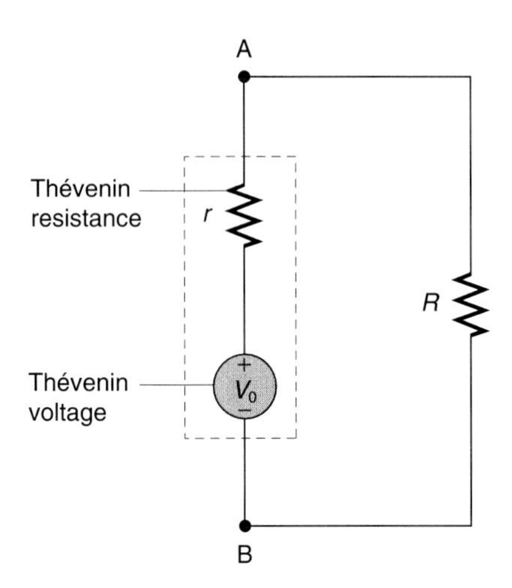

FIGURE 19.77 **The Thévenin equivalent of a complicated circuit.**

The potential difference of the ideal independent voltage source in the model (or equivalent circuit) is called the **Thévenin voltage** and the internal resistance r of the model is called the **Thévenin resistance**. The resulting single loop circuit depicted in Figure 19.77 is used frequently, and so this simple circuit is one you will encounter in many contexts. (See Problem 60.)

EXAMPLE 19.16

For the circuit indicated in Figure 19.78, what value of R maximizes the power absorbed by all the resistors?

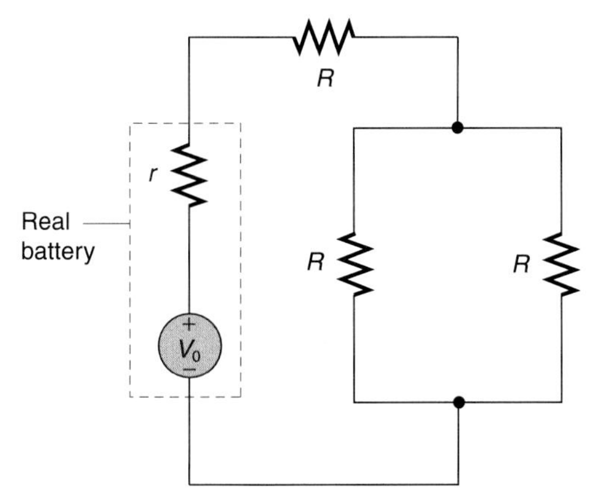

FIGURE 19.78

Solution

To maximize the power absorbed, the total equivalent resistance connected to the real battery must be equal to the internal resistance of the real battery. The parallel pair of resistors has an equivalent resistance equal to their product divided by their sum:

$$R_{eq} = \frac{RR}{R + R} = \frac{R}{2}$$

The $R/2$ equivalent resistance is in series with resistance R. The total resistance attached to the real battery therefore is

$$R + \frac{R}{2} = \frac{3R}{2}$$

According to the maximum power transfer theorem, the total equivalent resistance external to the real battery must be equal to the internal resistance r of the real battery, and so

$$\frac{3R}{2} = r$$

or

$$R = \frac{2}{3} r$$

19.15 BASIC ELECTRONIC INSTRUMENTS: VOLTMETERS, AMMETERS, AND OHMMETERS

Potential differences are measured with a **voltmeter**, currents are measured with an **ammeter**, and resistances are measured with an **ohmmeter**. The three instruments occasionally are packaged together as a single **multimeter** that can be used for measuring potential differences, currents, or resistances by selecting the appropriate function with a switch.

Such instruments are common in most laboratories, and it is important to have an understanding of their characteristics and how they are used. Ultimately, all three instruments use a sensitive current-detecting device. Historically, a galvanometer was used, consisting of a coil of wire suspended in a magnetic field, but other specialized devices now are employed to measure very small currents. We designate any such sensitive current-detecting device by the symbol shown in Figure 19.79, which was the historic symbol for a galvanometer.

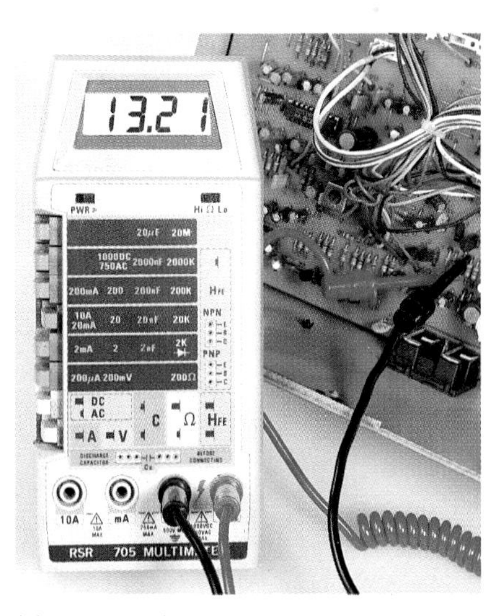

A common laboratory multimeter.

FIGURE 19.79 Symbol for a sensitive current-detecting device.

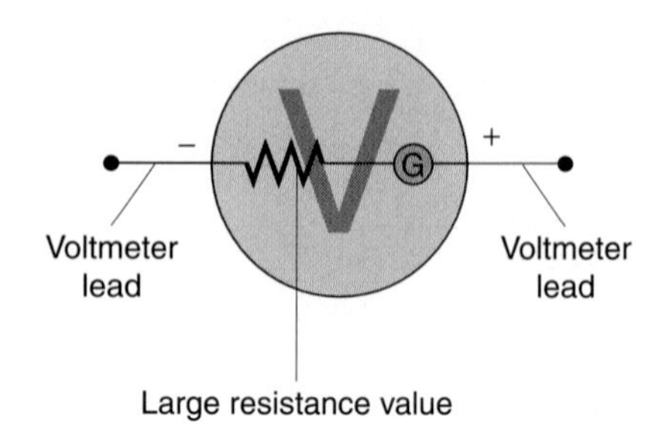

FIGURE 19.82 A voltmeter has a high effective resistance.

Voltmeters

A voltmeter is an instrument used to measure the potential difference between two nodes of a circuit. Such meters are symbolized as shown in Figure 19.80. The positive polarity lead of the meter typically is marked with a (+), is a red-colored terminal, or is red-colored wire. The negative polarity lead of the meter is either marked with a (–), is a black-colored terminal, or is a black-colored wire.

To measure the potential difference across a circuit element or between two nodes, the positive lead (the red wire) of the voltmeter is touched or connected to the high-potential side of the element [the (+) side of the circuit element], while the negative lead (the black wire) of the meter is touched or connected to the low-potential side of the element (–); see Figure 19.81. The meter indicates the value of the potential difference V between the two points. If the voltmeter leads are reversed, the meter will read $-V$ rather than V.

Notice in Figure 19.81 that a voltmeter is connected in *parallel* with the circuit element. This ensures that the potential difference across the voltmeter is the same as the potential difference across the circuit element.

Connecting a voltmeter to an electrical circuit means, of course, that the voltmeter becomes part of the circuit. The circuit is not the same circuit any more! To ensure that the effect of connecting the voltmeter to the circuit is minimal, we ideally want no current in the leads connecting the voltmeter to the circuit. This is accomplished by manufacturing the voltmeter to have a very large effective resistance. Inside the voltmeter, this may be accomplished by placing a high resistance in series with the sensitive current-measuring device, as shown in Figure 19.82. In this way, the current through the actual circuit is negligibly affected by the presence of the voltmeter in the circuit.

So the two things to remember about a voltmeter are the following:

1. It is connected in parallel with the circuit element in question and therefore has the same potential difference across it.
2. It has a very large effective resistance so as to affect the circuit minimally.

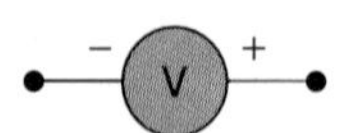

FIGURE 19.80 Symbol for a voltmeter.

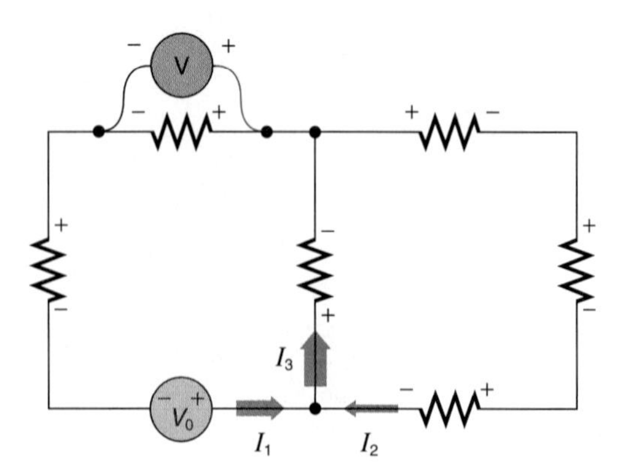

FIGURE 19.81 How to measure the potential difference across a circuit element with the voltmeter.

Ammeters

An ammeter is a device for measuring the current through some branch of a circuit. The symbol for an ammeter is shown in Figure 19.83. The positive polarity lead of the meter typically is marked with a (+) sign, is colored red, or is a red-colored wire. The negative polarity lead of the meter is either marked with a (–), is colored black, or is a black-colored wire.

To measure the current, the ammeter must be inserted in series so that all the current in the wire passes through the meter. The (+) side of the ammeter is placed at the higher-potential side of the branch, as shown in Figure 19.84.

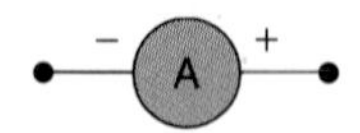

FIGURE 19.83 Symbol for an ammeter.

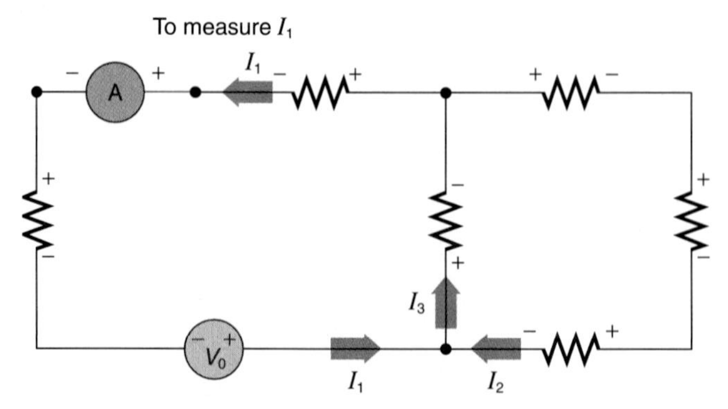

FIGURE 19.84 An ammeter is connected in series in the branch whose current is to be measured.

The original circuit is changed by inserting the ammeter into the circuit. To ensure that the ammeter does not significantly affect the value of the current in the branch of the circuit into which it will be placed, the ammeter must have a very small effective resistance. Then the potential difference across the ammeter itself will be negligible compared with that across the other circuit elements. Within the ammeter itself, a very small resistance is placed in parallel with the sensitive current-detecting device, as shown schematically in Figure 19.85. The small resistance ensures that the overall resistance of the ammeter itself is very small.

Hence an ammeter

1. is connected in series; and
2. must have very small effective resistance.

With the advent of microelectronics or circuits of very small size, it may be physically impossible to place an ammeter in the circuit path. The circuit may be hard-wired, which means that to place the ammeter in the circuit, wires must be cut and the meter inserted. In such situations, placing an ammeter in a circuit may be out of the question, if not just a pain in the neck. So what can be done to determine the current? All hope is not lost. If there is a resistor in the branch along which the current is to be measured, we can measure the potential difference across the resistor with a voltmeter. If the resistance is known (or can be determined from its color code), then the current can be found by using Ohm's law for the resistor, Equation 19.14:

$$V = IR$$

or

$$I = \frac{V}{R}$$

In this way, the current is determined without resorting to an ammeter.

Ohmmeters

An ohmmeter is depicted with the symbol shown in Figure 19.86. To measure the resistance of a resistor, the leads of the ohmmeter are attached to the leads of the resistor, as indicated in Figure 19.87.

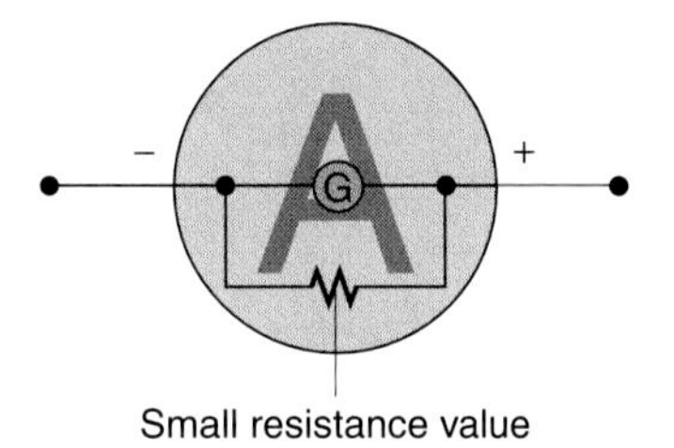

FIGURE 19.85 An ammeter has a low effective resistance.

The resistor must *not* be part of the circuit in which the resistor is used when such a measurement is made.

If an ohmmeter is used when the resistor is part of the rest of a circuit, as in Figure 19.88, the ohmmeter will indicate the equivalent resistance between the two terminals to which it is attached. The equivalent resistance is the parallel combination of the given resistance with that of the rest of the circuit. The ohmmeter then does not indicate the resistance of the desired resistor.

If the resistor cannot be easily removed from the circuit, its approximate value may be determined from its color code if it is so labeled. The value of a resistance also should be listed on the schematic circuit diagram if one is available

An ohmmeter measures resistance by using a battery to create a small current through the resistor whose resistance is to be measured; see Figure 19.89. The current in the sensitive current detector is inversely proportional to the value of the resistance.

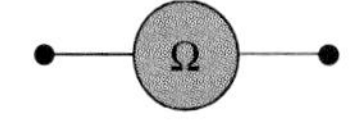

FIGURE 19.86 The symbol for an ohmmeter.

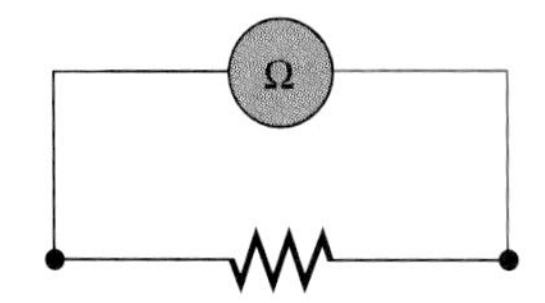

FIGURE 19.87 How to measure the resistance of a resistor with an ohmmeter.

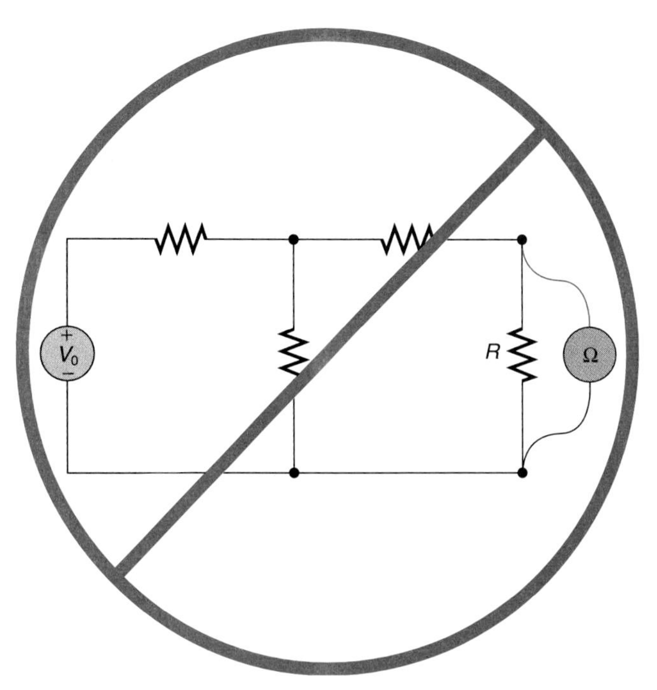

FIGURE 19.88 *Incorrect* way to measure the resistance of a resistor.

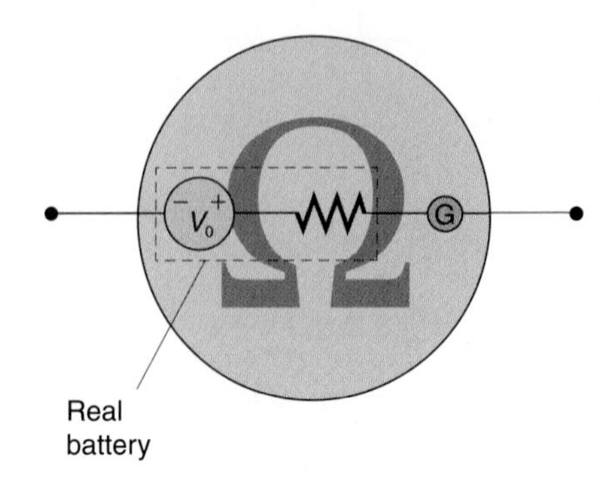

FIGURE 19.89 A model of an ohmmeter.

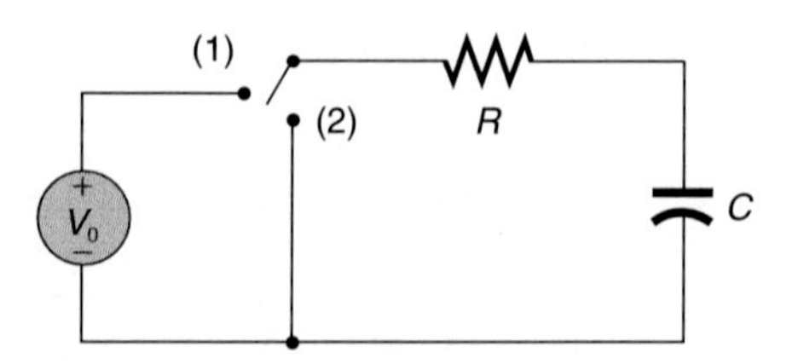

FIGURE 19.90 A series RC network (there is no closed path, so that it is not yet a circuit).

19.16 An Introduction to Transients in Circuits: A Series RC Circuit*

When you turn on the flash unit for your camera, there typically is a delay before the flash unit is ready to fire. Sometimes this delay can be quite annoying if you want to take a candid picture immediately. The delay is caused by the time it takes the battery to fully charge a capacitor used to accumulate the electrical potential energy needed for the flash lamp. This is an example of a transient effect in a circuit.

Transients in circuits are one-shot effects that occur, for example, when a circuit is first completed (turned on) or disconnected (turned off). As an example of such transients, we examine a single loop circuit consisting of an independent voltage source (an ideal battery), a resistor, and a capacitor, as shown in Figure 19.90. The arrangement is called a **series RC circuit**. A switch is included so that we can investigate several situations depending on which way the switch is positioned.

A Charging Capacitor

Let the capacitor in the *RC* circuit initially be uncharged and the switch open (as shown in Figure 19.90). The system is an electrical network, but not a circuit because there is no closed path. We want to see what happens when, at time $t = 0$ s, the switch is closed to position (1), thus creating the single loop circuit shown in Figure 19.91. In particular, we want to find the current in the circuit as a function of time.

At any instant t after the switch is closed, let $I(t)$ be the current in the circuit, directed as shown in Figure 19.91. According to our conventions, once the current direction is chosen, the resistor polarity is determined as indicated. At time t, let the capacitor have the charges $\pm q$ on its plates as shown (we take $q > 0$ C).

From Equation 18.1, the capacitance is (by definition)

$$C = \frac{|q|}{|V|}$$

so that the potential difference V across the capacitor is

$$V = \frac{q}{C}$$

with the polarity indicated in Figure 19.91.

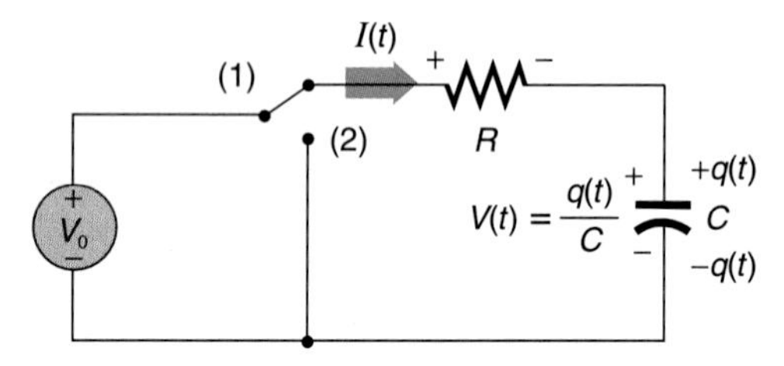

FIGURE 19.91 The switch is closed to position (1) when $t = 0$ s.

We apply the KVL to the single loop, beginning in the lower left corner of the circuit and traversing the loop in the clockwise sense:

$$-V_0 + IR + \frac{q}{C} = 0 \text{ V} \qquad (19.36)$$

But the current I is related to q since

$$I = \frac{dq}{dt}$$

We differentiate Equation 19.36 with respect to t, recognizing that the emf V_0 is a constant; we find

$$0 \text{ V/s} + R\frac{dI}{dt} + \frac{1}{C}\frac{dq}{dt} = 0 \text{ V/s}$$

Substituting I for

$$\frac{dq}{dt}$$

yields

$$R\frac{dI}{dt} + \frac{1}{C}I = 0 \text{ V/s}$$

or

$$\frac{dI}{I} = -\frac{1}{RC}dt$$

This is in a form that can be integrated. We let the initial current in the circuit be I_0 and the current at instant t be I. Then

$$\int_{I_0}^{I}\frac{dI}{I} = -\frac{1}{RC}\int_{0\text{ s}}^{t} dt$$

$$\ln I - \ln I_0 = -\frac{1}{RC}t$$

Taking the antilogarithms, we find

$$\frac{I}{I_0} = e^{-t/(RC)}$$

$$I(t) = I_0 e^{-t/(RC)} \tag{19.37}$$

We still have to figure out the value of the initial current I_0. At the instant when the switch is closed, the capacitor has no charge on it and thus no potential difference across it. Thus, applying the KVL to the loop at the initial instant when current I_0 is in the circuit (see Figure 19.92), we have, going around clockwise,

$$-V_0 + I_0R + 0\ \text{V} = 0\ \text{V}$$

Solving for I_0,

$$I_0 = \frac{V_0}{R}$$

Hence Equation 19.37 becomes

$$I(t) = \frac{V_0}{R} e^{-t/(RC)} \tag{19.38}$$

The current in the circuit $I(t)$ is plotted as a function of time in Figure 19.93.

When $t = RC$, the current is e^{-1} of its initial value. This time interval is called the **time constant** τ of the circuit. The time constant is used to characterize the decay of the current because, in this model, the current never actually reaches zero: notice that as t increases, the current I decreases *asymptotically* to zero. After a time equal to, say, five time constants (that is, when $t = 5\tau$), the current in the circuit is only e^{-5} of its initial value. Since $e^{-5} = 0.006\,738$, the current effectively is negligible after an interval of about 5τ; it is less than 1% of its initial value. We thus reach the important conclusion that the dc current (the steady, time-independent current) through a capacitor is *zero*.

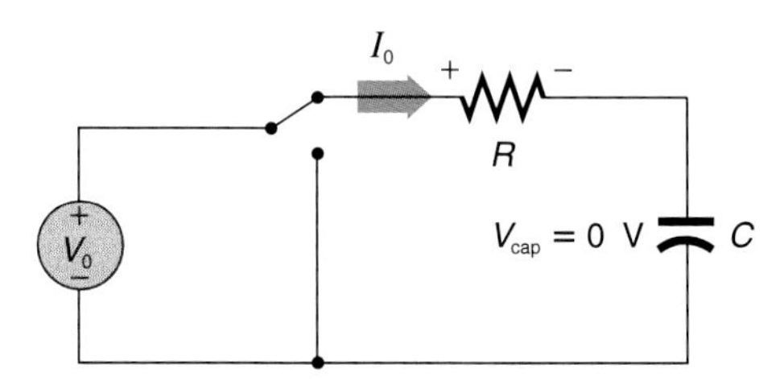

FIGURE 19.92 The situation when $t = 0$ s.

Once charged, a capacitor in a dc circuit acts like an open switch in the branch in which it is placed.

That makes sense, because a perfect capacitor has an infinitely good insulator between its two conducting plates. Capacitors are used to prevent dc currents from reaching certain parts of circuits; this property of a capacitor is exploited in many circuits that use transistors, which makes capacitors essential ingredients in modern electronics.

The initial slope of the graph of Equation 19.38—that is, the derivative of I with respect to t, evaluated when $t = 0$ s—has the value

$$-\frac{I_0}{\tau}$$

Hence in Figure 19.93 the line representing the initial slope of the curve intersects the time axis at the time τ, as shown in Figure 19.94. In other words, if the current decreased at this constant rate, the current would be zero at the time τ. But the current actually decreases exponentially, and by the time τ it is $e^{-1} = 0.368$ of its initial value.

The growth of the magnitude of the charge accumulating on each capacitor plate as a function of time in the charging process is found by integrating the current. That is, since

$$I = \frac{dq}{dt}$$

we have

$$dq = I\,dt$$

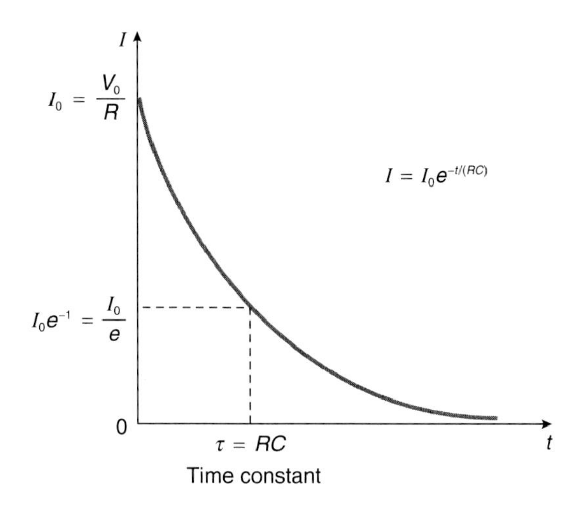

FIGURE 19.93 A graph of the current versus time.

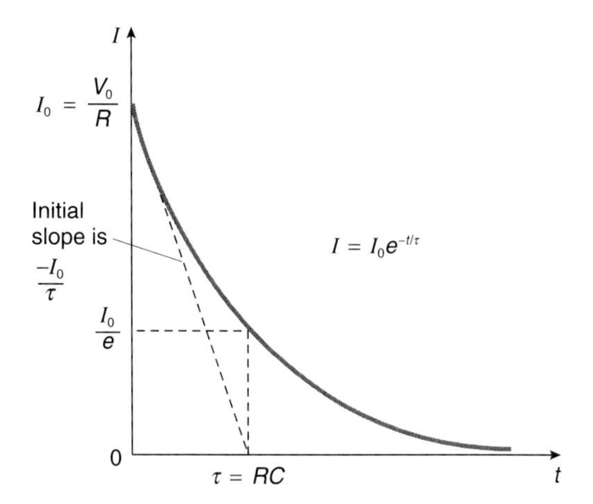

FIGURE 19.94 The initial slope of the current versus time graph is $-I_0/\tau$.

We use Equation 19.38 for I:

$$dq = \frac{V_0}{R} e^{-t/(RC)}\, dt$$

When $t = 0$ s, the initial charge on the capacitor is zero, and so the integral for the charge takes the form

$$\int_{0\text{ C}}^{q} dq = \frac{V_0}{R} \int_{0\text{ s}}^{t} e^{-t/(RC)}\, dt$$

Performing the integration, we obtain

$$q(t) = \frac{V_0}{R} \left[e^{-t/(RC)}(-RC) \right] \Bigg|_{0\text{ s}}^{t}$$

or

$$q(t) = V_0 C[1 - e^{-t/(RC)}] \qquad (19.39)$$

As $t \to \infty$ s, the magnitude of the charge on each plate of the capacitor approaches the value $Q_0 \equiv V_0 C$, and so we can rewrite Equation 19.39 as

$$q(t) = Q_0[1 - e^{-t/(RC)}]$$

This is graphed schematically in Figure 19.95.

The initial slope of this curve—that is, the derivative

$$\frac{dq}{dt}$$

evaluated when $t = 0$ s—is

$$\frac{Q_0}{\tau}$$

where $\tau = RC$, the time constant of the series RC circuit. Hence in Figure 19.95, the line tangent to the curve at $t = 0$ s would reach the value Q_0 at the time τ. The actual value of the charge on the capacitor at the time τ is

$$Q_0(1 - e^{-1}) = 0.632 Q_0$$

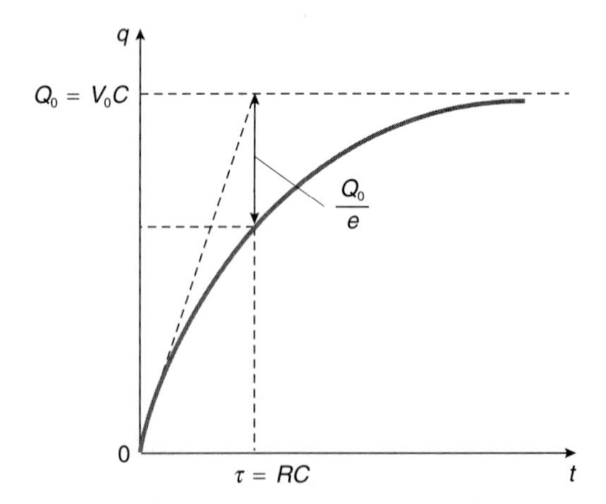

FIGURE 19.95 The magnitude of the charge on the capacitor plates as a function of time.

EXAMPLE 19.17

In Figure 19.93, how long does it take for current to decay to half its value? This time can be considered to be the half-life of the current in a charging series *RC* circuit. Express your result in terms of the time constant of the circuit.

Solution

The current in a charging series RC circuit is given by Equation 19.37:

$$I(t) = I_0 e^{-t/(RC)}$$

When the current is half its initial value, the equation becomes

$$\frac{I_0}{2} = I_0 e^{-t/(RC)}$$

$$\frac{1}{2} = e^{-t/(RC)}$$

Taking the natural logarithms of both sides, you get

$$\ln 1 - \ln 2 = -\frac{t}{RC}$$

Solving for t, you find

$$t = RC \ln 2$$

Since the time constant $\tau = RC$, you have

$$t = 0.693\tau$$

A Discharging Capacitor

Suppose we have waited long enough for the current in the circuit in Figure 19.96 to vanish. The capacitor in the circuit is fully charged and, since the capacitor acts like an open circuit for dc currents, there is zero current in the circuit.

Since no current is present, according to Ohm's law there is no potential difference across the resistor:

$$\begin{aligned} V_{res} &= IR \\ &= (0 \text{ A})R \\ &= 0 \text{ V} \end{aligned}$$

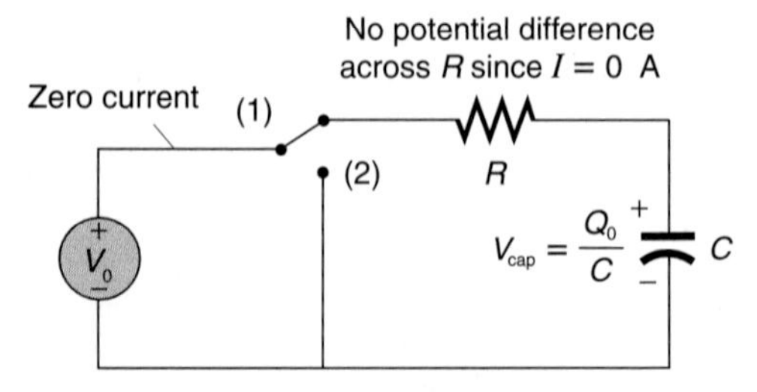

FIGURE 19.96 The capacitor is fully charged and the current is zero.

We apply the KVL to the single loop, going clockwise around the loop beginning in the lower left corner:

$$-V_0 + V_{cap} = 0 \text{ V}$$

so that

$$V_{cap} = V_0$$

The potential difference across the capacitor is equal to the potential difference across the independent voltage source.

Now we open the switch, as in Figure 19.97. We now have an electrical network; there is no circuit. Thus absolutely nothing changes! The capacitor is isolated and will not discharge since the charge is locked onto the capacitor; there is no electrical path for the charges to move from one plate to get to the other plate.

To discharge a capacitor, we need to move the switch to position (2), as shown in Figure 19.98, thus creating a (different!) circuit. We now use the same direction employed before for the current and apply the KVL to the single closed loop (the battery now is not part of the circuit); going clockwise around the new loop, we obtain

$$IR + \frac{q}{C} = 0 \text{ V} \tag{19.40}$$

For variety, let us see what happens to the charge q on the capacitor as a function of time. Since

$$I = \frac{dq}{dt}$$

Equation 19.40 becomes

$$R\frac{dq}{dt} + \frac{q}{C} = 0 \text{ V}$$

giving

$$\frac{dq}{dt} = -\frac{1}{RC}q$$

We rearrange the equation slightly:

$$\frac{dq}{q} = -\frac{1}{RC}dt$$

Now we integrate from $t = 0$ s (when q is Q_0, the charge on the fully charged capacitor) to time t (when the charge on the capacitor is q):

$$\int_{Q_0}^{q}\frac{dq}{q} = -\frac{1}{RC}\int_{0\text{ s}}^{t}dt$$

$$\ln q - \ln Q_0 = -\frac{t}{RC}$$

Simplifying and taking the antilogarithms, we get

$$q(t) = Q_0e^{-t/(RC)} \tag{19.41}$$

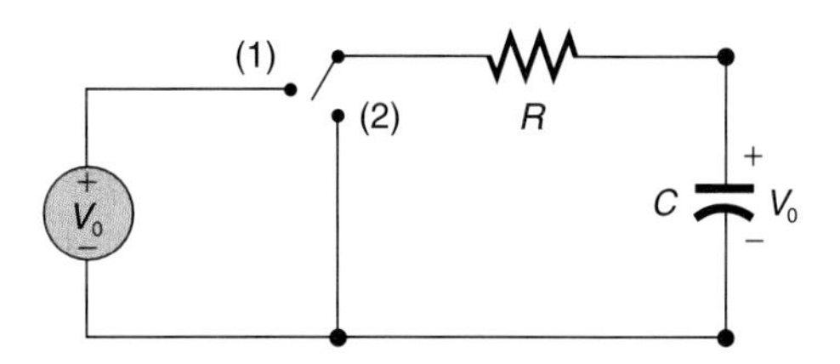

FIGURE 19.97 The switch is opened; nothing changes.

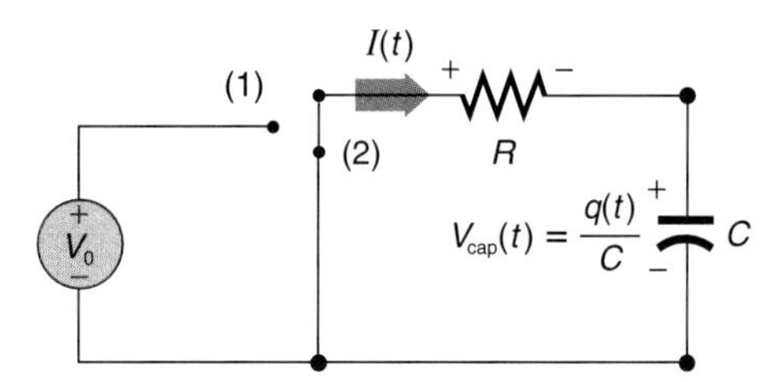

FIGURE 19.98 To discharge the capacitor, move the switch to position (2).

The initial magnitude of the charge on the capacitor is found from

$$C = \frac{Q_0}{V_0}$$

and so

$$q(t) = CV_0e^{-t/(RC)} \tag{19.42}$$

The charge on the capacitor decreases exponentially with time. The current in the circuit is

$$I = \frac{dq}{dt}$$
$$= CV_0e^{-t/(RC)}\left(-\frac{1}{RC}\right)$$

or

$$I(t) = -\frac{V_0}{R}e^{-t/(RC)} \tag{19.43}$$

The minus sign indicates that the current in the circuit actually is in the direction opposite to that shown in Figure 19.98. The current decreases in magnitude exponentially with time. The time constant still is $\tau = RC$. After a time of about 5τ, the current once again is negligible.

Note that if the resistance R is very small, as it would be if a simple wire were connected to the terminals of the charged capacitor, the time constant of the circuit is very small and the discharge of the capacitor takes place very quickly. This is what happens when you short-circuit the terminals of a charged capacitor, as we did in Section 19.1 when we first investigated the concept of electric current.

QUESTION 9

A capacitor is charged by connecting it to a real battery with internal resistance r. Do the final charge on the capacitor and the potential difference between its plates depend on the value of the internal resistance of the real battery? Explain your answer.

EXAMPLE 19.18

The capacitor in the series RC circuit of Figure 19.98 is discharging. At what time (in terms of the time constant τ) is the charge equal to one-half its initial value?

Solution

Use Equation 19.41,

$$q(t) = Q_0 e^{-t/(RC)}$$

Since $\tau = RC$, you have

$$q(t) = Q_0 e^{-t/\tau}$$

You want to know t when $q = Q_0/2$. Hence

$$\frac{Q_0}{2} = Q_0 e^{-t/\tau}$$

or

$$\frac{1}{2} = e^{-t/\tau}$$

Taking the natural logarithms of both sides, you find

$$-0.693 = -t/\tau$$

and so

$$t = 0.693\tau$$

EXAMPLE 19.19

For a discharging capacitor in a series RC circuit, at what time is the potential energy equal to one-half of its initial value? Express your answer in terms of the time constant τ.

Solution

The potential energy stored in the capacitor is found using Equation 18.12,

$$\text{PE} = \frac{1}{2}\frac{Q^2}{C}$$

Equation 19.41 describes how the charge on the discharging capacitor decreases with time, where $\tau = RC$. Substituting for the charge in the expression for the potential energy, you have

$$\text{PE} = \frac{1}{2C} Q_0^2 e^{-2t/\tau}$$

But

$$\frac{1}{2C} Q_0^2$$

is the initial potential energy PE_0 when $t = 0$ s, when the capacitor starts to discharge. So

$$\text{PE} = \text{PE}_0\, e^{-2t/\tau}$$

When the potential energy is half its initial value, this becomes

$$\frac{1}{2} = e^{-2t/\tau}$$

Taking the natural logarithms of both sides, you get

$$-0.693 = -2\frac{t}{\tau}$$

$$t = 0.347\tau$$

The potential energy decreases to half its value in half the time it takes the charge to decrease to half its value (see Example 19.18). The reason for this is that the potential energy depends on the square of the charge.

EXAMPLE 19.20

Find the time constant associated with the arrangement of resistors and capacitors shown in Figure 19.99.

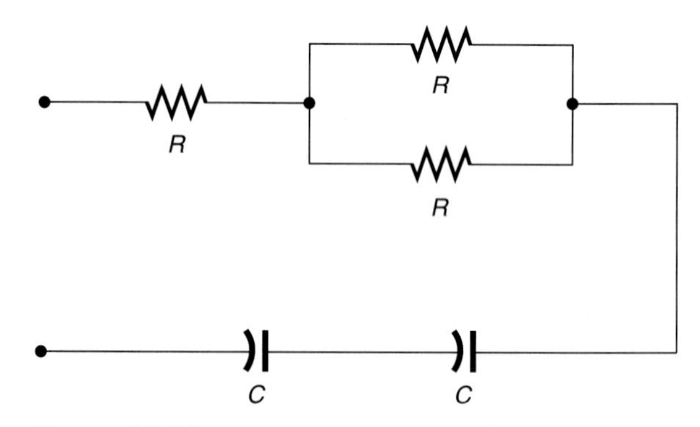

FIGURE 19.99

Solution

Combine the resistors and the capacitors into an equivalent resistor and an equivalent capacitor. The equivalent resistor has a value of $3R/2$ for its resistance and the equivalent capacitor has a value of $C/2$ for its capacitance. The result is a single resistor in series with a single capacitor, as shown in Figure 19.100.

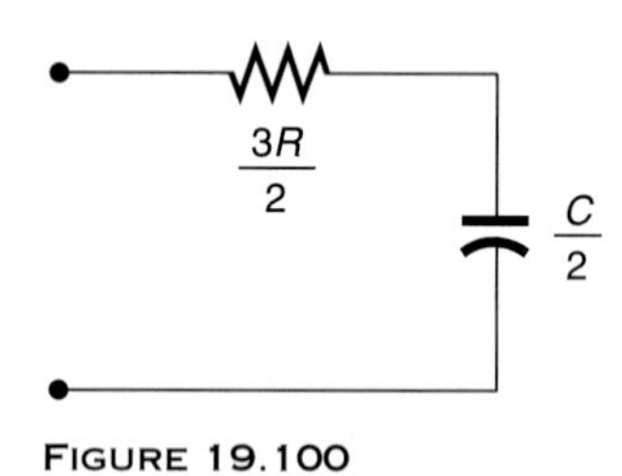

FIGURE 19.100

The time constant of the series combination is

$$\tau = R_{eq}C_{eq}$$
$$= \frac{3R}{2}\frac{C}{2}$$
$$= \frac{3}{4}RC$$

CHAPTER SUMMARY

An *electric current* represents charge in motion. The electric current I is the time rate of positive charge flow past a given location:

$$I \equiv \frac{dQ}{dt} \tag{19.1}$$

The current also can be expressed as

$$I = qnA\langle v \rangle \tag{19.3}$$

where n is the number of charge carriers per cubic meter, q is the charge on each charge carrier, $\langle v \rangle$ is the *average drift speed* (really the average velocity component), and A is the cross-sectional area through which the charges are moving (measured perpendicular to the average drift velocity).

In *ohmic materials*, the potential difference between the ends of a length ℓ of the material is proportional to the current through the material (which is *Ohm's law*):

$$V = IR \tag{19.14}$$

where R is the *resistance*

$$R \equiv \frac{\rho\ell}{A} \tag{19.13}$$

and A is the cross-sectional area of the material. The SI unit of resistance is the *ohm* (Ω). The constant ρ is the *resistivity* of the material and is measured in $\Omega\cdot$m using SI units.

Resistors connected in series have an equivalent resistance of

$$R_{eq} = R_1 + R_2 + R_3 + \cdots \quad \text{(resistors in series)} \tag{19.20}$$

Resistors connected in parallel have an equivalent resistance obtained from

$$\frac{1}{R_{eq}} = \frac{1}{R_1} + \frac{1}{R_2} + \frac{1}{R_3} + \cdots \quad \text{(resistors in parallel)} \tag{19.24}$$

Two resistors in parallel have an equivalent resistance that is the product of the resistances divided by their sum.

A graph of the current through a circuit element versus the potential difference across it is called the *characteristic curve* of the circuit element. The characteristic curve of an independent voltage source is a straight vertical line; that of a resistor is a line of positive slope whose value is the reciprocal of the resistance.

The electric power P absorbed by (transferred to) a circuit element is the product of the current I into its positive polarity terminal and the potential difference V across it:

$$P = IV \tag{19.26}$$

Resistors always absorb positive power since

$$P_{res} = I^2R \quad \text{(for a resistor } only\text{)} \tag{19.28}$$

Resistors convert the electric potential energy lost by charges passing through them into thermodynamic internal energy, increasing the temperature of the resistor. Heat transfer to the environment occurs because of this increase in temperature.

If the power absorbed by a circuit element is negative, the circuit element is increasing the electric potential energy of charges passing through it. Negative power absorbed does *not* imply the circuit element is cooling off by decreasing its thermodynamic internal energy.

An electrical *circuit* is a *network* with at least one closed path. Circuits are analyzed using the two Kirchhoff's laws. The *Kirchhoff current law* (KCL) states that the algebraic sum of the currents leaving any node of a circuit is zero. This is a statement of conservation of charge (per unit time). The *Kirchhoff voltage law* (KVL) states that the algebraic sum of the potential differences around any closed loop in a circuit is zero, where appropriate accounting must be made of the polarities of the various potential differences as they are encountered around the path. If a positive polarity terminal of a circuit element is encountered first, the potential difference is entered into the sum with a plus sign; if the negative polarity terminal is encountered first, the potential difference is entered into the sum with a minus sign. The KVL is a statement of conservation of energy (per coulomb of charge). We typically apply the KVL only to the *elementary loops* of a circuit.

The sum of the power absorbed by all circuit elements in a circuit is zero; this statement represents conservation of energy per unit time in the circuit as a whole.

A *real battery* is modeled as an independent voltage source (an ideal battery) in series with a resistance r known as the *internal resistance* of the real battery. Real batteries die because their internal resistance r gradually increases to huge values over the lifetime of the real battery.

The *maximum power transfer theorem* states that the maximum power absorbed by a single (or equivalent) load resistor R connected to the terminals of a real battery occurs when the resistance R has a value equal to the internal resistance r of the real battery.

A *voltmeter* is an instrument for measuring potential differences. It is a device of high effective resistance. A voltmeter is connected in parallel with the circuit element whose potential difference is to be measured. An *ammeter* is a device for measuring current; it is a low-resistance device that is connected in series in the branch of the circuit where the current is to be determined. An *ohmmeter* is an instrument for measuring resistance. To measure the resistance, the resistor must not be part of the circuit in which it is used.

When a capacitor is in series with a resistor, the charging or discharging of the capacitor are not accomplished instantaneously. The charge on the capacitor and the current follow exponential relationships that involve a characteristic time $\tau = RC$, known as the *time constant* of the series RC circuit.

A capacitor acts as an open switch in a dc circuit; the dc current is zero in any branch of a dc circuit that has a capacitor.

SUMMARY OF PROBLEM-SOLVING TACTICS

19.1 **(page 849)** As with the capacitors in series, be careful using Equation 19.24 in (1) adding the fractions correctly and (2) remembering that the sum of the reciprocals is the *reciprocal* of the equivalent resistance.

19.2 **(page 849)** The equivalent resistance of two resistors in parallel is the product of their resistances divided by the sum.

19.3 **(page 851)** Since resistors always become warm or hot when they have a nonzero current, resistors always absorb positive power (or zero power if the current through the resistor is zero).

19.4 **(page 851)** If you find the current through a resistor is negative, $-|I|$, Ohm's law leads to a negative potential difference, $-|V|$. The power absorbed by the resistor still is positive.

19.5 (page 851) Independent voltage sources absorb either positive or negative power depending on the direction of the current through them. Positive power is absorbed when the current *enters* the positive polarity terminal; negative power when the current *emerges* from the positive polarity terminal.

19.6 (page 851) If a current I is emerging from the positive polarity terminal of a circuit element, current $-I$ is entering the positive polarity terminal.

19.7 (page 855) The KCL really becomes useful only at nodes connecting three or more circuit elements.

19.8 (page 856) Write KVL equations only for each *elementary loop* in the circuit.

QUESTIONS

1. **(page 841); 2. (page 845); 3. (page 846); 4. (page 847); 5. (page 849); 6. (page 852); 7. (page 865); 8. (page 866); 9. (page 876)**

10. In a conductor, an electric field is necessary to maintain a constant drift speed for the electrons. The electric field provides the force on the charges constituting the current. Does this violate Newton's first law of motion? Prior to Newton, people thought a constant force was needed to maintain a constant speed. Are we harking back to pre-Newtonian physics with our model of currents in conductors? Explain.

11. Why does the gauge number of common wires increase for decreasing wire diameters as in Table 19.2? (Consult an electrical engineers handbook to see on what basis the gauge number is assigned. See also Problem 19.)

12. Construct an analogy between charge, current, and resistance on the one hand and cars and traffic on the other. What plays the role of charge? What plays the role of current? What plays the role of resistance?

13. Which has the greater resistance between opposite faces: a large cube of copper or a small cube of copper? Justify your answer.

14. If a copper wire and an aluminum wire of circular cross section and of the same length have the same resistance, what (if anything) can be said about the ratio of the diameters of the wires?

15. Is it correct to talk about the resistance of copper or the resistivity of copper? What is the distinction?

16. A contractor has a large electric drill for drilling through concrete walls that requires considerable current. The drill has a cord that is too short and you are sent to a hardware store to get an extension cord. You notice that the extension cords available have wire gauges from 10 to 18. To be safe, explain which gauge should you purchase.

17. The cockney accent in the East End district of London is noted for not pronouncing the "h's" at the beginning of a word. Thus "here" is pronounced as "ear," "he" is pronounced as a long-e sound, and so on. This gives rise to a number of interesting expressions since the word "home" then sounds like "ohm." Perhaps you can have fun deciphering Figure Q.17.*

FIGURE Q.17

18. The term brownout means that the electric company reduces the potential difference available at the outlets in our homes. Why and under what circumstances is this done?

19. Fuses are pieces of metal wire in a circuit that melt if the current exceeds a certain value. They are used to prevent excessive currents that may damage other circuit elements. Should a fuse be placed in series or parallel with the circuit element to be protected? Explain the reasoning behind your answer.

20. Two resistors, with $R_1 < R_2$, are connected in parallel with the combination connected to the terminals of an independent voltage source; see Figure Q.20. Which resistor absorbs the most power? Justify your answer.

21. Two resistors, with $R_1 < R_2$, are connected in series with the combination connected to the terminals of an ideal voltage

*The phrases are from a short note by Marshall Ellenstein in *The Physics Teacher*, 29, #6, September 1991, page 347.

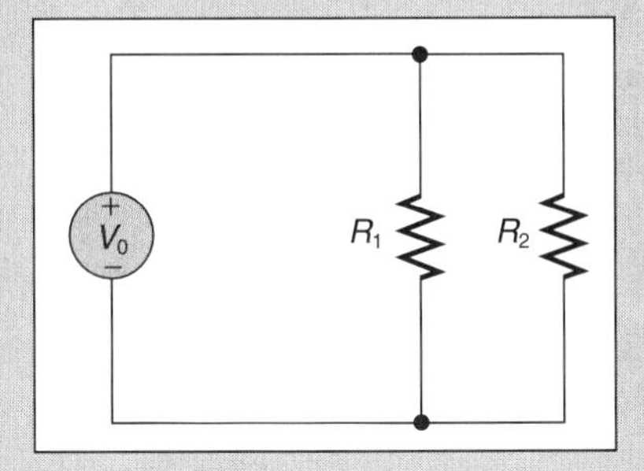

FIGURE Q.20

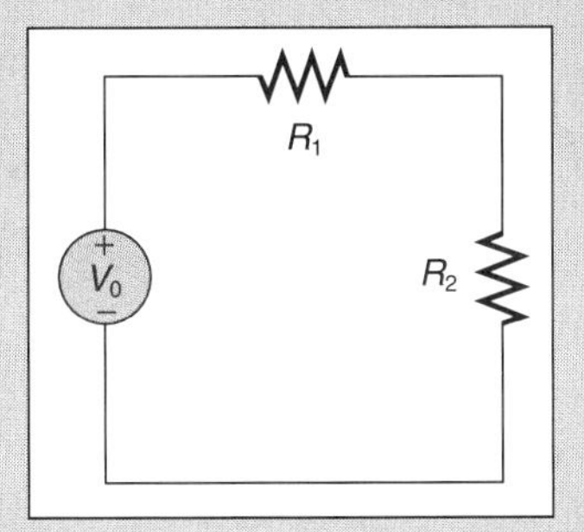

FIGURE Q.21

source; see Figure Q.21. Which resistor absorbs more power? Justify your answer.

22. A news report of an electrocution states that death occurred "when 14 000 V of electricity passed through the person." What is wrong with this statement?

23. *Future Shock* [title of a book by Alvin Toffler (Random House, New York, 1970)]. A live power line comes down during a violent storm and rests on your parked car while you are entertaining a friend inside the car. Are you both electrocuted? If not, should you get out of your car immediately? What should you do (other than scream and shout)? Explain.

24. If your car is hit by lightning, are you in any danger of electrocution? Is it better to have the windows up or down?

25. A small resistance and a large resistance each is connected in turn to a given independent voltage source. Which resistor absorbs more power? Justify your answer.

26. An electric company constructs a long transmission line from its generators to its customers. The resistance of the lines is symbolized by r. To minimize power absorption by the transmission lines themselves, should a large current and a small potential difference be used, or a large potential difference with a small current? Explain your reasoning.

27. Electrical appliances such as hot plates, portable electric ovens, and window-unit air conditioners that require large currents usually come with a warning not to operate them using a common extension cord. (See Figure Q.27.) Why is this admonition good advice? Many fires are caused by ignoring this warning, some with fatal consequences.

28. Carefully explain what is meant by each of the following electrical terms: (a) ground; (b) short circuit; (c) circuit; (d) characteristic curve.

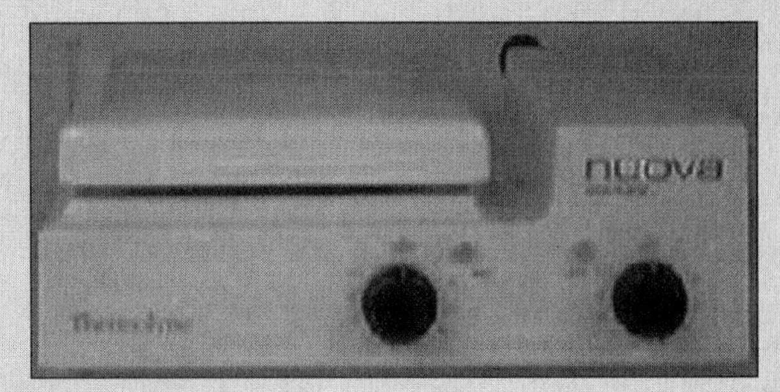

FIGURE Q.27

29. Carefully state the Kirchhoff circuit laws and indicate what fundamental law of physics underlies each circuit law.

30. Sketch a schematic characteristic curve for each of the following circuit elements. Clearly label appropriate axes. (a) An ideal battery; (b) a real battery; (c) an ohmic material; (d) a nonohmic material; (e) (optional) the characteristic curve of your professor.

31. A copper-clad pot is set on top of an electric stove burner that is glowing brightly from the electrical power absorbed by the heating element. Why does the copper bottom of the pot not short-circuit the heating element?

32. Incandescent lights frequently burn out when turned on rather than while they are on. See if you can suggest plausible reasons for why this is generally the case.

33. Call your local electric utility to determine the rate structure for using, say 2000 kW · h of energy per month. Determine the cost, excluding and then including any local taxes.

34. Given a battery and three different resistors, how many different circuits can be made using all four circuit elements in each? Sketch the circuits.

35. Three different resistors are to be connected together to produce various equivalent resistances. (a) How many different ways can they be connected together? (b) How should they be connected together to produce as large an equivalent resistance as possible? (c) How should the three be connected to produce the minimum equivalent resistance?

36. Two resistors have resistances such that $R_1 > R_2$. (a) If the two are connected in series to an independent voltage source, which resistor has the greater potential difference across it? (b) If the two are connected in parallel with the ideal battery, which has the greater current?

37. In many homes, it is possible to turn a room light on or off using either of two switches located at different points in the room. Switches used for this purpose are constructed as shown schematically in Figure Q.37. Design a circuit with a battery, two such switches, and a light bulb that does the job. If possible, test your circuit in a laboratory.

38. Several identical light bulbs are arranged in a circuit as shown in Figure Q.38. The switch labeled S is closed. (a) What happens to bulb C? (b) What happens to bulbs A and B?

39. James Thurber (1894–1961) once wrote: "She came naturally by her confused and groundless fears, for her own mother lived

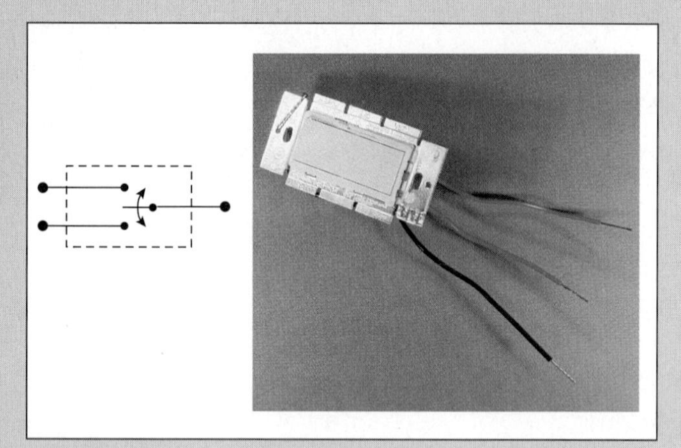

FIGURE Q.37

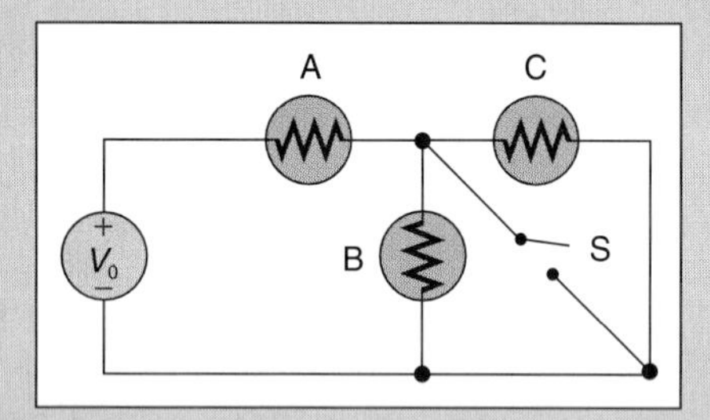

FIGURE Q.38

the latter years of her life in the horrible suspicion that electricity was dripping invisibly all over the house. It leaked, she contended, out of empty sockets as if the wall switch had been left on." [*My Life and Hard Times* (Harper and Brothers, New York, 1933), Chapter 2, "The Car We Had to Push"]. Why is this a groundless fear?

40. Perhaps you have noted that the brightness of a light bulb slowly decreases over its lifetime even though it is connected to an independent voltage source with a constant potential difference. What might account for this observation?

41. In choosing a new car (real) battery, you are offered a choice of various kinds, all of which have a potential difference between their terminals of 12 V. The saleswoman states the various choices differ in ampere-hour (A·h) rating. What is the meaning of the term ampere-hour?

42. You likely are familiar with the various common real battery sizes, called D-cell, C-cell, AA, AAA, and so on. Each has a potential difference between its terminals of about 1.5 V. Other than their physical dimensions and mass, what is the essential difference between them?

43. If you start your car with its headlights on, the lights dim until the motor starts. Suggest reasons for this observation.

44. Explain the distinction between an ammeter and a voltmeter: (a) when comparing the intrinsic electrical characteristics of the instruments; (b) when using the instruments to make a measurement in an electrical circuit.

45. You receive a bill from the electric company for 1500 kW·h. Are you being billed for *power* usage or *energy* usage? Explain.

46. (a) A multimeter M is attached to the resistor as indicated in Figure Q.46. (a) If the multimeter is set so it is a voltmeter, what does it indicate for the potential difference across the resistor? (b) The multimeter is switched to the ohmmeter setting. What will it indicate for the resistance?

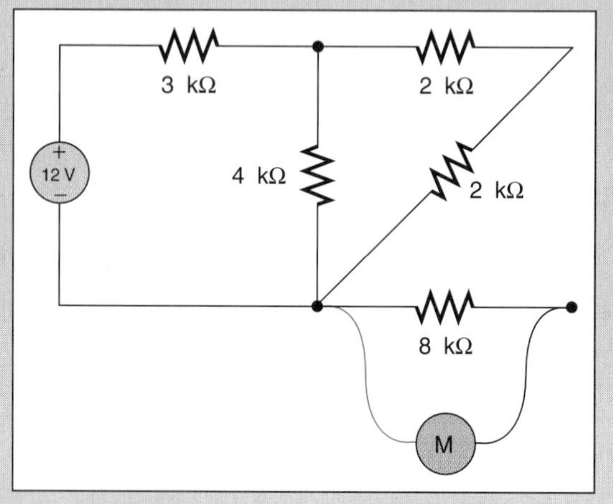

FIGURE Q.46

47. Look at the schematic diagram of an ohmmeter shown in Figure 19.89. Why is it *not* a good idea to leave a resistor connected to the terminals of an ohmmeter for long periods of time?

48. The time constant of a series *RC* circuit is $\tau = RC$. Show that the product of ohms and farads is seconds: $\Omega \cdot \text{F} = \text{s}$.

49. What is the power absorbed by a capacitor that is already fully charged?

50. Sketch the characteristic curve (with the dc current on the vertical axis) for a fully charged capacitor.

51. The charges on an isolated, real, charged capacitor gradually decrease with time because of the finite conductivity of most insulating materials separating the plates. The charges gradually leak across the gap separating the two conductors of the capacitor and the capacitor becomes neutralized. This is known as the *leakage current*. (If the capacitor is connected to an independent voltage source, the leaking charges are replenished by the source.) Construct a model of a real capacitor using an ideal capacitor and a resistance. Should the resistance be placed in series or parallel with the ideal capacitor to model the leakage current? Should the resistor have a high resistance or a low resistance? Explain the reasons for your choices.

52. Which graph in Figure Q.52 best represents the qualitative behavior of the charge on a capacitor as a function of time when the initially charged capacitor is connected in parallel with a resistor?

53. In most household and laboratory circuits, the duplex outlet plugs (think of them as independent voltage sources when wired up to the power grid) have *three* wires, one of which is called a ground wire. See Figure Q.53. What is the reason for such a third wire?

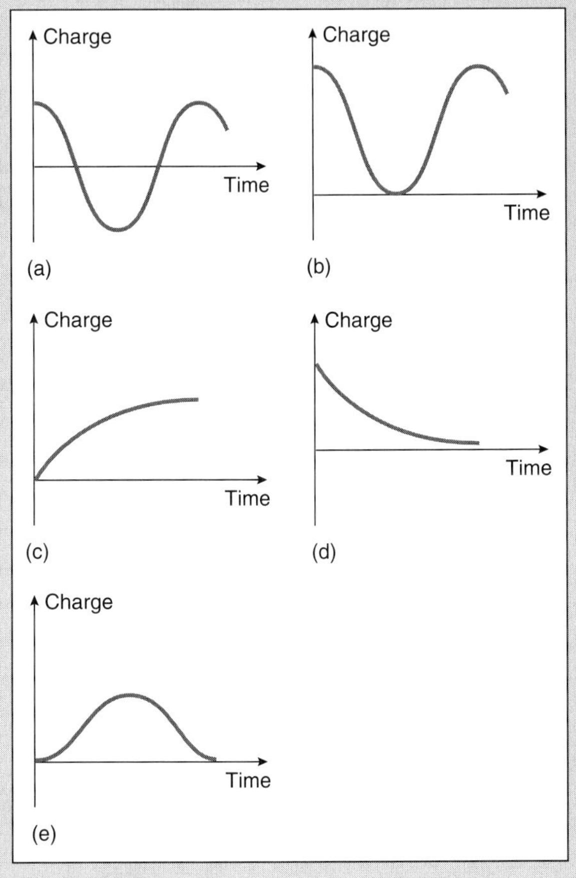

FIGURE Q.52

FIGURE Q.53

PROBLEMS

Sections 19.1 The Concept of Electric Current
19.2 Electric Current

1. The transport of electrons in large molecules has many significant implications. For example, light stimulates the transport of electrons through chlorophyll, the molecule intimately connected to photosynthesis. The conduction of up to 10^6 electrons per second along the axis of segments of the DNA double helix molecule is stimulated by the presence of ruthenium, which binds to ribose, one of the components of the helical DNA strands. What current does this electron transport represent? It is hoped that the development of ways to stimulate electron transport in DNA can lead to the detection of genetic defects and ways of detecting killer viruses, such as the HIV and Ebola viruses.

2. A charge q is in a circular orbit of radius r, moving at speed v. (a) What is the period of the motion? (b) What is the frequency of the motion? (c) What is the current?

•3. A cylindrically shaped beam of protons has a diameter of 2.00 mm and has 1.20×10^6 protons per cubic centimeter. The kinetic energy of each proton is 1.00 keV. (a) What is the speed of each proton in the beam? (b) What is the beam current in amperes?

•4. Protons in a 1.00 μA beam are traveling at a speed nearly equal to that of light. (a) How many protons are along each meter of the beam? (b) If the beam is 1.00 mm in diameter, what is the density of charge carriers (the number of protons per cubic meter)?

•5. (a) Calculate the drift speed in a 12 gauge shiny silver wire (see Table 19.2 on page 843) carrying a current of 1.50 A. Assume there is one charge carrier (an electron) per atom. The molar mass of silver is 108 g and its density is 10.5×10^3 kg/m^3. (b) How many hours does it take the charge carriers to travel one meter?

•6. A charge q is uniformly distributed along a straight line of length ℓ. The line rotates at angular speed ω about an axis through one end as shown in Figure P.6. What is the current past a fixed radial line?

•7. A closed, square insulating ring has sides of length ℓ and charge Q uniformly distributed along its perimeter. The square rotates about one edge at angular speed ω. What is the current past the fixed plane indicated in Figure P.7?

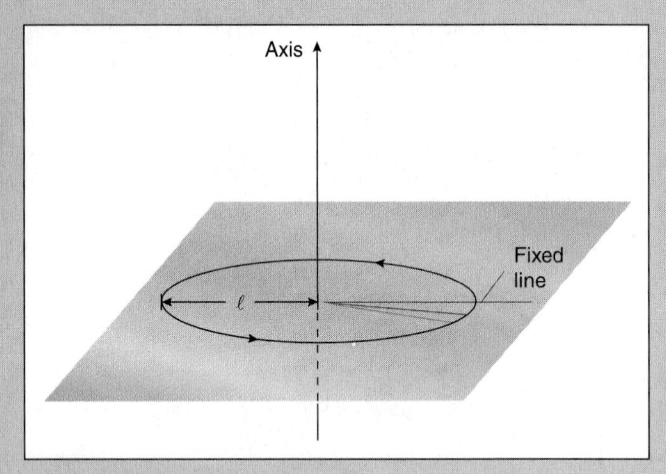

FIGURE P.6

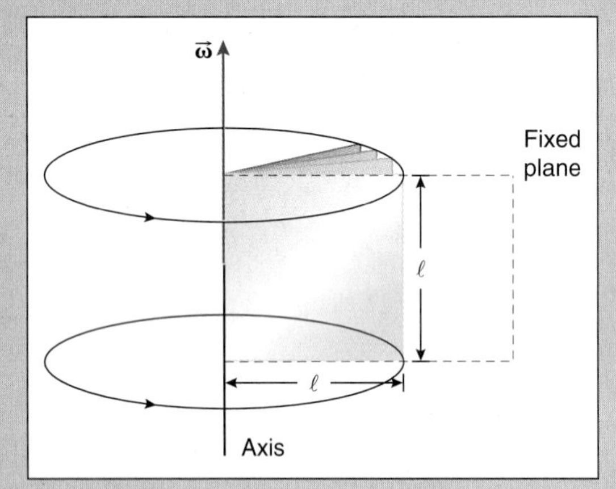

FIGURE P.7

•**8.** An insulating ring of radius a has a charge Q spread uniformly around its circumference. (a) The ring spins at constant angular speed ω about a diameter. Find the current past the line momentarily and occasionally coincident with half the circumference of the ring indicated in Figure P.8a. (b) If the ring spins at the same angular speed about an axis through the center of the ring and perpendicular to its plane as shown in Figure P.8b, what is the current past the indicated line?

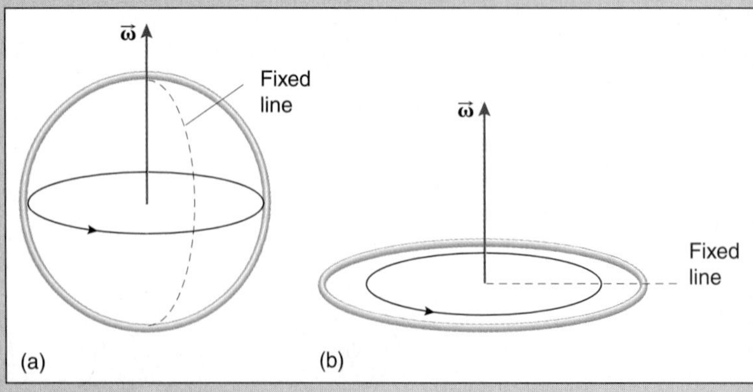

FIGURE P.8

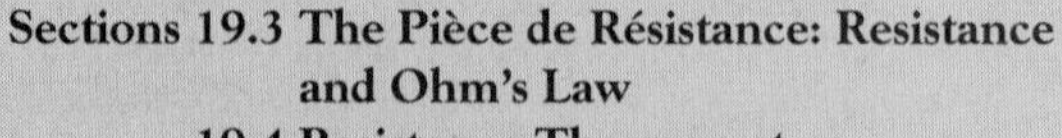

Sections 19.3 The Pièce de Résistance: Resistance and Ohm's Law
19.4 Resistance Thermometers
19.5 Characteristic Curves

9. Resistors are fabricated on integrated circuit chips by suitably tailoring the dimensions of a conducting layer such as gold. Let the resistor to be fabricated be a gold wire of rectangular cross section. If the resistor is 1.0 μm by 10.0 μm in cross section and 400 μm long, what is the resistance of the gold wire?

10. A wire of length ℓ and cross-sectional area A is melted and recast into a wire of length 3ℓ using the same mass of material. (a) What is the new cross-sectional area of the wire? (b) How is the resistivity of the wire affected by the change? (c) How is the resistance of the wire affected by the change?

•**11.** A copper cable consisting of ten strands of copper wire, each 1.00 mm in diameter, is one of several cables used to supply electrical service to a cow barn. The length of each cable is 200 m. What is the resistance of each cable?

•**12.** Consider two solid cubes of a material with resistivity ρ. One cube has sides of length ℓ, the other has sides of length 3ℓ. Which cube has the greater resistance between opposite faces, and by what factor is the resistance of this cube greater than the resistance of the other cube?

•**13.** The Alaska oil pipeline is approximately 1.27×10^3 km long, 120 cm in diameter, and made of steel with a thickness of about 1.0 cm. See Figure P.13. Steel has a resistivity of $1.80 \times 10^{-7}\ \Omega\cdot$m. What is the total resistance of the pipeline over its entire length?

•**14.** A copper wire is stretched uniformly (maintaining constant density and resistivity) so that it is ten times its original length (but of the same volume). By what factor is its resistance changed?

•**15.** A metal cube has a resistivity of $2.0 \times 10^{-8}\ \Omega\cdot$m and a resistance of 1.0 $\mu\Omega$ between opposite faces. (a) What is the length of a side of the cube? The cube now is compressed to form a square plate of the same density as before but of only one-tenth the original thickness. (b) What is the length of a side of the square plate? (c) What is the resistance between the two opposite square faces?

•**16.** Consult Table 19.2 on page 843. For wires of a given length, if you increase the wire gauge number by 2, by about what

FIGURE P.13

factor does the resistance of a wire increase? Does your result depend on the initial wire gauge?

•17. Two wires are made from the same material, are the same length, and carry equal currents. One wire is half the diameter of the other. Which has the larger average drift speed and by what factor?

•18. Assume the temperature coefficient of resistance of carbon (diamond) is independent of temperature and has the value $-0.5 \times 10^{-3}\ K^{-1}$ given in Table 19.1 for 20 °C. (a) At what temperature is its resistivity equal to zero? How does this compare with the melting temperature of diamond (~3.5×10^3 °C)? (b) What does the result of part (a) likely imply about the assumption of a constant temperature coefficient of resistivity?

•19. Consult Table 19.2 on page 843. For wires of a given length, show that if the wire gauge number is increased by 10, the resistance of the wire increases by about a factor of 10.

•20. A mass of 1.00 kg of copper (density 8.93×10^3 kg/m^3) is drawn into 12 gauge wire (see Table 19.2). (a) What is the length of the wire? (b) What is the resistance of this wire?

•21. For a wire of fixed length ℓ and made from a given material of density ρ_{mass}, show that the resistance of the wire is inversely proportional to its mass. What is the proportionality constant?

•22. As a physics student, you consider a cool amber liquid made from hops in ways different from the normal population. You measure the temperature of the liquid using a platinum resistance thermometer. The resistance thermometer has a resistance of 220.25 Ω at 25.0 °C. When immersed in the stein, the resistance of the thermometer is 199.99 Ω. (a) Determine the temperature of the liquid. (b) If the same thermometer is used to measure one's body temperature (assuming it is normal: 37.0 °C), what resistance will the thermometer indicate?

•23. On the same graph, accurately draw the characteristic curves of a 1.00 kΩ resistor and a 2.00 kΩ resistor when the potential difference across them varies from −10.0 V to 10.0 V.

Sections 19.6 Series and Parallel Connections Revisited 19.7 Resistors in Series and in Parallel

•24. An unusual copper wire is 4.00 m long. For 3.00 m of its length, the diameter is 3.00 mm and for the remaining 1.00 m of its length the diameter of the wire is only 1.00 mm. The wire is covered with a pretty red plastic insulation that is 0.90 mm thick. (a) The wire is connected to the terminals of an independent voltage source that provides 20.0 A current for the thick end of the wire. What current is in the thin end? (b) What is the resistance of the 3.00 m section of the wire? (c) What is the resistance of the 1.00 m section of the wire? (d) What is the total resistance of the wire? (e) What is the potential difference between the ends of the wire?

•25. For each of the resistor combinations shown in Figure P.25, find the resistance of the single equivalent resistor that can be placed between the terminals A and B.

•26. Find the equivalent resistance between the terminals A and B shown in Figure P.26.

•27. Find the single resistance that is equivalent to that of the resistor assortment of Figure P.27 between the terminals A and B.

Section 19.8 Electric Power

28. A resistor has a potential difference of 5.00 V across it and absorbs 15.0 W of power. Determine the current through the resistor and the value of its resistance.

29. Two light bulbs absorb 100 W and 200 W of power when used with a potential difference of 120 V across each of the filaments. Which filament has the greater resistance?

•30. Two resistors with resistances 5.0 Ω and 10.0 Ω are to be connected to an independent voltage source with an emf of 5.0 V. The resistors are first connected in series with the independent voltage source and then in parallel with it. (a) Determine the power absorbed by the two resistors for each arrangement. (b) Which arrangement results in the most total power absorbed by the two resistors?

•31. The 12.0 V battery (an independent voltage source) of your car is connected in series with the starter motor. The battery provides 90.0 A for 2.00 s to the starter motor. What power is absorbed by the motor?

•32. Show that the power absorbed by a resistor R can be written in each of the following equivalent ways:

$$P_{res} = \frac{V^2}{R}$$

or

$$P_{res} = I^2R$$

where V is the potential difference across the resistor and I is the

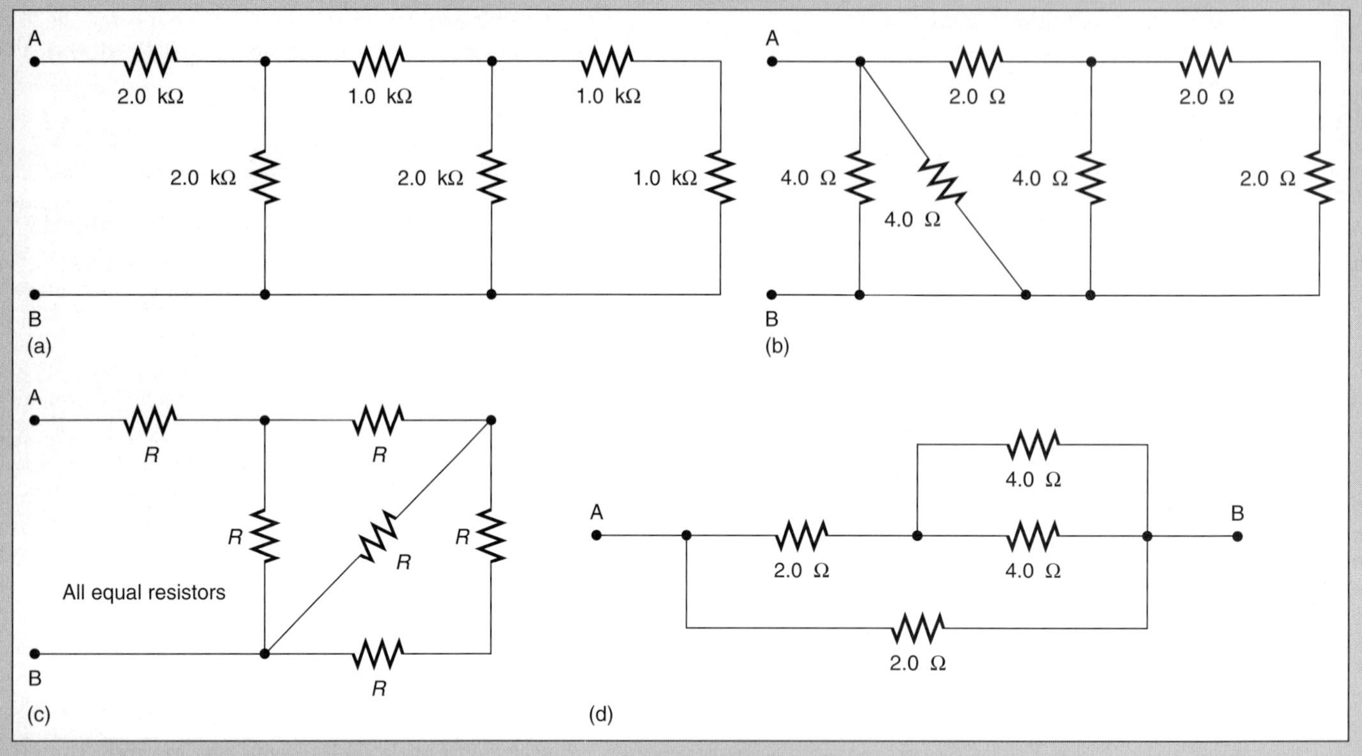

FIGURE P.25

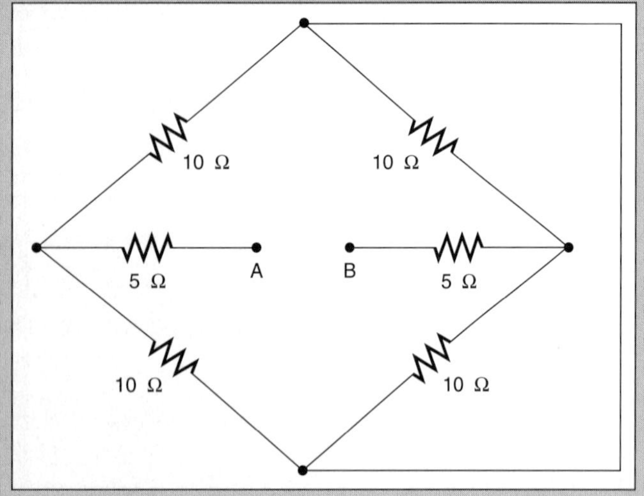

FIGURE P.26

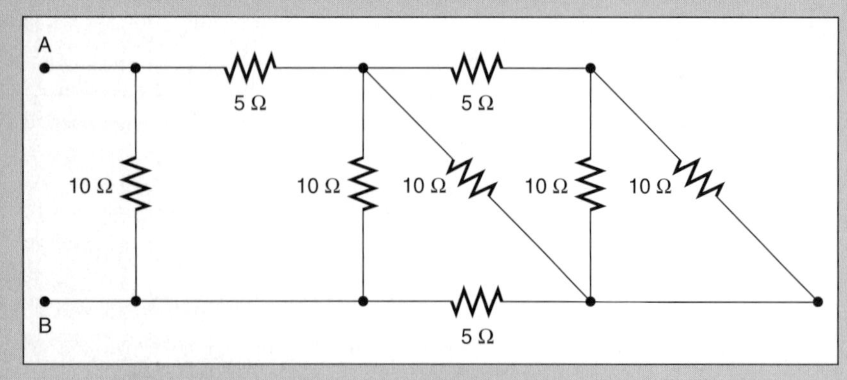

FIGURE P.27

current through it. The implication of the first equation is that, for a given potential difference V, the power absorbed by a resistor is inversely proportional to the resistance. The implication of the second result is that, for a given current I, the power absorbed by a resistor is directly proportional to R. Is anything wrong?

•**33.** If house wiring gets too hot, a fire may result. A wire is to carry 15 A and the power absorbed per meter of length is to be no greater than 1.0 W/m. (a) What diameter of copper wire would absorb 1.0 W/m? (b) What is the maximum gauge number of copper wire that should be used (see Table 19.2 on page 842)?

•**34.** Your electric toothbrush absorbs 5.0 W of power when used. If you use the brush for 5.0 minutes each day and a kilowatt-hour of energy costs $0.13, how much does it cost to use the cavity fighter for a year?

•**35.** If the price of a kilowatt-hour of energy is $0.13, what is the cost to use your 200 W study lamp for 6.0 hours every night for 9 months studying physics?

•**36.** A yellow bellied sapsucker (*Sphyrapicus varius*) perches on a power line that carries a 1.00 kA current. The feet of the bird are 5.0 cm apart. The resistance of the wire per meter of length is 5.0×10^{-5} Ω/m. Assume the wire is copper with a resistivity of 1.77×10^{-8} Ω·m. (a) What is the diameter of the wire? (b) What is the potential difference between the bird's feet? (c) What power is absorbed by the wire section between the feet of the bird?

Sections 19.9 Electrical Networks and Circuits
19.10 Electronics
19.11 Kirchhoff's Laws for Circuit Analysis
19.12 Electric Shock Hazards*
19.13 A Model for a Real Battery
19.14 Maximum Power Transfer Theorem

•**37.** A 100 W light bulb and a 60 W light bulb normally are connected in parallel, so if there is a potential difference of 120 V across each of them, they absorb 100 W and 60 W of power respectively. Suppose, instead, that you wire the two bulbs in series with a 120 V source placed across the combination. Determine the power absorbed by each and indicate which bulb is brightest.

•**38.** Typical house circuits at 120 V have fuses or circuit breakers rated for a maximum of 15 A. How many 1200 W toaster ovens can be used at full power on the same circuit before the fuse blows? First consider how the toaster ovens are connected to the 120 V source.

•**39.** Consider the circuit shown in Figure P.39. (a) What is the potential difference across the 15.0 Ω resistor? (Hint: Look at the relationship between the battery and the 15.0 Ω resistor. How are they connected together?) (b) What is the power absorbed by the 15.0 Ω resistor? (c) What is the current through the 10.0 Ω resistors? (d) What is the power absorbed by the battery? (e) What power is absorbed by each of the 10.0 Ω resistors?

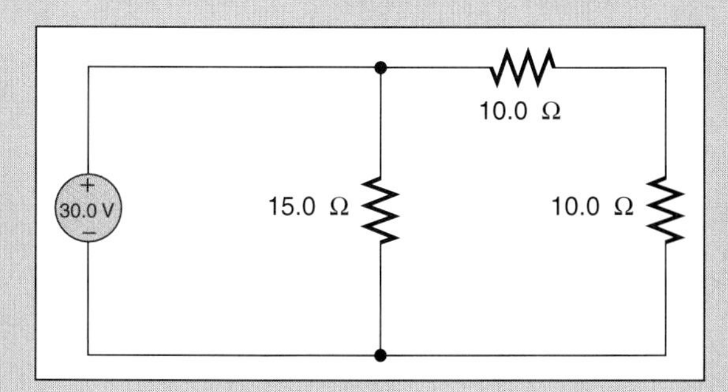

FIGURE P.39

•**40.** The following questions make reference to the circuit of Figure P.40. The resistors represent identical light bulbs, each with the same resistance *R*. Refer to the bulbs by number with your answers and explanations. (a) Which bulb(s) has the greatest current? Which bulb(s) absorbs the greatest power? Which is (are) the brightest? (b) Suppose bulb 3 is removed from the circuit. What happens to the brightness of bulbs 1 and 2? (c) Reinsert bulb 3. Now connect a wire from point A to point B. What happens to the brightness of bulbs 2 and 3? What happens to the brightness of bulb 1? (d) Remove the wire inserted in part (c). Connect a wire from point A to point C. What happens to the brightness of bulb 1? What happens to the brightness of bulbs 2 and 3?

This problem was inspired by similar problems in Arnold Arons, *A Guide to Introductory Physics Teaching* (John Wiley, New York, 1990), pages 184–187.

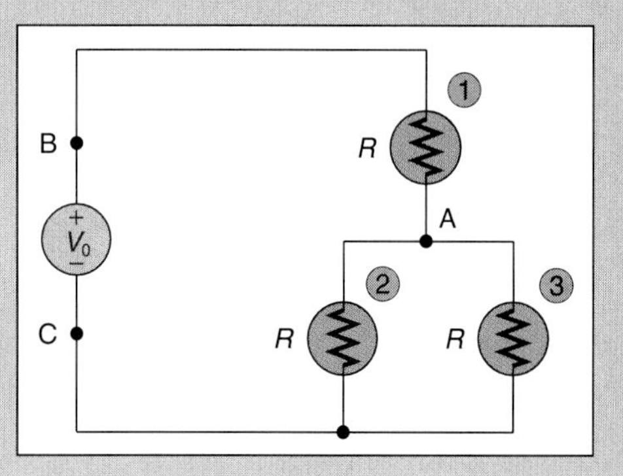

FIGURE P.40

•**41.** Examine the circuit of ideal batteries and resistors shown in Figure P.41. (a) Indicate clearly in a sketch the current directions you choose for each branch of the circuit to solve the problem. (b) Indicate appropriate polarities associated with each resistor. (c) Use Kirchhoff's laws to find the current through each circuit element. (d) Calculate the electric power absorbed by each circuit element. (e) Check to be sure that the sum of the power absorbed by all the circuit elements is zero.

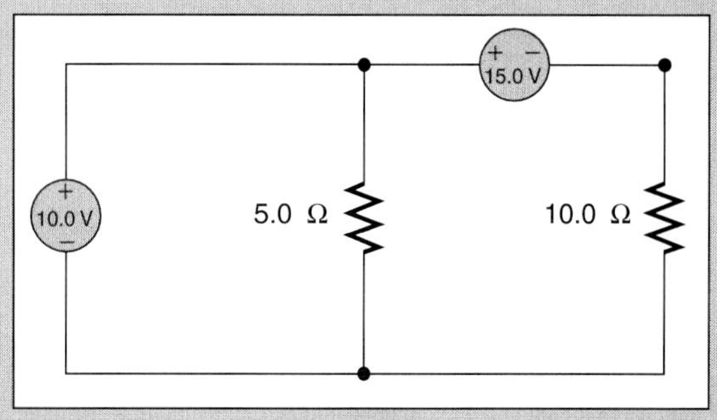

FIGURE P.41

•**42.** You decide to leisurely toast marshmallows by placing them over the 6.00 Ω resistor in the circuit shown in Figure P.42. (a) Indicate appropriate current directions for each branch of the circuit. (b) Mark the resistor polarities consistent with your choice of the current directions. (c) Find the current through each circuit element. (d) Find the power absorbed by each circuit element. (e) Show that the total power absorbed by all the circuit elements is zero.

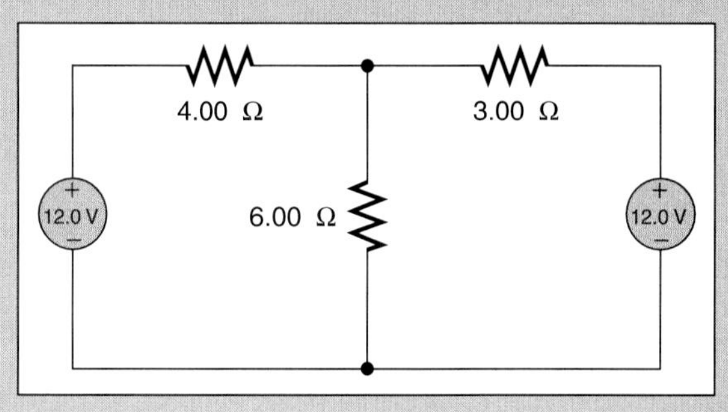

FIGURE P.42

•**43.** The circuit shown in Figure P.43 is known as a bridge circuit.* It is used extensively in applications such as temperature measurement. The purpose of the circuit is to measure the resistance of an unknown resistor R_x (such as the resistance of a resistance thermometer) in terms of known precision variable resistances R_2, R_3, and R_4. The values of R_2, R_3, and R_4 are adjusted until the ammeter indicates zero current; the bridge circuit then is said to be *balanced*. (a) When the ammeter indicates zero current, what must be the relationship between the current in R_x and the current in R_2? What about the relationship between the current in R_4 and the current in R_3? (b) When the ammeter indicates zero current, what is the potential difference between points A and B? (c) Use the result of part (a) to determine the relationship between the potential differences across R_x and R_4. What about the relationship between the potential differences across R_2 and R_3? (d) Using the preceding results, derive an expression for the unknown resistance R_x in terms of R_2, R_3, and R_4.

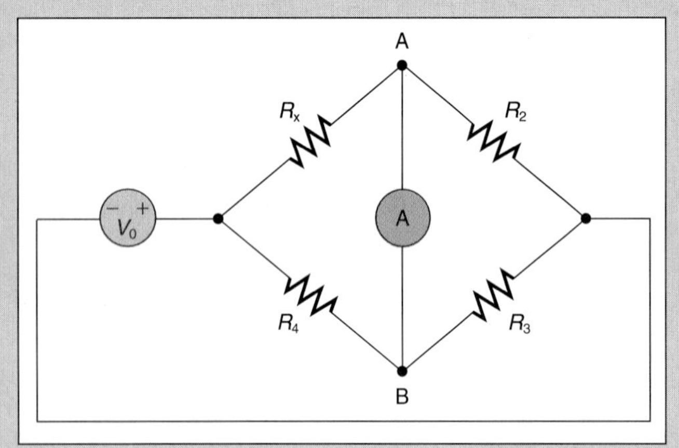

FIGURE P.43

•**44.** Yet another beautiful circuit (see Figure P.44). (a) Indicate clearly on a sketch the direction you choose for the currents in each distinct branch of the circuit. Label resistor polarities in accordance with your choice. (b) Write the KVL for each elementary loop of the circuit. Does this provide a sufficient number of equations to solve for the unknown currents? (c) Use the KCL at a significant node to secure yet another equation relating the unknown currents. (d) Solve the equations for the unknown currents. (e) Calculate the potential difference across each of the resistors. (f) Calculate the power absorbed by each circuit element and check that the total power absorbed by the circuit is zero.

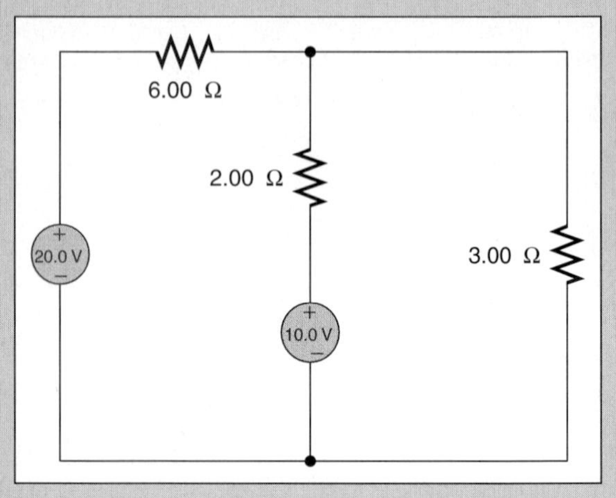

FIGURE P.44

•**45.** The circuit shown in Figure P.45 is constructed from identical resistances R for your greater enjoyment. The equivalent resistance is found by Zeus to be 13.0 Ω. (a) Determine the value of the resistance R used in the circuit. (b) Which of the resistors R will have the least current through it? (Specify the resistors by number.) (c) Find the least current through the resistors indicated in part (b). (d) Indicate how you would connect another single resistance R so as to make the total current through the battery the greatest.

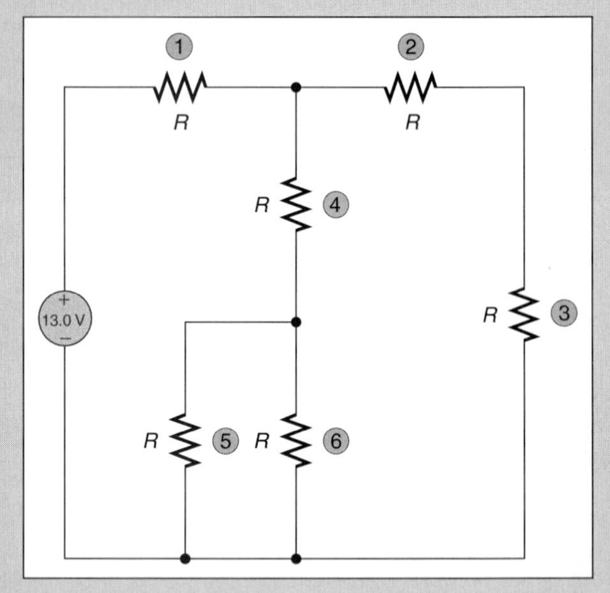

FIGURE P.45

•**46.** Analyze the circuit shown in Figure P.46 in the following manner: (a) Clearly indicate appropriate currents with labels and directions, indicating the resistor polarity markings accordingly. (b) Set up a number of equations sufficient to solve for the unknown currents delineated in part (a). (c) Solve these equations for the unknown currents using appropriate techniques. (d) Calculate the power absorbed by each circuit element and show that these powers sum to zero.

•**47.** The circuit shown in Figure P.47 is called a *voltage divider*. Show that the potential difference across resistance R_1 is

$$\frac{R_1}{R_1 + R_2} V_0$$

*This is not the same kind of bridge circuit you or your parents might enjoy with a deck of cards. This bridge circuit was popularized by Charles Wheatstone (1802–1875), although the circuit actually was invented in 1833 by Samuel Christie, only son of James Christie, founder of the well-known auction gallery. Wheatstone, a musician as well, patented the concertina in 1829.

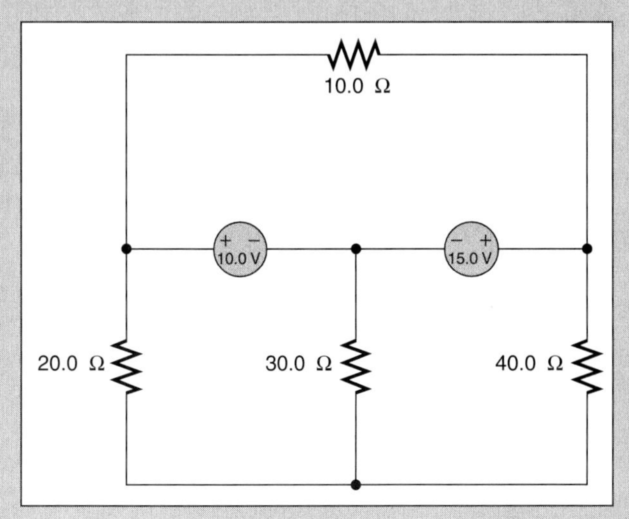

FIGURE P.46

and the potential difference across resistance R_2 is

$$\frac{R_2}{R_1 + R_2} V_0$$

Notice that these results can be summarized in the following way: the fraction of the potential difference V_0 that appears across a given resistance is the given resistance divided by the sum of the two resistances.

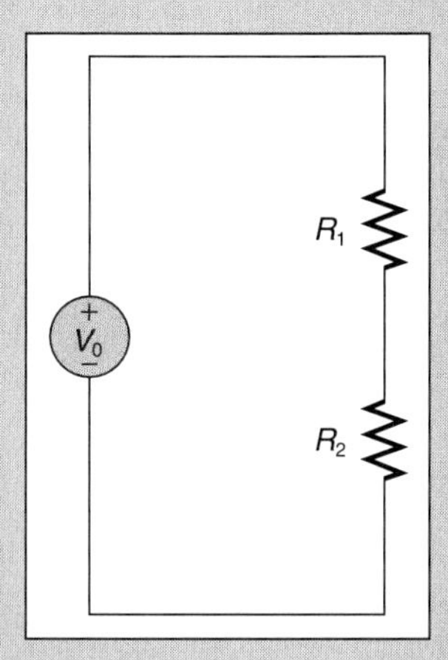

FIGURE P.47

•**48.** In the circuit depicted in Figure P.48, the current I_0 is split (or divided) between the two resistances R_1 and R_2. Show that the current in R_1 is

$$\frac{R_2}{R_1 + R_2} I_0$$

and that the current in R_2 is

$$\frac{R_1}{R_1 + R_2} I_0$$

Hence this circuit is called a *current divider*. Notice that the results can be summarized in the following way: if a current I_0 is split between two resistances, the fraction of the current in a given resistor is the *other* resistance divided by the sum of the two resistors.

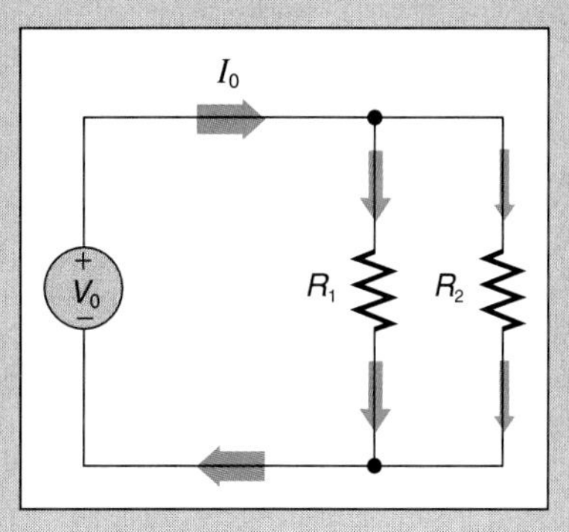

FIGURE P.48

•**49.** Consider the circuit shown in Figure P.49 (a) Determine the current in the 4.0 Ω resistor. (b) Determine the potential difference between points A and B. (c) Since there is a nonzero potential difference between points A and B, why is there no current in the 4.0 Ω resistor? (d) Calculate the power absorbed by the 8.0 V, 12.0 V, and 16.0 V independent voltage sources. (e) Calculate the total power absorbed by all the resistors.
This problem was inspired by Robert E. Viens, "A Kirchhoff's rules puzzler," *The Physics Teacher, 19*, #1, page 45 (January 1981).

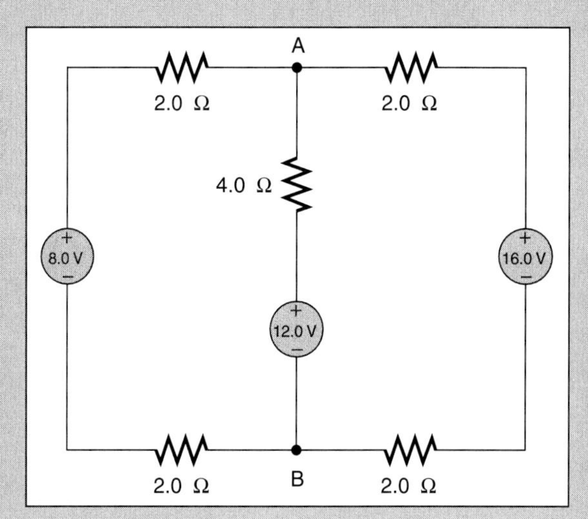

FIGURE P.49

•**50.** You are to design an automobile light circuit that has a 12.0 V battery, a fuse, a switch, two 50.0 W headlights, and two 10.0 W taillights. Each light needs 12.0 V for proper operation. The switch is to be placed so that all four lights are lit when the switch is closed and must ensure that if one light blows out (becomes an infinite resistance or open circuit), the remaining lights are unaffected. (a) Sketch a circuit that will accomplish this objective. (b) What current must the fuse be able to handle without melting?

•**51.** Two long house wires of uniform diameter have resistances R_2 = 1.0 Ω and R_3 = 1.0 Ω spread uniformly along their lengths. In Figure P.51, the resistance of the wires is lumped into one location for convenience on the circuit diagram. A heater with a resistance of 13.0 Ω is connected between the wires as indicated. An independent voltage source with an emf of 120 V is connected in series with a 0.30 Ω resistor that is a fuse. The fuse will melt (and break, or open, the circuit) if the power absorbed by the fuse exceeds 30 W. The power absorbed along the long house wires safely radiates away via heat transfer unless the currents become too large; this is the purpose of the fuse, among other things. (a) How much current will barely melt the fuse? (b) What current is at point A? (c) What current is at point B? (d) What current passes through the heater? (e) What is the potential difference across R_4? (f) If another 13.0 Ω heater were connected in parallel with the one shown, would the fuse melt? Justify your answer.

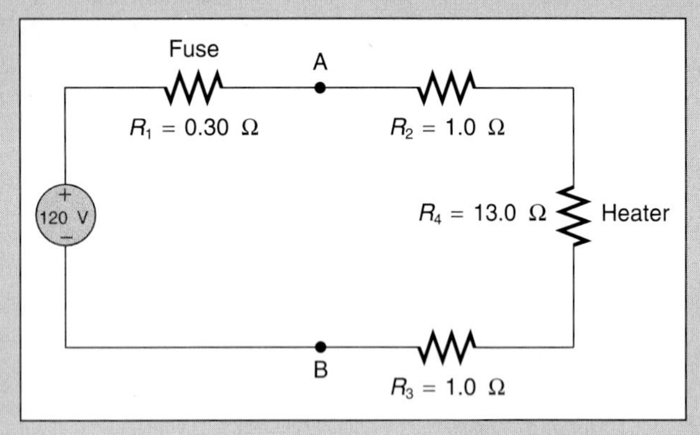

FIGURE P.51

•**52.** As an electrical engineer employed by the firm of Coppers, Gold, Silver, & Shorts you are asked to use a 12.0 V independent voltage source and three 8.0 Ω resistors to provide (in some part of the circuit) a current of 1.0 A and a potential difference of 8.0 V. Sketch a circuit that has these characteristics and indicate where the current is 1.0 A and between what two points the potential difference is 8.0 V.

•**53.** Consider the circuit shown in Figure P.53. (a) Simplify the circuit as much as possible (b) Choose a current direction emerging from the (+) terminal of the independent voltage source. Find this current. (c) What power is absorbed by the 8.0 Ω resistor? (d) What power is absorbed by the independent voltage source?

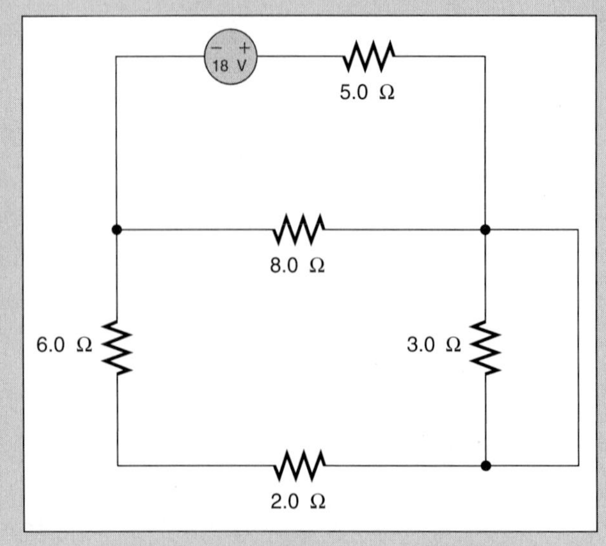

FIGURE P.53

•**54.** A 5.0 W flashlight bulb is powered by two fresh D-cell (real) batteries. The batteries die after 2.0 h of continuous use. (a) How many kilowatt-hours of energy were absorbed by the bulb? (b) If the batteries cost $5.00, what is the cost of one kilowatt-hour of energy supplied by such batteries?

•**55.** An ideal 6.00 V independent voltage source is used to model a real battery. The real battery gets warm when it delivers current to an external load resistance. When the current provided by the battery is 2.00 A, the power absorbed by the battery is found to be 1.16 W. (a) What is the internal resistance of the real battery? (b) When the real battery has a current of 2.00 A, what is its terminal potential difference? (c) Sketch a circuit that will deliver the maximum possible current from this battery. What is this current?

•**56.** Conditions are such that your effective body resistance is 50 kΩ. Assume a 50 mA current is sufficient to induce ventricular fibrillation (see Section 19.12). (a) What potential difference will induce this unfortunate consequence? (b) If your body resistance is lowered to 1.0 kΩ by foolishly having wet contact with wires from a voltage source, what potential difference will induce ventricular fibrillation? (c) Why does a venerable utility wirewoman attempt to do all work with rubber boots, leather climbing belt, and rubber gloves and with one hand at a time?

•**57.** Consider the circuit shown in Figure P.57. (a) If the resistance R is infinite, find: (i) the potential difference between the points A and B; (ii) the power absorbed by the battery. Express your answers in terms of V_0, R_0, and r. (b) If the resistance R is zero, find: (i) the potential difference between the points A and B; (ii) the current emerging from the battery terminal of positive polarity. Express your answers in terms of V_0, R_0, and r.

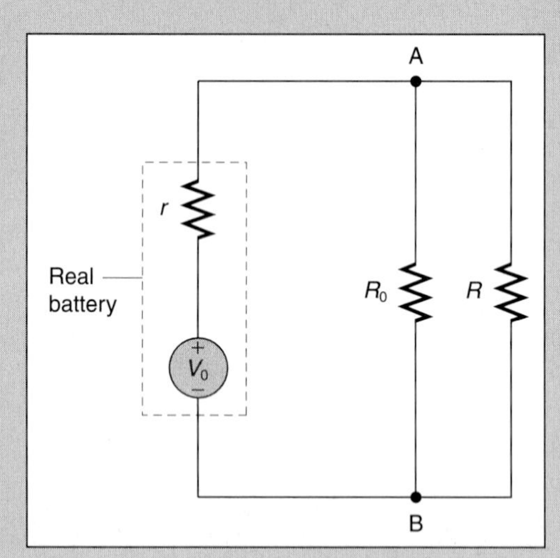

FIGURE P.57

•**58.** In the circuit shown in Figure P.58, what resistance R is needed to maximize the power absorbed by the combination of R and R_0? Express your result in terms of R_0 and r. What resistance R is needed to maximize the power absorbed by R itself?

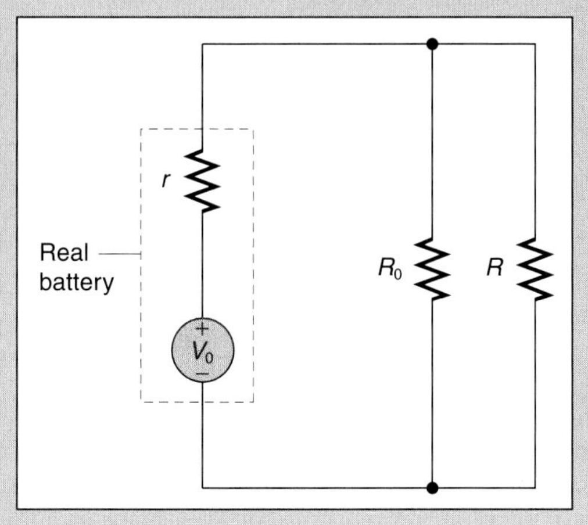

FIGURE P.58

•59. The condition

$$\frac{dP}{dR} = 0 \text{ W}/\Omega$$

is the condition for a maximum or a minimum for power absorbed by load resistor R in the circuit shown in Figure P.59. To show that the condition indeed is a maximum, what condition must the second derivative

$$\frac{d^2P}{dR^2}$$

satisfy? Show that this condition is met when $R = r$.

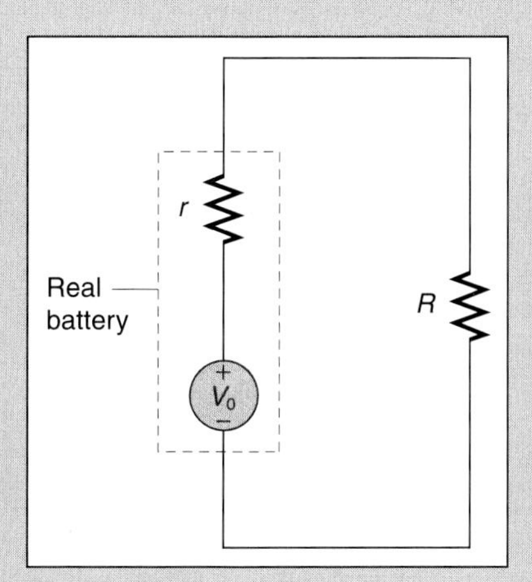

FIGURE P.59

•60. An independent voltage source maintains the same potential difference between its terminals no matter what the current is through the device. An *independent current source* is a circuit element that is the current analog of the independent voltage source (an ideal battery). An independent current source maintains a constant current I_0 through it no matter what the potential difference across its terminals. The characteristic curves of the two sources are shown in Figure P.60a, where I_0 is the fixed current supplied by the current source. An ideal current source is indicated with the symbol shown in Figure P.60b, where the arrow indicates the direction of the current. Consider an independent voltage source V_0 in series with a resistance r connected to a load resistor R as in Figure P.60c. The independent voltage source in series with the resistance r together make up what is called a Thévenin equivalent source. (a) Find the current I through the load resistance R as well as the potential difference V across the load resistor in terms of V_0 and r. (b) When the load resistor R has a resistance equal to zero, the circuit is said to be short-circuited between the terminals A and B. The current then is called the short circuit current I_{sc} between the terminals A and B. Find the short circuit current in terms of V_0 and r. (c) When the load resistance is infinite, the circuit is said to be open. When the load resistance is infinite, what is the current in the circuit? What is the potential difference between the points A and B when $R = \infty\ \Omega$? This potential difference is called the open circuit voltage (or the Thévenin voltage) between the terminals A and B. Now replace the Thévenin equivalent source with an independent current source whose current I_0 is equal to the short circuit current I_{sc} calculated in part (b). The current source is in parallel with the same resistance r used in Figure P.60c. The new arrangement is connected to the load resistance R. The new circuit is known as the *Norton equivalent circuit*. (d) Show that with the circuit of Figure P.60d, the current through the load resistance R and the potential difference across the load resistance R are the same as in part (a). Thus, as far as the load resistance is concerned, it makes no difference whether it is connected to (1) the voltage source in series with r or (2) the current source in parallel with r at terminals A and B (when the current source has a current equal to the short circuit current I_{sc}). Both arrangements result in the same current through R and the same potential difference across R. The two source arrangements (the Thévenin and Norton equivalent sources) are known as *source transformations* of each other. Such source transformations are used occasionally in more advanced courses in circuits to simplify the analysis of more complicated circuits.

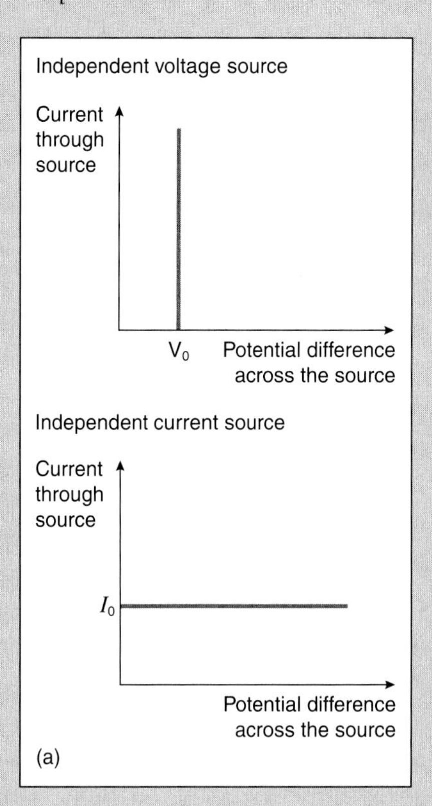

FIGURE P.60 (Continued on next page.)

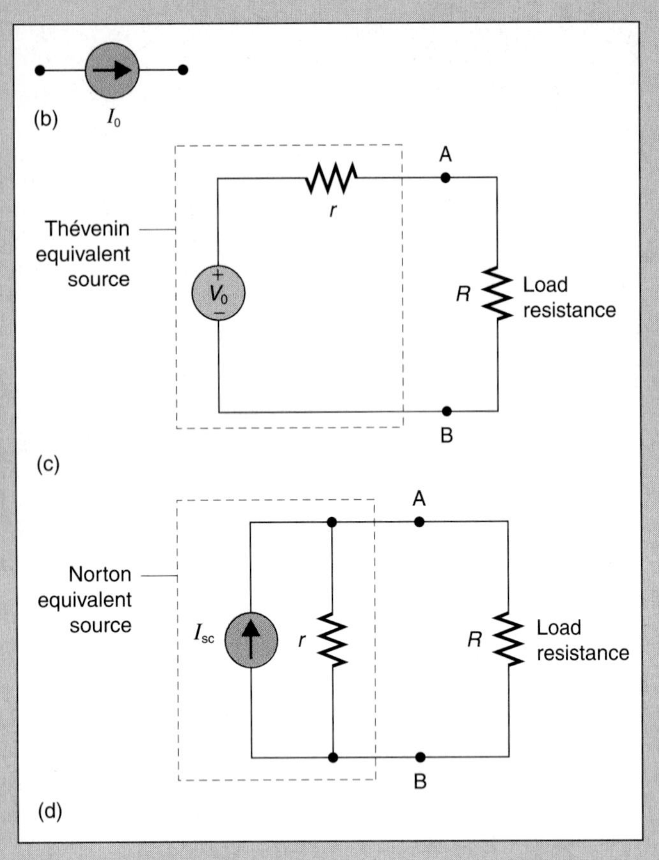

FIGURE P.60 *(continued)*

Section 19.15 Basic Electronic Instruments: Voltmeters, Ammeters, and Ohmmeters

61. You have set up the circuit shown in Figure P.61. The voltmeter indicates 5.00 V and the ammeter indicates 0.250 A. (a) Indicate the direction of the current through the resistor. (b) Determine the value of *R*. (c) Determine the power absorbed by the resistor. (d) What is the emf of the independent voltage source?

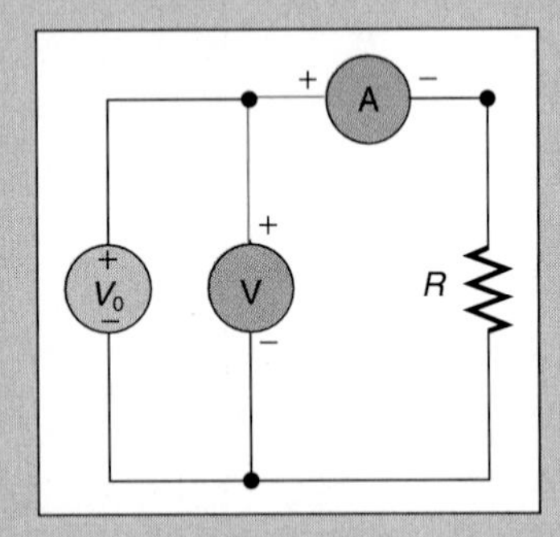

FIGURE P.61

•**62.** A real battery (Figure P.62) is modeled by an independent voltage source V_0 in series with a resistor *r*. The real battery is connected to a load resistor *R*, and a voltmeter is used to measure the potential difference across the load resistor as indicated. The following data are taken meticulously by engineer Milly Watt: (i) When the load resistance is 6.000 Ω, the voltmeter indicates 1.452 V; (b) When the load resistance is 4.000 Ω, the voltmeter indicates 1.429 V. Use this information to determine the internal resistance *r* in the model as well as the emf V_0 of the independent voltage source.

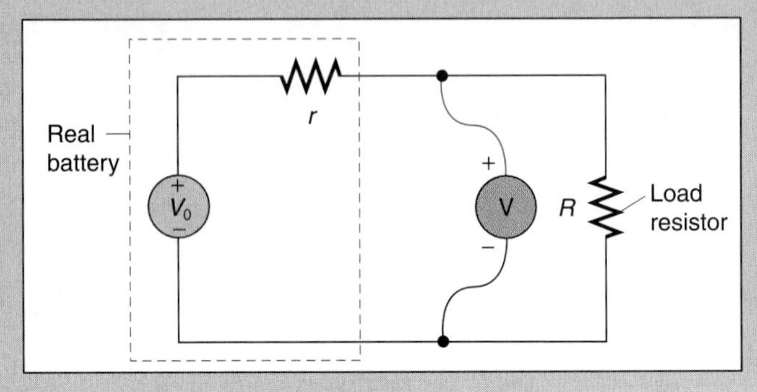

FIGURE P.62

•**63.** The potential difference V_0 and internal resistance *r* used to model a real battery can be found by connecting an adjustable resistance (called a *pot*) across the terminals of the real battery as shown in Figure P.63. The potential difference across *R* is measured for two different but known values of *R* (as determined with an ohmmeter when *R* is disconnected from the circuit). An experiment produces the following data: (i) When *R* = 1.00 kΩ, the potential difference across *R* is 6.00 V; (ii) When *R* = 10.0 Ω, the potential difference across *R* is 3.00 V. Use these data to determine V_0 and *r*.

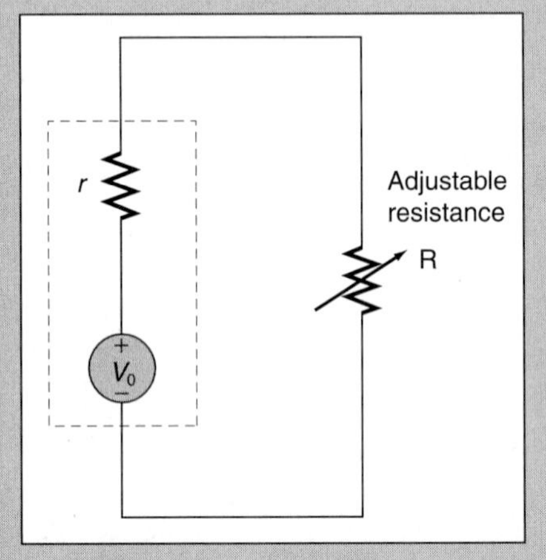

FIGURE P.63

•**64.** Alas. A student has connected the circuit shown in Figure P.64 and his lab instructor is not pleased. Indicate three problems associated with the way the circuit is wired.

•65. In the circuit shown in Figure P.65, voltmeter V_1 indicates 10.0 V when the switch is open. (a) What does voltmeter V_2 indicate? What do the ammeters A_1 and A_2 indicate? What is the emf V_0 of the independent voltage source? (b) The switch is closed and voltmeter V_1 indicates 8.0 V, V_2 indicates 6.0 V, A_1 indicates 0.50 A, and A_2 indicates 0.75 A. Use this information to find the numerical values of R_3, R_2, R_1, and *r*.

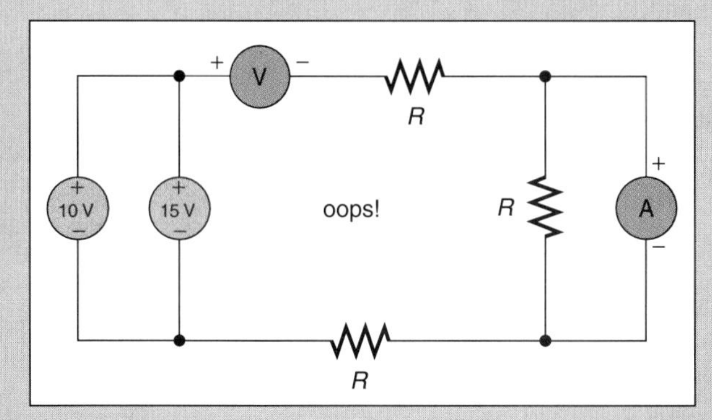

FIGURE P.64

Section 19.16 An Introduction to Transients in Circuits: A Series RC Circuit*

66. If you want a series *RC* circuit with a time constant of 60 s and have a capacitor with a capacitance of 100 μF on hand, what resistance is necessary? If you only have 100 kΩ resistors on hand, how do you secure the needed resistance?

•**67.** Show that when a resistor is connected across an isolated charged capacitor to form a single loop, series *RC* circuit, the electrical potential energy stored in the capacitor decays exponentially at a rate twice as fast as the current.

•**68.** Capacitors are used for energy storage, among other things. When collections of capacitors are used for such purposes, they are colloquially referred to as *capacitor banks*. Let there be 1.00 J of electrical potential energy stored in a charged capacitor that has a potential difference of 500 V between its plates. (a) Determine the capacitance of the capacitor. (b) The energy is released by discharging the capacitor via a series *RC* circuit. After about five time constants, the capacitor is effectively discharged. If you want to discharge the capacitor in part (a) so that the capacitor is effectively discharged in 0.100 ms, through what resistance should the capacitor be discharged? (c) What is the average power provided by the capacitor during the 0.100 ms interval?

•**69.** In a series *RC* circuit, at what time (in terms of the time constant τ) is the charge on a charging capacitor equal to one-half its final value? Does it take the same time for the capacitor to lose half its charge when discharging?

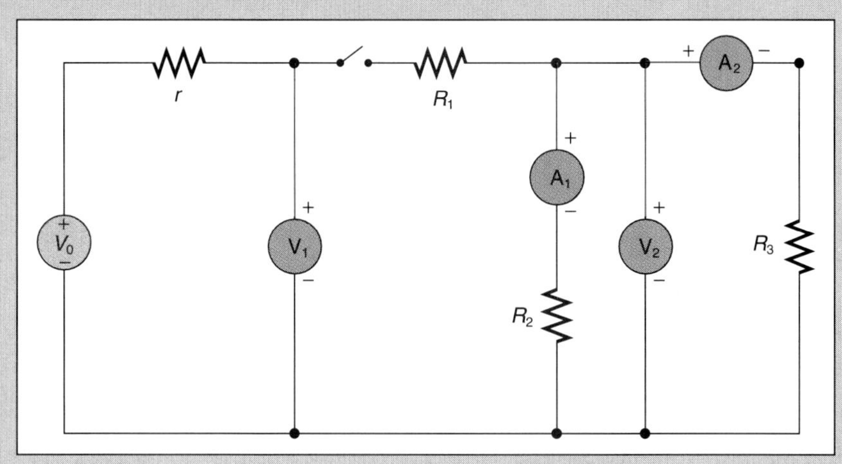

FIGURE P.65

•**70.** The capacitors in Figure P.70 are initially charged. When the switch is closed, find the time constant of the circuit.

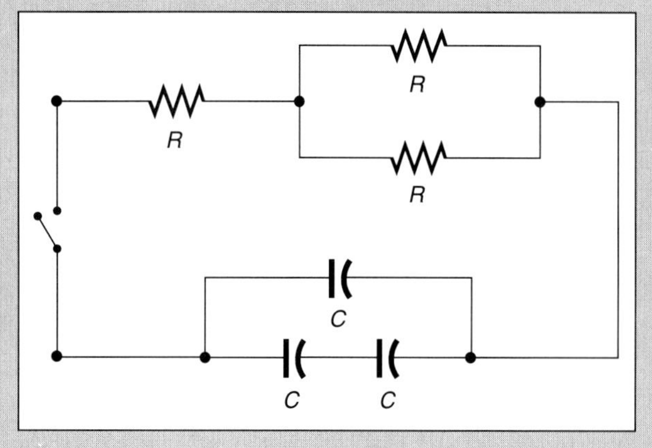

FIGURE P.70

•**71.** For a discharging capacitor in a series *RC* circuit, show that line representing the slope of a graph of the charge *q* versus *t*, evaluated when $t = \tau$, will intersect the time axis when $t = 2\tau$.

•**72.** The circuit of Figure P.72 has been turned on for a long time, so that you can ignore the transient behavior of the capacitor.

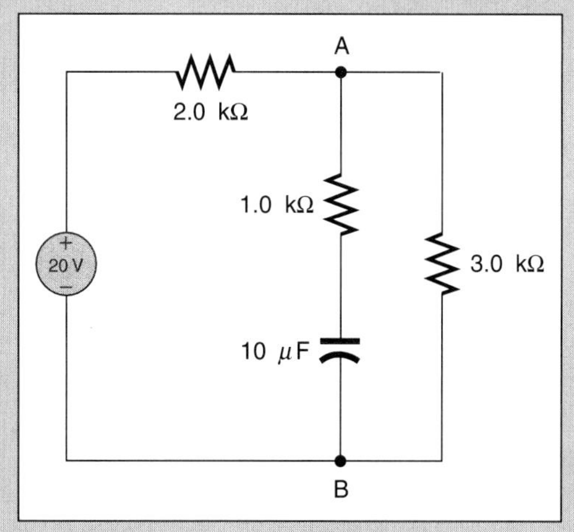

FIGURE P.72

(a) What is the dc current in the branch between nodes A and B that contains the capacitor? (b) What is the potential difference across the 1.0 kΩ resistor? (c) Determine the current through the 2.0 kΩ and 3.0 kΩ resistors. (d) What is the potential difference across the 3.0 kΩ resistor? (e) What is the potential difference across the capacitor? (f) Compute the power absorbed by each circuit element and verify that the sum of the power absorbed by the circuit is zero.

•73. A model of a *real capacitor* consists of a high resistance R, known as a *leakage resistor*, in parallel with an ideal capacitor (see Figure P.73a). The electrical path through R provides a way to account for the observation that all capacitors gradually discharge; the opposite charges on each conductor gradually are neutralized as electrons migrate through the material separating the two conductors of the capacitor. You have two *real* capacitors with leakage resistances R_1 and R_2 in series with an independent voltage source of emf V_0 as shown in Figure P.73b. Determine the potential difference across each capacitor in terms of V_0, R_1, and R_2.

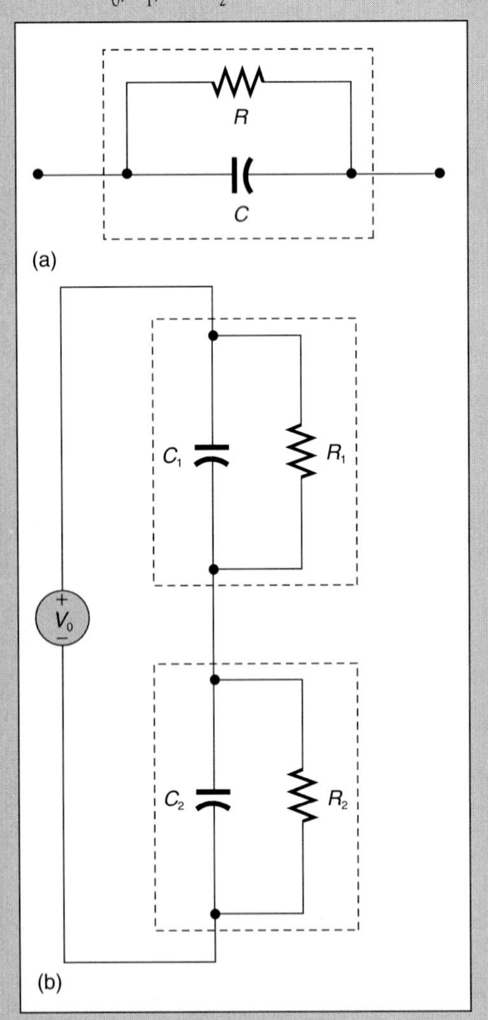

FIGURE P.73

74. (a) Calculate the power $P(t)$ absorbed by a charging capacitor in the series *RC* circuit (Figure P.74) as a function of time. Express your result in terms of R, V_0, and τ. (b) Integrate $P(t)$ from $t = 0$ s to $t = \infty$ s to show that the energy stored in the capacitor is

$$\frac{1}{2}CV_0^2$$

when it is fully charged.

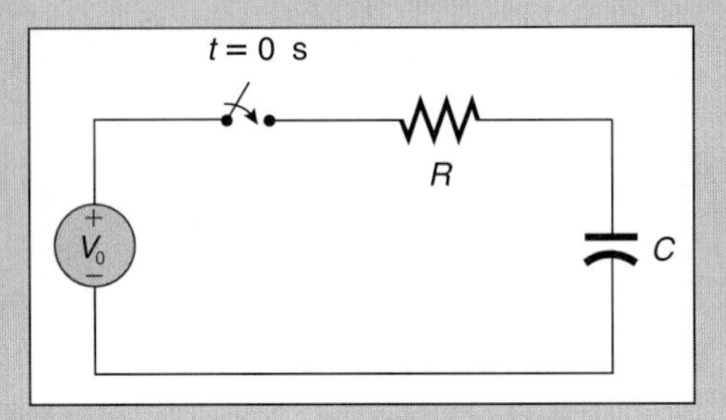

FIGURE P.74

75. Having earned a B.S. and an M.S., you are intent on getting the third degree in physics. Here it comes. A capacitor with capacitance C_1 is initially charged so that it has a potential difference V_0 between its plates. The capacitor then is disconnected from the independent voltage source and connected across the terminals of an uncharged capacitor with capacitance C_2 as indicated in Figure P.75. The resistance R models the actual resistance of the connecting wires. (a) Use charge conservation to show that the final potential difference V_f across each capacitor is

$$V_f = \frac{C_1}{C_1 + C_2}V_0$$

(b) Show that the final potential energy PE_f stored in the two-capacitor system is

$$PE_f = \frac{1}{2}\frac{C_1^2V_0^2}{C_1 + C_2}$$

(c) Show that the energy absorbed by the resistor in this process is

$$\frac{1}{2}\frac{C_1C_2}{C_1 + C_2}V_0^2$$

This problem is based on a paper by C. J. Macdonald, "Conservation and capacitance," *Physics Education*, 23, #4, page 202 (July 1988).

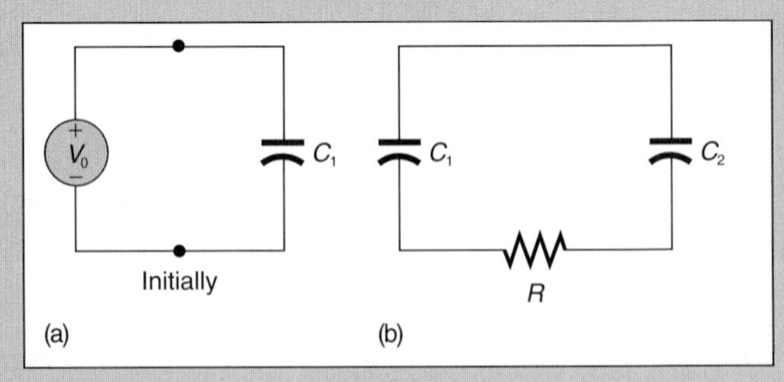

FIGURE P.75

Investigative Projects

A. Expanded Horizons

1. Investigate and report on the physical principles behind polygraphs (lie detectors).
James Allan Matte, *The Art and Science of the Polygraph Technique* (Thomas, Springfield, Illinois, 1980).
Eugene B. Black, *Lie Detectors: Their History and Use* (D. McKay, New York, 1977).

2. Investigate the underlying physics behind electrocardiography. You might even want to build your own electrocardiograph!
W. Bruce Fye, "A history of the origin, evolution, and impact of electrocardiography," *American Journal of Cardiography*, *73*, #13, pages 937–949 (15 May 1994).
H. Edward Roberts, "Electrocardiography," *Radio-Electronics*, *62*, #7, pages 31–40 (July 1991); #8, pages 44–49 (August 1991).

3. Electrical trauma still is an active area of research. The following reference is an interesting introduction to the subject; write a summary of it.
David L. Wheeler, "Understanding electrical trauma," *Chronicle of Higher Education*, *41*, #4, pages A10, A14 (21 September 1994).

4. Investigate the electrical characteristics of a defibrillator, used in attempts to restart a heart that has undergone ventricular fibrillation.
John R. Cameron and James G. Skofronick, *Medical Physics* (Wiley, New York, 1978), page 229.
Colin Brown, "Electric shock and the human body," *Physics Education*, *21*, #6, pages 350–353 (November 1986).

5. Investigate the physics and chemistry of real batteries and the progress being made toward the development of long-lasting batteries with small mass for applications in notebook PCs.
Karl V. Kordesch and Klaus Tomantschger, "Primary batteries," *The Physics Teacher*, *19*, #1, pages 12–21 (January 1981).
Mark Dewey, "Battery technology," *Radio-Electronics*, *62*, #1, pages 45–49 (January 1991).

6. Resistors can be connected in series or parallel, as we have seen. Other types of connections occur as well. Investigate the type of circuit connections known as Y–Δ connections and how they can be transformed into each other.
James W. Nilsson, *Electric Circuits* (3rd edition, Addison-Wesley, Reading, Massachusetts, 1990), pages 53–57.

7. Investigate how semiconducting materials are doped to become n-type or p-type materials.
James J. Brophy, *Semiconducting Devices* (McGraw-Hill, New York, 1964).

8. Discover how a semiconducting *pn junction* acts as a *diode*.
Edward Boyes, "Understanding the p-n junction," *Physics Education*, *25*, #1, pages 53–59 (January 1990).
Albert Paul Malvino, *Electronic Principles* (3rd edition, McGraw-Hill, New York, 1984), Chapters 1, 2.

9. Investigate how semiconducting materials are used to create a bipolar *transistor*. Indicate what is meant by the bipolar transistor terminals known as the *emitter*, the *collector*, and the *base*.
Albert Paul Malvino, *Electronic Principles* (3rd edition, McGraw-Hill, New York, 1984), Chapter 5.

10. Investigate what is meant by the term bipolar *transistor biasing* and how such biasing is accomplished in electronics. In particular, investigate the type of biasing known as voltage-divider (or ladder) biasing.
Albert Paul Malvino, *Electronic Principles* (3rd edition, McGraw-Hill, New York, 1984), Chapter 6.

11. Circuit breakers, like fuses, are used to protect electronic devices from excessive currents. The advantage of a circuit breaker is that after experiencing an excessive current it can simply be reset rather than replaced. Investigate and prepare an oral report on the distinction between the ways that a fuse and circuit breaker interrupt or open-switch a circuit. In desperation, some people insert pennies to replace fuses; explain why this practice is *quite* dangerous and can cause electrical fires.

12. Electric fences are common on many farms and ranches as well as for many backyard gardeners. Visit a manufacturer and/or retailer of such fences and determine what factors are involved in the design of such fencing, the potential differences used, and the currents that typically flow when an animal (which may be you!) inadvertently encounters the fence.

B. Lab and Field Work

13. Devise an experiment to measure the resistivity of the conducting jells used when taking electrocardiograms. Secure samples from a local physician or hospital. Newer devices do not use separate jells.

14. Design and perform an experiment to measure the characteristic curve of a light bulb. Determine if the light bulb is an ohmic device. Do your results vary greatly with the wattage rating of the bulb?

15. Devise and perform an experiment to measure the emf V_0 of the ideal battery and internal resistance r used to model a real battery.

16. Devise and test a procedure to measure the capacitance of a capacitor using a series *RC* circuit with a large known resistance *R*, an independent voltage source, a voltmeter, and a stopwatch.

17. Write or visit a state or federal prison that enforces the death penalty by electrocution. Determine what potential differences and currents are used to minimize the trauma of the condemned person during their electrocution.

18. Devise and perform an experiment to measure the characteristic curve of a semiconducting pn junction diode.

19. Design and test an apparatus similar to that shown in Figure 19.1 to demonstrate the concept of an electrical current.

C. Communicating Physics

20. Frequently you read about electricity in the newspaper. Journalistic accounts, however, occasionally are tinged with inaccurate science that manifests the scientific ignorance of the reporter. Statements such as "10 000 volts of electrical power," "a 12 000 volt current," "100 amperes of electrical energy," and other such nonsense are, unfortunately, not infrequent; you might even collect a few for the amusement of your class. The editor of your local or campus newspaper has hired you as a technical consultant to provide a 30-minute short-course for the edification the reporting staff on the proper use of electrical terminology. Prepare

such a short-course, designing appropriate visual aids, with a focus on the words power, current, and potential.

21. The electric eel, *Electrophorus electricus*, is able to stun or even kill nearby prey without electrocuting itself, a neat trick. Consult a biology professor at your university and see if the two of you can develop a model for explaining how this fishy trick is accomplished. Present the results of your model in the form of a 15-minute presentation to your physics class.

22. Prepare a brief oral or written report addressing the purpose of a surge protector and discussing a circuit diagram of such a device.

"BETTER REDO THAT
POWER CALCULATION"

CHAPTER 20

MAGNETIC FORCES AND THE MAGNETIC FIELD

We know that the lodestone has a wonderful power. . . . When I first saw it I was thunderstruck. . . . Who would not be amazed at the virtue of this stone. . .?

St. Augustine (A.D. 354–430)*

Magnets and magnetic effects have immense natural, technological, and cultural importance. Spectacular auroras arise when charged particles from the Sun collide with the atmosphere; the particles are funneled by the magnetic field of the Earth toward its magnetic poles. Magnetic fields in planetary, stellar, and galactic astronomy lead to almost unbelievably violent cosmic events. Without the simple magnetic compass, navigating across uncharted seas would have been vastly more dangerous and difficult; the Age of Discovery would have taken place with more trepidation, and over a longer time period than it did.

Birds use the magnetic field of the Earth (as well as clues from the Sun and stars) for navigation during long migrations. More immediately, the tapes you play for relaxation and amusement in your Walkman or stereo, the diskettes and devices you use and need to store large quantities of information, magnetic imaging technologies used for life-saving diagnostics in medicine, even the very process used to generate much of the electricity on which so much of our modern society depends—all are based on magnetism and magnetic effects. Thus magnetism is closely associated with a wide variety of technological and natural phenomena.

We first encountered the idea of a field in our study of the gravitational force in Chapter 6. The gravitational force is conveyed by a gravitational field $\vec{g}$. Likewise, in our study of electrical interactions in Chapter 16, we considered the electrical force to be conveyed by an electric field $\vec{E}$. The electric situation is very similar to the gravitational case, except that we have two types of charges (positive and negative) but only one type of mass (positive mass).

The study of magnetism introduces a third important field concept[†]: the magnetic field. It has to do with the *motion* of electric charges (currents). We shall see that while the magnetic field has similarities to the other fields, it also has some unique features that clearly distinguish it from the others and provide its *raison d'être*. In this chapter we see how the need for the magnetic field concept arose, how charges are affected by its presence, what causes this type of field, and what are the salient characteristics of this field itself.

*(Chapter Opener) *The City of God*, translated by Marcus Dods (The Modern Library, New York, 1950), Book XXI, Chapter 4, page 768. Lodestone literally means a stone that leads.

[†]Other fields are encountered in more advanced courses in physics.

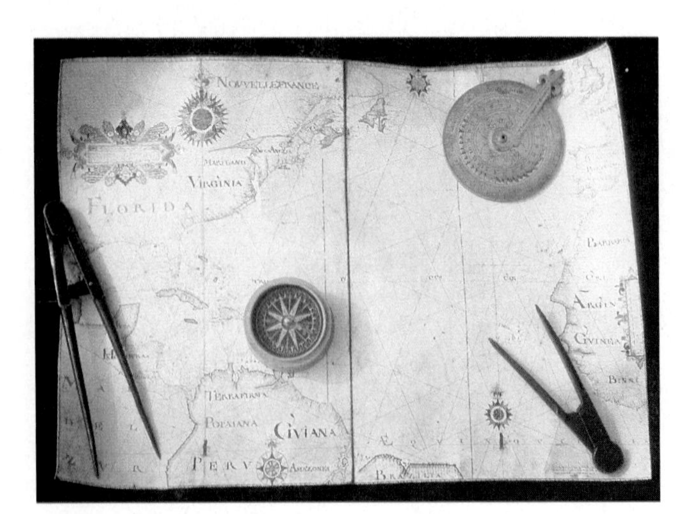

The compass aided the Age of Discovery.

20.1 The Magnetic Field

A wonder of this kind I experienced as a child of 4 or 5 years when my father showed me a compass. That this needle behaved in such a determined way did not at all fit into the kind of occurrences that could find a place in the unconscious world of concepts. . . . I can still remember—or at least believe I can remember—that this experience made a deep and lasting impression on me. Something deeply hidden had to be behind things.

Albert Einstein (1879–1955)[‡]

Magnets are always fun to play with. They fascinated both St. Augustine and a young boy named Albert Einstein.

Undoubtedly you know that magnets have two **poles**, designated north (N) and south (S). The designation arose from the navigational compass. The needle of such a compass is a small, thin magnet.

> The end of a compass needle that points generally in a northerly direction at most places on the Earth is defined to be the needle's **north magnetic pole** N (usually shortened to the **north pole** of the compass needle magnet).[§] The opposite end of the compass needle is defined to be its **south magnetic pole** S (or **south pole**).

In playing with two such permanent magnets, it is easy to determine that opposite poles, called *unlike* poles, attract each other (see Figure 20.1), while poles of the same kind, called *like* poles, repel each other (see Figure 20.2). These observations, coupled with the definition of a north magnetic pole, imply that the magnetic pole of the earth-magnet that is located in the

[‡]*Albert Einstein: Philosopher-Scientist*, edited by Paul Arthur Schilpp (Library of Living Philosophers, Evanston, Illinois, 1949), page 9.

[§]The *geographic* north and south poles of the Earth define the two points where the spin (rotational) axis of the Earth intersects its surface. The locations of the geographic poles are determined from astronomical observations and are distinct from the magnetic north and south poles of the Earth. The geographic poles exist because the Earth spins. Even if the Earth had no magnetic field, geographic north and south poles would exist.

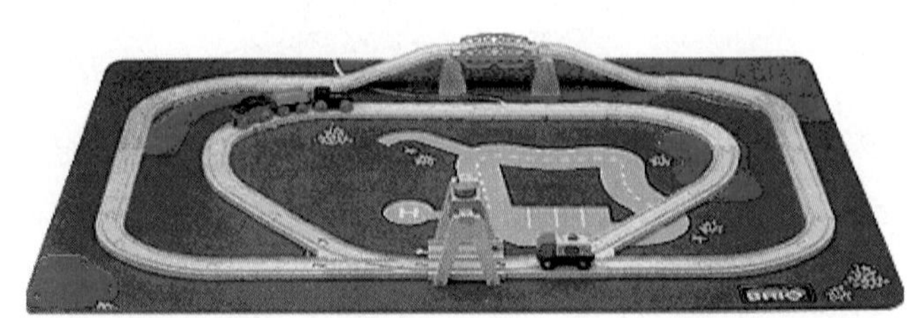

Magnets are used on many toys.

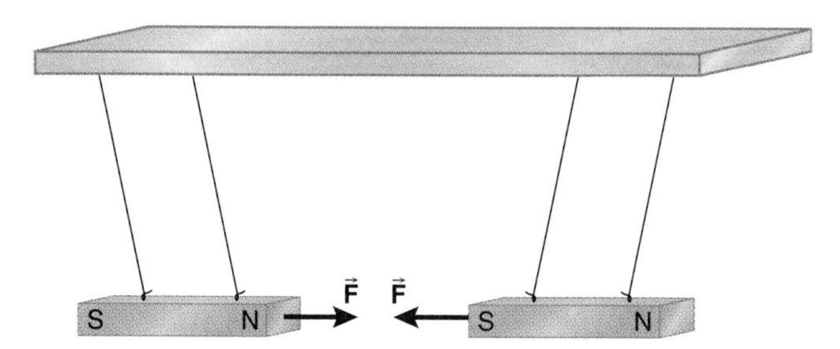

FIGURE 20.1 Unlike poles attract each other.

northern geographic hemisphere of the Earth must, therefore, technically be a *south* magnetic pole; see Figure 20.3.*

The magnetic forces of attraction and repulsion of magnetic poles on each other are very similar to the electrical force interactions between electric charges (with the north and south magnetic poles playing the roles of positive and negative charge). We shall see, however, that charges and magnetic poles are *not at all* the same thing.

If we take any bar magnet and sprinkle iron filings around it, a pattern emerges, as shown in Figure 20.4. The pattern indicates that the magnet affects the space in its vicinity. The pattern of lines traces out the magnetic field $\vec{B}$ of the magnet, which we imagine to be present in the space around the magnet, even without the presence of the iron filings. The direction of the magnetic field at any point is tangent to the field line at that point.

Such **magnetic field lines** are similar to the electric field lines we used to trace out the pattern of the electric field surrounding charges (see Section 16.7). By convention, when in the space outside the magnet material, the direction of the magnetic field is tangent to the field line directed *from* the north magnetic end of the magnet *to* its south magnetic end.† Just as with electric field, the number density of the magnetic field lines in a region is a measure of the magnitude of the magnetic field there. Since the field lines bunch closer together as you approach either pole end of the magnet, the magnitude of the magnetic field increases accordingly.

*Confoundingly, almost everyone (including geoscientists) calls this magnetic pole of the earth-magnet the north magnetic pole anyway. There is no confusion about the science, however.

†Within the magnet itself, the field lines run from the south polar end toward the north polar end, preserving their continuity from inside to their curving return pattern outside.

The pattern of magnetic field lines surrounding a magnet is similar to the pattern of electric field lines surrounding an electric dipole (see Figure 16.57). However, if we try to isolate the north and south poles of the magnet by, say, cutting the magnet in two (see Figure 20.5), we find that each piece has a dipolar pattern for the magnetic field surrounding it, albeit a weaker one.

Indeed, if we keep breaking the magnets into smaller and smaller pieces, we find that it is impossible to isolate either a north or a south magnetic pole; they *always* appear inseparably paired. Hence we say that magnets and the magnetic field always are *dipolar*.

This is quite unlike electric charge. Isolated positive and negative charges (electric **monopoles**) are quite common. Indeed the fundamental unit of charge e exists in positive and negative denominations on particles such as the proton and the electron. With magnets, *no magnetic monopoles ever have been discovered* despite fiendishly valiant and painstaking searches for them in all manner of materials from fossils and ancient rocks to cosmic rays. Magnetic poles always occur in north–south pairs (**magnetic dipoles**) that produce a dipolar magnetic field in the surrounding space. A monopolar magnetic field has never been observed; monopolar electric fields are common.

It is the effect of a magnetic field on charged particles that clearly indicates that magnetic poles and the magnetic field are not the same sort of beast as electric charge and the electric field. Experiments indicate the following:

1. A charge q placed at rest in a magnetic field experiences *zero* force. This single observation means the magnetic field is quite different from the electric field: a charge at rest in an electric field experiences a nonzero force $\vec{F}_{elec} = q\vec{E}$ (Equation 16.8).
2. If we move the charge along a magnetic field line, either parallel or antiparallel to the direction of the magnetic field, the moving charge again experiences zero force.
3. If the charge is moved at speed v at an angle θ ($\neq 0°$) with the direction of a uniform magnetic field, a nonzero magnetic force exists on the charge. The magnitude of the force is found to be proportional to v (for fixed θ) and proportional to $\sin\theta$ (for fixed v). The magnetic force on q is a *velocity-dependent force*, since it depends on both the speed *and* direction of the velocity vector.
4. If we vary the field magnitude, say, by performing experiments in regions with half or double the number density of magnetic field lines, the force is found to vary with the field

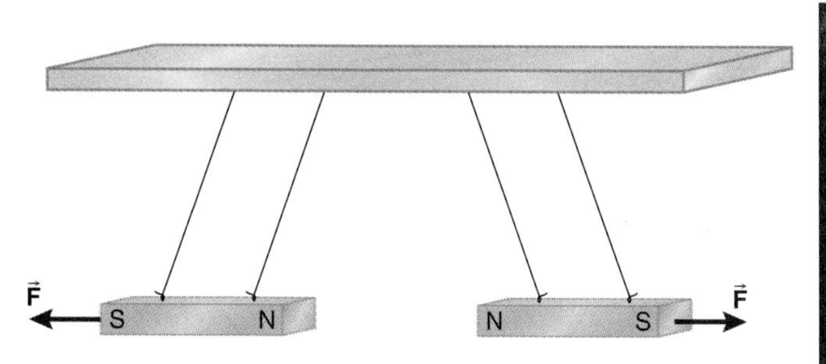

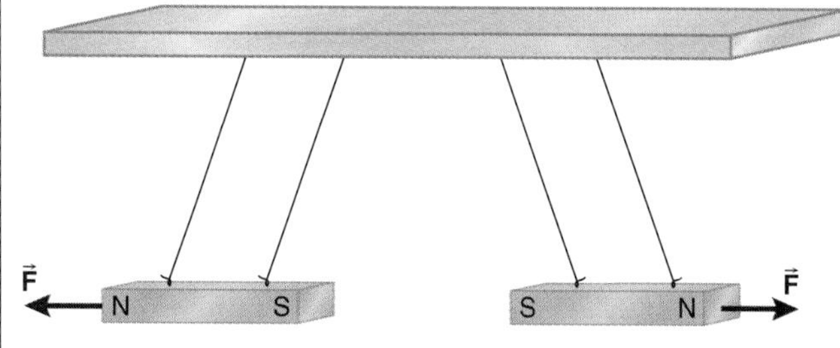

FIGURE 20.2 Like poles repel each other.

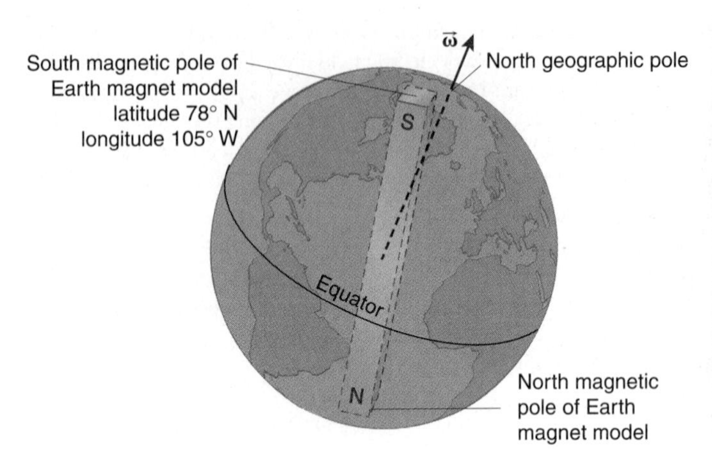

FIGURE 20.3 The magnetic pole of the Earth located in the northern hemisphere is a south magnetic pole because it attracts the north polar end of a compass needle.

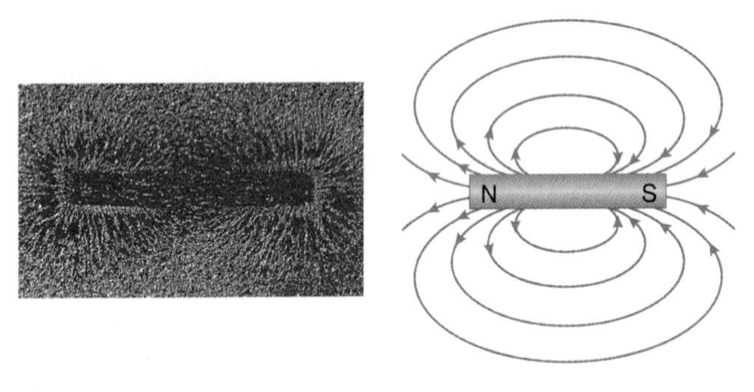

FIGURE 20.4 Iron filings sprinkled around a magnet reveal the presence of its magnetic field.

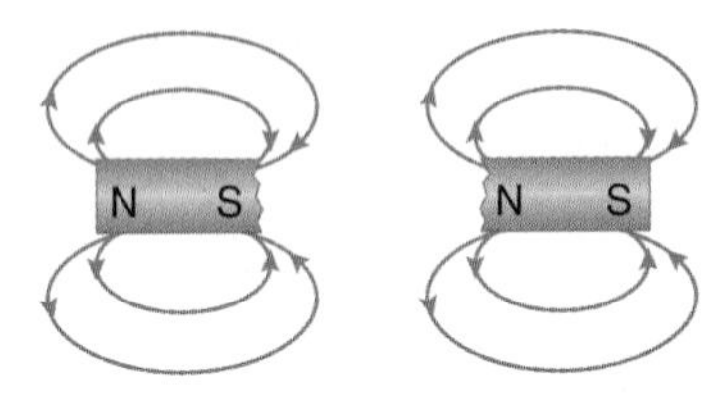

FIGURE 20.5 If we break a magnet in two, we get two dipolar magnets.

magnitude B (as determined from the number of magnetic field lines).

5. The direction of the force on q (determined from its acceleration) depends on the *sign* of the moving charge; charges of opposite sign experience forces in opposite directions. In every case the force (if nonzero) is perpendicular to *both* the velocity vector $\vec{v}$ of the charge and the field direction $\vec{B}$.

The idea of the polyphase induction electric motor occurred to Tesla as a classic "aha!" experience. He emigrated to the United States to develop the idea, arriving with a mere $0.04 in his pockets.

All five observations are summarized eloquently in the following expression for the magnetic force on a moving charge q:

$$\vec{F}_{\text{magnet on } q} = q\vec{v} \times \vec{B} \tag{20.1}$$

This illustrates yet again the usefulness and cogency of the vector product to express a wealth of information in physics. Equation 20.1 is used to define the magnitude of the magnetic field in SI units, since the units for force, charge, and velocity are known. The dimensions of the magnetic field must be those of force divided by the units for charge and velocity, or

$$\frac{\text{N}}{\text{C}\cdot\text{m/s}} = \text{N}\cdot\text{s/(C}\cdot\text{m)}$$

in SI units. This combination of SI units is defined to be a **tesla** (T), named in honor of Nikola Tesla (1856–1943), a Serbian engineer of Croatian birth who emigrated to the United States to develop his invention of alternating current electric motors.

Another common and convenient unit of magnetic field is the **gauss**, although it is not the SI unit. The gauss is a smaller unit than the tesla; the conversion factor between gauss and teslas is

$$10^4 \text{ gauss} \equiv 1 \text{ T}$$

We shall not use gauss in this text, but the unit is frequently used in geology and geophysics. The magnitude of the magnetic field of the Earth near its surface is quite weak: only about 10^{-4} T, or one gauss.

PROBLEM-SOLVING TACTIC

20.1 When using Equation 20.1, be sure to use the appropriate sign of the charge q. For positive charges, the force is in the direction of $\vec{v} \times \vec{B}$; for negative charges, the force is directed opposite to $\vec{v} \times \vec{B}$.

QUESTION 1

Are the forces shown in Figures 20.1 and 20.2 Newton's third law force pairs?

20.2 APPLICATIONS

Magnetic fields have many practical applications in the sciences; here we examine but a few.

A Velocity Selector

A combination of magnetic and electric fields can be used to sort charged particles according to their speeds. We consider a particle with charge q moving with constant velocity $\vec{\mathbf{v}}$. The particle enters a region of space where there exists a uniform magnetic field $\vec{\mathbf{B}}$ and a uniform electric field $\vec{\mathbf{E}}$ perpendicular to it (see Figure 20.6). The magnetic field is directed perpendicularly into the page, symbolized with the ×'s, representing the crossed tail feathers of a retreating arrow. The electric field is directed as shown in Figure 20.6. The velocity of the particles here is initially perpendicular to *both* the magnetic and electric fields.

With the coordinate system indicated in Figure 20.6, the magnetic field is

$$\vec{\mathbf{B}} = -B\hat{\mathbf{k}}$$

and the velocity is initially

$$\vec{\mathbf{v}} = v_0\hat{\mathbf{\imath}}$$

Thus the initial magnetic force on the charge is, from Equation 20.1,

$$\begin{aligned}\vec{\mathbf{F}}_{\text{magnet}} &= q\vec{\mathbf{v}} \times \vec{\mathbf{B}} \\ &= qv_0\hat{\mathbf{\imath}} \times (-B\hat{\mathbf{k}}) \\ &= qv_0B\hat{\mathbf{\jmath}}\end{aligned}$$

If q is positive, the charged particle experiences a magnetic force parallel to $\hat{\mathbf{\jmath}}$ for this choice of the coordinate system; if q is negative, the force on it is toward $-\hat{\mathbf{\jmath}}$.

The uniform electric field of magnitude E is directed so that the electrical force on the charge initially is opposite to the direction of the magnetic force on it. With an electric field in the downward direction, the electrical force on q is

$$\vec{\mathbf{F}}_{\text{elec}} = qE(-\hat{\mathbf{\jmath}})$$

The total force initially acting on the charge then is

$$\vec{\mathbf{F}}_{\text{magnet}} + \vec{\mathbf{F}}_{\text{elec}} = qv_0B\hat{\mathbf{\jmath}} - qE\hat{\mathbf{\jmath}}$$

Now we adjust the magnitude of the electric field (by varying the potential difference between the parallel plates causing the uniform electric field) until the total force acting on the charge is zero:

$$\begin{aligned}\vec{\mathbf{F}}_{\text{magnet}} + \vec{\mathbf{F}}_{\text{elec}} &= qv_0B\hat{\mathbf{\jmath}} - qE\hat{\mathbf{\jmath}} \\ &= 0 \text{ N}\end{aligned}$$

Hence, when the total force is zero, we have

$$qv_0B = qE \tag{20.2}$$

When the total force on the charge is zero, the charged particle continues to move merrily through the fields at a constant velocity and is not deviated by either the magnetic or electrical forces acting on it (since they vector sum to zero). Solving Equation 20.2 for v_0, we find

$$v_0 = \frac{E}{B} \tag{20.3}$$

This beautifully simple result can be put to practical use. Imagine a beam of charged particles, each with the same charge q, but having a variety or distribution of speeds, all moving in the same direction, as shown in Figure 20.7. For those particles with speeds *smaller* than the speed v_0 given by Equation 20.3, the electrical force is greater in magnitude than the magnetic force. Hence these particles are accelerated in the direction of the electrical force, away from the incident direction. For those particles with speeds *greater* than the speed v_0 given by Equation 20.3, the magnetic force is greater in magnitude than the electrical force. These particles also are accelerated away from the incident direction, in the direction of the magnetic force. *Only* those particles with the speed v_0 given by Equation 20.3 continue to move undeviated through the apparatus, since only particles with this speed have zero total force acting on them. By placing a baffle with a slit or hole at the end of the device, as in Figure 20.7, only the particles with the speed v_0 emerge from it.

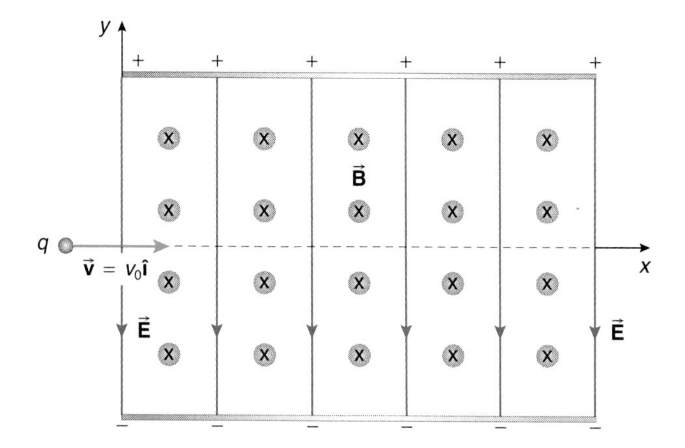

FIGURE 20.6 A charged particle with its velocity initially perpendicular to mutually perpendicular magnetic and electric fields.

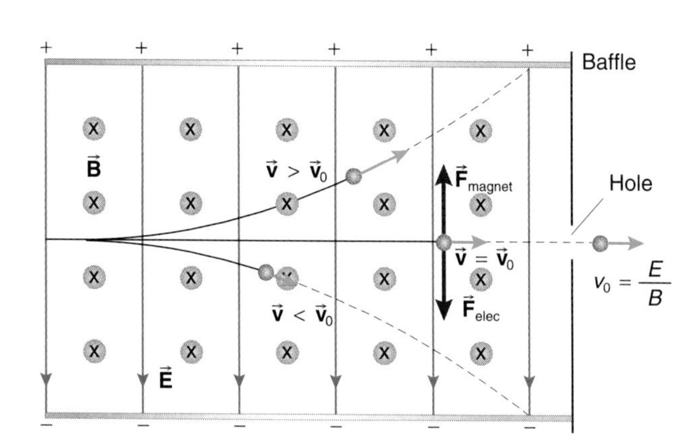

FIGURE 20.7 A velocity selector.

Since we can control the magnitude of the electric field easily, we can select the speed of the particles we want to emerge from the device. In other words, from the incident beam of particles with a dispersion of speeds, we have a way of selecting only those particles with a particular speed. This is a **velocity selector**. Such devices commonly are used to produce monoenergetic particles in experiments and devices that use charged particle beams.

Notice that the selected speed v_0 given by Equation 20.3

1. depends only on the magnitude of the fields;
2. is independent of the identical charge of the particles (as long as the charge is not zero); and
3. is independent of the mass of the particles.

QUESTION 2

The simultaneous presence of a magnetic and an electric field in a velocity selector is *not* an example of the principle of superposition of *fields*. Explain why. The velocity selector, however, *is* an example of the principle of superposition of *forces*. Explain why.

EXAMPLE 20.1

Ions, each with a charge $+e$, are injected appropriately into a velocity selector containing a magnetic field of magnitude 0.200 T and an electric field of magnitude 2.50×10^5 N/C.

a. What is the speed of the ions that emerge undeflected?
b. If the charged plates creating the electric field are separated by 2.00 cm, what is the potential difference between them?

Solution

a. The undeflected particles in the velocity selector have electric and magnetic forces on them of equal magnitude in opposite directions. Use Equation 20.3 to find the speed of the undeflected particles:

$$\begin{aligned} v_0 &= \frac{E}{B} \\ &= \frac{2.50 \times 10^5 \text{ N/C}}{0.200 \text{ T}} \\ &= 1.25 \times 10^6 \text{ m/s} \end{aligned}$$

b. Resurrecting ancient material from Chapter 17, the absolute value of the potential difference V between two oppositely charged plates is related to the magnitude of the field E and the plate separation d by

$$V = Ed$$

With the given field and plate separation, the potential difference is

$$\begin{aligned} V &= (2.50 \times 10^5 \text{ N/C})(2.00 \times 10^{-2} \text{ m}) \\ &= 5.00 \times 10^3 \text{ V} \end{aligned}$$

A Mass Spectrometer

The different isotopes of an element have different masses, but all behave indistinguishably in chemical reactions. Hence the discovery and separation of isotopes by purely chemical means is not possible. Here we see how it is possible to detect the presence of the various isotopes of a given element.

Send a particle of mass m and charge q into a region that has only a uniform magnetic field. The velocity vector is perpendicular to the field, as indicated in Figure 20.8. We want to determine the trajectory of the particle.

Upon entering the region of the field, the particle experiences a magnetic force given by Equation 20.1:

$$\vec{F}_{\text{magnet}} = q\vec{v} \times \vec{B}$$

The vector product indicates that the direction of the magnetic force on the particle is perpendicular to the velocity vector *and* the magnetic field vector. Thus, if the particle is positively charged, the force initially is directed as shown in Figure 20.9. If the particle is negatively charged, the force is in the opposite direction.

The direction of the velocity vector begins to change in response to the force; but the magnitude of the velocity vector is not affected, since the force is perpendicular to it. However, the magnetic force *always* is perpendicular to the velocity (and the field), and so the direction of the magnetic force also changes as the direction of the velocity vector changes. Recall that a force of constant magnitude that is always perpendicular to the velocity vector produces circular motion; hence the particle here is forced into a circular path, as shown in Figure 20.10.

Now we apply Newton's second law of motion to the particle. The magnetic force is the only force on the particle and produces an acceleration; here, it is a centripetal acceleration.

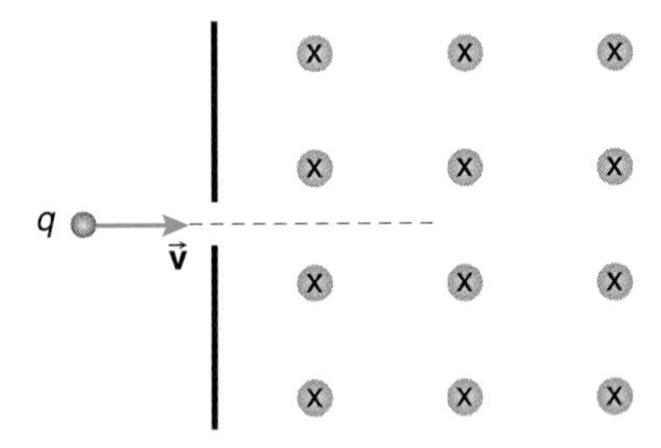

FIGURE 20.8 **A charged particle is sent into a region with a uniform magnetic field perpendicular to the velocity.**

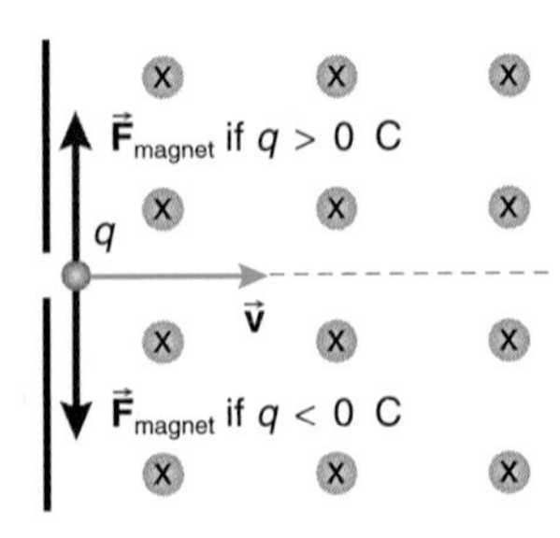

FIGURE 20.9 **The initial direction of the magnetic force on the charged particle.**

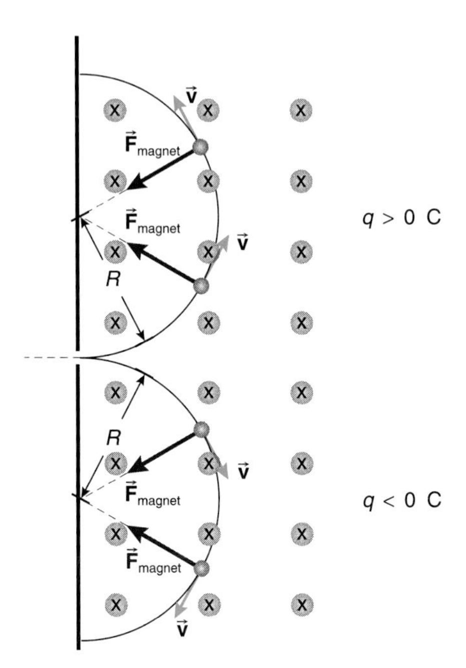

FIGURE 20.10 The trajectory of the charged particle is a circle.

The relationship between the magnitude of the total force and the magnitude of the acceleration is, as we know,

$$F = ma$$

Since the velocity vector is perpendicular to the field, the magnitude of the magnetic force is $F_{magnet} = |q|vB$. The magnitude of the centripetal acceleration is v^2/R, where R is the radius of the resulting circular path of the particle. Substituting for the magnitude of the force into the left-hand side of Newton's second law and for the magnitude of the acceleration on the right-hand side, we obtain

$$|q|vB = m\frac{v^2}{R}$$

The radius of the circular path of the particle thus is

$$R = \frac{mv}{|q|B} \tag{20.4}$$

and is called the **cyclotron radius**, but that is just technical lingo.

This result is of practical import. Imagine a beam of identically charged masses (atoms, molecules, or perhaps other charged particles such as electrons or protons) all moving with the same velocity incident upon the apparatus. The charged particles may represent an ion beam, that is, a collection of atoms of a particular element ionized so that the particles (ions) all have the same positive charge [say, one electron missing from each atom, giving each resulting atom (ion) a charge of $+e$]. By running the ions first through a velocity selector (described earlier), a beam of identically charged ions with identical velocities is formed. Feed this beam into the device pictured in Figure 20.8. According to our analysis, once inside the device the particles move in circular paths whose radii are given by Equation 20.4.

Now here is the punch line:

> If the beam is composed of ions of different isotopes of the same element, all with the same charge, each isotope moves in a circle with a different radius.

Equation 20.4 indicates that more massive isotopes have paths of larger radii (since the charge $|q|$ and speed v of all the ion particles are the same; they all are in the same magnetic field $\vec{\mathbf{B}}$ as well). The radius of the circular path thus depends only on the mass of the particle. So the device separates the isotopes according to their mass. Indeed the device also is capable of determining the relative abundance of the isotopes by comparing the relative depositions of material at the various terminations of the semicircular paths. Such a mass separator is called a **mass spectrometer** and was instrumental in the discovery of many isotopes of the elements. Such devices also can be used to separate uranium isotopes and enrich nuclear fuel, although this method is quite inefficient and expensive. Mass spectrometers also are used to detect small leaks in high-vacuum systems.

EXAMPLE 20.2

A beam of singly ionized atoms of carbon (each with charge $+e$) all have the same speed and enter a mass spectrometer. The ions accumulate in two different locations (see Figure 20.11), spaced 5.00 cm apart. The more abundant $^{12}_{6}C$ isotope traces a path of smaller radius, 15.0 cm. What is the atomic mass number of the other isotope in the beam?

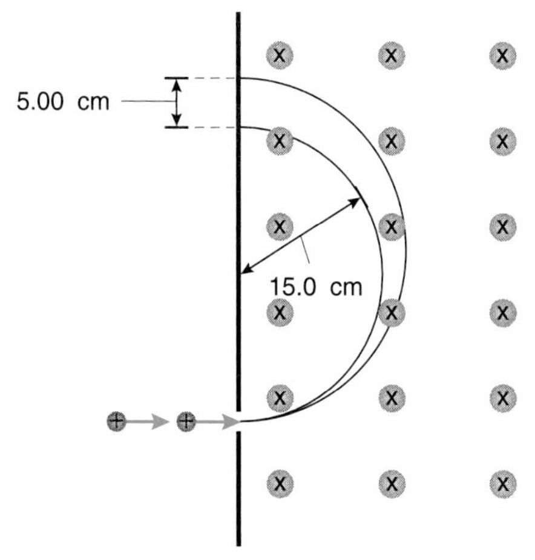

FIGURE 20.11

Solution

Let R_1 be the radius of the trajectory of the more abundant isotope $^{12}_{6}C$ and R_2 that of the less abundant, unknown isotope. The trajectory of the unknown isotope has a greater radius, and so the mass of the unknown isotope is greater than that of the $^{12}_{6}C$ isotope. Since the accumulations are separated by 5.00 cm, the geometry implies that

$$2R_2 - 2R_1 = 5.00 \times 10^{-2}\ \text{m}$$

Hence you have

$$R_2 - R_1 = 2.50 \times 10^{-2}\ \text{m}$$

Since $R_1 = 15.0 \times 10^{-2}$ m, the larger radius is

$$R_2 = 15.0 \times 10^{-2}\ \text{m} + 2.50 \times 10^{-2}\ \text{m}$$
$$= 17.5 \times 10^{-2}\ \text{m}$$

Write Equation 20.4 for each isotope:

$$R_1 = \frac{m_1 v}{|q|B} \qquad R_2 = \frac{m_2 v}{|q|B}$$

Taking the ratio of these expressions eliminates several unknown quantities:

$$\frac{R_2}{R_1} = \frac{m_2}{m_1}$$

Let A be the atomic mass number of the unknown isotope. The atomic mass number of the known isotope is 12. The masses of the isotopes are proportional to the mass number:

$$m_1 = \alpha(12) \qquad m_2 = \alpha A$$

where α is a constant that converts atomic mass numbers to kilograms. Hence the mass ratio is

$$\frac{m_2}{m_1} = \frac{A}{12}$$

and the ratio of the radii becomes

$$\frac{R_2}{R_1} = \frac{A}{12}$$

Substituting for the radii, you get

$$\frac{17.5 \times 10^{-2}\ \text{m}}{15.0 \times 10^{-2}\ \text{m}} = \frac{A}{12}$$

Solving for the unknown atomic mass number A, you find

$$A = 14$$

Thus the unknown isotope is ${}^{14}_{6}C$. This isotope of carbon is radioactive and is used frequently for dating organic archaeological artifacts, as we shall see in Chapter 26.

The Hall Effect

In 1879 Edwin Hall, working in Cambridge, Massachusetts, discovered an interesting effect that now bears his name. Its importance arises because it provides a way for determining the sign of the charge carriers in a current. The effect also can be used to measure precisely the magnitude of a magnetic field. There is no new physics involved in the so-called classical **Hall effect**: it is another, although sophisticated, application of the magnetic and electrical forces on charged particles. Yet ironically, the Hall effect had some surprises in store for physics almost a century after its discovery.

Consider a conducting rectangular bar of material carrying a current or stream of charged particles moving along the axis of the bar. The same current could be caused by positive carriers moving one way or by negative charges moving oppositely. Which is it? The bar is placed in a uniform magnetic field that is perpendicular to the direction of motion of the charged particles creating the current, as shown in Figure 20.12.

We consider two scenarios.

Scenario 1: Positive Charge Carriers

Say the current in the bar arises from positive charge carriers, each with charge $q > 0$ C. They each experience a magnetic force given by Equation 20.1. With the coordinate system in Figure 20.13, this is

$$\vec{\mathbf{F}}_{\text{magnet}} = q\vec{\mathbf{v}} \times \vec{\mathbf{B}}$$
$$= q\langle v\rangle\hat{\mathbf{i}} \times B\hat{\mathbf{k}}$$
$$= -q\langle v\rangle B\hat{\mathbf{j}}$$

The quantity $\langle v\rangle$ is the (single) component of the average drift velocity of the charge carriers. Under the influence of the magnetic force, the positive charges move downward and accumulate near the lower edge of the material, leaving behind an equivalent amount of negative charge near the upper edge of the rectangular bar. The charge separation produces an electric field transverse to the current, directed as in Figure 20.13.

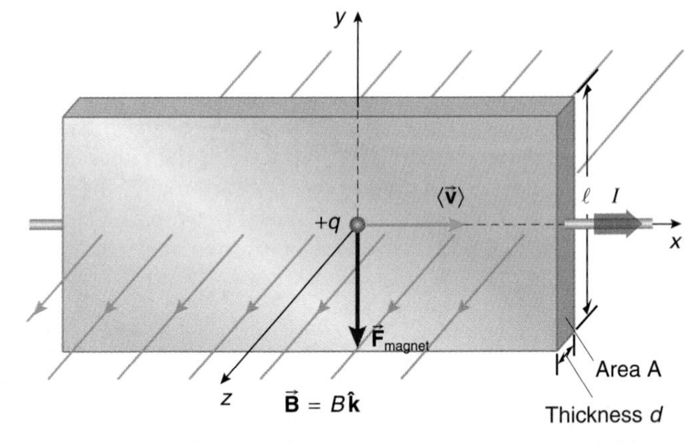

FIGURE 20.12 **Positive charge carriers moving in a material in a magnetic field.**

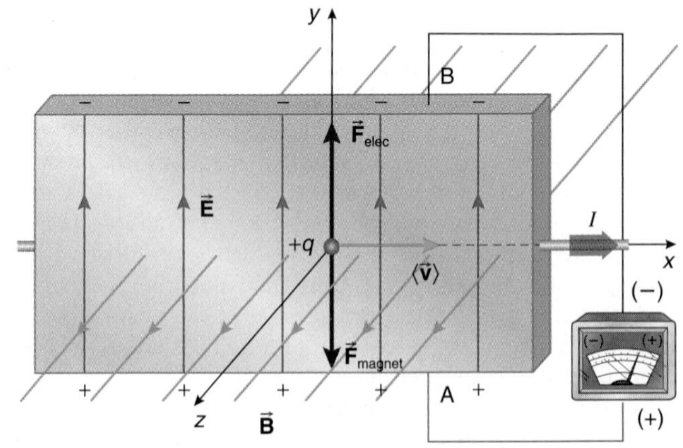

FIGURE 20.13 **Charge separation produces an electric field transverse to the current.**

As the charge accumulates, this electric field increases in magnitude and produces an electrical force on the positive charge carriers as well:

$$\vec{\mathbf{F}}_{\text{elec}} = qE\hat{\jmath}$$

The charge accumulation ceases when the magnitude of the opposing electrical force has the same magnitude as that of the magnetic force on each charge carrier. The direction of the electric field (toward $+\hat{\jmath}$) indicates that the lower edge of the bar is at a higher electric potential than the upper edge. If the potential difference between the edges of the bar is measured with a voltmeter (as in Figure 20.13) with the (+) lead of the voltmeter on the lower edge and the (−) lead on the upper edge, the voltmeter will indicate a positive potential difference across the bar.

Scenario 2: Negative Charge Carriers

If the current is caused by negative charge carriers $-|q|$, moving in the opposite direction, then the magnetic force on the negative charge carriers is (see Figure 20.14 for the coordinate system)

$$\begin{aligned}\vec{\mathbf{F}}_{\text{magnet}} &= (-|q|)(-\langle v\rangle\hat{\imath}) \times B\hat{\mathbf{k}} \\ &= -|q|\langle v\rangle B\hat{\jmath}\end{aligned}$$

The magnetic force on the charge carriers *again* is in the downward direction, as in Scenario 1.

Now it is the negative charge carriers that move downward to the lower edge, leaving behind equivalent positive charges near the top edge. The charge separation produces an electric field directed toward $-\hat{\jmath}$, as indicated in Figure 20.15.

The electric field produces an electrical force on the negative charge carriers:

$$\begin{aligned}\vec{\mathbf{F}}_{\text{elec}} &= (-|q|)(-E\hat{\jmath}) \\ &= +|q|E\hat{\jmath}\end{aligned}$$

The charge separation continues and the electric field (caused by the separated charges) grows in magnitude until the electrical force is equal in magnitude to the oppositely directed magnetic force on each charge carrier. The direction of the electric field indicates that the upper edge of the material is at a higher electric potential than the lower edge; this is the *reverse* of Scenario 1. If a voltmeter is connected (as shown in Figure 20.15) to the edges of the bar the same way as in Scenario 1, the voltmeter now will indicate a *negative* potential difference.

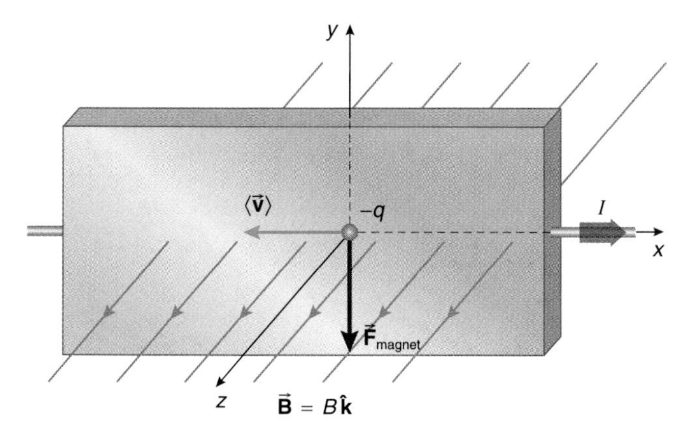

FIGURE 20.14 **Negative charge carriers moving in the opposite direction also can account for the same current.**

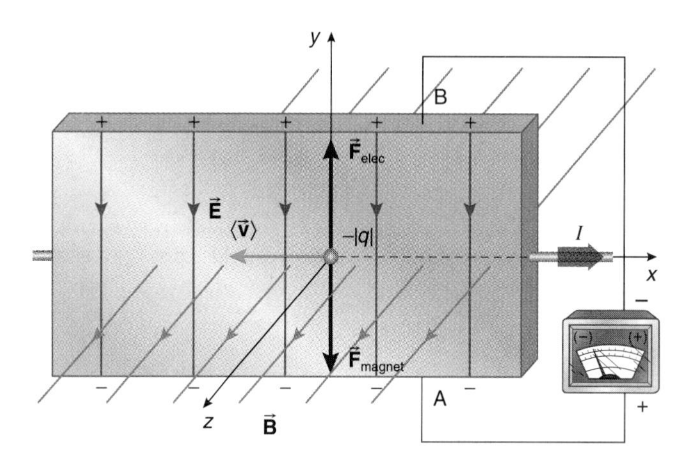

FIGURE 20.15 **The electric field produced by charge separation if the charge carriers are negative.**

Therefore, if the voltmeter when connected as in Figure 20.13 or 20.15 indicates a positive potential difference between the edges, the charge carriers are positive; if the voltmeter indicates a negative potential difference, the charge carriers are negative.* This is a neat, crucial experiment that says yes or no to an idea (the sign of the charge carriers) even before numbers are obtained.

The results of Hall effect experiments using ordinary metallic conductors indicate the charge carriers in them are negative (they are electrons). Nowadays, we also use semiconducting materials that are contaminated with impurity elements (a process called doping the pure semiconducting material) to create semiconductors in which the dominant charge carriers are either negative (n-type semiconducting material) or positive (p-type semiconducting materials). The Hall effect in doped semiconductors is used to distinguish experimentally between n-type and p-type materials.

An expression for the potential difference between the edges of the bar in the Hall effect can be obtained by equating the magnitudes of the electric and magnetic forces on a charge carrier:

$$\begin{aligned}|q|E &= |q|\langle v\rangle B \\ E &= \langle v\rangle B \qquad (20.5)\end{aligned}$$

The potential difference between the edges is found from Equation 17.9. Using Scenario 1,

$$V_{\text{B}} - V_{\text{A}} = -\int_{\text{A}}^{\text{B}} \vec{\mathbf{E}} \cdot d\vec{\mathbf{r}}$$

*Of course, if the voltmeter is connected to the edges the other way around, a negative potential difference indicates positive carriers and a positive potential difference indicates negative charge carriers. You have to be careful to connect the voltmeter appropriately with respect to the geometry!

where we integrate between the edges (see Figure 20.13). Performing the integration,

$$\begin{aligned} V_B - V_A &= -\int_A^B E\hat{\jmath} \cdot dy\,\hat{\jmath} \\ &= -E\ell \end{aligned}$$

where ℓ is the separation of the edges of the bar. Let $V_{Hall} \equiv V_A - V_B$ be the **Hall voltage**. In this way V_{Hall} is the value the voltmeter reads in Figures 20.13 and 20.15. Thus

$$V_{Hall} = E\ell$$

and Equation 20.5 becomes

$$\frac{V_{Hall}}{\ell} = \langle v \rangle B$$

$$V_{Hall} = \langle v \rangle \ell B \tag{20.6}$$

This can be put in a more useful form by eliminating the drift speed via its relation to the current (Equation 19.3):

$$I = nqA\langle v \rangle$$

where n is the number of charge carriers per unit volume and A is the cross-sectional area of the wire (here the bar). Solving Equation 19.3 for $\langle v \rangle$,

$$\langle v \rangle = \frac{I}{nqA}$$

Substituting for $\langle v \rangle$ into Equation 20.6, we find the Hall voltage is

$$V_{Hall} = \frac{I\ell B}{nqA} \tag{20.7}$$

Since the cross-sectional area A of the bar is the product of ℓ and the thickness d, the Hall voltage can be written as

$$V_{Hall} = \frac{IB}{nqd} \tag{20.8}$$

Note that the Hall voltage is measured across the width ℓ; d is the *other* cross-sectional dimension of the bar.

Semiconducting materials can be fabricated with precisely controlled values for the charge carrier density n. Indeed Equation 20.8 is used to measure n using a known magnetic field. Conversely, once n is determined we can turn the problem around. Let a known current pass through an arrangement like Figure 20.13, now called a **Hall probe**. Currents, of course, are quite easily measured with a sensitive ammeter. Knowing n, q, d, and I means that a measurement of the Hall voltage V_{Hall} permits a determination of the magnitude of the magnetic field. Such Hall probes are one way in which the magnitude of a magnetic field is measured with precision in the laboratory.

The Hall effect still is a topic of current interest and research in physics. Unexpected surprises lurk in many corners of the sciences. In 1980 it was discovered that in very thin sheets of material, the Hall voltage exhibits a steplike behavior as the magnetic field is increased. But Equation 20.7 predicts a smooth, linear relationship between V_{Hall} and B_0. The steplike behavior was quite surprising and unanticipated. The steps in the Hall voltage are a rare example of a macroscopic manifestation of quantum mechanical effects. The so-called *quantum Hall effect* won its discoverer (Klaus von Klitzing) the 1985 Nobel prize in physics. The quantized Hall effect is used to measure very accurately an important constant of physics known as the fine structure constant; but this is getting a bit over our heads for an introductory course. The point is that the Hall effect certainly is not a dead subject, even though it was discovered well over a century ago.

EXAMPLE 20.3

A copper Hall probe of thickness 125 μm is placed appropriately in a magnetic field (as in Figure 20.13). A 25.0 A current is in the strip. A Hall voltage of − 11 μV is measured across the 2.0 cm width of the strip. What is the magnitude of the magnetic field?

Solution

In Example 19.1 on page 841, you found the number of charge carriers per cubic meter for copper is

$$n = 8.49 \times 10^{28} \text{ electrons/m}^3$$

The charge on each carrier is $q = -e$, since they are electrons. Solving Equation 20.8 for the magnitude of the magnetic field, you find

$$\begin{aligned} B &= \frac{nqdV_{Hall}}{I} \\ &= [(8.49 \times 10^{28} \text{ electrons/m}^3)(-1.602 \times 10^{-19} \text{ C/electron}) \\ &\quad \times (125 \times 10^{-6} \text{ m})(-11 \times 10^{-6} \text{ V})] \,/\, 25.0 \text{ A} \\ &= 0.75 \text{ T} \end{aligned}$$

This is a strong magnetic field and illustrates how large the tesla unit for the magnetic field really is. For comparison, the magnetic field of the Earth is only about 10^{-4} T.

20.3 MAGNETIC FORCES ON CURRENTS

Electric currents are charges in motion. If the current is in a wire located in a region where there is a magnetic field, the wire will experience a force that is merely a manifestation of the basic magnetic interaction on the moving charged particles within it.

Imagine a wire segment of length $d\ell$ and cross-sectional area A carrying a current I in the direction indicated in Figure 20.16. The magnetic field at the location of the segment of wire $d\ell$ is $\vec{\mathbf{B}}$. Let n be the number of charge carriers per unit volume, $\langle \vec{\mathbf{v}} \rangle$ the

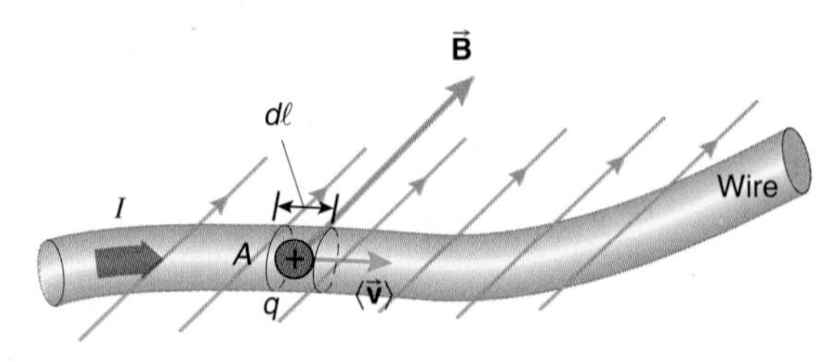

FIGURE 20.16 A current-carrying wire in a magnetic field.

average drift velocity of the charge carriers, and q the charge on each; the current is (from Equation 19.3)

$$I = nqA\langle v\rangle$$

The average magnetic force on a charge carrier in the wire segment is given by

$$q\langle\vec{\mathbf{v}}\rangle \times \vec{\mathbf{B}}$$

where $\vec{\mathbf{B}}$ is the value of the magnetic field at the position of the charge q in the segment $d\ell$.

The number of charge carriers in the length $d\ell$ is the product of the number of charge carriers per unit volume (n) and the volume of the little segment ($A\,d\ell$). Thus the total differential magnetic force $d\vec{\mathbf{F}}_{\text{magnet}}$ on the segment of wire is

$$d\vec{\mathbf{F}}_{\text{magnet}} = nA\,d\ell\, q\langle\vec{\mathbf{v}}\rangle \times \vec{\mathbf{B}} \qquad (20.9)$$

Now we switch vector horses: instead of using the vector drift velocity $\langle\vec{\mathbf{v}}\rangle$, we use the magnitude $\langle v\rangle$ of the drift velocity and account for the direction via the length segment $d\ell$. That is, let a vector $d\vec{\ell}$ point in the direction of the drift velocity; this is the same direction as the conventional current (positive charge flow). With this change, we rewrite Equation 20.9 in the equivalent form

$$d\vec{\mathbf{F}}_{\text{magnet}} = nqA\langle v\rangle\, d\vec{\ell} \times \vec{\mathbf{B}}$$

In this way we can identify $nqA\langle v\rangle$ with the current I (Equation 19.3). Hence the differential force on this segment of the wire is

$$d\vec{\mathbf{F}}_{\text{magnet}} = I\, d\vec{\ell} \times \vec{\mathbf{B}}$$

To find the total force on the whole wire, we simply integrate over the length of the wire:

$$\vec{\mathbf{F}}_{\text{magnet}} = \int_{\text{wire}} I\, d\vec{\ell} \times \vec{\mathbf{B}}$$

(well, maybe not so simply since this is a vector integration!). Since the current in a wire is constant over its length (charge conservation), I can be brought outside the integral.

The magnetic force on a current-carrying wire is

$$\vec{\mathbf{F}}_{\text{magnet}} = I\int_{\text{wire}} d\vec{\ell} \times \vec{\mathbf{B}} \qquad (20.10)$$

If the wire is straight and the field is constant along the length of the wire segment, as in Figure 20.17, then the constant field can be factored out from the integral:

$$\vec{\mathbf{F}}_{\text{magnet}} = \left(I\int_{\text{wire}} d\vec{\ell}\right) \times \vec{\mathbf{B}}$$

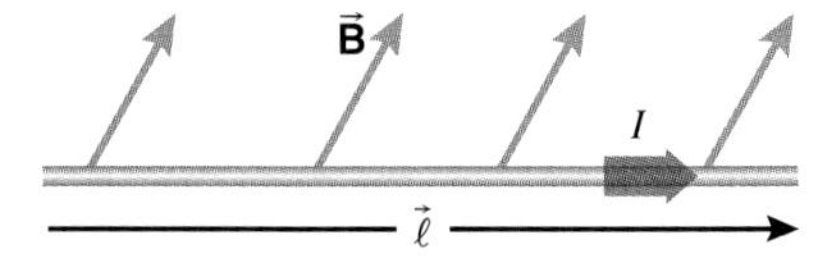

FIGURE 20.17 **A straight wire in a magnetic field that is constant along its length.**

The integration over the straight line segment is the directed length of that segment. Hence the magnetic force on the wire is

$$\vec{\mathbf{F}}_{\text{magnet}} = I\vec{\ell} \times \vec{\mathbf{B}} \qquad \text{(straight wire segment of length } \ell, \ \vec{\mathbf{B}} \text{ constant along its length)} \qquad (20.11)$$

EXAMPLE 20.4

A wire of total length ℓ_0 carrying a current I passes through a region permeated by a uniform magnetic field of magnitude B, as shown in Figure 20.18. Only a straight portion of wire, of length ℓ, finds itself in the field.

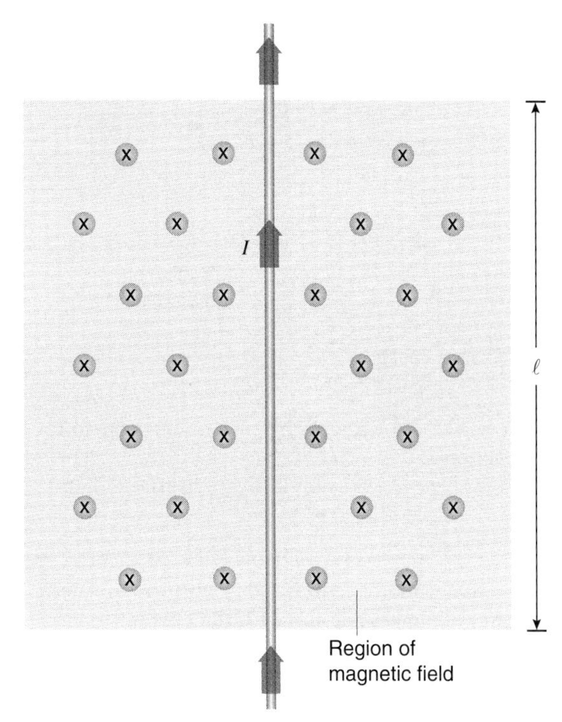

FIGURE 20.18

a. Calculate the force on the wire.
b. What happens if the direction of the current is reversed?

Solution

a. You need to consider only the portion of the wire in the region of the field. The portion of the wire not in the field has zero magnetic force acting on it. Since the wire is straight and the field is uniform over the portion of the wire in the field, you can use Equation 20.11,

$$\vec{\mathbf{F}}_{\text{magnet}} = I\vec{\ell} \times \vec{\mathbf{B}}$$

The angle between $\vec{\ell}$ and $\vec{\mathbf{B}}$ is 90°; hence the magnitude of the magnetic force on the wire is

$$F_{\text{magnet}} = I\ell B \sin 90° = I\ell B$$

The direction of the vector product is the direction of the magnetic force, found by the vector product right-hand rule. The force is shown in Figure 20.19.

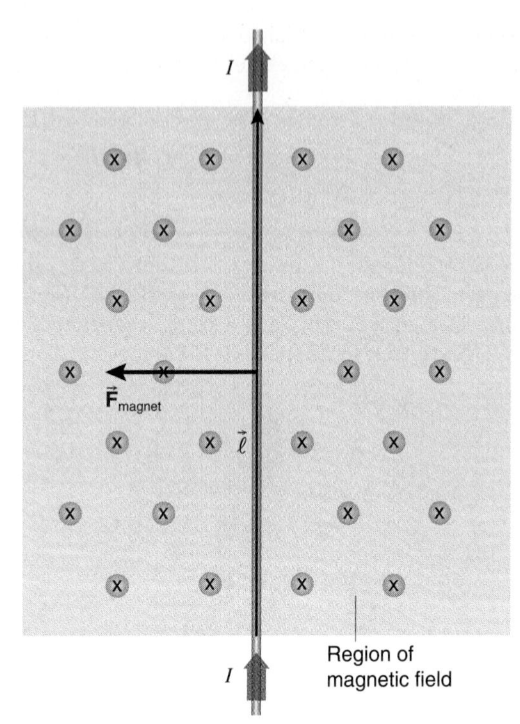

FIGURE 20.19

b. Reversing the direction of the current amounts to reversing the direction of the vector $d\vec{\ell}$; the effect of this is to reverse the direction of the force—that is, $\vec{F}_{magnet}$ is to the right.

EXAMPLE 20.5

A closed current loop is entirely in a uniform magnetic field. What is the total magnetic force on the current loop?

Solution

The force on the loop is found from Equation 20.10:

$$\vec{F}_{magnet} = I \int_{wire} d\vec{\ell} \times \vec{B}$$

Since the magnetic field is constant over the entire loop, you can factor it out from the integral:

$$\vec{F}_{magnet} = I\left(\int_{wire} d\vec{\ell}\right) \times \vec{B}$$

For a closed loop, the integral of $d\vec{\ell}$ around the whole loop is zero since for every $d\vec{\ell}$ pointing in one direction, there is another section of the loop with $d\vec{\ell}$ pointing in the opposite direction and each pair vector sums to zero. Thus the total force on a closed current loop in a uniform field is zero.

20.4 WORK DONE BY MAGNETIC FORCES

A magnetic field that does not vary with time is called a *static* magnetic field. The force it exerts on moving charges is a static magnetic force.

The magnetic force on a charge q we found to be (Equation 20.1)

$$\vec{F}_{magnet} = q\vec{v} \times \vec{B}$$

The differential work dW_{magnet} done by this force on the charge when the particle undergoes a differential change $d\vec{r}$ in its position vector is, from the definition of work, Equation 8.1;

$$\begin{aligned} dW_{magnet} &= \vec{F}_{magnet} \cdot d\vec{r} \\ &= (q\vec{v} \times \vec{B}) \cdot d\vec{r} \end{aligned}$$

Recall that the velocity $\vec{v}$ is the time rate of change of the position vector:

$$\vec{v} = \frac{d\vec{r}}{dt}$$

or

$$d\vec{r} = \vec{v}\,dt$$

This means the differential change in the position vector is parallel to the velocity vector. The differential work done by the magnetic force is, therefore,

$$dW_{magnet} = (q\vec{v} \times \vec{B}) \cdot \vec{v}\,dt$$

Since the magnetic force is $q\vec{v} \times \vec{B}$, the force always is perpendicular to the velocity vector of the charge. Hence the *scalar product* of the magnetic force with $\vec{v}$ is always *zero*.

Therefore the differential work done by the static magnetic force always is zero:

$$dW_{magnet} = 0 \text{ J}$$

and, therefore, W_{magnet} also is zero.

PROBLEM-SOLVING TACTIC

20.2 The static magnetic force does no work on a moving charge and, by the CWE theorem, cannot by itself change the kinetic energy of the charge.

EXAMPLE 20.6

An electron moving in and perpendicular to a uniform, static magnetic field executes a circular orbit with a radius equal to the cyclotron radius R (Equation 20.4). What is the work done by the magnetic force on the electron as it executes one orbital circumference?

Solution

The electron has a force acting on it and moves through a distance equal to the circumference. However, the magnetic force is at all times perpendicular to the velocity of the electron. The work done is zero. You should *not* be fooled into thinking the work done by the magnetic force is $F_{magnet}(2\pi R)$! Furthermore, even if the static magnetic field is not uniform and the path is not circular, no work is done on the moving charge by the field.

20.5 Torque on a Current Loop in a Magnetic Field

Electric motors are used in everything from the toy cars and trains we enjoyed as kids (and may still enjoy!) to essential appliances such as refrigerators, washing machines, elevators, and subways. Such motors all are based on a remarkable observation: a current loop in a magnetic field can experience a torque. Couple the current loop appropriately to an axle and we have the beginnings of an elemental motor. Here we see how the torque on a loop of current arises.

Imagine a rectangular current loop of area A and dimensions ℓ_1 and ℓ_2 in a uniform magnetic field $\vec{B}$ that makes an angle θ with a line perpendicular to the plane of the loop, as shown in Figure 20.20. For convenience in describing the orientation of the loop with respect to the field, we introduce an area vector associated with the current loop. The magnitude of the area vector is equal to the area of the loop $A = \ell_1\ell_2$. The direction of the area vector is perpendicular to the plane of the loop in the sense of the following right-hand rule:

Curl the fingers of your right hand around the perimeter of the current loop in the same sense as the current; then the extended thumb of your right hand indicates the direction of the area vector of the loop. The area vector $\vec{A}$ makes an angle θ with the direction of the magnetic field, as indicated in Figure 20.20.

Example 20.5 showed that the *total* magnetic force on *any* current loop in a *uniform* magnetic field is zero, regardless of the orientation of the loop. While the total force on the loop is zero, there may nonetheless be a nonzero torque on the loop. This torque has significant practical applications in electric motor technology. To find the torque, we first need to find the force on each segment, and then the torque of each force about the center.

Since the field is uniform over each segment of the loop, we can use Equation 20.11 to find the force on each of the four straight segments of the loop:

$$\vec{F}_{\text{magnet}} = I\vec{\ell} \times \vec{B}$$

We use the coordinate system in Figure 20.20 (and Figures 20.21–20.25). The magnetic field everywhere is $\vec{B} = B\hat{k}$.

We start with segment (1) in Figure 20.21. The directed current segment $\vec{\ell}_1$ here is

$$\vec{\ell}_1 = \ell_1\hat{j}$$

and is perpendicular to the magnetic field, as shown in Figure 20.21. The magnetic force on this segment is

$$\begin{aligned}\vec{F}_1 &= I\vec{\ell}_1 \times \vec{B}\\ &= I\ell_1\hat{j} \times (B\hat{k})\\ &= I\ell_1B\hat{i}\end{aligned}$$

which is also shown in Figure 20.21.

For the corresponding segment (3) on the opposite side of the loop (see Figure 20.22), the directed current segment is $\vec{\ell}_3 = -\ell_1\hat{j}$, and so the magnetic force is

$$\begin{aligned}\vec{F}_3 &= I\vec{\ell}_3 \times \vec{B}\\ &= I(-\ell_1\hat{j}) \times (B\hat{k})\\ &= -I\ell_1B\hat{i}\end{aligned}$$

The directed wire segment (2) in Figure 20.23 is a bit more complicated. From the geometry of Figure 20.23, we can write

$$\vec{\ell}_2 = \ell_2[(-\cos\theta)\hat{i} - (\sin\theta)\hat{k}]$$

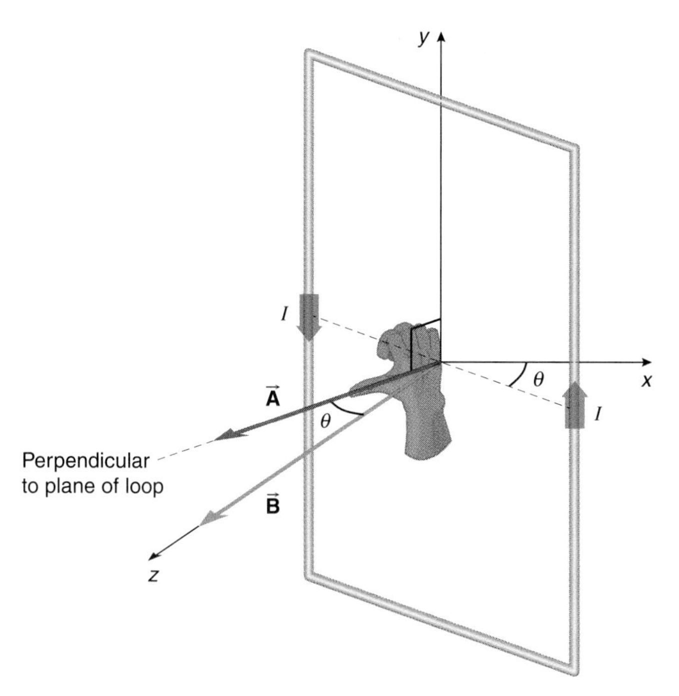

FIGURE 20.20 A current loop with area vector $\vec{A}$ in a magnetic field $\vec{B}$.

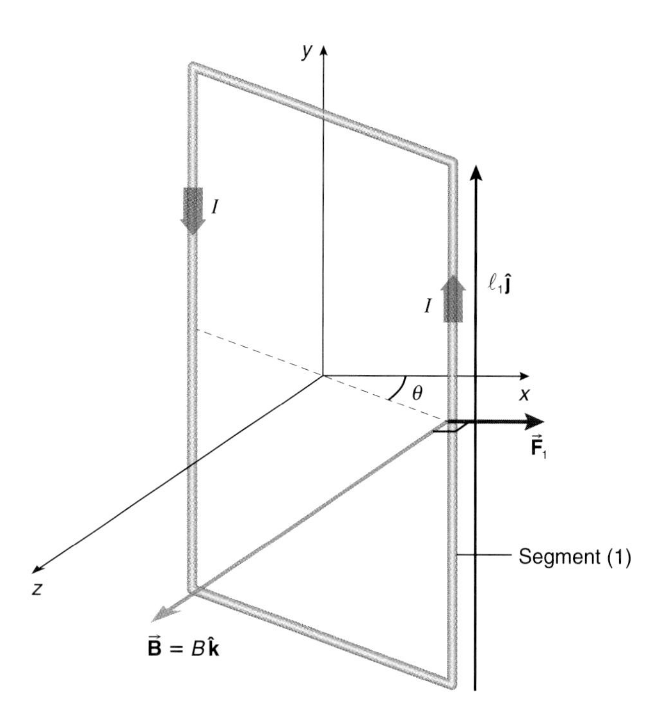

FIGURE 20.21 The force on segment (1).

Thus the magnetic force on this segment is

$$\begin{aligned}\vec{\mathbf{F}}_2 &= I\vec{\ell}_2 \times \vec{\mathbf{B}} \\ &= I\ell_2[(-\cos\theta)\hat{\mathbf{i}} - (\sin\theta)\hat{\mathbf{k}}] \times B\hat{\mathbf{k}} \\ &= (I\ell_2 B\cos\theta)\hat{\mathbf{j}}\end{aligned}$$

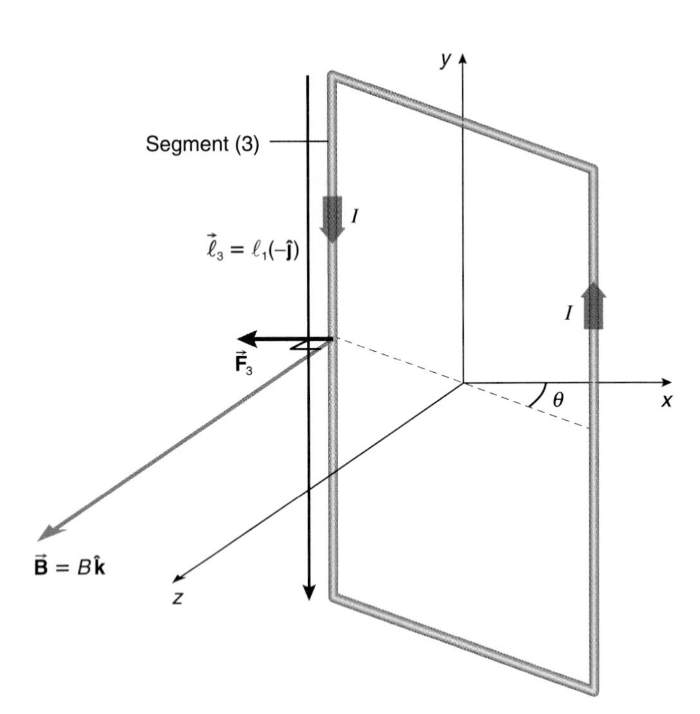

FIGURE 20.22 The force on segment (3).

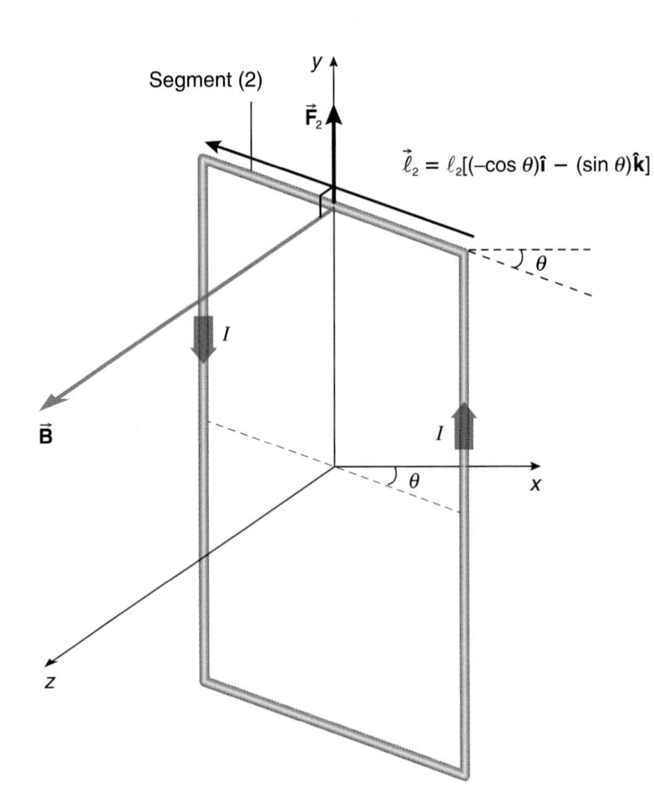

FIGURE 20.23 The force on segment (2).

The current in segment (4) in Figure 20.24 is directed opposite to the current in segment (2). Thus the directed wire segment is (see Figure 20.24)

$$\vec{\ell}_4 = \ell_2[(\cos\theta)\hat{\mathbf{i}} + (\sin\theta)\hat{\mathbf{k}}]$$

The force on this segment is

$$\begin{aligned}\vec{\mathbf{F}}_4 &= I\vec{\ell}_4 \times \vec{\mathbf{B}} \\ &= I\ell_2[(\cos\theta)\hat{\mathbf{i}} + (\sin\theta)\hat{\mathbf{k}}] \times B\hat{\mathbf{k}} \\ &= (-I\ell_2 B\cos\theta)\hat{\mathbf{j}}\end{aligned}$$

We note incidentally that the total force on the current loop is the vector sum of the forces on each segment. We find

$$\vec{\mathbf{F}}_1 + \vec{\mathbf{F}}_2 + \vec{\mathbf{F}}_3 + \vec{\mathbf{F}}_4 = 0\ \text{N}$$

as we knew from Example 20.5.

Now we can find the torque about the center of the loop. The points of applications of the forces we can take to be the centers of each segment since the field is uniform along each length. To find the torque of each force, we write the position vectors from the center of the loop to the points of application of the forces (see Figure 20.25).

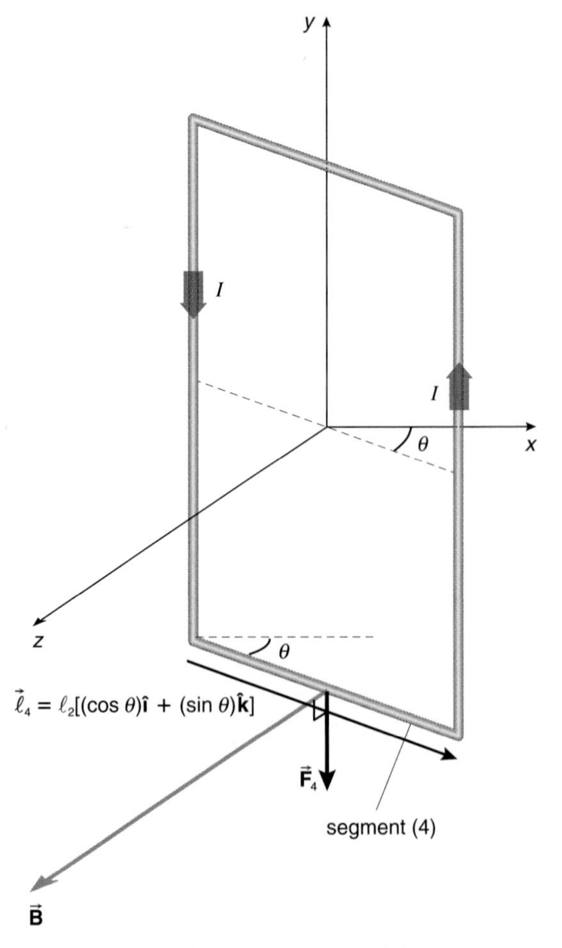

FIGURE 20.24 The force on segment (4).

Note in Figure 20.25 that the lines of action of the forces $\vec{F}_2$ and $\vec{F}_4$ pass through the point about which we are taking the torque, the center of the loop. These forces thus contribute zero torque. Only $\vec{F}_1$ and $\vec{F}_3$ produce a nonzero torque, and so we only need expressions for $\vec{r}_1$ and $\vec{r}_3$. These position vectors are

$$\vec{r}_1 = \frac{\ell_2}{2}\left[(\cos\theta)\,\hat{\imath} + (\sin\theta)\,\hat{k}\right]$$

$$\vec{r}_3 = \frac{\ell_2}{2}\left[(-\cos\theta)\,\hat{\imath} - (\sin\theta)\,\hat{k}\right]$$

The torques of the respective forces are found from the general equation for the torque of a force $\vec{\tau} = \vec{r} \times \vec{F}$:

$$\vec{r}_1 \times \vec{F}_1 = \frac{\ell_2}{2}\left[(\cos\theta)\,\hat{\imath} + (\sin\theta)\,\hat{k}\right] \times I\ell_1 B\,\hat{\imath} = \left(\frac{I}{2}\ell_1\ell_2 B\sin\theta\right)\hat{\jmath}$$

$$\vec{r}_3 \times \vec{F}_3 = \frac{\ell_2}{2}\left[(-\cos\theta)\,\hat{\imath} - (\sin\theta)\,\hat{k}\right] \times (-I\ell_1 B)\hat{\imath} = \left(\frac{I}{2}\ell_1\ell_2 B\sin\theta\right)\hat{\jmath}$$

The total torque on the entire current loop is the vector sum of the torques of $\vec{F}_1$ and $\vec{F}_3$; we find

$$\vec{\tau} = (I\ell_1\ell_2 B\sin\theta)\hat{\jmath}$$

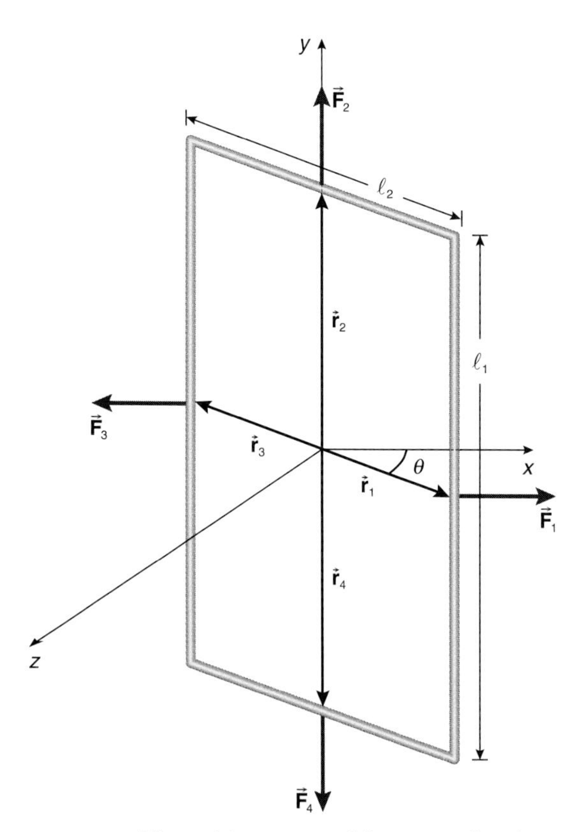

FIGURE 20.25 The position vectors of the points of application of the forces on the four segments.

Since $\ell_1\ell_2$ is the area A of the loop, we can express the torque as

$$\vec{\tau} = (IAB\sin\theta)\hat{\jmath} \tag{20.12}$$

Whew! A lot of work, but we finally did it. The total torque is along $\hat{\jmath}$; so, if the loop is free to turn, it will begin to spin about the $\hat{\jmath}$ direction. If you align the thumb of your right hand in the direction of the torque, the right-hand fingers indicate the sense that the loop will spin.

Notice from Figure 20.20 that the area vector $\vec{A}$ can be written as

$$\vec{A} = (-A\sin\theta)\hat{\imath} + (A\cos\theta)\hat{k}$$

The magnetic field is $B\hat{k}$. Hence the vector product of $\vec{A}$ with $\vec{B}$ is

$$\vec{A} \times \vec{B} = [(-A\sin\theta)\hat{\imath} + (A\cos\theta)\hat{k}] \times B\hat{k} = (AB\sin\theta)\hat{\jmath}$$

Compare this result with Equation 20.12 for the torque. We see that the torque on the current loop can be written elegantly as

$$\vec{\tau} = I\vec{A} \times \vec{B} \tag{20.13}$$

It is customary to lump the current I with the area vector $\vec{A}$ to form a new vector associated with the current loop. The new vector is defined to be

$$\vec{\mu} \equiv I\vec{A} \tag{20.14}$$

and is called the **magnetic dipole moment** of the current loop.

This may seem like strange terminology right now, but this is only because we have not yet explored how magnetic fields arise in the first place. We get to that in due course.

Hence we write the expression for the torque on the current loop in its final form:

$$\vec{\tau} = \vec{\mu} \times \vec{B} \tag{20.15}$$

When the magnetic dipole moment is parallel or antiparallel to $\vec{B}$, the vector product is zero and we have zero torque. When the magnetic dipole moment is perpendicular to $\vec{B}$, the torque has maximum magnitude.

Equation 20.15 looks surprisingly like the expression we found (Equation 16.14 on page 730) for the torque of an electric dipole moment $\vec{p}$ in an electric field $\vec{E}$:

$$\vec{\tau} = \vec{p} \times \vec{E}$$

We also discovered (Equation 17.26 on page 787) that an electric dipole moment in an electric field has a potential energy given by

$$PE = -\vec{p} \cdot \vec{E}$$

Analogously, the potential energy of a magnetic dipole moment in a magnetic field is*

$$PE = -\vec{\mu} \cdot \vec{B} \tag{20.16}$$

*We make the analogy without proof for the sake of brevity. A formal derivation is not easy to find in recent books on electromagnetism; but one can be found in William Taussig Scott, *The Physics of Electricity and Magnetism* (Wiley, New York, 1959), pages 297 and 335.

A current loop (magnetic dipole moment) turns under the action of the torque in order to lower its potential energy. The potential energy is minimized when the magnetic dipole moment vector is parallel to the direction of the magnetic field; the potential energy is maximized with the magnetic dipole moment antiparallel to the field.

Although we derived Equations 20.15 and 20.16 using a rectangular loop, they are valid for planar current loops of *any* shape. The magnetic dipole moment of the loop is the product of the current in the loop and its area vector, directed in the sense of the right-hand rule described at the beginning of this section.

If the idea of a current loop in a magnetic field seems rather esoteric, or far-fetched, recall that the torque on such a current loop is the basic physical principle governing the operation of electric motors, as we mentioned at the beginning of this section.* The effect has enormous technological applications. There also are applications at the atomic level. An electron orbiting the nucleus of an atom can be viewed as a tiny current loop (with the direction of the current opposite to the direction of motion of the electron, because it is a negative charge that is moving). Thus an electron in an atom creates an orbital magnetic dipole moment. The electron and other fundamental particles such as the proton and neutron also have intrinsic magnetic dipole moments associated with their spin. When a collection of such magnetic dipole moments is placed in an external magnetic field, the magnetic dipole moments align with the direction of the field to minimize their potential energies. In this way submicroscopic systems with magnetic dipole moments can be aligned. Many medical imaging technologies exploit this behavior of atomic and nuclear magnetic dipole moments.

PROBLEM-SOLVING TACTIC

20.3 If a wire carrying a current is wrapped into a coil with n turns, each with the same area A, the magnetic dipole moment of the coil is n times that of a single turn. That is, the magnetic moment of a coil is

$$\vec{\mu} = nI\vec{A} \qquad (20.17)$$

The area A can be circular, rectangular, or whatever. The common direction of $\vec{\mu}$ and $\vec{A}$ is determined from the same right-hand rule: wrap the fingers of your right hand around the coil in the sense of the (positive) current, and the extended right-hand thumb indicates the direction of both $\vec{\mu}$ and $\vec{A}$. Motors have coils with many turns of wire to increase the magnetic moment and therefore the torque on the coil if it carries a current in a magnetic field.

EXAMPLE 20.7

A current of 15.0 A is in a circular coil of radius 5.00 cm with 50 turns of wire (see Figure 20.26). The plane of the coil makes an angle of 30.0° with a uniform magnetic field of magnitude 0.150 T.

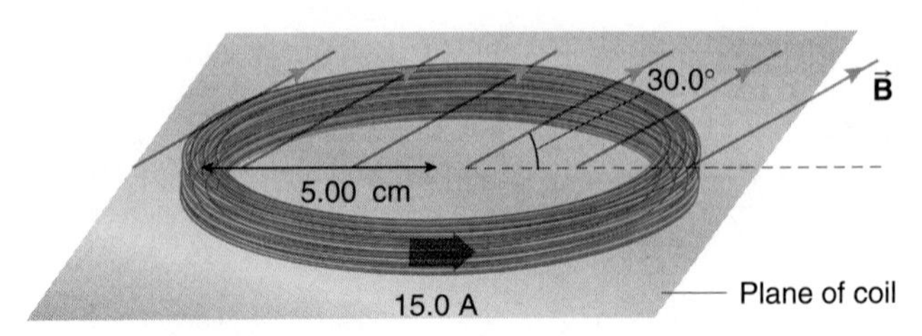

FIGURE 20.26

a. Find the magnitude of the magnetic dipole moment of the coil.
b. Determine the magnitude of the torque on the coil. Indicate in a sketch the direction that the coil will spin if it is free to move.
c. Determine the potential energy of the coil in its given orientation.

Solution

a. According to Problem-Solving Tactic 20.3, the magnitude of the magnetic dipole moment of the coil is

$$\begin{aligned}\mu &= nIA \\ &= (50)(15.0\ \text{A})[\pi(5.00 \times 10^{-2}\ \text{m})^2] \\ &= 5.89\ \text{A}\cdot\text{m}^2\end{aligned}$$

b. The magnitude of the torque is found from Equation 20.15:

$$\tau = \mu B \sin\theta$$

where θ is the angle between the magnetic dipole moment and the magnetic field. The direction of $\vec{\mu}$ is determined from the right-hand rule; it is shown in Figure 20.27. The angle is between $\vec{\mu}$ and $\vec{B}$ is seen to be 60.0°. Hence the torque has magnitude

$$\begin{aligned}\tau &= (5.89\ \text{A}\cdot\text{m}^2)(0.150\ \text{T})\sin 60.0° \\ &= 0.765\ \text{N}\cdot\text{m}\end{aligned}$$

The direction of $\vec{\tau}$ and the sense the coil will spin are indicated in Figure 20.27.

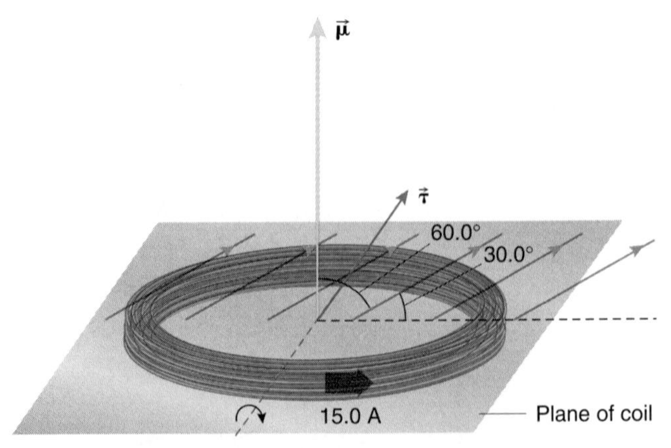

FIGURE 20.27

*One also needs to reverse the current direction in the loop periodically to keep the loop turning; the reversals are accomplished by means of a split ring commutator. See Investigative Project 5 in Chapter 22.

c. The potential energy of the coil in its given configuration is found from Equation 20.16:

$$\begin{aligned} \text{PE} &= -\vec{\mu} \cdot \vec{B} \\ &= -\mu B \cos\theta \\ &= -(5.89\ \text{A}\cdot\text{m}^2)(0.150\ \text{T})\cos 60.0° \\ &= -0.442\ \text{J} \end{aligned}$$

STRATEGIC EXAMPLE 20.8

An electron is in a circular orbit about the nucleus of an atom, as indicated in Figure 20.28 (not to scale!). Find a relationship between the orbital magnetic dipole moment $\vec{\mu}$ and the orbital angular momentum $\vec{L}$ of the electron about the center of its orbit.

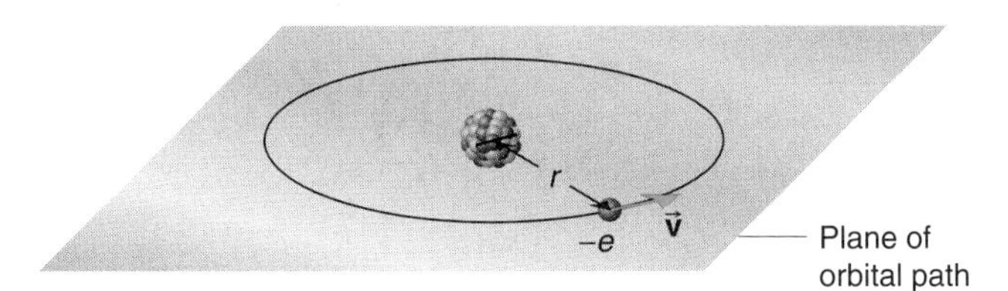

FIGURE 20.28

Solution

Let v be the speed of the electron and r the radius of its orbit. The orbiting electron creates a current opposite to its velocity, because the convention for current is the direction of positive charge motion; see Figure 20.29.

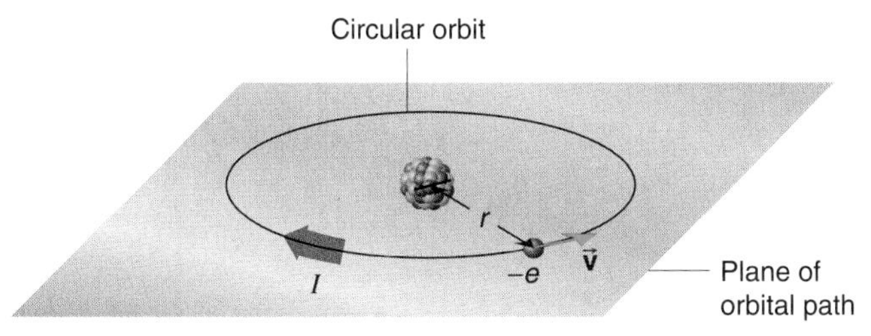

FIGURE 20.29

The current is the number of coulombs that pass any given point along the circumferential path during one second. Hence the current is the fundamental unit of charge e times the frequency ν of the orbital motion (the number of times the electron completes an orbit during one second). That is,

$$I = e\nu$$

The period T of the motion is the inverse of the frequency, so that

$$I = \frac{e}{T} \tag{1}$$

The period is the time to complete one revolution. The electron travels the distance of the circumference $2\pi r$ at the speed v; so the time it takes (the period) is

$$T = \frac{2\pi r}{v}$$

Substituting for T into equation (1) for the current, you find

$$I = \frac{ev}{2\pi r}$$

The magnitude of the magnetic dipole moment of the orbiting electron (a current loop!) is, by definition, the product of the current and the area of the loop (Equation 20.14):

$$\begin{aligned} \mu &= IA \\ &= \frac{ev}{2\pi r}\pi r^2 \\ &= \frac{evr}{2} \end{aligned} \tag{2}$$

The direction of the magnetic dipole moment is perpendicular to the loop in the sense of the right-hand rule; see Figure 20.30.

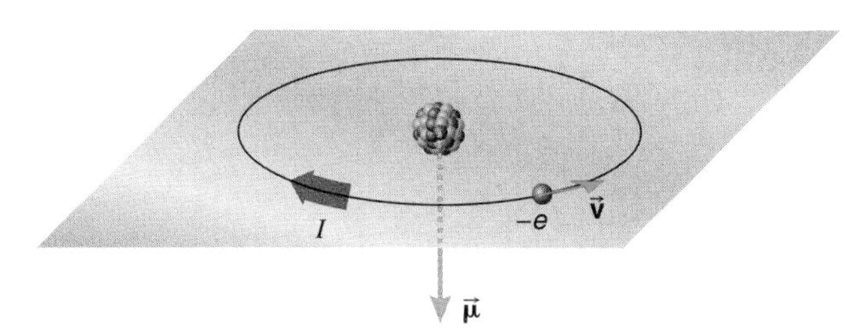

FIGURE 20.30

The orbital angular momentum of the electron is (see Chapter 10)

$$\vec{L}_{\text{orbit}} = \vec{r} \times \vec{p}$$

where $\vec{r}$ is the position vector of the electron and $\vec{p}$ is its momentum; see Figure 20.31. In circular motion $\vec{r}$ is perpendicular to $\vec{p}$, and so the magnitude of the orbital angular momentum is

$$\begin{aligned} L_{\text{orbit}} &= rp\sin 90° \\ &= rp \\ &= mvr \end{aligned} \tag{3}$$

where m is the mass of the electron. The direction of the orbital angular momentum is determined from the vector product right-hand rule.

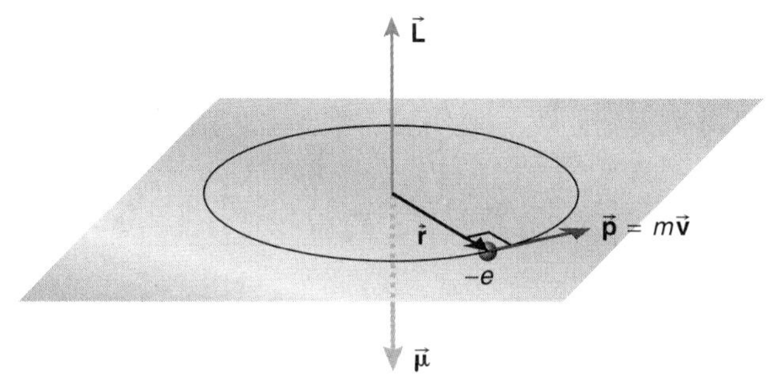

FIGURE 20.31

The orbital magnetic dipole moment and the orbital angular momentum vectors point in opposite directions! Comparing equation (2) with equation (3) shows the two vectors are related by

$$\vec{\mu} = -\frac{e}{2m}\vec{L} \tag{4}$$

20.6 THE BIOT–SAVART LAW

From our study of the gravitation, we saw that a mass m experiences a gravitational force when in a gravitational field $\vec{g}$. The gravitational field itself is caused, in turn, by other masses M. Think of this relationship schematically as

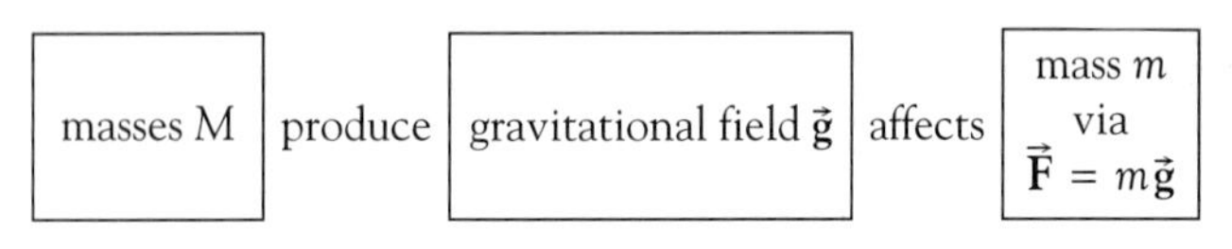

Likewise a charge q in an electric field $\vec{E}$ experiences an electrical force. The electric field itself is caused by other charges Q:

charges Q	produce	electric field $\vec{E}$	affects	charge q via $\vec{F} = q\vec{E}$

We have seen that magnetic fields exert magnetic forces on charges in motion (currents):

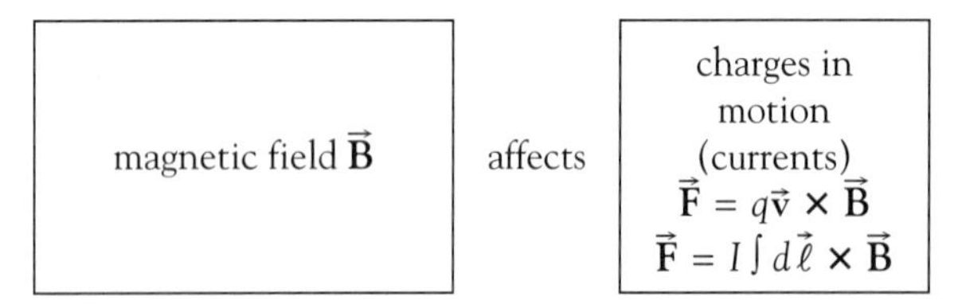

What produces a magnetic field? A glance at the gravitational and electrical cases provides a strong clue. It should come as no surprise that since magnetic fields affect charges in motion (currents), the magnetic field itself is caused by other charges in motion (other currents):

charges in motion: current	produces	magnetic field $\vec{B}$	affects	other charges in motion: (currents)

Here we want to find the relationship between a current (charge in motion) and the magnetic field it produces. Historically, the relationship was not a trivial thing to deduce; it was figured out for the first time during the 19th century (and without the benefit of vector notation!). Indeed much experimentation and head scratching went on before the relationship became apparent.

Part of the reason for this was the nature of the problem. With the gravitational and electrical cases, it was relatively easy to find or calculate the appropriate field of pointlike masses or charges (as well as spherically symmetric mass or charge distributions). If the mass or charge distribution was not one of these geometries, it was more difficult to calculate the gravitational or electrical fields: vector integrations are needed, as we saw in Chapters 6 and 16.

Unfortunately, for the magnetic field there are no macroscopic pointlike currents or spherically symmetric currents; current distributions almost always have messier geometries. To calculate the magnetic field of a current virtually always involves a somewhat intimidating vector integration over the path of the current. Physics, like life, is not always easy.

On a microscopic level, certain particles such as the electron, proton, and neutron (and others) have permanent intrinsic magnetic dipole moments. Just as an electric dipole produces an electric field, a magnetic dipole produces a magnetic field. A magnetic dipole moment has a magnitude that is equal to the current in the current loop times an area, but it is by no means clear what the current or the area is for these particles that create their permanent intrinsic magnetic dipole moment. We simply know the particles have an intrinsic magnetic dipole moment because we can measure it. To the extent that these particles are pointlike particles, the magnetic field their magnetic dipole moments produce can be considered to be caused by very small or pointlike sources of a magnetic field. The fields they produce, however, always are *dipolar* fields: isolated magnetic monopoles never have been detected. These microscopic dipolar elements were, however, unknown during the early 19th century.

By careful observations of the oscillations of a suspended magnet near a long, current-carrying wire, the French scientists Jean-Baptiste Biot (1774–1862) and Félix Savart (1791–1841) experimentally discovered the relationship between a current and the magnetic field it produces in the surrounding space. The relationship now is known as the **Biot–Savart law.**

Consider a small, differential length $d\vec{\ell}$ of a current-carrying wire, as shown in Figure 20.32. The vector $d\vec{\ell}$ is directed in the same sense as the current I. The current element $I\,d\vec{\ell}$ produces a differential bit of magnetic field $d\vec{B}$ at a point P located a distance r away. Let $\hat{r}$ be a unit vector pointing *from* the current element *to* the point P.

Biot and Savart found that the magnetic field at P produced by the current element is expressed (using our modern notation) by

$$d\vec{B} = \frac{\mu_0}{4\pi} I \frac{d\vec{\ell} \times \hat{r}}{r^2} \tag{20.18}$$

where μ_0 is a scalar constant we discuss in more detail later.* Its role is to specify how successful (i.e., effective) the current element is in producing the magnetic field.

To find the magnetic field caused by the entire wire, we integrate over the wire:

$$\vec{B} = \frac{\mu_0}{4\pi} I \int_{\text{wire}} \frac{d\vec{\ell} \times \hat{r}}{r^2} \tag{20.19}$$

This expression may look formidable to you. Actually the law is quite similar to the vector expression for the gravitational field of an extended mass (Equation 6.45 on page 258),

$$\vec{g} = G\int_{\substack{\text{mass}\\\text{distribution}}} \frac{dM}{r^2}\hat{r}$$

*Do not confuse the magnetic dipole moment $\vec{\mu}$, a vector quantity, with the constant μ_0, which is a scalar. The use of the Greek letter μ in two different contexts in magnetism is a regrettable but common notation in each instance. Some texts use $\vec{m}$ for the magnetic dipole moment vector, but the situation is no better since m also is used for mass (a scalar).

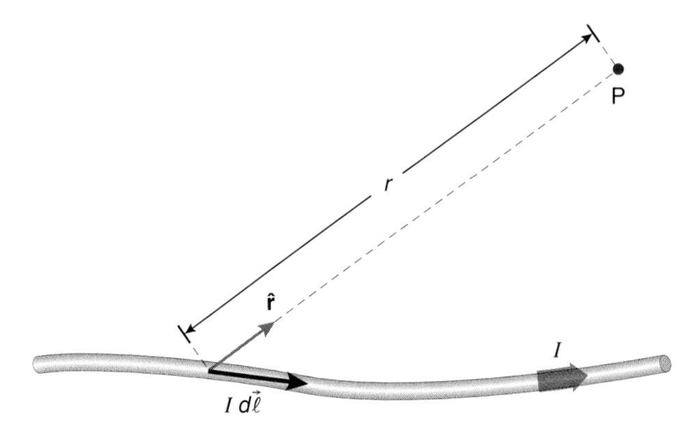

FIGURE 20.32 **A current element $I\,d\vec{\ell}$ produces a differential amount of magnetic field $d\vec{B}$ at P.**

and to the electric field of an extended charge distribution (Equation 16.16 on page 733),

$$\vec{E} = \frac{1}{4\pi\varepsilon_0}\int_{\substack{\text{charge}\\ \text{distribution}}} \frac{dQ}{r^2}\,\hat{r}$$

Specifically the similarities are as follows:

1. Each has a pointlike source term that causes the field:

   gravitational case (mass): dM
   electrical case (charge): dQ
   magnetic case (current element): $I\,d\vec{\ell}$

2. Each expression for the field has a dimensional constant that reflects how effective the source is in producing the field:

   gravitational case: G

   electrical case: $\dfrac{1}{4\pi\varepsilon_0}$

   magnetic case: $\dfrac{\mu_0}{4\pi}$

3. Each is an inverse square law, where r is the distance between the source term and the point where the field is to be calculated.
4. In each case the unit vector $\hat{r}$ is directed *from* the pointlike source of the field *to* the point where the field is to be calculated.
5. Each involves a vector integration over the extended source.

The only additional complication for the magnetic field is the vector product in the numerator of the Biot–Savart law.

The constant in the Biot–Savart law,

$$\frac{\mu_0}{4\pi}$$

reflects what was done with the electrical constant. The factor of 4π is inserted into the constant here so that a factor of 4π will *not* appear in more useful equations later on. The quantity μ_0 has a fancy name: the **permeability of free space**. (Recall that ε_0 was the *permittivity of free space*; this terminology may be a bit confusing to you at first.) The quantity

$$\frac{\mu_0}{4\pi}$$

has the (exact) numerical value

$$\frac{\mu_0}{4\pi} \equiv 10^{-7}\ \text{T}\cdot\text{m/A}$$

with the indicated SI units. These units of μ_0 can be deduced from the Biot–Savart law itself.

The magnetic fields produced by some common current distributions are given in Table 20.1.

EXAMPLE 20.9

A circular wire loop of radius R carries a current I.

a. Use the Biot–Savart law to find the magnetic field at the center of the loop.
b. For a loop of radius 5.0 cm, what current yields a magnetic field of magnitude 1.0×10^{-4} T at the center of the loop (which is about the magnitude of the magnetic field of the Earth).

Solution

a. The Biot–Savart law (Equation 20.19) states that

$$\vec{B} = \frac{\mu_0}{4\pi} I \int_{\text{wire}} \frac{d\vec{\ell} \times \hat{r}}{r^2}$$

For a current element $I\,d\vec{\ell}$ on the loop, the vector $\hat{r}$ points from the current element to the point where you want to calculate the field: the center of the loop; see Figure 20.33. The vector product $d\vec{\ell} \times \hat{r}$ points upward along the axis of the loop. Choose an origin at the center of the loop and let the z-axis be perpendicular to the plane of the loop.

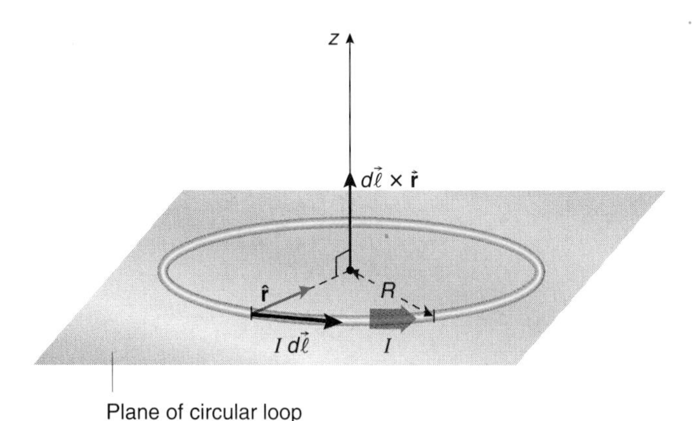

FIGURE 20.33

The angle between $d\vec{\ell}$ and $\hat{r}$ is 90° and is the same for every current element as you go around the loop. Indeed the direction of the vector product also is unchanged as you run

TABLE 20.1 Some Commonly Used Magnetic Fields

(1) At the center of circular loop of radius R (see Example 20.9):

$$\vec{\mathbf{B}}_{\text{center}} = \frac{\mu_0}{4\pi}\frac{2\pi I}{R}\hat{\mathbf{k}}$$

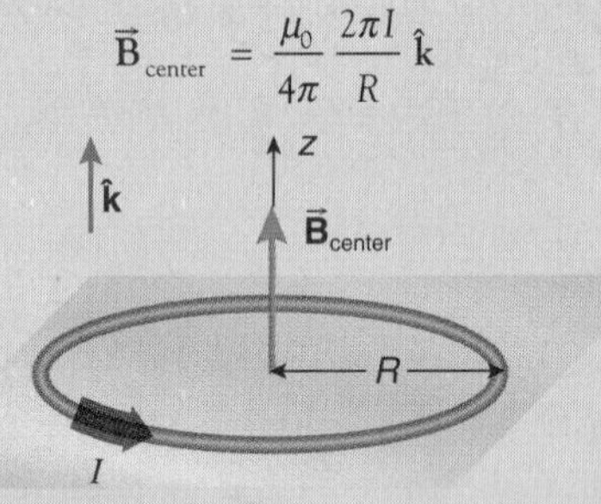

(2) On the axis of a circular current loop (see Example 20.10):

$$\vec{\mathbf{B}}_{\text{axis}} = \frac{\mu_0}{4\pi}\frac{I(2\pi R^2)}{(R^2 + z)^{3/2}}\hat{\mathbf{k}}$$

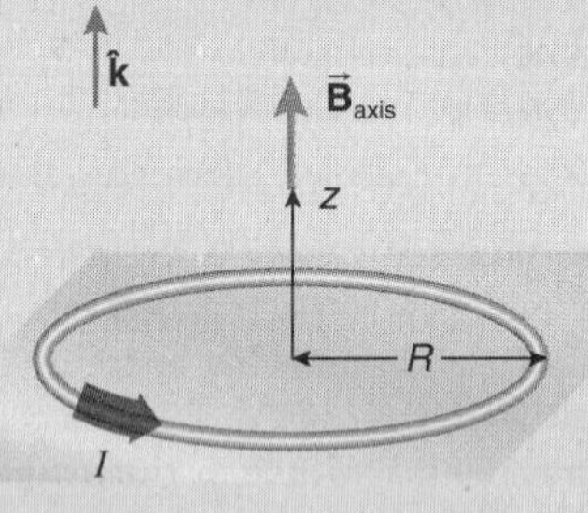

(3) For a circular coil of n loops, all of the same radius, multiply the preceding results by n.

(4) A distance d from an infinite wire (see Example 20.11):

$$B = \frac{\mu_0}{4\pi}\frac{2I}{d}$$

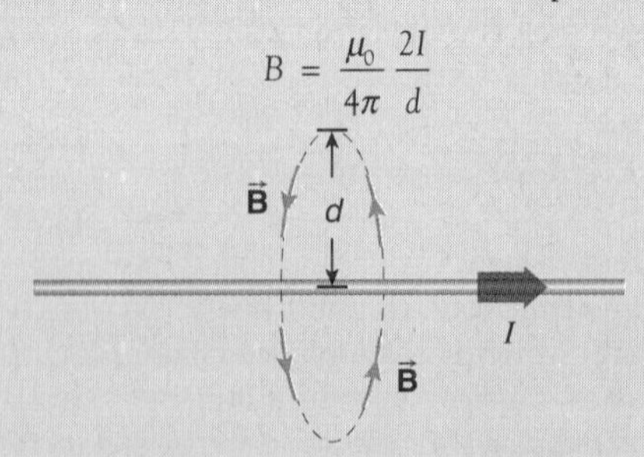

(5) Inside a long solenoid having n turns per meter of its length, each carrying current I, far from its ends (see Example 20.13):

$$B = \mu_0 n I$$

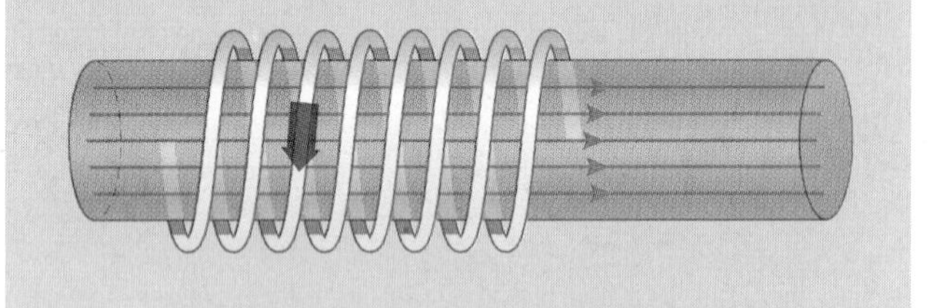

around the loop. The magnetic field, therefore, points along $\hat{\mathbf{k}}$. Hence the Biot–Savart law reduces to

$$\vec{\mathbf{B}}_{\text{center}} = \frac{\mu_0}{4\pi} I \int_{\text{wire}} \frac{d\ell\,(1)\sin 90^\circ}{r^2}\hat{\mathbf{k}}$$

Since the distance r between each current element and the place where you want to find the field is constant and equal to the radius R of the circle, the integration reduces to

$$\vec{\mathbf{B}}_{\text{center}} = \frac{\mu_0}{4\pi}\frac{I}{R^2}\int_{\text{wire}} d\ell\,\hat{\mathbf{k}}$$

The integral of $d\ell$ around the loop is the circumference of the loop: $2\pi R$. Thus the field at the center of the loop is

$$\vec{\mathbf{B}}_{\text{center}} = \frac{\mu_0}{4\pi}\frac{I}{R^2}2\pi R\,\hat{\mathbf{k}}$$
$$= \frac{\mu_0}{4\pi}\frac{2\pi I}{R}\hat{\mathbf{k}} \quad (1)$$

Notice that the direction of the field can be obtained by grasping the wire with your right hand with the thumb in the direction of the current; the curled fingers of your right hand then indicate the way the field surrounds the wire and threads the loop.

The magnetic field at other points in the plane of the loop (whether inside or outside the loop) is much more difficult to calculate.

b. The magnitude of the magnetic field at the center of the loop is

$$B_{\text{center}} = \frac{\mu_0}{4\pi}\frac{2\pi I}{R} \quad (2)$$

Solve this expression for I:

$$I = \frac{4\pi}{\mu_0}\frac{RB_{\text{center}}}{2\pi}$$

Remember to use SI units. Substituting $R = 5.0 \times 10^{-2}$ m and $B_{\text{center}} = 1.0 \times 10^{-4}$ T, you find

$$I = \frac{(5.0 \times 10^{-2}\ \text{m})(1.0 \times 10^{-4}\ \text{T})}{(10^{-7}\ \text{T}\cdot\text{m/A})2\pi}$$
$$= 8.0\ \text{A}$$

STRATEGIC EXAMPLE 20.10

Find the magnetic field at a point P located a distance z from the center along the axis of a circular current loop of radius R and current I; see Figure 20.34.

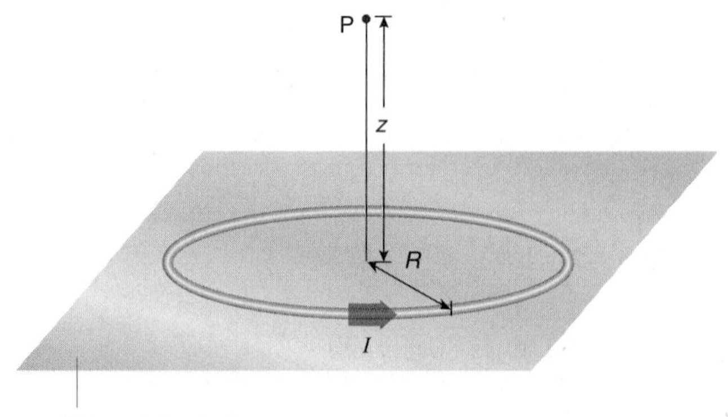

Plane of circular loop

FIGURE 20.34

Solution
Use the Biot–Savart law, Equation 20.19:

$$\vec{B} = \frac{\mu_0}{4\pi} I \int_{\text{wire}} \frac{d\vec{\ell} \times \hat{r}}{r^2}$$

The unit vector $\hat{r}$ points from the current element $I\,d\vec{\ell}$ to the point where you want to compute the field, and r is the distance between the current element and the field point; see Figure 20.35. The angle between $I\,d\vec{\ell}$ and $\hat{r}$ is 90° for every current element along the wire ring. The distance r also is the same for every current element around the ring. This is beginning to seem like Example 20.9 all over again. But one thing is different: the direction of the vector product $d\vec{\ell} \times \hat{r}$ is not along the axis, as shown in Figure 20.35.

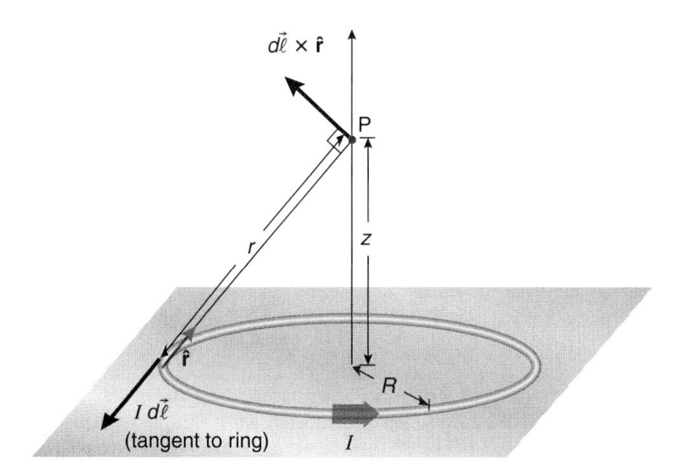

FIGURE 20.35

As you integrate around the loop, the various differential contributions to the magnetic field sweep out a cone-shaped megaphone, pictured in Figure 20.36. The components perpendicular to the axis vector sum pairwise to zero. Only the components along the z-axis survive the integration; the z-component of $d\vec{\ell} \times \hat{r}$ is $|d\vec{\ell} \times \hat{r}| \cos\theta$. Hence the total surviving magnetic field at the point P is

$$\vec{B}_{\text{axis}} = \frac{\mu_0}{4\pi} I \int_{\text{wire}} \frac{d\ell\,(1)\sin 90°}{r^2} \cos\theta\,\hat{k}$$

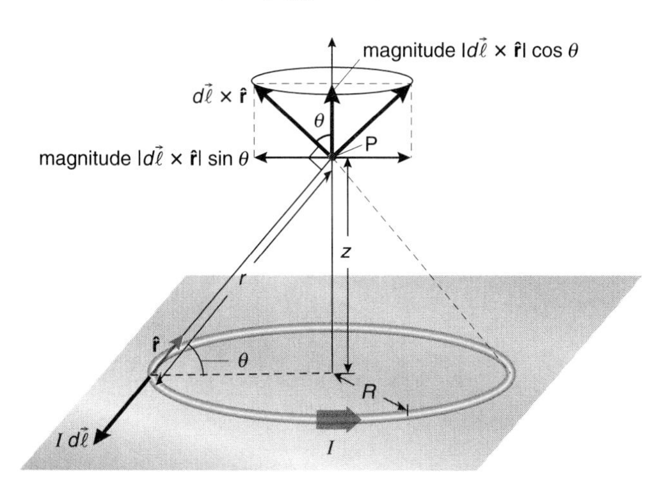

FIGURE 20.36

Since θ and r are the same for every current element around the ring,

$$\vec{B}_{\text{axis}} = \frac{\mu_0}{4\pi} \frac{I}{r^2} \cos\theta \int_{\text{wire}} d\ell\,\hat{k}$$

The integration around the wire loop is the circumference $2\pi R$. From the geometry of Figure 20.36, you see

$$\cos\theta = \frac{R}{r}$$

Hence the magnetic field is

$$\vec{B}_{\text{axis}} = \frac{\mu_0}{4\pi} \frac{I}{r^2} \frac{R}{r} 2\pi R\,\hat{k}$$

The Pythagorean theorem gives

$$r = (R^2 + z^2)^{1/2}$$

and so the field is

$$\vec{B}_{\text{axis}} = \frac{\mu_0}{4\pi} \frac{I\,(2\pi R^2)}{(R^2 + z^2)^{3/2}}\,\hat{k} \qquad (1)$$

Notice that when $z = 0$ m you are at the center of the current loop and the result is the same as equation (1) of Example 20.9. It had better be! Note also that the field is in the same direction for both positive and negative values of z, as shown in Figure 20.37. Grasp the wire with your right hand with the thumb in the direction of the current; your fingers thread the loop in the direction of the field along the axis of the loop.

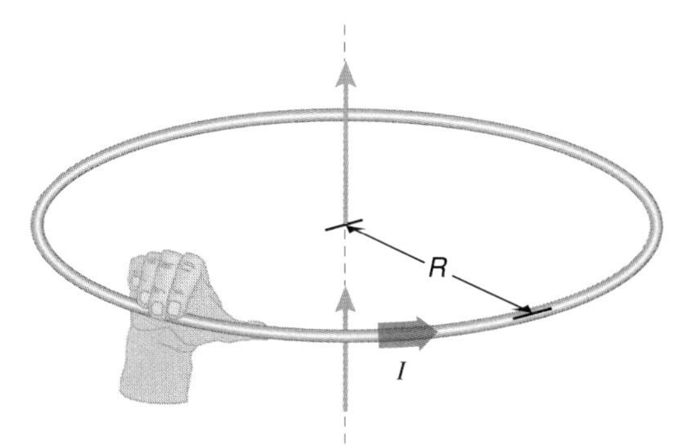

FIGURE 20.37

STRATEGIC EXAMPLE 20.11

a. Find the magnetic field at a point P located a distance d from a long (infinite) wire carrying a current I.

b. If the wire carries a current of 15.0 A, at what distance from the wire is the magnitude of the magnetic field 1.0×10^{-4} T? This is approximately the magnitude of the magnetic field of the Earth.

Solution

a. Begin with the Biot–Savart law, Equation 20.19:

$$\vec{B} = \frac{\mu_0}{4\pi} I \int_{\text{wire}} \frac{d\vec{\ell} \times \hat{r}}{r^2}$$

Choose the x-axis to be along the wire, as indicated in Figure 20.38. By the right-hand rule, the vector product $I\,d\vec{\ell} \times \hat{\mathbf{r}}$ is directed along the z-axis for every current element along the wire. You can see this explicitly by noting that the current element $I\,d\vec{\ell}$ is $I\,dx\,\hat{\imath}$ and, geometrically, the vector $\hat{\mathbf{r}}$ can be written as

$$\hat{\mathbf{r}} = (\cos\theta)\hat{\imath} + (\sin\theta)\hat{\jmath}$$

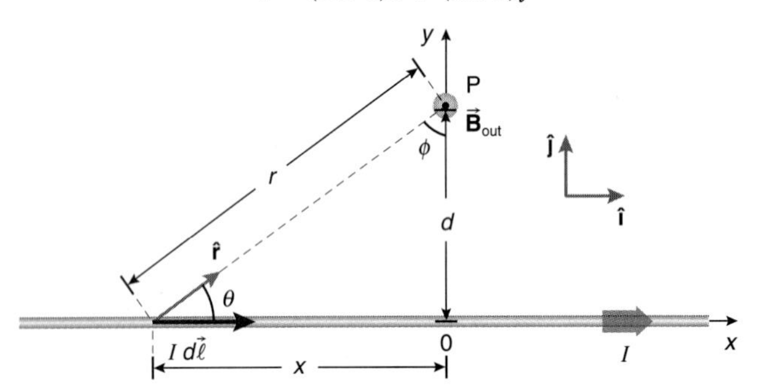

FIGURE 20.38

Hence the vector product is

$$\begin{aligned} I\,d\vec{\ell} \times \hat{\mathbf{r}} &= I\,dx\,\hat{\imath} \times [(\cos\theta)\hat{\imath} + (\sin\theta)\hat{\jmath}] \\ &= (I\,dx\sin\theta)\hat{\mathbf{k}} \end{aligned}$$

The Biot–Savart law becomes

$$\vec{\mathbf{B}} = \frac{\mu_0}{4\pi} I \int_{\text{wire}} \frac{dx\sin\theta}{r^2}\,\hat{\mathbf{k}}$$

Geometrically, $r^2 = d^2 + x^2$ and

$$\sin\theta = \frac{d}{r} = \frac{d}{(d^2 + x^2)^{1/2}}$$

With these substitutions in the expression for $\vec{\mathbf{B}}$, the field is

$$\begin{aligned} \vec{\mathbf{B}} &= \frac{\mu_0}{4\pi} I \int_{\text{wire}} \frac{dx\,d}{(d^2 + x^2)^{3/2}}\,\hat{\mathbf{k}} \\ &= \frac{\mu_0}{4\pi} Id \int_{\text{wire}} \frac{dx}{(d^2 + x^2)^{3/2}}\,\hat{\mathbf{k}} \end{aligned}$$

You have to integrate over the entire wire from $x = -\infty$ m to $x = \infty$ m. Since x appears in the integrand to an even power, the integral is symmetric in x, and so you can take twice the integral from 0 m to ∞ m. The *magnitude* of the field thus is

$$B = \left(\frac{\mu_0}{4\pi} Id\right) 2 \int_{0\text{ m}}^{\infty\text{ m}} \frac{dx}{(d^2 + x^2)^{3/2}} \qquad (1)$$

The integral has a form that is ripe for trigonometric substitution methods. In particular, substitute

$$x = d\tan\phi \qquad (2)$$

(see Figure 20.38 for the geometric interpretation of ϕ). For the limits of integration, when $x = 0$ m, $\phi = 0°$ and when $x = +\infty$ m, $\phi = 90°$. To find dx, take differentials of equation (2):

$$dx = d\sec^2\phi\,d\phi$$

With the substitution $x = d\tan\phi$, the denominator of the integrand becomes

$$\begin{aligned} (d^2 + x^2)^{3/2} &= (d^2 + d^2\tan^2\phi)^{3/2} \\ &= [d^2(1 + \tan^2\phi)]^{3/2} \\ &= (d^2\sec^2\phi)^{3/2} \\ &= d^3\sec^3\phi \end{aligned}$$

Making these substitutions into equation (1) for the magnitude of the magnetic field, you find

$$\begin{aligned} B &= \frac{\mu_0}{4\pi} 2Id \int_{0°}^{90°} \frac{d\sec^2\phi\,d\phi}{d^3\sec^3\phi} \\ &= \frac{\mu_0}{4\pi}\frac{2I}{d} \int_{0°}^{90°} \cos\phi\,d\phi \end{aligned}$$

Perhaps setting up the problem was difficult, but at least the integration is quite easy to do! You finally get

$$\begin{aligned} B &= \frac{\mu_0}{4\pi}\frac{2I}{d}\sin\phi\,\Big|_{0°}^{90°} \\ &= \frac{\mu_0}{4\pi}\frac{2I}{d} \end{aligned} \qquad (3)$$

for the magnitude of the magnetic field at a distance d from an infinite straight wire.

The direction of the field at P is perpendicularly out of the page in Figure 20.38. Notice that the problem has cylindrical symmetry: no matter where the point P is chosen around the wire (as long as it is a distance d from the wire), the field is perpendicular to the wire. Thus the magnetic field lines around the wire are circles, centered on the wire, as shown in Figure 20.39. If you grasp the wire with your right hand with the thumb in the direction of the current, your curled fingers indicate the directional sense in which the magnetic field points around the wire.

b. To find the distance from the wire at which the magnitude of the magnetic field is 1.0×10^{-4} T, solve equation (3) for d and substitute appropriate numerical values:

$$\begin{aligned} d &= \frac{\mu_0}{4\pi}\frac{2I}{B} \\ &= \frac{(10^{-7}\ \text{T}\cdot\text{m/A})\,[2(15.0\ \text{A})]}{1.0 \times 10^{-4}\ \text{T}} \\ &= 3.0 \times 10^{-2}\ \text{m} \\ &= 3.0\ \text{cm} \end{aligned}$$

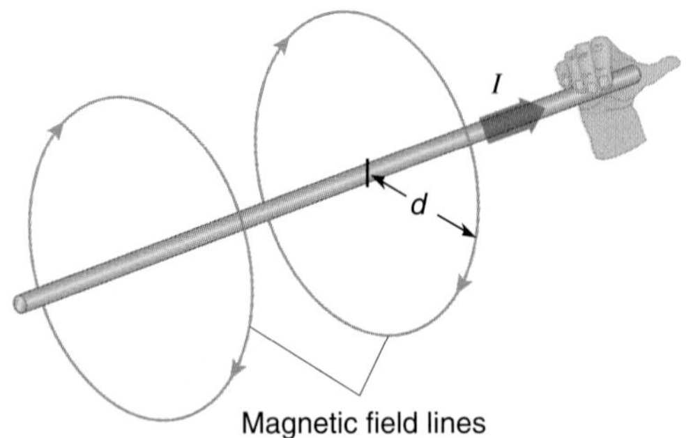

FIGURE 20.39

20.7 Forces of Parallel Currents on Each Other and the Definition of the Ampere

Two or more wires exert magnetic forces on each other just as two or more charges exert electrical forces on each other. Each wire finds itself in the magnetic field produced by the other wire(s). Here we consider the special situation of two parallel, infinitely long, current-carrying wires separated by a distance d, as shown in Figure 20.40.

Wire (1) has current I_1 and this current produces a magnetic field in the space surrounding the wire. With the coordinate system indicated in Figure 20.40, the magnetic field of wire (1) at the position of wire (2) is (using the results of Example 20.11 and the right-hand rule to determine the direction of the field)

$$\vec{\mathbf{B}}_{1 \text{ of wire (1) at the position of wire (2)}} = \frac{\mu_0}{4\pi}\frac{2I_1}{d}\hat{\mathbf{k}} \qquad (20.20)$$

Wire (2) has current I_2 and finds itself in the magnetic field of wire (1). The magnetic field of wire (1) is constant along the straight length of wire (2); therefore, we can use Equation 20.11 for the force on a length ℓ of wire (2):

$$\vec{\mathbf{F}}_{2 \text{ on length } \ell \text{ of wire (2) due to wire (1)}} = I_2\vec{\ell} \times \vec{\mathbf{B}}_1$$

With the coordinate system in Figure 20.40, this becomes

$$\begin{aligned}\vec{\mathbf{F}}_{2 \text{ on length } \ell \text{ of wire (2) due to wire (1)}} &= I_2\ell\hat{\imath} \times B_1\hat{\mathbf{k}} \\ &= -I_2\ell B_1\hat{\jmath}\end{aligned}$$

Substituting for the magnitude of $\vec{\mathbf{B}}_1$ using Equation 20.20, we have

$$\begin{aligned}\vec{\mathbf{F}}_{2 \text{ on length } \ell \text{ of wire (2) due to wire (1)}} &= -I_2\ell\frac{\mu_0}{4\pi}\frac{2I_1}{d}\hat{\jmath} \\ &= -\frac{\mu_0}{4\pi}\frac{2I_1I_2}{d}\ell\hat{\jmath} \qquad (20.21)\end{aligned}$$

In a similar fashion, wire (2) produces a magnetic field at the position of wire (1). Using the results of Example 20.11 and the right-hand rule to determine the direction of the field, the magnetic field of wire (2) at the position of wire (1) is

$$\vec{\mathbf{B}}_{2 \text{ of wire (2) at the position of wire (1)}} = \frac{\mu_0}{4\pi}\frac{2I_2}{d}(-\hat{\mathbf{k}}) \qquad (20.22)$$

Wire (1), carrying current I_1, finds itself in the magnetic field of wire (2). The field is constant along the length of wire (1), and so we can use Equation 20.11 to find the force on a length ℓ of wire (1):

$$\vec{\mathbf{F}}_{1 \text{ on length } \ell \text{ of wire (1) due to wire (2)}} = I_1\vec{\ell} \times \vec{\mathbf{B}}_2$$

With the coordinate system in Figure 20.40, this becomes

$$\begin{aligned}\vec{\mathbf{F}}_{1 \text{ on length } \ell \text{ of wire (1) due to wire (2)}} &= I_1\ell\hat{\imath} \times (-B_2\hat{\mathbf{k}}) \\ &= I_1\ell B_2\hat{\jmath}\end{aligned}$$

Substituting for the magnitude of $\vec{\mathbf{B}}_2$ using Equation 20.22, we have

$$\begin{aligned}\vec{\mathbf{F}}_{1 \text{ on length } \ell \text{ of wire (1) due to wire (2)}} &= I_1\ell\frac{\mu_0}{4\pi}\frac{2I_2}{d}\hat{\jmath} \\ &= \frac{\mu_0}{4\pi}\frac{2I_1I_2}{d}\ell\hat{\jmath} \qquad (20.23)\end{aligned}$$

Notice that in accordance with Newton's third law, the force of wire (1) on a length ℓ of wire (2) is equal in magnitude but opposite in direction to the force of wire (2) on a length ℓ of wire (1). Indeed, once we found the force of wire (1) on wire (2), there was no need to go back to the beginning to calculate the force of wire (2) on wire (1); we simply could have used Newton's third law to find the other force! We took the scenic route to gain a little more practice in working with magnetic fields and forces.

Notice that the forces that parallel wires exert on each other are attractive if the currents are parallel (see Figure 20.41).

If the currents are antiparallel, the forces on the wires are repulsive, as shown in Figure 20.42.

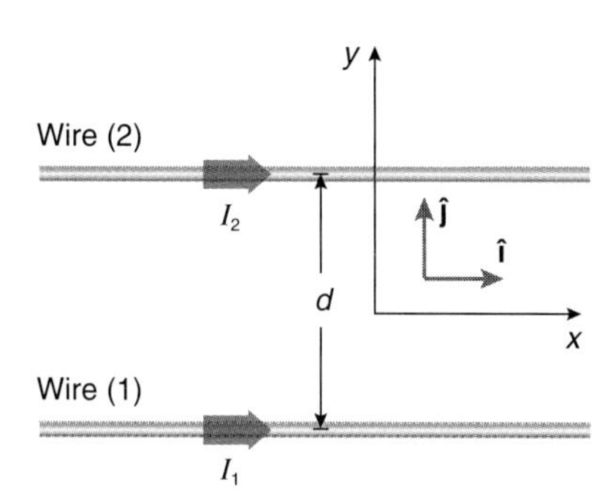

FIGURE 20.40 Two infinite and parallel current-carrying wires.

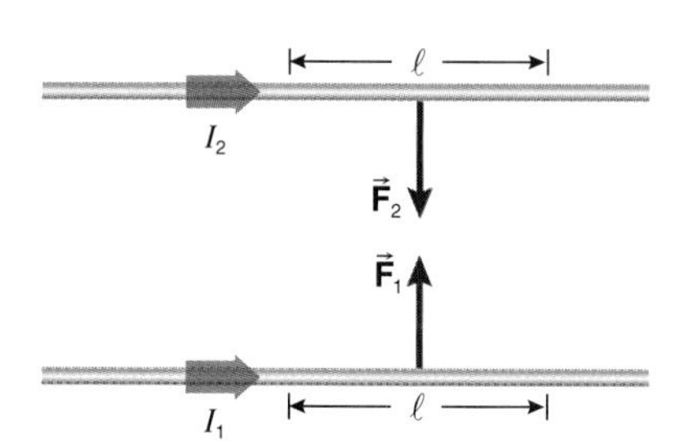

FIGURE 20.41 Parallel currents attract each other.

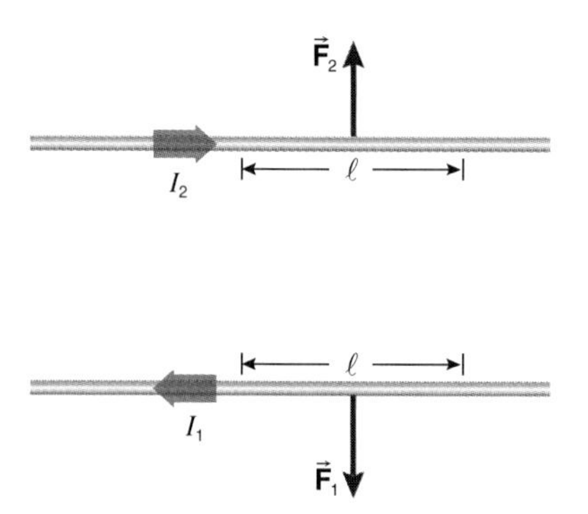

FIGURE 20.42 Antiparallel currents repel each other.

Two infinitely long, parallel, straight current-carrying wires exert forces on a length ℓ of either wire of magnitude

$$F = \frac{\mu_0}{4\pi}\frac{2I_1I_2}{d}\ell \quad (20.24)$$

from either Equation 20.21 or 20.23. The SI unit of current (the ampere) is defined from Equation 20.24. Let the wires be exactly one meter apart, so that $d \equiv 1$ m. Consider the force on exactly one meter of length of the wires, so that $\ell \equiv 1$ m. Let the currents be of equal magnitude: $I_1 = I_2 = I$. Adjust the size of the equal currents I until the magnitude of the force on each wire is exactly 2×10^{-7} N. The magnitude of the current that yields a force of this magnitude per meter of length is defined to be one ampere. The equation for the magnitude of the force then becomes

$$2 \times 10^{-7}\ \text{N} \equiv (10^{-7}\ \text{T}\cdot\text{m/A})\frac{2I^2}{1\ \text{m}}(1\ \text{m})$$

Solving for I, we obtain, exactly,

$$I = 1\ \text{A}$$

Since a coulomb is an ampere-second, this procedure also defines the SI unit of charge.

QUESTION 3

Convince yourself that the antiparallel currents in Figure 20.42 produce repulsive forces on each other.

20.8 GAUSS'S LAW FOR THE MAGNETIC FIELD*

We developed Gauss's law for the two other fields we have studied, the gravitational field $\vec{g}$ and the electric field $\vec{E}$. The law involves the flux of the field vector through a *closed surface*.

For the gravitational case, we discovered in Equation 6.55 that the flux of the gravitational field through a closed surface S is equal to -4π times the mass enclosed within the surface:

$$\int_{\text{clsd surface}} \vec{g} \cdot d\vec{S} = -4\pi M_{\text{within S}}$$

The specific location of the mass within the closed surface was irrelevant; what mattered was simply whether the mass was within the surface or not. If there was no mass within the Gaussian surface, the flux of the gravitational field through that closed surface was zero.

For the electric field, we found that the flux of the electric field through a closed surface was equal to the total (net) charge enclosed by the surface divided by the constant ε_0 (Equation 16.18):

$$\int_{\text{clsd surface}} \vec{E} \cdot d\vec{S} = \frac{\text{net charge enclosed by the surface}}{\varepsilon_0}$$

If there is zero net (total) charge within the Gaussian surface, then the flux of the electric field through that surface is zero. It makes no difference where the charges are within the surface, only that they *are* within it. In particular, if the charges within the surface all are electric dipoles, as shown in Figure 20.43,

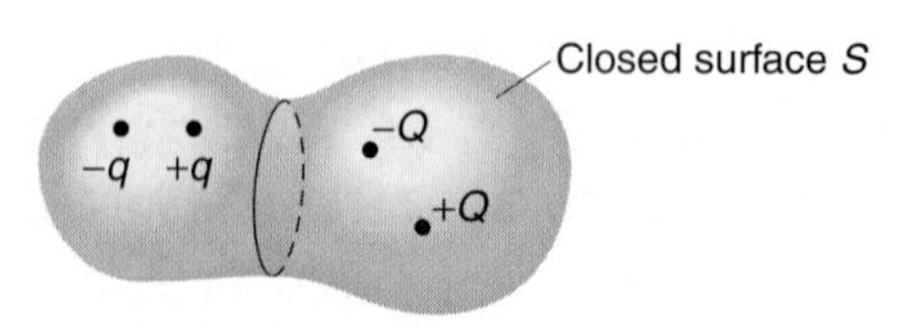

FIGURE 20.43 A collection of electric dipoles produces zero flux of the electric field through a surrounding closed surface.

then the flux of the total electric field vector (resulting from the vector superposition of the individual fields of the individual charges) through the surface is zero.

Gauss's law for the magnetic field likewise involves the flux of the magnetic field vector through a closed surface:

$$\int_{\text{clsd surface}} \vec{B} \cdot d\vec{S}$$

The flux again must fit our observations. We mentioned as we began our study of magnetism that isolated magnetic poles (magnetic monopoles) evidently do not exist: the magnetic field is a *dipolar* field. The poles of the field always are paired. If there are magnets within the surface, the magnets are always dipolar. Thus, in strict analogy with the electric dipole case, we can say the following.

The flux of the magnetic field through *any* closed surface must always be zero:

$$\int_{\text{clsd surface}} \vec{B} \cdot d\vec{S} = 0\ \text{T}\cdot\text{m}^2 \quad (20.25)$$

This is **Gauss's law for the magnetic field.**

Gauss's law for the magnetic field has several implications.

Gauss's law for the magnetic field is, in essence, a statement about the nonexistence of magnetic monopoles (isolated, unpaired north or south magnetic poles).

Mass monopoles exist; electric monopoles exist; but magnetic monopoles evidently do *not* exist. The existence of monopoles is necessary to have a nonzero term on the right-hand side of Gauss's law.

Gauss's law for the magnetic field implies that the magnetic field lines form topologically closed loops.

Magnetic field lines do not begin or end anywhere: they always form closed loops. In gravitation, the field lines representing the field begin at infinity and end at (point toward) the (positive) mass monopoles; there are no negative mass monopoles. Electric field lines begin on positive charge (point away from positive charge) and end on negative charge (point toward negative charge).

Gauss's law for the magnetic field is the second of the four fundamental equations of electromagnetism known collectively as the *Maxwell equations*. Recall that the first of the Maxwell equations was Gauss's law for the electric field (Section 16.11).

20.9 MAGNETIC POLES AND CURRENT LOOPS

We mentioned as we began our study of magnetism that magnets have poles designated N (north) and S (south). Yet we found that it is moving charge (a current) that is the source of the magnetic field. In this section we reconcile these views and see how to determine the north and south poles of a current loop.

The magnetic field produced by a traditional bar magnet is shown in Figure 20.44. The magnetic field produced by a current loop is shown in Figure 20.45.

The magnetic field lines in both cases form closed loops in compliance with Gauss's law for the magnetic field. Outside the bar magnet, the field direction is from the north to the south pole; within the magnet, the field is from the south to the north. The right-hand rule is used to determine the direction of the field in the vicinity of the current loop. The current loop has a north pole side and a south pole side.

You might argue that a current loop is one thing and a permanent magnet is quite another. But the distinction is not apparent, even at the microscopic level. If we keep breaking the bar magnet into smaller and smaller pieces, creating smaller and smaller magnets, eventually we get down to the atomic level where we find that the magnet is really nothing more than a current loop consisting of an electron orbiting a nucleus. A macroscopic permanent magnet is a superposition of many small, generally aligned, submicroscopic current loops represented by the electronic orbital motions inside atoms (the current directions are opposite to the motion of the electrons).

So the idea of magnetic poles is nothing more than an extension of the idea of the two sides of a current loop. The magnetic dipole moment of the current loop is parallel to the axis of a bar magnet we imagine it to represent: the head of the magnetic dipole moment vector represents the north end and the tail of the magnetic dipole moment vector represents the south end.

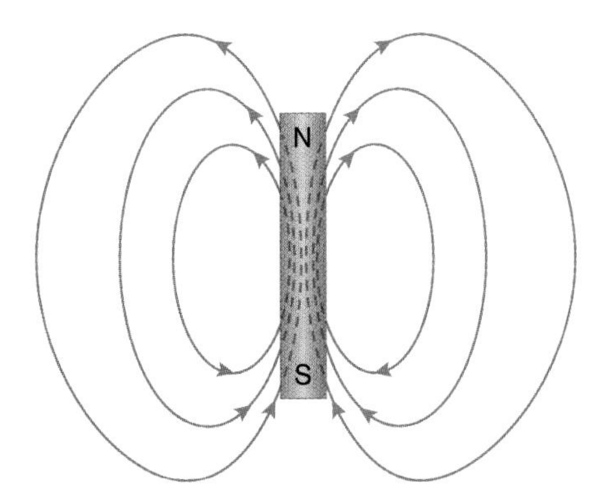

FIGURE 20.44 The magnetic field near a bar magnet.

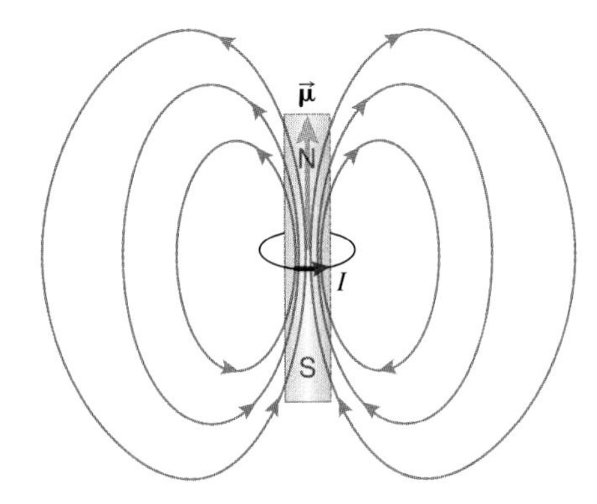

FIGURE 20.45 The magnetic field of a current loop can be imagined as like that of a bar magnet.

20.10 AMPERE'S LAW*

The electrical force on a charge is related to the electric field (caused by other charges) by the now well-worn equation

$$\vec{F}_{elec} = q\vec{E}$$

Just like the gravitational force, the static electrical force is a conservative force. This means that the work done by the static electrical force around any *closed* path is zero:

$$q\int_{\text{clsd path}} \vec{E} \cdot d\vec{r} = 0 \text{ J}$$

Hence we have

$$\int_{\text{clsd path}} \vec{E} \cdot d\vec{r} = 0 \text{ V} \qquad (20.26)$$

In other words, the integral of the static (time-independent) electric field around a closed path is zero.

What about the integral of the magnetic field around a closed path?† That is, we want to determine the value of

$$\int_{\text{clsd path}} \vec{B} \cdot d\vec{r}$$

Here we have to be careful. The quantity $\vec{B} \cdot d\vec{r}$ does not represent some physical quantity, and certainly not work. Although the static magnetic *force* does no work on a moving charge, we *cannot* conclude that the path integral of the magnetic *field* around a closed path is zero, since the magnetic field and the magnetic force point in different directions: they are perpendicular to each other. We are just curious about what this analogous line integral amounts to. To see what the path integral of the magnetic field is around a closed contour, we look at a specific example and then generalize from what we learn.

For this purpose, then, consider the relatively simple magnetic field of an infinitely long current-carrying wire (Example 20.11). The magnetic field of the wire at a distance d from the wire is of magnitude

$$B = \frac{\mu_0}{4\pi}\frac{2I}{d}$$

†We use the term closed path or, equivalently, closed contour rather than the term closed loop here because the path or contour can be a mathematical path rather than a physical entity such as a current loop. We reserve the term loop for current loops.

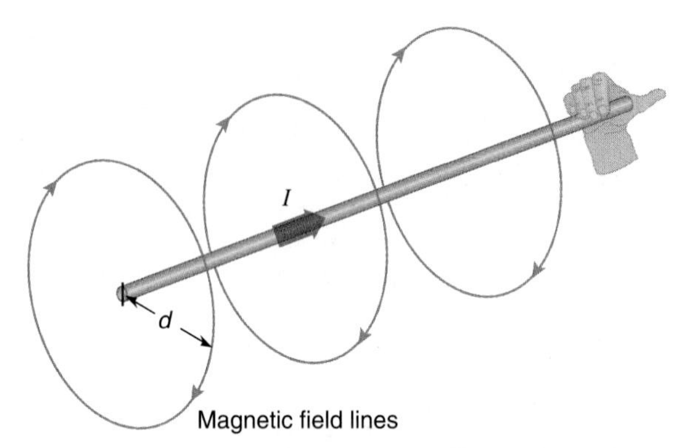

FIGURE 20.46 Integrate the magnetic field around a circle centered on the wire.

The direction of the field is found from the right-hand rule: grab the wire with your right hand with the thumb in the direction of the current. Your curled fingers then circle the wire in the sense of the field; see Figure 20.46.

Case 1: A Circular Path

We integrate the field clockwise around a closed circular path of radius d when viewed along the direction of the current, as shown in Figure 20.46. The magnetic field $\vec{\mathbf{B}}$ is parallel to $d\vec{\mathbf{r}}$ at all points along the path, and so the scalar product in the integrand of the path integral becomes

$$\int_{\text{clsd path}} \vec{\mathbf{B}} \cdot d\vec{\mathbf{r}} = \int_{\text{clsd path}} B\, dr \cos 0° = \int_{\text{clsd path}} B\, dr$$

The magnitude of the field also is constant all along the chosen path, so that

$$\int_{\text{clsd path}} \vec{\mathbf{B}} \cdot d\vec{\mathbf{r}} = B \int_{\text{clsd path}} dr$$

The integral of dr around the entire path is the circumference of the path: $2\pi d$. Thus

$$\int_{\text{clsd path}} \vec{\mathbf{B}} \cdot d\vec{\mathbf{r}} = B(2\pi d)$$

Substituting for the magnitude of the field, we find

$$\int_{\text{clsd path}} \vec{\mathbf{B}} \cdot d\vec{\mathbf{r}} = \frac{\mu_0}{4\pi}\frac{2I}{d}2\pi d$$
$$= \mu_0 I$$

Clearly the result for the static magnetic field is not like the static electric field result (Equation 20.26). Something new is happening.

If we integrate the field around the circular path in a direction opposite to the field, the scalar product in the integrand then is

$$\int_{\text{clsd path}} \vec{\mathbf{B}} \cdot d\vec{\mathbf{r}} = \int_{\text{clsd path}} B\, dr \cos 180°$$
$$= -\int_{\text{clsd path}} B\, dr$$

and the result is negative:

$$\int_{\text{clsd path}} \vec{\mathbf{B}} \cdot d\vec{\mathbf{r}} = -\mu_0 I$$

Are these results peculiar to this example or to this path? We check out another path.

Case 2: A Path with Circular and Radial Segments

We choose the path indicated in Figure 20.47, which consists of four segments: (1) a circular segment of radius d; (2) a radial segment extending from radius d to radius r; (3) a circular segment at radius r; and (4) a radial segment from radius r back to radius d, thus completing this closed path.

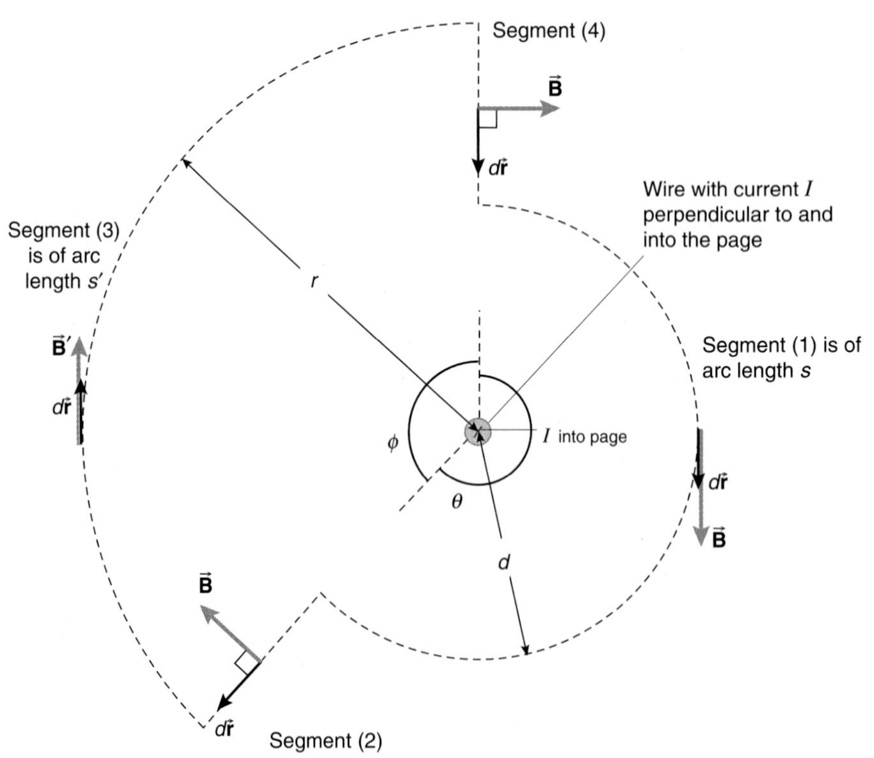

FIGURE 20.47 A more complicated contour around the wire.

The path integral of $\vec{\mathbf{B}} \cdot d\vec{\mathbf{r}}$ over the whole path breaks down into the integral over each of the four segments:

$$\int_{\text{whole clsd path}} \vec{\mathbf{B}} \cdot d\vec{\mathbf{r}} = \int_{\text{segment (1)}} \vec{\mathbf{B}} \cdot d\vec{\mathbf{r}} + \int_{\text{segment (2)}} \vec{\mathbf{B}} \cdot d\vec{\mathbf{r}} + \int_{\text{segment (3)}} \vec{\mathbf{B}} \cdot d\vec{\mathbf{r}} + \int_{\text{segment (4)}} \vec{\mathbf{B}} \cdot d\vec{\mathbf{r}} \tag{20.27}$$

Along the radial segments (2) and (4), the magnetic field $\vec{\mathbf{B}}$ is perpendicular to $d\vec{\mathbf{r}}$, making their scalar product zero in the integrands along these segments. No contribution comes from the radial segments. Along segment (1), the magnetic field is parallel to $d\vec{\mathbf{r}}$, and so

$$\int_{\text{segment (1)}} \vec{\mathbf{B}} \cdot d\vec{\mathbf{r}} = \int B\,dr$$

Since the circular segment is centered on the wire, B is constant along this path, giving

$$\int_{\text{segment (1)}} \vec{\mathbf{B}} \cdot d\vec{\mathbf{r}} = B\int dr$$
$$= Bs$$

where s is the arc length of segment (1). This arc length s is related to the angle θ; indeed, from the definition of a radian, the angle θ in radians is

$$\theta = \frac{s}{d}$$

so that

$$s = \theta d$$

Thus the line integral along this segment is

$$\int_{\text{segment (1)}} \vec{\mathbf{B}} \cdot d\vec{\mathbf{r}} = Bs$$
$$= B\theta d$$

The magnitude of $\vec{\mathbf{B}}$ along this segment is

$$B = \frac{\mu_0}{4\pi}\frac{2I}{d}$$

Thus the line integral along this segment is

$$\int_{\text{segment (1)}} \vec{\mathbf{B}} \cdot d\vec{\mathbf{r}} = B\theta d$$
$$= \frac{\mu_0}{4\pi}\frac{2I}{d}\theta d$$
$$= \frac{\mu_0}{4\pi}2I\theta$$

Now for segment (3). Along this segment the magnetic field is different in magnitude than along segment (1). Call the magnetic field $\vec{\mathbf{B}}'$ along segment (3). We use the same reasoning as for segment (1) to evaluate the line integral. The field is parallel to $d\vec{\mathbf{r}}$; the magnitude B' of the field also is constant along this segment. So the path integral along this segment is

$$\int_{\text{segment (3)}} \vec{\mathbf{B}}' \cdot d\vec{\mathbf{r}} = \int_{\text{segment (3)}} B'\,dr$$
$$= B'\int_{\text{segment (3)}} dr$$

The integral of dr along this segment is the arc length s'. But the angle ϕ is

$$\phi = \frac{s'}{r}$$

so that

$$\int_{\text{segment (3)}} \vec{\mathbf{B}}' \cdot d\vec{\mathbf{r}} = B's'$$
$$= B'\phi r$$

The magnitude of $\vec{\mathbf{B}}'$ is

$$B' = \frac{\mu_0}{4\pi}\frac{2I}{r}$$

and so

$$\int_{\text{segment (3)}} \vec{\mathbf{B}}' \cdot d\vec{\mathbf{r}} = B'\phi r$$
$$= \frac{\mu_0}{4\pi}\frac{2I}{r}\phi r$$
$$= \frac{\mu_0}{4\pi}2I\phi$$

Substituting the results for each segment into Equation 20.27 for the complete path, we find

$$\int_{\text{clsd path}} \vec{\mathbf{B}} \cdot d\vec{\mathbf{r}} = \frac{\mu_0}{4\pi}2I\theta + 0\ \text{T}\cdot\text{m} + \frac{\mu_0}{4\pi}2I\phi + 0\ \text{T}\cdot\text{m}$$
$$= \frac{\mu_0}{4\pi}2I(\theta + \phi)$$

But geometrically $\theta + \phi$ is 2π. Thus we obtain

$$\int_{\text{clsd path}} \vec{\mathbf{B}} \cdot d\vec{\mathbf{r}} = \mu_0 I$$

which is the same result we obtained for the circular contour in Case 1!

Case 3: Paths with Many Circular and Radial Segments

The arguments of Case 2 can be extended to more complex paths encircling the wire such as a closed path consisting of alternating circular and radial segments, as in Figure 20.48.

No contribution to the path integral is obtained from the radial segments (along which the field is perpendicular to $d\vec{\mathbf{r}}$), and the sum of the contributions from the circular segments once again will sum to just $\mu_0 I$. Amazing. It is also apparent that if the plane of the path is not perpendicular to the wire, path segments extending parallel to the wire axis would always be perpendicular to $\vec{\mathbf{B}}$ and therefore make no contribution to the line integral.

Case 4: Any Closed Path Around the Wire

An arbitrary closed path encircling the wire can be approximated by a series of alternating radial and circular segments, as shown in Figure 20.49.

As the size of these segments becomes smaller and smaller and their number increases indefinitely, the segmented path approaches the actual path as a limit. Thus, for any arbitrary closed path encircling the wire, we still get

$$\int_{\text{clsd path}} \vec{\mathbf{B}} \cdot d\vec{\mathbf{r}} = \mu_0 I$$

Notice that it makes no difference where the closed path lies relative to the current going through it or, from the opposite

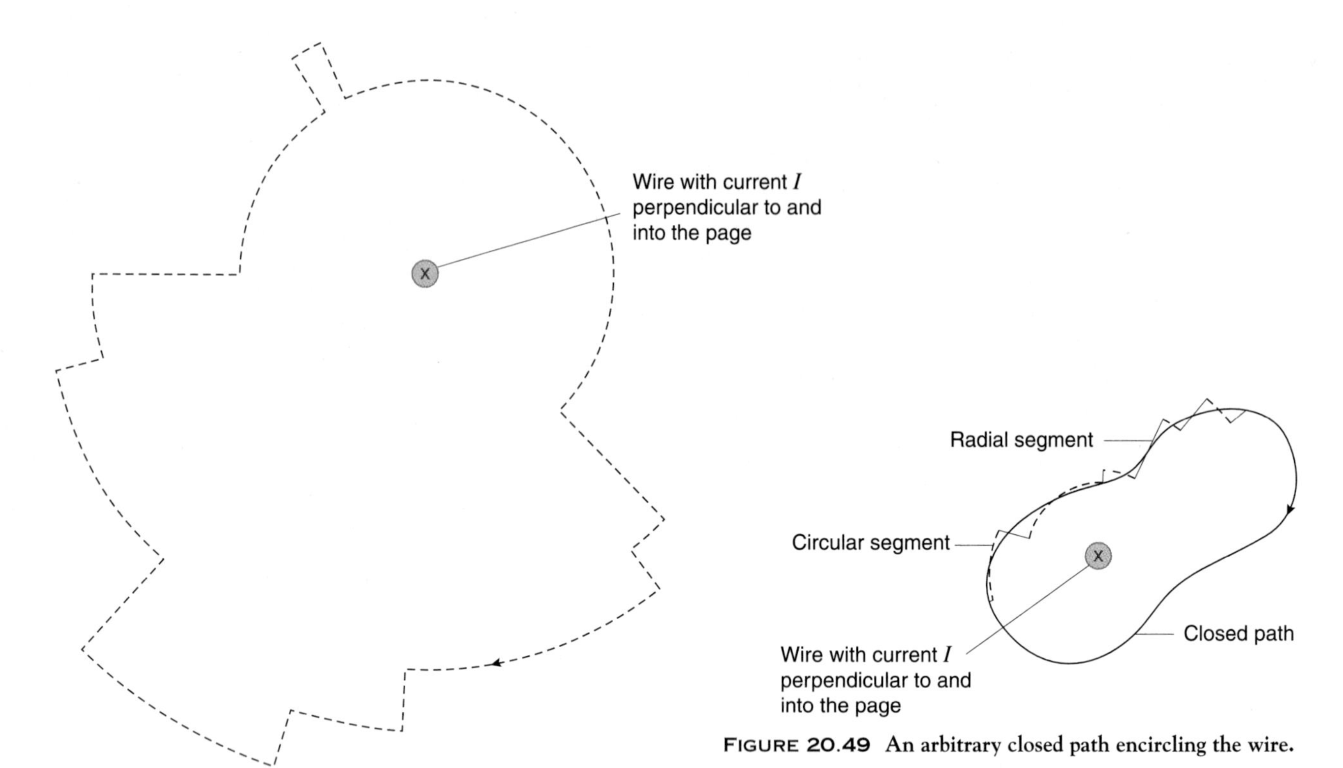

FIGURE 20.48 A complicated path consisting of radial and circular segments.

FIGURE 20.49 An arbitrary closed path encircling the wire.

viewpoint, where the current is located within a given path. The essential thing is that the current *threads through* the area bounded by the closed path chosen.

We need to explain more precisely just what is meant by the term: "current threading the path (or contour)." The term means that the current pierces *any surface* that has the contour of the path as a boundary (see Figure 20.50), like open soap bubble surfaces before they leave the rim of the ringlike contour used to blow them. The path (contour) need not be planar. Another way to think of the path is as the *perimeter* of various hat-shaped surfaces; there is a wide variety of hat surfaces for a given hat size (perimeter), as a visit to a hat department in a department store easily will confirm! To thread the path, the current must pierce the bubble surface or hat. The same path can bound many different bubble-shaped or hat-shaped surfaces.

Strictly speaking, the current must pierce the hat-shaped surface an *odd* number of times (see Figure 20.51).*

Hat shaped surfaces with the contour as a boundry

I

Integration path or contour

FIGURE 20.50 Hat-shaped surfaces with the contour as boundary.

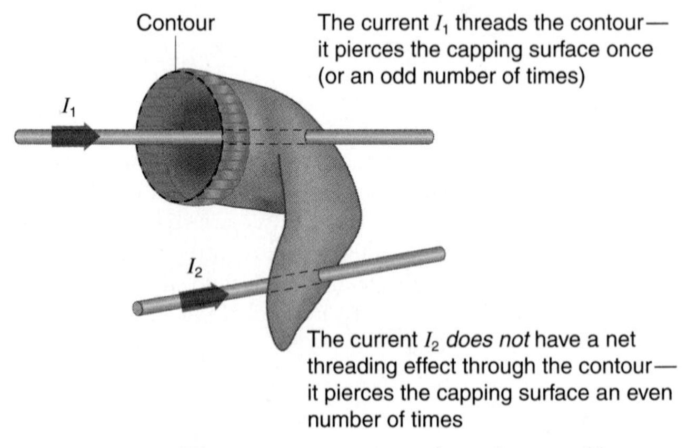

FIGURE 20.51 The current must pierce the surface an odd number of times.

*When the same current pierces the surface an even number of times, the net threading effect is zero.

Case 5: A Path Not Threaded by the Current

What happens if the current does *not* thread the path around which we integrate the field, as in Figure 20.52?

There are four segments to the path depicted. No contribution is obtained along the radial segments because the field is perpendicular to $d\vec{\mathbf{r}}$ at each point along these segments. Along segment (1) we will get

$$\frac{\mu_0}{4\pi}\frac{2Is}{d}$$

or, since $s = \theta d$,

$$\frac{\mu_0}{4\pi}2I\theta$$

as we did with case (2). But along segment (3), the field and $d\vec{\mathbf{r}}$ are directed opposite to each other. The scalar product in the integrand is

$$\begin{aligned}\vec{\mathbf{B}} \cdot d\vec{\mathbf{r}} &= B\,dr\cos 180^\circ \\ &= -B\,dr\end{aligned}$$

The result we get will be negative. That is, along segment (3) the result of the integration is

$$\int_{\text{segment (3)}} \vec{\mathbf{B}} \cdot d\vec{\mathbf{r}} = -\frac{\mu_0}{4\pi}2I\frac{s'}{r}$$

or, since $s' = \theta r$,

$$= -\frac{\mu_0}{4\pi}2I\theta$$

The sum of the contributions from the four segments [only segments (1) and (3) contribute nonzero terms] is

$$\begin{aligned}\int_{\text{clsd path}} \vec{\mathbf{B}} \cdot d\vec{\mathbf{r}} &= \frac{\mu_0}{4\pi}2I\theta - \frac{\mu_0}{4\pi}2I\theta \\ &= 0\ \text{T}\cdot\text{m}\end{aligned}$$

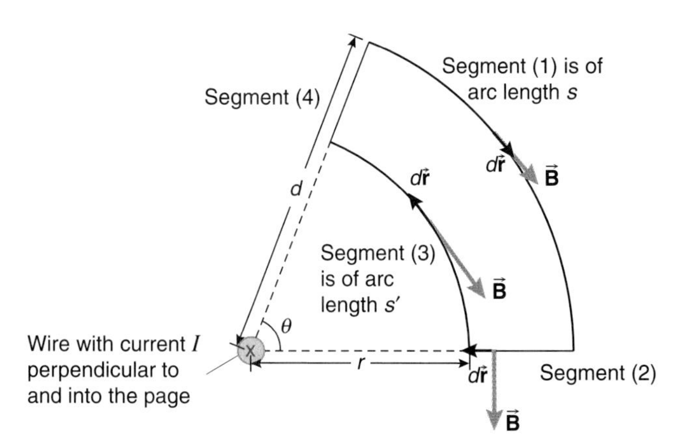

FIGURE 20.52 Zero current threads this contour.

Any closed path not encircling the current can be broken up into alternating radial and circular segments as we did for paths circling the current. But if the path does not encircle the current, the grand sum of all the contributions from the segments inevitably will be zero.

Now we summarize all these results into a single statement. The path integral of the static magnetic field around any closed contour,

$$\int_{\text{clsd path}} \vec{\mathbf{B}} \cdot d\vec{\mathbf{r}}$$

we place on the left-hand side of an equation. The integration is performed by going around the path in either direction.

1. If the current threads the path in such a direction that the magnetic field produced by the current is directed around the wire in the *same* sense as the direction of the path integration, then the path integral gives a contribution of $+\mu_0 I$.
2. If the current threads the path so that the magnetic field produced by the current is directed around the wire in the *opposite* sense to the direction of the path integration, the path integral gives a contribution of $-\mu_0 I$.
3. If the current does not thread the path of the integration, nothing (zero) is contributed to the path integral.

All the verbiage can be elegantly condensed into the following mathematical statement, another example of how we use mathematics to cogently codify much information:

$$\int_{\text{clsd path}} \vec{\mathbf{B}} \cdot d\vec{\mathbf{r}} = \mu_0 I_{\text{net current threading the path}} \qquad (20.28)$$

This is known as **Ampere's law**.

Although we derived this result using the magnetic field of an infinitely long wire, the result is a very general one and applies to *any static magnetic field*.

If multiple (steady) currents are present simultaneously, as in Figure 20.53, the path integral of the total magnetic field around a closed path is equal to μ_0 times the algebraic sum of the currents threading the path.

If you wrap the fingers of your right hand around the closed path in the direction of the path integration, those currents directed through the path in the direction of your thumb appear on the right-hand side of Ampere's law with a plus sign; those in the opposite direction appear with a minus sign.

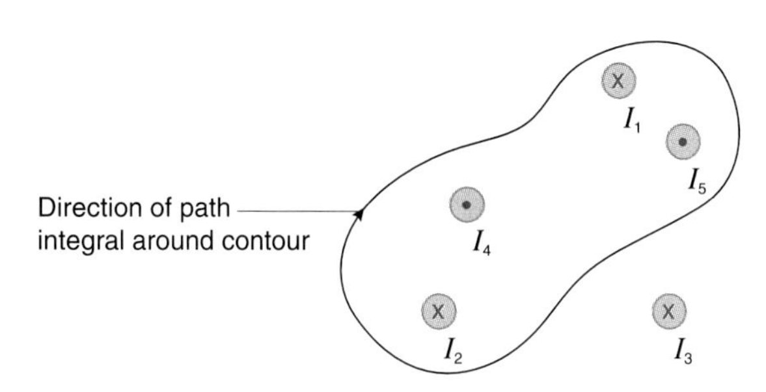

FIGURE 20.53 Multiple currents threading the path.

Thus, in Figure 20.53, currents I_1 and I_2 appear with a plus sign; currents I_4 and I_5 appear with a minus sign. Current I_3 does not thread the path and so does not contribute to the right-hand side of Ampere's law. Ampere's law for the path in Figure 20.53 thus reads

$$\int_{\text{clsd path}} \vec{\mathbf{B}} \cdot d\vec{\mathbf{r}} = \mu_0(I_1 + I_2 - I_4 - I_5)$$

To perform the actual path integration on the left-hand side of this equation typically is very difficult; but the result of the integration is easy to find because the right-hand side of Ampere's law is the simple matter of an algebraic sum of the currents threading the path. In a few cases, where the magnetic field is constant along the sections of a chosen closed path, Ampere's law often can provide a breathtaking way to calculate the field quickly (see Examples 20.12 and 20.13).

QUESTION 4

Compare and contrast Gauss's law for the static electric field with Ampere's law for the static magnetic field.

EXAMPLE 20.12

Use Ampere's law to find the magnetic field at a distance d from an infinite straight wire carrying current I.

Solution

This was effectively done with Case 1, but let's summarize the argument here. By symmetry, the magnitude of the magnetic field at a fixed distance d from the wire is constant. Hence choose a circular contour of radius d, centered on the wire coincident with a magnetic field line. Integrate around the contour in the same sense as the direction of the magnetic field, as in Figure 20.54.

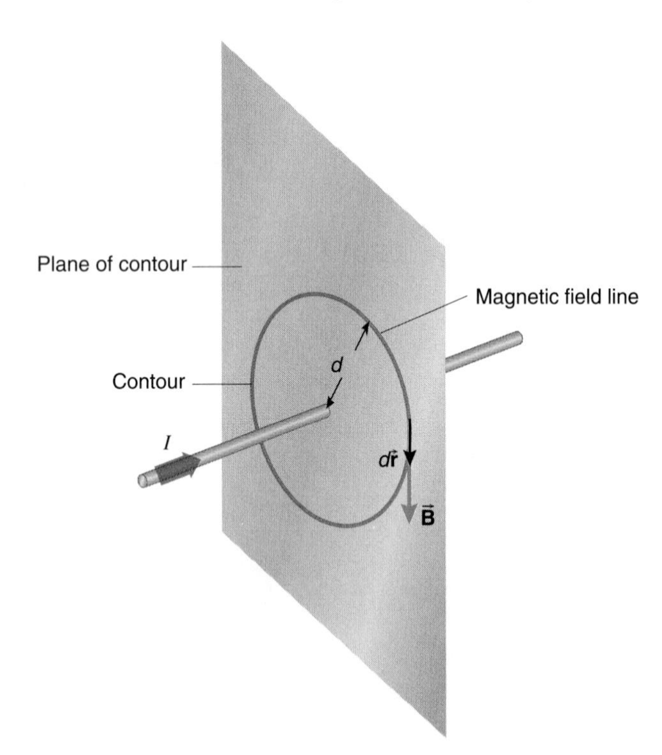

FIGURE 20.54

Note that the current threading the contour is the current I in the wire. Ampere's law here is

$$\int_{\substack{\text{circular}\\\text{contour}}} \vec{\mathbf{B}} \cdot d\vec{\mathbf{r}} = \mu_0 I$$

The magnetic field is parallel to each differential $d\vec{\mathbf{r}}$ along the contour; hence $\vec{\mathbf{B}} \cdot d\vec{\mathbf{r}} = B\,dr$, and you have

$$\int_{\substack{\text{circular}\\\text{contour}}} B\,dr = \mu_0 I$$

Since the magnetic field has a constant magnitude along the chosen contour, the integral becomes

$$B\int_{\substack{\text{circular}\\\text{contour}}} dr = \mu_0 I$$

The integral of dr around the circular contour is the circumference of the contour: $2\pi d$. Hence Ampere's law yields

$$B(2\pi d) = \mu_0 I$$

Solving for B, you find the magnetic field has a magnitude given by

$$B = \frac{\mu_0}{2\pi}\frac{I}{d}$$

In order to keep

$$\frac{\mu_0}{4\pi}$$

as a unit (since its value is 10^{-7} T·m/A), rewrite the field magnitude as

$$B = \frac{\mu_0}{4\pi}\frac{2I}{d}$$

You obtained the same result in Example 20.11 using a more complicated vector integration over the current elements along the wire.

STRATEGIC EXAMPLE 20.13

A long wire wound into a helical shape whose length is typically much greater than its diameter is known as a *solenoid*; see Figure 20.55.

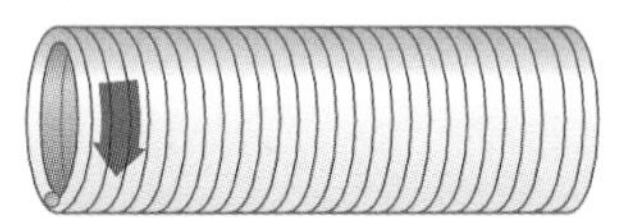

FIGURE 20.55

When the wire carries a current I, a magnetic field is created within the solenoid. Use Ampere's law to calculate the field inside the solenoid, far from either end.

Solution

First consider a solenoid with the individual loops of the wire somewhat separated from each other, as in Figure 20.56. For points very close to each turn, the magnetic field will be approximately that of a long wire and will circle the wire as shown.

Notice that in the space between the loops, neighboring loops have fields between them that are in opposite directions.

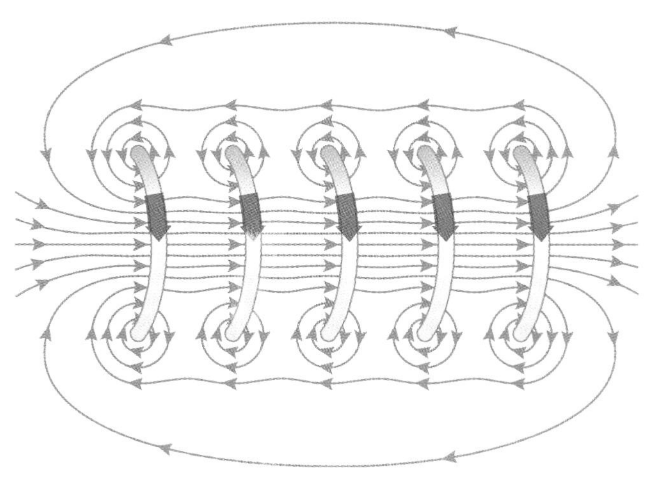

FIGURE 20.56

If the solenoid is tightly wound, as in Figure 20.57, the close-in fields between the loops essentially vector sum to zero. The resulting magnetic field inside the solenoid then is quite uniform and is directed along the axis of the solenoid.

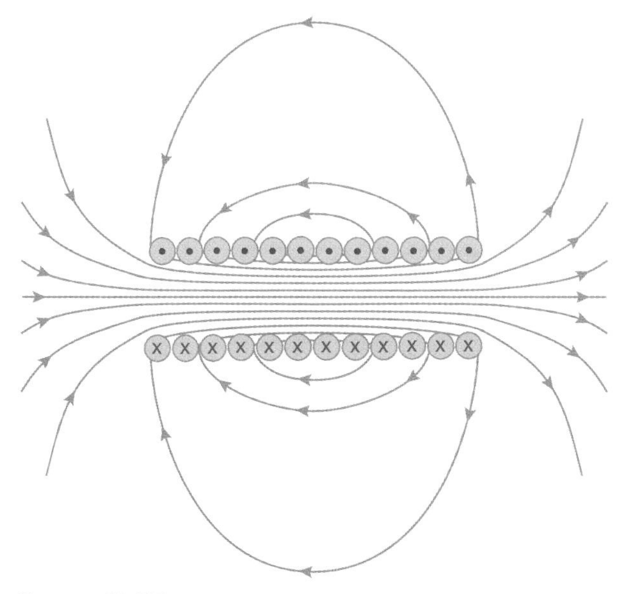

FIGURE 20.57

The magnetic field at points outside the solenoid is quite small compared with the field inside the solenoid. For points near the center of a long solenoid, you can consider the field outside the solenoid to be essentially zero and the field inside the solenoid to be directed along the axis and uniform across the cross-sectional area of the solenoid. These approximations become exact for a close-wound solenoid of infinite length. In practice, as long as the length of the solenoid is much greater than its diameter and you are far from the ends, these approximations are quite good.

Use Ampere's law to calculate the magnetic field inside the solenoid. Ampere's law states that

$$\int_{\text{clsd path}} \vec{\mathbf{B}} \cdot d\vec{\mathbf{r}} = \mu_0 I_{\text{net current threading the path}}$$

To find the magnetic field, you need to choose a path or contour around which to integrate the magnetic field. How do you go about choosing an appropriate path? To explicitly evaluate the left-hand side of the law, you need to pick a path that makes the integration easy. Such paths are those along which the magnetic field is constant over part or all of the path. The shape of the path sometimes reflects the symmetry of the particular situation. A couple of choices might seem appropriate.

Choice 1: A circular path centered on the axis of the solenoid, as shown in Figure 20.58. The magnetic field certainly is constant over the entire contour. The only problem with this choice is that $d\vec{\mathbf{r}}$ along the contour is everywhere perpendicular to the magnetic field. Thus the scalar product $\vec{\mathbf{B}} \cdot d\vec{\mathbf{r}}$ in the integrand on the left-hand side of Ampere's law is zero everywhere. The left-hand side of Ampere's law thus is zero for this contour. Notice that the right-hand side also is zero because no current threads the contour. This shows $0\ \text{T}\cdot\text{m} = 0\ \text{T}\cdot\text{m}$, which is nice and true, but not too informative. So this choice of a contour is not appropriate.

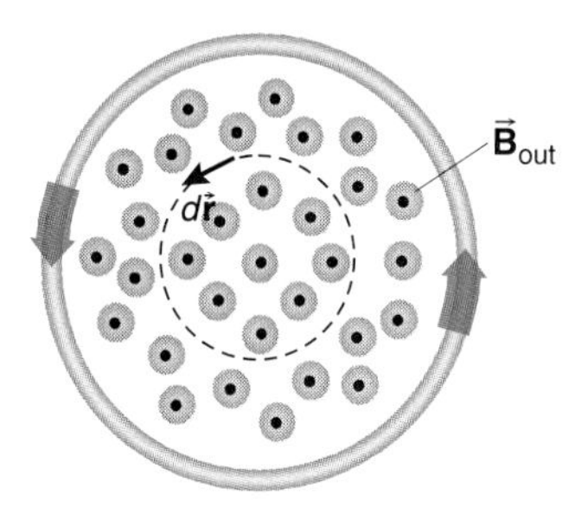

FIGURE 20.58

Choice 2: A rectangular path within the solenoid, as indicated in Figure 20.59. There are four segments of the rectangular path.

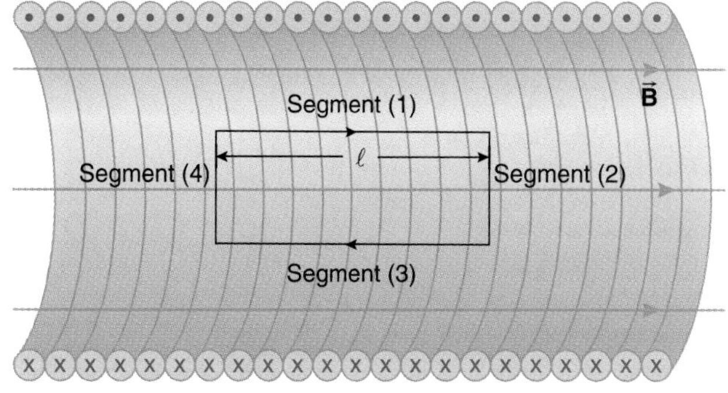

FIGURE 20.59

Along segments (2) and (4), the magnetic field is perpendicular to $d\vec{\mathbf{r}}$, so that the scalar product $\vec{\mathbf{B}} \cdot d\vec{\mathbf{r}}$ is zero along these segments. Along segment (1), $\vec{\mathbf{B}}$ is parallel to $d\vec{\mathbf{r}}$, so that the integral of $\vec{\mathbf{B}} \cdot d\vec{\mathbf{r}}$ along this segment is

$$\int_{\text{segment (1)}} \vec{\mathbf{B}} \cdot d\vec{\mathbf{r}} = \int_{\text{segment (1)}} B\,dr \cos 0° = \int_{\text{segment (1)}} B\,dr$$

The magnitude of the field is constant along this segment, and so you have

$$\int_{\text{segment (1)}} \vec{\mathbf{B}} \cdot d\vec{\mathbf{r}} = B \int_{\text{segment (1)}} dr = B\ell$$

Along segment (3), $\vec{\mathbf{B}}$ is antiparallel to $d\vec{\mathbf{r}}$, and so along this segment you have

$$\int_{\text{segment (3)}} \vec{\mathbf{B}} \cdot d\vec{\mathbf{r}} = \int_{\text{segment (3)}} B\,dr \cos 180° = -\int_{\text{segment (3)}} B\,dr$$

Once again the field is constant over the length of the segment, so that

$$\int_{\text{segment (3)}} \vec{\mathbf{B}} \cdot d\vec{\mathbf{r}} = -B \int_{\text{segment (3)}} dr = -B\ell$$

Hence the left-hand side of Ampere's law is

$$B\ell + 0\ \text{T}\cdot\text{m} - B\ell + 0\ \text{T}\cdot\text{m} = 0\ \text{T}\cdot\text{m}$$

The right-hand side of Ampere's law also is zero since the contour has no current threading the path. Once again this is an inappropriate path. So . . . these choices illustrate that to get anything out of Ampere's law you need to have some current threading the closed path chosen!

Choice 3: Choose a rectangular path that cuts through the side of the solenoid, as indicated in Figure 20.60.

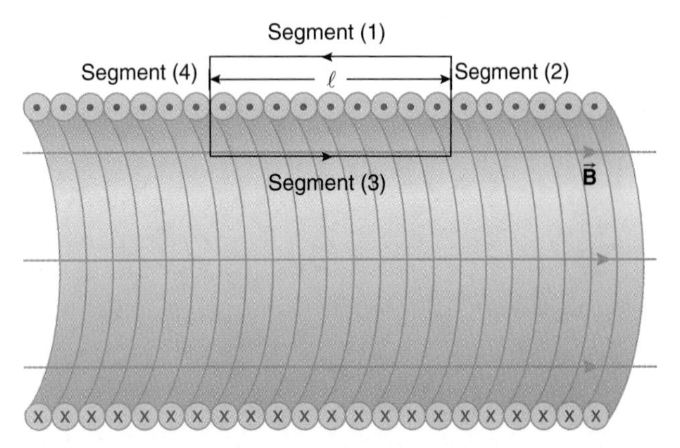

FIGURE 20.60

Now at least you have some current threading the path, so that the right-hand side of Ampere's law will not be zero. To evaluate the right-hand side of Ampere's law, you need to know how many wires thread the path; each wire carries a current I. Let n be the number of turns of wire that are wound on each meter of the solenoid. The number of turns of wire in ℓ meters then is just $n\ell$. If you integrate around the contour in the counterclockwise sense in Figure 20.60, each of the currents is a positive current on the right-hand side according to the convention established with Ampere's law. Thus the right-hand side of Ampere's law is

$$\mu_0 n\ell I$$

To evaluate the left-hand side of the law, integrate the magnetic field around all four segments of the path. Along segments (2) and (4), the magnetic field is either perpendicular to $d\vec{\mathbf{r}}$ (for points on these segments inside the solenoid) or zero (for points outside the solenoid). Thus there is no contribution to the path integral from these segments. Likewise, the magnetic field is zero along segment (1) because you are outside the solenoid and the field is approximately zero there.

Along segment (3), the magnetic field is parallel to $d\vec{\mathbf{r}}$ and constant over the length ℓ of the path. The integration along this segment then is

$$\int_{\text{segment (3)}} \vec{\mathbf{B}} \cdot d\vec{\mathbf{r}} = \int_{\text{segment (3)}} B\,dr = B \int_{\text{segment (3)}} dr = B\ell$$

So the entire left-hand side of Ampere's law is just $B\ell$. Equate the left-hand side of the law with the right-hand side:

$$\int_{\text{clsd path}} \vec{\mathbf{B}} \cdot d\vec{\mathbf{r}} = \mu_0 I_{\text{net current threading the path}}$$

$$B\ell = \mu_0 n\ell I$$

Solving for the magnitude of the magnetic field, you find

$$B = \mu_0 n I \qquad (1)$$

EXAMPLE 20.14

Using Ampere's law, find the magnetic field produced by the current in a long wire wrapped around a torus (or doughnut), as shown in the cross-sectional view of Figure 20.61.

Solution

Like the solenoid (Example 20.13), the magnetic field is directed along (around) the toroid. An appropriate closed contour along which to evaluate the left-hand side of Ampere's law is a circle of radius r, as indicated in Figure 20.62.

The symmetry implies that the magnitude of the magnetic field is constant along the length of the path. Perform the path integration in the same directional sense as the field. Thus the left-hand side of Ampere's law becomes

$$\int_{\substack{\text{circular}\\\text{contour}}} \vec{\mathbf{B}} \cdot d\vec{\mathbf{r}} = \int_{\substack{\text{circular}\\\text{contour}}} B\,dr \cos 0° = B(2\pi r)$$

The right-hand side of Ampere's law involves summing the currents threading the contour. The number of current-carrying wires threading the contour is equal to the total number of turns of wire on the entire torus; call this number N. Then the total current threading the contour is NI. Equating the left- and right-hand sides of Ampere's law, you get

$$B(2\pi r) = \mu_0 NI$$

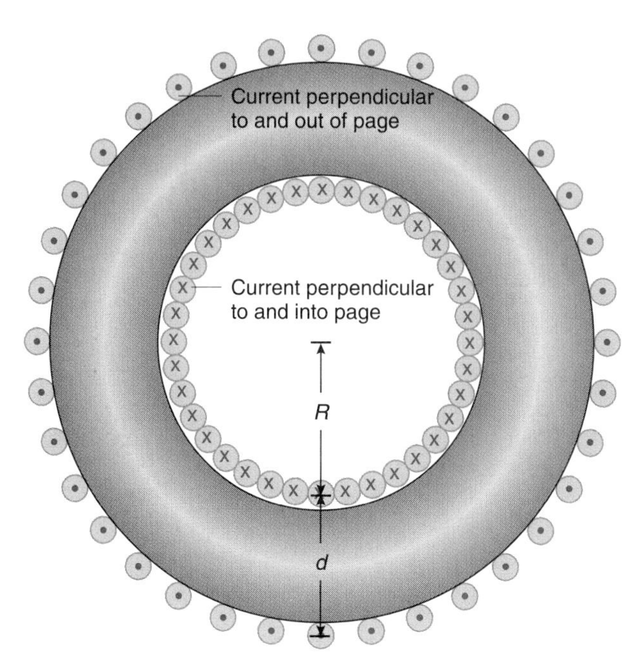

FIGURE 20.61

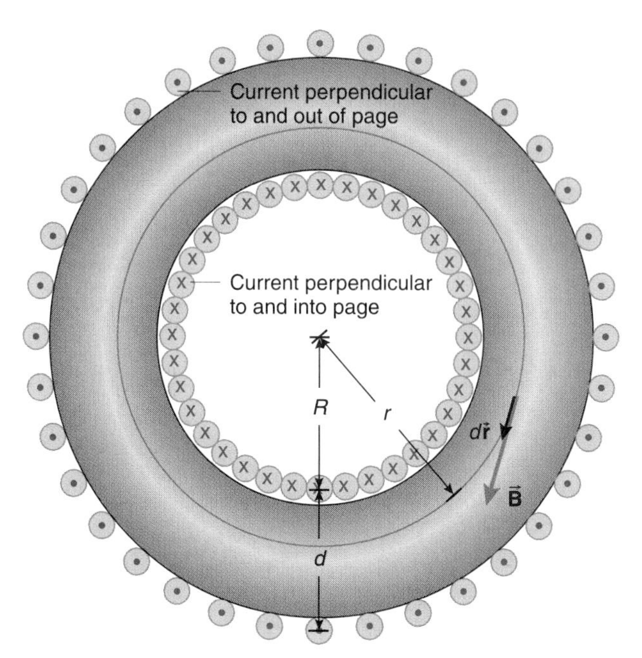

FIGURE 20.62

Solving for the magnitude of the magnetic field, you find

$$B = \frac{\mu_0 NI}{2\pi r} \tag{1}$$

The magnitude of the magnetic field is not constant over the cross-sectional area of the torus because of the dependence on the radius r. For a torus with a very large radius R when compared with the diameter d of the ring, the magnitude of the field approaches that of a long solenoid (and is essentially constant over the diameter of the cross section). When R is large, then,

$$\frac{N}{2\pi r} \approx \frac{N}{2\pi R} \approx n = \text{number of turns per meter}$$

and equation (1) for the toroidal geometry becomes

$$B = \mu_0 nI \tag{2}$$

the same as that for a long solenoid (Example 20.13).

20.11 THE DISPLACEMENT CURRENT AND THE AMPERE–MAXWELL LAW*

The magnetic energy, as developed in the mariner's needle, is, as all know, essentially one with the electricity beheld in heaven. . . .

Herman Melville (1819–1891)†

In Section 19.16 we considered what happens when a source of emf (an independent voltage source) is connected to a series combination of a resistor and a capacitor, as in Figure 20.63. With an initially uncharged capacitor and the switch open, there is no current and so no magnetic field is present. When the switch is closed, a current exists for a brief interval in order to charge the capacitor. Such a current produces a magnetic field. Eventually, when the capacitor becomes fully charged, there is no more current and the magnetic field disappears. This brief magnetic field thus is time dependent because the current is a function of time.

Let's look more closely at the charging process, the temporary current, the magnetic field produced by this time-dependent current, and the effect of the time-dependent current and magnetic field on Ampere's law (Section 20.10). This sounds like quite a task! Specifically, consider the charging of the parallel plate capacitor shown in Figure 20.64, extracted from the circuit of Figure 20.63 for clarity. Ampere's law states that the path integral of the magnetic field around a closed contour is equal to μ_0 times the current threading the path. By threading the path we mean piercing any hat-shaped surface that has the path as its boundary or perimeter. We choose the path indicated in Figure 20.64.

During the charging process, a time-dependent conduction current exists in the wire, and this current is equal to the time

†*Moby Dick*, Chapter 124, "The Needle." First published in 1851 (University of California Press, Berkeley, 1979).

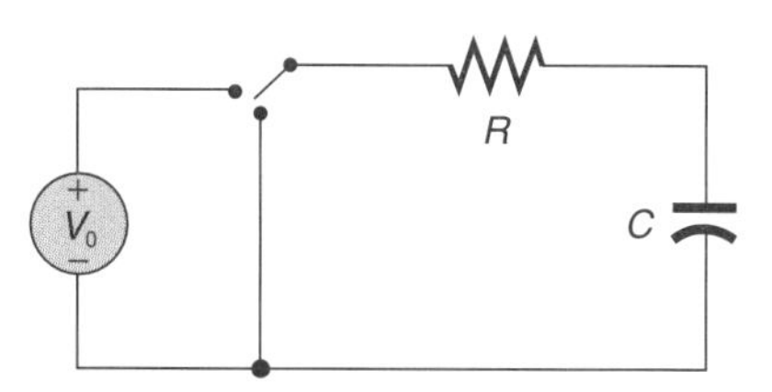

FIGURE 20.63 A series *RC* circuit.

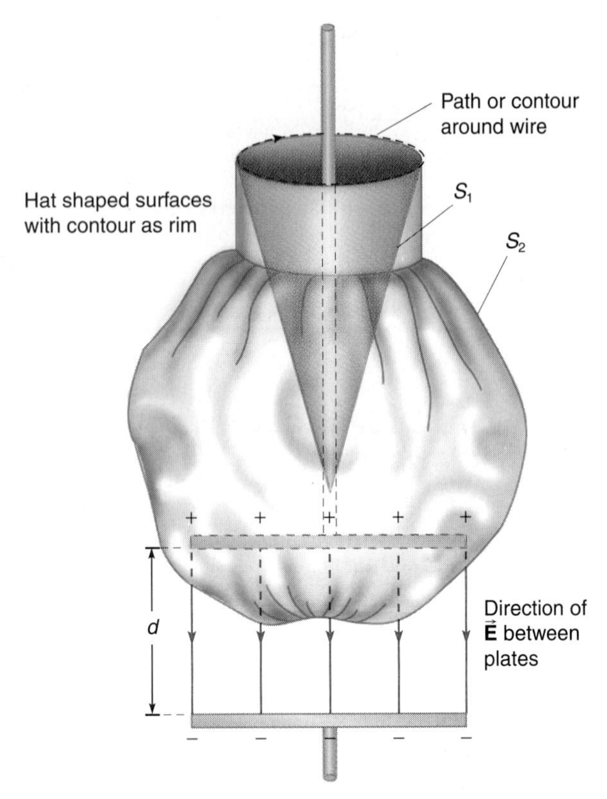

FIGURE 20.64 A path around the wire for Ampere's law.

rate at which the charge $Q(t)$ on the capacitor is changing. From the general definition of the current, we have

$$I = \frac{dQ(t)}{dt}$$

This current pierces the dunce cap–shaped surface S_1 indicated in Figure 20.64, and so the right-hand side of Ampere's law is not zero. The left-hand side of Ampere's law therefore is not zero, indicating the presence of a magnetic field surrounding the wire. Nothing new here.

But now we examine the chef's hat–shaped surface S_2 in Figure 20.64, which passes between the plates of the capacitor and has the same path (contour) for its perimeter as the surface S_1. Ampere's law states that the threading current is the current through *any* surface with the contour or path as a boundary or perimeter. The plates of a capacitor are electrically insulated from one another, so that no conduction current can pass between the plates of the capacitor.

We *know* the left-hand side of Ampere's law is not zero (from the previous argument using the dunce cap surface S_1), but the conduction current does not pierce the chef's hat surface S_2. This fact means that the right-hand side of Ampere's law apparently is zero while the left-hand side is not! There is something funny going on here. Something needs to be done (i.e., cooked up) to the right-hand side of Ampere's law to eliminate the paradoxical result with the chef's hat surface S_2. It was James Clerk Maxwell (1831–1879) who first realized what had to be done to eliminate this apparent inconsistency on the right-hand side of Ampere's law. Here is Maxwell's idea.

When the capacitor has a charge Q on its plates, the potential difference V between the plates is found from the definition of the capacitance:

$$C = \frac{Q}{V}$$

We rearrange this slightly:

$$Q = CV$$

Since the capacitance of the capacitor is fixed (and only a function of geometry and the material between the plates, here assumed to be a vacuum), the time-varying charge means the potential difference between the plates also changes with time:

$$\frac{dQ}{dt} = C\frac{dV}{dt} \qquad (20.29)$$

The potential difference V between the plates can be expressed in terms of the magnitude E of the electric field between the plates and the plate separation d (Example 17.1):

$$V = Ed$$

We substitute this into Equation 20.29, obtaining (for fixed plate separation d)

$$\frac{dQ}{dt} = C\frac{dE}{dt}d \qquad (20.30)$$

The capacitance of a parallel plate capacitor is found from Example 18.3:

$$C = \varepsilon_0 \frac{A}{d}$$

where A is the area of a plate of the capacitor. We make this substitution for C into Equation 20.30:

$$\begin{aligned}\frac{dQ}{dt} &= \varepsilon_0 \frac{A}{d}\frac{dE}{dt}d \\ &= \varepsilon_0 A \frac{dE}{dt}\end{aligned}$$

Since the area of the plates is fixed, we can write this as

$$\frac{dQ}{dt} = \varepsilon_0 \frac{d}{dt}(EA) \qquad (20.31)$$

Maxwell realized that the quantity EA was the *flux of the electric field* through the chef's hat surface S_2. Although S_2 extends beyond the plates of the capacitor, the electric field (ideally) is nonzero only between the plates of the capacitor (neglecting edge effects), so that the flux of $\vec{E}$ through the surface S_2 arises only from that portion of S_2 that lies between the capacitor plates. Let Φ_{elec} be the flux of the electric field vector through this surface. Equation 20.31 then is written as

$$\frac{dQ}{dt} = \varepsilon_0 \frac{d\Phi_{elec}}{dt} \qquad (20.32)$$

Maxwell regarded the right-hand side of this equation as a new type of current I_D through S_2. Its dimensions and units were right. Namely, he defined the right-hand side of Equation 20.32 to be a **displacement current** I_D:

$$I_D \equiv \varepsilon_0 \frac{d\Phi_{elec}}{dt} \qquad (20.33)$$

The name is a bit unfortunate, for it has nothing to do with displacements of any kind. Historically, Maxwell called it the displacement current under the erroneous assumption that it had to do with a real displacement of charges rather than fields; the term is another instance of historical inertia in the notation and terminology of physics. Equation 20.32 indicates that the displacement current through S_2, in quantity and directional sense, is equal to the ordinary conduction current

$$\frac{dQ}{dt}$$

that is present through S_1. The conduction current through S_1 arises from the physical transport of charge; the displacement current through S_2 arises from the time dependence of the flux of the electric field between the plates of the capacitor.

The right-hand side of Ampere's law should include *both* conduction and displacement currents (if appropriate):

$$\int_{\text{clsd path}} \vec{\mathbf{B}} \cdot d\vec{\mathbf{r}} = \mu_0 (I + I_D)_{\text{threading the path}} \qquad (20.34)$$

This generalization is called the **Ampere–Maxwell law**.

For our example, through surface S_1 there is a conduction current I but no displacement current I_D. On the other hand, through surface S_2 there is a displacement current I_D, equal in quantity and direction to the conduction current, but no conduction current I.

The crux of Maxwell's argument was the realization that magnetic fields are produced via two distinct mechanisms:

1. by electric charges in motion (conduction current); and
2. by time-varying electric fields (via the displacement current).

The Ampere–Maxwell law was a surprising breakthrough in physics and is the third of the four fundamental equations of electromagnetism known as the Maxwell equations.*

Returning to the specific example of the charging parallel plate capacitor, we apply the Ampere–Maxwell law to a circular contour located between the capacitor plates (with the hat-shaped area A' taken to be the plane of the contour, as shown in Figure 20.65. There is no conduction current in this region, and so the Ampere–Maxwell law involves only the displacement current:

$$\int_{\text{clsd path}} \vec{\mathbf{B}} \cdot d\vec{\mathbf{r}} = \mu_0 I_D$$

*The other two Maxwell equations that we have studied so far are Gauss's law for the electric field and Gauss's law for the magnetic field. We have one more Maxwell equation to go to complete the set; we study this additional equation in Chapter 21.

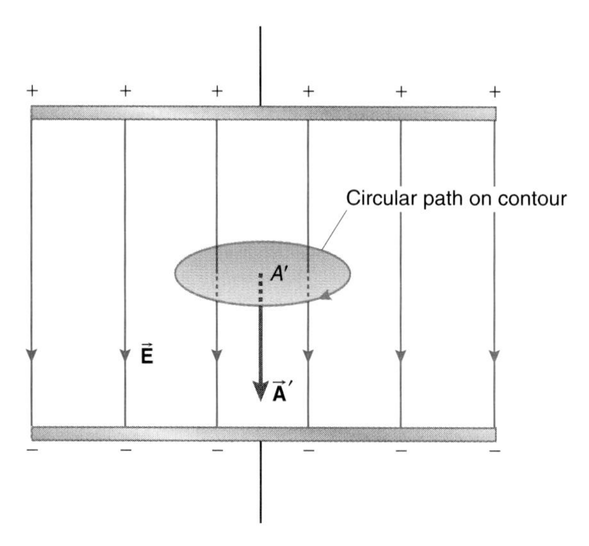

FIGURE 20.65 Apply the Ampere–Maxwell law to the indicated contour.

Using Equation 20.33 for the displacement current, this becomes

$$\int_{\text{clsd path}} \vec{\mathbf{B}} \cdot d\vec{\mathbf{r}} = \mu_0 \varepsilon_0 \frac{d\Phi_{elec}}{dt} \qquad (20.35)$$

where the electric flux is taken through the area A' bounded by the contour. Since the electric field between the plates of a parallel plate capacitor is the same everywhere between the plates, the flux of the electric vector through the area of the contour is

$$\Phi_{elec} = \vec{\mathbf{E}} \cdot \vec{\mathbf{A}}' = EA'$$

The direction of the area vector $\vec{\mathbf{A}}'$ is determined by wrapping the fingers of your right hand around the contour in the direction taken around the path; thus, $\vec{\mathbf{A}}'$ is parallel to the field direction. Substituting this expression for the electric flux into Equation 20.35, we have

$$\int_{\text{clsd path}} \vec{\mathbf{B}} \cdot d\vec{\mathbf{r}} = \mu_0 \varepsilon_0 \frac{d}{dt}(EA')$$

Since A' does not depend on the time, it comes through unscathed from the differentiation:

$$\int_{\text{clsd path}} \vec{\mathbf{B}} \cdot d\vec{\mathbf{r}} = \mu_0 \varepsilon_0 A' \frac{dE}{dt}$$

The displacement current is directed across the entire region between the capacitor plates; if the capacitor is charging, the electric field between the capacitor plates is increasing with time and the derivative

$$\frac{dE}{dt}$$

is positive. The displacement current (in Figure 20.65) is directed down (from the top plate to the bottom plate), but recognize that the displacement current occurs throughout the region between the plates. Place the thumb of your right hand along the direction of the displacement current; your right-hand fingers then indicate the sense of the magnetic field caused by the displacement current (see Figure 20.66).

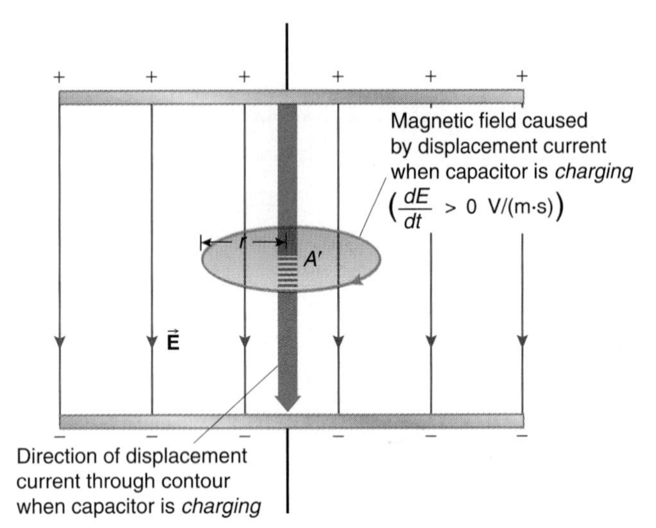

FIGURE 20.66 The direction of the displacement current and the magnetic field it creates for a charging capacitor.

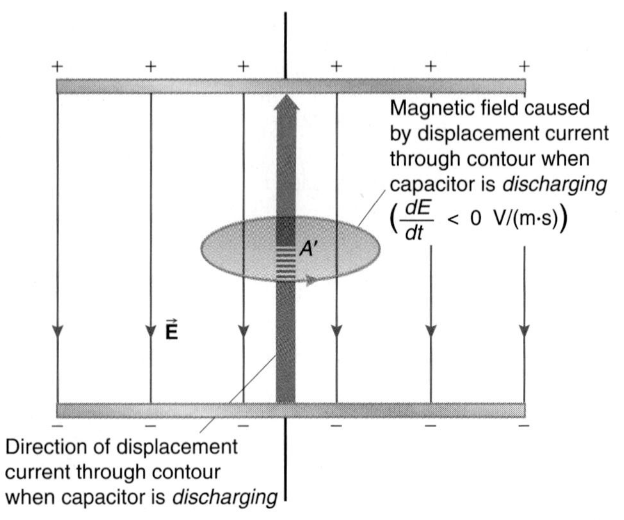

FIGURE 20.67 The direction of the displacement current and the magnetic field it creates for a discharging capacitor.

The cylindrical symmetry implies that the magnetic field has a constant magnitude along the circular contour. If we integrate around the contour in the same direction as the magnetic field, the Ampere–Maxwell law becomes

$$\int_{\text{clsd path}} \vec{\mathbf{B}} \cdot d\vec{\mathbf{r}} = \mu_0 I_D$$

$$B(2\pi r) = \mu_0 \varepsilon_0 A' \frac{dE}{dt}$$

Since $A' = \pi r^2$, we have

$$B(2\pi r) = \mu_0 \varepsilon_0 \pi r^2 \frac{dE}{dt}$$

$$B = \frac{\mu_0}{2} \varepsilon_0 r \frac{dE}{dt} \qquad (20.36)$$

This explicitly indicates how a changing electric field produces a magnetic field. Note, incidentally, that this magnetic field increases linearly with r, reflecting the fact that as we choose greater r, more displacement current is within the contour because the electric flux through the area of the contour increases.

Equation 20.36 also could be obtained by saying the displacement current is I_D and then using the expression for the magnetic field of a long wire, equation (3) of Example 20.11:

$$B = \frac{\mu_0}{4\pi} \frac{2 I_D}{r} \qquad (20.37)$$

The displacement current I_D through the contour is

$$\begin{aligned} I_D &= \varepsilon_0 \frac{d\Phi_{\text{elec}}}{dt} \\ &= \varepsilon_0 \frac{d}{dt}(EA') \\ &= \varepsilon_0 A' \frac{dE}{dt} \\ &= \varepsilon_0 \pi r^2 \frac{dE}{dt} \end{aligned}$$

Substituting for I_D into Equation 20.37 gives the same result as Equation 20.36.

This example indicates that a displacement current causes a magnetic field just as effectively as does a normal conduction current of moving charges.

If the capacitor is discharging, then the electric field between the capacitor plates is decreasing with time, and so the derivative

$$\frac{dE}{dt}$$

is negative. The displacement current then is directed upward through the contour (from the bottom plate to the top plate), as shown in Figure 20.67. The magnetic field generated by the displacement current then is in the direction opposite to that when the capacitor was charging.

Thus there is a magnetic field generated by the displacement current between the plates of the capacitor; this field is in one direction when the capacitor is charging and in the reverse direction when the capacitor is discharging, reflecting the different direction of the displacement current.

Also note that the magnetic field induced by the displacement current is perpendicular to the (changing) electric field that causes it.

When the capacitor is fully charged (or fully discharged), the conduction current in the wires is zero, the displacement current between the plates of the capacitor is zero, and there is no magnetic field.

QUESTION 5

For the parallel plate capacitor, the displacement current I_D between the capacitor plates is equal in quantity and direction to the conduction current I in the wire feeding the plates. Explain why the right-hand side of the Ampere-Maxwell law should *not* be either $2\mu_0 I$ or $2\mu_0 I_D$.

20.12 MAGNETIC MATERIALS*

When we studied electric phenomena and electric fields, we began with electrical effects in vacuum and described the electrical properties of materials by changing the permittivity of free space (a vacuum) ε_0 to the permittivity of the material ε. The ratio of the permittivities was the dielectric constant κ of the material:

$$\kappa = \frac{\varepsilon}{\varepsilon_0}$$

A similar thing is done for the magnetic properties of most materials. The permeability of free space μ_0 changes to the **permeability of the material** μ. The ratio of the permeabilities is known as the **relative permeability**:

$$\kappa_{\mathrm{m}} = \frac{\mu}{\mu_0}$$

(Do not confuse the permeability μ with the magnitude of the magnetic dipole moment vector $\vec{\boldsymbol{\mu}}$.)

Materials are classified magnetically according to the approximate size of the relative permeability (see Table 20.2):

1. **Diamagnetic materials** have κ_{m} slightly less than 1.
2. **Paramagnetic materials** have κ_{m} slightly greater than 1.
3. **Ferromagnetic materials** have κ_{m} significantly greater than 1.

In Example 20.8, we saw that an orbiting electron in an atom is a tiny current loop. The circulating electron thus has an orbital magnetic dipole moment. The electron also has an intrinsic magnetic dipole moment associated with its *spin* (if the spin is imagined literally as a rotating electron, the spin of the charged electron also produces a very tiny current loop). The total magnetic moment of an atom consists of an appropriate (quantum mechanical) vector sum of the orbital and spin magnetic moments of the electrons in the atom.

The protons and neutrons in the nucleus also have magnetic moments, but these are much smaller than the magnetic moments of the electrons because of the much greater mass of those nuclear particles. The (quantum mechanical) vector sum of the magnetic moments of the protons and neutrons in a nucleus produces a (small) nuclear magnetic dipole moment whose value depends on the specific element and isotope. The physics of the interaction of these nuclear magnetic moments with external magnetic fields is the basis for important chemical and medical technologies such as nuclear magnetic resonance (NMR) and imaging techniques for noninvasive (i.e., nonsurgical) examination of soft tissues. Traditional x-ray imaging typically is inappropriate for soft tissues unless the tissue is doped with a strong x-ray absorber.

TABLE 20.2 Relative Permeability of Selected Materials

Material	κ_{m}
Diamagnetic materials	
Copper (Cu)	$1 - 9.4 \times 10^{-6}$
Gold (Au)	$1 - 2 \times 10^{-5}$
Mercury	$1 - 3.2 \times 10^{-5}$
Lead	$1 - 1.7 \times 10^{-5}$
Paramagnetic materials	
Aluminum (Al)	$1 + 2.1 \times 10^{-5}$
Magnesium (Mg)	$1 + 5 \times 10^{-5}$
Platinum	$1 + 2.9 \times 10^{-4}$
Sodium (Na)	$1 + 2 \times 10^{-5}$
Ferromagnetic materials	
Iron (Fe)	2×10^2 to 6×10^3
78 Permalloy (78% Ni, 22% Fe)	4×10^3 to 10^5
Supermalloy (5% Mo, 79% Ni, 16% Fe)	10^5 to 10^6

Source: Dale R. Corson and Paul Lorrain, *Introduction to Electromagnetic Fields and Waves* (W. H. Freeman, San Francisco, 1962), page 284. *CRC Handbook of Chemistry and Physics* 78th edition (CRC Press, Boca Raton, Florida, 1997), pages 12-116 and 12-119.

We will not go into how these various magnetic moments are combined, but such a summation can lead to two results:

1. The atom as a whole has a nonzero, permanent magnetic dipole moment. Paramagnetic and ferromagnetic materials are of this type.
2. The atom does not have a permanent magnetic dipole moment. Diamagnetic materials are of this type.

The presence of a nonzero magnetic moment in paramagnetic and ferromagnetic materials means that if the material is placed in an external magnetic field, the magnetic moments will tend to align *parallel* to the direction of the field to minimize their potential energies. The alignment is not complete because of random motions due to thermal energy (internal energy). The important factor is the size of the thermal energy per particle (proportional to kT) relative to the magnitude of the potential energy of its magnetic dipole moment in the external field ($\mathrm{PE} = -\vec{\boldsymbol{\mu}} \cdot \vec{\mathbf{B}}$). The specific degree of the alignment depends on the particular material and the temperature.

The distinction between paramagnetic and ferromagnetic materials arises when the external field is removed. For paramagnetic materials, the individual magnetic moments of the atoms and molecules become randomly oriented after the external field is removed, and so the total magnetic moment of the material (the vector sum of the magnetic dipole moments of the particles) is zero.

For ferromagnetic materials, the magnetic dipole moments do not become completely randomized with the removal of the external field and a residual total magnetic moment remains: a permanent magnet is produced. A permanent magnet can have its magnetic dipoles completely randomized by increasing the temperature of the material, by sufficiently jarring the material (for instance, by repetitively dropping it on the floor or striking the material several times with a hammer), or by degaussing (using a coil with an opposing magnetic field to neutralize a given field),† thus making the total magnetic moment zero once again.

Diamagnetic materials have zero magnetic moment. When placed in a magnetic field, a nonzero magnetic moment is *induced* in the material (much the way electric polarization is induced via electric fields). The induced magnetic moment is directed *antiparallel* to the applied field, because opposite magnetic poles attract each other.

A detailed examination of the magnetism of materials is left for more advanced courses in electromagnetism.

20.13 The Magnetic Field of the Earth*

The Earth has a permanent magnetic field. A crude model of the field imagines a huge bar magnet embedded within the Earth, as

†If you have a tape player, frequent degaussing of the recording and playback heads is recommended.

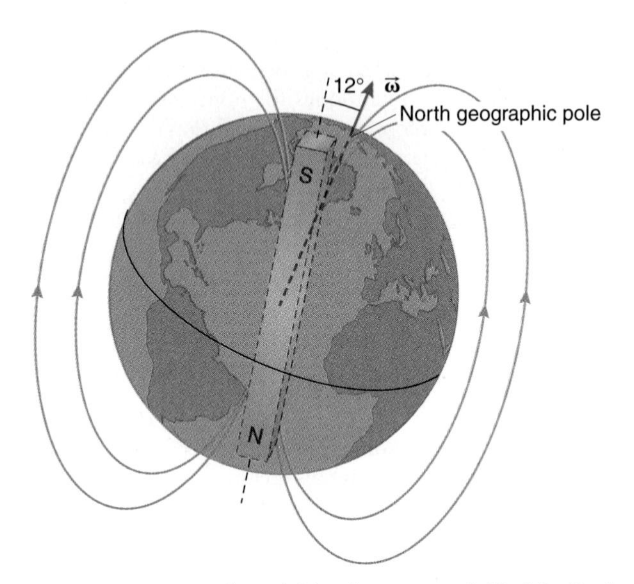

FIGURE 20.68 **A crude model for the magnetic field of the Earth.**

shown in Figure 20.68. The magnetic axis of the Earth (a line between its south and north magnetic poles) is inclined to its spin axis by about 12°.

A compass needle is a small magnetic dipole that orients itself parallel to the horizontal component of the magnetic field of the Earth. The direction indicated by a line from the center of a compass to the north pole of the compass needle is called **magnetic north**. A line indicating the direction to the north geographic pole is **true north** (or geographic north). The angle between the magnetic and true north directions at a given location is the **magnetic declination**. A compass mounted in the vertical plane aligned along the magnetic north–south direction is called a **dip needle**; it indicates the angle that the magnetic field of the Earth makes with the horizontal direction.

The origin of the magnetic field of the Earth still is a matter of much interest and research. There are only a few aspects of the magnetic field that are known:

1. The field certainly is *not* really caused by a huge, embedded permanent magnet such as in Figure 20.68. Although the core of the Earth consists of much iron and other ferromagnetic material capable of sustaining a permanent magnetic dipole moment on a bulk scale under typical temperatures at the surface of the Earth, the high temperatures in the interior of the Earth prohibit the formation of a permanent magnetic field; the material is not ferromagnetic at these high temperatures.
2. The magnetic field likely is caused by convection currents of ionic material within the outer parts of the fluid core of the Earth. It is thought that these convection currents are closely tied to the rotation of the Earth. The discovery that the fluid outer core of the Earth spins slightly faster than the mantle has added considerable excitement to this research. Slowly rotating planets like Venus have weaker magnetic fields than the Earth; faster rotating planets like Jupiter have stronger magnetic fields than the Earth. The detailed mechanisms that generate the magnetic field of the Earth still are quite elusive and complex; in other words, we do not really have the answer to the question! The problem is difficult to probe. The *rule of difficulty* is at work once again: if you know the answer, the problem is easy; if you do not know the answer, the problem is hard.
3. The magnetic field of the Earth has undergone *reversals* of magnetic polarity every few million years. What instigates these polarity reversals also is not known. The reversal occurs quite rapidly on a geologic time scale. When reversing direction, the magnetic poles wander relatively quickly (within about 10 000 years) from one hemisphere to the other.*

The discovery of the field reversals was made during the 20th century while exploring the interesting geology near the center of the mid-ocean ridge in the Atlantic. Such ridges exist on the seabed of other oceans as well. The mid-Atlantic ridge feature is a region where hot magma (lava) slowly wells up from the interior of the Earth and cools as it approaches the surface of the Earth (at the bottom of the ocean). Within hot magma, minute particles with a magnetic dipole moment orient themselves parallel to the magnetic field of the Earth at the time. The magnetic orientation of the particles thus is locked into the rocks when they solidify and provides a record of the magnetic field direction of the Earth.

Since the magnetic field of the Earth reverses itself, the present magnetic field at a given location may be greater or smaller than the average magnetic field due to the parallel or antiparallel alignment of the magnetic material in the underlying rocks. A sensitive instrument (a magnetometer) can measure the difference between the actual magnetic field at a given location and the average field. This difference is known as the **magnetic anomaly**. A distinct pattern of alternating magnetic anomalies exists around the regions of the mid-ocean ridges, as shown in Figure 20.69. In particular, if the magnetic anomaly is measured along a line perpendicular to the ridge, a pattern of bands or strips of alternating anomaly is apparent that is quite symmetric about the ridge. The age of the rocks increases with distance from the ridge.

The pattern is consistent with a slow, conveyor belt type of transport of material away from the ridge, indicative of sea-floor spreading, as shown in Figure 20.70. The recently cooled surface rocks record the recent direction of the magnetic field vector and, as they are slowly pushed away from their source ridge, they display a pattern of the past. The existence of these bands is firm proof of the ideas of plate tectonics in geology.

*There also is recent evidence (from analyzing ancient volcanic lava flows at Steens Mountain in southern Oregon, latitude 42.5° N, longitude 118.5°W) of very rapid changes in the direction of the magnetic field, with the poles moving as much as 6° in a single day. See R. S. Coe, M. Prévot, and P. Camps, "New evidence for extraordinarily rapid change of the geomagnetic field during a reversal," *Nature*, 374, #6524, pages 687–692 (20 April 1995).

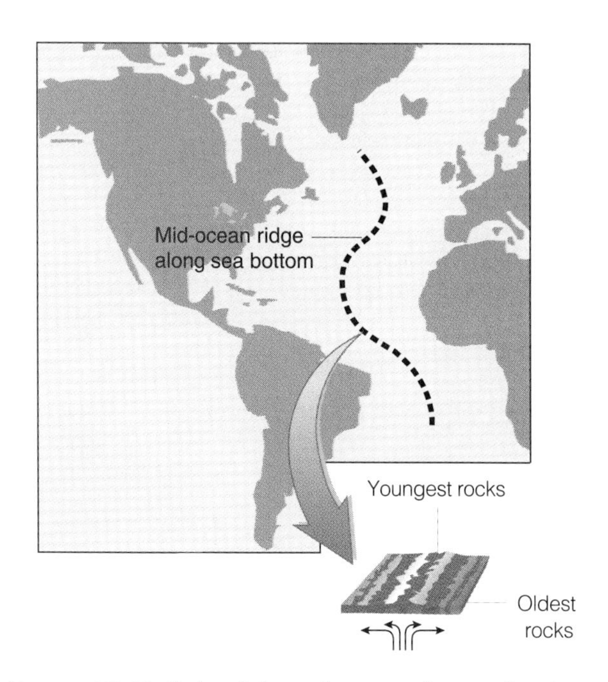

FIGURE 20.69 Strips of alternating magnetic anomaly exist near the ridge.

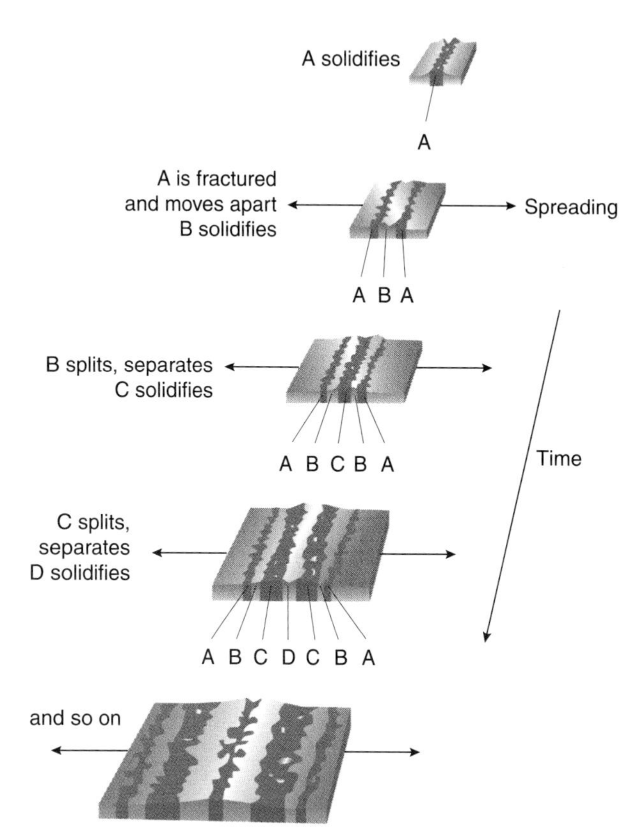

FIGURE 20.70 Formation of the strips of alternating magnetic anomaly.

CHAPTER SUMMARY

The end of a compass needle that points generally in a northerly direction at most places on the surface of the Earth is defined to be a *north* magnetic pole; the opposite end of the needle is a *south* magnetic pole. Like magnetic poles repel each other, unlike poles attract. Magnetic poles always are paired; isolated magnetic poles (magnetic monopoles) evidently do not exist.

Magnets produce a magnetic field $\vec{B}$ in the surrounding space. The SI unit of magnetic field is the *tesla* (T).

A particle with charge q, moving at velocity $\vec{v}$ in a magnetic field $\vec{B}$, experiences a magnetic force $\vec{F}$ given by

$$\vec{F} = q\vec{v} \times \vec{B} \tag{20.1}$$

Static magnetic forces do no work on the charge because the force always is perpendicular to the velocity (and thus to the differential change in the position vector of the particle).

A current I in a magnetic field $\vec{B}$ experiences a force given by

$$\vec{F} = I\int_{\text{wire}} d\vec{\ell} \times \vec{B} \tag{20.10}$$

where the current element $I\,d\vec{\ell}$ is directed in the same sense as the current. If the wire is straight and the magnetic field is constant along its entire length, then Equation 20.10 reduces to

$$\vec{F} = I\vec{\ell} \times \vec{B} \qquad \text{(straight wire of length } \ell, \ \vec{B} \text{ constant along its length)} \tag{20.11}$$

A loop of current I with area A has a magnetic dipole moment $\vec{\mu}$ given by

$$\vec{\mu} \equiv I\vec{A} \tag{20.14}$$

where the common direction of $\vec{\mu}$ and $\vec{A}$ is given by the following right-hand rule: wrap the fingers of your right hand around the loop in the directional sense of the current. Then your extended right-hand thumb indicates the direction of $\vec{\mu}$ and $\vec{A}$.

A magnetic dipole moment in a magnetic field $\vec{B}$ experiences a torque $\vec{\tau}$ given by

$$\vec{\tau} = \vec{\mu} \times \vec{B} \tag{20.15}$$

and has a potential energy

$$\text{PE} = -\vec{\mu} \cdot \vec{B} \tag{20.16}$$

The magnetic field produced by a current I is found from the *Biot–Savart law*:

$$\vec{B} = \frac{\mu_0}{4\pi} I \int_{\text{wire}} \frac{d\vec{\ell} \times \hat{r}}{r^2} \quad (20.19)$$

where $\hat{r}$ is a unit vector pointing from the current element $I\,d\vec{\ell}$ to the point where the field is to be found, and r is the distance between the current element and the field point. The constant μ_0 is the permeability of free space; it satisfies

$$\frac{\mu_0}{4\pi} = 10^{-7}\ \text{T}\cdot\text{m/A}$$

The ampere, the SI unit of current, is defined in the following way: two infinite parallel wires each carrying one ampere, when spaced one meter apart, exert a force on each other of magnitude exactly 2×10^{-7} N per meter of length.

Gauss's law for the magnetic field states that the flux of the magnetic field $\vec{B}$ through any closed surface always is zero:

$$\int_{\text{clsd surface S}} \vec{B} \cdot d\vec{S} = 0\ \text{T}\cdot\text{m}^2 \quad (20.25)$$

Gauss's law for the magnetic field is a statement that magnetic monopoles do not exist; the magnetic field is dipolar.

Ampere's law states that the path integral of the magnetic field around a closed path is equal to μ_0 times the net current threading the contour:

$$\int_{\text{clsd path}} \vec{B} \cdot d\vec{r} = \mu_0 I_{\text{net current threading the path}} \quad (20.28)$$

The currents should include displacement currents when and where they exist, in which case the law is called the *Ampere–Maxwell law* (see Equation 20.34). Currents are considered positive if they thread the path according to the following right-hand rule: wrap the fingers of your hand around the contour in the same sense as $d\vec{r}$; your extended right-hand thumb indicates the positive sense for currents threading the path.

A time-varying flux of the electric field produces a displacement current I_D given by

$$I_D \equiv \varepsilon_0 \frac{d\Phi_{\text{elec}}}{dt} \quad (20.33)$$

which also creates a magnetic field (via the Biot–Savart law).

To calculate magnetic effects in materials, use the permeability of the material μ rather than the permeability of free space μ_0.

Summary of Problem-Solving Tactics

20.1 (page 898) When using Equation 20.1, $\vec{F} = q\vec{v} \times \vec{B}$, be sure to use the appropriate sign of the charge q. For positive charges, the force is in the direction of $\vec{v} \times \vec{B}$; for negative charges, the force is directed opposite to $\vec{v} \times \vec{B}$.

20.2 (page 906) The static magnetic force does no work on a moving charge and, by the CWE theorem, cannot by itself change the kinetic energy of the charge.

20.3 (page 910) If a wire carrying a current is wrapped into a coil with n turns, each with the same area A, the magnetic dipole moment of the coil is n times that of a single turn. The common direction of $\vec{\mu}$ and $\vec{A}$ is determined from the same right-hand rule: wrap the fingers of your right-hand around the coil in the sense of the (positive) current, and the extended right-hand thumb indicates the direction of both $\vec{\mu}$ and $\vec{A}$.

Questions

1. **(page 898)**; 2. **(page 900)**; 3. **(page 918)**; 4. **(page 924)**; 5. **(page 930)**

6. How do you think the ancient Greeks and Chinese first discovered magnets over two millennia ago? Were they capable of distinguishing between magnetism and electricity?

7. Suggest some experiments that indicate that the interaction between two magnets is not an electrostatic interaction.

8. Indicate in a sketch several magnetic field lines that illustrate: (a) a situation where the magnetic field $\vec{B}$ is constant; (b) a situation where the magnetic field is not constant.

9. The diagrams in Figure Q.9 trace lines of a static electric field and/or a static magnetic field. Indicate which represent electric field lines, which represent magnetic field lines, and which could be either.

10. In electrostatics, the electric field is either parallel or antiparallel to the electrical force on a charged particle. Explain why it is *not* a good idea to define the direction of the magnetic field to be either parallel or antiparallel to the direction of the magnetic force on a charged particle moving in the field.

11. Consider the equation $\vec{F} = q\vec{v} \times \vec{B}$. (a) Which vector quantities always are perpendicular to each other? (b) Which are not necessarily perpendicular to each other?

12. Charged particles from the Sun strike the Earth more frequently near its magnetic poles than elsewhere. Explain why. Collisions of such particles with the atmosphere produce auroras.

13. The electric and magnetic forces on the undeflected ions in a velocity selector have equal magnitudes and opposite directions. Are these forces a Newton's third law force pair? Explain.

14. According to Newton's third law, all forces occur in pairs. What is the force that is the third law counterpart to the magnetic force on a charged particle moving in a magnetic field?

15. A charged particle q is moving in a circle in a uniform magnetic field. An electric field is turned on in the same direction as the magnetic field. Describe and sketch the subsequent trajectory of the particle. Consider both cases: $q > 0$ C and $q < 0$ C.

16. An electron begins initially traveling horizontally west in the United States (to enjoy the scenery) in a region where the magnetic field of the Earth points true north. Describe the initial direction of the force on the electron.

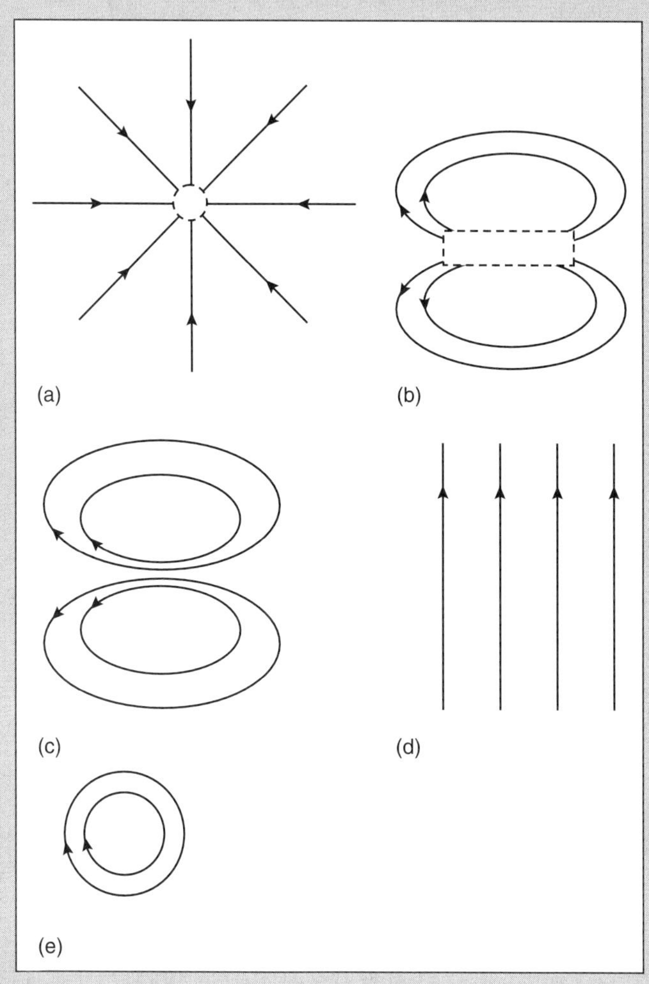

FIGURE Q.9

17. A charge q is moving at velocity $\vec{v}$ in a magnetic field, yet experiences zero magnetic force. Indicate in a sketch and with a short explanation how can this happen.

18. A charged particle moving with velocity $\vec{v}$ enters a region where there exists either an electric or a magnetic field. The force on the particle is perpendicular to $\vec{v}$. Can you conclude definitively that the field is magnetic rather than electric? Explain.

19. If a charge is at rest in a magnetic field, can it be set in motion with a static magnetic field?

20. Describe an experiment to determine whether a given end of a bar magnet is a north magnetic pole or a south magnetic pole.

21. You suspect the existence of a magnetic field in some region of space. Describe an experiment to verify the existence (or nonexistence) of the magnetic field in the region.

22. Describe the similarities and the differences between static electric and magnetic fields and the effects they have on charged particles.

23. A field exists in a region of space but you are uncertain about whether it is a gravitational, an electric, or a magnetic field. Given a particle of mass m with positive charge q, describe a series of experiments that will enable you to tell if the field is gravitational, electric, or magnetic as well as the direction of the field.

24. A wire carrying a current has zero total charge. How, then, can a magnetic field exert a force on the wire?

25. A wire carries a current. The same current results if you assume that positive charge flows one way or negative charge in the opposite direction. Do the magnitude and direction of the magnetic force on the wire depend on the sign of the charge carrier assumed to be responsible for the current? Explain.

26. A charged particle moves in a magnetic field but the magnetic force on the particle is zero. What can you conclude about the orientation of the velocity of the particle with respect to the magnetic field direction?

27. The magnetic field lines of two parallel, long, current-carrying wires are shown in Figure Q.27, but without directional arrows for the currents. (a) Are the current directions in the wires parallel or antiparallel to each other? (b) Indicate the current directions in the wires.

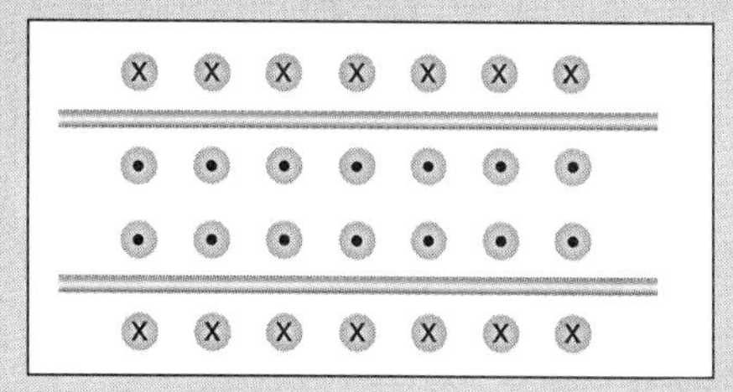

FIGURE Q.27

28. Bring a small, weak bar magnet near a TV, PC monitor, or oscilloscope (while they are turned on!). Explain why the display on the screen is distorted.

29. A moving electron and a proton enter a region with their velocities perpendicular to a uniform magnetic field. For each of the following cases, determine which particle (if either) is deflected into a circle of greater radius: (a) the speeds of the particles are the same; (b) the kinetic energies of the particles are the same; (c) the momenta of the particles are the same.

30. Explain why the following statement is true: A static magnetic field cannot change the *speed* of a charged particle.

31. Explain why the following statement usually is true: A static magnetic field can change the *velocity* of a charged particle. Under what circumstances is the statement false?

32. Describe a situation in which a magnetic dipole moment experiences: (a) zero torque in a magnetic field; (b) nonzero torque in a magnetic field.

33. A jumbo jet is flying horizontally over the Atlantic Ocean near Newfoundland in the direction east as indicated by a magnetic compass. The magnetic field of the Earth makes an angle θ with the horizontal. (a) What is the direction of the horizontal component of the magnetic field of the Earth? (b) What is the direction (up or down?) of the vertical component of the magnetic field? (c) Will the magnetic force cause an accumulation of electrons in the nose, tail, or left or right wing tip (left and right determined facing forward)?

34. A *bubble chamber* is a device in which the paths of tiny charged particles in a fluid can be photographed. A magnetic field is used to deflect the particles from their incident directions. The paths of an electron and positron (a particle with the same mass as an

electron but with charge $+e$) are shown in Figure Q.34. (a) If the magnetic field points toward you perpendicular to the page, which is the path for $-e$ and which for $+e$? (b) Notice that the paths are spiral and not circular. What does this imply about the speed of the particles as a function of time? Develop a hypothesis that may account for the spiral paths.

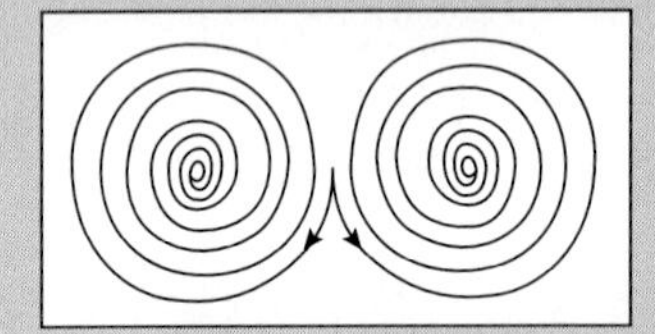

FIGURE Q.34

35. Two long wires carry oppositely directed currents of equal magnitude. To make the magnetic field of the wires a minimum at large distances from the wires, is it better to arrange the wires as in Figure Q.35a or Q.35b, or does it make any difference at all?

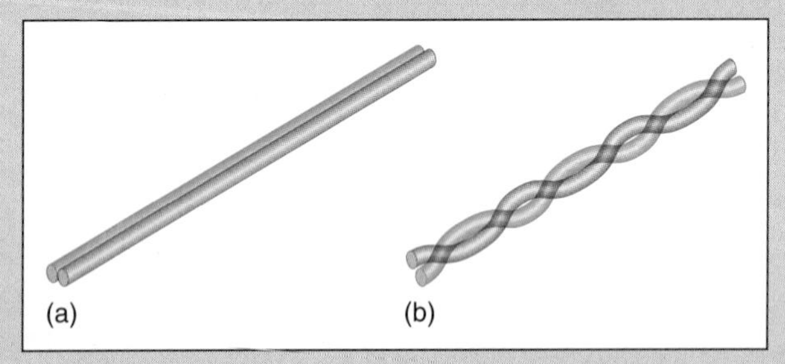

FIGURE Q.35

36. Your instructor gives you a piece of wire of length ℓ to conduct a given current. What shape of loop will give you the maximum magnitude of magnetic dipole moment? Is it better to use a single loop or a large number of small coils?

37. Take a highly flexible wire of small mass and lay it on a frictionless surface as in Figure Q.37. Close the switch. Describe and explain what happens to the flexible wire.

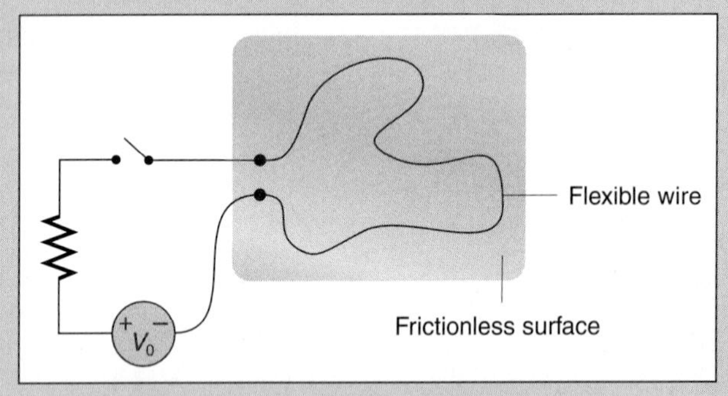

FIGURE Q.37

38. A hollow copper water pipe carries an electrical current I along its length that is uniformly distributed around the circumference of the pipe. What is the magnetic field within the hollow tube of the pipe? What is the magnetic field exterior to the pipe?

39. Why can either pole of a magnet attract an unmagnetized paper clip?

40. Pictured in Figure Q.40 are the magnetic field lines going into the page • and out of the page x produced by various current distributions. In each case, sketch the placement of the wire and the direction of the current that produces the indicated field. A few words of explanation also may be appropriate.

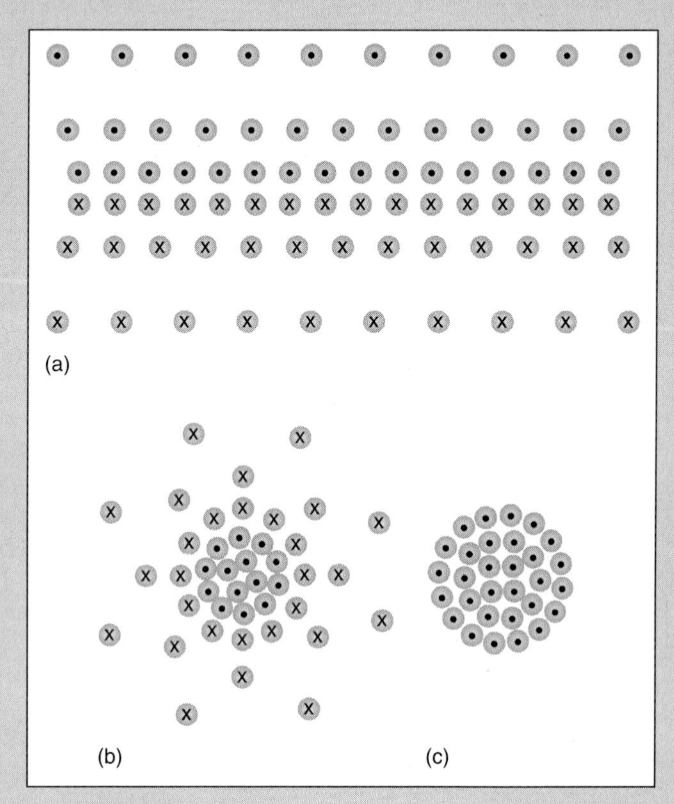

FIGURE Q.40

41. A permanent magnet with the polarities indicated in Figure Q.41 was discovered by Michael Davis ["A magnetic tripole," *The Physics Teacher*, 14, #1, page 34 (January 1976)]. Sketch how a current-carrying wire could be wound into a coil to produce such a tripolar arrangement.

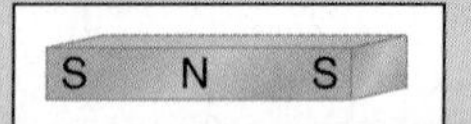

FIGURE Q.41

42. The magnetic axis of the Earth is inclined to its rotational axis by about 12°. Is the magnetic declination angle 12° at all locations on the surface of the Earth? Explain.

PROBLEMS

Sections 20.1 The Magnetic Field
20.2 Applications

1.–6. The wallpaper-like patterns in Figures P.1–P.6 represent various directions for a uniform magnetic field. Charged particles are incident into each region of the field with velocity vectors directed as indicated by the arrows on each charge. Sketch the *initial* direction of the magnetic force that acts on each charged particle.

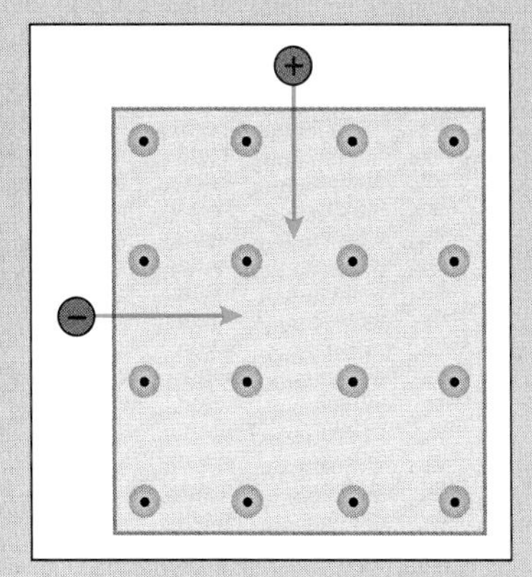

FIGURE P.1

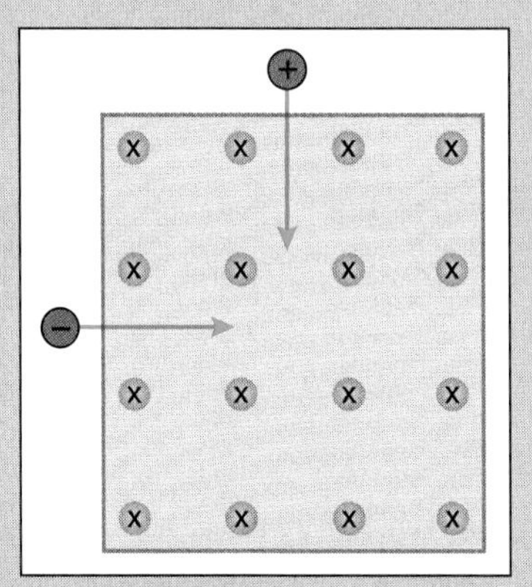

FIGURE P.2

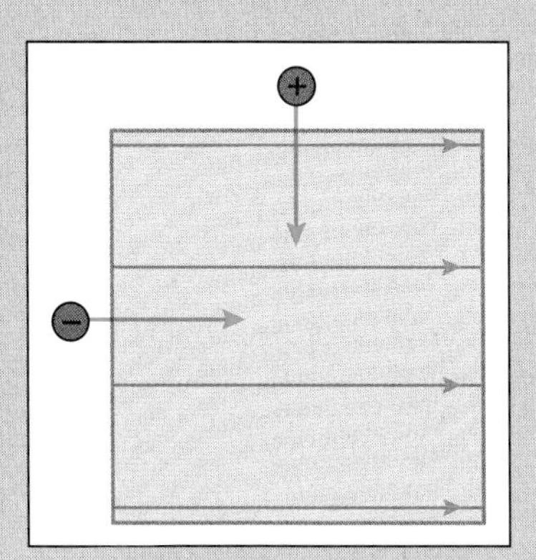

FIGURE P.3

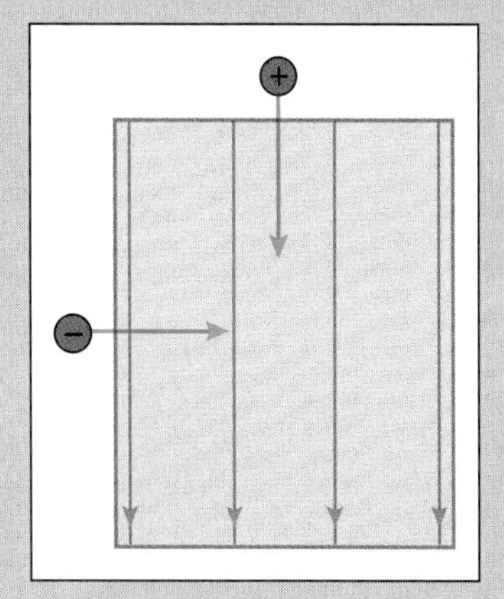

FIGURE P.4

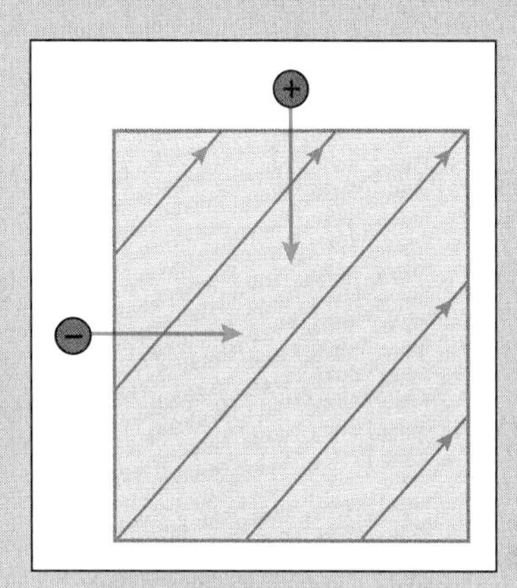

FIGURE P.5

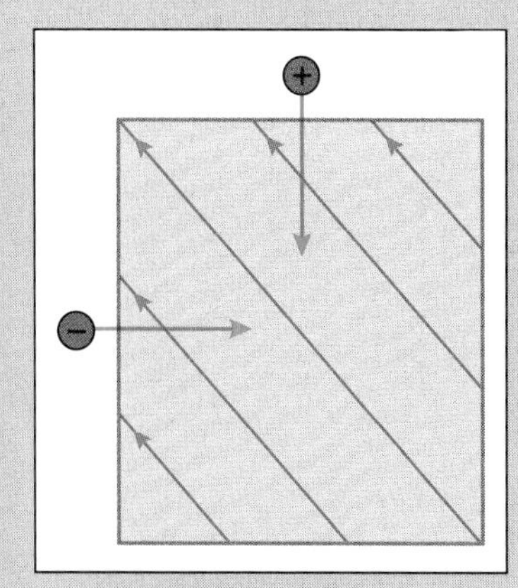

FIGURE P.6

7. An electron is moving at speed 5.0×10^6 m/s in the magnetic field of magnitude 6.0×10^{-2} T indicated in Figure P.7. (a) Find the magnitude and direction of the initial magnetic

force on the electron. (b) What is the magnitude of the acceleration of the electron?

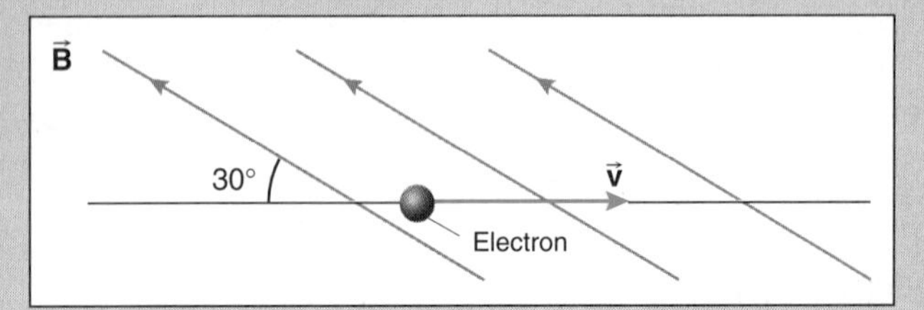

FIGURE P.7

•8. In the beginning, there was a field; and the field was magnetic and uniform; and it was good. Yea, and the field was directed perpendicularly out of the page as indicated in Figure P.8. And charged particles were sent forth into the field with various velocities (all perpendicular to the field direction); and forces and accelerations on these particles resulted. Indeed, the paths of the particles in the field were noted and recorded and none went astray but each to its own place. And each particle was numbered accordingly. (a) At some point along each of the paths, sketch the direction of the magnetic force (if any) acting on the particle. (b) Indicate clearly which particles (if any) were possessed with negative charge, which were possessed with positive charge, and which were possessed with no charge. (c) If all the particles have the same absolute magnitude of charge (if they have nonzero charge) and the same speed, which of the particles has the smallest mass? If this cannot be determined from the sketch, indicate that this is the case and why.

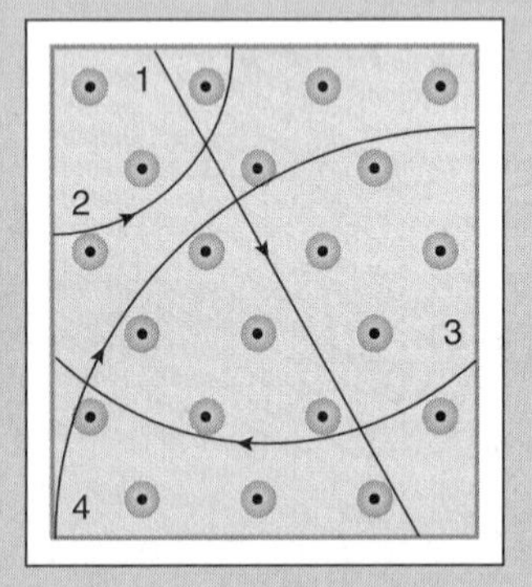

FIGURE P.8

• 9. An electron has velocity $(5.0 \text{ km/s})\hat{\imath} - (6.0 \text{ km/s})\hat{\jmath}$ in a magnetic field. The force on the electron is found to be $(3.84 \times 10^{-19}\text{ N})\hat{\imath} + (3.20 \times 10^{-19}\text{ N})\hat{\jmath} + (2.40 \times 10^{-19}\text{ N})\hat{k}$. The magnetic field lacks an x-component. Find the magnetic field.

•10. An electron, a proton, and a neutron each has a kinetic energy of 1.00 keV. They are projected into a semi-infinite region with a uniform magnetic field of magnitude 50.0 mT directed as indicated in Figure P.10. (a) Determine the speed of each particle. (b) Will the kinetic energy of each particle change in the region of the magnetic field? Explain your answer. (c) Schematically indicate the path that each particle follows (be sure to indicate which path is appropriate for each particle!). (d) Determine the radii of the path followed by each particle.

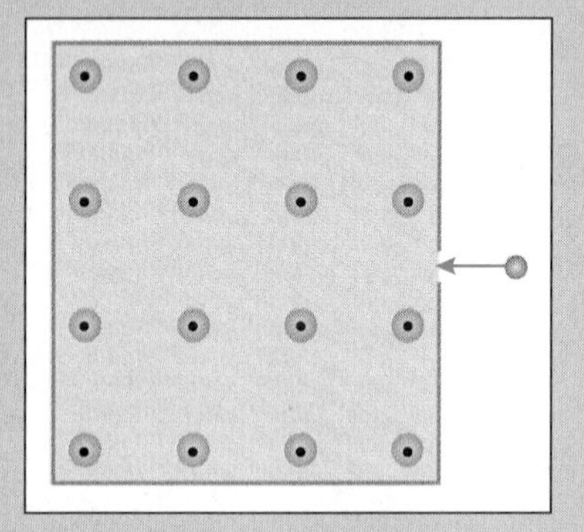

FIGURE P.10

•11. $\vec{B}$ careful. A charged bee ($q = -2.50 \times 10^{-3}$ C) is streaking at a speed of 22.0 m/s across a uniform magnetic field of magnitude 25 mT as indicated in Figure P.11. What are the magnitude and direction of the magnetic force on the bee?

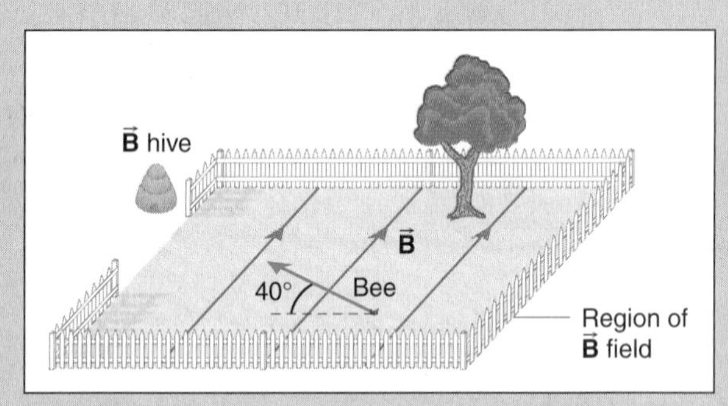

FIGURE P.11

•12. An electron moves at constant velocity in a region where there is an electric field of magnitude 500 N/C and a magnetic field of magnitude 0.150 T. (a) What is the angle between the electric and magnetic fields? (b) What is the minimum value for the speed of the electron? (c) Will a proton moving with the same minimum velocity also move with a constant velocity through the field region?

•13. A happy particle of mass 2.00 mg has charge $-10.0\ \mu$C and finds itself in a region where the only force on it is a magnetic force. The magnitude of the magnetic field at the location of the particle is 15.0 mT and the speed of the particle is 2.00 km/s. The angle between the directions of the magnetic field and the velocity vector of the particle is 75.0°. (a) Sketch the situation. (b) Determine the magnitude of the magnetic force acting on the particle. Specify the direction of the force. (c) Determine the magnitude of the acceleration of the particle.

•14. Magnetic fields are used to change the direction of charged particle beams. (a) A beam of electrons has a speed of 9.50×10^6 m/s. A uniform magnetic field, oriented perpendicular to the velocity of the electrons, is to be used to turn the electron beam through a 90° angle along a circular arc of radius 5.00×10^{-2} m. What is the magnitude of the magnetic field used? (b) If a beam of protons with the same speed is used in the same apparatus, what is the radius of the circular arc along which they move?

•**15.** An electron is moving nonchalantly with a velocity of $(3.00\ \text{km/s})\hat{\imath} + (4.00\ \text{km/s})\hat{\jmath}$ in a uniform magnetic field directed along the z-axis. The force on the electron is measured to be of magnitude 2.40×10^{-16} N, making an angle of 143.13° with $\hat{\imath}$. (a) Make a schematic diagram of the situation. (b) Determine the magnitude of the magnetic field and express the field in Cartesian vector form.

•**16.** A tiny particle of mass m with a charge q traveling at speed v enters a semi-infinite region with a uniform magnetic field of magnitude B directed as indicated in Figure P.16. Show that the time t the particle is in the region with the field is independent of the speed v and is given by

$$t = \frac{\pi m}{|q|B}$$

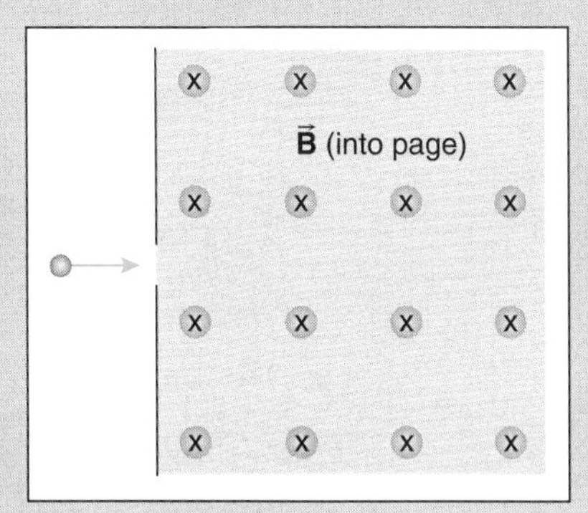

FIGURE P.16

•**17.** A velocity selector has a magnetic field of magnitude 0.15 T and an electric field of magnitude 2.00×10^4 N/C. A beam of negatively charged morons (each with charge $q = -3e$) enters the field as shown in Figure P.17. (a) If the electric field in the velocity selector is directed as indicated in Figure P.17, sketch the directions of the magnetic and electrical forces acting on each moron. (b) Indicate on the same sketch the orientation of the magnetic field inside the velocity selector. (c) What is the speed of the morons that emerge undeflected from the velocity selector?

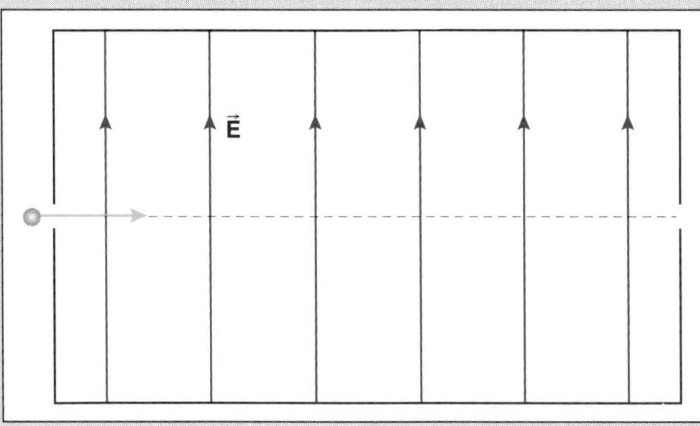

FIGURE P.17

•**18.** In a mass spectrograph, positive ions of mass m and charge q are accelerated from rest through a potential difference V and then encounter a uniform magnetic field perpendicular to their motion as indicated in Figure P.18. (a) In a sketch, indicate the trajectories of the ions in the region of the magnetic field. (b) Use the CWE theorem to find an expression (in terms of m, q, and V) for the speed of the ions as they exit the region in which they are accelerated. (c) Beginning with Newton's second law, show that the charge-to-mass ratio q/m of the ions is

$$\frac{q}{m} = \frac{8V}{B^2 d^2}$$

where d is the distance from the entrance slit of the magnetic field region to the place where the ions strike the wall of the device.

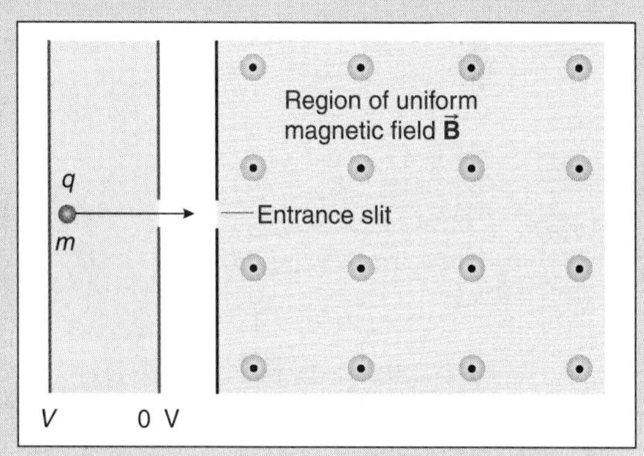

FIGURE P.18

•**19.** A semiconductor with a rectangular cross section carries a current I along the direction $\hat{k}$ as shown in Figure P.19. A magnetic field is parallel to $\hat{\imath}$. (a) If the charge carriers are predominantly negative (an n-type semiconductor), across which pair of faces of the conductor should the Hall voltage be measured? Which face is at the higher potential? (b) Repeat part (a) if the charge carriers are predominantly positive (a p-type semiconductor).

20. A particle with mass m and charge q traveling at speed v enters a region with a uniform magnetic field. The velocity vector makes an angle α with the field direction as indicated in Figure P.20. The trajectory of the particle is a helix whose axis is the direction of the magnetic field vector. (a) Show that the radius r of the helix is

$$r = \frac{mv \sin \alpha}{|q|B}$$

(b) Show that the time t for one revolution of the particle around a single turn of the helix is

$$t = \frac{2\pi m}{|q|B}$$

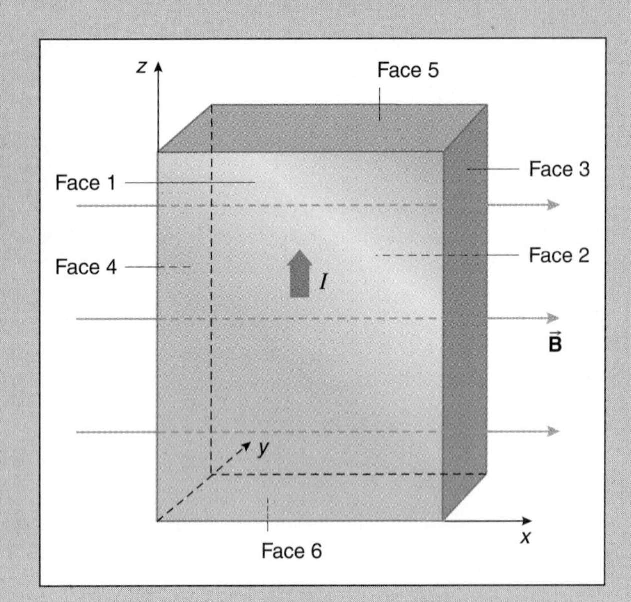

FIGURE P.19

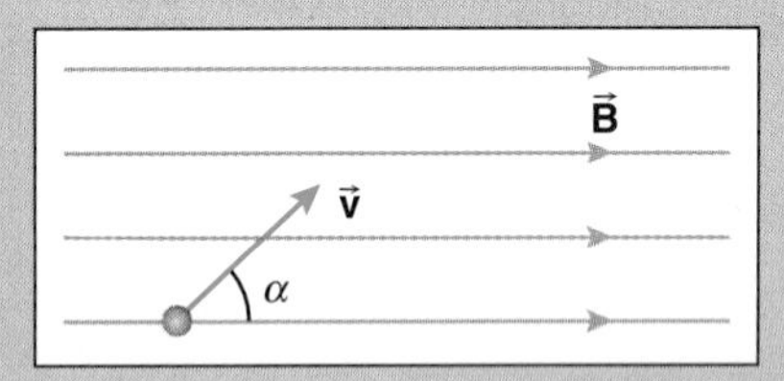

FIGURE P.20

(c) Show that the distance d between successive turns of the helix measured parallel to the axis, known as the pitch of the helix, is

$$d = \frac{2\pi mv \cos \alpha}{|q|B}$$

(d) What effect does the *sign of the charge* have on the spiral motion? Charged particles in space (electrons, ions, cosmic rays) encountering magnetic fields in space (such as that due to the Earth) are confined to such spiral motion. When such particles collide with the atmosphere of the Earth, *auroras* result.

21. Electrons of mass m and speed v pass through a small slit and enter a region of uniform magnetic field as shown in Figure P.21. The velocity vectors of the electrons all lie within a small angle $\theta << 1$ rad of the direction of the field, so the electrons enter the field region initially moving away from each other. Show that the magnetic forces on the electrons will confine the beam to a diameter d given by

$$d \approx \frac{2mv\theta}{eB}$$

where θ is in radians. This illustrates how a magnetic field can be used to keep a charged particle beam confined. The same principle is applicable to beams of other charged particles.

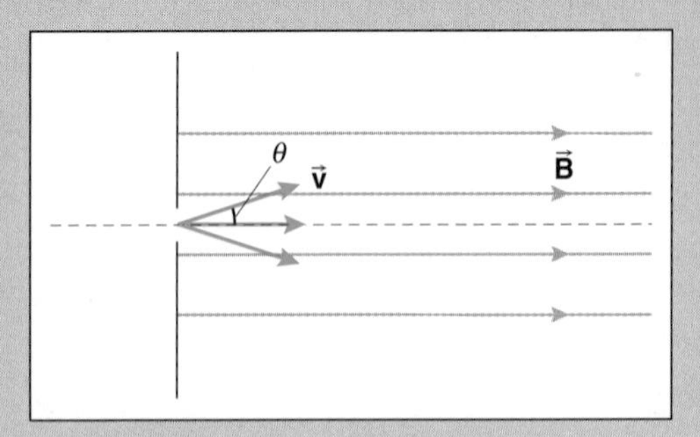

FIGURE P.21

Sections 20.3 Magnetic Forces on Currents
20.4 Work Done by Magnetic Forces
20.5 Torque on a Current Loop in a Magnetic Field

22. A wire of infinite length carries a current of 15.0 A in a uniform magnetic field of magnitude 0.55 T as indicated in Figure P.22. (a) Find the force on a one meter length of the wire. (b) If the angle that the wire makes with the magnetic field is tripled, what happens to the direction of the force? Is the magnitude of the force tripled? Explain your answer.

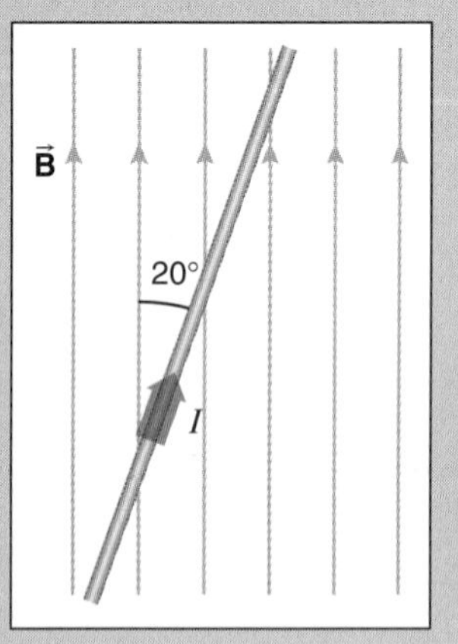

FIGURE P.22

•23. The wire shown in Figure P.23 carries a current I in a uniform magnetic field $\vec{B}$ directed as indicated. Show that the total magnetic force on the wire is zero.

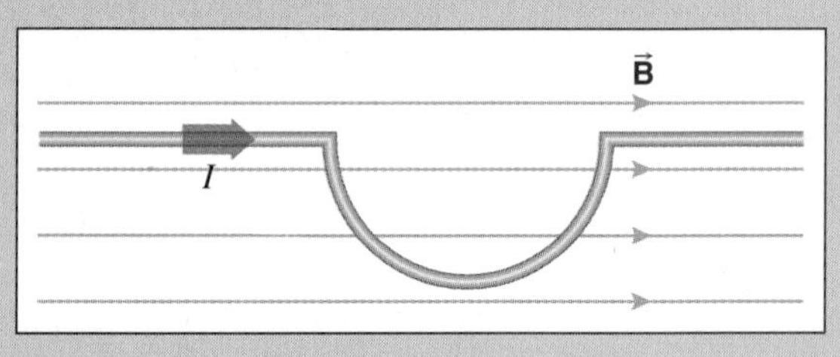

FIGURE P.23

•24. A semicircular wire carries a current I in a uniform magnetic field $\vec{B}$ as shown in Figure P.24. Show that the total magnetic force on the wire is $-2IBR\hat{\jmath}$.

•25. A circular current loop of radius 5.0 cm carries a current of 1.50 A. The loop is situated in a uniform magnetic field of magnitude 0.60 T directed as indicated in Figure P.25. (a) What is the magnitude of the magnetic dipole moment of the current loop?

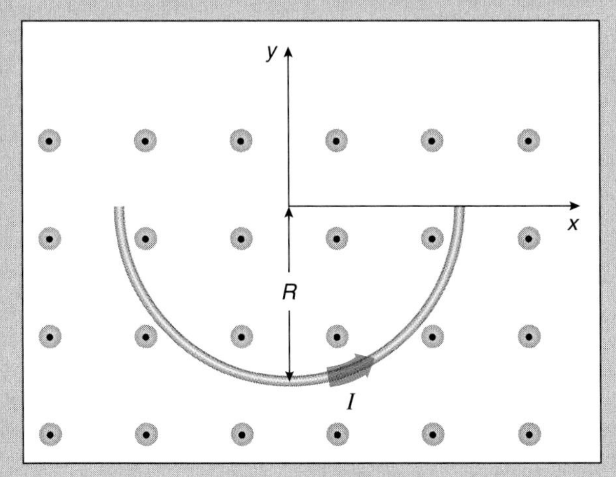

FIGURE P.24

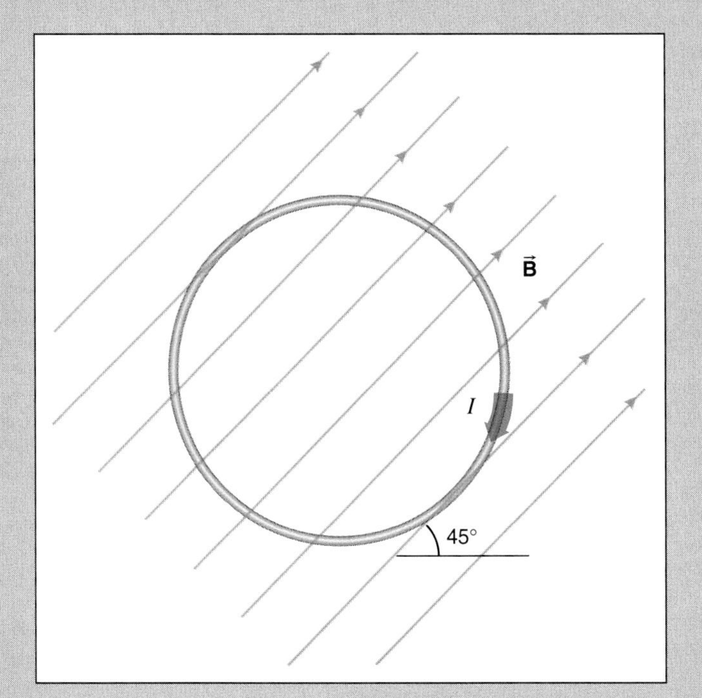

FIGURE P.25

(b) Calculate the magnitude of the torque on the current loop. (c) Indicate clearly in a sketch the axis about which the loop will rotate, as well as the rotational sense of the motion under the influence of the magnetic torque. (d) What is the potential energy of the current loop in the orientation indicated?

•**26.** The current loop in Figure P.26 consists of two semicircles of different radii. (a) Calculate the magnitude of the magnetic dipole moment of the current loop. (b) Specify two directions for a uniform magnetic field that produce a total torque on the loop of 0 N · m. (c) Specify a direction of a magnetic field that produces the maximum magnitude of torque on the current loop. (d) Calculate the magnitude of the maximum torque on the loop.

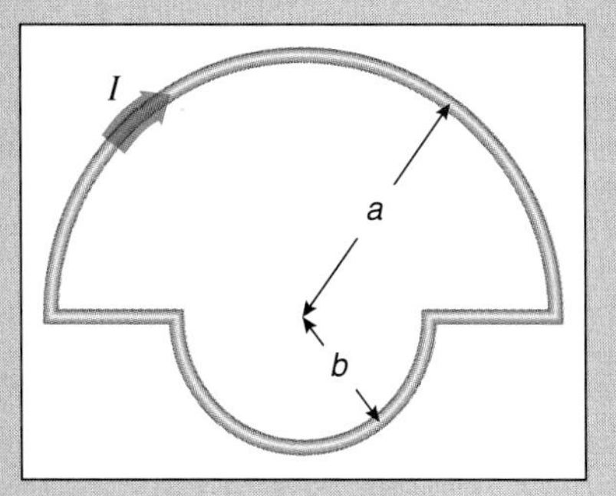

FIGURE P.26

•**27.** A constant magnetic field of magnitude 1.0 mT is directed horizontally. Three independent, flat current loops are suspended to pivot about a vertical symmetry axis that touches the loop at two points. Each current loop carries a current of 15.0 A and each loop is oriented so the plane of the loop makes an angle of 30° with the direction of the magnetic field. The torques on all the loops are identical. One of the loops is a square with sides of 20.0 cm. (a) Find the magnitude of the torque on the loops. (b) Another of the loops is an equilateral triangle with sides ℓ. Find ℓ. (c) The third loop is a circle of radius R. Find R.

•**28.** Calculate the magnitude of the magnetic dipole moment of the current loop indicated in Figure P.28.

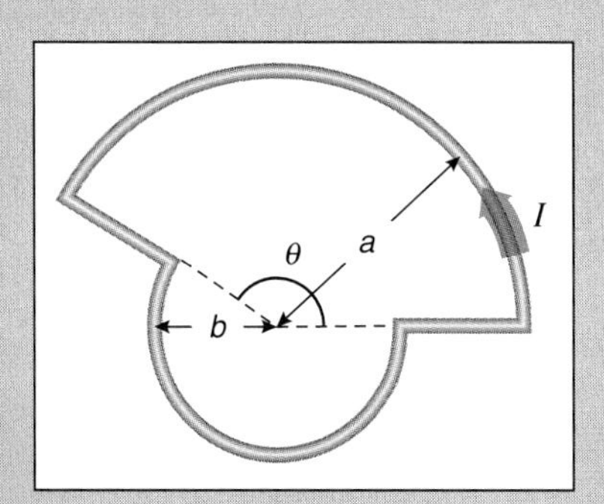

FIGURE P.28

•29. Two parallel, frictionless rails, separated by distance ℓ and inclined at angle θ to the horizontal, are bridged by a conducting bar of mass m, initially at rest, that is free to move; see Figure P.29. The system lies in a uniform magnetic field $\vec{B}$ directed vertically upward on the surface of the Earth. Neglect the magnetic field of the Earth. When $t = 0$ s, an appropriate source establishes the indicated current I in the system. (a) Determine the magnitude of the magnetic force on m. (b) Draw a second law force diagram indicating all the forces acting on m. (c) Find an expression for the velocity of m as a function of time. (d) What value of B keeps the conducting bar in equilibrium?

•**30.** Two circular current loops have radii r_1 and r_2 as shown in Figure P.30. Each loop carries the same current I. (a) If the

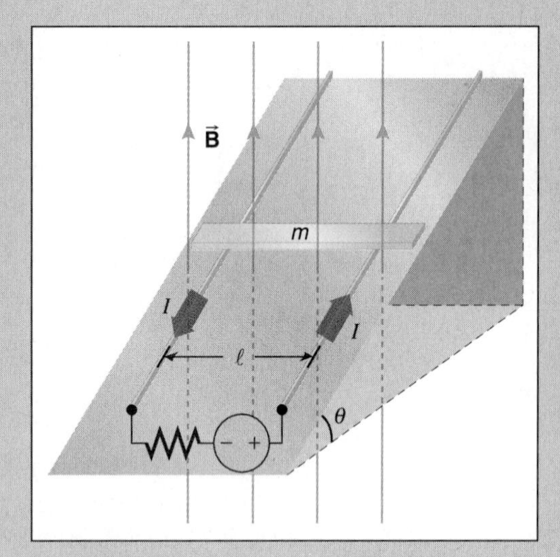

FIGURE P.29

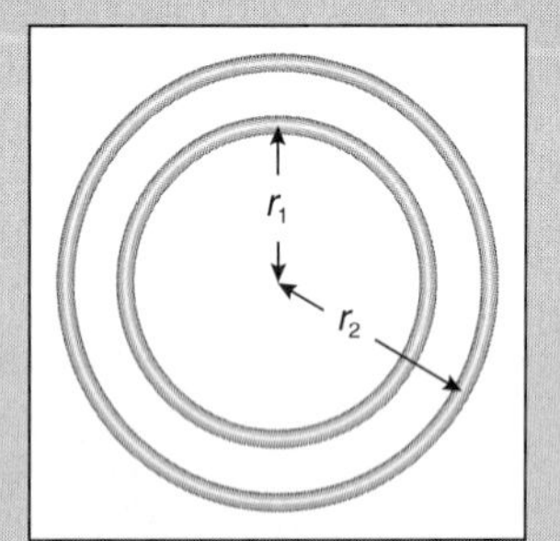

FIGURE P.30

current in each loop is clockwise, determine the magnitude of the magnetic dipole moment of the system of two loops. (b) If the current in the outer loop is clockwise while that of the inner loop is counterclockwise, determine the magnitude of the magnetic dipole moment of the system of two loops.

•**31.** A rectangular current loop carrying a current I is placed in a uniform magnetic field as shown in Figure P.31. (a) Determine the total force on the current loop. (b) Determine the magnetic dipole moment of the current loop. (c) Determine the total torque on the current loop.

•**32.** A rectangular current loop carrying a current I is placed in a uniform magnetic field as shown in Figure P.32. (a) Determine the total force on the current loop. (b) Determine the magnetic dipole moment of the current loop. (c) Determine the total torque on the current loop.

•**33.** Specify the area of a planar coil with 100 turns of wire carrying 25.0 A that can have a maximum torque of magnitude 20.0 N·m when placed in a magnetic field of magnitude 0.200 T.

•**34.** A flat, square piece of wood 20.0 cm on a side lies on a flat table. An external, horizontal field of magnitude 1.2 T is present (see Figure P.34). Around the perimeter of the wood square are wound 100 turns of wire carrying a current I. The total mass of the wood and loop system is 2.00 kg. If the current I is large enough, the square will rise up about one of its edges when the torque of the magnetic force on the loop exceeds the torque of the weight of the system taken about the same edge.

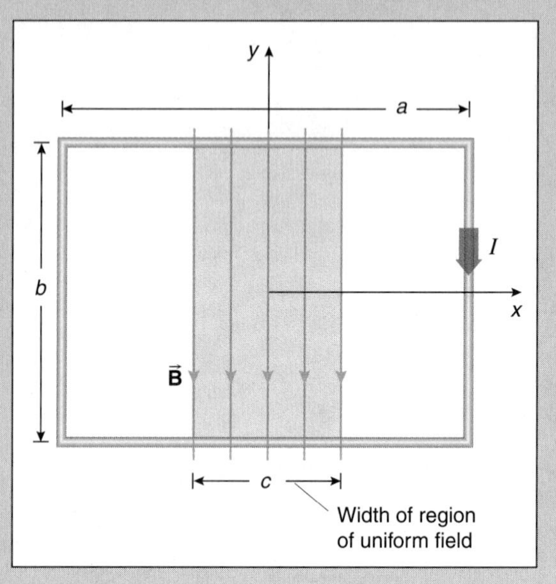

FIGURE P.31

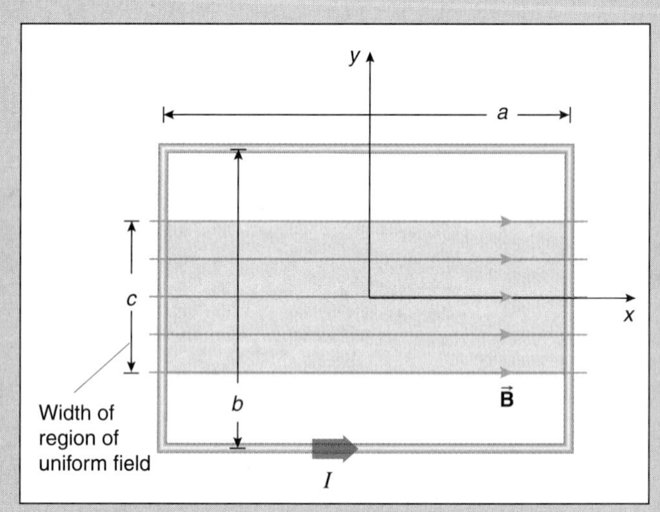

FIGURE P.32

(a) Indicate in a diagram which of the four edges of the square remains in contact with the table when the current I becomes large enough to make the square begin to rotate. (b) Calculate the minimum value for the current I needed to make the square begin to rise about an edge. (c) By what factor would the answer in (b) change if the 20.0 cm square changed to 40.0 cm, with the mass and field staying the same?

•**35.** Electromagnetic rail guns can accelerate small masses to very high speeds extraordinarily quickly. They have the potential to be used for antimissile defense systems as well as for launching materials from the lunar surface to spacecraft in lunar orbit. A rail gun consists of two parallel conducting rails, oriented perpendicular to a strong magnetic field as shown in Figure P.35. Consider the rails to be horizontal. Immediately to the rear of the projectile is a conducting fuse that vaporizes and forms a conducting gas when a large current is initiated briefly in the parallel conductors. The current in the ionized gas experiences a magnetic force that accelerates the gas and the projectile along the rails. Let the constant current pulse be 15 MA and the distance between the

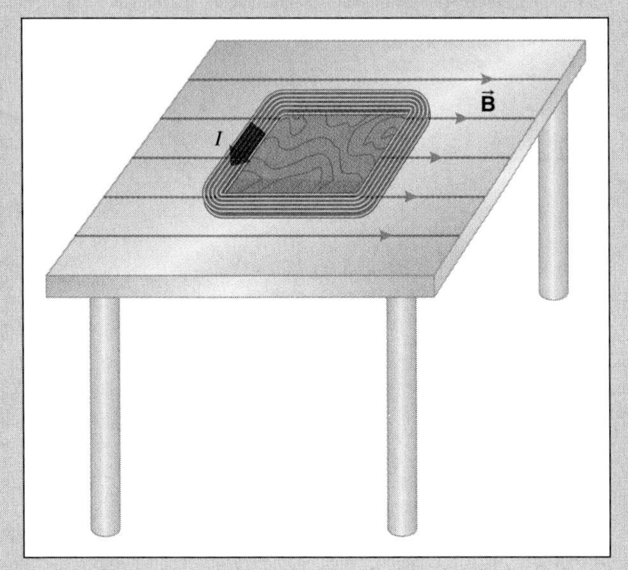

FIGURE P.34

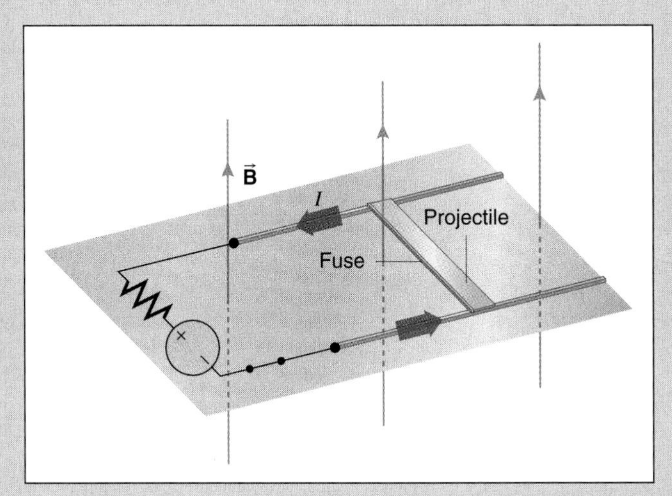

FIGURE P.35

conducting rails be 2.0 cm; the rails are 4.0 m in length. The magnetic field has a magnitude of 2.6 T. (a) Determine the magnitude of the magnetic force on the conducting gas. (b) Assume the force on the gas is transferred without diminution to a projectile of mass 0.10 kg. Neglect any friction between the projectile and rails. Determine the magnitude of the acceleration of the projectile and the launch speed.

36. A charge q is distributed uniformly on the surface of a sphere of radius R rotating about a diameter with angular velocity $\vec{\omega}$. Show that the magnetic dipole moment $\vec{\mu}$ of the sphere is

$$\vec{\mu} = \frac{1}{3}qR^2\vec{\omega}$$

Section 20.6 The Biot–Savart Law

37. Show that the SI units $A \cdot m^2 \cdot T$ are equivalent to J.

38. A long, straight, 12 gauge copper wire (diameter 2.05 mm) carries a current of 15.0 A. Use the results of Example 20.11 to find the magnitude of the magnetic field at the surface of the wire.

•**39.** A very long wire carrying 25.0 A has a semicircular deformation as indicated in Figure P.39. Poor thing. (a) What contribution does the straight segment of wire to the left of the semicircle make to the magnetic field at point P? (b) What is the contribution to the magnetic field at P by the straight segment of wire to the right of the semicircle? (c) What is the total magnetic field at point P? (d) *Describe* the difficulties encountered in using the Biot–Savart law to calculate the magnetic field at a point that is *not* at the center of the semicircular loop. (Do not make the calculation.)

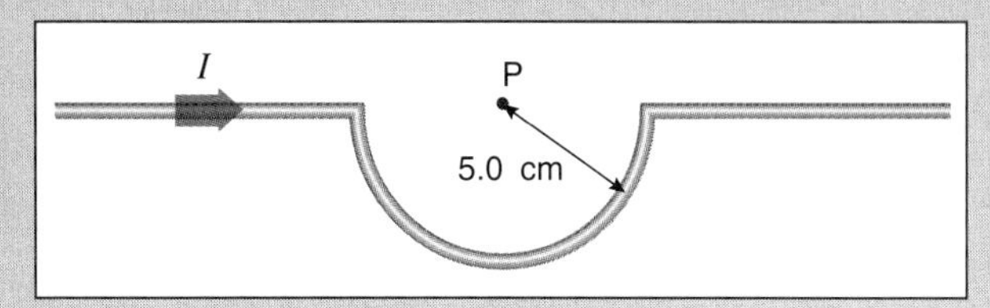

FIGURE P.39

•**40.** A long wire makes a semicircular U-turn of radius R about point 0 in Figure P.40. If the wire carries a current I, use the Biot–Savart law to find the magnetic field at point 0.

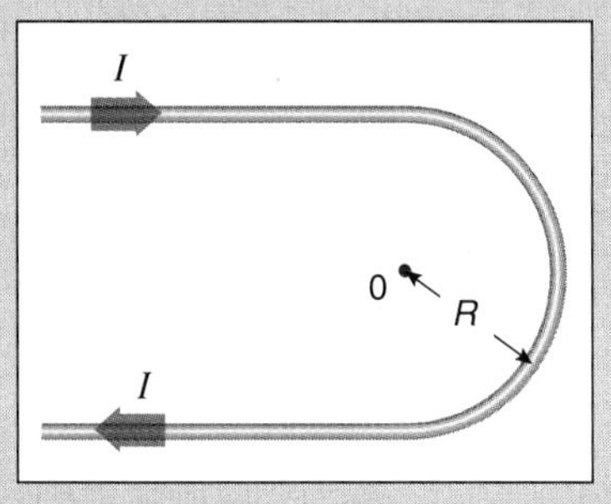

FIGURE P.40

•41. A long wire splits into two identical semicircular segments as shown in Figure P.41. Find the magnetic field at the center of the circle.

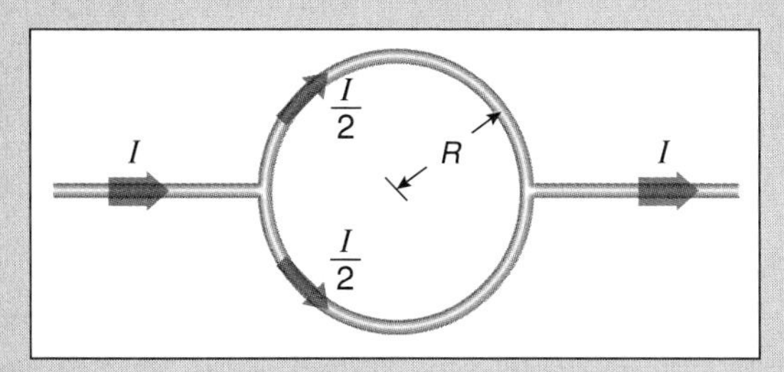

FIGURE P.41

•**42.** A long wire splits into two identical triangular segments as shown in Figure P.42. Find the magnetic field at point P.

•**43.** The electron in the hydrogen atom orbits the nuclear proton in a circular orbit of radius 0.529×10^{-10} m with a frequency of 6.58×10^{15} Hz. (a) Sketch the situation; include the orbital direction of the electron and the direction of the current its

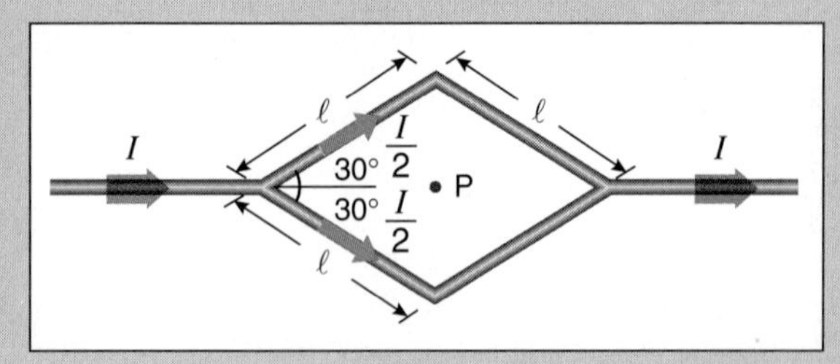

FIGURE P.42

motion represents. (b) What is the magnitude of the magnetic field produced by the electron at the position of the nuclear proton? Indicate the direction of the field in the sketch in part (a). (c) The intrinsic magnetic dipole moment of the proton has a magnitude of

$$\mu_{\text{proton}} = 1.41 \times 10^{-26}\ \text{A}\cdot\text{m}^2$$

The magnetic dipole moment of the proton aligns itself parallel to the magnetic field produced by the circulating electron. What work needs to be done (with no change in kinetic energy) to turn the magnetic dipole moment of the proton so it is antiparallel to the magnetic field? Express your result in eV.

•**44.** A straight wire segment of length ℓ carries a current I (see Figure P.44). (a) Use the Biot–Savart law to show that the magnitude of the magnetic field at a distance z from the wire along its perpendicular bisector is

$$B = \frac{\mu_0}{4\pi}\frac{2I}{z}\frac{\ell}{(\ell^2 + 4z^2)^{1/2}}$$

(b) Indicate the direction of the field in a sketch. (c) Show that as $\ell \to \infty$ m, this expression for the magnitude of the field approaches that of an infinite wire (the result of Example 20.11).

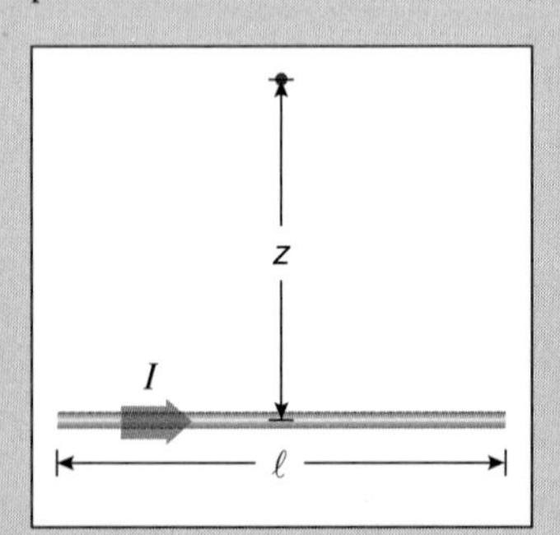

FIGURE P.44

•**45.** (a) Use the result of Problem 44 to show that magnitude of the magnetic field at the center of a square current loop with sides ℓ, carrying current I (see Figure P.45), is

$$B = \frac{\mu_0}{4\pi}8\sqrt{2}\frac{I}{\ell}$$

(b) Specify the direction of the field.

•**46.** (a) Use the Biot–Savart law to show that the magnitude of the magnetic field at a distance z along a perpendicular from the end of a straight wire segment of length ℓ (see Figure P.46), carrying current I, is

$$B = \frac{\mu_0}{4\pi}\frac{I}{z}\frac{\ell}{(\ell^2 + z^2)^{1/2}}$$

(b) Specify the direction of the field.

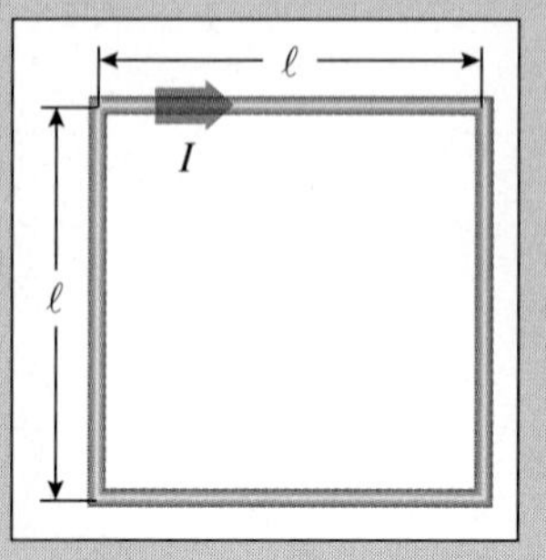

FIGURE P.45

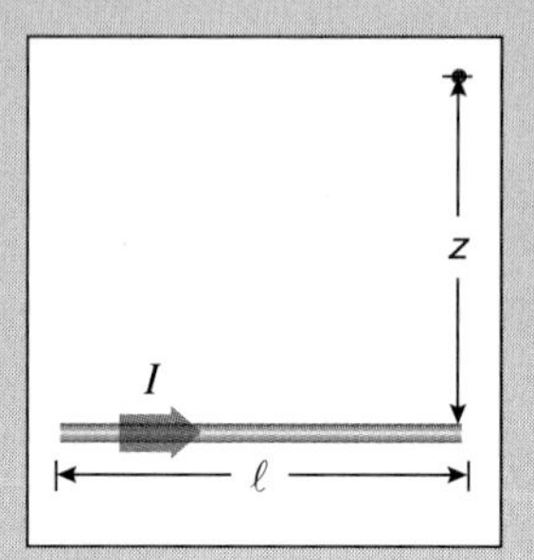

FIGURE P.46

•**47.** Use superposition, the results of Problem 46, and suitable changes of notation to redo Problem 44.

•**48.** Two infinitely long straight wires lie in the same plane and carry the currents depicted in Figure P.48. (a) Find the magnetic field at the point P located 25 cm from the intersection of the wires along the bisector of the acute angle between them. (b) Find the magnetic field at the point S, located 25 cm from the intersection of the wires along the bisector of the obtuse angle between the wires.

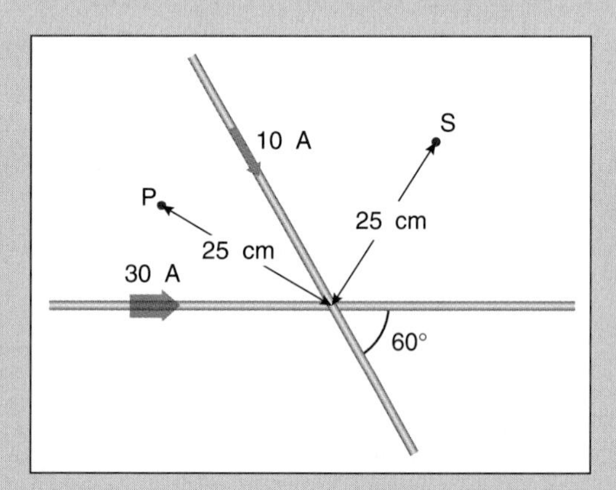

FIGURE P.48

•**49.** A proton is moving with speed 5.0×10^7 m/s near a long, straight wire carrying a current of 15 A as shown in Figure P.49. (a) What is the magnitude of the magnetic field of the wire at the position of the proton? (b) Determine the magnitude of the force on the proton. Show the direction of the force in a sketch. (c) What is the magnitude and direction of the force of the proton on the wire? (d) What is the speed of the proton after it has moved 5.0 cm from the position shown in Figure P.49?

•**50.** An otherwise infinite, straight, wire has two concentric loops of radii a and b carrying equal currents in opposite directions as

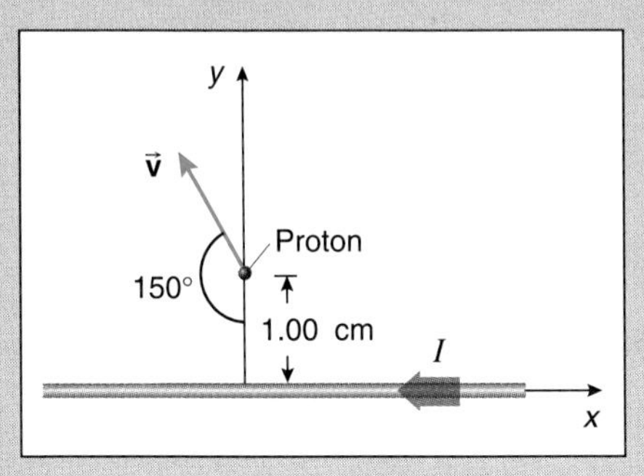

FIGURE P.49

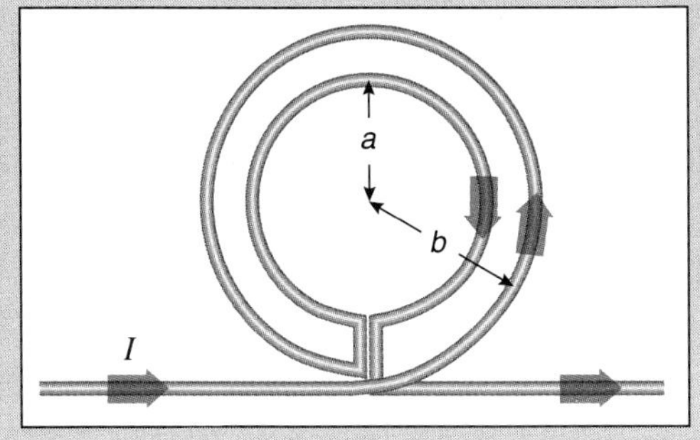

FIGURE P.50

shown in Figure P.50. Show that the magnitude of the magnetic field at the common center of the loops is zero if the radii have the ratio

$$\frac{a}{b} = \frac{\pi}{\pi + 1} \approx 0.7585$$

The idea for this problem comes from Kenneth W. Ford, *Classical and Modern Physics* (Xerox College Publishing, Lexington, Massachusetts, 1973), page 812.

•**51.** A circular wire ring of diameter 20 cm carries a current of 5.0 A directed as indicated in Figure P.51. A moving electron just happens to be passing through the neighborhood. When the electron is at the center of the circular ring and moving at speed 2.50×10^6 m/s in the direction indicated in the sketch, find the acceleration of the electron.

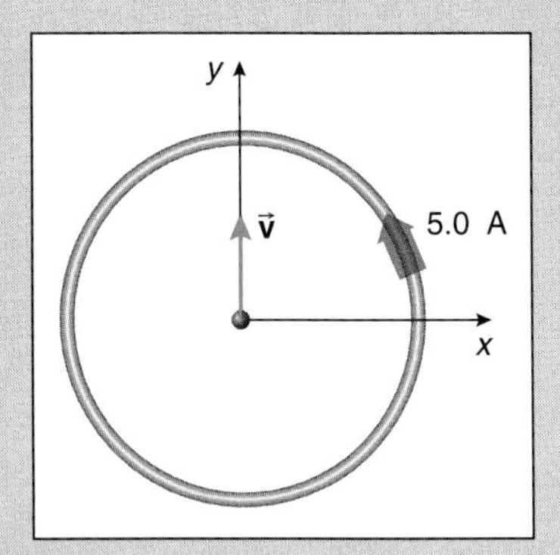

FIGURE P.51

•**52.** (a) For the two infinitely long wires shown in Figure P.52, determine the magnitude and direction of the magnetic field of the top wire at the position of the bottom wire. (b) What is the magnitude and direction of the magnetic field of the bottom wire at the position of the top wire? (c) What is the force of the top wire on 2.00 m of the length of the bottom wire? (d) What is the force of the bottom wire on 2.00 m of the top wire?

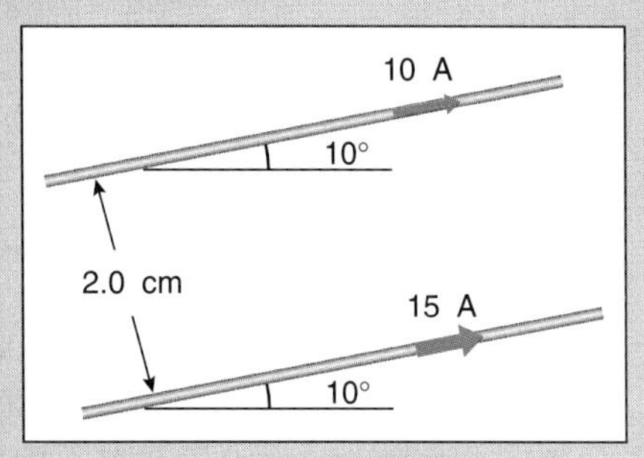

FIGURE P.52

•**53.** The rectangular current loop in Figure P.53 finds itself in the pleasant company of an infinite, straight wire also carrying current. (a) Determine the magnetic field of the infinite wire at the position of the side of the rectangular loop most distant from the infinite wire. (b) Determine the magnetic force of the infinite wire on the side of the rectangle most distant from the infinite wire. (c) Determine the magnetic field of the infinite wire at the position of the side of the rectangular loop closest to the infinite wire. (d) Determine the magnetic force of the infinite wire on the side of the rectangle closest to the infinite wire. (e) By considering symmetrically placed segments of the rectangular wire perpendicular to the infinite wire, show that the total magnetic force on these perpendicular segments is zero. (f) Calculate the total magnetic force on the rectangular current loop due to the magnetic field of the infinite wire. (g) What is the total force exerted on the infinite wire by the (complicated!) magnetic field of the current loop? (Hint: It is not necessary to do an intricate calculation; think of Newton's laws.)

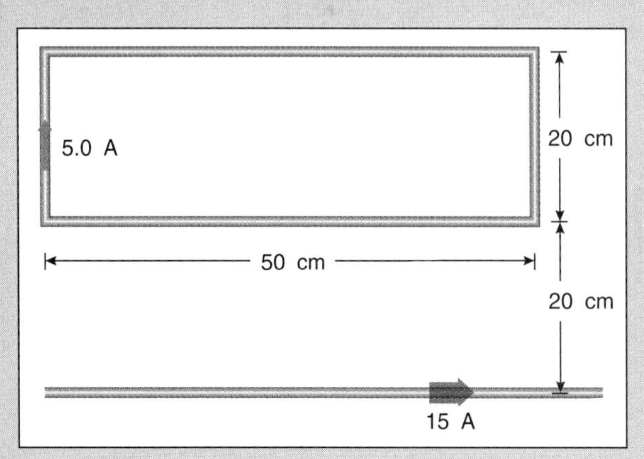

FIGURE P.53

•**54.** An infinitely long, straight wire lies along the z-axis of a Cartesian coordinate system and carries a current of 10.00 A toward increasing values of z. (a) Determine the magnetic field of the wire at the point P described by the coordinates: $x = 3.000$ m, $y = 4.000$ m, $z = 5.000$ m. Express $\vec{B}$ in Cartesian form. (b) An electron is at point P, cavorting along at a speed of 2.000×10^6 m/s parallel to the current in the wire. Determine the magnetic force acting on the electron when it is at this location; express the force in Cartesian form. (c) Show explicitly that $\vec{F} \cdot \vec{v} = 0\ \text{N·m/s}$ and $\vec{F} \cdot \vec{B} = 0\ \text{N·T}$. What is the meaning of these vanishing scalar products? (d) Will the force in (b) continue to act in the same direction at later times? Explain why or why not.

•**55.** A wire carrying current I has the shape indicated in Figure P.55. Beginning with the Biot–Savart law, find the magnetic field at the point P at the center of the circular arcs.

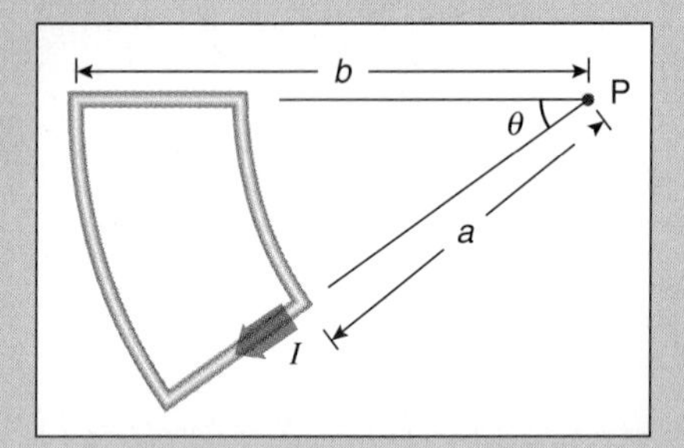

FIGURE P.55

•**56.** The two infinite, insulated wires indicated in Figure P.56 lie in the same plane and carry equal currents I, directed as shown. The four positions P_1, P_2, P_3, and P_4 are located symmetrically and are at a distance d from the point where the wires cross over each other. (a) Which points have the greatest value for the magnitude of the magnetic field? What is the magnitude of the field at those points? (b) Which points have the smallest value of the magnitude of the magnetic field? What is the magnitude of the field at those points?

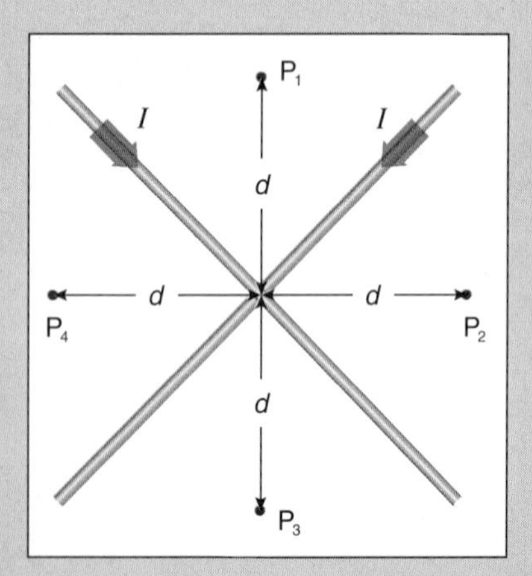

FIGURE P.56

•**57.** A coaxial cable (see Figure P.57) consists of a long straight inner wire of diameter 1.0 mm surrounded by a thin, cylindrical conducting shell of radius 8.0 mm symmetrically placed around the central wire. The cable carries a 3.0 A current via the inner conductor to a resistor some distance away; the current returns through the outer sheath. (a) Find the magnitude of the magnetic field close to the surface of the inner wire. Indicate its direction in a sketch. (b) Find the magnitude of the magnetic field at a distance of 20 mm radially away from the axis of the system.

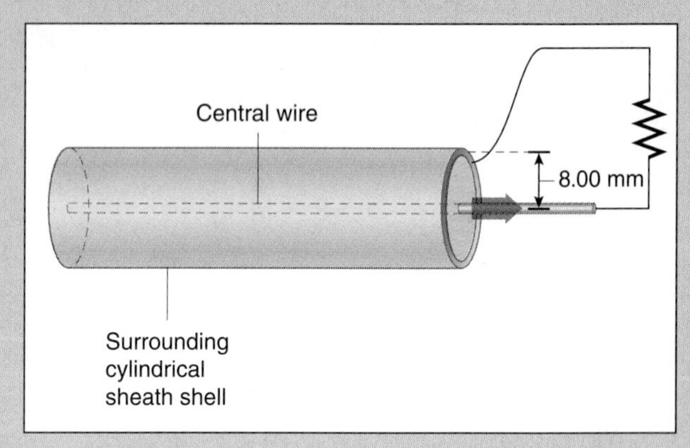

FIGURE P.57

•**58.** (a) Use the results of Example 20.10 to find the magnitude of the magnetic field along the axis of a circular current loop (with current I) of radius R at a point P a distance $z = R/2$ from the plane of the loop. (See Figure P.58a). (b) If instead of a single current loop, there are N loops essentially at the same location, what is the magnitude of the magnetic field at the location in part (a)? (c) If a second identical coil of N loops is arranged as in Figure P.58b, show that the magnitude of the magnetic field at point P is

$$B = \frac{\mu_0}{4\pi} \frac{32\pi}{5\sqrt{5}} \frac{NI}{R}$$

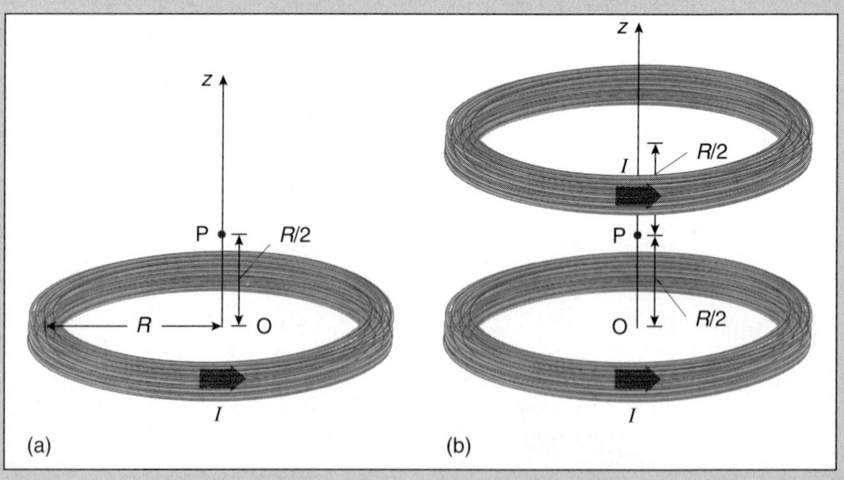

FIGURE P.58

Such an arrangement of twin coils, separated by a distance equal to the radius of the coils, is known as a *Helmholtz coil.* The arrangement commonly is used in the laboratory to produce a magnetic field in the region near the centrally located point P that is quite uniform. For distances along the axis that are up to 10% of the radius of the coil, the magnetic field varies from the value calculated above by less than 1 part in 5000 (0.02%).

•**59.** Two parallel wires, separated by a distance d, are oriented coming out of the page in Figure P.59. Each wire is of infinite length and carries current I. The point P lies equidistant from each wire. Find the magnitude and direction of the magnetic field at point P: (a) if the current in both wires is along $\hat{\mathbf{k}}$; (b) if the current in the left wire is along $\hat{\mathbf{k}}$ and that in the right wire is along $-\hat{\mathbf{k}}$.

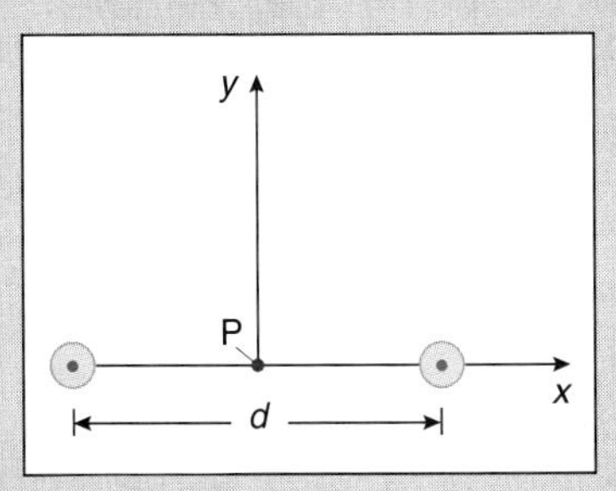

FIGURE P.59

•**60.** Two parallel, infinitely long wires, separated by a distance d, carry parallel currents I and βI, where β is a purely numerical multiple (see Figure P.60). (a) Find the location(s) of the lines along which the total magnetic field is 0 T. (b) Use the result of part (a) to show that if $\beta = 1$, the location of the line is midway between the two wires.

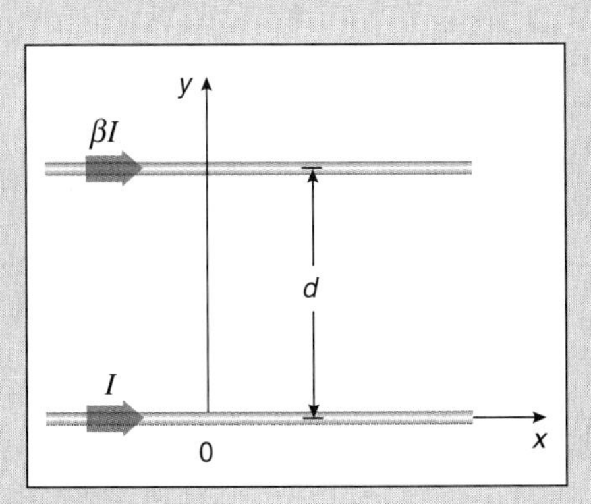

FIGURE P.60

61. A wire snakes along an arbitrary curved path in the x–y plane between the origin and point P; see Figure P.61. A uniform magnetic field is parallel to $\hat{\mathbf{k}}$. An appropriate source (not shown) establishes a current I in the wire. (a) By considering a series of infinitesimal steps parallel to $\hat{\mathbf{\imath}}$ and $\hat{\mathbf{\jmath}}$, show that the total force on the wire is the same as that on a straight wire between the origin and point P. (b) Show that if the wire is a complete loop, so that P is coincident with the origin, the total force on the wire vanishes.

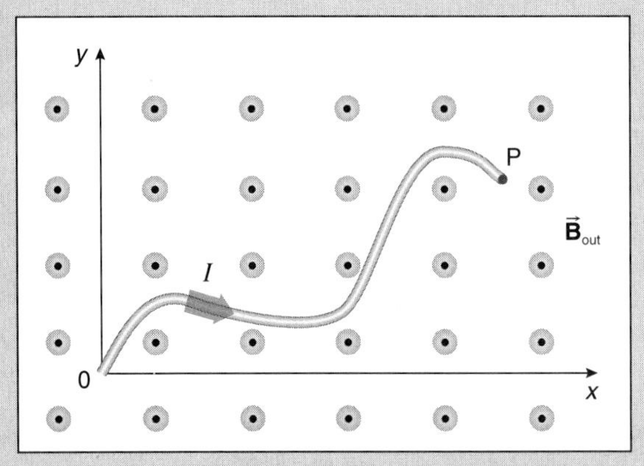

FIGURE P.61

62. A wire with current I follows the y-axis from $y = +\infty$ m to the origin, then the x-axis out to $x = +\infty$ m, as shown in Figure P.62. Show that the magnetic field in the first quadrant (where $a > 0$ m and $b > 0$ m) is

$$\vec{\mathbf{B}} = \frac{\mu_0}{4\pi} I \left[\frac{1}{a} + \frac{1}{b} + \frac{a}{b\left(a^2 + b^2\right)^{1/2}} + \frac{b}{a\left(a^2 + b^2\right)^{1/2}} \right] \hat{\mathbf{k}}$$

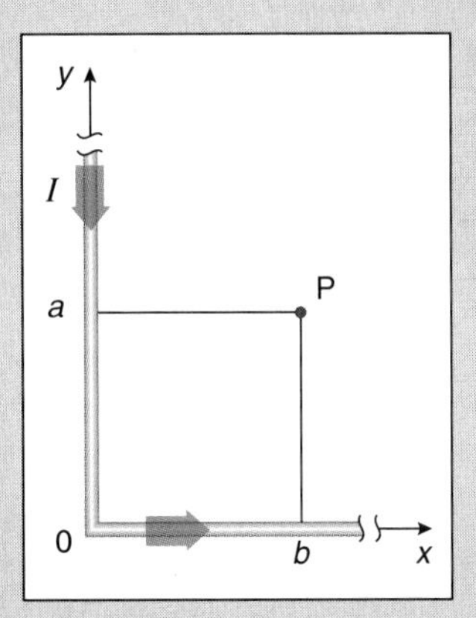

FIGURE P.62

63. The magnetic field of the Earth has a south magnetic pole in the northern geographic hemisphere and a north magnetic pole in the southern geographic hemisphere. The Earth also rotates in an eastward sense, taking one day to complete a rotation. For simplicity, assume that the line between the magnetic poles is coincident with the rotational axis of the Earth. (a) Imagine that the magnetic field of the Earth is produced by an excess charge distributed uniformly over the surface of the Earth. The spin of the Earth means that the excess charges form current loops. In order to produce the observed polarity of the magnetic field of the Earth, must the excess charge of the Earth be positive or negative? Indicate your reasoning. (b) Imagine (unrealistically) the excess charge Q to be localized in an equatorial ring of radius essentially equal to the radius of the Earth (6370 km). Given that the rotational period of the Earth is $T = 23$ h 56 min, what is the effective current represented by this equatorial distribution of charge? (c) The order of magnitude of the magnetic field of the Earth is about 10^{-4} T. Make the assumptions and use the results of part (b). What must be the magnitude of the charge Q to produce a magnetic field of magnitude $\approx 10^{-4}$ T at the center of the equatorial current loop?

64. A charge q moves at constant velocity $v_0\hat{\imath}$ along the x-axis as shown in Figure P.64. The charge passes the origin when $t = 0$ s. The moving charge is, of course, a current and thus produces a magnetic field. At any given point in space, this magnetic field varies with time. Take a point P to be located at coordinate y along the y-axis. (a) Give arguments to demonstrate that the magnetic field at P is directed along $\hat{k}$ at all times. (b) Show that the magnetic field at P varies with time as

$$\vec{B}(t) = \frac{\mu_0}{4\pi}\frac{qv_0y}{(y^2 + v_0^2t^2)^{3/2}}\hat{k}$$

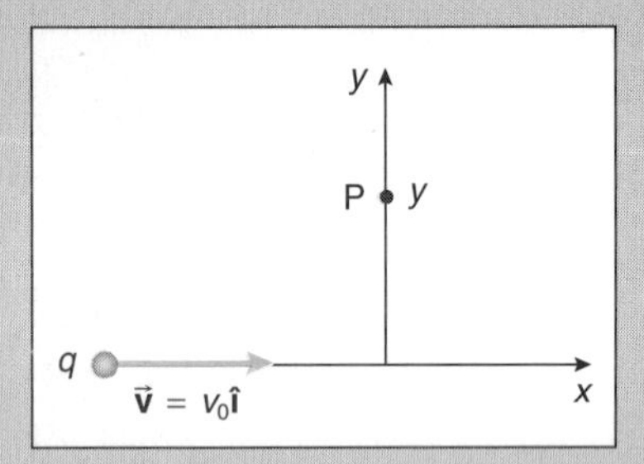

FIGURE P.64

Sections 20.7 Forces of Parallel Currents on Each Other and the Definition of the Ampere
20.8 Gauss's Law for the Magnetic Field*
20.9 Magnetic Poles and Current Loops

•**65.** You discover an electron promenading around in a circle (see Figure P.65) under the action of a magnetic force. (a) In a sketch, indicate the direction of the magnetic force on the electron. (b) On the same sketch, indicate the direction of the magnetic field causing the circular motion of the electron. (c) Beginning with Newton's second law, find the relationship between the radius r of the circle, the magnetic field magnitude B, and the intrinsic properties of the electron (charge, mass, etc.). (d) Calculate the work done by the magnetic force as the electron traverses one orbital path. (e) A great distance away (so you can effectively ignore the circulating electron), a proton is found at rest in the same field. Describe what happens to the proton.

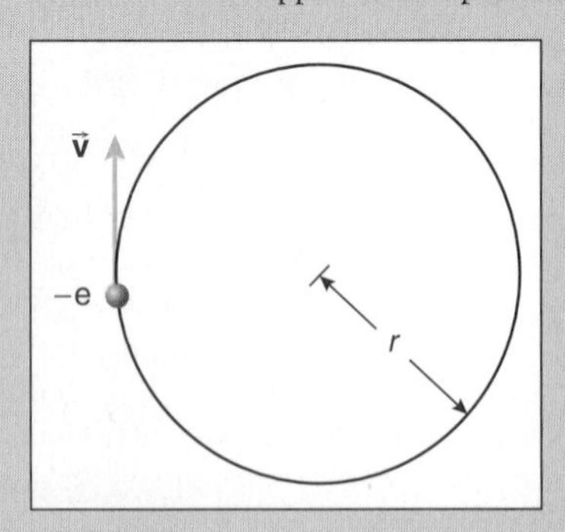

FIGURE P.65

•**66.** A square current loop 15.0 cm on a side and carrying 20.0 A clockwise in the x–y plane (see Figure P.66) finds itself in a uniform magnetic field $\vec{B} = (4.00\text{ T})\hat{\imath} + (5.00\text{ T})\hat{k}$. (a) What is the magnetic dipole moment of the current loop? (b) What is the torque on the current loop? (c) Will the current loop rotate? (d) What is the potential energy of the current loop in its initial orientation? Is this the minimum value of the potential energy of the current loop? (e) Calculate the flux of the magnetic field through the current loop when it is in its initial orientation.

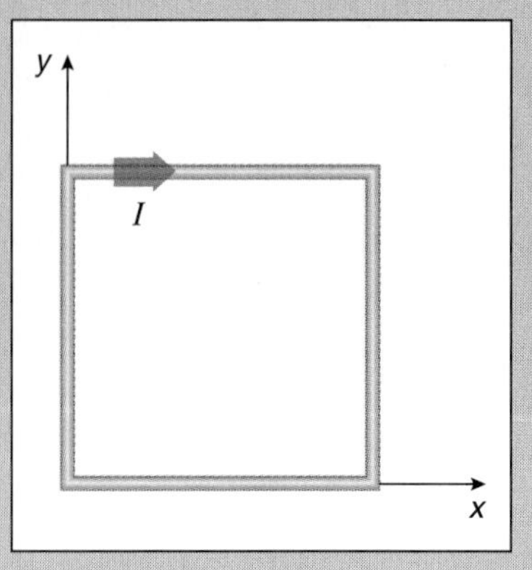

FIGURE P.66

•**67.** An infinite straight wire carrying a current I produces a magnetic field surrounding the wire. (a) Is the magnetic field uniform over the entire rectangular area depicted in Figure P.67a? (b) Consider the area vector of the rectangle to be in the same direction as the magnetic field at its location. Show that the magnetic flux through the rectangular area is

$$\Phi = \frac{\mu_0}{4\pi}2I\ell\ln\left(\frac{b}{a}\right)$$

(c) What is the magnetic flux through a circular area of radius R centered on and perpendicular to the wire as in Figure P.67b?

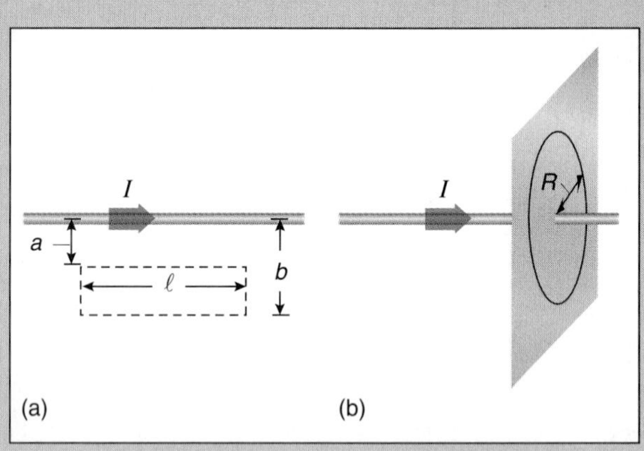

FIGURE P.67

•**68.** (a) A closed Gaussian surface surrounds the positive charge of an electric dipole as shown in Figure P.68a. What is the flux of the electric field through the Gaussian surface? Sketch the electric field lines in the vicinity of the dipole. (b) A closed Gaussian surface surrounds the north pole end of a bar magnet as shown in Figure P.68b. What is the flux of the magnetic field through the closed surface? Sketch the magnetic field lines around and through the magnet. Why is this result so different from the electrical dipole case in (a)?

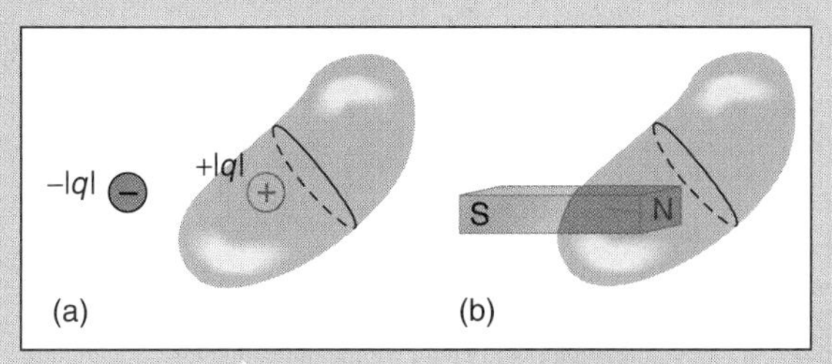

FIGURE P.68

•69. An electron in a TV tube moves with speed 7.50×10^6 m/s into a region with a uniform magnetic field of magnitude 50.0 mT. (a) What is the angle between the velocity of the electron and the magnetic field that produces the maximum magnitude of force on the electron? What is this force magnitude? (b) If the magnitude of the force is only 0.25 the maximum, what is the angle between the velocity of the electron and the magnetic field direction? (c) What is the initial kinetic energy of the electron in eV? (d) What is the kinetic energy of the electron as it leaves the region with the magnetic field?

Sections 20.10 Ampere's Law*
20.11 The Displacement Current and the Ampere–Maxwell Law*
20.12 Magnetic Materials*
20.13 The Magnetic Field of the Earth*

70. Many parallel wires pierce the page as indicated in Figure P.70. Each wire carries a current of 1.5 A either into or out of the page (as indicated). (a) What is the path integral of the magnetic field taken clockwise around the indicated closed contours? (b) What is the path integral if it is taken counterclockwise around each path?

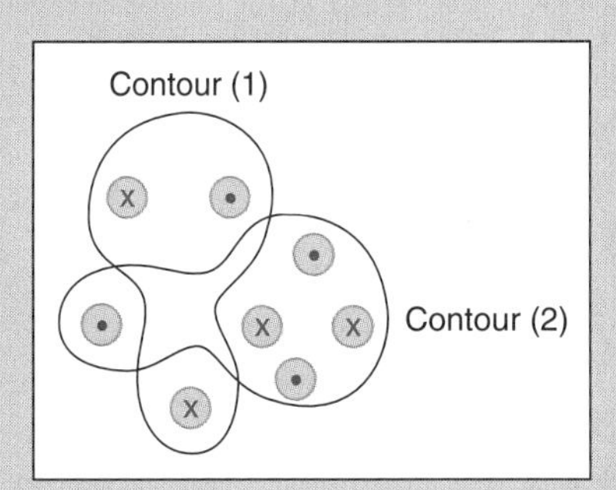

FIGURE P.70

•71. Design a long solenoid to produce a magnetic field of magnitude 50.0 mT along its axis with a current no greater than 15.0 A.

•72. A long, hollow, cylindrical conductor with inner radius a and outer radius b carries a current I uniformly distributed over the cross-sectional area of the conductor (see Figure P.72). (a) Use Ampere's law to show that the magnitude of the magnetic field at a radius r from the axis of the conductor: (a) is zero if $r < a$; (b) is

$$\frac{\mu_0}{4\pi}\frac{2I}{r}\frac{r^2 - a^2}{b^2 - a^2}$$

if $b < r < a$; (c) is equal to the magnitude of the field of an infinite wire,

$$\frac{\mu_0}{4\pi}\frac{2I}{r}$$

if $r > b$. (d) When $a = 1.00$ cm, $b = 1.20$ cm, and $I = 25.0$ A, make an accurate graph of the magnitude of B versus r over the domain 0 cm $< r <$ 2.00 cm.

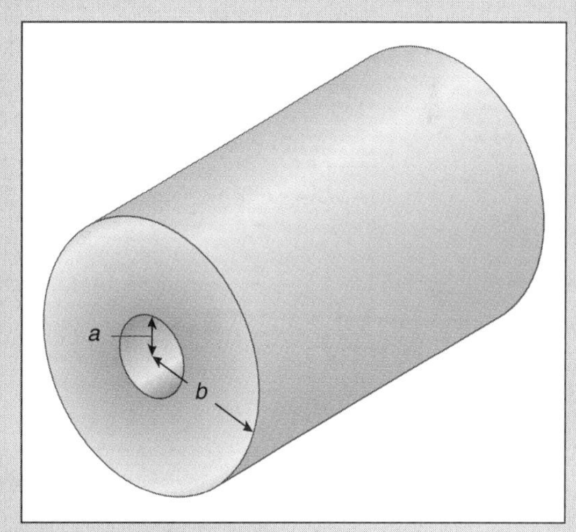

FIGURE P.72

•73. A uniform magnetic field in a region is shown in Figure P.73. The field line diagram implies there is zero field to the right and left of the indicated field lines on the far right and left of the figure, respectively. Using the indicated closed path and

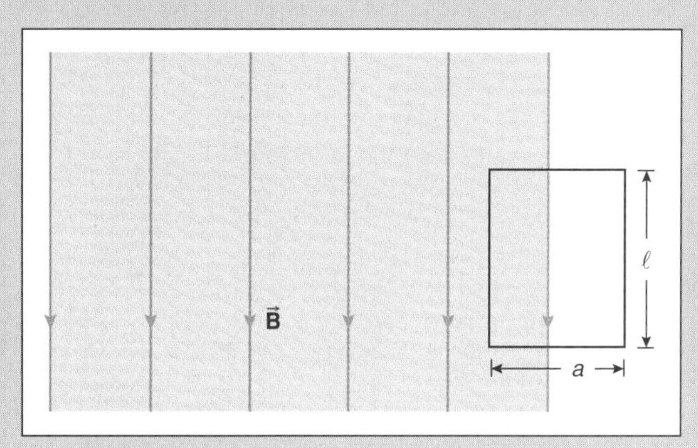

FIGURE P.73

Ampere's law, show that the magnetic field *cannot* drop abruptly to zero as implied by the field line diagram. Real magnets have so-called *fringing fields* to the right and left of the indicated field lines, so the magnetic field gradually approaches zero, rather than abruptly.

•**74.** Three infinite wires, each carrying a current of 5.0 A, are indicated in Figure P.74. What are the magnitude and direction of the magnetic field at the point $x = 2.0$ cm, $y = 6.0$ cm produced by these currents?

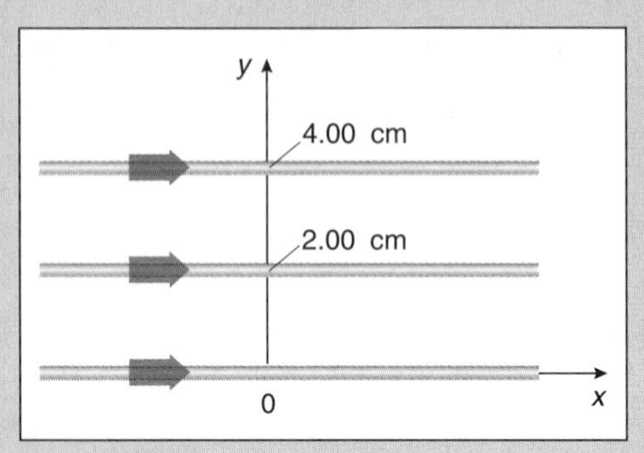

FIGURE P.74

INVESTIGATIVE PROJECTS

A. Expanded Horizons

1. Birds use a number of factors to sense direction when migrating over long distances, among them the magnetic field of the Earth. Investigate and report on the status of research on magnetic effects affecting navigation in migratory birds.
Thomas Alerstam, *Bird Migration* (Cambridge University Press, New York, 1990).
Chris Mead, *Bird Migration* (Facts on File, New York, 1983).
Kenneth P. Able and Mary A. Able, "Daytime calibration of magnetic orientation in a migratory bird requires a view of skylight polarization," *Nature*, *364*, #6437, pages 523–525 (5 August 1993).
Kenneth P. Able and Mary A. Able, "Calibration of the magnetic compass of a migratory bird by celestial rotation," *Nature*, *347*, #6291, pages 378–380 (27 September 1990).

2. You might enjoy reading about and summarizing the life of Nikola Tesla for it is an interesting case study of ambition, invention, and failure. He died in poverty as a complete recluse, living in the anonymity of a single-occupancy, resident hotel in New York City.
John J. O'Neill, *Prodigal Genius: The Life of Nikola Tesla* (Washburn, New York, 1944).
Stephen S. Hall, "Tesla: a scientific saint, wizard, or carnival salesman?" *Smithsonian*, *17*, #3, pages 120–134 (June 1986).
Bill Lawren, "Rediscovering Tesla," *Omni*, *10*, #6, pages 64–68 (March 1988).
Margaret Cheney, *Tesla: Man Out of Time* (Barnes & Noble, New York, 1993).
Ben Johnston, editor, *My Inventions: The Autobiography of Nikola Tesla* (Barnes & Noble, New York, 1995)

3. As you know, electricity is used by various biological species for defense and to stun prey (see Investigative Project 8 in Chapter 16). Consult an appropriate faculty member in the biology department at your university to see if there are ways in which biological species exploit magnetism in ways other than for navigation on long migrations (see Investigative Project 1).

4. The characteristics, origin, and evolution of the magnetic field of the Earth still are active areas of research. Survey the problem from both a historical and contemporary viewpoint.
David W. Strangway, *History of the Earth's Magnetic Field* (McGraw-Hill, New York, 1970).
Jeremy Bloxham and David Gubbins, "The evolution of the Earth's magnetic field," *Scientific American*, *261*, #6, pages 68–75 (December 1989).
J. A. Jacobs, "The Earth's magnetic field and reversals," *Endeavor, New Series Volume 19*, #4, pages 166–171 (December 1995).
J. A. Jacobs, "The Earth's magnetic field," *Contemporary Physics*, *36*, #4, pages 267–277 (July 1995).
Ray Ladbury, "Geodynamo turns toward a stable magnetic field," *Physics Today*, *49*, #1, pages 17–18 (January 1996).

5. The navigational compass has a long and uncertain history that encompasses Chinese, Arabic, and European sources. Investigate its history and write a brief report about its origins and development.
Encyclopaedia Britannica (11th edition, 1910), volume 6, pages 806–809.
H. L. Hitchins and W. E. May, *From Lodestone to Gyro-compass* (Hutchinson, London, 1955).
A. Crichton Mitchell, "Chapters in the history of terrestrial magnetism," *Terrestrial Magnetism and Atmospheric Electricity*, *37*, pages 105–146 (1932); *42*, pages 241–280 (1937); *44*, pages 77–80 (1939); and *51*, pages 323–351 (1946).

6. Detail the search for hypothetical magnetic monopoles.
Alfred S. Goldhaber and W. Peter Trower, "Resource Letter MM-1: Magnetic Monopoles," *American Journal of Physics*, *58*, #5, pages 429–439 (May 1990).
Henry J. Frisch, "Quest for magnetic monopoles," *Nature*, *344*, #6268, pages 706–707 (19 April 1990).

7. Electricity and magnetism have inspired several attempts at constructing perpetual motion machines, none of which can elude the consequences of the laws of thermodynamics. A look at these attempts may be interesting:
Arthur W. J. G. Ord-Hume, *Perpetual Motion: The History of an Obsession* (St. Martin's Press, New York, 1977), Chapter 5, pages 83–93; also see the bibliography on pages 224–227.

8. Consult appropriate sources in your science library to make a table of the approximate value of the surface magnetic fields on the planets of the solar system, the Sun, and exotic stars such as white dwarfs and neutron stars.

9. For the appropriate orientations, a current experiences a magnetic force in a magnetic field. Imagine a liquid conductor such as mercury or sodium in a pipe. Design a system with a current through the conducting liquid and a magnetic field that produces a magnetic force of maximum magnitude that serves to pump the liquid along the pipe. Such *electromagnetic pumps* are used in some nuclear reactors to facilitate heat transfer. Mercury and sodium are very dangerous substances, and so such pumps must be very carefully constructed to avoid dangerous leaks and adverse environmental consequences.

B. Lab and Field Work

10. Your physics department likely has a small permanent bar magnet. Tape the magnet with its axis along an east–west direction on top of a large sheet of paper on a wooden table, well away from ferromagnetic materials. With a dime-store magnetic compass of small diameter, devise and execute a procedure for tracing out the magnetic field lines surrounding the magnet. The field lines represent the vector superposition of the magnetic field of the bar magnet with the local horizontal component of the magnetic field of the Earth. Determine the locations of the two places where the total magnetic field is zero.

11. Carefully devise and perform a series of experiments on a small charge (such as the charge that can be placed on a small, tethered, conducting ball) in a magnetic field to illustrate the facts on which Equation 20.1 is based.

12. For historical perspective, investigate the methods and experiments used by Biot and Savart to formulate the Biot– Savart law.
    Herman Erlichson, "The experiments of Biot and Savart concerning the force exerted by a current on a magnetic needle," *American Journal of Physics*, 66, #5, pages 385–391 (May 1998).

13. For whom the bell tolls. Your life as a student likely is regulated by the periodic ringing of class bells. Your professor likely can provide you with an example of such an electrically activated bell. Explore the construction of such a bell. Sketch a circuit representation of the bell system and explain how a current provided to the bell makes it ring repetitively.

14. Magnetic stirrers are used frequently in chemistry and biology laboratories. Investigate and describe the principles underlying their operation.

15. If your physics department has a Hall probe for measuring the magnitude of magnetic fields, explore the magnitude of the magnetic field between the pole faces of a permanent magnet or within a pair of Helmholtz coils (see Problem 58).

16. Devise and perform an experiment to measure the magnitude and direction of the magnetic field of the Earth at your location.

17. Discover how a degausser is able to eliminate the permanent magnetic field of a material. (See Section 20.12.)

C. Communicating Physics

18. Your former high school physics teacher has invited you back to give a talk about fields to a group of high school physics students. For this audience, create a 30-minute talk about the field concept. Illustrate the talk with a clear discussion of the distinctions among electric, magnetic, and gravitational fields. A few hands-on demonstrations to accompany the talk would enhance its interest for the audience.

19. Write a short paragraph about the origin of the word *lodestone*. Consult the *Oxford English Dictionary*.

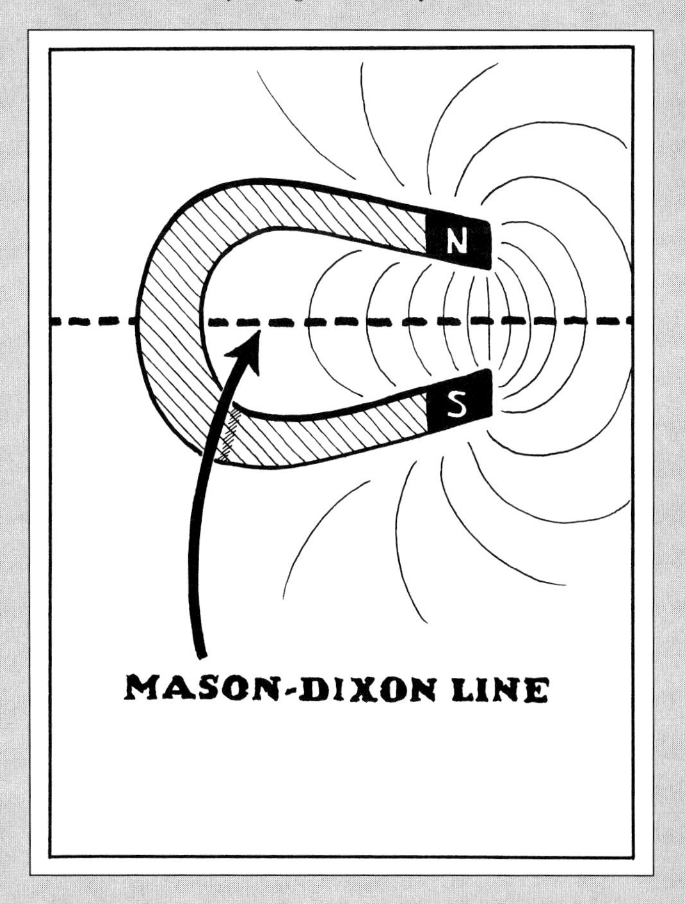

Chapter 21

Faraday's Law of Electromagnetic Induction

And yet it is Faraday's spark which now shines upon our coasts, and promises to illuminate our streets, halls, quays, squares, warehouses, and perhaps at no distant day, our homes.

John Tyndall (1820–1893)*

Without Faraday's law, you likely would be using beeswax and kerosene to study by candle and lamp light. Chemical batteries or Wimshurst machines would be needed to operate every electrical appliance. Electricity would be prohibitively expensive.[†] Generating electricity cheaply for common and geographically widespread use relies on Michael Faraday's remarkable discovery of the law of electromagnetic induction.

There is an interesting story about Faraday (1791–1867), likely apocryphal,[‡] that goes like this. When the British Chancellor of the Exchequer William Gladstone (1809–1898) asked Faraday what earthly good ever could come of his discovery of electromagnetic induction, Faraday is said to have replied: "Why sir, there is every probability that you will soon be able to tax it!" And we thought tax-and-spend was just a recent phenomenon!

21.1 Faraday's Law of Electromagnetic Induction

In the last chapter (Section 20.11) we discovered that the time-varying flux of the electric field produces a magnetic field via the displacement current. The fields are mutually perpendicular (orthogonal). What about the converse: does a time-varying flux of the magnetic field produce an electric field?

One of the most technologically significant discoveries of the 19th century occurred when Michael Faraday realized in 1831 that this is indeed the case. Since the displacement current was invented by Maxwell about 30 years after Faraday's work, the originality of Faraday's discovery is all the more remarkable. It also is humbling to know that the most sophisticated mathematics ever used by Faraday in his technical papers and discoveries never went beyond simple algebra and arithmetic. Faraday was a true experimental genius, who insisted on using and exploiting the richness of the English language to the fullest. He was a master wordsmith. It was James Clerk Maxwell (1831–1879) who first used sophisticated mathematics to cogently describe the wealth of experimental information gleaned by Faraday from nature.

We easily can perform two experiments similar to those undertaken by Faraday.

Experiment 1: A Magnet and a Loop of Wire

Hold a permanent magnet near a closed loop or coil of wire, as in Figure 21.1. A very sensitive galvanometer (or current detector) is connected in series with the wire to detect the presence and direction of any, even a very small, electric current in the wire loop. The wire loop represents an electric circuit although there are no batteries or other sources of energy in the circuit. The magnetic field of the magnet produces a magnetic flux through the coil of wire, since the lines of the field thread the coil.[§] The following qualitative observations then are made:

1. If the magnet is held in a fixed position, so that the magnetic flux through the loop is constant (independent of time), the galvanometer indicates there is zero current in the wire loop.
2. If the magnet is moved toward the loop, as shown in Figure 21.1, thus increasing the magnetic flux through the loop of wire, a current is detected in the wire as long as the magnetic flux through the loop is changing with time. The faster the magnet is moved, the greater the time rate of change of the magnetic flux, and the greater the current observed in the loop.
3. If the magnet is moved away from the loop, as in Figure 21.2, thus decreasing the magnetic flux through the loop, the current observed in the wire is in the opposite direction. Once again: the faster the magnet is moved, the greater the change in the magnetic flux through the loop, and the greater the current observed in the wire.

We can summarize these results by saying that a current is observed in the loop as long as the magnetic flux through the loop is changing with time. Indeed, the greater the time rate of change of the magnetic flux, the greater the current in the loop: they are directly proportional. If the magnetic flux through the loop is constant (not changing with time), no current is observed in the loop. The current in the loop is in one direction when the magnetic flux through the loop is increasing with time, and in the opposite direction when the magnetic flux decreases with time. In this experiment the changing magnetic flux is associated with the motion of the magnet.

*(Chapter Opener) John Tyndall, *Fragments of Science* (6th edition, D. Appleton, New York, 1897), Volume 2, Chapter XVI, "The Electric Light," page 452. First edition published in 1871.

[†]It costs about a kilodollar ($1000) to provide a kilowatt-hour of energy using batteries!

[‡]See the letters by I. Bernard Cohen, *Nature*, 157, #3981, pages 196–197 (16 February 1946) and by R. A. Gregory, *Nature*, 157, #3984, page 305 (9 March 1946), which cast doubt on the authenticity of this story. The quotation is taken from the letter by Cohen.

[§]Remember that the flux is a quantitative measure of the extent to which the field lines thread the loop.

Faraday came from an impoverished family and his formal education was essentially nonexistent. A deeply religious man, Faraday had a strong sense of community and a deep affection for children (though he had none of his own). Apprenticed as a bookbinder, his interest in electricity was sparked when he chanced upon an article about it while rebinding a volume of the *Encyclopaedia Britannica*.

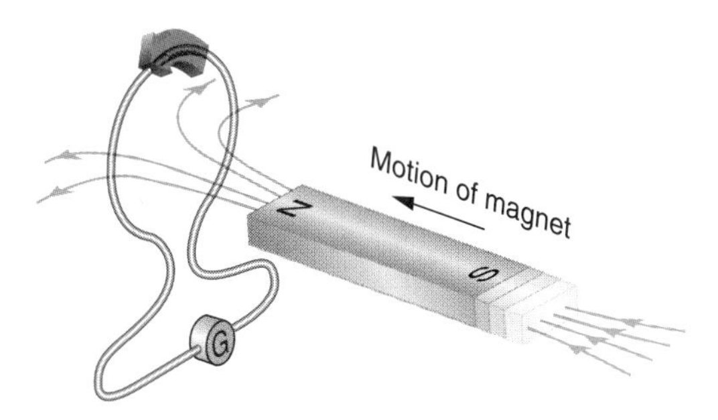

FIGURE 21.1 If the magnet is moved toward the coil, a current is detected in the coil. The curvature of the magnetic field lines through the coil here (and in similar drawings that follow below and on pages 961–962) is exaggerated to clarify the perspective.

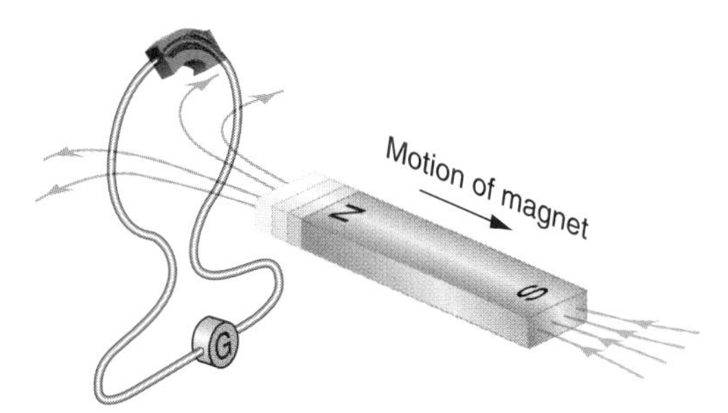

FIGURE 21.2 If the magnet is moved away from the coil, a current is detected in the opposite direction.

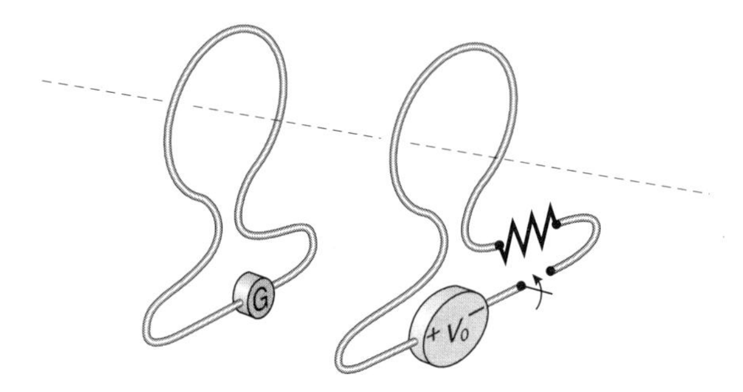

FIGURE 21.3 Two loops or coils, one with a battery and switch.

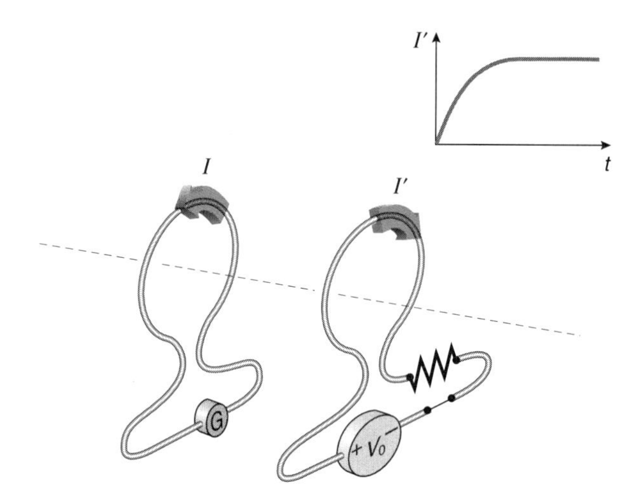

FIGURE 21.4 A current I is detected by the galvanometer only while the current I' in the other coil is increasing with time.

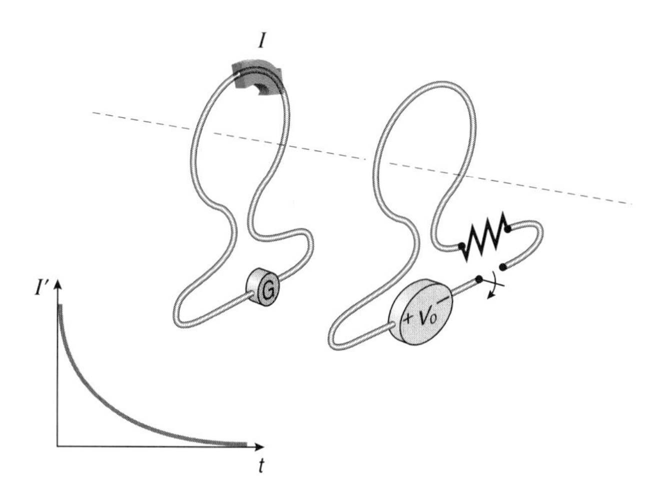

FIGURE 21.5 A current I is detected in the opposite direction when the current I' in the other coil decreases with time.

Experiment 2: Two Loops of Wire, a Battery, and a Switch

In this experiment no permanent magnet is used. Rather, the coil and galvanometer system used in Experiment 1 is placed near another loop or coil of wire connected to a battery and a switch, as shown in Figure 21.3.

The following observations then are made:

1. With the switch open in the circuit with the battery, there is no current in the wire loop on the right.* No magnetic fields are present. The magnetic flux through the wire loop connected to the galvanometer is zero. No current is detected by the galvanometer in the wire loop.
2. Now we close the switch in the circuit on the right, thus establishing a current in it (see Figure 21.4). The loop on the right can be arranged so that the direction of the magnetic field produced by the current in the wire is the same as that of the magnet in Experiment 1. During the very brief interval during which the current increases in the circuit on the right, the magnetic field of this current also increases and briefly creates an increasing magnetic flux through the loop connected to the galvanometer. The galvanometer detects a current in the wire loop on the left *only* during this brief interval during which the magnetic field of the current in the circuit on the right is changing with time.

   The direction of the current detected by the galvanometer is in the same sense as when the magnetic flux through the loop in Experiment 1 was increasing with time. When the current in the circuit on the right reaches a constant value, making the magnetic flux through the loop on the left constant, the galvanometer detects no current in the loop.
3. The switch in the circuit on the right now is opened (see Figure 21.5), and the current in it falls to zero during a very brief time interval as the contact is broken. The magnetic flux through the loop on the left now decreases quickly with

*The network on the right actually is not a circuit when the switch is open!

time and the galvanometer detects a current in the opposite direction to the current detected in #2. The current detected by the galvanometer is in the same direction as the current in Experiment 1 when the magnetic flux through the loop was decreasing with time. When the current in the circuit on the right reaches zero (and becomes constant), no current is detected by the galvanometer in the loop on the left.

We summarize these results as follows. During those intervals in which the magnetic flux through the loop with the galvanometer changes with time, a current is detected in the loop. The current is in one direction when the magnetic flux is increasing with time, and in the opposite direction when the magnetic flux through the loop is decreasing with time. In this experiment, the change in the magnetic flux is not caused by the motion of a magnet (as in Experiment 1) but by a change in the magnetic field itself.

A current exists in the wire loop with the galvanometer as long as the magnetic flux through the loop is changing with time. Such a current is called an **induced current**, since it is induced or produced not by conventional batteries but by a changing magnetic flux through the loop.

Electric currents, of course, are charges in motion. Electric charges move in response to electrical forces caused by electric fields* via Equation 16.8:

$$\vec{\mathbf{F}}_{\text{elec}} = q\vec{\mathbf{E}}$$

Hence we can say that a changing magnetic flux through a loop induces (i.e., produces) an electric field—called an **induced electric field**—that causes the charges to move and produce the electric current detected by the galvanometer.

We shall have more to say about this induced electric field shortly; it is an electric field of a very different character than the electric fields we have encountered previously.

The experimental genius of Faraday was complemented by equal theoretical insight into the real significance of his discoveries. Faraday realized that the presence of the wire loop with the galvanometer served only to reveal or manifest the induced electric field. That is, charges present in the conducting wire loop detect the presence of the induced electric field; remove the wire loop and we are left with the mere *mathematical* outline of the closed path. Nonetheless, even if the conductor is absent, the induced electric field (caused by the changing magnetic flux) *still* is present in space.†

To investigate the nature of this induced electric field more thoroughly, we consider a more symmetrical situation. A circular wire loop of planar area A is located in a region of space where there exists a uniform magnetic field $\vec{\mathbf{B}}$; see Figure 21.6.

We let the area vector associated with the area of the wire loop be directed parallel to the field, so that we have a reference direction associated with the loop. With the thumb of your right hand along the direction of the area vector, the fingers of your right hand define the positive sense around the loop; this sense is counterclockwise in Figure 21.6.

*Recall that a magnetic field exerts zero force on a charge at rest. Only an electrical force can begin to move a charge at rest.

†This is just like what happens when a charge Q establishes a static electric field in the surrounding space (that can act on another charge *q if it is present* in the field).

Now, as Faraday did, we perform three experiments: (1) If the magnetic field does not change, there is no induced current observed in the loop. (2) If the magnetic field increases in magnitude, thus increasing the magnetic flux through the loop, the direction of the induced current in the loop is observed to be in the negative sense (here, clockwise around the loop in Figure 21.7). (3) If the magnitude of the magnetic field is decreasing, thus decreasing the magnetic flux through the loop, the direction of the induced current in the loop is in the positive sense (here, counterclockwise around the loop in Figure 21.8).

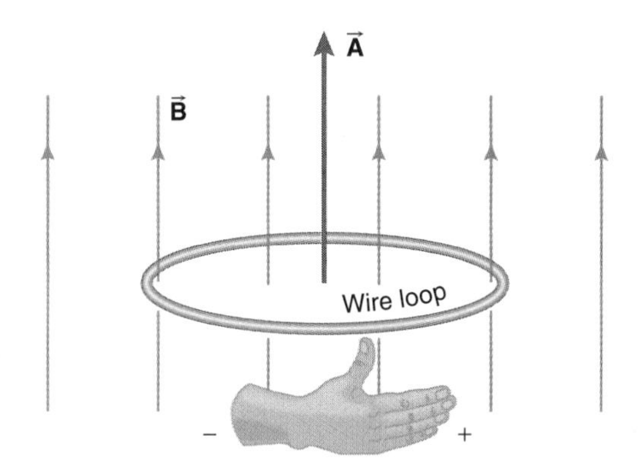

FIGURE 21.6 A circular loop wire loop in a uniform magnetic field.

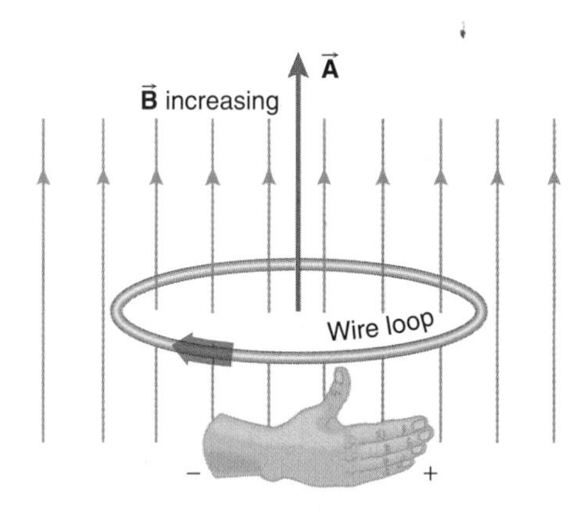

FIGURE 21.7 Increase the magnetic field and the induced current is in the negative sense as the change in flux takes place.

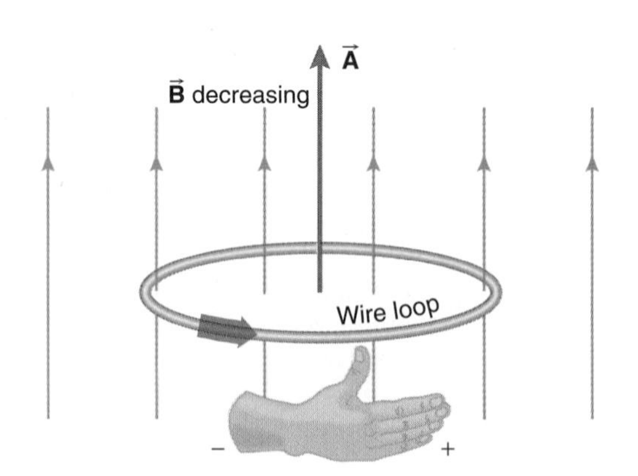

FIGURE 21.8 Decrease the magnetic field and the induced current is in the positive sense as the change in flux takes place.

What is the direction of the induced electric field? Since the charges are moving around the loop, we surmise that the induced electric field also is directed around the loop in the same sense as the current (the direction of positive charge motion). You might wonder if the field might be directed in a radial sense; this is not the case because a radial electric field would not cause charge to move around the loop. So the direction of the electric field induced by the changing magnetic flux through the loop is around the circumference of the loop in the direction of the current induced. This is a new type of electric field, quite unlike the static electric field encountered in our study of electrostatic phenomena in Chapter 16. The electric field lines representing the induced electric field form *closed contours*!

The fact that the induced electric field lines form closed loops (just as all magnetic field lines do) means that the flux of the induced electric field through any closed surface is zero (just as the flux of any magnetic field through a closed surface is zero). Gauss's law for the electric field still holds and need not be modified: this new induced electric field has no flux through a closed surface, and so the only electric field that is significant in Gauss's law for the electric field is the *static* electric field arising from stationary electric charges. *Static* electric field lines do *not* form closed loops. The static electric field lines begin on positive charge and end on negative charge.

The induced electric field produces an electrical force on a charge q. Let's calculate the work done by the electrical force in taking a charge q around a circular closed path when the magnetic field through it is increasing in magnitude. This work will be

$$W_{elec} = \int_{\text{clsd path}} \vec{\mathbf{F}}_{elec} \cdot d\vec{\ell}$$

Since $\vec{\mathbf{F}}_{elec} = q\vec{\mathbf{E}}_{induced}$, we have

$$W_{elec} = q\int_{\text{clsd path}} \vec{\mathbf{E}}_{induced} \cdot d\vec{\ell}$$

The induced electric field is parallel to $d\vec{\ell}$ around the whole path, and so the scalar product simplifies:

$$W_{elec} = q\int_{\text{clsd path}} E_{induced}\, d\ell$$

The circular symmetry indicates that the magnitude of the induced electric field is constant in magnitude along the path, yielding

$$\begin{aligned} W_{elec} &= qE_{induced}\int_{\text{clsd path}} d\ell \\ &= qE_{induced}(2\pi r) \end{aligned}$$

where r is the radius of the circular loop. The work done by the electrical force due to the induced electric field around a closed path is *not zero*.

The electrical force produced by the induced electric field is *not* a conservative force! The induced electric field is not like an electrostatic electric field caused by charges at rest (static electric fields *are* conservative).

The electric work done per unit charge around the closed path is

$$\frac{W_{elec}}{q} = E_{induced}(2\pi r) \qquad (21.1)$$

Faraday discovered that the work done per unit charge by the induced electric field around a closed path is equal to the negative of the time rate of change of the magnetic flux through the same path:

$$\int_{\text{clsd path}} \vec{\mathbf{E}} \cdot d\vec{\ell} = -\frac{d\Phi_{\text{magnetic through the enclosed area}}}{dt} \qquad (21.2)$$

This is **Faraday's law of electromagnetic induction.**

We can immediately generalize the circular path to one of any shape.

The dimensions of the left-hand side of Equation 21.2 are those of the electric field times a distance; using SI units, we have:

$$(\text{N/C})(\text{m}) = \text{J/C} = \text{V}$$

or the same as the electric potential. Notice that the left-hand side of Faraday's law is a path integral of the electric field (around a *closed path*) and thus is similar to the path integral (*not* necessarily around a closed path), Equation 17.9, used to define the (static) electric potential difference between two points. The similarity is the reason that the left-hand side of Faraday's law is called an **induced emf.**

The induced emf is, by definition, the path integral

$$\text{induced emf} \equiv \int_{\text{clsd path}} \vec{\mathbf{E}} \cdot d\vec{\ell} \qquad (21.3)$$

Faraday imagined the motion of charges arising from the induced emf to be motivated or caused by what he picturesquely called an "electromotive force." We use the term *emf* rather than electromotive force because the emf is *not* a force as we now define the word in physics. The work done by the electric field arising from Faraday's law also is *not* to be associated with a potential energy (as it was for electrostatic electric fields) since potential energies *only* can be associated with conservative forces. Unlike the electrostatic force, the electrical force stemming from the induced electric field in Faraday's law is *not* a conservative force: it does finite (nonzero) work around a closed path.

Faraday's law (Equation 21.2) also is commonly written as

$$\text{induced emf} = -\frac{d\Phi_{magnet}}{dt} \qquad (21.4)$$

Faraday's law of electromagnetic induction is an important, very general law of electromagnetism and is the last of the four fundamental equations of electromagnetism known collectively as the Maxwell equations.

Returning to our specific circular example, the magnetic flux of the uniform magnetic field through the area bounded by the path here is

$$\begin{aligned} \Phi_{magnet} &= \int \vec{\mathbf{B}} \cdot d\vec{\mathbf{A}} \\ &= BA \end{aligned}$$

Since the flux is changing because the field is changing, we have

$$\frac{d\Phi_{magnet}}{dt} = A\frac{dB}{dt}$$

The right-hand side of Faraday's law (Equation 21.2) then is

$$-A\,\frac{dB}{dt}$$

The left-hand side of Faraday's law is the induced emf, which we found to be given by Equation 21.1:

$$E_{\text{induced}}(2\pi r)$$

Thus Faraday's law for this specific case becomes

$$E_{\text{induced}}(2\pi r) = -A\,\frac{dB}{dt}$$

Since $A = \pi r^2$ for the circle, we have, for the induced field,

$$E_{\text{induced}} = -\frac{r}{2}\frac{dB}{dt}$$

which shows explicitly how a changing magnetic field can give rise to an induced electric field. You also can note from Figures 21.7 and 21.8 that the induced electric field (directed around the circumference of the circular path) is perpendicular to the changing magnetic field that causes it.

Nature, along with our inventiveness, is neatly symmetric:

- changing magnetic fields give rise to induced electric fields via Faraday's law;
- changing electric fields give rise to magnetic fields via the displacement current and the Ampere–Maxwell law.

These electric and magnetic fields are mutually perpendicular to each other.

PROBLEM-SOLVING TACTIC

21.1 Notice that Faraday's law, as expressed by Equation 21.4, indicates that the induced emf is equal to the negative of the slope of a graph of Φ_{magnet} versus time.

QUESTION 1

A small, planar wire loop is placed in a magnetic field directed along k̂. Sketch an orientation for the loop that produces zero magnetic flux through it. Sketch an orientation for the loop that maximizes the magnetic flux through it.

STRATEGIC EXAMPLE 21.1

The magnetic flux through a single current loop varies with time as indicated in Figure 21.9. Make a graph of the corresponding induced emf in the current loop as a function of time.

Solution

Problem-Solving Tactic 21.1 indicates that the induced emf is equal to the *negative* of the slope of the graph of Φ_{magnet} versus time. The induced emf thus is the negative of the slope of the graph shown in Figure 21.9. The slopes of the various segments are found in the following way.

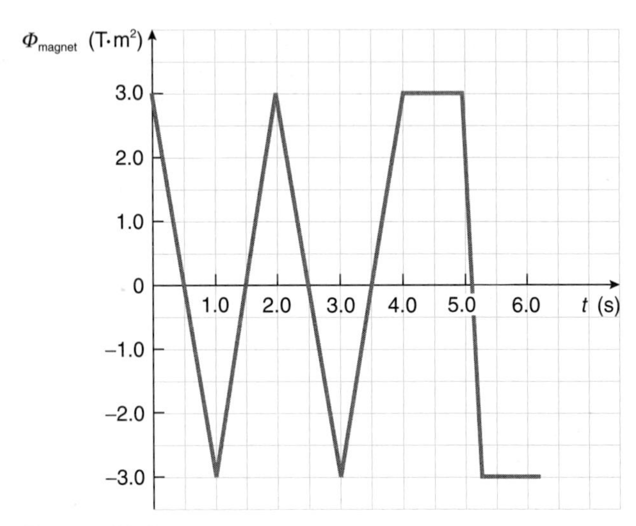

FIGURE 21.9

a. During the interval from $t = 0$ s to $t = 1.0$ s, the slope of the graph is constant and equal to

$$\text{slope} = \frac{-3.0\ \text{T}\cdot\text{m}^2 - (3.0\ \text{T}\cdot\text{m}^2)}{1.0\ \text{s}} = -6.0\ \text{T}\cdot\text{m}^2/\text{s}$$

The induced emf is the negative of the slope:

$$\begin{aligned}\text{induced emf} &= -(\text{slope})\\ &= -(-6.0\ \text{T}\cdot\text{m}^2/\text{s})\\ &= 6.0\ \text{T}\cdot\text{m}^2/\text{s}\\ &= 6.0\ \text{V}\end{aligned}$$

b. During the time interval from $t = 1.0$ s to $t = 2.0$ s, the slope of the graph is

$$+6.0\ \text{T}\cdot\text{m}^2/\text{s}$$

and the induced emf is

$$\begin{aligned}\text{induced emf} &= -(+6.0\ \text{V})\\ &= -6.0\ \text{V}\end{aligned}$$

c. The slope of the graph during the interval from $t = 2.0$ s to $t = 3.0$ s is the same as from $t = 0$ s to $t = 1.0$ s, and so the induced emf is the same as what you calculated in part (a):

$$+6.0\ \text{V}$$

d. The slope during the interval from $t = 3.0$ s to $t = 4.0$ s is the same as in part (b), and so the induced emf is the same as in (b):

$$-6.0\ \text{V}$$

e. During the interval from $t = 4.0$ s to $t = 5.0$ s, the slope of the graph is zero; the magnetic flux is not changing with time during this interval, and so the induced emf is 0 V.

f. From $t = 5.0$ s to $t = 5.25$ s the slope of the graph is

$$\text{slope} = \frac{-3.0\ \text{T}\cdot\text{m}^2 - (3.0\ \text{T}\cdot\text{m}^2)}{0.25\ \text{s}} = -24\ \text{T}\cdot\text{m}^2/\text{s}$$

and so the induced emf is

$$\text{induced emf} = -(\text{slope}) = +24\ \text{V}$$

g. For times $t > 5.25$ s, the magnetic flux is constant; the slope is zero, and the induced emf is 0 V during this interval.

These results are summarized in Figure 21.10.

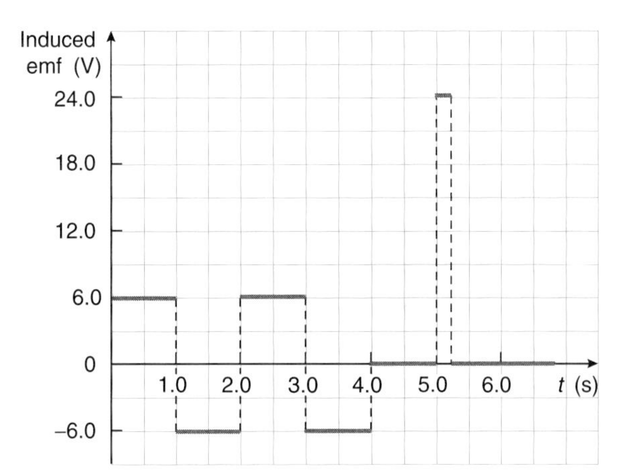

FIGURE 21.10

STRATEGIC EXAMPLE 21.2

A metal rod of length ℓ is rotated at angular velocity $\vec{\omega}$ about an end in a uniform magnetic field $\vec{B}$ that is parallel to $\vec{\omega}$, as shown in Figure 21.11. Find the absolute value of the induced emf and indicate which end of the rod is at the higher electric potential.

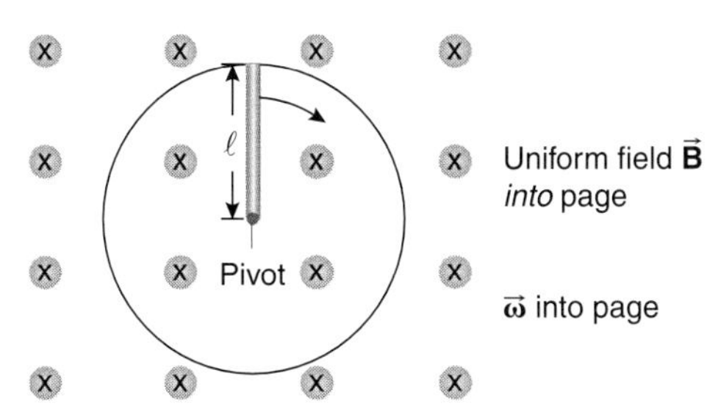

FIGURE 21.11

Solution

Electrons in the conductor are free to move. Since the electrons find themselves moving in a magnetic field, the electrons experience a magnetic force given by Equation 20.1:

$$\vec{F}_{\text{magnet}} = q\vec{v} \times \vec{B} = (-e)\vec{v} \times \vec{B}$$

This force is directed *toward* the pivot. Since $\vec{v}$ is perpendicular to $\vec{B}$, the force has a magnitude of

$$F_{\text{magnet}} = evB$$

The magnetic force produces charge separation in the conducting rod, with the tip of the rod becoming *positive* and the pivot end becoming *negative*.

Method 1

The separation of charge will not occur indefinitely, however, because the separated charges produce an *electric field* directed from the tip of the rod toward the pivot. This electric field produces an electrical force on the electrons that is in the direction opposite to the magnetic force. When the two forces are of equal magnitude, no further charge separation occurs. The result is a potential difference (an induced emf) between the pivot and the tip of the rod, with the outer tip at a higher potential. This effect is similar to the Hall effect discussed in Section 20.2. At equilibrium the electric and magnetic forces are equal in magnitude (but opposite in direction, so the total force is zero):

$$eE = evB$$

so that

$$E = vB$$

For circular motion $v = r\omega$; hence

$$E = r\omega B$$

This electric field is directed toward the pivot, so that

$$\vec{E} = -\omega B r \hat{r}$$

The potential difference between the ends of the rod is found from the definition of the potential difference, Equation 17.9:

$$V_f - V_i = -\int_i^f \vec{E} \cdot d\vec{r}$$

Integrate from the pivot to the tip:

$$V_{\text{tip}} - V_{\text{pivot}} = -\int_{0\ \text{m}}^{\ell} (-\omega B r \hat{r}) \cdot dr\,\hat{r} = \omega B \int_{0\ \text{m}}^{\ell} r\,dr = \frac{\omega B \ell^2}{2}$$

This is the emf induced between the ends of the rod.

Method 2

To calculate the magnitude of the induced emf, determine the flux of the magnetic field through a pie-shaped segment of the circle of angle θ, as shown in Figure 21.12. We express θ in radians.

The area S of the pie-shaped segment is $\theta/2\pi$ times the area of the circle itself:

$$S = \frac{\theta}{2\pi}\pi\ell^2 = \frac{\theta\ell^2}{2}$$

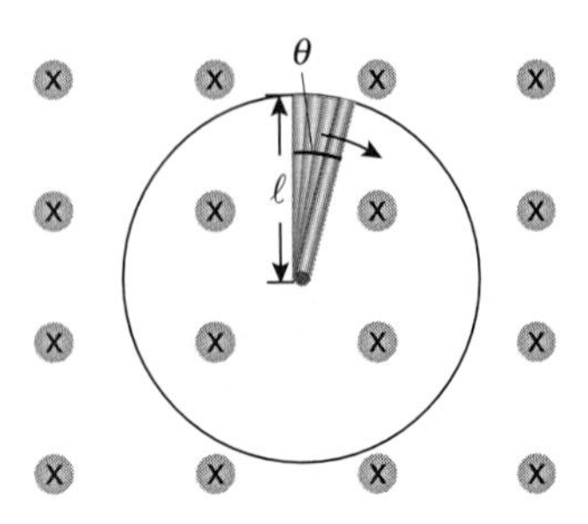

FIGURE 21.12

The flux Φ of the uniform magnetic field through the pie-shaped segment is

$$\Phi = \int \vec{B} \cdot d\vec{S}$$

The differential area vector $d\vec{S}$ can be taken in the same direction as $\vec{B}$ (since the area does not enclose a volume, you have a choice about the direction for the area vector). Thus the scalar product reduces to just $B\,dS$:

$$\Phi = \int B\,dS$$

Since the magnetic field is uniform, the flux becomes

$$\begin{aligned} \Phi &= BS \\ &= B\frac{\theta \ell^2}{2} \end{aligned}$$

The absolute value of the induced emf is

$$|\text{induced emf}| = \left|\frac{d\Phi}{dt}\right|$$

The only time-dependent quantity is the angle θ. Thus

$$|\text{induced emf}| = \frac{B\ell^2}{2}\left|\frac{d\theta}{dt}\right|$$

But

$$\left|\frac{d\theta}{dt}\right|$$

is the angular speed ω. The absolute value of the induced emf then is

$$|\text{induced emf}| = \frac{\omega B \ell^2}{2}$$

The outer tip of the rod has the higher potential.

21.2 Lenz's Law

We need to find an easy way to determine the direction of the induced electric field caused by a changing magnetic flux. If a conducting loop is coincident with the closed path in Faraday's law, positive charges will flow around the conducting loop in the same sense as the induced electric field (or negative charges will flow in the opposite direction), producing an induced current. So the way to find the direction of the field is to find a rule for determining the direction of the induced current* if a conducting path is provided. We emphasize again that the induced (nonconservative) electric field is present with or without the presence of a conducting path; a conducting path merely serves to detect the presence of the induced field by providing charges a chance to move in response to it, thereby producing an induced current.

Not long after Faraday discovered the law of electromagnetic induction, Heinrich Emil Lenz (1804–1865) formulated a convenient rule for determining the directional sense of an induced current around a closed conducting path coincident with the closed path in Faraday's law. The rule now is called **Lenz's law**. It is important to realize that Lenz's law contains nothing that is not already implicit in Faraday's law. Lenz's law is merely a rule to help us easily figure out the directional sense of the induced electric field and current (since positive charge flows in the same sense as the induced field) that arise from a changing magnetic flux. Regrettably, the word *law* occasionally is invoked too casually in physics.

Lenz's law states that the induced current always will be directed so as to *oppose the change* in the magnetic flux that is taking place.

The induction "tries" to preserve the status quo and resists *any* change, much like a two-year-old child; in its own way, the induced current is quite contrarian.

PROBLEM-SOLVING TACTIC

21.2 **To see how Lenz's law actually is used is best done by carefully examining a number of specific examples.** See the examples that follow. Study them thoughtfully and with good cheer. Once you get a feeling for the examples, you will be able to figure out the directions of induced currents and induced electric fields with confidence.

STRATEGIC EXAMPLE 21.3

Return to the first experiment performed when we introduced Faraday's law. Part of this experiment involved moving a permanent magnet *toward* a loop or coil of wire connected to a galvanometer (used to detect the presence of the induced current and its direction); see Figure 21.13. Use Lenz's law to determine the direction of the current induced in the loop.

Solution

When the magnet is moved toward the loop, the magnetic flux through the loop increases. According to Lenz's law, the induced current is directed so as to oppose the change in the flux. You can determine the direction of the induced current in two ways.

*Remember that the conventional direction taken for current is that associated with the motion of positive charge.

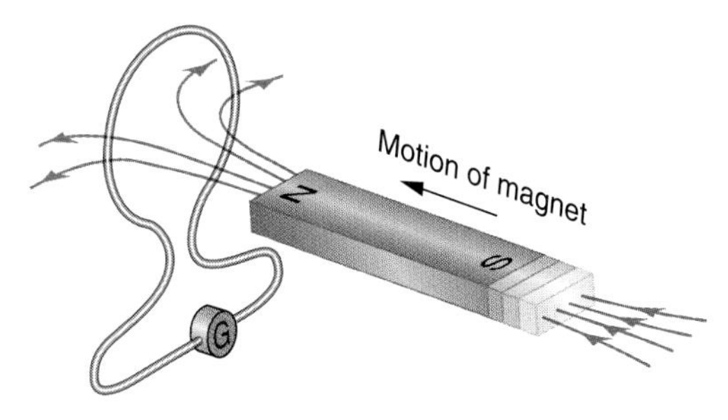

FIGURE 21.13

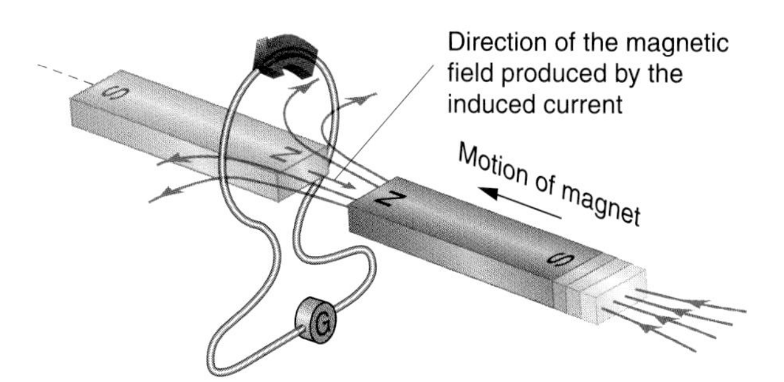

FIGURE 21.15

Method 1
Since the flux is increasing because the field is increasing through the loop, the direction of the induced current will be so that the *magnetic field it creates* is directed *opposite* to the field of the approaching magnet. Since the magnetic field of the magnet is directed through the loop from right to left and is increasing in magnitude (thus increasing the flux through the loop), the magnetic field of the induced current will be directed left to right through the loop to oppose the increasing field (thus decreasing the magnetic flux through the loop).

Grab the loop with your right hand so that your fingers thread the loop from left to right, as in Figure 21.14. The direction of the induced current will be in the direction of your extended right-hand thumb.

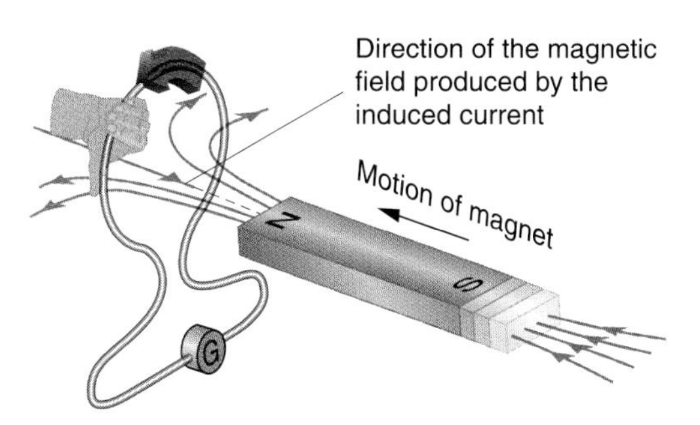

FIGURE 21.14

Method 2
Since the north end of the magnet is approaching the loop, the *magnetic field of the induced current* will be directed to oppose this motion. That is, the direction of the induced current will be such that the magnetic field it creates has a north magnetic pole facing the approaching north pole of the moving magnet; see Figure 21.15. The two like poles then will tend to repel each other, thus opposing the change (the approach of the north pole of the permanent magnet). In other words, the induced current opposes the change taking place. The direction of the induced current must be as indicated in Figure 21.15 to make the side of the loop facing the approaching magnet a north pole.

EXAMPLE 21.4

If the permanent magnet in Example 21.3 now is drawn away from the loop, as shown in Figure 21.16, use Lenz's law to determine the direction of the current induced in the wire loop.

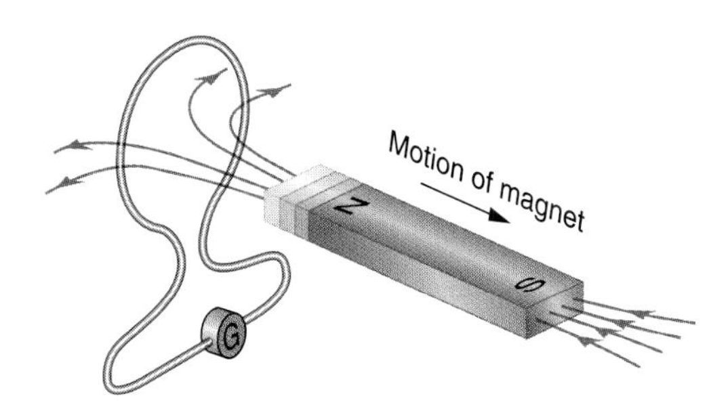

FIGURE 21.16

Solution
The motion of the magnet away from the loop means the magnetic flux through the loop is decreasing with time. According to Lenz's law, the induced current flows in the loop in a direction to oppose this change. You can determine this direction in two equivalent ways.

Method 1
The flux is decreasing through the loop because the magnetic field through the loop is decreasing in magnitude. The *magnetic field caused by the induced current* will be directed so as to reinforce the waning field of the permanent magnet. That is, the field of the induced current will be directed in the *same* direction as the field of the permanent magnet.

This means that you grasp the loop with your right hand so that the fingers thread the loop from right to left, parallel to the field of the permanent magnet through the loop, as shown in Figure 21.17. The extended thumb of your right hand then indicates the direction of the current.

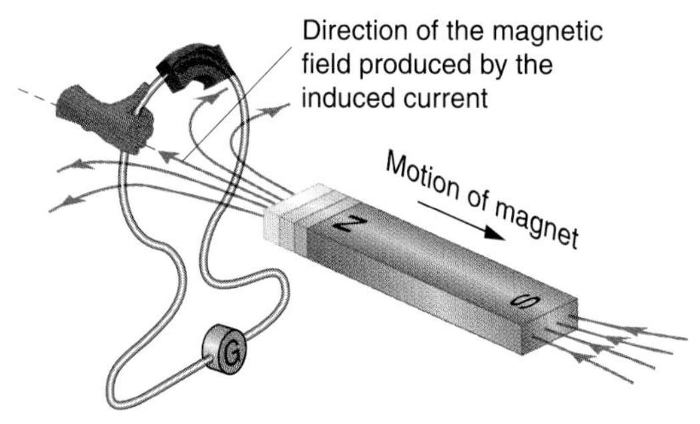

FIGURE 21.17

Method 2

Since the north pole of the permanent magnet is moving away from the loop, the magnetic field of the induced current will oppose this change. So the induced current develops a south magnetic pole on the side facing the departing north pole of the permanent magnet; see Figure 21.18. In this way, the two unlike poles attract each other and the induced current fights or opposes the change (the departure of the north pole of the permanent magnet).

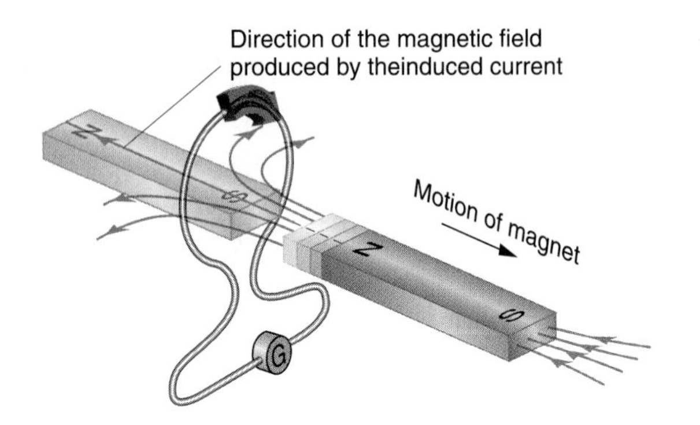

FIGURE 21.18

EXAMPLE 21.5

Close the switch in the circuit shown in Figure 21.19. This was the second experiment in Section 21.1. Determine the direction of the induced current in the other wire loop using Lenz's law.

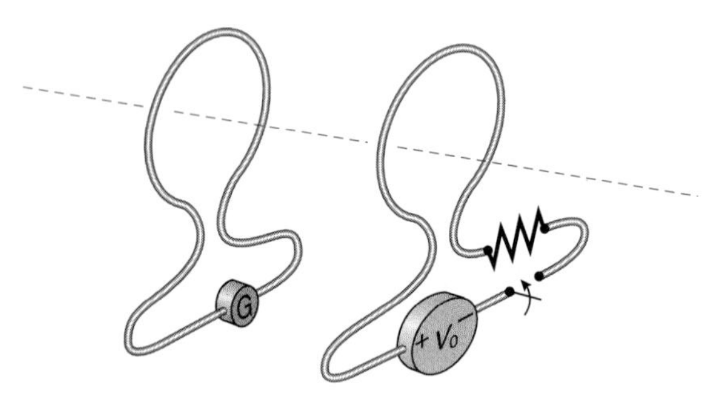

FIGURE 21.19

Solution

During a brief interval, the current in the right-hand circuit increases with time. The magnetic field created by this current, therefore, also increases with time, so that the flux of the magnetic field of this current through the left loop increases. Lenz's law states that the current induced in the loop is directed so as to oppose this change. Since the magnetic field is increasing through the left loop, the magnetic field of the induced current will be directed to fight or oppose the change. The induced current does this by moving in a direction that creates a magnetic field directed opposite to that of the increasing field.

Grasp the loop so that the fingers of your right hand thread the loop from left to right; your extended thumb is directed as indicated in Figure 21.20. This is the direction of the induced current.

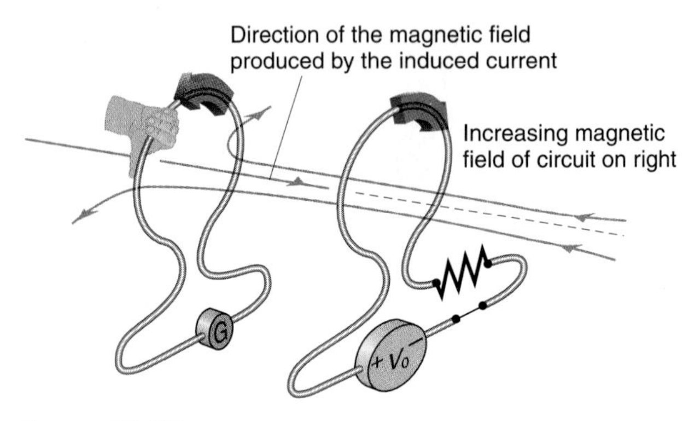

FIGURE 21.20

EXAMPLE 21.6

When the switch is opened in the circuit on the right of Figure 21.21, determine the direction of the current induced in the wire loop.

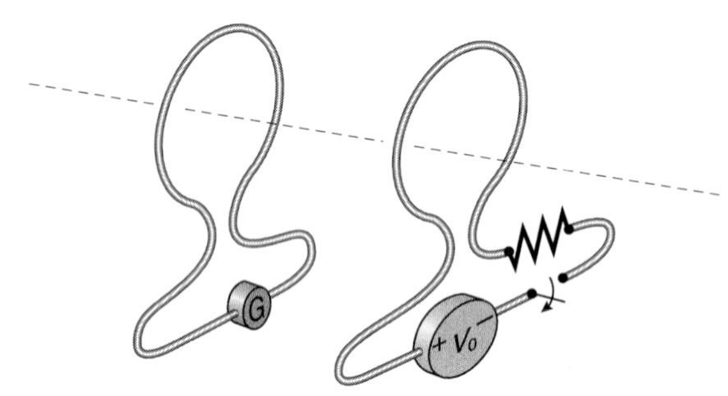

FIGURE 21.21

Solution

As the current falls precipitously to zero, so does the magnetic field created by this current. The magnetic flux of this field through the left loop thus decreases. Lenz's law states that the induced current is directed to oppose the change. Hence the field of the induced current through the left loop must be directed parallel to the field of the falling current in the circuit on the right.

Grasp the loop with the fingers of your right hand threading the loop from right to left (parallel to the existing decreasing field), as in Figure 21.22. The extended thumb of your right hand is directed as indicated. This is the direction of the induced current.

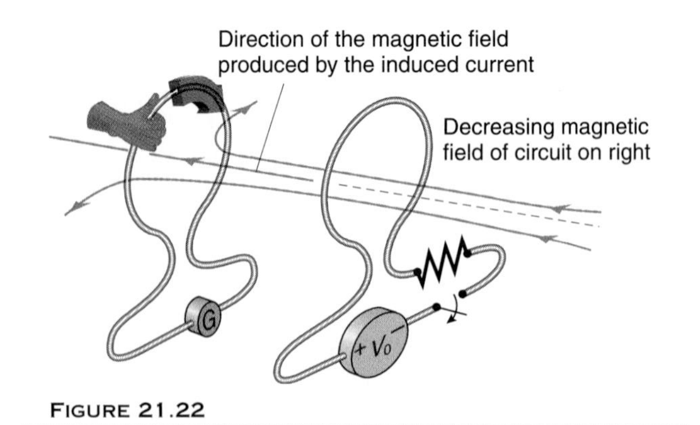

FIGURE 21.22

EXAMPLE 21.7

A flat, closed, continuous coil has 50 turns of wire with a total resistance of 0.20 Ω. The area of the coil is 60 cm². The coil is positioned in a uniform magnetic field $\vec{B}$ as shown in Figure 21.23. The magnitude of the magnetic field increases at a constant rate from 0 T to 2.0 T during 1.50 s. Find

a. the emf induced in the coil;
b. the size of the current induced in the coil; and
c. the direction of the induced current.

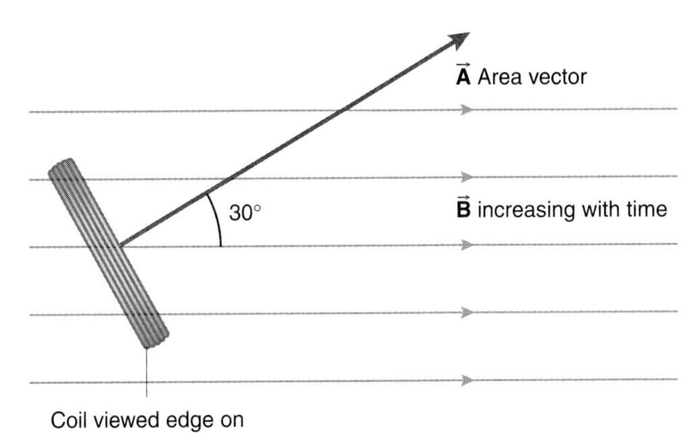

FIGURE 21.23

Solution

a. To calculate the induced emf you need to find the rate at which the magnetic flux through the coil is changing with time. Since the magnetic field is uniform over the area of the coil, the magnetic flux through *each turn* of the coil is

$$\begin{aligned}\Phi_{\text{single turn}} &= \vec{B} \cdot \vec{A} \\ &= BA\cos\theta\end{aligned}$$

where θ is the angle between the magnetic field and the area vector of a loop of the wire; here, $\theta = 30°$. The total magnetic flux through the coil then is

$$\Phi_{\text{total}} = NBA\cos\theta$$

where N is the number of turns in the coil; here, $N = 50$. The induced emf is given by Equation 21.4:

$$\begin{aligned}\text{induced emf} &= -\frac{d\Phi_{\text{total}}}{dt} \\ &= -\frac{d}{dt}(NBA\cos\theta)\end{aligned}$$

Here the only thing that is changing with time is the magnitude of the magnetic field; the induced emf thus is

$$\text{induced emf} = (-NA\cos\theta)\frac{dB}{dt}$$

The rate at which the magnetic field is changing with time is

$$\begin{aligned}\frac{dB}{dt} &= \frac{2.0\ \text{T} - 0\ \text{T}}{1.50\ \text{s}} \\ &= 1.3\ \text{T/s}\end{aligned}$$

With the information provided—$N = 50$, $A = 60\ \text{cm}^2 = 60 \times 10^{-4}\ \text{m}^2$, and $\theta = 30°$—the induced emf is found to be

$$\begin{aligned}\text{induced emf} &= -50(60 \times 10^{-4}\ \text{m}^2)(\cos 30°)(1.3\ \text{T/s}) \\ &= -0.34\ \text{V}\end{aligned}$$

The minus sign is a manifestation of Lenz's law.

b. The size of the current induced in the coil can be found using Ohm's law:

$$|\text{induced emf}| = IR$$

or

$$\begin{aligned}I &= \frac{0.34\ \text{V}}{0.20\ \Omega} \\ &= 1.7\ \text{A}\end{aligned}$$

c. The direction of the induced current is determined using Lenz's law. The direction of the induced current will oppose the change in flux. Since the magnetic flux through the coil is increasing with time, the induced current produces a magnetic field opposite to the direction of the applied magnetic field. Use your right hand, and grasp the loops of the coil so that your fingers are directed through the coil in the direction opposite to the direction of the applied field, as shown in Figure 21.24. The thumb of the right hand then indicates the direction of the induced current around the coil loops.

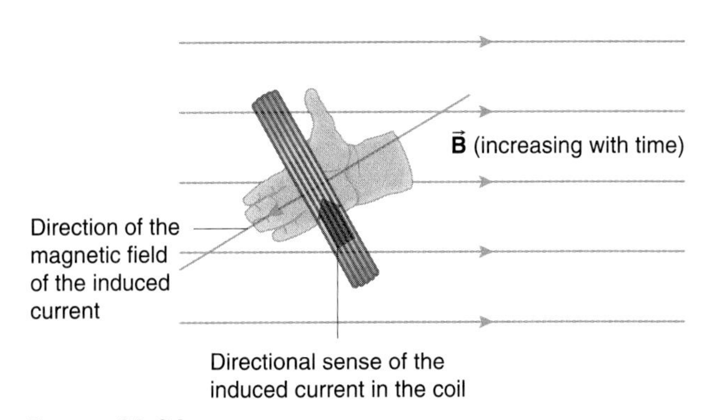

FIGURE 21.24

21.3 AN AC GENERATOR

The phenomenon described by Faraday's law is used for the very practical purpose of generating an alternating (oscillating) emf (an ac independent voltage source). This is the way most electric utilities as well as the alternator in your car generate a source of emf for many useful and convenient electrical applications.

We consider a planar loop of wire with an area vector $\vec{A}$ in a uniform magnetic field $\vec{B}$, as shown in Figure 21.25. The magnetic flux through the loop is

$$\begin{aligned}\Phi_{\text{magnet}} &= \int \vec{B} \cdot d\vec{S} \\ &= BA\cos\theta \qquad (21.5)\end{aligned}$$

where θ is the angle between the area vector $\vec{A}$ and the magnetic field $\vec{B}$. Now let some external agent rotate the loop about its symmetry axis at a constant angular velocity $\vec{\omega}$.

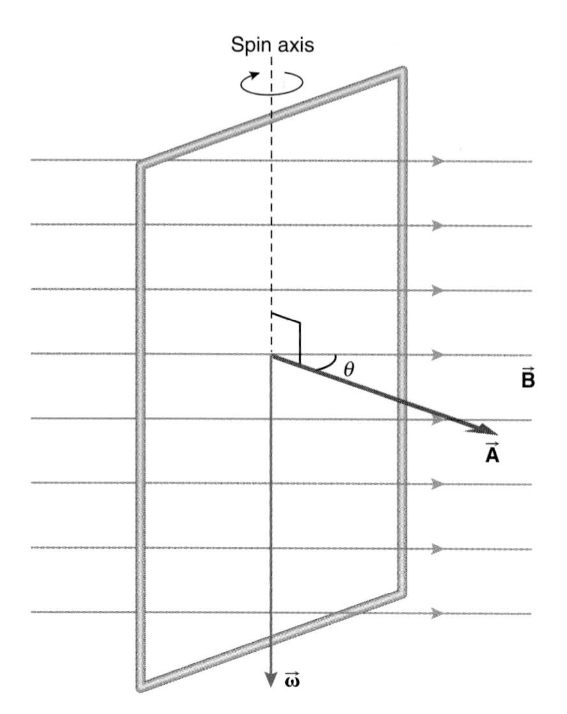

FIGURE 21.25 A wire loop is rotated in a uniform magnetic field.

From our study of uniform circular motion (Chapter 4), if $\theta = 0$ rad when $t = 0$ s, the angle θ at instant t is

$$\theta = \omega t$$

This means that the magnetic flux through the loop (Equation 21.5) is

$$\Phi_{\text{magnet}} = BA\cos(\omega t) \tag{21.6}$$

Since the magnetic flux through the loop is time dependent, Faraday's law implies that an emf is induced in the loop:

$$\begin{aligned}\text{induced emf} &= -\frac{d\Phi_{\text{magnet}}}{dt} \\ &= -\frac{d}{dt}\left[BA\cos(\omega t)\right]\end{aligned}$$

Since the magnitude of the magnetic field and the area of the loop are constant, the differentiation yields

$$\begin{aligned}\text{induced emf} &= -BA[-\sin(\omega t)]\omega \\ &= BA\omega\sin(\omega t)\end{aligned} \tag{21.7}$$

This induced emf oscillates sinusoidally with time, as shown in Figure 21.26.

If instead of using a single loop, a coil of wire of N loops (of identical area A) is used, the flux through the coil is N times that through a single loop. Therefore the induced emf in the coil then is also N times that of the single loop:

$$\text{induced emf} = NBA\omega\sin(\omega t) \tag{21.8}$$

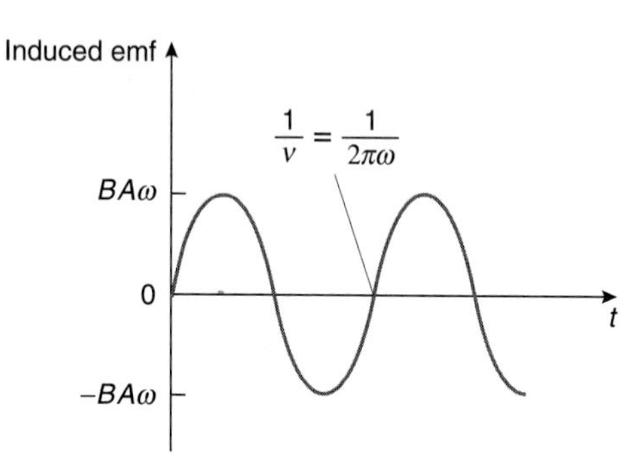

FIGURE 21.26 The emf induced in the wire loop is a sinusoidal function of time.

This is the way that the commercial electrical industry "generates electricity"; Faraday's law is at the very heart of the enterprise!

The rotating coil can be packaged into a box and the ends of the wire run to the outside world. The circuit symbol for a sinusoidal ac independent voltage source (an **ac generator** or an ideal ac battery) is shown in Figure 21.27. The term ac literally means *a*lternating current, but we use ac generically to mean oscillating with time.

When associated with a generator, we call the induced emf the **source voltage** $V_{\text{source}}(t)$, and treat it as the ac potential difference associated with the source:

$$V_{\text{source}}(t) \equiv \text{induced emf}$$

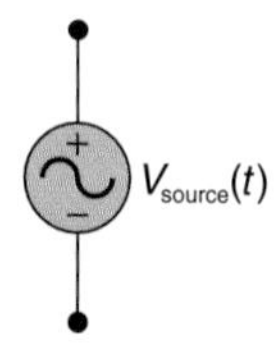

FIGURE 21.27 Circuit symbol for a sinusoidal independent voltage source.

Faraday's law is applied to create an ac independent voltage source in these large commercial generators.

An induced current exists only if the ends of the coil are connected together to an external electrical network to form an electrical circuit.

The electrical network in the United States and the rest of North and South America is based on a standard frequency $\nu = 60$ Hz; the frequency is controlled quite precisely. The coil rotates in the magnetic field 60 times during each second. Since each revolution of the coil represents 2π rad, the angular speed ω (and angular frequency) is related to the frequency ν by

$$\begin{aligned}\omega &= 2\pi\nu \\ &= (2\pi \text{ rad})(60 \text{ Hz}) \\ &\approx 377 \text{ rad/s}\end{aligned}$$

In Europe, Japan, Australia, Asia, and Africa, the standard frequency is 50 Hz, in which case $\omega \approx 314$ rad/s.

Of course, it requires work to rotate the current-generating coil; Lenz's law sees to that! The external agent used to rotate the coil in the magnetic field characterizes the electrical generating plant. Typically these mechanical sources are of three types:

1. Hydroelectric: Falling water is directed against the blades of a turbine that is coupled to the generator coils.
2. Steam: High-pressure steam is directed against the blades of a turbine coupled to the generator coils. The steam is produced by heating water using various types of fuels or energy sources: coal, oil, nuclear energy, geothermal energy, or even wood and garbage.
3. Wind: Wind is used to turn a propeller (turbine) that is coupled to the rotating coils of the generator.

EXAMPLE 21.8

You are commissioned to design an ac generator whose coil is to be turned at 60.0 Hz in a strong magnetic field of 0.150 T. The peak emf of the generator must be at least 170 V. A square coil with sides 10.0 cm is used. What is the number of turns of wire needed in the coil?

Solution

The emf induced in the rotating coil is given by Equation 21.8:

$$V_{\text{source}}(t) = NBA\omega\sin(\omega t)$$

The peak emf V_0 is the coefficient of $\sin(\omega t)$. So

$$V_0 = NBA\omega$$

Hence the number of turns in the coil is

$$N = \frac{V_0}{BA\omega}$$

The angular frequency ω is

$$\begin{aligned}\omega &= 2\pi\nu = (2\pi \text{ rad})(60.0 \text{ Hz}) \\ &= 377 \text{ rad/s}\end{aligned}$$

Making the substitutions, you find

$$\begin{aligned}N &= \frac{170 \text{ V}}{(0.150 \text{ T})(0.100 \text{ m})(0.100 \text{ m})(377 \text{ rad/s})} \\ &= 301\end{aligned}$$

The coil should have 301 turns of wire.

21.4 SUMMARY OF THE MAXWELL EQUATIONS OF ELECTROMAGNETISM

The four *Maxwell equations* are the pillars of electromagnetism. We have studied them one by one as we progressed through the various electric and magnetic phenomena of the previous chapters. It is useful now to bring the equations together here for convenience, for reflection, and indeed, for celebration. The Maxwell equations are the following:

1. Gauss's law for the electric field:

$$\int_{\text{clsd surface S}} \vec{\mathrm{E}} \cdot d\vec{\mathrm{S}} = \frac{Q_{\text{net enclosed by S}}}{\varepsilon_0} \qquad (21.9)$$

   Gauss's law is a consequence of the inverse-square nature of Coulomb's law for the electrical force interaction between pointlike charges. Recall that there are *two distinct types* of electric fields: (a) the static electric field produced by electric charges; this electric field produces a conservative force on charges; and (b) the induced electric field created by a changing magnetic flux; electrical forces produced by this field are *not* conservative. Only the static electric field need be considered in Gauss's law: the flux of the induced electric field through a closed surface is *zero* because the induced electric field lines are closed loops!

2. Gauss's law for the magnetic field:

$$\int_{\text{clsd surface S}} \vec{\mathrm{B}} \cdot d\vec{\mathrm{S}} = 0 \text{ T}\cdot\text{m}^2 \qquad (21.10)$$

Unlike Faraday, Maxwell came from a privileged family and published his first paper at age 14! He was the first professor of experimental physics at Cambridge University and established the famous Cavendish Laboratory there.

This is a statement about the nonexistence of magnetic monopoles; magnets are dipolar. Magnetic field lines form closed contours.

3. The Ampere–Maxwell law:

$$\int_{\text{clsd path}} \vec{\mathbf{B}} \cdot d\vec{\ell} = \mu_0(I + I_D) = \mu_0 I + \mu_0 \varepsilon_0 \frac{d\Phi_{\text{elec}}}{dt} \quad (21.11)$$

where I and I_D are the conduction and displacement currents, respectively, enclosed by the path (or equivalently, piercing any hat-shaped surface that has the path as a perimeter). This law is a statement that magnetic fields are caused by electric conduction currents and/or by a changing electric flux (via the displacement current).

4. Faraday's law of electromagnetic induction:

$$\int_{\text{clsd path}} \vec{\mathbf{E}} \cdot d\vec{\ell} = -\frac{d\Phi_{\text{magnet}}}{dt} \quad (21.12)$$

This is a statement about how changes in magnetic flux produce (nonconservative) electric fields.

The Maxwell equations as we have written them are in what is called their *integral form*. There also is an equivalent *differential form* for the equations that we will neither write nor pursue, since it involves aspects of differential vector calculus we have not used in this text. A subsequent physics course in electromagnetic theory will develop the useful differential forms for the Maxwell equations in all their rightful glory.

Here we will only point ahead to two of the more remarkable aspects of the Maxwell equations:

1. Maxwell was able to show (as we will in Section 21.5) that electric and magnetic fields can *propagate* themselves through space according to the classical wave equation. The speed c of these traveling *electromagnetic fields* in a vacuum was found by Maxwell to be related to the familiar electric and magnetic constants simply by

$$c = \frac{1}{(\mu_0 \varepsilon_0)^{1/2}} \quad (21.13)$$

where ε_0 is the permittivity of free space and μ_0 is the permeability of free space. When Maxwell evaluated the numerical value for the speed of these waves by substituting the numerical values for ε_0 and μ_0, a remarkable result was secured:

$$\frac{1}{(\mu_0 \varepsilon_0)^{1/2}} = \frac{1}{\left[(4\pi \times 10^{-7}\ \text{T·m/A}) \dfrac{1}{4\pi \times 9.00 \times 10^{9}\ \text{N·m}^2/\text{C}^2} \right]^{1/2}} = 3.00 \times 10^{8}\ \text{m/s}$$

Maxwell recognized this result as very close to the measured speed of light. He then made a great conceptual leap: the waves indeed *were* light.

Light is an **electromagnetic wave** that travels in a vacuum with a speed c that is given by

$$c = \frac{1}{(\mu_0 \varepsilon_0)^{1/2}}$$

This equation is one of the great equations of physics, stemming as it does from the electromagnetic theory of Maxwell; the equation unifies three seemingly disparate fields of physics: electricity, magnetism, and optics.

2. The other remarkable feature of the Maxwell equations is that the later revolution created by relativity (see Chapter 25) had *no effect* on the equations. The equations are, as we say, relativistically correct (a term less disparaging than "politically correct"!). The equations do not need to be modified because of the revolutionary discoveries made by Einstein in 1905 regarding our notions of space, time, and energy. Newtonian mechanics, on the other hand, is fundamentally altered in the relativistic domain (as we have alluded to on occasion and will explore in detail in Chapter 25). In fact, what Einstein did was to take the incompatibility of the Newtonian laws with the Maxwellian equations of electromagnetism and resolve it in favor of Maxwell.

21.5 ELECTROMAGNETIC WAVES*

We all know what light is; but it is not as easy to tell what it is.
Samuel Johnson (1709–1784)†

In a vacuum, there is no matter, no charge, and no conduction current; displacement currents exist when the electric flux changes with time. In a vacuum, the four Maxwell's equations (Equations 21.9–21.12 of Section 21.4) take on a symmetric form:

Gauss's law for the electric field:

$$\int_{\text{clsd surface } S} \vec{\mathbf{E}} \cdot d\vec{\mathbf{S}} = 0\ \text{V·m} \quad (21.14)$$

Gauss's law for the magnetic field:

$$\int_{\text{clsd surface } S} \vec{\mathbf{B}} \cdot d\vec{\mathbf{S}} = 0\ \text{T·m}^2 \quad (21.15)$$

Ampere–Maxwell law for the magnetic field:

$$\int_{\text{clsd path}} \vec{\mathbf{B}} \cdot d\vec{\ell} = \mu_0 \varepsilon_0 \frac{d\Phi_{\text{elec}}}{dt} \quad (21.16)$$

Faraday's law for the induced electric field:

$$\int_{\text{clsd path}} \vec{\mathbf{E}} \cdot d\vec{\ell} = -\frac{d\Phi_{\text{magnet}}}{dt} \quad (21.17)$$

A time-varying electric flux gives rise to a magnetic field and a time-varying magnetic flux gives rise to an electric field. The circularity of this argument suggests the possibility of *self-sustaining*,

†*Boswell's Life of Johnson*, edited by George Birkbeck Hill (Oxford at the Clarendon Press, Oxford, 1934), Volume III, page 38. Entry for 12 April 1776.

perhaps propagating, time-dependent, electric and magnetic fields; we call them electromagnetic waves.

The plethora of electromagnetic waves of various kinds likely is familiar to you: radio waves, microwaves, visible light, even x-rays are common examples. With them arise some significant questions:

1. How do we know that these various kinds of light really *are* electromagnetic waves?
2. How are such waves produced?
3. Are the waves longitudinal or transverse?
4. What is oscillating in an electromagnetic wave?

We finally have all the tools at our disposal to answer these important questions. Here we see how the Maxwell equations in a vacuum lead to a wave equation for propagating electric and magnetic fields, as well as how Maxwell made the connection between these propagating electromagnetic disturbances and light, alluded to in the previous section.

Recall (from Chapter 12) that a classical wave disturbance $\Psi(x, t)$, propagating at speed v in either direction parallel to the x-axis, satisfies the classical wave equation (Equation 12.7):

$$\frac{\partial^2\Psi}{\partial x^2} - \frac{1}{v^2}\frac{\partial^2\Psi}{\partial t^2} = 0$$

Here we show that Maxwell's equations in a vacuum lead to wave equations for the space- and time-varying electric and magnetic fields. We also show how the fields manage to propagate themselves through a vacuum without the necessity of nearby charges or currents: each field is generated by a time variation in the other field, as implied by the Ampere–Maxwell law and Faraday's law in a vacuum.

To accomplish these objectives, we consider a simple and familiar system: a parallel plate capacitor with plates of *finite* area (see Figure 21.28). If the charges on the capacitor vary with time, perhaps periodically reversing themselves, the electric field in the vicinity of the capacitor also must change its magnitude and reverse its direction accordingly. A component of the electric field in the vicinity of the capacitor, say, in the plane midway between the capacitor plates, must be a function of spatial position and time.* We call it $E_y(x, t)$, as in Figure 21.28.

The time-varying electric field gives rise to a displacement current

$$I_D = \varepsilon_0 \frac{d\Phi_{\text{elec}}}{dt}$$

The displacement current in turn gives rise to a magnetic field in the vicinity, whose direction is determined from the now familiar right-hand rule: right-hand thumb in the direction of the current means the right-hand fingers indicate the directional sense of the magnetic field caused by the current. The displacement current is directed from one capacitor plate to the other, and so the magnetic field in the midplane has a component $B_z(x, t)$, as shown in Figure 21.29.

Notice that the electric and magnetic fields are perpendicular to each other: the electric field has a component $E_y(x, t)$ while the magnetic field has a component $B_z(x, t)$. As the charges vary on

*Other components also vary with space and time, but we focus on this component for simplicity.

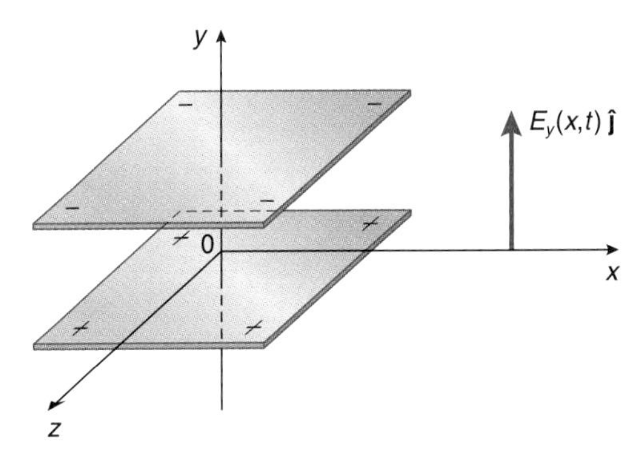

FIGURE 21.28 If the charges on the capacitor vary with time, the electric field in the vicinity is a function of position x and time t.

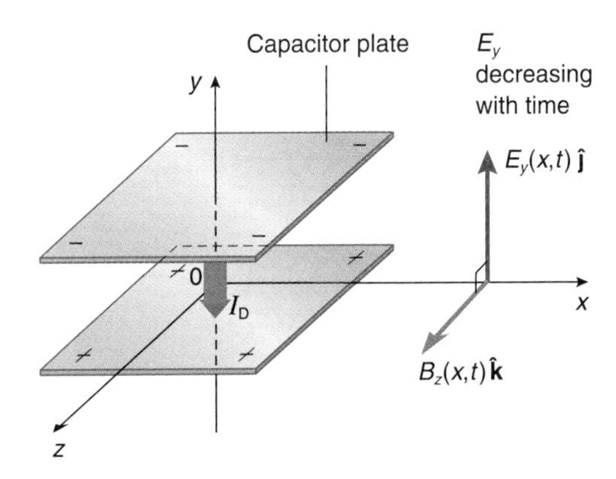

FIGURE 21.29 The displacement current produces a magnetic field.

the capacitor, the fields in the vicinity also must change accordingly. You might ask, quite reasonably, if the changes in the fields some distance away from the capacitor occur instantaneously, or whether there is a finite speed associated with the propagation of the changes. We seek to answer that query. The mutual orthogonality of the electric field, the magnetic field, and the direction of the possible propagation of their changes has important implications that we are now quite able to explore.

Do changes in the fields propagate parallel to the x-axis? Imagine a small, differential rectangle in the x–y plane of width dx and small length ℓ, as shown in Figure 21.30. We apply Faraday's law (Equation 21.17) to this rectangle:

$$\int_{\text{clsd path}} \vec{\mathbf{E}} \cdot d\vec{\ell} = -\frac{d\Phi_{\text{magnet}}}{dt}$$

We integrate the electric field around the perimeter of the differential rectangle, say in the counterclockwise sense; Φ_{magnet} is the flux of the magnetic field through the differential area. Along both the narrow widths dx of the differential rectangle, the electric field is perpendicular to $d\vec{\ell}$ (see Figure 21.31), and so the scalar product in the integrand vanishes on the short sides.

Along the left side of the rectangle where the component of the electric field is E_y, $d\vec{\ell}$ is antiparallel to the field (see Figure 21.32), and so the contribution to the integral is $-E_y\ell$. Along the other long side of the differential rectangle, the component of the electric

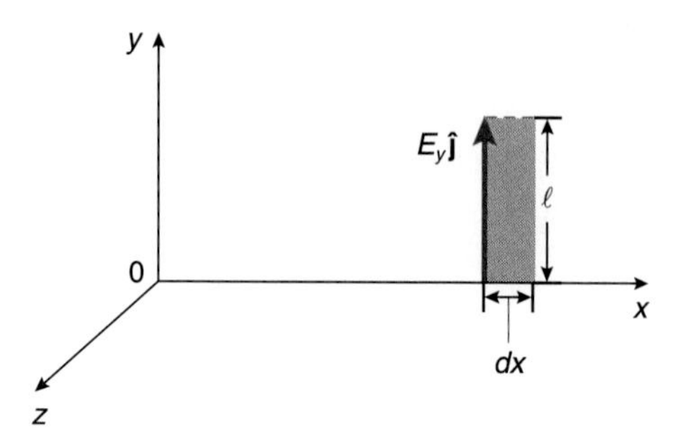

FIGURE 21.30 Apply Faraday's law to the indicated differential rectangle.

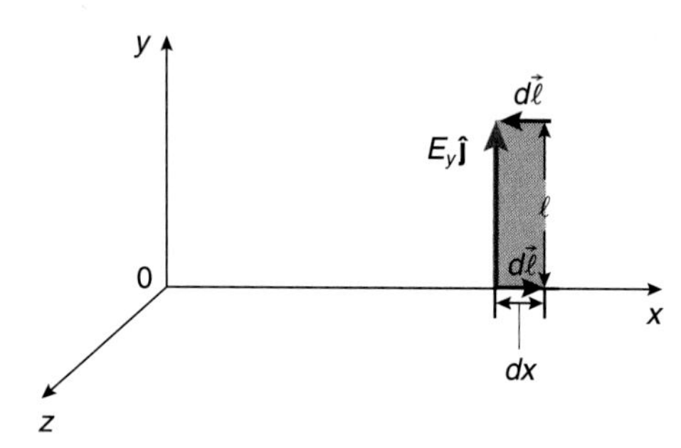

FIGURE 21.31 Along the short sides of the rectangle, $d\vec{\ell}$ is perpendicular to the electric field.

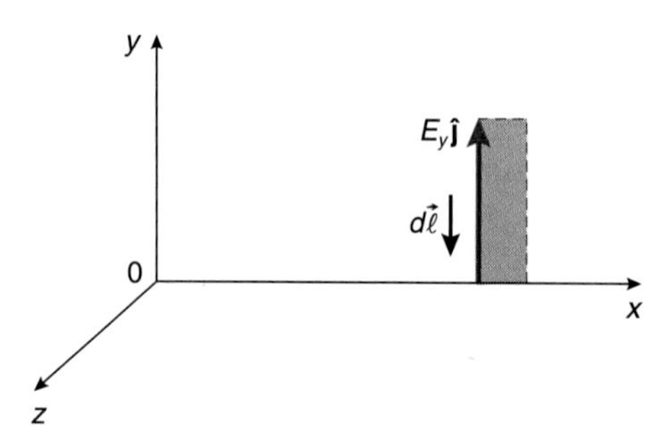

FIGURE 21.32 Along the left side of the rectangle, $d\vec{\ell}$ is antiparallel to the electric field.

field is $E_y + dE_y$ and the field is parallel to $d\vec{\ell}$ (see Figure 21.33); thus, the scalar product yields a contribution of $(E_y + dE_y)\ell$.

Hence the path integral of the electric field around the differential rectangle (the left-hand side of Faraday's law) is

$$-E_y\ell + (E_y + dE_y)\ell + 0 \text{ J/C} + 0 \text{ J/C}$$

or $\ell\, dE_y$.

The magnetic field is perpendicular to the differential area. Since we integrated counterclockwise around the path, the right-hand rule indicates the differential area vector is parallel to the magnetic field (see Figure 21.34). The flux of the magnetic field through the area thus is $B_z\ell\, dx$.

Hence Faraday's law (Equation 21.17) becomes

$$\ell\, dE_y = -\frac{d}{dt}(B_z\ell\, dx)$$

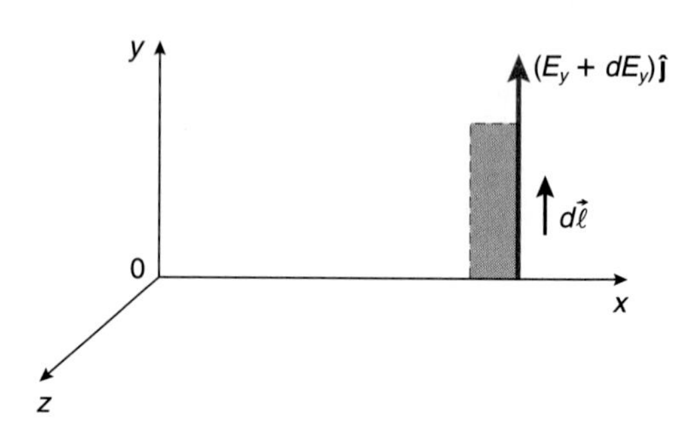

FIGURE 21.33 Along the right side of the rectangle, $d\vec{\ell}$ is parallel to the electric field.

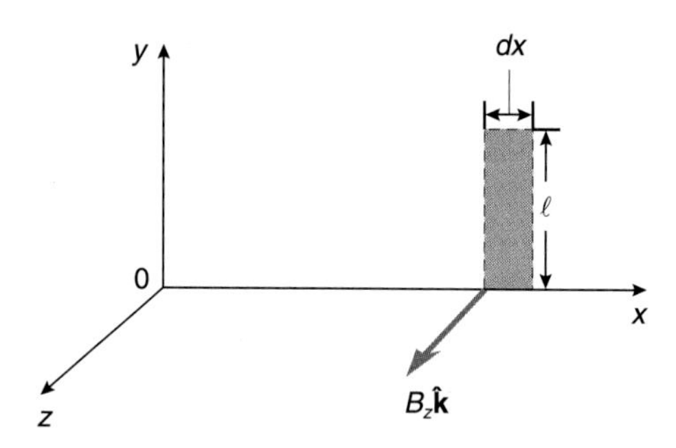

FIGURE 21.34 The magnetic field is perpendicular to the differential rectangle.

or

$$\frac{dE_y}{dx} = -\frac{dB_z}{dt}$$

Since both fields are functions of x and t, the derivatives really should be written properly as partial derivatives:

$$\frac{\partial E_y}{\partial x} = -\frac{\partial B_z}{\partial t} \tag{21.18}$$

The rate at which the electric field changes with position is the negative of the rate at which the magnetic field changes with time.

We now perform a similar calculation, applying the Ampere–Maxwell law in a vacuum (Equation 21.16) to a differential rectangle in the x–z plane (see Figure 21.35):

$$\int_{\text{clsd path}} \vec{\mathbf{B}} \cdot d\vec{\ell} = \mu_0\varepsilon_0 \frac{d\Phi_{\text{elec}}}{dt}$$

We integrate counterclockwise around the differential rectangle. Along the short sides of the rectangle, the magnetic field is perpendicular to $d\vec{\ell}$ (see Figure 21.36), and so no contribution to the integral arises there.

Along the left side where the magnetic field component is B_z, $d\vec{\ell}$ is parallel to the field (see Figure 21.37), and so we get a contribution $B_z\ell$ to the path integral. Along the other long side, the value of the magnetic field component is $B_z + dB_z$ and the field is antiparallel to $d\vec{\ell}$ (see Figure 21.38); thus we get a contribution $-(B_z + dB_z)\ell$ to the integral.

The path integral of the magnetic field around the differential rectangle, the left-hand side of the Ampere–Maxwell law in a vacuum, yields $-dB_z\,\ell$.

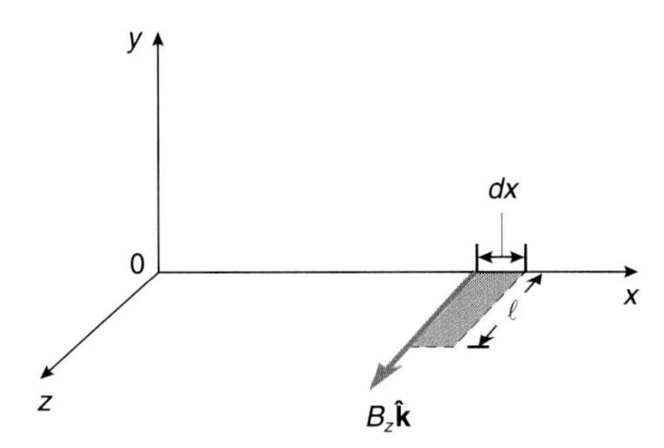

FIGURE 21.35 Apply the Ampere–Maxwell law to a differential rectangle in the x–z plane.

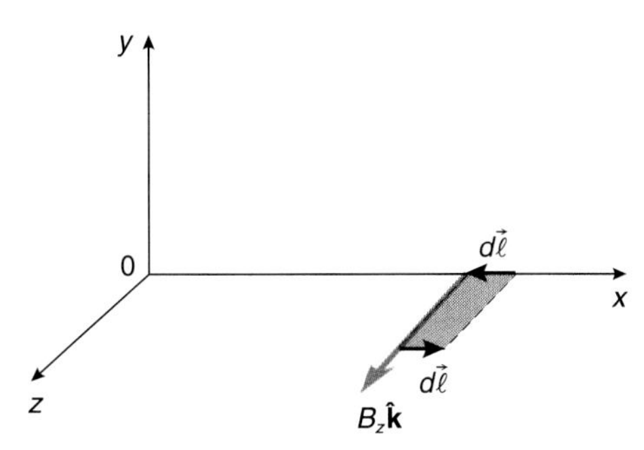

FIGURE 21.36 Along the short sides of the rectangle, $d\vec{\ell}$ is perpendicular to the magnetic field.

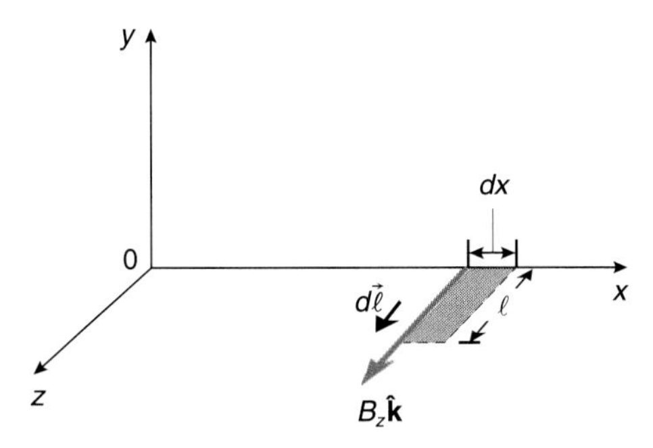

FIGURE 21.37 Along the left side of the rectangle, $d\vec{\ell}$ is parallel to the magnetic field.

The electric field is perpendicular to the differential rectangle in the x–z plane (see Figure 21.39). Since we integrated counterclockwise around the path, the differential area vector is parallel to the electric field. Hence the flux of the electric field through the area is $E_y \ell\, dx$.

The Ampere–Maxwell law in a vacuum (Equation 21.16) then becomes

$$-dB_z\, \ell = \mu_0 \varepsilon_0 \frac{d}{dt}(E_y \ell\, dx)$$

or

$$-\frac{dB_z}{dx} = \mu_0 \varepsilon_0 \frac{dE_y}{dt} \tag{21.19}$$

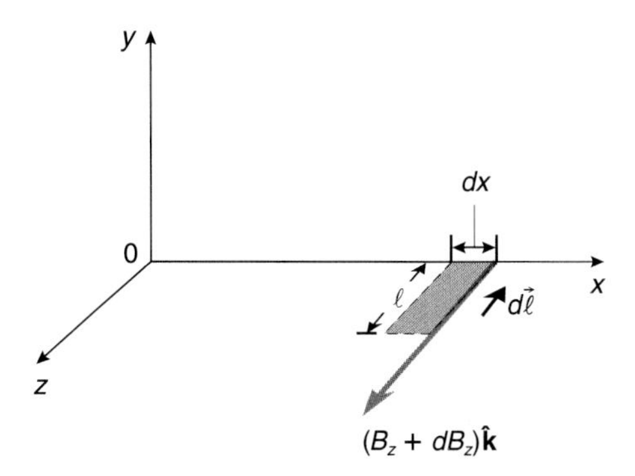

FIGURE 21.38 Along the other long side, $d\vec{\ell}$ is antiparallel to the magnetic field.

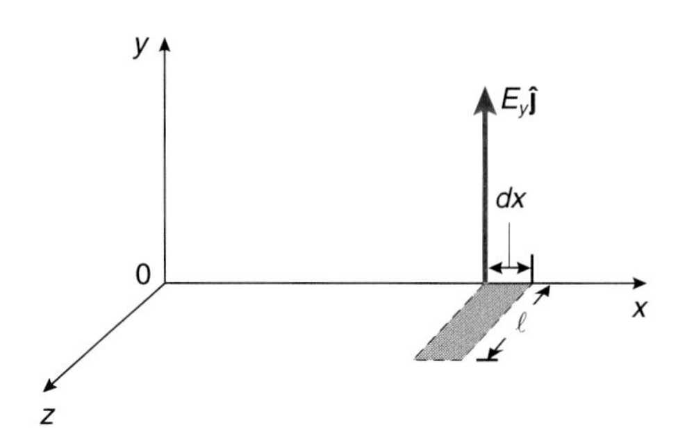

FIGURE 21.39 The electric field is perpendicular to the rectangle in the x–z plane.

Once again, since the fields are functions of both space and time, the derivatives properly are partial derivatives; thus, we write Equation 21.19 as

$$-\frac{\partial B_z}{\partial x} = \mu_0 \varepsilon_0 \frac{\partial E_y}{\partial t} \tag{21.20}$$

The spatial variation in the magnetic field is proportional to the time rate of change of the electric field.

Mathematically, when taking a partial derivative with respect to t, we keep x constant; likewise, when taking a partial derivative with respect to x, we keep t constant. Therefore it makes no difference whether you first differentiate a function with respect to x and then with respect to t, or first with respect to t and then with respect to x:

$$\frac{\partial^2 E_y}{\partial t\, \partial x} = \frac{\partial^2 E_y}{\partial x\, \partial t} \qquad \frac{\partial^2 B_z}{\partial t\, \partial x} = \frac{\partial^2 B_z}{\partial x\, \partial t} \tag{21.21}$$

We exploit this fact in the following way. Take Equation 21.18 and differentiate it with respect to x:

$$\frac{\partial^2 E_y}{\partial x^2} = -\frac{\partial^2 B_z}{\partial x\, \partial t} \tag{21.22}$$

Likewise differentiate Equation 21.20 with respect to t:

$$-\frac{\partial^2 B_z}{\partial t\, \partial x} = \mu_0 \varepsilon_0 \frac{\partial^2 E_y}{\partial t^2} \tag{21.23}$$

Combining Equations 21.22 and 21.23 using the partial derivative property in Equation 21.21, we find

$$\frac{\partial^2 E_y}{\partial x^2} = \mu_0 \varepsilon_0 \frac{\partial^2 E_y}{\partial t^2}$$

After transposing, we discover that the electric field component satisfies a partial differential equation that has the form of the classical wave equation (Equation 12.7):

$$\frac{\partial^2 E_y}{\partial x^2} - \mu_0 \varepsilon_0 \frac{\partial^2 E_y}{\partial t^2} = 0 \text{ N/(C·m}^2) \qquad (21.24)$$

In like manner, if we differentiate Equation 21.18 with respect to t and Equation 21.20 with respect to x, and combine them using Equation 21.21, we discover the following:

The magnetic field component also satisfies a partial differential equation of the form of the classical wave equation:

$$\frac{\partial^2 B_z}{\partial x^2} - \mu_0 \varepsilon_0 \frac{\partial^2 B_z}{\partial t^2} = 0 \text{ T/m}^2 \qquad (21.25)$$

A comparison of Equations 21.24 and 21.25 with the general form for the classical wave equation (Equation 12.7) indicates that the speed of propagation of such both waves of electric and magnetic fields is

$$\frac{1}{(\mu_0 \varepsilon_0)^{1/2}} \qquad (21.26)$$

Substituting the numerical values of μ_0 and ε_0, we find that the speed of these propagating electric and magnetic disturbances is numerically equal to the experimentally determined speed of light. It was from this equality that Maxwell made the conceptual leap and concluded that these propagating electromagnetic disturbances indeed *are* what we know as light.

If the charges move sinusoidally with time, the electromagnetic waves also are sinusoidal (far from the source) and we represent the electric and magnetic fields of the electromagnetic wave with sinusoidal wave functions like those we studied in Chapter 12:

$$E_y(x, t) = E_0 \cos(kx - \omega t) \qquad (21.27)$$

$$B_z(x, t) = B_0 \cos(kx - \omega t) \qquad (21.28)$$

As we noted before, the electric field, the magnetic field, and the direction of propagation all are mutually orthogonal (perpendicular); hence, light is a *transverse* electromagnetic wave.

Indeed, the vector product $\vec{E} \times \vec{B}$ indicates the direction of propagation.*

*In a subsequent course on electromagnetism, you will discover that this vector product, called the *Poynting vector*, has a magnitude proportional to the *intensity* of the wave: the power per square meter (oriented perpendicular to the direction of propagation) transmitted by the electromagnetic wave.

Also note that since the magnetic field arises because of *changes* in the electric field, and vice versa, oscillating changes in the fields arise fundamentally from the *acceleration of charges*.†

From this we see that the classical electromagnetic theory of Maxwell predicts that *accelerations of charges produce electromagnetic waves* that travel at the speed of light; they *are* light.‡

It is this principle that is exploited for generating radio waves, for example. Then, when such electromagnetic waves impinge on a conducting wire (or antenna), the electric field of the wave there causes electrons in the conductor to move back and forth, forming a small alternating current. The current is subsequently amplified and changed, using various electronic tricks, to produce a visual display (TV) or sound (radio).

Example 21.9

The antenna for the public radio station WVTF in Roanoke, Virginia, emits electromagnetic waves with a frequency of 89.1 MHz. What is the corresponding wavelength of the electric and magnetic field waves in a vacuum?

Solution

In a vacuum, electromagnetic waves travel with a speed $c = 3.00 \times 10^8$ m/s. The speed c, wavelength λ, and frequency ν of a wave are related by Equation 12.10:

$$c = \nu\lambda$$

The wavelength thus is

$$\lambda = \frac{c}{\nu} = \frac{3.00 \times 10^8 \text{ m/s}}{89.1 \times 10^6 \text{ Hz}} = 3.37 \text{ m}$$

The frequency of this radio station is near the lower end of the FM band of radio frequencies. Wavelengths of FM radio waves are on the order of meters.

†For example, if the charges undergo simple harmonic oscillation, the charges have nonzero acceleration (except at the instant when the charges pass through their equilibrium position).

‡We shall see in Chapter 26 that the emission of light by atoms is fundamentally different because of quantum mechanical considerations that have no classical analog.

Maxwell is celebrated on this stamp from San Marino. Now where in the world is San Marino?

21.6 Self-Inductance*

We leave the travel of electromagnetic waves for now (though we consider light again in Chapters 23 and 24) to introduce a new thought: the magnetic analog of a capacitor. Actually, the introduction hardly is necessary, for we have been learning about it for some time: a coil of wire. The coil may consist of a single loop or, more likely, multiple turns of wire such as in a solenoid, a toroidal coil, or another such arrangement of wire loops. Such coils of wire are called **inductors** and are represented schematically by the ideal circuit symbol shown in Figure 21.40.

The **self-inductance** L of an inductor (a coil) is defined as the ratio of the total magnetic flux Φ_{magnet} through the coil to the current I in the coil:

$$L \equiv \frac{\Phi_{\text{magnet}}}{I} \tag{21.29}$$

The current I in the coil is the current that produces the magnetic field threading the coil, which is why L is called the *self*-inductance. The dimensions of inductance are those of magnetic flux ($\text{T}\cdot\text{m}^2$ in SI units) divided by current (A in SI units). The SI unit combination $\text{T}\cdot\text{m}^2/\text{A}$ is called a **henry** (H), named for Joseph Henry (1797–1878):

$$\text{T}\cdot\text{m}^2/\text{A} \equiv \text{H} \tag{21.30}$$

Henry apparently discovered electromagnetic induction before Faraday but did not publish his findings.†

Analogously to the capacitance, the inductance is a function of the geometric arrangement of the coils of wire and the material (if any) within the coil itself. In other words, although the defining equation for the inductance (Equation 21.29) involves the ratio of the total magnetic flux through the coil to the current producing the flux (via the magnetic field), the ratio is independent of the specific values of Φ_{magnet} and I. We can see the parallels with the capacitance, since C is defined as the ratio Q/V, where the value of the ratio is independent of both Q and V. The capacitance depends only on the geometry (shape) of the capacitor and the material between its conductors (as we saw in Chapter 18).

We rearrange Equation 21.29 to get

$$\Phi_{\text{magnet}} = LI \tag{21.31}$$

If the current I in the coil changes with time, then the magnetic flux through the coil also changes with time. Faraday's law then implies that there will be an emf induced in the coil:

$$\text{induced emf} = -\frac{d\Phi_{\text{magnet}}}{dt}$$

†It is customary in the sciences to give credit to the persons who first publish a finding. The merits of this custom have been vigorously debated over the years, causing many feuds over patents and the priority of discovery. The priority issue for Faraday and Henry is by no means settled among scientific historians. See Investigative Project 15.

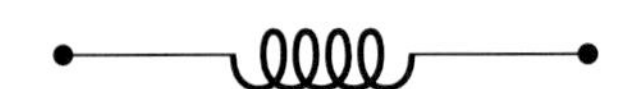

Figure 21.40 Circuit symbol for an ideal inductor.

Substituting for the magnetic flux using Equation 21.31, we obtain

$$\text{induced emf} = -L\frac{dI}{dt} \tag{21.32}$$

Imagine a coil such as that shown in Figure 21.41. The current I in the coil creates a magnetic field directed along the axis of the coil. Let the current I be increasing with time; hence, the magnetic flux through the coil also increases with time. Lenz's law states that a current is induced in the coil in a direction that opposes the change taking place. Since the flux is increasing with time, the direction of the induced current (created in response to the induced emf) tends to decrease the flux through the loop by diminishing the current I. Thus the induced current direction opposes that of I, and the polarity of the induced emf also does, as shown in Figure 21.42. If the coil is wound the other way (see Figure 21.43) and I increases with time, we still have the same directions of opposition for the induced current and also for the polarity for the induced emf in the coil.

We now do something that may seem peculiar at first. We define a potential difference V to be

$$V \equiv -\text{induced emf} \tag{21.33}$$

Then, according to Equation 21.32,

$$V = L\frac{dI}{dt} \tag{21.34}$$

When

$$\frac{dI}{dt}$$

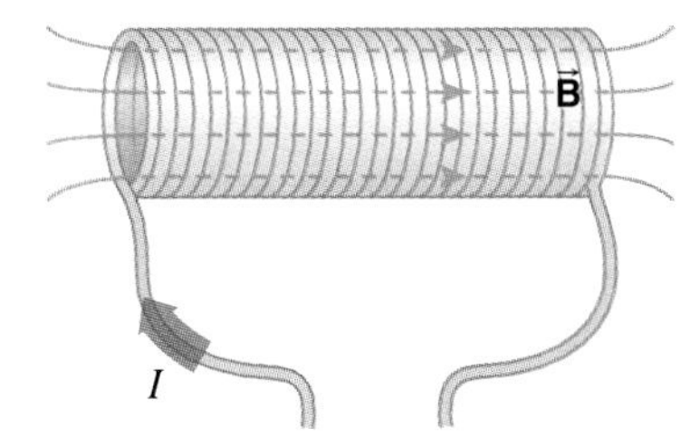

Figure 21.41 A current in a coil produces a magnetic field through the coil.

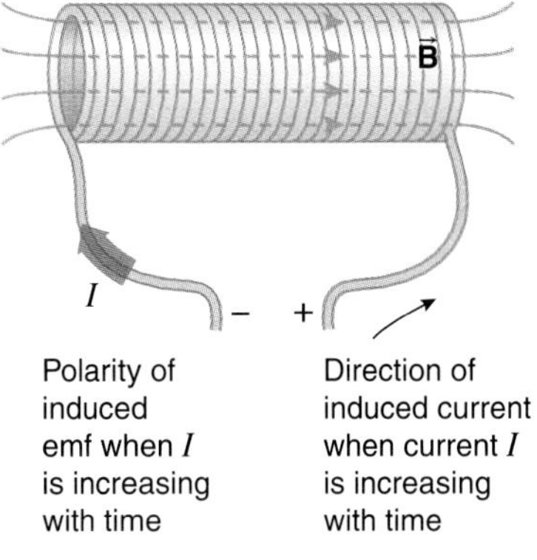

Figure 21.42 Direction of the induced current and polarity of the induced emf in the coil when the current in it increases with time.

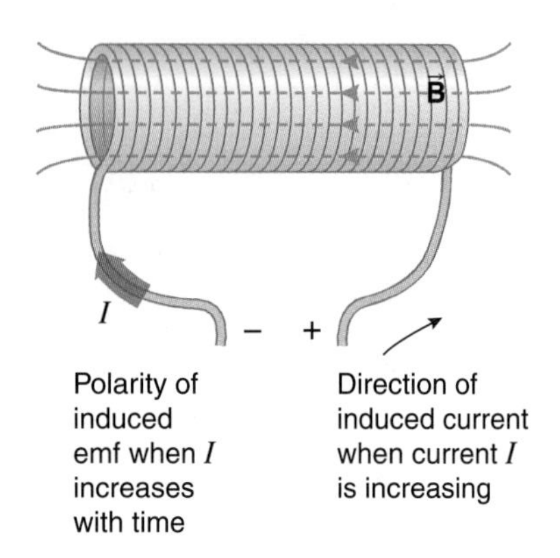

FIGURE 21.43 For a coil wound the other way, the direction of the induced current and polarity of the induced emf are the same as for the coil in Figure 21.42.

is positive (meaning that I is increasing with time), V has the polarity indicated in Figure 21.44.

The introduction of V to denote the opposition is done for consistency with the sign conventions on other circuit elements such as resistors and capacitors. In other words, we treat V as if it were a potential difference between two points (the two ends of the coil) rather than the negative of an induced emf. An inductor actually behaves like a peculiar battery or voltage source with the current going *into* its positive polarity terminal (unlike a normal battery or voltage source, where we usually think of the current as emerging *from* the positive polarity terminal). The emf is actually distributed continuously along the length of the wire of the coil; with V we just look at the two ends of the wire of the coil.

In this way we can treat the inductor as a common circuit element (an electrical device of some kind with two wires coming out of it). Why are we playing a bit of a semantic and physical shell game? From the viewpoint of physics, one should think of an inductor as an example of Faraday's and Lenz's laws and an induced emf. But from the more practical engineering viewpoint when using inductors in circuits, it is useful and convenient to treat V as if it were just an ordinary potential difference between the terminals of the inductor. It really is not, but doing so simply recognizes the convenient and conventional way inductors are treated in circuits and electronics. By using V instead of the induced emf, we can treat an inductor as a common circuit element: once we choose a current direction, an inductor circuit element is labeled with (+) and (−) polarity markings in accordance with the direction we choose for the current, with the current going into the terminal (lead) with the (+) polarity sign.

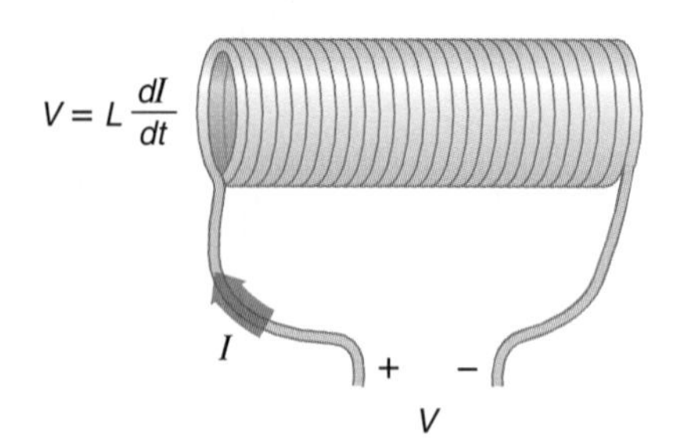

FIGURE 21.44 Polarity of the potential difference V, defined by Equations 21.33 and 21.34.

Recall that for a resistor, once we chose the current direction, we labeled the polarity of the potential difference across the resistor so the current enters R at the (+) polarity marking, as indicated in Figure 21.45. For a capacitor, if the current charging the capacitor is directed as shown in Figure 21.46, the potential difference across the capacitor has the indicated polarity. Notice that if the *potential difference* V is not changing with time, the current through the capacitor is zero; the capacitor then acts like an *open circuit* for dc (time-independent voltages and currents).

Now similarly for an inductor, if the current is directed through the inductor as shown in Figure 21.47, then there is a potential difference V across the inductor with the indicated polarity. In this way, the three circuit elements are treated similarly in circuits when applying the Kirchhoff voltage and current laws (see Sections 21.8 and 21.10).

PROBLEM-SOLVING TACTICS

21.3 Be careful to keep clear the distinction between the induced emf in an inductor (equal to the negative time rate of change of the magnetic flux) and the potential difference

$$V = -\text{ induced emf } = L\frac{dI}{dt}$$

introduced so that an inductor can be treated with the same sign conventions as resistors and capacitors. The induced emf and the potential difference V are the negatives of each other.

21.4 For dc situations, an inductor acts as a short circuit. Since the potential difference across an inductor is proportional to the time rate of change of the current, if the current does not change with time (i.e., the current is constant: a dc current), there then is no potential difference across the inductor and it acts like a short circuit. The inductor then can be mentally replaced by an ideal wire.

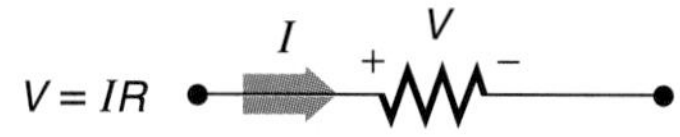

$V = IR$ (I, $+$ V $-$)

FIGURE 21.45 Polarity convention for a resistor.

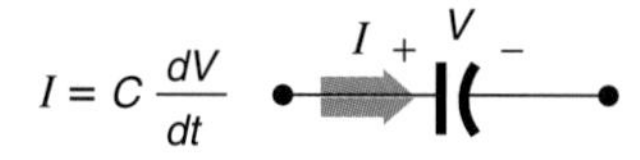

$I = C\frac{dV}{dt}$ (I, $+$ V $-$)

FIGURE 21.46 Polarity convention for a capacitor.

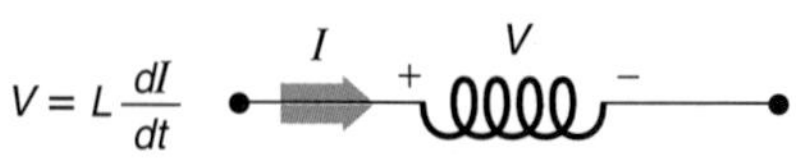

$V = L\frac{dI}{dt}$ (I, $+$ V $-$)

FIGURE 21.47 Polarity convention for an inductor.

QUESTION 2

You have a specified length of wire. (a) Describe a geometric configuration for the wire that makes its self-inductance large. (b) Describe a geometric configuration that minimizes its self-inductance.

EXAMPLE 21.10

Calculate the self-inductance of a long solenoid of length ℓ, with n coils per meter of its length, each with cross-sectional area A. Assume the magnetic field lines are confined to the region within the coil over its entire length.

Solution

The self-inductance is the ratio of the magnetic flux through the coil to the current in the coil (Equation 21.29):

$$L = \frac{\Phi_{\text{magnet}}}{I}$$

The magnitude of the magnetic field inside a long solenoid, far from its ends, is (from Example 20.13)

$$B = \mu_0 nI$$

where n is the number of turns of wire per meter of its length. Since the magnetic field is uniform across the cross-sectional area of the solenoid, the flux through a single coil of the solenoid is BA, where A is the cross-sectional area.

The number of loops of wire in a length ℓ of the solenoid is $n\ell$. Thus the total magnetic flux through a length ℓ of the solenoid is $n\ell$ times the flux though a single coil:

$$\begin{aligned}\Phi_{\text{magnet}} &= n\ell BA \\ &= n\ell\mu_0 nIA \\ &= n^2\ell A\mu_0 I\end{aligned}$$

The self-inductance of a length ℓ of the long solenoid then is

$$\begin{aligned}L &= \frac{\Phi_{\text{magnet}}}{I} \\ &= \mu_0 n^2 A\ell\end{aligned}$$

From this result we see that the self-inductance depends on geometric factors associated with the coil; L is independent of the magnetic flux through the coil and the current in it. The self-inductance of a right-handed coil is identical to that of its mirror image with left-handed windings, because Lenz's law of opposition to a changing current must apply equally to both.

From a practical viewpoint, the self-inductance of a coil is rarely calculated. It is measured experimentally (by various techniques; see Problem 42). The same also could be said of capacitances and resistances: they are rarely calculated from their defining equations except when designing or fabricating them.

21.7 SERIES AND PARALLEL COMBINATIONS OF INDUCTORS*

Just as for other circuit elements, it occasionally is useful to combine inductors into series or parallel combinations to secure different equivalent inductances. For reasons that will become apparent in Section 21.11, we assume here that the magnetic field of each inductor produces zero flux through any other inductor present.

Inductors in Series

A collection of inductors in series is shown in Figure 21.48. Circuit elements in series have the same current through them. Hence for a series connection of inductors, we have, for each inductor,

$$\begin{aligned}V_1 &= L_1\frac{dI}{dt} \\ V_2 &= L_2\frac{dI}{dt} \\ V_3 &= L_3\frac{dI}{dt}\end{aligned} \tag{21.35}$$

We replace the series with a single equivalent inductance L_{eq} with the same potential difference V between the terminals A and B, as indicated in Figure 21.49. Thus we must have

$$V = L_{\text{eq}}\frac{dI}{dt} \tag{21.36}$$

The potential difference V is the sum of those in Figure 21.48:

$$V = V_1 + V_2 + V_3$$

We substitute for V on the left-hand side of this equation using Equation 21.36 and use Equation 21.35 for the terms on the right-hand side. The same current is in each inductor in series. The result is

$$L_{\text{eq}}\frac{dI}{dt} = L_1\frac{dI}{dt} + L_2\frac{dI}{dt} + L_3\frac{dI}{dt}$$

or

$$L_{\text{eq}} = L_1 + L_2 + L_3 \qquad \text{(inductors in series)}$$

The preceding argument applies for any number of inductors in series.

Inductors in series combine like resistors in series or capacitors in parallel:

$$L_{\text{eq}} = L_1 + L_2 + L_3 + \cdots \tag{21.37}$$

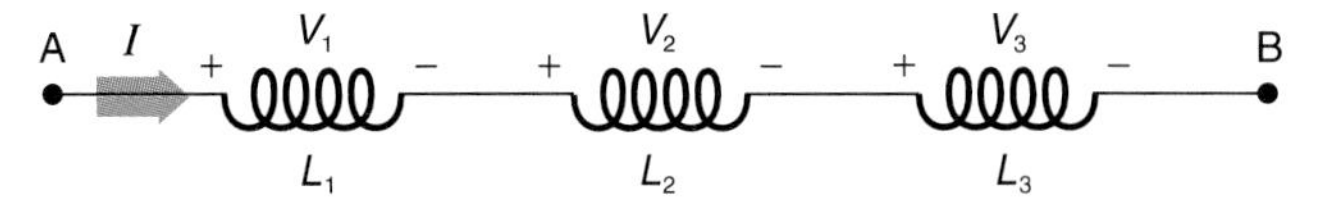

FIGURE 21.48 Inductors in series.

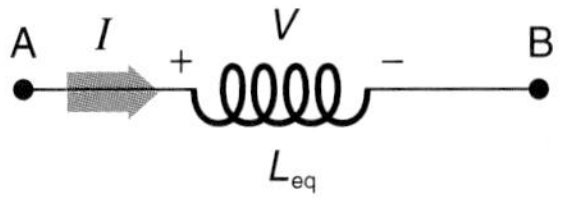

FIGURE 21.49 The equivalent inductor with inductance L_{eq}.

Inductors in Parallel

A collection of inductors in parallel is shown in Figure 21.50. Since circuit elements in parallel are connected to the same two nodes, the same potential difference exists across each circuit element. Thus, for the parallel combination of inductors pictured, we have

$$V = L_1 \frac{dI_1}{dt}$$
$$V = L_2 \frac{dI_2}{dt} \quad (21.38)$$
$$V = L_3 \frac{dI_3}{dt}$$

We simplify the situation by replacing the collection with a single equivalent inductor L_{eq} having the same potential difference V between its terminals, as shown in Figure 21.51, where

$$V = L_{eq} \frac{dI}{dt} \quad (21.39)$$

The Kirchhoff current law implies that

$$I = I_1 + I_2 + I_3$$

Differentiating this equation with respect to t, we get

$$\frac{dI}{dt} = \frac{dI_1}{dt} + \frac{dI_2}{dt} + \frac{dI_3}{dt}$$

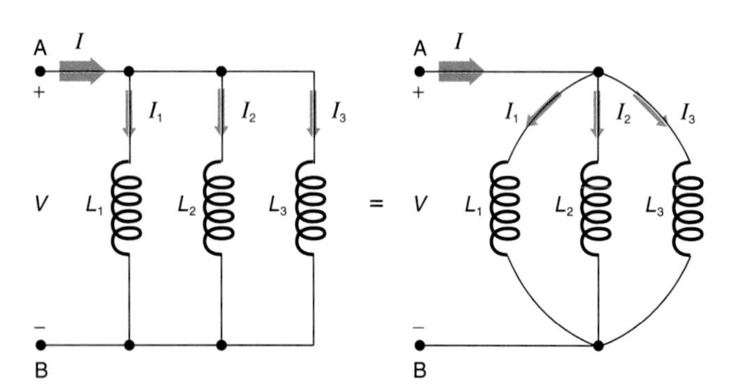

FIGURE 21.50 A parallel collection of inductors.

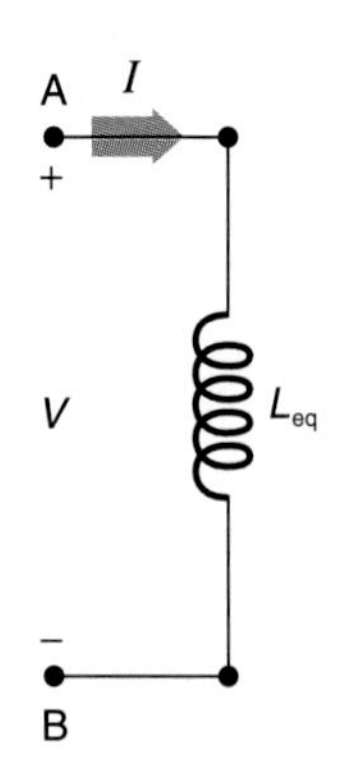

FIGURE 21.51 The equivalent inductor has inductance L_{eq}.

We use Equation 21.39 to substitute for the derivative on the left-hand side of this equation, and Equation 21.38 for the terms on the right-hand side; this yields

$$\frac{V}{L_{eq}} = \frac{V}{L_1} + \frac{V}{L_2} + \frac{V}{L_3}$$

or

$$\frac{1}{L_{eq}} = \frac{1}{L_1} + \frac{1}{L_2} + \frac{1}{L_3} \quad \text{(inductors in parallel)}$$

The preceding argument applies for any number of inductors in parallel.

> Inductors in parallel combine like resistors in parallel or capacitors in series:
>
> $$\frac{1}{L_{eq}} = \frac{1}{L_1} + \frac{1}{L_2} + \frac{1}{L_3} + \cdots \quad (21.40)$$

21.8 A Series LR Circuit*

Faraday's law ensures that changes in currents cannot be accomplished instantaneously. In this section, we examine the transient behavior of a **series *LR* circuit** consisting of an independent voltage source V_0, an inductor L, and a resistor R connected in series, as shown in Figure 21.52. The series *LR* circuit is analogous to the series *RC* circuit we treated in Chapter 19.

Case 1: Current upon Closing a Switch

With the switches open, there is zero current. We want to determine how the current varies with time after switch (1) is closed when $t = 0$ s (see Figure 21.53). Switch (2) is left open. We choose the direction for the current I shown in Figure 21.53. The choice of a current direction forces us to label the resistor and inductor polarities accordingly, with the current going into the

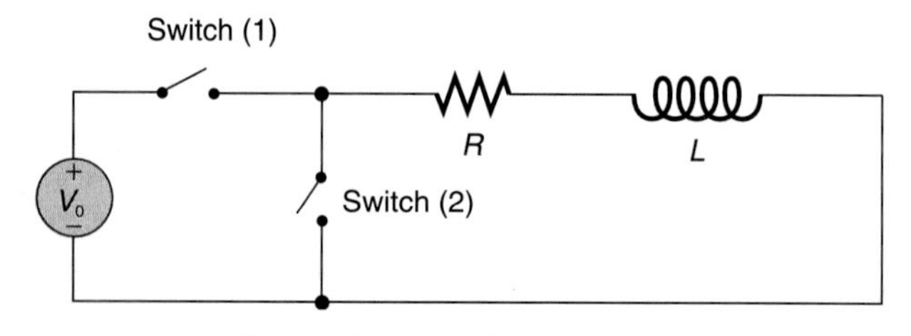

FIGURE 21.52 A series *LR* network (note there is no closed path, so it is not yet a circuit).

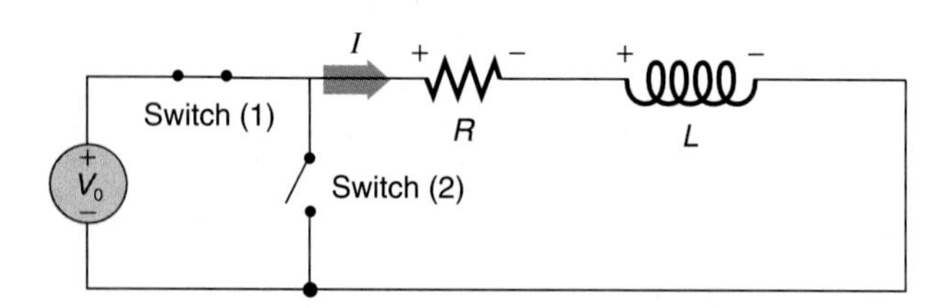

FIGURE 21.53 Switch (1) is closed when $t = 0$ s.

positive polarity terminal of both, as indicated. From Ohm's law, the potential difference across the resistor is IR. The potential difference across the inductor is

$$L\frac{dI}{dt}$$

according to the scheme we instituted in Section 21.6.

We apply the Kirchhoff voltage law (KVL) the same way we have on many occasions in the past (Chapter 19). We traverse the circuit in the clockwise sense, beginning in the lower left corner:

$$-V_0 + IR + L\frac{dI}{dt} = 0 \text{ V} \tag{21.41}$$

Rearranging the terms slightly, we have

$$\frac{dI}{dt} + \frac{R}{L}I = \frac{V_0}{L} \tag{21.42}$$

This is another example of a differential equation, an equation involving an unknown variable (I), and its derivatives, here just the first derivative

$$\frac{dI}{dt}$$

We have encountered differential equations before (for example, when studying simple harmonic motion in Chapter 7), but not one of this type. Since you likely have not yet had a formal course in how to solve these equations, we will have to solve this equation using the seat of our pants, your growing knowledge of calculus, and a bit of (mathematically legitimate) sleight of hand. What we do to solve this differential equation may not occur to you, at first, as reasonable. In other words, you may not be able to solve this differential equation on your own just yet, but you should be able to follow what happens here without feeling that any mathematical wool is being pulled over your eyes.

To begin, we change variables. This is the step that may not appear obvious to you at all. In particular, let a parameter x be defined as*

$$x \equiv I - \frac{V_0}{R} \tag{21.43}$$

Since both V_0 and R are fixed, differentiation with respect to time yields

$$\frac{dx}{dt} = \frac{dI}{dt}$$

We substitute for I and its derivative in Equation 21.42:

$$\frac{dx}{dt} + \frac{R}{L}\left(x + \frac{V_0}{R}\right) = \frac{V_0}{L}$$

Simplifying slightly, we have

$$\frac{dx}{dt} + \frac{R}{L}x = 0 \text{ V/H}$$

*Here we use x as an unknown quantity, *not* the Cartesian coordinate x.

Rearranging the terms yields

$$\frac{dx}{x} = -\frac{R}{L}\,dt$$

which now is in a form that can be integrated. Indeed, the purpose of the change in variables was to achieve just this objective.

We integrate from when the switch is closed at $t = 0$ s to instant (and elapsed time) t, with corresponding limits on x (i.e., from x_0 to x):

$$\ln x - \ln x_0 = -\frac{R}{L}t$$

or

$$\ln\left(\frac{x}{x_0}\right) = -\frac{R}{L}t$$

Taking the antilogarithms, we get

$$\frac{x}{x_0} = e^{-(R/L)t}$$

or

$$x = x_0 e^{-(R/L)t} \tag{21.44}$$

In Equation 21.43, when $t = 0$ s the current $I = 0$ A, and so

$$x_0 = -\frac{V_0}{R}$$

Substituting for x and x_0 in Equation 21.44, we find

$$I - \frac{V_0}{R} = -\frac{V_0}{R}e^{-(R/L)t}$$

Solving for I yields

$$I(t) = \frac{V_0}{R}\left[1 - e^{-(R/L)t}\right] \tag{21.45}$$

A few reassuring things to note about this solution for the current:

1. Notice that when $t = 0$ s the current is $I = 0$ A, which of course was the initial condition of the circuit.
2. After a long time, as $t \to \infty$ s, the exponential term approaches 0, and so the current in the circuit eventually becomes equal to V_0/R. In other words, after a long time the current becomes constant, and the inductor *has no effect* (it acts as a short circuit).

When I becomes constant, there is no potential difference across the inductor. An inductor has an effect *only* while the current is *changing* with time. For the steady-state situation, there is no potential difference across an inductor since the current no longer is changing with time.

Indeed, this can be seen from Equation 21.34:

$$V = L\frac{dI}{dt}$$

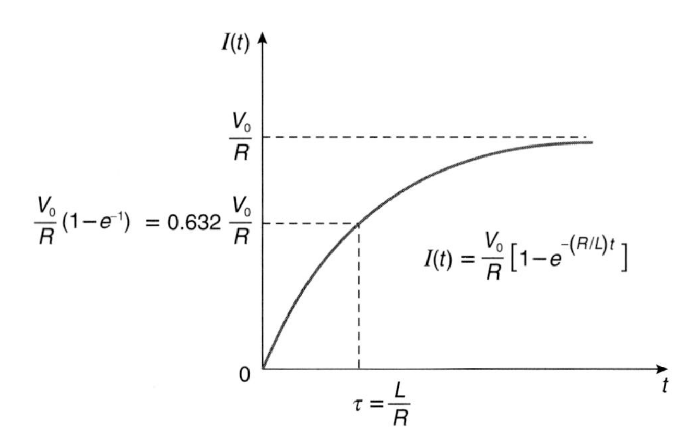

FIGURE 21.54 The current versus time in a series LR circuit.

When I is constant ($= V_0/R$), its derivative is zero, making $V = 0$ V.

3. The exponent for e in Equation 21.45 must always be dimensionless. Thus the quantity L/R must have the SI units of seconds.

This characteristic time is called the **time constant** τ of the circuit:

$$\tau \equiv \frac{L}{R} \qquad (21.46)$$

After a time t equal to five time constants ($t = 5\tau$), the exponential term is then $e^{-5} = 0.006\,74$, and the current I is within 1% of its final value of V_0/R.

A graph of the current versus time is shown in Figure 21.54.

Since *every* circuit consists of at least one loop of wire, there is some (small) self-inductance associated with every circuit, which justifies the behavior of the current we presumed in Experiment 2 in Section 21.1 (see Figure 21.4).

Case 2: Disconnecting the Independent Voltage Source

Wait a sufficient time so the current in Case 1 has reached its steady-state value V_0/R. What happens when the switch (2) is closed just when switch (1) is opened (see Figure 21.55)?

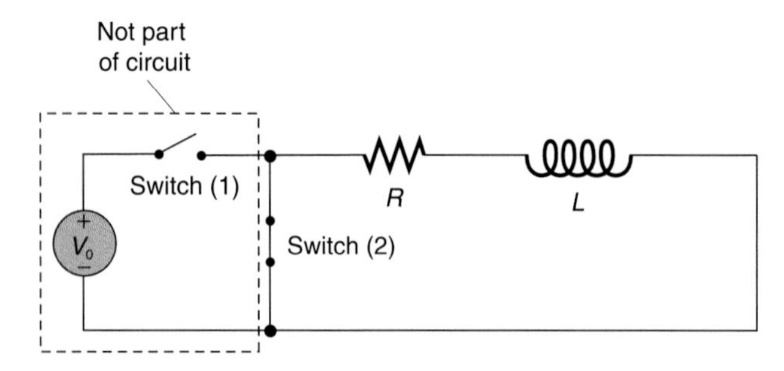

FIGURE 21.55 After a long time, switch (2) is closed while switch (1) is opened.

The result of this operation is to disconnect the independent voltage source V_0 from the circuit. The voltage source V_0 is part of a network but the *circuit* now consists of only the resistor and the inductor. Initially (i.e., when $t = 0$ s here) the current in the resistor and inductor is V_0/R. What happens to the current as a function of time? It cannot instantaneously vanish: Lenz's law prohibits it. Once again we label the resistor and inductor polarities in accordance with the direction chosen for the current, as in Figure 21.56.

We apply the KVL to the circuit, going around clockwise beginning in the lower left corner of Figure 21.56:

$$IR + L\frac{dI}{dt} = 0 \text{ V}$$

Fortunately, this differential equation is easier to solve than the one in Case 1! We rearrange the terms to collect those involving I:

$$\frac{dI}{I} = -\frac{R}{L}\,dt$$

This can be integrated immediately. Integrating on the left-hand side from I_0 ($= V_0/R$) to I and on the right-hand side from $t = 0$ s to t yields

$$\int_{I_0}^{I} \frac{dI}{I} = -\frac{R}{L}\int_{0\text{ s}}^{t} dt$$

$$\ln I - \ln I_0 = -\frac{R}{L}t$$

We combine the logarithms:

$$\ln\left(\frac{I}{I_0}\right) = -\frac{R}{L}t$$

Now we take antilogarithms:

$$\frac{I}{I_0} = e^{-(R/L)t}$$

or

$$I(t) = I_0 e^{-(R/L)t}$$

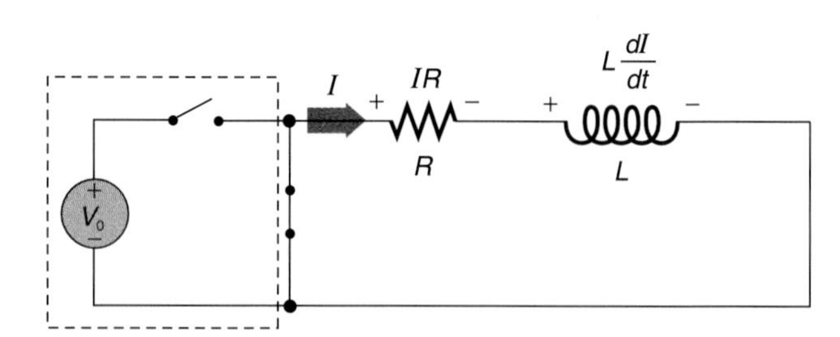

FIGURE 21.56 The polarities are marked consistent with direction chosen for I.

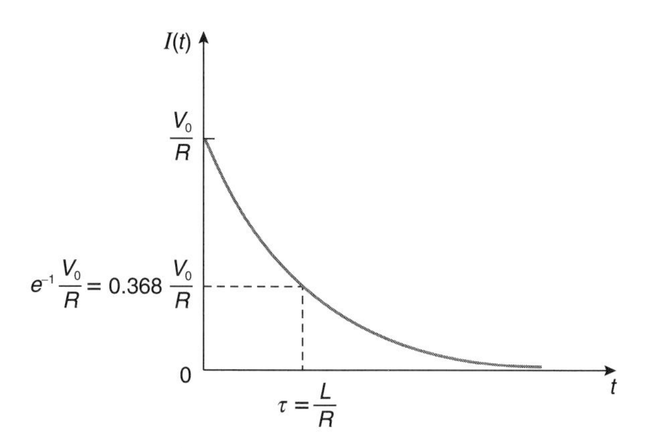

FIGURE 21.57 **The current decreases exponentially with time.**

Substituting for I_0, we obtain

$$I(t) = \frac{V_0}{R}e^{-(R/L)t} \tag{21.47}$$

This indicates that the current in the circuit decreases exponentially with time, as shown in Figure 21.57.

Since the exponent must be dimensionless, L/R must have the dimensions of time. Indeed, we have the same time constant that we had for Case 1:

$$\tau = \frac{L}{R}$$

After a time t equal to five time constants, the exponential term is $e^{-5} = 0.006\,74$ and the current I is less than 1% of its initial value.

STRATEGIC EXAMPLE 21.11

Consider the series *LR* network shown in Figure 21.58.

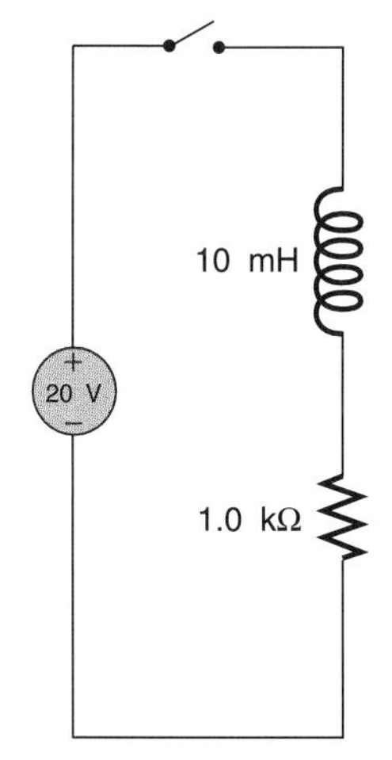

FIGURE 21.58

a. At the instant the switch is closed, determine the following quantities: (i) the current in the resistor; (ii) the current in the inductor; (iii) the potential difference across the resistor; (iv) the potential difference across the inductor; and (v) the rate at which the current through the inductor is changing with time.

b. After many time constants (i.e., for $t \to \infty$ s), determine the same quantities.

c. Show explicitly that the KVL is satisfied in the circuit at any instant t.

d. Calculate the time constant of the circuit.

e. Make detailed plots of the potential difference across the resistor and inductor as functions of time.

Solution

a. The inductor and resistor are in series and so must have the same current through them at any instant. The switch initially was open and the current was zero. At the instant the switch is closed, the initial value of the current must also be zero. This follows from Equation 21.45 (case 1) for the current through an inductor in a series *LR* circuit:

$$I(t) = \frac{V_0}{R}\left[1 - e^{-(R/L)t}\right]$$

which when $t = 0$ s reduces to

$$I(0 \text{ s}) = \frac{V_0}{R}(1 - 1)$$
$$= 0 \text{ A}$$

Since R and L are in series, *both* the resistor and inductor have zero current when $t = 0$ s.

Since the resistor has no current when $t = 0$ s, the potential difference across the resistor at this instant is zero (from Ohm's law: $V = IR$). With no potential difference across the resistor when $t = 0$ s, the KVL, taken around the loop (say in the clockwise sense) implies that the potential difference across the inductor must be 20 V at this instant, with the polarity indicated in Figure 21.59. The KVL must be satisfied *at any instant*, since it represents a statement about energy conservation (you will soon see this again explicitly).

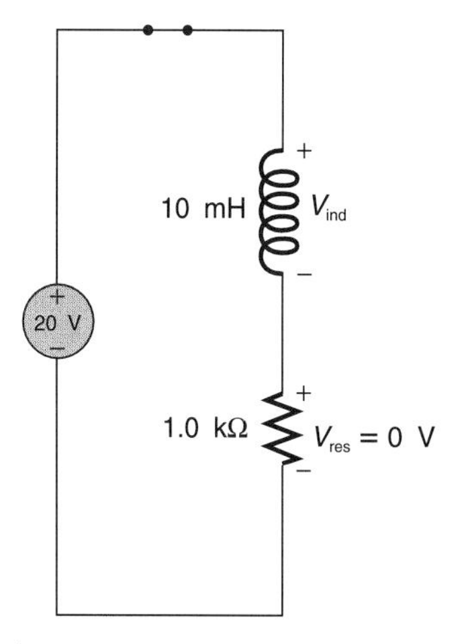

FIGURE 21.59 **The situation when $t = 0$ s.**

The potential difference V across the inductor depends on how fast the current in the inductor is changing with time. That is, from Equation 21.34,

$$V = L\frac{dI}{dt}$$
$$20\ \text{V} = (10 \times 10^{-3}\ \text{H})\frac{dI}{dt}$$

Solving for

$$\frac{dI}{dt}$$

you find

$$\frac{dI}{dt} = \frac{20\ \text{V}}{10 \times 10^{-3}\ \text{H}} = 2.0 \times 10^{3}\ \text{A/s}$$

This is the initial slope when $t = 0$ s of a graph of I versus t through the inductor. Since the current in the inductor and the resistor are the same (since these circuit elements are in series here), this also represents the rate at which the current through the resistor is changing with time when $t = 0$ s.

You also could obtain this result using Equation 21.45 for the current through the inductor:

$$I(t) = \frac{V_0}{R}\left[1 - e^{-(R/L)t}\right]$$

Differentiating with respect to t, you find

$$\frac{dI}{dt} = \frac{V_0}{R}(-1)\,e^{-(R/L)t}\left(-\frac{R}{L}\right) = \frac{V_0}{L}e^{-(R/L)t} \qquad (1)$$

Notice that when $t = 0$ s you get the previous result:

$$\frac{dI}{dt} = \frac{V_0}{L} = \frac{20\ \text{V}}{10 \times 10^{-3}\ \text{H}} = 2.0 \times 10^{3}\ \text{A/s}$$

b. After many time constants have elapsed, the potential difference across the inductor is zero because the current no longer is changing with time. That is, using equation (1) and letting $t = \infty$ s means that the exponential vanishes, and so

$$\frac{dI}{dt} = 0\ \text{A/s}$$

The current through the inductor then is constant and from Equation 21.45 equal to

$$\frac{V_0}{R} = \frac{20\ \text{V}}{1.0 \times 10^{3}\ \Omega} = 20\ \text{mA}$$

This is equal to the current through the resistor since the two circuit elements are in series. Using Ohm's law, the potential difference across the resistor is

$$V_{\text{res}} = IR = (20 \times 10^{-3}\ \text{A})(1.0 \times 10^{3}\ \Omega) = 20\ \text{V}$$

The inductor behaves as if it was shorted out.

c. To show that the KVL is satisfied at any time t, we need to find the potential differences across the resistor and inductor at any time t. The potential difference $V_{\text{res}}(t)$ across the resistor is

$$V_{\text{res}}(t) = I(t)R$$

Using Equation 21.45 for $I(t)$, you get

$$V_{\text{res}}(t) = V_0[1 - e^{-(R/L)t}] \qquad (2)$$

The potential difference across the inductor at any time is

$$V_{\text{ind}}(t) = L\frac{dI}{dt}$$

which, using equation (1) for $\frac{dI}{dt}$, is

$$V_{\text{ind}}(t) = L\frac{V_0}{L}e^{-(R/L)t} = V_0 e^{-(R/L)t} \qquad (3)$$

The polarities are indicated in Figure 21.60.

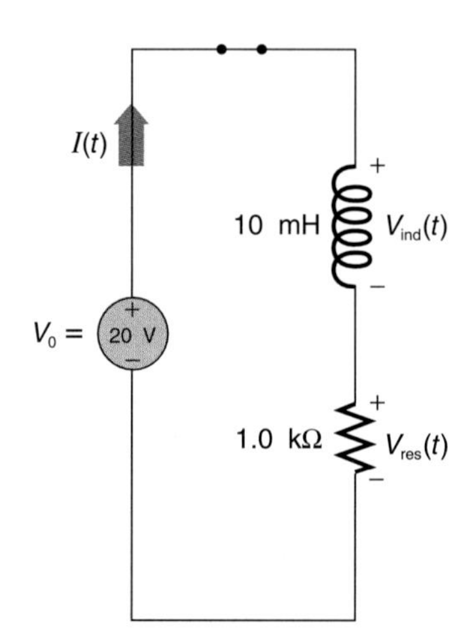

FIGURE 21.60

Now invoking the KVL and marching around the loop (in the CW sense beginning in the lower left corner), you have

$$-V_0 + V_{\text{ind}}(t) + V_{\text{res}}(t) = 0\ \text{V}$$

Using equations (3) and (2), this becomes

$$-V_0 + V_0 e^{-(R/L)t} + V_0[1 - e^{-(R/L)t}] = 0\ \text{V}$$
$$0\ \text{V} = 0\ \text{V}$$

The KVL is indeed satisfied at any instant.

d. The time constant of the circuit is

$$\tau = \frac{L}{R} = \frac{10 \times 10^{-3}\ \text{H}}{1.0 \times 10^{3}\ \Omega} = 1.0 \times 10^{-5}\ \text{s}$$

e. The potential difference across the resistor is

$$V_{res}(t) = V_0[1 - e^{-(R/L)t}] = (20\ \text{V})[1 - e^{-(1.0 \times 10^5\ \text{s}^{-1})t}]$$

The potential difference across the inductor is

$$V_{ind} = V_0 e^{-(R/L)t} = (20\ \text{V})e^{-(1.0 \times 10^5\ \text{s}^{-1})t}$$

Both are plotted in Figure 21.61.

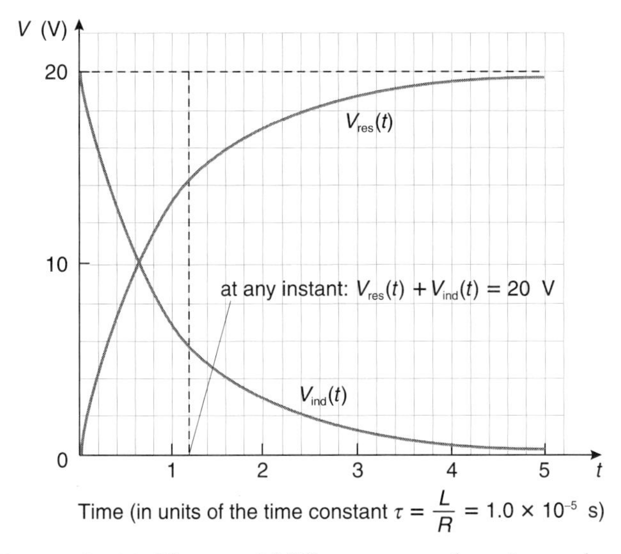

FIGURE 21.61 The potential difference across the resistor and inductor as a function of time.

21.9 ENERGY STORED IN A MAGNETIC FIELD*

When we investigated capacitors in Chapter 18, we discovered that electrical potential energy is stored by the capacitor in the electric field between its plates. Here we discover that inductors store energy in their magnetic field, illustrating once again that an inductor is the magnetic analog of a capacitor.

Recall the situation investigated in Case 1 of the previous section. This involved a description of how the current in a series *LR* circuit increases when a switch is closed to complete the circuit. By applying the KVL (which, in closet form, is a consequence of conservation of energy), we found that (from Figure 21.53 and Equation 21.41)

$$-V_0 + IR + L\frac{dI}{dt} = 0\ \text{V}$$

If we multiply this equation by the current I, each term has an interesting physical interpretation:

$$-IV_0 + I^2R + IL\frac{dI}{dt} = 0\ \text{W} \qquad (21.48)$$

Recall the sign convention we introduced for the power in Chapter 19: the power absorbed by (transferred to) a circuit element is the potential difference across the circuit element times the current into its positive polarity terminal.

1. The potential difference across the battery is V_0 and the current into its positive polarity terminal is $-I$. Hence the power absorbed by the battery is $-IV_0$, the first term of Equation 21.48. This power is negative because the battery actually is providing power to the circuit.
2. The second term is the power absorbed by the resistor. The power absorbed by the resistor is positive (as it should be), as resistances will always warm up.
3. The third term must also represent power: the power absorbed by the inductor. Once again, the current I is going into the positive terminal of the inductor. Since power is the time rate of change of the energy, the inductor is absorbing (securing) energy U at a rate

$$P = IV$$

$$\frac{dU}{dt} = IL\frac{dI}{dt} \qquad (21.49)$$

or

$$dU = IL\,dI$$

This expression can be integrated to find the energy U stored by the inductor when a current I is passing through the inductor.

The energy stored by the inductor is

$$U = \frac{1}{2}LI^2 \qquad (21.50)$$

when the inductor has a current I passing through it. Notice that the energy does not depend on

$$\frac{dI}{dt}$$

but just on I itself. In other words, the inductor stores energy (as long as $I \neq 0$ A) even when the current is *not* changing with time.

A few important things to glean from Equations 21.49 and 21.50:

The power absorbed by an inductor,

$$P = \frac{dU}{dt} = IL\frac{dI}{dt}$$

depends on the rate at which the current is changing with time.

If the current is increasing with time,

$$\frac{dI}{dt} > 0\ \text{A/s}$$

the power absorbed by the inductor is positive. If the current is decreasing with time,

$$\frac{dI}{dt} < 0 \text{ A/s}$$

the power absorbed is negative, indicating that the inductor actually is providing power (actually returning energy from the field to the electrons). So the power absorbed by an inductor can be positive or negative depending on the sign of

$$\frac{dI}{dt}$$

the time rate of change of the current. If the current is not changing with time,

$$\frac{dI}{dt} = 0 \text{ A/s}$$

and the inductor absorbs zero power.

Table 21.1 illustrates the parallels between a capacitor and an inductor. Recall that a capacitor stores its energy in the electric field between its plates. For an inductor, the energy is stored in its magnetic field. We illustrate this with a long solenoid. A long solenoid has a self-inductance L given by (see Example 21.10)

$$L = \mu_0 n^2 A\ell$$

where A is the cross-sectional area, n is the number of turns per meter, and ℓ is the length of the solenoid. The magnitude of the magnetic field inside a long solenoid, far from its ends, is (from Example 20.13)

$$B = \mu_0 nI$$

where I is the current in the solenoid. The energy stored in the magnetic field of the solenoid is (Equation 21.50):

$$U = \frac{1}{2}LI^2$$

Substituting for L and for I from the two previous equations, we find

$$U = \frac{1}{2}\mu_0 n^2 A\ell\left(\frac{B}{\mu_0 n}\right)^2$$

$$= \frac{1}{2}\frac{A\ell B^2}{\mu_0}$$

But $A\ell$ is the volume of the solenoid. Thus the energy stored per unit volume (assuming the magnetic field is zero outside the solenoid, an assumption valid for an infinite solenoid) is

$$\frac{U}{A\ell} = \frac{1}{2}\frac{B^2}{\mu_0}$$

The quantity

$$\frac{1}{2}\frac{B^2}{\mu_0} \quad (21.51)$$

is known as the **magnetic energy density**.

Notice that the magnetic energy density depends on the square of the magnetic field. The result here is analogous to the result we obtained for the energy density of an electric field, Equation 18.13:

$$\frac{1}{2}\varepsilon_0 E^2$$

Although we derived the magnetic energy density using a long solenoid and the electric energy density using a parallel plate capacitor, the results for the energy densities are very general.

Whatever the electric and/or magnetic field is at a particular location, the energy densities associated with them are proportional to the square of the magnitude of the field.

Example 21.12

What magnitude magnetic field has an energy density of 1.0 J/m³?

Solution

The energy density of the magnetic field is given by Equation 21.51. Use the given energy density:

$$1.0 \text{ J/m}^3 = \frac{B^2}{2\mu_0}$$

Solve for B^2:

$$B^2 = 2\mu_0(1.0 \text{ J/m}^3)$$

Substituting for μ_0 and solving for B, you obtain

$$B = 1.6 \text{ mT}$$

21.10 A Parallel LC Circuit*

We have investigated the transient behavior of series RC and LR circuits. In those circuits we found that the potential difference across the capacitor in a series RC circuit, or the current in a series LR circuit, increased or decreased exponentially with time with a characteristic time constant ($\tau = RC$ or $\tau = L/R$ as the case may be). We now turn to another possible paired combination of these circuit elements: a **parallel LC circuit**. A few surprises await us!

We begin with the circuit shown in Figure 21.62. Initially, switch (1) is closed and switch (2) is open. This means that initially the inductor is not part of the circuit. The capacitor is in parallel

Table 21.1 Similarities Between a Capacitor and an Inductor

	Power absorbed	*Energy stored*
Capacitor	$VC\frac{dV}{dt}$	$\frac{1}{2}CV^2$
Inductor	$IL\frac{dI}{dt}$	$\frac{1}{2}LI^2$

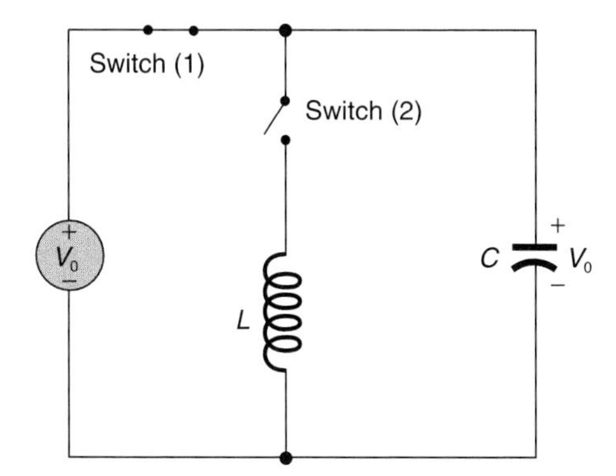

FIGURE 21.62 An initial circuit.

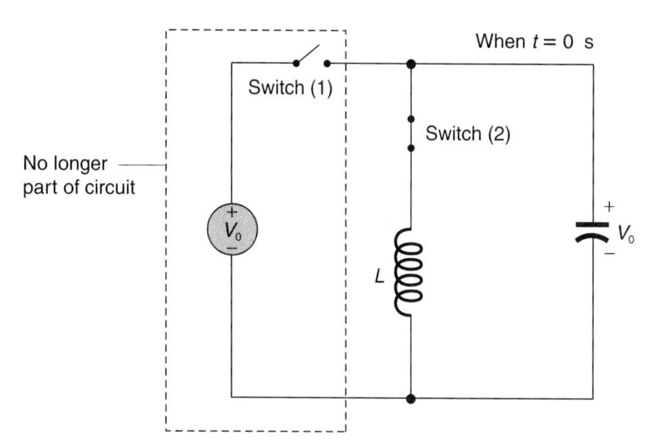

FIGURE 21.63 When $t = 0$ s, the circuit involves just the capacitor and inductor.

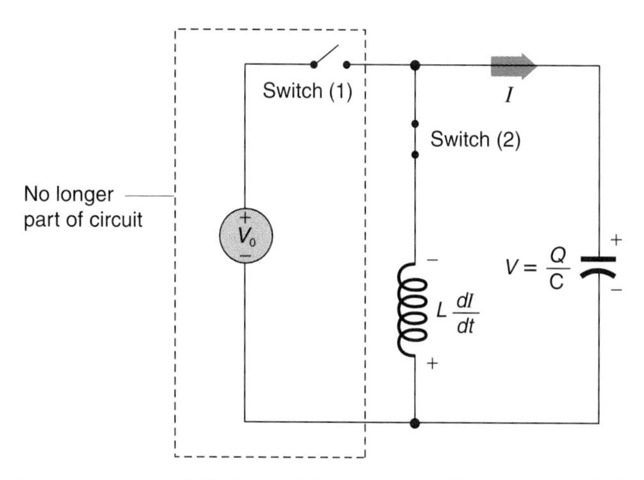

FIGURE 21.64 Polarity markings are assigned consistent with the direction chosen for I.

with the independent voltage source V_0, since the two circuit elements are connected to the same two nodes (here they also are in series!). Since the capacitor and independent voltage source are in parallel, the two circuit elements have the same potential difference across them; thus the capacitor has a potential difference V_0 between its plates with the polarity indicated in Figure 21.62.

The charge initially on the capacitor is found from the definition of the capacitance:

$$C = \frac{Q_0}{V_0}$$

so that

$$Q_0 = CV_0$$

The initial energy stored in the capacitor is found from Equation 18.12:

$$\frac{1}{2}\frac{Q_0^2}{C}$$

When $t = 0$ s, we close switch (2) and simultaneously open switch (1), securing the circuit shown in Figure 21.63. This takes the voltage source V_0 out of the circuit and places the inductor and capacitor in parallel (also in series, since again we have a one-loop, two-element circuit). We want to discover how the charge Q on the capacitor varies with time as it discharges through the inductor.

Since the capacitor initially has the polarity indicated in Figure 21.64, we choose the current I to be in the direction indicated, consistent with the polarity markings on the capacitor. With this choice for the current direction, we are forced to mark the polarity of the inductor as indicated.*

A differential equation for Q can be found by applying the KVL to the LC loop in Figure 21.64. We go clockwise around the loop, beginning in the lower left corner of the loop:

$$L\frac{dI}{dt} + \frac{Q}{C} = 0 \text{ V}$$

*The actual potential difference across the inductor has the reverse polarity since the actual current is opposite to the direction chosen; we choose this polarity to be consistent with our convention for assigning polarity based on the direction of the current.

Then we use

$$I = \frac{dQ}{dt}$$

in the first term to obtain

$$L\frac{d^2Q}{dt^2} + \frac{1}{C}Q = 0 \text{ V}$$

or

$$\frac{d^2Q}{dt^2} + \frac{1}{LC}Q = 0 \text{ A/s} \qquad (21.52)$$

The differential equation for Q (Equation 21.52) is an old friend; it is the differential equation for *simple harmonic oscillation*!

The equation for Q has the same form as the differential equation that describes the oscillation of a mass on a Hooke's law spring (Equation 7.11):

$$\frac{d^2x}{dt^2} + \omega^2 x = 0 \text{ m/s}^2$$

The only difference here is that the differential equation is for the charge Q, not for the position x of a mass on the end of a spring. For the harmonic oscillator in Chapter 7, we found the solution to Equation 7.11 to be

$$x(t) = A\cos(\omega t + \phi) \tag{21.53}$$

By analogy, we write the solution to Equation 21.52 as

$$Q(t) = Q_0\cos(\omega t + \phi) \tag{21.54}$$

The coefficient of Q in the differential equation (Equation 21.52) once again is ω^2.

Thus the angular frequency of the charge oscillations is

$$\omega = \frac{1}{(LC)^{1/2}} \tag{21.55}$$

We know that when $t = 0$ s the charge on the capacitor was Q_0. Setting $t = 0$ s in Equation 21.54 means that the left-hand side must be Q_0, so that

$$Q_0 = Q_0\cos(0 \text{ rad} + \phi)$$

This implies that $\cos\phi = 1$, and $\phi = 0$ rad. Thus our solution for Q (Equation 21.54) reduces to

$$Q(t) = Q_0\cos(\omega t)$$

Since

$$I = \frac{dQ}{dt}$$

a differentiation shows that the current in the circuit also varies sinusoidally:

$$I(t) = -Q_0\omega\sin(\omega t) \tag{21.56}$$

The initial value of the current, when $t = 0$ s, thus is 0 A.

Graphs of the charge Q and the current I as functions of time are shown in Figure 21.65. These graphs correspond to the position and velocity component of a mass undergoing one-dimensional simple harmonic oscillation, released from position x_0 when $t = 0$ s.

The energy in the circuit oscillates between the capacitor and the inductor. The energy in the capacitor at any time is

$$\frac{1}{2}\frac{Q^2}{C} = \frac{1}{2}\frac{Q_0^2}{C}\cos^2\omega t$$

while that in the inductor is

$$\frac{1}{2}LI^2 = \frac{1}{2}LQ_0^2\omega^2\sin^2\omega t$$

At any instant t the sum of these energies is

$$\frac{1}{2}\frac{Q_0^2}{C}\cos^2\omega t + \frac{1}{2}LQ_0^2\omega^2\sin^2\omega t$$

But since

$$\omega^2 = \frac{1}{LC}$$

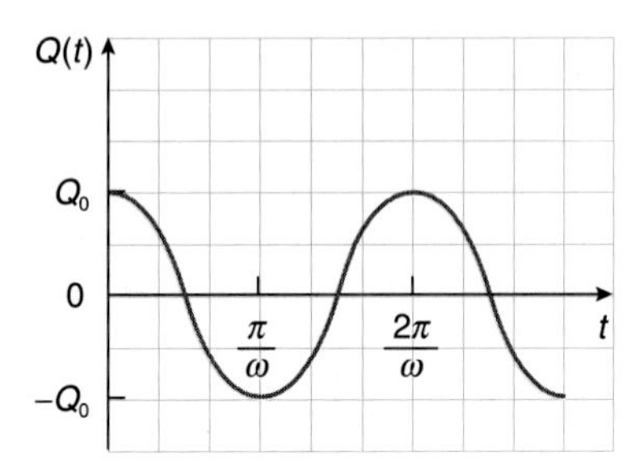

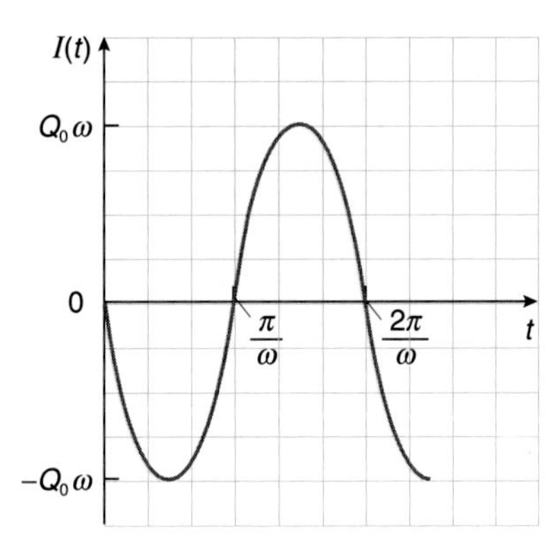

FIGURE 21.65 Graphs of the charge and current oscillations as functions of time.

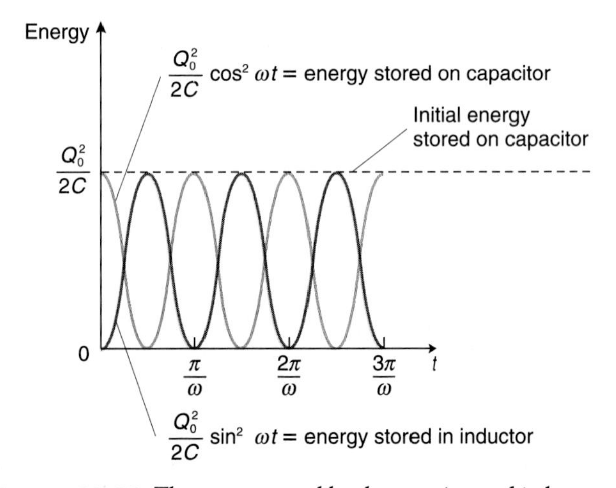

FIGURE 21.66 The energy stored by the capacitor and inductor as functions of time.

the energy sum becomes

$$\frac{1}{2}\frac{Q_0^2}{C}(\cos^2\omega t + \sin^2\omega t) = \frac{1}{2}\frac{Q_0^2}{C}$$

which is just the initial energy initially stored on the capacitor. Energy is conserved at every instant, as indicated in Figure 21.66.

In terms of our mechanical example, the energy stored in the capacitor is analogous to the potential energy stored in the spring, and the energy present in the inductor when charge is moving through it is analogous to the kinetic energy of the moving mass.

An *ideal* simple harmonic oscillator oscillates indefinitely. In a *real*, macroscopic harmonic oscillator, frictional effects eventually damp out the motion with the initial mechanical energy eventually transferred to the internal energy of its environment. Similar effects occur in an LC circuit. For an *ideal* LC circuit, the charge oscillations continue indefinitely. However, a *real* LC circuit inevitably has some resistance in the circuit, if only from the wire that forms the inductor. The resistor in an LC circuit plays the role of the friction in a damped harmonic oscillator. The charge oscillations eventually die out. The resistor absorbs positive power no matter which way the current is directed through it. Thus the initial electrical energy in the circuit eventually appears as an energy transfer to the resistor, increasing its internal energy and passing the energy to the internal energy of the environment via heat transfer.

EXAMPLE 21.13

You are asked to design a parallel LC circuit that can oscillate with a frequency of 540 kHz, which is in the AM radio band of frequencies. A 15 mH inductor is available. Find the capacitance needed.

Solution

The angular frequency of oscillation ω is given by Equation 21.55, which after squaring is

$$\omega^2 = \frac{1}{LC}$$

The angular frequency ω and the frequency ν are related by

$$\omega = 2\pi\nu$$

Hence you have

$$4\pi^2\nu^2 = \frac{1}{LC}$$

Solving for C, you obtain

$$\begin{aligned} C &= \frac{1}{4\pi^2\nu^2 L} \\ &= \frac{1}{4\pi^2(540 \times 10^3 \text{ Hz})^2 (15 \times 10^{-3} \text{ H})} \\ &= 5.8 \times 10^{-12} \text{ F} \\ &= 5.8 \text{ pF} \end{aligned}$$

21.11 MUTUAL INDUCTANCE*

Here we see how remote circuits can magnetically induce emfs in another circuit via Faraday's law through **mutual inductance**. Since every circuit can be imagined geometrically as at least one complete conducting path or coil, the effects of mutual inductance, however small or large, are present in every circuit unless it is a truly isolated, one-loop circuit. The effects of mutual inductance are hazardous to some sensitive electronic devices (such as heart pacemakers) but provide the *raison d'être* for others such as the transformers we consider in the next section.

In Section 21.6 we investigated what happens when the magnetic flux through a coil changes with time: an emf is induced in the coil. In particular, in Section 21.6 the flux through the coil was due to the magnetic field caused by the current in the same coil. We described the effect by means of an essentially geometric factor known as the self-inductance L of the coil.

Here we want to extend these ideas to two coils in close proximity. We find, perhaps remarkably as Faraday discovered, that changes in one coil affect the other, even though they have no common conducting link. We consider a number of situations.

Case 1: A Closed Circuit near an Open Circuit

Imagine coil (1) with current I_1 with a nearby second coil that is an open circuit (see Figure 21.67); the second coil is a network but not a circuit.

The magnetic field of current I_1 in the first coil produces some magnetic flux Φ_{21} through the second coil. We define the **mutual inductance** M_{21} to be the ratio of the magnetic flux through the second coil (caused by the magnetic field of the current in the first coil) to the current in the first coil:

$$M_{21} \equiv \frac{\Phi_{21}}{I_1} \qquad (21.57)$$

This definition somewhat resembles the definition of the self-inductance (Equation 21.29). Hence, in SI units, the mutual inductance is expressed in henries (like the self-inductance) and also turns out to be a function of the geometric size and arrangement of the coils and the materials (i.e., the medium) between them.

If current I_1 changes with time, the magnetic flux Φ_{21} also changes with time. Faraday's law (Equation 21.4) then implies that an induced emf is produced in the second coil:

$$\text{induced emf}_2 = -\frac{d\Phi_{21}}{dt}$$

We use Equation 21.57 for the magnetic flux:

$$\text{induced emf}_2 = -M_{21}\frac{dI_1}{dt}$$

Just as we did for the emfs induced by self-inductance, we define V_2 to be the *negative* of this induced emf, so that

$$V_2 \equiv M_{21}\frac{dI_1}{dt} \qquad (21.58)$$

The mutual inductance M_{21} relates the changes in the current in the first coil to the resulting effects in the second coil.

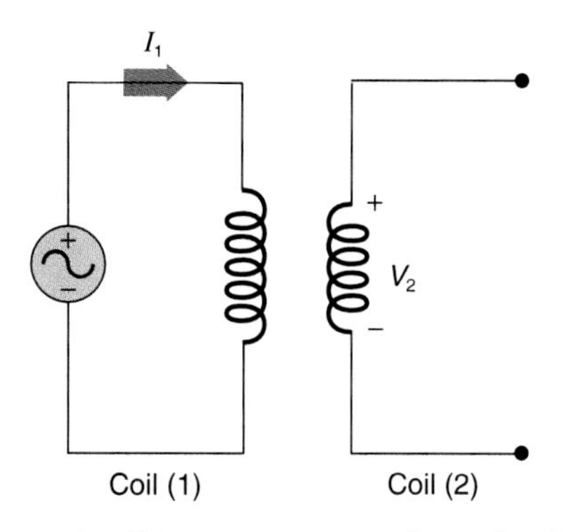

FIGURE 21.67 A coil in proximity to another coil with a current.

Case 2: A Reciprocal Arrangement

Now let the first coil be an open circuit and let a current I_2 be in the second coil, as shown in Figure 21.68.

In this case, current I_2 creates a magnetic field and this magnetic field produces a magnetic flux Φ_{12} through the first coil. Now define the mutual inductance M_{12} to be the ratio of the magnetic flux Φ_{12} to the current I_2:

$$M_{12} \equiv \frac{\Phi_{12}}{I_2} \tag{21.59}$$

Once again, the mutual inductance M_{12} depends only on geometric factors and the material medium between the coils.

If the current I_2 changes with time, the magnetic flux Φ_{12} changes with time, and an induced emf is produced in the first coil according to Faraday's law:

$$\text{induced emf}_1 = -\frac{d\Phi_{12}}{dt}$$

Using Equation 21.59, we have

$$\text{induced emf}_1 = -M_{12}\frac{dI_2}{dt}$$

Call V_1 the negative of this induced emf, so that

$$V_1 \equiv M_{12}\frac{dI_2}{dt} \tag{21.60}$$

The mutual inductance M_{12} relates the changes in the second coil to the resulting effects in the first coil.

How are the two coefficients of mutual inductance, M_{21} and M_{12}, related to each other? They both depend on geometric factors such as the number of turns and the area of each coil, and on the material medium between the coils. Since the geometry and the medium between the coils are the same in both situations, it is certainly plausible that

$$M_{21} = M_{12}$$

In fact, this is exactly the case, although our argument here is not a rigorous proof. In view of this equality, we write

$$M_{21} = M_{12} \equiv M \tag{21.61}$$

and call M mutual inductance between the coils.

Case 3: Both Circuits Closed

The situation is more complicated when currents exist in both coils, as shown in Figure 21.69.

Now we have to consider not only the mutual effects, via the mutual inductance, but also the self effects, via the self-inductance. The total induced emf in each coil therefore has two contributions: one from the self-inductance (of the particular coil) and one from the mutual inductance (from the other coil). The total potential difference therefore must also have two contributions. That is,

$$V_1 = L_1\frac{dI_1}{dt} + M\frac{dI_2}{dt} \tag{21.62}$$

where L_1 is the self-inductance of the first coil. Likewise,

$$V_2 = L_2\frac{dI_2}{dt} + M\frac{dI_1}{dt} \tag{21.63}$$

where L_2 is the self-inductance of the second coil. Things get a bit involved! The detailed implications of these effects between the coils we leave for a more advanced course in electromagnetic theory and electrical circuits.

The implication of Equations 21.58 and 21.60 is that it is possible to remotely induce an emf in a circuit by means of a changing magnetic flux through a coil instigated by changes in another circuit.

Unfortunately, mutual inductance is the cause of electronic interference in circuits by time-varying magnetic fields such as those caused by motion of a circuit through a nonuniform magnetic field, or the time-varying magnetic fields of distant lightning, auroras, and even electric machinery.

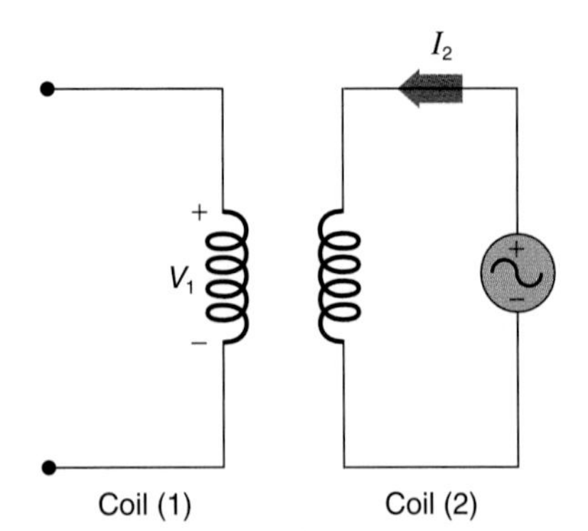

FIGURE 21.68 A reciprocal arrangement of the coils.

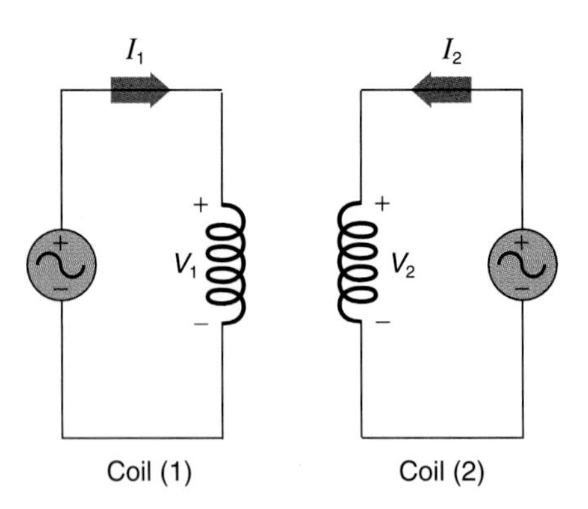

FIGURE 21.69 Currents in both coils.

21.12 An Ideal Transformer*

Transformers are used for many practical purposes; you likely had a small one for your electric train as a child. If you have a low-voltage lighting system in your home, you also will find a transformer in each fixture or a single transformer for the whole collection of lighting fixtures. Open up your stereo amplifier or PC and you also will find transformers. The large, typically cylindrical devices that you see hanging from utility poles all around town are transformers, as are many of the large devices in utility substations. What is the function of a transformer and on what principle of physics are they based?

An ideal transformer has two coils of wire, electrically isolated from each other, and arranged so that all the magnetic field lines produced by the currents in each coil completely thread the other coil, with no leakage of magnetic field lines. In practice, this idealization can be approximated by either (1) tightly wrapping the coils around each other, as in Figure 21.70, or (2) linking the coils with a highly ferromagnetic material such as iron, as in Figure 21.71.

An ideal transformer thus has the following characteristic: the magnetic flux through each *individual* turn (or loop) of wire in both coils is the same. Let Φ be the magnetic flux through a *single* turn of wire; Φ is the same through *every* loop of wire in the entire arrangement, regardless of which coil contains the individual loop. If the two coils are of different cross-sectional area, no matter; then the magnetic field compensates accordingly, so that the flux through the individual loop is the same as that through every other loop. For example, if the second coil has a larger cross-sectional area, then the number of lines of magnetic field per square meter is fewer, and so the total flux through the loop is the same as that through any other loop. The important thing to realize is that Φ is the magnetic flux caused by the total magnetic field arising from whatever currents are in both coils.

Several transformers on a utility pole. The devices on the lower wooden crossbar are fuses.

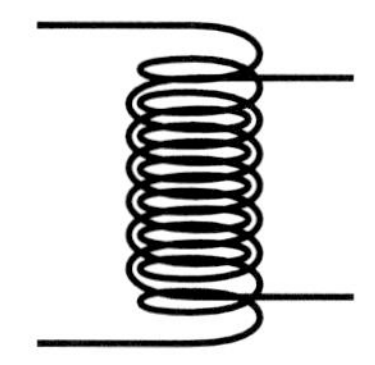

FIGURE 21.70 One coil wrapped around another coil.

The total magnetic flux through the first coil is the number of turns of wire (or loops) N_1 in this coil times the magnetic flux Φ through a single coil:

$$\text{total magnetic flux through coil (1)} = N_1\Phi$$

Likewise the total magnetic flux through the second coil, with N_2 turns, is

$$\text{total magnetic flux through coil (2)} = N_2\Phi$$

If the currents are changing with time, the magnetic flux through the coils changes with time. Faraday's law implies there then is an emf induced in each coil equal to the negative time derivative of the total magnetic flux through the coil. Since the number of turns of wire in each coil is fixed, we have

$$\begin{aligned}\text{induced emf in coil (1)} &= -\frac{d}{dt}\left(N_1\Phi\right)\\ &= -N_1\frac{d\Phi}{dt}\\ \text{induced emf in coil (2)} &= -\frac{d}{dt}\left(N_2\Phi\right)\\ &= -N_2\frac{d\Phi}{dt}\end{aligned}$$

Once again, we introduce potential differences V_1 and V_2 equal to the negative of the respective emfs:

$$V_1 = N_1\frac{d\Phi}{dt} \tag{21.64}$$

$$V_2 = N_2\frac{d\Phi}{dt} \tag{21.65}$$

However, since

$$\frac{d\Phi}{dt}$$

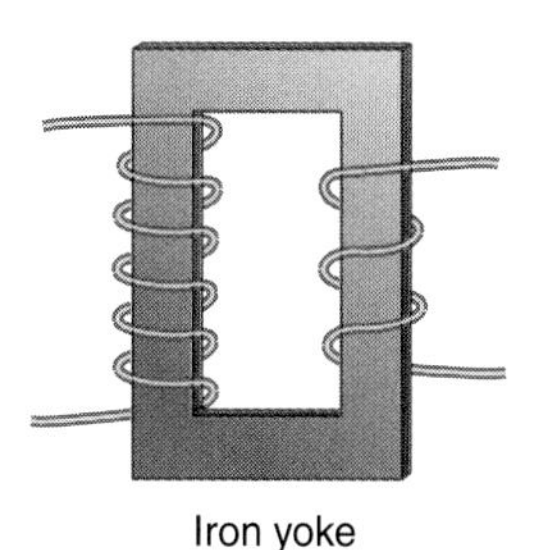

FIGURE 21.71 Two coils linked with an iron yoke.

is the *same* for every loop of wire in the entire system, we can divide Equations 21.64 and 21.65 to obtain

$$\frac{V_1}{V_2} = \frac{N_1}{N_2} \tag{21.66}$$

The ratio of the potential differences has the same value as the ratio of the number of coils.

The beauty of the ideal transformer is that we can avoid the more complicated Equations 21.62 and 21.63 we investigated in Section 21.9 involving the self and mutual inductances of the coils themselves (Case 3). The ratio of the potential differences is simply set by the number of turns (coils). The treatment of *real* transformers, however, is more complex than this ideal scenario; we defer their consideration to engineering courses in electrical circuits. Fortunately, with almost all real transformers, the ideal transformer is a reasonable first approximation.

The input coil of a transformer customarily is called the **primary** coil; the output coil of the arrangement is known as the **secondary** coil.

If the output potential difference V_2 is greater than the input V_1, the transformer is called a **step-up transformer.** From Equation 21.66, we can deduce that $N_2 > N_1$ for a step-up transformer.

If the output potential difference V_2 is less than the input V_1, then the transformer is called a **step-down transformer.** Equation 21.66 implies $N_2 < N_1$ for a step-down transformer.

By convention, the polarity of V_1 and V_2 is indicated by means of a *dot convention* next to a symbolic representation of the transformer (see Figure 21.72). Dots indicate the terminals with corresponding polarity so the details of the winding need not be examined.

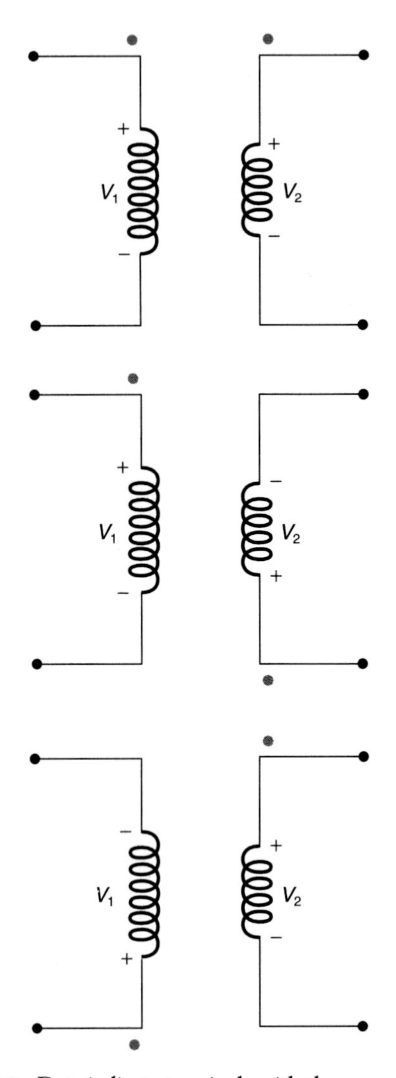

FIGURE 21.72 **Dots indicate terminals with the same polarity.**

It might seem as if we get something for nothing in a transformer. We can change ac potential differences up or down, but energy conservation prevails, as it must. In an ideal transformer, no electrical energy is gained or lost to the system. The sum of the electrical power absorbed by the primary and secondary coils in an ideal transformer must be zero. Note in Figure 21.73 that the directions chosen for the currents I_1 and I_2 are *into* the terminals marked (+) for both V_1 and V_2 respectively. Thus the power absorbed by the primary is

$$I_1V_1$$

with our usual sign convention for the power: the current must be into the (+) polarity terminal. The power absorbed by the secondary is

$$I_2V_2$$

with the same sign convention. The sum of the power absorbed is zero by energy conservation:

$$I_1V_1 + I_2V_2 = 0 \text{ W}$$

We solve for I_2:

$$I_2 = -\frac{V_1}{V_2} I_1$$

Using Equation 21.66 for the ratio V_1/V_2, we have

$$I_2 = -\frac{N_1}{N_2} I_1$$

or equivalently,

$$N_2I_2 = -N_1I_1 \tag{21.67}$$

The minus sign indicates that if I_1 has the direction shown in Figure 21.73, the current I_2 actually is in the direction opposite to that shown in Figure 21.73. We use these directional and polarity conventions for the currents and potentials indicated in Figure 21.73 because they are the ones used throughout electrical engineering.

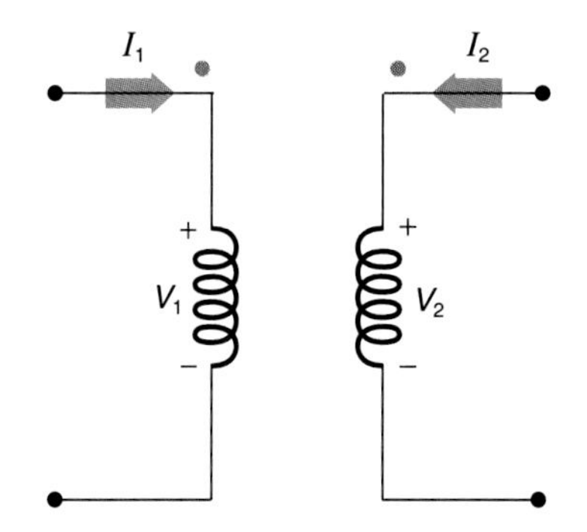

FIGURE 21.73 Conventional directions for the currents.

In a *step-up* transformer, the output potential difference V_2 is greater than the input V_1, but the output current I_2 is less than the input I_1 (in absolute value) in accordance with energy conservation.

Conversely, in a *step-down* transformer, the output potential difference V_2 is less than the input V_1, and the output current I_2 is greater (in absolute value) than the input I_1.

Transformers are used ubiquitously in electronics and electrical power transmission for changing (i.e., transforming) ac potential differences from one value to another.

A transformer will not transform dc (steady) potential differences.

You should be able to explain why!

Transformers are used in the electrical utility industry to facilitate transmission of electrical power over long distances with small resistive power losses. Such transmission involves lengths of wire many kilometers long. Such lengths of wire, of course, have a significant electrical resistance, say R. The power absorbed by a resistance R is proportional to the square of the current: I^2R (Equation 19.28). So, to minimize the power loss caused by the resistance of the transmission wires, the current should be as small as possible.

With a step-up transformer at the generating plant, the output potential difference is large but the corresponding current is small. Thus power transmission over long distances is done at a high potential difference (upward of a megavolt) and small current. Once the wires reach your town, the potential difference is lowered (typically to about 15 kV) with a large step-down transformer, and the current correspondingly increases. The final stage before entering your home is another step-down transformer to secure common household potential differences of 220 V and 120 V.* Transformers have long service lifetimes since they are completely passive devices; there are no moving parts.

Electric arc welding machines use step-down transformers to decrease the potential difference while increasing the current to maximize resistive power absorption (and heating) of steel.

EXAMPLE 21.14

A common step-down transformer has an input potential difference of 14.5 kV and an output of 120 V.

a. Determine the ratio of number of turns of wire in the primary coil to that in the secondary.

b. If the maximum current in the secondary is 300 A, what is the maximum current in the primary?

Solution

a. The input and output potential differences are related to the number of turns in the coils by Equation 21.66:

$$\frac{V_1}{V_2} = \frac{N_1}{N_2}$$

$$\frac{14.5 \times 10^3 \text{ V}}{120 \text{ V}} = \frac{N_1}{N_2}$$

$$\frac{N_1}{N_2} = 121$$

Notice that the secondary coil has many fewer coils since this is a step-down transformer.

b. The currents in the coils are related to the number of turns by Equation 21.67:

$$N_2 I_2 = -N_1 I_1$$

Solving for I_1, you find

$$\begin{aligned} I_1 &= -\frac{N_2}{N_1} I_2 \\ &= -\frac{1}{121}(300 \text{ A}) \\ &= -2.48 \text{ A} \end{aligned}$$

The maximum current in the primary is 2.48 A. What is the significance of the minus sign in view of Figure 21.73?

*A neat trick is used to provide 220 V. You might ask your professor about it.

Chapter Summary

Faraday's law of electromagnetic induction states that an *induced emf* is produced by and is directly proportional to the time rate of change of magnetic flux:

$$\text{induced emf} = -\frac{d\Phi_{\text{magnet}}}{dt} \qquad (21.4)$$

The induced emf itself is the path integral of a (nonconservative) electric field $\vec{E}$ around the closed contour through which the magnetic flux is taken:

$$\text{induced emf} \equiv \int_{\text{clsd path}} \vec{E} \cdot d\vec{\ell} \qquad (21.3)$$

Lenz's law is a rule for determining the direction of the induced electric field and any induced current caused by it. The induced field and current always are directed so as to oppose the change in the magnetic flux that is taking place.

A coil with N turns of wire, each of area A, rotating at angular velocity $\vec{\omega}$ perpendicular to the area vector $\vec{A}$ in a uniform magnetic field $\vec{B}$ (also perpendicular to $\vec{\omega}$)produces an emf given by

$$V_{\text{induced}}(t) = NBA\omega \sin(\omega t) \qquad (21.8)$$

The four Maxwell equations of electromagnetism, together with the equations for the magnetic and electrical forces, describe all electromagnetic phenomena. The force equations (from Chapters 16 and 20) are

$$\vec{F}_{\text{elec}} = q\vec{E} \qquad (16.8)$$

$$\vec{F}_{\text{magnet}} = q\vec{v} \times \vec{B} \qquad (20.1)$$

The Maxwell equations are

1. Gauss's law for the electric field:

$$\int_{\text{clsd surface S}} \vec{E} \cdot d\vec{S} = \frac{Q_{\text{net enclosed by S}}}{\varepsilon_0} \qquad (21.9)$$

2. Gauss's law for the magnetic field:

$$\int_{\text{clsd surface S}} \vec{B} \cdot d\vec{S} = 0\ \text{T}\cdot\text{m}^2 \qquad (21.10)$$

3. The Ampere–Maxwell law:

$$\int_{\text{clsd path}} \vec{B} \cdot d\vec{\ell} = \mu_0(I + I_D) \qquad (21.11)$$

where I is the conduction current threading the closed path and I_D is the displacement current. The displacement current is defined as

$$I_D \equiv \varepsilon_0 \frac{d\Phi_{\text{elec}}}{dt}$$

where Φ_{elec} is the flux of the electric field through any hatlike surface that has the closed path as a boundary.

4. Faraday's law of electromagnetic induction:

$$\int_{\text{clsd path}} \vec{E} \cdot d\vec{\ell} = -\frac{d\Phi_{\text{magnet}}}{dt} \qquad (21.12)$$

where Φ_{magnet} is the flux of the magnetic field through any surface that has the closed path as a boundary.

Among the consequences of Maxwell's equations in a vacuum are the existence of electromagnetic waves that propagate with a speed equal to

$$\frac{1}{(\mu_0\varepsilon_0)^{1/2}} \qquad (21.26)$$

This speed is identical to the measured speed of light c. These electromagnetic waves *are* light in the broadest sense of the term, encompassing γ-rays, x-rays, ultraviolet light, visible light, infrared light, microwaves, and radio waves.

A coil of wire, otherwise called an inductor, is the magnetic analog of a capacitor. An inductor has a self-inductance L defined as the ratio

$$L \equiv \frac{\Phi_{\text{magnet}}}{I} \qquad (21.29)$$

where Φ_{magnet} is the flux of the magnetic field through the coil caused by the current I in the coil. If a time-varying current exists in the coil, there is a potential difference V across the coil given by

$$V = L\frac{dI}{dt} \qquad (21.34)$$

The potential difference V is the negative of the emf induced in the coil by the time-varying magnetic flux through the coil. The potential difference V is introduced so that an inductor can be treated as another circuit element with the same sign conventions previously introduced for resistors and capacitors. (See Figure 21.74 for the polarity convention for an inductor.)

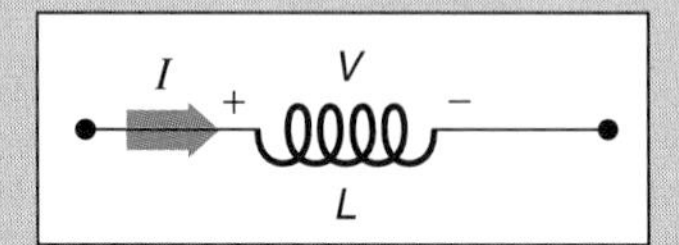

Figure 21.74 Polarity convention for an inductor.

Inductors in series can be combined into a single equivalent inductor:

$$L_{\text{eq}} = L_1 + L_2 + L_3 + \cdots \qquad \text{(series connection of inductors)} \qquad (21.37)$$

Inductors in parallel also can be combined into a single equivalent inductor:

$$\frac{1}{L_{\text{eq}}} = \frac{1}{L_1} + \frac{1}{L_2} + \frac{1}{L_3} + \cdots \qquad \text{(parallel connection of inductors)} \qquad (21.40)$$

A magnetic field stores energy. The energy density associated with a magnetic field is

$$\frac{1}{2}\frac{B^2}{\mu_0} \qquad (21.51)$$

The mutual inductance M of a pair of coils is the ratio of the flux of the magnetic field Φ_{21} through a coil (2) to the current in the *other* coil (1) which created the field:

$$M = \frac{\Phi_{21}}{I_1} \qquad (21.57)$$

If the current in the other coil (1) varies with time, the time-varying flux of its magnetic field through coil (2) induces an emf in the second coil. The potential difference V_2 (the negative of the induced emf) across the given coil (2) is

$$V_2 \equiv M \frac{dI_1}{dt} \tag{21.58}$$

A similar relationship exists for the other coil.

A transformer has a primary coil with N_1 coils and a secondary with N_2 coils. It is ideal if it does not gain or lose electrical energy per unit time. The potential differences across its primary and secondary coils are related to the numbers of turns:

$$\frac{V_1}{V_2} = \frac{N_1}{N_2} \tag{21.66}$$

(See Figure 21.75 for the polarity and current conventions for a transformer.) The respective currents in the primary and secondary also are related to their numbers of turns:

$$N_2 I_2 = -N_1 I_1 \tag{21.67}$$

An ideal transformer absorbs zero total electrical power.

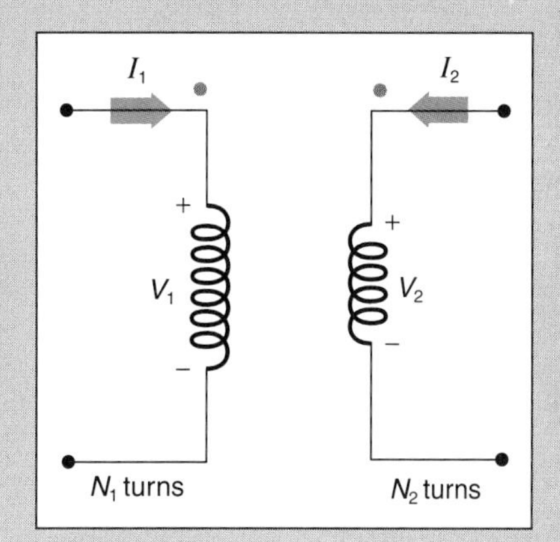

FIGURE 21.75 Polarity and current conventions for an ideal transformer.

SUMMARY OF PROBLEM-SOLVING TACTICS

21.1 (page 958) Notice that Faraday's law, as expressed by Equation 21.4, indicates that the induced emf is equal to the *negative of the slope* of a graph of Φ_{magnet} versus time.

21.2 (page 960) To see how Lenz's law actually is used is best done by carefully examining a number of specific examples.

21.3 (page 972) Be careful to keep clear the distinction between the induced emf in an inductor (equal to the *negative* time rate of change of the magnetic flux) and the potential difference

$$V = -\text{induced emf} = L\frac{dI}{dt}$$

introduced so that an inductor can be treated with the same sign conventions as resistors and capacitors . The induced emf and the potential difference V are the negatives of each other.

21.4 (page 972) For dc situations, an inductor acts as a short circuit.

QUESTIONS

1. **(page 958); 2. (page 973)**
3. The Earth has a magnetic field. What is the total flux of the magnetic field of the Earth through the surface of the Earth?
4. Your friends are confused. As their tutor, explain the distinction between the terms magnetic flux and magnetic field. Is either or both a scalar quantity? Can you say a magnetic field exists at a *point* in space? Can the same be said for the magnetic flux?
5. Two conducting loops have a magnet located along their common axis as indicated in Figure Q.5. The magnet is moved toward loop (1). Indicate the directions of the currents induced in each loop.

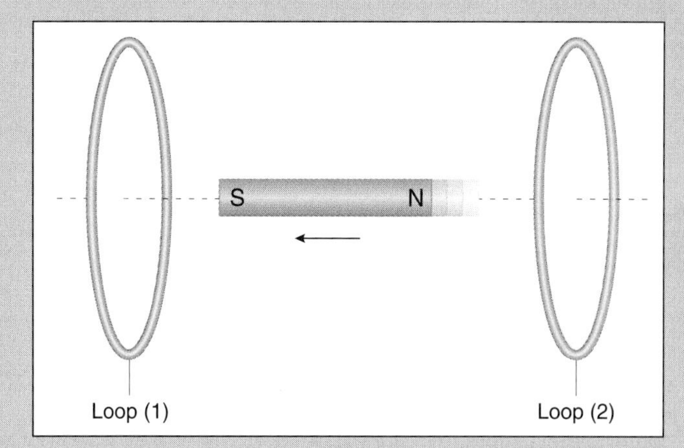

FIGURE Q.5

6. Figure Q.6 shows the magnetic field inside a very long solenoid (an end-on view). The magnetic field is decreasing with time. For each of the indicated wire loops, indicate the direction of any current induced in the loop.
7. In Figure Q.6 (previous question) the magnetic field does not vary with time. The largest wire loop is moved in such a way that every point of the loop perimeter always remains outside the region with the nonzero magnetic field. Explain why there is no current induced in the largest loop.

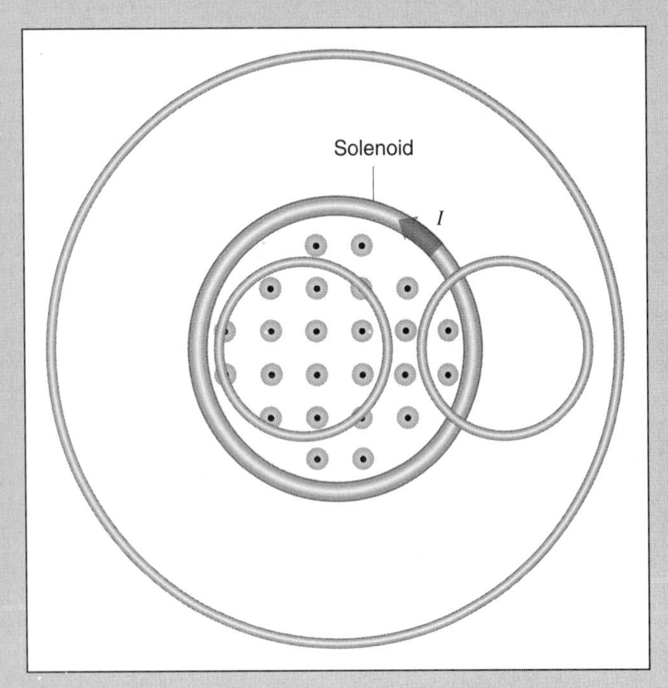

FIGURE Q.6

8. Small gasoline engines, such those used for chain saws, lawn mowers, weed whackers, and small outboard motors, generate a spark for combustion by means of a *magneto*, a permanent magnet attached to the flywheel. A stationary coil of wire is connected to the spark plug. When the magnet passes the coil, a spark is produced across the spark plug gap. Explain how this happens.

9. A bicycle wheel spins about its axis in a constant magnetic field $\vec{B}$ that is in the same direction as the angular velocity vector. Indicate the polarity of the emf induced along a spoke.

10. A single metal ring of diameter 10 cm spins steadily about a diameter with a frequency of 5.0 Hz. A uniform magnetic field of magnitude 0.100 T is perpendicular to the spin axis. What is the orientation of the ring when the induced emf is (instantaneously) zero?

11. In nuclear magnetic resonance (NMR) imaging technology, carefully controlled, oscillating, magnetic fields are used in complex ways to form visible images of soft tissues such as the brain. In the past neurosurgeons occasionally used metal clamps to block cerebral aneurysms. Subsequently, it was found that NMR images in the vicinity of the metal clamps were not as clear as in soft tissues further from the clamps. Explain how Faraday's law gives rise to currents in the clamps that produce magnetic fields that can interfere with the NMR fields, thus spoiling the image. Plastic clamps now are used.

12. AM radio waves have frequencies on the order of 1000 kHz. What is the corresponding order of magnitude of their electromagnetic wavelengths?

13. A scientific charlatan claims sound is an electromagnetic wave. What experimental evidence can you cite to prove that sound is *not* an electromagnetic wave?

14. Briefly state the four Maxwell equations in your own words.

15. Seismographs are used to detect and record vibrations from earthquakes and explosions (deliberately set to probe for underground reservoirs of oil or to analyze subsurface rock features). One way of detecting ground vibrations is to use a permanent magnet and a coil of wire. Explain how such a detector might function.

16. A planar loop of wire is in a uniform magnetic field oriented perpendicular to the plane of the loop. The loop then is pulled at constant velocity in a direction in the plane of the loop. Is an emf induced in the loop? If the magnetic field is oriented at an angle θ to the plane of the loop and the experiment is repeated, can an emf be induced in the loop? If no emf is induced under these circumstances, explain what (if anything) can be done in the uniform field so an emf *is* induced in the coil.

17. A wire segment AC is moved near the magnet shown in Figure Q.17 (a) Is an emf induced in the wire? If so, proceed with the following questions. If not, you are home free. (b) Which end of the wire segment has the positive polarity? (c) Given the same magnet, describe two things you could do to increase the magnitude of the induced emf. (d) What should you do to reverse the polarity of the induced emf?

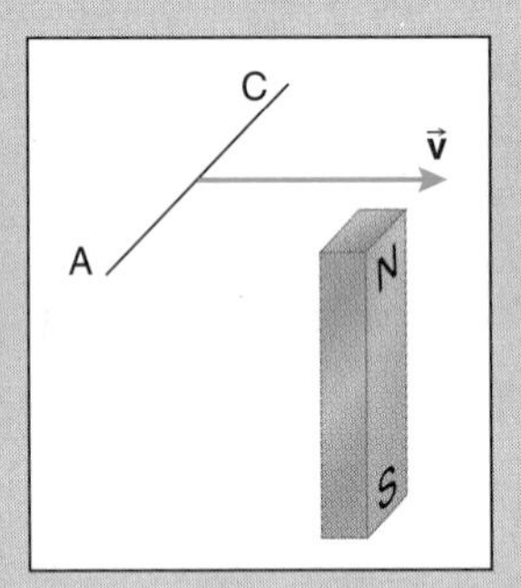

FIGURE Q.17

18. A conducting bar moves along conducting rails as shown in Figure Q.18. Is the direction of the induced current clockwise or counterclockwise? Experimentally, it is found that a constant force is necessary to move the bar at constant velocity; explain why. Does this violate Newton's second law of motion? Explain.

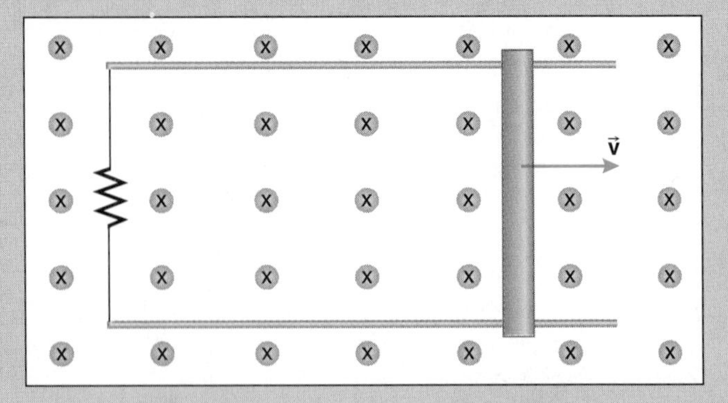

FIGURE Q.18

19. A short bar magnet is dropped through a coil of wire whose plane is horizontal as shown in Figure Q.19. Make a qualitative graph of the emf induced in the coil as a function of time.

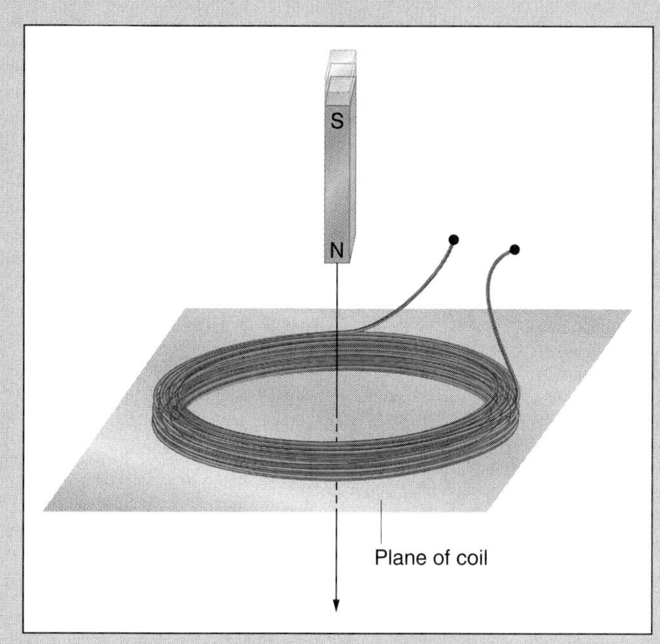

FIGURE Q.19

20. The potential difference V across an inductor is

$$V = L\frac{dI}{dt}$$

Imagine a graph of $V(t)$ versus t, beginning when $t = 0$ s. At an instant t on the graph, what is the geometric interpretation of the current $I(t)$?

21. A metal washer rests on top of a vertical coil as shown in Figure Q.21. The coil has a large self-inductance. When the switch is closed the current increases with time. If the time rate of change of the current is large enough, the metal washer flies vertically off the coil. Explain why.

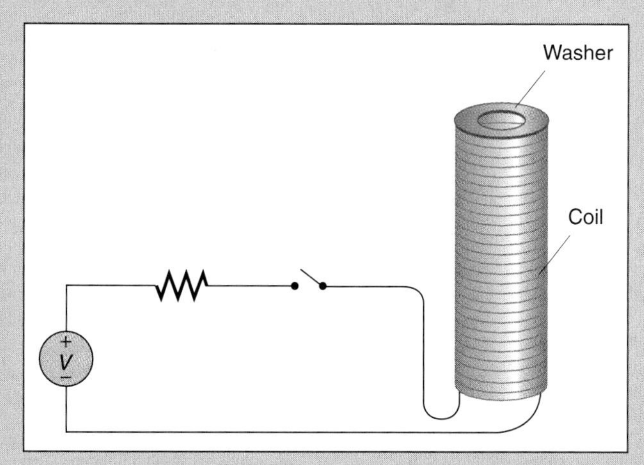

FIGURE Q.21

22. The current through the inductor indicated in Figure Q.22 is increasing with time. Indicate the polarity of the potential difference

$$V = L\frac{dI}{dt}$$

Repeat if the current is decreasing with time.

FIGURE Q.22

23. An *LR* circuit with a long time constant τ has a constant current. A switch suddenly opens the circuit. Explain why a spark or arc likely is seen as the switch opens. Is the spark likely to be seen if the time constant is very small?

24. An induction stove has a conducting coil imbedded just below a ceramic cooktop surface. An oscillating current exists in the coil. A metal pot with finite resistivity is placed on the stove. Explain how Faraday's law leads to an increase in the temperature of the pot, thus cooking the food within it. The ceramic surface itself may not get very hot in the process.

25. A newspaper account about the safety of induction stoves states that "induction elements work by creating magnetic friction (instant heat) in the pot or pan."* Critique this journalistic account.

26. An independent voltage source V_0 is in a series *LR* network. The switch is suddenly closed and the current eventually reaches a constant value. The source suddenly is doubled from V_0 to $2V_0$. What is the effect of this latter change on the value of the time constant of the circuit?

27. A coil of wire is rotated at constant angular velocity about a symmetry axis parallel to a uniform magnetic field, as indicated in Figure Q.27. Is there an emf induced in the coil?

Roanoke Times & World News (Roanoke, Virginia), 4 June 1995, page C-1.

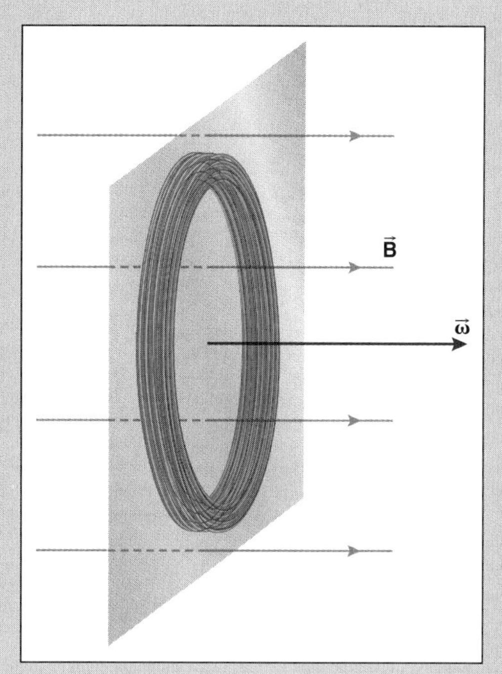

FIGURE Q.27

28. A coil of wire is rotated at constant angular velocity about a symmetry axis perpendicular to a uniform magnetic field, as indicated in Figure Q.28. Is there an emf induced in the coil?

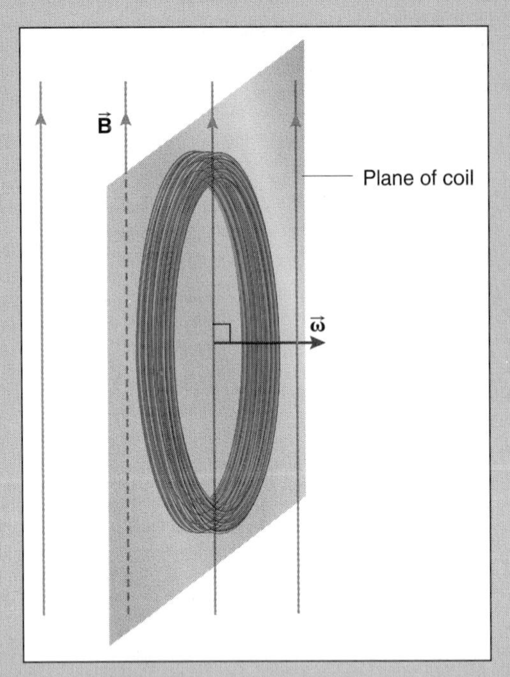

FIGURE Q.28

29. A coil of wire is rotated at constant angular velocity about a symmetry axis in the plane of the coil and perpendicular to a uniform magnetic field, as indicated in Figure Q.29. (a) Is the absolute value of the emf induced in the coil constant throughout one complete rotation of the coil? (b) If the coil forms a circuit, does the induced emf produce an induced current in the same direction in the coil throughout one complete rotation of the coil?

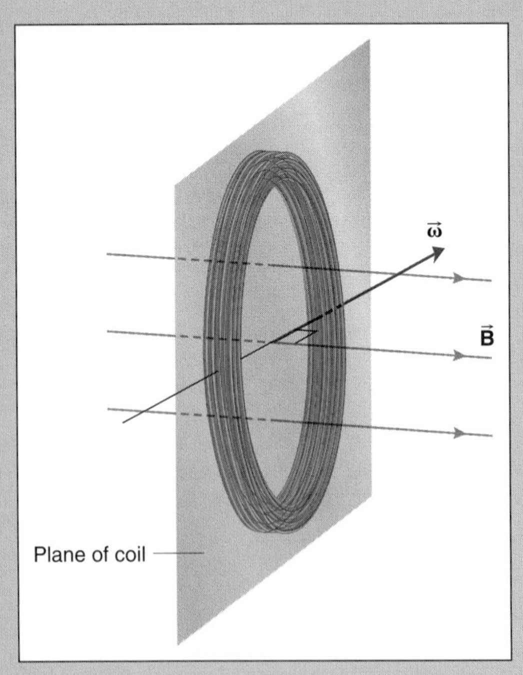

FIGURE Q.29

30. In your physics lab, you drop a magnet through a solid ring of copper, as in Figure Q.30. Neglecting air resistance, is the magnitude of the acceleration of the magnet constant? Does it make a difference whether the magnet falls north pole first or south pole first?

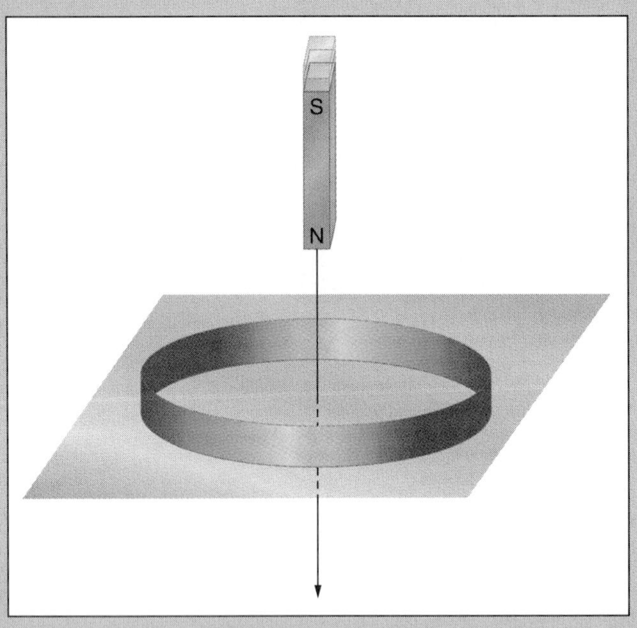

FIGURE Q.30

31. Explain why a small induced emf typically develops between the wing tips of an airplane in flight. Under what circumstances will the induced emf be zero?

32. Explain why a transformer cannot step up or step down dc independent voltage sources such as common batteries.

33. You have two planar coils of wire separated by a small, fixed distance. (a) Describe an orientation for the coils that maximizes their mutual inductance. (b) What orientation minimizes their mutual inductance?

34. Many battery-operated devices such as video cameras and portable PCs typically operate with low-voltage dc independent voltage sources (9 V or other values). An adaptor permits you to use a nearby standard household 120 V ac source, thereby enabling you to save the battery for when true portability is needed. Among other electronic components, what might be inside the adapter?

35. The utility company in your town has distribution lines at about 14.5 kV AC that are stepped down to the much smaller common household potential differences by a step-down transformer near each collection of several homes. The company has a suggestion box and makes cash rewards to employees who invent ways to save money. The suggestion box receives two suggestions for cost cutting: (a) Eliminate the step-down transformers and have each house use the 14.5 kV source directly. (b) Eliminate all the step-down transformers and have the distribution lines throughout town at the smaller value potential differences. As a member of the R&D division, diplomatically explain why each suggestion is not worthy of a monetary award.

36. A transformer is enclosed in a case with four wire leads emerging form it. (a) What can you do to discover whether

two leads represent the opposite ends of one coil or the ends of two different coils? (b) Describe an experiment you can perform to determine the ratio of the number of turns in the two coils and which coil has the larger number of turns.

37. A single coil of wire is wound with two layers of wire. When a current passes through the coil, the current in the two layers is in opposite directional senses, the inner wrapping clockwise, the outer counterclockwise (or vice versa). Explain why this arrangement minimizes the self-inductance of the coil.

38. A home is served by an ac underground electric cable, but the location of the cable under the property is not known. Suggest a device that can be used to detect the nearby presence of such a cable, thus locating the approximate path of the cable through the property.

PROBLEMS

Sections 21.1 Faraday's Law of Electromagnetic Induction
21.2 Lenz's Law
21.3 An ac Generator

1. Show that the SI units for the time rate of change of the magnetic flux, $T \cdot m^2/s$, are equivalent to volts (V).

2. When the switch in the circuit on the left in Figure P.2 is opened, indicate the direction of the momentary induced current in the resistor in the circuit on the right.

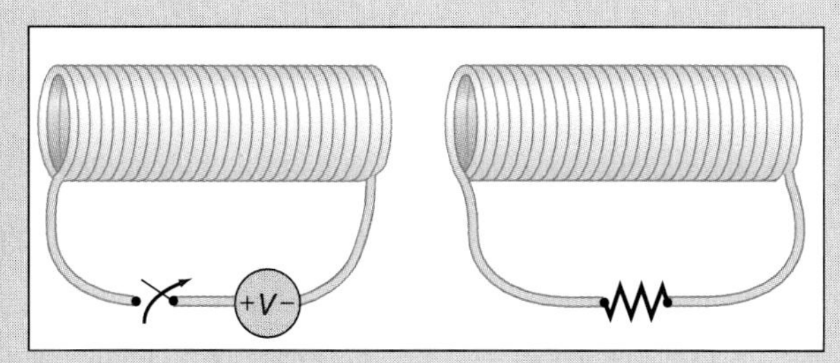

FIGURE P.2

3. While the resistance R in the circuit on the left in Figure P.3 is increased, indicate the direction of the current induced in the circuit on the right.

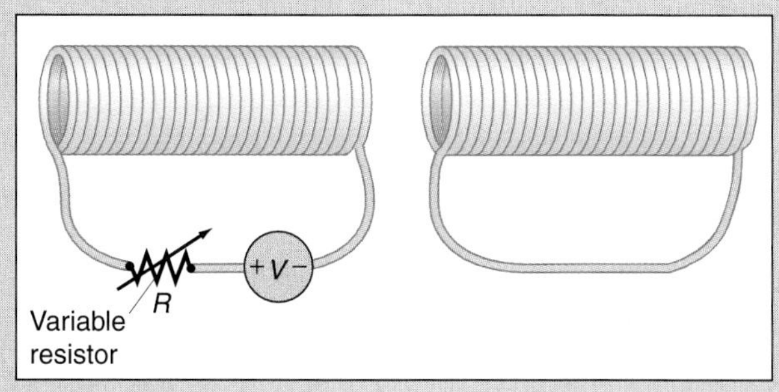

FIGURE P.3

4. If the metal rod in Example 21.2 has a length of 1.00 m and is spun in a magnetic field whose magnitude is 50 mT, what angular speed ω produces an emf of 1.00 V between the ends of the rod? Express your result in both rad/s and rev/s.

5. An ac generator has a source voltage given by

$$(311 \text{ V}) \sin[(314 \text{ rad/s})t]$$

(a) What is the maximum value of the output? (b) What is the frequency of the generator? This is the common, ac sinusoidal independent voltage source available at electrical outlets in Europe, Africa, Asia, Japan, and Australia.

6. An ac generator has a source voltage given by

$$(170 \text{ V}) \sin[(377 \text{ rad/s})t]$$

(a) What is the maximum value of the output? (b) What is the frequency? This is the common, ac sinusoidal independent voltage source available at electrical outlets in North and South America.

•7. A circular wire loop is held in the horizontal plane as indicated Figure P.7. A magnet is dropped through the loop so that the N pole of the magnet initially faces the loop. You observe the event from the vantage point of a perch above the loop. (a) As the magnet approaches the loop, is the current induced in the loop in the clockwise or counterclockwise sense? (b) As the magnet recedes from the loop (after passing through it), is the induced current in the loop in the clockwise or counterclockwise sense?

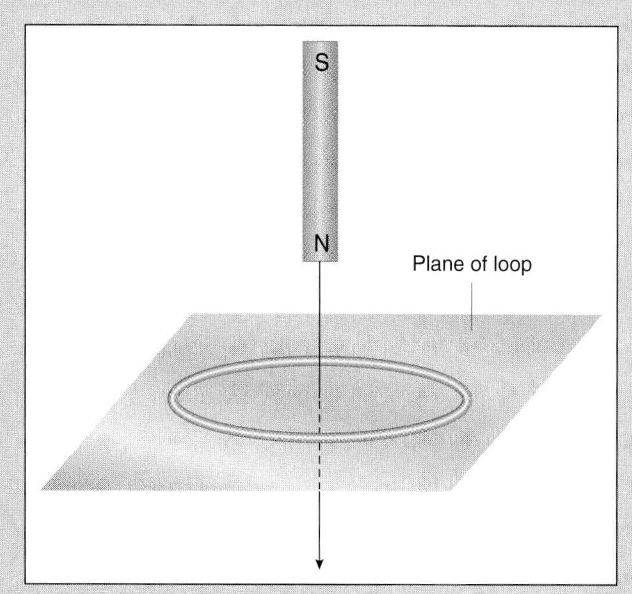

FIGURE P.7

•8. The magnetic flux through a circular current loop of radius 10 cm changes with time according to the graph in Figure P.8. To gain a few brownie points, make an accurate graph of the emf induced around the loop as a function of time.

•9. A straight conducting rod of length ℓ is moving at constant velocity $\vec{v}$ in a region where there is a uniform magnetic field $\vec{B}$ as indicated in Figure P.9. (a) Electrons in the moving rod experience a magnetic force and are free to move in response to it. In this way, electric charge accumulates on the ends of

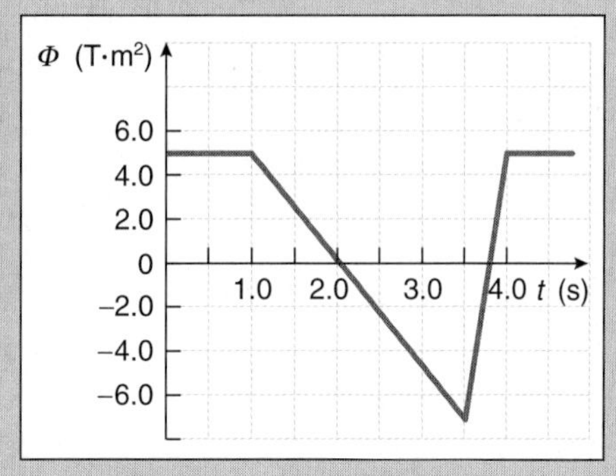

FIGURE P.8

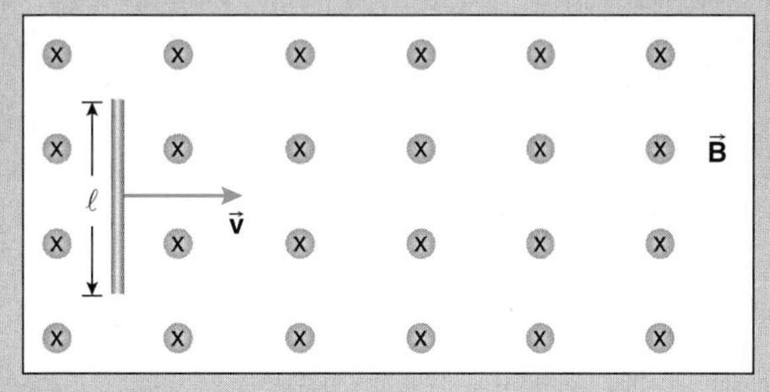

FIGURE P.9

the rod. This accumulation will not continue indefinitely, however. What limits the further accumulation of charge? (b) Which end of the rod accumulates negative charge and which end positive charge? (c) Determine the absolute value of the potential difference between the ends of the rod. (Assume a constant electric field along the length of the rod.) (d) During a time interval Δt, what area is swept out by the rod? (e) Find

$$\frac{|\Delta \Phi|}{|\Delta t|}$$

where $|\Delta \Phi|$ is the absolute value of the magnetic flux swept out by the rod during the interval Δt. Compare this result with the answer you found in part (c).

•**10.** You are not one to miss out on a good physics lecture, and so are racing to class after an early round of golf. You are streaking horizontally in your sports car at 120 km/h. The magnetic field of the Earth has a magnitude of 0.80×10^{-4} T, inclined at 60° to the surface in the vicinity. Your car is 1.5 m wide and 4.0 m long. (a) Will both the horizontal and vertical components of the magnetic field of the Earth contribute to an induced emf between the door handles of the car? Indicate your reasoning. (b) Determine the sign and magnitude of the induced emf between the opposite side door handles of your car if they are separated by 1.5 m. (c) If you connect a sensitive voltmeter between the door handles, which handle is at the higher potential?

•**11.** A U-shaped wire is bridged by a small metal rod AB of length ℓ as indicated in Figure P.11. For your enjoyment, a uniform magnetic field $\vec{\mathbf{B}}$ is directed out of the page. (a) Calculate the magnetic flux through the loop formed by the wire and rod. (b) Now move the bar at constant velocity $\vec{v}$ to the right as indicated. Any way you can, find the emf induced in the loop as the rod is moving. (c) Indicate the direction of the current induced in the loop. (d) The induced current in the movable rod is a current-carrying wire moving in a magnetic field. Calculate the direction of the magnetic force on the induced current in the movable bar. Note that this force opposes the force you need to exert on the rod to move it; this means that *you* have to do work on the rod to keep it moving, even at a constant velocity. Does this violate Newton's second law?

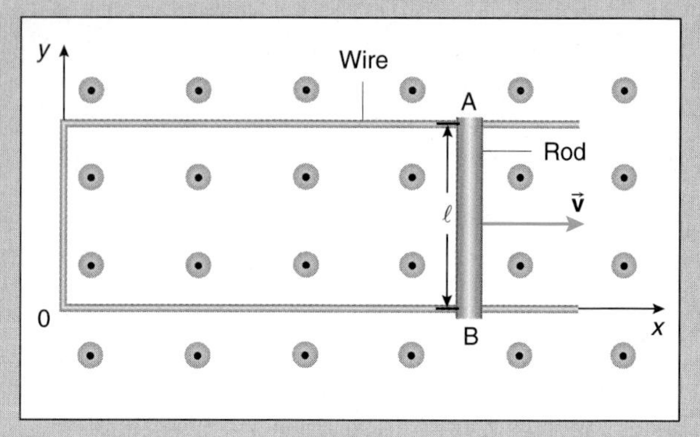

FIGURE P.11

•**12.** A movable metal rod can slide freely on two parallel frictionless conducting rails as in Figure P.12. An independent voltage source produces a current in the circuit, so that the movable rod is a current-carrying wire in a magnetic field. (a) Determine the direction of the magnetic force on the movable rod. (b) The magnetic flux through the circuit is changing with time because of the motion of the rod. Calculate the induced emf and indicate the direction of the induced current in the circuit. Note that this induced current opposes that provided by the voltage source. What are the implications of this phenomenon? Consult the following reference if needed: Mario Iona, "Why Johnny can't learn physics from textbooks I have known," *American Journal of Physics*, 55, #4, pages 299–307 (April 1987); see page 305 in particular.

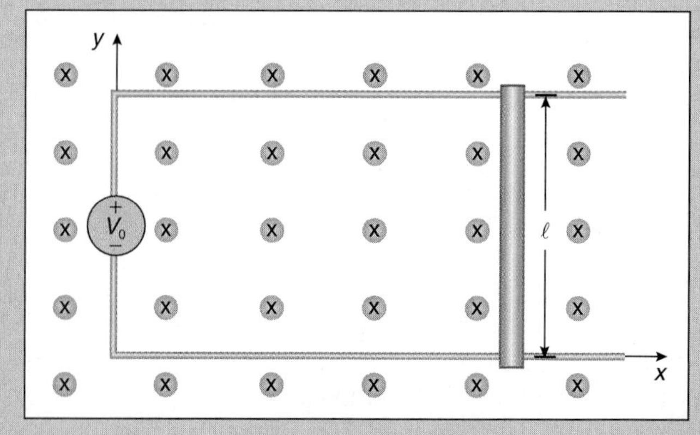

FIGURE P.12

•**13.** A magnetic field $\vec{\mathbf{B}}(t) = B_0 e^{-\alpha t}\hat{\mathbf{k}}$ is perpendicular to a circular path of radius r, which makes an angle θ with the x–y plane, as indicated in Figure P.13. (a) Calculate the flux of the magnetic field through the path as a function of time. (b) Find the emf induced around the path. If a wire is placed coincident with the path, indicate the direction of the current induced in the wire.

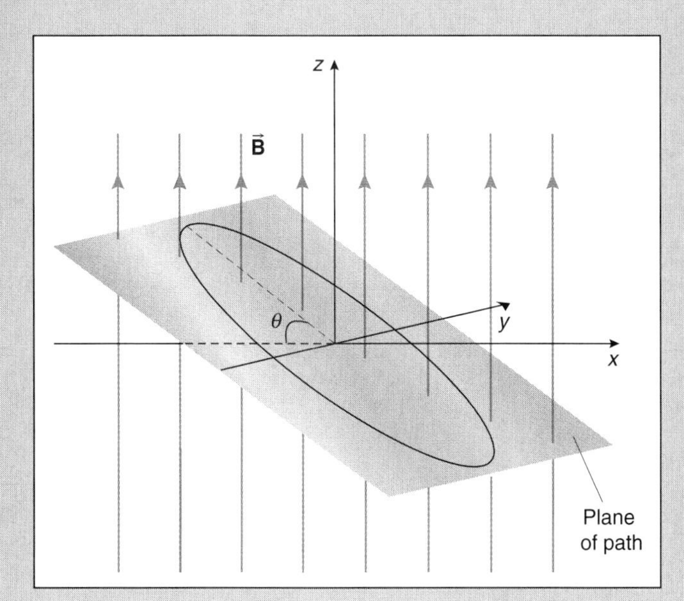

FIGURE P.13

•**14.** A solenoid has 2000 turns of wire and is 125 cm long and 10.0 cm in diameter; see Figure P.14. A circular wire loop of diameter 5.0 cm lies along the axis of the solenoid near the middle of its length as shown. (a) If the current in the solenoid initially is 4.0 A, find the magnetic flux through the smaller loop. (b) If the current in the solenoid is switched off and falls to zero in 2.0 s, calculate the average value of the emf induced in the smaller loop. (c) Indicate the direction of the current induced in the smaller loop as the magnetic field in the solenoid decreases to zero.

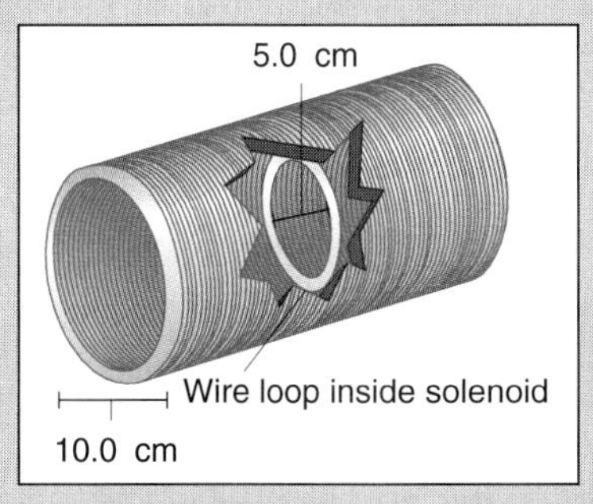

FIGURE P.14

•**15.** A jumbo jet is flying horizontally over the United States at 900 km/h. The vertical component of the magnetic field of the Earth in the vicinity is on the order of 5×10^{-5} T (its precise value depends on the geographic location; the vertical component of the field is greatest near the magnetic poles and zero at the magnetic equator). (a) The wingspan of the jet is 64 m. Estimate the magnitude of the emf induced between the wing tips of the jet as it flies. (b) Recall that the south magnetic pole of the magnetic field of the Earth is in the northern geographic hemisphere. (Why?) In the vicinity of the south magnetic pole of the Earth, which wing tip [port (left) or starboard (right)] of the jet is at the higher electric potential? Justify your answer. (c) How would your answer to part (b) be different over Antarctica? (d) The length of the jet is 71 m. Will there also be an induced emf between the nose and tail? Justify your answer.

•**16.** A helicopter has blades 4.5 m in length, rotating in the horizontal plane at 8.0 Hz. The magnitude of the vertical component of the magnetic field of the Earth in the vicinity is 0.5×10^{-4} T. What is the magnitude of the emf induced between the tip of each blade and the rotor hub?

•**17.** The high-speed TGV train travels horizontally at a speed of 180 km/h in a region where the vertical component of the magnetic field of the Earth is about 0.5×10^{-4} T. A steel axle connects the wheels, separated by the standard western rail gauge of 1.36 m. (a) What emf is induced between the wheels as a result of the motion of the train? (b) Explain why you do not need to know the value of the horizontal component of the magnetic field. (c) Explain why there can be a steady emf present along a highly conducting steel axle.

•**18.** The rectangular wire loop shown in Figure P.18 is moving at constant velocity $\vec{v}$ and enters a semi-infinite region with a uniform magnetic field of magnitude B_0 directed as indicated. The loop begins to enter the field when $t = 0$ s. (a) Make a graph of the magnetic flux through the loop as a function of time for $t \geq 0$ s. Indicate appropriate times on the graph. (b) Make another graph depicting the emf induced in the wire loop for times $t \geq 0$ s.

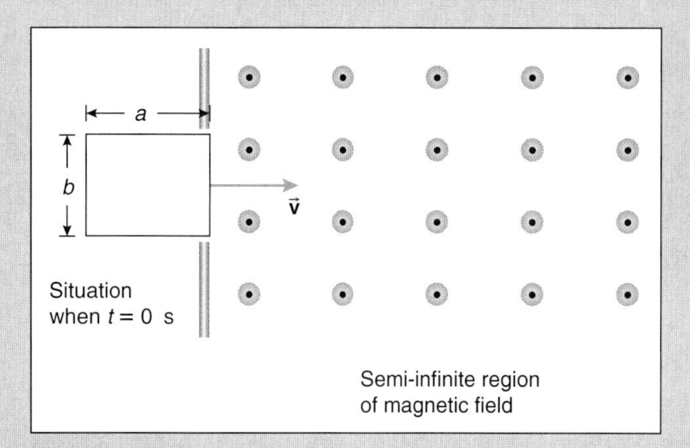

FIGURE P.18

•**19.** You are home on the range in Wyoming. A continuous barbed wire fence encloses a square 10 km on a side with one gate open on the north side of the enclosure. The downward vertical component of the magnetic field of the Earth at the location of the huge ranch is 0.50×10^{-4} T. Suppose the magnetic field of the Earth increases temporarily by 1.0% during a 20 s interval, as it might do on an unusual occasion. (a) If the gate is always open, is the east or west side of the gate at the higher electric potential during the increase? (b) Find the induced emf. (c) During the (different) interval over which the magnetic field later decreases to its original value, must the induced emf be as large as it was in part (b)? Explain.

•**20.** A circular coil of radius 5.0 cm with N turns of wire (all of equal area) is rotated about a diameter at a frequency of 60.0 Hz in a uniform magnetic field of magnitude 0.150 T. The axis of rotation is perpendicular to the direction of the magnetic field. How many turns of wire are needed so that the peak value of the induced emf in the coil is 30 V?

•21. You need to make an ac generator with a sinusoidal output of 12.0 V amplitude and frequency 50.0 Hz. A 10 cm by 20 cm rectangular coil of wire with 300 turns of wire is available. (a) At what angular frequency should the coil be rotated? (b) What magnitude magnetic field is needed?

•22. A rectangular coil 5.0 cm wide and 10.0 cm long has 150 turns of wire. The coil is turned vigorously at constant frequency ν about a symmetry axis (see Figure P.22) perpendicular to the magnetic field of the Earth (of magnitude ~0.60×10^{-4} T). In order to induce an emf of peak absolute value 1.0 V, what frequency is needed?

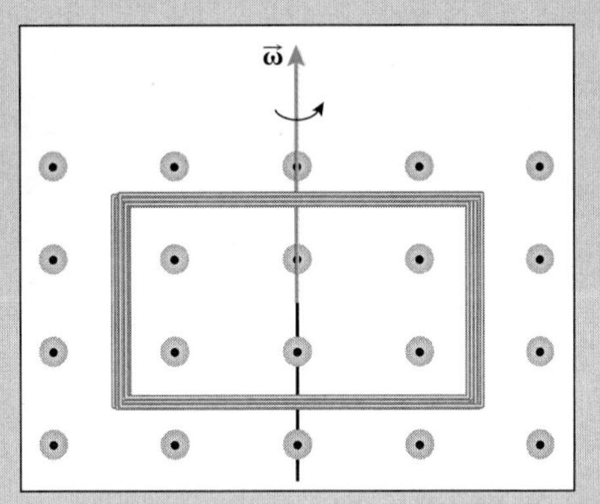

FIGURE P.22

•23. An electrical generator consists of 160 turns of wire wrapped around a frame of area 0.020 m^2, rotating at a frequency of 10.0 Hz about a symmetry axis perpendicular to a magnetic field of magnitude 0.150 T as shown in Figure P.23. The ends of the coil are connected via slip rings to an external 200 Ω resistor. (a) Calculate the maximum current through the resistor. (b) How many times each second is there zero instantaneous power absorbed by the resistor?

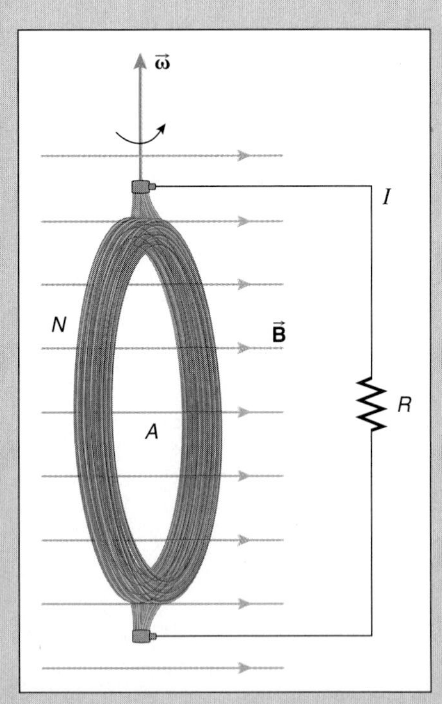

FIGURE P.23

•24. A thin wire of mass m and length ℓ has a resistance R. The wire can slide freely, with no friction, along twin vertical rails (of negligible resistance) as shown in Figure P.24. A horizontal, uniform magnetic field of magnitude B is perpendicular to the rails and wire. Neglect the magnetic field of the Earth. The wire falls under the action of the gravitational force near the surface of the Earth. Show that the falling wire attains a terminal speed equal to

$$\frac{mgR}{B^2\ell^2}$$

Note that the power and the work done on the wire by the gravitational force are positive. The wire also absorbs positive electrical power because of its finite resistance. Discuss the problem from the standpoint of energy considerations.

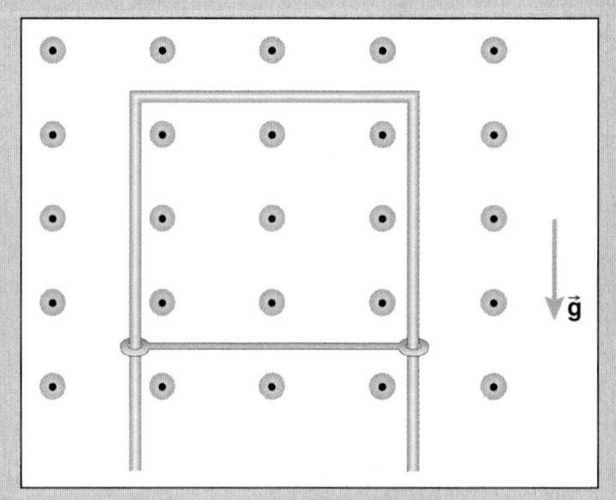

FIGURE P.24

Sections 21.4 Summary of the Maxwell Equations of Electromagnetism* 21.5 Electromagnetic Waves*

25. Calculate the wavelength of the radio waves associated with your favorite radio station.

26. Compare the wavelengths of the radio waves associated with the frequencies 890 kHz in the AM radio band and 89.0 MHz on the FM band.

27. What are the frequencies corresponding to electromagnetic waves with the following wavelengths: (a) 1.00 km; (b) 1.00 m; (c) 1.00 cm; (d) 1.00 mm; (e) 1.00 μm; (f) 1.00 nm.

Sections 21.6 Self-Inductance* 21.7 Series and Parallel Combinations of Inductors*

28. Show that the SI units for μ_0, usually written as T·m/A, also can be expressed as H/m.

29. An inductor has a self-inductance of 150 mH. Determine the potential difference V across the terminals of the inductor (as defined in Section 21.6) when the current in the inductor: (a) is 0.200 A and is increasing at a rate of 60 A/s; (b) is 0.200 A and is decreasing at a rate of 50 A/s; (c) is zero and increasing at a rate of 40 A/s.

30. You have an inductor with an inductance of 75 mH. At what rate must the current change in the inductor to secure a potential difference V across the inductor equal to 3.0 V? Should you have the current increasing or decreasing with time?

•**31.** An inductor has an inductance of 125 mH. A graph of the current in the inductor as a function of time is shown in Figure P.31. Plot the corresponding emf induced in the inductor *and* the potential difference V across the inductor (as defined in Section 21.6) as functions of time.

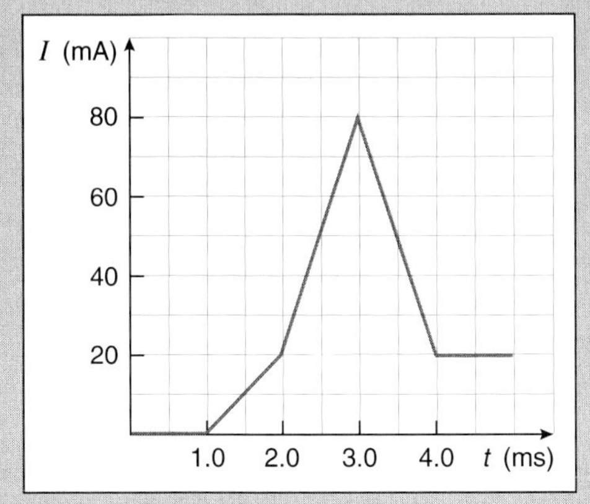

FIGURE P.31

•**32.** When $t = 3.0$ s, the current in a 60 mH inductor is 120 mA and is increasing at a rate of 25 mA/ms. (a) What is the potential difference V across the inductor at this instant? (b) What is the instantaneous power absorbed by the inductor at this time?

•**33.** A coaxial cable consists of an inner conductor of radius R_1 and an outer conductor of radius R_2, as shown in Figure P.33. The currents in the conductors are of equal magnitude but in opposite directions. (a) Calculate the magnetic flux through a rectangular section of length ℓ whose plane contains the axis of the cable (see Figure P.33). (b) Use the result of part (a) to show that the self-inductance per meter of length is

$$\frac{\mu_0}{4\pi} 2 \ln\left(\frac{R_2}{R_1}\right)$$

(c) Calculate the self-inductance per meter of RG58/U coaxial cable, which has an inner conductor of radius 0.41 mm and an outer conductor of radius 1.7 mm. Assume the material separating the conductors has the magnetic properties of a vacuum. This type of coaxial cable frequently is used (with so-called BNC connectors) in scientific laboratories.

•**34.** Two parallel wires, each of radius R, are separated by distance $d >> R$, as shown in Figure P.34. The wires carry equal currents in opposite directions. Wire of this type is used frequently to connect your TV to its aerial. Assume the material separating the conductors has the magnetic properties of a vacuum. (a) Show that the magnetic flux of the currents through a rectangular area of length ℓ, indicated in Figure P.34, is

$$\Phi = \frac{\mu_0}{4\pi} 4I\ell \ln\left(\frac{d-R}{R}\right)$$

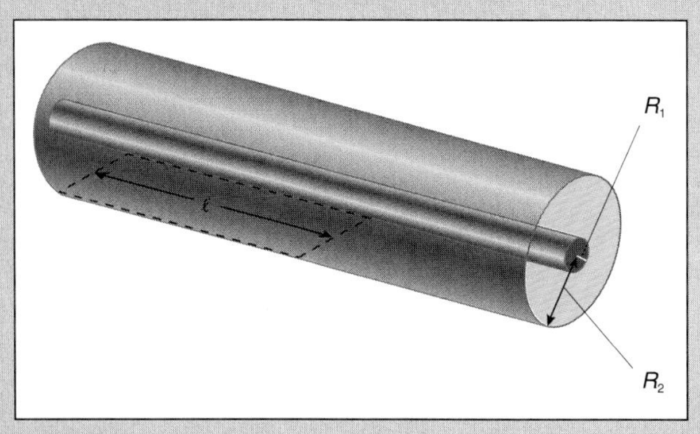

FIGURE P.33

(b) Show that the self-inductance per meter length of the wire is

$$\frac{\mu_0}{4\pi} 4 \ln\left(\frac{d-R}{R}\right)$$

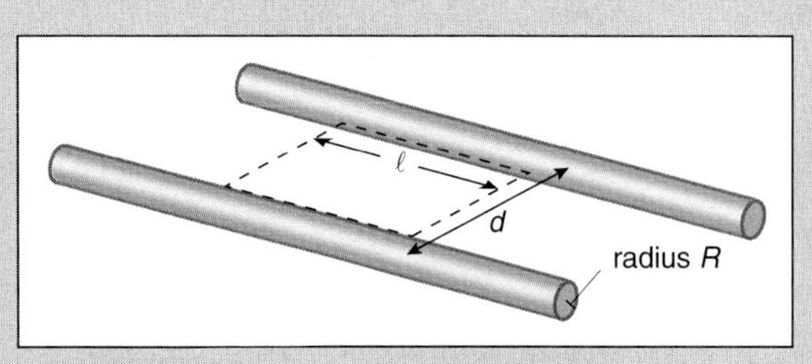

FIGURE P.34

•**35.** A toroidal solenoid (see Example 20.14) of average radius R contains N loops of wire. Each circular loop is of radius r as indicated in Figure P.35. Show that the self-inductance is approximately

$$L \approx \frac{\mu_0}{4\pi} 2\pi \frac{N^2 r^2}{R}$$

if $R >> r$. (With $R >> r$, the field within the toroid is essentially uniform).

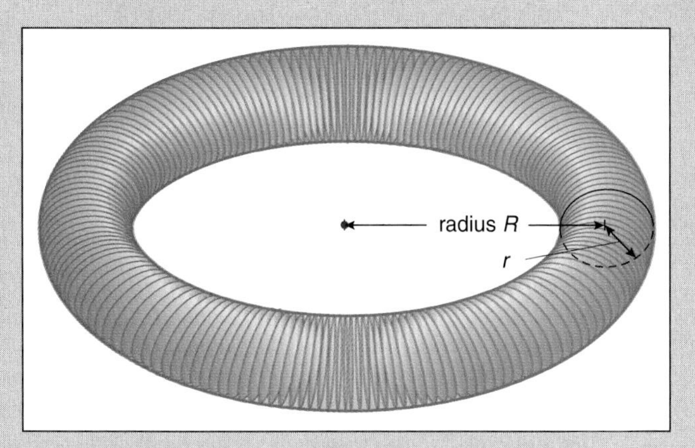

FIGURE P.35

•**36.** The potential difference V across a 150 mH inductor as a function of time is shown in Figure P.36. Assume the initial value for the current in the inductor is 0 A. (a) Find and graph an expression for the current $I(t)$ as a function of time. (b) What is the current when $t = 2.0$ ms? When $t = 4.0$ ms? (c) How can you check your answers to part (b) by using the *graph* of $V(t)$ versus t? (Hint: Think of the geometric interpretation of I on the graph of $V(t)$ versus t.)

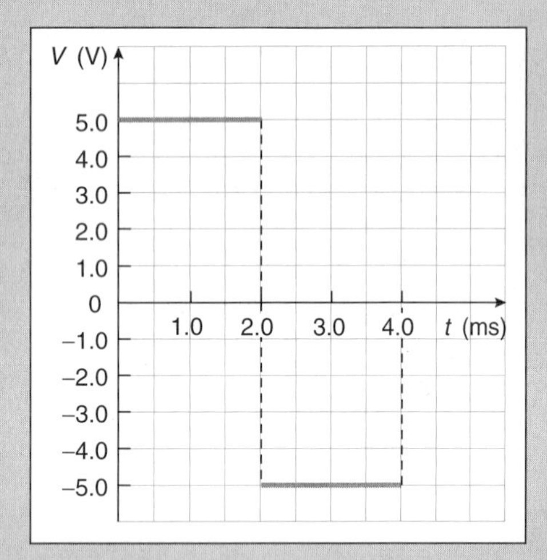

FIGURE P.36

•**37.** The potential difference across a 150 mH inductor as a function of time is shown in Figure P.37. Assume the initial value for the current in the indictor is 0 A. (a) Find and graph an expression for the current $I(t)$ as a function of time. (b) What is the current when $t = 2.0$ ms? When $t = 4.0$ ms? (c) How can you check your answers to part (b) by using the *graph* of $V(t)$ versus t? (Hint: Think of the geometric interpretation of I on the graph of $V(t)$ versus t.)

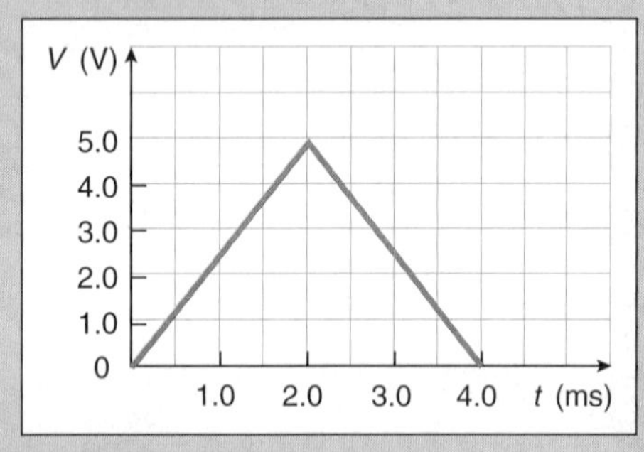

FIGURE P.37

•**38.** Find the equivalent inductance for each collection of inductors in Figure P.38. Assume each inductor produces negligible magnetic flux through any of the other inductors.

•**39.** Given three inductors each with self-inductance L: (a) What is the maximum inductance that can be made using all three? (b) What is the minimum inductance that can be made using all three? (c) What other inductances can be made using all three? Sketch the connections. Assume there is no magnetic flux through any inductor from the other inductors.

Sections 21.8 A Series *LR* Circuit*
21.9 Energy Stored in a Magnetic Field*
21.10 A Parallel *LC* Circuit*

•**40.** When a series LR circuit is turned on, the current reaches 25% of its final value in 1.10 s. Find the time constant of the circuit.

•**41.** You turn on a series LR circuit. How many time constants must you wait until the current is 99.44% of its final value?

•**42.** To measure the self-inductance of an inductor, you connect it in series with a 100 Ω resistor and a constant 15.0 V independent voltage source as shown in Figure P.42. (a) What is the steady-state potential difference across the resistor and across the inductor? (b) What is the steady-state current in the circuit? (c) When $t = 0$ ms, switch (1) is opened while switch (2) is closed. You measure the potential difference across the resistor as a function of time using an oscilloscope. From the oscilloscope trace, you secure the following data:

Time (ms)	*Potential difference across the resistor* (V)
0.00	15.00
1.00	7.71
2.00	3.96
3.00	2.03
4.00	1.05

Plot the data as a function of time. (d) From the graph, determine the time constant of the circuit. (e) What is the inductance L?

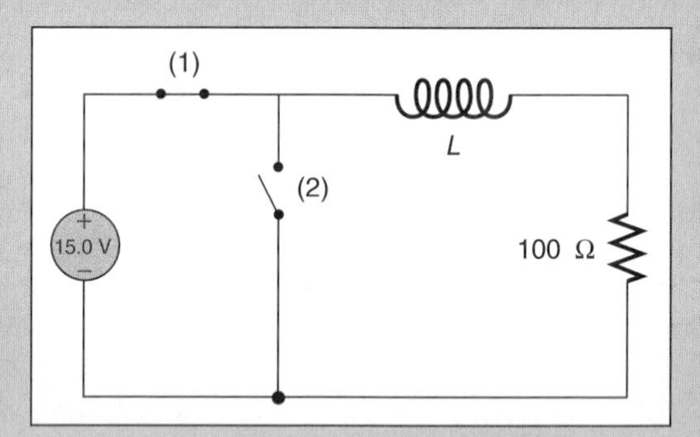

FIGURE P.42

•**43.** A series LR circuit has an inductance of 50 mH and a resistance of 0.25 Ω connected to a 15.0 V dc battery by a switch. The switch is closed when $t = 0$ s. (a) What is the current in the resistor when $t = 0.20$ s? (b) What is the potential difference V across the inductor when $t = 0.20$ s? (c) At what time does the current reach half its long-term value?

•**44.** When $t = 0$ s, a series LR circuit is connected to the independent voltage source shown in Figure P.44. (a) What is the time constant of the circuit? (b) What is the initial time rate of change of the current? (c) What is the time rate of change of the current when $t = \tau$? (d) What is the steady-state current?

•**45.** A 12.0 V independent voltage source is connected when $t = 0$ s to a series LR circuit with $R = 150$ Ω and $L = 250.0$ mH. (a) What is the time constant of the circuit? (b) What is the steady-state current in the circuit? (c) How much time does it take for the current to reach 50% of its steady-state value?

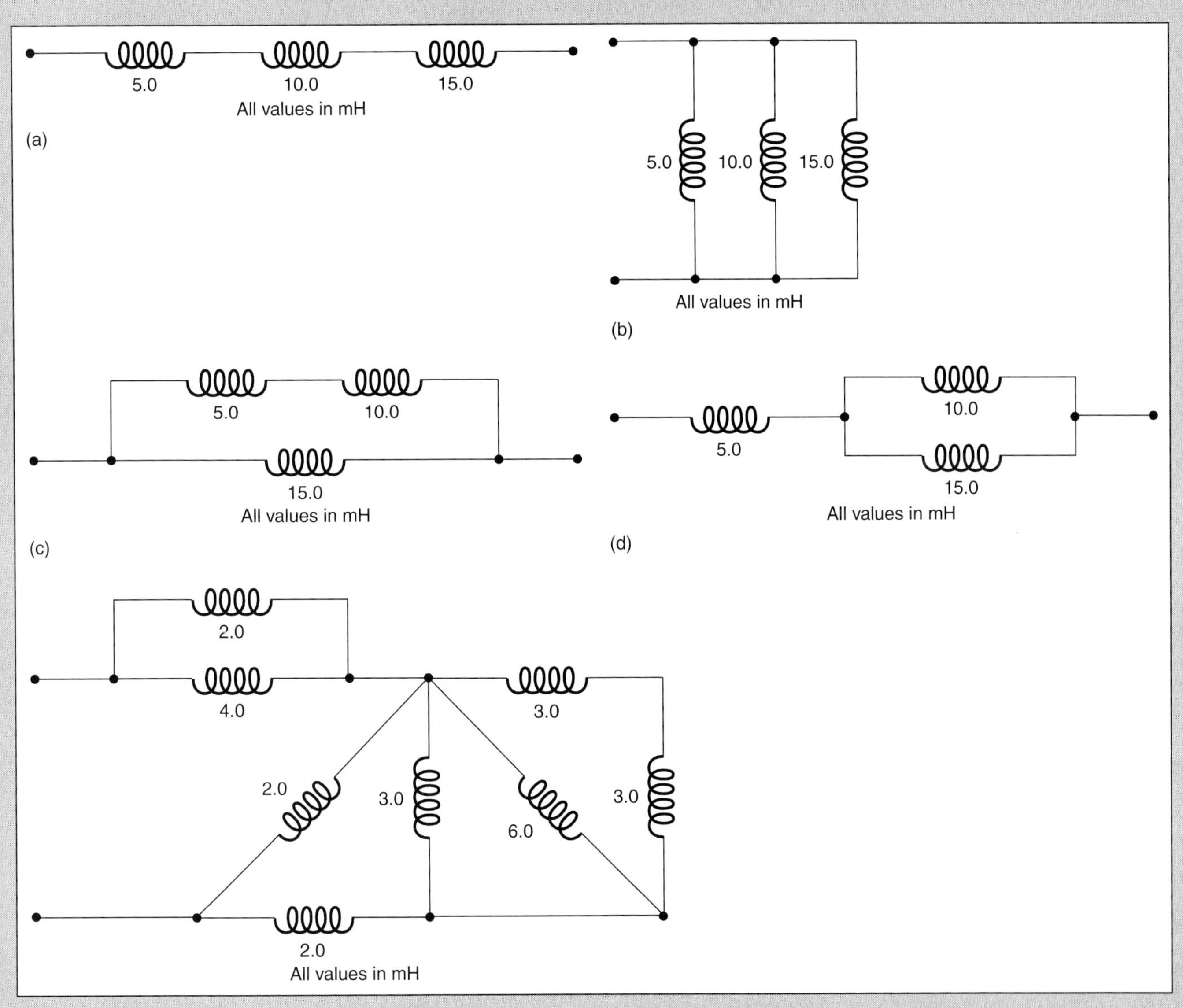

FIGURE P.38

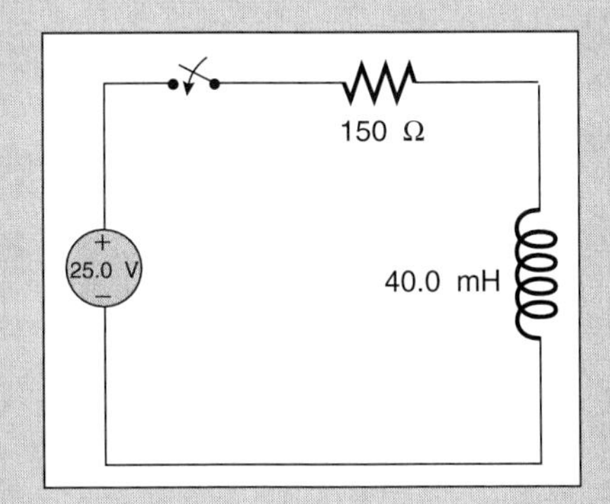

FIGURE P.44

•**46.** If it takes 2.0 ms for the current in an *LR* circuit to reach half its maximum value, what is the time constant of the circuit?

•**47.** (a) Calculate the energy stored in an 800 mH inductor that carries a constant current of 1.50 A. (b) To double the energy stored, by what factor must the current increase? (c) For a 100 μF capacitor to store the same amount of energy, what potential difference must exist across the capacitor?

•**48.** In the series *LR* circuit indicated in Figure P.48, the switch is closed when $t = 0$ s. (a) Find an expression for the energy $U(t)$ stored in the inductor as a function of time in terms of L, R, and V_0. (b) Let $V_0 = 10.0$ V, $R = 100\ \Omega$, and $L = 150$ mH. Make accurate graphs of $I(t)$ and $U(t)$ for $0 \text{ s} \leq t \leq 0.010$ s. In what ways are the graphs similar? In what ways are they different?

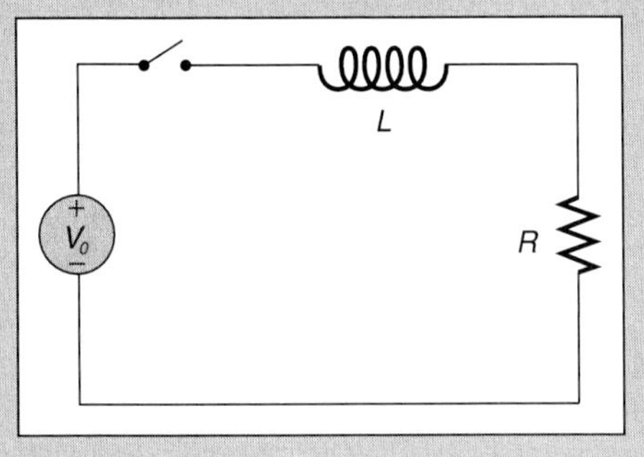

FIGURE P.48

•49. A device (not important to this question) provides a current to the two terminals of a circuit element in a sealed box as shown in Figure P.49a. The current varies with time as indicated in Figure P.49b. The potential difference V measured across the terminals of the box varies with time as shown in Figure P.49c. Negative values for the current and potential difference mean the current and polarity markings are opposite to the illustrated direction and/or polarity shown in Figure P.49a. (a) Explain why these graphs indicate that the circuit element is *not* a resistor. (b) Explain why these graphs indicate that the circuit element is *not* a capacitor. (c) Are the graphs consistent with the device being an inductor? Why or why not? (d) Calculate the inductance. (e) Complete the graph of the potential difference versus time for times greater than 2.0 s, indicating proper polarity and magnitude.

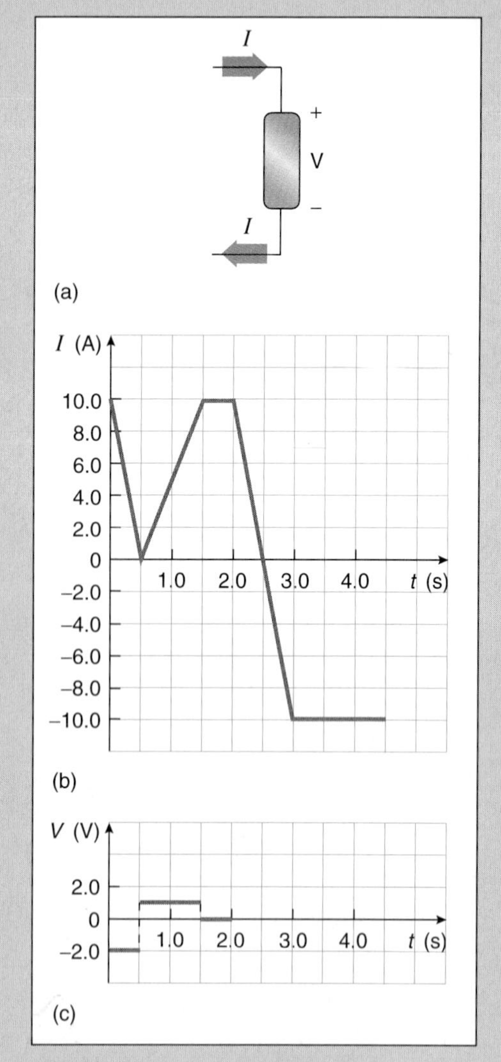

FIGURE P.49

•50. Use the KVL in the LC circuit depicted in Figure 21.63 to show that the current I also satisfies the differential equation for simple harmonic oscillation:

$$\frac{d^2I}{dt^2} + \frac{1}{LC}I = 0 \text{ A/s}^2$$

•51. The magnitude of the magnetic field of the Earth near its surface is on the order of 0.6×10^{-4} T. What is the energy density of this magnetic field?

•52. What is the oscillation frequency of a parallel LC circuit with a capacitance of 1.00 μF and an inductance of 100 mH?

•53. AM radio frequencies range from about 500 kHz to about 1600 kHz, a range called the AM band. The tuning circuit of an AM receiver typically is a parallel LC circuit with a fixed inductor and a capacitor whose capacitance can be varied to permit the circuit to oscillate at the frequency corresponding to a given radio station. If a receiver has a fixed inductor with $L = 0.33$ mH, over what range of capacitances must the capacitor be able to vary?

54. Consider the circuit shown in Figure P.54. (a) When $t = 0$ s the switch is closed. At that instant determine: (i) the current in the inductor; (ii) the current in the 100 Ω resistor; (iii) the current in the 200 Ω resistor; (iv) the current in the switch; (v) the potential difference across the 100 Ω resistor; (vi) the potential difference across the 200 Ω resistor; (vii) the potential difference across the inductor; and (viii) the time rate at which the current is changing through the inductor. (b) After many time constants (effectively for $t \to \infty$ s), determine the quantities listed in part (a).

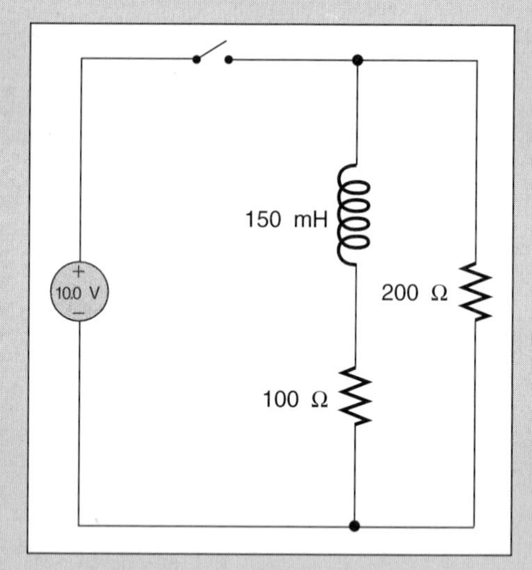

FIGURE P.54

55. In the series LR circuit of Figure P.55 the current has reached its steady-state value. (a) What energy is stored in the magnetic field of the inductor? (b) When $t = 0$ s, switch (1) is opened while switch (2) is closed and the current decays to zero. Calculate the energy absorbed by the resistor between $t = 0$ s and $t = \infty$ s. Compare this with the answer to part (a).

56. Two wires are connected to an unknown circuit element(s) inside a closed box (see Figure P.56). A 6.0 V dc independent voltage source is connected to its leads when $t = 0$ s. The current in the circuit gradually increases from 0 A when $t = 0$ s to 1.0 A when $t = 0.50$ s, and continues to grow until it reaches a steady value of 3.0 A. It is noted that the current does *not* increase linearly with time. The current does not oscillate but grows continuously as described. (a) Does the box contain solely a resistor? Explain your reasoning. (b) Does

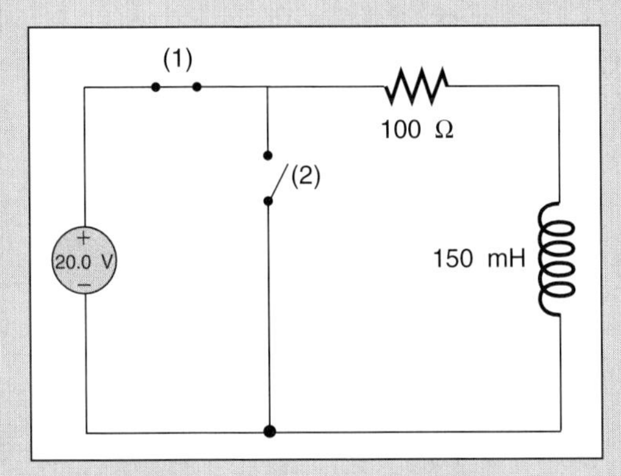

FIGURE P.55

the box contain only a capacitance? Explain your reasoning. (c) Does the box contain only an inductance? Explain your reasoning. (d) What circuit elements are inside the box and how are they connected? Sketch the complete circuit. (e) Determine the expression $I(t)$ for the current as a function of time.

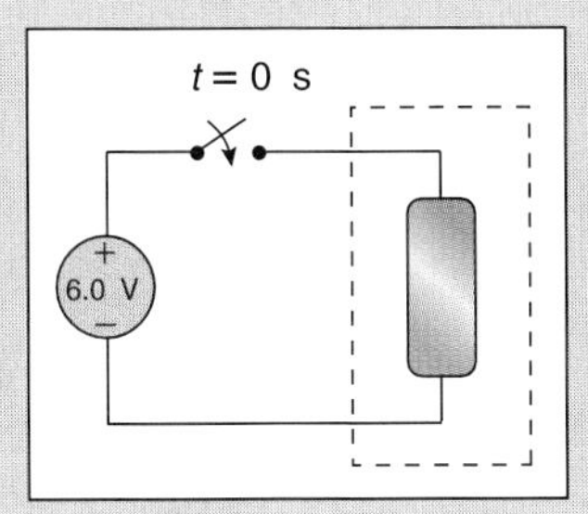

FIGURE P.56

Sections 21.11 Mutual Inductance* 21.12 An Ideal Transformer*

57. A step-down transformer has an input of 14 kV and an output of 220 V. What is the ratio of the number of turns in the primary to that in the secondary?

58. You have just purchased a low-voltage lighting system that operates at 24 V ac. The independent voltage source available in your home is 120 V ac. Specify the ratio of the number of turns in the primary to those in the secondary for a suitable transformer.

59. Your electric train set operates at 12 V ac, whereas the source in your home is 120 V ac. What is the ratio of the number of turns in the primary to those in the secondary of a suitable transformer?

•60. A solenoid has 250 turns of wire wrapped along its length of 25 cm. The diameter of the solenoid is 1.5 cm. In the middle of the solenoid is a smaller coil of diameter 1.0 cm with 50 turns of wire along its 2.0 cm length. The two coils are coaxial. A current of 15 A is in the larger solenoid with no current in the smaller coil. (a) Determine the magnetic flux through the smaller coil. (b) Determine the mutual inductance of the pair of coils.

•61. For each transformer shown in Figure P.61, imagine the current in the primary coil to be directed from terminal (1) to terminal (2) and increasing with time. Indicate which wire lead of the secondary has the positive polarity of the emf and which lead of the secondary has the positive polarity of the potential difference V.

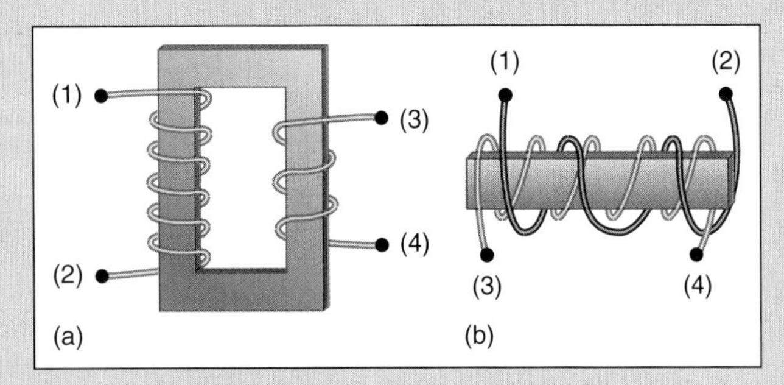

FIGURE P.61

•62. Transformers can be used to change the effective value of a resistance in an ac circuit. To see how this is done, connect a resistor R across the terminals of the secondary coil of a transformer as shown in Figure P.62. Show that the ratio of V_1 to I_1 in the primary coil is

$$\frac{V_1}{I_1} = \left(\frac{N_1}{N_2}\right)^2 R$$

Thus the effective value of the resistance R in the primary circuit is amplified by the factor

$$\left(\frac{N_1}{N_2}\right)^2$$

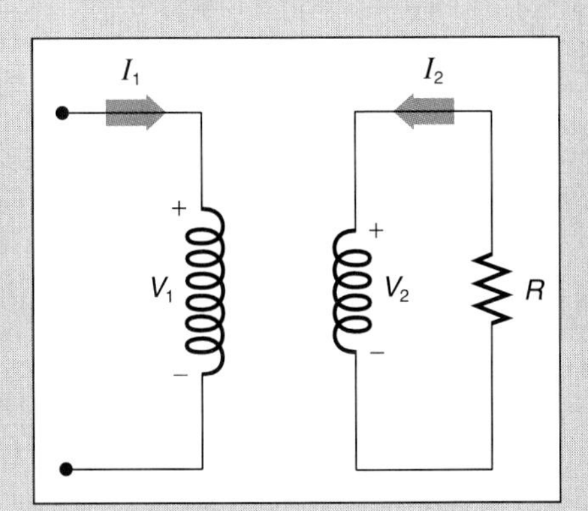

FIGURE P.62

•63. (Refer to the previous problem.) An electrocardiograph (EKG) is used to measure electric potentials associated with currents that traverse the heart muscles. The EKG is modeled as an oscillating independent voltage source in series with a large resistor (40 kΩ). An audio speaker with an effective resistance of 8.0 Ω is to be connected via a transformer to the device (see Figure P.63). To maximize power transfer from the EKG to the speaker, the resistance of the speaker must be amplified by the transformer to equal that of the EKG. Specify the ratio of turns in the primary to that in the secondary for an appropriate transformer.

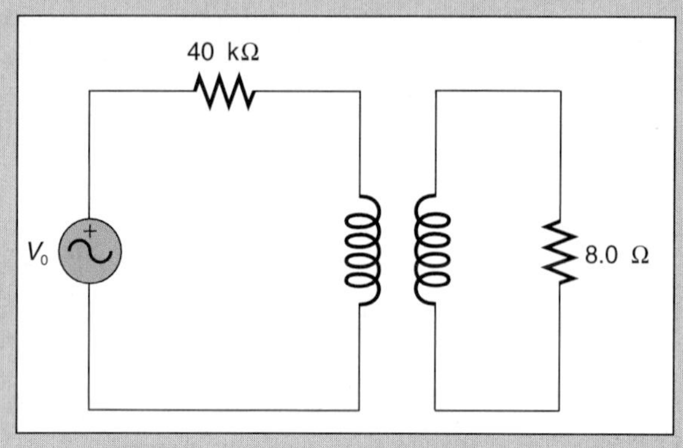

FIGURE P.63

INVESTIGATIVE PROJECTS

A. Expanded Horizons

1. Hints of a subtle connection between light and electricity were gleaned by Francis Hauksbee (c. 1666–1713) early in the 18th century. A study of his work on electroluminescence and static electricity is a good illustration of the careful observational research made by our scientific forebears. Hauksbee was an artisan (apprenticed as a draper) with little formal education but a keen and observant mind. This was recognized by Isaac Newton, who invited Hauksbee to meetings of the Royal Society to present demonstrations. Hauksbee's work led Dufay and Franklin to the discovery of two types of electric charge (see Chapter 16). Investigate and try to duplicate Hauksbee's experiments.
   Duane Roller and Duane H. D. Roller, "Francis Hauksbee," *Scientific American*, 189, #2, pages 64–69 (August 1953); additional reference on page 100.
   Dictionary of Scientific Biography (Scribner, New York, 1972), volume VI, pages 169–175.
2. If you are interested in learning more about Michael Faraday and his famous and ingenious experiments, see the following references.
   John Meurig Thomas, *Michael Faraday and the Royal Institution* (Adam Hilger, Bristol, 1991).
   Herbert Kondo, "Michael Faraday," *Scientific American*, 189, #4, pages 90–99 (October 1953); additional references on page 122.
3. Heinrich Rudolph Hertz (1857–1894) was a renowned experimental scientist and the first person (at the ripe age of only 30) to create and detect electromagnetic radiofrequency waves in the laboratory (1887). His many ingenious experiments, performed over a very short lifetime, are a rich mine, worthy of study. Investigate and report on his experiments and their significance.
   Joseph P. Mulligan, "Heinrich Hertz and the development of physics," *Physics Today*, 42, #3, pages 50–57 (March 1989).
   Joseph P. Mulligan, "Max Planck and the 'black year' of German physics," *American Journal of Physics*, 62, #12, pages 1089–1097 (December 1994).
   Jed Z. Buchwald, *The Creation of Scientific Effects: Heinrich Hertz and Electric Waves* (University of Chicago Press, Chicago, 1994).
4. The biological effects of electromagnetic fields and waves are of subject of some controversy. You might find it interesting to narrow this topic and investigate it in some detail. For this purpose, the following resource guide will be useful.
   David Hafemeister, "Resource Letter BELFEF-1: Biological effects of low-frequency electromagnetic fields," *American Journal of Physics*, 64, #8, pages 974–981 (August 1996).
5. Although they lack moving parts, real transformers such as those on utility poles hum at audible frequencies when in operation. Investigate why (perhaps by calling your local electric utility company) and report on your findings to your classmates.

B. Lab and Field Work

6. Design an experiment to detect and measure the variation in the acceleration of a magnet falling vertically through a coil of wire. Test to see if it makes any difference which pole of the magnet falls first.
7. Design and perform an experiment to measure the self-inductance of a coil of wire. See Problem 42. Insert a bar of iron into the coil and remeasure its self-inductance.
8. Design and perform an experiment to demonstrate oscillations in an LC circuit. Investigate how changing the value of C affects the frequency of the oscillations.
9. Perhaps your professor can arrange for you to use a research magnet in your physics department. After determining the magnitude of the magnetic field available (perhaps with a Hall probe), design and build a hand-cranked ac generator with a peak output of, say, 10 V.
10. Visit and consult with a cardiac surgeon and/or a manufacturer of heart pacemakers to learn more about how electrical energy is supplied to them. On another note, some cordless electric toothbrushes transfer electrical energy using Faraday's law.
    Peter P. Tarjan and Alan D. Bernstein, "An engineering overview of cardiac pacing," *IEEE Engineering in Medicine and Biology Magazine*, 3, #2, pages 10–14 (1984).
11. Visit your local electric utility company to learn more about the variety of common transformers used in the electrical distribution system in your locale. Detail the common potential differences encountered as inputs and outputs and, from these, determine the ratio of primary to secondary turns in the various

transformers. You also might inquire about how switching large ac potential differences is accomplished without initiating large sparks or discharges.

12. Call, write, or e-mail the R&D department of a manufacturer of induction stoves to learn more about their performance with the goal of simulating a *Consumer Reports* type article about such stoves.
13. Visit an automotive shop to investigate the use of induction lights to set the timing of gasoline engines. Demonstrate their use for your classmates.

C. Communicating Physics

14. Most people are little aware of Faraday, let alone of his law of electromagnetic induction, despite the fact that it is "Faraday's spark" that, in large measure, has made a modern, electronic, technological society possible. (See the chapter-opening quotation by John Tyndall on page 953.) Certainly we all complain when the power goes off because of storms! For an audience without technical training (such as most readers of a regional newspaper), write a science feature article and/or essay detailing how Faraday's law maintains and is responsible for a high standard of living.
15. In Section 21.6 we briefly mentioned that Joseph Henry apparently discovered electromagnetic induction before Faraday but did not publish the finding; credit is given to Faraday. Is this fair? Should credit be given to the person who first makes the observation or who first publishes the finding? Write a one page essay detailing your opinion about the controversy. There is risk in publication, because the observations and results are exposed to peer scrutiny and, perhaps, negative criticism and ridicule. Even today, questions about the priority of discovery plague the sciences, mathematics, and engineering. The matter is not purely academic. Huge amounts of money are involved if a discovery has important and patentable implications.

Chapter 22
Sinusoidal ac Circuit Analysis

. . . electricity from the square root of minus one.

Vladimir Karapetoff (1876–1948)*

A century ago, one of the great technological debates was whether the electrical distribution system (the world-wide-web of the day) should be ac or dc.[†] The debate was yet to be settled in 1895 when George Vanderbilt constructed a summer retreat called the Biltmore, still the largest private residence in the United States, in Asheville, North Carolina (see chapter opening photo). A technophile with huge financial resources inherited from his grandfather, the shipping and railroad baron Cornelius Vanderbilt, George hedged his bets and had the 220-room mansion built to accommodate both systems.[‡]

Nationally, the choice eventually was settled in favor of ac distribution, in large measure because of the inventive genius of one man: Charles Proteus Steinmetz (1865–1923).[§] A German immigrant to the United States and an electrical engineer of the first rank, Steinmetz realized that the then purely mathematical idea of complex variables could be used to great practical advantage to simplify the analysis of ac circuits. He explained how to do so in a nine-volume, practical treatise. Among other things, this application of pure mathematics to the sciences and engineering illustrates the thesis that the pure math of today is often the physics of tomorrow.

Steinmetz spent a 30-year career working for the relatively new General Electric Company in Schenectady, New York.[#] He came to be known as the Wizard of Schenectady, able to perform amazing calculations in his head. His use of complex variable theory for the practical and quick solution of problems in ac circuits was a precursor of the importance complex variables also were found to play in the development of quantum mechanics during the early decades of the 20th century. For the same reasons, then, we use the algebra of complex variables here to analyze ac circuits; the algebra makes ac circuits much easier to solve than other methods and is essential for the study of quantum mechanics.

We first begin with a short review of the algebra of complex variables. If you already are conversant with this algebra, proceed to Section 22.3. There we see how the algebra is applied to ac circuits with independent voltage sources that vary sinusoidally with time. Sinusoidal, independent voltage sources at the frequency of 60 Hz ($\omega \approx 377$ rad/s) are available from common electrical outlets in North and South America. Sinusoidal voltage sources with a frequency of 50 Hz ($\omega \approx 314$ rad/s) are the European, African, Asian, and Japanese standard.

Sinusoidally varying sources form the basis for the analysis of more complicated periodic (and nonperiodic) time variations, thanks to a branch of mathematics known as Fourier analysis. Fourier analysis enables us to consider *any* time variation as a

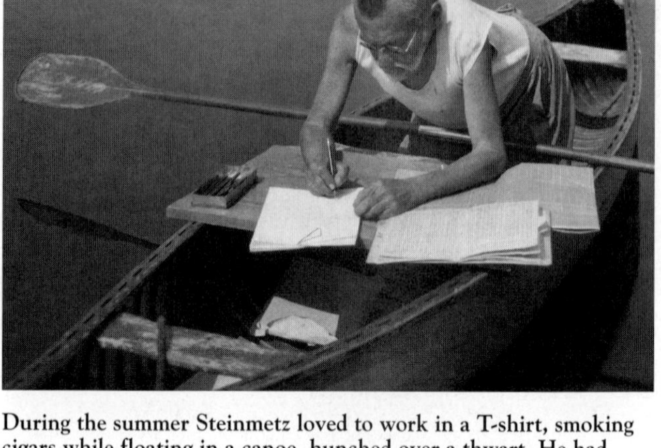

During the summer Steinmetz loved to work in a T-shirt, smoking cigars while floating in a canoe, hunched over a thwart. He had little need for large books of mathematical tables (well before the advent of electronic, scientific calculators). Among other mathematical feats, he could find logarithms of numbers to five significant figures in his head.

suitable infinite series or integral of sinusoidal time variations at different frequencies (see Section 12.21). Thus, if we know how to treat a single sinusoidal source in an ac circuit, we have the basic building block for more advanced ac circuit analysis.

22.1 Representations of a Complex Variable

Complex variables make use of the imaginary number $i \equiv \sqrt{-1}$.[**]

A **complex variable** z consists of two parts: its **real part** x and its **imaginary part** y. Here, z, x, and y represent *algebraic variables; they do not necessarily represent the Cartesian coordinates.*[††] The variables z, x, and y may represent any physical quantity but must all be dimensionally consistent (i.e., dimensionally the same). In other words, in this introduction to complex variables, think of z, x, and y as algebraic variables, familiar to you from the halcyon days of your algebra and calculus courses. When complex variables are used in physics and engineering, symbols more suggestive of the physical quantity to be represented are employed (as we will see later in this chapter). In particular, complex variables will be used to represent ac independent voltage sources, potential differences, currents, and other electrical parameters. For now, we think of z, x, and y in typically generic, mathematical terms.

Complex variables are written in a number of equivalent representations, each with its own particular usefulness. It is

*(Chapter Opener) *Cornell Daily Sun*, 29 October 1923; in an obituary for Charles Proteus Steinmetz.

†One only has to see how difficult it is to settle on a unique and universal operating system for computers to realize that such great technological debates continue even today.

‡Some computers today are built to accommodate multiple operating systems; history repeats itself.

§Others contributed to the development of ac circuit theory as well, but it was Steinmetz who, with passion, made the treatment understandable to a wide audience of engineers.

#He also was a great teacher, serving without pay at the nearby, well-respected, Union College. Steinmetz, a dedicated socialist, lived his creed and never pursued wealth for its own sake.

**The letter comes from the first letter of *i*maginary. In electrical engineering, the letter $j \equiv \sqrt{-1}$ is used, which was the notation of Steinmetz, while i is reserved for ac current. We use I for *any* current, and so will use the standard mathematical notation for the imaginary number as is done in quantum mechanics: $i \equiv \sqrt{-1}$. We apologize to any electrical engineers among you, and pray for your understanding of the difficulties associated with choosing appropriate notation.

††The letter z is the traditional notation for a general complex variable in mathematics. Likewise its real part traditionally is called x and its imaginary part y.

important for you to be able to change from one representation to another with facility since, as we shall see, various arithmetic and algebraic operations are performed more easily in one representation than another.

Rectangular Form

The complex variable z in **rectangular form** is written as

$$z = x + iy \tag{22.1}$$

We designate the real part of the complex variable $z = x + iy$ as

$$\text{Re } z \equiv x \tag{22.2}$$

and its imaginary part, the coefficient of i, as

$$\text{Im } z \equiv y \tag{22.3}$$

The imaginary part of a complex variable is itself a *real variable*.

Both the real and imaginary parts of a complex variable are *real variables*. The complex nature of z is explicitly and totally contained in the imaginary number i.

From a physical standpoint, the real and imaginary parts of a complex variable must have the same dimensions, since i is a pure number.

EXAMPLE 22.1

Specify the real and imaginary parts of the complex number

$$z = 4.5 - i(2.2)$$

Solution

The real part is equal to 4.5 and the imaginary part is the coefficient of i, or -2.2:

$$\text{Re } z = 4.5 \qquad \text{Im } z = -2.2$$

Graphical Representation of a Complex Number: The Complex Plane

The **complex plane** is a two-dimensional plane that plots the real part of a complex number along the horizontal axis and the imaginary part along the vertical axis.

A complex number thus is a point *on* the complex plane; equivalently, we also say *in* the complex plane. The **graphical representation** of a complex number pictures its location in the complex plane; see Figure 22.1.

If a line is drawn from the origin of the complex plane to the point representing the complex number as in Figure 22.1, the length of this line is the **magnitude** r of the complex number z. Note that the magnitude can be found from the real and imaginary parts using the Pythagorean theorem:

$$r = (x^2 + y^2)^{1/2} \tag{22.4}$$

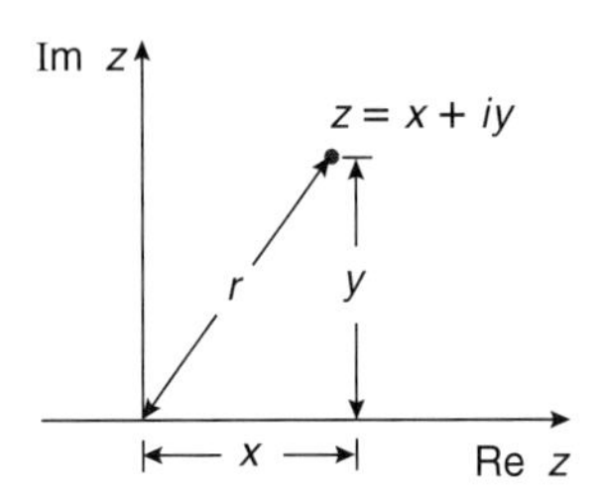

FIGURE 22.1 Graphical representation of a complex number in the complex plane.

The magnitude of the complex number z is the positive square root of the sum of the squares of its real and imaginary parts. The magnitude of a complex number is always a real number greater than or equal to zero. Physically, r is not necessarily a distance (it is only a distance if x and y are distances).

You may note that a complex number is similar to a two-dimensional vector. The magnitude of the complex number is found using the same procedure as that used to find the magnitude of a two-dimensional vector.

EXAMPLE 22.2

Plot the complex number $z = -2.0 + i(3.0)$ on the complex plane and determine its magnitude.

Solution

The real part of the complex number is -2.0 and its imaginary part is 3.0. Its location in the complex plane is shown in Figure 22.2.

The magnitude of the complex number is found from Equation 22.4:

$$\begin{aligned} r &= (x^2 + y^2)^{1/2} \\ &= [(-2.0)^2 + (3.0)^2]^{1/2} \\ &= 3.6 \end{aligned}$$

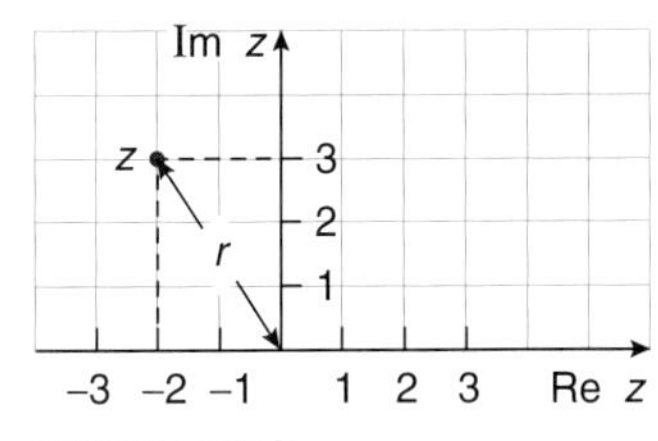

FIGURE 22.2

Polar Form

A complex number also can be represented in **polar form** by specifying its magnitude r and the angle θ that locates the complex number in the complex plane, as shown in Figure 22.3. The polar form is written symbolically as

$$z = r\angle\theta \tag{22.5}$$

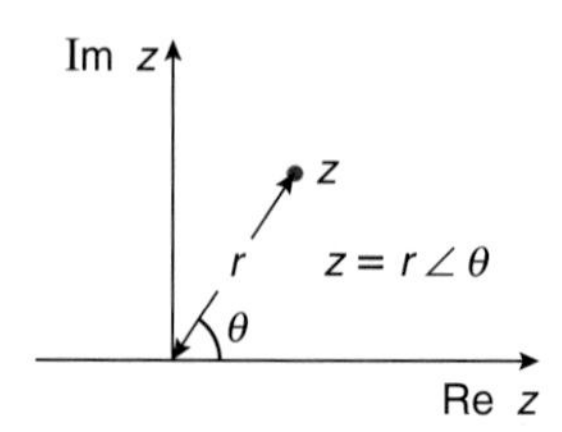

FIGURE 22.3 Polar representation of a complex number in the complex plane.

The angle θ can be expressed either in degrees or radians; for consistency, *we will always use radians.*

From the geometry, the real part of the complex number is

$$x = r \cos \theta \quad (22.6)$$

and its imaginary part is

$$y = r \sin \theta \quad (22.7)$$

The angle θ in the polar expression for the complex number is the angle whose tangent is the imaginary part divided by the real part of the complex number:

$$\tan \theta = \frac{\text{Im } z}{\text{Re } z} = \frac{y}{x} \quad (22.8)$$

Using Equations 22.6, 22.7, and 22.8, we have a way to convert from polar to rectangular form:

$$\begin{aligned} z &= r \angle\theta \\ &= r \cos \theta + ir \sin \theta \end{aligned} \quad (22.9)$$

If all this seems to have a vaguely familiar ring, you are not mistaken. The algebra of complex numbers is very similar to two-dimensional vector analysis. We never need to extend this on into three dimensions, because there are only two parts to a complex number.

PROBLEM-SOLVING TACTIC

22.1 When converting between rectangular and polar forms, be sure your calculator is in radian mode for performing trigonometric operations. The reason for this is because we choose to express the angles in radians, not degrees.

EXAMPLE 22.3

Convert the complex number $z = 4.00 + i(3.00)$ to polar form.

Solution

From Equation 22.4, the magnitude of the complex number is the (positive) square root of the sum of the squares of its real and imaginary parts:

$$\begin{aligned} r &= (x^2 + y^2)^{1/2} \\ &= [(4.00)^2 + (3.00)^2]^{1/2} \\ &= 5.00 \end{aligned}$$

The polar angle is found from Equation 22.8

$$\begin{aligned} \tan \theta &= \frac{y}{x} \\ &= \frac{3.00}{4.00} \\ &= 0.750 \\ \theta &= \tan^{-1}(0.750) \\ &= 0.644 \text{ rad} \end{aligned}$$

Thus the polar form of $4.00 + i(3.00)$ is

$$5.00 \angle(0.644 \text{ rad})$$

Exponential Form

Yet another way to represent a complex number makes use of the amazing Euler identity.* The identity states that

$$e^{i\theta} = \cos \theta + i \sin \theta \quad (22.10)$$

The proof of this identity stems from comparing the series expansions of the exponential, the sine, and the cosine; the series expansions of the left- and right-hand sides of the Euler relation are seen to be identical.†

We multiply both sides of the Euler identity by the magnitude r of the complex number:

$$re^{i\theta} = r \cos \theta + ir \sin \theta \quad (22.11)$$

The right-hand side of Equation 22.11 we recognize as the rectangular form of the complex number (from Equation 22.9). The left-hand side of Equation 22.11, $re^{i\theta}$, is the **exponential form** of the complex number z.

The angle in the exponent of the exponential form *will always be expressed in radians* because we express the angle in the polar form in like fashion.

In summary, a complex variable z has three distinct representations:

rectangular form:	$z = x + iy$
polar form:	$z = r \angle\theta$
exponential form:	$z = re^{i\theta}$

*Leonhard Euler (1707–1783), a great mathematician and scientist, gave us many of our modern mathematical notations, among them i for $\sqrt{-1}$, e for the base of natural logarithms, and $f(x)$ for a function of x.

†These series expansions are

$$e^{i\theta} = 1 + (i\theta) + \frac{(i\theta)^2}{2!} + \frac{(i\theta)^3}{3!} + \frac{(i\theta)^4}{4!} + \cdots$$

$$\cos \theta = 1 - \frac{\theta^2}{2!} + \frac{\theta^4}{4!} - \frac{\theta^6}{6!} + \cdots$$

$$\sin \theta = \theta - \frac{\theta^3}{3!} + \frac{\theta^5}{5!} - \frac{\theta^7}{7!} + \cdots$$

QUESTION 1

Discuss the ways the use of the word *imaginary* for $i = \sqrt{-1}$ is simultaneously appropriate and inappropriate.

EXAMPLE 22.4

Find the exponential form for the complex number $4.00 + i(3.00)$.

Solution

In Example 22.3 you found the polar form for this complex number; so you know

$$r = 5.00 \quad \text{and} \quad \theta = 0.644 \text{ rad}$$

The exponential form is thus

$$re^{i\theta} = 5.00e^{i(0.644 \text{ rad})}$$

STRATEGIC EXAMPLE 22.5

a. Graph the complex number $z = i$ in the complex plane.
b. Find the polar form for this complex number.
c. Find the exponential form for this complex number.

Solution

a. In the complex plane, the complex number $z = i$ is located one unit along the imaginary axis, as shown in Figure 22.4

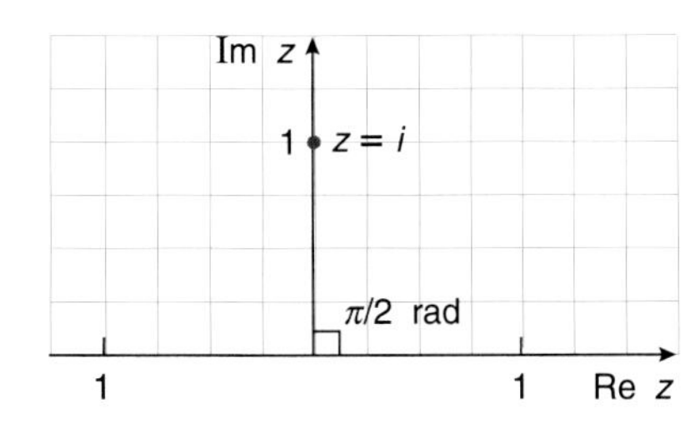

FIGURE 22.4

b. The complex number i has a real part equal to 0 and imaginary part equal to 1. Its magnitude is thus

$$\begin{aligned} r &= [(0)^2 + (1)^2]^{1/2} \\ &= 1 \end{aligned}$$

From Equation 22.8, the polar angle has an infinite tangent. Hence

$$\theta = \frac{\pi}{2} \text{ rad}$$

Hence the complex number $z = i$ has the polar form

$$1 \angle\left(\frac{\pi}{2} \text{ rad}\right)$$

c. The exponential form of the complex number i thus is

$$e^{i(\pi/2 \text{ rad})}$$

This is an important example that will be used frequently!

EXAMPLE 22.6

a. Determine the exponential form of the complex number

$$5.00 \angle\left(\frac{\pi}{6} \text{ rad}\right)$$

b. Determine the real and imaginary parts of this complex number.

Solution

a. The given complex number is in polar form $r \angle\theta$. The exponential form is

$$re^{i\theta} = 5.00e^{i(\pi/6 \text{ rad})}$$

b. From the Euler identity, Equation 22.10, we have

$$\begin{aligned} re^{i\theta} &= r\cos\theta + ir\sin\theta \\ &= 5.00\cos\left(\frac{\pi}{6}\text{ rad}\right) + i\left[5.00\sin\left(\frac{\pi}{6}\text{ rad}\right)\right] \\ &= 4.33 + i(2.50) \end{aligned}$$

The real part is 4.33 and the imaginary part is 2.50.

The techniques of this example will be used frequently later in the chapter.

22.2 ARITHMETIC OPERATIONS WITH COMPLEX VARIABLES

The importance of the ability to add, subtract, multiply, and divide complex numbers will become apparent when we use them for ac circuit analysis.

Addition and Subtraction of Complex Variables

The addition of two complex variables,

$$z_1 = x_1 + iy_1$$

and

$$z_2 = x_2 + iy_2$$

produces another complex variable. Operationally,

$$z_1 + z_2 = (x_1 + iy_1) + (x_2 + iy_2)$$

or rearranging,

$$z_1 + z_2 = (x_1 + x_2) + i(y_1 + y_2) \tag{22.12}$$

The real parts are added together, and likewise for the imaginary parts.

Subtraction proceeds along similar lines:

$$z_1 - z_2 = (x_1 + iy_1) - (x_2 + iy_2)$$

or rearranging,

$$z_1 - z_2 = (x_1 - x_2) + i(y_1 - y_2) \qquad (22.13)$$

The respective real parts are subtracted, as are the imaginary parts.

PROBLEM-SOLVING TACTIC

22.2 Addition and subtraction of complex variables can only be performed when they are in rectangular form. If the complex variables are in polar or exponential form, they *must* be converted to rectangular form in order to perform the addition or subtraction. The sum or difference then can be converted back to polar or exponential form if desired. Addition and subtraction cannot be performed when the complex numbers are in polar or exponential forms.

EXAMPLE 22.7

Find the sum of the complex numbers

$$3 + i(4)$$

and

$$2 - i(2)$$

Solution

The real and imaginary parts are separately added to obtain

$$(3 + 2) + i[4 + (-2)] = 5 + i(2)$$

EXAMPLE 22.8

Find the sum of the complex numbers

$$4.00 \angle\left(\frac{\pi}{6}\ \text{rad}\right)$$

and

$$3.00 \angle\left(\frac{\pi}{4}\ \text{rad}\right)$$

and convert it into polar and exponential forms.

Solution

You first need to convert the complex numbers into rectangular form:

$$4.00 \angle\left(\frac{\pi}{6}\ \text{rad}\right) = 4.00 \cos\left(\frac{\pi}{6}\ \text{rad}\right) + i\left[4.00 \sin\left(\frac{\pi}{6}\ \text{rad}\right)\right]$$
$$= 3.46 + i(2.00)$$

Likewise,

$$3.00 \angle\left(\frac{\pi}{4}\ \text{rad}\right) = 3.00 \cos\left(\frac{\pi}{4}\ \text{rad}\right) + i\left[3.00 \sin\left(\frac{\pi}{4}\ \text{rad}\right)\right]$$
$$= 2.12 + i(2.12)$$

The sum is then written in rectangular form:

$$[3.46 + i(2.00)] + [2.12 + i(2.12)] = 5.58 + i(4.12)$$

Now convert the sum $5.58 + i(4.12)$ back to polar form. From Equation 22.4 you have

$$r = [(5.58)^2 + (4.12)^2]^{1/2}$$
$$= 6.94$$

The angle is found from Equation 22.8:

$$\tan\theta = \frac{y}{x} = \frac{4.12}{5.58} = 0.738$$

$$\theta = 0.636\ \text{rad}$$

Thus the polar form of the sum is

$$6.94 \angle(0.636\ \text{rad})$$

The exponential form for the sum is

$$6.94 e^{i(0.636\ \text{rad})}$$

Multiplication of Complex Variables

Multiplication of two complex variables is accomplished most easily when they are in exponential or in polar form. For example, let

$$z_1 = r_1 e^{i\theta_1}$$

and

$$z_2 = r_2 e^{i\theta_2}$$

The product is

$$z_1 z_2 = r_1 e^{i\theta_1} r_2 e^{i\theta_2}$$

Rearranging and exploiting the properties of exponentials, we find

$$\boxed{z_1 z_2 = r_1 r_2 e^{i(\theta_1 + \theta_2)}} \qquad (22.14)$$

The product has a magnitude equal to the product of the magnitudes and the angle in the exponential form is simply the sum of the polar angles.

This means that multiplication in polar form also is operationally easy. We multiply the magnitudes and add the polar angles:

$$z_1 z_2 = (r_1 \angle\theta_1)(r_2 \angle\theta_2)$$
$$= r_1 r_2 \angle(\theta_1 + \theta_2) \qquad (22.15)$$

A product also can be performed when the complex numbers are in rectangular form, but the result is algebraically more involved:

$$z_1 z_2 = (x_1 + iy_1)(x_2 + iy_2)$$
$$= (x_1 x_2 - y_1 y_2) + i(x_1 y_2 + x_2 y_1) \qquad (22.16)$$

PROBLEM-SOLVING TACTIC

22.3 For the multiplication of two complex numbers in rectangular form, it occasionally is easier first to convert them to polar or exponential form, then do the multiplication. The result can be converted back into rectangular form if desired.

EXAMPLE 22.9

Find the product of the two complex numbers

$$3.00 \angle\left(\frac{\pi}{4} \text{ rad}\right)$$

and

$$4.00 \angle\left(\frac{\pi}{6} \text{ rad}\right)$$

Express the result in both polar and exponential forms.

Solution

In polar form you have

$$(3.00)(4.00) \angle\left(\frac{\pi}{4} \text{ rad} + \frac{\pi}{6} \text{ rad}\right) = 12.0 \angle\left(\frac{5\pi}{12} \text{ rad}\right)$$

The exponential form is

$$12.0\, e^{i(5\pi/12 \text{ rad})}$$

EXAMPLE 22.10

Find the product of the complex numbers

$$3.00 + i(4.00) \qquad \text{and} \qquad 2.00 - i(3.00)$$

Express the result in rectangular form.

Solution

Method 1

Perform the product directly:

$$[3.00 + i(4.00)][2.00 - i(3.00)]$$
$$= (3.00)(2.00) + (3.00)[-i(3.00)]$$
$$+ [i(4.00)](2.00) + [i(4.00)][-i(3.00)]$$
$$= 18.0 - i(1.00)$$

Method 2

First convert the complex numbers to polar form:

$$3.00 + i(4.00) = 5.00 \angle(0.927 \text{ rad})$$
$$2.00 - i(3.00) = 3.61 \angle(-0.983 \text{ rad})$$

Then perform the product:

$$(5.00)(3.61) \angle[0.927 \text{ rad} + (-0.983 \text{ rad})]$$
$$= 18.1 \angle(-0.056 \text{ rad})$$

Now convert this to rectangular form:

$$18.1 \angle(-0.056 \text{ rad}) = 18.1 \cos(-0.056 \text{ rad})$$
$$+ i[18.1 \sin(-0.056 \text{ rad})]$$
$$= 18 - i(1.0)$$

which is identical (except with two significant figures) to the product found from their rectangular forms.

Division of Complex Numbers

Division of complex variables also is most easily performed when they are in exponential or polar form.

In exponential form, the division is

$$\frac{z_1}{z_2} = \frac{r_1 e^{i\theta_1}}{r_2 e^{i\theta_2}}$$
$$= \frac{r_1}{r_2} e^{i(\theta_1 - \theta_2)} \qquad (22.17)$$

We divide the magnitudes and subtract the polar angles to find the quotient. This result also means that division in polar form is quite straightforward:

$$\frac{z_1}{z_2} = \frac{r_1 \angle\theta_1}{r_2 \angle\theta_2}$$
$$= \frac{r_1}{r_2} \angle(\theta_1 - \theta_2) \qquad (22.18)$$

Division of complex numbers in rectangular form also can be accomplished by making use of complex conjugates (see the next subsection).

PROBLEM-SOLVING TACTIC

22.4 To divide two complex numbers in rectangular form, it occasionally is easier first to convert them to polar or exponential form, then do the division. The result can then be put back into rectangular form again if desired.

EXAMPLE 22.11

Divide the two complex numbers

$$5.00 \angle\left(\frac{\pi}{4} \text{ rad}\right)$$

and

$$2.00 \angle\left(\frac{\pi}{6} \text{ rad}\right)$$

Solution
You divide the magnitudes and subtract the polar angles to obtain

$$\frac{5.00}{2.00}\angle\left(\frac{\pi}{4}\text{ rad}-\frac{\pi}{6}\text{ rad}\right)=2.50\angle\left(\frac{\pi}{12}\text{ rad}\right)$$

EXAMPLE 22.12

Divide 3.00 + i(4.00) by 2.00 − i(3.00).

Solution
First, convert each complex number to polar form:

$$3.00+i(4.00)=5.00\angle(0.927\text{ rad})$$
$$2.00-i(3.00)=3.61\angle(-0.983\text{ rad})$$

Perform the division with their polar forms:

$$\frac{5.00}{3.61}\angle\left[0.927\text{ rad}-(-0.983\text{ rad})\right]=1.39\angle(1.910\text{ rad})$$

This can be put back into rectangular form if needed:

$$\begin{aligned}1.39\angle(1.910\text{ rad}) &= 1.39\cos(1.910\text{ rad})\\ &\quad + i[1.39\sin(1.910\text{ rad})]\\ &= -0.463+i(1.31)\end{aligned}$$

Complex Conjugation

The **complex conjugate** of a complex variable z is often useful. The complex conjugate of z is designated as z^*. The complex conjugate is found by replacing i with $-i$ wherever it is found in z. Thus the complex conjugate of

$$z = x + iy$$

is

$$z^* = x - iy$$

Notice that the complex conjugate of the complex conjugate is the original complex number:

$$\begin{aligned}(z^*)^* &= (x-iy)^*\\ &= x-(-i)y\\ &= x+iy\\ &= z\end{aligned}$$

In exponential form, if $z = re^{i\theta}$, then its complex conjugate is

$$z^* = re^{-i\theta}$$

This implies that in polar form if

$$z = r\angle\theta$$

then the complex conjugate is

$$z^* = r\angle(-\theta)$$

Notice that the product of a complex variable and its complex conjugate gives the square of the magnitude of the complex variable:

$$\begin{aligned}zz^* &= re^{i\theta}re^{-i\theta}\\ &= r^2e^0\\ &= r^2\end{aligned}$$

QUESTION 2

For a complex number z plotted on the complex plane, where is its complex conjugate z^* located?

EXAMPLE 22.13

Find the complex conjugate of $z = 5 + i(6)$.

Solution
Replace i with $-i$:

$$z^* = 5 - i(6)$$

EXAMPLE 22.14

Find the complex conjugate of $z = 2 - i(3)$.

Solution
Replace i with $-i$ to obtain

$$\begin{aligned}z^* &= 2-(-i)3\\ &= 2+i(3)\end{aligned}$$

EXAMPLE 22.15

Find the complex conjugate of $z = 5e^{i(\pi/6\text{ rad})}$.

Solution
Replace i with $-i$ to obtain

$$z^* = 5e^{-i(\pi/6\text{ rad})}$$

EXAMPLE 22.16

Find the complex conjugate of $z = 2e^{-i(\pi/4\text{ rad})}$.

Solution
Replace i with $-i$ to obtain

$$z^* = 2e^{i(\pi/4\text{ rad})}$$

EXAMPLE 22.17

Find the complex conjugate of

$$z = 4\angle\left(\frac{\pi}{3}\text{ rad}\right)$$

Solution
Change the sign of the polar angle to obtain

$$z^* = 4\angle\left(-\frac{\pi}{3}\text{ rad}\right)$$

Rationalization

We all love to rationalize. But rationalization means something special in the mathematics of complex variables, illustrating yet again how common words often are used differently in technical contexts. **Rationalization** is a procedure to collect in the numerator all the i's in a complex-valued expression. Rationalization also is a way that the division of two complex numbers expressed in rectangular form can be performed directly.

The process is best examined with an example. We consider the following quotient of two complex numbers in rectangular form:

$$\frac{2.00 + i(3.00)}{3.00 - i(4.00)}$$

We want to get all the i's into the numerator. This is accomplished by multiplying *both* the numerator and denominator by the *complex conjugate of the denominator*; this procedure multiplies the original expression by unity (in a very special form) and leaves the value of the original quotient unchanged.

The complex conjugate of the denominator here is $3.00 + i(4.00)$. So we multiply the expression by

$$\frac{3.00 + i(4.00)}{3.00 + i(4.00)}$$

In this way, we obtain a real denominator and a different complex numerator:

$$\frac{2.00 + i(3.00)}{3.00 - i(4.00)}\left[\frac{3.00 + i(4.00)}{3.00 + i(4.00)}\right]$$

Performing the multiplications (in rectangular form), we get

$$\begin{aligned}\frac{6.00 - 12.00 + i(8.00 + 9.00)}{9.00 + i(12.0) - i(12.0) + 16.0} &= \frac{-6.00 + i(17.00)}{25.0}\\ &= \frac{-6.00}{25.0} + i\left(\frac{17.00}{25.0}\right)\\ &= -0.240 + i(0.680)\end{aligned}$$

This puts the original expression into the standard rectangular form for a complex number. The real part of

$$\frac{2.00 + i(3.00)}{3.00 - i(4.00)}$$

thus is -0.240 and the imaginary part is 0.680.

Another way to rationalize is to put both the numerator and denominator of the complex fraction into polar form and perform the division. So, for example, to rationalize

$$\frac{2.00 + i(3.00)}{3.00 - i(4.00)}$$

we express the numerator in polar form:

$$\begin{aligned}2.00 + i(3.00) &= 3.61\,\angle\tan^{-1}\left(\frac{3.00}{2.00}\right)\\ &= 3.61\,\angle(0.983\ \text{rad})\end{aligned}$$

and the denominator in polar form:

$$\begin{aligned}3.00 - i(4.00) &= 5.00\,\angle\tan^{-1}\left(\frac{-4.00}{3.00}\right)\\ &= 5.00\,\angle(-0.927\ \text{rad})\end{aligned}$$

Thus

$$\frac{2.00 + i(3.00)}{3.00 - i(4.00)} = \frac{3.61\,\angle(0.983\ \text{rad})}{5.00\,\angle(-0.927\ \text{rad})}$$

Performing the division in polar form, we obtain

$$\frac{3.61}{5.00}\angle[0.983\ \text{rad} - (-0.927\ \text{rad})] = 0.722\,\angle(1.910\ \text{rad})$$

The result can then be expressed in rectangular form:

$$\begin{aligned}0.722\,\angle(1.910\ \text{rad}) &= 0.722\cos(1.910\ \text{rad})\\ &\quad + i[0.722\sin(1.910\ \text{rad})]\\ &= -0.240 + i(0.681)\end{aligned}$$

This differs slightly from the previous result because of rounding errors.

22.3 Complex Potential Differences and Currents: Phasors

With a background in complex variables now under your belt, we now see how useful they are for analyzing sinusoidal ac circuits. We first encountered oscillatory behavior in Chapter 7 when we examined the dynamics of simple harmonic oscillation. As we chose to there, here we also represent all sinusoidal oscillations using cosine functions. Thus sinusoidally oscillating potential differences and currents will be written as

$$V(t) = V_0\cos(\omega t + \theta) \qquad \text{and} \qquad I(t) = I_0\cos(\omega t + \phi)$$

Since the product ωt is in *radians*, the **phase angles** θ and ϕ also should be expressed in radians in the arguments of the cosines.* Recall that any sine function can be represented as a cosine function by subtracting $\pi/2$ rad from the argument of the sine function, an identity Example 22.18 will use.

You likely recall that working with trigonometric functions necessitates intimate knowledge of a raft of trigonometric identities that you had the pleasure (?) of proving during your trig course in high school. Working with trigonometric functions is cumbersome from an algebraic viewpoint. Not impossible, mind you, but awkward.

*Electrical engineers sometimes express the phase angle in degrees, while keeping the product ωt in radians. To avoid potential confusion, we choose not to follow that convention and always express both ωt *and* the phase angle in radians.

We express all oscillating potential differences or currents in the form of cosine functions to exploit the Euler identity, Equation 22.10, to *invent* complex potential differences and currents whose *real parts* are the actual potential differences and currents.

That is, for a real oscillating potential difference

$$V(t) = V_0 \cos(\omega t + \theta)$$

we construct a *complex potential difference*

$$\begin{aligned} \mathbf{V}(t) &\equiv V_0 e^{i(\omega t + \theta)} \\ &= V_0 \cos(\omega t + \theta) + iV_0 \sin(\omega t + \theta) \end{aligned} \qquad (22.19)$$

Notice that only the *real part* of the complex potential difference $\mathbf{V}(t)$ is the actual (i.e., real) oscillating potential difference. The imaginary part of the complex potential difference just comes along for a mathematical ride. But by bringing the imaginary part along, we will see that trigonometric identities can be avoided in favor of exponentials. We will shortly see that the calculations in ac circuits then are no more difficult conceptually that those for dc circuits. This was the great contribution of Steinmetz to the simplification of ac circuit analysis. The beauty and utility of suitable notation is quite remarkable.

For a real oscillating current

$$I(t) = I_0 \cos(\omega t + \phi)$$

we likewise invent a *complex current*

$$\begin{aligned} \mathbf{I}(t) &\equiv I_0 e^{i(\omega t + \phi)} \\ &= I_0 \cos(\omega t + \phi) + iI_0 \sin(\omega t + \phi) \end{aligned} \qquad (22.20)$$

The real part of the complex current is the actual oscillating current. The imaginary part of the complex current comes along for the ride.

We do this because the complex potential differences and currents are much easier to work with than trigonometric functions. Exponential functions are easier to multiply, divide, and even differentiate than trigonometric functions. More significantly, we will see that ac circuit equations deduced from the Kirchhoff voltage law (KVL) and the Kirchhoff current law (KCL) turn out to be *algebraic* equations for the complex potential differences and currents. On the other hand, if only the real sinusoidal voltages and currents are used, the circuit equations turn out to involve both integrals and derivatives: integro-differential equations—a horrible mathematical quagmire. Algebraic equations are much easier to solve!

The complex potential difference

$$\mathbf{V}(t) = V_0 e^{i(\omega t + \theta)}$$

has the polar representation

$$\mathbf{V}(t) = V_0 \angle(\omega t + \theta)$$

The complex potential difference $\mathbf{V}(t)$ is called a **potential difference phasor.*** In the complex plane, the complex potential difference phasor is represented by a length V_0 rotating counterclockwise at the angular frequency ω, as shown in Figure 22.5. At any instant, the real part of the complex potential difference is the horizontal projection of V_0:

$$V_0 \cos(\omega t + \theta)$$

which is the real oscillating potential difference.

Likewise a complex current

$$\mathbf{I}(t) = I_0 e^{i(\omega t + \phi)}$$

has a polar representation

$$\mathbf{I}(t) = I_0 \angle(\omega t + \phi)$$

The complex current $\mathbf{I}(t)$ is called a **current phasor**. Just like the complex potential difference, the complex current can be thought of as a length I_0 rotating counterclockwise in the complex plane at the angular frequency ω, as shown in Figure 22.6. The projection of I_0 along the horizontal axis is the real part of $\mathbf{I}(t)$ and is the real current

$$I_0 \cos(\omega t + \phi)$$

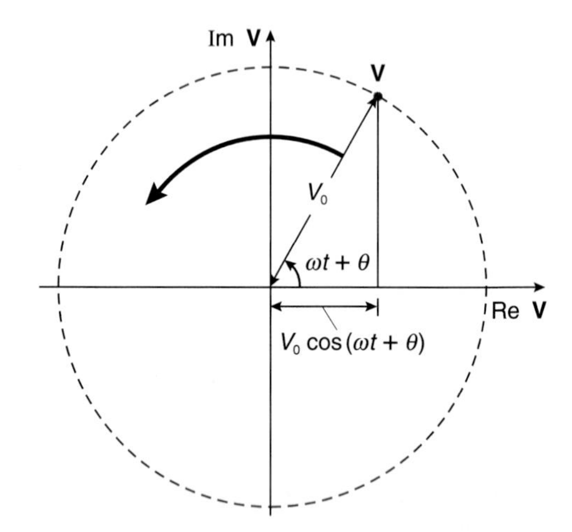

FIGURE 22.5 The potential difference phasor spins counterclockwise at angular frequency ω in the complex plane.

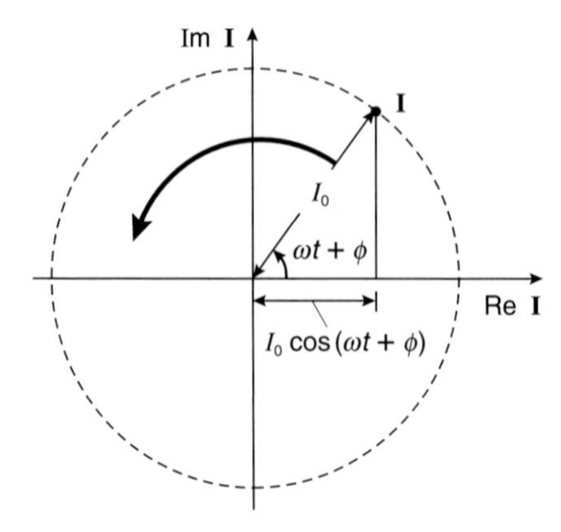

FIGURE 22.6 The current phasor spins counterclockwise at angular frequency ω in the complex plane.

*Sorry, Star Trekkers, electrical engineers long ago invented the term phasor. The Star Trek phasors are figments of Hollywood imaginations (pun intended). Groan.

EXAMPLE 22.18

A potential difference varies sinusoidally as given by the equation $V(t) = V_0 \sin(\omega t + 0.157 \text{ rad})$. Express this in the form

$$V(t) = V_0 \cos(\omega t + \theta)$$

Solution

To convert sines to cosines, take the cosine of the argument of the sine, minus $\pi/2$ rad. That is, take

$$\begin{aligned} V(t) &= V_0 \cos\left[(\omega t + 0.157 \text{ rad}) - \left(\frac{\pi}{2} \text{ rad}\right)\right] \qquad (1) \\ &= V_0 \cos(\omega t - 1.414 \text{ rad}) \end{aligned}$$

The double angle cosine formula from trigonometry verifies that equation (1) is equivalent to the original expression:

$$\begin{aligned} V(t) &= V_0\left[\cos(\omega t + 0.157 \text{ rad}) \cos\left(\frac{\pi}{2} \text{ rad}\right)\right. \\ &\quad \left. + \sin(\omega t + 0.157 \text{ rad}) \sin\left(\frac{\pi}{2} \text{ rad}\right)\right] \\ &= V_0 \sin(\omega t + 0.157 \text{ rad}) \end{aligned}$$

TABLE 22.1 The Real Current and Real Potential Difference Relationships

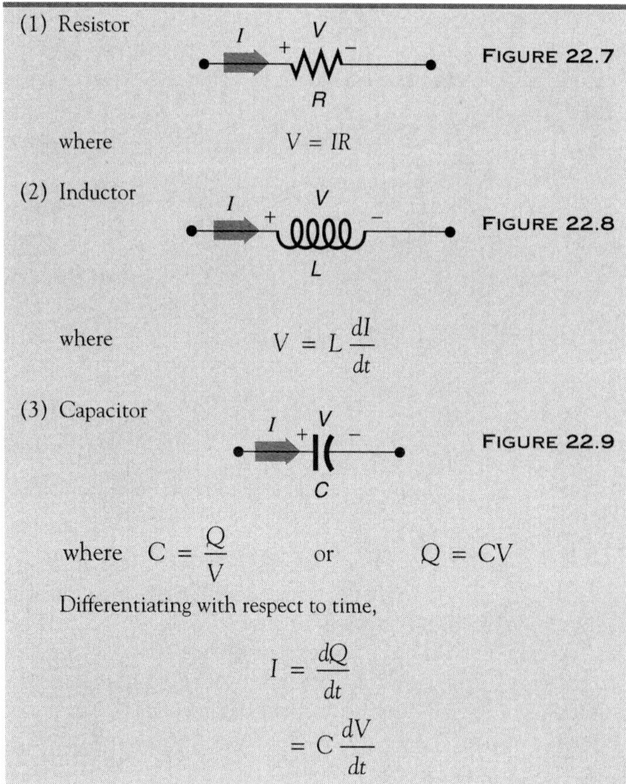
(1) Resistor

FIGURE 22.7

where $V = IR$

(2) Inductor

FIGURE 22.8

where $V = L\dfrac{dI}{dt}$

(3) Capacitor

FIGURE 22.9

where $C = \dfrac{Q}{V}$ or $Q = CV$

Differentiating with respect to time,

$$\begin{aligned} I &= \frac{dQ}{dt} \\ &= C\frac{dV}{dt} \end{aligned}$$

22.4 THE POTENTIAL DIFFERENCE AND CURRENT PHASORS FOR RESISTORS, INDUCTORS, AND CAPACITORS

In dc circuits and in the transient analysis we investigated in Chapters 19 and 21, the relationship between the (real!) current I through a circuit element and the (real) potential difference V across it were as indicated in Table 22.1.

We next want to determine the relationship between the complex current phasor through these circuit elements and the corresponding complex potential difference phasor across each of them.

Resistor

Take the complex potential difference phasor

$$\mathbf{V}(t) = Ve^{i(\omega t + \theta)} \qquad (22.21)$$

and a complex current phasor

$$\mathbf{I}(t) = Ie^{i(\omega t + \phi)} \qquad (22.22)$$

and substitute them into Ohm's law:

$$\mathbf{V}(t) = R\mathbf{I}(t) \qquad (22.23)$$

In polar form this is

$$V \angle(\omega t + \theta) = RI \angle(\omega t + \phi)$$

Since we have $V = IR$ in step at all times, the phase angle θ of the potential difference must be the same as that of the current, ϕ. We say the current and the potential difference are **in phase** with each other since $\phi = \theta$. In the complex plane, the current and potential difference phasors rotate *together* (the two phasors are on top of each other at all times, as shown in Figure 22.10).

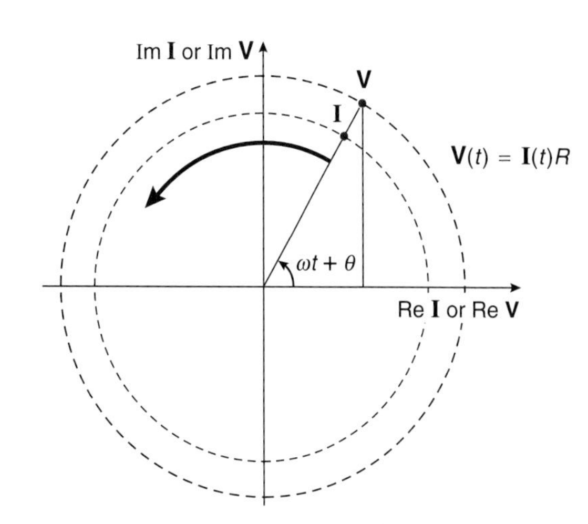

FIGURE 22.10 **For a resistor, the potential difference and current phasors are in phase and rotate together in the complex plane.**

The phasor magnitudes are related by

$$V = IR$$

In other words, Ohm's law expresses for a resistor both the relationship between the current and potential difference phasors and also their magnitudes. Think of the relationship between the current and potential difference phasors through a resistor in the way shown in Figure 22.11, where

$$\mathbf{V}(t) = \mathbf{I}(t)R \qquad (22.24)$$

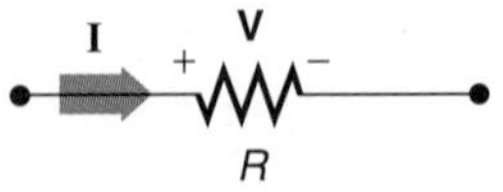

FIGURE 22.11 The potential difference and current phasors for a resistor.

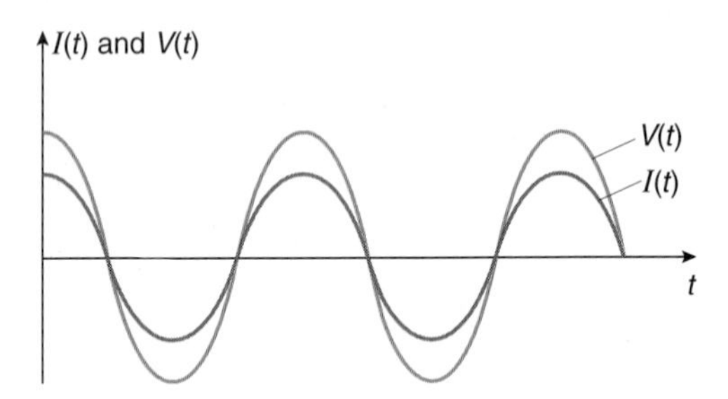

FIGURE 22.12 The real current in a resistor and the real potential difference across it reach their peak values simultaneously.

The real parts (and the imaginary parts) of the potential difference and current phasors have the same phase angle since the real (actual) potential difference across the resistor and the real (actual) current through it are in phase, reaching their respective peak values at the same instant, as indicated in Figure 22.12.

Inductor

Let the complex current

$$\mathbf{I}(t) = Ie^{i(\omega t + \phi)}$$

pass through an inductor. The potential difference across the inductor is related to the current through it (see Table 22.1) by:

$$V(t) = L\frac{dI(t)}{dt} \tag{22.25}$$

Now we can begin to see the advantage of using the complex quantities. We substitute the complex current and voltage phasors and something remarkable happens. Thus

$$\begin{aligned}\mathbf{V}(t) &= L\frac{d\mathbf{I}(t)}{dt}\\ &= L\frac{dIe^{i(\omega t+\phi)}}{dt}\\ &= Li\omega Ie^{i(\omega t+\phi)}\\ &= i\omega L\mathbf{I}(t) \qquad (22.26)\end{aligned}$$

What is remarkable here is that this is an *algebraic* relationship between the current and potential difference phasors, thanks to the unique properties of the exponential. The algebraic relationship between the phasors is precisely the motivation for introducing the complex currents and potential differences. If we use the real current and potential difference, the relationship between them in an inductor involves a derivative (Equation 22.25).

Equation 22.26 has some important ramifications. We write the current phasor in polar form:

$$\mathbf{I}(t) = I\angle(\omega t + \phi)$$

and the potential difference phasor in polar form:

$$\mathbf{V}(t) = V\angle(\omega t + \theta)$$

Also we convert the complex number i to polar form (see Example 22.5):

$$i = 1\angle\left(\frac{\pi}{2}\text{ rad}\right)$$

Substituting these forms into Equation 22.26, we find

$$V\angle(\omega t + \theta) = \left[1\angle\left(\frac{\pi}{2}\text{ rad}\right)\right]\left[\omega LI\angle(\omega t + \phi)\right]$$

The product of the two complex numbers on the right-hand side means their phase angles add; hence

$$V\angle(\omega t + \theta) = \omega LI\angle\left(\omega t + \phi + \frac{\pi}{2}\text{ rad}\right)$$

From this we can conclude that the phase angles for the current and potential difference are *not* the same in an inductor but differ by $\pi/2$ rad (= 90°):

$$\theta = \phi + \frac{\pi}{2}\text{ rad}$$

Therefore the location of the potential difference phasor in the complex plane is $\pi/2$ rad (= 90°) *ahead* of the current phasor; we say ahead, since the phase angles increase in a counterclockwise sense in the complex plane. Another way of expressing the same thing is to say the current phasor *lags* the potential difference phasor by $\pi/2$ rad in an inductor. As the two phasors rotate in the complex plane, they maintain this $\pi/2$ rad separation; see Figure 22.13.

The magnitudes of the potential difference and current phasors are related by

$$V = \omega LI$$

but the phasors always are at right angles to each other in the complex plane.

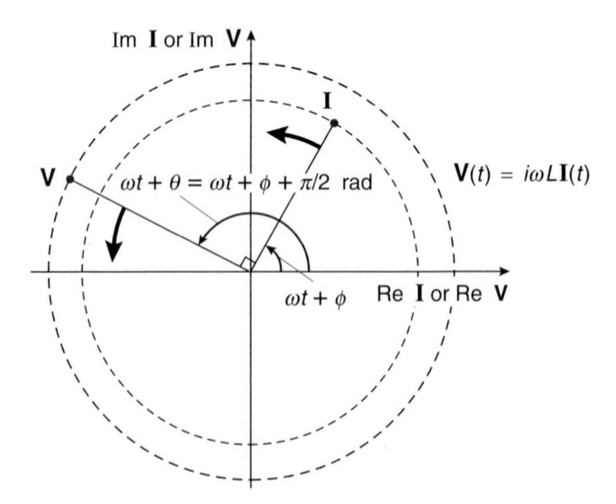

FIGURE 22.13 The potential difference phasor always is $\pi/2$ rad ahead of the current phasor for an inductor as they both rotate in the complex plane.

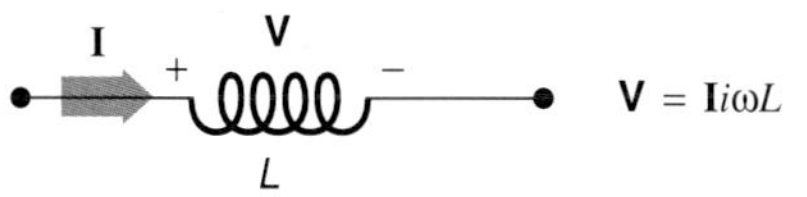

I V + − L V = I*i*ω*L*

FIGURE 22.14 The potential difference and current phasors for an inductor.

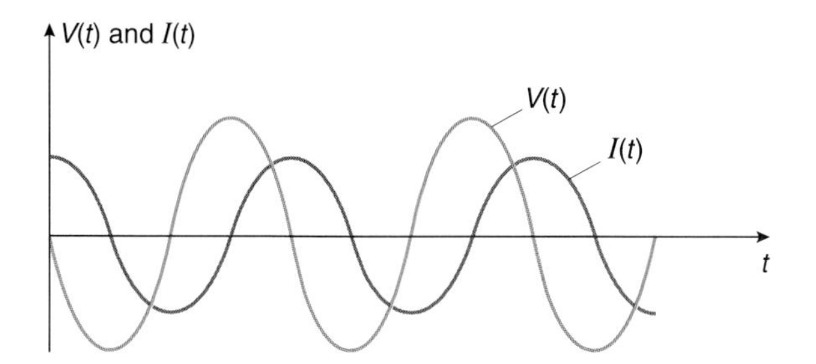

FIGURE 22.15 In an inductor, the real potential difference reaches its peak value ahead of (before) the real current.

Hence for an inductor in an ac circuit, the relationship between the current and potential difference phasors is as indicated in Figure 22.14 and Equation 22.26:

$$\mathbf{V} = \mathbf{I}i\omega L$$

The real parts of the potential difference and current phasors, which represent the real (actual) potential difference across the inductor and the real (actual) current through it, also have a phase difference of $\pi/2$ rad. The potential difference reaches its peak value before (i.e., ahead of) the current, as shown in Figure 22.15.

Capacitor

Let the complex potential difference across a capacitor be

$$\mathbf{V}(t) = Ve^{i(\omega t + \theta)}$$

According to Table 22.1, the current through the capacitor is

$$I(t) = C\frac{dV(t)}{dt}$$

We use the complex potential difference and current phasors and perform the differentiation; we find

$$\begin{aligned}\mathbf{I}(t) &= C\frac{d\mathbf{V}}{dt}\\ &= C\frac{dVe^{i(\omega t+\theta)}}{dt}\\ &= CVi\omega e^{i(\omega t+\theta)}\\ &= i\omega C\mathbf{V}(t) \qquad (22.27)\end{aligned}$$

Once again, notice that the relationship between the current and potential difference phasors is algebraic rather than the differential relationship between the real current and potential difference across the capacitor.

We write the potential difference phasor in polar form:

$$\mathbf{V}(t) = V\angle(\omega t + \theta)$$

the current phasor in polar form:

$$\mathbf{I}(t) = I\angle(\omega t + \phi)$$

and the complex number i in polar form:

$$i = 1\angle\left(\frac{\pi}{2}\text{ rad}\right)$$

Substituting these forms into Equation 22.27, we find

$$I\angle(\omega t + \phi) = \left[1\angle\left(\frac{\pi}{2}\text{ rad}\right)\right]\left[\omega CV\angle(\omega t + \theta)\right]$$

The product of the complex numbers in polar form on the right-hand side means the polar angles are added, yielding

$$I\angle(\omega t + \phi) = \omega CV\angle\left(\omega t + \theta + \frac{\pi}{2}\text{ rad}\right)$$

We see the phase angles are connected by the following relation:

$$\phi = \theta + \frac{\pi}{2}\text{ rad}$$

The current phasor in a capacitor is $\pi/2$ rad ahead of the potential difference phasor in the complex plane. Equivalently, the potential difference phasor lags the current phasor by $\pi/2$ rad, as shown in Figure 22.16.

The phase relationship between the potential difference and current phasors in a capacitor is just the reverse of that in an inductor. As the phasors rotate in the complex plane, they maintain the $\pi/2$ rad separation in angle.

We solve Equation 22.27 for $\mathbf{V}(t)$ so it has the same form as for the other circuit elements:

$$\mathbf{V}(t) = \frac{\mathbf{I}(t)}{i\omega C} \qquad (22.28)$$

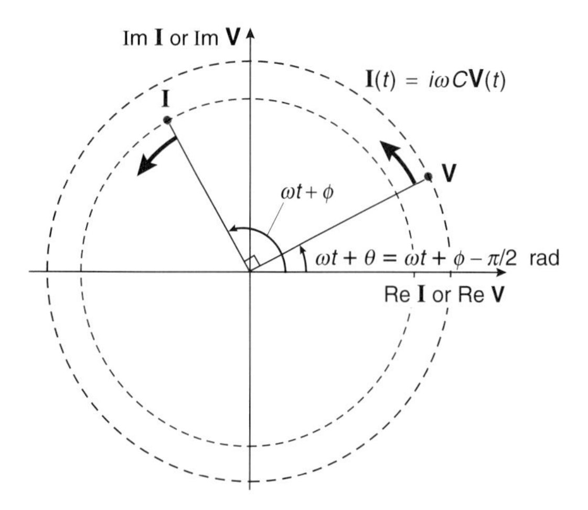

FIGURE 22.16 The current phasor always is $\pi/2$ rad ahead of the potential difference phasor in a capacitor as they both rotate in the complex plane.

Thus, for a capacitor, the relationship between the current and potential difference phasors is as indicated in Figure 22.17, where

$$\mathbf{V} = \mathbf{I}\frac{1}{i\omega C}$$

The real parts of the potential difference and current phasors also have a $\pi/2$ rad difference in phase. Hence the real (actual) current through the capacitor and the real (actual) potential difference across it differ in phase by $\pi/2$ rad, with the current reaching its peak value ahead of (i.e., before) the potential difference, as shown in Figure 22.18.

For each circuit element the relationship between the phasors is algebraic and has the general form:

$$\boxed{\mathbf{V} = \mathbf{I}Z} \qquad (22.29)$$

where for a

$$\text{resistor} \quad Z_R = R \qquad (22.30)$$

$$\text{inductor} \quad Z_L = i\omega L \qquad (22.31)$$

$$\text{capacitor} \quad Z_C = \frac{1}{i\omega C} \qquad (22.32)$$

Equation 22.29 is called the **ac version of Ohm's law**.

Equations 22.29–22.32 make the relationship between the phasors **V** and **I** look like Ohm's law for each of the three circuit elements. This is wonderful!

The quantity Z is called the **impedance** of the circuit element. Using SI units, the impedance is expressed in ohms (Ω). For a resistor, the impedance is just its resistance. For capacitors and inductors, the impedance is a (purely imaginary) complex number that depends on the angular frequency ω.

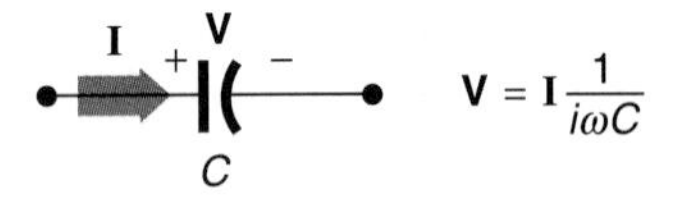

$$\mathbf{V} = \mathbf{I}\frac{1}{i\omega C}$$

FIGURE 22.17 The potential difference and current phasors for a capacitor.

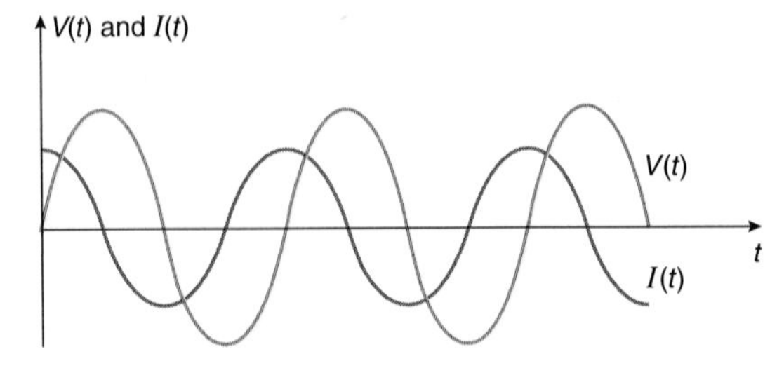

FIGURE 22.18 In a capacitor, the current reaches its peak value ahead of (before) the real potential difference.

The impedance of an inductor

$$Z_L = i\omega L$$

occasionally is written as

$$Z_L = iX_L \qquad (22.33)$$

where $X_L \equiv \omega L$ and is called the **inductive reactance**. The inductive reactance is a *real* number, measured in ohms.

The impedance of a capacitor

$$Z_C = \frac{1}{i\omega C}$$

is rationalized by multiplying the numerator and denominator by $-i$ to obtain

$$Z_C = \frac{1}{i\omega C}\left(\frac{-i}{-i}\right)$$

$$= -\frac{i}{\omega C}$$

This is rewritten as

$$Z_C = -iX_C \qquad (22.34)$$

where

$$X_C \equiv \frac{1}{\omega C}$$

X_C is known as the **capacitive reactance**.

A resistor has an impedance that is a *real* number; for capacitors and inductors, the impedance is purely *imaginary*. For combinations of these circuit elements, the equivalent impedance, in general, is a *complex number* (with real and imaginary parts).

The impedance is *not* a phasor and does not rotate in the complex plane, since there is no ωt term associated with it; there is an ωt term for the current and potential difference phasors.

Notice that the impedance of a capacitor is *infinite* when $\omega = 0$ rad/s, which is a dc (time-independent) situation. Capacitors, therefore, act like open switches for dc currents, since no dc current can pass through them.

Any branch of a circuit that contains a capacitor has a (steady-state) dc current equal to zero. Capacitors therefore can be used to isolate or confine dc currents to only certain sections of a complicated circuit. This property of a capacitor is used extensively in transistor circuits to bias transistors properly. To **bias** a transistor is to provide a dc potential difference to ensure its proper operation.

For sufficiently large angular frequencies, the impedance of a capacitor *approaches zero*; at these high angular frequencies, then, the capacitor acts like a *short circuit* for ac currents; this characteristic also is used extensively in transistor circuits.

As a function of angular frequency, the impedance of an inductor behaves in just the reverse manner as that of a capacitor. For $\omega = 0$ rad/s (a dc situation), the impedance of an inductor is *zero*, and so the inductor acts like a *short circuit* for dc currents. If the angular frequency is very high, however, the impedance of an inductor becomes very large and it increasingly behaves like an *open switch* for ac currents (i.e., very small ac current exists in the inductor).

EXAMPLE 22.19

Find the impedance and the capacitive reactance of a 10.0 μF capacitor at a frequency of 60.0 Hz.

Solution

The 60.0 Hz frequency corresponds to an angular frequency of $\omega = 2\pi\nu = 377$ rad/s. From Equation 22.32, the impedance of the capacitor is

$$\begin{aligned} Z_C &= \frac{1}{i\omega C} \\ &= \frac{1}{i(377 \text{ rad/s})(10.0 \times 10^{-6} \text{ F})} \\ &= \frac{265\ \Omega}{i} \end{aligned}$$

Rationalizing, you find

$$\begin{aligned} Z_C &= \frac{265\ \Omega}{i}\left(\frac{-i}{-i}\right) \\ &= -i(265\ \Omega) \end{aligned}$$

Notice that the impedance is a *complex number*.

The capacitive reactance X_C is a *real number* defined from Equation 22.34 to be

$$Z_C \equiv -iX_C$$

Comparing this with the impedance just calculated shows that the capacitive reactance is

$$X_C = 265\ \Omega$$

EXAMPLE 22.20

Calculate the impedance and inductive reactance of a 25.0 mH inductor at a frequency of 60.0 Hz.

Solution

From Equation 22.31 the impedance of an inductor is

$$Z_L = i\omega L$$

The angular frequency corresponding to 60.0 Hz is

$$\begin{aligned} \omega &= 2\pi\nu \\ &= 377 \text{ rad/s} \end{aligned}$$

You are given $L = 25.0$ mH; thus the impedence is

$$\begin{aligned} Z_L &= i(377 \text{ rad/s})(25.0 \times 10^{-3} \text{ H}) \\ &= i(9.43\ \Omega) \end{aligned}$$

The inductive reactance X_L is a *real number* defined from Equation 22.33:

$$Z_L = iX_L$$

which means that the inductive reactance here is

$$X_L = 9.43\ \Omega$$

22.5 SERIES AND PARALLEL COMBINATIONS OF IMPEDANCES

The beauty of the complex variable approach to ac circuits is becoming apparent. The relationship between the current phasor through each of the primary circuit elements (resistors, capacitors, and inductors) and the potential difference phasor across them has the form of the ac version of Ohm's law given by Equation 22.29:

$$V = IZ$$

where Z is the impedance of the circuit element. We now exploit this to combine impedances in series and in parallel.

Impedances in Series

Impedances in series combine like resistors did in series.

That is, for the series of impedances shown in Figure 22.19, we can replace the series combination with a single equivalent impedance Z_{eq}:

$$Z_{eq} = Z_1 + Z_2 + Z_3 + \cdots \tag{22.35}$$

Since the impedances are complex numbers, a complex number addition must be performed to find Z_{eq}.

Impedances in Parallel

Impedances in parallel combine like resistors did in parallel.

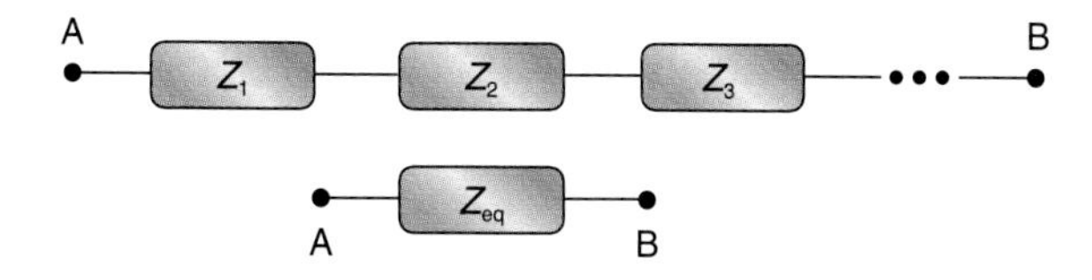

FIGURE 22.19 Impedances in series and their equivalent.

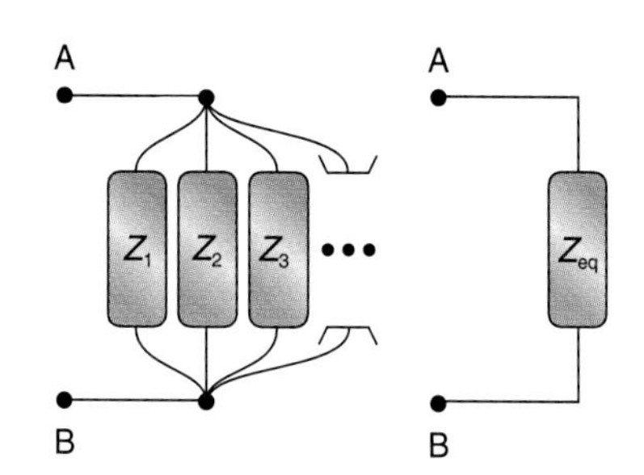

FIGURE 22.20 Impedances in parallel and their equivalent.

For the parallel combination of impedances shown in Figure 22.20, we can replace the collection with an equivalent impedance Z_{eq} found from

$$\frac{1}{Z_{eq}} = \frac{1}{Z_1} + \frac{1}{Z_2} + \frac{1}{Z_3} + \cdots \qquad (22.36)$$

Once again, these are complex number arithmetic operations.

PROBLEM-SOLVING TACTIC

22.5 For two impedances in parallel, you can make use of the shortcut we found for resistors in parallel: the equivalent impedance is the product of the two impedances divided by their sum:

$$Z_{eq} = \frac{Z_1 Z_2}{Z_1 + Z_2} \qquad (22.37)$$

Remember this rule is valid *only for two* parallel impedances at a time.

EXAMPLE 22.21

A circuit is a series combination of a 1.000 kΩ resistor, a 10.0 μF capacitor, a 0.100 H inductor, and a 60.0 Hz ac independent voltage source. Find the equivalent impedance of the series combination of the resistor, capacitor, and inductor.

Solution

First find the impedances of the individual circuit elements. The angular frequency is $\omega = 2\pi\nu = 377$ rad/s. The impedance of the resistor is

$$Z_R = R = 1.000\ \text{k}\Omega$$

The impedance of the capacitor is

$$\begin{aligned} Z_C &= \frac{1}{i\omega C} \\ &= \frac{1}{i(377\ \text{rad/s})(10.0 \times 10^{-6}\ \text{F})} \\ &= \frac{265\ \Omega}{i} \end{aligned}$$

Rationalizing, you find

$$Z_C = -i(265\ \Omega)$$

The impedance of an inductor is

$$\begin{aligned} Z_L &= i\omega L \\ &= i(377\ \text{rad/s})(0.100\ \text{H}) \\ &= i(37.7\ \Omega) \end{aligned}$$

The equivalent impedance of the series combination is the sum

$$\begin{aligned} Z_{eq} &= Z_R + Z_C + Z_L \\ &= 1000\ \Omega - i(265\ \Omega) + i(37.7\ \Omega) \\ &= 1000\ \Omega - i(227\ \Omega) \end{aligned}$$

EXAMPLE 22.22

In a circuit operating at 60.0 Hz, a 10.0 μF capacitor is in parallel with a 0.100 H inductor. Find the equivalent impedance of the parallel combination.

Solution

From Example 22.21, the impedance of these circuit elements at the frequency of 60.0 Hz is

$$Z_C = -i(265\ \Omega)$$

and

$$Z_L = i(37.7\ \Omega)$$

Since the two circuit elements are in parallel, the equivalent impedance is their product divided by their sum (Equation 22.37):

$$\begin{aligned} Z_{eq} &= \frac{Z_C Z_L}{Z_C + Z_L} \\ &= \frac{[-i(265\ \Omega)][i(37.7\ \Omega)]}{-i(265\ \Omega) + i(37.7\ \Omega)} \\ &= -\frac{44.0\ \Omega}{i} \end{aligned}$$

Rationalizing, you obtain

$$\begin{aligned} Z_{eq} &= -\frac{44.0\ \Omega}{i}\left(\frac{-i}{-i}\right) \\ &= i(44.0\ \Omega) \end{aligned}$$

22.6 COMPLEX INDEPENDENT AC VOLTAGE SOURCES

An ideal, ac, sinusoidal, independent voltage source produces a potential difference between its two terminals expressed as

$$V_{source}(t) = V_0 \cos(\omega t) \qquad (22.38)$$

where ω is the angular frequency of the source and V_0 is its *amplitude*. We take the phase angle of the voltage source to be zero.

A *complex ac voltage source*, known as a **voltage source phasor**, has the actual ac voltage source as its real part:

$$\mathbf{V}_{source}(t) = V_0 \cos(\omega t) + iV_0 \sin(\omega t) \qquad (22.39)$$

In exponential form, the complex source is

$$\mathbf{V}_{source}(t) = V_0 e^{i(\omega t)} \qquad (22.40)$$

and in polar form, the source is

$$\mathbf{V}_{source}(t) = V_0 \angle(\omega t) \qquad (22.41)$$

An ac source is symbolized in circuit diagrams as in Figure 22.21.

The terminals nominally have polarity markings (+) and (−) and indicate the polarity of the potential difference between the two terminals when $t = 0$ s.

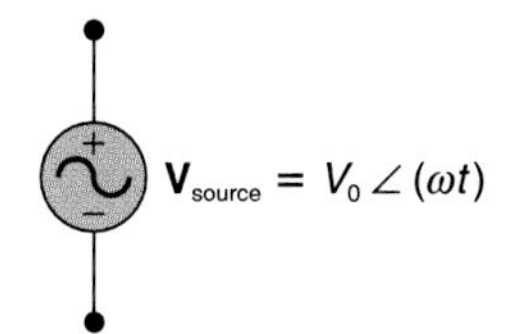

FIGURE 22.21 An ac voltage source phasor symbol.

22.7 Power Absorbed by Circuit Elements in ac Circuits

As we have seen, power is an important electrical concept. The instantaneous power $P(t)$ absorbed by (transferred to) a circuit element is the product of the potential difference $V(t)$ across it and the current $I(t)$ into its (nominally) positive terminal:

$$P(t) = V(t)I(t) \quad (22.42)$$

If the potential difference and current are oscillating sinusoidally with time, we have

$$V(t) = V_0 \cos(\omega t + \theta) \quad (22.43)$$

$$I(t) = I_0 \cos(\omega t + \phi) \quad (22.44)$$

where V_0 and I_0 are the peak values of the potential difference across and the current through the circuit element. Substituting these quantities into Equation 22.42, we find the instantaneous power is

$$P(t) = V_0 I_0 \cos(\omega t + \theta)\cos(\omega t + \phi) \quad (22.45)$$

In many cases, we are not interested so much in the instantaneous power absorbed as in the *average* power absorbed over a complete cycle (or many complete cycles) of the oscillation. To find the average power absorbed, we resurrect a trigonometric identity for the product of the cosines of two angles:

$$\cos\alpha\cos\beta = \frac{1}{2}\left[\cos(\alpha + \beta) + \cos(\alpha - \beta)\right] \quad (22.46)$$

Applying this identity to the product of the two cosines in Equation 22.45, we obtain

$$\begin{aligned} P(t) &= V_0 I_0 \frac{1}{2}\Big[\cos(\omega t + \theta + \omega t + \phi) \\ &\qquad + \cos(\omega t + \theta - \omega t - \phi)\Big] \\ &= V_0 I_0 \frac{1}{2}\left[\cos(2\omega t + \theta + \phi) + \cos(\theta - \phi)\right] \quad (22.47) \end{aligned}$$

The first term in the brackets in Equation 22.47 is time dependent; the second term is independent of time. Since the first term oscillates with time as a cosine, its time-averaged value over a period (or multiple periods) is zero, as shown in Figure 22.22.

The second term in Equation 22.47, while also a cosine, is independent of time and does not oscillate. Thus the average power $\langle P\rangle$ absorbed by the circuit element is

$$\langle P\rangle = \frac{1}{2} V_0 I_0 \cos(\theta - \phi) \quad (22.48)$$

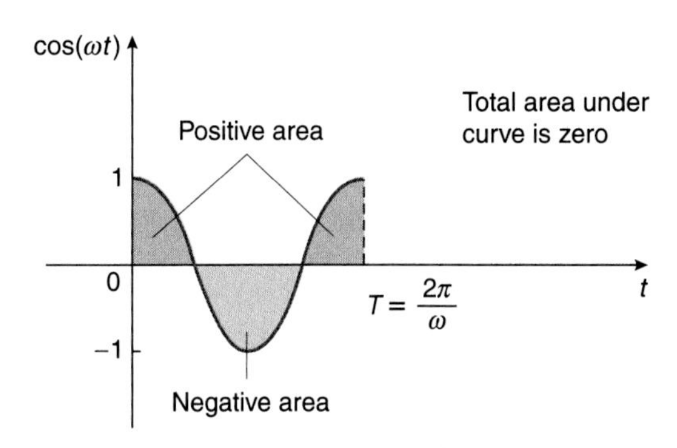

FIGURE 22.22 The time-averaged value of an oscillating cosine (or sine) function over a period (or multiple periods) is zero.

The cosine term here involves the phase difference $\beta \equiv \theta - \phi$ between the potential difference across the circuit element and the current through it. The cosine of this phase difference is called the **power factor:**

$$\cos(\theta - \phi) \equiv \cos\beta \equiv \text{power factor} \quad (22.49)$$

Hence the average power absorbed by a circuit element is

$$\boxed{\langle P\rangle = \frac{1}{2} V_0 I_0 \cos\beta} \quad (22.50)$$

The peak values are the magnitudes of the potential difference and current phasors.

The average power absorbed by a circuit element also can be expressed in terms of the potential difference and current phasors associated with the circuit element in the complex plane. Consider the expression

$$\frac{1}{2}\operatorname{Re}(\mathbf{V}\mathbf{I}^*) \quad (22.51)$$

The potential difference phasor in exponential form is

$$\mathbf{V} = V_0 e^{i(\omega t + \theta)}$$

and the current phasor is

$$\mathbf{I} = I_0 e^{i(\omega t + \phi)}$$

The complex conjugate of the current phasor then is

$$\mathbf{I}^* = I_0 e^{-i(\omega t + \phi)}$$

Substituting into Equation 22.51, we find

$$\frac{1}{2}\operatorname{Re}\left[V_0 e^{i(\omega t+\theta)} I_0 e^{-i(\omega t+\phi)}\right] = \frac{1}{2}\operatorname{Re}\left[V_0 I_0 e^{i(\theta-\phi)}\right]$$

We take the real part of the complex exponential:

$$\frac{1}{2} V_0 I_0 \cos(\theta - \phi) \equiv \frac{1}{2} V_0 I_0 \cos\beta$$

which is identical to Equation 22.50 for the average power absorbed by the circuit element. Hence an alternative expression for the average power absorbed by a circuit element is

$$\langle P\rangle = \frac{1}{2}\operatorname{Re}(\mathbf{V}\mathbf{I}^*) \quad (22.52)$$

Frequently, we rearrange Equation 22.50 for the average power:

$$\langle P \rangle = \frac{V_0}{\sqrt{2}} \frac{I_0}{\sqrt{2}} \cos \beta \tag{22.53}$$

The peak values divided by $\sqrt{2}$ are known as the **effective values** of the potential difference and current. They also are known as **rms values** (for *r*oot *m*ean *s*quare values). That is,

$$V_{rms} = \frac{V_0}{\sqrt{2}} \tag{22.54}$$

$$I_{rms} = \frac{I_0}{\sqrt{2}} \tag{22.55}$$

The average power then is

$$\langle P \rangle = V_{rms} I_{rms} \cos \beta \tag{22.56}$$

The oscillatory independent voltage source at most common electrical outlets in the United States is 120 V; this is the rms value V_{rms}. The peak value V_0 is greater by the factor of $\sqrt{2}$.*

On a practical note, multimeters that measure ac potential differences and/or ac currents read the rms values of these quantities (V_{rms} and/or I_{rms}), *not* the peak values (V_0 and/or I_0).

For resistors, the phase difference β between the potential difference and the current is 0 rad. Thus the power factor is

$$\cos \beta = 1$$

and the average power absorbed by a resistor is

$$\langle P \rangle_{res} = \frac{1}{2} V_0 I_0 = V_{rms} I_{rms} \tag{22.57}$$

For capacitors, the phase difference β between the potential difference and the current is always $-\pi/2$ rad. The power factor is

$$\cos\left(-\frac{\pi}{2} \text{ rad}\right) = 0$$

and the average power absorbed by the ideal capacitor is zero:

$$\langle P \rangle_{cap} = 0 \text{ W}$$

The *instantaneous* power is *not* zero at every instant but the *average* power absorbed *is* zero. This simply means that the capacitor absorbs positive power over half a cycle and negative power over the other half, and the average over each period is zero.

For inductors, the phase difference between the potential difference and current in an inductor is $\pi/2$ rad, so that the power factor is

$$\cos\left(\frac{\pi}{2} \text{ rad}\right) = 0$$

and the average power absorbed by an inductor is zero:

$$\langle P \rangle_{ind} = 0 \text{ W}$$

As in capacitors, the instantaneous power in inductors is not zero at every instant but the average power over any period is zero.

These are characteristics of *ideal* circuit elements. *Real* capacitors and inductors are modeled using ideal circuit elements (see Chapter 19 and 21). Real capacitors and inductors always have some resistance, and it is this resistance that absorbs positive electrical power and accounts for the observation that real capacitors and inductors become warm in use.

We now have all the mathematical and conceptual hardware to analyze a plethora of ac circuits. We will examine just a few to become familiar with the techniques. The important thing to realize is that the methods used closely resemble those employed for dc circuits. The complex variable approach to ac circuits makes this possible.

22.8 A Filter Circuit

To illustrate the concepts we have learned about ac circuits, we apply them to a special class of circuits of some technological import. Most stereo receivers permit you to selectively eliminate high or low frequencies and create the musical mix most enjoyable to your ears. There is a wide variety of circuits, called **filter circuits**, that let certain frequencies pass relatively unimpeded and filter out or eliminate one or another range of frequencies.

Here we examine the simplest kind of filter circuit: an ac independent voltage source whose single frequency can be varied, a resistor with resistance R, and a capacitor with capacitance C, all connected in series, as shown in Figure 22.23.

The sinusoidal ac independent voltage source varies as

$$V_0(t) = V_0 \cos(\omega t) \tag{22.58}$$

We want to find the ac current in the circuit and the ac potential differences across both the resistor and the capacitor.

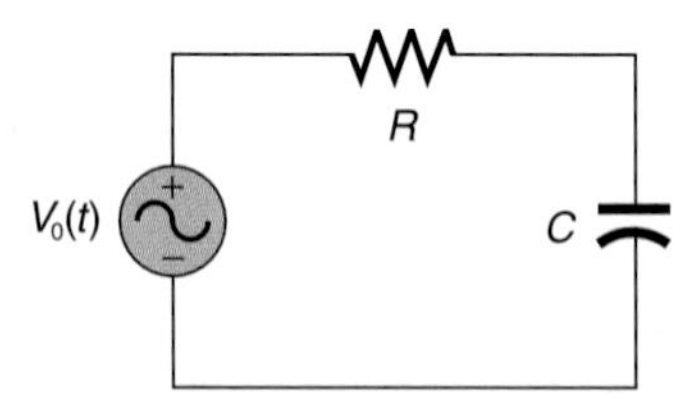

FIGURE 22.23 A series *RC* filter circuit.

*In Europe and other countries the rms voltage source is 220 V, so the peak value is $\sqrt{2}$ (220 V) = 311 V. This is the reason that voltage adapters are needed when using U.S. made appliances in Europe to protect the appliances from too high a potential difference. The adapters are really transformers to step down the potential difference to 120 V rms (or 170 V peak). In Japan the rms voltage is 100 V, less than in the United States. No voltage adapters are needed for U.S. appliances (although the appliances will not operate as effectively because of the reduced voltages and power). The frequency used in Europe and Japan is 50 Hz. In both Europe and Japan, one also needs plug adapters since the topological configuration of electrical plugs varies from country to country.

Finding the Current Phasor

To solve the circuit, we proceed in the following way.

Step 1: We convert the circuit to the complex domain. To accomplish this:

a. We change the independent ac voltage source to a complex voltage source phasor. For a real source voltage $V_0 \cos(\omega t)$, the complex source voltage phasor is

$$\mathbf{V}_{\text{source}} = V_0 \angle(\omega t) \tag{22.59}$$

whose real part is the real voltage source.

b. We indicate the impedances of the resistor and capacitor next to their circuit symbols, as shown in Figure 22.24. The impedance of the resistor is just R; the impedance of the capacitor is $-iX_C$ where X_C is the capacitive reactance

$$X_C = \frac{1}{\omega C}$$

Step 2: We indicate a direction for the current phasor **I**. The choice of this direction is yours to make but we customarily choose it to be directed out of the nominally positive terminal of the ac voltage source, just as for dc circuits. Once the direction for the current phasor is chosen, the nominal polarities of the impedances are determined and can be marked accordingly: the current phasor goes into the (+) polarity terminal of each impedance and leaves at the (−) polarity terminal, again just as it did for dc circuits. The circuit diagram then has the appearance shown in Figure 22.25.

The potential difference phasor across each of the impedances is found from the ac version of Ohm's law, Equation 22.29:

$$\mathbf{V} = \mathbf{I}Z$$

Hence the potential difference phasor across the resistor is

$$\mathbf{V}_R = \mathbf{I}R \tag{22.60}$$

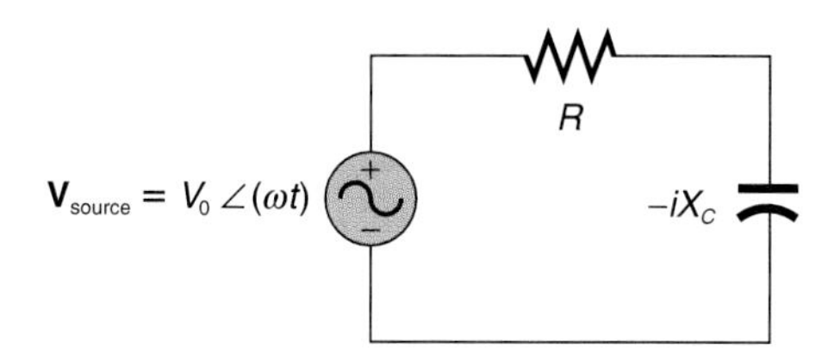

FIGURE 22.24 The circuit in the complex domain.

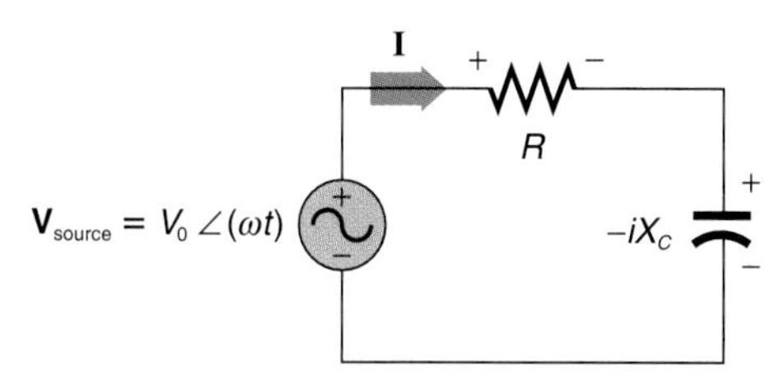

FIGURE 22.25 The phasor current direction is indicated and the polarities of the impedances are marked accordingly.

Across the capacitor, the potential difference phasor is

$$\mathbf{V}_C = \mathbf{I}(-iX_C) \tag{22.61}$$

with the polarities indicated, consistent with the direction chosen for the current phasor.

Step 3: We use the KVL to go around the loop, using the same conventions we established for dc circuits:

- If, in going around the loop, we first encounter a (+) polarity sign for a circuit element, the potential difference phasor for that circuit element is entered into the KVL equation with a plus (+) sign.
- If, in going around the loop, we first encounter a (−) polarity sign for the circuit element, the potential difference phasor for that circuit element is entered into the KVL equation with a minus (−) sign.

As for dc circuits, we can go around the loop in either direction. We go around the loop of Figure 22.25 clockwise and begin in the lower left corner of the loop. Then the KVL here becomes

$$-\mathbf{V}_{\text{source}} + \mathbf{I}R + \mathbf{I}(-iX_C) = 0 \text{ V} \tag{22.62}$$

Step 4: We solve the equation for the current phasor; we use Equation 22.62:

$$\mathbf{I}(R - iX_C) = \mathbf{V}_{\text{source}}$$

The current phasor thus is

$$\mathbf{I} = \frac{\mathbf{V}_{\text{source}}}{R - iX_C} \tag{22.63}$$

We need to put Equation 22.63 into polar form. The denominator is a complex number in rectangular form; we convert the denominator to polar form:

$$R - iX_C = \left[R^2 + (-X_C)^2\right]^{1/2} \angle\tan^{-1}\left(\frac{-X_C}{R}\right) \tag{22.64}$$

Since $\tan(-\alpha) = -\tan\alpha$,

$$\tan^{-1}\left(-\frac{X_C}{R}\right) = -\tan^{-1}\left(\frac{X_C}{R}\right)$$

We define ϕ to be

$$\phi \equiv \tan^{-1}\left(\frac{X_C}{R}\right) \tag{22.65}$$

Thus Equation 22.64 becomes

$$R - iX_C = (R^2 + X_C^2)^{1/2} \angle(-\phi)$$

Substituting this into the expression for the current phasor (Equation 22.63), we obtain

$$\mathbf{I} = \frac{\mathbf{V}_{\text{source}}}{(R^2 + X_C^2)^{1/2} \angle(-\phi)} \tag{22.66}$$

Division by complex numbers in polar form means we can move the $\angle(-\phi)$ in the denominator of Equation 22.66 into the numerator by changing the sign of the angle. Thus

$$\mathbf{I} = \frac{\mathbf{V}_{\text{source}} \angle\phi}{(R^2 + X_C^2)^{1/2}} \tag{22.67}$$

Substituting the source voltage phasor in polar form (Equation 22.59) yields

$$\mathbf{I} = \frac{V_0 \angle(\omega t) \angle\phi}{(R^2 + X_C^2)^{1/2}}$$

When multiplying complex numbers in polar form, the angles add, which means that the current phasor becomes

$$\mathbf{I} = \frac{V_0}{(R^2 + X_C^2)^{1/2}} \angle(\omega t + \phi) \tag{22.68}$$

To find the real current $I(t)$ as a function of time, we take the real part of the current phasor:

$$\begin{aligned} I(t) &= \mathrm{Re}\,\mathbf{I} \\ &= \frac{V_0}{(R^2 + X_C^2)^{1/2}} \cos(\omega t + \phi) \end{aligned} \tag{22.69}$$

The angle ϕ is given by Equation 22.65.*

Now that we know the current, we can find the potential differences across the capacitor and the resistor by using the ac Ohm's law for each impedance.

The Potential Difference Across the Capacitor

The phasor representing the potential difference across the capacitor $\mathbf{V}_C$ is found by taking the product of the impedance of the capacitor with the current phasor through it:

$$\mathbf{V}_C = \mathbf{I}Z_C$$

Using $Z_C = -iX_C$ and Equation 22.68 for the current phasor, we have

$$\mathbf{V}_C = \frac{V_0 \angle(\omega t + \phi)}{(R^2 + X_C^2)^{1/2}} (-iX_C) \tag{22.70}$$

*If the angular frequency of the source is zero, we have a dc situation. When $\omega = 0$ rad, we say we are at the *dc limit*. Since a capacitor acts as an open circuit for dc current, when $\omega = 0$ rad/s, we expect I to be zero. Does Equation 22.69 for $I(t)$ yield 0 A when $\omega = 0$ rad/s? Let's check. First, we have to evaluate the angle ϕ using Equation 22.65:

$$\phi = \tan^{-1}\left(\frac{X_C}{R}\right)$$

The capacitive reactance is

$$X_C = \frac{1}{\omega C}$$

and when $\omega = 0$ rad/s, $X_C = \infty\ \Omega$. Thus ϕ is the angle whose tangent is infinite:

$$\begin{aligned} \phi &= \tan^{-1}(\infty) \\ &= \frac{\pi}{2} \text{ rad} \end{aligned}$$

Equation 22.69 for the real current then is

$$I(t) = \frac{V_0}{(R^2 + X_C^2)^{1/2}} \cos\left(0 \text{ rad} + \frac{\pi}{2} \text{ rad}\right)$$

The cosine term is zero *and* the coefficient of the cosine is zero (since $X_C = \infty\ \Omega$). Thus the current is indeed zero, as it should be for the dc situation.

Since the complex number $-i$ in polar form is

$$1\angle\left(-\frac{\pi}{2} \text{ rad}\right)$$

Equation 22.70 becomes

$$\mathbf{V}_C = \frac{V_0 \angle(\omega t + \phi)}{(R^2 + X_C^2)^{1/2}} \left[1\angle\left(-\frac{\pi}{2} \text{ rad}\right)\right] X_C$$

Once again, the product rule for complex numbers in polar form means the angles in the numerator are added together, yielding

$$\mathbf{V}_C = X_C \frac{V_0 \angle(\omega t + \phi - \pi/2 \text{ rad})}{(R^2 + X_C^2)^{1/2}} \tag{22.71}$$

Notice incidentally that the phase difference between the potential difference phasor and the current phasor is $-\pi/2$ rad (as it always will be for a capacitor).

The real potential difference across the capacitor $V_C(t)$ is the real part of its potential difference phasor $\mathbf{V}_C$:

$$\begin{aligned} V_C(t) &= \mathrm{Re}\,\mathbf{V}_C \\ &= X_C \frac{V_0}{(R^2 + X_C^2)^{1/2}} \cos\left(\omega t + \phi - \frac{\pi}{2} \text{ rad}\right) \end{aligned} \tag{22.72}$$

The actual potential difference across the capacitor (Equation 22.72) and actual current (Equation 22.69) also have a phase difference of $-\pi/2$ rad.

We look at the magnitude of the peak potential difference across the capacitor as a function of angular frequency. We call this $V_{C\,\text{peak}}(\omega)$; this is the coefficient of the cosine in Equation 22.72:

$$V_{C\,\text{peak}}(\omega) = \frac{X_C V_0}{(R^2 + X_C^2)^{1/2}}$$

We can simplify this algebraically by taking the X_C in the numerator and putting it in the denominator as $1/X_C$:

$$V_{C\,\text{peak}}(\omega) = \frac{V_0}{\frac{1}{X_C}(R^2 + X_C^2)^{1/2}}$$

Putting X_C into the square root, we obtain

$$V_{C\,\text{peak}}(\omega) = \frac{V_0}{\left(\frac{R^2}{X_C^2} + 1\right)^{1/2}}$$

Since

$$X_C = \frac{1}{\omega C}$$

the peak potential difference across the capacitor is

$$V_{C\,\text{peak}}(\omega) = \frac{V_0}{\left[1 + (\omega RC)^2\right]^{1/2}} \tag{22.73}$$

This expression shows how the peak value of the potential difference across the capacitor varies with the angular frequency of the source. In particular, we look at two limiting cases.

Case 1: If the angular frequency is zero, then

$$V_{C\,peak} = V_0$$

which is the amplitude of the voltage source. This is just what we would expect, because when $\omega = 0$ rad/s we have a dc situation. Zero current is present in the circuit since a capacitor acts as an open circuit for dc. There then is zero potential difference across the resistor and so the capacitor effectively is connected in parallel to the dc battery (an ac voltage source at zero frequency).

Case 2: For very large values of the angular frequency of the source, Equation 22.73 indicates that the peak value of the potential difference across the capacitor approaches zero.

As the angular frequency increases, the impedance of the capacitor decreases. At very large ω, the capacitor essentially has zero impedance and acts more and more like just a piece of wire (with zero potential difference across it). We say the capacitor shorts out for large ω.

A schematic sketch of the peak potential difference across the capacitor as a function of ω is shown in Figure 22.26.

We consider the voltage source to be an input and the potential difference across the capacitor to be an output, as indicated in Figure 22.27.

Since the peak potential difference across the capacitor (the output voltage) is appreciable only for low (i.e., small) angular frequencies, this output is known as a **low pass filter.**

The term arises because only for *low* angular frequencies is there a significant potential difference passed from the input (the voltage source) to the output (the potential difference across the capacitor).

The ratio of the peak output potential difference to the peak input potential difference is the **voltage gain** of the circuit. Here, the voltage gain is (using Equation 22.73)

$$\text{voltage gain} \equiv \frac{V_{C\,peak}}{V_0} = \frac{1}{\left[1 + (\omega RC)^2\right]^{1/2}} \quad \text{(low pass filter)} \qquad (22.74)$$

At zero angular frequency, the voltage gain is equal to 1; as the frequency increases, the voltage gain falls, approaching 0 at very high frequencies. At very high frequencies, the impedance of the capacitor approaches zero, so that the capacitor acts like a short circuit (a wire) with essentially no potential difference across it.

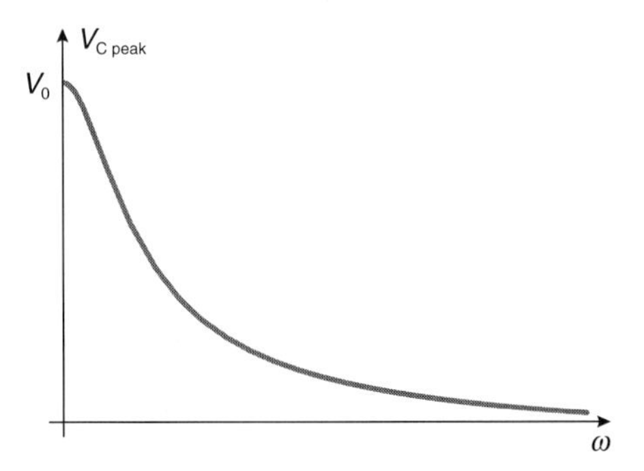

FIGURE 22.26 The peak potential difference across the capacitor as a function of ω.

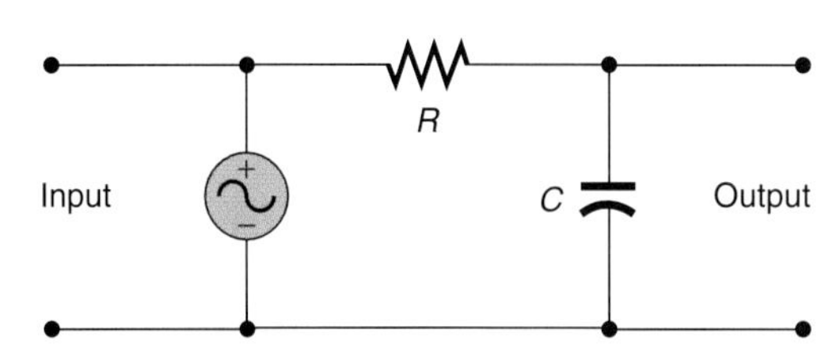

FIGURE 22.27 A low pass filter.

The Potential Difference Across the Resistor

The phasor representing the potential difference across the resistor is the product of the current phasor **I** times the impedance of the resistor:

$$\mathbf{V}_R = \mathbf{I}R$$

The current phasor is the same as that through the capacitor since the two circuit elements are in series. That is, the current phasor is that given by Equation 22.68. Thus

$$\mathbf{V}_R = \frac{V_0 \angle(\omega t + \phi)}{(R^2 + X_C^2)^{1/2}} R = \frac{RV_0 \angle(\omega t + \phi)}{(R^2 + X_C^2)^{1/2}}$$

The real potential difference $V_R(t)$ across the resistor is the real part of this potential difference phasor:

$$V_R(t) = \text{Re}\,\mathbf{V}_R = \frac{RV_0}{(R^2 + X_C^2)^{1/2}} \cos(\omega t + \phi) \qquad (22.75)$$

Notice that the real current through the resistor (Equation 22.69) and the real potential difference across the resistor (Equation 22.75) have the same phase angle: they are in phase. This is *always* the case in a resistor, as we discovered in Section 22.4.

Now we look at the magnitude of the peak potential difference across the resistor $V_{R\,peak}$ as a function of the angular frequency ω of the source. The peak potential difference is the coefficient of the cosine term in Equation 22.75:

$$V_{R\,peak}(\omega) = \frac{RV_0}{(R^2 + X_C^2)^{1/2}} \qquad (22.76)$$

We substitute for the capacitive reactance to see how $V_{R\,peak}$ depends on the angular frequency. Since

$$X_C = \frac{1}{\omega C}$$

the peak potential difference across the resistor is

$$V_{R\,peak}(\omega) = \frac{RV_0}{\left[R^2 + \left(\frac{1}{\omega C}\right)^2\right]^{1/2}}$$

After some algebra (that infamous expression), we find

$$V_{R\,\text{peak}}(\omega) = \frac{\omega RCV_0}{\left[1 + (\omega RC)^2\right]^{1/2}} \qquad (22.77)$$

We look at the low and high angular frequency limits.

Case 1: For angular frequencies approaching zero, the peak potential difference across the resistor approaches zero. When $\omega = 0$ rad/s, the current in the circuit is zero, and there should then be no potential difference across the resistor. This is the dc limit.

Case 2: For very high angular frequencies, the denominator of Equation 22.77 approaches

$$[1 + (\omega RC)^2]^{1/2} \approx [(\omega RC)^2]^{1/2} = \omega RC$$

Hence the peak potential difference across the resistor approaches

$$V_{R\,\text{peak}} \approx \frac{\omega RCV_0}{\omega RC} = V_0$$

For high angular frequencies, the impedance of the *capacitor* becomes negligible; we say the capacitor shorts out and acts more and more like just a common wire. In this high angular frequency limit, most of the source potential difference appears across the resistor since the Kirchhoff voltage law must be satisfied at all times in the circuit.

Next let's look at the complement of the circuit by trading the positions of R and C (they are in series so their order is immaterial). Consider the output voltage to be the potential difference across the resistor, as in Figure 22.28. The input is the ac voltage source.

Since the potential difference across the resistor (the output) is appreciable only for high values of the angular frequency, this arrangement is called a **high pass filter**.

It is only for high frequencies that a significant potential difference appears across the output resistor; see Figure 22.29.

For the high pass filter, the voltage gain is, using Equation 22.77,

$$\text{voltage gain} \equiv \frac{V_{R\,\text{peak}}(\omega)}{V_0} = \frac{\omega RC}{\left[1 + (\omega RC)^2\right]^{1/2}} \qquad \text{(high pass filter)} \qquad (22.78)$$

Two things to note about our solution of this ac circuit:

1. The nominal polarities of the potential differences across the capacitor and the resistor are indicated in Figure 22.25.

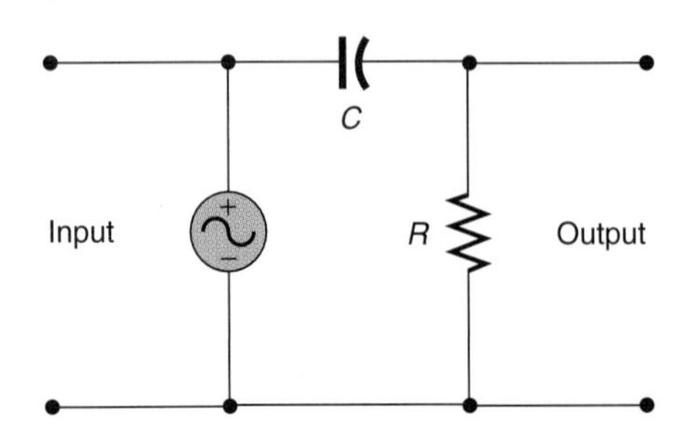

FIGURE 22.28 A high pass filter.

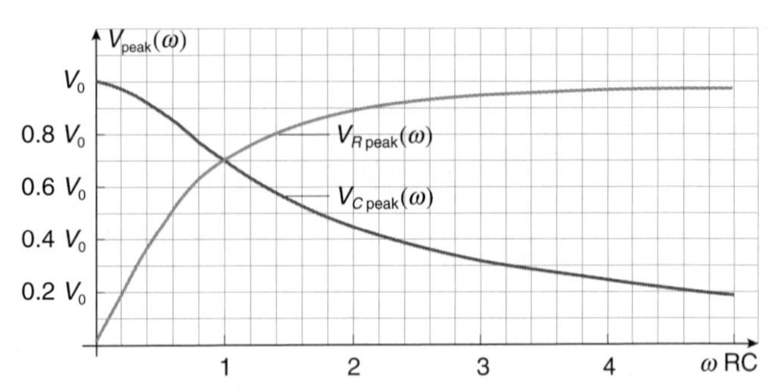

FIGURE 22.29 Peak output potential differences in an RC circuit as a function of ωRC.

At any instant t, the KVL around the loop must be satisfied (energy conservation!). That is,

$$-V_0(t) + V_R(t) + V_C(t) = 0 \text{ V} \qquad (22.79)$$

This is not at all obvious from the expressions for the actual potential differences (Equations 22.58, 22.72, and 22.75); but it is indeed the case (as you are asked to show in Problem 50). The phasor approach guarantees that Equation 22.79 is the case (if we do the mathematics correctly) since we used the KVL to find the voltage phasors.

2. At any *nonzero* angular frequency ω, the sum of the *amplitudes* (peak values) of the oscillating potential differences across the capacitor and the resistor (the peak values) does *not* equal the amplitude of the source. That is,

$$V_0 \neq \frac{\omega RCV_0}{\left[1 + (\omega RC)^2\right]^{1/2}} + \frac{V_0}{\left[1 + (\omega RC)^2\right]^{1/2}}$$

The reason for inequality is that the respective potential differences are *not* all in phase: they do not reach their peaks simultaneously.

PROBLEM-SOLVING TACTIC

22.6 To solve an ac circuit, convert the circuit to the complex domain and employ the same techniques and conventions as in dc circuit analysis. Use impedances, source voltage phasors, current phasors, and potential difference phasors. The real currents and potential differences are the real parts of the complex current and potential difference phasors.

22.9 A Series RLC Circuit

One day looking at the formulas in some book or other, I discovered a formula for the frequency of a resonant circuit which was $f = 1/[2\pi(LC)^{1/2}]$, where L is the inductance and C was the capacitance of the circuit. And there was pi, and where was the circle? You laugh, but I was serious then. Pi was a thing with circles, and here is pi coming out of an electric circuit. Where was the circle?

Richard Feynman (1918–1988)*

*Richard P. Feynman, "What is science?" *The Physics Teacher, 7*, #6, pages 313–320 (September 1969); the quote is on page 315.

When you tune your radio or select a TV channel, a **series *RLC* circuit** may be employed to select the frequency at which a resistor has maximum potential difference—that is, absorbs maximum average power. In order to tune in only one station at a time, the circuit also must respond only to a narrow range (or band) of frequencies around the desired one. A series *RLC* circuit accomplishes these objectives. As its name implies, a series *RLC* circuit consists of a resistor R, an inductor L, and a capacitor C in series with an ac voltage source, as shown in Figure 22.30.

Our focus will be to obtain an expression first for the potential difference across the resistor, as this is amplified and detected by other electronic components (that we will not investigate). The average power absorbed by the resistor is a maximum when the potential difference across it is a maximum (since for a resistor $P = V^2/R$). We know from Section 22.7 that inductors and capacitors each absorb zero average power over each cycle or multiple of it. Hence the average power absorbed by the circuit is the sum of the power absorbed by the resistor and by the source. In keeping with energy conservation and our sign convention for the power, the sum of the average power absorbed by these two circuit elements over a cycle (or multiple of cycles) must be zero:

$$\langle P \rangle_{\text{res}} + \langle P \rangle_{\text{source}} = 0 \text{ W}$$

A resistor always absorbs positive power, so the source absorbs negative power, indicating that it is generating the power for the circuit.

Eventually we calculate the expression for the average power absorbed by the resistor, which is proportional to the square of the potential difference across it. To do this, we need to find expressions for the current through the resistor and the potential difference across it. To accomplish this objective, we proceed as we did with the filter circuit of the previous section.

Step 1: We convert the ac independent voltage source to the complex domain by using a voltage source phasor. Since the source voltage is

$$V(t) = V_0 \cos(\omega t) \tag{22.80}$$

the voltage source phasor is

$$\mathbf{V} = V_0 e^{i\omega t} \quad \text{(exponential form)} \tag{22.81}$$

$$= V_0 \angle(\omega t) \quad \text{(polar form)} \tag{22.82}$$

Also we use the impedance of the resistor, capacitor, and inductor. Converted to the complex domain, the circuit now appears as in Figure 22.31.

Step 2: We simplify the circuit by combining circuit elements that are in series or parallel. In the present case the impedances of the resistor, inductor, and capacitor are in series with each other. The total impedance of the series combination is the sum of the three impedances:

$$Z_{\text{total}} = R + i\omega L + \frac{1}{i\omega C}$$

After rationalizing the impedance of the capacitor, the total impedance becomes

$$Z_{\text{total}} = R + i\omega L - i\frac{1}{\omega C}$$

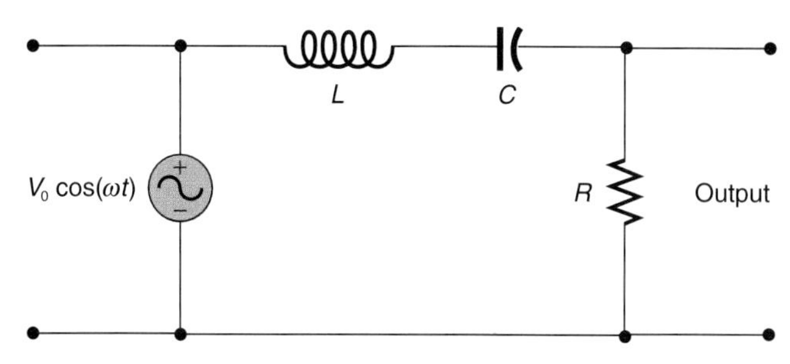

FIGURE 22.30 A series *RLC* circuit.

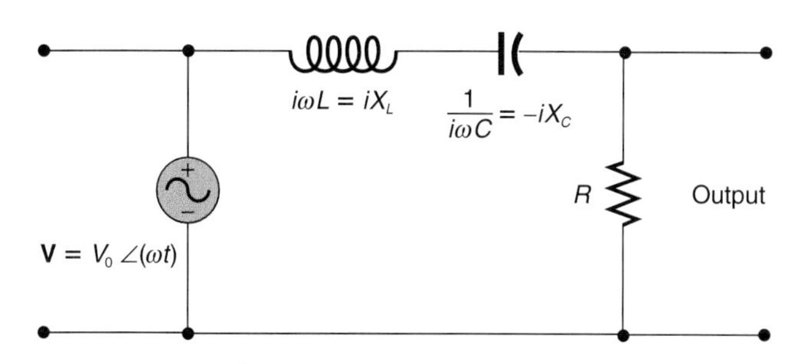

FIGURE 22.31 The series *RLC* circuit in the complex domain.

We use the inductive reactance $X_L = \omega L$ and the capacitive reactance

$$X_C = \frac{1}{\omega C}$$

to write the total impedance more cleanly as

$$Z_{\text{total}} = R + i(X_L - X_C) \tag{22.83}$$

The circuit now appears as shown in Figure 22.32.

Step 3: We choose a direction for the current phasor **I** and mark the polarity of each impedance in accordance with this choice. The choice for the direction of the phasor current is out of the nominally positive terminal of the ac source. The impedance Z_{total} then must be labeled with the polarity indicated in Figure 22.33.

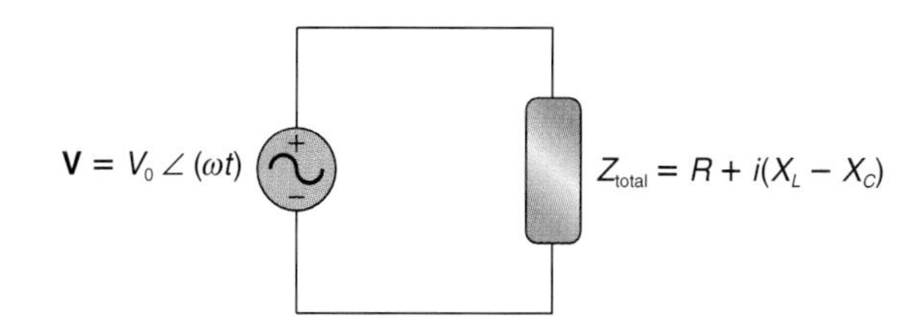

FIGURE 22.32 The simplified circuit in the complex domain.

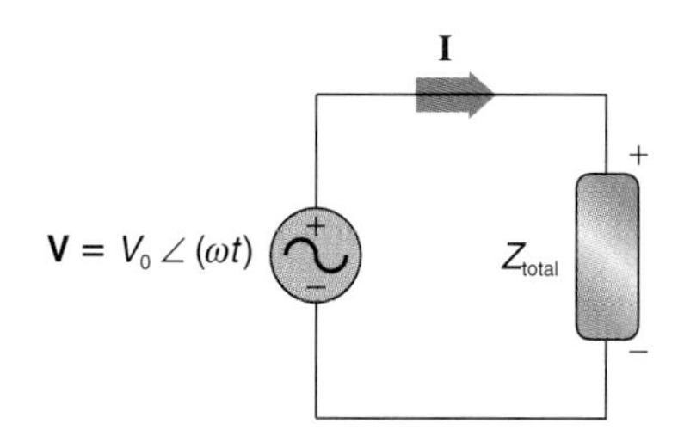

FIGURE 22.33 The direction for the current phasor and the corresponding polarity of the impedance.

Step 4: We use the KVL to form the loop equations. Going around the loop in the clockwise sense and beginning in the lower left corner of the loop in Figure 22.33, the KVL results in the following equation:

$$-\mathbf{V} + \mathbf{I}Z_{\text{total}} = 0 \text{ V} \tag{22.84}$$

There is only one loop in the circuit, so the KVL yields only one equation. But we have only one unknown current phasor, and so this single equation is sufficient to solve the problem.

Just as with dc circuit techniques, if the application of the KVL to each elementary loop of the circuit does not result in a sufficient number of equations to solve for the unknown currents, the KCL is then employed at one or more significant nodes until the total number of equations (from the KVL and KCL) is equal to the number of unknown current phasors in the problem.

Step 5: We solve the equation(s) for the unknown current(s). (In the present case, there is only one equation and one unknown current.) We solve Equation 22.84 for **I**, yielding

$$\mathbf{I} = \frac{\mathbf{V}}{Z_{\text{total}}}$$

Substituting for Z_{total} using Equation 22.83, we find

$$\mathbf{I} = \frac{\mathbf{V}}{R + i(X_L - X_C)} \tag{22.85}$$

That is the end of the physics, but we have to express **I** in proper polar form to extract its real part, which is the real current. The denominator is a complex number in rectangular form. We express the denominator in polar form:

$$R + i(X_L - X_C) = [R^2 + (X_L - X_C)^2]^{1/2} \angle\theta$$

where

$$\theta = \tan^{-1}\left(\frac{X_L - X_C}{R}\right)$$

We make this substitution into the denominator of Equation 22.85 and write the voltage source phasor in polar form; the current phasor then becomes

$$\mathbf{I} = \frac{V_0 \angle(\omega t)}{\left[R^2 + (X_L - X_C)^2\right]^{1/2} \angle\theta}$$

Division of complex numbers in polar form means that we put the angle in the denominator into the numerator by changing its sign:

$$\mathbf{I} = \frac{V_0 \angle(\omega t) \angle(-\theta)}{\left[R^2 + (X_L - X_C)^2\right]^{1/2}}$$

The angles in the numerator then are combined according to the product rule for complex numbers in polar form: the angles just add. Thus the current phasor is

$$\mathbf{I} = \frac{V_0 \angle(\omega t - \theta)}{\left[R^2 + (X_L - X_C)^2\right]^{1/2}} \tag{22.86}$$

To find the phasor representing the potential difference across the resistance, we multiply the current phasor **I** by the impedance of the resistor, which is just R:

$$\begin{aligned} \mathbf{V}_{\text{res}} &= \mathbf{I}R \\ &= \frac{RV_0 \angle(\omega t - \theta)}{\left[R^2 + (X_L - X_C)^2\right]^{1/2}} \end{aligned} \tag{22.87}$$

Notice that the current and potential difference phasors for the resistor (Equations 22.86 and 22.87) are in phase, as they must be for any resistor.

To find the average power absorbed by the resistor, we use Equation 22.50, modified for the notation here:

$$\langle P \rangle = \frac{1}{2} V_{\text{res}} I_{\text{res}} \cos\beta$$

where V_{res} is the amplitude of the oscillating potential difference across the resistor, I_{res} is the amplitude of the oscillating current through the resistor, and $\cos\beta$ is the power factor (the cosine of the phase angle difference between the potential difference and the current). The phase angle difference is zero for a resistor, and so the power factor is unity:

$$\cos\beta = \cos(0 \text{ rad}) = 1 \qquad \text{(for any resistor)}$$

The amplitude of the potential difference across the resistor is the magnitude of the potential difference phasor of Equation 22.87:

$$V_{\text{res}} = \frac{RV_0}{\left[R^2 + (X_L - X_C)^2\right]^{1/2}}$$

Likewise the amplitude of the current through the resistor is the magnitude of the current phasor of Equation 22.86:

$$I_{\text{res}} = \frac{V_0}{\left[R^2 + (X_L - X_C)^2\right]^{1/2}}$$

Thus the average power absorbed by the resistor is

$$\langle P \rangle = \frac{1}{2} \frac{RV_0^2}{R^2 + (X_L - X_C)^2} \tag{22.88}$$

This also is the average power absorbed by the series *RLC* combination, because the capacitor and inductor absorb no average power.

We want to determine the angular frequency ω that makes $\langle P \rangle$ as large as possible. From Equation 22.88, $\langle P \rangle$ is greatest when $X_L = X_C$. The dependence of the average power on the angular frequency can be made explicit by substituting for the inductive and capacitive reactances. In particular, the term $X_L - X_C$ in the denominator of Equation 22.88 can be rewritten as

$$\begin{aligned} X_L - X_C &= \omega L - \frac{1}{\omega C} \\ &= \frac{\omega^2 LC - 1}{\omega C} \end{aligned} \tag{22.89}$$

Recall that when we examined the transient behavior of an *LC* circuit (in Section 21.10), we discovered that the quantity

$$\frac{1}{(LC)^{1/2}}$$

was an angular frequency. We define this quantity to be the **natural resonant angular frequency** ω_0 of the circuit:

$$\omega_0 \equiv \frac{1}{(LC)^{1/2}} \qquad (22.90)$$

The reason for this terminology will become apparent shortly. With this definition, Equation 22.89 can be rewritten, after some algebraic manipulation, as

$$X_L - X_C = \frac{L}{\omega}(\omega^2 - \omega_0^2) \qquad (22.91)$$

Substituting this expression into Equation 22.88 for the average power results in

$$\langle P \rangle = \frac{1}{2}\frac{RV_0^2}{R^2 + \frac{L^2}{\omega^2}(\omega^2 - \omega_0^2)^2}$$

With a bit more algebraic manipulation, this can be reexpressed as

$$\langle P \rangle = \frac{1}{2}\frac{\omega^2 RV_0^2}{R^2\omega^2 + L^2(\omega^2 - \omega_0^2)^2} \qquad (22.92)$$

which is about as tidy (!) as the expression for the average power can be made.

We seek the angular frequency that maximizes the average power in, the current through, and potential difference across *R* because the latter is amplified by subsequent electronic circuitry. The condition $X_L = X_C$ maximizes the power, which implies the desired angular frequency ω is equal to the natural resonant angular frequency of the circuit ω_0.

There are several aspects of Equation 22.92 that merit comment.

If the average power is plotted as a function of the angular frequency of the source, the peak of the curve occurs when $\omega = \omega_0$.

This can be seen from Equation 22.92 directly: when $\omega = \omega_0$, the denominator is smallest, so the average power then is greatest. Equivalently, we could take the derivative of $\langle P \rangle$ with respect to ω and then set the result equal to zero to locate the extremum (see Problem 46). A numerical value for ω_0 is determined from the specific values of *L* and *C* using Equation 22.90. Since ω_0 is the natural resonant angular frequency, the power is a maximum when the angular frequency of the source is equal to the natural resonant angular frequency. If the angular frequency of the source is too large or too small, the resistor does not absorb as much power.

When $\omega = \omega_0$ in a series *RLC* circuit, we have an electronic example of **resonance**.* The circuit is analogous to the forced, simple harmonic oscillator that we studied in Chapter 7. If the frequency of the driving force on a child's swing is too fast or too slow, the swing absorbs little energy; if the driving frequency is equal to the natural oscillation frequency of the swing, the energy (and amplitude) of the swing is greatest.

The average power absorbed by the resistor when the angular frequency of the source is equal to ω_0 can be found by substituting $\omega = \omega_0$ into Equation 22.92. The result is

$$\langle P \rangle_{max} = \frac{1}{2}\frac{V_0^2}{R} \qquad (22.93)$$

Since

$$V_{rms} = \frac{V_0}{\sqrt{2}}$$

the peak power also can be written as

$$\langle P \rangle_{max} = \frac{V_{rms}^2}{R} \qquad (22.94)$$

For fixed *L* and *C* (and therefore for fixed ω_0), the numerical value of the resistance affects the shape of the graph of $\langle P \rangle$ versus ω. The smaller the resistance *R*, the more sharply peaked the curve.

This effect is illustrated in Figure 22.34 for two values of *R*. To make the graphs, the following values were used for the parameters of the circuit:

amplitude of the source voltage $V_0 = 10.0$ V
inductance $L = 100$ mH
capacitance $C = 1000$ pF

With these values for *L* and *C* the resonant angular frequency of the circuit is

$$\omega_0 = \frac{1}{(LC)^{1/2}} = 1.00 \times 10^5 \text{ rad/s}$$

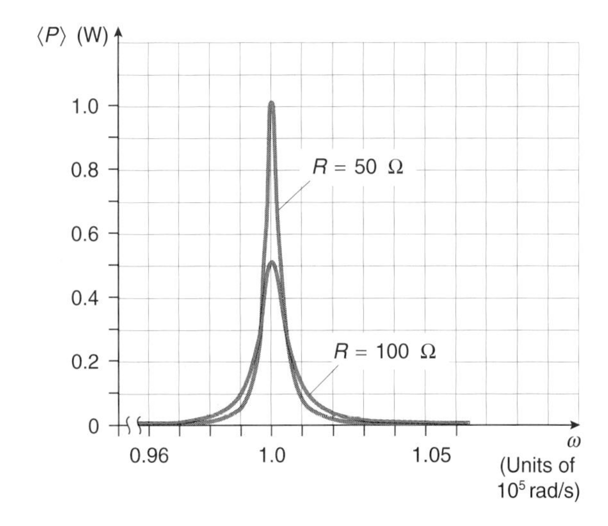

FIGURE 22.34 The value of *R* affects the shape of the average power versus ω curve.

*Operationally, for a fixed source angular frequency (say a radio station), we tune the circuit to the desired angular frequency by adjusting the value of the capacitor in the circuit (which is the tuning knob on our radios and TVs).

The graphs are for $R = 50.0\ \Omega$ and $100\ \Omega$.

The effect of R on the shape of the average power curve also can be explored analytically.

Consider the two angular frequencies ω_1 and ω_2 on either side of ω_0 at which the average power falls to *half* the peak average power, as shown in Figure 22.35. The values of these angular frequencies are found in the following way. At these angular frequencies, the average power is half the peak power or

$$\frac{1}{2}\langle P\rangle_{\max} = \frac{1}{2}\frac{1}{2}\frac{V_0^2}{R}$$

Using this on the left-hand side of Equation 22.92, we obtain

$$\frac{1}{2}\frac{1}{2}\frac{V_0^2}{R} = \frac{1}{2}\frac{\omega^2 R V_0^2}{R^2\omega^2 + L^2(\omega^2 - \omega_0^2)^2} \qquad (22.95)$$

Now we solve this equation for ω. This solution for the angular frequencies at half power is a bit tricky algebraically, but eventually we find that Equation 22.95 reduces to

$$L^2(\omega^2 - \omega_0^2)^2 = R^2\omega^2$$

There are both positive and negative choices when we take the square root:

$$L(\omega^2 - \omega_0^2) = \pm R\omega$$

Taking the positive term on the right-hand side yields

$$L(\omega^2 - \omega_0^2) = +R\omega$$

or

$$\omega^2 - \frac{R}{L}\omega - \omega_0^2 = 0\ \text{rad}^2/\text{s}^2$$

Taking the negative term on the right-hand side yields

$$L(\omega^2 - \omega_0^2) = -R\omega$$

or

$$\omega^2 + \frac{R}{L}\omega - \omega_0^2 = 0\ \text{rad}^2/\text{s}^2$$

Each of these equations has two roots:

$$\omega_2 = \frac{\frac{R}{L} \pm \sqrt{\left(-\frac{R}{L}\right)^2 + 4\omega_0^2}}{2}$$

$$\omega_1 = \frac{-\frac{R}{L} \pm \sqrt{\left(\frac{R}{L}\right)^2 + 4\omega_0^2}}{2}$$

To avoid negative angular frequency solutions, we must reject the – sign choice for both cases. The positive solutions are

$$\omega_2 = \frac{R}{2L} + \frac{1}{2}\sqrt{\left(-\frac{R}{L}\right)^2 + 4\omega_0^2} \qquad (22.96)$$

$$\omega_1 = -\frac{R}{2L} + \frac{1}{2}\sqrt{\left(\frac{R}{L}\right)^2 + 4\omega_0^2} \qquad (22.97)$$

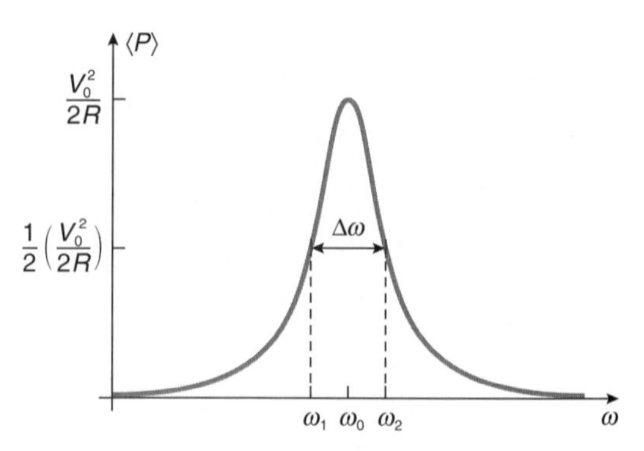

FIGURE 22.35 The frequencies ω_1 and ω_2 at which the average power absorbed falls to half the peak.

The difference between these half-power angular frequencies is called the **angular frequency bandwidth** $\Delta\omega$ of the circuit:

$$\Delta\omega \equiv \omega_2 - \omega_1 \qquad (22.98)$$

$$= \frac{R}{L} \qquad (22.99)$$

Equation 22.99 shows that for fixed ω_0 (i.e., fixed L and C), the half-power width is directly proportional to R; the smaller R is, the smaller the half-power angular frequency bandwidth, as confirmed in Figure 22.34.

The ratio between the resonant angular frequency ω_0 and the angular frequency bandwidth $\Delta\omega$ is called the **quality factor** Q of the circuit:

$$Q \equiv \frac{\omega_0}{\Delta\omega} \qquad (22.100)$$

Substituting for ω_0 using Equation 22.90 and for $\Delta\omega$ using Equation 22.99, the quality factor can be expressed in terms of the circuit components:

$$Q = \frac{\frac{1}{(LC)^{1/2}}}{\frac{R}{L}}$$

$$= \left(\frac{L}{CR^2}\right)^{1/2} \qquad (22.101)$$

A circuit with a large value of Q is one with a small (narrow) angular frequency bandwidth; such a condition is necessary to avoid cross talk, the simultaneous reception of two different radio or TV frequencies.

The quality factors of the circuits whose power curves are plotted in Figure 22.34 are found using Equation 22.101. Using those values of L and C, the quality factor of the circuit is

$$Q = \frac{1.00 \times 10^4\ \Omega}{R}$$

For the $100\ \Omega$ resistor (using the given values for L and C),

$$Q_{100} = 100$$

and for the $50.0\ \Omega$ resistor,

$$Q_{50} = 200$$

indicating that as the resistance is *decreased* the quality factor of the circuit *increases* (for fixed L and C).

Notice in Figure 22.34 that the average power absorbed is more sharply peaked at the resonant angular frequency ω_0 for the 50.0 Ω resistor than for the larger, 100 Ω resistor, reflecting the difference in the quality factors of the circuit with the two different resistors.

Since the resonant angular frequency ω_0 depends on C (see Equation 22.90), changing the value of the capacitance shifts the angular frequency at which the peak in the power curve occurs. If the capacitance is increased, Equation 22.90 shows that the resonant angular frequency decreases. Note that changing C also affects the quality factor Q of the circuit (see Equation 22.101); increasing C decreases the quality factor Q. The angular frequency bandwidth $\Delta\omega$ is unaffected by changing the value of the capacitance (see Equation 22.99).

After our leisurely tour through the complex algebraic thicket of the series *RLC* circuit, you should realize that it is the basic element behind tuned communication systems, so it has a wide range of applications!

CHAPTER SUMMARY

A *complex variable* z can be represented in *rectangular form* as

$$z = x + iy \tag{22.1}$$

where $i \equiv \sqrt{-1}$. The *real part* of z is x and its *imaginary part* is y:

$$\text{Re}\, z \equiv x \tag{22.2}$$

$$\text{Im}\, z \equiv y \tag{22.3}$$

Both the real and imaginary parts are real variables. Here z, x, and y represent general algebraic variables, *not necessarily Cartesian coordinates*.

The *polar form* of a complex variable z has the form

$$z = r\angle\theta \tag{22.5}$$

where r is the *magnitude* of the complex number, also known occasionally as its *amplitude*:

$$\begin{aligned} r &= [(\text{Re}\, z)^2 + (\text{Im}\, z)^2]^{1/2} \\ &= (x^2 + y^2)^{1/2} \end{aligned} \tag{22.4}$$

and

$$\tan\theta = \frac{\text{Im}\, z}{\text{Re}\, z} = \frac{y}{x} \tag{22.8}$$

The *exponential form* of the complex variable z is

$$z = re^{i\theta}$$

The *Euler identity* is useful for changing from the exponential form to the rectangular form of a complex variable:

$$re^{i\theta} = r\cos\theta + ir\sin\theta \tag{22.11}$$

Two complex variables $z_1 = x_1 + iy_1$ and $z_2 = x_2 + iy_2$ can be added and subtracted in rectangular form:

$$z_1 \pm z_2 = (x_1 \pm x_2) + i(y_1 \pm y_2) \qquad (22.12, 22.13)$$

The product of two complex variables in polar form, $z_1 = r_1\angle\theta_1$ and $z_2 = r_2\angle\theta_2$, is

$$z_1 z_2 = r_1 r_2 \angle(\theta_1 + \theta_2) \tag{22.15}$$

while their quotient is

$$\frac{z_1}{z_2} = \frac{r_1}{r_2}\angle(\theta_1 - \theta_2) \tag{22.18}$$

The *complex conjugate* z^* of a complex variable z is found by replacing i with $-i$ everywhere. Thus, if

$$z = x + iy = re^{i\theta} = r\angle\theta$$

then its complex conjugate is

$$z^* = x - iy = re^{-i\theta} = r\angle(-\theta)$$

The magnitude r of a complex variable z can be found from

$$r = (zz^*)^{1/2}$$

Sinusoidal, ac potential differences $V(t) = V_0\cos(\omega t + \theta)$, currents $I(t) = I_0\cos(\omega t + \phi)$, and ac independent voltage sources are represented by complex *phasors* whose real parts are the actual (i.e., real) potential differences, currents, and sources, respectively:

$$\mathbf{V}(t) \equiv V_0 e^{i(\omega t + \theta)} = V_0\angle(\omega t + \theta) \tag{22.19}$$

$$\mathbf{I}(t) \equiv I_0 e^{i(\omega t + \phi)} = I_0\angle(\omega t + \phi) \tag{22.20}$$

$$\mathbf{V}_{\text{source}}(t) \equiv V_0\angle(\omega t) \tag{22.41}$$

In resistors, inductors, and capacitors, the potential difference phasor and the current phasor are related by the *ac version of Ohm's law*:

$$\mathbf{V} = \mathbf{I}Z \tag{22.29}$$

where Z is the *impedance* of the circuit element:

$$Z_R = R \quad \text{for a resistor} \tag{22.30}$$

$$Z_L = i\omega L \equiv iX_L \quad \text{for an inductor} \tag{22.31}$$

$$Z_C = \frac{1}{i\omega C} \equiv -iX_C \quad \text{for a capacitor} \tag{22.32}$$

The quantity $X_L \equiv \omega L$ is the *inductive reactance* and

$$X_C \equiv \frac{1}{\omega C}$$

is the *capacitive reactance*.

The potential difference phasor across a resistor and the current phasor through it are in phase. In an inductor, the phase difference between the potential difference phasor and the current phasor is $\pi/2$ rad. For a capacitor, the phase difference between the potential difference phasor and the current phasor is $-\pi/2$ rad. The same phase relationships exist between the real potential differences and the real currents in the various circuit elements.

Impedances in series are added to find an equivalent impedance:

$$Z_{eq} = Z_1 + Z_2 + Z_3 + \cdots \tag{22.35}$$

The equivalent impedance for impedances in parallel is

$$\frac{1}{Z_{eq}} = \frac{1}{Z_1} + \frac{1}{Z_2} + \frac{1}{Z_3} + \cdots \tag{22.36}$$

If V_0 is the amplitude of the sinusoidal potential difference across a circuit element and I_0 is the amplitude of the sinusoidal current into its nominally positive terminal, the average power $\langle P\rangle$ absorbed by (transferred to) the circuit element is

$$\langle P\rangle = \frac{1}{2} V_0 I_0 \cos\beta \qquad (22.50)$$

$$= V_{rms} I_{rms} \cos\beta \qquad (22.56)$$

where $\cos\beta$ is the *power factor*, the cosine of the phase difference between the potential difference and the current. The *rms values* (also called *effective values*) are related to the peak values by

$$V_{rms} = \frac{V_0}{\sqrt{2}} \qquad (22.54)$$

$$I_{rms} = \frac{I_0}{\sqrt{2}} \qquad (22.55)$$

For a resistor, the power factor is equal to 1. For inductors and capacitors, the power factor is 0; therefore, both inductors and capacitors absorb *zero* average power.* The average power absorbed also can be expressed as

$$\langle P\rangle = \frac{1}{2} \mathrm{Re}(\mathbf{VI}^*) \qquad (22.52)$$

In the complex domain, sinusoidal ac circuits are solved with the KVL and KCL using conventions and techniques like those used for dc circuit analysis.

*Remember these are *ideal* circuit elements.

SUMMARY OF PROBLEM-SOLVING TACTICS

22.1 **(page 1008)** When converting between rectangular and polar forms, be sure your calculator is in radian mode for performing trigonometric operations.

22.2 **(page 1010)** Addition and subtraction of complex variables can only be performed when they are in rectangular form.

22.3 **(page 1011)** For the multiplication of two complex numbers in rectangular form, it occasionally is easier first to convert them to polar or exponential form, then do the multiplication.

22.4 **(page 1011)** To divide two complex numbers in rectangular form, it occasionally is easier first to convert them to polar or exponential form, then do the division.

22.5 **(page 1020)** For two impedances in parallel, you can make use of the shortcut we found for resistors in parallel: the equivalent impedance is the product of the two impedances divided by their sum.

22.6 **(page 1026)** To solve an ac circuit, convert the circuit to the complex domain and employ the same techniques and conventions as in dc circuit analysis.

QUESTIONS

1. **(page 1009)**; 2. **(page 1012)**

3. Considering Lenz's law from Chapter 21, explain why the inductive reactance should increase with increasing angular frequency.

4. Explain why the capacitive reactance should decrease with increasing angular frequency.

5. Explain why a capacitor acts as a short circuit at high angular frequencies and as an open circuit at low angular frequencies.

6. Explain why an instructor, er. . . no, an inductor acts as a short circuit at low angular frequencies and an open circuit at high angular frequencies.

7. A light bulb (a resistor) is connected in series with a capacitor and a sinusoidal ac independent voltage source with a variable angular frequency ω as in Figure Q.7. For small values of ω, will the light bulb be brightly lit or dim? For large values of ω, will the bulb be brightly lit or dim? Explain your reasoning in each case.

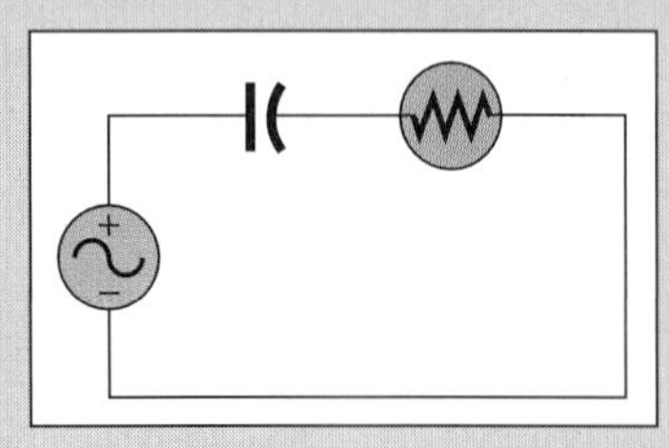

FIGURE Q.7

8. A light bulb (a resistor) is connected in series with an inductor and a sinusoidal ac independent voltage source with a variable angular frequency ω as in Figure Q.8. For small values of ω, will the light bulb be brightly lit or dim? For large values of ω, will the bulb be brightly lit or dim? Explain your reasoning in each case.

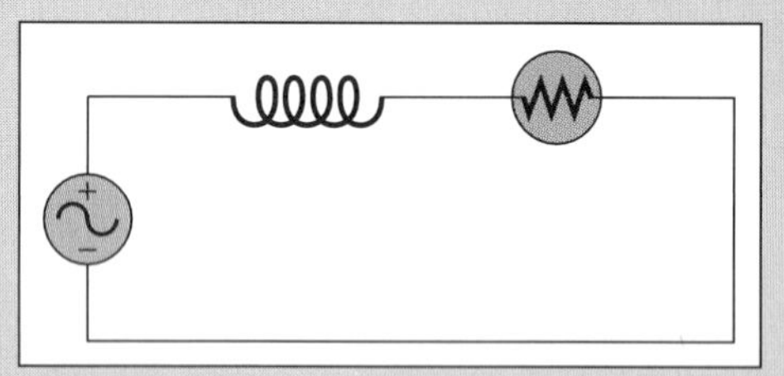

FIGURE Q.8

9. Can an ohmmeter be used to measure the impedance of a resistor? Of a capacitor? Of an inductor?

10. Two sinusoidal ac independent voltage sources have the same amplitude but different angular frequencies:

$$V_0 \cos(\omega_1 t) \qquad V_0 \cos(\omega_2 t)$$

Do they have equal rms amplitudes?

11. Fluorescent lights use an inductor to limit the glow discharge current through the bulb. Why is this usage more energy efficient than using a resistor to limit the current?

12. The angular frequency in a series *RLC* circuit is doubled. By what factor is the impedance of the resistor changed? Answer the same question for the inductor, the capacitor, and the total impedance.

13. At resonance, what is the total impedance of a series *RLC* circuit?

14. One of the *RL* circuits in Figure Q.14 is a high pass filter and the other is a low pass filter. By considering the qualitative behavior of the impedance of an inductor as a function of the angular frequency ω, deduce which circuit is the high pass filter and which is the low pass filter. Explain your reasoning.

15. Compared with acoustic frequencies, is radio static high frequency or low frequency? To eliminate such static should a low pass or high pass filter be used?

16. In the schematic diagram in Figure Q.16, explain why the capacitor prevents dc currents in each circuit from passing into the other circuit but permits the passage of ac currents. Such circuits are said to be dc isolated and ac capacitively coupled. Capacitors are used frequently in transistor circuits in just this way.

17. In the schematic diagram of Figure Q.17, explain why node A is effectively an electrical ground for sufficiently large angular frequencies. Capacitors used in this way are called bypass capacitors and are used frequently in transistor circuits.

18. The sum of the peak values of the potential differences across the resistor, inductor, and capacitor in an *RLC* circuit is usually greater than the peak value of the source voltage. Does this contradict the KVL? Explain why or why not.

19. Since the ac current and ac potential difference across a light bulb (a resistor) vary rapidly with time, does the temperature of the filament and the consequent light output of the bulb vary accordingly? What might limit the size of any temperature variation?

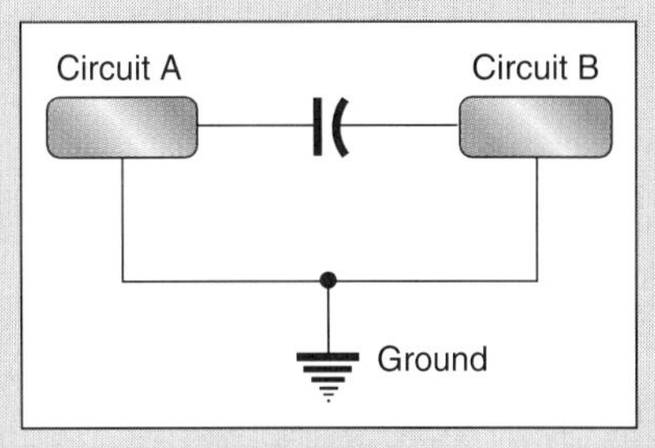

FIGURE Q.16

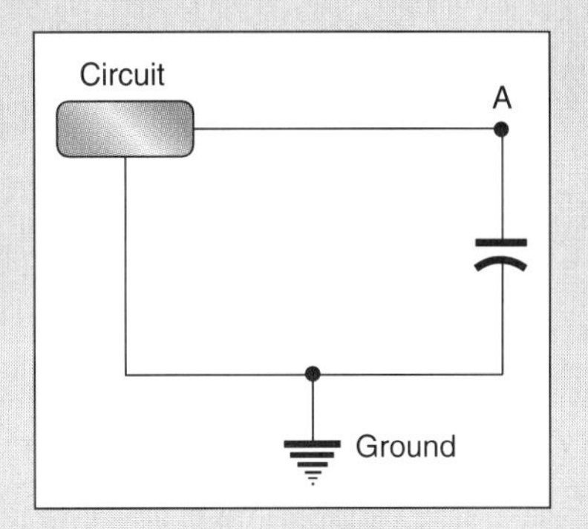

FIGURE Q.17

20. What is the phase angle θ of a series *RLC* circuit (on page 1028) when the inductive reactance is equal to the capacitive reactance? Is this the condition for resonance?

21. Why are capacitors used more frequently than inductors in integrated circuits?

22. If a source consists of *both* a dc emf and an ac oscillatory emf as well, is it permissible to consider the effects of each separately in determining currents and potential differences in a circuit and then add the results? Explain your answer. This procedure is used frequently in analyzing transistor circuits.

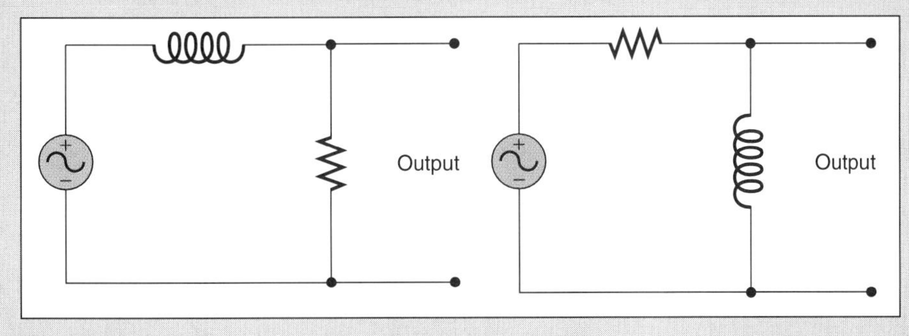

FIGURE Q.14

PROBLEMS

Sections 22.1 Representations of a Complex Variable
22.2 Arithmetic Operations with Complex Variables

•1. Two complex numbers are given by

$$z_1 = 3 + i(4) \qquad \text{and} \qquad z_2 = 2 - i(5)$$

Find: (a) the magnitude of each complex number; (b) the polar form of each complex number; and (c) the exponential form of each complex number. (d) Locate each complex number on the complex plane.

•2. Perform the following arithmetic operations on

$$z_1 = 3 + i(4) \qquad \text{and} \qquad z_2 = 2 - i(5)$$

Express each result in rectangular, polar, and exponential form. (a) $z_1 + z_2$; (b) $z_1 - z_2$; (c) $z_1 z_2$; (d) z_1/z_2. (e) Locate the results of (a)–(d) on the complex plane.

•3. For the two complex numbers

$$z_1 = -2 + i(3) \qquad \text{and} \qquad z_2 = 3 - i(4)$$

find: (a) the magnitude of each complex number; (b) the polar form of each complex number; (c) the exponential form of each complex number; and (d) the location of each complex number on the complex plane.

•4. Perform the following arithmetic operations on the complex numbers

$$z_1 = -2 + i(3) \qquad \text{and} \qquad z_2 = 3 - i(4)$$

Express each result in rectangular, polar, and exponential form. (a) $z_1 + z_2$; (b) $z_1 - z_2$; (c) $z_1 z_2$; (d) z_1/z_2. (e) Locate the results of (a)–(d) on the complex plane.

•5. Find the simplest form for the following complex expressions: (a) $(1 - i)^4$; (b) $(\sqrt{2} - i) - i(1 - i\sqrt{2})$;

$$\text{(c)}\ \frac{5}{(1 - i)(2 - i)(3 - i)}.$$

•6. Rationalize the following expressions:

$$\text{(a)}\ \frac{2}{2 + i};\ \text{(b)}\ \frac{i(5)}{1 - i(2)}.$$

•7. Evaluate the following: (a) $[2 - i(3)]^*$; (b) $[4e^{-i(\pi/4\ \text{rad})}]^*$; (c) $[3.0\ \angle(\pi/6\ \text{rad})]^*$.

•8. By using the polar form of the complex number $z = -1 + i$, show that

$$(-1 + i)^7 = -8(1 + i)$$

•9. Show that the addition and subtraction of two complex variables is equivalent to the addition and subtraction of a pair of two-dimensional vectors whose components are the real and imaginary parts of the complex variables.

•10. For two complex variables

$$A = A_x + iA_y \qquad \text{and} \qquad B = B_x + iB_y$$

show that if A_x and A_y are the two Cartesian components of a two-dimensional vector $\vec{A} = A_x\hat{\imath} + A_y\hat{\jmath}$, and B_x and B_y are the two Cartesian components of another two-dimensional vector $\vec{B} = B_x\hat{\imath} + B_y\hat{\jmath}$: (a) then $\text{Re}(AB^*)$ is the scalar product $\vec{A} \cdot \vec{B}$; and (b) $\text{Im}(AB^*)$ is the single nonzero component of the vector product $\vec{B} \times \vec{A}$.

•11. (a) Show that

$$\frac{1}{2}(1 + i)^2 = i$$

Hence $\sqrt{i}$ can be represented by the complex number

$$\frac{1}{\sqrt{2}}(1 + i)$$

This ensures that there are no "supercomplex" numbers given by multiple roots of i itself; multiple roots of i can be expressed in terms of i itself. (b) By writing i in exponential form, show that the exponential form of $\sqrt{i}$ is

$$\sqrt{i} = e^{i(\pi/4\ \text{rad})}$$

•12. (a) After studying Problem 11, show that

$$(-i)^{1/2} = \frac{1}{\sqrt{2}}(1 - i)$$

(b) Show that

$$(-i)^{1/2} = e^{-i(\pi/4\ \text{rad})}$$

Sections 22.3 Complex Potential Differences and Currents: Phasors
22.4 The Potential Difference and Current Phasors for Resistors, Inductors, and Capacitors
22.5 Series and Parallel Combinations of Impedances
22.6 Complex Independent ac Voltage Sources
22.7 Power Absorbed by Circuit Elements in ac Circuits

13. Show that the inductive reactance has the SI units of ohms.

14. Show that the capacitive reactance has the SI units of ohms.

15. Make an accurate graph of the resistance of a 10.0 kΩ resistor as a function of frequency from 0 Hz to 1.0 kHz.

•16. Make an accurate graph of the capacitive reactance of a 0.010 μF capacitor as a function of frequency ν from 0 Hz to 1.0 kHz.

•17. Make an accurate graph of the inductive reactance of a 0.500 H inductor as a function of frequency ν from 0 Hz to 1.0 kHz.

•18. At what frequency will a 50.0 μF capacitor have a capacitive reactance of 53.05 Ω?

•19. At what frequency will a 150 mH inductor have an inductive reactance of 47.12 Ω?

•20. Given an inductor with inductance L and a capacitor with capacitance C, at what angular frequency is the inductive reactance equal to the capacitive reactance?

•21. For what range of frequencies will a 50 μF capacitor have a capacitive reactance less than 10.0 Ω?

•22. For what range of frequencies will a 50 mH inductor have an inductive reactance less than 10.0 Ω?

•23. Find the impedance of the following circuit elements at a frequency of 1.00 kHz: (a) a resistance with R = 10.0 kΩ;

(b) a capacitor with $C = 0.0100\ \mu F$; and (c) an inductor with $L = 0.500$ H. (d) What is the capacitive reactance of the capacitor at the given frequency? (e) What is the inductive reactance of the inductor at the indicated frequency?

•**24.** (a) The three circuit elements in Problem 23 are connected in series. Find the equivalent impedance. Express your result in both rationalized rectangular form and polar form. (b) If the three circuit elements are connected in parallel, find the equivalent impedance. Express your result in both rationalized rectangular form and polar form.

•**25.** (a) An inductor L and a capacitor C are connected in series. At what frequency (in Hz) will the equivalent impedance of the pair be zero? (b) Is there a (finite) nonzero frequency at which the *parallel* combination of the pair will produce an equivalent impedance equal to zero? Justify your answer.

•**26.** A 150 Ω resistor is in series with a 0.250 H inductor and a capacitor, with the collection connected to an ac independent voltage source at 60.0 Hz. What capacitance results in a total impedance of 150 Ω?

•**27.** A parallel combination of a 1.00 kΩ resistor and a 250 mH inductor is in series with a capacitor C, as shown in Figure P.27. The combination is connected to a source with a frequency of 1.00 kHz. What value of C produces a total impedance whose real part is ten times greater than its imaginary part?

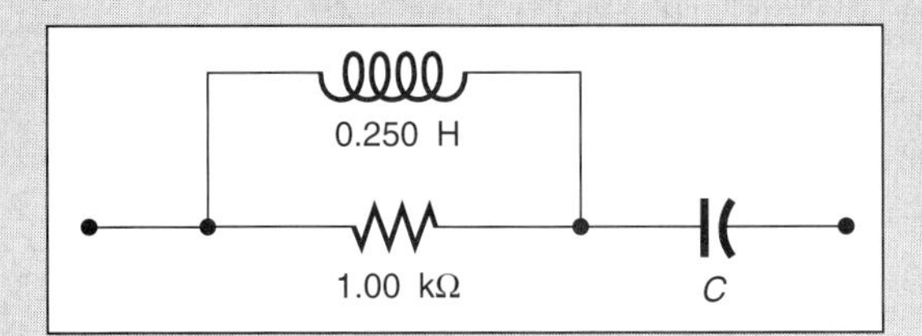

FIGURE P.27

•**28.** Show that the time-averaged value of $\cos(\omega t)$, averaged over a period T of the oscillation, is zero. That is, show that

$$\langle\cos(\omega t)\rangle = \frac{1}{T}\int_{0\text{ s}}^{T}\cos(\omega t)\,dt = 0$$

29. Three circuit elements are at your disposal:

$$R = 1.00\ \text{k}\Omega$$
$$C = 100\ \mu\text{F}$$
$$L = 500\ \text{mH}$$

(a) Using a single log–log graph grid,* plot
(i) the logarithm (base 10) of the capacitive reactance X_C;
(ii) the logarithm (base 10) of the inductive reactance X_L; and
(iii) the logarithm (base 10) of the resistance

as functions of the logarithm (base 10) of the angular frequency ω from 1.00 rad/s to 1.00×10^4 rad/s. The logarithm of the angular frequency should be the abscissa. (b) Explain why each graph is a straight line. (c) At what angular frequency is the inductive reactance equal to 1.00 kΩ? What frequency is this in Hz? (d) At what angular frequency is the capacitive reactance equal to 1.00 kΩ? What frequency is this in Hz? (e) At what angular frequency is the inductive reactance equal to the capacitive reactance? What frequency is this in Hz?

*Such graph paper *automatically* takes the logarithms of the numbers plotted. If you are unfamiliar with the use of log–log graph paper, consult your instructor.

Sections 22.8 A Filter Circuit
22.9 A Series *RLC* Circuit

•**30.** For the sake of your professor, consider the circuit indicted in Figure P.30 (a) What is the frequency ν of the source? (b) What is the impedance of the resistor? (c) What is the independent voltage source phasor? Express your result in both rationalized rectangular and polar form. (d) Choose a direction for the current phasor and label the resistor polarity accordingly. (e) Use the KVL to find the current phasor. Express your result in both rationalized rectangular and polar form. (f) What is the potential difference phasor across the resistor? Express your result in both rationalized rectangular and polar form. (g) What is the real current $I(t)$ through the resistor? (h) What is the real potential difference $V(t)$ across the resistor? (i) What is the phase difference between the current through the resistor and the potential difference across it? (j) What is the peak value of the current through the resistor? (k) What is the rms value of the current in the resistor? (l) What is the peak value of the potential difference across the resistor? (m) What is the rms value of the potential difference across the resistor? (n) What average power is absorbed by the resistor?

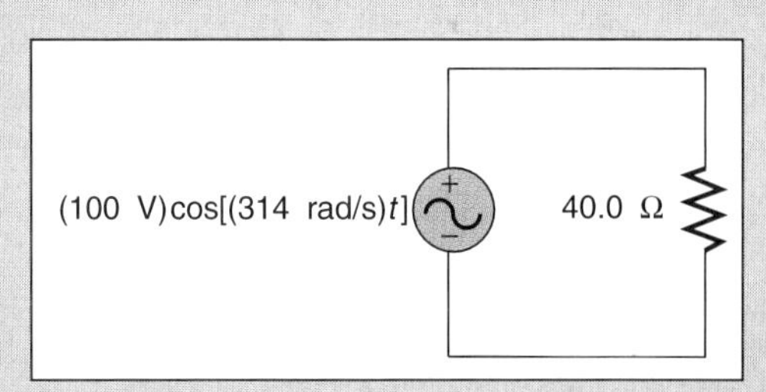

FIGURE P.30

•**31.** An ac ammeter and an ac voltmeter are used in Problem 30 to measure the current through the resistor and the potential difference across it. (a) Indicate how they should be connected to the circuit. (b) What will be the reading indicated by each meter?

•**32.** A resistor is connected to an ac independent voltage source as shown in Figure P.32. The peak value of the ac current through the resistor is I_0 when the source frequency is ν. If the frequency of the source is changed to 2ν, what is the value of the peak current through the resistor?

•**33.** Find the dc current I_{dc} through a resistor R that will produce the same power absorbed as an ac current with a peak value of I_0.

•**34.** Consider the circuit indicated in Figure P.34. (a) What is the frequency ν of the source? (b) What is the impedance of the capacitor? (c) What is the independent voltage source phasor? Express your result in both exponential and polar form. (d) Choose a direction for the current phasor and label the capacitor polarity accordingly. (e) Use the KVL to find the

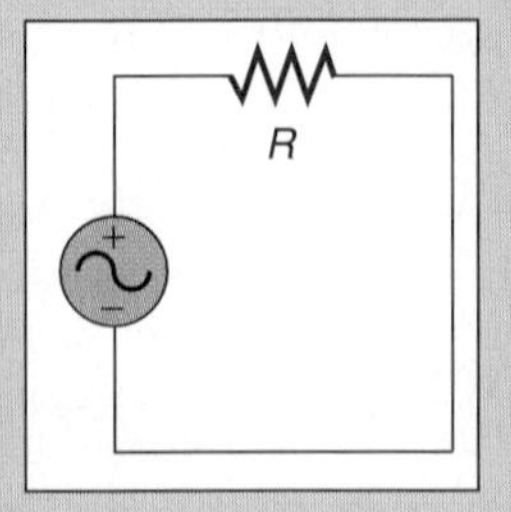

FIGURE P.32

current phasor. Express your result in both polar and exponential form. (f) What is the potential difference phasor across the capacitor? Express your result in both polar and exponential form. (g) What is the real current $I(t)$ through the capacitor? (h) What is the real potential difference $V(t)$ across the capacitor? (i) What is the phase difference between the potential difference across the capacitor and the current through it? (j) What is the peak value of the current through the capacitor? (k) What is the rms value of the current in the capacitor? (l) What is the peak value of the potential difference across the capacitor? (m) What is the rms value of the potential difference across the capacitor? (n) What is the power factor associated with the capacitor? (o) What average power is absorbed by the capacitor?

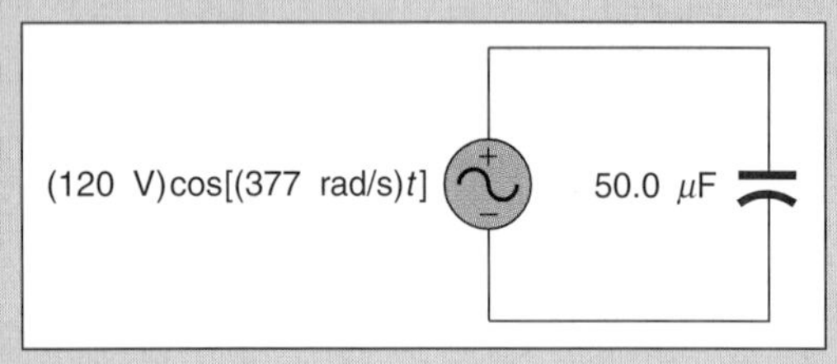

FIGURE P.34

•**35.** An ac independent voltage source described by

$$(20.0\ \text{V})\cos[(377\ \text{rad/s})t]$$

is connected to a 100 μF capacitor as shown in Figure P.35. (a) What is the impedance of the capacitor? (b) What is the capacitive reactance X_C? (c) What is the *peak* value of the current through the capacitor? (d) At what time $t \geq 0$ s does the current in the capacitor first reach its peak value?

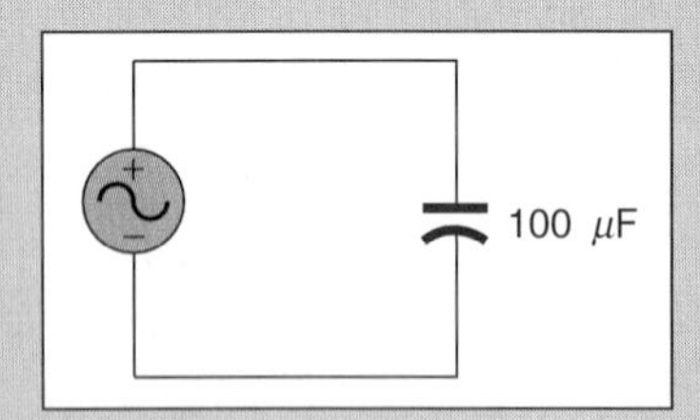

FIGURE P.35

•**36.** A capacitor is connected to an ac independent voltage source as shown in Figure P.36. The peak value of the ac current through the capacitor is I_0 when the source frequency is ν. If the frequency of the source is changed to 2ν, what is the new value of the peak current through the capacitor?

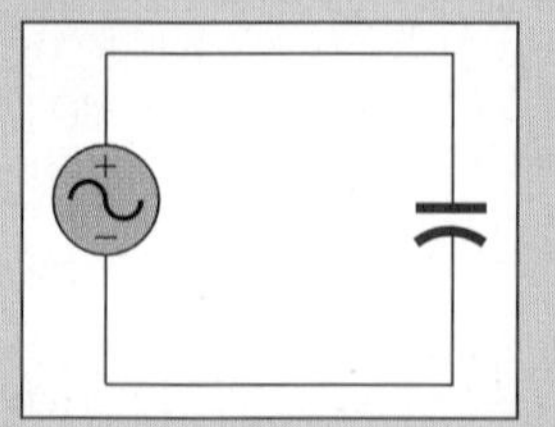

FIGURE P.36

•**37.** Consider the circuit indicated in Figure P.37 (a) What is the frequency ν of the source? (b) What is the impedance of the inductor? (c) What is the independent voltage source phasor? Express your result in both exponential and polar form. (d) Choose a direction for the current phasor and label the inductor polarity accordingly. (e) Use the KVL to find the current phasor. Express your result in both polar and exponential form. (f) What is the potential difference phasor across the inductor? Express your result in both polar and exponential form. (g) What is the real current $I(t)$ through the inductor? (h) What is the real potential difference $V(t)$ across the inductor? (i) What is the phase difference between the potential difference across the inductor and the current through it? (j) What is the peak value of the current through the inductor? (k) What is the rms value of the current in the inductor? (l) What is the peak value of the potential difference across the inductor? (m) What is the rms value of the potential difference across the inductor? (n) What is the power factor associated with the inductor? (o) What average power is absorbed by the inductor?

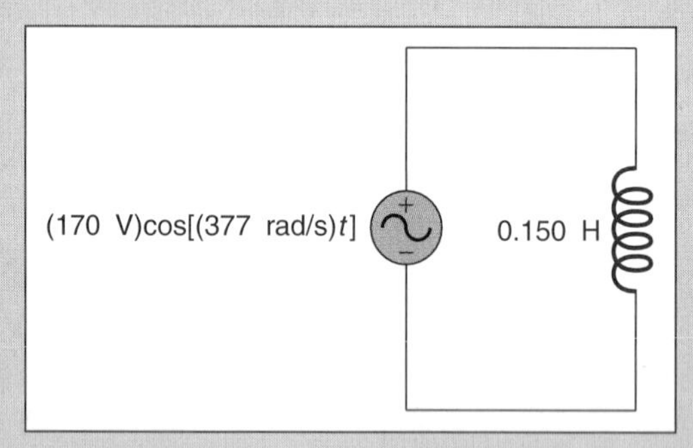

FIGURE P.37

•**38.** An inductor is connected to an ac independent voltage source as shown in Figure P.38. The peak value of the ac current through the inductor is I_0 when the source frequency is ν. If the frequency of the source is changed to 2ν, what is the new value of the peak current through the inductor?

FIGURE P.38

•**39.** An ac independent voltage source described by

$$(20.0 \text{ V}) \cos[(377 \text{ rad/s})t]$$

is connected to a 50 mH inductor as shown in Figure P.39. (a) What is the impedance Z_L of the inductor? (b) What is the inductive reactance X_L? (c) What is the peak value of the current through the inductor? (d) At what time $t \geq 0$ s does the current in the inductor first reach its peak value?

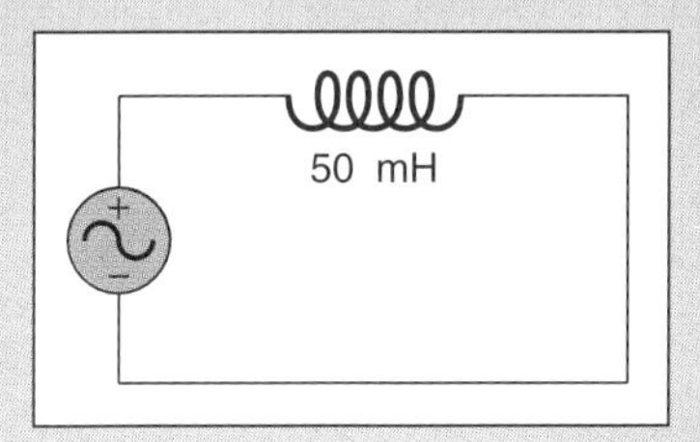

FIGURE P.39

•**40.** A series *LR* circuit is driven by an ac independent voltage source as shown in Figure P.40. (a) Redraw the circuit in the complex domain indicating the impedances of the circuit elements as well as the polar form of the complex source voltage phasor $\mathbf{V}_{\text{source}}$. (b) Choose the direction for the complex current phasor $\mathbf{I}$ so it emerges from the nominally positive terminal of the ac voltage source. Use the KVL to show that the current phasor has the form

$$\mathbf{I} = \frac{\mathbf{V}_{\text{source}}}{R + i\omega L}$$

(c) Show that the current phasor in polar form is

$$\mathbf{I} = \frac{V_0}{\left[R^2 + (\omega L)^2\right]^{1/2}} \angle(\omega t - \theta)$$

where

$$\tan\theta = \frac{\omega L}{R}$$

(d) Show that the real current $I(t)$ in the circuit is

$$I(t) = \frac{V_0}{\left[R^2 + (\omega L)^2\right]^{1/2}} \cos(\omega t - \theta)$$

(e) Find the potential difference phasors across both the resistor and the inductor and use them to show that the real potential differences across the resistor and the inductor are

$$V_{\text{res}}(t) = \frac{V_0 R}{\left[R^2 + (\omega L)^2\right]^{1/2}} \cos(\omega t - \theta)$$

and

$$V_{\text{ind}}(t) = \frac{V_0 \omega L}{\left[R^2 + (\omega L)^2\right]^{1/2}} \cos\left(\omega t - \theta + \frac{\pi}{2}\text{ rad}\right)$$

Notice that the potential difference across the resistor is in phase with the current through it, while the potential difference across the inductor and the current through it are $\pi/2$ rad out of phase. (f) What is the power factor of the inductor? (g) What is the average power absorbed by the inductor? (Think! A detailed calculation is not needed!) (h) What is the power factor of the resistor? (i) What is the average power absorbed by the resistor? (j) What is the power factor of the source? (k) What is the average power absorbed by the independent voltage source?

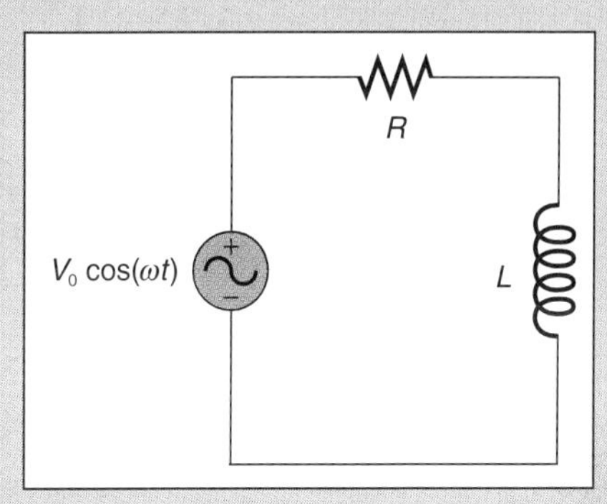

FIGURE P.40

•**41.** For the low pass filter of Figure P.41, what value of the angular frequency produces a peak potential difference across the capacitor equal to half the amplitude of the source voltage? That is, find the value of ω for which

$$V_{C\,\text{peak}}(\omega) = \frac{V_0}{2}$$

Under these circumstances, what is the voltage gain?

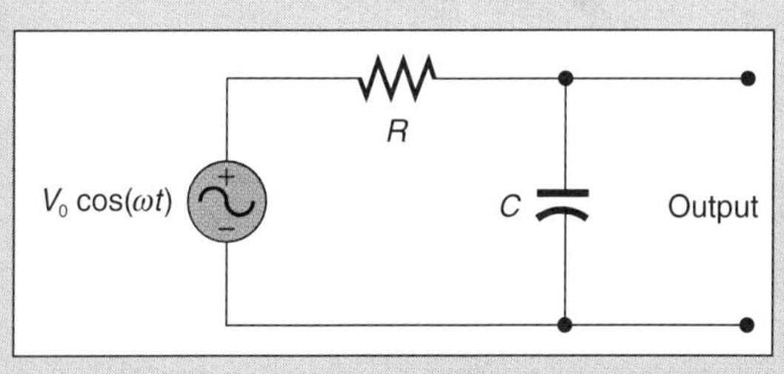

FIGURE P.41

•**42.** For what value of the angular frequency is the peak potential difference across the resistor in a high pass *RC* filter equal to half the source voltage? That is, find the value of ω for which

$$V_{R\,\text{peak}}(\omega) = \frac{V_0}{2}$$

What is the voltage gain under these circumstances?

•**43.** (a) In Equation 22.65 on page 1023, for what value of ω is the phase angle $\pi/2$ rad? For what value of ω is the phase angle $\pi/4$ rad? (b) Sketch a graph of the phase angle ϕ in Equation 22.65 as a function of the angular frequency ω. The angle ϕ represents the phase angle difference between the current in the circuit and the source voltage. Notice that as the angular frequency increases and the impedance of the capacitor decreases, the current and source voltage are increasingly in phase. For very large ω, the capacitor shorts out and the circuit is purely resistive with the current in phase with the source voltage.

•**44.** A high pass filter has $R = 1.00\ k\Omega$ and $C = 0.0500\ \mu F$ as shown in Figure P.44. A graph of the logarithm (base 10) of the voltage gain (Equation 22.78 on page 1026) versus the logarithm (base 10) of the frequency ν ($= 2\pi\omega$) is called a *Bode plot.* (The logarithm of the voltage gain is the ordinate and the logarithm of the frequency is the abscissa.) (a) Construct a Bode plot for this high pass filter over the frequency domain from 10 Hz to 100 kHz. (b) Estimate the frequency at which the voltage gain is 0.50.

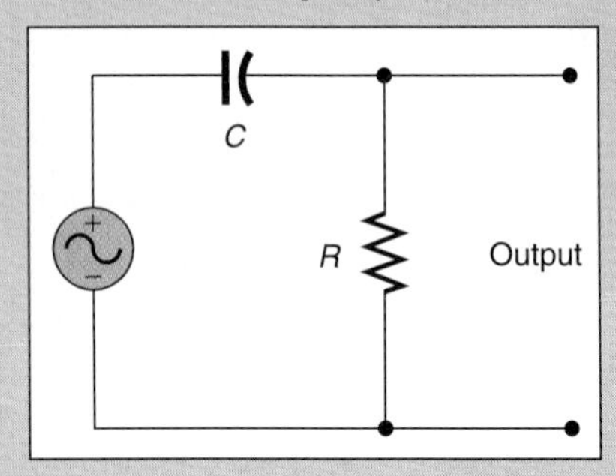

FIGURE P.44

•**45.** A low pass filter circuit has $R = 1.00\ k\Omega$ and $C = 0.0500\ \mu F$ as shown in Figure P.45. A graph of the logarithm (base 10) of the voltage gain (Equation 22.74 on page 1025) versus the logarithm (base 10) of the frequency ν ($= 2\pi\omega$) is called a *Bode plot.* (The logarithm of the voltage gain is the ordinate and the logarithm of the frequency is the abscissa.) (a) Construct a Bode plot for this low pass filter over the frequency domain from 10 Hz to 100 kHz. (b) Estimate the frequency at which the voltage gain is 0.50.

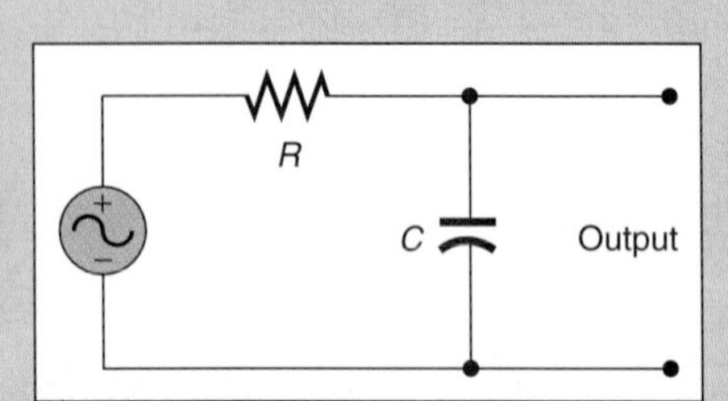

FIGURE P.45

•**46.** Show that the peak of the average power absorbed by the resistor in a series *RLC* circuit, Equation 22.92 on page 1029, occurs where $\omega = \omega_0$ by taking the derivative

$$\frac{d\langle P\rangle}{d\omega}$$

and setting it equal to zero.

•**47.** In a series *RLC* circuit, a variable capacitor is used to tune the circuit over the frequencies of the AM radio band, from about 500 kHz to 1600 kHz. (a) If the inductor has an inductance $L = 650\ \mu H$, what must be the range of the variable capacitor to be able to tune in the spectrum of frequencies in the AM radio band? The radio stations of the AM radio band are separated from each other by 10 kHz in frequency. The bandwidth of the tuning circuit must be smaller than this to ensure that when you tune in one station, you do not pick up neighboring stations (which would result in cross talk, listening to two stations simultaneously!). (b) If the bandwidth of the circuit is to be 2.0 kHz, what is the Q of the circuit when tuned to a station at 500 kHz? What does this Q imply about the value of the resistance R in the circuit? (c) For a bandwidth of 2.0 kHz, what is the Q of the circuit when tuned to a station at the other end of the AM band at 1600 kHz? What does this Q imply about the size of the resistance in the circuit?

‡**48.** The potential difference across a resistor with resistance R has the time dependence shown in Figure P.48. (a) Determine the power absorbed by the resistor during the interval from when $t = 0$ s to $t = T/2$. (b) Determine the power absorbed by the resistor from when $t = T/2$ to $t = T$. (c) Calculate the average power absorbed by the resistor during the interval from when $t = 0$ s to $t = T$. (d) Determine the average potential difference across the resistor over a complete cycle from when $t = 0$ s to $t = T$. (e) If the average potential difference across the resistor over one cycle now is applied continuously (in a dc way) to the resistor, what is the power absorbed by the resistor?

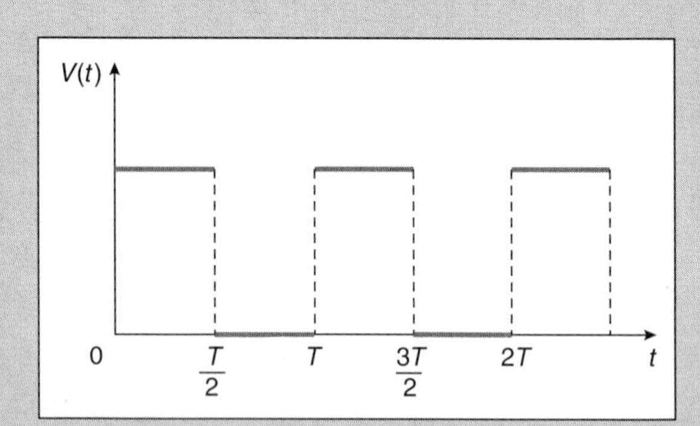

FIGURE P.48

‡**49.** Figure P.49 depicts a *parallel RLC* circuit. Notice that since the resistor, capacitor, and inductor are in parallel with the ac independent voltage source, all the circuit elements have the same potential difference across them, equal to the potential difference across the source. (a) Convert the circuit to the complex domain by indicating the polar form of the source voltage phasor and the impedances of the resistor, the capacitor, and the inductor. Use X_C for the capacitive reactance of the capacitor and X_L for the inductive reactance of the inductor. (b) Show that the polar form for the current phasors in the resistor, the capacitor, and the inductor are (using the current directions indicated in Figure P.49)

$$\mathbf{I}_{\text{res}} = \frac{V_0}{R}\angle(\omega t)$$

$$\mathbf{I}_{\text{ind}} = \frac{V_0}{X_L}\angle\left(\omega t - \frac{\pi}{2}\ \text{rad}\right)$$

$$\mathbf{I}_{\text{cap}} = \frac{V_0}{X_C}\angle\left(\omega t + \frac{\pi}{2}\ \text{rad}\right)$$

(c) Show that the equivalent impedance Z_{eq} of the parallel *RLC* combination satisfies

$$\frac{1}{Z_{eq}} = \frac{1}{R} + i\left(\frac{1}{X_C} - \frac{1}{X_L}\right)$$

(d) Use the expression for the equivalent impedance found in part (c) to show that the peak value of the current in the source is

$$V_0\left[\left(\frac{1}{R}\right)^2 + \left(\frac{1}{X_C} - \frac{1}{X_L}\right)^2\right]^{1/2}$$

(e) What angular frequency ω maximizes the peak value of the total current from the source? This is the resonant angular frequency for the parallel *RLC* circuit. How does the resonant angular frequency for the *parallel RLC* circuit compare with the resonant angular frequency for the *series RLC* circuit for the same values of *R*, *L*, and *C*?

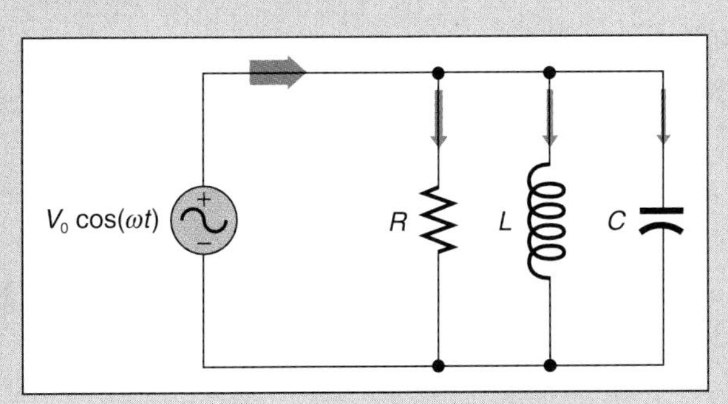

FIGURE P.49

50. For the *RC* filter circuit discussed in Section 22.8, show explicitly that the real potential differences across the voltage source, the resistor, and the capacitor satisfy the KVL at any instant *t*. See Figure P.50. That is, show that

$$-V(t) + V_R(t) + V_C(t) = 0 \text{ V}$$

where

$$V(t) = V_0 \cos(\omega t)$$

$$V_R(t) = \frac{\omega RC V_0}{\left[1 + (\omega RC)^2\right]^{1/2}} \cos(\omega t + \phi)$$

and

$$V_C(t) = \frac{V_0}{\left[1 + (\omega RC)^2\right]^{1/2}} \cos\left(\omega t + \phi - \frac{\pi}{2} \text{ rad}\right)$$

Recall from Equation 22.65 that

$$\tan \phi = \frac{1}{\omega RC}$$

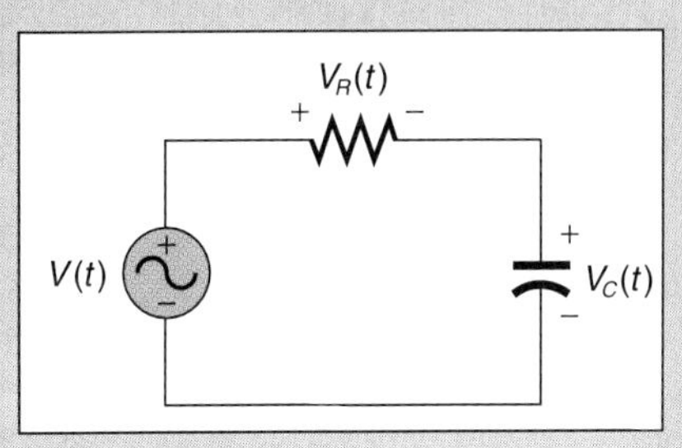

FIGURE P.50

51. The electrical grid in the United States is a *three-phase* grid because the ac independent voltage sources provided consist of three sources that can be represented by the following expressions:

$$V_1(t) = V_0 \cos(\omega t)$$

$$V_2(t) = V_0 \cos\left(\omega t - \frac{2\pi}{3} \text{ rad}\right)$$

$$V_3(t) = V_0 \cos\left(\omega t - \frac{4\pi}{3} \text{ rad}\right)$$

Show that the potential difference between *any two* of these sources has a peak value of $\sqrt{3}V_0$ and oscillates sinusoidally at angular frequency ω. Hint:

$$\cos\alpha - \cos\beta = -2\sin\left(\frac{\alpha+\beta}{2}\right)\sin\left(\frac{\alpha-\beta}{2}\right)$$

Investigative Projects

A. Expanded Horizons

1. The invention of new math is part of the creative process enjoyed by mathematicians; such math occasionally finds use in physics and engineering. Although it is somewhat removed from physics, investigate the properties of the so-called *hallucinatory numbers*, also called the *perplex numbers*, based on a number *h* whose absolute value is minus one: $|h| = -1$.
Paul Fjelstad, "Extending special relativity via the perplex numbers," *American Journal of Physics*, 54, #5, pages 416–422 (May 1986).
V. Majernik, "The perplex numbers are in fact the binary numbers," *American Journal of Physics*, 56, #8, page 763 (August 1988).

2. Charles Proteus Steinmetz and Albert Einstein were equally fascinating to the public during the early decades of the 20th century. Steinmetz made significant contributions to lightning research and even wrote a book about special relativity. Someone (perhaps in jest) long ago suggested the frequency unit not be named after Heinrich Hertz, but after C. P. Steinmetz, for then the unit could be abbreviated cps,

a cycle per second, from his initials! If you are interested in learning more about Steinmetz and perhaps discovering the reasons that he subsequently has become all but unknown to the general public (unlike Einstein), you might enjoy reading and comparing each of the following biographies, one written just after his death, the other after the passage of some time.
Ronald R. Kline, *Steinmetz: Engineer and Socialist* (The Johns Hopkins University Press, Baltimore, 1992).
John Winthrop Hammond, *Charles Proteus Steinmetz* (The Century Co., New York, 1924).

3. Investigate the electrical effects generated in body tissue with a view toward understanding the electricity and electronics of an electrocardiogram.
John R. Cameron and James G. Skofronick, *Medical Physics* (Wiley, New York, 1978), Chapter 9.
Robert O. Becker and Andrew A. Marino, *Electromagnetism and Life* (State University of New York Press, Albany, 1982).

4. Investigate the applications of high- and low-frequency currents and potential differences in medicine.
John R. Cameron and James G. Skofronick, *Medical Physics* (Wiley, New York, 1978), Chapter 11.

5. Investigate the differences between ac and dc electrical motors.
Richard A. Honeycutt, *Electromechanical Devices* (Prentice-Hall, Englewood Cliffs, New Jersey, 1986), Chapters 5 and 6, pages 111–163.

6. Investigate the differences between ac generators and alternators.
Richard A. Honeycutt, *Electromechanical Devices* (Prentice-Hall, Englewood Cliffs, New Jersey, 1986), Chapter 4, pages 83–110.

B. Lab and Field Work

7. Your physics department likely has an independent voltage source of variable frequency (occasionally called a signal or function generator) and an oscilloscope for measuring time-dependent potential differences. Use a 1.0 kΩ resistor and a 0.50 μF capacitor to design and investigate the frequency dependence of the voltage gain of both a high pass and a low pass filter. See Problems 44 and 45.

8. Design and perform an experiment to investigate the frequency dependence of the phase difference between a sinusoidal ac source and the resulting current in a series *RC* circuit. Your instructor may be able to assist you in devising a method for displaying and measuring the phase difference on an oscilloscope.

9. Given an inductor with an unknown inductance and a capacitor with an unknown capacitance, devise and perform an experimental procedure to measure their reactances and thereby determine the values of the inductance and capacitance.

C. Communicating Physics

10. Write a paper investigating, comparing, and contrasting the advantages and disadvantages of transmitting electrical energy over long distances by dc and by ac methods. In the early part of the 20th century, why did ac methods win out over dc methods? Will the practical development of relatively low cost, high temperature superconducting materials (which have zero resistance for conduction currents) change the terms of the debate? Write the paper for a nontechnically educated audience.
James P. Rybak, "AC or DC?" *Popular Electronics, 11*, #9, pages 42–48 (September 1994).
George Westinghouse Commemoration (The American Society of Mechanical Engineers, New York, 1937).

11. Amazingly, the invention of the electric chair was associated with the competition between ac and dc means of transmitting electricity. Thomas Edison (1847–1931) favored dc transmission while George Westinghouse (1846–1914), among others, favored ac. Edison invented an ac electric chair and had the method of execution adopted by New York State, primarily to illustrate the danger associated with ac electricity and to discredit the technology. Ironically, the ploy failed. This amazing story is a fascinating blend of physics, engineering, politics, and the worst elements of human nature. You might find it an interesting topic to investigate and report to your class.
"Edison's Miracle of Light" (WGBH Educational Foundation show, *American Enterprise*, 23 October 1995, Transcript 802).

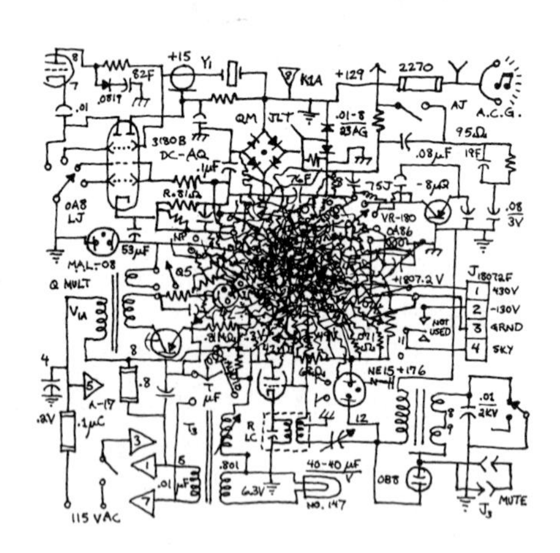

115 VAC
MUTE
52. Calculate the potential difference across and the power absorbed by each circuit element.

Chapter 23
Geometric Optics

But soft! what light through yonder window breaks?
William Shakespeare (1564–1616)*

The nature of light and the way it passes through and interacts with matter are rich veins of physics that have led to many significant discoveries, revolutionary ideas, and technological advances. Newton thought light was a stream of particles and his theory of the mechanics of particles was able to account for both reflection and refraction. In Chapter 21 (Sections 21.4 and 21.5), we saw how the four fundamental equations of electromagnetism led James Clerk Maxwell to conclude light was electromagnetic waves.† Later, in Chapter 26, we shall see that light has aspects of *both* particles and waves, a seemingly contradictory situation. In the 20th century the speed of light also assumed an important role in relativity theory (Chapter 25).

But investigations into light and optical instruments had begun millennia before Newton and Maxwell, who published their work in the 17th and 19th centuries. Indeed, reflections from water surfaces likely were noted curiously by prehistoric peoples. Mirrors of polished metal were manufactured by hand in ancient Egypt as well as other ancient middle eastern cultures. The hieroglyph for *life*, called the *ankh*, ☥, is thought to be a stylized rendition of the shape of ancient Egyptian hand mirrors. Today, infants and even some of our pets seem intrigued by their reflections in mirrors. Birds have been known to attack their reflections, with typically frustrating results!

Transparency of certain materials (for example, water and crystals) is a fact of nature. The discovery of glass and glass making arose in many of the early middle eastern civilizations of the Fertile Crescent. The great Greek-Alexandrian mathematician, astronomer, and scientist Ptolemy‡ (~A.D. 150) wrote a long treatise on the refraction or bending of light at the interface between two transparent materials, even deducing a relationship between the directions of the incident and refracted rays that was correct for small angles.

The invention of lenses arose from unknown sources in the early Middle Ages, likely by North African Islamic Arabs, and resulted in the invention of spectacles to correct vision problems. The word lens come from the Latin *lens*, meaning "lentil bean," because of their similar shapes. Ibn-al-Haithan (965–1038) discovered the relationship between object and image distances in lenses.

The likely accidental invention of the telescope§ about 1608 and the microscope# about the same time began the application of the optical principles of reflection and refraction to more sophisticated instrumentation that continues to this day.

In this chapter we first explore the domains of optics and the empirical laws of reflection and refraction. We will see how these laws are used to understand the image-forming properties of useful optical devices such as mirrors and lenses, as well as more complicated optical systems consisting of several optical elements.

The ankh hieroglyph ☥, meaning "life," appears on the base of the statue at the Ramasseum near Luxor, Egypt, that inspired Percy Bysshe Shelley's famous poem *Ozymandias*.

Little known outside the Middle East, Ibn-al-Haithan (spellings vary) made significant contributions to optics at the turn of the last millennium.

Manufactured optical instruments such as cameras, microscopes, and telescopes have been and continue to be of immense practical use.** Our eye (we have two!) is one of the most remarkable optical instruments known. It is humbling to realize that eyes developed naturally through biological evolution during aeons in which the scientific and inventive prowess of our intellect were nonexistent. Nature is more creative and inventive than we are.

23.1 THE DOMAINS OF OPTICS

To begin our investigations into light, we assume that the directions of motion of light are represented by directed straight lines called **rays**. Rays do not travel, although quite frankly the language we associate with them might imply that they do! Rays really indicate *paths* of light. Rays are perpendicular to the wave crests in a wave model.

*(Chapter Opener) *Romeo and Juliet*, Act II, Scene ii.

†The Maxwell equations further are able to account for the laws of reflection and refraction. There also were much earlier experiments (by Young in 1800) which implied that light had wavelike properties. We discuss Young's experiments in Chapter 24.

‡Ptolemy also codified the then dominant geocentric theory of the solar system, known as the Ptolemaic theory.

§The discovery was made independently by three Dutch spectacle makers: Hans Lippershey, Zacharias Jansen, and James Metius.

#By Robert Hooke (1635–1703) (and others), whom we also know for Hooke's law; see Chapter 7.

**The development of sophisticated CCD cameras, electron microscopes, and new-technology telescopes still are based, fundamentally, on the basic principles we study in this chapter.

Light rays indicate the paths that light follows. A thin laser beam can be considered as a ray such as in this laboratory.

The study of light can be organized into three broad domains: geometric optics, physical optics, and quantum optics. These domains are not strictly disjoint. The complicated transitions between them are continuous, not sharp; but, for simplicity, we consider them as distinct. We distinguish the domains as follows:

- **geometric optics**: employs only rays
- **physical optics**: employs rays and waves
- **quantum (photon) optics**: employs rays, waves, and energy bundles called photons

The distinction between geometric and physical optics is delineated by comparing the wavelength of the light under study with the characteristic dimensions of the devices or systems interacting with the light.* The distinction between physical and quantum optics will be based on whether the wave or the (yet to be studied) particle aspects of light are manifest.

*We briefly touched on these ideas in Section 12.17.

To see how the dimensions of devices interacting with light are used to distinguish between the geometric and wave domains, consider the following analogy. Imagine a series of ocean wave crests or swells methodically moving across the surface of the sea. Let the wave have a characteristic wavelength λ and speed v. The wave enters a large bay as shown in Figure 23.1.

If we neglect the complicated edge effects near the shore, the wave crests go directly into the bay; they eventually crash into the beach giving much pleasure to local surfers. However, suppose you just invested in a very expensive sailboat and want to moor the boat in the bay. This could be quite hazardous to the boat during severe storms because the boat is unprotected from the ravages of large wave crests entering the wide bay. Your congresswoman obligingly uses her influence to get the Army Corps of Engineers to construct a breakwater across the mouth of the bay as shown in Figure 23.2. The breakwater narrows the entrance of the bay to a fraction of its former size. The incident ocean wave crests now are presented with a different geometry: a much smaller entrance to the harbor. The wave crests now fail to go straight into the harbor in a single direction, but rather spread out as they come through the opening, traveling in *many* different directions. Your boat is safe, though the surfing is ruined. The amplitude of the wave crests in the harbor is greatest in the straight-through direction, but there are wavelets of smaller amplitude propagating in directions quite different from the incident direction of the ocean swell.

This situation can be summarized in the following way. When the aperture of the bay is much greater than the wavelength of the waves, the wave crests go straight into the bay, propagating in a single direction (neglecting the effects at the edges of the wide bay aperture). This is the geometric approximation or limit. On the other hand, when the size of the aperture is on the order of the wavelength or less, the wave crests spread out in the region behind the opening, moving in many different directions. This spreading out of a wave (diffraction) as it passes through an opening or around an obstacle occurs when the size of the opening (or obstacle) is comparable to or smaller than the wavelength of the waves incident on it.

The same ideas apply to light:

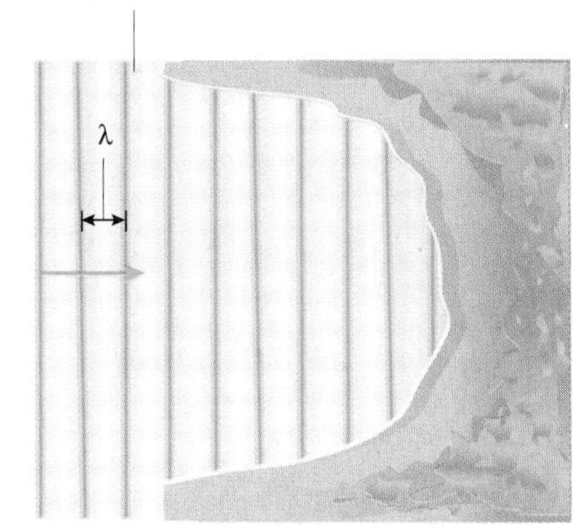

FIGURE 23.1 Waves enter a large bay.

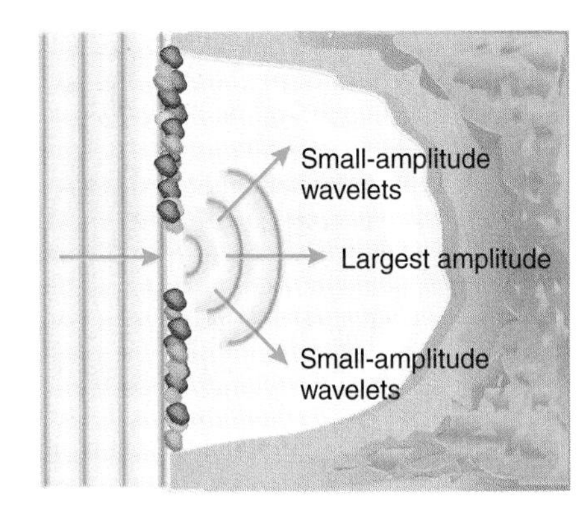

FIGURE 23.2 **A breakwater across the mouth of the bay causes diffraction.**

If its wavelength is much less than the size of the opening or obstacle presented to the light, then the geometric limit is appropriate and each incident light ray subsequently travels in a unique direction or single ray. Well-defined shadows exist in regions where there are no rays. This is the domain of geometric optics.

If its wavelength is on the order of the size of the opening through which the light passes, then the wave limit is appropriate and we must account for diffraction effects. This is the domain of physical optics, which we consider in Chapter 24.

The geometric limit of light is appropriate when the size of its wavelength is negligible, such as when visible light passes through the openings in these clouds.

Although we shall be concerned principally with *visible* light, these ideas also apply equally to other portions of the electromagnetic spectrum. For example, visible light passes straight through large apertures such as door and window openings, even small keyholes, with well-defined shadow regions; the geometric limit is appropriate and we can conclude that the wavelength of visible light is much smaller than the dimensions of these apertures. Thus visible light travels in straight lines by or through these openings, and the ray approximation is an adequate description of the behavior.

On the other hand, if you carry a radio through a door opening or around to the back of your house, the antenna does not lose the station; hence, radio waves must diffract around such openings or obstacles,* indicating that radio waves have wavelengths comparable to or exceeding the dimensions of common objects in our environment. The geometric limit, therefore, is not appropriate for radio waves interacting with such obstacles or apertures. The diffraction of radio waves around objects of macroscopic size such as small hills is why one does not have to be along a clear line of sight to a radio transmitter in order to receive its signal.

In this chapter we study geometric optics, and so we need be concerned here only with the interaction of rays of light with reflecting surfaces (mirrors) and refracting surfaces (abrupt boundaries between transparent media) that are large compared with the wavelength. Wavelengths are occasionally mentioned in this chapter for their familiar use in describing different colors of light, but the observational evidence that light is a wave can be postponed to serve as the real core of the following chapter on physical optics.

*This assumes the materials are not transparent to radio waves. In fact, many nonconducting materials are transparent to radio waves, but you likely are able to see the point we are trying to make here.

23.2 The Inverse Square Law for Light

There is one glory of the sun, and another of the moon, and another glory of the stars; for one star differeth from another star in glory.
The New Testament, 1 Corinthians 15:41

Go outside on a clear night and there are few of us so crass that we are not impressed with or humbled by the magnificence of the starry sky. Some stars are quite bright, others are barely detectable to the eye. Stars differ in brightness not only because of their various intrinsic luminosities (their power output) but also because they are at many different distances from us. These common and well-known conclusions (which are suggested by, but are not at all obvious from simply looking at the night sky) raise questions about what we mean by dim or bright and how the light collected by, say, our eyes or telescopes varies with their distance from the light source.

The more distant you are from a pointlike light source, the dimmer it is. A pointlike star, after all, is a distant sun, but it provides precious little nighttime illumination because of its vast distance from us. Consider such a pointlike or spherical source of light, as in Figure 23.3, radiating equally in all directions, much like a distant star or the relatively nearby Sun. Let L be a measure of the total power of the light sent out in all directions by the source, which is the amount of energy emitted per second (in watts), called the **luminosity** of the source.†

The wattage of a light bulb is a measure of its luminosity (at all wavelengths, not necessarily just in the visible portion of the electromagnetic spectrum). The light bulb source is only pointlike when viewed from distances r at which its physical size is negligible. Certainly all the power must pass through any imagined sphere of radius r centered on the pointlike source (as in Figure 23.3). Since $4\pi r^2$ is the area of the sphere, the power per unit area on the sphere, called the **intensity** I, thus is

$$I \equiv \frac{L}{4\pi r^2} \tag{23.1}$$

†The luminosity of the Sun is 3.83×10^{26} W, a rather bright light bulb!

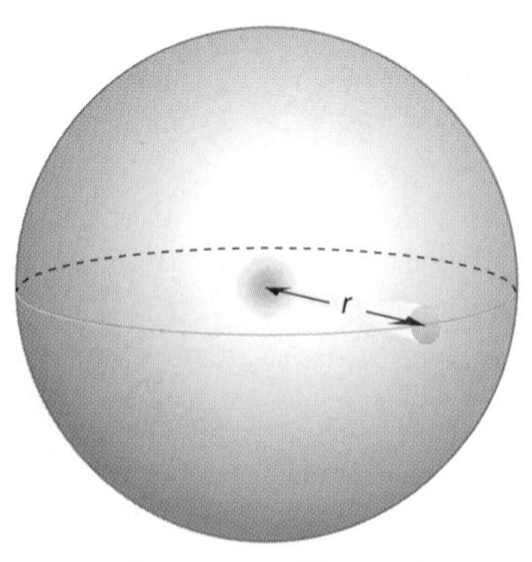

FIGURE 23.3 **A pointlike source of light radiates equally in all directions.**

The intensity received from the pointlike or spherical source decreases as the inverse square of the distance r, an **inverse square law**.

The inverse square law for light intensity makes no distinction as to whether the light energy is associated with waves or particles. The law applies equally well to *both* viewpoints and cannot be used to distinguish between them; other experiments must do that.

Recall that in Chapter 12 (page 548), we defined the intensity of *any* wave in a similar way: the average power transmitted through one square meter oriented perpendicular to the direction the wave is travelling. The intensity is proportional to the square of the amplitude of the wave.

The **brightness** B is a measure of the power collected by, say, your eye or a telescope of aperture area A oriented perpendicular to the rays from the source:

$$B \equiv IA \qquad (23.2)$$

Hence using a telescope with a larger aperture than your eye increases the brightness of a star, although the intensity of the light incident on both is the same, since the telescope is at the same distance from the source as your eye. Likewise, you see light sources as brighter with a dilated pupil than with it contracted, something you may have experienced uncomfortably during and after an opthamological exam.

From the meaning of the intensity we can deduce that the amplitude of the *spherical wave* diverging from a pointlike or spherical source must decrease as $1/r$. That is, the amplitude of a spherical wave is proportional to the inverse first power of the distance r from the source.

EXAMPLE 23.1

The Sun has a luminosity of 3.83×10^{26} W. What is the intensity of solar light on the Earth, located 1.496×10^8 km from the Sun? This intensity is called the *solar constant*.

Solution

Since the light energy from the Sun is spread equally over a sphere centered on the Sun, the intensity is the luminosity divided by the area of a sphere with a radius equal to the distance of the Earth from the Sun. Use Equation 23.1:

$$\begin{aligned} I &= \frac{L}{4\pi r^2} \\ &= \frac{3.83 \times 10^{26}\ \text{W}}{4\pi(1.496 \times 10^{11}\ \text{m})^2} \\ &= 1.36 \times 10^3\ \text{W/m}^2 \end{aligned}$$

It is this considerable intensity that motivates the use of solar energy; it is free and nonpolluting as well.

23.3 THE LAW OF REFLECTION

The phenomenon of reflection commonly is observed by us all. Indeed, for better or worse, a reflection from a mirror is one of the first apparitions that greets us each and every morning.

The path of a narrow flashlight beam or laser beam can be represented as a single ray. If the ray encounters a smooth reflecting surface as in Figure 23.4, the direction of the ray is changed abruptly at the surface. A line perpendicular to the surface at the point where the incident ray strikes the surface is called a normal line, or more simply, the **normal** to the surface. The angle θ that the **incident ray** makes with the normal line is the **angle of incidence**. The angle θ' that the **reflected ray** makes with the normal line is the **angle of reflection**.

Simple observations and measurements readily indicate two things:

1. The incident ray, the normal line, and the reflected ray all lie in the same plane.
2. The angle of incidence is equal to the angle of reflection:

$$\theta = \theta' \qquad (23.3)$$

These two observations constitute the **law of reflection**.

Note that a graph of θ' versus θ is linear and makes a 45° angle with each axis, as shown in Figure 23.5.

The first observation is important because it is conceivable (but wrong) that the reflected ray might not lie in the plane of the incident ray and the normal line, even if the two angles were equal ($\theta = \theta'$), as indicated in Figure 23.6.

The place where the incident ray strikes the reflecting surface must be **locally smooth**. This means the differential area where the ray contacts the surface must be locally flat over distances much larger than the wavelength of the incident light,

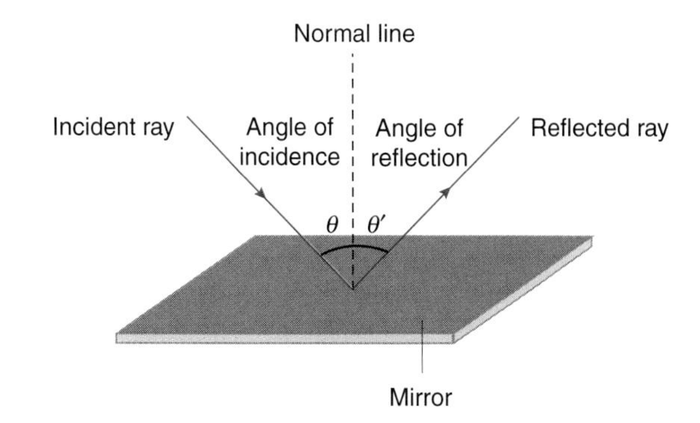

FIGURE 23.4 Reflection of a ray from a surface.

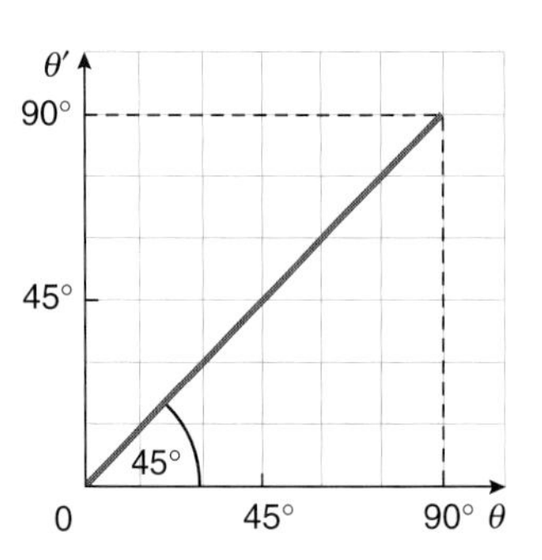

FIGURE 23.5 A graph of the angle of reflection versus the angle of incidence is a straight line with a 45° slope.

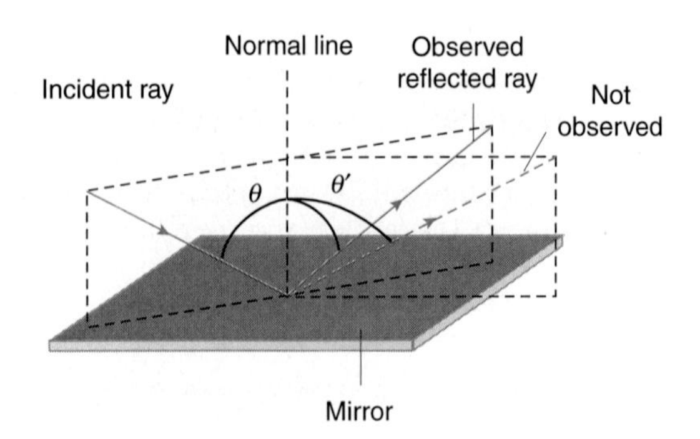

FIGURE 23.6 The reflected ray, normal line, and incident ray all lie in the same plane.

as shown in Figure 23.7. Any deviations from being locally flat are much smaller than the wavelength of the light under consideration. This type of reflection is known as **specular reflection*** and is the type of reflection that principally concerns us.

The local area is **locally rough** if the deviations from a plane in the topography of the local differential surface area are on the order of or larger than the wavelength of the incident light. A ray reflected from locally rough surfaces produces many rays going off at many different angles and directions, producing **diffuse reflection**; see Figure 23.8.

The appearance of the reflection of approaching headlights from a dry pavement surface illustrates diffuse reflection; if the pavement is wet, the reflection is more specular.

A surface that is locally rough for visible light can be locally smooth for longer wavelengths such as microwaves and radio waves. For example, a wire mesh screen is an effective locally smooth surface and a good specular reflector for microwaves and radio waves. The holes in the metal screen windows in microwave ovens permit the passage of visible light, but the screen acts as a reflecting surface for microwaves, since their wavelength is on the order of centimeters in size.

An elementary experiment indicates that light rays are reversible: if you reverse the direction of a ray, the light retraces its path, as shown in Figure 23.9.

Sophisticated experiments indicate that if the reflecting surface is moving with a significant nonzero velocity component along the normal line, the angle of incidence no longer is equal to the angle of reflection. If the law of reflection is to hold, the velocity component v of the reflecting surface along the normal line must be negligible compared with the speed c of light: $v << c$.† Similar experiments also indicate that if the reflecting surface has a velocity *parallel* to the surface itself, the law of reflection still applies.

An unusual mirror called a **corner cube reflector** consists of three plane mirrors that meet at right angles with each other. The terminology comes about because the system of three mirrors would form the corner of a cube. A ray of light reflected from a corner cube mirror system is antiparallel to the incident ray regardless of the angle that the incident ray makes with the first normal line. A corner cube reflector is a generalization to three mirrors of the situation examined in Example 23.3. The device turns the ray around and sends it back from where it came, along a parallel, though slightly displaced, ray.

These devices have many applications:

1. A satellite known as LAGEOS (an acronym for *LA*ser *GEO*dynamic *S*atellite) orbits the Earth. The satellite is a passive device that bristles with corner cube reflectors over its spherical surface. By measuring the time it takes a laser pulse to go from a point on the Earth to the satellite and back, it is possible to calculate the distance to the satellite very accurately. If this is done simultaneously from two different locations on the surface of the Earth, it is possible to triangulate and obtain an accurate measurement of the distance between the two widely separated points on the Earth.‡ Such distance determinations are so accurate that the data confirm the theory of plate tectonics: the continents are slowly shifting their positions relative to one another.

*The word comes from the Latin *specularis*, meaning "of or pertaining to mirrors."

†A treatment of reflection from high-speed mirrors can be found in R. W. Ditchburn, *Light* (Interscience, New York, 1964), pages 437–438.

‡For other ways in which precise locations are determined on the Earth, see the article by Thomas A. Herring, "The global positioning system," *Scientific American*, 274, #2, pages 44–50 (February 1996).

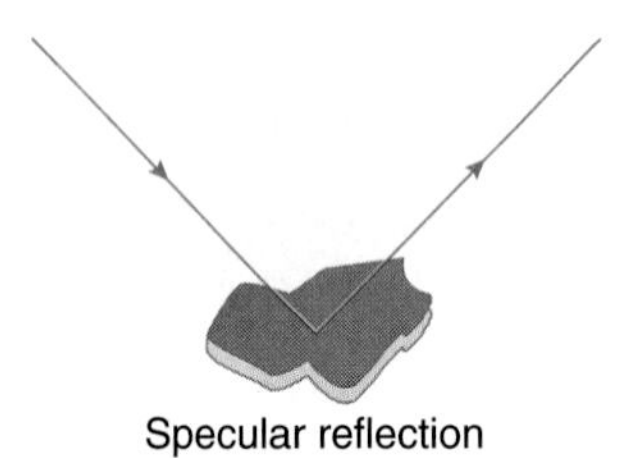

FIGURE 23.7 The differential area from which the incident ray reflects must be locally smooth for the reflected ray to follow the law of reflection.

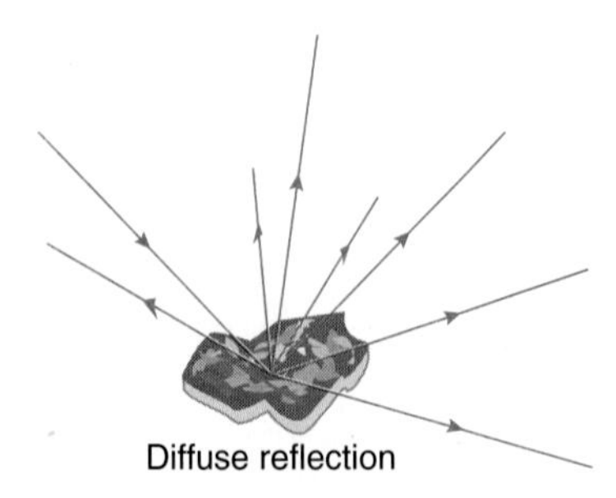

FIGURE 23.8 Diffuse reflection from locally rough surfaces.

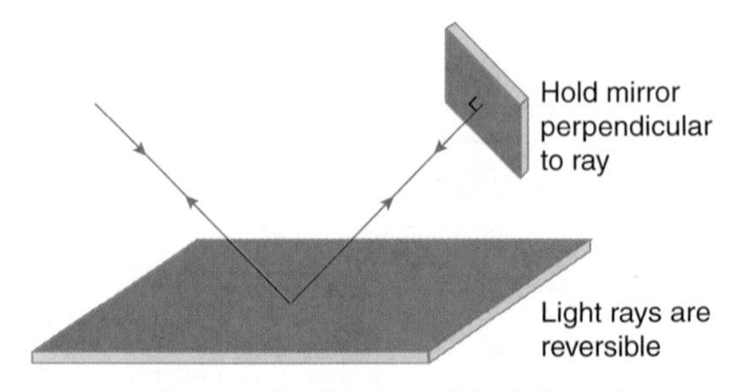

FIGURE 23.9 Reverse the direction of the light and it retraces its path.

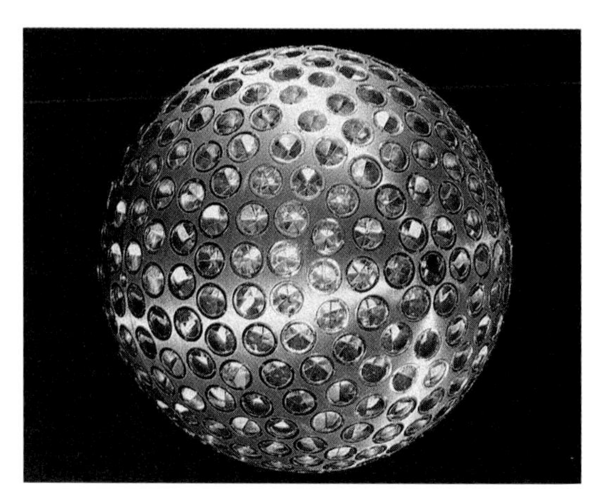

The LAGEOS satellite is covered with corner cube reflectors.

A corner cube reflector array was placed on the lunar surface during the Apollo Moon exploration voyages.

2. The first Apollo flights to the Moon set up and left behind small arrays of corner cube reflectors pointed in the general direction of the Earth. The round-trip travel times of laser pulses from the Earth to these corner cube arrays are used to measure the distance between a point on the Earth and the array on the Moon with extraordinary precision (an uncertainty of less than a centimeter). These experiments have been under way for many years now and confirm that the distance between the Moon and the Earth is increasing at the rate of 3.8 centimeters per year. The cause for this slow increase in the Earth–Moon distance is the friction between our ocean tides and the continental land masses (see Problem 53 in Chapter 10).
3. Most modern surveying and leveling equipment routinely uses corner cube reflectors.

QUESTION 1

Describe several surfaces that produce specular reflection and several that produce diffuse reflection for visible light; repeat the question for radio waves (which have much longer wavelengths than visible light).

EXAMPLE 23.2

A light ray is incident at angle θ on the reflecting surface shown in Figure 23.10. The reflecting surface now is turned through an angle ϕ as shown in Figure 23.11. Show that the *change* in the angle between the incident and reflected ray is 2ϕ, independent of the original incidence angle θ.

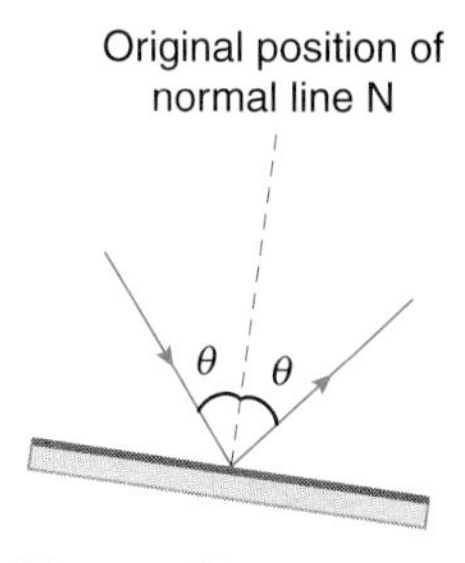

FIGURE 23.10

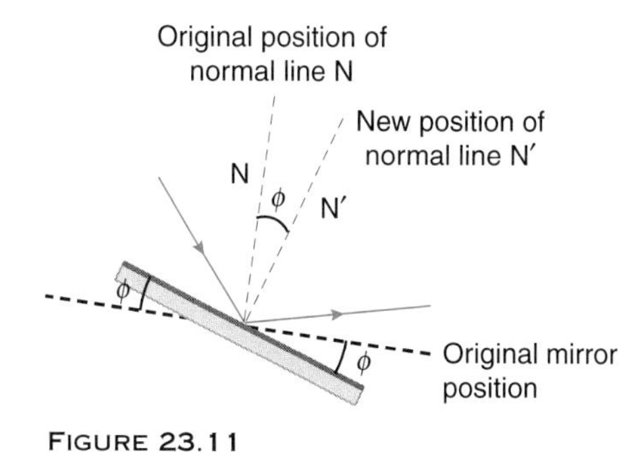

FIGURE 23.11

Solution

For the initial position of the mirror (Figure 23.10), the angle between the reflected ray and the incident ray is 2θ, according to the law of reflection.

When the mirror is rotated through the angle ϕ, the normal line also rotates through the same angle ϕ. The angle of incidence now is $\theta + \phi$, as shown in Figure 23.12.

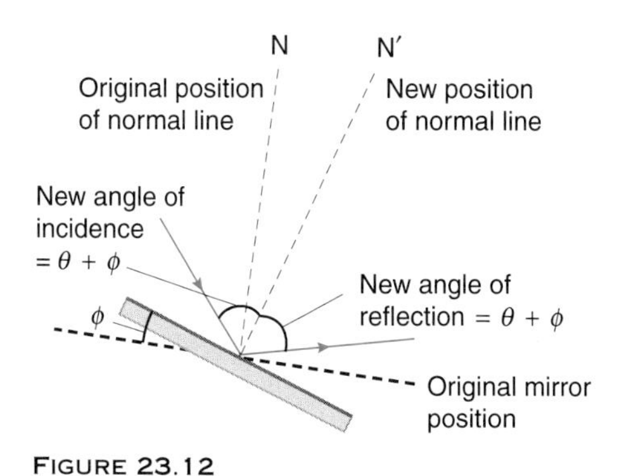

FIGURE 23.12

By the law of reflection, the angle of reflection also now is $\theta + \phi$. The angle between the new reflected ray and the incident ray then is $2(\theta + \phi)$. Hence the *change* in the angle that the reflected ray makes with the incident ray is

$$2(\theta + \phi) - 2\theta = 2\phi$$

This device is called an *optical lever*. The classic Cavendish experiment, used to determine the numerical value of the universal gravitational constant G (see Figure 6.3), uses an optical lever to double the small angular deflection of the mirror.

QUESTION 2

Is the change in the direction of the reflected ray in Example 23.2 independent of the axis about which the mirror is rotated through the angle ϕ?

EXAMPLE 23.3

A ray is incident at angle θ on one of two mirrors that make a 90° angle with each other as shown in Figure 23.13. Determine the direction of the final reflected ray with respect to the incident direction. Such an arrangement of mirrors is called a *corner mirror*. You see corner mirrors frequently in the clothing sections of department stores as well as in mirrored rooms such as the Hall of Mirrors in the palace at Versailles, near Paris.

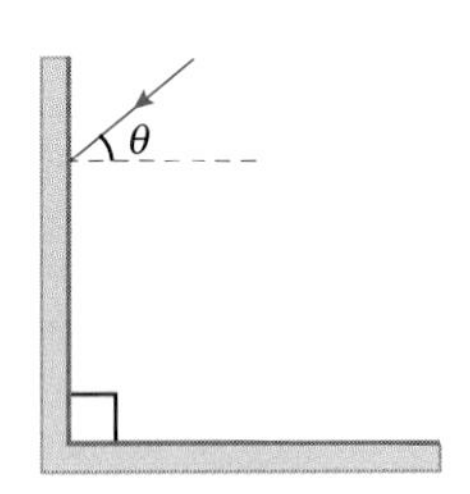

FIGURE 23.13

Solution

The law of reflection is applied at each reflection. Let β be the angle the final reflected ray makes with the horizontal direction, as indicated in Figure 23.14.

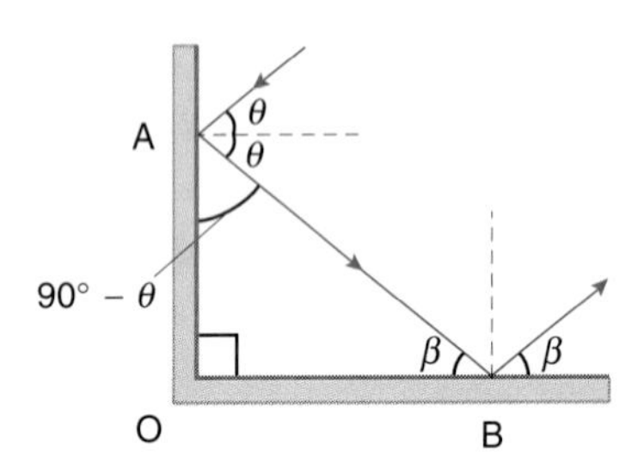

FIGURE 23.14

The angles of triangle AOB in Figure 23.14 are indicated. Since the sum of the angles of any triangle must be 180°, you have

$$\beta + 90° + (90° - \theta) = 180°$$

Solving for β, you obtain

$$\beta = \theta$$

Therefore the reflected ray is *antiparallel* to the incident ray whatever the angle of incidence happens to be. The corner mirror simply reverses the direction of the incident ray; the reflected ray is not coincident with the incident ray but is always antiparallel to it.

23.4 THE LAW OF REFRACTION

Countless wonders are performed by nature in accordance with the laws of these refractions;

Roger Bacon (c. 1214–1294)*

You have no doubt noticed that when a light ray strikes an abrupt boundary between two homogeneous, isotropic, and transparent media, such as that between air and water, the direction of the light ray changes permanently at the interface (unless the light ray approaches along a perpendicular to the surface). Such bending of light rays is called **refraction**.†

We must be careful to distinguish between the words

- **homogeneous**, meaning uniform composition and structure throughout the material; and
- **isotropic**, meaning identical or invariant in all directions.

The distinction is illustrated in Figure 23.15 for various patterns. Water, glass, and gases in equilibrium are examples of homogeneous and isotropic materials. Many crystals (such as quartz and calcite) are homogeneous but not isotropic (**anisotropic**).

Opus Majus, translated by Robert Belle Burke (University of Pennsylvania Press, Philadelphia, 1928), Volume I, page 131.

†The word comes from the Latin *refringere*, meaning "to break."

(a)

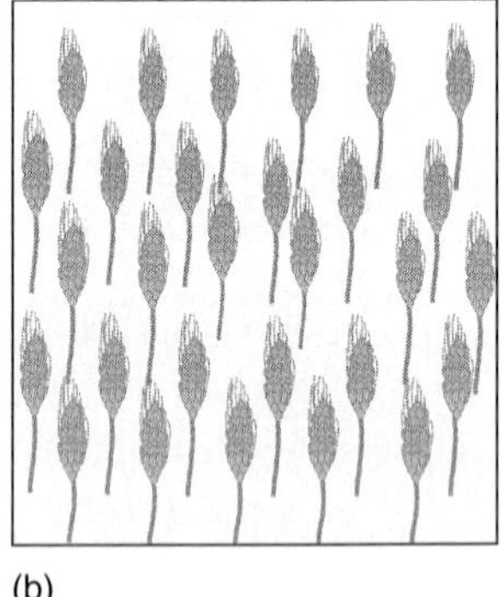

(b)

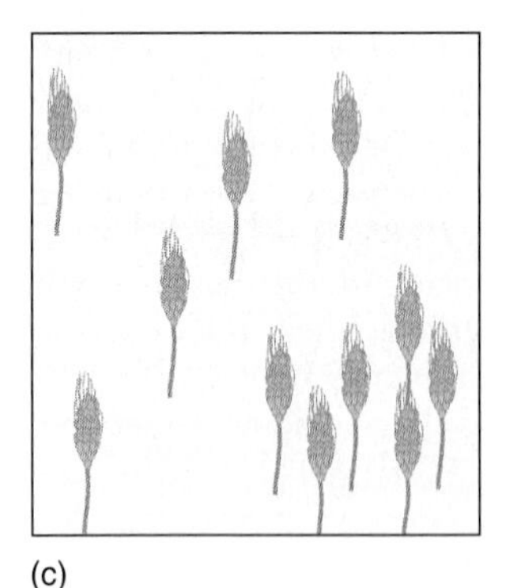

(c)

FIGURE 23.15 **Patterns that are (a) homogeneous and isotropic, (b) homogeneous and anisotropic, and (c) inhomogeneous and anisotropic.**

Refraction is the reason that a leaning, straight soda straw in a glass of water looks bent when viewed through the level surface (the second and lower image of the straw in water is formed by refraction at the curved surface of the container).

Refraction was noted in ancient Greece as well as in Islamic countries and Europe during the early Middle Ages. Experiments performed by Willebrord Snel* (1580–1626) led to the discovery of the quantitative relationship between the angle of incidence and the angle of refraction, occasionally called **Snel's law.**†

As with reflection, the angle of incidence θ_1 is taken in reference to a normal line to the interface, as in Figure 23.16. The ray in the second medium is the **refracted ray**. The **angle of refraction θ_2** is the angle between the refracted ray and the normal line. We also can notice that some light *reflects* (usually weakly) from these abrupt boundaries between transparent materials, but here we are concerned with the transmitted (refracted) ray and its change of direction.

*Many references spell his name with two l's: Snell; but Snel used only one letter l in the spelling of his last name. The author feels we should use Snel's name the way he spelled it.

†In France, Snel's law is called Huygens's law, after Christian Huygens (1629–1695). The priority of the discovery still is in dispute.

In Figure 23.16, clearly θ_2 is not equal to θ_1. To determine the relationship between the angles, we can measure and plot θ_2 for various incident angles θ_1 on a given transparent interface (see Figure 23.17). While θ_2 is proportional to θ_1 for small angles (but not equal to it), a graph of θ_2 versus θ_1 is clearly not simply linear, especially for larger angles of incidence, as shown in Figure 23.17 for an air-to-water interface.

Geometrically, if we draw a circle of any radius r centered on the point where the ray refracts, as in Figure 23.18, experiments indicate that, although θ_1/θ_2 is a changing ratio, the ratio of the ray components *parallel* to the interface, x_1/x_2, shown in Figure 23.18, stays constant at all angles. Hence a graph of $\sin\theta_2 = x_2/r$ versus $\sin\theta_1 = x_1/r$ *is* linear, as shown in Figure 23.19.

Hence the ratio of the sine of the angle of incidence to the sine of the angle of refraction is constant for the given pair of media. Change one of the media and the ratio still is a constant, but with a different numerical value.

If the incidence medium is a vacuum, the constant determined from the ratio of the sine of the angle of incidence θ_{vacuum}

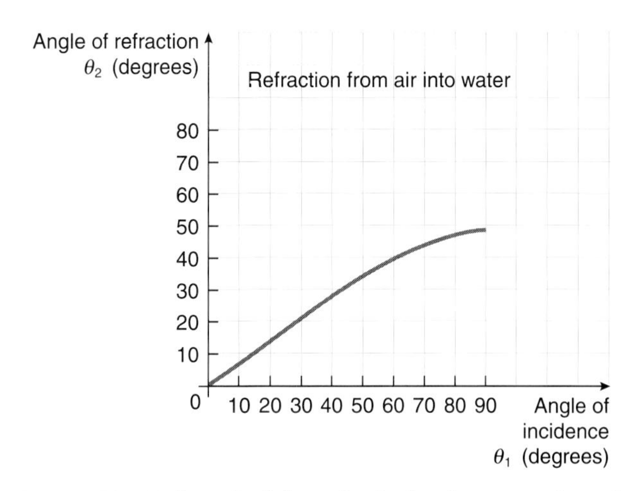

FIGURE 23.17 A graph of the angle of refraction versus the angle of incidence for air-to-water refraction.

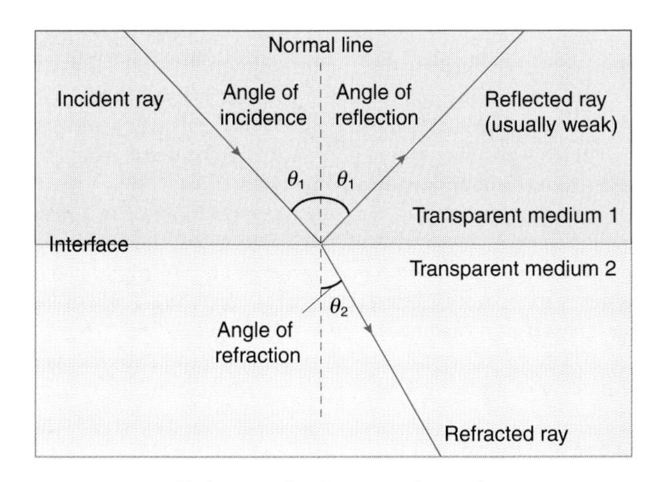

FIGURE 23.16 Refraction (with some reflection) at a transparent interface.

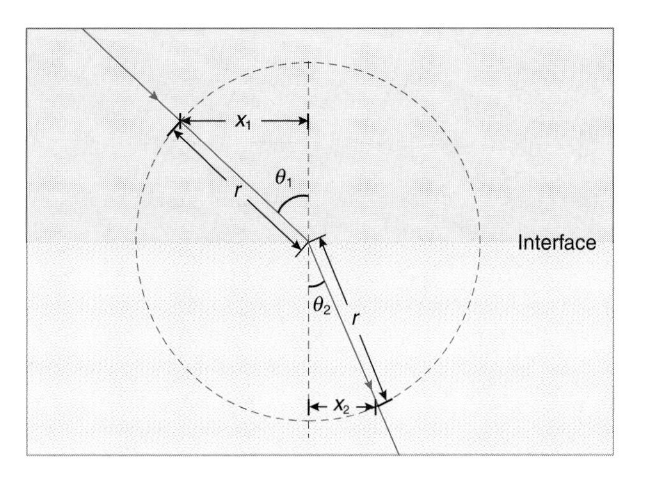

FIGURE 23.18 The ratio of the distances x_1/x_2 is constant for all angles of incidence.

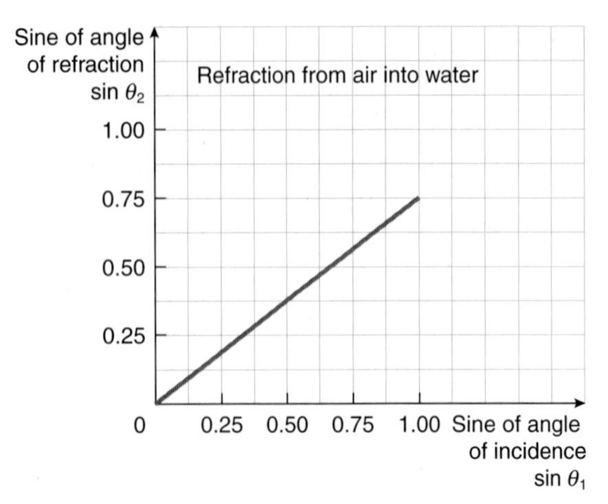

FIGURE 23.19 A graph of the sine of the angle of refraction versus the sine of the angle of incidence is linear.

to the sine of the angle of refraction θ_{refract} is a characteristic of the material and defined to be its **index of refraction**[†]:

$$n \equiv \frac{\sin \theta_{\text{vacuum}}}{\sin \theta_{\text{refract}}} \tag{23.4}$$

The index of refraction of a vacuum thus is equal to exactly 1. Table 23.1 lists the index of refraction of some common transparent materials. When refraction occurs at the interface between two materials with indices of refraction n_1 and n_2, the ratio of the sine of the angle of incidence θ_1 to the sine of the angle of refraction θ_2 is equal to the ratio of n_2 to n_1:

$$\frac{\sin \theta_1}{\sin \theta_2} = \frac{n_2}{n_1} \tag{23.5}$$

Rearranging this slightly, we obtain

$$\boxed{n_1 \sin \theta_1 = n_2 \sin \theta_2} \tag{23.6}$$

Equation 23.6, coupled with the observation that the incident ray, the refracted ray, and the normal line all lie in the same plane, as shown in Figure 23.20, constitute the **law of refraction** for homogeneous, isotropic, and transparent materials.

Furthermore, experiments prove that the light retraces its path if its direction is reversed.

Boundaries such as those that exist between

- air and a water surface;
- air and glass;
- water and glass; and
- glass and transparent plastics

TABLE 23.1 Approximate Values of the Index of Refraction*

Substance	*Index of refraction*
Solids	
Amber	1.55
Diamond	2.42
Glass (depends on the type)	~1.5 (about 3/2)
Plastic (depends on the type)	~1.3–1.4
Water (ice)	1.30
Liquids	
Benzene	1.50
Carbon tetrachloride	1.46
Ethanol	1.36
Methanol	1.33
Water (liquid)	1.33 (about 4/3)
Gases	
Air (atmospheric pressure)	1.000 29 (usually taken to be 1.000)
Many minerals have relatively high indices of refraction:	
Hutchinsonite	3.1
Proustite	3.1

*Precise values depend on the temperature and the wavelength of light used.

Source: *Handbook of Chemistry and Physics* (Chemical Rubber Publishing Company, Cleveland, Ohio)

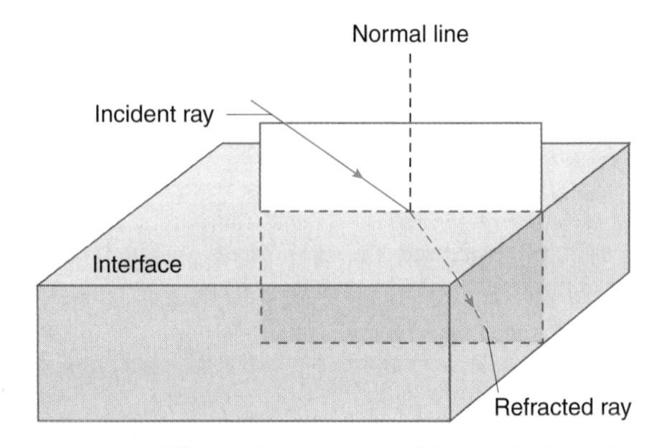

FIGURE 23.20 The incident ray, normal line, and refracted ray all lie in the same plane.

typify abrupt boundaries. The boundary between the atmosphere of a planet and outer space is not abrupt and the refraction of light entering or leaving such an atmosphere results in ray paths that change gradually (continuously), as indicated in Figure 23.21.[‡] Such atmospheres are inhomogeneous (the density decreases with height) and anisotropic (differences in the vertical direction differ from those in horizontal directions).

Nonetheless, as long as the light path is confined to a region small if compared with the atmospheric height, the atmosphere typically can be considered homogeneous and isotropic. It is the variation of index of refraction with density and temperature that gives rise to the phenomenon of mirages.

While most crystals are homogeneous, some are isotropic while others are not. Thus one has to be careful to consider

[†]In Chapter 24 we shall see that the index of refraction also is equal to the ratio of the speed of light in vacuum to its speed in the material.

[‡]If a ray is incident tangentially on the atmosphere of the Earth from the vacuum of space, the total change in the direction of the ray is only about 0.5°.

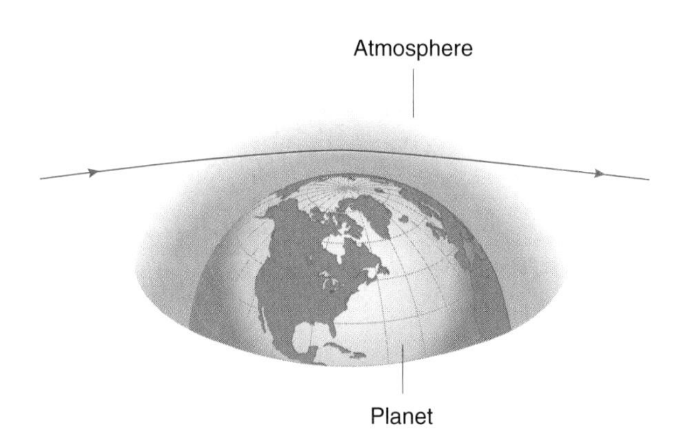

FIGURE 23.21 Refraction of light entering a planetary atmosphere is not abrupt but gradual because of the variation of the density of the atmosphere with height. The thickness of the atmosphere is exaggerated here for clarity.

whether to apply the same law of refraction to everything that is transparent to light. Certain crystals follow the law of refraction; in other crystals, the law needs to be supplemented to account for the anisotropic nature of the material. We consider the description of the refraction of light in these more complex crystals in the next chapter.

PROBLEM-SOLVING TACTIC

23.1 When using the law of refraction,

$$n_1 \sin \theta_1 = n_2 \sin \theta_2 \qquad (23.6)$$

you will find it convenient and less confusing to let the medium in which the incident ray lies always be medium 1 (the first medium) and let the medium in which the refracted ray lies be medium 2 (the second medium).

QUESTION 3

What is the meaning of the slope of a graph of sin θ_1 (as ordinate) versus sin θ_2 (as abscissa)?

EXAMPLE 23.4

A light ray in air is incident at an angle of 40.0° to the normal to a surface of water, as shown in Figure 23.22. Determine the *deviation angle* ϕ of the ray.

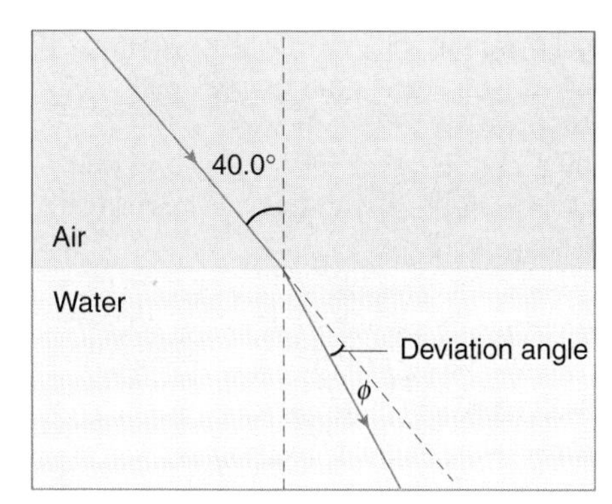

FIGURE 23.22

Solution

Apply the law of refraction to the ray, Equation 23.6:

$$n_1 \sin \theta_1 = n_2 \sin \theta_2$$

The indices of refraction of the materials are found in Table 23.1. The incident ray is in air, so that $n_1 = 1.00$ and $\theta_1 = 40.0°$. The refracted ray is in water, so that $n_2 = 1.33$. Making these substitutions into the law of refraction, you find

$$1.00 \sin 40.0° = 1.33 \sin \theta_2$$
$$\sin \theta_2 = 0.483$$
$$\theta_2 = 28.9°$$

The deviation angle is the difference between the angles of incidence and refraction:

$$\theta_1 - \theta_2 = 40.0° - 28.9°$$
$$= 11.1°$$

QUESTION 4

If, as pictured at the beginning of this section, a straw is inclined to the normal by the angle 40.0° in Example 23.4, is the kink in the straw the same as the 11.1° deviation angle?

23.5 TOTAL INTERNAL REFLECTION

If you have an aquarium, you may have noticed that when you view the tank from certain angles, with your eye well away from a normal to the surface, a fish is not visible through the top surface when it is swimming in certain regions of the tank. Here we see how this disappearing act comes about.

A material with a higher index of refraction than another is known as the more **optically dense** of the two, quite independent of the real relative density of the materials in kg/m^3.* The medium with the lower index of refraction is called the less optically dense of the pair. An interesting effect occurs when light passes from a more optically dense medium into a less optically dense material, say from water into air, as in Figure 23.23.

We let n_1 be the index of refraction of the optically more dense medium (here, the water), since it contains the incident ray. We have

$$n_1 > n_2$$

The law of refraction at the interface is given by Equation 23.6:

$$n_1 \sin \theta_1 = n_2 \sin \theta_2$$

Since the light is traveling into the less optically dense medium, the refracted ray is bent farther away from the normal line; that is, since $n_1 > n_2$, we must have $\theta_2 > \theta_1$.

Now we increase the angle of incidence θ_1, as in Figure 23.24. The angle of refraction θ_2 also increases. If we keep increasing θ_1, eventually a situation is reached where the angle of refraction θ_2

*Nonetheless, there is a strong correlation between mass density in kg/m^3 and optical density.

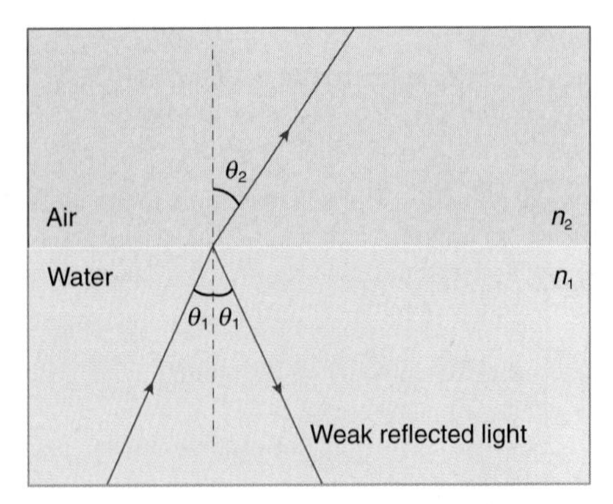

FIGURE 23.23 Light incident from water into air.

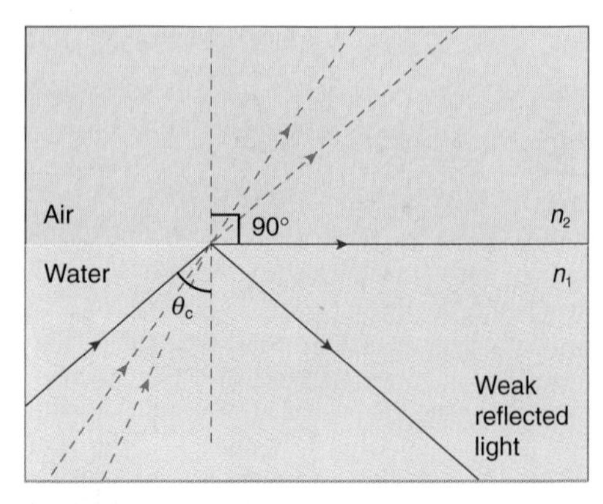

FIGURE 23.25 When the light is incident at the critical angle, an angle of refraction of 90° results.

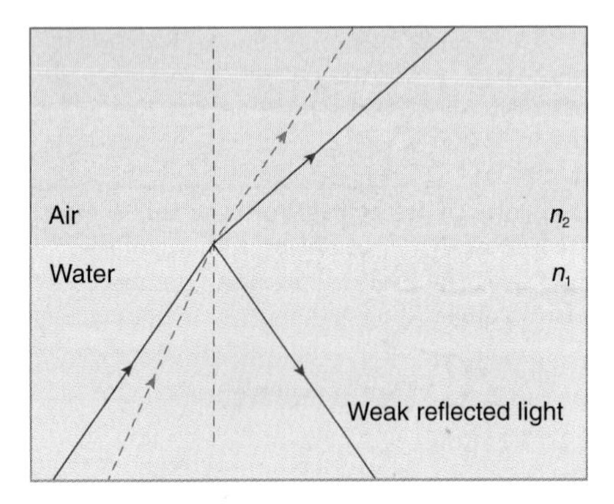

FIGURE 23.24 Increase the angle of incidence and the angle of refraction increases as well.

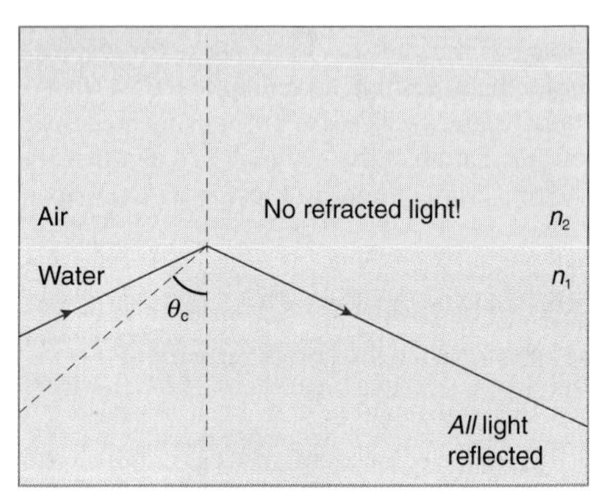

FIGURE 23.26 Total internal reflection.

becomes 90°, as shown in Figure 23.25. The angle of incidence that produces an angle of refraction of 90° is called the **critical angle** θ_c; the refracted ray is along the interface itself as shown in Figure 23.25.

When the angle of incidence is equal to the critical angle, the law of refraction, Equation 23.6, becomes

$$n_1 \sin \theta_c = n_2 \sin 90° \tag{23.7}$$

Hence the critical angle is found from

$$\sin \theta_c = \frac{n_2}{n_1} \qquad (n_1 > n_2) \tag{23.8}$$

What happens if the angle of incidence is greater than the critical angle in the more optically dense medium? We have "run out of" angle of refraction, since it cannot be greater than 90°. When the angle of incidence is greater than the critical angle, experimentally we find that the transmission ceases and the weak reflected light at the interface jumps to a strong value that persists at all greater angles of incidence. The light is totally internally reflected at the interface back into the more optically dense medium, as shown in Figure 23.26.

For incidence angles greater than the critical angle, the transparent interface acts like a good *mirror*; reflection is all that occurs, and so the effect is called **total internal reflection.**

In fact, there is a small "evanescent" disturbance that trickles into the second medium, decreasing exponentially over just a few wavelengths, but there is no net energy transfer into the second medium.*

The phenomenon of total internal reflection is one that has many technological applications, only two of which are mentioned here:

1. Total internal reflection is the basic physical principle that underlies the transmission of light in fiber optic communication systems. Light pulses propagate along the fiber and follow the path of the fiber optic cable, because each time the light is incident on the walls of the fiber the angle of incidence is greater than the critical angle of the fiber. Thus the light is totally internally reflected back into the fiber. Successive total internal reflections along the fiber ensure that the light follows the fiber no matter what its slowly curving shape may be.
2. Light pipes are used in medical instruments for noninvasive (i.e., nonsurgical) examination of the upper and lower intestinal tract and the respiratory tract. Such light pipes also are used in arthroscopic surgery.

*This observation is confirmed by a detailed application of the Maxwell equations of electromagnetism at the interface. The evanescent disturbance is an excellent classical example of an analogous phenomenon called *tunneling* in quantum mechanics.

The small optical fiber cables on the left can replace the huge array of conventional copper cables on the right for voice and data transmission because of the high frequency associated with visible light. Total internal reflection keeps light confined to an optical fiber in optical communication systems.

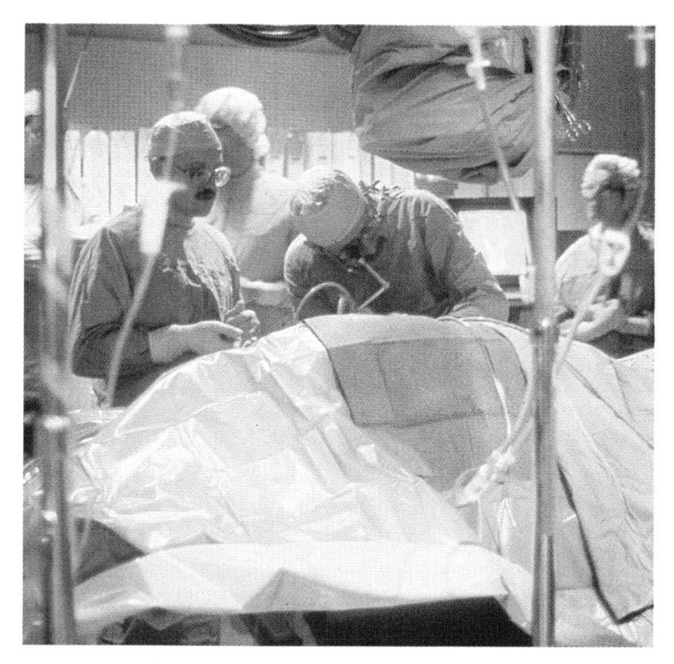

Light pipes that use total internal reflection enable surgeons to view the interior of the body with small incisions.

PROBLEM-SOLVING TACTIC

23.2 **In using Equation 23.8 for the critical angle, the material with the greater index of refraction (in which the incident ray lies) is in the denominator.** If you confuse the indices of refraction of the materials and place the larger index of refraction in the numerator, the resulting sine of the critical angle is greater than 1, a mathematically impermissible result. This should indicate to you that something is awry. Most calculators even will flag an error on their display if you try to find the angle whose sine is greater than 1.

EXAMPLE 23.5

Calculate the critical angle for a water–air interface.

Solution

From Table 23.1, the index of refraction of water is 1.33 while that for air is 1.00. When light is incident at the critical angle, the angle of refraction is 90°. Hence the law of refraction becomes

$$1.33 \sin \theta_c = 1.00 \sin 90°$$
$$\sin \theta_c = 0.752$$
$$\theta_c = 48.8°$$

23.6 DISPERSION

Rainbows are beautiful. They are caused by the passage of sunlight through water droplets. When sunlight also passes through the prismatic crystals of a chandelier, we may also observe small rainbow-like segments of color on the walls or table. Here we see that such beautiful effects arise because of a variation of the index of refraction of materials with wavelength.

The visible region of the electromagnetic spectrum consists of light with a rather small range of wavelengths: from about 400 nm to about 700 nm in a vacuum (or air). This small range of wavelengths is very important to us since it is the region to which our eyes are sensitive.* There is a correspondence between color and wavelength. White light is composite, consisting of all wavelengths of the visible spectrum (and typically other wavelengths in the near infrared and near ultraviolet regions as well).

If a parallel beam of white light is incident with a nonzero angle of incidence on a transparent interface, a spectrum of colors is produced as shown in Figure 23.27. We say the incident light mixture of wavelengths has been *dispersed* into its component colors or wavelengths.

If the white light is incident on a prism, two refractions take place in the passage through the prism: one at the first air–glass boundary, the second at the glass–air interface on another side of the prism, as indicated in Figure 23.28. We assume that the index of refraction of the air is the same as in a vacuum—that is, $n_1 = 1.00$ (Table 23.1 indicates that this is a good approximation). At the first boundary, all the wavelengths have the same angle of incidence θ_1. If we let n be the index of refraction of the prism, then the law of refraction, Equation 23.6, becomes

$$1.00 \sin \theta_1 = n \sin \theta_2 \qquad (23.9)$$

*Not coincidentally, this is the region where the Sun produces most of its output of electromagnetic waves. Our eyes evolved to take advantage of the region of the spectrum where the most light is produced.

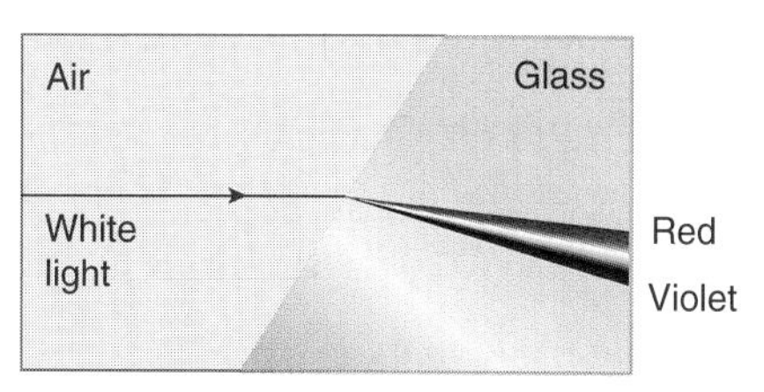

FIGURE 23.27 The refraction of white light produces a spectrum of colors.

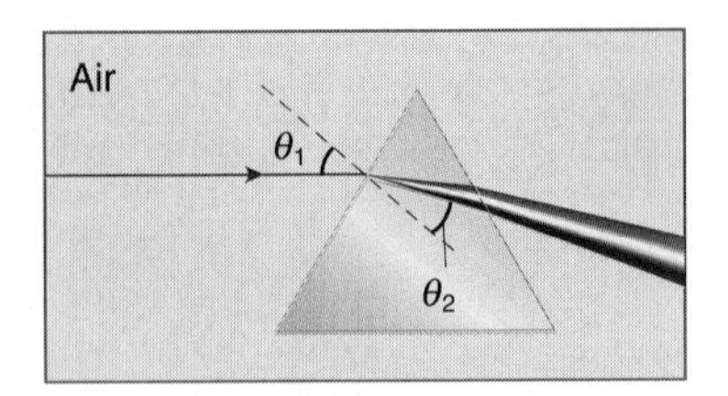

FIGURE 23.28 A prism can be used to produce a spectrum of white light.

Figure 23.28 indicates that the various colors (wavelengths) have *different* angles of refraction θ_2. Since the left-hand side of Equation 23.9 is the same for all the wavelengths, but θ_2 is different for each wavelength, the index of refraction n of the prism must be a function of the wavelength: $n(\lambda)$.

The variation of the index of refraction with wavelength† is known as **dispersion**.

Figure 23.29 is a graph of the variation of n with wavelength for various types of glass and quartz; the variations in these and other materials typically are slight numerically, but significant physically. The wavelengths indicated along the abscissa of Figure 23.29 are the vacuum wavelengths λ of the light.‡

Dispersion is the underlying physical principle behind an important instrument known as a prism spectrometer. Each chemical element has a characteristic **emission spectrum** of specific wavelengths corresponding to various transitions of the electrons in the atom; the spectrum is like a natural bar-coding for the element. If the light from a collection of various atoms then is incident on a prism such as in Figure 23.28, the prism separates the light into its constituent wavelengths. By measuring the angles through which the light is deviated by the prism, it is possible to determine the specific wavelengths emitted by the light source. From a table of the known wavelengths emitted by various atoms, the particular element responsible for the emission of the light then can be identified.

†It is more proper to say that dispersion is variation of the index of refraction with the frequency, because that does not change when light passes from one medium to another. But the wavelength (in air) is the favored variable for colors, probably because it is much easier to measure than the (very high) frequency of visible light.

‡In Chapter 24, we show that the wavelength λ_n of the light in a medium with an index of refraction n is given by

$$\lambda_n = \frac{\lambda}{n}$$

This equation is *not* what is meant by dispersion, since it indicates how a particular wavelength λ of light in a vacuum changes to λ_n in a medium with an index of refraction n. Dispersion means that the index of refraction itself varies with the particular wavelength, $n(\lambda)$, usually in a small but complicated way.

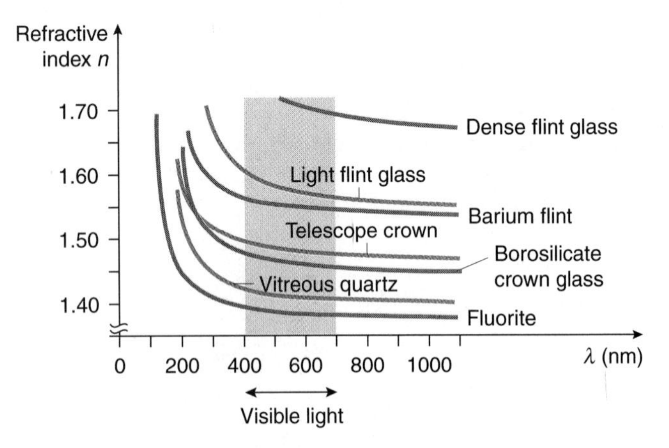

FIGURE 23.29 The variation of the index of refraction with wavelength in various kinds of glasses and quartz.

QUESTION 5

Since violet light is deviated through a greater angle than red light when both are incident at the same angle on a prism (see Figure 23.28), is the index of refraction of glass for red light greater than or less than the index of refraction for violet light?

EXAMPLE 23.6

A certain sample of flint glass has an index of refraction equal to 1.571 for red light (656 nm) and 1.594 for violet light (434 nm). If white light is incident from air at an incidence angle of 35.0°, what is the angular separation of the red and violet rays in the refracted beam?

Solution

Apply the law of refraction to each ray to determine the angle of refraction of each.

For the red ray:

$$\begin{aligned} n_1 \sin\theta_1 &= n_2 \sin\theta_2 \\ 1.000 \sin 35.0° &= 1.571 \sin\theta_2 \\ \theta_2 &= 21.4° \end{aligned}$$

For the violet ray:

$$\begin{aligned} n_1 \sin\theta_1 &= n_2 \sin\theta_2 \\ 1.000 \sin 35.0° &= 1.594 \sin\theta_2 \\ \theta_2 &= 21.1° \end{aligned}$$

The angular separation of the red and violet rays is the difference between the two angles of refraction, only 0.3°.

23.7 RAINBOWS*

Rain, rain, and sun! a rainbow in the sky!

Alfred, Lord Tennyson (1809–1892)§

Among the most beautiful of natural atmospheric effects are rainbows. As we alluded to at the beginning of the previous section, they are caused by dispersion in appropriately placed water drops. Sunlight is essentially white light: a mixture of all colors (wavelengths) in the visible spectrum. The particular white light ray involved in the formation of a rainbow enters a droplet and is refracted and dispersed into its component colors, as indicated at point (1) in Figure 23.30; some light also is reflected at the air–water interface, but this light is not involved in the formation of rainbows.

§ *Idylls of the King*, The Coming of Arthur, line 402. *The Poetic and Dramatic Works of Alfred Lord Tennyson*, Cambridge edition (Houghton Mifflin, Boston, 1898), page 309.

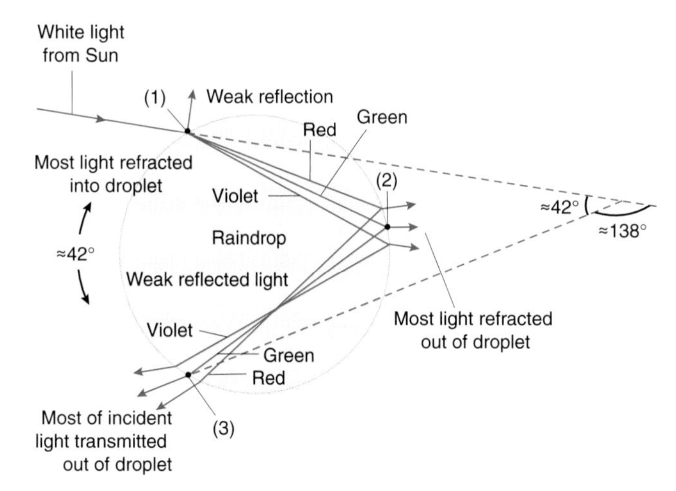

FIGURE 23.30 Path of a ray through a raindrop involved in the primary rainbow.

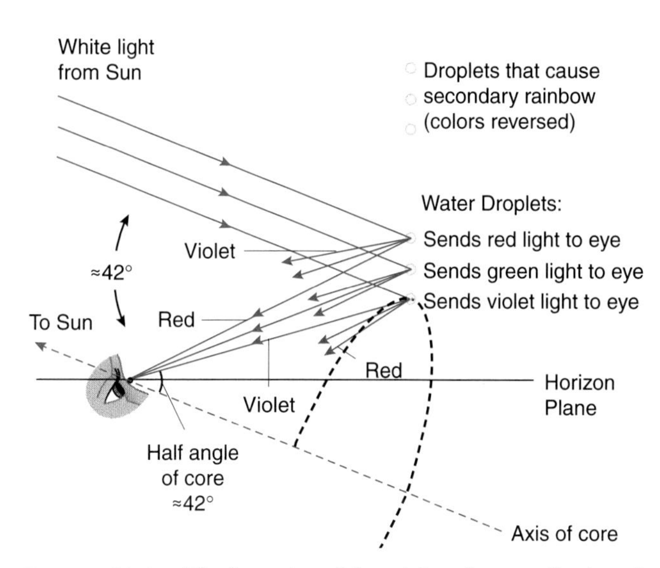

FIGURE 23.31 The formation of the rainbow from a collection of water droplets.

When the dispersed beam of light rays encounters the rear surface of the raindrop at point (2) in Figure 23.30, most of the light is refracted out of the droplet; but some light is reflected from the water–air surface. It is this small amount of light reflected from the back of the droplet that produces the rainbow; rainbows are dim because most of the light is transmitted at the back of the raindrop, not reflected.*

The reflected light is refracted and dispersed additionally at its third encounter with the water–air interface, at point (3) in Figure 23.30. The deviation angle between the incident and emergent beams is about 138°, so the acute angle is about 42° (see Problem 32).

An observer situated to receive red light from the drop also will receive red light from raindrops if they are located on a cone whose vertex is the eye of the observer, whose axis is the extension of the line from the Sun to the eye, and whose half angle is about 42° (see Figure 23.31). A circular arc of light results, centered on the antisolar point opposite the Sun. Droplets located lower than those sending the red light to the eye send the other colors of the **primary rainbow** to the eye. The red arc (at an angular radius of about 42.5°) appears outside the arcs of the other colors, with violet on the inside edge (at an angular radius of about 40°). There is some overlapping of the colors because of the finite angular size of the Sun (about 0.5°).

The rainbow continues to exist as long as new drops replenish those falling through the imaginary cones on the observer's line of sight. An observer on a level plain (whether in Spain or Kansas) sees only a segment of the circular arc (and only along portions of the cone that contain raindrops). Partial rainbows are seen if raindrops exist along only portions of the cone. Observers viewing waterfalls from high cliffs can see more of a circular arc from a rainbow formed by water droplets from the spray of the waterfall.

*Some references state (erroneously) that the light is totally internally reflected from the back of the water droplet. This is not the case. Note that each individual dispersed ray inside the water droplet makes equal angles with the two normal lines to the surfaces at (1) and (2) in Figure 23.30. If we reverse the direction of the ray, it is apparent that the angle of incidence is less than the critical angle, so that no total internal reflection occurs at the back surface. The author thanks Professor Richard Ditteon (at Rose-Hulman Institute of Technology) for bringing this common misconception to his attention.

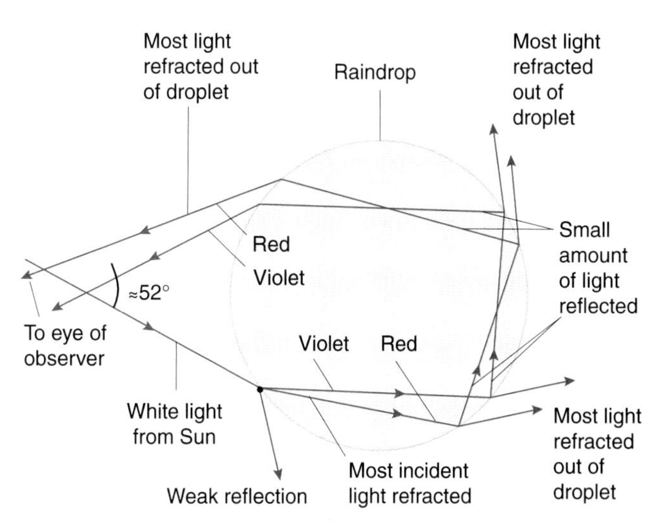

FIGURE 23.32 Light ray paths that form the secondary rainbow.

A **secondary rainbow** also can be seen occasionally. The secondary rainbow is caused by light that undergoes *two* reflections before emerging from the droplet. In this case the appropriate white light ray enters the droplet on the bottom and leaves on the top, as shown in Figure 23.32. Since only a small amount of light is reflected with each encounter with the water–air interface, the secondary rainbow is *much* dimmer than the primary rainbow and, therefore, even more difficult to see.

The angle between the incident and emergent beams is about 52°. Because of the extra reflection, the colors of the secondary rainbow are the reverse of those in the primary rainbow: red is now on the inside of the bow (at an angular radius of about 50°) and violet on the outside (at about 54°). The width of the secondary rainbow is about twice that of the primary rainbow, another reason secondary rainbows appear fainter and are more difficult to see than primary rainbows.

The primary (brighter) and secondary (dimmer) rainbows. Note the reversal of the color sequence in the two rainbows.

23.8 Objects and Images

When you look into a mirror, you see an image of yourself. Camera lenses also form images of scenes on film. The lens of a video camera forms an image on an electronic detector. We want to investigate the image-forming properties of optical devices such as mirrors and lenses. To do this, we need to know what is meant by the terms object and image.

When a given scene or **object** is placed in front of an optical device, we think of each point on the object as an **object point**. Each object point is a source of many diverging light rays in many different directions from the point, as shown in the left side of Figure 23.33. It does not matter whether the object point is diffusely reflecting these rays or generating them.

A pointlike object consists of a single object point source of light rays, as the name implies. An **extended object** consists of many different object points, as in the right side of Figure 23.33. If we know what happens to the rays from a single object point, the generalization to an extended object is not difficult.

Some of the light rays from an object point may enter a somewhat wide optical device, as in Figure 23.34. We are not interested in those rays from the object point that miss or do not enter the optical device. The entering rays are processed by the device in some fashion, which may involve reflections, refractions, or some combinations of these processes. The processed rays then leave the optical device (see Figure 23.34) and reveal where the image is located.

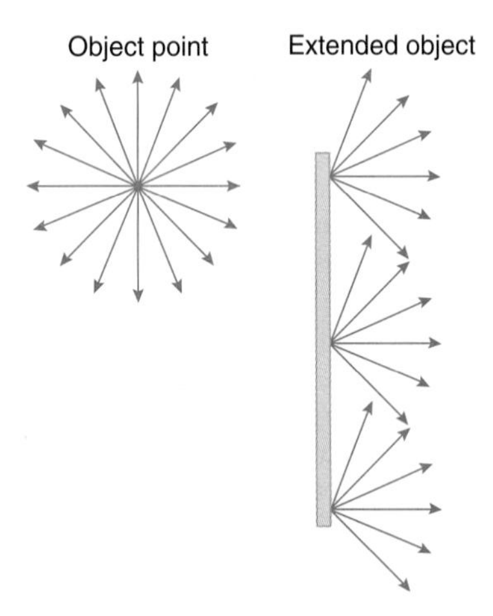

FIGURE 23.33 An object point and an extended object consisting of many object points.

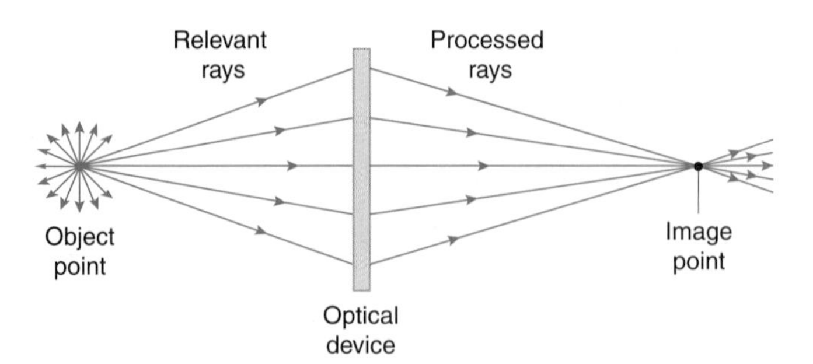

FIGURE 23.34 The processed rays leave the optical device and reveal the location of the image; here the image is real.

If the exiting light rays intersect, or even appear to intersect, at some point, that point is the **image point** of the object point. If these rays physically converge at the image point, the image is a **real image** (the rays *really* intersect there). If the exiting rays only *appear* to intersect at the image point, the image point is a **virtual image**.*

For various optical devices such as mirrors, refracting surfaces, and lenses, our goals are the following:

1. to find a relationship between the location of an object and the location of its corresponding real or virtual image;
2. for extended objects, to determine if the resulting extended image is larger, smaller, or the same size as the object—that is, to determine the magnification; and
3. to determine if an extended image is upright or inverted.

To accomplish these goals, it is convenient to introduce a sign convention associated with distances in the problem.

23.9 The Cartesian Sign Convention

A sign convention is used to facilitate computations of the positions and sizes of images and the assessment of the nature of the image (real or virtual) and its magnification and orientation. Distances and sizes take on positive or negative values depending on certain assumed rules or conventions.

The sign convention we adopt is the **Cartesian sign convention**, known for its simplicity and named for its association with the standard Cartesian coordinate system.† A standard geometry is used.

*The word virtual means *not in actual fact*. Unfortunately, the current term virtual reality makes mincemeat out of this distinction.

†This sign convention was recommended by the International Commission on Optics many years ago; see "International Commission of Optics," edited by S. S. Ballard, *Journal of the Optical Society of America*, *41*, #2, pages 140–141 (February 1941). Despite its simplicity, the Cartesian sign convention has yet to be universally adopted in the literature. You need to be aware of this confusion in the literature. We will do our part to further the cause of the Cartesian sign convention.

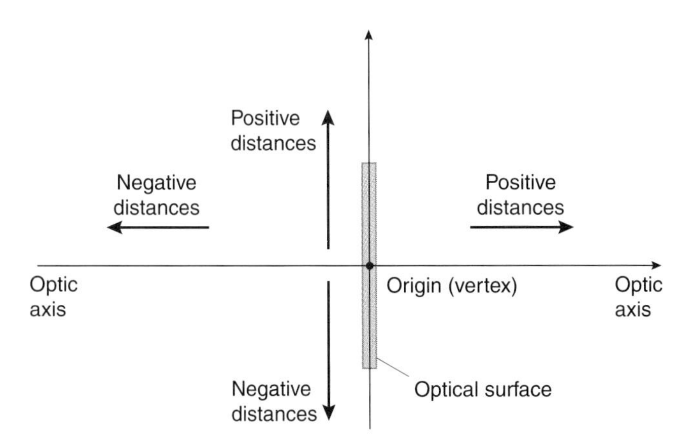

FIGURE 23.35 The Cartesian sign convention.

The object is placed to the left of the optical surface (a mirror or refracting surface). The light incident on the optical device from the object initially is traveling from left to right.

The center of the optical surface, called its **vertex**, is taken to be the origin of a Cartesian coordinate system. The horizontal Cartesian coordinate axis is called the **optic axis**. Distances measured to the right of the origin along the optic axis (i.e., measured to the right of the optical surface) are *positive* distances, since they are along the positive axis of the standard Cartesian coordinate system; see Figure 23.35. Distances measured to the left of the origin along the optic axis (i.e., to the left of the optical surface) are *negative* distances, since they are along the negative Cartesian coordinate axis. Likewise, distances measured upward from the horizontal Cartesian axis also are positive; distances measured downward from the horizontal axis are negative.

That is all there is to the sign convention for distances. Just imagine the vertex of the optical surface to be the origin of a standard Cartesian coordinate system and everything else follows quite readily from there. The same Cartesian sign convention for distances and magnification will be used throughout our discussion of geometric optics.

There also is a sign convention for the magnification.

The **magnification** m is defined as the ratio of the size of the image to the size of the object. If the magnification m is positive, the image of the object is **upright** (right-side-up), meaning that the image has the same vertical orientation as the object. If the magnification is negative, the image is **inverted** (up-side-down) with respect to the vertical orientation of the object.

Assessing whether left and right are preserved or reversed depends on how the image is viewed by the observer (see Investigative Project 9).

23.10 IMAGE FORMATION BY SPHERICAL AND PLANE MIRRORS

Mirror, mirror, on the wall, who is fairest of them all?
From the well-known fairy tale, *Snow White*

When you look	*kool uoy nehW*
into a mirror	*rorrim a otni*
it is not	*ton si ti*
yourself you see,	*,ees uoy flesruoy*
but a kind	*dnik a tub*
of apish error	*rorre hsipa fo*
posed in fearful	*lufraef ni desop*
symmetry.	*.yrtemmys*
	—ekidpU nhoJ

John Updike (1932–)*

Here we see how simple polished or coated surfaces, such as a bathroom mirror, a curved makeup mirror, or the curved rearview mirror on the passenger side of your car, use the law of reflection to produce an image and how that image differs from the object.

A **spherical mirror** is a reflecting surface whose shape is part of (a section of) a spherical surface. The radius of the spherical surface is the **radius of curvature** R of the mirror, as shown in Figure 23.36. The **center of curvature** C of the mirror is the center of the sphere of which the mirror surface is part.

Rather than treat a **plane mirror** as something different, we can and do consider it as a special case of a spherical mirror: one with an infinite radius of curvature.

There are two types of spherical mirrors, concave and convex mirrors, shown in Figure 23.37. According to the Cartesian sign convention, with its origin at the mirror, a concave mirror has a negative radius of curvature and a convex mirror has a positive radius of curvature.

Consider an object point in front of a convex spherical mirror, as in Figure 23.38. A line from the object point through the center of curvature C of the mirror defines the optic axis, as indicated in Figure 23.38. The point O on the mirror surface is the vertex of the mirror, and is the origin of the Cartesian coordinate system used to specify the signs of the appropriate distances. The distance of the object from the mirror is the **object distance** s. According to the

**The New Yorker*, 30 November 1957, page 200.

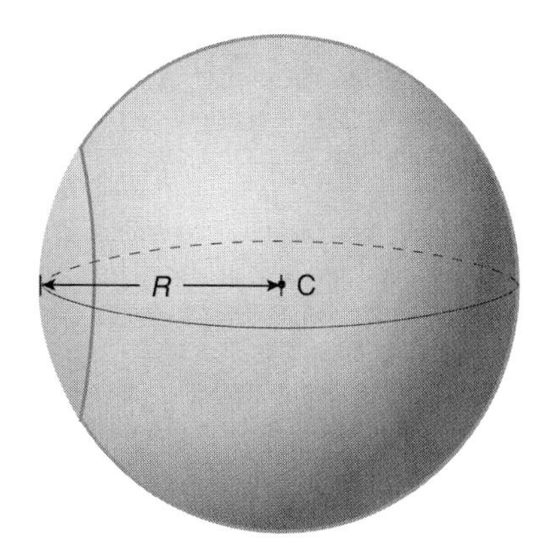

FIGURE 23.36 A spherical mirror is a section of a spherical surface.

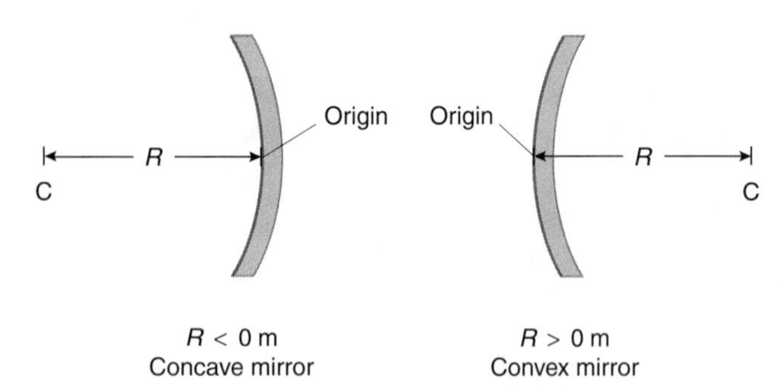

FIGURE 23.37 A concave mirror and a convex mirror.

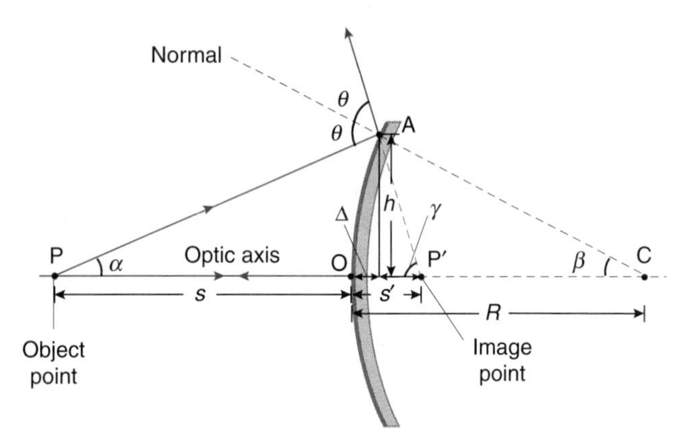

FIGURE 23.38 The formation of the image of point P at P′ by a convex mirror.

Cartesian sign convention, the object distance s here is *negative*. That is, here s itself has a negative value. The radius of curvature R of the convex mirror is positive, since the center of curvature of this mirror and the length R are to the right of the origin.

To locate the image point, we examine a few rays diverging from the object point to see what happens to them after they are processed (here, reflected) by the mirror surface. We look at two rays in particular:

1. A ray from the object point at P directed along the optic axis; see Figure 23.38. This incident ray hits the mirror at point O, the mirror vertex. The normal to the mirror surface at this point is the optic axis itself. The angle of incidence is zero and so, applying the law of reflection, the angle of reflection also is zero. The ray merely retraces its path, as shown in Figure 23.38.
2. Another ray from the object point P strikes the mirror at point A in Figure 23.38. The normal to the mirror surface at point A is a line from the center of curvature to point A. The angle of incidence is θ and, applying the law of reflection to this ray, the angle of reflection also is θ as shown. The location of the image point is found by examining the two rays after they have been processed (reflected) by the mirror. The two extended reflected rays appear to have an intersection at point P′; this, therefore, is the image point. The image here is virtual because the two rays do not physically meet at P′; the rays only *appear* to have come from P′. Let s' be the distance of the image from the origin, called the **image distance**. Here, since the image is located to the right of the origin, s' is a positive distance according to the Cartesian sign convention.

Derivation of the Mirror Equation

We want to find a relationship between the object distance s, the image distance s', and the radius of curvature R of the mirror. We define the angles α, β, and γ as in Figure 23.38. The angle θ is an exterior angle of triangle PAC. The exterior angle of a triangle is equal to the sum of the two remote interior angles.* Thus

$$\theta = \alpha + \beta \tag{23.10}$$

Likewise, the angle 2θ is an exterior angle of triangle PAP′. Applying the same geometric theorem, we find

$$2\theta = \alpha + \gamma \tag{23.11}$$

Substituting for θ from Equation 23.10, we find

$$2(\alpha + \beta) = \alpha + \gamma \tag{23.12}$$

Rearranging this slightly, we get

$$\gamma - \alpha = 2\beta \tag{23.13}$$

Now we make the assumption that the incident rays from the object point that are processed by the mirror make small angles with the optic axis, and so never are located far from the optic axis. This is called the **paraxial ray approximation** (meaning the rays are almost *par*allel to the optic *axis*). This assumption is a reasonably good first approximation in most cases.

If the angles α, β, and γ are small, we can use the **small angle approximation** to relate them to the distances. For small angles, the angle (in radians) is approximately equal to the tangent of the angle (it is also approximately equal to the sine of the angle). For angles up to about 10° (= 0.17 rad), the difference between the angle (in radians!), the tangent of the angle, and the sine of the angle is less than 1%. Notice in Figure 23.39 that graphs of θ in radians, $\tan\theta$, and $\sin\theta$ all are close together for small θ.

With the small angle approximation and the geometry of Figure 23.38, we have

$$\beta \approx \tan\beta = \frac{h}{R - \Delta} \tag{23.14}$$

$$\gamma \approx \tan\gamma = \frac{h}{s' - \Delta} \tag{23.15}$$

$$\alpha \approx \tan\alpha = \frac{h}{-s + \Delta} \tag{23.16}$$

In Equation 23.16 for the tangent of α, the quantity $-s$ is used in the denominator because s itself is negative according to the Cartesian sign convention; thus, $-s$ is positive.

If the mirror is thin† and the angles are small, the distance Δ in the denominator is negligible compared with

---

*This theorem from plane geometry follows because the sum of the angles of a plane triangle must be 180°. Hence, in the triangle pictured, we must have

$$(180° - \theta) + \alpha + \beta = 180°$$

and so

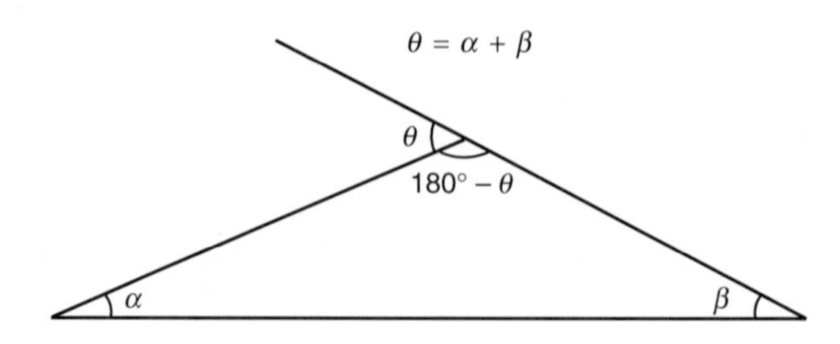

† A thin mirror is only a small section of the sphere from which its surface was cut.

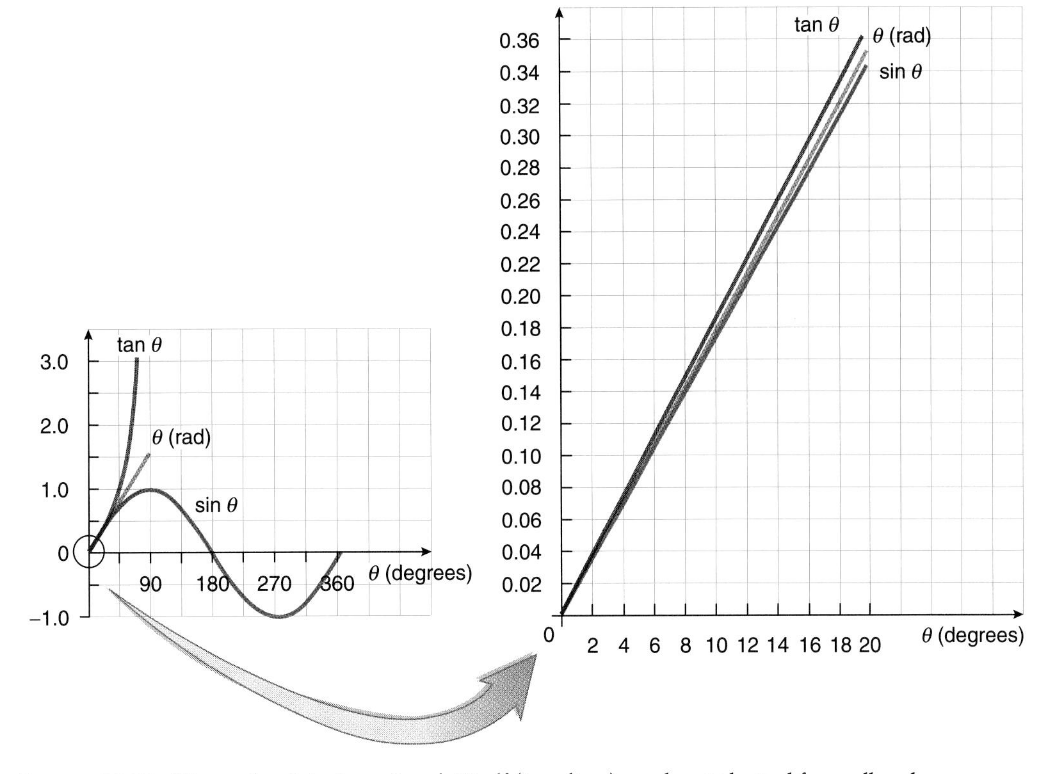

FIGURE 23.39 The graphs of sin θ, tan θ, and θ itself (in radians) are almost identical for small angles.

other distances such as s, s', and R. Hence Equations 23.14–23.16 become

$$\beta \approx \frac{h}{R} \tag{23.17}$$

$$\gamma \approx \frac{h}{s'} \tag{23.18}$$

$$\alpha \approx \frac{h}{-s} \tag{23.19}$$

We use the expressions in Equations 23.17–23.19 for the angles in Equation 23.13. The result is

$$\frac{h}{s'} - \frac{h}{-s} = 2\frac{h}{R} \tag{23.20}$$

We divide by h and perform some minor algebraic housekeeping. The result is the **mirror equation** for spherical mirrors:

$$\frac{1}{s} + \frac{1}{s'} = \frac{2}{R} \tag{23.21}$$

The Focal Point and Focal Length

One convenient refinement is made to the mirror equation.

> If the object point P is located infinitely far away from the mirror, then $s = -\infty$ m and the position of the image is called the **focal point** F of the mirror. The distance of the focal point from the mirror is the **focal length** f of the mirror; the focal length is given a sign in accordance with the Cartesian sign convention.

If we substitute $s = -\infty$ m into the mirror equation (Equation 23.21), we find

$$\frac{1}{-\infty \text{ m}} + \frac{1}{s'} = \frac{2}{R}$$

Since the image distance for an object infinitely far away is defined to be the focal length f, we put $s' = f$ and obtain

$$0 \text{ m}^{-1} + \frac{1}{f} = \frac{2}{R}$$

$$f = \frac{R}{2} \tag{23.22}$$

> The focal length of a spherical mirror is half its radius of curvature.

The focal point of the mirror is located halfway between the center of curvature and the mirror itself. Light rays from an object point at increasing distances from the mirror become increasingly close to parallel as they reach the mirror, as shown in Figure 23.40. Incident parallel rays after reflection intersect (or appear to intersect) at the focal point (the image point for an object point infinitely far away), as in Figure 23.41.

> An important point: the image of an object located a *finite* object distance from the mirror is *not* at the focal point! An image is at the focal point *only* when the object is infinitely far away from the mirror.

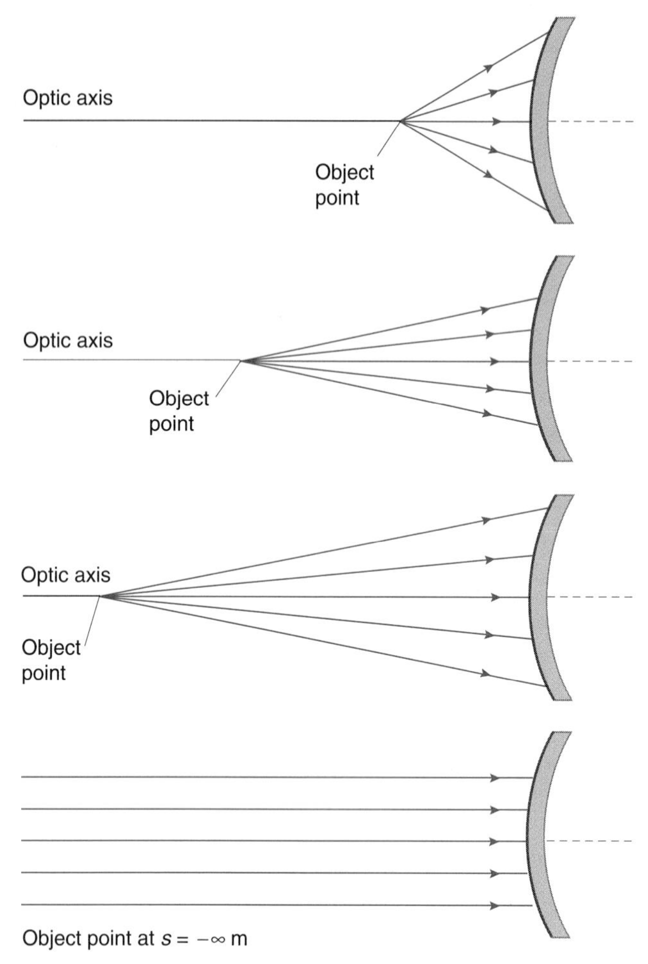

FIGURE 23.40 Light rays from an increasingly distant object point approach the optical surface increasingly close to parallel.

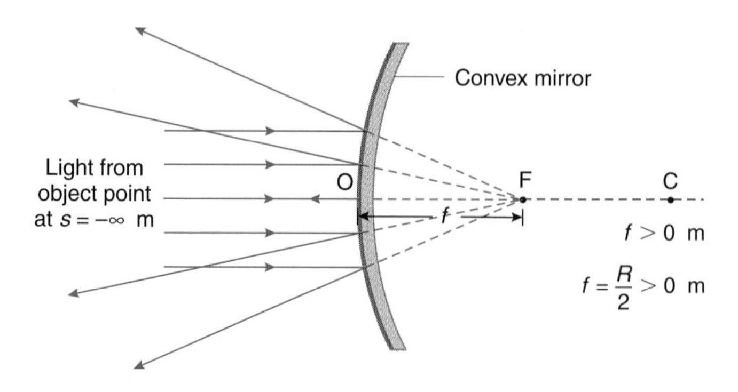

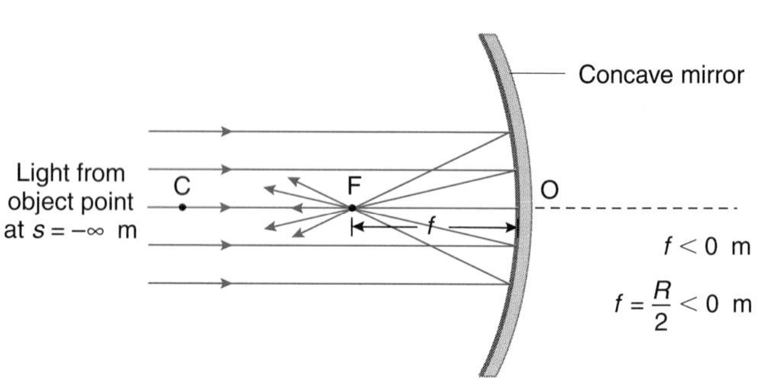

FIGURE 23.41 Parallel incident rays intersect (or appear to intersect) at the focal point after reflection from the spherical mirror.

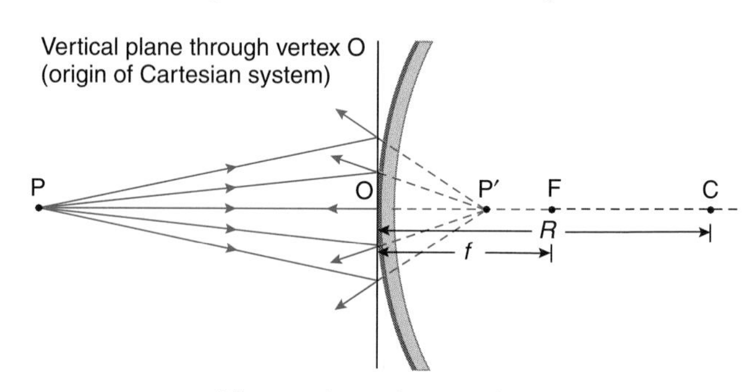

FIGURE 23.42 All paraxial rays from an object point intersect (or appear to intersect) after reflection at its image point. Here they appear to intersect at P′.

All paraxial rays from an object point that are processed (reflected) by the mirror either intersect or appear to intersect at the corresponding image point.

Because we consider only paraxial rays, we henceforth draw the reflections of the rays as if they occur at the vertical plane through the vertex of the mirror (see Figure 23.42) rather than at the indicated mirror surface itself. The curvature of the mirror is exaggerated in the figures for clarity.

Using Equation 23.22, we rewrite the mirror equation (Equation 23.21) in its most versatile form:

$$\frac{1}{s} + \frac{1}{s'} = \frac{2}{R} = \frac{1}{f} \quad \text{(spherical mirror equation)} \qquad (23.23)$$

Although we derived the mirror equation (Equation 23.23) using a convex mirror, a similar derivation for a concave mirror leads to the same result. Hence Equation 23.23 can be applied to *any* spherical mirror, be it convex or concave. We just must be careful to apply the appropriate signs to the distances according to the Cartesian sign convention.

You can use units of convenience in Equation 23.23 as long as you use the *same* units for all the distances.

If the image is located behind the mirror, then the image must be virtual, since light never really gets in back of the mirror. On the other hand, if the image is located in front of the mirror, the image is real, since the reflected rays really will pass through the image in that case. Thus for any mirror, if the image distance $s' > 0$ m, the image is virtual; if $s' < 0$ m, the image is real. These relationships are not worth memorizing, since they can be determined by thinking about where the convergent light that forms the image really is located.

PROBLEM-SOLVING TACTIC

23.3 When using the general mirror equation (Equation 23.23), recognize that it involves the reciprocals of the various distances. Do not forget to take a final reciprocal with your calculator to find s, s', f, or R.

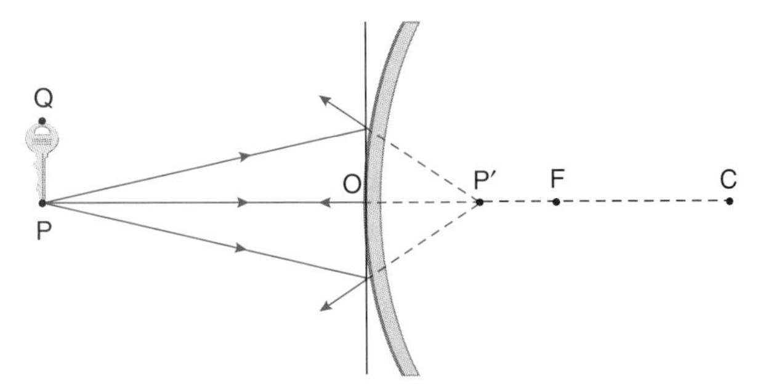

FIGURE 23.43 An object point P on the optic axis has its image point P′ on the optic axis.

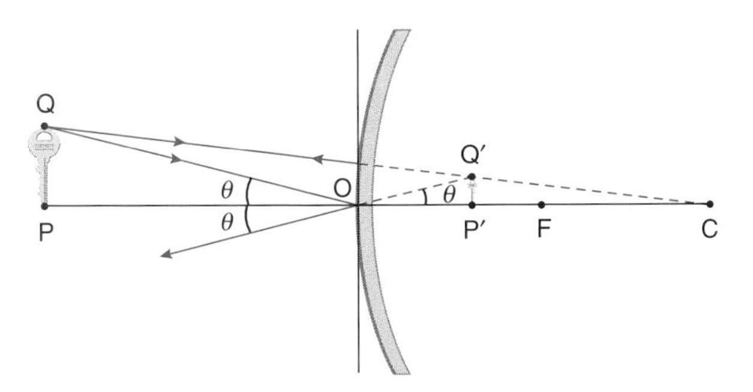

FIGURE 23.44 An object point Q off the optic axis has its image point Q′ off the optic axis.

An Extended Object

If an *extended object* is placed in front of the mirror, the location of the image is found by applying the law of reflection to the rays from each point on the object. In particular, the object point P on the optic axis has an image point P′ on the optic axis located by using the general mirror equation (Equation 23.23), as in Figure 23.43.

The point Q at the top of the object (see Figure 23.44) has an image point Q′ that can be located geometrically in the following way. A ray from the top of the object directed toward the center of curvature merely reflects back on itself since it strikes the mirror along a line normal to the mirror at this point. Another ray from the object point Q to the origin O makes an angle θ with the optic axis. By the law of reflection at this point, the reflected ray also makes an angle θ with the optic axis, as indicated in Figure 23.44. The image point Q′ is at the point of intersection from which the reflected rays appear to have come from. The extended image of the extended object is located between the points P′ and Q′.

The image in Figure 23.44 is virtual, since the light never is really in or behind the space in back of the mirror. Since the points on the extended object all are essentially the same distance s from the mirror (if the mirror is thin), the locations of all the image points also are essentially the same distance s' from the mirror as is the image point on the optic axis.

If we block or cover a portion of the mirror, as in Figure 23.45, the image is *dimmer*, since fewer rays from each object point now are processed by the mirror; but *the entire extent of the image still is formed*. Every part of the mirror is involved in forming every part of the image.

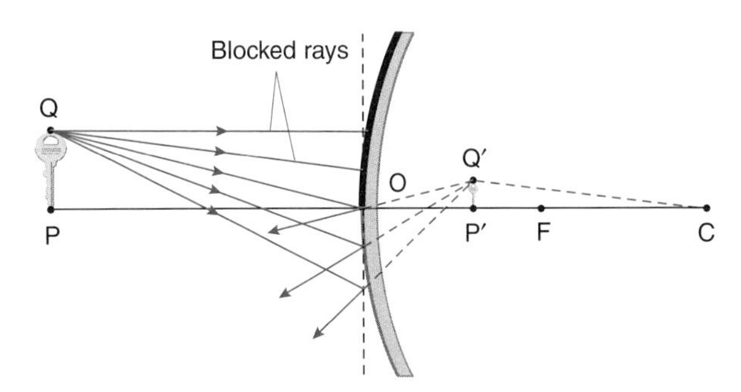

FIGURE 23.45 If we block a portion of the mirror, the entire image still is formed, but is dimmer.

Thus, as long as some light from each object point reaches the mirror, each image point is formed; the entire extent of the image is formed. Covering part of the mirror simply reduces the amount of light processed by the mirror; some light from each object point still is processed, and so the entire extent of the image nevertheless is formed.

If half of the *object* is blocked, the image is halved as well. In this instance, no light from the blocked part of the object reaches the mirror, and so the blocked part of the object is irrelevant and does not appear as part of the image.

Magnification

To determine the magnification, consider Figure 23.44. The magnification indicates the size of the image relative to the size of the object. Notice that triangles PQO and P′Q′O are similar triangles. This means, of course, that their corresponding sides are proportional. The distances PQ and P′Q′ in Figure 23.44 both are positive according to the Cartesian sign convention. Since s is negative, $-s$ is positive.

Hence the magnification is

$$m = \frac{\text{image size}}{\text{object size}} = -\frac{s'}{s} \quad \text{(mirror magnification)} \qquad (23.24)$$

So for Figure 23.44, Equation 23.24 for the magnification yields a positive number, indicating that the image is upright according to the sign convention for the magnification.

Equation 23.24 for the magnification can be applied to *any* spherical mirror. Moreover, if the appropriate signs are used for the object and image distances s and s', the sign of m automatically will indicate whether the image is upright (if $m > 0$) or inverted ($m < 0$). Wonderful!

The law of reflection is independent of the medium in which the light is traveling. Thus the mirror and magnification equations (Equations 23.23 and 23.24) can be applied whatever the transparent medium surrounding the mirror happens to be. The mirror and magnification equations are the same regardless of whether the mirror is immersed in air, water, gin, or cranberry juice.

QUESTION 6

Why is a plane mirror considered to have an infinite radius of curvature rather than a radius of curvature of zero?

EXAMPLE 23.7

You stand some distance $|s|$ in front of a plane mirror. Determine the location of your image, its magnification, and whether it is real or virtual.

Solution

A plane mirror is a spherical mirror with an infinite radius of curvature. Thus the general mirror equation (Equation 23.23) becomes

$$\frac{1}{s} + \frac{1}{s'} = \frac{2}{\infty \text{ m}}$$
$$= 0 \text{ m}^{-1}$$

So

$$s' = -s \qquad (1)$$

The object distance s is negative according to the Cartesian sign convention (see Figure 23.46). Since the object distance s

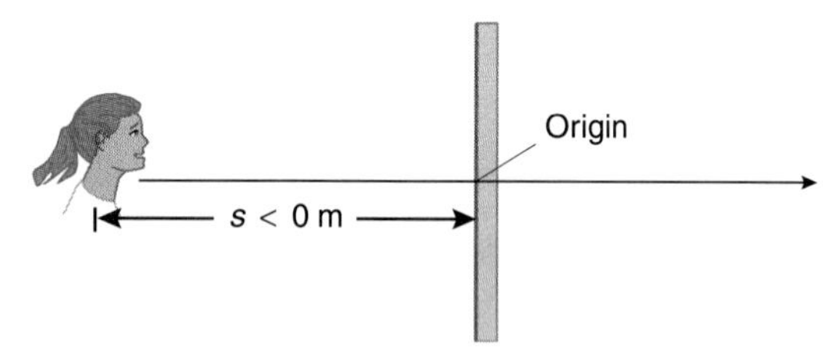

FIGURE 23.46

is negative, equation (1) indicates that the image distance s' is positive; the image therefore is located to the right of the mirror and is as far from the mirror as the object. The image location is behind the mirror; no light from the object really reaches this region. Therefore the image is virtual. But you can surely see it!

The magnification of a plane mirror is found using the general magnification equation (Equation 23.24):

$$m = -\frac{s'}{s}$$

Since you found $s' = -s$, the magnification is

$$m = -\frac{(-s)}{s}$$
$$= +1$$

The image is upright (since $m > 0$) and the same size as the object. It certainly would make life more difficult if a plane mirror gave an inverted image!

STRATEGIC EXAMPLE 23.8

A lipstick container 6.0 cm in height is placed 100 cm from a concave spherical cosmetic mirror cut from a sphere of radius 50 cm. Determine

a. the location of the image;
b. the nature of the image (real or virtual);
c. the magnification and size of the image; and
d. the focal length of the mirror.

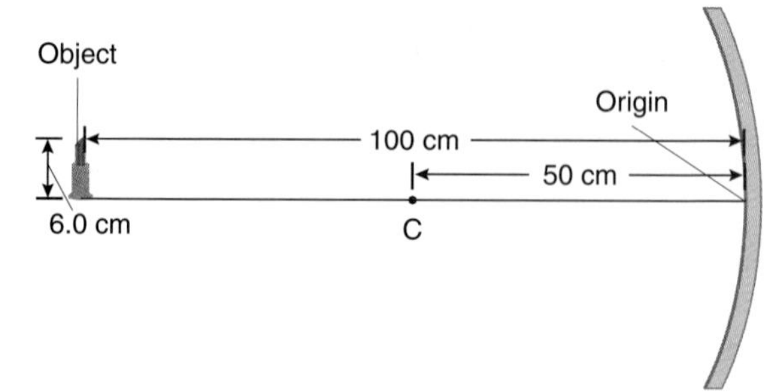

FIGURE 23.47

Solution

a. The situation is depicted with the standard geometry, shown in Figure 23.47. The origin for the Cartesian sign convention is at the vertex of the mirror. Since the object is located to the left of the origin, the object distance s is negative:

$$s = -100 \text{ cm}$$

The mirror is concave, so that the center of curvature of the mirror surface also is to the left of the origin. Thus the radius of curvature of the mirror R is negative:

$$R = -50 \text{ cm}$$

Use the general mirror equation, Equation 23.23:

$$\frac{1}{s} + \frac{1}{s'} = \frac{2}{R}$$

$$\frac{1}{-100 \text{ cm}} + \frac{1}{s'} = \frac{2}{-50 \text{ cm}}$$

Solving for s', you find

$$\frac{1}{s'} = -\frac{2}{50 \text{ cm}} + \frac{1}{100 \text{ cm}}$$
$$= -\frac{3}{100 \text{ cm}}$$

Problem-Solving Tactic 23.3 cautions you to remember to take the final reciprocal:

$$s' = -\frac{100 \text{ cm}}{3}$$
$$= -33 \text{ cm}$$

b. Since the image distance is negative, the image is located to the *left* of the origin (to the left of the mirror). Since the reflected light actually traverses this region, the image is *real*.

c. The magnification is found from the general mirror magnification equation (Equation 23.24):

$$m = -\frac{s'}{s}$$
$$= -\frac{-33 \text{ cm}}{-100 \text{ cm}}$$
$$= -0.33$$

Since the magnification is negative, the image is inverted. The image size is about 1/3 that of the object, or

$$-0.33(6.0 \text{ cm}) = -2.0 \text{ cm}$$

in height, extending below the optic axis according to the Cartesian sign convention.

d. The focal length of the mirror is half the radius of curvature of the mirror (Equation 23.22):

$$\begin{aligned} f &= \frac{R}{2} \\ &= \frac{-50 \text{ cm}}{2} \\ &= -25 \text{ cm} \end{aligned}$$

The focal point thus is 25 cm to the *left* of the origin (i.e., to the left of the mirror).

EXAMPLE 23.9

A pretty but deadly 75 cm long coral snake (*Micrurus fulvius*) is stretched out along the optic axis of a convex mirror of radius of curvature 150 cm. Its head is 50 cm from the mirror vertex with its tail further away.

a. Sketch the situation using the standard geometry.
b. Determine the locations of the images of the head and tail of the snake.
c. Is the length of the image of the snake longer or shorter than its actual length?

Solution

a. The situation appears as in Figure 23.48.
b. Use the general mirror equation (Equation 23.23) to locate the images:

To locate the image of the head	*To locate the image of the tail*
$s = -50 \text{ cm}$	$s = -125 \text{ cm}$
$R = 150 \text{ cm}$	$R = 150 \text{ cm}$
$\frac{1}{s} + \frac{1}{s'} = \frac{2}{R}$	$\frac{1}{s} + \frac{1}{s'} = \frac{2}{R}$
$\frac{1}{-50 \text{ cm}} + \frac{1}{s'} = \frac{2}{150 \text{ cm}}$	$\frac{1}{-125 \text{ cm}} + \frac{1}{s'} = \frac{2}{150 \text{ cm}}$
$s' = 30 \text{ cm}$	$s' = 47 \text{ cm}$

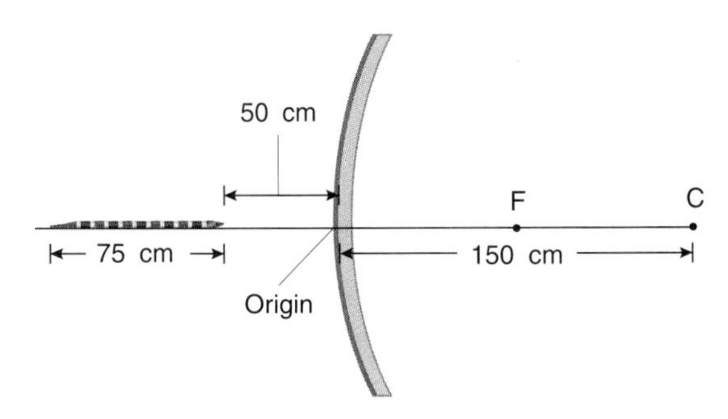

FIGURE 23.48

c. The length of the virtual image of the snake is

$$47 \text{ cm} - 30 \text{ cm} = 17 \text{ cm}$$

and so is shorter than the actual snake.

23.11 RAY DIAGRAMS FOR MIRRORS

Ray diagrams are useful geometric constructions that serve as a check on calculations made with the mirror and magnification equations. The diagrams also serve as a useful check on whether the image is real or virtual. In addition, ray diagrams provide an alternative method for solving mirror (and lens) problems.

Ray diagrams begin with an extended object placed in front of a mirror, as in Figure 23.49. The image of the object point P on the optic axis also will lie on the optic axis. Hence we just need to determine where the image of the point Q at the top of the object lies. A perpendicular line from the image of point Q to the optic axis will indicate the extent of the image of the extended object.

Many rays diverge from point Q and are processed by the mirror. *All* the processed rays intersect at the image point Q′. To locate the image point Q′, then, we need consider only a few rays. Two rays will suffice, but ray diagrams typically use three or four, just as a check on the construction. Of the bevy of rays diverging from the object point Q, we choose four special rays, called **principal rays**, whose paths are easy to trace geometrically. Ray diagrams also are called **principal ray diagrams** as a result. Since we consider only paraxial rays, the plane of effective reflection is coincident with a plane perpendicular to the optic axis passing through the vertex point of the mirror, the origin. We draw rays as if they reflect from this plane.

The privileged rays that warrant specific attention (because their paths are easy to trace) are the following:

1. A ray from the object point directed toward the center of curvature of the mirror, shown as ray (1) in Figure 23.49. After reflection, this ray simply retraces its path because it strikes the mirror along the normal line to the mirror.
2. A ray from the object point parallel to the optic axis, shown as ray (2) in Figure 23.49. *Any* ray parallel to the optic axis will pass through (or *appear* to pass through) the focal point; here, this ray after reflection *appears* to pass through the

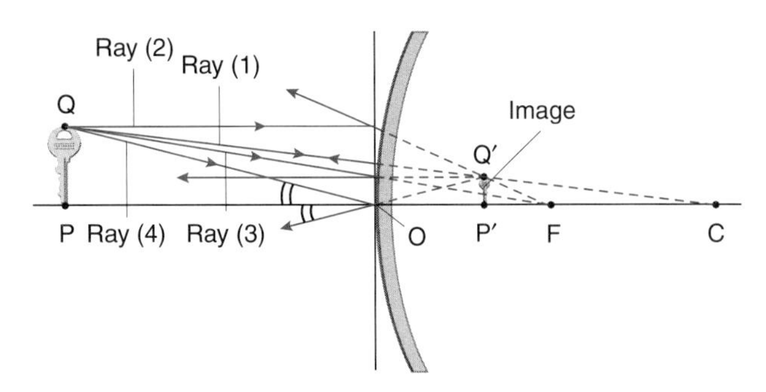

FIGURE 23.49 The principal rays for a mirror ray diagram.

focal point, located halfway between the center of curvature and the mirror itself.

3. A ray from the object point directed toward the focal point F, indicated as ray (3) in Figure 23.49. Since rays are reversible, this ray will emerge after reflection traveling parallel to the optic axis.
4. A ray from the object point to the vertex of the mirror, shown as ray (4) in Figure 23.49. This ray will have equal angles of incidence and reflection above and below the optic axis. This reflected ray also will appear to come from Q′.

Any two of these rays are sufficient to locate the position of the image. Once the position of the image is found, two other things can be assessed easily:

a. The nature of the image (real or virtual) is determined from where the image is relative to the mirror itself. If the image is in front of the mirror, the image is real; if the image is in back of the mirror, the image is virtual.
b. We also are able to see if the magnification is positive or negative and greater or less than unity.

EXAMPLE 23.10

A Baltimore oriole (*Icterus galbula*), a bird, not a baseball player, is perched in front of a concave spherical mirror such that $|s| > |R|$.

a. Draw the corresponding ray diagram to locate the image of the bird.
b. From your ray diagram, indicate whether the magnification is positive or negative, and whether the image is smaller or larger than the object.
c. Determine whether the image is real or virtual.

Solution

a. The situation is depicted schematically in Figure 23.50. Principal ray (1) to the center of curvature reflects back on itself. Principal ray (2), incident parallel to the optic axis, passes through the focal point F. Principal ray (3), incident through the focal point, emerges parallel to the optic axis. Principal ray (4) makes equal angles of incidence and reflection with the optic axis at the mirror vertex.

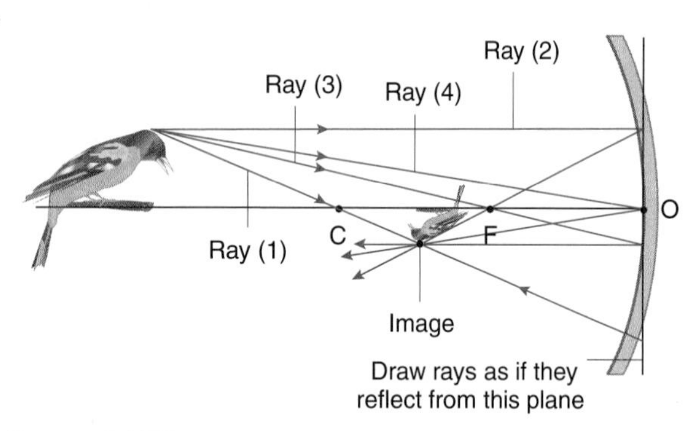

FIGURE 23.50

b. The image is inverted, so the magnification m is negative. The image also is smaller than the object, so $|m| < 1$.
c. The four rays actually pass through the image point, so the image is real.

23.12 REFRACTION AT A SINGLE SPHERICAL SURFACE

One side of either a water-filled, thin-walled spherical fish bowl or a solid crystal ball is an example of a single spherical refracting surface. The law of refraction relates the directions of the incident and refracted rays through the interface. We want to investigate the imaging properties of such single, spherically shaped, refracting surfaces. As with mirrors, we want to locate the image, determine whether it is real or virtual, and find the magnification. Since simple lenses are composed of two refracting surfaces, the work we do here soon will enable us to analyze lenses (in Section 23.13).

Derivation of the Single Surface Refraction Equation

An object point P is located at object distance s from a spherical refracting surface with a radius of curvature R. We formulate the problem with the standard geometry, as in Figure 23.51. Let the medium in which the object lies have an index of refraction n_1 and the second medium have index n_2. The Cartesian sign convention is used for all distances, of course. The sign of R depends on where the center of curvature is located with respect to the surface. For the system pictured in Figure 23.51, the radius of curvature R is positive and the object distance s is negative.

To locate the image of the object point, we need to see where at least two rays, originally diverging from the object point, intersect (or appear to intersect) after they have been processed (here, refracted) by the surface. In particular, we consider the following two rays:

1. A ray from the object point directed along the optic axis, ray (1) in Figure 23.51. This ray strikes the refracting surface at the origin O. We apply the law of refraction to the ray. Since the ray is directed along the normal to the surface at O, the angle of incidence of this ray is zero. Thus the angle of refraction also is zero and the ray proceeds undeviated into the second medium.
2. For a second ray, consider the ray from the object point that strikes the refracting surface at point A; this is ray (2) in Figure 23.51. Once again, we apply the law of refraction to the ray. The normal to the surface where the ray is refracted is a line from the center of curvature to the point A. The angle of incidence is θ_1 and the angle of refraction is θ_2. The two refracted rays intersect at point P′ and, thus, this is the image

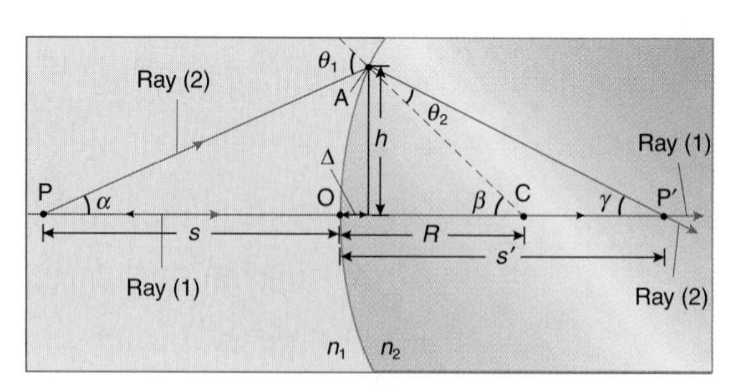

FIGURE 23.51 Geometry for deriving the single surface refraction equation.

point. The law of refraction (Equation 23.6) applied to ray (2) refracted at point A states that

$$n_1 \sin \theta_1 = n_2 \sin \theta_2$$

We restrict ourselves to paraxial rays from the object point, so that the angles θ_1 and θ_2 are small. The small angle approximation means that we can replace the sines in the law of refraction with the angles themselves (in radians). In other words, Equation 23.6 becomes*

$$n_1\theta_1 = n_2\theta_2 \tag{23.25}$$

We introduce the angles α, β, and γ, as shown in Figure 23.51. The angle θ_1 is an exterior angle of triangle PAC. Thus θ_1 is equal to the sum of the two remote interior angles:

$$\theta_1 = \alpha + \beta \tag{23.26}$$

Likewise β is an exterior angle of triangle P′AC, and so

$$\beta = \theta_2 + \gamma \tag{23.27}$$

using the same geometric theorem. Substituting for θ_1 and θ_2 in Equation 23.25 using Equations 23.26 and 23.27, we obtain

$$n_1(\alpha + \beta) = n_2(\beta - \gamma)$$

We rearrange this slightly:

$$n_1\alpha + n_2\gamma = (n_2 - n_1)\beta \tag{23.28}$$

Since the angles α, β, and γ are small, we approximate the angles with their tangents (the small angle approximation once again):

$$\gamma \approx \tan \gamma = \frac{h}{s' - \Delta} \tag{23.29}$$

$$\beta \approx \tan \beta = \frac{h}{R - \Delta} \tag{23.30}$$

$$\alpha \approx \tan \alpha = \frac{h}{-s + \Delta} \tag{23.31}$$

For the tangent of α, we use $-s$ in the denominator because s itself is negative; thus, $-s$ is positive.

As we did for mirrors, we assume the refracting surface is a small section of the sphere defining its curvature. Thus we neglect the small distance Δ in comparison with s, s', and R, and Equations 23.29–23.31 become

$$\gamma \approx \frac{h}{s'} \tag{23.32}$$

$$\beta \approx \frac{h}{R} \tag{23.33}$$

$$\alpha \approx \frac{h}{-s} \tag{23.34}$$

*It was this small angle form for the law of refraction that was discovered by Ptolemy during the second century A.D., as we mentioned in the introduction to this chapter. For small angles, the angle of refraction is directly proportional to the angle of incidence.

Using Equations 23.32–23.34 for the angles in Equation 23.28, we find

$$n_1 \frac{h}{-s} + n_2 \frac{h}{s'} = (n_2 - n_1)\frac{h}{R}$$

Hence the single surface refraction equation is

$$-\frac{n_1}{s} + \frac{n_2}{s'} = \frac{n_2 - n_1}{R} \tag{23.35}$$

(single surface refraction equation)

Although we derived Equation 23.35 for the case of a convex surface, a similar derivation shows that it is also applicable to concave spherical refracting surfaces, including a flat surface (in which case $R = \infty$ m). We must use the Cartesian sign convention for all the distances in the single surface refraction equation (Equation 23.35), including R.

Whether the image is real or virtual is assessed using common sense (a dangerous term!).

If the image is located to the right of the origin (positive image distances), then the image must be real since the refracted rays really exist in this region and will converge to the image point. On the other hand, if the image distance is negative, the image is to the left of the origin (the refracting surface). Since the refracted light rays really are on the other side of the surface, the image is virtual; the refracted rays never really intersect at a virtual image point, they only appear to intersect there. Thus a negative image distance here means a virtual image.

Magnification

To determine the magnification caused by a single surface refraction, we need to examine what happens to an extended object. The object point P in Figure 23.52 is imaged at P′ on the optic axis; the location of P′ is determined using the single surface refraction equation (Equation 23.35). The object point Q is imaged at point Q′. This means that all the rays diverging from object point Q converge on the image point Q′. Once again, because of the small angle approximation, we draw the refractions as if they occur at the plane perpendicular to the optic axis through the vertex of the surface (the origin of the Cartesian system).

We look at just one particular ray from Q to Q′, the ray directed from Q to the center of curvature C, shown in Figure 23.53. This ray is undeviated upon refraction because it is directed along the normal to the surface at the point where it strikes the surface.

Notice that triangles PQC and P′Q′C are similar. The corresponding sides of these triangles are proportional. In particular, the magnification is the ratio between the image and object heights:

$$m = \frac{\text{image height}}{\text{object height}} \tag{23.36}$$

The object height is PQ, and this is a positive distance according to the Cartesian sign convention since it is above the

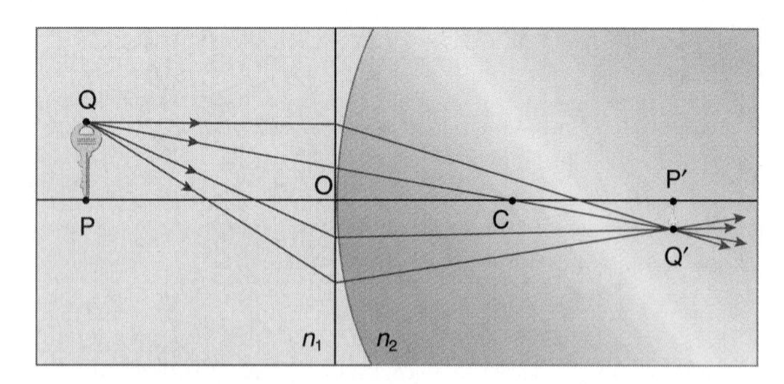

FIGURE 23.52 Formation of the image of an extended object.

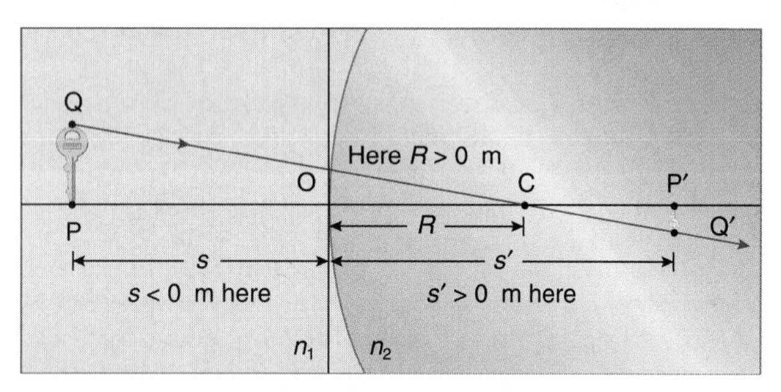

FIGURE 23.53 Consider one particular ray along the normal line; it is not deviated at the surface.

optic axis. The image height is P′Q′ and, since it is below the optic axis, it is negative according to the Cartesian sign convention. So m itself will be negative according to Equation 23.36. This is consistent with the sign convention for m: a negative magnification means the image is inverted with respect to the object.

It is more convenient to put the magnification in terms of the object and image distances. To do this, we use the proportionality of the sides of similar triangles PQC and P′Q′C:

$$m = \frac{\text{image height}}{\text{object height}}$$

$$= -\frac{s' - R}{-s + R} \tag{23.37}$$

Some explanation of the signs is needed here. The height of the image in Figure 23.53 is negative according to the Cartesian sign convention. In Figure 23.53, the distances s' and R are both positive and $s' > R$, so that $s' - R$ is positive; thus, we substitute $-(s' - R)$ for the image distance. The height of the object in Figure 23.53 is positive. Since the object is to the left of the origin, the object distance s is negative; hence, $-s$ is positive and $-s + R$ also is positive and is the appropriate side of the similar triangle. The magnification comes out negative, indicating that the image is inverted in Figure 23.53.

A more convenient equation for the magnification emerges if we eliminate R from Equation 23.37 by using the single surface refraction equation (Equation 23.35). To accomplish this goal, we solve the single surface refraction equation for R. This yields

$$R = \frac{ss'(n_2 - n_1)}{-n_1 s' + n_2 s}$$

Substituting this expression for R into the magnification equation (Equation 23.37), we obtain

$$m = -\frac{s' - \dfrac{ss'(n_2 - n_1)}{-n_1 s' + n_2 s}}{-s + \dfrac{ss'(n_2 - n_1)}{-n_1 s' + n_2 s}}$$

And this is supposed to be simpler! Not yet. But if we grind through a simplification of this equation (putting things over common denominators and so on), a rather remarkable series of cancellations takes place (check it out!) and this rather horrid looking magnification equation reduces to the following elegant form:

The magnification for single surface refraction is

$$m = \frac{n_1 s'}{n_2 s} \tag{23.38}$$

(magnification for single surface refraction)

Equation 23.35 for the location of the image and Equation 23.38 for the magnification are applicable to refraction at any spherical refracting surface (and for a flat or plane refracting surface, in which case $R = \infty$ m). Remember that the Cartesian sign convention must be used for all the distances in the equations.

Just as was the case for mirrors, if part of the spherical surface is blocked in some way, as in Figure 23.54, the entire image of an extended object still is formed; the image is dimmer because fewer light rays reach the image.

If more than one refracting surface is present, we take each refracting surface in the order in which it is encountered by the light. A succession of single surface refraction problems then must be solved individually. The location of the image formed by each surface is calculated using the single surface refraction equation (Equation 23.35); the magnification of each surface is found using Equation 23.38.

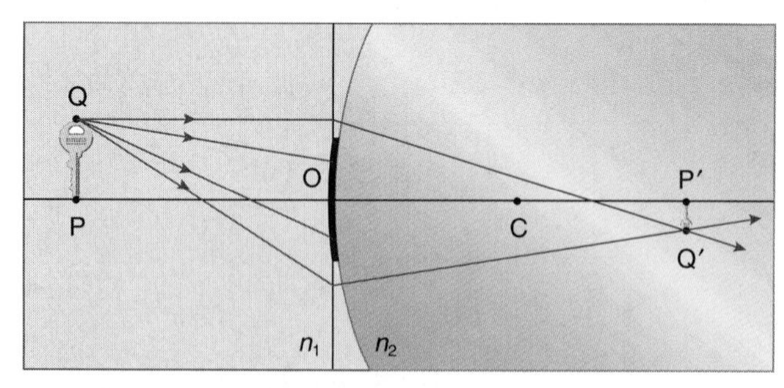

FIGURE 23.54 The entire image is formed, even if part of the refracting surface is blocked.

PROBLEM-SOLVING TACTICS

23.4 When considering refraction (or, for that matter, reflection) of the light from an object by two (or more) surfaces in succession, the image formed by the first surface becomes the object for the second surface. This fact does *not* necessarily mean the image distance of the first surface becomes the object distance of the second surface. For the purpose of assigning appropriate signs and numerical values for the *distances*, the origin of the Cartesian system is always taken to be at the particular refracting surface then under consideration. *Distances are measured from the particular surface under consideration, not from any previous surface.* Numerical values of distances (and their signs) must be adjusted to account for the shift in origin as you move from surface to surface in the problem (see Example 23.12).

23.5 The indices of refraction n_1 and n_2 in Equations 23.35 and 23.38 are the indices of the first medium and the second medium respectively; these change as you progress from surface to surface. That is, if we have two refractions:

Air to glass for the first refraction	*and then*	*Glass to water for the second refraction*
$n_1 = 1.00$		$n_1 = 1.50$
$n_2 = 1.50$		$n_2 = 1.33$

23.6 The total magnification of the entire system of surfaces is the product of the magnifications produced by the individual surfaces.

$$m_{\text{total}} = m_1 m_2 m_3 \cdots$$

where m_1, m_2, m_3, . . . are the magnifications produced by the first, second, third, and successive refracting (or reflecting) surfaces. The sign of the total magnification indicates whether the final image is upright or inverted with respect to the original object.

QUESTION 7

How could you define a focal length f for a single refracting surface? How is the focal length related to the indices of refraction n_1 and n_2 and the radius of curvature R? What conditions on the indices of refraction and the radius of curvature lead to $f > 0$ m? To $f < 0$ m? If an object is at a finite distance from the refracting surface, is the image located at the focal point of the surface?

EXAMPLE 23.11

A heads-up Sacagawea dollar coin lies at the bottom of the shallow end of a swimming pool at a depth of 100 cm, as shown in Figure 23.55.

a. To an observer looking straight down at the coin through the calm surface at a height of 200 cm above the surface, what is the apparent depth of the pool?

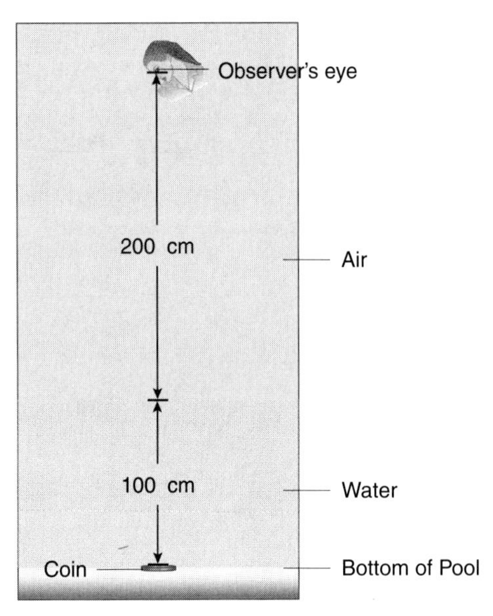

FIGURE 23.55

b. Is the image real or virtual?
c. What is the magnification of the coin?
d. Does the image appear right-side-up or inverted?

Solution

a. A flat surface is treated as a spherical surface of infinite radius. The single surface refraction Equation 23.35 becomes, with $R = \infty$ m,

$$-\frac{n_1}{s} + \frac{n_2}{s'} = \frac{n_2 - n_1}{R} = 0 \text{ m}^{-1} \qquad (1)$$

Change your perspective of the pool geometry to put the problem into the standard geometry, shown in Figure 23.56. Since the object (the coin) is in the water and the light is refracted into the air, you must take

$$n_1 = 1.33$$
$$n_2 = 1.00$$

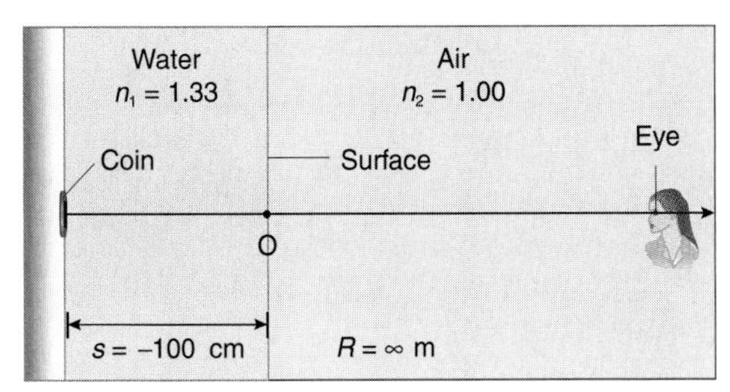

FIGURE 23.56

The object distance is $s = -100$ cm. The height of the observer's eye above the surface of the pool is irrelevant; you want the location of the image of the *coin*; it is this *image* that is seen by the eye (the image becomes the object for the eye). Substituting these numerical values into equation (1), you obtain

$$-\frac{1.33}{-100\ \text{cm}} + \frac{1.00}{s'} = 0\ \text{m}^{-1}$$

Solving for s', you find

$$s' = -75.2\ \text{cm}$$

b. The image is virtual because it is on the same side of the interface as the object and is 75.2 cm below the surface. The water thus appears to be 75.2 cm deep, shallower than it actually is.

c. The magnification of the coin is found from Equation 23.38:

$$m = \frac{n_1 s'}{n_2 s} = \frac{1.33\,(-75.2\ \text{cm})}{1.00\,(-100\ \text{cm})} = 1.00$$

d. Since the magnification is 1.00, the virtual image of the coin is upright and the same size as the object.

QUESTION 8

Does the upright image in Example 23.11 mean merely heads-up in this example (with the writing backward), or that the portrait of Sacagawea has the same orientation as on the object, with the writing in proper direction?

STRATEGIC EXAMPLE 23.12

The 2.0 cm diameter gaudy ring of a charlatan is placed 100 cm in front of a huge, spherical, glass crystal ball of radius 25 cm with an index of refraction equal to 1.50, shown in Figure 23.57. Determine the position of the final image of the ring and its magnification and size.

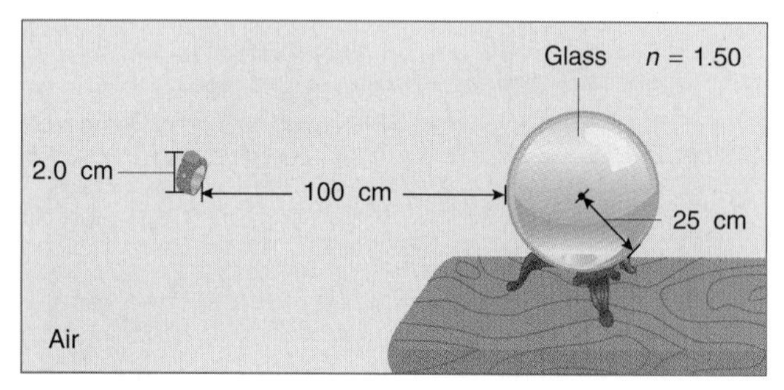

FIGURE 23.57

Solution

Light rays from the object (the ring) are refracted by the crystal ball once as they enter the ball and then on the other side when leaving it. You consider only paraxial rays close to the optic axis, so the single surface refraction equation can be used. There are *two* successive single surface refraction problems to solve.

Refraction at the First Surface

Since the object is in air and the ball is glass, the indices of refraction are

$$n_1 = 1.00 \qquad n_2 = 1.50$$

The object is located 100 cm to the left of the first refracting surface. Since the refracting surface is the origin, the Cartesian sign convention indicates that the object distance is negative and the radius of curvature of the first refracting surface is positive (see Figure 23.58):

$$s = -100\ \text{cm} \qquad R = +25\ \text{cm}$$

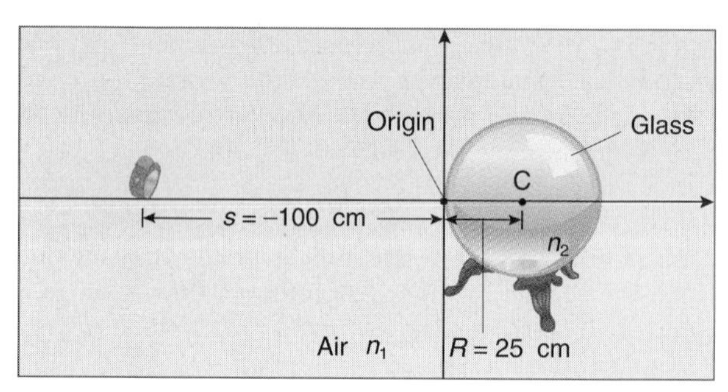

FIGURE 23.58

Substitute these values into the single surface refraction equation (Equation 23.35):

$$-\frac{n_1}{s} + \frac{n_2}{s'} = \frac{n_2 - n_1}{R}$$

$$-\frac{1.00}{-100\ \text{cm}} + \frac{1.50}{s'} = \frac{1.50 - 1.00}{+25\ \text{cm}}$$

Solving for s', you find

$$s' = 150\ \text{cm}$$

The image is located 150 cm to the right of the first refracting surface. The magnification m_1 caused by this first refraction is found using Equation 23.38:

$$m_1 = \frac{n_1 s'}{n_2 s} = \frac{1.00\,(150\ \text{cm})}{1.50\,(-100\ \text{cm})} = -1.00$$

Refraction at the Second Surface

Following Problem-Solving Tactic 23.4, for the refraction at the second surface you must shift the origin of the Cartesian system to the vertex of the second refracting surface. Following Problem-Solving Tactic 23.5, in the single surface refraction equation, the first medium now is the glass, so $n_1 = 1.50$; the second medium (into which the light is going) is the air, so $n_2 = 1.00$. The problem appears as in Figure 23.59.

The image formed by the first surface becomes the object for the second surface. The image formed by the first surface is located

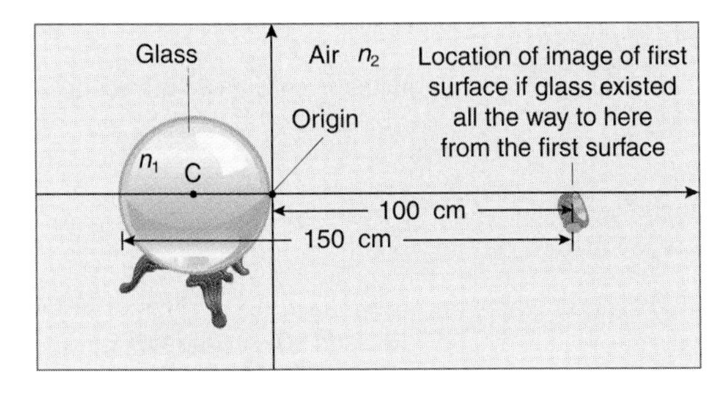

FIGURE 23.59

150 cm to its right or 100 cm to the right of the second refracting surface. Hence the object distance for the refraction at the second surface is

$$s = +100 \text{ cm}$$

This is a *virtual object*! The light forming the image of the first surface is refracted by the second surface before it can actually converge to form the image created by the first surface.

The radius of curvature for the second surface is negative since its center of curvature lies to the left of the origin at the second surface:

$$R = -25 \text{ cm}$$

Making these substitutions into the single surface refraction equation, Equation 23.35, you get

$$-\frac{n_1}{s} + \frac{n_2}{s'} = \frac{n_2 - n_1}{R}$$

$$-\frac{1.50}{100 \text{ cm}} + \frac{1.00}{s'} = \frac{1.00 - 1.50}{-25 \text{ cm}}$$

Solving for s', you find

$$s' = +29 \text{ cm}$$

The image is located 29 cm to the right of the second refracting surface. The final image is real because the light refracted by the second surface actually reaches this point. The magnification m_2 caused by the second surface is

$$\begin{aligned} m_2 &= \frac{n_1 s'}{n_2 s} \\ &= \frac{1.50\,(29 \text{ cm})}{1.00\,(100 \text{ cm})} \\ &= 0.44 \end{aligned}$$

Following Problem-Solving Tactic 23.6, the total magnification of the original object is the product of the magnifications of each refraction:

$$\begin{aligned} m_{\text{total}} &= m_1 m_2 \\ &= (-1.00)(0.44) \\ &= -0.44 \end{aligned}$$

The minus sign indicates that the final image is inverted with respect to the original object. The height of the final image is

$$\begin{aligned} \text{image size} &= m_{\text{total}}(\text{object size}) \\ &= -0.44(2.0 \text{ cm}) \\ &= -0.88 \text{ cm} \end{aligned}$$

The minus sign also indicates that the image is located below the optic axis, in keeping with the Cartesian sign convention.

23.13 THIN LENSES

Those of us, including the author, who need eyeglasses or contact lenses to see clearly, certainly appreciate the gift that refracting surfaces give us every waking moment! Without our eyeglasses, the world we see would appear as an impressionistic blur of light and color. While artistically beautiful, that view is annoying—and downright dangerous when driving. Eyeglasses are thin lenses, and here we see how they form an image.

A **thin lens** consists of two closely spaced refracting surfaces. The distance between the surfaces (the thickness of the lens) is considered negligible compared with object distances, image distances, and the radii of curvature of the two refracting surfaces.*

To find the location of the image formed by such a lens, we apply the single surface refraction equation successively to each surface. In particular, in Figure 23.60, let

n_1 = the index of refraction of the first medium (where the object lies)

n_2 = the index of refraction of the material out of which the lens is constructed

n_3 = the index of refraction of the medium on the other side of the lens from the object

Derivation of the Thin Lens Equation

We examine the refraction at each of the two surfaces.

Refraction at the First Surface

Applying the single surface refraction equation (Equation 23.35) here, we find

$$-\frac{n_1}{s} + \frac{n_2}{s_1'} = \frac{n_2 - n_1}{R_1} \tag{23.39}$$

*Our artistic rendition of thin lenses makes them look overly plump on the scale of our sketches.

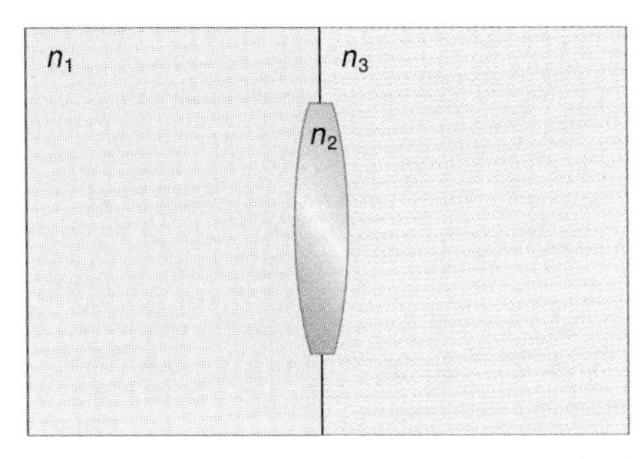

FIGURE 23.60 A thin lens with different media on both sides.

where s is the object distance, s_1' is the image distance of the first surface refraction, and R_1 is the radius of curvature of the first refracting surface. The magnification produced by the first surface refraction is given by Equation 23.38; here, it has the form

$$m_1 = \frac{n_1 s_1'}{n_2 s} \tag{23.40}$$

Refraction at the Second Surface

The single surface refraction equation here becomes

$$-\frac{n_2}{s_2} + \frac{n_3}{s'} = \frac{n_3 - n_2}{R_2} \tag{23.41}$$

where s_2 is the object distance of the second surface, s' is the image distance of the second surface, and R_2 is the radius of curvature of this surface. The magnification produced by the second surface refraction is

$$m_2 = \frac{n_2 s'}{n_3 s_2} \tag{23.42}$$

The position s' of the image formed by the second surface is the location of the final image.

The image of the first surface becomes the object for the second surface. Since the lens is thin, the two surfaces are essentially at the same location and have the *same origin* for the Cartesian sign convention. Thus, because the lens is thin,

$$s_1' = s_2 \tag{23.43}$$

Two terms in Equations 23.39 and 23.41, rewritten here, are identical (the circled terms below are the same):

$$-\frac{n_1}{s} + \frac{n_2}{s_1'} = \frac{n_2 - n_1}{R_1} \tag{23.44}$$

$$-\frac{n_2}{s_2} + \frac{n_3}{s'} = \frac{n_3 - n_2}{R_2} \tag{23.45}$$

If we add Equations 23.44 and 23.45 together, we are left with an equation that involves only the original object distance s and the final image distance s':

$$-\frac{n_1}{s} + \frac{n_3}{s'} = \frac{n_2 - n_1}{R_1} + \frac{n_3 - n_2}{R_2} \tag{23.46}$$

This eliminates the intermediate step, since we really do not care where the image formed by the first surface lies. We just want to locate the final image.

Equation 23.46 is the most general form of the thin lens equation; it is useful if the media on either side of the lens are not the same, such as with contact lenses.

Magnification

The total magnification resulting from the two refractions is

$$\begin{aligned} m &= m_1 m_2 \\ &= \frac{n_1 s_1'}{n_2 s}\frac{n_2 s'}{n_3 s_2} \end{aligned} \tag{23.47}$$

Since the lens is thin, the image distance of the first surface s_1' is the same as the object distance for the second surface s_2. Thus Equation 23.47 becomes

$$\begin{aligned} m &= \frac{n_1 s_2}{n_2 s}\frac{n_2 s'}{n_3 s_2} \\ &= \frac{n_1}{n_3}\frac{s'}{s} \end{aligned} \tag{23.48}$$

A Widely Encountered Special Case

In many instances the media are the same on both sides of the lens. Then $n_3 = n_1$ and Equation 23.46 simplifies to

$$-\frac{n_1}{s} + \frac{n_1}{s'} = \frac{n_2 - n_1}{R_1} + \frac{n_1 - n_2}{R_2}$$

We rearrange this slightly:

$$-\frac{1}{s} + \frac{1}{s'} = \frac{n_2 - n_1}{n_1}\left(\frac{1}{R_1} - \frac{1}{R_2}\right) \tag{23.49}$$

When $n_3 = n_1$, the magnification Equation 23.48 also simplifies to

$$m = \frac{s'}{s} \tag{23.50}$$

Focal Point and Focal Length

As we saw in Figure 23.40, if an object point is placed infinitely far away from the lens, the incident rays are parallel. The image then is at the *focal point* of the lens; the distance of the focal point from the lens is called the *focal length*. We use Equation 23.49 with an infinite object distance, letting the image distance be the focal length f:

$$-\frac{1}{-\infty \text{ m}} + \frac{1}{f} = \frac{n_2 - n_1}{n_1}\left(\frac{1}{R_1} - \frac{1}{R_2}\right)$$

$$\frac{1}{f} = \frac{n_2 - n_1}{n_1}\left(\frac{1}{R_1} - \frac{1}{R_2}\right) \tag{23.51}$$

Equation 23.51 is known as the **lens maker's equation** because it indicates what possible radii to use when manufacturing a lens with a given focal length and material for use in a given medium.

The focal length depends on three things:

a. the curvature of the two refracting surfaces of the thin lens (R_1 and R_2);
b. the index of refraction n_2 of the material out of which the lens is made; and
c. the medium in which the lens is used (n_1).*

Lenses with positive focal lengths are called **converging lenses**, as in Figure 23.61; lenses with negative focal lengths are called **diverging lenses**.

With the lens maker's equation (Equation 23.51) the thin lens equation (Equation 23.49) reduces to the more elegant form:

$$-\frac{1}{s}+\frac{1}{s'}=\frac{1}{f} \quad \text{(thin lens equation; medium is the same on both sides of lens)} \tag{23.52}$$

The magnification is given by Equation 23.50:

$$m=\frac{s'}{s}$$

Just as with mirrors and single refracting surfaces, if part of a thin lens is blocked, the entire image still is present; the image is simply dimmer, since less light reaches it than before. This is the principle behind an iris diaphragm. Such a device controls the amount of light through a lens, but does not affect the size or position of the image. It is good that this is the case: otherwise, in bright sunlight we would have tunnel vision! The size of the iris merely controls the amount of light reaching the retina of the eye or the film of a camera.

*The dependence of the focal length on the medium in which the lens is used is why, among other reasons, your eyes and eyeglasses do not perform properly underwater without a face mask.

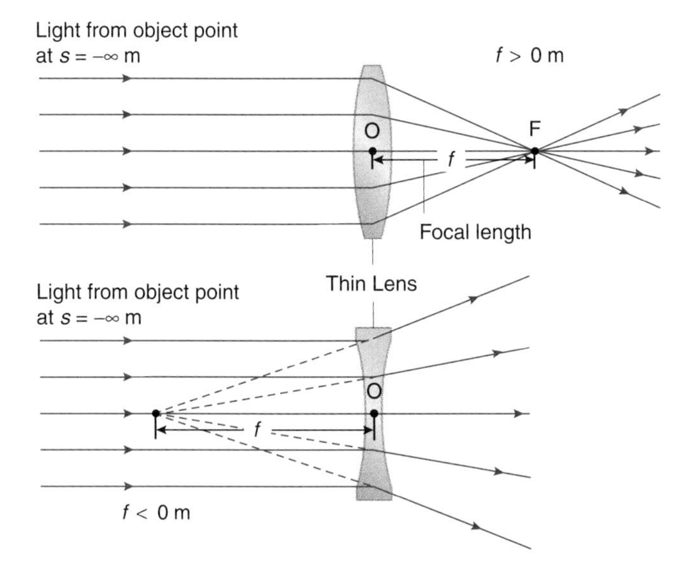

FIGURE 23.61 **A converging lens has a positive focal length, while a diverging lens has a negative focal length.**

If the image distance is positive, the image is to the right of the lens in the standard geometry; this image is real. On the other hand, if the image distance is negative ($s' < 0$ m), then the image is to the left of the lens, even though the refracted rays are on the right side of the lens. Thus the refracted rays do not really pass through the image point if $s' < 0$ m and the image is virtual.

PROBLEM-SOLVING TACTIC

23.7 Be sure to distinguish between the equations for lenses (Equations 23.52 and 23.50) and those for mirrors (Equation 23.23 and 23.24); they are not the same. However, the Cartesian sign convention is used for *both* mirrors and lenses, as well as for single refracting surfaces. The mirror reflects light, but not the Cartesian sign convention, and that makes the equations for lenses and mirrors different.

EXAMPLE 23.13

Determine whether the lenses shown in Figure 23.62 are converging or diverging.

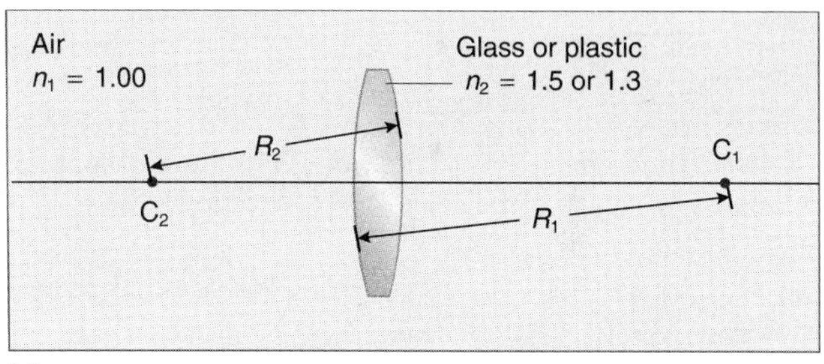

(a)

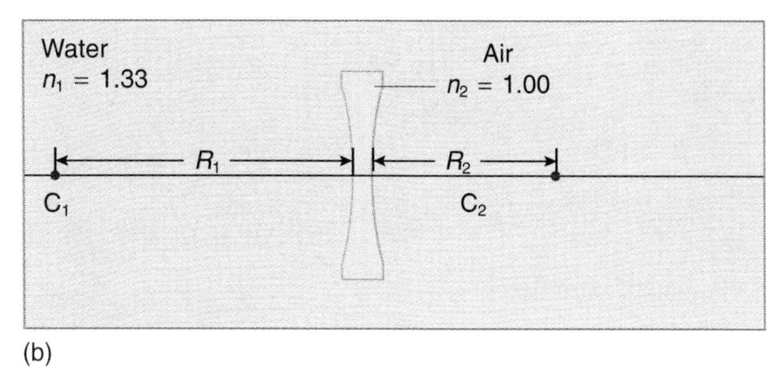

(b)

FIGURE 23.62

Solution

a. Use the lens maker's equation (Equation 23.51):

$$\frac{1}{f}=\frac{n_2-n_1}{n_1}\left(\frac{1}{R_1}-\frac{1}{R_2}\right)$$

Here $n_2 > n_1$, so that the term $n_2 - n_1$ is positive. The radii of curvature $R_1 > 0$ m and $R_2 < 0$ m; thus, the term with the radii also is positive. The focal length f is therefore greater than zero, and so the lens is a converging lens.

b. Use the lens maker's equation (Equation 23.51). In this case $n_2 < n_1$, so $n_2 - n_1 < 0$. The radii term in the lens maker's equation also is negative, since $R_1 < 0$ m and $R_2 > 0$ m. So the focal length is positive and the air lens in water is converging.

STRATEGIC EXAMPLE 23.14

A lucky four-leaf clover is placed 30 cm in front of the two-lens system shown in Figure 23.63. Determine the location of the final image, its magnification, and whether it is upright or inverted and real or virtual.

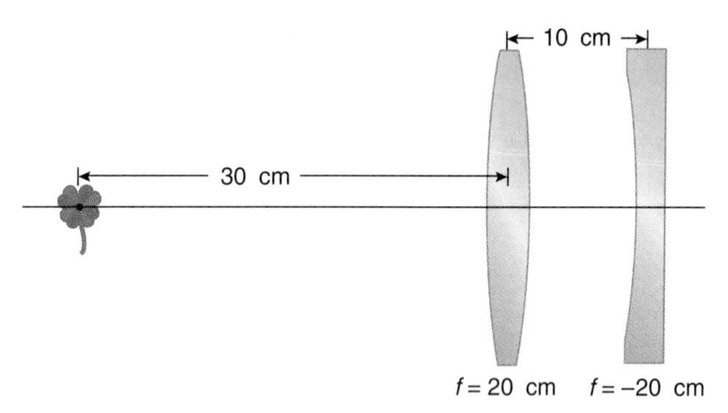

FIGURE 23.63

Solution

Take the lenses one at a time, beginning with the first lens encountered by rays from the object, the converging lens.

The First Lens

Place the origin of the Cartesian system at the vertex of the first lens. Then, for this lens, you have $s = -30$ cm and $f = +20$ cm. Use the thin lens equation (Equation 23.52) to determine the position of the image of the first lens:

$$-\frac{1}{s} + \frac{1}{s'} = \frac{1}{f}$$

$$-\frac{1}{-30\ \text{cm}} + \frac{1}{s'} = \frac{1}{20\ \text{cm}}$$

Solving for s', you find

$$s' = 60\ \text{cm}$$

The image of the first lens is 60 cm to its right.

The magnification produced by the first lens is found using Equation 23.50:

$$\begin{aligned} m_1 &= \frac{s'}{s} \\ &= \frac{60\ \text{cm}}{-30\ \text{cm}} \\ &= -2.0 \end{aligned}$$

The Second Lens

The image of the first lens becomes the object for the second lens. You must transfer the origin for the Cartesian sign convention to the position of the second lens and adjust for this change in determining how far the image of the first lens is from the second lens (Problem-Solving Tactic 23.4). Hence the object distance for the second lens is $s = +50$ cm, *not* 60 cm. This is an example of a *virtual object*.

Use the thin lens equation to determine the position of the image of the second lens:

$$-\frac{1}{s} + \frac{1}{s'} = \frac{1}{f}$$

$$-\frac{1}{+50\ \text{cm}} + \frac{1}{s'} = \frac{1}{-20\ \text{cm}}$$

Solving for s', you find

$$s' = -33\ \text{cm}$$

The image formed by the second lens is the final image. This image is 33 cm to the left of the second lens because s' is negative. The light refracted by the second lens is to its right; this light does not physically pass through the image location. Therefore the final image is virtual.

The magnification produced by the second lens is found using Equation 23.50:

$$\begin{aligned} m_2 &= \frac{s'}{s} \\ &= \frac{-33\ \text{cm}}{50\ \text{cm}} \\ &= -0.66 \end{aligned}$$

The total magnification is the product of the magnifications produced by each lens:

$$\begin{aligned} m_{\text{total}} &= m_1 m_2 \\ &= (-2.0)(-0.66) \\ &= 1.3 \end{aligned}$$

The final image thus is upright (since $m_{\text{total}} > 0$) and enlarged (since $|m_{\text{total}}| > 1$).

23.14 Ray Diagrams for Thin Lenses

We introduced ray diagrams with spherical mirrors as a useful check on the calculations made to locate the position of the image and determine its nature (real or virtual) and relative size. Ray diagrams also can be drawn for thin lenses to accomplish the same purposes. Since the lens is thin and because of the paraxial ray approximation, we draw ray diagrams as if all the refraction occurs at a plane perpendicular to the optic axis passing through the center (vertex) of the lens.

The focal point located with the lens maker's equation (Equation 23.51),

$$\frac{1}{f} = \frac{n_2 - n_1}{n_1}\left(\frac{1}{R_1} - \frac{1}{R_2}\right)$$

is the **primary focal point** of a lens. For a converging lens, $f > 0$ m and the primary focal point is to the right of the lens in the standard geometry; see Figure 23.64a. For a diverging lens, $f < 0$ m and the primary focal point is to the left of the lens in the standard geometry; see Figure 23.64b.

The significance of the primary focal point is as follows. *Any* ray incident on the lens parallel to the optic axis (say, from an object point at $s = -\infty$ m), when refracted by the lens, passes through (or appears to pass through) the primary focal point, as shown in Figure 23.64a and 23.64b.

Each lens also has a **secondary focal point** F_2; this is the location for an object that leads to an infinitely distant image. If the medium is the same on both sides of the lens, the secondary focal point is located the same distance (in absolute value) from the lens as the primary focal point, but is on the *opposite* side of the lens as the primary focal point.

The significance of the secondary focal point is as follows. *Any* ray passing through (or appearing to pass through) the secondary focal point emerges from the lens traveling parallel to the optic axis, as shown in Figure 23.65. If we reverse the direction of the light (by using a plane mirror facing and perpendicular to the rays), the light will retrace its path. The secondary focal point is useful when sketching ray diagrams to locate the position of an image, as we will see.

As with mirrors, we use an extended object and locate the position of the image of an object point not lying on the optic axis, say point Q in Figure 23.66. Of the many rays diverging from point Q that are refracted by the lens, we consider just three principal rays whose paths through the lens are easy to trace. Two rays are sufficient to locate the image, but the third ray serves as a useful check of the ray diagram itself.

The three rays used in a ray diagram for thin lenses are the following:

1. A ray from object point Q through the center of the thin lens (the origin for the Cartesian sign convention), as indicated by ray (1) in Figure 23.66. The central region of a thin lens is approximately a very thin parallel plate. Problem 29 shows that a ray passing through a parallel plate emerges parallel to its incident direction, though translated. Since the lens is thin, the thickness of the essentially

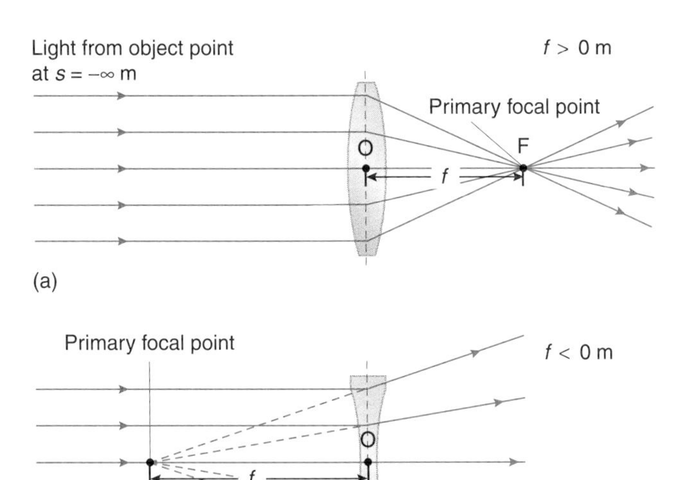

FIGURE 23.64 The primary focal point F. (a) Converging lens. (b) Diverging lens.

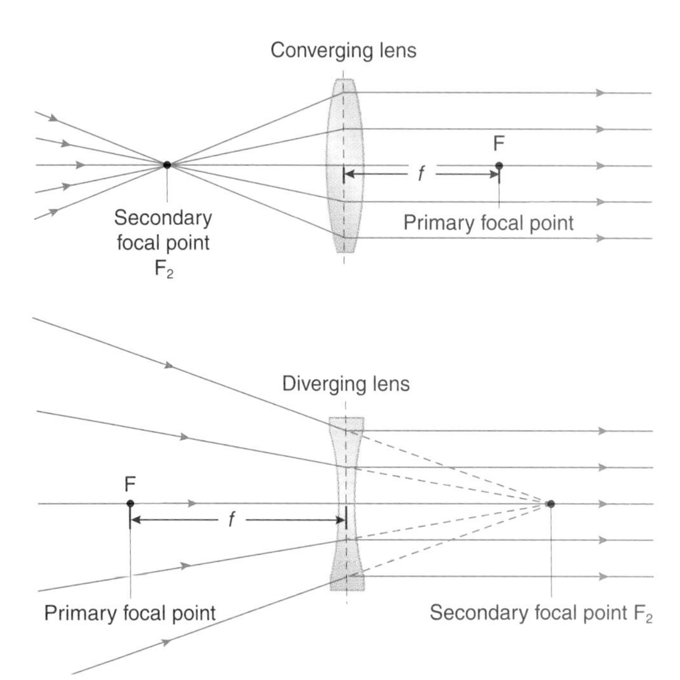

FIGURE 23.65 Secondary focal points F_2 of converging and diverging lenses.

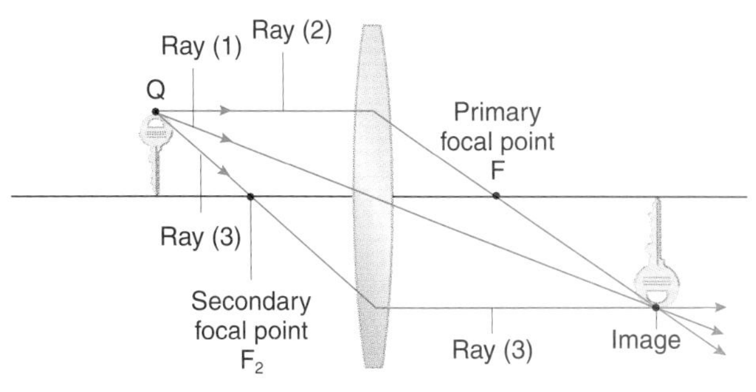

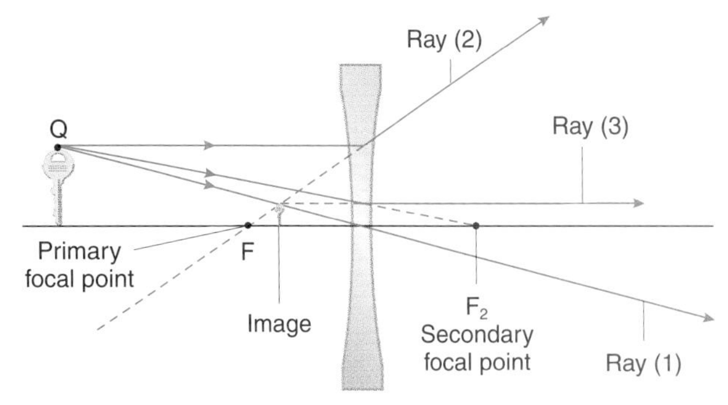

FIGURE 23.66 Ray diagram for a converging (a) and for a diverging lens (b).

parallel plate at the center of the lens is thin and the translation of the ray is negligible. Thus this ray passes essentially straight through the lens.

2. A ray from object point Q parallel to the optic axis, ray (2) in Figure 23.66. This ray passes through (or appears to pass through) the primary focal point of the lens. Any ray parallel to the optic axis passes through (or appears to pass through) the primary focal point when refracted by the lens; the single ray from object point Q that approaches the lens parallel to the optic axis is no exception.
3. A ray from object point Q through (or toward) the secondary focal point emerges from the lens parallel to the optic axis; this is ray (3) in Figure 23.66.

The place where the three rays intersect (or appear to intersect) after they are refracted by the lens locates the position of the image. Once the image is located geometrically using the three principal rays, the nature of the image can be distinguished:

- whether it is real or virtual;
- whether it is upright or inverted;
- whether it is larger or smaller than the object.

EXAMPLE 23.15

A small bracket fungus (*Polyporus applanatus*) is placed 10 cm in front of a converging lens of primary focal length 20 cm. Draw a thin lens ray diagram to locate the image and determine whether it is upright or inverted, and real or virtual.

Solution

Draw a diagram approximately to scale. Principal ray (1) passes through the vertex of the lens undeviated, as in Figure 23.67. Principal ray (2), incident parallel to the axis, passes through the primary focal point. Principal ray (3) acts *as if* it came through the secondary focal point, and emerges from the lens parallel to the optic axis.

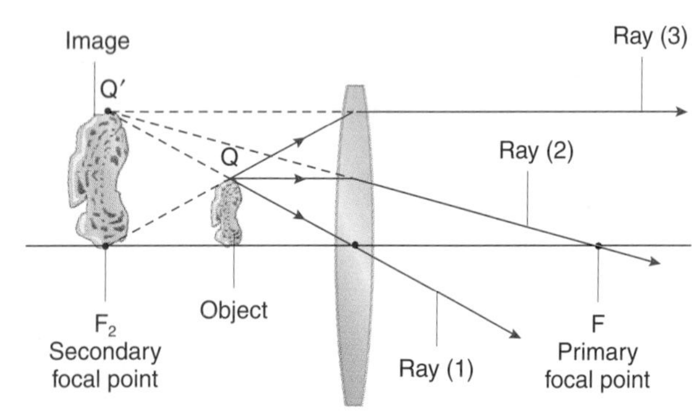

FIGURE 23.67

It is apparent that the three principal rays do not intersect on the right-hand side of the lens; rather they appear to intersect at point Q′ on the left-hand side of the lens. The image, therefore, is virtual. The diagram also shows that the image is upright and enlarged.

23.15 OPTICAL INSTRUMENTS

Where the telescope ends, the microscope begins. Which of the two has the grander view? Choose.

Victor Hugo (1802–1885)*

Here we examine a few optical instruments of particular interest and import in the sciences. These instruments also form the basis of more complex optical systems. A wide variety of more sophisticated optical instruments continually are being invented, illustrating that applications of geometric optics are limited only by our imaginations.

A Simple Magnifier

A magnifying glass may be the first seemingly magical optical instrument you remember from your childhood. A simple magnifier is a converging lens used in a particular way. If we are given a magnifying glass and are asked to demonstrate its use, we bring an object close to the glass and look through it to see a magnified image of the object. How close? To use the lens as a magnifier, the object is placed between the converging lens and its secondary focal point. Then an enlarged virtual image is produced, as shown in the ray diagram of Figure 23.68.

The magnification is calculated using the magnification equation for a thin lens (Equation 23.50):

$$m = \frac{s'}{s}$$

Since both the object and image distances are negative here, the magnification is positive, indicating that the virtual image is upright. This confirms what you already knew: when using a magnifying glass, the object does not have to be turned up-side-down to see the image right-side-up!

There is another way that the magnification of a simple magnifier is specified that involves comparing the *angular size* of the object and image: the **angular magnification** $m_{\angle}$.

The angular size of the object θ_o is defined and measured when the object is exactly 25 cm away from the eye, an arbitrarily chosen standard distance called the **standard eye**

Les Misérables, St. Denis, Book III, The House in the Rue Plumet, Chapter 3, "Foliic ac Frondibus" (A. L. Burt, New York, 1862) page 163.

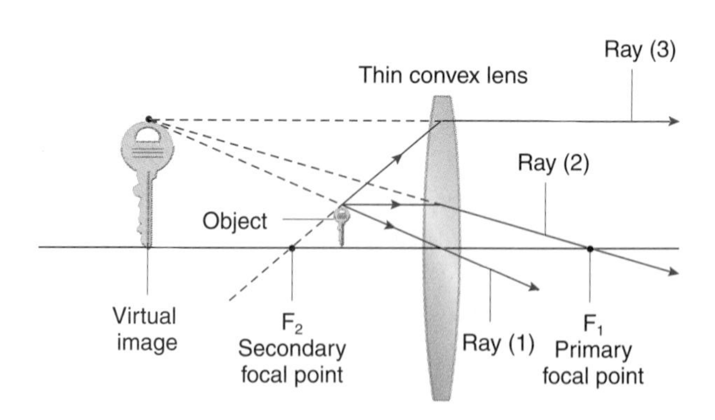

FIGURE 23.68 A simple magnifier.

near point,* as shown in Figure 23.69. From the geometry of Figure 23.69, we have (using the small angle approximation)

$$\theta_o = \frac{h}{25\ \text{cm}}$$

The angular size of the image θ_i when it is at the standard eye near point is found using triangle QOP in Figure 23.70 (using the small angle approximation again):

$$\theta_i = \frac{h}{-s} \tag{23.53}$$

where the term $-s$ is used in the denominator since s itself is negative.

The thin lens equation (Equation 23.52) is solved for s with the image distance $s' = -25$ cm:

$$-\frac{1}{s} + \frac{1}{-25\ \text{cm}} = \frac{1}{f} \tag{23.54}$$

All the distances must be in centimeters since the image distance is expressed in centimeters. Solving Equation 23.54 for $-s$, we find

$$-s = \frac{(25\ \text{cm})f}{25\ \text{cm} + f} \qquad (s \text{ and } f \text{ in centimeters}$$

Using this expression for $-s$ in Equation 23.53 for the angular size of the image, we obtain

$$\theta_i = \frac{h(25\ \text{cm} + f)}{(25\ \text{cm})\, f}$$

The angular magnification is the ratio of the angular size of this image at the standard eye near point to that of the object when it is at the standard eye near point:

$$m_\angle \equiv \frac{\theta_i}{\theta_o} \tag{23.55}$$

$$= \frac{\dfrac{h(25\ \text{cm} + f)}{(25\ \text{cm})\, f}}{\dfrac{h}{25\ \text{cm}}}$$

$$= \frac{25\ \text{cm}}{f} + 1 \qquad (f \text{ in centimeters}) \tag{23.56}$$

Another way to use the lens as a magnifying glass is to place the object at the secondary focal point of the lens; the virtual image then is at $-\infty$ m, as shown in Figure 23.71. With the final image at $-\infty$ m, the ciliary muscles that control the shape of the eye lens are relaxed when we view the image through the lens. In this case the angular size of the image, determined from Figure 23.71, is (using the small angle approximation)

$$\theta_i = \frac{h}{f} \tag{23.57}$$

The angular magnification, Equation 23.55, then becomes

$$m_\angle = \frac{\dfrac{h}{f}}{\dfrac{h}{25\ \text{cm}}}$$

$$= \frac{25\ \text{cm}}{f} \qquad (f \text{ in centimeters}) \tag{23.58}$$

This magnification is slightly less than the magnification with the *image* at the standard eye near point (Equation 23.56); however, Equation 23.58 is chosen to typify the magnification because the typical eye is then most relaxed. In adjusting

*The actual near point of the eye depends on the individual.

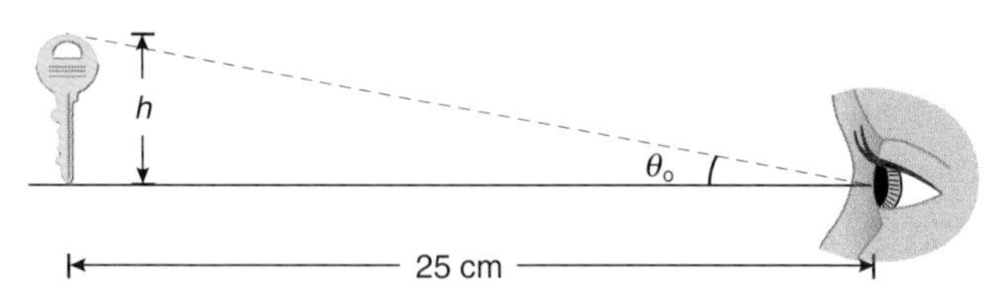

FIGURE 23.69 The angular size θ_o of the object when placed at the standard eye near point.

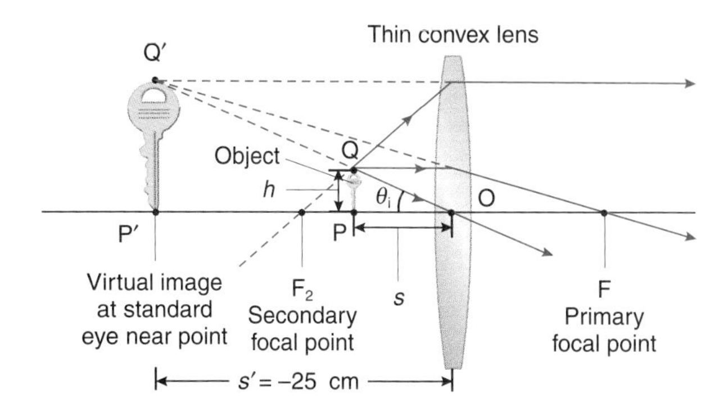

FIGURE 23.70 The angular size θ_i of the image when it is at the standard eye near point.

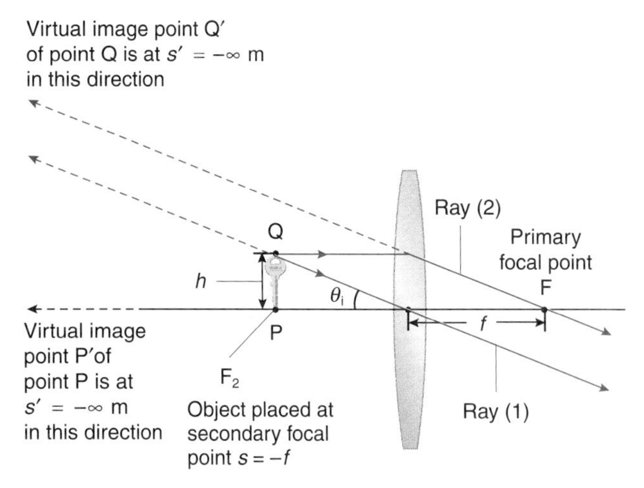

FIGURE 23.71 With the object at the secondary focal point, the image is at $-\infty$ m.

an object distance for clear focus, you should move the object in from too great an object distance and stop at the first clearness of image in order to avoid eye muscle strain during use of the magnifier.

QUESTION 9

Why can you *not* draw principal ray (3) in Figure 23.71?

A Pinhole Camera

One of the more remarkable cameras has an optical element that consists of nothing more than a pinhole. The aperture of a pinhole is very small, of course. Only a small amount of light passes into the lighttight box containing the film. This means that long exposure times may be necessary. The object must be stationary during the exposure, so it is difficult to obtain pictures of moving objects with a pinhole camera.

The ray diagram for an ideal pinhole camera consists of a *single* ray from each object point to each image point, as shown in Figure 23.72. Other than this single ray through the pinhole, no other rays reach a given image point. Thus the image of the object is found at *any* and *every* image distance on the positive distance side of the pinhole.

This means that the image is fairly sharp wherever the film is placed. The position of the film or screen determines the magnification using the magnification Equation 23.50:

$$m = \frac{s'}{s}$$

As the image distance and magnification increase, however, the image becomes dimmer because the same small amount of light through the pinhole is spread over a larger area.

Since a given object can be registered on the film at many image distances, we can turn the problem around and say that for a fixed image distance (corresponding to where the film is placed), objects at many different object distances are equally well imaged on the film simultaneously. This quality is called good **depth of field**. Objects close to and far from the pinhole camera are equally well recorded on the film, as when taking a picture of your friends against a distant mountain background.

We can easily illustrate the principle of pinhole image formation by performing a simple experiment. We form an image of the Sun on a piece of paper with a pinhole in an index card, as in Figure 23.73. As the screen is moved closer to the pinhole, the image size decreases and the brightness of the image increases. This technique is a common method for safely observing the Sun to detect sunspots or the partial phases of solar eclipses.

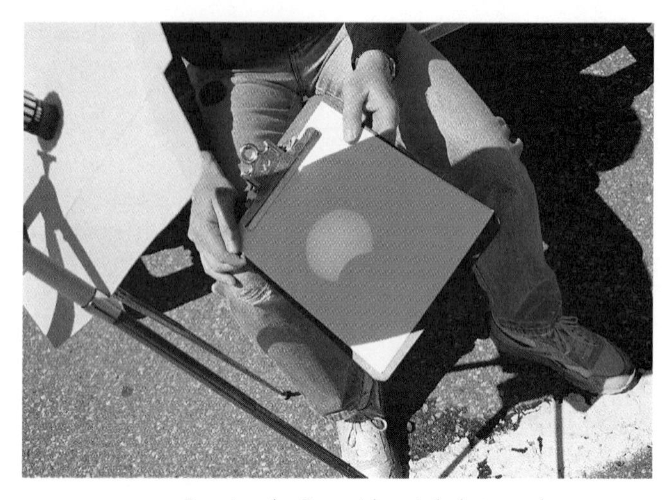

FIGURE 23.73 Imaging the Sun with a pinhole is a safe way to view a solar eclipse.

The principle of pinhole projection also is the reason you squint if you are near- or farsighted and do not use your eyeglasses. By squinting, you are trying to make the aperture of the eye small. The small aperture or pinhole over the eye lens reduces the cone of ray convergence that would tend to blur the poorly focused images near and far. The depth of field is thus improved. Try it. If you are highly nearsighted, just look at the (bright) world through a pinhole and you will be surprised at how much more clearly you can see. Just don't drive at night with pinholes!

Standard Cameras

The purpose of the lens of a camera is to form a real image of an object on the film, as indicated in Figure 23.74.

The lens must be a converging lens; a single diverging lens cannot form a real image of real objects, no matter where they are placed (see Problem 61). Of course, a shutter is provided to determine when light is allowed onto the film, and various other bells and whistles are added to increase the flexibility of use (and

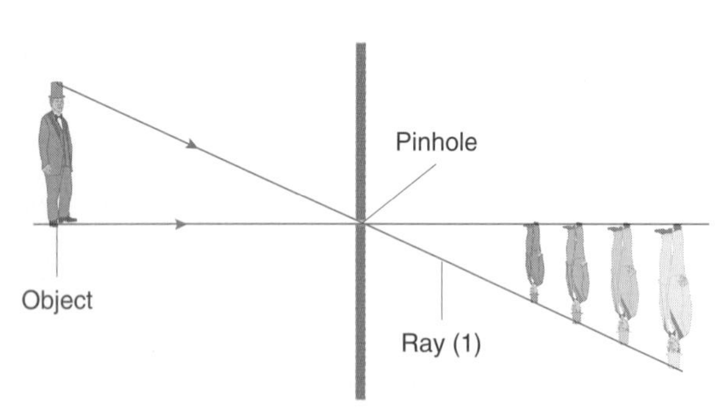

FIGURE 23.72 Ray diagram for a pinhole aperture.

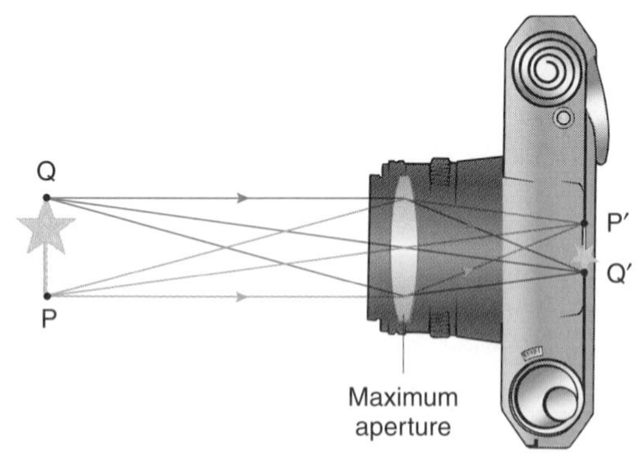

FIGURE 23.74 A camera lens forms a real image on the film.

cost). The essential imaging element is the converging lens. It can be much wider than a pinhole because the refracting lens gathers a wide bundle of rays from each object point and converges them to a single point on the image plane.

An iris diaphragm is used on some camera lenses to regulate the amount of light reaching the film as well as to control the depth of field. The smaller the aperture size, the more the effective part of the lens is like a pinhole, and the greater the depth of field. In other words, to take a picture of your compatriots against a lovely but distant background scene, use a small aperture.

Two numbers are used to specify a camera lens: its focal length (typically quoted in millimeters) and the ***f*/number** of the lens.* The *f*/number of the lens is defined to be the focal length divided by the maximum aperture (diameter) of the lens:

$$f/\text{number} \equiv \frac{\text{focal length}}{\text{maximum aperture}} \tag{23.59}$$

The *f*/number is dimensionless. The slash / can be a bit confusing at first. When a lens is said to be *f*/1.4, it means that the *f*/number is 1.4. In other words:

$$1.4 = \frac{\text{focal length}}{\text{maximum aperture}}$$

If the focal length of the lens is 50 mm, the maximum aperture of this lens thus is

$$\begin{aligned}\text{maximum aperture} &= \frac{50 \text{ mm}}{1.4} \\ &= 36 \text{ mm}\end{aligned}$$

Smaller apertures (meaning larger *f*/numbers) can be obtained with an iris diaphragm. These larger *f*/numbers are called effective *f*/numbers since they determine effectively how much of the entire lens is being used. For a lens *specified* as *f*/1.4, an iris diaphragm can be used to secure *larger* (effective) *f*/numbers of 2.8, 5.6, 11, and so forth, but not *f*/numbers smaller than the one in the lens specification. That is, the iris cannot be used to make the *f*/number of this lens 1.2. The *f*/number on the spec sheet of the lens is the *smallest* *f*/number the lens can attain (corresponding to the maximum aperture). Thus an iris diaphragm can be used to give an *f*/1.4 lens an effective *f*/number of *f*/5.6; that is, the effective aperture of the lens has been reduced by the diaphragm, so that

$$5.6 = \frac{\text{focal length}}{\text{effective aperture}}$$

The aperture determines the amount of light entering the camera. The amount of light reaching the film is proportional to the area of the lens opening; the area, in turn, is proportional to the square of the diameter or the square of the aperture (apart from numerical factors). Thus the amount of light reaching the lens is inversely proportional to the square of the *f*/number.

To have good depth of field, one wants to shoot with a high effective *f*/number, since then the aperture of the camera is small and approaches the imaging properties of a pinhole. Inexpensive, even disposable cameras have lenses of small, fixed aperture and, consequently, a relatively high *f*/number to obtain reasonably good depth of field and eliminate the need for focusing hardware.

*The term also is known as the *focal ratio*; we will not use this terminology.

The Eyes of Vertebrates

The eyes of vertebrates (see Figure 23.75) are certainly some of the more remarkable optical instruments that exist.† The eye produces an inverted (!) real image of objects on the retina. The brain then performs another (essentially electronic) inversion of the image to produce the image we perceive in its proper orientation. The retina has two types of cells: **rods**, which specialize in detecting low levels of light; and **cones**, which specialize in color vision and detecting bright light conditions.

Most of the refraction of light rays from an object occurs at the first surface encountered by the rays from the object: the **cornea**. However, for simplicity, we draw the refraction in Figures 23.76 and 23.77 as if the eye lens was the principal culprit. With the eye, *the image distance is fixed* because of anatomical considerations (the eye is essentially of fixed size). (It would be quite amazing if our eye lens poked in and out of our heads to focus, like a camera lens!) For objects at various distances from the eye, then, the focal length of the eye lens must change to ensure that the image lies on the retina. The eye lens tinkers with the refraction to produce a sharp image on the retina. The eye lens is a smart lens: **ciliary muscles** adjust the curvature of the lens, thus changing its focal length (see Equation 23.46) to ensure that the image of the object lies on the retina.

An **iris** regulates the amount of light entering the eye, automatically expanding the aperture (called the **pupil**) of the eye under dim lighting conditions or decreasing the aperture under bright lighting conditions. We do not usually talk about the *f*/number of the eye because both the focal length of the eye lens and its aperture can change. When you look at something, the eye turns so that the image of the object being examined falls on a central region of the retina known as the **fovea**, containing almost exclusively cones.

†Various life forms have different kinds of eyes. For example, insects typically have an eye known as a *compound eye* (see Investigative Project 10).

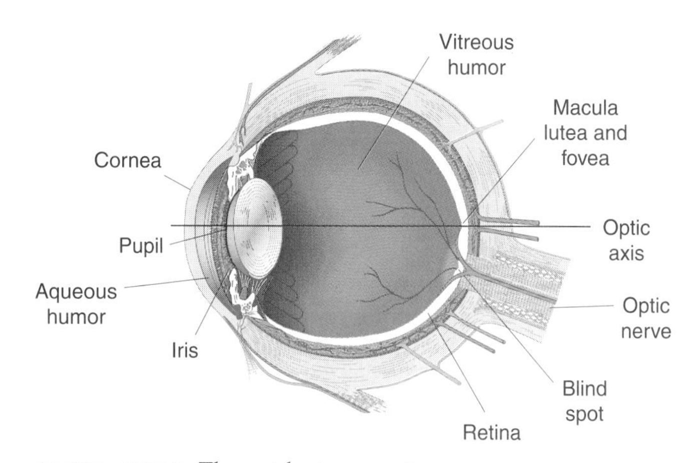

FIGURE 23.75 The vertebrate eye anatomy.

As many of you likely know from personal experience, not all eyes perform well. Various types of vision defects exist and can be corrected with supplementary optical hardware: a fancy word for eyeglasses or contact lenses. We consider some of the more common vision problems and how they are corrected.

Myopia (Nearsightedness)

The operational problem with **myopia** is an inability to see distant objects clearly. The focal length of the eye lens is too small; even when the ciliary muscles around the lens are relaxed, they are unable to let the focal length of the eye lens be long enough to image distant objects on the retina. For a myopic eye, the image of distant objects is always located between the eye lens and the retina, on the *near* side of the retina (see Figure 23.76). A corrective lens is added to make the final image of the lens and eye optical system appear on the retina. We can see what type of lens is needed from the following analysis.

The most distant (farthest) object that can be seen clearly by a myopic person is called the **far point** d_{far} (consider d_{far} always to be a positive number). A normal eye would have a far point of infinity. The term far point is a bit ironic, for in many cases it actually is not very far from the eye, especially for highly myopic eyes. Indeed, far points of only 15 cm are not uncommon.

The purpose of using a lens in front of the eye is to place the image of distant objects at the far point of the eye. The image provided by the lens becomes the object for the eye; if the object for the eye appears to be within its far point, the eye can see the image clearly (meaning that the final image is on the retina). The lens thus creates a close *virtual image* of a distant object. The virtual image of the lens then becomes a *virtual object* for the eye.

What lens can place the image of a distant object at the far point of the eye? We use the thin lens equation (Equation 23.52) to determine this focal length:

$$-\frac{1}{s} + \frac{1}{s'} = \frac{1}{f}$$

Let the distant object be at $s = -\infty$ m; we want s' to be equal to the far point—that is, $s' = -d_{far}$, where the minus sign indicates that the image is virtual (since d_{far} was taken to be intrinsically positive). We make these substitutions into Equation 23.52:

$$-\frac{1}{-\infty \text{ m}} + \frac{1}{-d_{far}} = \frac{1}{f}$$

Solving for the focal length, we find

$$f = -d_{far}$$

A diverging lens ($f < 0$ m) whose focal length is equal in absolute value to the far point is needed to correct myopia.

Optometrists and ophthalmologists quote the focal length not in meters or centimeters but in **diopters** (dp). The diopter value of a lens is the reciprocal of the focal length, when the focal length is expressed in *meters*:

$$\text{diopter} \equiv \frac{1}{f_{\text{in meters}}} \tag{23.60}$$

The diopter value is indicated on your eyeglass prescription (if you need it!). Thus a myopic person with a far point of 15 cm needs a corrective lens of focal length $f = -15$ cm or

$$\frac{1}{-0.15 \text{ m}} = -6.7 \text{ dp}$$

Hyperopia (Farsightedness)

Hyperopia is an inability to see objects clearly that are as close to the eye as perhaps 25 cm. Distant objects are seen clearly, but the simple task of reading is difficult because the material must be placed far from the eye in order to be seen clearly. In this case the eye produces an image of a nearby object that is behind the retina, on its *far* side. The image is not on the retina itself, as shown in Figure 23.77.

The closest that a person with hyperopia can bring an object and still see it clearly is called the **near point**. That is, the eye can

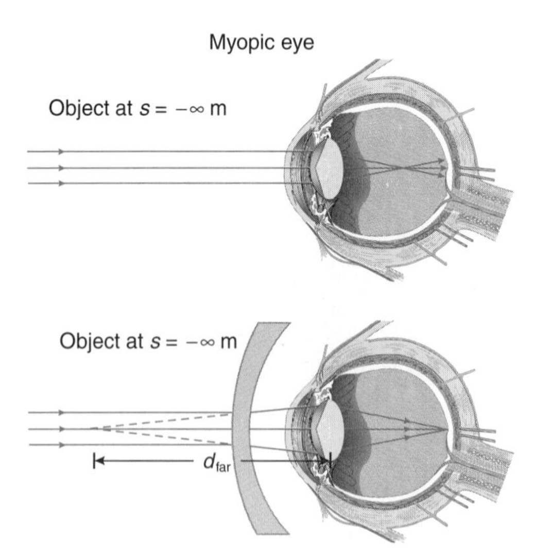

FIGURE 23.76 A myopic eye and a lens to correct for myopia.

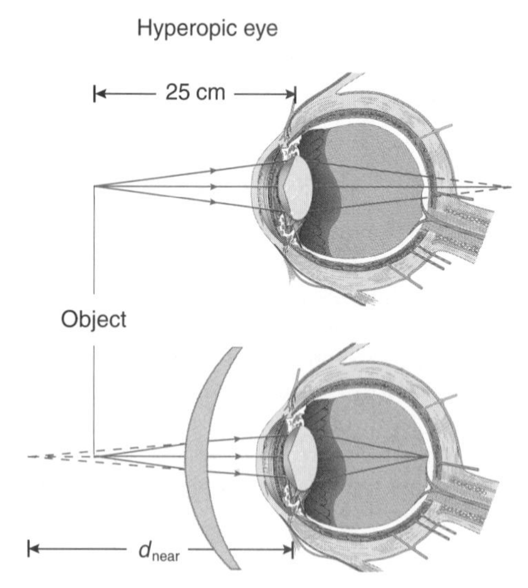

FIGURE 23.77 The hyperopic eye and a lens to correct for hyperopia.

see clearly only if the object it is viewing is located *further* from the eye than its near point.

We can see how corrective eyeglasses can be prescribed. Let d_{near} be the closest distance from the unaided eye that reading material can be held and still seen clearly. The function of corrective lenses is to create a virtual image of the reading material essentially at $-d_{near}$; this virtual image then becomes a virtual object for the eye. The eye thus views a virtual object that is far away, and the image produced by the eye is sharp on the retina. We use the lens equation (Equation 23.52) to see what type of lens is needed:

$$-\frac{1}{s} + \frac{1}{s'} = \frac{1}{f}$$

The object for the lens is the reading material, placed a convenient distance d away (consider d to be always a positive distance); therefore, according to the Cartesian sign convention, $s = -d$ and the image distance is to be $s' = -d_{near}$. Making these substitutions in Equation 23.52, we find

$$-\frac{1}{-d} + \frac{1}{-d_{near}} = \frac{1}{f}$$

Since $d < d_{near}$, the focal length is positive. Thus a converging lens is needed to correct for hyperopia.

Astigmatism

Astigmatism is a more complex vision problem caused by variations in the curvature of the corneal surface. We will not study this in greater detail except to mention that astigmatism is corrected with a lens that has different curvatures along two perpendicular axes in the plane of the lens.

Other Vision Problems

A **cataract** is a clouding of the eye lens, rendering it translucent rather than transparent. It is corrected only by surgically removing the natural eye lens and implanting an artificial substitute.

Glaucoma is an abnormally high pressure in the fluid within the eye, which if uncorrected (via drugs or surgery) can irreversibly rupture blood vessels within the eye, causing the death of retinal cells (the rods and cones) and permanent blindness.

A Simple Microscope

A simple microscope is a two-lens system designed to make small objects appear larger. Both lenses are converging lenses of short focal length, as we will see. The lens closest to the object is called the **objective lens**. The object is placed at a distance from the objective lens that is slightly greater than the secondary focal distance, as shown in Figure 23.78.

The objective lens then forms a real, enlarged image with a magnification m_o given by

$$m_o = \frac{s'}{s} \quad (23.61)$$

The object distance s (<0 m) is essentially the negative of the focal length of the objective lens:

$$s \approx -f_o$$

where $f_o > 0$ m for a converging lens. Since the eyepiece focal length also is short, the image distance s' is essentially the length of the microscope tube ℓ:

$$s' \approx \ell$$

The magnification of the objective, Equation 23.61, thus is essentially

$$m_o \approx -\frac{\ell}{f_o} \quad (23.62)$$

Since ℓ and f_o are both positive, the magnification of the objective is negative, indicating (as Figure 23.78 shows) that the image of the objective is inverted.

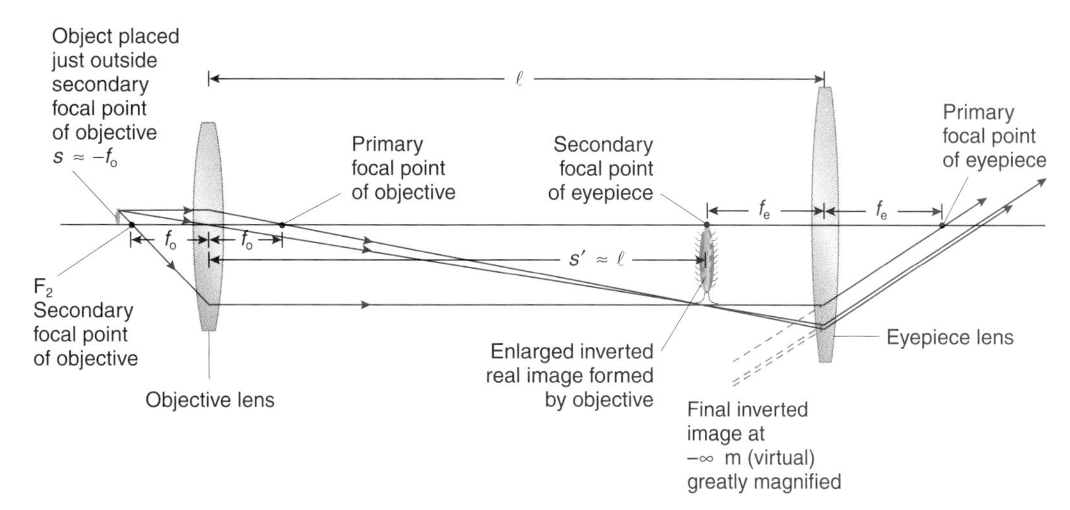

FIGURE 23.78 A microscope.

The image of the objective lens becomes the object for the **eyepiece lens**, which acts as a simple magnifier. Thus the eyepiece produces an angular magnification $m_{\angle e}$ given by Equation 23.58:

$$m_{\angle e} = \frac{25 \text{ cm}}{f_e}$$

where f_e is the eyepiece focal length ($f_e > 0$ cm, since it is a converging lens).

The total magnification of the microscope is the product of the magnifications:

$$\begin{aligned} m_{total} &= m_o m_{\angle e} \\ &\approx -\frac{\ell}{f_o}\frac{25 \text{ cm}}{f_e} \end{aligned} \tag{23.63}$$

(all distances and focal lengths *must* be in centimeters)

This expression for the magnification of a microscope indicates why the objective and eyepiece lenses are chosen to have small focal lengths.

> The magnification of a microscope is inversely proportional to the product of the focal lengths.

Since ℓ, f_o, and f_e are all positive, the minus sign indicates that the final image is inverted with respect to the original object. The inversion typically is not a problem, since an inverted amoeba (*Amoeba proteus*) looks to all the world like another one that is upright.

Most microscopes have the magnification of the objective element and eyepiece elements inscribed on the barrels of the lenses. A product of the numbers yields the total magnification.

An Astronomical Telescope

A refracting telescope consists of two converging lenses. In this case the objective lens (the one closest to the object) has a long focal length (for reasons that will become apparent) and a large aperture (to collect as much light as possible). A telescope, of course, is designed to look at distant objects and make them appear closer. Hence the object distance is effectively infinite: $s = -\infty$ m.

The thin lens equation (Equation 23.52) indicates that the image distance of the objective lens is the primary focal length. Since the object distance is effectively infinite, the usual equation for the magnification of a lens (Equation 23.50),

$$m = \frac{s'}{s}$$

is inappropriate. With $s = -\infty$ m, the magnification is zero. But the image still has finite size! So we revert to defining an angular magnification. The angular size of the object θ_o is the angle subtended by the object without optical aid (i.e., without the telescope).

Rays from the object point Q approach the objective as parallel rays making the angle θ_o with the optic axis as indicated in Figure 23.79. One of these rays [ray (1)] passes undeviated through the center of the objective. Another ray [ray (3)] passes through the secondary focal point of the objective and emerges parallel to the optic axis.

The extended image P′Q′ of the extended object PQ is in the **focal plane**, a plane perpendicular to the optic axis at the primary focal point of the objective. The rays then enter the eyepiece lens, placed so that the final image is a virtual image at $-\infty$ m. In other words, the eyepiece is placed so that the image formed by the objective lens is at the secondary focal point of the eyepiece lens.

The ray incident on the eyepiece parallel to the optic axis passes through the primary focal point of the eyepiece as shown in Figure 23.79. Using the small angle approximation, the angle subtended by the image θ_i is found using triangle FAO′:

$$\theta_i = \frac{-h}{f_e} \tag{23.64}$$

The distance h is negative (according to the Cartesian sign convention), so $-h$ is positive.

Using the small angle approximation again, the angle subtended by the object itself, θ_o, is found from

$$\theta_o = \frac{-h}{f_o} \tag{23.65}$$

where $-h$ is positive since h itself is negative.

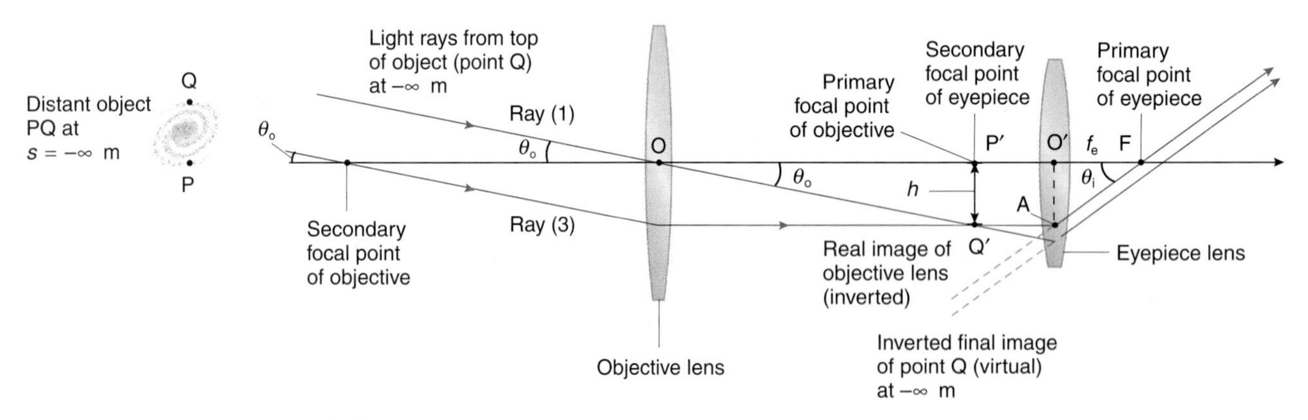

FIGURE 23.79 An astronomical telescope.

The absolute magnitude of the angular magnification of the telescope is the ratio of these angles; a minus sign is introduced since the final image is inverted:

$$m_{\angle} = -\frac{\theta_i}{\theta_o}$$

Substituting for the angles using Equations 23.64 and 23.65, we find

$$m_{\angle} = -\frac{f_o}{f_e} \tag{23.66}$$

The magnification of a telescope is the negative of the ratio of the focal length of the objective to that of the eyepiece.

It is difficult, not to mention expensive, to change the magnification of a telescope by changing its objective element because of its large aperture. So, to change the magnification of a telescope, we switch to an eyepiece of different focal length. To increase the magnification, a shorter focal length eyepiece is used.

Despite the Madison Avenue hype, never purchase a telescope on the basis of its magnification. You can get any useful magnification with any telescope with a suitable choice for an eyepiece. In practice, astronomers rarely work with magnifications greater than several hundred; increasing the magnification beyond that only increases blur because of refractive turbulence within the atmosphere along the line of sight. The critical factor to consider (other things aside, such as cost, mechanical performance, and beauty) for an astronomical telescope is aperture, for the area of the objective determines the amount of light collected by the telescope.

Like biologists, astronomers are not bothered by a final image that is inverted: a star, planet, or galaxy looks essentially the same either way. For terrestrial use, however, an inverted final image is awkward, if not inconvenient. Various methods are used to produce a final upright image in a terrestrial telescope:

a. A *field lens* is inserted between the eyepiece and the objective. The purpose of this lens is to invert the inverted image of the objective, so that the simple magnifier eyepiece looks at an upright image.
b. A diverging eyepiece lens is used in conjunction with the objective lens, in place of the traditional converging eyepiece lens, as in Figure 23.80. Such a two-lens system produces a right-side-up final image. This is the design employed in most opera glasses and terrestrial spyglasses.
c. Traditional binocular telescopes (binoculars) use a prism to invert the image of the objective before the eyepiece magnifies the result. Binoculars have two numbers associated with them: for example, 7 × 35, 7 × 50, or 8 × 35. The first number (7, 7, or 8 here) is the angular magnification; the second (35, 50, or 35 here) is the diameter of the objective element in millimeters.

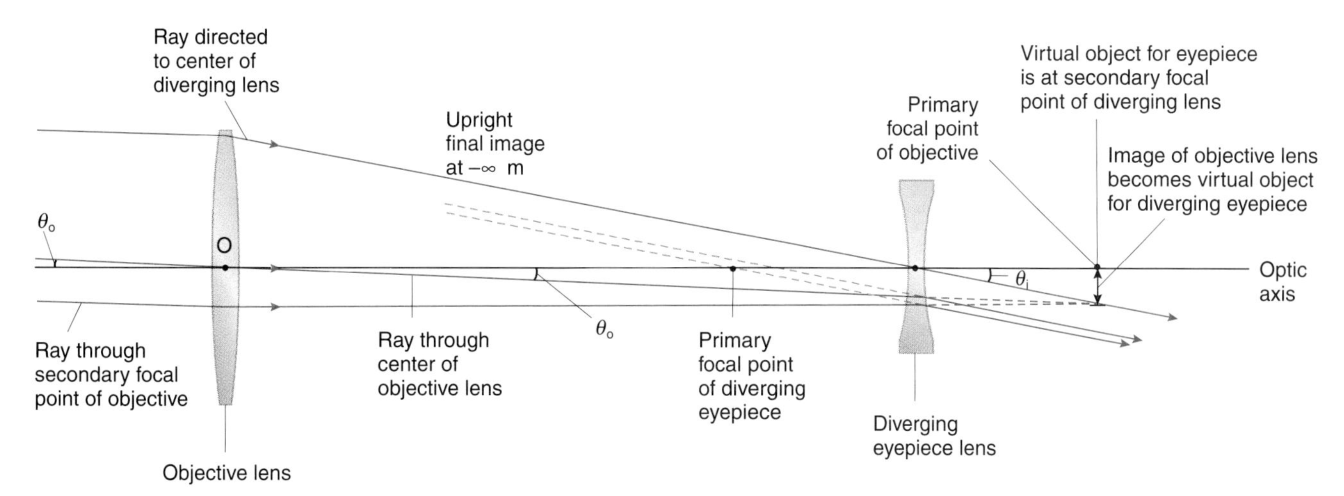

FIGURE 23.80 A terrestrial spyglass telescope produces an upright final image.

Chapter Summary

Light *rays* are perpendicular to light wavefronts and indicate the path of light. *Geometric optics* is appropriate if light waves interact with apertures or obstacles that are much larger than the wavelength of the light.

The *intensity* of light is the average power transmitted through one square meter oriented perpendicular to the direction the light is travelling. The light intensity I from a pointlike source of *luminosity* L decreases as the inverse square of the distance r from the source:

$$I = \frac{L}{4\pi r^2} \tag{23.1}$$

In SI units, the luminosity L is expressed in watts, the distance r in meters, and the intensity I in watts per square meter.

The *law of reflection* states that the angle of incidence of a light ray is equal to the angle of reflection from a mirror surface; the incident ray, the reflected ray, and the normal line are in a common plane. The angles are measured with respect to a normal to the surface at the point of reflection.

The *law of refraction* at an interface between two transparent media states that the angle of incidence θ_1 of a light ray and its angle of refraction θ_2 are related by

$$n_1 \sin\theta_1 = n_2 \sin\theta_2 \tag{23.6}$$

where n_1 and n_2 are the indices of refraction of the two respective media; the incident ray, refracted ray, and normal line lie in one plane.

When a light ray passes from a more *optically dense* medium with index of refraction n_1 to a less optically dense medium with index of refraction n_2, *total internal reflection* occurs for angles of incidence greater than the critical angle θ_c. There then is no refracted ray. The critical angle is found from

$$\sin\theta_c = \frac{n_2}{n_1} \qquad (n_1 > n_2) \tag{23.8}$$

Dispersion occurs because the index of refraction n of a material is a function of the wavelength of light: $n = n(\lambda)$.

Optical surfaces may cause rays diverging from an object point to reconverge. The point of intersection is called the image point of the object point. If the light rays really pass through the image point, the image is *real*; if the light rays only appear to pass through the image point, the image is *virtual*.

The *Cartesian sign convention* is used for assigning positive and negative signs in geometric optics. The standard geometry assumes light from an object initially travels toward an optical device from left to right. An origin is placed at the vertex of the optical device, with Cartesian coordinates increasing horizontally to the right and vertically upward. Hence distances measured from the origin horizontally to the right or vertically upward are positive; those measured from the origin horizontally to the left or vertically down are negative. A positive value for the magnification indicates that the image is *upright* (the same vertical orientation as the object); a negative magnification shows the image is *inverted* (up-side-down).

When an object is placed at object distance s in front of a spherical mirror of radius R, the image is located at distance s' from the mirror, according to

$$\frac{1}{s} + \frac{1}{s'} = \frac{2}{R} = \frac{1}{f} \qquad \text{(spherical mirror equation)} \tag{23.23}$$

where f is the focal length of the mirror, equal to half its radius of curvature. The magnification m is found from

$$m = -\frac{s'}{s} \qquad \text{(mirror magnification)} \tag{23.24}$$

The Cartesian sign convention is used to assign positive or negative values to all distances. A ray diagram that traces the path of principal rays is useful for checking calculations.

Refraction at a single spherical surface of radius R is governed by the single surface refraction equation:

$$-\frac{n_1}{s} + \frac{n_2}{s'} = \frac{n_2 - n_1}{R} \qquad \text{(single surface refraction)} \tag{23.35}$$

The magnification is found from

$$m = \frac{n_1 s'}{n_2 s} \qquad \text{(single surface refraction magnification)} \tag{23.38}$$

The Cartesian sign convention is used to assign positive or negative values to all distances.

The image formed by the first refracting surface encountered by light rays becomes the object for the next surface encountered. The Cartesian origin is transferred to each successive surface in turn, with distances adjusted in value and sign to account for the transposition of the origin. The total magnification of the final image is the product of the magnifications produced by each optical surface.

A *thin lens* is one whose thickness is negligible compared with the radii of curvature of its surfaces and the object and image distances. A thin lens with the same medium on both sides has the object and image distances related by

$$-\frac{1}{s} + \frac{1}{s'} = \frac{1}{f} \qquad \text{(thin lens equation)} \tag{23.52}$$

where f is the focal length of the lens. The focal length f is a function of the radii of curvature of the lens surfaces, the index of refraction n_2 of the material from which the lens is made, and the index of refraction n_1 of the medium in which the lens is used:

$$\frac{1}{f} = \frac{n_2 - n_1}{n_1}\left(\frac{1}{R_1} - \frac{1}{R_2}\right) \tag{23.51}$$

This relationship is called the *lens maker's equation*. We use the Cartesian sign convention for thin lenses as well.

Focal lengths occasionally are quoted in *diopters* (dp). The diopter value of the focal length is the *reciprocal* of the focal length in *meters*.

The magnification of a thin lens is

$$m = \frac{s'}{s} \tag{23.50}$$

A thin lens ray diagram is useful for checking calculations.

Note that the mirror equations for distances and magnification (Equations 23.23 and 23.24) are *different* from the thin lens equations (Equations 23.52 and 23.50). This is the price paid for using the same Cartesian sign convention for both mirrors and lenses.

When a simple magnifier of focal length $f > 0$ cm produces a virtual image at $-\infty$ m, the *angular magnification* is

$$m_\angle = \frac{25 \text{ cm}}{f} \qquad (f \text{ in centimeters}) \qquad (23.58)$$

The magnification of a microscope that has an objective lens of focal length $f_o > 0$ cm and an eyepiece lens of focal length $f_e > 0$ cm is inversely proportional to the product of the two focal lengths:

$$m \approx -\frac{\ell}{f_o}\frac{25 \text{ cm}}{f_e} \qquad \text{(all distances in centimeters)} \qquad (23.63)$$

where ℓ is the length of the tube of the microscope.

The magnification of a telescope with an objective element of focal length $f_o > 0$ cm and an eyepiece lens of focal length f_e is

$$m_\angle = -\frac{f_o}{f_e} \qquad (23.66)$$

Summary of Problem-Solving Tactics

23.1 **(page 1051)** When using the law of refraction,

$$n_1 \sin \theta_1 = n_2 \sin \theta_2 \qquad (23.6)$$

you will find it convenient and less confusing to let the medium in which the incident ray lies always be medium 1 (the first medium) and let the medium in which the refracted ray lies be medium 2 (the second medium).

23.2 **(page 1053)** In using Equation 23.8 for the critical angle, the material with the greater index of refraction (in which the incident ray lies) is in the denominator.

23.3 **(page 1060)** When using the general mirror and lens equations (Equations 23.23 and 23.52), recognize that they involve the reciprocals of the various distances. Do not forget to take a final reciprocal with your calculator.

23.4 **(page 1067)** When considering refraction (or, for that matter, reflection) of the light from an object by two (or more) surfaces in succession, the image formed by the first surface becomes the object for the second surface. This fact does *not* necessarily mean the image distance of the first surface becomes the object distance of the second surface. For the purpose of assigning appropriate signs and numerical values for the *distances*, the origin of the Cartesian system is always taken to be at the particular surface then under consideration. *Distances are measured from the particular surface under consideration, not from any previous surface.*

23.5 **(page 1067)** The indices of refraction n_1 and n_2 in Equations 23.35 and 23.38 are the indices of the first medium and the second medium respectively; these change as you progress from surface to surface.

23.6 **(page 1067)** The total magnification of the entire system is the product of the magnifications produced by the individual surfaces or devices.

23.7 **(page 1071)** Be sure to distinguish between the equations for lenses (Equations 23.52 and 23.50) and those for mirrors (Equation 23.23 and 23.24); they are not the same.

Questions

1. **(page 1047)**; **2.** **(page 1048)**; **3.** **(page 1051)**; **4.** **(page 1051)**; **5.** **(page 1054)**; **6.** **(page 1061)**; **7.** **(page 1067)**; **8.** **(page 1068)**; **9.** **(page (1076)**

10. Is a blind person incapable of detecting *any* electromagnetic waves, or only portions of the electromagnetic spectrum? Explain.

11. Historians study the past. On a clear night, when you look out into space, you literally see history in real time. Explain why.

12. The oscillating electric field of an electromagnetic wave emanating from a pointlike source decreases as the inverse of the radial distance r from the source. In what significant way does this radial dependence differ from the static electric field of a pointlike charge?

13. Starry, starry night .... Clearly explain the distinction between a star that is too faint to see and a star that is invisible because of its color.

14. When you drive your car through an old steel-truss bridge (see Figure Q.14), still common in some areas, radio reception is noticeably poorer than just before entering the bridge or just after leaving it. Does this happen because of diffraction, reflection, or what?

FIGURE Q.14

15. Explain the difference between the appearance of the Moon as seen reflected from the dead-calm surface of a pond and that seen reflected from a surface with small ripples.

16. When driving your car at night, is it more difficult to see the road if the pavement is wet or dry? Explain your answer.

17. Look into a corner cube reflector. If one is unavailable, make one with three small cosmetic mirrors. Is your eyelid on the bottom or top of the image? Explain the observation.

18. Does the law of reflection describe diffuse reflection?

19. Look at the Moon. How can you tell that its surface is locally rough and not locally smooth like a spherical mirror? Why are some of the illuminated areas noticeably darker than others?

20. In Example 23.2, explain why the term *optical lever* is appropriate.

21. When you brush your hair in front of a mirror, is the image of your hair real or virtual?

22. Since acoustic waves cannot exist in a vacuum, how could you define an index of refraction for acoustic waves in various material media such as gases, liquids, and solids?

23. Explain why the refracted ray bends toward the normal line when entering a more optically dense material and away from the normal when entering a less optically dense material.

24. An optical system produces a magnification of −1 for a heads-up penny. Is the image tails-up? Explain.

25. If you place a screen (or camera film) at the location of a virtual image, does the image deliver energy to the screen?

26. When you watch a movie, is the image on the screen real or virtual?

27. Will an air prism in glass (see Figure Q.27) disperse white light into its component colors? If so, sketch the path of red and violet rays through the prism.

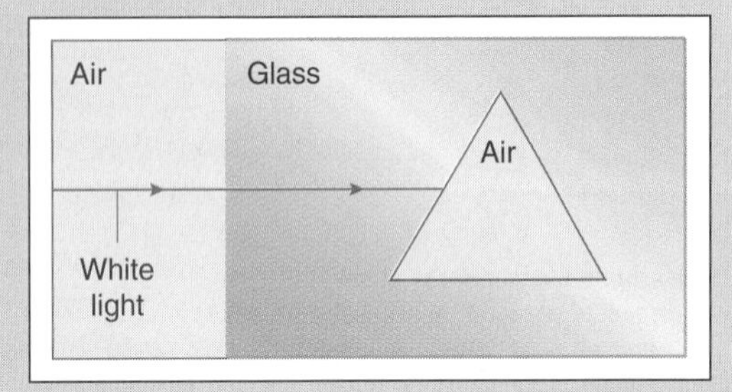

FIGURE Q.27

28. Do other types of waves, such as surface water waves and sound waves, reflect and refract?

29. In Example 23.6, will the angular separation of the red and violet rays increase, decrease, or remain the same if the angle of incidence of the white light increases?

30. Explain the operation of a two-way mirror.

31. Diamond, with its relatively large index of refraction ($n = 2.42$), is cut so rays of light that are refracted into the top face of the gem reemerge through the top face, thus giving diamonds their optical brilliance. Does the light reemerge because the diamond is cut into a corner cube configuration and silvered, or because of total internal reflection of the light within the diamond?

32. What is the focal length of a plane mirror?

33. If a clear solid crystal is placed in a clear liquid with the same index of refraction, is the crystal easily visible? Explain. If strong materials could be made with the same index of refraction as that for most radar wavelengths in air, they would have the ultimate properties for stealth technology in military aircraft.

34. Explain why the total magnification of a series of optical surfaces is given by the product of the individual magnifications, *not* by their sum.

35. An observer looking vertically down into a pool sees a coin lying on the bottom of the pool. The image appears closer to the observer than the coin really is, but the magnification is 1. (See Example 23.11.) Should not the image appear larger than the coin at the bottom of the pool? (Does a diamond grow in value when you view it at close range?)

36. Pure water is colorless and transparent, yet you can easily see water drops if you spill several onto a glass table. Explain.

37. Determine whether the lenses pictured in Figure Q.37 are converging or diverging.

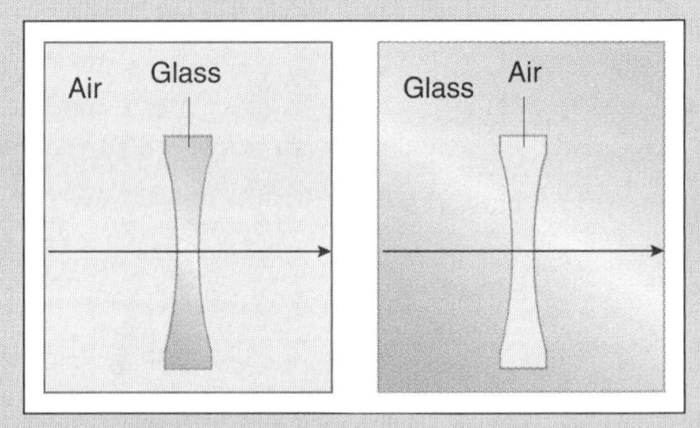

FIGURE Q.37

38. When used in air, a particular concave mirror and a particular converging lens have the same absolute value for their focal lengths. Each is now submerged in water. Which, if any, of their focal lengths change in absolute value?

39. If you look at the eyes of a myopic person when they have their eyeglasses on, do their eyes appear smaller or larger than when the eyeglasses are off? Explain your answer with the assistance of a ray diagram.

40. If you look at the eyes of a hyperopic person when they have their eyeglasses on, do their eyes appear smaller or larger than when the eyeglasses are off? Explain your answer with the assistance of a ray diagram.

41. Many telescopes use a mirror for an objective element to form a real image at its focal point. The eyepiece then magnifies the real image formed by the objective. Explain why it is difficult in practice (though not in theory) to use a mirror for the objective element of a microscope.

42. As an engineer working for the laser development company of Light, Powers, and Watt, you are asked to design a beam expander that will take the thin pencil of rays from a laser and expand it into a wide beam of parallel light as indicated in Figure Q.42. Specify what to place inside the beam expander to accomplish the job, indicating any pertinent spacing of optical components. (The solution is not unique.)

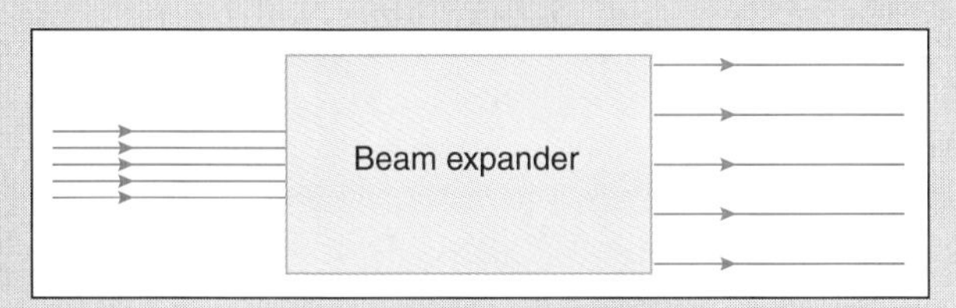

FIGURE Q.42

43. In William Golding's famous novel *Lord of the Flies*,* the boy mockingly called Piggy was quite myopic and his glasses were used "as burning glasses" to start a fire by focusing the rays of the Sun onto dry tinder. If only one lens was used, what is wrong with this description? Devise a way both lenses could be combined to form a "burning glass."

The inconsistency of this literary passage first was pointed out by Sam Prytulak, "Interdisciplinary application," *The Physics Teacher*, *29*, #3, page 135 (March 1991).

44. An inventor has submitted a patent application for an unusual device: a brightly polished, solid steel sphere that concentrates the rays of the Sun on a small target and sets it afire. The application has been sent to you for review. Should a patent be granted? Use diagrams or calculations to support your decision.

45. An object is placed in front of a mirror or a lens and a real image is formed of magnification m. If the object now is placed where the image was located, where is the new image of the object located? What is the value of the new magnification?

46. Shaving and makeup mirrors are concave. With your face close to such a mirror, what happens to your image as you gradually back away? When sufficiently far from the mirror, your image appears in a different orientation. How far is your face from the mirror relative to its radius of curvature when the change takes place?

47. You see a fish some distance from a river bank in a calm pool. Is the actual location of the fish above or below the image you see? What implications does this have for spear fishing?

48. On the front of an ambulance, the word is spelled backward with the letters also reversed:

ƎƆИA⅃UᗺMA

Explain why.

49. How should the poem by John Updike at the beginning of Section 23.10 on page 1057 really be printed to indicate accurately the process of reflection?

50. Telephoto lenses foreshorten images. Explain why. (See Figure Q.50.)

51. The passenger side rear view mirror on your car is a convex mirror while that on the driver's side is a plane mirror. What are the advantages and disadvantages of each as a rearview mirror?

52. Examine a good camera lens and tabulate its *f*/numbers. Beginning with the lowest *f*/number, how are the *f*/numbers related to the aperture diameter of the lens? To its area?

53. When you change the *f*/number setting on your camera lens by one full "stop" (or setting), by how much do you change the amount of light reaching the film?

54. How does the depth of field depend on the effective *f*/number of a lens?

55. Perhaps you or your parents wear bifocal eyeglasses. Does this term mean a single lens with two focal points, or what? What vision problems do bifocal eyeglasses attempt to correct?

56. Most astronomical telescopes of large aperture use concave mirrors for the objective. What advantage in cost and optical performance do mirrors have compared with lenses as the objective element?

57. The specifications of most binoculars quote a number known as the field of view. Find out what is meant by this term.

58. If you stand under a leafy tree during a partial solar eclipse, you may observe hundreds of images of the partially eclipsed Sun on the ground. What forms these images? Why do you not normally notice similar images of the uneclipsed Sun?

59. A slide projector has a converging lens that projects a real, inverted image of the slide onto a screen. Imagine you have a slide of the words

THIS END UP ↑

There are four different ways you can put the slide into the projector: the "proper way," turning the slide left for right, turning the slide top for bottom, and again reversing it left for right. Sketch the four possible ways the image can appear on the screen.

*William Golding, *Lord of the Flies* (Coward-McCann, New York, 1955), pages 45 and 202.

FIGURE Q.50

PROBLEMS

Sections 23.1 The Domains of Optics
23.2 The Inverse Square Law for Light

1. The solar constant, 1.36 kW/m^2, represents the intensity of solar light, the amount of energy per second received from the Sun on a one square meter area perpendicular to the direction to the Sun, at the distance of the Earth from the Sun (1.496×10^8 km). What is the intensity of sunlight at the position of a space probe in the vicinity of Saturn, located 9.54 times as far from the Sun as the Earth?

•2. The intensity of sunlight is about 1.36 kW/m^2 at the distance of the Earth from the Sun (1.496×10^8 km). (a) Use only this data to calculate the power output of the Sun in watts, known as the solar luminosity. (b) Determine the value of the solar constant at the spatial average distance of the planet Pluto from the Sun (about 40 times the distance of the Earth from the Sun).

•3. Light spreads out in all directions from a small pointlike source in air and passes points A, B, C, and D located distances 1.00 m, 2.00 m, 8.00 m, and 9.00 m from the source respectively. (a) Find the ratio of the intensity of the light at point A to that at point B. (b) Find the ratio of the intensity of the light at point C to that at point D. (c) Why are the ratios calculated in parts (a) and (b) the same or different? (d) How would the answers to parts (a)–(c) be affected if the source of the light and the points A, B, C, and D were in water?

•4. Over the course of a year, the distance of the Earth from the Sun varies from 1.47×10^8 km on about 3 January to 1.52×10^8 km on about 6 July. (a) By about what percentage does the energy received from the Sun vary between January and July? (b) Is this variation the cause of the seasons?

•5. The luminosity of the Sun is 3.83×10^{26} W. How much solar energy is contained in the light in 1.00 cubic kilometer of space at a distance from the Sun equal to that of the Earth (1.496×10^8 km)?

⁑6. Consider a universe uniformly sprinkled with stars throughout its volume.* Let the number density of stars be $\langle n \rangle$, the number of stars per unit volume. Let $\langle L \rangle$ be average luminosity of a star. (a) Imagine a thin spherical shell of such stars centered on the Earth. The radius of the shell is r and its thickness is Δr (where $\Delta r << r$) (see Figure P.6). Each star in the shell is essentially the same distance (r) away from the Earth. Show that the amount of light from each star in the shell collected by an aperture of area A on the Earth is

$$\frac{\langle L \rangle}{4\pi r^2} A$$

(b) Show that the total light collected by the aperture from the stars in the shell is

$$\frac{\langle n \rangle \langle L \rangle A}{4\pi} \Delta r\, \Omega$$

where Ω is the solid angle of the shell intercepted by the area A (see Figure P.6). Note about this result is *independent* of the radius of the shell. This means that we get the *same* amount of light from *every* shell surrounding the Earth. If the universe were of infinite extent, the amount of light collected from the infinite number of shells surrounding the Earth also would be infinite and the night sky would be brighter than the surface of the Sun! This calculation and conclusion is the *dark night sky* paradox. Clearly, the calculation produces an amazing result, contradicted by the obvious observation that the night sky is dark. For a complete treatment of the dark night sky paradox, as well as its resolution, see the wonderful book by Edward R. Harrison, *Cosmology* (Cambridge University Press, Cambridge, England, 1981), Chapter 12, pages 249–265. Also see Chapter 1, Problem 74 of this book.

*Stars actually are clumped into vast arrays called galaxies that each contain several hundred billion stars. There are estimated to be about 10^{12} such galaxies.

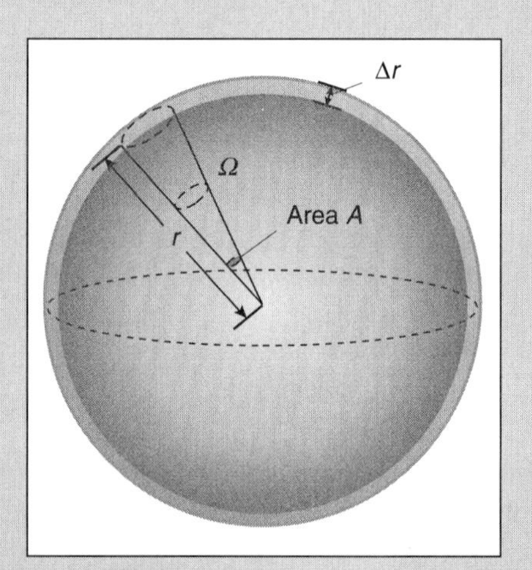

FIGURE P.6

Section 23.3 The Law of Reflection

7. Newton thought light was a stream of particles. Imagine such particles of light approaching a rigid wall (reflecting surface) at angle θ (see Figure P.7). Each particle makes an elastic collision with the wall and rebounds at angle θ'. By conserving the component of the momentum of the particle parallel to the surface, show that $\theta = \theta'$, thus proving a part of the law of reflection. Why is the component of the particle momentum perpendicular to the wall *not* conserved?

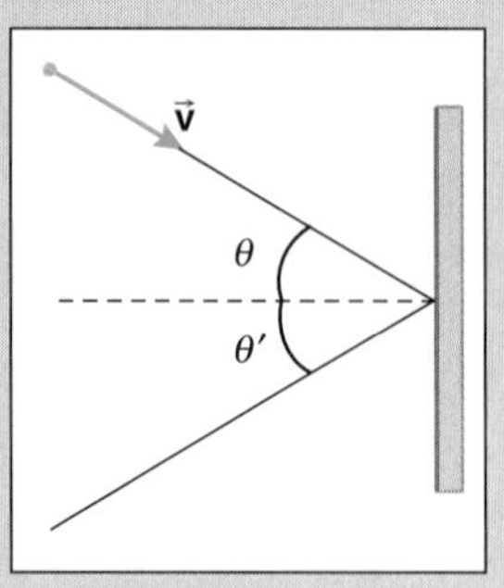

FIGURE P.7

•8. Two plane mirrors are inclined at an angle ϕ to each other as indicated in Figure P.8. An incident ray is reflected from each of the mirrors as shown. (a) Show that $\phi = \theta + \beta$. (b) Show that the exiting ray makes an angle of $180° - 2\phi$ with the incident ray and that this is the case regardless of the angle of incidence (as long as the ray is reflected from both mirrors).

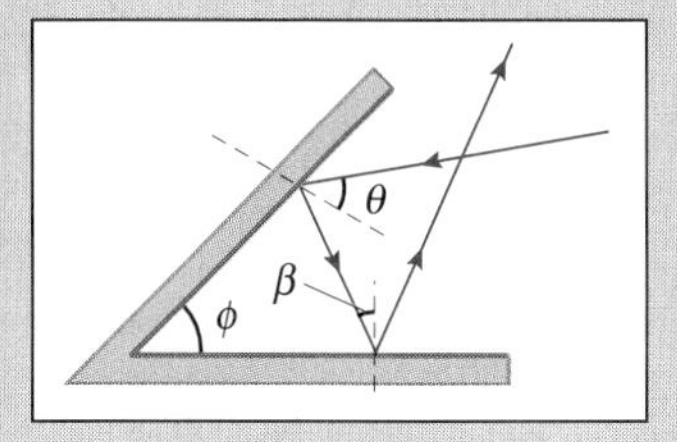

FIGURE P.8

•9. Two parallel rays are incident on two plane mirrors that make an angle ϕ with each other, as shown in Figure P.9. Show that the two reflected rays make an angle 2ϕ with each other, independent of the angles of incidence that the two parallel rays make with the two mirrors.

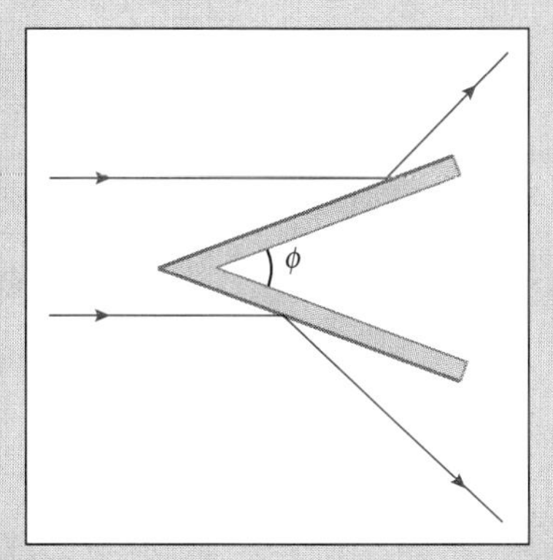

FIGURE P.9

10. A paraboloid of revolution is formed by rotating a parabola about its axis of symmetry. Prove that any ray of light parallel to the axis of such a parabolic reflector is reflected to the focus, as shown in Figure P.10. Such parabolic reflectors are used in astronomical telescopes to collect light. Car headlamps have a light source placed at the focus of a parabolic reflector, so that the rays emerge parallel to the axis of the parabola.

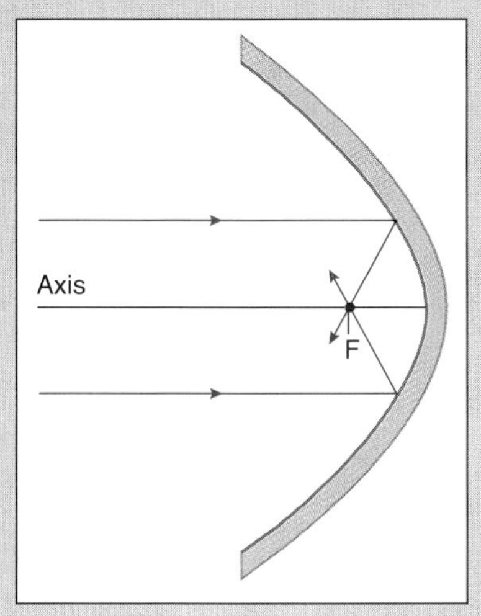

FIGURE P.10

11. A long reflecting cavity has a cross section that is an ellipse. Prove that a ray in the plane of the ellipse, emitted from one focus, is reflected to the other focus, as shown in Figure P.11. Such elliptical reflecting cavities are used in some lasers.

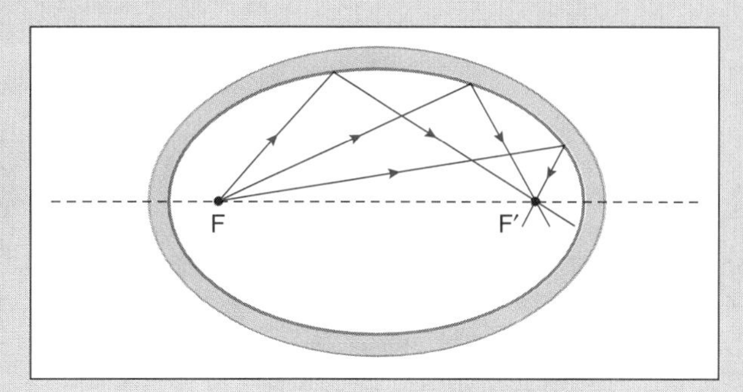

FIGURE P.11

12. A hyperbola of revolution is formed by rotating a hyperbola about its axis of symmetry, as in Figure P.12. Show that a ray emitted from one focus of such a reflector is reflected so that it appears to come from the focus of its opposite twin.

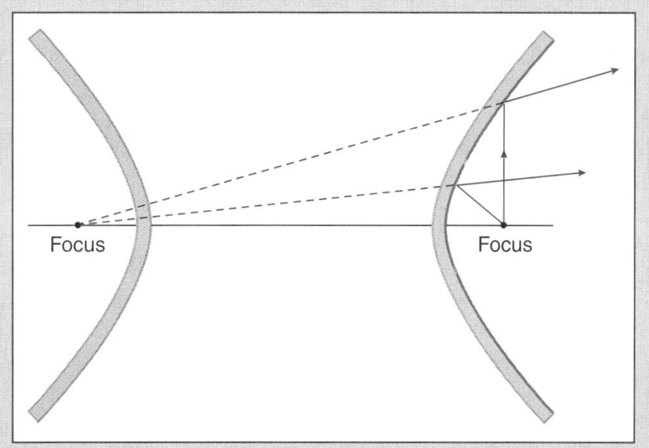

FIGURE P.12

13. A hollow cylindrical reflecting surface has diameter D and length ℓ as in Figure P.13. One end of the reflector has a cap with a small hole through which light from a light source can enter the reflecting pipe. An eye at point P views concentric circular rings of light caused by multiple reflections of light from the sides of the tube. (a) Show that the ring observed closest to the axis of the tube is formed by light from the hole making a reflection at a distance $\ell/2$ from the end of the tube with the hole. (b) Show that the next ring out from the axis

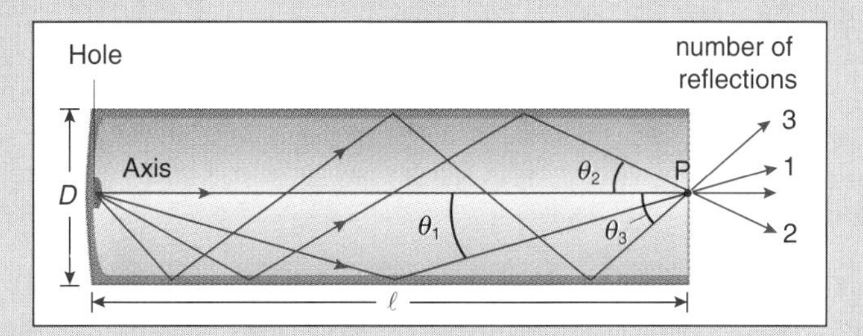

FIGURE P.13

of the tube is formed by a ray making *two* reflections from the sides of the tube, the first one at a distance $\ell/4$ from the end of the tube with the hole. (c) Show that the third ring from the axis of the tube is formed by three reflections from the sides of the tube, the first at a distance of $\ell/6$ from the end of the tube with the hole. (d) Show that the angular radius θ_m of the mth ring of light is found from

$$\tan\theta_m = \frac{Dm}{\ell}$$

where m is the number of reflections that the ray makes with the sides of the tube in propagating to point P.
This problem was inspired by Laurence A. Marschall and Emma Beth Marschall, "Reflections in a polished tube," *The Physics Teacher*, *21*, #2, page 105 (February 1983).

Sections 23.4 The Law of Refraction
23.5 Total Internal Reflection
23.6 Dispersion
23.7 Rainbows*

14. (a) On the same piece of graph paper, make an accurate graph of the angle of refraction versus the angle of incidence for light incident from air onto a water surface and from air onto a glass surface. (b) Make similar graphs of the sine of the angle of refraction versus the sine of the angle of incidence. (c) What is meaning of the slope of the graphs in part (b)?

15. Determine the critical angle for a diamond ($n = 2.42$) immersed in water ($n = 1.33$).

•16. A ray of light is incident in air on a transparent material with an index of refraction n, as shown in Figure P.16. Determine the angle of incidence that makes the angle of refraction *half* the angle of incidence.

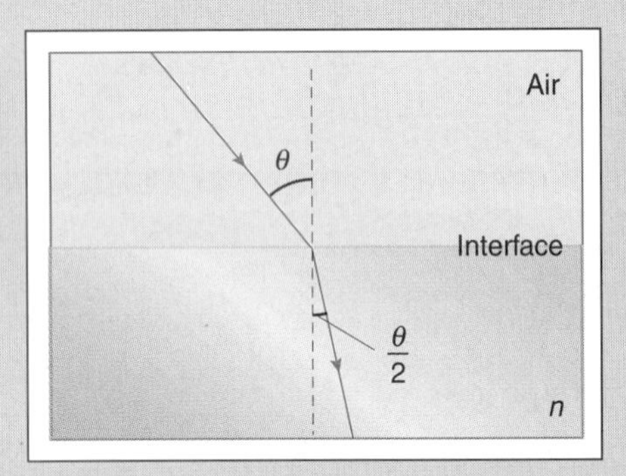

FIGURE P.16

•17. A laser beam of cross-sectional diameter d is incident at angle θ on a glass plate with index of refraction n, as shown in Figure P.17. What is the width of the beam in the glass?

•18. A ray of light is incident as shown in Figure P.18 on a 30°–60°–90° glass prism ($n = 1.50$), so that when *inside* the prism the ray is traveling parallel to the hypotenuse. (a) Determine the angle of incidence at which the incident ray struck the short face of the prism. (b) Find the direction of the ray after it leaves the prism.

•19. An expensive spotlight is located at the bottom of a gold-plated swimming pool of depth 2.00 m (see Figure P.19).

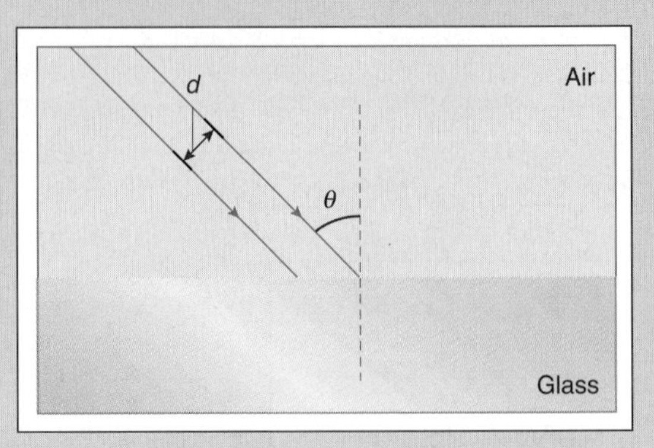

FIGURE P.17

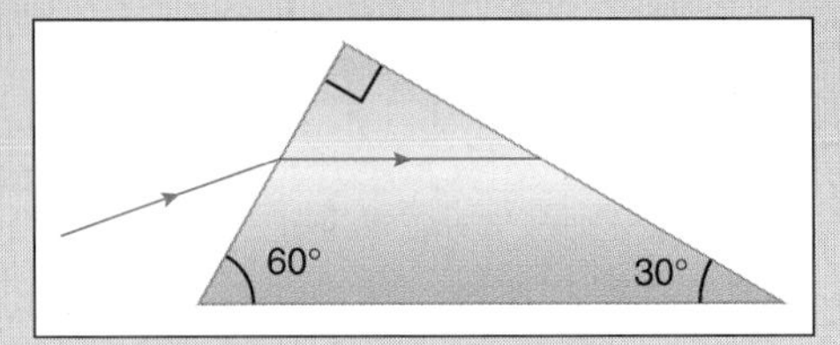

FIGURE P.18

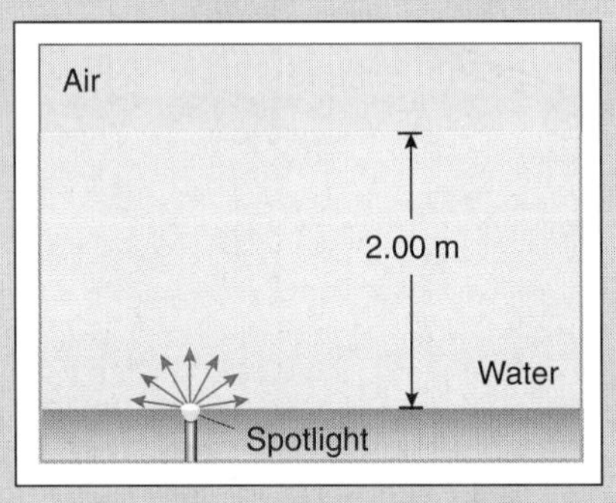

FIGURE P.19

Determine the *diameter* of the circle from which light emerges from the tranquil surface of the pool.

•20. A ray of white light is incident at an angle of 60° on a glass plate of thickness 5.0 cm as shown in Figure P.20. The index of refraction for red light is 1.500 and that for violet is 1.510. Determine the width of the emerging beam of colors.

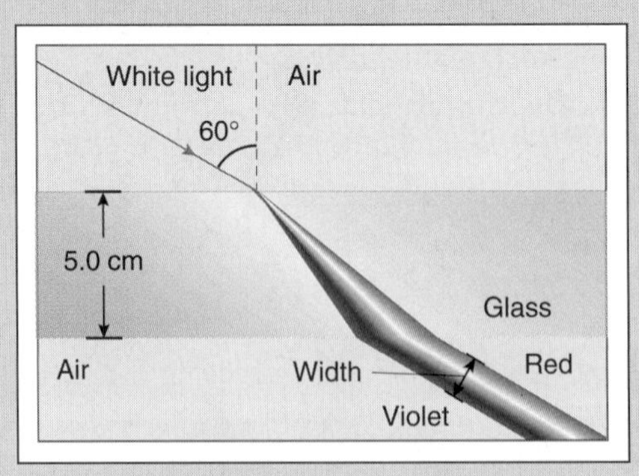

FIGURE P.20

•21. The red light from a helium–neon laser has a wavelength of 632.8 nm. If the light is incident on a plate of glass as indicated in Figure P.21, what is the *deviation angle* ϕ?

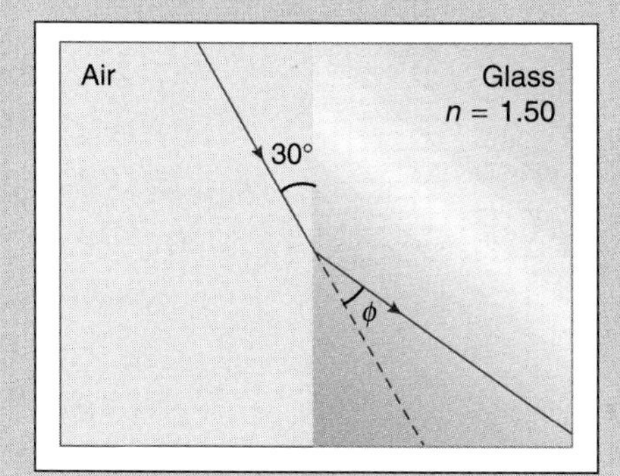

FIGURE P.21

•22. A ray of light is incident on the transparent prism in Figure P.22. (a) Determine what index of refraction of the prism makes the light incident on the hypotenuse just barely totally reflect from the prism–air interface. (b) If the prism is made of material with an index *greater* than that calculated in part (a), what will happen to the path of the light? (c) If the prism is made of material with an index *less* than that calculated in part (a), what will happen to the path of the light?

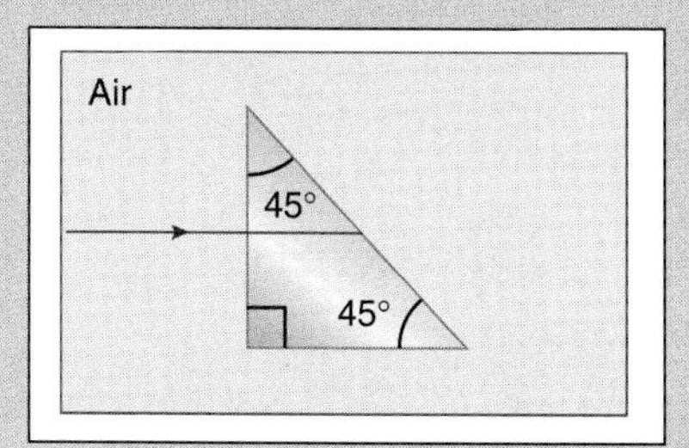

FIGURE P.22

•23. Your cat is attentively watching a tasty ichthyological specimen from a vantage point level with the surface of the water and to the right of the opaque side of a water garden pond, as indicated in Figure P.23. The cagey fish, however, remarkably knows some physics. If the fish is at a depth of 5.0 cm, where relative to the edge of the pond can the fish swim so your cat is unable to see it?

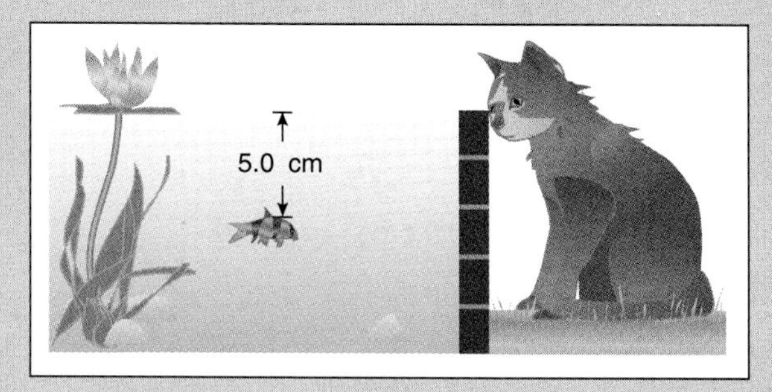

FIGURE P.23

•24. In Figure P.24, when viewed from S, point P appears to be at point P′ along the perpendicular from P to the interface. The original version of the law of refraction formulated by Willebrord Snel (c. 1621) stated that the ratio of the distance OP to OP′ was a constant for all rays incident on the interface between two given transparent materials. Show that this statement leads to the current version of the law of refraction:

$$\frac{\sin \theta_1}{\sin \theta_2} = \text{constant}$$

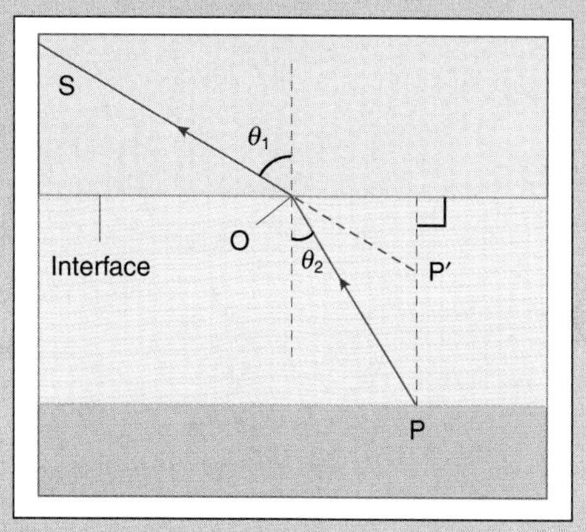

FIGURE P.24

•25. Precise measurements of the index of refraction are made using a prism with an apex angle A as shown in Figure P.25. A light ray is directed through the prism. The incident light is refracted twice and undergoes a deviation angle δ, which is a minimum when the ray inside the prism travels parallel to the base of the prism. When this is the case, show that the index of refraction of the prism (for the wavelength of the light used in the experiment) is given by

$$n = \frac{\sin\left[\frac{1}{2}(A + \delta)\right]}{\sin\left(\frac{A}{2}\right)}$$

where δ is the angle of minimum deviation.

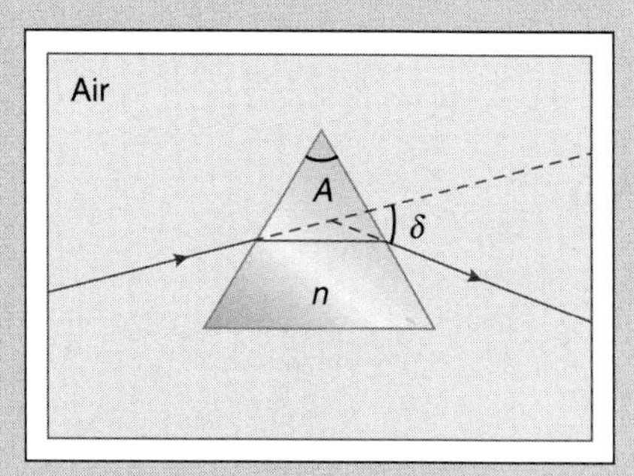

FIGURE P.25

•26. To calm troubled waters, a sheet of glass 2.0 cm thick is placed in full contact over water as indicated in Figure P.26. (a) What is the critical angle for the glass–air interface? (b) At what angle of incidence θ must light approach the water–glass interface so that the light is incident on the glass–air interface at the critical angle?

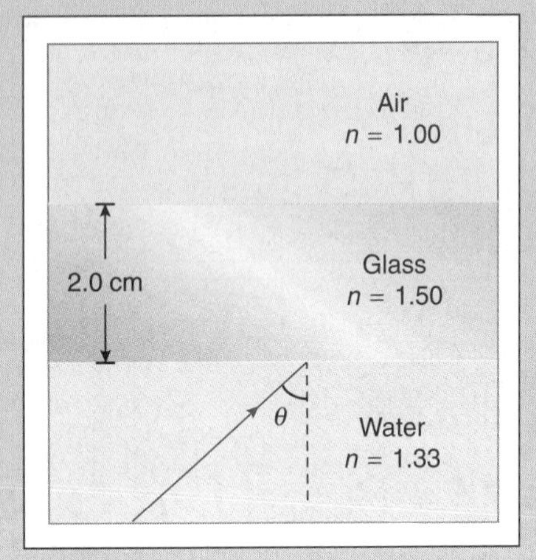

FIGURE P.26

•27. An optical fiber of index of refraction n is surrounded by air. A light ray enters the end of the fiber as shown in Figure P.27. (a) Show that the largest value of ϕ permitted, if the ray in the fiber is incident on the wall at the critical angle for the fiber–air interface, is

$$\sin\left(\frac{\phi}{2}\right) = (n^2 - 1)^{1/2}$$

The angle ϕ defines what is called the *acceptance angle* of an optical fiber. (b) To make the acceptance angle small, would you use a fiber with a high or low index of refraction? Explain your reasoning.

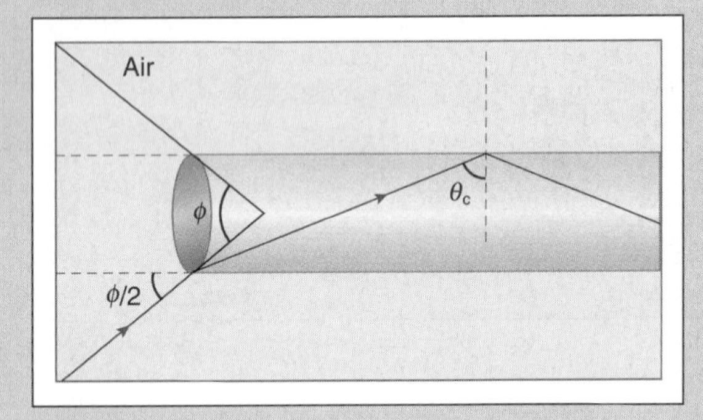

FIGURE P.27

•28. An old abandoned quarry has a vertical cylindrical shaft 20.0 m in diameter and 20.0 m deep. The hole is filled with water completely to ground level. To prevent children from falling into the water, a 1.5 m high fence surrounds the pool 3.0 m from the edge of the water. The water is perfectly calm. (a) Where under water can you lurk and not be seen by any person 2.0 m tall standing at the fence? (b) Where in the pool can you be located in order to see the entire fence height in all directions?

29. A parallel plate of thickness d and index of refraction n_2 is surrounded by a medium with index of refraction n_1. A ray of light is incident on the parallel plate with an angle of incidence θ_1, as indicated in Figure P.29. (a) Show that the emerging ray is parallel to the incident ray. (b) Show that the lateral shift ℓ of the ray is

$$\ell = (d \sin\theta_1)\left[1 - \frac{n_1 \cos\theta_1}{(n_2^2 - n_1^2 \sin^2\theta_1)^{1/2}}\right]$$

Note when $\theta_1 = 0°$, there is no lateral shift. Why? Also, when $n_2 = n_1$, there is no lateral shift, regardless of the angle of incidence. Why?

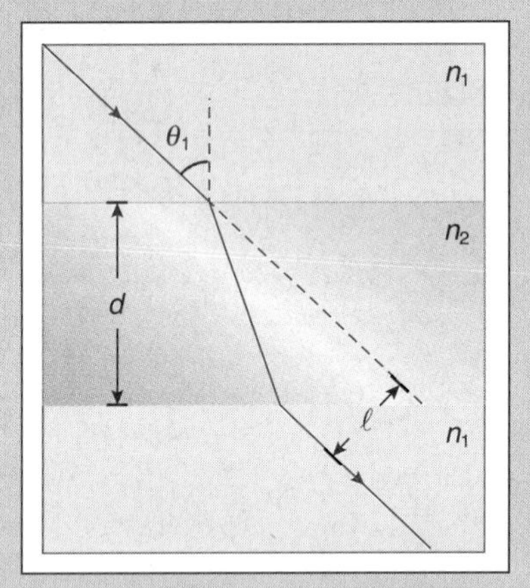

FIGURE P.29

30. Light passes from point A (coordinates $x = 0$ m, $y = \ell$) to point B (coordinates $x = \ell$, $y = -\ell$) by refracting somewhere at the surface between two transparent media at the point x as indicated in Figure P.30. Forget the law of refraction and let x vary. (a) Using the coordinate system provided, find an expression for the total time t it takes for light to travel from A to B via point x. (b) From the relation involving x, find the special value of x that makes the elapsed time from A to B a *minimum*. (c) Relate the result of part (b) to trigonometric functions of the angle of incidence θ_1 and the angle of refraction θ_2 to show that the law of refraction emerges from this minimization of the time. This is known as *Fermat's principle*.

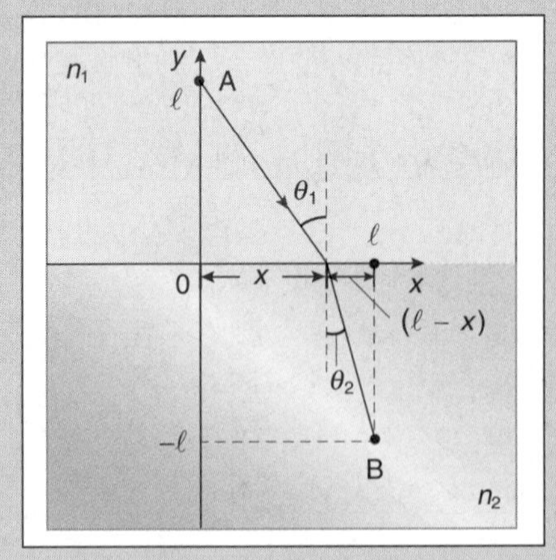

FIGURE P.30

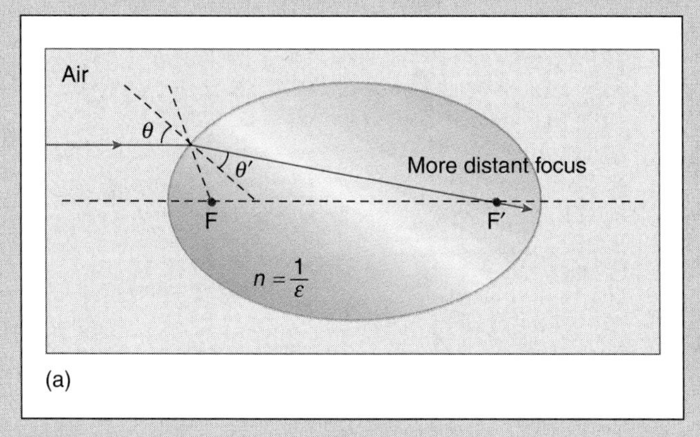

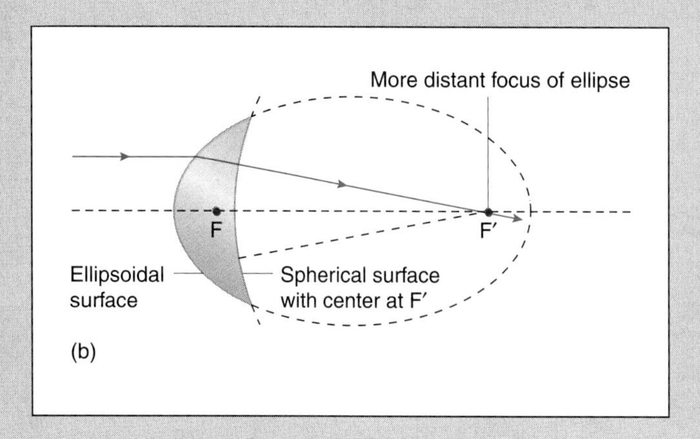

FIGURE P.31

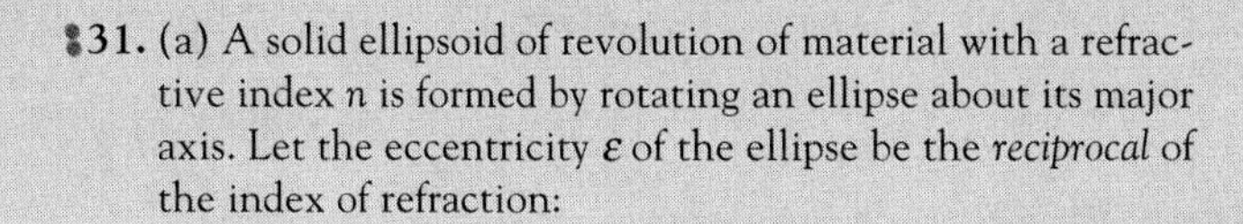

31. (a) A solid ellipsoid of revolution of material with a refractive index n is formed by rotating an ellipse about its major axis. Let the eccentricity ε of the ellipse be the *reciprocal* of the index of refraction:

$$\varepsilon = \frac{1}{n}$$

Show that for such an ellipse a ray of light incident parallel to the major axis is refracted so that it passes through the more distant focus (see Figure P.31a). (b) If such an ellipsoid is truncated to form a lens by the surface of a sphere whose center is the "more distant" focus (see Figure P.31b), the light refracted at the ellipsoidal surface will proceed undeviated through the spherical surface and continue to the focus. (Why?) Such an arrangement has been suggested as a novel solar collector.

Part (a) was inspired by David Mountford, "Refraction properties of conics," *The Mathematical Gazette*, 68, pages 134–137 (1984). For the application in part (b), see A. Tan, "A novel refracting solar collector," *Physics Education*, 22, #3, page 141 (May 1987).

32. The path of a light ray involved in the formation of the primary rainbow is illustrated in Figure P.32. The figure depicts the path of a single wavelength with a refractive index n. The incident light is incident with an *impact parameter* b on a spherical water drop of radius R. The angle of incidence at point A is θ_1 and the angle of refraction is θ_2. At point A, the ray is deviated by an angle α. Some of the light is reflected from the back surface of the drop at point B, and deviated through an angle β at this point. Finally, the ray is refracted out of the drop at point C and deviated through the angle γ at this point. (a) The total angle of deviation of the light ray is the sum of the deviation angles $\alpha + \beta + \gamma \equiv \phi$. Show that

$$\phi = 180° + 2\theta_1 - 4\theta_2$$

(b) Examine the geometry at the point A and show that

$$\sin\theta_1 = \frac{b}{R}$$

(c) Use the law of refraction at point A to show that

$$\sin\theta_2 = \frac{b}{nR}$$

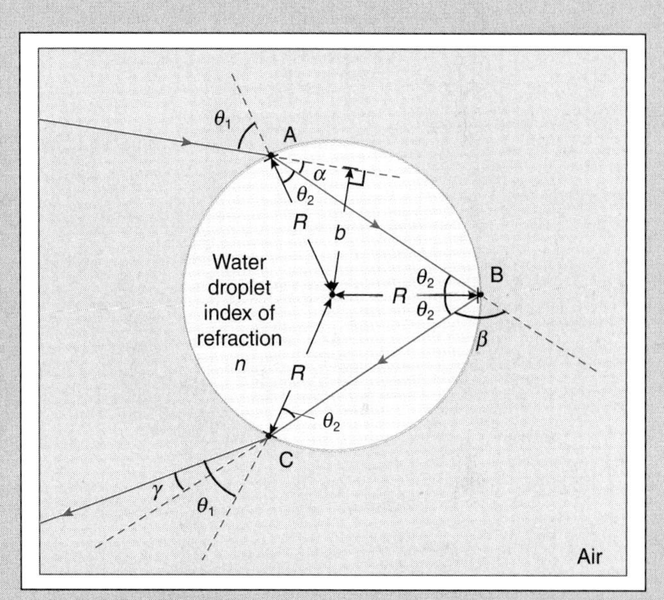

FIGURE P.32

(d) Let the ratio of the impact parameter b to the radius of the drop R be defined as

$$\psi \equiv \frac{b}{R}$$

a dimensionless parameter. Then the results of parts (a), (b), and (c) can be expressed as

$$\phi = 180° + 2\sin^{-1}(\psi) - 4\sin^{-1}\left(\frac{\psi}{n}\right)$$

Show that the deviation angle ϕ has a *minimum* when

$$\psi = \left(\frac{4 - n^2}{3}\right)^{1/2}$$

(e) Use the result of part (d) to show that if the spherical drop is water (with $n = 1.33$), then the minimum deviation angle occurs for

$$\psi = \frac{b}{R} = 0.862$$

and the minimum deviation angle then is

$$\phi = 138°$$

The acute angle between the incident and exiting rays then is

$$180° - 138° = 42°$$

Because of dispersion (the variation of n with wavelength), the various colors emerge with slightly different angles from the drop and give rise to the rainbow.

This problem was inspired by Colin M. Cartwright, "Rainbows," *Physics Education, 27*, #3, pages 155–158 (May 1992).

Sections 23.8 Objects and Images
23.9 The Cartesian Sign Convention
23.10 Image Formation by Spherical and Plane Mirrors
23.11 Ray Diagrams for Mirrors

•33. A hummingbird (*Calypte anna*) is hovering 50 cm in front of a spherical garden mirror of diameter 20.0 cm. (See Figure P.33.) Hmmm. (a) Determine the location of the image of our feathered friend. (b) Determine the magnification. (c) Is the image real or virtual? (d) Is the image upright or inverted?

FIGURE P.33

•34. A puffin (*Fratercula arctica*) is in front of a concave spherical mirror of radius R. Its image is real and the same size. Where is the puffin relative to the mirror?

•35. For a fixed focal length mirror, plot the image distance s' versus the object distance $s < 0$ m assuming: (a) $f = 25$ cm; (b) $f = -25$ cm. Let $-50 \text{ cm} \leq s < 0$ cm.

•36. A basketball player of height 2.2 m is standing 3.0 m in front of a convex spherical mirror of radius of curvature 4.0 m. (a) What is the focal length of the mirror? (b) Locate the image. (c) Determine the size of the image. (d) Is the image real or virtual? (e) Is the image upright or inverted? (f) Draw a neat ray diagram to confirm the calculations and conclusion drawn in parts (a)–(e).

•37. Dr. Ruth Canal is a dentist who is designing a small dental mirror. When placed 2.0 cm from a molar, the mirror will provide an image that is the same size as the tooth and (of course) upright. (a) Is the image real or virtual? (b) Determine the radius of curvature for the mirror.

•38. A horseshoe crab (*Limulus polyphemus*) in sea water finds itself 3.0 m in front of a convex, spherical mirror of radius of curvature 1.0 m. (a) Locate the position of the crabby image. (b) Is the image real or virtual? (c) What is the magnification? (d) Is the image upright or inverted? Does the crab care? (e) Draw a neat ray diagram to confirm your calculations.

•39. You are preening yourself for the annual Fancy Dress Ball in front of a large plane mirror as indicated in Figure P.39. Show geometrically that the vertical height of the mirror needed to see yourself is completely half your height. This means that a full-length mirror (properly mounted on a wall) need only be half your height. You can never view yourself in entirety in a cosmetic, pocket size mirror.

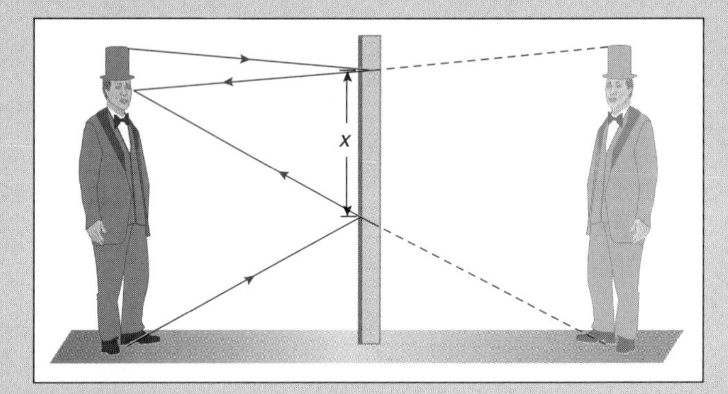

FIGURE P.39

•40. The Wicked Witch of the West is rushing along a normal line toward a plane mirror at speed v. At what speed is her image approaching her?

•41. If the witch in the previous problem approaches the mirror at speed v at an angle of incidence θ, at what speed is her image approaching her?

•42. A convex mirror of focal length +2.00 m is used to form an image of an impala (*Aepyceros melampus*) located 6.0 m from the mirror. (a) Find the location of the image. (b) Determine the magnification. (c) Is the image upright or inverted? (d) Draw a ray diagram to confirm your calculations.

•43. A tasty lollipop is located 200 cm in front of a concave spherical mirror with a focal length of absolute value 50 cm. (a) Locate the image of the sucker. (b) Determine the magnification. (c) Is the image real or virtual? (d) Is the image upright or inverted? (e) Construct a *neat* ray diagram to confirm your calculations and conclusions in parts (a)–(d).

•44. A large, circular mirror has a focal length of +2.0 m and a diameter of 0.50 m. (a) Is the mirror concave or convex? (b) What is the radius of curvature of the mirror? (c) A chowder clam (*Venus mercenaria*) is placed 1.0 m in front of the mirror. Where is the image, what is its magnification, and is the image real or virtual? (d) Draw a ray diagram to confirm the aspects of the image found in part (c). (e) If the entire system is submerged in crystal clear water, how are the answers to (b), (c), and (d) affected? Explain your reasoning.

•45. The "Optic Mirage" toy (Edmund Scientific Co.) produces very realistic real images of objects placed at the vertex of the mirror at the bottom of the two-mirror array. (See Figure P.45.) The two concave mirrors have radii of curvature of identical

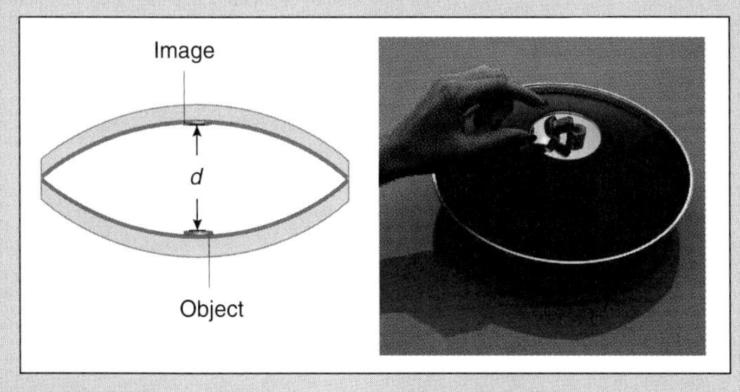

FIGURE P.45

absolute magnitude $|R|$. Show that if the separation d of the mirror vertices is *either* $|R|/2$ or $3|R|/2$, a real image of a heads-up coin is formed at the mirror vertex at which the coin is *not* located. Will the image show heads-up or tails-up? Determine the magnification of the object for each mirror separation.
The idea for this problem originated with the following article: Andrzej Sieradzan, "Teaching geometrical optics with the 'Optic Mirage,'" *The Physics Teacher*, 28, #8, pages 534–536 (November 1990).

•**46.** Two plane mirrors face each other a distance d apart as shown in Figure P.46. You stand a distance x in front of the mirror on the right, admiring a new addition to your wardrobe. (a) How many images are there? (b) Are the images real or virtual? (c) What is the separation of successive images in each mirror? (d) If you have a patch over your left eye, which images have the patch over the left eye of the image and which have the patch over the right eye of the image, or do they all have the patch over the same eye (in which case, which eye?)?

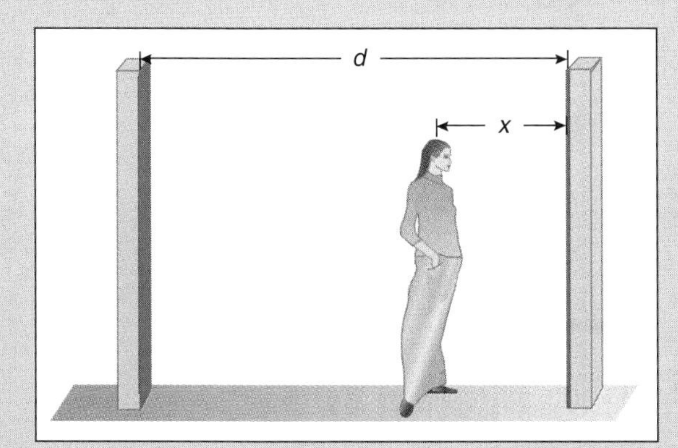

FIGURE P.46

•**47.** Two plane mirrors are in contact with each other along one edge and form an angle of 40° between their faces as shown in Figure P.47. A small American firefly (*Photuris pennsylvanica*) light source is hovering midway between the two mirrors. In a carefully drawn figure, geometrically locate all the images of this marvelous insect. How many are there?

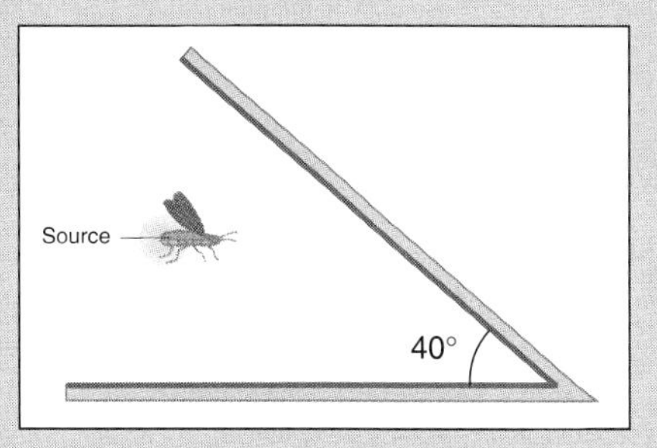

FIGURE P.47

•**48.** A magician uses a clear plate of glass and a large concave mirror in a black enclosure (see Figure P.48) to project a real image of a disembodied head. The plate glass reflects and also transmits light. The center of curvature of the mirror is at C in Figure P.48. (a) The flat glass forms a reflection image. Where is this image and how is it oriented? (b) The curved mirror uses the image of the flat glass plate as its object to form a final real image. Where is the final image, what is its size, and what is its orientation?

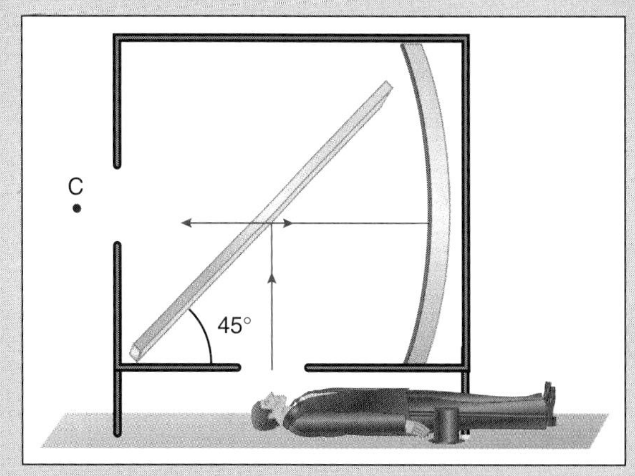

FIGURE P.48

•**49.** Four lollipops of different shapes are in a stand in front of a plane mirror as shown in Figure P.49. Sketch the total image.

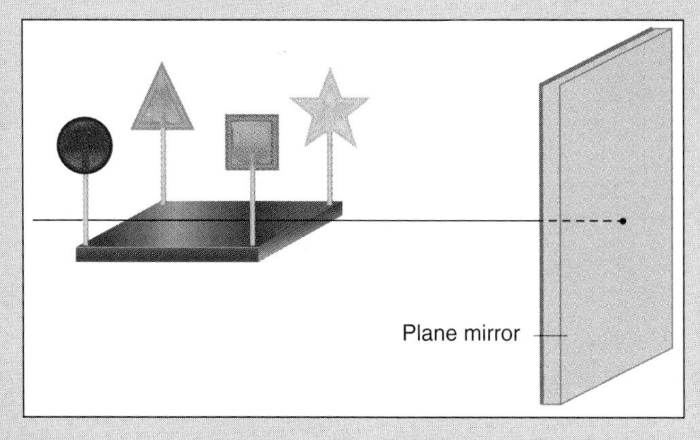

FIGURE P.49

•50. A flat semicircle with one quadrant shaded is placed in front of two plane mirrors oriented as shown in Figure P.50. Sketch the orientation and shading of the three images.

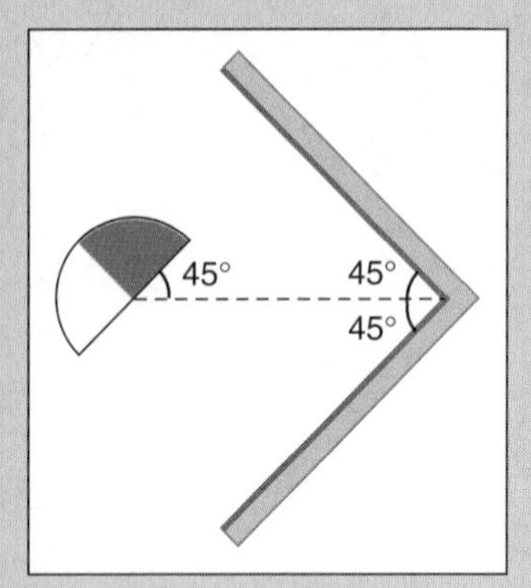

FIGURE P.50

⁑51. Telescopic mirrors typically are shaped not as sections of a spherical surface but as paraboloids of revolution. Several large new-technology telescope mirrors were made by a new process of spinning a mass of molten glass, a process known as spin casting.* The spinning molten glass, as we will see in this problem, assumes a surface that is a paraboloid of revolution. The melted glass then is cooled slowly while spinning so that it solidifies into the desired paraboloid shape. To see that a spinning surface assumes the shape of a parabola of revolution and to see how the focal length of a mirror surface of this shape is set by the angular speed of the rotation, consider the following: (a) Imagine a small mass element m of the mirror as it is spinning at constant angular speed ω at a radius x from the axis as in Figure P.51a. Two forces act on the mass element: its weight and the normal force of the surface acting perpendicular to the surface arising from the rest of the glass in the melt. Use Newton's second law to show that

$$\tan\theta = \frac{x\omega^2}{g}$$

where θ is the angle that the tangent to the surface makes with the horizontal direction at the location of the mass. (b) Notice that with the coordinate system in Figure P.51a,

$$\tan\theta = \frac{dy}{dx}$$

Use this relation from analytic geometry and the result of part (a) to show that

$$y = \frac{\omega^2}{2g}x^2$$

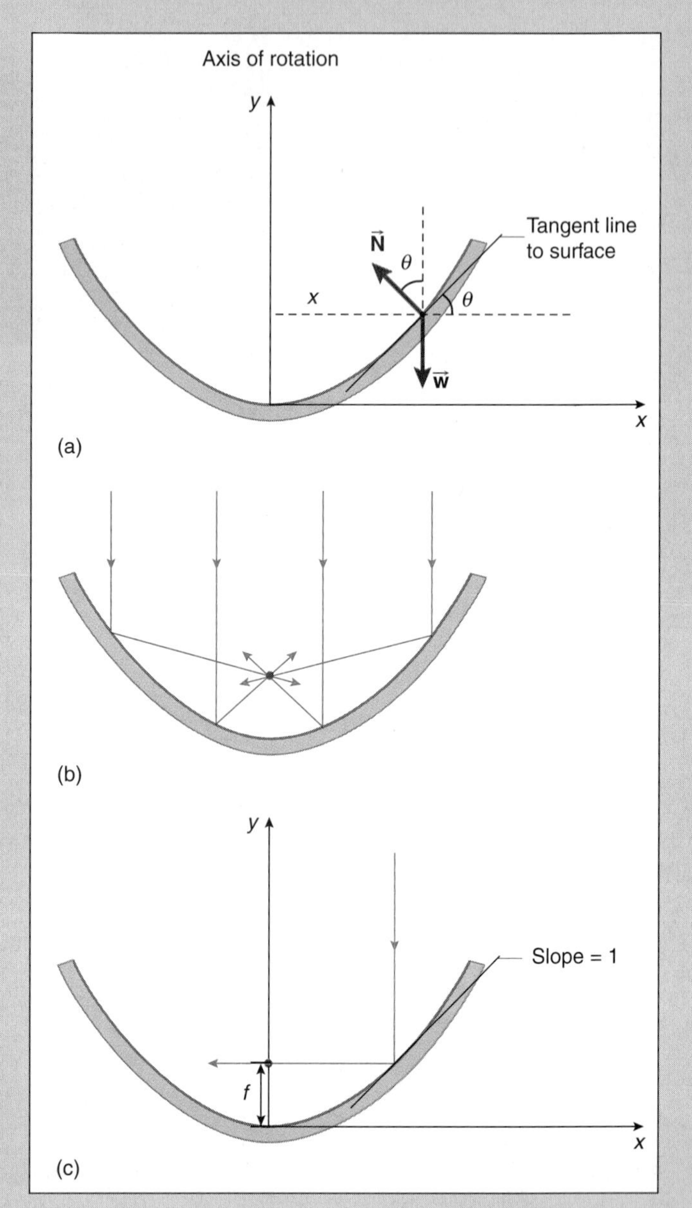

FIGURE P.51

This shows that the shape of the surface is parabolic. (c) Any light ray incident parallel to the axis of the parabola will go through the focus (the focal point!) of the parabola (see Figure P.51b). Consider the special parallel ray incident on the surface where the slope of the surface is equal to unity (Figure P.51c). Show that for this ray

$$x = \frac{g}{\omega^2}$$

(d) Use the result of part (c) and the equation of the parabola from part (b) to show that the focal length f of the paraboloid surface is

$$f = \frac{g}{2\omega^2}$$

*See the following articles: Mark Dragovan and Don Alvarez, "Making a mirror by spinning a liquid," *Scientific American, 270*, #2, pages 116–117 (February 1994); Ben Iannotta, "Spinning images from mercury mirrors," *New Scientist, 147*, #1986, pages 38–41 (15 July 1995); and Ermanno F. Borra, "Liquid mirrors," *Scientific American, 270*, #2, pages 76–81 (February 1994).

(e) Calculate the spin rate in revolutions per minute needed to make a mirror with a focal length of 5.00 m.

Section 23.12 Refraction at a Single Spherical Surface

•**52.** Your pet angelfish (*Pterophyllum scalare*) is located 10.0 cm from the edge of a thin-walled fish tank as shown in Figure P.52. Neglect the effect of the thin glass wall. (a) Determine the location of the image of the fish. (b) Determine the magnification. (c) Is the image real or virtual? (d) Is the image upright or inverted? (e) Why is the thin wall optically unimportant?

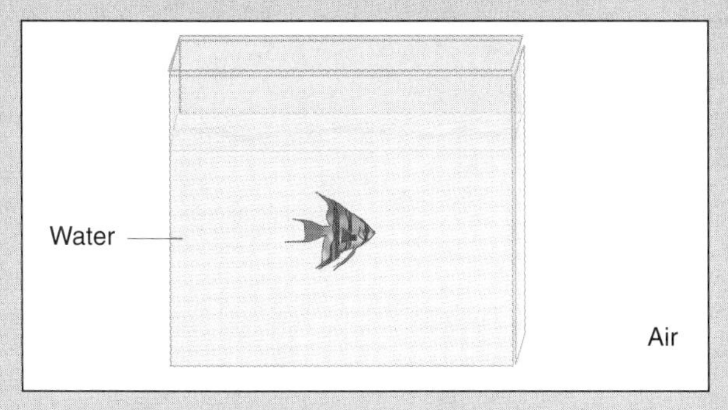

FIGURE P.52

•**53.** A small olive is suspended 3.0 cm from the bottom of a large glass filled to the brim with peculiarly flavored water, as shown in Figure P.53. (a) What is the apparent depth of the olive as seen from a point directly above the glass? (b) What is the magnification of the olive? (c) Is the image real or virtual? Upright or inverted?

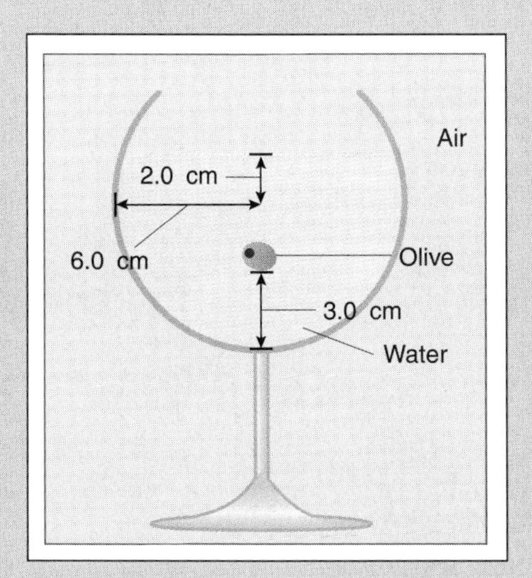

FIGURE P.53

•**54.** A glass crystal ball of radius 10 cm is used by a local wizard to divine your grade for the course. (See Figure P.54.) How far is the focal point of the sphere from its center?

•**55.** The warm and inviting waters of Tahiti beckon you from the rigors of winter. While boating in a calm cove of depth 4.0 m, you spy a pirate treasure poking up slightly from the bottom (Figure P.55). What is the apparent depth of the treasure when viewed along the normal line?

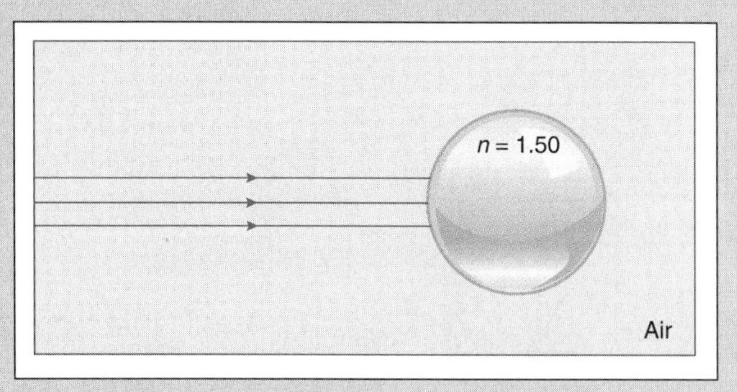

FIGURE P.54

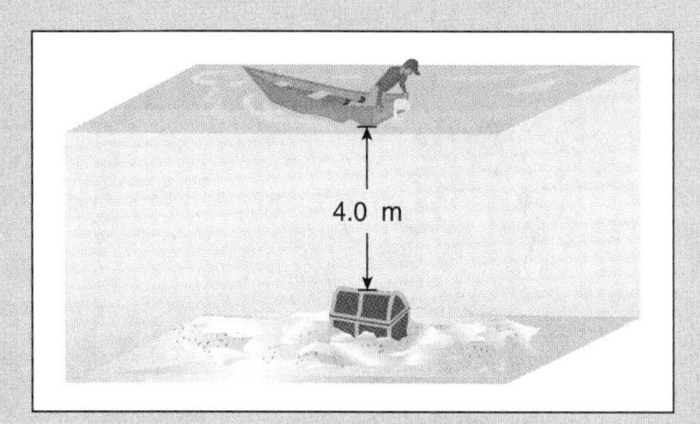

FIGURE P.55

•**56.** While scuba diving (using a face mask with a very thin glass plate) in the tranquil and warm waters off the French Riviera, you stretch your arms straight out in front of you. (a) Will your arms appear to you to be shorter, longer, or the same size than they really are? (b) If your arms really are 70 cm in length, how long do they appear to be? (c) You notice a small fish touching your fingertips. If the fish swims at a speed v toward the face mask, will its observed *image* move at speed v? Faster? Slower?

•**57.** A layer of crystal clear ice 1.00 m thick ($n = 1.30$) floats on the surface of a pond. A brown trout (*Salmo trutta*) lies motionless at the boundary between the ice and water. (a) Sketch the path of the light ray from the trout that spends the shortest time in the ice before passing into the air. (b) Sketch the path of the light ray from the trout that spends the greatest time in the ice before passing into the air. (c) How thick does the ice appear when viewed from above? (d) How thick does the ice appear when viewed from the water underneath?

•**58.** A penny is located 15.0 cm from the surface of a glass crystal ball of radius 5.00 cm with $n = 1.500$. (a) Determine the location of the final image of the penny. (b) Is the image real or virtual? (c) Determine the magnification and orientation of the penny.

•**59.** A parallel beam of laser light is incident on a sphere (surrounded by air). The light is brought to a focus at the rear surface of the sphere as indicated in Figure P.59. What is the index of refraction of the sphere? Does your result depend on the radius of the sphere?

⁝**60.** A spinning vertical cylinder of water forms a curved surface that is approximately spherical (actually it is a paraboloid; see

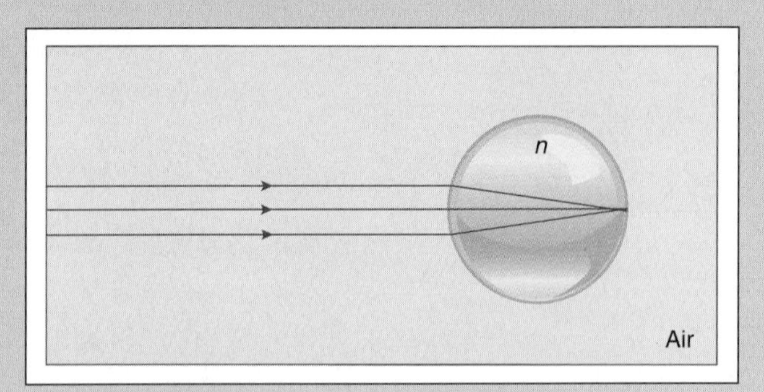

FIGURE P.59

Problem 51) and of radius of curvature 0.40 m. A bright electric spark occurs in the water at a depth of 0.80 m. (a) Where does the spark appear to be located as seen by a person above, looking along the vertical axis of the cylinder? (b) We shall see in Chapter 24 that the index of refraction of light is the ratio of the speed of light in vacuum to that in the material. The speed of the *sound* of the spark in water is approximately four times the speed of sound in air. The sound wave follows the standard refraction laws. Using the speed of sound in air as a reference speed, calculate the *sonic* index of refraction for water relative to air. (c) The sound from the spark is transmitted from the water into the air. Where is the acoustic image of the sound source?

Sections 23.13 Thin Lenses
23.14 Ray Diagrams for Thin Lenses

Assume all lenses are thin unless otherwise stated.

•61. Use the thin lens equation (Equation 23.52) to show that a diverging lens can never form a real image of a real object no matter where the object is placed.

•62. For a fixed focal length lens, sketch a plot of s' versus s assuming: (a) $f = 25$ cm; (b) $f = -25$ cm. Assume a domain for s of $-100 \text{ cm} < s < 0$ cm.

•63. A lens has radii of curvature of its first and second surfaces of 20 cm and 25 cm, respectively, as indicated in Figure P.63. The lens is made from recycled physic elixir bottles with an index of refraction of 1.50. The lens is surrounded by air. (a) Find the focal length of the lens. (b) Express the focal length in diopters. (c) Is the lens converging or diverging? (d) An uninteresting object is placed 40 cm from the lens with the standard geometry. Find the location of its equally uninteresting image. Is the image real or virtual? Determine the magnification. Is the image upright or inverted? Verify your calculations with a ray diagram.

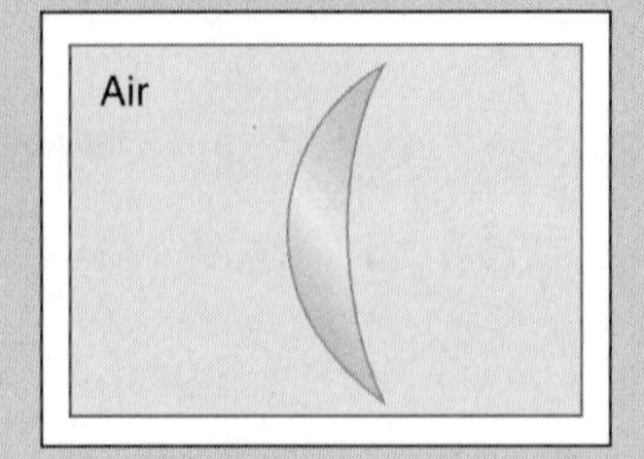

FIGURE P.63

•64. A small snipe (*Capella gallinago*) finds itself 150 cm in front of a diverging lens. Time for a snipe hunt. (a) If the absolute value of the image distance is 50 cm, find the focal length of the lens. (b) Determine the magnification. (c) Is the image real or virtual? (d) Is the image upright or inverted? (e) Draw a neat ray diagram to locate the position of the image. (f) If the upper half of the lens is now painted black so that no light can pass through this portion of the lens, describe what happens (if anything) to the image.

•65. A small gremlin from Academia finds itself 150 cm in front of a two-lens optical system as indicated in Figure P.65. For light entertainment, determine: (a) the location of the final image; and (b) the total magnification. (c) Is the final image real or virtual? Upright or inverted?

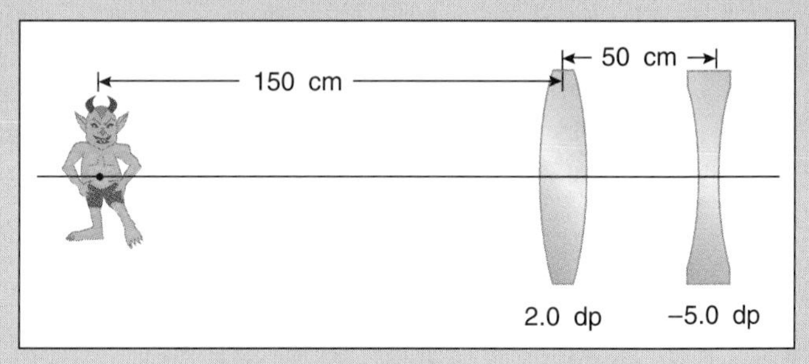

FIGURE P.65

•66. Two thin lenses of focal lengths f_1 and f_2 are placed in contact with each other; the combination still is thin. Show that the equivalent focal length f of the combination is

$$f = \frac{f_1 f_2}{f_1 + f_2}$$

•67. A glass convex lens ($n = 1.50$) has radii of curvature for both of its surfaces of absolute magnitude $|R|$. (See Figure P.67.) (a) Determine the focal length of the lens when used in air. (b) In terms of the focal length in air, determine the focal length of the lens when it is immersed in a medium with an index of refraction n.

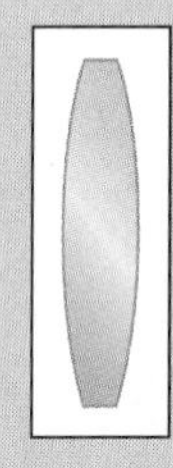

FIGURE P.67

•68. Santa Claus has given you a converging lens with a 10.0 cm focal length. The instructions ask you to place a candy cane 50.0 cm from a screen. (a) Determine where to place the lens so that a real, enlarged image of the candy cane appears on the screen. What is the magnification? Is the image upright or inverted? (b) Determine where to place the lens such that a real, reduced image of the candy cane appears on the screen. What is the magnification? Is the image upright or inverted?

•69. A lens of focal length f_{lens} is placed directly in front of and in contact with a mirror of focal length f_{mirror}. The combination is considered to be thin. What is the effective focal length of the combination?

•70. Most single-lens-reflex (SLR) cameras come with a standard lens of focal length 50.0 mm. The lens is used to form an image of an Oxford gargoyle, located 10.00 m from the lens. (a) Determine the image distance and the magnification. (b) The lens is replaced with another lens of focal length 200 mm. Determine the image distance and the magnification of the gargoyle using this lens. (c) What is the *ratio* of the magnification with the 200 mm lens to that with the standard 50.0 mm lens? This is the reason that camera lenses with focal length longer than the standard 50.0 mm lens are known as *telephoto lenses*. Compare this ratio with the ratio of the focal lengths. (d) If the lens is switched to one with a focal length of 28.0 mm, determine the image distance and the magnification of the gargoyle. Compare the magnification of the 28.0 mm lens with that of the 50.0 mm lens. This is the reason that lenses with focal lengths shorter than the standard 50.0 mm focal length lens are known as *wide angle lenses*.

•71. A candelabra is located 5.00 m from a screen. Determine the focal length and location of a lens needed to form a real, inverted image that is three times the size of the candelabra.

•72. A converging lens is placed at the center of curvature of a concave mirror as indicated in Figure P.72. The absolute magnitude of the focal lengths of both the lens and mirror is 25.0 cm. A ladybug (*Anatis quindecimpunctatum*) is placed 50.0 cm in front of the lens. Determine the location, magnification, and nature of the final image of the bug (real or virtual; upright or inverted).

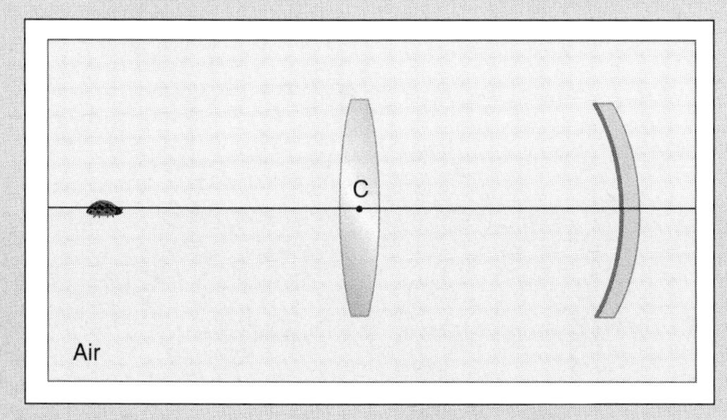

FIGURE P.72

•73. A projector with a lens of 10.0 cm focal length is to produce an image of slide (22.0 mm × 34.0 mm) on a screen (150 cm × 150 cm). The image fills as much of the screen as possible. (a) What is the magnification? (b) Is the image inverted? Explain how to place the slide to yield an image that is properly oriented. (c) How far is the slide from the lens? (d) What is the distance from the lens to the screen?

•74. A diverging lens with a focal length of −20.0 cm and a mirror lie 10.0 cm apart as shown in Figure P.74. The radius of curvature of the mirror is not known. Parallel light rays, from an object infinitely afar, enter the lens as indicated. You note that the final image is located 20.0 cm to the left of the lens. (a) Determine the radius of curvature of the mirror. (b) Is the mirror concave or convex? (c) Is the final image real or virtual?

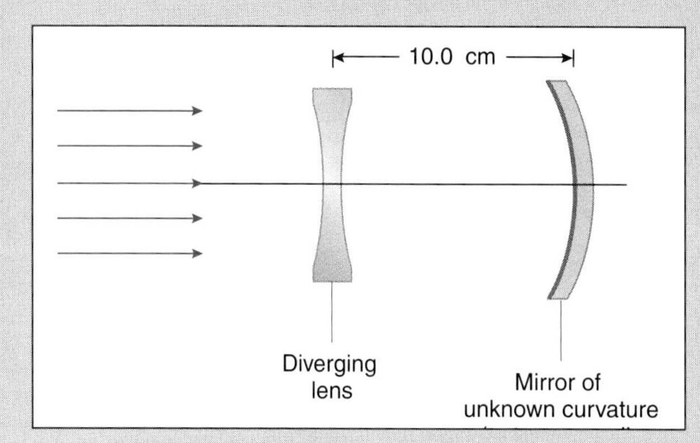

FIGURE P.74

•75. A beam of parallel light is incident on a thin lens of focal length $f = +20.0$ cm. Some distance from the lens is a glass crystal ball ($n = 1.50$) of radius 10.0 cm. The light is brought to a focus at the center of the ball. What is the distance between the center of the ball and the lens?

•76. A black widow spider (*Latrodectus mactans*) is 60.0 cm in front of the first lens of a two-lens system as indicated in Figure P.76. (a) Find the position of the final image relative to the second lens. (b) Determine the total magnification. (c) Is the final image real or virtual? Upright or inverted with respect to the orientation of the original spider object?

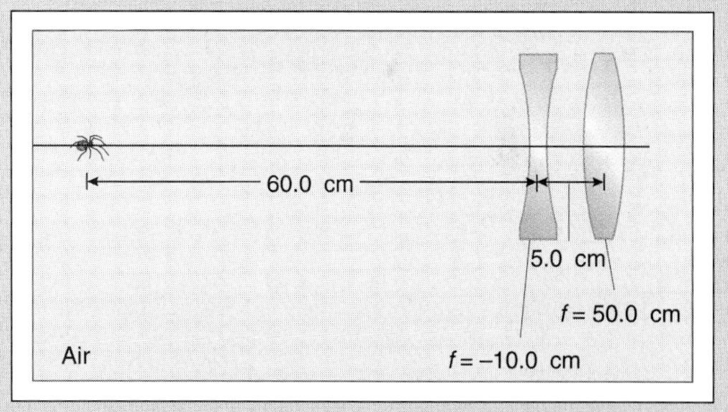

FIGURE P.76

•77. Four lollipops of different shapes are in a small stand 50.0 cm in front of a convex lens of focal length 25.0 cm. See Figure P.77. Consider all the lollipops to be essentially the same distance from the lens. Sketch the image of the suckers.

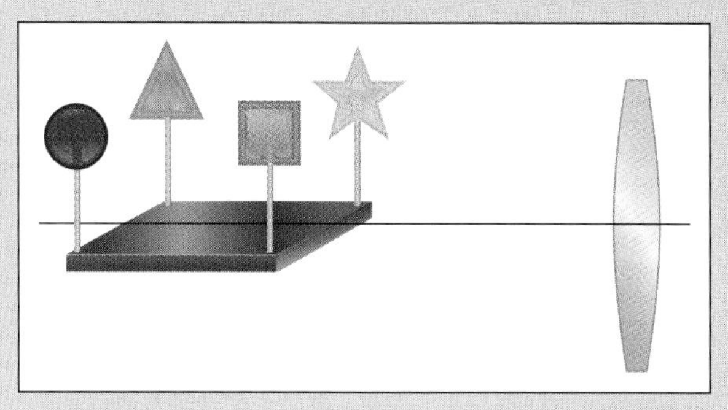

FIGURE P.77

•78. An object is placed at the secondary focal point of a thin converging lens, and the lens is followed (at any distance) by a plane mirror as indicated in Figure P.78. Show that the final image is located at the same position as the object but is inverted. Describe how you can use this result to experimentally find the focal length of a converging lens.

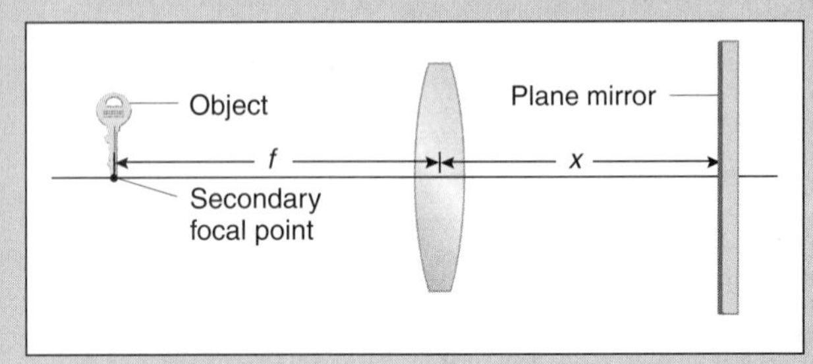

FIGURE P.78

•79. A romantic, lighted candle is placed 40 cm in front of a diverging lens. The light passes through the diverging lens and on to a converging lens of focal length 10 cm that is 5 cm from the diverging lens. The final image is real, inverted, and 20 cm beyond the converging lens. Find the focal length of the diverging lens.

•80. The diameter of the Sun subtends an angle of 0.50° when viewed from the Earth. The intensity of direct sunlight is 1.36 kW/m^2 on the surface of the Earth. A 5.0 cm diameter lens forms a hot image of the Sun that is 1.0 cm in diameter on a screen. (a) How do you know that this is a real image? (b) What light power goes through the lens? (c) What light power reaches the hot image? (d) What is the intensity of the light in the image W/m^2? Your result is much greater than the intensity of direct solar light. Does this result violate energy conservation? Explain. (e) What is the focal length of the lens? Is the lens converging or diverging?

•81. Three thin converging lenses each of focal length 60 cm are placed close together so that the combination still can be considered a thin lens. An object is placed 30 cm in front of the system. Find the location of the image.

:82. (a) For a converging lens of focal length f (> 0 m), show that the distance d (> 0 m) between a real object and its real image ($d = -s + s'$) is equal to

$$d = -\frac{s^2}{s+f}$$

(b) Show that the distance d is a *minimum* when $s = -2f$.

:83. (a) Beginning with the thin lens equation, show that if the object distance s changes with time, the image distance s' changes with time according to

$$\frac{ds'}{dt} = m^2 \frac{ds}{dt}$$

(b) Beginning with the mirror equation, show that the corresponding relationship is

$$\frac{ds'}{dt} = -m^2 \frac{ds}{dt}$$

Section 23.15 Optical Instruments

84. A myopic lawyer has a far point of 25 cm. What is the diopter value of suitable corrective eyeglasses?

85. You have misplaced your eyeglasses and are bumbling around the physics building. Your far point is 15 cm. Specify the diopter value of suitable corrective eyeglasses.

86. A telescope with an objective lens of focal length 200.0 cm is provided with an assortment of eyepiece lenses of focal lengths 6.0 mm, 25 mm, and 40 mm. (a) What is the greatest absolute value of magnification that can be achieved? (b) What is the smallest absolute value of magnification that can be achieved?

•87. The film in a pinhole camera is moved from a distance d (from the pinhole) to a distance $2d$. (a) What happens to the magnification of the image? (b) By how much is it necessary to change the exposure time if you need the same total amount of light per unit area on the film during the exposure?

•88. A microscope with an eyepiece lens of focal length 2.00 cm and an objective lens of focal length 1.00 cm is used to examine a section of the brain of a sacrificed rat. The distance between the lenses of the instrument is 22.0 cm and the final image is virtual and at infinity. (a) How far from the objective lens is the object? (b) What is the magnification produced by the objective lens? (c) What is the total magnification produced by the microscope?

•89. You are admiring the views of Mt. Phobos and Mt. Deimos. The twin peaks are 1.0 km apart and 5.0 km distant along the perpendicular bisector of the line joining the peaks. You whip out your 35 mm SLR camera with its 50.0 mm focal length lens to record the panorama. What is the distance separating the twin peaks on the film?

•90. A camera lens with an aperture of 20 mm and a focal length of 50.0 mm is used to photograph your class for an upcoming issue of the alumni magazine. The exposure time is 10 ms. (a) With the 20 mm aperture, what is the *f*/number used for this exposure? (b) What diameter of lens opening would produce the same exposure on the film in a time of 40 ms? What is the corresponding effective *f*/number for this exposure?

•91. Two 1.0 cm focal length converging lenses are used to form a microscope with a magnification of −350 with a final virtual image infinitely far from the eye. Find the approximate separation of the lenses.

•92. A hyperopic patient can see objects clearly only if they are farther than 2.00 m away. Distant vision is good. The patient wishes to read a book when it is 25 cm away. (a) Calculate the focal length of the eyeglasses necessary to provide this correction. (b) Where is the image formed by the eyeglasses? Is the image real or virtual? (c) What is the magnification caused by the eyeglasses? (d) Why is the eye now able to see the image in (b) clearly? (Explain with no numbers!) (e) Is the final image on the retina real or virtual? Upright or inverted? With a magnification $|m| > 1$ or < 1?

•93. (a) Take appropriate measurements to determine the following: when viewed at arm's length, about what angle does the four-finger width of your hand fill? Express your result in radians and degrees. (b) The Moon has a diameter of 3.48×10^6 m and is 3.84×10^8 m away. What is the angular size of the Moon? Express your result in radians and in degrees.

A giant telescope has an objective element that is a concave mirror 2.00 m in diameter with a radius of curvature of −24.00 m. (c) What is the focal length of the mirror? (d) What angle in radians (and degrees) does the image of the Moon subtend as measured from the mirror? (e) Is the image of the Moon real or virtual? (f) What is the linear width of the image of the Moon as photographed on film that is placed at the focal plane of the mirror? (g) If the mirror aperture were *square* in shape instead of circular, what would be the shape of the image of the Moon? Explain your reasoning.

•**94.** A microscope consists of two lenses of 15 mm focal length that are 20.0 cm apart. The device is used to examine a cootie that is of diameter 0.10 mm. (a) How far from the objective lens must the cootie be placed for most comfortable viewing through the eyepiece (i.e., for a final image that is virtual and infinitely far away)? (b) What is the angular extent of the final image?

•**95.** A tree 15.0 m high is 1.000 km away from a two-lens system consisting of a converging lens of focal length 12.0 cm followed by a diverging lens of focal length −2.0 cm. The separation of the lenses is 10.3 cm. (a) Find the location of the image of the first lens. (b) Find the angular size of the image of the first lens. (c) Find the object distance for the second lens. (d) Find the image distance of the second lens. (e) Find the magnification provided by the second lens. (f) Find the height of the final image. This optical system is a compact telephoto lens.

•**96.** A bug-a-boo is located 25.0 cm in front of a −10.0 dp lens. Another +5.0 dp lens is located 25.0 cm from the first lens as indicated in Figure P.96. (a) Locate the position of the final image of the system. (b) Determine the total magnification. (c) Is the final image real or virtual? (d) Is the final image upright or inverted?

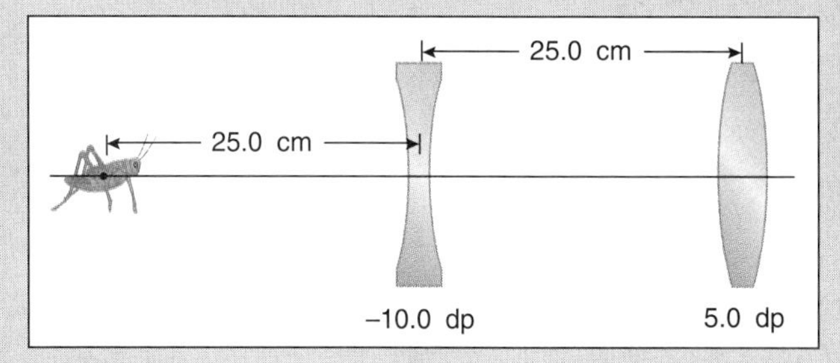

FIGURE P.96

•97. It is fun to look through a telescope backward—that is, with the light from a distant source first entering, say, a 2.0 cm focal length converging lens, followed by a 20.0 cm focal length converging lens. (a) Calculate the magnification of this arrangement and describe the image. (b) How could this backward arrangement be used to look at an object close to you on the table so that the final image has a magnification of absolute magnitude greater than one?

INVESTIGATIVE PROJECTS

A. Expanded Horizons

1. You might enjoy reading and reporting about some of the revolutionary applications of optical fibers in both communications and medicine.

W. S. Boyle, "Light-wave communications," *Scientific American*, 237, #2, pages 40–48, 140 (August 1977).
Harry Rheam, "Lightwave communications," *Science Teacher*, 60, #5, pages 26–29 (May 1993).
Emmanuel Desurvire, "Lightwave communication: the fifth generation," *Scientific American*, 266, #1, pages 114–121 (January 1992).
Alastair M. Glass, "Fiber optics," *Physics Today*, 46, #10, pages 34–38 (October 1993).
Gary Stix, "The last frontier," *Scientific American*, 270, #3, pages 105–106 (March 1994).
Abraham Katzir, "Optical fibers in medicine," *Scientific American*, 260, #5, pages 120–125 (May 1989).
Abraham Katzir, *Lasers and Optical Fibers in Medicine* (Academic Press, San Diego, 1993).

2. The magnification of many microscopes can be increased by placing a small drop of oil on the microscope slide and bringing the objective lens into contact with the oil drop. Investigate these so-called *oil immersion* techniques in microscopy and determine how the droplet is able to increase the magnification.

Francis A. Jenkins and Harvey E. White, *Fundamentals of Optics* (4th edition, McGraw-Hill, New York, 1976).
Max Born and Emil Wolf, *Principles of Optics* (6th edition, Cambridge University Press, Cambridge, England, 1997), pages 253–254.

3. Investigate the cause of the *green flash* in atmospheric optics that occurs occasionally as the Sun sets or rises.

Roger W. Sinnott, "The green flash," *Sky and Telescope*, 83, #2, pages 200–203 (February 1992).
Robert Gannon, "Stalking the elusive green flash," *Focus*, 38, #3, pages 10–11 (Fall 1988).
Sky and Telescope, 87, #2, pages 110–111 (February 1994).
D. J. K. O'Connell, "The green flash," *Scientific American*, 202, #1, pages 112–122, 189–190 (January 1960).
Robert Greenler, *Rainbows, Halos and Glories* (Cambridge University Press, Cambridge, England, 1989).

4. What causes the ring seen occasionally around the Moon? What is the angular size of the ring? Does it have colors like the rainbow?

Walter Tape, *Atmospheric Halos* (American Geophysical Union, Washington, D.C., 1994).
Bartley L. Cardon, "An unusual lunar halo," *American Journal of Physics*, 45, #4, pages 331–335 (April 1977).
Robert Greenler, *Rainbows, Halos and Glories* (Cambridge University Press, Cambridge, England, 1989).
David K. Lynch and William Livingston, *Color and Light in Nature* (Cambridge University Press, Cambridge, England, 1995).

5. Investigate how mirrors occasionally are used for deception by magicians.

Michael J. Ruiz and Terry L. Robinson, "Mirrors in magic," *The Physics Teacher*, 25, #4, pages 206–212 (April 1987).
Derek B. Swinson, "Magic mirrors—front and back," *The Physics Teacher*, 32, #6, page 329 (September 1994).
Marshall Ellenstein, "Magic and physics," *The Physics Teacher*, 20, #2, pages 104-106 (February 1982).

6. Investigate the optical path of light through a spherical rain drop in the formation of a *secondary rainbow*. The region

between the primary and secondary rainbows is noticeably darker than that within the primary or outside the secondary rainbows; the darker region between the rainbows is known as the *Alexander dark band*. Investigate its cause.
Colin M. Cartwright, "Rainbows," *Physics Education*, *27*, #3, pages 155–158 (May 1992).

7. Investigate the use of materials with a spatially varying index of refraction (so-called *gradient index materials*) in the fabrication of lenses.
Neil Morton, "Gradient refractive index lenses," *Physics Education*, *19*, #2, pages 86–90 (March 1984).
Erich W. Marchand, *Gradient Index Optics* (Academic Press, New York, 1978).
James Evans and Mark Rosenquist, "'$F = ma$' optics," *American Journal of Physics*, *54*, #10, pages 876–883 (October 1986).
James Evans, "The ray form of Newton's law of motion," *American Journal of Physics*, *61*, #4, pages 347–350 (April 1993).
K. C. Mamola, Wilhelm F. Mueller, and Bruce J. Regittko, "Light rays in gradient index media: a laboratory exercise," *American Journal of Physics*, *60*, #6, pages 527–529 (June 1992).

8. Investigate the geometric optics associated with the appearance of mirages.
Alistair B. Fraser and William H. Mach, "Mirages," *Scientific American*, *234*, #1, pages 102–111 (January 1976).
Walter Tape, "The topology of mirages," *Scientific American*, *252*, #6, pages 120–129 (June 1985).
David S. Falk, Dieter R. Brill, and David G. Stork, *Seeing the Light* (Harper and Row, New York, 1986), pages 58–62.
Alistair B. Fraser, "Theological optics," *Applied Optics*, *14*, #4, pages A92–A93 (April 1975).
E. Khular, K. Thyagarajan, and A. K. Ghatak, "A note on mirage formation," *American Journal of Physics*, *45*, #1, pages 90–92 (January 1977).
G. P. Sastry, "Teaching mirages," *American Journal of Physics*, *46*, #7, page 765 (July 1978).

9. You look in a plane mirror and see that it reverses right for left but does not reverse up for down. Should a flat mirror know up–down from right–left? Investigate what is meant by this description and whether in a strict sense it is true or not.
J. Ken Gee, "The myth of lateral inversion," *Physics Education*, *23*, #5, pages 300–301 (September 1988).

10. The human eye is one of the more remarkable optical instruments. Investigate the physiology of the eye. Compare the invertebrate eye with the compound eyes of many insects such as the common (and annoying) house fly (*Musca domestica*).
Paul L. Pease, "Resource letter CCV-1: Color and color vision," *American Journal of Physics*, *48*, #11, pages 907–917 (November 1980); this contains many references to the subject.

11. Stars are classified according to their brightness with a *stellar magnitude scale*. For example, the Sun has an apparent visual magnitude of −26, while the bright star Sirius (α Canis Majoris) is −1.5. Polaris (α Ursae Minoris) has a magnitude of about 2, while the dimmest stars visible to the naked eye have magnitudes of about 6. Consult an astronomy text and discover the basis for the magnitude scale. If two stars differ in magnitude by 1, what is the ratio of their apparent brightness? What is the distinction between the apparent visual magnitude of a star and its absolute magnitude? Prepare a short report on your findings.

B. Lab and Field Work

12. Design and perform an experiment to measure the angle of reflection for various angles of incidence of a light ray on a mirror.

13. Design and perform an experiment to measure the angle of refraction for various angles of incidence for a light ray incident from air into a water-filled, thin-glass-walled fish tank.

14. To prevent fraud, an association of jewelry stores has asked you to design and test a simple experimental procedure to easily distinguish between a real diamond and an imitation diamond (zirconium). Design such a test. Several instruments for this purpose actually exist. Discover the principles underlying their operation.

15. The variation of the index of refraction with wavelength (dispersion) for many materials is approximately given by an empirical equation developed by Augustin Louis Cauchy (1789–1857):

$$n(\lambda) = A + \frac{B}{\lambda^2}$$

Your physics department likely has a prism spectrometer and a light source (such as a mercury lamp) that emits several well-known precise wavelengths. Design and perform an experiment to measure accurately the index of refraction of the glass prism for each wavelength; see Problem 25. Fit your data to Cauchy's equation by plotting n versus $1/\lambda^2$ to determine the value of the constants A and B (and their units, if any).

16. Measure the focal length of a converging lens. A diverging lens does not form a real image of a real object. Devise an experimental technique to measure the focal length of a diverging lens using a converging lens of known focal length.

17. If a transparent material such as glass is immersed in a liquid with the same index of refraction (an experimental technique called *index matching*), the glass appears to disappear. Take a pyrex beaker and place it in a large glass beaker. Under a fume hood, add a solution consisting of equal amounts of benzene and carbon tetrachloride and watch the pyrex apparently disappear as the fluid level rises. Some vegetable oils (such as Wesson oil) can be used instead of the chemical solution.
William R. Gregg, "An old physics demonstration—redone more safely," *The Physics Teacher*, *31*, #1, page 40 (January 1993).

18. Visit a ophthalmologist who specializes in the surgical procedure called *radial keratotomy*, which corrects myopia and hyperopia. Determine how the procedure corrects for these vision defects. Learn the risks and long-term prognoses for it.
Raymond Munna, *As I See It: Radial Keratotomy Before, During and After Surgery* (Granite, Metairie, Louisiana, 1985).

19. Investigate the optical characteristics and operation of a *zoom lens*. In particular, determine the *minimum* number and types of thin lenses needed to construct a zoom lens. Construct a simple zoom lens on an optical bench to demonstrate its operation.
Michael J. Ruiz, "Camera optics," *The Physics Teacher*, *20*, #6, pages 372–380 (September 1982).

20. The index of refraction has a slight temperature dependence. Scavenge your physics department for an appropriate experimental arrangement to perform an experiment to measure the variation of the index of refraction of glass or water as a function of temperature.

C. Communicating Physics

21. Light has been used as an image by writers and poets in many varied literal and allegorical contexts from biblical times to the present. Choose one of your favorite writers and/or works and explore in an essay how the word *light* is used in the context of the work. Does the author have an understanding of light in a physical context?

22. Rainbows are common in artistic works, and even on clothing. In many cases the order of the colors in such artistic rainbows is not correct. Search for and collect several examples of such incorrect rainbows.

23. Investigate the physics and chemistry of color mixing in the paint industry and prepare a short report of your findings.
    Kurt Nassau, *The Physics and Chemistry of Color: The Fifteen Causes of Color* (John Wiley, New York, 1983).

Chapter 24
Physical Optics

We see how we may determine their forms [referring to the planets and stars], their distances, their bulk, and their motions, but we can never know anything of their chemical . . . [composition].

Auguste Compte (1798–1857)*

As any politician knows, never say never. Auguste Compte was proved wrong shortly after his now infamous comment on what is impossible for us to know in astronomy. (see chapter opening quotation on page 1103). By inventing techniques for precisely measuring the wavelengths of light, physicists shortly realized that each chemical element emits characteristic wavelengths, thereby enabling astronomers to probe the composition of light sources anywhere in the visible universe.

Behold, I will send my messenger, and he shall prepare the way...
The Old Testament, Malachi 3:1

Light is the messenger. What we know about the stars and the universe as a whole depends almost entirely on our ability to squeeze as much information as possible from the thin thread of light available to us from such distant sources.[†] In the previous chapter, we saw how light can be manipulated by mirrors and lenses to form images. Here we see how we can exploit the wave nature of light to invent new and useful devices and instruments to glean much additional information about stars, galaxies, and more local materials and phenomena of technological and practical import.

Heretofore only alluded to, we now first present convincing experimental evidence of the wave nature of light that was discovered long before Maxwell developed his theory of electromagnetic waves. We investigate several aspects of physical optics, the study of phenomena that manifest the wave aspects of light: interference, diffraction, and polarization.

24.1 Existence of Light Waves

Look at a distant, bright, and small light source such as a star at night through two closely spaced, narrow, and parallel slits. It is not a hallucination when you see *many* bright spots, not one. From such a simple observation we can deduce that light must spread out considerably from these narrow double slits and, in overlapping, add to give bright and dark regions. The multiple spots are a sure sign of overlapping waves, like overlapping ripples on a pond. Small apertures elucidate the wave nature of visible light, as we alluded to in Section 23.1.

24.2 Interference

. . . and light was against light . . .
The Old Testament, 1 Kings 7:4

A superposition of waves may give rise to variations in the resulting amplitude of the total wave disturbance, known as **interference**; the interference of surface water waves is common. We first encountered interference effects in Chapter 12 when we superimposed

- two waves of the same frequency traveling in opposite directions, producing standing waves; and
- two waves of slightly different frequency traveling in the same direction, producing beats.

Similar effects also are seen with light waves, but only with some experimental and technical finesse.

Here we examine other aspects of interference, restricting ourselves to *sinusoidal* waves that meet the following conditions:

1. The waves must be of the same physical type; in this chapter the sinusoidal waves are light waves, but one also can observe the same effects using sound, water, or other waves.
2. The waves must have the same frequency, and so are described by similar sinusoidal wavefunctions; from an equivalent viewpoint, we say the waves, traveling similarly, have identical sinusoidal wavelengths. A wave with a definite sinusoidal wavelength is called a **monochromatic wave.**[‡]
3. The sources of the waves must be **coherent**, meaning the sinusoidal waves have *a phase difference that is independent of time*. Recall that the phase of a sinusoidal wave is the argument of the sinusoid describing the wave. That is, for two sinusoidal waves of the same frequency, traveling toward increasing values of x, the wavefunctions are

$$\Psi_1(x, t) = A\cos(kx - \omega t)$$
$$\Psi_2(x, t) = A\cos(kx - \omega t + \delta)$$

The phase of each wave is the angle of its respective cosine: $kx - \omega t$ for Ψ_1 and $kx - \omega t + \delta$ for Ψ_2. The **phase difference** δ is the difference between the individual phases of the two waves. Notice that the phase of each wave *is* time dependent, but the phase difference δ between the two waves may or may not be. For the two waves to be coherent, the phase difference δ must be independent of time. By convention, we express the phase of a wave (and any phase difference δ) in radians (rad), *not* degrees.

As we considered in Chapters 12, a line or surface connecting points of constant phase on a wave is called a **wavefront**; for a water wave, a wavefront is imagined as the line along each crest of the wave (or equivalently, each trough), as indicated in Figure 24.1.

Coherent wave sources typically are obtained from a single wave by one of two methods.

1. Each wavefront of the wave is divided by a series of slits or holes into multiple coherent wave sources, a process called **wavefront division**. We see examples of how this is done in Sections 24.3 and 24.8.
2. Each wavefront of the wave is divided into a collection of smaller-amplitude wavefronts by reflection and/or transmission, a process called **amplitude division**. We see an example of how this is accomplished in Section 24.11.

What happens when the waves from the multiple coherent sources arrive at the same place at the same time (i.e., are

*(Chapter Opener) *Cours de philosophie positive*, Book II, Astronomy (1835), from *Auguste Compte and Positivism*, edited by Gertrud Lenzer (Harper and Row, New York, 1975), page 130.

[†]There are other messengers, too, from the distant recesses of the universe, but they are more difficult to decipher: cosmic rays and neutrinos. Neutrinos are extremely difficult to detect because they interact *extraordinarily* weakly with matter.

[‡]Note the careful wording here. A *nonsinusoidal*, periodic wave with a definite wavelength, such as

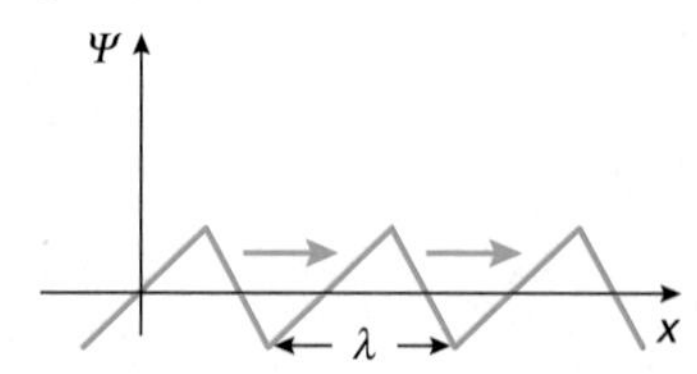

actually contains many sinusoidal frequencies even though it periodically repeats itself. Such nonsinusoidal waves can be treated as collections of sinusoidal waves through *Fourier analysis* but, for simplicity, we do not consider them in this text, except as we did briefly in Section 12.21.

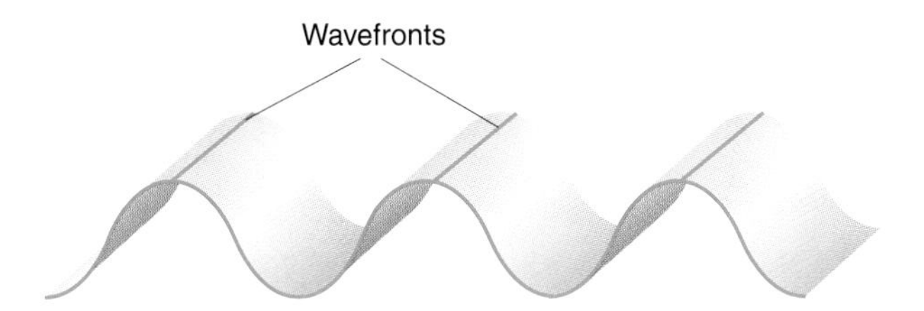

FIGURE 24.1 Wavefronts are lines or surfaces of constant phase.

superimposed)? For light and many other waves, experiments indicate that the algebraic sum of the individual wave disturbances at the same point accurately describes the resulting disturbance; this is the principle of linear superposition. That is, for two disturbances $\Psi_1(x, t)$ and $\Psi_2(x, t)$, the resulting disturbance is

$$\Psi(x, t) = \Psi_1(x, t) + \Psi_2(x, t)$$

When two coherent, monochromatic waves at the same location superimpose and interfere, the resulting wave disturbance depends on their phase difference δ.

If the phase difference between the waves is 0 rad, or an *integral multiple* of 2π rad, the waves are said to be **in phase**. Graphs of the two waves at a fixed x or a fixed t will show the waves to be in step with each other, the maxima occurring at the same times or places, as shown in Figure 24.2.

When the waves are in phase, the interference is called **constructive interference** and the amplitude of the resulting disturbance is the sum of the amplitudes of the individual disturbances. The intensity of the resulting wave is proportional to the square of the resulting amplitude. The resulting disturbance has a maximum intensity when the waves are in phase with each other. If the two interfering waves are of equal amplitude A, the resulting amplitude with constructive interference is 2A, and the resulting intensity is proportional to $(2A)^2 = 4A^2$, which is *four times* the intensity of each of the individual interfering waves.

If the phase difference δ between the waves is π rad, or any *odd integral multiple* of π rad, the waves are said to be completely **out of phase** with each other. Graphs of the two waves as functions of either space or time will show that the maximum of one wave occurs at the place or time of the minimum of the other wave, as indicated in Figure 24.3.

When the waves are completely out of phase, the interference is called completely **destructive interference**, and the amplitude of the resulting disturbance is the difference in the amplitudes of the individual disturbances. If the two interfering waves are of equal amplitude, the amplitude of the resulting disturbance is zero for completely destructive interference.

One way to produce a phase difference between two coherent waves is to let them travel different distances from their sources to where they are superimposed. The **path difference** causes a phase difference.

A path difference of one wavelength corresponds to a phase difference of 2π rad, because a shift in position of one wavelength along the wave changes its phase by a complete cycle of the cosine—that is, by 2π rad.

Hence, for a path difference Δx, the resulting phase difference δ_{path} in radians is

$$\delta_{\text{path}} = \frac{\Delta x}{\lambda}(2\pi \text{ rad}) \tag{24.1}$$

The factor

$$\frac{\Delta x}{\lambda}$$

is the number of wavelengths corresponding to the path difference; each wavelength of path difference represents a phase difference of 2π rad.

QUESTION 1

Does constructive interference violate energy conservation? Does completely destructive interference?

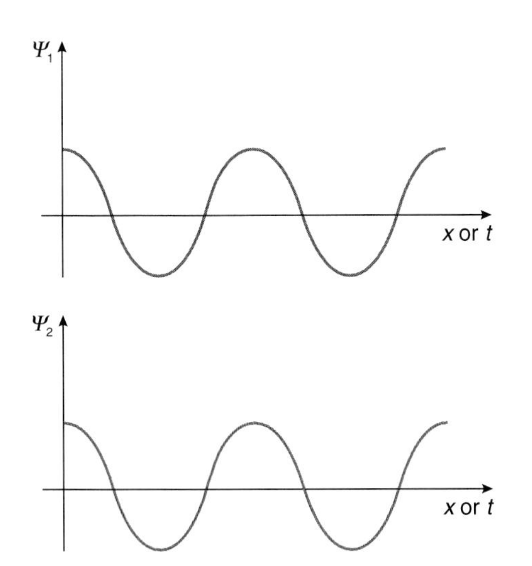

FIGURE 24.2 Two waves in phase: constructive interference results.

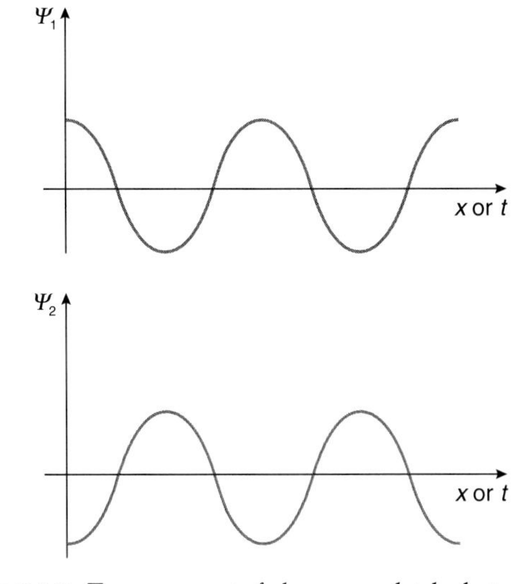

FIGURE 24.3 Two waves out of phase: completely destructive interference results.

24.3 YOUNG'S DOUBLE SLIT EXPERIMENT

. . . whenever two portions of the same light arrive at the eye by different routes, either exactly or very nearly in the same direction, the light becomes most intense when the difference of their routes is any multiple of a certain length, and least intense in the intermediate state of the interfering portions; and this length is different for light of different colours.

Thomas Young (1773–1829)*

In 1801 Thomas Young, a medical doctor, scientist, and one of the first to successfully translate Egyptian hieroglyphs (talk about a Renaissance man!), described the results of a now famous experiment that bears his name. **Young's double slit experiment** is one of the classic experiments of physics; it was the first convincing demonstration that light has wavelike characteristics, though the electromagnetic nature of the wave still awaited Maxwell's work some six decades later. Despite its apparent simplicity, the double slit experiment has implications that reverberate into the modern philosophy of quantum mechanics, in which the complementary particle-like (photon) aspects of light must be taken into account. The full significance and importance of Young's experiment is difficult to appreciate in this first encounter; we return to it again briefly in Chapter 27.

The geometry of the experiment is simple. Parallel wavefronts of a monochromatic wave (from a distant pointlike source) are incident on two identical, narrow slits, each of width a, separated by center-to-center distance d, as shown in Figure 24.4.† The slit width a and their separation d are on the order of the wavelength of the incident monochromatic light.

Since each wavefront arrives at the two slits at the same instant, the two slits form coherent sources, secured by wavefront division. There is a time-independent phase difference equal to 0 rad between the two slit sources. The coherent slit sources also have equal amplitudes. The diffracted wavelets (recall Section 23.1) emanating from the two slits interfere with each other where the wavelets overlap. We want to examine the interference effects on a screen parallel to the plane of the slits. We return to other effects of the diffraction in Section 24.7.

How the diffracted waves from the two slits interfere at any point depends only on the path difference from the slits to the point in question. In particular, the point O in Figure 24.5 is equidistant from each slit; the path difference between the waves from each slit thus is 0 m and the two waves from the slits are in phase with each other at O. Constructive interference results and the point O appears bright. We say there is a **bright fringe** at point O.

We consider arrival points on the screen on one or the other side of point O, as indicated in Figure 24.6. From a point P of superposition on the screen, we draw an arc of constant radius through the nearer slit to show the extra path Δx from the more distant slit.

As P moves away from O on the screen, the path difference Δx between the waves from the two sources increases from zero. Eventually the path difference will be half a wavelength. A path difference of $\lambda/2$ corresponds to a phase difference of (from Equation 24.1)

$$\begin{aligned}\delta &= \frac{\Delta x}{\lambda}(2\pi \text{ rad})\\ &= \frac{\lambda/2}{\lambda}(2\pi \text{ rad})\\ &= \pi \text{ rad}\end{aligned}$$

The two wavelets reaching the screen at this location then are completely out of phase and therefore interfere completely destructively. We say there is a **dark fringe** at that location.

If we select points farther away from point O, the path difference keeps increasing and eventually becomes equal to one wavelength. A path difference of λ corresponds to a phase difference of 2π (according to Equation 24.1):

$$\begin{aligned}\delta &= \frac{\Delta x}{\lambda}(2\pi \text{ rad})\\ &= \frac{\lambda}{\lambda}(2\pi \text{ rad})\\ &= 2\pi \text{ rad}\end{aligned}$$

The waves are now back in phase with each other again: the constructive interference results in the formation of a bright fringe. Every time the path difference between the two waves is an integral number of wavelengths, the interference is constructive and a bright fringe appears on the screen.

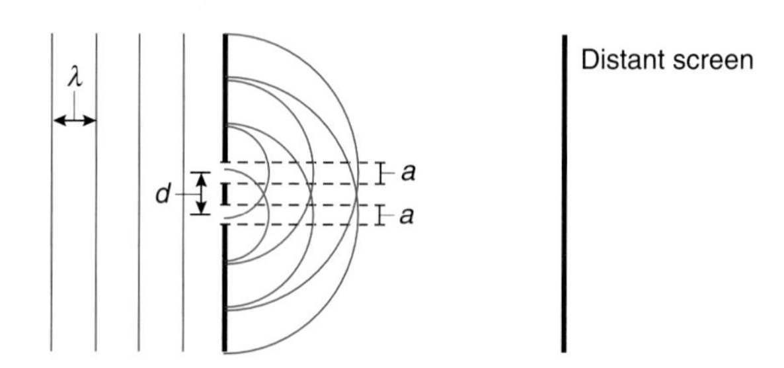

FIGURE 24.4 Geometry for Young's double slit experiment.

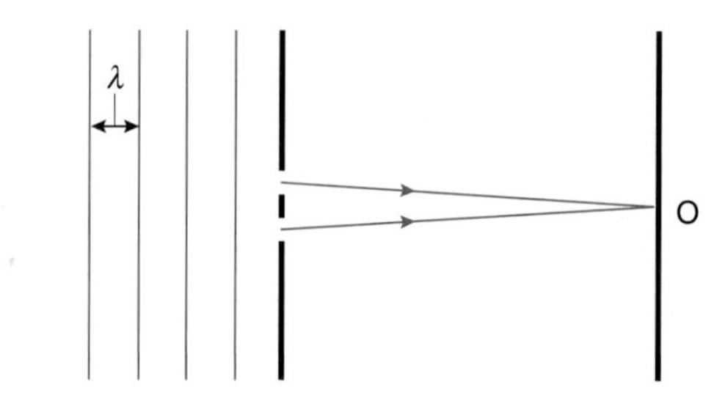

FIGURE 24.5 Zero path difference for waves interfering at point O.

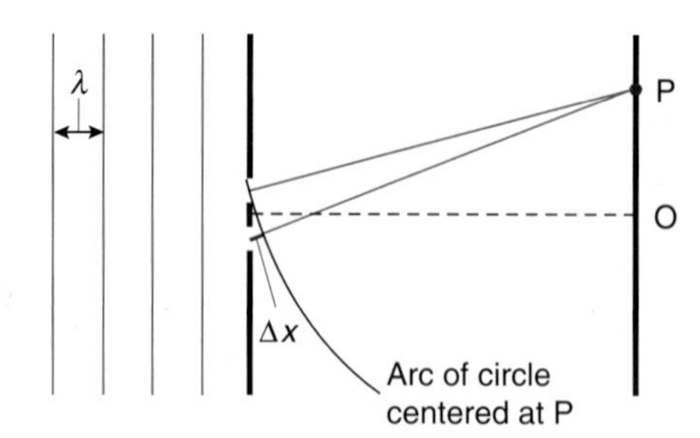

FIGURE 24.6 Nonzero path difference Δx for waves interfering at point P.

*Miscellaneous Works of the Late Thomas Young, M.D., F.R.S. . . ., edited by George Peacock (J. Murray, London, 1855), Volume I, page 170.

†More conveniently, we also can put the pointlike source at the secondary focal point of a converging lens, in which case the incident wavefronts (remember rays are perpendicular to wavefronts) on the double slit are parallel.

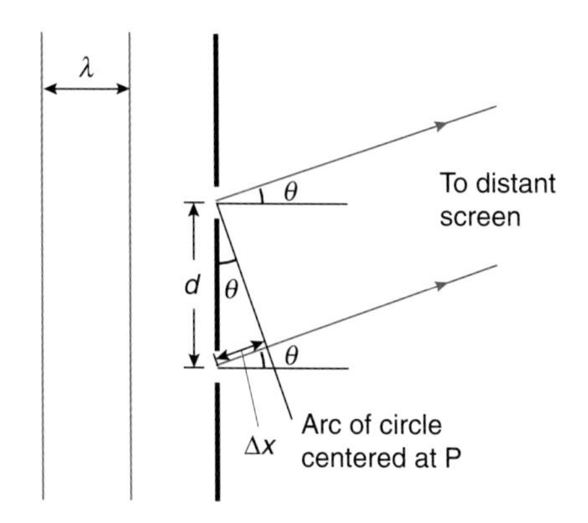

FIGURE 24.7 Magnified view of Young's double slit arrangement for a distant screen geometry.

For a *distant* screen, the small arc of the circle from P to the slits approaches a straight line, as indicated in Figure 24.7.* The path difference then can be expressed trigonometrically in terms of the separation of the slits d and the angle θ from the straight-through direction to any point P on the screen. The path difference is

$$\Delta x = d \sin \theta$$

When the path difference is an integral number of wavelengths, we have constructive interference and a bright fringe on the screen.

Thus the condition for constructive interference on the distant screen is

$$d \sin \theta = m\lambda \tag{24.2}$$

(condition for constructive interference: bright fringe)

where m is an integer that can take on the values $0, \pm 1, \pm 2, \ldots$.

Each value of m corresponds to a particular bright fringe, as shown in Figure 24.8. Positive values of m correspond to the bright fringes for positive θ (measured counterclockwise from the straight-through direction) using the point O′ halfway between the slits as an origin. The value $m = 0$ is the central bright fringe in the straight-through direction. Negative values of m correspond to negative values for θ.

The absolute value of m is known as the **order of interference**, or equivalently, the **order number**. Thus $|m|$ indicates how many bright fringes we are away in either direction from the central bright fringe at the straight-through direction. Since $\sin \theta \approx \theta$ rad for small values of θ, we also have described the observation that the multiple fringes appear uniformly spaced on a distant screen near the center of the pattern.

*Experimentally, we can place a converging lens beyond the slits and place the screen at the focal point of the lens. Optically, the screen then is effectively an infinite distance from the slits.

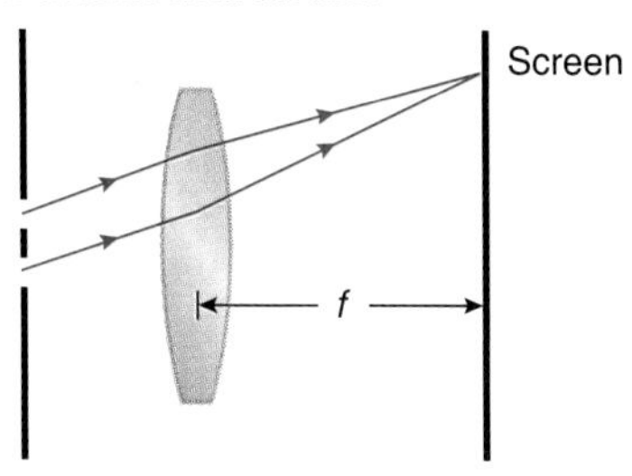

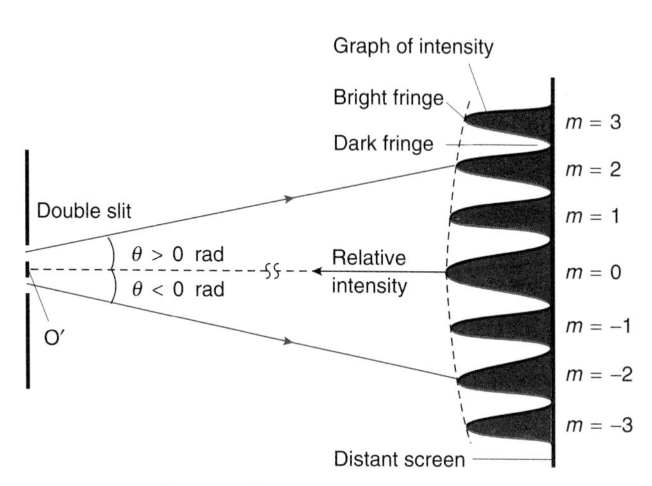

FIGURE 24.8 Pattern of fringes in Young's double slit experiment.

Notice from Equation 24.2 that for a fixed wavelength λ and fixed order number m, if the separation between the slits d now is decreased, the angle θ locating the interference maximum corresponding to any given order number m must increase. So the smaller the slit separation, the greater the angular separation of the interference peaks.

QUESTION 2

The same note from a single sound source is fed to two loudspeakers separated by a distance d. (a) Are the two acoustic sources coherent? (b) Would you expect to hear loud and quiet acoustic interference fringes as you scan your ear across a line parallel to the line between the speakers?

EXAMPLE 24.1

Two parallel slits, each 0.10 mm wide, are separated by 0.50 mm and are illuminated by a parallel beam of laser light with a wavelength of 633 nm. A screen is placed 4.00 m from the slits. What is the separation of the bright interference fringes on the screen?

Solution

Since the distance to the screen is much greater than the distance d between the slits, use Equation 24.2 to find the locations of the interference maxima. In particular, the first-order ($|m| = 1$) fringe is located at an angle θ, where

$$d \sin \theta = 1\lambda$$

$$\sin \theta = \frac{\lambda}{d} = \frac{633 \times 10^{-9} \text{ m}}{0.50 \times 10^{-3} \text{ m}} = 1.3 \times 10^{-3}$$

Since the sine of θ is so small, the sine is also virtually equal to θ in radians, according to the small angle approximation. Hence the angle is

$$\theta = 1.3 \times 10^{-3} \text{ rad}$$

Imagine a circle centered on the slits with a radius equal to the distance to the screen, as in Figure 24.9. For small angles θ, the arc length is approximately equal to y, and so you have

$$\theta = \frac{y}{\ell}$$

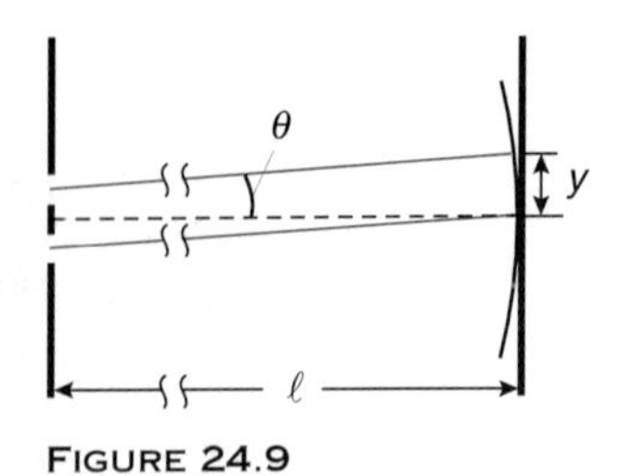

FIGURE 24.9

where y is the distance on the screen between the fringes for order numbers $m = 0$ and $m = 1$. Hence you have

$$\begin{aligned} y &= \theta\ell \\ &= (1.3 \times 10^{-3}\ \text{rad})(4.00\ \text{m}) \\ &= 5.2 \times 10^{-3}\ \text{m} \end{aligned}$$

Since the fringes are equally spaced for small angles, the separation between the fringes observed on the screen is 5.2 mm.

24.4 Single Slit Diffraction

When monochromatic light passes through a single slit whose aperture is on the order of the size of the wavelength, the light diffracts, or spreads out, on the other side of the slit. As we have seen, diffraction is characteristic of wave phenomena in general. Here we examine the diffraction of light as it passes through a single slitlike aperture of width a; the results, though, are equally applicable to other types of wave diffraction.

The intensity distribution of the diffracted light on a distant screen is indicated in Figure 24.10. The observed distribution of the light intensity on the screen indicates that the light is not uniformly spread out or diffracted: certain regions on the screen have no light, while others have some. Thus the amplitude of the diffracted light depends on the angle θ from the straight-through direction in Figure 24.10.

To account for this we resort to an old and what might seem to be odd geometric construction, called **Huygens's principle**. It was first used in the 17th century by Christian Huygens (1629– 1695) to explain how waves propagate from one place to another. Huygens's idea was the following. Imagine a wavefront of any shape at some position in space at a particular time, as in Figure 24.11. According to Huygens's principle, each point on a wavefront acts as a source of a wavelet that propagates outward from the point; the points of the wavefront are coherent emitters of the wavelets. During a time interval Δt, each wavelet of light travels a distance $c\ \Delta t$ from its source point, as shown in Figure 24.11. The new position of the wavefront is determined by the envelope* of the wavelets from the individual points on the original position of the wavefront.

We apply this to the single slit. Using Huygens's principle, we may usefully regard the points on the wavefront filling the slit as a collection of coherent sources of wavelets propagating into the region beyond the slit, as shown in Figure 24.12. Their envelope creates the wavefront anywhere beyond the slit.

Arriving and superimposing at any given point on the screen are the contributions from the multitude of coherent pointlike sources arrayed across the small width of the slit. To visualize the path difference to a point P on the screen, we draw an arc of constant radius centered on the screen point of superposition P through the nearer slit edge, as in Figure 24.13.

If the distance ℓ to the screen is much larger than the tiny slit width a, the extra path Δx from the farther slit edge is part of a small right triangle,† blown up for clarity in Figure 24.14. From the right triangle in Figure 24.14 and definition of the sine, the path difference Δx is found to be

$$\Delta x = a \sin \theta$$

If the screen point is at the center (straight-through direction), the contributions from paired point-like sources placed symmetrically on either side of the straight-through direction arrive in phase and we have a maximum amplitude and intensity at this position.

*A geometrical *envelope* is a curve (or surface) that is tangent to a set of curves (or surfaces).

†To avoid having to put the screen far away to make $\ell >> a$, we alternatively can put a converging lens near the slit aperture and place the screen at the focal plane of the lens. Then those rays that are superimposed on reaching each point on the screen are indeed exactly parallel on leaving the slit aperture, as if $\ell = \infty$ m.

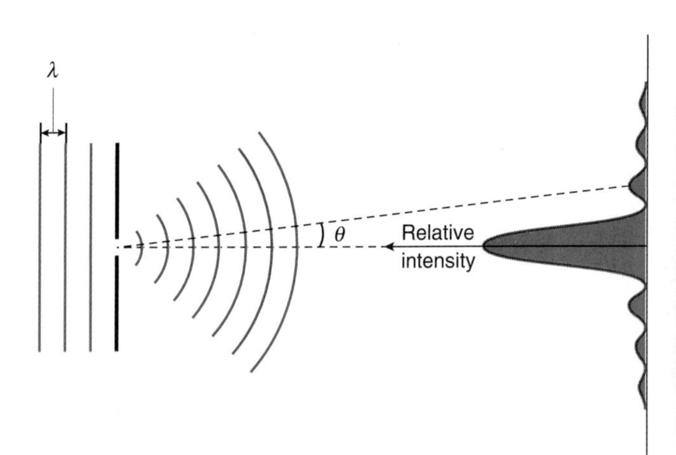

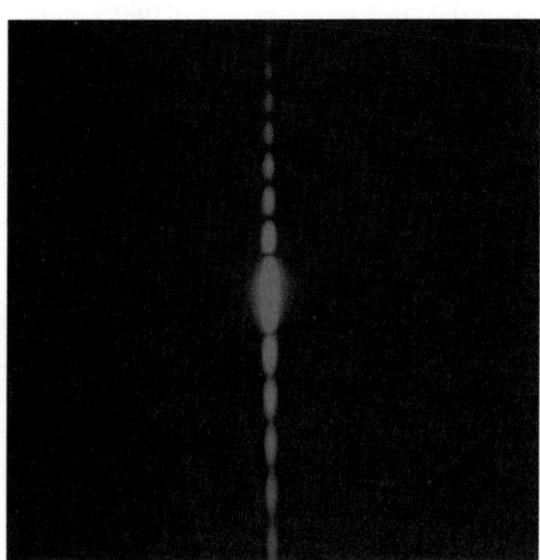

FIGURE 24.10 Diffraction pattern formed by a single slit.

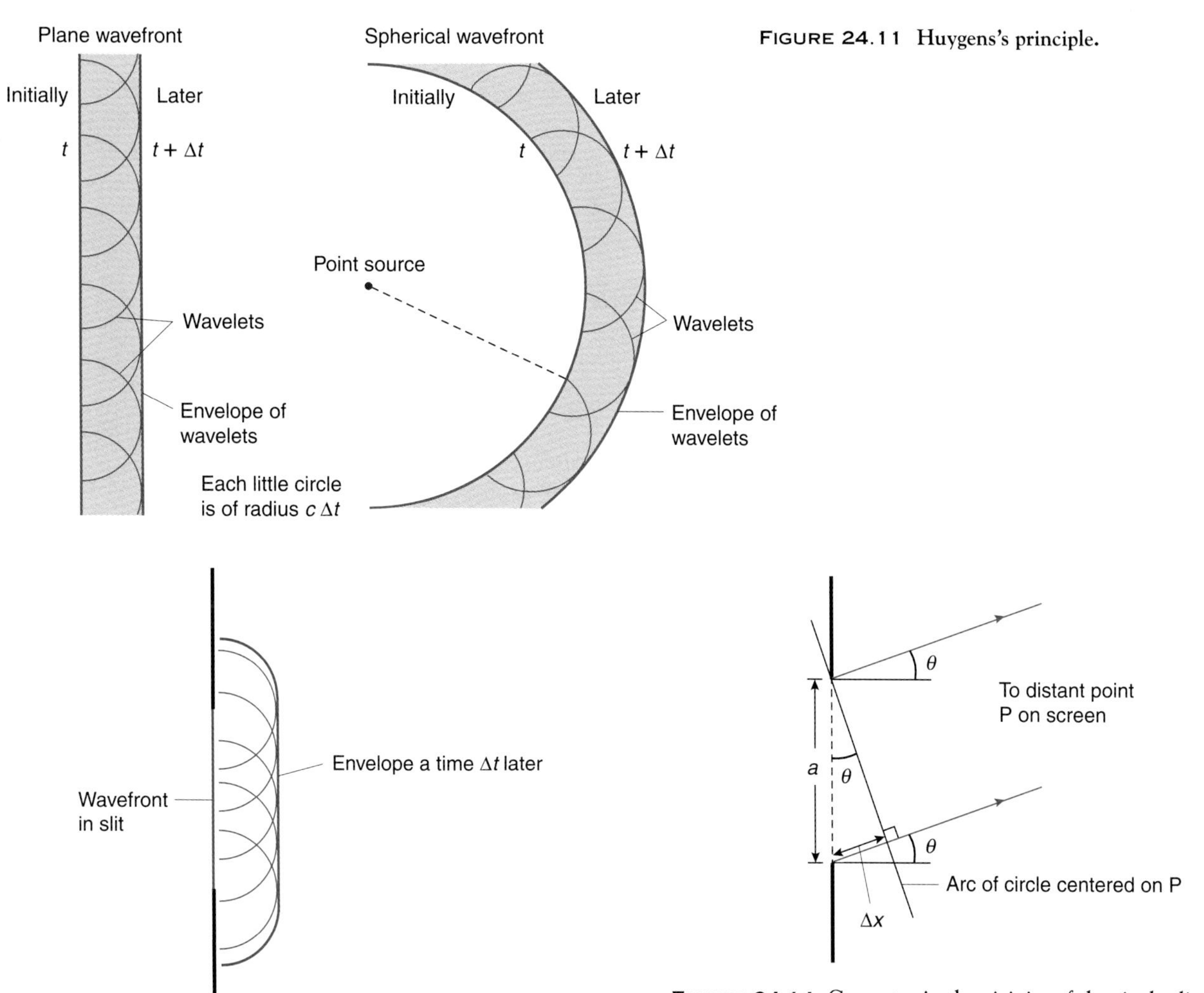

FIGURE 24.11 Huygens's principle.

FIGURE 24.12 Huygens's principle applied to a single slit (magnified view).

FIGURE 24.14 Geometry in the vicinity of the single slit for a distant screen.

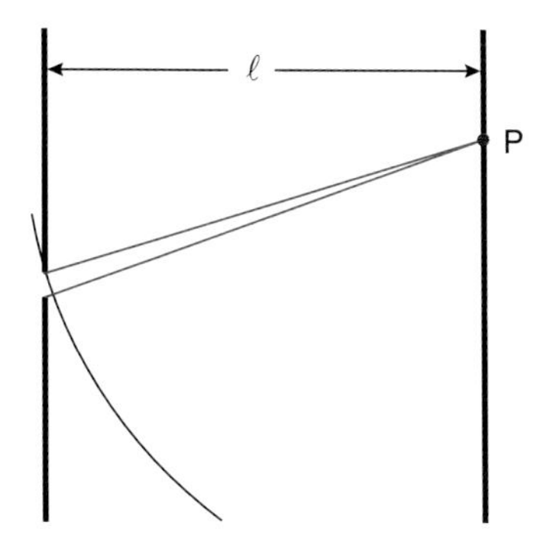

FIGURE 24.13 A circle centered on P helps to delineate the path difference.

The minima at the sides of the central diffraction maximum come about in the following way. Take, for example, the first minimum to the side. The wavelets from the coherent point sources distributed across the aperture of the slit now travel different distances to the screen. As we move away from the central maximum on the screen, eventually we reach a position where the path difference to the point P on the screen between the point source at the near edge (top) of the aperture and the point source at the *middle* of the slit is half a wavelength, as shown in Figure 24.15. The half wavelength path difference for the pair corresponds to a phase difference of π rad (from Equation 24.1). The two wavelets from these two points then interfere destructively on the screen and contribute nothing.

Now let's consider the pair of point sources in the aperture just below the first pair. The wavelets from these sources also have a difference in path to point P of half a wavelength (and a phase difference of π rad), and so these two wavelets *also* interfere destructively. As we consider similarly paired points across the full aperture, *every* pair adds to give zero amplitude. The sum of any number of zeros is still zero. Thus we have pairwise destructive interference from all the sources in pairs half the aperture apart, distributed across the entire aperture.

The $\lambda/2$ path difference between a point at the top and a point at the *middle* of the slit also means that the path difference between the points at the top and at the *bottom* of the slit is a full wavelength, as shown in Figure 24.15. Geometrically, this path difference between the extremes of the slit is related to the width a of the slit and the angle θ_1 locating the first minimum. The path difference in Figure 24.15 is $a \sin \theta_1$.

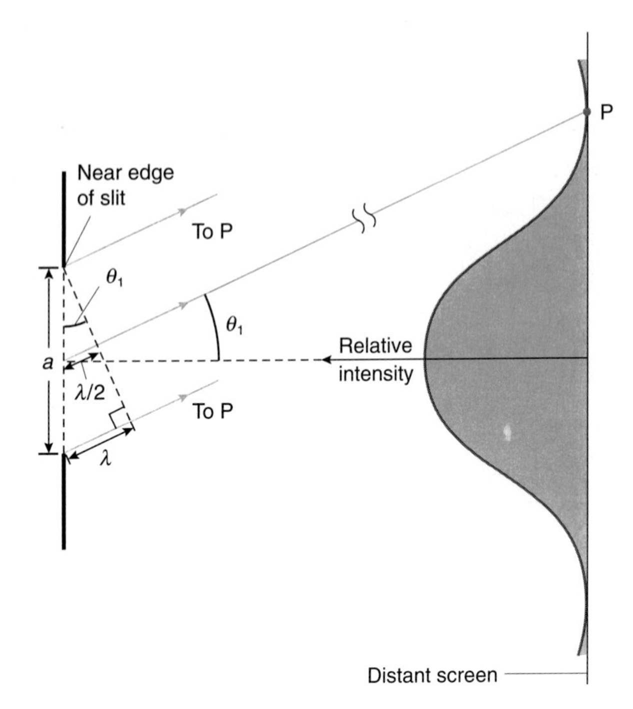

FIGURE 24.15 Path difference to the screen of $\lambda/2$ between the near edge and middle of the slit.

So the *first minimum* of the diffraction pattern occurs where

$$a \sin \theta_1 = \lambda \quad (24.3)$$

(first minimum, single slit diffraction)

Other diffraction minima exist when the path difference $a \sin \theta$ between opposite edges of the slit is any integral multiple m of the wavelength:

$$a \sin \theta = m\lambda \qquad (m = \text{integer}) \quad (24.4)$$

(single slit diffraction minima)

For small angles, $\sin \theta \approx \theta$ in radians, so the minima are equally spaced. In between the minima are maxima, but their locations are harder to specify than the minima, except for the central diffraction maximum, and we will not consider them further. These other maxima are progressively weaker than the central maximum.

Notice in Equation 24.3 that for a fixed wavelength, if the slit width a is made smaller, the angle θ_1 locating the first minimum of the diffraction pattern must increase, and so the broad central maximum increases in width.

The *narrower* the slit, the *greater* the total angular width $2\theta_1$ of the complete central diffraction peak.

You can see this by looking through a crack between your fingers at a distant pointlike street light.

Conversely, the greater the width of the slit, the smaller the angle θ_1 to the first minimum of the diffraction pattern. As the slit width a becomes much greater than the wavelength ($a >> \lambda$), the angle θ_1 approaches zero and we enter the geometric limit of Chapter 23, where we can essentially ignore the wave nature of the light and treat the light as propagating in straight lines (rays).

PROBLEM-SOLVING TACTIC

24.1 Do not confuse the order number m locating the interference *maxima* for Young's double slit experiment in Equation 24.2 with the integer m locating the single slit diffraction *minima* in Equation 24.4. While both m's are integers, each represents a different effect!

EXAMPLE 24.2

Parallel light of wavelength 633 nm is incident on a slit of width 0.12 mm.

a. Find the total angular width of the central diffraction maximum on a screen located 3.50 m from the slit.

b. What is the width of the central diffraction maximum on the screen in centimeters?

Solution

a. Since the distance to the screen is much larger than the width of the slit, you can use Equation 24.3 to find the angle θ_1, which is half the total angular width of the central diffraction maximum:

$$a \sin \theta_1 = \lambda$$

$$\sin \theta_1 = \frac{\lambda}{a} = \frac{633 \times 10^{-9}\ \text{m}}{0.12 \times 10^{-3}\ \text{m}} = 5.3 \times 10^{-3}$$

Since the sine of the angle is so small, the small angle approximation implies the angle is

$$\theta_1 = 5.3 \times 10^{-3}\ \text{rad}$$

The total angular width of the pattern is

$$2\theta_1 = 1.1 \times 10^{-2}\ \text{rad} = 0.63°$$

b. The extent y of the central diffraction maximum on the screen is determined by making use of the definition of the radian measure of an angle, as in Figure 24.16. For small angles, the distance y is approximately equal to the arc length of a circle of radius ℓ centered on the slit. Hence

$$2\theta_1 = \frac{y}{\ell}$$

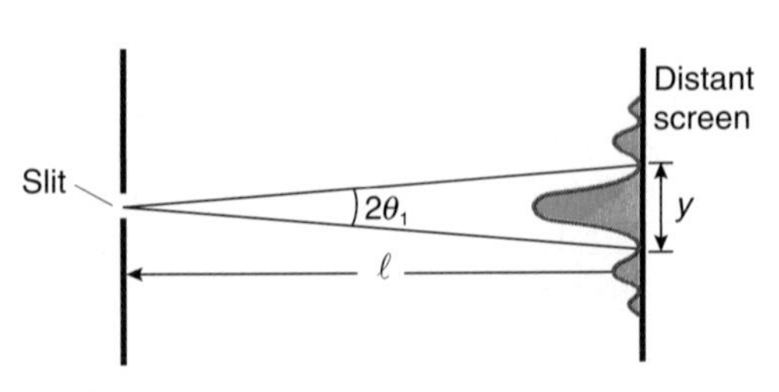

FIGURE 24.16

Therefore you have

$$y = 2\theta_1 \ell$$
$$= (1.1 \times 10^{-2} \text{ rad})(3.50 \text{ m})$$
$$= 3.9 \times 10^{-2} \text{ m}$$
$$= 3.9 \text{ cm}$$

24.5 DIFFRACTION BY A CIRCULAR APERTURE

Circular apertures are common to lenses and our eyes, and so it is of more than passing interest to know something about diffraction by such a geometry.

If monochromatic light is incident on a circular aperture of *diameter a*, rather than on a slit, diffraction of the light through the aperture produces a circularly symmetric pattern on a distant screen, indicated in Figure 24.17. The pattern consists of a central circular bright spot surrounded by a series of increasingly dimmer concentric circles of light.

The concentric dark rings are not equally spaced (as the fringes are for diffraction by a single slit), because the wavelet sources across the diameter have varying lengths (circumferences) and importance (as opposed to the equal segments across a slit), and the effective circle width is somewhat less than the diameter. A detailed mathematical analysis of the locations of the secondary peaks and the minima of the diffraction pattern of a circular aperture is beyond the scope of an introductory course in physics. However, it is of practical import to know the location of the first minimum of the diffraction pattern of such a circular aperture, as we will see in Section 24.6.

The first minimum of the circular diffraction pattern is located at an angle θ_1 found from the following relation*:

$$a \sin \theta_1 = 1.220\lambda \qquad (24.5)$$

(first diffraction minimum for a circular aperture)

where a is the *diameter* of the circular aperture.

This is very similar to the equation that locates the first minimum of the diffraction pattern of a single slit of width a (Equation 24.3):

$$a \sin \theta_1 = \lambda \qquad \text{(first diffraction minimum for a single slit)}$$

The distinction between the two equations is the effective value $a/1.220$ for the circular aperture. The number 1.220 arises from a special class of functions, called Bessel functions, that invariably are associated with circular geometries in physics; Bessel functions are analogous to the familiar special functions you know as the sines and cosines.

Notice, as for a single slit, that for a fixed wavelength, the *smaller* the diameter a of the circular aperture, the *greater* the angle θ_1 to the first minimum of the diffraction pattern. In other words, the smaller the aperture, the greater the angular width of the central diffraction peak. Conversely, the greater the size of the aperture, the smaller the angular width of the central diffraction peak and the less apparent the effects of diffraction.

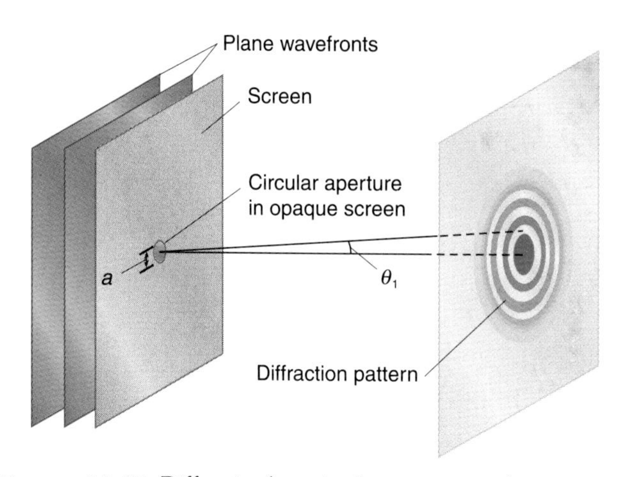

FIGURE 24.17 Diffraction by a circular aperture produces a pattern of concentric rings.

PROBLEM-SOLVING TACTIC

24.2 For circular apertures, be sure to use Equation 24.5 to locate the angle to the first minimum of the diffraction pattern, not Equation 24.3 for slit apertures.

EXAMPLE 24.3

Light with a wavelength of 633 nm is incident on a pinhole of diameter 0.30 mm.

a. What is the angular width of the central diffraction peak on a screen 4.0 m from the pinhole?
b. What is the linear extent of the central diffraction peak on the screen?

Solution

a. The angular width of the central diffraction peak is $2\theta_1$, where θ_1 is the angle from the straight-through direction to the first minimum on either side of the peak. The angle θ_1 is found from Equation 24.5:

$$a \sin \theta_1 = 1.220\lambda$$

$$\sin \theta_1 = \frac{1.220\lambda}{a}$$

The diameter of the pinhole is $a = 0.30 \text{ mm} = 3.0 \times 10^{-4} \text{ m}$. Substituting for a and λ, you obtain

$$\sin \theta_1 = \frac{1.220\,(633 \times 10^{-9} \text{ m})}{3.0 \times 10^{-4} \text{ m}}$$
$$= 2.6 \times 10^{-3}$$

Using the small angle approximation, we have

$$\theta_1 = 2.6 \times 10^{-3} \text{ rad}$$

The angular width of the central peak is

$$2\theta_1 = 5.2 \times 10^{-3} \text{ rad}$$
$$= 0.30°$$

*Note that we did not derive this. Alas, we apologize; to derive the factor 1.220 is not simple.

b. The linear extent y of the central peak on the distant screen is found from the definition of an angle in radians, as shown in Figure 24.18. Since y is approximately equal to the arc length of a circle of radius ℓ centered on the hole, you have

$$2\theta_1 = \frac{y}{\ell} \qquad (\theta_1 \text{ in radians})$$

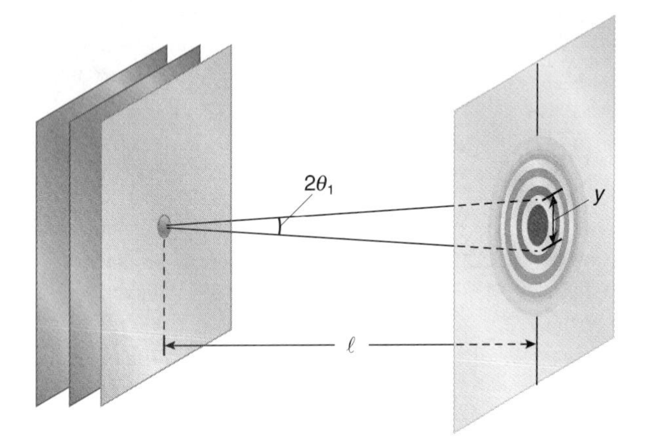

FIGURE 24.18

Solving for y, you find

$$\begin{aligned} y &= 2\theta_1\ell \\ &= 2(2.6 \times 10^{-3} \text{ rad})(4.00 \text{ m}) \\ &= 2.1 \times 10^{-2} \text{ m} \\ &= 2.1 \text{ cm} \end{aligned}$$

24.6 Resolution

When you see a very distant car approaching at night, it is hard to distinguish the individual headlights; they appear as one source. We say they are **unresolved.** When the car is closer, it is easy to tell there are two headlights; we say they then are **resolved.*** Now look at the starry sky. Almost half the stars visible to the naked eye at night really are not single stars like the Sun but binary stars, two stars gravitationally bound to and orbiting each other. A telescope easily reveals the binary character of many stars that the naked eye perceives as one.† What determines whether we can distinguish whether a source is single or binary? For a telescope, does this ability depend on the magnification? At the other extreme, what determines the smallest details you can discern in a microscope? Here we answer these important questions about **resolution**, the ability to separate and distinguish fine details.

Independent binary sources, such as two headlights, two stars, or two amoebas, are **incoherent sources**, unlike a pair of double slits illuminated by monochromatic light. The phase of the light from each source has no correlation with that of the light from the other source; the light from each is emitted by independent atoms and so their phase difference is not constant in time.

We saw in the previous section how, when monochromatic light traverses a circular aperture of diameter a, a diffraction pattern of concentric rings is produced. This pattern appears in the focal plane of the eye, or the focal plane of the objective element of telescope, where the Huygens's wavelets are superimposed. The angular position of the first minimum of the diffraction pattern of a circular aperture is located using Equation 24.5:

$$a \sin \theta_1 = 1.220\lambda$$

Two incoherent, monochromatic, pointlike, distant sources each produce their own diffraction pattern.

Since the sources are incoherent, the waves from one source do not consistently interfere with those from the other. Let the angle between the sources be ϕ. The angular separation of the two diffraction patterns is equal to the angle ϕ between the sources, as shown in Figure 24.19.

If the sources are well separated in angle, the diffraction patterns on the distant screen or focal plane are quite distinct from each other and it is easy to tell that two sources of light are present. Under these circumstances, we say the two sources are *well resolved.*

Certainly if the angle between the two sources is zero, the two diffraction patterns are superimposed on top of one another and we cannot tell that there are really two sources producing the pattern on the screen. The two sources are *unresolved.*

The question then arises: what is the minimum separation angle between the sources that allows us to barely resolve or distinguish that there are really two diffraction patterns present, indicating the presence of the two sources? As the angle between the sources increases, the diffraction patterns caused by each become further separated as well. When the angle ϕ is large enough, a dip appears in the combined diffraction pattern that indicates two patterns are present, not one, as in Figure 24.20. In the 19th century. Lord Rayleigh [John William Strutt (1842–1919)] proposed a convenient criterion, now called the **Rayleigh criterion**, for determining the minimum angular separation angle for resolution.

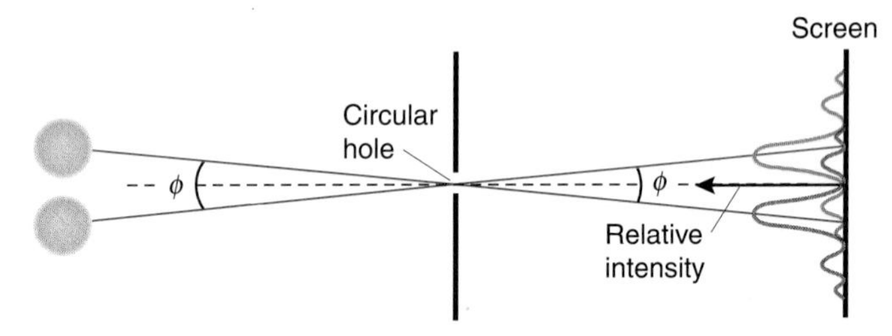

FIGURE 24.19 Two incoherent sources produce two independent diffraction patterns; here the two are well resolved.

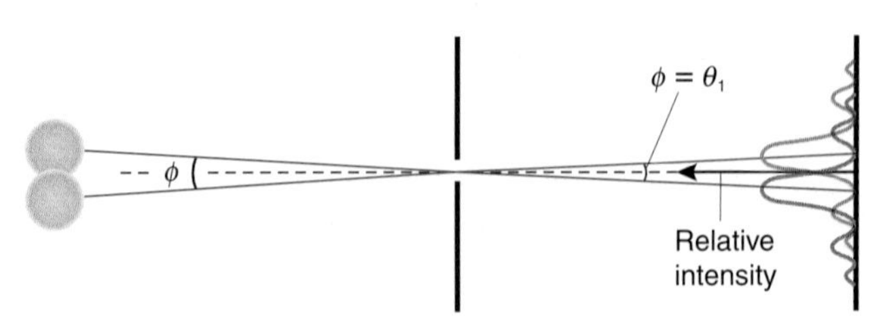

FIGURE 24.20 Two sources barely resolved according to the Rayleigh criterion.

*The word comes from one of the meanings of the word resolve: to decide.
†The orbital periods for such visual binaries are quite long, even centuries, so the orbital motion is not readily apparent in a telescope.

When the central peak of one diffraction pattern is located at the position of the first minimum of the *other* diffraction pattern, we say the two sources are *barely resolved.*

The angle ϕ between the sources then is equal to the angle θ_1, the angle that locates the first minimum of either diffraction pattern.

Mathematically, for a circular aperture, the sources are barely resolved if their angular separation ϕ is equal to the angle θ_1 in Equation 24.5:

$$a \sin \theta_1 = 1.220\lambda$$

Since the angle θ_1 is almost always very small, the small angle approximation can be used, and we have

$$a\theta_1 = 1.220\lambda \qquad (\theta_1 \text{ in radians}) \text{ (circular aperture)} \qquad (24.6)$$

For stars and headlights, which emit many different wavelengths, to determine the angle in the Rayleigh resolution criterion, use an *effective wavelength*, typically in the yellow region of the spectrum ($\lambda \approx 550$ nm) where our eyes are most sensitive.

The eyes of cats and other felines, some other mammals, and a few reptiles like boa constrictors have vertical slit apertures. For these life forms, it is apparently more important to resolve vertical details than horizontal. If the aperture is a slit, the minimum resolution angle is found from Equation 24.3, $a \sin \theta_1 = \lambda$, which for small angles is

$$a\theta_1 = \lambda \qquad (\theta_1 \text{ in radians}) \text{ (slit aperture)} \qquad (24.7)$$

Consider a telescope with a circular aperture. The aperture diameter of the objective lens or mirror determines the resolution of the instrument. If the angular separation of two sources is less than θ_1, where θ_1 is determined from the Rayleigh criterion and λ is the effective wavelength of the incident light, then the two sources cannot be distinguished or resolved. Problems such as this occur quite frequently in astronomy. In order to see binary stars as indeed separate sources of light, a telescope of sufficient aperture is needed: one whose Rayleigh criterion minimum resolution angle θ_1 is *smaller* than the angular separation ϕ of the stars.

The resolution is *not* determined by the magnification. This can be easily confirmed by looking at a binary star through a telescope, such as Mizar (ζ Ursae Majoris), the second star from the end of the handle of the Big Dipper (Ursa Major), shown in Figure 24.21.* You will see that Mizar is a close binary in the telescope. Now place a piece of cardboard over the aperture of the telescope with a hole about the diameter of the eye pupil (about 1 cm); you will no longer be able to see the two stars resolved through the telescope, regardless of the magnification you are using! Thus resolution depends on the aperture, *not* the magnification.

As another example, the eye has a circular aperture and it is the size of its aperture that determines the ability of the eye to distinguish two sources as being separate.† The function of the cornea and eye lens is merely to bring the distant screen to a more reasonable finite distance: the screen is the retina. If the separation of the two sources is less than the angle determined from the Rayleigh criterion, then the eye cannot distinguish the presence of the two sources and they are unresolved.

*Mizar has a dim, nearby, naked eye companion, called Alcor. Fix your attention only on Mizar.

†For the eye, there is a biological factor to consider as well: the angular size of the cone cells (measured from the pupil) on the fovial region of the retina. See Example 24.4.

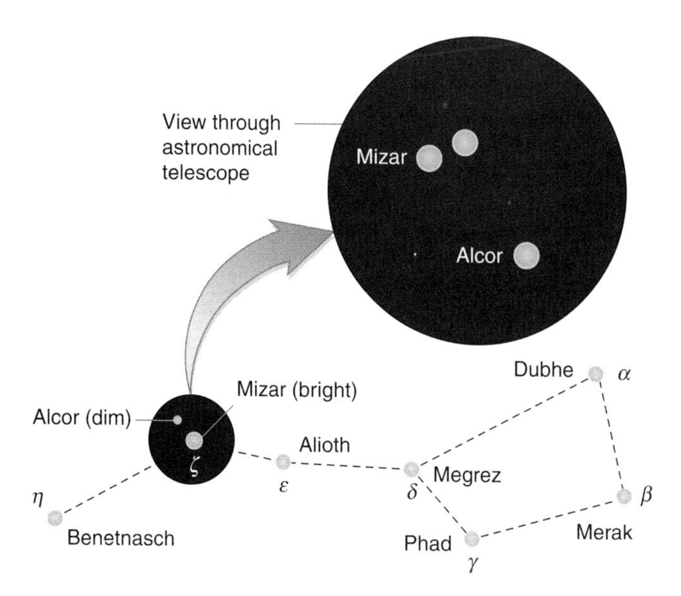

FIGURE 24.21 The location of Mizar in the constellation Ursa Major.

The same considerations apply to microscopes. The detail that can be distinguished or resolved is essentially the ability to tell if two object points are distinct or not. Such resolution ultimately depends on the aperture of the objective lens of the instrument and the wavelength of the light used to view the object. Using smaller-wavelength light enhances the resolution, since for a fixed aperture, decreasing λ decreases the minimum resolution angle θ_1. Electron microscopes use electron wavelengths that are about 1000 times smaller than visible light wavelengths, so such microscopes have extraordinary resolution.

It is diffraction that sets a theoretical limit on the resolution of optical instruments. The resolution is determined by the aperture of the instrument and the wavelength, *not the magnification.*

The *smaller* the angle θ_1 found from the Rayleigh criterion, the *better* the resolution. That is, the easier it is to distinguish the presence of the two point sources. Good resolution means that the angle θ_1 to the first minimum of the diffraction pattern is small, which happens when the circular aperture (or the slit width) is large. So for good resolution, the angle θ_1 between the center and first minimum of each diffraction pattern produced by two incoherent sources is smaller than the separation angle between the two source centers.

QUESTION 3

A slit has a width equal to the diameter of a circular aperture. Which has the better resolution?

EXAMPLE 24.4

What is the minimum angle of resolution for the human eye as determined from the Rayleigh criterion if the effective wavelength of the light is 550 nm? Consider the maximum aperture of the eye when fully dilated to be about 7.0 mm.

Solution
Use Equation 24.5:

$$a \sin \theta_1 = 1.220\lambda$$

The resolution angle θ_1 is small, so the small angle approximation is appropriate:

$$a\theta_1 = 1.220\lambda \qquad (\theta_1 \text{ in radians})$$

$$\theta_1 = \frac{1.220\lambda}{a}$$

$$= \frac{1.220\,(550 \times 10^{-9}\ \text{m})}{7.0 \times 10^{-3}\ \text{m}}$$

$$= 9.6 \times 10^{-5}\ \text{rad} \qquad \text{(about 20 arc seconds)}$$

The actual resolution limit of the eye is about ten times this result, or about three arc minutes (60 arc seconds ≡ 1 arc minute), because of the size of the cone cells on the fovea region of the retina. The cone cells subtend an angle of about one arc minute from the pupil. Each resolved diffraction pattern must form on a separate cone cell with an unaffected cone cell in between them, and so the resolution limit of the eye actually is about three arc minutes (but varies greatly with the individual).

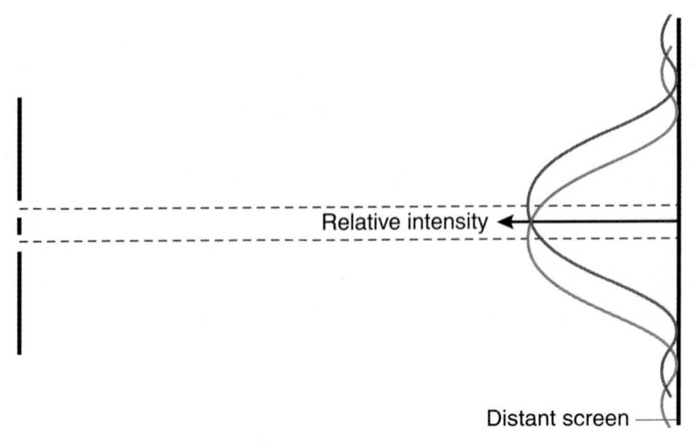

FIGURE 24.22 **Two slits each produce a diffraction pattern on a distant screen when illuminated one at a time; the two patterns are very slightly separated from one another.**

24.7 The Double Slit Revisited

This is not a Hollywood sequel.* We must revisit the double slit to account for the simultaneous interference effects between the slits and the diffraction effects of each slit.

Each slit, acting alone, produces a diffraction pattern on a distant screen. If one slit is covered, the incident monochromatic, coherent light only passes through the other slit, and the illuminated slit produces a single slit diffraction pattern on the screen. When each slit is illuminated separately, the individual diffraction patterns of each slit are very slightly separated from each other, as shown in Figure 24.22.

The individual diffraction patterns cannot be observed on the screen simultaneously because they appear only when one or the other slit is illuminated, but not both. Since the slit separation is small, the separation of the individual diffraction patterns is negligible. But when both slits are illuminated by monochromatic light, the slit sources are coherent and the light from one slit now can also interfere with the light from the other slit. A double slit interference pattern, with equally spaced dark and bright fringes (for small angles), is seen on the distant screen.

The intensity of the interference peaks, however, is modulated by the **envelope** of the single slit diffraction pattern, as shown in Figure 24.23.

In other words, the light coming from the slits, which forms the bright fringes of the double slit interference pattern, is reduced because each slit sends out less light to interfere constructively at off-center directions of its own broad diffraction pattern. Diffraction effects are apparent from the modulation of the intensity of the interference peaks. The diffraction effects of each single slit are manifested like the Cheshire Cat: the diffraction disappears with only its "smile" remaining.†

We now can play with a few variations.

1. If the slit separation d is fixed and the slit width a is decreased, the interference peaks remain in the same locations, since they are determined by the fixed slit separation d. But since the slit width a is decreased, the location of the first minimum of the diffraction pattern appears at a larger angle θ_1 according to Equation 24.3:

$$a \sin \theta_1 = \lambda$$

With the right-hand side of this equation constant (for a fixed wavelength), decreasing a means that $\sin \theta_1$ must increase, and so θ_1 itself must increase. The result is that more interference peaks now appear under the expanded size of the envelope of the central diffraction peak, as indicated in Figure 24.24.

2. Now we fix the slit width a, but decrease the slit separation d. The envelope of the diffraction pattern now stays fixed in size, since the slit width is kept constant. However, as the slit separation d decreases, the angular position of a fixed order number m of the interference peaks must increase according to Equation 24.2:

$$d \sin \theta = m\lambda$$

With the right-hand side of this equation fixed (for a fixed order number and fixed wavelength), if d decreases, then $\sin \theta$ must increase—and so θ itself must increase—for that

*Perhaps fortunately, no one has been able to create a credible screenplay for a potential TV show called *L.A. Physics* analogous to *L.A. Law*. Unfortunately, scientists and engineers invariably are depicted by Hollywood as mad or deranged. See Investigative Project 16.

†With apologies to Lewis Carroll and his *Alice's Adventures in Wonderland*.

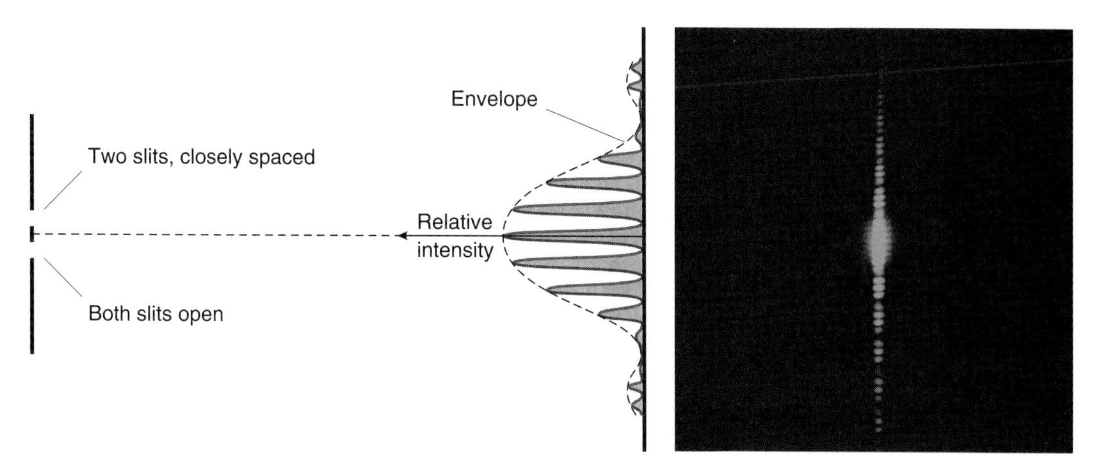

FIGURE 24.23 Double slit interference and diffraction. (Photo is not to the same scale as graph.)

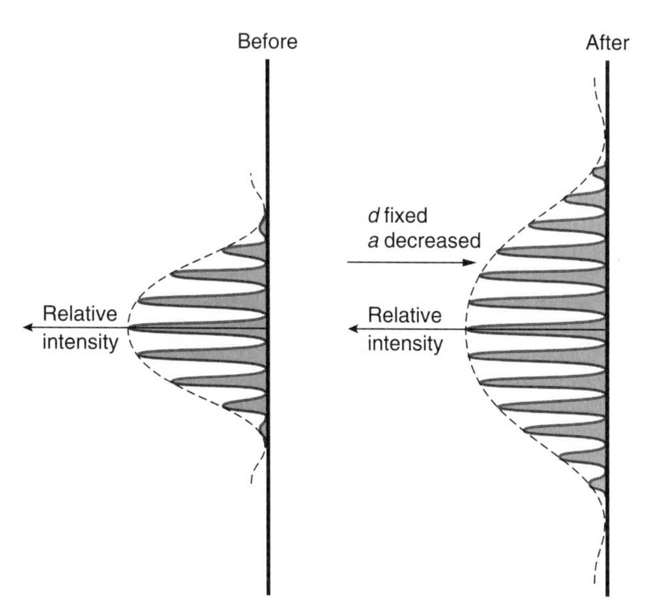

FIGURE 24.24 Decrease the slit width for fixed slit separation; more interference peaks appear under the expanded angular size of the central diffraction envelope.

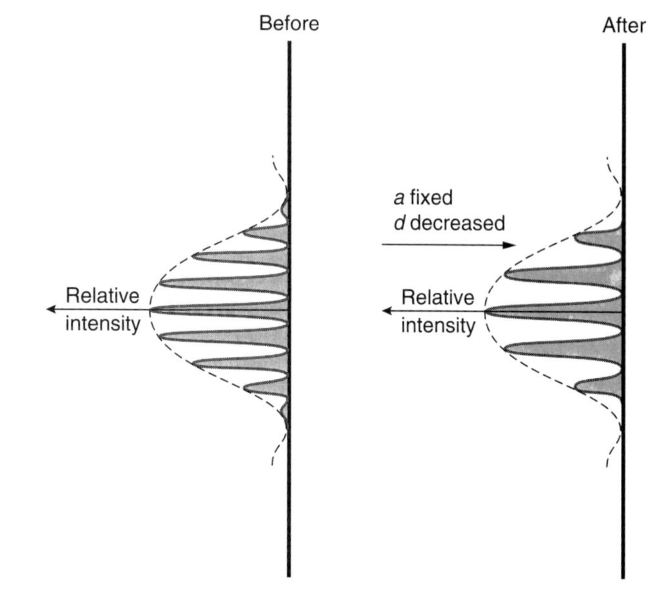

FIGURE 24.25 Decrease the slit separation with fixed slit width; the interference peaks get further apart and the diffraction envelope remains the same.

order. This results in fewer interference peaks under the same size central diffraction envelope, as shown in Figure 24.25.

EXAMPLE 24.5

Two slits, each of width 0.20 mm, are separated by 0.70 mm and illuminated with light of wavelength 633 nm in a Young's double slit experiment. How many bright interference fringes appear under the central diffraction envelope?

Solution

The interference fringes are located using Equation 24.2:

$$d \sin \theta = m\lambda$$

where d is the slit separation. The first minimum of the central diffraction peak is found from Equation 24.3:

$$a \sin \theta_1 = \lambda$$

where a is the slit width. You can find the order number that appears at $\theta = \theta_1$ by substituting θ_1 into Equation 24.2:

$$d \sin \theta_1 = m\lambda$$

Now dividing by Equation 24.3, you find

$$\frac{d}{a} = m$$

With the given values for d and a, this is

$$m = \frac{0.70 \text{ mm}}{0.20 \text{ mm}} = 3.5$$

Since the order number must be an integer, this result means that the seven bright fringes corresponding to $m = 0$,

± 1, ± 2, and ± 3 are within the angle $\pm \theta_1$, while those with higher absolute values of m are not. Note that the result is independent of λ. Why?

24.8 Multiple Slits: The Diffraction Grating

A single slit produces the single slit diffraction pattern of Figure 24.10. The double slit produces the double slit interference pattern of Figure 24.23, in which the diffraction pattern of a single slit manifests itself as the modulation of the intensity of the interference fringes. What happens to the interference and diffraction patterns if many more additional identical, equally spaced slits are presented to the incident light? Such a multiple arrangement of N slits is a **diffraction grating**.

Since the N slits each are of width a, the diffraction patterns of each slit, taken individually, are identical in shape (but very slightly separated in space by the slit separation). If the slit width a is very small, the resulting diffraction patterns of the individual slits have large angular widths. The envelopes of all the individual single slit diffraction patterns are essentially coincident, since the total width Nd of the entire array of slits typically is not appreciable compared with the width of any single diffraction central maximum.

The locations of the interference maxima of a double slit are determined from Equation 24.2:

$$d \sin \theta = m\lambda$$

where the order of interference m is an integer (or 0). This equation means that for constructive interference the path difference $d \sin\theta$ from the two slits to a point on a distant screen is an integral number of wavelengths ($= m\lambda$). The addition of more identical slits with the same slit separation d does not change this argument: the path difference between adjacent slits still is $d \sin \theta$ as indicated in Figure 24.26, and the resulting intensity is increased.

When the path difference between adjacent slits is an integral number of wavelengths, $m\lambda$, constructive interference again occurs at the same locations as with a double slit with the same slit separation.

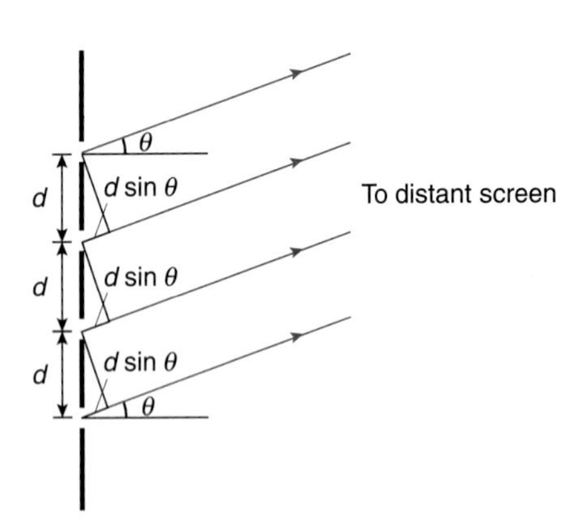

Figure 24.26 The path difference $d \sin \theta$ between adjacent slits of a diffraction grating is the same.

The locations of the interference maxima for a multiple slit diffraction grating thus are found using

$$d \sin \theta = m\lambda \qquad (24.8)$$
(diffraction grating maxima)

which is known as the **grating equation**. The slit separation d is called the **grating spacing**.

Does *anything* different happen in going from a double slit to a multiple slit array of identical slits with the same slit separation? Indeed.

As the number of slits increases, the angular widths of the interference maxima decrease. In other words, the places where the interference is constructive become more sharply defined, as shown in Figure 24.27.

The maxima occur for values of θ determined from the grating equation, Equation 24.8. Even small variations from these specific values of θ cause the waves from the various slits to interfere in essentially a completely destructive manner. Why is this? With a large number of slits, as θ varies even slightly from the places where the path difference between adjacent slits is an integral number of wavelengths, the wavelet from any given slit finds a wavelet from another slit that is out of phase with it and the two amplitudes cancel.

For example, if the path difference between *adjacent* slits is, say, 1.1 wavelengths, then two slits spaced 5 slits apart will have waves that have a path difference of 0.5 wavelength and so are out of phase with each other. So slits that are $5d$ apart interfere destructively; we get a pairwise cancellation of waves on the screen from slits separated by five slit separations. Only the few slits left over from all this pairwise cancellation contribute a small amount of light to the region between the interference

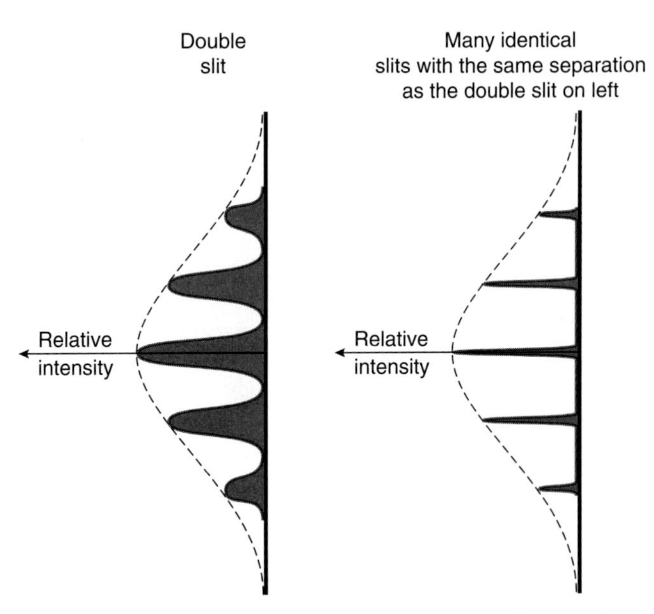

Figure 24.27 Increasing the number of slits with the same slit separation sharpens the interference maxima.

maxima. Very narrow and bright fringes remain for the orders of any given wavelength.

We can use the technique we first employed for locating the first minimum of a single slit diffraction pattern to determine the angular width of the central *interference maximum* for a diffraction grating (the maximum corresponding to order number $m = 0$). To find the minima of the single slit diffraction pattern, we broke the slit into many coherent wave sources (using Huygens's principle). We then examined the interference of the myriad of coherent sources. We discovered that the location of the first minimum is found from Equation 24.3:

$$a \sin \theta_1 = \lambda$$

where a is the width of the collection of Huygens's sources (the width of the slit).

A diffraction grating with N slits illuminated also consists of a myriad (actually N) of coherent wave sources as well. Thus the location of the first minimum of the central *interference maximum* of these N coherent sources is found by replacing the slit width a in Equation 24.3 with the width of the *illuminated* grating: Nd. Thus the location θ_{min} of the first minimum of the central interference fringe is

$$Nd \sin \theta_{min} = \lambda$$

Since the angle θ_{min} is small, we use the small angle approximation to write this as

$$Nd\theta_{min} = \lambda$$

or

$$\theta_{min} = \frac{\lambda}{Nd} \tag{24.9}$$

The angular width of the interference maximum is $2\theta_{min}$. Equation 24.9 shows explicitly that as the number of slits increases greatly, the angular width of the interference maximum decreases drastically and the maxima become very narrow ("sharp"). Note that Equation 24.9 for a single slit ($N = 1$) gives the expected result (where d then is equal to a).

Although we derived Equation 24.9 for the central interference fringe, the result is applicable to *any* of the m interference maxima of the diffraction grating. We just use the projected width of the illuminated grating, $Nd \cos \theta$ (shown in Figure 24.28), as the effective width of the aperture and obtain the line width about the mth maximum:

$$\theta_{min} = \frac{\lambda}{Nd \cos \theta} \tag{24.10}$$

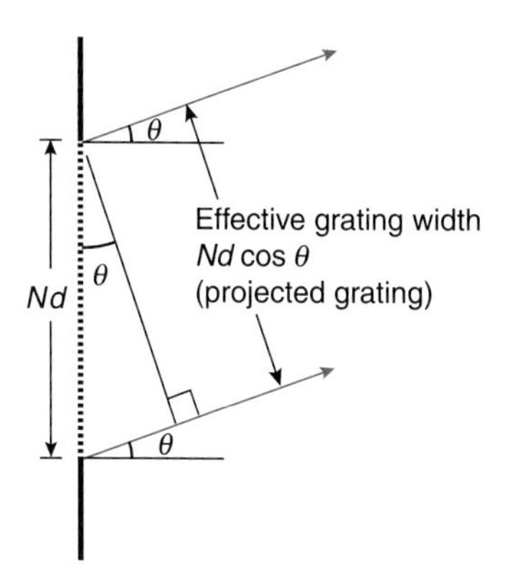

FIGURE 24.28 The projected width of a grating.

The quantity N is the number of slits of a grating that are *illuminated* by the incident light. N is different from the total number of slits in the grating if only a portion of the total number of slits actually are illuminated.

EXAMPLE 24.6

Light waves from a sodium vapor lamp (like the common yellow street light) are incident along the normal to a diffraction grating with 12.0 × 103 slits per centimeter. The first-order interference fringe is found to be located at an angle of 45.0° from the incident, straight-through direction. Determine the wavelength of the light.

Solution

You need to know the grating spacing d. There are 12.0×10^3 slits per centimeter, or 1.20×10^6 slits per meter. The grating spacing is the distance d between slits, which is the *inverse* of the number per meter. Hence

$$d = 8.33 \times 10^{-7} \text{ m}$$

Use the grating equation (Equation 24.8):

$$d \sin \theta = m\lambda$$

Solving for λ, you obtain

$$\lambda = \frac{d \sin \theta}{m}$$

The order number m is $m = 1$. Making the appropriate substitutions, you find

$$\lambda = \frac{(8.33 \times 10^{-7} \text{ m}) \sin(45.0°)}{1} = 5.89 \times 10^{-7} \text{ m} = 589 \text{ nm}$$

24.9 RESOLUTION AND ANGULAR DISPERSION OF A DIFFRACTION GRATING

Illuminate a diffraction grating with light from a source of wavelength λ_1. The interference maxima are located at angles determined by the grating equation, Equation 24.8:

$$d \sin \theta_1 = m\lambda_1 \tag{24.11}$$

(Here the angle θ_1 is the angle at which wavelength λ_1 has order number m, *not* the angular position of the first minimum of the diffraction peak of a single slit.) If light of *another* wavelength λ_2

is incident simultaneously on the diffraction grating, its interference maxima are found from

$$d \sin \theta_2 = m\lambda_2 \tag{24.12}$$

Thus an interference pattern of bright fringes is established for *each* wavelength present in the incident light, as shown in Figure 24.29.

Examples of such multiple-wavelength light sources include gas discharge lamps such as sodium vapor lamps (common yellowish street lights), mercury vapor lamps (common bluish street lights), hydrogen vapor lamps, and neon signs. The atoms in each source act independently (incoherently), but each light wave from an individual atom forms multiple coherent sources at the grating by wavefront division.

We want to determine how close the wavelengths λ_1 and λ_2 can be and still reveal the two interference patterns formed by the grating as separate. In other words, if the difference between the wavelengths $\Delta\lambda = \lambda_2 - \lambda_1$ is too small, the two interference patterns are essentially superimposed on each other and we cannot infer the presence of two different wavelengths. We are interested in determining the *resolution* of the diffraction grating: its ability to distinguish between two incident wavelengths.

Both wavelengths produce an interference fringe at $\theta = 0$ rad corresponding to order number $m = 0$. So we need to consider order numbers m whose absolute values are greater than zero to have any hope of distinguishing between the interference patterns. For the same order number m, the interference fringes for each wavelength are found at

$$d \sin \theta_1 = m\lambda_1 \quad \text{and} \quad d \sin \theta_2 = m\lambda_2$$

To see how changing λ changes θ, we use the general grating equation (Equation 24.8)

$$d \sin \theta = m\lambda$$

and take its derivative with respect to λ (remembering that the grating spacing d is constant for a given grating; the order number m also is constant, since we are considering the *same* order interference maxima for both wavelengths):

$$(d \cos \theta)\frac{d\theta}{d\lambda} = m$$

$$\frac{d\theta}{d\lambda} = \frac{m}{d \cos \theta} \tag{24.13}$$

This expression enables us to find the small angular separation $d\theta$ of two closely spaced wavelengths (the difference between the wavelengths is $d\lambda$). Equation 24.13 is the way in which θ changes with wavelength λ, and is known as the **angular dispersion** of the grating. Notice that the angular dispersion

- is directly proportional to the order number; the angular separation of two wavelengths increases as the interference order number increases; and
- is inversely proportional to the grating spacing d; the smaller the slit separation d, the greater the angular separation of the wavelengths.

If the angles are small, the cosine is near unity, so the angular dispersion then is approximately

$$\frac{d\theta}{d\lambda} \approx \frac{m}{d} \tag{24.14}$$

We say the two interference patterns of two different but nearly equal wavelengths are barely resolved if the interference maximum for one wavelength falls on the first interference minimum of the other wavelength: this is analogous to the Rayleigh criterion invoked in considering resolution for single aperture diffraction patterns. This resolution criterion means that the small angular difference $d\theta$ in the locations of the interference maxima is the angle θ_{min} in Equation 24.9:

$$\theta_{min} = \frac{\lambda}{Nd}$$

Making this substitution for $d\theta$ in Equation 24.14 and writing $\Delta\lambda$ for the small difference in wavelength, we find

$$\frac{\lambda}{Nd} \approx \frac{m}{d}\Delta\lambda \tag{24.15}$$

We can use either λ_1 or λ_2 for the wavelength in this equation since the two wavelengths are virtually the same. Rearranging Equation 24.15 slightly, we find

$$\frac{\lambda}{\Delta\lambda} = Nm \tag{24.16}$$

The expression $\lambda/\Delta\lambda$ is known as the **resolving power** of the diffraction grating. A grating with a large resolving power can discern between two wavelengths that are not very different from each other.

Although we derived Equation 24.16 for the resolving power of a grating making use of the small angle approximation, Equation 24.16 is legitimate even for large angles (although we have not proved this explicitly).

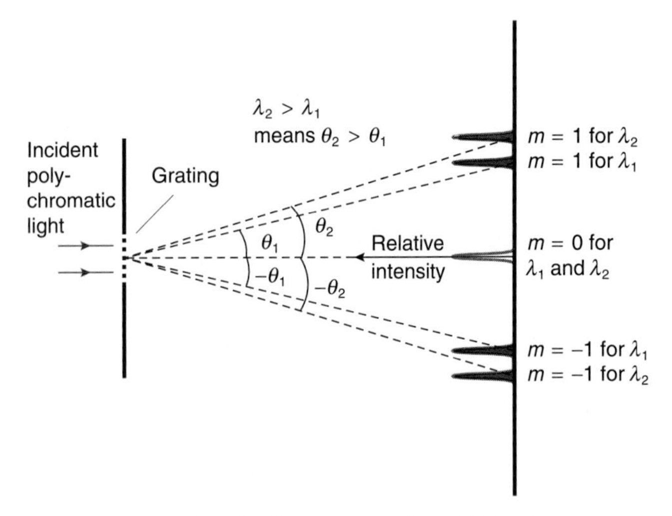

FIGURE 24.29 **Each incident wavelength establishes its own interference pattern.**

Notice that the resolving power of a grating depends on two factors:

1. The order number m. The resolution increases for larger order numbers.
2. The number of slits N illuminated in the grating. The greater N, the greater the resolution. This is the principal justification for increasing the number of illuminated slits in a grating.

Diffraction gratings are extraordinarily useful instruments. If light with a collection of many different wavelengths is incident on a diffraction grating, it separates the light into its component wavelengths according to the grating equation, Equation 24.8. Each wavelength appears at a different angle θ for a given order number m. The collection of wavelengths present in a source is called its spectrum.* Each chemical element has its own unique spectrum. Thus the determination of the wavelengths present in a source (known as spectroscopy) provides information about the chemical nature of the source, a technique as useful in astronomy as in forensics. A prism also can be used to separate or distinguish the various wavelengths present in a light source (thanks to dispersion—the variation of index of refraction with wavelength), but diffraction gratings are preferred because they are much easier to use, have greater angular dispersion, and have better resolution.

Spectroscopy is an important tool of many fields of physics and astronomy. Spectroscopy is the primary tool of astronomy, for very little comes to us from distant stars and galaxies other than light. Indeed it was spectroscopy that penetrated one of the "impossible problems" of the 19th century: the chemical composition of the stars (cf. the quote by Auguste Compte opening this chapter).

The first precision diffraction gratings were manufactured in the latter part of the 19th century by Henry Rowland at The Johns Hopkins University. In the best spirit and interest of science, he provided them at cost to many research laboratories around the world, likely forgoing a fortune in royalties for the invention. His gratings are as valuable today as then because they have no electronic or moving parts—quite a legacy!

STRATEGIC EXAMPLE 24.7

A grating with 2000 slits spaced equally over a 1.000 cm distance is used to analyze the spectrum of the element mercury. Among other wavelengths, mercury emits light at 576.959 nm and 579.065 nm. What is the angular separation of the two wavelengths in the second-order spectrum?

Solution

Since there are 2000 slits in each centimeter of the grating, the distance d between the slits of the grating (the grating spacing) is

$$\begin{aligned} d &= \frac{1.000 \text{ cm}}{2000} \\ &= 5.000 \times 10^{-4} \text{ cm} \\ &= 5.000 \times 10^{-6} \text{ m} \end{aligned}$$

*The word comes from the Latin *spectrum*, meaning "an appearance."

The second-order spectrum corresponds to order number $m = 2$. If $m = -2$ we get similar results.

Method 1

Use the grating equation (Equation 24.8)

$$d \sin \theta = m\lambda$$

to find the angles at which each wavelength appears in the second order:

$$\sin \theta = \frac{m\lambda}{d}$$

A subtraction of the resulting angles then yields the angular separation of the two wavelengths.

Thus the angle at which the 576.959 nm wavelength appears is

$$\begin{aligned} \sin \theta_{576.959 \text{ nm}} &= \frac{2(576.959 \times 10^{-9} \text{ m})}{5.000 \times 10^{-6} \text{ m}} \\ &= 0.2308 \end{aligned}$$

$$\theta_{576.959 \text{ nm}} = 13.34°$$

The 579.065 nm wavelength appears at

$$\begin{aligned} \sin \theta_{579.065 \text{ nm}} &= \frac{2(579.065 \times 10^{-9} \text{ m})}{5.000 \times 10^{-6} \text{ m}} \\ &= 0.2316 \end{aligned}$$

$$\theta_{579.065 \text{ nm}} = 13.39°$$

The angular separation of the two wavelengths in the second order thus is

$$\theta_{579.065 \text{ nm}} - \theta_{576.959 \text{ nm}} = 13.39° - 13.34° = 0.05°$$

Note that to use this method, you need to know many significant figures for the wavelengths and grating spacing to be able to perform the subtraction. Method 2 is a better way.

Method 2

Since the two wavelengths are very closely spaced you can *differentiate* the grating equation (Equation 24.8) with respect to λ to see how the angle θ varies with wavelength. This is the angular dispersion, Equation 24.13; slightly rearranged, you have

$$d\theta = \frac{m}{d \cos \theta} d\lambda \tag{1}$$

Let the small angle and wavelength differences be $\Delta\theta$ and $\Delta\lambda$. You found the slit separation $d = 5.000 \times 10^{-6}$ m. The order number is $m = 2$. The difference between the wavelengths $\Delta\lambda$ is

$$\begin{aligned} \Delta\lambda &= 579.065 \text{ nm} - 576.959 \text{ nm} \\ &= 2.106 \text{ nm} \\ &= 2.106 \times 10^{-9} \text{ m} \end{aligned}$$

For the angle θ you can use the angle at which *either* wavelength appears:

Using $\theta_{576.959\ \text{nm}} = 13.34°$ in equation (1), you have

$$\Delta\theta = \frac{2(2.106 \times 10^{-9}\ \text{m})}{(5.000 \times 10^{-6}\ \text{m})\cos(13.34°)} = 8.658 \times 10^{-4}\ \text{rad} = 0.049\,61°$$

Using $\theta_{579.065\ \text{nm}} = 13.39°$ in equation (1), you have

$$\Delta\theta = \frac{2(2.106 \times 10^{-9}\ \text{m})}{(5.000 \times 10^{-6}\ \text{m})\cos(13.39°)} = 8.659 \times 10^{-4}\ \text{rad} = 0.049\,61°$$

The angular separation is 0.049 61°. Notice that you secure a result to four significant figures using the angular dispersion equation, whereas with the grating equation you only secured a result to one significant figure. Notice also that it makes little difference which wavelength is used in the equation for the angular dispersion (as long as the wavelengths are nearly the same).

EXAMPLE 24.8

Among other wavelengths, a sodium vapor lamp emits two slightly different wavelengths: 588.995 nm and 589.592 nm. It is these wavelengths that give sodium vapor lamps their predominantly yellowish pall. How many slits are needed to barely resolve these wavelengths in the first-order spectrum?

Solution

Equation 24.16 determines the resolution:

$$\frac{\lambda}{\Delta\lambda} = Nm$$

The wavelength difference is

$$\Delta\lambda = 589.592\ \text{nm} - 588.995\ \text{nm} = 0.597\ \text{nm}$$

You can use either wavelength for λ in Equation 24.16 since their difference is small. The wavelengths are to be resolved in the first order, so that $m = 1$. Substituting into Equation 24.16, you find

$$\frac{589\ \text{nm}}{0.597\ \text{nm}} = N \times 1$$

$$N = 987$$

A grating with a minimum of about 1000 illuminated slits is needed.

24.10 THE INDEX OF REFRACTION AND THE SPEED OF LIGHT

The law of refraction (and the law of reflection), dear to us from geometric optics in Chapter 23, can be derived from the Maxwell equations of electromagnetism. From this derivation (which we leave for a more advanced course on electromagnetism) comes another meaning for the index of refraction of a medium.

The index of refraction is the ratio of the speed of light in vacuum to its speed in the transparent medium:

$$n = \frac{\text{speed of light in vacuum}}{\text{speed of light in the medium}} = \frac{c}{v} \tag{24.17}$$

We discussed in Section 21.5 how the speed of light in vacuum is related to the permittivity of free space ε_0 and permeability of free space μ_0, the electrical and magnetic properties of free space (a vacuum). The speed of light in a material also is related to the electrical and magnetic properties of the medium: its permittivity ε and its permeability μ. The reason the electrical and magnetic properties of the materials come into play is that the electric and magnetic fields of the light wave in the two media must each be properly matched on both sides of the interface between the media. This means that the oscillations of the fields of the electromagnetic wave must be at the same frequency in each material. If this were not true, we would find intolerable numbers of oscillations piling up on one side, waiting to be transmitted.

Therefore the frequency of the light is the same in both media.

It is from this matching of the fields that the law of refraction arises.

Since the index of refraction of visible light in transparent materials is a number greater than 1 (see Table 23.1), Equation 24.17 implies that the speed of light waves in a material medium is *less* than that in a vacuum. The reason has to do with the process of light transmission through such materials. That process involves the absorption and subsequent emission of light photons (energy bundles or particle-waves of light) by atoms. The photons of light travel between the atoms at the vacuum speed of light, but the absorption and emission processes each take a bit of time. The result is that the effective speed of the light through the material is less than if the material were not present. The process is similar to having annoying toll booths spaced every 10 km along a superhighway; they slow down the effective speed of the traffic.

The product of the frequency and the wavelength is equal to the speed of the wave. Thus for light in a vacuum we have

$$c = \nu\lambda \tag{24.18}$$

where ν is the frequency of the light and λ is its wavelength in vacuum. In a medium where the speed of light is v_1, we have a similar relation:

$$v_1 = \nu\lambda_1 \qquad (24.19)$$

where λ_1 is the wavelength of the light in the medium. The frequency of the light in the medium is the *same* as in the vacuum. Dividing Equation 24.18 by 24.19, we obtain

$$\frac{c}{v_1} = \frac{\lambda}{\lambda_1}$$

The left-hand side of this equation is the index of refraction n_1 of the medium (from Equation 24.17). Thus

$$n_1 = \frac{\lambda}{\lambda_1}$$

After slight rearrangement, this becomes

$$\lambda_1 = \frac{\lambda}{n_1} \qquad (24.20)$$

Experiments indicate that the index of refraction of visible light in transparent materials is greater than 1 (see Table 23.1).

Hence Equation 24.20 indicates that the wavelength of such light in a material medium is *shorter* (smaller) than the wavelength in vacuum.

EXAMPLE 24.9

The index of refraction of water is about 1.33 (from Table 23.1). Determine the speed of light in water.

Solution
Equation 24.17 relates the speed of light in a material to the index of refraction:

$$n = \frac{c}{v}$$

So

$$v = \frac{c}{n} = \frac{3.00 \times 10^8 \text{ m/s}}{1.33} = 2.26 \times 10^8 \text{ m/s}$$

This speed still is *quite* fast!

EXAMPLE 24.10

Light with a vacuum wavelength of 632.8 nm enters glass with an index of refraction of 1.50. What is the wavelength of the light in glass?

Solution
The wavelength of light when in a material medium is shorter than that in vacuum by a factor equal to the index of refraction, according to Equation 24.20:

$$\lambda_{\text{glass}} = \frac{\lambda_{\text{vacuum}}}{n_{\text{glass}}} = \frac{632.8 \text{ nm}}{1.50} = 422 \text{ nm}$$

24.11 THIN-FILM INTERFERENCE*

No light, but rather darkness visible

John Milton (1608–1674)†

You may have noticed that fine lenses for cameras, microscopes, and telescopes have transparent coatings on their optical surfaces. What purpose do the coatings serve, or are they purely decorative and protective? Quite apart from this, perhaps while taking a relaxing bubble bath, you have noted that the bubbles have interesting rainbow-like, swirled, colored bands. Less romantically, when filling up at a gas station on a rainy day, you may have noted the delicate rainbow-like colored bands on puddles, a clear indication of some gasoline polluting the water surface. Here we shall see that these apparently disparate observations all are manifestations of **thin-film interference** of light.

Aside from a path difference, there is another way to secure a phase difference. When a wave is reflected, a *phase change* of π rad may, or may not, occur for the reflected wave, as we saw in Section 12.10. Under what circumstances do electromagnetic waves experience such a phase change of π rad when reflected? Experiments indicate two circumstances lead to such phase changes:

a. Reflection from a conducting surface (a metal). This is analogous to the rope wave reflection off a rigid wall, since the conductor cannot sustain an electric field.
b. Reflection from an optically more dense material ($n_2 > n_1$). This situation is like a rope pulse reflecting from a section of

†*Paradise Lost*, Book I, line 63.

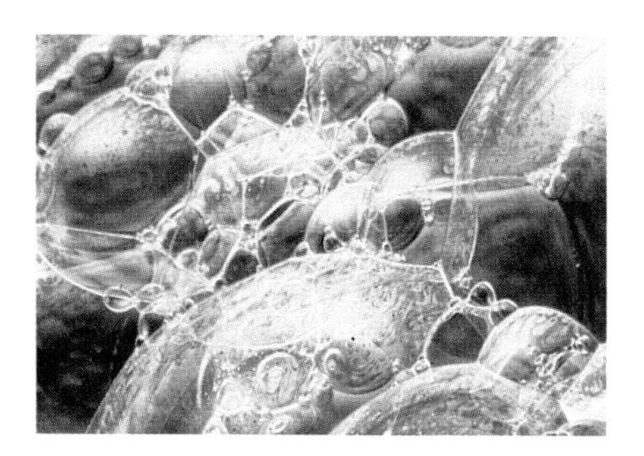

Thin-film interference is apparent in soap bubbles.

rope with a greater mass per unit length than that of the incident pulse. If $n_2 < n_1$, no phase change occurs in the reflected wave. We shall see in the next subsection how these conclusions are sustained by experimental observations.

When light passes the boundary between two transparent media, some light is reflected at the boundary. The amount of light reflected depends on a number of factors, among them the indices of refraction of the two media and the angle of incidence. For light incident along the normal line from air to glass, about 4% of the incident light is reflected back into the air at the boundary; this increases to almost 100% reflection for angles of incidence approaching 90°. At such angles, the glass acts much like a mirror surface.* Here we examine only light incident normally on the boundaries between transparent materials.

We consider, in particular, monochromatic light incident on a thin film of material separating two other transparent media, as shown in Figure 24.30. Some light is reflected at the first interface and some from the second interface.

We are interested in the interference of the two reflected waves when they superimpose in the first material along the normal line. The two reflected waves are monochromatic and also coherent because they arise from the same monochromatic incident light wave via *amplitude division*. Once the wave reflected from the second surface gets back to the wave reflected from the first surface, the two waves interfere, since they are superimposed along the same normal line.

To determine the nature of the interference of the reflected waves, we need to consider *two* factors that contribute to their phase difference: a path difference and phase changes upon reflection.

We consider each in turn.

*Most surfaces reflect most light when the angle of incidence approaches 90°. You can observe this with even a sheet of paper. If you hold the paper up before a light bulb so the light is incident at a large angle to the normal, you will see that much of the light from the bulb is reflected from the paper surface.

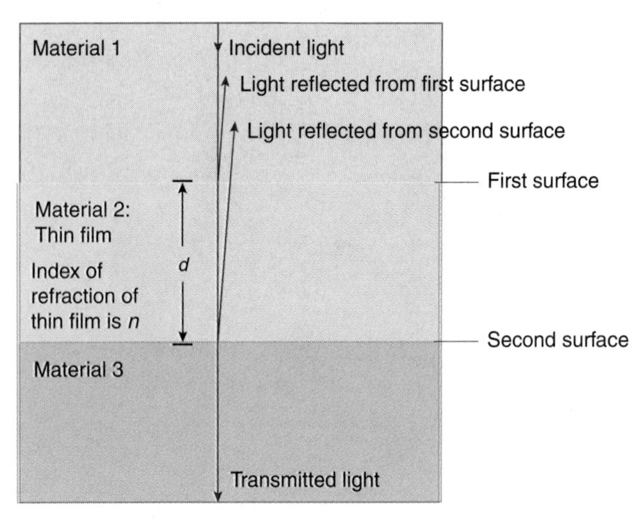

FIGURE 24.30 Light incident along the normal to a thin film separating two other transparent media.

Path Difference

The wave reflected from the lower interface travels an additional distance $2d$ before it is superimposed and interferes with the wave reflected from the top surface, where d is the thickness of the film. Hence the path difference Δx between the two reflected rays is $2d$. From Equation 24.1, this path difference corresponds to a phase difference of

$$\delta_{\text{path}} = \frac{2d}{\lambda_{\text{film}}}(2\pi \text{ rad}) \qquad (24.21)$$

PROBLEM-SOLVING TACTIC

24.3 When investigating thin-film interference, note that the wavelength of the light in the film is *different* from that in vacuum. It is the wavelength of the light in the film that is involved in assessing the phase difference associated with the path difference. So be careful to use the wavelength of the light in the film in Equation 24.21. In particular, we saw from Equation 24.20 that

$$\lambda_{\text{film}} = \frac{\lambda_{\text{vacuum}}}{n}$$

Phase Changes upon Reflection

We also need to assess whether there are any phase changes involved in the two reflections. *This assessment must be done on a case-by-case basis*, taking into consideration the specific indices of refraction of the three media involved. Three situations may arise; the first two are the following:

1. *Neither* reflected wave experiences a phase change upon reflection.
2. *Both* reflected waves experience a phase change upon reflection.

Because the *same* phase shift has occurred to *both* reflected waves in either of these two scenarios, the phase changes upon reflection thus are irrelevant in assessing the status of the interference of the two reflected rays. No *difference* in phase results between the two reflected waves from this cause in these instances. For either of these cases, the interference of the two waves is determined *solely* from the path difference between the two reflected waves.

In particular, if the thickness of the film is such that the phase difference arising from the path difference is a multiple of 2π rad, the interference of the two waves is completely constructive and the reflected light is bright:

$$\frac{2d}{\lambda_{\text{film}}}(2\pi \text{ rad}) = m(2\pi \text{ rad}) \qquad m = \text{integer}$$
$$\text{(constructive interference)} \qquad (24.22)$$

On the other hand, if the thickness of the film is such that the resulting phase difference is an *odd multiple* of π rad, the resulting interference of the two reflected rays is completely destructive. If m is an integer, then $2m + 1$ is an odd integer. Hence the condition for destructive interference is

$$\frac{2d}{\lambda_{\text{film}}}(2\pi \text{ rad}) = (2m + 1)(\pi \text{ rad}) \qquad m = \text{integer}$$
$$\text{(destructive interference)} \qquad (24.23)$$

The third scenario is the following:

3. One of the reflected waves experiences a phase change of π rad upon reflection and the other wave does not.

It makes no difference which one experiences the phase change; the important thing is that one reflected wave does while the other one does not. If this is the case, then the previous conclusions are reversed. That is, the condition for destructive interference is

$$\frac{2d}{\lambda_{\text{film}}}(2\pi \text{ rad}) = m(2\pi \text{ rad}) \qquad m = \text{integer}$$
$$\text{(destructive interference)} \qquad (24.24)$$

and the condition for constructive interference is

$$\frac{2d}{\lambda_{\text{film}}}(2\pi \text{ rad}) = (2m + 1)(\pi \text{ rad}) \qquad m = \text{integer}$$
$$\text{(constructive interference)} \qquad (24.25)$$

In other words, if the path difference implies a phase difference between the waves of a multiple of 2π rad, initially leading us to think the waves are in phase and yielding constructive interference, the phase change upon reflection of one but not the other wave means the two will be out of phase and the interference actually is destructive. Conversely, if the path difference implies a phase change of an odd multiple of π rad, initially leading us to think the waves are out of phase and interfering destructively, the phase change upon reflection of one wave but not the other means the reflected waves actually are in phase and will interfere constructively.

Combining all the scenarios for phase difference, we write the total phase difference as

$$\delta = \delta_{\text{path}} + \delta_{\text{bdy}} \qquad (24.26)$$
$$= \begin{cases} \dfrac{\Delta x}{\lambda_{\text{film}}}(2\pi \text{ rad}) + 0 \text{ rad} & \text{(for similar reflections)} \\ \dfrac{\Delta x}{\lambda_{\text{film}}}(2\pi \text{ rad}) + \pi \text{ rad} & \text{(for different reflections)} \end{cases}$$

PROBLEM-SOLVING TACTIC

24.4 There is no magic bullet formula to memorize that can treat all situations of thin-film interference. Each situation must be assessed individually to determine (a) the path difference and its resulting phase difference and (b) the phase changes (if any) upon reflection and whether they are relevant or not. The two pieces of information then must be interwoven to determine if the superposition of the reflected waves is constructive or destructive.

Thin-film *destructive* interference is the principle behind coating optical surfaces such as lenses to *minimize* reflections; more light then is transmitted by the lens. Since an uncoated air–glass interface reflects approximately 4% of the light incident along the normal, it does not take many such surfaces to significantly attenuate the incident light, particularly in optical systems consisting of many lenses. However, if each lens surface is coated with an appropriate thickness of a transparent material (such as magnesium fluoride [MgF_2]), the reflected light can be minimized with a resulting gain in the transmission of the lens system. All fine optical lens systems are so coated.

Thin-film *constructive* interference in the reflected light (at various wavelengths) also explains the colored bands seen when gasoline is on water. Each colored band indicates constructive interference in the light reflected from the air–gasoline and gasoline–water interfaces for that color. Each colored band thus traces out a film of constant thickness of gasoline. It is similar for soap bubbles.

If we place a slightly convex glass surface on a flat glass surface as shown in Figure 24.31, and view the reflected light from the glass–air and the air–glass interfaces, we observe a dark spot at the point of contact. At the point of contact, the rays reflected from the glass–air and air–glass interfaces have zero path difference. The interference nonetheless is seen to be destructive. Since the path difference is zero, the two waves are out of phase because one reflected wave experienced no phase change upon reflection (the wave from the glass-to-air reflection) while the other reflected wave had a phase change of π rad (the one from the air-to-glass reflection). This confirms the conclusions we drew earlier about phase changes upon reflection.

Note that the pattern of the constructive and destructive interference in the reflected light consists of a series of concentric

A thin film coating on a lens minimizes reflections and enhances transmission.

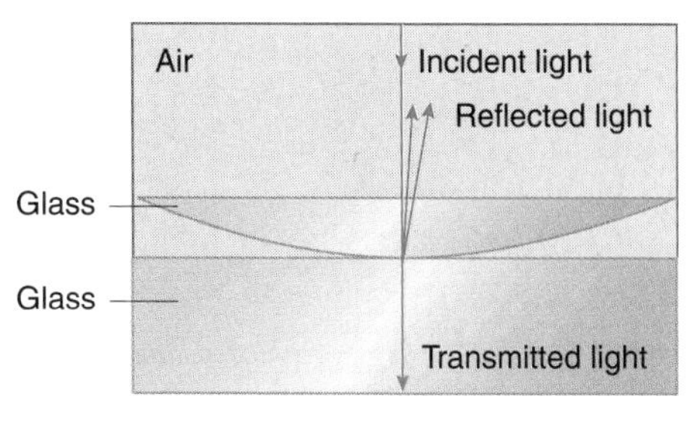

FIGURE 24.31 **Newton's rings in reflected light produce a dark spot at the point of contact.**

circles, called **Newton's rings**, that trace out contours of constant thickness of the air gap between the surfaces.

If we view the transmitted light, the central spot is bright, as shown in Figure 24.32. There is no path difference between the two transmitted waves, and no phase difference from reflection since the transmitted light is not reflected. The transmitted light also consists of a series of concentric bright and dark rings.

Thin-film interference is used extensively to test the shape of various optical surfaces. The constructive and destructive interference conditions map contours of constant thickness. A convex surface on a flat surface produces concentric rings of constructive and destructive interference fringes centered on the point of contact. If the surface is not flat, or the convex surface is not symmetric about its optic axis, the rings are distorted in shape. See Figure 24.33.

If two optical flats are inclined slightly to each other to form an air wedge between them, the contours of constant thickness should be straight lines; any apparent deviations of the interference fringes from this geometry indicate that one or the other flat (or both) really is not flat.

FIGURE 24.32 Newton's rings in transmitted light produce a bright spot in the center.

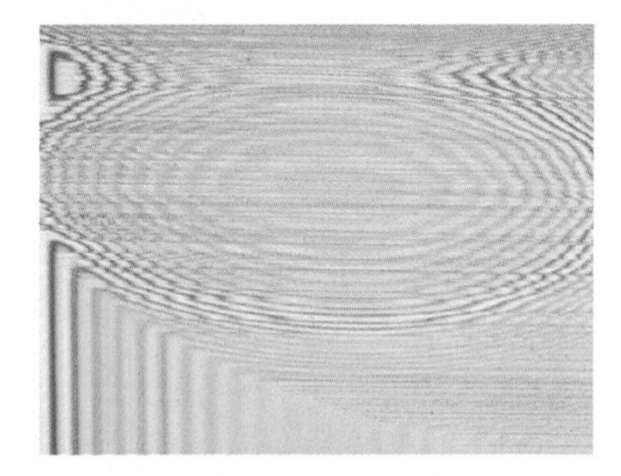

FIGURE 24.33 Distortions in fringe shapes reveal deviations from spherical or flat geometries of the surfaces.

QUESTION 4

If both surfaces of a lens with index of refraction $n > 1$ in air are coated with a material of index n' to minimize reflections, are the coatings of the same thickness? Explain.

EXAMPLE 24.10

Light with a vacuum wavelength 633 nm is incident normally on a thin-film coating with an index of refraction $n = 1.35$ over glass, as in Figure 24.34. What is the minimum thickness d for the film needed to minimize reflected light at normal incidence?

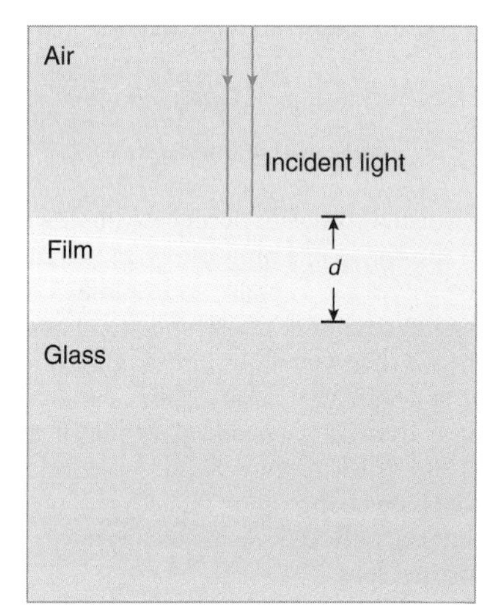

FIGURE 24.34

Solution

Since the index of refraction of the film is greater than that of the air, the wave reflected from the air-to-film interface experiences a change of phase of π rad upon reflection. Likewise, the wave reflected from the second interface (film-to-glass) also experiences a phase change of π rad upon reflection, because the index of refraction of the glass ($n \approx 1.5$) is greater than that of the film. Since *both* reflected rays experience a phase change of π rad upon reflection, the phase changes upon reflection are *irrelevant* for determining the total phase difference between the two reflected waves.

The path difference between the two reflected rays is, therefore, the sole factor that determines the character of the interference. The path difference is $2d$. Since you want to make the two reflected waves interfere destructively, the phase difference δ_{path} corresponding to the path difference $2d$ should be an odd integral multiple of π rad. To keep the thickness of the film to a minimum, the phase difference is made the first odd multiple of π, just π itself. Thus Equation 24.21 becomes

$$\delta_{\text{path}} = \pi \text{ rad} = \frac{2d}{\lambda_{\text{film}}}(2\pi \text{ rad})$$

Solving for d, you obtain

$$d = \frac{\lambda_{\text{film}}}{4}$$

The wavelength in the film is related to the vacuum wavelength by Equation 24.20:

$$\lambda_{\text{film}} = \frac{\lambda_{\text{vacuum}}}{n}$$

where n is the index of refraction of the film. Hence the minimum film thickness needed is

$$\begin{aligned} d &= \frac{\lambda_{\text{vacuum}}}{4n} \\ &= \frac{633 \text{ nm}}{4(1.35)} \\ &= 117 \text{ nm} \end{aligned}$$

Example 24.11

The edge of a thin sheet of tissue paper is placed between two optical flats to form an air wedge, as shown in Figure 24.35. Light of vacuum wavelength 589 nm is incident normally on the arrangement and 41 parallel, bright fringes of constructive interference are seen in the reflected light. A bright fringe is seen essentially coincident with the edge of the paper. Find the thickness of the sheet of paper.

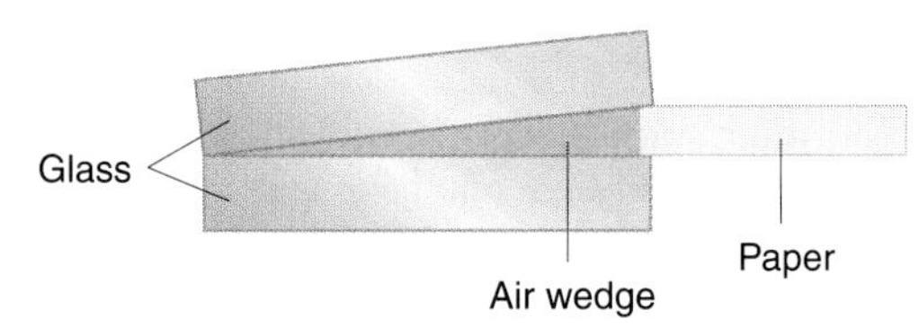

FIGURE 24.35

Solution

Here the interference occurs between the wave reflected from the top of the air wedge and the wave reflected from the bottom of the air wedge, as shown in Figure 24.36.

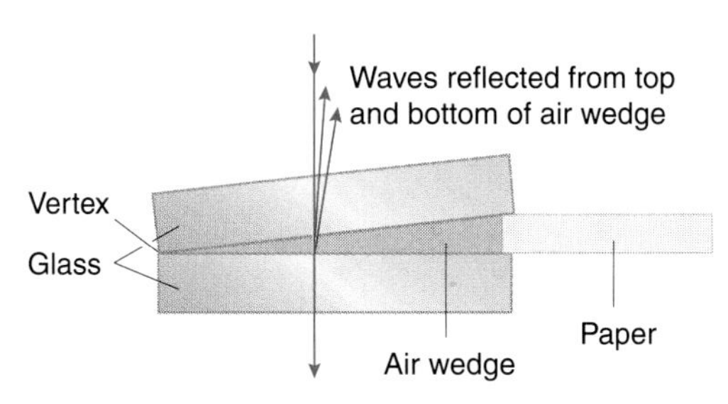

FIGURE 24.36

The top reflection (at the glass-to-air wedge interface) does not experience a phase change of π rad upon reflection, because the index of refraction of air (the second medium) is less than that of glass (the first medium). The bottom reflection (at the air wedge-to-glass interface) does experience a phase change upon reflection, since the index of refraction of the glass (now the second medium) is greater than that of the air (now the first medium). Thus you have a situation in which one of the reflected waves experiences a phase change upon reflection while the other does not.

At the vertex of the air wedge, the path difference between the waves is zero. But the two reflected waves here are nevertheless completely out of phase because of the phase change of one of the waves upon reflection. Hence the two reflected waves interfere completely destructively at this location and a dark fringe is seen at the wedge vertex. As you move out from the vertex, when the phase difference corresponding to the path difference $2d$ (twice the thickness d of the wedge at that location) becomes equal to an odd multiple of π rad, the two reflected waves will interfere constructively. Thus, from Equation 24.25, a bright interference fringe occurs whenever

$$\frac{2d}{\lambda_{\text{film}}}(2\pi \text{ rad}) = (2m + 1)(\pi \text{ rad})$$

where m is an integer. The first bright fringe from the vertex corresponds to $m = 0$, the second bright fringe corresponds to $m = 1$, and so forth. Thus the 41st bright fringe (where the paper is located) corresponds to $m = 40$. Thus the thickness d of the air wedge at the location of the paper must satisfy

$$\begin{aligned} \frac{2d}{\lambda_{\text{film}}}(2\pi \text{ rad}) &= [2(40) + 1](\pi \text{ rad}) \\ &= 81\pi \text{ rad} \end{aligned}$$

The thin film in this situation is the air wedge. Since the index of refraction of the air is essentially the same as the index of refraction of the vacuum, $\lambda_{\text{film}} = \lambda_{\text{vacuum}} = 589$ nm. Hence the thickness d is

$$\begin{aligned} d &= \frac{81}{4}\lambda_{\text{film}} \\ &= \frac{81}{4}(589 \text{ nm}) \\ &= 1.19 \times 10^4 \text{ nm} \\ &= 1.19 \times 10^{-5} \text{ m} \end{aligned}$$

24.12 Polarized Light*

There are two distinct types of waves: longitudinal and transverse. We first encountered them in Chapter 12. The distinction is made on the basis of the direction of the "jiggling" or oscillatory motion associated with the wave motion. If the jiggling of the waves is parallel and antiparallel to the direction of propagation of the wave, the wave is longitudinal. If the jiggling is perpendicular to the direction of propagation, the wave is transverse. Sound is a longitudinal wave in fluids but can have longitudinal and transverse characteristics in solids. We showed how the Maxwell equations indicate that light is a transverse electromagnetic wave in Section 21.5. Here we see how that can be determined experimentally.

We shall consider many different ways to produce what we shall term "polarized" light in subsequent sections of this chapter. Passing over the many natural processes for the moment, take your polarizing sunglasses and borrow a second pair from a friend in order to make a few observations and conclusions about the nature of light waves. Look at any light source through these polarizing samples held together. Rotate one about the axis of viewing, while keeping the second pair fixed. You see the intensity vary between bright and dark through every 90° of rotation.

That is all the experimenting; now ponder. If light waves were longitudinal (jiggling along the axis of travel, like sound waves in air), there is *no way* they could distinguish the transverse angle of rotation of the transmitter of intensity. But they obviously do. The light waves must have a sense of direction angle *around* the axis, and we now have proof they are *transverse* in character. Light is a transverse (and electromagnetic) wave in vacuum and in most media: the electric and magnetic field oscillations of a light wave are perpendicular to the direction of propagation, as we saw in Chapter 21.*

To describe the **polarization** of the wave is to geometrically describe what is happening to the orientation of the oscillation as a function of time. By convention, it is the oscillations of the electric field vector of a light wave (rather than the associated perpendicular magnetic field vector) that we choose to define the polarization of a light wave. The plane of the magnetic field vector oscillations is perpendicular to the plane of the electric field oscillations.

Only transverse waves can exhibit polarization.

We fix a position in space (i.e., we fix x) and follow the oscillation of the electric field vector associated with the light wave as a function of time. If the oscillations of the transverse wave at a particular location are confined to a line (see Figure 24.37), we say the wave is **plane polarized.**†

Other types of polarized light can be created by judicious superposition of plane polarized light. For example, consider two coherent, plane polarized light waves with the following characteristics:

- the waves are of equal amplitude;
- both waves are traveling along the $+x$-axis;
- each wave is plane polarized, but with the directions of the oscillations at right angles to each other; and
- the waves are $\pi/2$ rad out of phase.

We write the electric vector of one of the waves as

$$\vec{\mathbf{E}}_1 = E_0 \cos(kx - \omega t)\hat{\mathbf{j}}$$

The other wave is described by

$$\vec{\mathbf{E}}_2 = E_0 \cos\left(kx - \omega t + \frac{\pi}{2} \text{ rad}\right)\hat{\mathbf{k}} \qquad (24.27)$$

We use the trigonometric formula for the cosine of the sum of two angles to express the second wave as

$$\vec{\mathbf{E}}_2 = -E_0 \sin(kx - \omega t)\hat{\mathbf{k}}$$

The superposition of the two waves is

$$\vec{\mathbf{E}}_1 + \vec{\mathbf{E}}_2 = E_0[\cos(kx - \omega t)\hat{\mathbf{j}} - \sin(kx - \omega t)\hat{\mathbf{k}}]$$

We look at the origin ($x = 0$ m) and see what happens to the electric vector as a function of time:

$$\begin{aligned}\vec{\mathbf{E}}_1 + \vec{\mathbf{E}}_2 &= E_0[\cos(-\omega t)\hat{\mathbf{j}} - \sin(-\omega t)\hat{\mathbf{k}}]\\ &= E_0[\cos(\omega t)\hat{\mathbf{j}} + \sin(\omega t)\hat{\mathbf{k}}] \qquad (24.28)\end{aligned}$$

The resulting electric vector traces out a circle in the y–z plane at the angular frequency ω, schematically shown in Figure 24.38. With the thumb of your right hand pointing along the direction of propagation, the curled fingers of the right hand indicate the direction of the circular motion of the electric vector given by Equation 24.28. Such light is said to be **right circularly polarized**, since the right hand relates the direction of the circular motion to the direction of propagation.

By changing the sign of the $\pi/2$ rad phase difference in Equation 24.27 for $\vec{\mathbf{E}}_2$, the electric vector of the resulting superposition of the two waves executes circular motion in the opposite sense. Such light is called *left circularly polarized* light, since the fingers of the left hand indicate the sense of the motion of the electric vector if the thumb of the left hand is pointed along the direction of propagation of the waves. If the amplitudes of the two waves are different but the other conditions (listed

*Light can assume both a transverse *and* longitudinal character under some special circumstances, such as when microwaves propagate down hollow conducting waveguides. We confine ourselves to purely transverse light waves.

†Plane polarized light also is known as linearly polarized light.

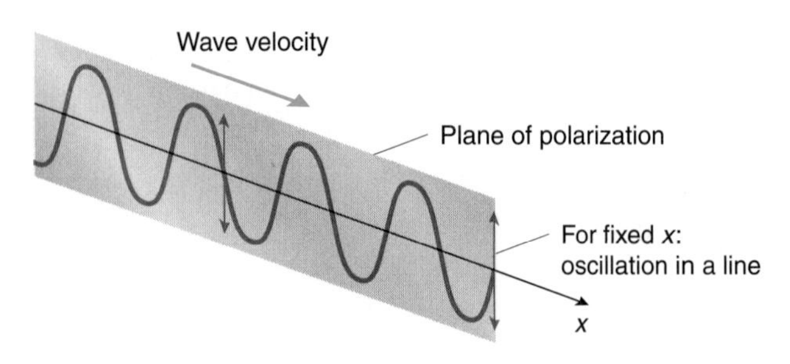

FIGURE 24.37 Plane polarized light.

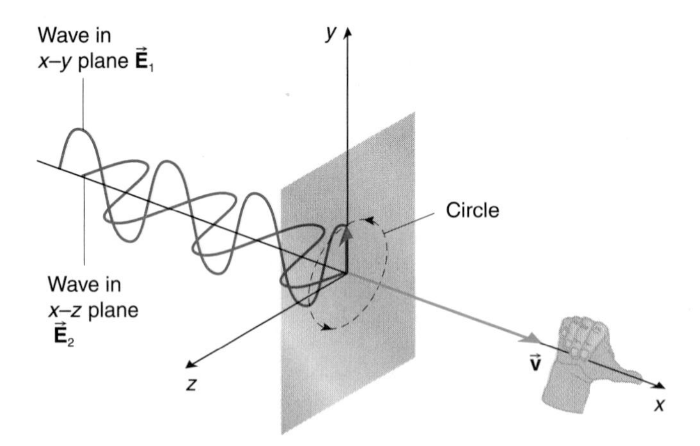

FIGURE 24.38 Right circularly polarized light.

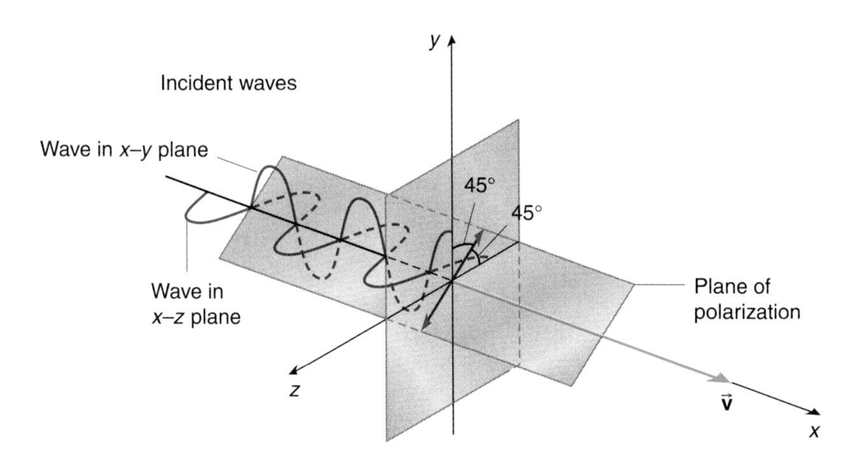

FIGURE 24.39 Plane polarized light.

previously) on the two waves are the same, the motion of the resulting electric vector traces out an ellipse; thus, we can have left and right **elliptically polarized light**.

We now consider two plane polarized light waves, differing in phase by π rad, but oscillating in planes at right angles to each other. The two waves are

$$\vec{\mathbf{E}}_1 = E_0 \cos(kx - \omega t)\hat{\mathbf{j}}$$

and

$$\vec{\mathbf{E}}_2 = E_0 \cos(kx - \omega t - \pi \text{ rad})\hat{\mathbf{k}}$$

We use the trigonometric formula for the cosine of the sum of two angles to obtain

$$\begin{aligned}\vec{\mathbf{E}}_2 &= E_0[\cos(kx - \omega t) \cos \pi + \sin(kx - \omega t) \sin \pi]\hat{\mathbf{k}} \\ &= -E_0 \cos(kx - \omega t)\hat{\mathbf{k}}\end{aligned}$$

The superposition of the two waves then is

$$\begin{aligned}\vec{\mathbf{E}}_1 + \vec{\mathbf{E}}_2 &= E_0 \cos(kx - \omega t)\hat{\mathbf{j}} - E_0 \cos(kx - \omega t)\hat{\mathbf{k}} \\ &= E_0 \cos(kx - \omega t)(\hat{\mathbf{j}} - \hat{\mathbf{k}})\end{aligned}$$

We fix our attention at the origin (where $x = 0$ m) and see what happens to the electric vector as a function of time; we have

$$\vec{\mathbf{E}}_1 + \vec{\mathbf{E}}_2 = E_0 \cos(\omega t)(\hat{\mathbf{j}} - \hat{\mathbf{k}}) \qquad (24.29)$$

This is *plane polarized light* with the plane of oscillation at 45° to the y- or z-axis, as shown in Figure 24.39.

The important point to note is that the two waves *do not interfere destructively* despite the π rad (= 180°) phase difference between them. The two waves are coherent because they have a constant phase difference between them. But the two waves do not interfere because the oscillations are perpendicular to each other! If the two oscillations were polarized in the *same* plane, they would interfere destructively: notice that this can be accomplished (mathematically) by simply changing the unit vector $\hat{\mathbf{k}}$ to the unit vector $\hat{\mathbf{j}}$ in Equation 24.29, in which case the superposition becomes zero at all times (i.e., destructive interference).

> Interference between coherent sources can occur only when the sources have the same polarization.

Most light originates from atoms. The light wave emitted during a very short interval from an individual atom is polarized. The next time it emits light, the polarization direction may be different. The atoms in a typical light source also all act independently of each other.† The resulting incoherent light from the entire collection of atoms within the source thus consists of countless waves with planes of polarization that are randomly oriented with respect to each other, as schematically illustrated in Figure 24.40. Such a continuum of polarization directions constitutes **unpolarized light**. Light from typical light sources is unpolarized light.

It is convenient to model unpolarized light in another way. We impose two mutually perpendicular directions on the sea of different planes of polarizations associated with unpolarized light. Then we resolve the amplitude of each oscillation into two components along the two mutually perpendicular directions chosen. For a given wave from a particular atom, these mutually perpendicular oscillations are in phase and coherent with each other. However, the *collective* oscillations along each perpendicular direction are incoherent, since they represent oscillations from many different and independently acting atoms in the source. Thus the resultant field oscillations along each mutually perpendicular direction are incoherent with each other. So we model unpolarized light by two mutually perpendicular, collective, incoherent oscillations, as shown in Figure 24.41.

†Lasers are *not* typical light sources because the atoms act in concert with each other. Many lasers emit polarized light.

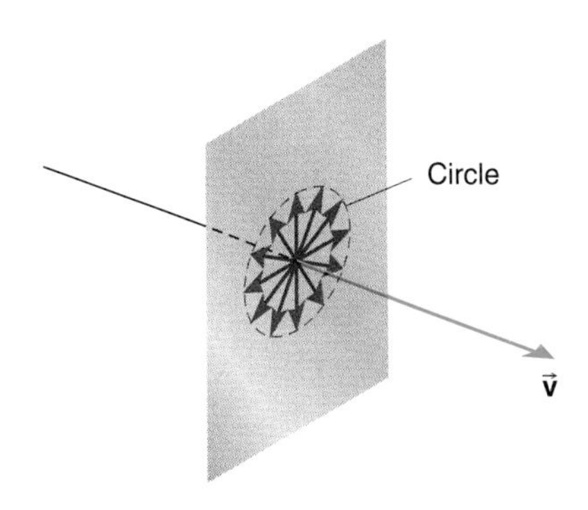

FIGURE 24.40 Unpolarized light consists of incoherent oscillations in every direction perpendicular to the velocity of the light.

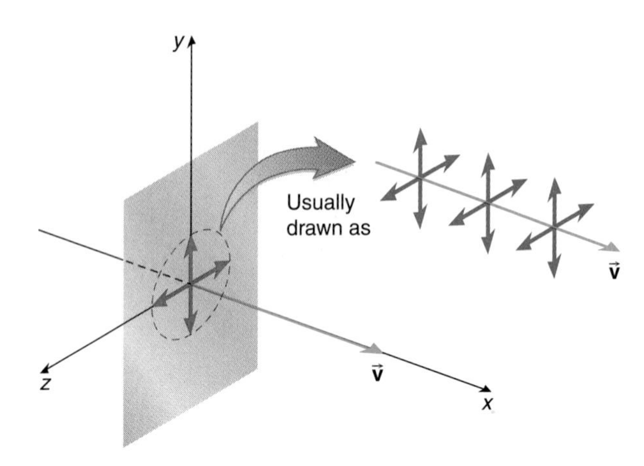

FIGURE 24.41 **Unpolarized light is modeled with two mutually perpendicular, incoherent oscillations.**

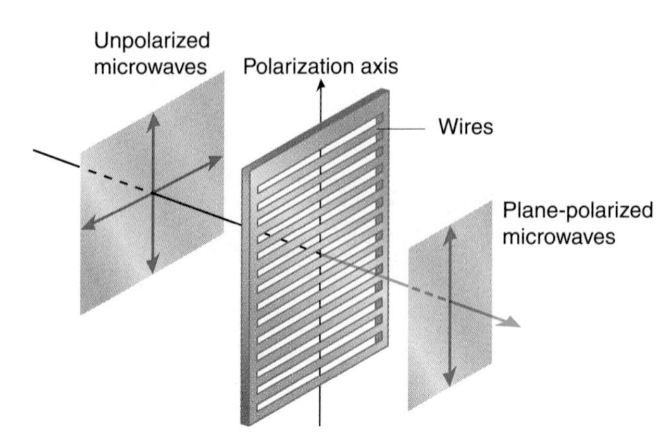

FIGURE 24.42 **Unpolarized microwaves are polarized by sending them through a set of parallel conducting wires. The axis of a microwave polarizer is perpendicular to the wires in the plane of the polarizer.**

QUESTION 5

Explain why longitudinal waves cannot be said to have their oscillations confined to a single plane and, therefore, why the concept of polarization is appropriate only for transverse waves.

24.13 POLARIZATION BY ABSORPTION*

How can the unpolarized light from a typical light source be polarized? That is, how can one of the two incoherent, mutually perpendicular plane polarizations used to model unpolarized light be selected or separated from the other? Several techniques are used; we explore one here and several others subsequently.

Light in the most general sense consists of the entire electromagnetic spectrum. For the moment, we consider the microwave region of the spectrum. Unpolarized microwaves can be polarized by sending the transverse waves through a grid of parallel conducting wires, as in Figure 24.42. We choose the two perpendicular directions used to represent the unpolarized incident beam to be parallel and perpendicular to the wires. The polarized waves with an electric vector parallel to the conducting wires are absorbed by the wires. The oscillatory field parallel to the wires transfers energy to the electrons that can move along the wires; it is the polarization direction perpendicular to the wires that is transmitted.

So the wire grid acts as a **polarizer**, a device for producing polarized microwaves. The **axis** of a polarizer is the direction parallel and antiparallel to the plane of polarization of the transmitted waves. The axis of a polarizer is not a unique line but simply a direction or a whole collection of lines oriented parallel to each other. Thus, for the wire grid polarizer of microwaves, the axis of the polarizer is a direction in the plane of the polarizer perpendicular to the direction of the wires, as shown in Figure 24.42.*

*This is quite unlike a transverse wave on a rope passing through a picket fence. For the rope, a transverse oscillation parallel to the pickets is transmitted and a transverse wave oscillating perpendicular to the direction of the pickets is absorbed.

In 1932 one of the more prolific American inventors of the 20th century, Edwin H. Land (1909–1991), discovered how to polarize visible unpolarized light via a mechanism similar to the wire grid polarizer for microwaves just outlined.† He developed a conducting polymeric material (an iodine-impregnated form of polyvinyl alcohol) whose long molecules conduct readily along their length but poorly across their very thin width. The molecules (analogous to the wires in the microwave polarizer) were oriented parallel by stretching the material in nitrocellulose sheets as they solidified; the molecules thereby served as a conducting medium analogous to the wires in the microwave example. Unpolarized incident light with electric vectors oriented along the oriented molecules is absorbed and that perpendicular to the long axis of the molecules is transmitted. Thus the axis of the polarizer once again is perpendicular to the long axis of the molecules, similar to the case for the wires and microwaves. Land called the material he created *Polaroid* and founded the now famous company bearing that name to manufacture it.‡

Thus unpolarized light can be polarized by selective absorption in passing it through a polarizing sheet such as a Polaroid®. Half the light intensity is absorbed by the polarizing material and half is transmitted with a plane of polarization parallel to the axis of the material.

Recall that the intensity is related to the energy of the wave, which in turn is proportional to the square of the wave amplitude.

Materials that produce polarized light by selective absorption are known as **dichroic** materials. Certain natural crystals (such as the mineral tourmaline) also have the prop-

†For an interesting historical and personal account of the development of polarizing sheets, see Edwin H. Land, *Journal of the Optical Society of America, 41*, #12, pages 957–963 (December 1951).

‡Polaroid cameras and Polaroid film are registered trademarks that have nothing to do with polarized light or polaroid material. The cameras and film were other inventions of Edwin H. Land and the Polaroid Corporation, provoked by an impatient question from his young daughter about why camera pictures could not appear "right away."

erty of selectively absorbing one of the two perpendicular oscillations representing unpolarized light. Such crystals are called dichroic crystals.

24.14 Malus's Law*

If unpolarized light is incident on a sheet of polarizing (dichroic) material, the transmitted light is plane polarized in a direction parallel to the transmission axis of the polarizer, as in Figure 24.43.

The polarizer transmits half of the incident light while the other half is absorbed by the sheet. If the linearly polarized light then is incident on a second polarizing sheet, as in Figure 24.44, the second sheet examines or analyzes the incoming light and can tell us something about the nature of the polarized light. The second polarizing sheet thus is called an **analyzer**.

The amount of light transmitted by the analyzer depends on the angle θ between its transmission axis and the direction of the oscillation of the incident plane polarized light. In particular, let E_0 be the amplitude of the incident plane polarized light; see Figures 24.44 and 24.45. The amplitude E_0 is resolved into two components:

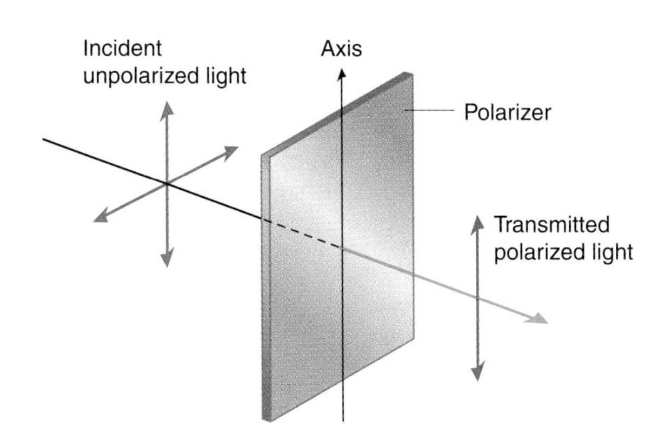

FIGURE 24.43 Production of polarized light by a polarizer.

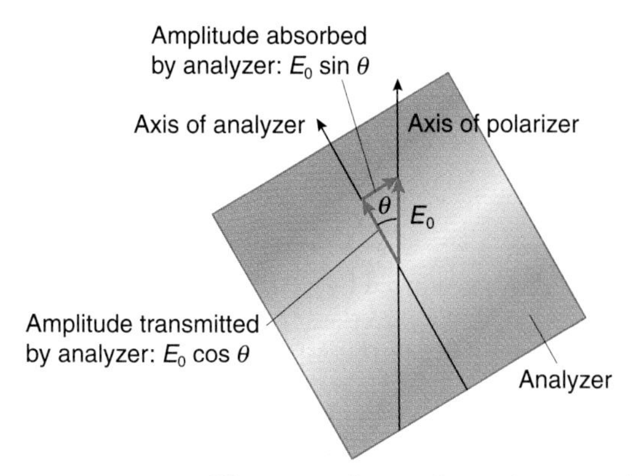

FIGURE 24.45 The action of an analyzer; the view is looking back along the light in Figure 24.44.

1. One component, $E_0 \sin\theta$, perpendicular to the axis of the analyzer. This component is absorbed by the analyzer and is not transmitted.
2. Another component, $E_0 \cos\theta$, parallel to the axis of the analyzer. This component is transmitted by the analyzer.

Thus the amplitude of the transmitted light is $E_0 \cos\theta$. The intensity transmitted is proportional to the square of the amplitude transmitted:

$$I_{\text{transmitted}} = K(E_0 \cos\theta)^2 = KE_0^2 \cos^2\theta$$

where K is a proportionality constant. But E_0^2 is proportional to the intensity of the light incident on the analyzer, $I_{\text{incident}} = KE_0^2$.

Thus

$$I_{\text{transmitted}} = I_{\text{incident}} \cos^2\theta \qquad (24.30)$$

This relationship between the incident and transmitted intensities was discovered in the early 19th century by the French physicist Étienne Louis Malus (1775–1812) and is, therefore, called **Malus's law**.

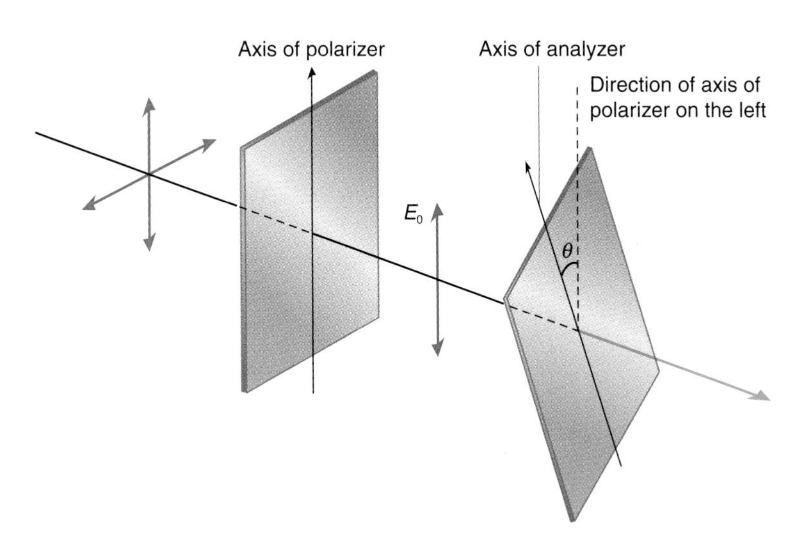

FIGURE 24.44 Polarizer and analyzer.

QUESTION 6

At what angle must the axis of the analyzer be oriented with respect to incident plane polarized light so that *no* light is transmitted by the analyzer?

EXAMPLE 24.12

Plane polarized light is produced by a polarizer. An analyzer is to be oriented at an angle θ to the plane of polarization of the incident polarized light so the intensity transmitted by the analyzer is 25% of the incident intensity. What is the angle θ required?

Solution

The intensity transmitted by the analyzer is determined from Malus's law, Equation 24.30:

$$I_{\text{transmitted}} = I_{\text{incident}} \cos^2\theta$$

Since $I_{\text{transmitted}}$ is to be 25% of I_{incident}, you have

$$0.25 I_{\text{incident}} = I_{\text{incident}} \cos^2\theta$$

$$0.25 = \cos^2\theta$$

Solving for θ, you obtain

$$\theta = 60°$$

24.15 POLARIZATION BY REFLECTION: BREWSTER'S LAW*

Perhaps you have noticed how Polaroid® sunglasses or a polarizing filter on a camera reduce reflected light (glare) from surfaces such as lakes. Here we see how light is polarized by reflection.

When light is incident on a transparent interface, some of the light is refracted into the second medium while some light is reflected back into the first medium. The proportion of the light reflected depends on several factors:

a. the angle of incidence: as it increases, the amount of the light reflected increases;
b. the numerical values of the indices of refraction of the two media: the greater the difference between the indices of refraction, the greater the reflection; and
c. the state of polarization of the incident light.

Let unpolarized light be incident on the interface at a nonzero angle of incidence, as in Figure 24.46. Consider the unpolarized beam to be composed of two mutually perpendicular, incoherent, plane polarized oscillations with

a. oscillations parallel to the interface (perpendicular to the plane of incidence encompassing the incident ray and the normal to the surface); and
b. oscillations in the plane of incidence (parallel to the plane of incidence, the plane containing the incident ray and the normal line).

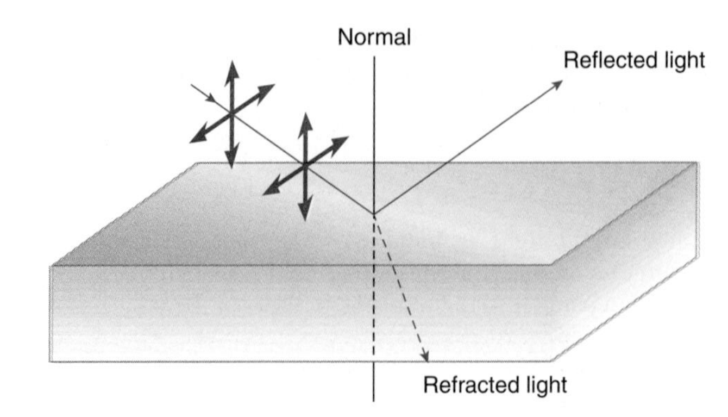

FIGURE 24.46 Unpolarized light incident on a transparent interface.

The following observations can be made about the reflected and refracted light by using a polarizing sheet as an analyzer of the light:

1. If we rotate the analyzer in the refracted beam,† the transmitted intensity varies but never to zero intensity, indicating unequal mixtures of the two types of oscillations. Such an unequal mixture of the two mutually perpendicular, incoherent oscillations is called **partially polarized** light. Hence, for angles of incidence greater than 0°, the refracted light is partially polarized.
2. Repeating the same procedure in the reflected beam, we find it, too, is partially polarized. By noting the orientation of the axis of the analyzer (which can be assessed independently if not marked on the analyzer), we find the reflected light has more of the oscillations perpendicular to the plane of incidence; the refracted light has more of the oscillations parallel to the plane of incidence, as in Figure 24.47.

At a particular angle of incidence θ_B, the reflected light is *completely* plane polarized and consists entirely of the oscillation perpendicular to the plane of incidence, as shown in Figure 24.48.

†For solid materials, this may be difficult to do! Consider using a liquid material such as water as the medium into which the ray is refracted.

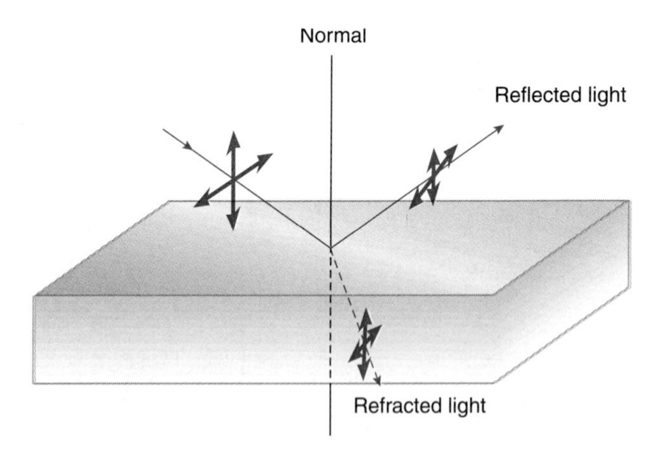

FIGURE 24.47 The reflected and transmitted light are each partially polarized.

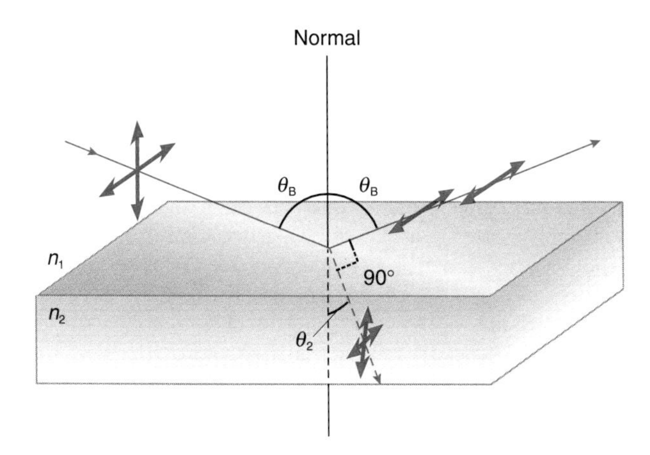

FIGURE 24.48 **The Brewster angle of incidence θ_B yields completely plane polarized reflected light with oscillations perpendicular to the plane of incidence (and parallel to the interface).**

The refracted light still is partially polarized, because not all of the oscillations perpendicular to the plane of incidence are reflected; even at this special angle of incidence, only *some* of that oscillation is reflected. The refracted beam thus always contains some of both oscillations. At this special angle of incidence θ_B, the angle between the reflected and refracted beam is found to be 90°. These observations were first made by David Brewster (1781–1868) and are known as **Brewster's law.***

> The special angle of incidence that results in a reflected beam that is completely plane polarized is called the **Brewster angle.** The refracted beam makes a 90° angle with this reflected beam.

An expression for the Brewster angle can be found using the law of refraction. When the angle of incidence is equal to the Brewster angle θ_B, the law of refraction (Equation 23.6) becomes

$$n_1 \sin \theta_B = n_2 \sin \theta_2$$

*All of these results can be derived from the Maxwell equations of electromagnetism.

The angle of refraction θ_2 can be expressed in terms of the Brewster angle θ_B by noting geometrically in Figure 24.48 that

$$\theta_B + 90° + \theta_2 = 180°$$
$$\theta_2 = 90° - \theta_B$$

Substituting for θ_2 in the law of refraction, we find

$$\begin{aligned} n_1 \sin \theta_B &= n_2 \sin(90° - \theta_B) \\ &= n_2(\sin 90° \cos \theta_B - \cos 90° \sin \theta_B) \\ &= n_2 \cos \theta_B \end{aligned}$$

We rearrange this slightly to yield

$$\frac{\sin \theta_B}{\cos \theta_B} = \frac{n_2}{n_1}$$

$$\tan \theta_B = \frac{n_2}{n_1} \qquad (24.31)$$

Thus the Brewster angle can be found from the indices of refraction.

Perhaps the most widespread use of Brewster's law is in laser technology. The windows of the discharge tubes in gas lasers are inclined to the axis of the tube by the Brewster angle to selectively enhance the polarization of the emitted laser light. Such windows are called **Brewster windows.**

Light reflected from other dielectric surfaces also is partially polarized, with the dominant polarization being the oscillation parallel to the surface or (equivalently) perpendicular to the plane of incidence. Such reflected light often is called *glare*. Most of the light reflected from a surface has an oscillation parallel to the surface. For the case of a horizontal surface, Polaroid® sunglasses and polarizing camera filters reduce glare by selectively blocking (absorbing) mostly the reflected oscillations that are horizontal; the transmission axis of the polarizing material is oriented vertically and thus absorbs most of the reflected oscillations, as in Figure 24.49.

Take a pair of Polaroid® sunglasses and examine strongly reflected light such as that from a water surface or even a well-polished floor; if the sunglasses are rotated, the transmitted light from the reflection is a minimum when the glasses are in a conventional orientation (or up-side-down). The reflected light is

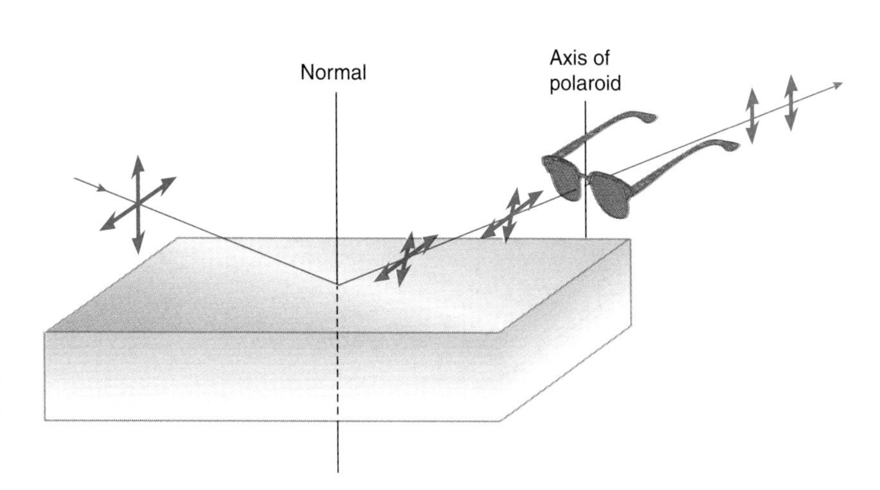

FIGURE 24.49 **Polaroid® sunglasses and filters block (i.e., absorb) some reflected light. If the reflected light is at the Brewster angle, all of the light will be blocked.**

transmitted strongly when the glasses are rotated 90° from their conventional orientation. Polarizing filters for cameras are essentially sunglasses that eliminate polarized glare from reflections when properly oriented. Such filters also are used by anglers.

QUESTION 7

What specific observations lead you to the conclusion that the refracted beam has more of the polarization oriented in the plane of incidence, while the reflected light has more of the oscillation perpendicular to the plane of incidence?

EXAMPLE 24.13

a. What is the Brewster angle for an air–water interface?
b. What is the Brewster angle for a water–air interface?

Solution

a. Use Brewster's law in the form of Equation 24.31:

$$\tan \theta_B = \frac{n_2}{n_1} = \frac{1.33}{1.00}$$

$$\theta_B = 53.1°$$

b. The role of the indices of refraction is reversed, and so Brewster's law becomes

$$\tan \theta_B = \frac{n_2}{n_1} = \frac{1.00}{1.33}$$

$$\theta_B = 36.9°$$

This is less than the critical angle for a water–air refraction, so that there is both a reflected and transmitted beam.

Note that the two angles sum to 90°. Why should they?

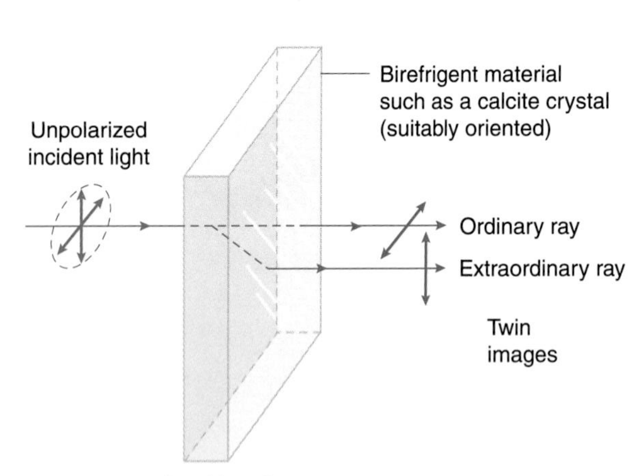

FIGURE 24.50 Double refraction.

24.16 POLARIZATION BY DOUBLE REFRACTION*

Certain anisotropic crystals (such as calcite and many others) exhibit a remarkable type of refraction. When an unpolarized light ray is incident on the crystal in certain directions, *two* refracted rays result! See Figure 24.50. Such **double refraction** also is known, equivalently, as **birefringence**. One of the refracted rays follows the ordinary law of refraction, and so this ray is known as the **ordinary ray** (or O ray). The other refracted ray is called the **extraordinary ray** (or E ray), since it does not follow the law of refraction. Indeed, the extraordinary ray usually goes off in a direction not even in the plane of incidence!

The extraordinary ray experiences an index of refraction that *varies* with the direction the incident ray makes with respect to microscopic symmetry axes of the crystal structure.

For our purposes, the most important aspect of double refraction is that the two refracted rays are completely plane polarized along two mutually perpendicular directions.

That is, the O ray polarization direction is perpendicular to the E ray polarization direction. Since the beams typically propagate in different directions within the crystal,[†] the two polarized rays can easily be separated in space.

Producing polarized light via double refraction has several advantages over the other methods for producing polarized light:

1. With double refraction, half the incident unpolarized light goes into each of the O and E rays. Polarization by reflection is less efficient, since typically only a small percentage of the light is reflected at the Brewster angle.

[†]If the crystal is cut and illuminated appropriately, the O and E rays will propagate in the same direction within the crystal; for some orientations within the crystal, they propagate in the same direction at the same speed, and in other crystal orientations, they propagate in the same direction at different speeds. This latter property is exploited to make circularly polarized light from linearly polarized light in a device known as a quarter-wave plate (see Investigative Project 12). The analysis of such anisotropic crystals is quite interesting and leads to many practical optical applications.

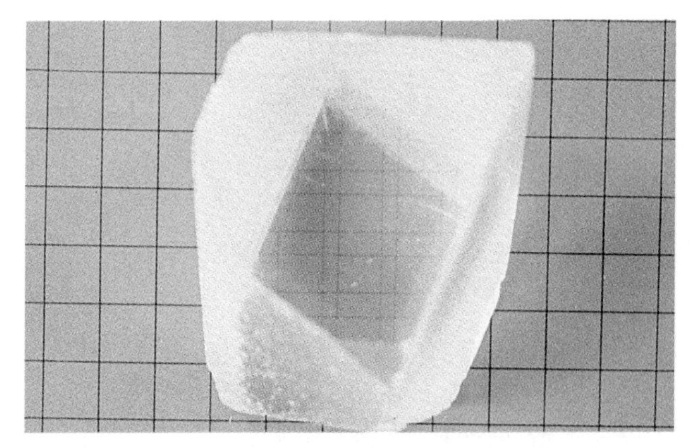

A properly oriented calcite crystal will show twin images. If you rotate the crystal, the image formed by the extraordinary rays rotates about that formed by the ordinary rays.

2. With polarization by double refraction, little of the light is absorbed by the transparent crystal, so it is more suited to high-power applications. Polarization by absorption such as with polarizing sheets only can be used effectively if the amount of light energy to be absorbed does not appreciably warm the polarizer. Excessive energy transfer can melt the material! High-power lasers easily can melt polarizing sheets because of the high energy the film may be called on to absorb.

24.17 POLARIZATION BY SCATTERING*

Take your Polaroid® sunglasses, hold them in front of your eyes, and look at the blue sky, preferably at a region about 90° from the Sun. Rotate the sunglasses and you find the light transmitted by them varies in intensity, but not to zero; hence, the sky light is partially polarized. If the sky light were completely plane polarized, there would be an orientation at which all the skylight was blocked and the view through the sunglasses would look black. The polarization of sky light arises because of **scattering**.

When a beam of light is incident on transparent matter (be it solid, liquid, or gas), some of the light is redirected or scattered from the original direction. The redirected light is called scattered light and the process is called scattering. It is scattered light that makes light rays visible in dusty conditions such as when chalk-filled erasers are clapped together along the path of a laser beam, or when sunbeams are seen streaming through the atmosphere. The scattered light in such circumstances is partially polarized.

We can qualitatively model these scattering processes in the following way. The oscillating electric field vectors of the incident light are always in a plane perpendicular to the ray and cause electrons in atoms or molecules to oscillate and produce electric fields in the same directions. The scattered light thus generated by these particles has electric fields that are the components perpendicular to the direction of viewing. Since sunlight is unpolarized, it has equal amounts of field in all directions normal to its incident ray. Viewed along an angle to the incident ray, its induced accelerations and fields in the scattering medium have larger components normal to the incident ray than parallel to it.

Light scattering by the atmosphere and dust particles makes this laser beam visible from the side.

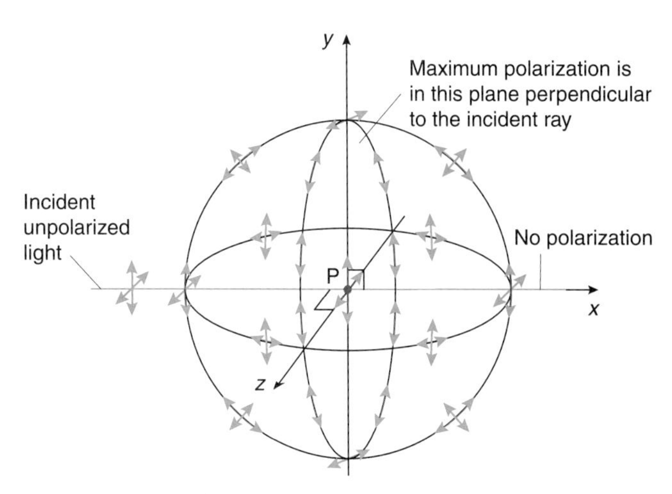

FIGURE 24.51 **Polarization of light scattered at various orientations relative to the incident ray.**

So the scattered light rays are not polarized if viewed straight ahead or back along the incident ray, but the scattered light at 90° is maximally polarized perpendicular to the rays of the Sun, as shown in Figure 24.51.

Atmospheric light scattering is only partially polarized because of several effects:

1. Multiple scattering takes place. In other words, the light reaching you from the sky has been scattered several times; this complicates the analysis.
2. The scattering molecules are anisotropic, so that the electrons do not necessarily experience oscillations in the same direction as an incident electric field vector of the light wave.
3. The degree of polarization of the scattered light depends on the size of the scattering particle. The smaller the scattering particle, the greater the polarization of the scattered light.

Nevertheless, sunlight scattered from a clear sky shows enough polarization to be well reduced by a properly oriented polarizing filter on your camera or by Polaroid® sunglasses.

24.18 RAYLEIGH AND MIE SCATTERING*

Why is the clear, daytime sky blue? Why are clouds white? What causes the Sun to be red when rising or setting? Here we see that such effects are caused by light scattered by the constituents of the atmosphere.

Light scattering from particles much smaller than the wavelength of the incident light is known as **Rayleigh scattering**, after the 19th-century English physicist Lord Rayleigh, who first quantitatively investigated such scattering.† The molecules of our atmosphere satisfy this condition very well for visible light wavelengths. Thus incident light from the Sun is scattered in the

†Recall from Section 24.6 that Lord Rayleigh also formulated the Rayleigh criterion for resolution.

atmosphere, giving our sky its characteristic blue color for reasons we shall soon see.* On the other hand, the Moon and some other planets (such as Mercury) and other moons lack atmospheres; on these celestial objects the sky between the stars appears black because there are no particles present to scatter the incident solar radiation.

Rayleigh scattering has a dependence on the wavelength of the incident light that accounts for the blue color of the sky. The dependence on wavelength can be deduced from a dimensional analysis of the scattering problem.† After pondering the problem, Rayleigh decided that the amplitude of the scattered light likely could depend on several factors:

- the total content or volume V of the scattering particle;
- the relatively large wavelength λ of the incident light;
- the distance r from the scattering particle to the observer; and
- the speed of light.

The ratio of the amplitude of the scattered light to the amplitude of the incident light will, of course, be dimensionless. So for this ratio we seek to formulate a dimensionless combination of the potential contributing factors listed here. The first three factors, V, λ , and r all involve the dimension of length, whereas the speed of light c has the dimensions of length per time. Since c is the *only* factor listed that involves a time dimension, the speed of light must in fact be irrelevant to the amplitude ratio (none of these other factors can cancel the time dimension, since they involve the length dimension only).

Therefore the ratio of the amplitudes must depend on V, λ, and r. Certainly as the particle volume vanishes, the scattering should vanish as well; thus it seems reasonable to suppose (at least to a first approximation) that the scattered wave amplitude is directly proportional to the volume of the scattering particle. The inverse square law governing the intensity of propagating waves (see Section 23.2) implies that the scattered amplitude received should be inversely proportional to the distance r between the scattering particle and the observer: in this way the scattered light intensity decreases as $1/r^2$, in keeping with the inverse square law. So the ratio of the scattered to incident amplitudes should go in part something like

$$\frac{\text{scattered amplitude}}{\text{incident amplitude}} \propto \frac{V}{r}$$

The right-hand side of this relationship is not yet dimensionless. However, if we put the remaining factor, the wavelength λ, into the denominator as λ^2, we can cause the right-hand-side to be dimensionless:

$$\frac{\text{scattered amplitude}}{\text{incident amplitude}} \propto \frac{V}{\lambda^2 r} \qquad (24.32)$$

Thus the amplitude of the scattered light is proportional to the inverse square of the wavelength.

The intensity of the scattered light is proportional to the square of the scattered amplitude, and so the scattered light intensity must therefore be proportional to the inverse *fourth* power of the wavelength. That is,

$$\text{scattered intensity} \propto \frac{1}{\lambda^4} \qquad (24.33)$$

omitting the other contributing factors. Therefore short wavelengths (blues) are scattered more efficiently than longer wavelengths (reds).

Both sunlight and our visual sensitivity are maximum around yellow and weaken rapidly toward the violet end of the spectrum. These trends, when multiplied by the scattering ratio which strongly favors short wavelengths, gives us a blue sky as a compromise. Thus the clear sky looks blue and is most blue 90° from the direction to the Sun; see Figure 24.52. When you observe a sunrise or sunset, the incident light traverses a very long atmospheric path (see Figure 24.52) compared with

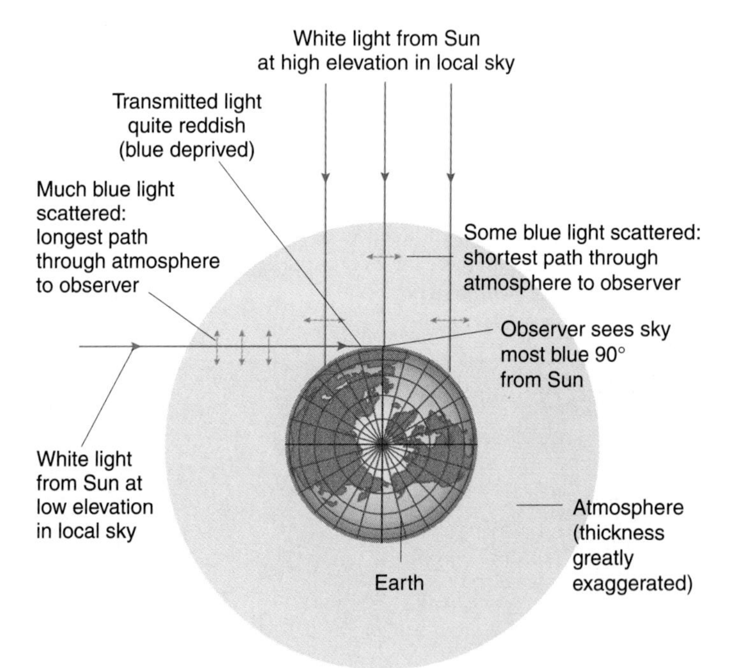

FIGURE 24.52 Light scattering in the atmosphere produces the blue sky and red sunrises and sunsets.

*A fine suspension of iron oxide particles gives the atmosphere on Mars a reddish hue.

†This argument parallels that first made by Lord Rayleigh in his epochal treatment of the scattering problem: John William Strutt [Lord Rayleigh], "On the light from the sky: its polarization and colour," *Philosophical Magazine*, *41*, pages 107–120 (1871).

the path at, say, noontime. More of the blue light is scattered out of the incident light and the resulting light reaching our eyes is blue deficient: sunsets and sunrises are red. Physics is amazing stuff!

The scattering by larger particles, on the order of the wavelength of the incident light, is called **Mie scattering** [after Gustav Mie (1868–1957), who first began investigating this type of scattering quantitatively in 1908†]. The wavelength dependence of such scattering is more complex than Rayleigh scattering; but as the particle size increases further, the scattering approaches diffuse reflection. This is the reason that clouds appear white: the scattering from the water particles that make up clouds becomes diffuse reflection, essentially independent of wavelength.

A special circumstance bears mentioning, however. If the scattering particles are between one and two wavelengths in size and are essentially the same uniform size, then Mie scattering predicts that the longer wavelengths are scattered more than the shorter wavelengths, just the opposite of Rayleigh scattering. If clouds of dust in the atmosphere meet these conditions, it is possible to observe the rising or setting Sun (or Moon) as blue or even green in color! Sightings of such colored Moons and Suns were quite common after the cataclysmic volcanic eruption of Krakatoa in the 19th century.‡ More recent sightings also have occurred.§ The rarity of such sightings may be the origin of the colloquial expression "once in a blue moon." *Calendrical* "blue moons" refer to the second full moon of a calendar month, which occur about every three years (although 1999 had two calendrical blue moons). Such calendrical blue moons do not, of course, appear blue in the sky!

†Gustav Mie, "Beiträge zur Optik trüber Medien, speziell kolloidaler Metallösungen," *Annalen der Physik, 25,* #3, pages 377–445 (1908).

‡The sightings are summarized and discussed in F. A. Rollo Russell and E. Douglas Archibald, "On the unusual optical phenomena of the atmosphere, 1883–1886, including twilight effects, coronal appearances, coloured suns, etc." which appeared as a section of the Report of the Krakatoa Committee of the Royal Society, *The Eruption of Krakatoa and Subsequent Phenomena*, edited by G. J. Symons (Trubner, London, 1888), Section I. (c), "The blue, green, and other coloured appearances of the sun and moon," pages 199–218.

§In 1950 a blue-colored Sun was sighted over eastern Canada and the northeastern United States caused by smoke from extensive forest fires in western Canada. Reports of this sighting are found in Rudolf Penndorf, "On the phenomenon of the colored sun, especially the 'blue sun' of September 1950," Geophysical Research Paper #20, Air Force Cambridge Research Center Technical Report 53-7 (April 1953); R. Wilson, "The blue sun of 1950 September," *Monthly Notices of the Royal Astronomical Society of Canada, 11,* pages 447–489 (1951); and William Paul and R. V. Jones, "Blue sun and moon," *Nature, 168,* page 554 (29 September 1951).

24.19 OPTICAL ACTIVITY*

Certain materials and solutions are able to rotate the plane of polarization of the light passing through them as shown in Figure 24.53. This phenomenon is called **optical activity.**

Optical activity occurs in certain crystals (e.g., quartz) when plane polarized incident light is incident along certain directions with respect to symmetry axes associated with the particular crystal structure. Some liquids (e.g., turpentine) and solutions (e.g., sugar solutions) also exhibit optical activity. The extent of the rotation of the plane of polarization depends on the distance ℓ traveled through the crystal or liquid and the concentration c of the solutions.# In solutions, the angle of rotation ϕ of the plane of polarization is written empirically as

$$\phi = \alpha \ell c \tag{24.34}$$

where α is a proportionality constant known as the **specific optical rotating power** and is a function of both wavelength and temperature. The dependence on the wavelength of the light leads to **rotary dispersion**, analogous to the variation of the index of refraction with wavelength. Thus linearly polarized incident white light is dispersed into various planes of polarization by optically active materials, as shown in Figure 24.54.

The directional sense of the rotation of the plane of polarization determines the type of optical activity in the sample. If the rotation of the plane is in a right-handed sense, the material is said to be **dextrorotatory**. Right-handed means that if the thumb of the right hand is pointing in the direction of propagation of the light, the curled fingers indicate the sense of the progressive rotation

#Do not confuse the concentration with the speed of light, also designated by *c*.

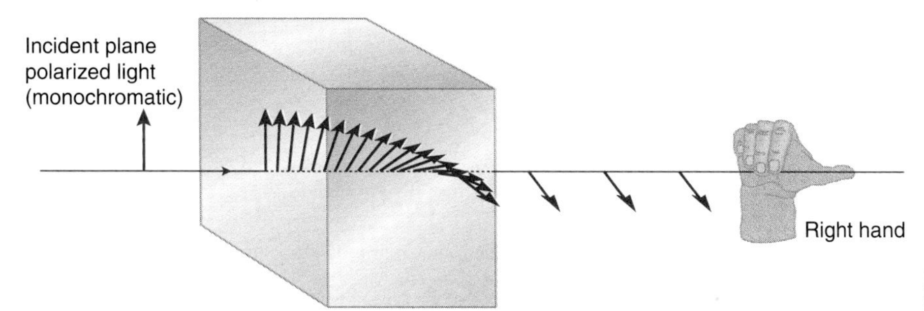

FIGURE 24.53 Optical activity. For clarity, only half of the transverse oscillation is shown.

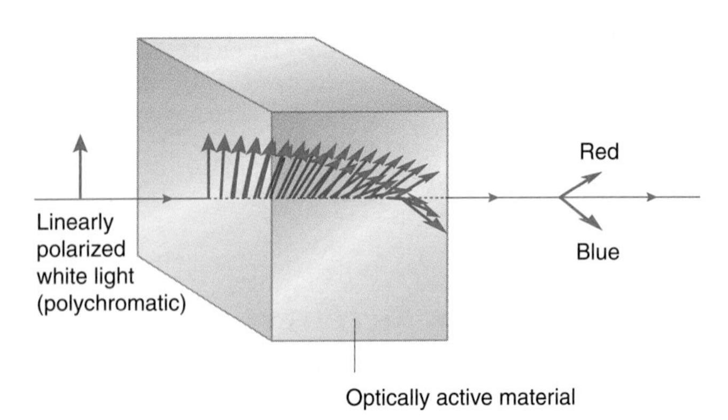

FIGURE 24.54 **Rotary dispersion of plane polarized white light. For clarity, only half of the transverse oscillation is shown.**

of the plane of polarization; see Figure 24.53.* If the rotation of the plane of polarization is in the opposite (left-handed) sense, the material is **levorotatory**.

Optical activity is used in many of the sciences including geology (crystalline minerals), chemistry (concentrations of solutions, isomeric structures of molecules), biology (biologically important molecules with spiral structures such as DNA and RNA), structural engineering (for optical stress analysis), electrical engineering (liquid crystals), and physics for a wide variety of purposes and applications.

*Regrettably, these conventions for the directional senses are not universal, although the one we choose is the most widely used. You need to be sure what convention is being employed when consulting other references. As we have seen before, many conventions in optics have yet to be made standards, used by all, once again illustrating the inherent conservative nature of science as well as the human inertia associated with change.

CHAPTER SUMMARY

Physical optics is the study of phenomena related to the wave nature of light.

A *wavefront* is a line or surface of constant phase in a propagating disturbance and is perpendicular to a ray. Two or more light sources are said to be *coherent* if the phase difference between them is independent of time. Coherent sources of light (or other waves) can be made by *wavefront division* (with two or more slits), or by *amplitude division* (such as in thin-film interference). Light from coherent sources will interfere when superimposed at the same place at the same time. The nature of the superposition depends on the *phase difference* between the waves. *Constructive interference* occurs if the phase difference between the waves is an integral multiple of 2π rad. *Destructive interference* occurs if the phase difference is an odd integral multiple of π rad.

A phase difference can arise from a *path difference* between waves going from coherent sources to the point of their superposition. A path difference of one wavelength corresponds to a phase difference of 2π rad. The phase difference δ_{path} arising from a path difference Δx is

$$\delta_{\text{path}} = \frac{\Delta x}{\lambda}(2\pi \text{ rad}) \qquad (24.1)$$

If monochromatic light is incident from a distant source on two parallel slits separated by a center-to-center distance d, constructive interference fringes are seen on a distant screen when the path difference $d \sin\theta$ between the wavelets from the slits is an integral number of wavelengths:

$$d \sin\theta = m\lambda \qquad (m = \text{integer}) \qquad (24.2)$$
(double slit, constructive interference)

where the angle θ is measured from the incident direction. The integer $|m|$ is called the *order number* of interference. This is *Young's double slit* experiment.

When monochromatic light from a distant source diffracts through a single slit of width a, the first intensity minimum of the diffraction pattern on a distant screen (measured from the incident direction) is located at an angle θ_1 given by

$$a \sin\theta_1 = \lambda \qquad (24.3)$$
(first diffraction minimum, slit aperture)

The quantity $a \sin\theta_1$ is the path difference between opposite sides of the slit to the first minimum on the distant screen.

When monochromatic light from a distant source diffracts through a *circular* aperture of diameter (aperture) a, the first minimum of the diffraction pattern is located at an angle θ_1 found from

$$a \sin\theta_1 = 1.220\lambda \qquad (24.5)$$
(first diffraction minimum, circular aperture)

The *Rayleigh criterion* states that two incoherent sources are *barely resolved* if the central diffraction peak from one source is located at the first minimum of the pattern formed by the other source. This angular separation of the sources then is θ_1, which for a circular aperture is found from Equation 24.5, and for a slit aperture is found from Equation 24.3. Good resolution means that the angle θ_1 between the center and first minimum of each diffraction pattern produced by two incoherent sources is smaller than the separation angle between the two source centers.

The intensity of interference fringes in a double (or multiple) slit experiment is modulated by the envelope of the single slit diffraction pattern.

A *diffraction grating* has many parallel, identical, equally spaced slits. Interference maxima occur at those locations where

$$d \sin\theta = m\lambda \qquad \text{(diffraction grating maxima)} \qquad (24.8)$$

where d is the *grating spacing* (the center-to-center distance between adjacent slits) and m is the order number. If the number of slits N is large, the interference peaks are narrow and sharply defined. The *resolution* of a diffraction grating, its ability to discern two nearly equal wavelengths whose difference is $\Delta\lambda$, is directly proportional to the order number m and the number of illuminated slits N of the grating:

$$\frac{\lambda}{\Delta\lambda} = Nm \tag{24.16}$$

where λ is either of the nearly equal wavelengths.

The *angular dispersion* of a diffraction grating is

$$\frac{d\theta}{d\lambda} = \frac{m}{d \cos \theta} \tag{24.13}$$

Diffraction gratings are widely used to measure the wavelength of light.

The index of refraction of a material is equal to the ratio of the speed of light c in vacuum to that in the material, v:

$$n = \frac{c}{v} \tag{24.17}$$

The wavelength λ_1 of light in a medium that has an index of refraction n_1 is shorter than in a vacuum (the frequency of the light is unchanged):

$$\lambda_1 = \frac{\lambda}{n_1} \tag{24.20}$$

where λ is the wavelength in vacuum.

Light reflected from a metal (a conducting surface), or from an optically more dense material (a medium with an index of refraction greater than that of the medium from which the light is incident), experiences a *phase change* of π rad upon reflection.

In cases of thin-film interference, the phase difference between waves arises from a path difference and from phase changes that may occur from reflections. The relevance of phase changes upon reflection must be assessed on a case-by-case basis.

Light is a transverse wave that can be plane polarized, meaning the oscillations of the electric vector of the light are confined to a plane that includes the ray. Unpolarized light arises from sources whose atoms emit light with planes of polarization in random orientations. For convenience, unpolarized light is imagined to be composed of two mutually perpendicular, incoherent, planes of polarization.

Unpolarized light can be polarized in one of several ways.

A sheet of polarizing material selectively transmits the light polarized parallel to its transmission axis, and absorbs the polarization perpendicular to it. If plane polarized light of amplitude E_0 is incident on a polarizing sheet with an angle θ between the axis of the polarizer and the plane of oscillation of the polarized light, the amplitude of the light transmitted is $E_0 \cos \theta$. Hence the light intensity (which is proportional to the square of the amplitude) transmitted by the polarizer follows *Malus's law*:

$$I_{\text{transmitted}} = I_{\text{incident}} \cos^2\theta \tag{24.30}$$

Reflected light typically is *partially polarized*, with a preponderance of the oscillations perpendicular to the plane of incidence. If the light is incident at the *Brewster angle* θ_B on a refractive medium, the reflected and transmitted rays are 90° apart and the reflected light is completely plane polarized with its (oscillating and propagating) electric field perpendicular to the plane of incidence. The Brewster angle is found from *Brewster's law*:

$$\tan \theta_B = \frac{n_2}{n_1} \tag{24.31}$$

Certain crystals produce *two* refracted rays, a phenomenon called *double refraction*. The *ordinary ray* follows the law of refraction while the *extraordinary ray* does not. The ordinary and extraordinary rays have plane polarizations perpendicular to each other.

The scattering of light by small particles (such as the atoms and molecules that make up the atmosphere) also produces partially polarized light.

The scattering of light by particles whose size is much smaller than the wavelength is called *Rayleigh scattering*. The intensity of Rayleigh scattering is proportional to the inverse fourth power of the wavelength, which accounts for the blue sky and red sunrises and sunsets. Light scattered by particles whose size is on the order of the wavelength is called *Mie scattering*, and accounts, under special circumstances, for the rare observation of blue- and green-colored Suns and Moons. The water particles that comprise clouds are comparatively large and scatter light independent of wavelength (approaching diffuse reflection), so that clouds appear white.

Certain crystals and solutions can rotate the plane of polarization of incident plane polarized light, a phenomenon called *optical activity*. The effect is slightly dependent on the wavelength of the incident light, leading to *rotary dispersion*.

SUMMARY OF PROBLEM-SOLVING TACTICS

24.1 **(page 1110)** Do not confuse the order number m locating the interference *maxima* for Young's double slit experiment in Equation 24.2 with the integer m locating the single slit diffraction *minima* in Equation 24.4.

24.2 **(page 1111)**For circular apertures, be sure to use Equation 24.5 to locate the angle to the first minimum of the diffraction pattern, not Equation 24.4 for slit apertures.

24.3 **(page 1122)** When investigating thin-film interference, note that the wavelength of the light in the film is *different* from that in vacuum. It is the wavelength of the light in the film that is involved in assessing the phase difference associated with the path difference. So be careful to use the wavelength of the light in the film in Equation 24.21.

24.4 **(page 1123)** There is no magic bullet formula to memorize that can treat all situations of thin-film interference. Each situation must be assessed individually to determine (a) the path difference and its resulting phase difference and (b) the phase changes (if any) upon reflection and whether they are relevant or not.

QUESTIONS

1. (page 1105); 2. (page 1107); 3. (page 1113); 4. (page 1124); 5. (page 1128); 6. (page 1130); 7. (page 1132)
8. When 76 trombones play the same note, is the sound a collection of 76 coherent or incoherent sources of acoustic waves?
9. Describe some simple observations that clearly indicate that the wavelength of visible light is quite short while that of sound is much longer.
10. Glass prisms change the direction of violet light more than that of red light. However, with double slit interference, the least deviated color is violet. Which of these two colors is shown by these facts to have the greater wavelength, and through what reasoning?
11. For a fixed slit separation d and slit width a, what happens to the appearance of the double slit pattern if light of a longer wavelength is first used and next changed to light of shorter wavelength? Draw sketches to illustrate the relative appearance of the two interference patterns.
12. Which color of visible light provides the best resolution in a microscope? In a telescope?
13. In a single slit diffraction pattern, the angular width of the central peak ($2\theta_1$) depends on several factors: the width of the slit and the wavelength of the light. Sketch rough graphs indicating how the angular width depends on each of these factors (with the other factor held constant).
14. A close pair of binary stars is more easily resolved if their color is predominantly blue rather than red. Explain why.
15. A single slit of width a is a distance $d/2$ above a plane mirror as indicated in Figure Q.15. Monochromatic light of wavelength λ is incident normally on the slit. A screen S is located far from the arrangement. (a) The mirror produces a virtual image of the slit. Is the virtual image *coherent* with the slit itself? (b) Is the pattern on the screen a double slit interference pattern or a pair of single slit diffraction patterns? (c) If the pattern is an interference pattern, is the fringe closest to the mirror surface bright or dark? Explain. This arrangement is known as *Lloyd's mirror*.

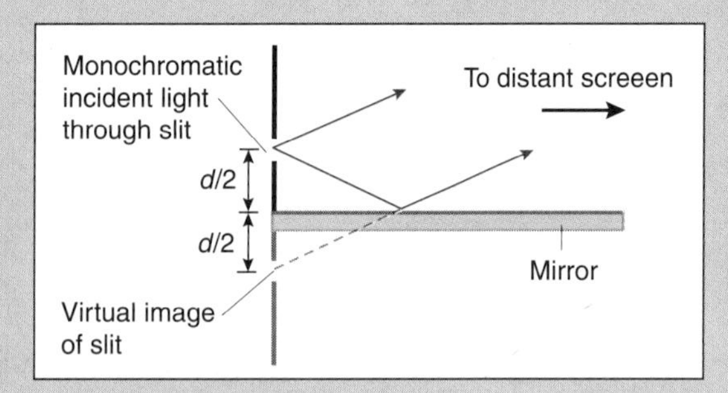

FIGURE Q.15

16. Take out one of your compact discs and observe white light reflected from the CD at a large angle of incidence. Explain why you see a spectrum of colors.
17. Can a panther with a vertical slit pupil in each eye judge vertical details such as tall grass better than horizontal details such as level tree branches? Explain with a convincing discussion.
18. Explain with the assistance of a diagram the appearance of a single slit diffraction pattern formed by *white light* incident from a distant source.
19. Explain the advantages of using a diffraction grating with: (a) many slits; (b) closely spaced slits.
20. Which visible light color experiences the greatest deflection when refracted through a prism? Which experiences the greatest deflection when using a diffraction grating?
21. To analyze the spectrum of a light source, a diffraction grating typically is preferred over a prism. Discuss several reasons for this.
22. A light source emits several distinct wavelengths of light. The light is incident on a diffraction grating. You note that the second order of one wavelength is coincident in angular position with the third order of another wavelength. What conclusions can you draw from this observation?
23. Your professor gives you a diffraction grating in the laboratory but has no information about its small grating spacing d. Describe a simple experiment that you can perform to measure the grating spacing.
24. Watch a water wave come in and reflect from a rigid sea wall. Does the reflected wave experience a phase change of π rad?
25. A red traffic light is seen by an underwater swimmer. Don't ask how. What color does the light appear to be to the swimmer? Explain.
26. Is the resolution of a circular aperture improved if the aperture is used under water instead of in air? Explain.
27. Is Equation 24.20 what is meant by the term dispersion?
28. Young's double slit experiment is performed in air and then in water with the same pair of slits and light source. Explain the difference (if any) in the interference pattern observed on a distant screen.
29. Discuss the feasibility of creating a thin-film antireflection coating on military aircraft to enhance their "stealth" capability of invisibility to radars (that typically have wavelengths of a few centimeters).
30. Does red or violet light travel more slowly in glass, or do they both travel at the same speed?
31. You have a rather attractive red bathing suit in air (red ≈ 660 nm). Will the color of the suit change when viewed under water?
32. You likely used a soap and glycerin solution and a bubble-blower loop as a child. Resurrect the toy. Notice the light reflected from the soap film on the loop. If the film is very thin at some location, explain why you see reflections from this location or region as black. (See Investigative Project 3.)
33. Describe an observation that clearly demonstrates that light is a transverse wave.

34. By means of a few simple experiments, see if the human eye can distinguish light according to various orientations of its plane of linear polarization. Some flying insects and birds can detect polarization and use such information for navigational purposes, since skylight is partially polarized.

35. A polarizer is placed with its axis parallel to one slit of two parallel slits in a Young's double slit experiment, while another polarizer is placed on the other with its axis perpendicular to the slit. The double slits are simultaneously illuminated with monochromatic, unpolarized light from a slit source. What is seen on a distant screen and why?

36. Describe how a single polarizer can be used to show how the Rayleigh scattered blue sky light on a sunny day is partially polarized.

37. What advantage would there be if the windshield of your car were covered with a polarizing material? What should be the orientation of the axis of the polarizing material to gain greatest advantage? Cite one significant disadvantage of covering the windshield with such a material.

38. You have a sheet of polarizing material but there is no information to indicate its axis. With only an unpolarized light source available, describe an experiment you can perform to determine the orientation of the axis of the polarizer.

39. The difference in the scattering of light by small and large particles (relative to the wavelength of the incident light) is well illustrated by your driving experiences with auto windshields and either (a) headlights from oncoming cars at night or (b) driving "into the Sun" during the day. If the windshield is littered with leaves (particles large compared with the wavelength of visible light), the incident light is more reflected (perhaps diffusely) than scattered. This is called *back scattering* (back to the source). Thus when you sit in the driver's seat your eye is shaded by the large particles. On the other hand, if the particles on the windshield are small, such as with road grit or dust, your eye sees the windshield ablaze with extra light. The scattering then is predominantly in the *forward* direction of the light with little back scattering. The rings of Saturn (located about 9.5 times further from the Sun than the Earth) are clearly visible from the Earth using even a moderately sized telescope. On the other hand, the ring around Jupiter (about 5 times farther from the Sun than the Earth) is invisible from the Earth and only was discovered by a spacecraft that had already *passed* the giant planet (on its way to Saturn), while looking *back* toward Jupiter. What can be said about the likely sizes of the particles in the two very different ring systems?

PROBLEMS

Sections 24.1 Existence of Light Waves
24.2 Interference
24.3 Young's Double Slit Experiment

1. Two narrow slits with 0.14 mm between their centers are illuminated by monochromatic light of wavelength 488 nm. What is the angle between the zeroth- and first-order fringes? Express your result in both radians and degrees.

2. A laser with light of wavelength 632.8 nm illuminates a double slit in a Young's double slit experiment. On a screen 2.00 m from the slits, interference fringes are observed that are separated by 1.0 cm. (a) Determine the angle between the zeroth- and first-order fringes. Express your result in both radians and degrees. (b) Determine the separation d of the two slits.

3. In a Young's double slit experiment using light of wavelength 488 nm, the angular separation of the interference fringes on a distant screen is measured to be 1.0°. What is the separation of the two slits used in the experiment?

4. Monochromatic light of unknown wavelength from a distant source is incident along the normal line to the twin slits of a Young's double slit experiment. The slits are separated by 0.50 mm. A screen located 6.50 m from the slits has interference fringes separated by 7.7 mm. What is the incident wavelength?

•5. The double slit experiment of Young is one of the classic experiments of physics. Light with a wavelength of 632.8 nm is incident on two slits separated by 0.10 mm. A screen 1.50 m by 1.75 m is located 5.00 m from the slits. (a) Calculate the separation of the interference fringes on the screen in centimeters. (b) If the slit separation is doubled, what happens to the separation of the fringes on the screen?

•6. Two light waves begin with identical phases from different places, propagate outward, and arrive at the same location. Each wave has amplitude A, wavelength 500 nm, and speed c, but one wave must travel 600 nm farther to arrive there and superimpose with the other wave. (a) What is the frequency of the resulting disturbance? (b) What is the phase difference (in radians) between the waves when they arrive?

•7. Each of two coherent light waves has an amplitude of 10 V/m and a frequency of 5.0×10^{14} Hz. The waves are identical except for a constant relative phase difference ϕ; the waves now are superimposed. (a) What is the wavelength? Is this within the range of visible light wavelengths? (b) If the phase difference is 0.60 rad, what is the combined amplitude? (c) If this phase difference is caused by a difference in path Δx traveled, find Δx. (d) If this phase difference is caused by a difference Δt in starting time, find Δt.

•8. Two synchronized radio wave transmitters are located 100 m apart along a north–south line and emit radio waves with a wavelength of 50.0 m. (Synchronized simply means that the peaks of the waves emitted occur at the same instant of time.) (a) What is the frequency of the radio waves? (b) Are the sources coherent? Explain why or why not. (c) Consider the full 360° horizontal plane of angles measured from the center point between the transmitters. Find all directions to distant locations for which the radio wave intensity is a maximum and sketch these directions on a map including the transmitters.

•9. The great operatic tenor Luciano Pavarotti is singing a perfect note with a frequency of 262 Hz (middle C). The speed of sound in air is 343 m/s. Some distance away are two slitlike openings in a wall, spaced 2.0 m between their centers, as shown in Figure P.9. (a) What is the wavelength of the sound waves? (b) At what minimum nonzero angle θ from the straight-through direction beyond the openings should you locate yourself to hear his voice loud and clear?

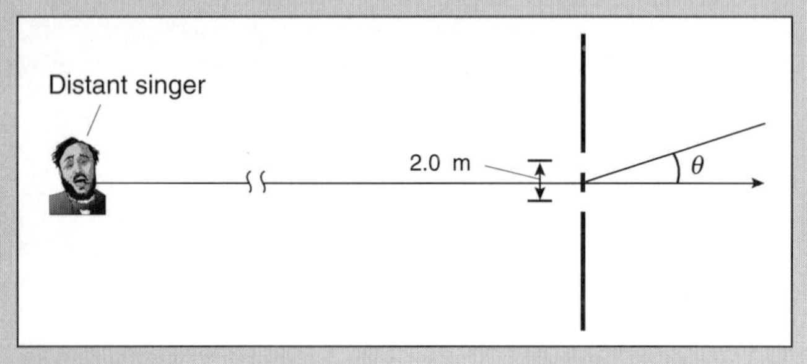

FIGURE P.9

10. Monochromatic light of wavelength λ is incident at an angle ϕ on a pair of slits separated by a distance d as indicated in Figure P.10. Determine the angles θ at which constructive interference takes place on a distant screen.

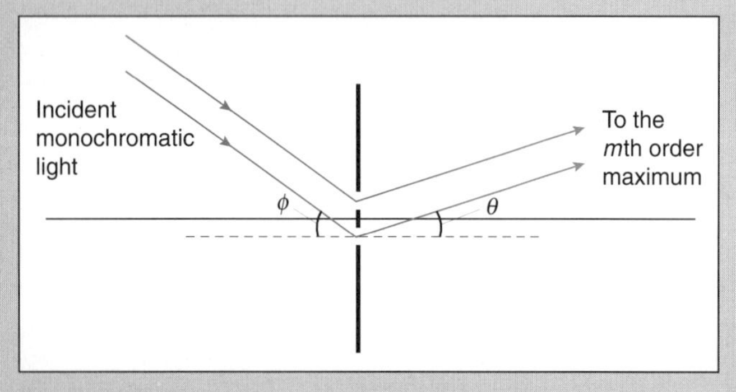

FIGURE P.10

Sections 24.4 Single Slit Diffraction
24.5 Diffraction by a Circular Aperture
24.6 Resolution
24.7 The Double Slit Revisited

11. Find the angle at which the first diffraction minimum is formed when light of wavelength 632.8 nm is sent through: (a) a slit of width 0.10 mm; (b) a circular aperture with a diameter of 0.10 mm.

•12. The separation of the headlights on your car is 110 cm. (a) At what distance are the headlights just barely resolved by a skeptical eye with an aperture of 7.0 mm? Use 550 nm as the effective wavelength of the light emitted by the headlights. Assume the resolution of the eye is diffraction limited, rather than determined by cone cell size on the retina. (b) On the retina of each eye, will there be two diffraction patterns or an interference pattern from two coherent sources?

•13. A distant monochromatic point source of wavelength 600 nm is viewed through a circular hole; the circular central image has an angular radius of 1.0×10^{-3} rad. (a) What is the diameter of the hole? (b) How would you change the diameter of the aperture to produce a smaller image?

•14. Calculate the minimum angular resolution angle for a human eye whose aperture is 7.0 mm for an effective wavelength of 550 nm. Express your result in arc seconds. Assume the resolution of the eye is limited by diffraction rather than by cone cell size. If our eyes had *slit* apertures with a 2.0 mm width instead of circular apertures, would the resolution of the eye be better, worse, or the same? Explain your reasoning.

•15. Estimate the angular separation of two stars that can be barely resolved by a telescope of aperture 10 cm. Express your result in both radians and arc seconds. Use 550 nm as the effective wavelength of the light. Neglect distortion caused by the atmosphere.

•16. The binary star Albireo in the constellation Cygnus (the swan) consists of two stars with an angular separation of 35 arc seconds and is among the most beautiful binary stars to observe. What is the minimum aperture needed for a telescope to resolve the pair? Use 550 nm as the effective wavelength of the light. Neglect distortion caused by the atmosphere.

•17. The Great Wall in China is approximately 5 m wide. Show that it is impossible for an astronaut on the Moon to resolve the thickness of the Great Wall with the unaided eye even when the Moon is at its perigee location in its orbit (its distance of closest approach, 3.63×10^5 km). Use a reasonable estimate of the aperture of the eye of the astronaut.

•18. Georges Seurat (1859–1891) was a French neo-impressionist painter most noted for his pointillist paintings. Aspects of his life were the focal point of a play, *Sunday in the Park with George*, by Stephen Sondheim and James Lapine. Seurat's paintings consist of a myriad of individual colored dots (see Figure P.18 for an example of his work). He must have gone mad painting the millions of dots on his canvases! The dots are on the order of several millimeters in diameter, say, 3.0 mm for the purposes of this problem. If your eye is sufficiently far from the painting, you cannot resolve the dots. Assume an eye with an aperture of 4.0 mm in a moderately lit gallery. For two adjacent red dots (with an effective wavelength of 600 nm), how far from the painting must you stand so you are barely able to resolve the dots if the resolution of the eye is diffraction limited? If two adjacent dots are of blue pigment (with an effective wavelength of 450 nm), what is the distance at which the dots can be barely resolved?

FIGURE P.18

•**19.** Big Brother is watching. A spy satellite at an altitude of 120 km is checking up on your diligence in doing physics homework. The satellite uses a lens with an aperture of 40 cm. Assume light with an effective wavelength of 550 nm. What is the separation of two objects on the ground that can be barely resolved by the lens?

•**20.** The Lilliputians in Jonathan Swift's *Gulliver's Travels* were said to be of a height of "somewhat under six Inches" (say, about 10 cm) and to have eyes that could "see with great Exactness" and "Sharpness."* (a) If we scale the height (say, 2.00 m) and eye diameter (7.0 mm) of a normal human accordingly, what is the aperture of the eye of a Lilliputian? (b) Compare the resolution limit of a Lilliputian eye with that of a normal human, assuming (incorrectly) the resolution of the eye is determined by the Rayleigh criterion. What does this suggest about Swift's knowledge of physics? Of course, Swift published *Gulliver's Travels* in the 18th century, well before Young discovered the wave aspects of light (in 1801), so Swift can be forgiven!
For an amusing article bearing on this question, see David Piggins, "Lilliputian eyes and vision," *Perception*, *7*, pages 609–610 (1978).

•**21.** A telescope with an aperture of 10 cm is used to view Mt. Hood from a distance of 150 km. Assume visible light has an effective wavelength of 550 nm, optimal focusing, and no distortions caused by atmospheric effects or turbulence. (a) Will details on the mountain be seen better with the unaided eye or with the telescope? Explain your answer. (b) What is the closest distance apart on the mountain of two bright objects that, in principle, could be distinguished through the telescope? (c) An eyepiece lens with half the focal length is substituted for the original eyepiece in the telescope. What effect does this change have on the magnification? What effect does this change have on your answer to part (b)?

•**22.** (a) Look at the fine detail of a poster on a distant wall, make appropriate length measurements, and determine the smallest angle your eye can resolve at an effective wavelength of 550 nm. (b) Have a friend estimate the diameter of your pupil† (or do it by looking in a mirror) and compare your measured angle in part (a) with the theoretical angular resolution limit of your eye.

•**23.** A telescope lens of diameter D is used to form an image of a distant star. The image is a diffraction disk in the focal plane of the objective. The aperture of the telescope is decreased to $D/2$. (a) By what factor does the amount of light collected by the telescope decrease? (b) What happens to the size of the diffraction disk of the star? (c) By what factor does the intensity of the light of the central image of the star change?

•**24.** A reconnaissance aircraft at an elevation of 30 km manages to photograph a railroad yard at a military installation. The railroad ties, spaced 40 cm apart, are just barely resolved. Assume no atmospheric distortion. Calculate the minimum diameter of the (circular) lens used to secure the photograph. Assume the effective wavelength of the light is 550 nm.

*Jonathan Swift, *Gulliver's Travels*, Chapter 6 (Doubleday, Garden City, New York, 1945).

†Be careful to do this safely: no rulers in the eye!

Sections 24.8 Multiple Slits: The Diffraction Grating
24.9 Resolution and Angular Dispersion of a Diffraction Grating

25. Light of wavelength 488 nm is incident normally on a grating with a grating spacing of 2.0×10^{-6} m. At what angle from the normal line are the first and second orders of interference located?

•**26.** A source of light emits two different wavelengths. One of them is known to be 488 nm but the other wavelength is not known. Careful experimentation by Milly Hertz has determined that the fourth-order interference maximum of the 488 nm wavelength is at the same angular position as the third-order interference maximum of the unknown wavelength. What is the unknown wavelength?

•**27.** You have pirated a sodium vapor lamp to investigate its spectrum. A diffraction grating with 10 000 slits per centimeter is used to separate the light into its component wavelengths. Sodium has two prominent emission wavelengths at 589.0 nm and 589.6 nm that cast an ugly yellow pall over much of our environment at night. (a) Calculate the angular separation of the two wavelengths in the first-order spectrum. (b) What is the minimum number of slits of the grating that must be illuminated by the light to barely resolve the doublet?

•**28.** You have been assigned the task of designing a grating for use in an astronomical research satellite. Lucky you! The grating is to examine the range of visible light from about 400 nm to 680 nm. The grating must be able to resolve (barely) wavelengths differing by only 0.010 nm in the first order. The first-order spectrum of visible light is to occupy an angular range of 30°. Describe the specifications of the grating, showing the calculations used to determine each parameter specified.

•**29.** The eyes of your fellow students in an 8 A.M. class have become glazed over with a thick coating of lecture verbiage in which there occurs complete destructive interference of too many obfuscations. But lo! White light (400 nm to 700 nm) is incident normally on a grating with 10 000 slits per centimeter. Calculate the angular spread of the resulting beautiful first-order spectrum.

•**30.** For cultural enrichment as well as to demonstrate your algebraic prowess, show that the angular dispersion of a grating (Equation 24.13) can be written in the following form:

$$\frac{d\theta}{d\lambda} = \frac{m}{\left[d^2 - (m\lambda)^2\right]^{1/2}}$$

•**31.** Show that the angular dispersion of a diffraction grating, Equation 24.13, can be expressed as

$$\frac{d\theta}{d\lambda} = \frac{\tan\theta}{\lambda}$$

•**32.** A diffraction grating consists of 500 parallel slits per millimeter. The grating is illuminated along the normal to the grating by a parallel beam of monochromatic light. The first bright line fringes are found at 18.44° on either side of the normal. (a) Find the grating spacing d. (b) Find the wavelength of the light.

•33. Two wavelengths of light are incident normally on a diffraction grating. One of the wavelengths is known to be 656 nm but the other is unknown. However, it is found that the third order of the unknown wavelength appears at the same angle as the second order of the 656 nm wavelength. (a) With this information, is it necessary to know the grating spacing d to determine the unknown wavelength? (b) If the answer to part (a) is no, calculate the unknown wavelength. Otherwise you can take the day off.

•**34.** Two wavelengths are incident normally on an arrangement of parallel slits. The intensity distribution for each of the colors on a distant screen is plotted in Figure P.34 as a function of sin θ. The first maximum for the longer, 600 nm wavelength occurs when sin $\theta = 0.010$. (a) Is it possible to determine the number of slits through which the light passed? If so, what is the number of slits? (b) What is the second wavelength if its interference maxima lie as illustrated relative to the 600 nm pattern? (c) What is the slit spacing d?

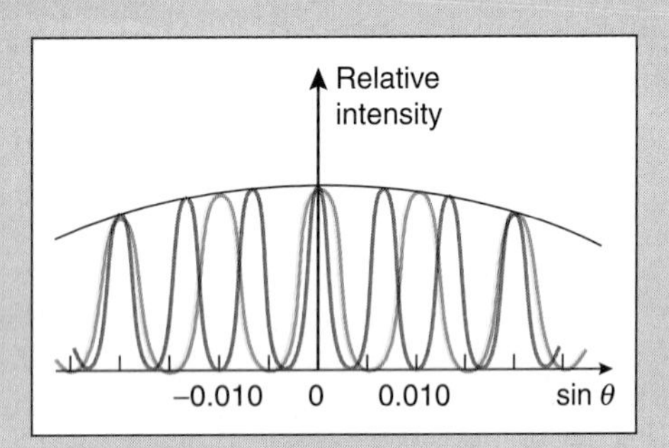

FIGURE P.34

35. A diffraction grating is illuminated at normal incidence by light with a frequency of 5.0×10^{14} Hz from a distant source. The first-order interference maximum occurs at 20° on either side of the straight-through direction. (a) How many slits are in one centimeter of the grating? (b) How many interference fringes, including the center and both sides, are formed by this grating? (c) If the overall width of the grating illuminated were doubled, keeping the same slit separation, what effect would this have on the interference pattern? (d) If the whole system were lowered into pure water ($n = 1.33$), at what angle would the first-order interference appear?

36. Four identical radio transmitter antennas each consist of a vertical wire that radiates radio waves of wavelength λ that propagate "important messages from our sponsors" horizontally in all directions into the hearts and minds of the population. The antenna array forms a square whose edges are equal to the wavelength (see Figure P.36). If all the antennas radiate in phase, in what horizontal directions will the combined radiation be strongest at great distances from the antennas?

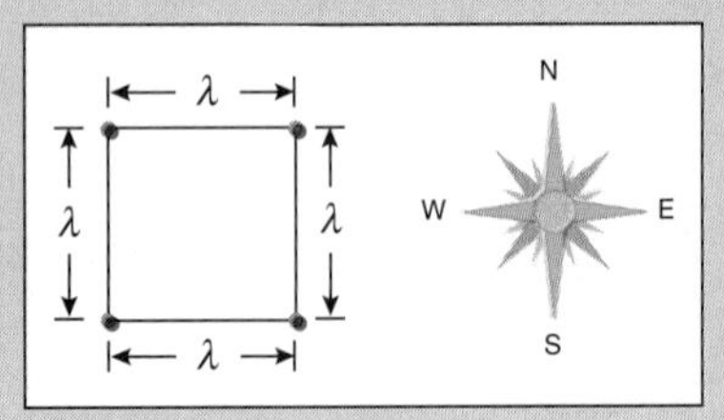

FIGURE P.36

Sections 24.10 The Index of Refraction and the Speed of Light
24.11 Thin-Film Interference*

37. The red light from a helium–neon laser has a wavelength of 632.8 nm. (a) What is the wavelength of this light in glass that has an index of refraction of 1.50? (b) What is the speed of the light in this glass?

38. Light is incident normally on a system of three glass lenses in air. About 4% of the incident light is reflected from each air–glass or glass–air interface. What percentage of the incident light is transmitted by the system?

•**39.** A long group of runners in single file are spaced 2.0 m apart when each is running at a speed of 5.0 m/s on grass. When they reach a sandy surface, they slow to 2.0 m/s, and when they reach pavement, they run at 6.0 m/s. No runners are added and none is lost as they cruise along. (a) How many runners per second (frequency) pass you on the grass? (b) How many runners per second (frequency) pass you on the sand? (c) How many runners per second (frequency) pass you on the pavement? Do any runners vanish? (d) What is their spacing (wavelength) on sand? (e) What is their spacing (wavelength) on pavement?

•**40.** Just to pass the time, the steady traffic flow along a highway is watched at many locations. No cars enter or leave the highway, but their speeds over three different sections are 20 m/s, 40 m/s, and 30 m/s. It is found that the cars traveling at 30 m/s are spaced 30 m apart between centers. (a) At what rate (frequency in cars/s) do cars go past three observing points located at the three specified regions of the highway? (b) What is the spacing (wavelength) between car centers in the other two regions? (c) Which quantity—speed, wavelength, or frequency—is the same all along the highway? Does this make physical sense in terms of piling up of cars over time? (d) What does this problem have to do with the passage of light through a succession of transparent media?

•41. Light is incident on a glass sphere of radius 1.00 cm and index of refraction $n = 1.50$. (a) Sketch the path of the light ray that spends the maximum amount of time within the glass sphere. (b) Calculate this maximum time. (c) Considering only paths of transmission, sketch the path of the light ray that spends the least amount of time inside the glass. (d) Calculate the minimum time. (The answer is not zero.)

•**42.** For the surfaces indicated in Figure P.42, light is incident from the left. What is the least thickness x for a coating that will minimize reflected light with an effective wavelength (in air) of 632.8 nm?

•**43.** A glass-covered mirror has a thin film (with $n_{\text{film}} = 1.60$) to minimize light reflected at normal incidence from the first glass surface, as shown in Figure P.43. The effective wavelength of the light is 550 nm in air. What is the minimum thickness for the coating?

•**44.** A piece of paper is used to form an air wedge between two optical flats as indicated in Figure P.44. Light of wavelength 633 nm is incident normally. In the reflected light, 25 bright interference fringes are observed from the vertex of the wedge to the paper, with the 25th bright fringe coincident with the edge of the paper. A dark fringe appears at the vertex of the air wedge. (a) Why is the fringe at the vertex dark? (b) Calculate the thickness of the piece of paper. (c) Calculate the small

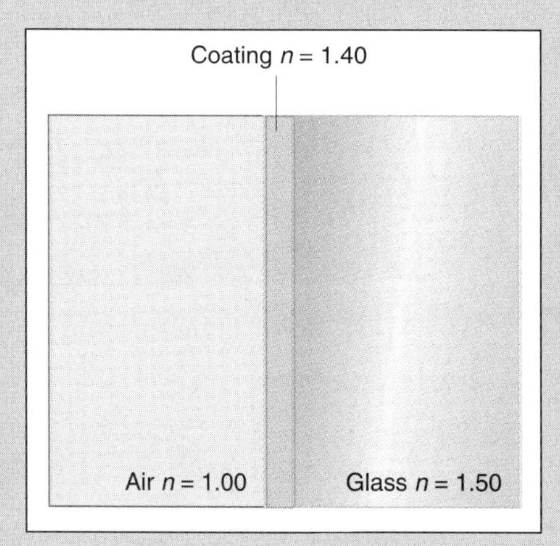

FIGURE P.42

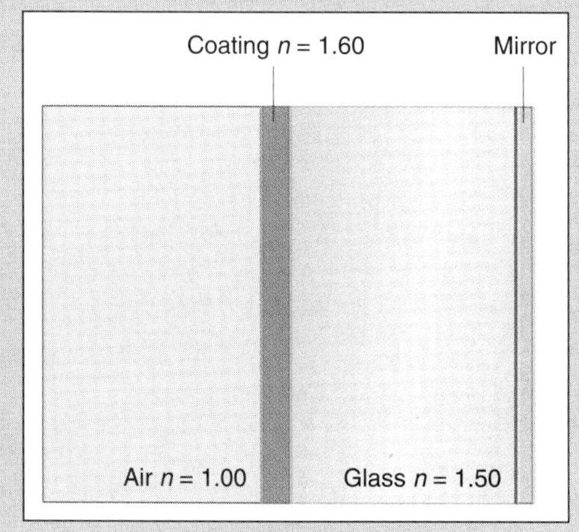

FIGURE P.43

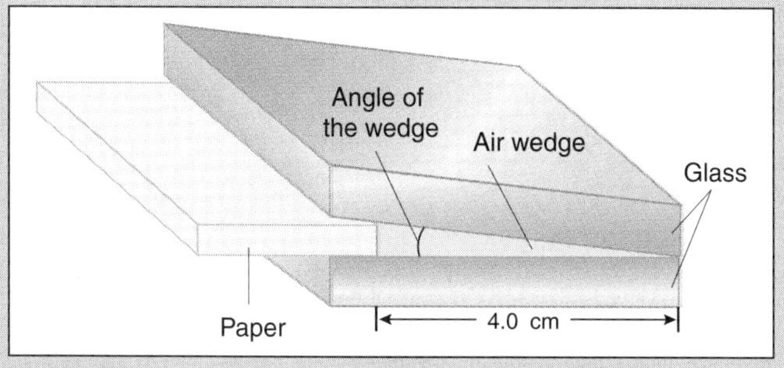

FIGURE P.44

apex angle of the wedge. (d) If the air wedge now is filled with brine ($n = 1.31$), what is the order number of the bright fringe located nearest the edge of the paper?

•45. You have just received delivery of two expensive circular optical flats (serial numbers 1 and 2) from the firm Glass, Waters, and Mirage, Inc. Your supervisor asks you to test the flats by placing them each in turn on another piece of glass that is known to be flat. Light of wavelength 632.8 nm illuminates the combination at normal incidence and the patterns indicated in Figure P.45 are seen in the reflected light. (a) Which of the newly received flats really is flat? (b) What is the approximate shape of the other newly received surface?

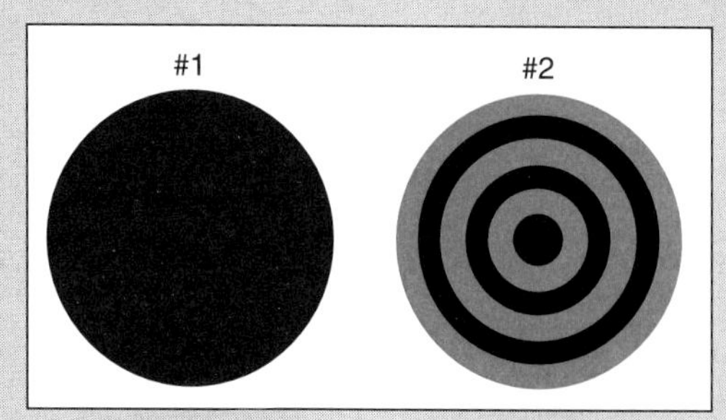

FIGURE P.45

•**46.** A thin oil film ($n = 1.30$) lies on level water and is illuminated normally by white light. The reflection at normal incidence in air lacks the wavelengths 400 nm (violet) and 666 nm (red) but has all the rest of the visible range of wavelengths. (a) What color does the reflection appear? (b) What is the thickness of the film?

•**47.** A thin film of water appears on a glass surface. The water is illuminated by white light at normal incidence. Excruciating measurements find that the thickness of the water film is 8.00×10^{-7} m. What wavelengths are missing in the reflected light in the visible light range of wavelengths (about 400 nm–700 nm)?

•**48.** A thin transparent layer of magnesium fluoride ($n = 1.3$) is applied to *both* sides of a glass plate (or lens) to reduce reflections and to increase transmission. Assume normal incidence. (a) Sketch the layered arrangement and label the reflections, indicating which reflected beams undergo a phase change upon reflection. (b) If the first coating is 317 nm thick, what (vacuum) wavelength in the visible portion of the spectrum does not reflect from the front side of the lens? (c) If the other coating also is 317 nm thick, does the same color also not reflect from the *back* side of the system after going through the glass? (d) What are the nearest (vacuum) wavelengths less than and greater than the one removed that would have *maximum* reflection?

⁑49. A plano-convex lens of index of refraction n is placed on an optical flat as indicated in Figure P.49. The radius of the spherical surface is R. The arrangement is illuminated along the normal with light of wavelength λ. The interference in the light reflected from the two surfaces forming the air gap produces circular fringes known as Newton's rings (although they were apparently first observed by Robert Hooke). (a) Show that the radius r of the mth bright fringe from the center is given approximately by

$$r \approx \left[(2m + 1)\frac{\lambda}{2}R\right]^{1/2}$$

(b) Show that the radius of the mth *dark* fringe from the center is given approximately by

$$r \approx (m\lambda R)^{1/2}$$

⁑**50.** The detector of a radio telescope is located on the Earth's equator 100 m above the sea atop a cliff (see Figure P.50). The telescope is used to detect the radio emissions with a 10.0 cm

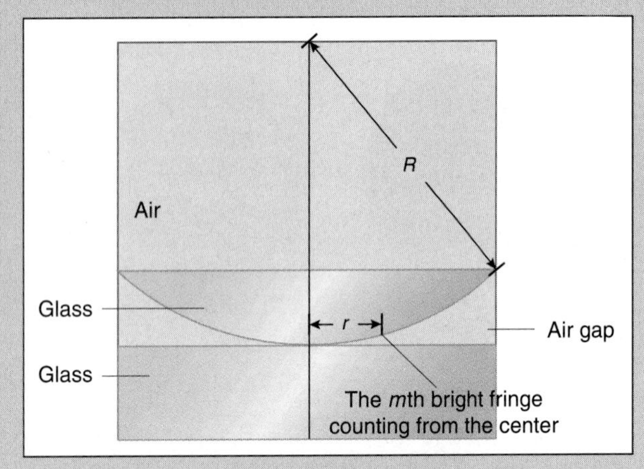

FIGURE P.49

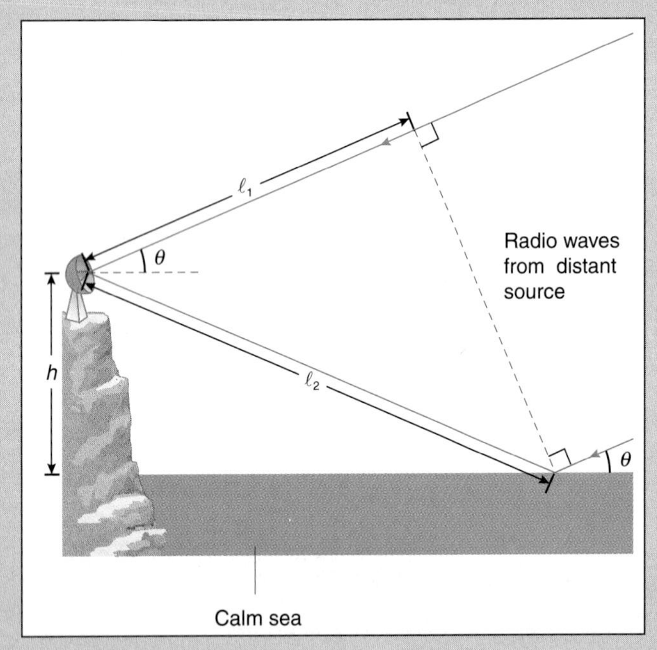

FIGURE P.50

wavelength from a distant star source as it sets in the west. The sea is very calm (i.e., flat). A reflection of the source from the water means that the telescope detects radio waves from *two* sources (the actual source and its image) with a phase difference of π rad between them. The separation of the source and its image decreases with time as the source sets. Neglect atmospheric refraction of the radio waves. (a) Show that the *path difference* between radio waves from the image and those directly from the radio source is (see Figure P.50)

$$\ell_2 - \ell_1 = 2h \sin \theta$$

where h is the height of the detector above the sea. (b) If the path difference is an integral number of wavelengths, will the interference between the double source be constructive or destructive? (c) At the instant the source sets, will the radio telescope detect a constructive interference peak or a destructive interference valley? (d) Sketch the pattern of interference detected by the radio telescope as a function of time as the source approaches the horizon, indicating clearly the instant the source sets. (e) Through how many degrees per hour does the Earth rotate? (f) Calculate the time interval in seconds between the interference peaks detected by the radio telescope as the source sets. (g) If the telescope were located at, say, latitude 37° N, how would this location qualitatively affect the answer to part (e) assuming all other information provided is the same?

Sections 24.12 Polarized Light*
24.13 Polarization by Absorption*
24.14 Malus's Law*
24.15 Polarization by Reflection: Brewster's Law*
24.16 Polarization by Double Refraction*
24.17 Polarization by Scattering*

51. What is the Brewster angle for glass with an index of refraction of 1.6 when used in air? What is the angle of refraction for the transmitted beam in this case?

52. Determine the Brewster angle for light incident on a glass cube ($n = 1.6$) immersed in water ($n = 1.33$).

•**53.** Unpolarized light is incident on a spherical glass marble (with $n = 1.50$) as indicated in Figure P.53. Indicate the directions (relative to the incident direction) at which the light reflected from the marble is completely plane polarized.

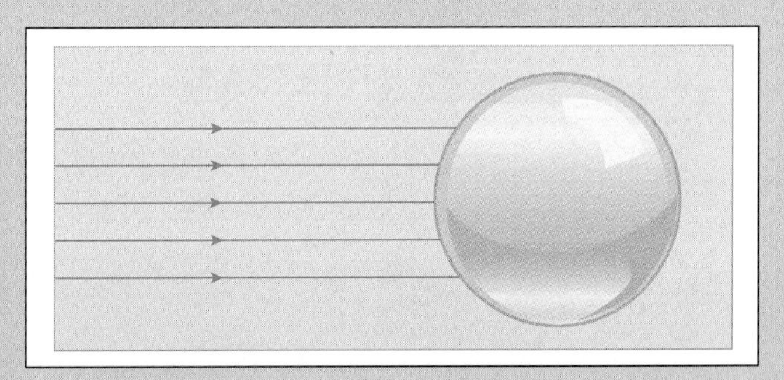

FIGURE P.53

•**54.** A cool loon (*Gavia immer*) floats on a calm lake and surveys the full reflection of the clear bright sky. In what directions down from its eye and around the horizon will the sky image be darkest? Explain.

•**55.** An unpolarized beam of light passes through two polarizers with a 30° angle between their transmission axes. (a) What is the ratio of the amplitude of the light transmitted through both polarizers to the amplitude of the light between the polarizers? (b) What is the ratio of the intensity of the light transmitted through both polarizers to the intensity between the polarizers?

•**56.** Two polarizing sheets are arranged so that initially the light transmitted through the combination is a maximum. (a) Through what angle should the second sheet be rotated so that the transmitted intensity decreases by 50%? (b) Does it make any difference which sheet is rotated through the angle calculated in part (a) if the only goal is to achieve a 50% decrease in the light intensity of the transmitted beam?

•57. Brewster's law states that the angle θ_B at which the reflected beam is completely polarized is related to the indices of refraction of the two media via

$$\tan \theta_B = \frac{n_2}{n_1}$$

where n_1 in the index of refraction of the medium in which the incident ray lies and n_2 is the index of refraction of the medium in which the refracted ray lies. (a) Show that if $n_2 > n_1$, then $\theta_B > 45°$. (b) Show that if $n_2 < n_1$, then $\theta_B < 45°$. (c) If the greatest value of the index of refraction is about 4.50, then find the smallest possible value for the Brewster angle. Describe the experimental situation that yields this minimum value for the Brewster angle. Is the light totally internally reflected?

•**58.** Given $n_1 > n_2$, show that the critical angle for total internal reflection always is greater than the Brewster angle at the same interface.

•**59.** Two pieces of polarizing material (labeled #1 and #2) are oriented so that their transmission axes are perpendicular to each other. Unpolarized light of intensity I_0 is incident on the pair. (a) What is the intensity of the light between the two sheets? (b) What is the intensity of the light after the second sheet? (c) A third polarizing sheet (#3) is now placed between the others so that its transmission axis is oriented at an angle θ relative to the transmission axis of the first sheet; see Figure P.59. What is the intensity of the light between the third and second sheets? (d) What is the intensity of the light after sheet #2?

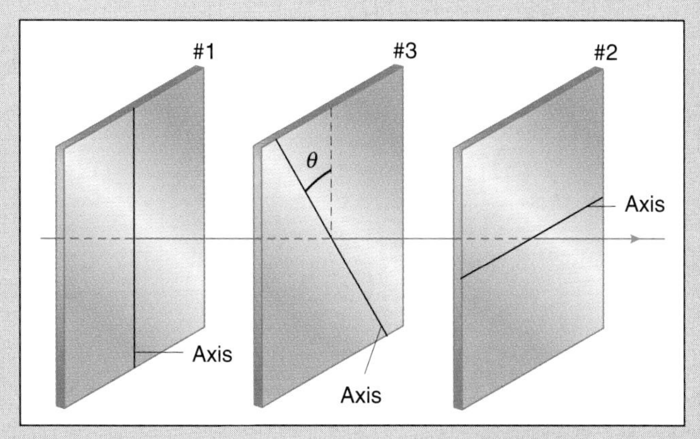

FIGURE P.59

•**60.** Unpolarized light is incident on three polarizing sheets whose transmission axes are oriented at 30° angles with respect to each other as illustrated in Figure P.60. Let I_0 be the intensity of the incident unpolarized light. What is the intensity of the light emerging from the third sheet?

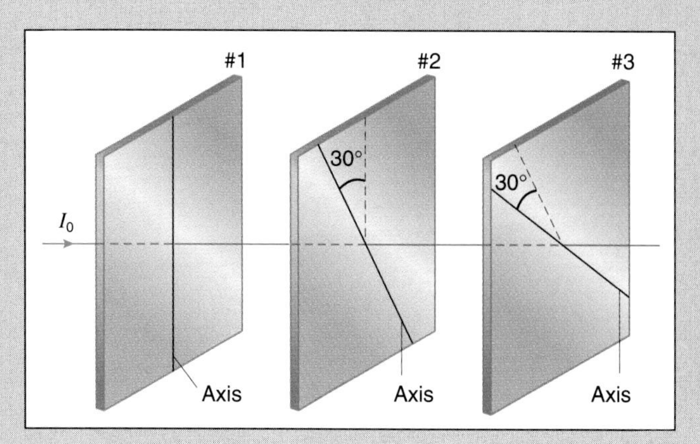

FIGURE P.60

•61. There you are, on the shores of a calm pond, romantically watching the reflection of the full Moon as it silently rises above the eastern horizon. Then you remember a bit of physics: when the elevation of the Moon is just right, the light from the reflected Moon will be completely polarized! To impress your date, and to make manifest the social utility of a knowledge of physics, you explain the situation and take out your Polaroid® sunglasses to demonstrate the effect at the appropriate time. What is the Moon's elevation angle above the horizon when its reflected light is completely polarized?

•**62.** You are trying to photograph the street scene illustrated in Figure P.62. The vertical store windows, whose normal lines are oriented at about 45° to your line of sight, reflect an annoying bright glare. You wish to reduce this glare relative to the general scene. (a) How should the axis of a polarizing filter be oriented in front of the lens to reduce the glare? (b) By what factor will the presence of the filter reduce the light from the bricks, wood, and other objects in the scene?

FIGURE P.62

•**63.** On the days of the March and September equinoxes (~20 March and ~22 September), the Sun sets due west. A vertical side window (with $n = 1.5$) of a car is oriented so that the reflected light of the setting Sun is completely plane polarized. In what four possible directions can the car be pointing? (Express your answer in degrees measured clockwise from north.)

INVESTIGATIVE PROJECTS

A. Expanded Horizons

1. Investigate the relationship between light, color, and the nature of color vision by the eye.
Paul L. Pease, "Resource letter CCV-1: Color and color vision," *American Journal of Physics*, 48, #11, pages 907–917 (November 1980); this contains many references to the subject.
Marcel G. J. Minnaert, *Light and Color in the Outdoors* (Springer-Verlag, New York, 1993).

2. Investigate the uses of lasers in various aspects of medicine.
Alan L. McKenzie and John A. S. Carruth, "Lasers in surgery and medicine," *Physics in Medicine and Biology*, 29, #6, pages 619–642 (June 1984).
Robert A. Weale, "Physics and ophthalmology," *Physics in Medicine and Biology*, 24, #3, pages 489–504 (May 1979).

3. Soap bubbles exhibit many interesting optical and physical effects. Investigate the physics of soap bubbles.
Cyril Isenberg, *The Science of Soap Films and Soap Bubbles* (Dover, New York, 1992).
Charles V. Boys, *Soap Bubbles* (Anchor, Garden City, New York, 1959).
David Lovett, *Demonstrating Science with Soap Films* (Institute of Physics Publishing, Philadelphia, 1994).
Coran Ramme, "Videotaping the lifespan of a soap bubble," *The Physics Teacher*, 33, #9, pages 558–561 (December 1995).
Feredoon Behroozi and Dale W. Olson, "Colorful demos with a long-lasting soap bubble," *American Journal of Physics*, 62, #9, pages 856–857 (September 1994).

4. See how polarized light is used for investigating the distribution of strain in materials. Such techniques have been used by engineers to study what elements of Gothic cathedrals carry the most stress, an important consideration when undertaking repairs or remodeling.
Frank L. Pedrotti and Leno S. Pedrotti, *Introduction to Optics* (Prentice-Hall, Englewood Cliffs, New Jersey, 1987), pages 378–380.

5. Investigate the function of Brewster windows in laser cavities.
Frank L. Pedrotti and Leno S. Pedrotti, *Introduction to Optics* (Prentice-Hall, Englewood Cliffs, New Jersey, 1987), pages 371–372.

6. Determine why it is generally *not* possible to see interference effects in *thick* films.
James Trefil, "Thick film interference," *The Physics Teacher*, 21, #2, pages 119–121 (February 1983).
Joseph C. Amato, Roger E. Williams, and Hugh Helm, "An inexpensive, easy to build Fabry-Perot interferometer and its use in the introductory laboratory," *American Journal of Physics*, 59, #11, pages 992–994 (November 1991).

7. Diffraction gratings come in many varieties; we have investigated the action of a plane, *transmission* grating since the light actually passes through the grating. Other types of gratings include *reflection gratings*, *concave spherical gratings*, and so-called *blazed gratings*. Investigate the characteristics and applications of such gratings.
Frank L. Pedrotti and Leno S. Pedrotti, *Introduction to Optics* (Prentice-Hall, Englewood Cliffs, New Jersey, 1987), pages 417, 419–423.

8. There are many electrical and magnetic effects associated with light. Effects associated with electric fields are known as *electro-optic effects*; those associated with magnetic fields are called *magneto-optic effects*. A few of these effects are listed here and would make fine topics for further reading and research:

Electro-optic effects	*Magneto-optic effects*
Stark effect	Zeeman effect
Kerr electro-optic effect	Cotton–Mouton effect
Electric double refraction	Voigt effect
	Faraday effect
	Kerr magneto-optic effect

These specialized effects cause ripples across many of the sciences and fields of engineering. Our survey of physical optics only touches the proverbial tips of many icebergs.

B. Lab and Field Work

9. For some fundamental experiments into the speed of light, which had unforeseen implications for the development of relativity, Albert A. Michelson (1852–1931) invented a new interference device, now called a *Michelson interferometer*. The device subsequently was used for precise measurements of the meter and also for the measurements of the diameters of large but distant stars. The device creates interference fringes from the amplitude division of a light wave. Investigate the construction of a Michelson interferometer and how it creates interference fringes. Your physics department likely has such an interferometer in its optics laboratory. Use it with an appropriate light source to form such interference fringes; also use it to measure precisely small changes in the position of one of the mirrors on the device. Research the historical importance of the interferometer (in the so-called *Michelson–Morley* series of experiments) in showing that light, unlike other waves, needs no medium in which to propagate.
Francis A. Jenkins and Harvey E. White, *Fundamentals of Optics* (4th edition, McGraw-Hill, New York, 1976).

10. Another interferometer with a wide range of technical applications for the precise measurements of wavelengths in light scattering experiments is the *Fabry–Perot interferometer*, which like the Michelson interferometer (see Investigative Project 9) employs amplitude division of light to create interference fringes. Investigate how a Fabry–Perot interferometer is constructed and how its interference fringes are formed. Determine what is meant by the terms *finesse* and *free spectral range* when associated with this instrument. If your physics department has a Fabry–Perot interferometer available, use it to form interference fringes of an appropriate light source.
Francis A. Jenkins and Harvey E. White, *Fundamentals of Optics* (4th edition, McGraw-Hill, New York, 1976).

11. Take a sheet of polarizing material and use it as an analyzer. Look through it at the liquid crystal display of your calculator, while rotating the analyzer. What do you observe and what conclusions can you draw from your observation? Investigate the principles and uses of liquid crystals in electronic displays.
Renate J. Ondris-Crawford, Gregory P. Crawford, and J. William Doane, "Liquid crystals, the phase of the future," *The Physics Teacher*, 30, #6, pages 332–339 (September 1992).
June E. Ball, "Liquid crystals," *Physics Education*, 15, #2, pages 108–109 (March 1980).

Anthony J. Nicastro, "Demonstrations of some optical properties of liquid crystals," *The Physics Teacher*, *21*, #3, pages 181–182 (March 1983).
Carl H. Hayn, "Liquid crystal displays," *The Physics Teacher*, *19*, #4, pages 256–257 (April 1981).
Renate J. Ondris-Crawford, Gregory P. Crawford, and J. William Doane, "Resource letter LC-1: Liquid crystals: physics and applications," *American Journal of Physics*, *63*, #9, pages 781–788 (September 1995); this paper has a lengthy list of references to the literature on liquid crystals.
Edward F. Carr and James P. McClymer, "A laboratory experiment on interference of polarized light using a liquid crystal," *American Journal of Physics*, *59*, #4, pages 366–367 (April 1991).

12. Doubly refracting (or birefringent) materials are fabricated into special devices known as *quarter-wave plates*. Discover what a quarter-wave plate is and demonstrate several applications of such plates.
Francis A. Jenkins and Harvey E. White, *Fundamentals of Optics* (4th edition, McGraw-Hill, New York, 1976).
Grant R. Fowles, *Introduction to Modern Optics* (2nd edition, Holt, Dover, New York, 1989).

13. Obtain a double refracting crystal (such as calcite) from your instructor. Place the crystal over an X you have made on a piece of paper to form a double image of the X as in Figure 24.50. Rotate the crystal and determine which image is formed from the ordinary ray and which from the extraordinary ray. Now view the images through a polarizer. Rotate the polarizer slowly, describe what happens, and explain your observation.

14. Investigate how chemists use the phenomenon of optical activity to determine the concentration of certain solutions. Design and perform an experiment to determine the concentration of a sugar solution using optical activity.
Thomas M. Lowry, *Optical Rotatory Power* (Dover, New York, 1964).
Francis A. Jenkins and Harvey E. White, *Fundamentals of Optics* (4th edition, McGraw Hill, New York, 1976), pages 572–573, 584–587.

15. Investigate, demonstrate, and explain the optical properties and characteristics of compact discs (CDs).
Christian Nöldeke, "Compact disc diffraction," *The Physics Teacher*, *28*, #7, pages 484–485 (October 1990).
Haym Kruglak, "Diffraction demonstration with a compact disc," *The Physics Teacher*, *31*, #2, page 104 (February 1993).
Thomas D. Rossing, "The compact disc audio system," *The Physics Teacher*, *25*, #9, pages 556–563 (December 1987).

C. Communicating Physics

16. Despite the significant contributions of scientists and engineers to our material well-being, health, and technological prowess, the TV and movie industries generally depict engineers and scientists as mad or deranged nerds, lacking a social conscience and social skills. What might account for such consistently unfavorable depictions of scientific personnel in the entertainment industry? Do you think such attitudes may be a reflection of the unfortunate experience of the tragic poet Sylvia Plath when she took a college physics course? Her comment: *The day I went into physics class it was death*. [*The Bell Jar* (New York, Harper and Row, 1967), Chapter 3.]

CHAPTER 25
THE SPECIAL THEORY OF RELATIVITY

Nothing puzzles me more than time and space; and yet nothing puzzles me less, for I never think about them.

Charles Lamb (1775–1834)*

Like Charles Lamb, Einstein also was puzzled deeply by space and time, but unlike Lamb, he *did* think about them, and seriously too. As a result, our ideas about space, time, light, and energy never will be the same. The theories of relativity are beautiful constructions of the human intellect; you will experience much real intellectual satisfaction from an understanding of them.

There is a certain mysterious aura surrounding relativity in our culture. Perhaps it stems from the popular view that equates the very name of Einstein with genius†; or perhaps it comes from the perceived difficulty of the theories. In any case, the theories of relativity invented by Einstein certainly are among the greatest of the many revolutions in physics during the 20th century. Relativity also begot enormous social and political ramifications as well, foremost among them fearful and awesome nuclear weapons.

Just what is meant by the word **relativity**? The term refers to the relative motion of two reference frames. Einstein created two theories of relativity: the *special theory* (published in 1905), which we consider in significant detail in this chapter; and the more difficult *general theory* (published during 1914–1916), to which we only briefly allude. There also is classical *Galilean relativity* that stems from the now familiar kinematics and dynamics of Galileo and Newton.

25.1 Reference Frames

The world of events forms a four-dimensional continuum.
Albert Einstein (1879–1955) and Leopold Infeld (1898–1968)‡

A **reference frame** consists of (1) a coordinate system and (2) a set of synchronized clocks distributed throughout the coordinate grid and at rest with respect to it.

*(Chapter Opener) Letter to Thomas Manning, 2 January 1810, in *The Complete Works and Letters of Charles Lamb* (The Modern Library, New York, 1935), letter CLXV, pages 775–777; the quote is on page 776.

†We all likely have referred to a gifted individual as "an Einstein."

‡*The Evolution of Physics* (Simon & Schuster, New York, 1938), page 219.

One of the most recognizable faces of the 20th century, indeed, of human history!

For example, we can consider a reference frame to be, say, a Cartesian coordinate system and a set of clocks distributed at any and every location throughout the coordinate system, as shown in Figure 25.1.

The clocks all are synchronized, so that they indicate the same time and tick at the same rate. The clocks are like all the timepieces that you and your classmates are wearing; they all indicate the same standard time (or daylight saving time)§ and progress at the same rate. We think of the clocks as having a stopwatch feature, so that time intervals can be measured conveniently. A reference frame thus has three spatial coordinates, such as x, y, and z, and one time coordinate, t; from a mathematical viewpoint, a reference frame is a four-dimensional space–time affair.

Historical events specify places and dates. The tragic assassination of President John F. Kennedy occurred just outside the Texas School Book Depository in Dallas at 12:30 P.M. on 22 November 1963; the Civil War Battle of Antietam occurred on 17 September 1862 in Maryland near Antietam Creek, a tributary of the Potomac River. Events in physics are no different, but hopefully are less tragic. As we did in Chapter 3 (long ago!), we define an **event** as something that happens at a particular place and instant: where and when. To specify where an event occurs means choosing a coordinate system, say a Cartesian coordinate system. A set of synchronized clocks is needed to specify when. Thus we use a reference frame to specify an event. Different reference frames describe the same event in different ways.

Among other things, relativity is concerned with how an event described in one reference frame is related to its description in another reference frame.

In particular, relativity specifies how the coordinates and times of events (and other physical quantities such as momentum, energy, and electric and magnetic fields) measured or specified in one reference frame are related to the coordinates, time, and corresponding physical quantities in another reference frame.

Given any pair of reference frames:

1. the reference frames may be moving at constant velocity (including zero velocity) with respect to each other; or

§The clocks in different time zones on the Earth all tick at the same rate but indicate times that differ from each other by an hour (or a half hour, in some instances). Nonetheless, we can set them all to the same time, much like simultaneously started stopwatches. The clocks then fulfill the condition for the clocks in a reference frame: the same time and the same tick-rate.

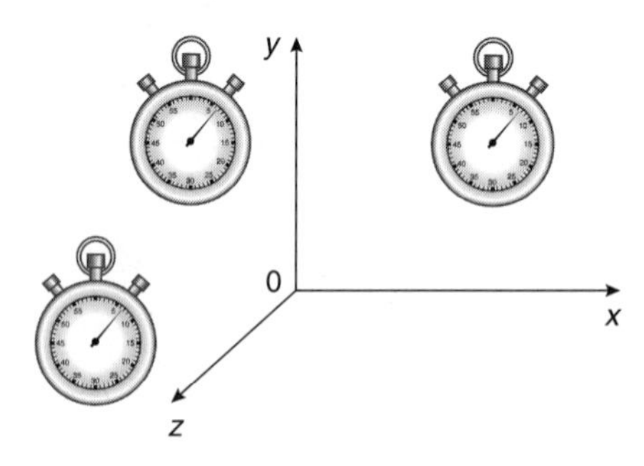

FIGURE 25.1 A reference frame is a coordinate system and a collection of synchronized clocks at rest.

2. one reference frame may be accelerating (which includes rotations) with respect to the other.

Inertial reference frames are the ones in which Newton's first law is true (see Chapter 5). Reference frames that are not inertial are called **noninertial reference frames**; accelerated reference frames are noninertial.

The **special theory of relativity** is concerned with the relationship between events and physical quantities specified in different inertial reference frames (and only these *special* reference frames). The **general theory of relativity** is concerned with the relationship between events and physical quantities in *any* (i.e., *general*) reference frames.

The equations that indicate how the four space and time coordinates (and physical quantities) specified in one reference frame are related to the corresponding quantities specified in another reference frame are called **transformation equations**. You already are familiar with transformation equations in a mathematical context when changing coordinate systems (say from rectangular to polar coordinates, or vice versa), and so the concept of such transformations should not be intimidating.

The **standard geometry** we use for the special theory of relativity has two inertial reference frames called S and S′, with their *x*- and *x*′-coordinate axes collinear (see Figure 25.2). Imagine collections of clocks (stopwatches) distributed at rest throughout each respective frame; the clocks all are set to 0 s when the two origins coincide.

If we imagine ourselves at rest in reference frame S, then frame S′ is moving to the right at any chosen constant velocity $v\hat{\imath}$, measured using clocks and rulers in S.

Changing horses, we imagine ourselves at rest in S′; then frame S is moving to the left with a velocity $-v\hat{\imath}'$, as in Figure 25.3, measured using clocks and rulers in S′. Note that the *relative speed* v (a positive scalar!) of the two systems is the same; they simply move in opposite directions with respect to each other.

QUESTION 1

Reference frames attached to (a) a plane flying with a constant velocity in nonturbulent air, (b) a ship traveling with a constant velocity in calm seas, or (c) your car moving at constant velocity across the level plains of Kansas all are examples of inertial reference frames. A reference frame attached to a spinning merry-go-round, an airplane taking off or landing, or your car while braking are examples of noninertial reference frames. Describe a specific experiment to tell whether a given reference frame is an inertial reference frame.

25.2 CLASSICAL GALILEAN RELATIVITY

Classical **Galilean relativity** relates the space and time coordinates and other physical quantities in two inertial reference frames when the relative speed v of the two frames is much less than the speed of light c—that is, when $v << c$.

This is the realm of our everyday common experience. The equations that relate the space and time coordinates and other physical quantities, such as velocity components, between the two standard inertial reference frames, when $v << c$, are the **Galilean transformation equations**.

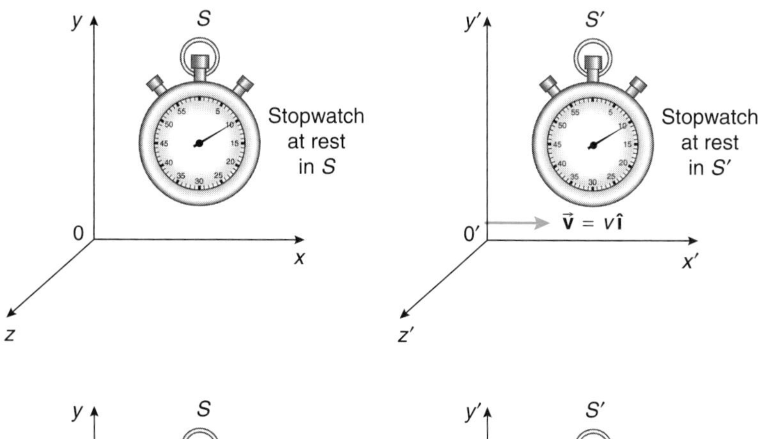

FIGURE 25.2 Frame S′ moves to the right as seen from frame S in the standard geometry.

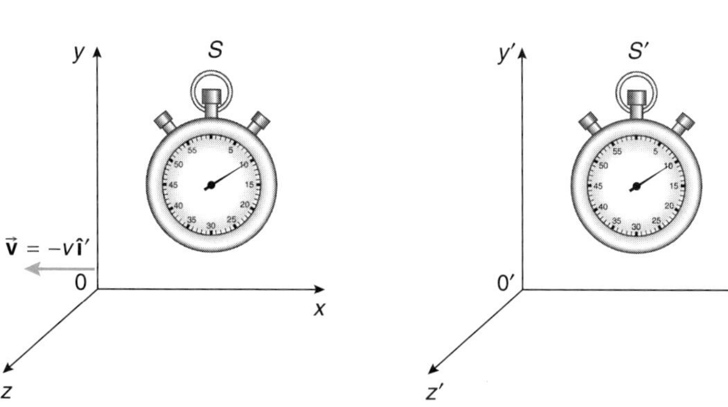

FIGURE 25.3 Frame S moves to the left as seen from frame S′ in the standard geometry.

The Galilean Time Transformation Equation

In classical physics time is a *universal* measure of the chronological ordering of events and the time interval between them.*

Watches in fast sports cars, airplanes, spacecraft, and oxcarts tick at the same rate as those at rest on the ground; this is a common, almost trivial observation, usually unstated and taken for granted. The time interval between two events and the rate at which time passes are independent of the speed of the moving clock; they are the same everywhere. In classical physics all the clocks distributed throughout any two inertial reference frames tick at the same rate and, once synchronized, can be arranged to indicate the same time instants: that is,

$$t' = t \tag{25.1}$$

The Galilean Spatial Coordinate Transformation Equations

The origins of the two inertial frames having the standard geometry coincide when $t = t' = 0$ s; see Figure 25.4. Therefore an object at position x' in frame S′ has the coordinate x in frame S, where

$$x = x' + vt \tag{25.2}$$

or, solving for x',

$$x' = x - vt \tag{25.3}$$

Note the slight asymmetry of these equations. The difference in the sign of the vt term arises because frame S′ is moving to the *right* as seen by frame S, while frame S is moving to the *left* when viewed from frame S′ in the standard geometry, so the velocity components of the frames have opposite signs. The other coordinates are identical; that is,

$$y' = y \tag{25.4}$$

$$z' = z \tag{25.5}$$

*We shall see in special relativity that this *cannot* be the case (Section 25.4). In special relativity the temporal interval between events depends on the inertial reference frame.

The Galilean space and time transformation equations are Equations 25.1 and 25.3–25.5, summarized here:

$$x' = x - vt \tag{25.3}$$

$$y' = y \tag{25.4}$$

$$z' = z \tag{25.5}$$

$$t' = t \tag{25.1}$$

The transformation equations for some other physical quantities can be found from Equations 25.1 and 25.3–25.5. For example, we consider the velocity and acceleration component transformations in the next two subsections.

The Galilean Velocity Component Transformation Equations

When you run horizontally at 3 m/s and toss a ball at 5 m/s in the same direction, the velocity component of the ball with respect to the ground is the algebraic sum of the velocity components: 8 m/s. We can see how this comes about from the Galilean transformation equations for position.

Imagine a particle moving to the right in frame S′ along the x'-axis with a velocity component

$$u_x' = \frac{dx'}{dt'}$$

in frame S′;† see Figure 25.5. The velocity component u_x of the particle from the viewpoint of the frame S is

$$u_x = \frac{dx}{dt}$$

The two velocity components can be related using the transformation Equation 25.3:

$$\begin{aligned} u_x &= \frac{dx}{dt} \\ &= \frac{d(x' + vt)}{dt} \end{aligned}$$

†We apologize for the slight change in notation, but in this section (and later in Sections 25.11 and 25.16), it is convenient to use the letter $\vec{\mathbf{u}}$ for velocity of a particle here to distinguish it clearly from the relative speed v of the two inertial reference frames.

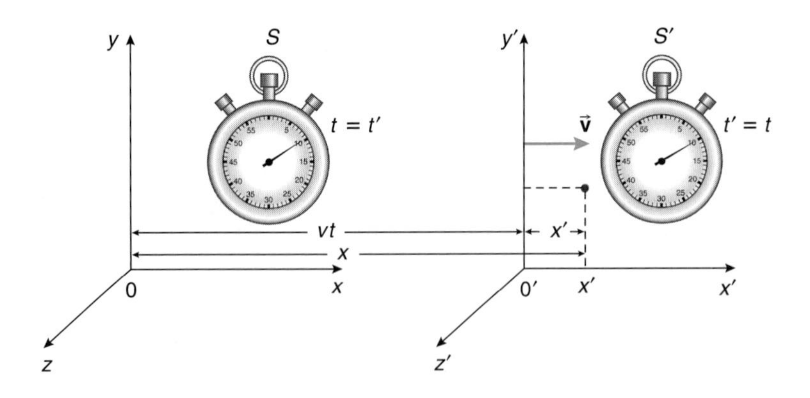

FIGURE 25.4 The view from frame S, when $t = t' > 0$ s.

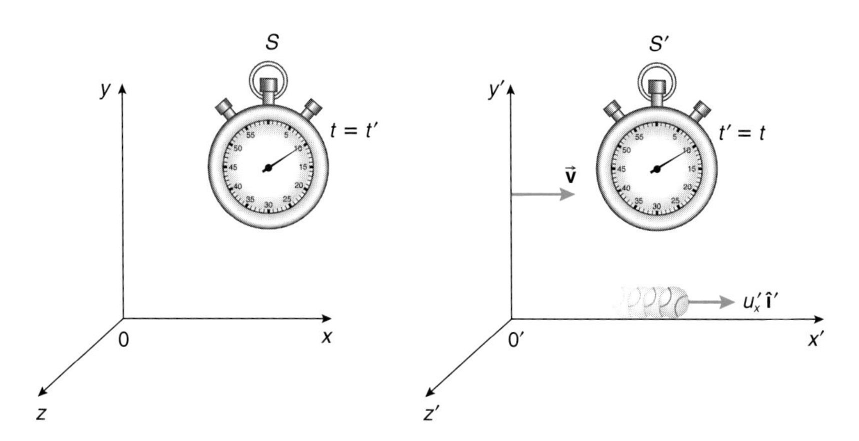

FIGURE 25.5 A particle moving in reference frame S'.

Recall that the relative speed v of the two inertial frames is constant. Hence we have

$$u_x = \frac{dx'}{dt} + v$$

Since $t' = t$ according to Galilean relativity (Equation 25.1), we have

$$u_x = \frac{dx'}{dt'} + v$$

But

$$\frac{dx'}{dt'} = u'_x$$

is the velocity component of the object in the frame S'.

Thus we have

$$u_x = u'_x + v \qquad (25.6)$$

Equation 25.6 agrees with our commonsense ideas about the way the velocity components should add. Indeed, Equation 25.6 is a modified version of the relative velocity addition equation we played with back in Chapter 4 (Equation 4.20).

We imagine a particle moving in the S' frame along the y'-axis, as in Figure 25.6. Its velocity component is

$$u'_y = \frac{dy'}{dt'}$$

as measured in frame S'. In reference frame S, the corresponding velocity component is

$$u_y = \frac{dy}{dt}$$

But since $t = t'$ and $y = y'$, the two velocity components are equal:

$$u_y = \frac{dy}{dt} = \frac{dy'}{dt'} = u'_y \qquad (25.7)$$

The velocity components along a direction *perpendicular* to the motion are the same in the two standard inertial reference frames.

The Galilean Acceleration Component Transformation Equations

We imagine a particle in frame S' that has an acceleration component

$$a'_x = \frac{d^2x'}{dt'^2}$$

The acceleration component of the particle as measured in frame S is

$$a_x = \frac{d^2x}{dt^2}$$

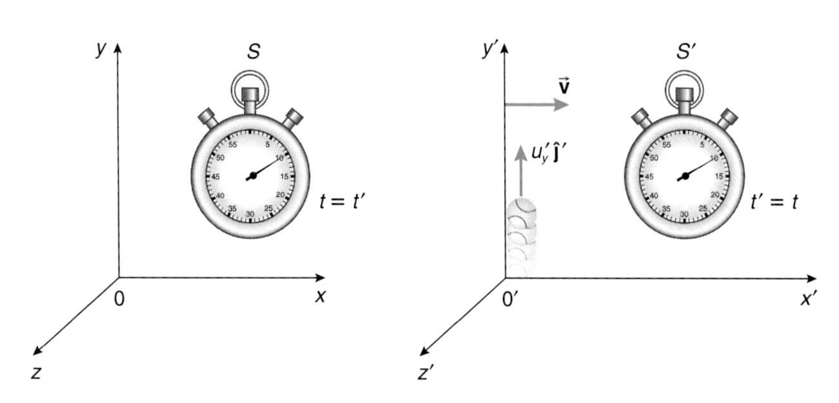

FIGURE 25.6 A particle moving along the y'-axis in frame S'.

We use Equation 25.2 for x and Equation 25.1 for t, remembering that the velocity component v is constant:

$$\begin{aligned} a_x &= \frac{d^2x}{dt^2} \\ &= \frac{d^2(x' + vt)}{dt^2} \\ &= \frac{d^2x'}{dt'^2} + 0 \text{ m/s}^2 \\ &= a_x' \end{aligned} \qquad (25.8)$$

In a similar fashion, we can show that

$$a_y = a_y'$$
$$a_z = a_z'$$

The acceleration components are the same in the two inertial reference frames. (We showed this once before in Section 4.4.)

It is the equality of the accelerations in the two inertial reference frames that restricts the application of Newton's laws of motion to such inertial reference frames.

25.3 The Need for Change and the Postulates of the Special Theory

The relativity theory arose from necessity, from serious and deep contradictions in the old theory from which there seemed no escape. The strength of the new theory lies in the consistency and simplicity with which it solves all these difficulties, using only a very few convincing assumptions.

Albert Einstein and Leopold Infeld*

So what's the deal? What is the need for changing the Galilean ideas of relativity? None of the Galilean results indicates that there is any subtlety about nature or its workings that warrants changing the Galilean transformation equations between the two inertial reference frames. Where $v << c$, experiments verify the validity of the equations within the common kinematic world of dynamics; welcome back to the world of 19th-century kinematics and dynamics! The problem is that we have little direct experience with objects or reference frames moving at speeds comparable to that of light. However, electromagnetism provides some subtle clues that all is not right with Galilean relativity.

The Maxwell theory predicts the existence of electromagnetic waves that were shown to travel at a speed equal to the known speed of light; they *are* light. The theory indicates that the speed c of such waves in vacuum stems from the permittivity of free space ε_0 and the permeability of free space μ_0 (see Sections 21.4 and 21.5):

$$c = \frac{1}{(\mu_0\varepsilon_0)^{1/2}}$$

Note that there is no dependence on the speed v of the source of the waves in this equation!

The Evolution of Physics (Simon & Schuster, New York, 1938), page 203.

Now we imagine an experiment that precisely measures the speed of light. Measuring the speed of light presents significant technological challenges, but we ignore them to focus on simply the results of such experiments. The experiment is performed twice, once with light from a stationary source, where the result obtained is c. The reference frame in which we perform this experiment we call S, as shown in Figure 25.7.

The experiment then is performed again when the source (at rest in frame S′) is moving at speed v in the same direction as the light is moving, as in Figure 25.8. The measurements again are made using clocks and rulers at rest in reference frame S.

According to Galilean relativity (Equation 25.6), the second experiment should yield $c + v$ for the speed of light, since the velocity components are both positive. In fact, the second experiment indicates that the speed of light is *still* c, the identical value obtained when the source is at rest! Such experiments have been performed many times in many ingenious different ways.† The results are always the same.

The measured speed of light in vacuum has the same numerical value c = 299 792 458 m/s, independent of the speed of the source *or* that of the observer, which contradicts the predictions of Galilean relativity.

†The most famous of these experiments is known as the Michelson–Morley experiment; see Investigative Project 2. Other experiments also confirm it: atoms moving at many different speeds emit light with the same speed c. A binary star, with one component approaching and the other receding from us, emits light with the same speed c.

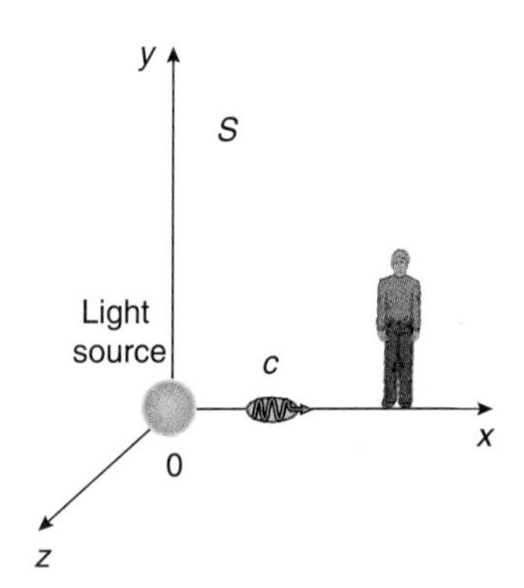

Figure 25.7 Measure the speed of light in S with the source at rest in S. The result is c.

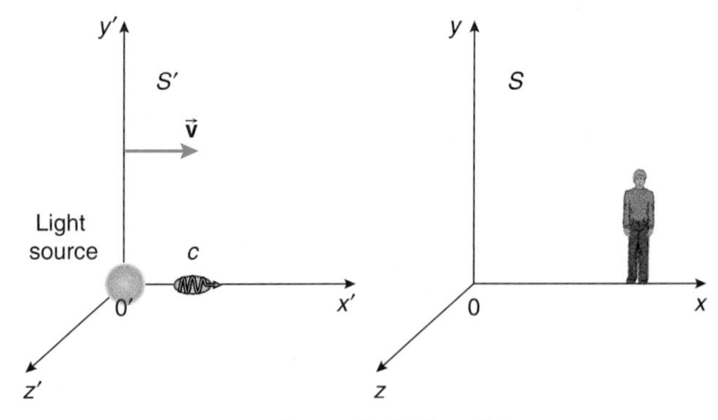

Figure 25.8 Measure the speed of light in S from a moving source that is at rest in S′. The result *also* is c.

These results are peculiar, even baffling, for they defy common sense, that is, the prediction of Galilean relativity. Since physics ultimately rests on experiment, the speed-of-light experiments indicate that Maxwell must be right and the speed of light must not depend on the speed of the source or observer.*

Therefore the Galilean transformation equations between two inertial reference frames need to be modified to accommodate the observed invariance of the speed of light. The modifications needed to make this accommodation truly revolutionized physics. The new transformation equations, called the **Lorentz transformation equations** (which we derive in Section 25.7), together with their implications and consequences, constitute the special theory of relativity.

The entire special theory stems from only two postulates, one rooted in experiment, the other stemming from aesthetics, even natural philosophy as well as experiment. The first bold postulate encompasses all measurements, early and modern, of the speed of light and the prediction of c by the Maxwell theory of electromagnetism.

Postulate 1: The speed of light in a vacuum has the same numerical value c when measured in any inertial reference frame, independent of the motion of the source and/or observer.

Despite the confounding nature of this statement, it reflects the way nature is; physics must take nature on its own terms.

The second postulate of the theory makes an aesthetic argument about the apparent experimental equivalence of all inertial reference frames.

Postulate 2: The fundamental laws of physics must be the same in all inertial reference frames.

Certainly we can apply the same laws of physics to the gas atoms, electric charges, or even frisbees in an airplane cabin flying at constant velocity as we can when it is at the terminal gate. In so many words, the second postulate states that if the laws of physics are codified into a sacred text (this one!?) in an inertial frame S, the *same* sacred text can be used in *any* other inertial reference frame S′. Whatever laws of physics are discovered in one inertial frame can be used in any other inertial frame. If this were not the case, new physics texts would have to be written for each inertial reference frame and the task of learning physics would be compounded immensely; that is a recipe for total confusion. So we insist that the laws of physics, whatever they may be, have the same form for all inertial observers.

Descriptions of what happens as a result of the laws of physics may differ from one inertial reference frame to another, but the underlying fundamental physical principles and laws are the same. For example, while running at constant velocity across the room, let your professor toss a ball vertically upward according to her. You see the ball follow a parabolic trajectory. She says it traveled in a straight line vertically up and down. But you both begin your description with the same laws of physics.

The entire edifice of the special theory of relativity rests on the foundation of these two postulates. Without either, the entire theory crumbles.

On these two commandments hang all the law and the prophets.
The New Testament, Matthew 22:40

QUESTION 2

The effects of special relativity are only apparent for speeds that are significant fractions of the speed of light. Based on your experience, estimate the order of magnitude of the speed of the fastest material thing you have actually seen with your naked eye.

25.4 TIME DILATION

In classical physics it was always assumed that clocks in motion and at rest have the same rhythm, that rods in motion and at rest have the same length. If the velocity of light is the same in all CS [coordinate systems], if the relativity theory is valid, then we must sacrifice this assumption. It is difficult to get rid of deep-rooted prejudices, but there is no other way.
Albert Einstein and Leopold Infeld†

Before we tackle the problem of finding the new transformation equations between the inertial reference frames S and S′, let us explore several interesting and peculiar consequences of the special theory of relativity. If we accept the postulates of the theory, we are hooked and cannot avoid their consequences.

We construct several identical, special clocks to measure time. They all run the same way. Each clock consists of a pulsed laser, a mirror located a distance ℓ_0 from the laser, and, adjacent to the laser, a detector and trigger mechanism to stimulate the next emission of a pulse of light from the laser, as shown in Figure 25.9.

The operation of each light clock is described by three events:

1. emission of the pulse from the laser;
2. reflection of the pulse from the mirror; and
3. detection of the pulse (and emission of the next pulse at the same instant).

The time interval for the laser pulse to travel to the mirror and back to the detector is the total round-trip distance traveled, $2\ell_0$, divided by the speed of light c. So the fundamental time interval τ_0, the tick rate of each clock at rest, is

$$\tau_0 = \frac{2\ell_0}{c} \tag{25.9}$$

This time interval is called the **rest time interval** or **proper time interval** of the clock because it is measured in the frame in which the clock is at rest.

To characterize the three events of each clock, place the laser and detector at rest in inertial reference frame S′ with the laser and detector at its origin, as in Figure 25.10. Let this clock at rest in S′ move at constant velocity (together with S′) along the x-axis of another inertial reference frame S.

*Historically, there is confusion as to what experiments (if any) motivated Einstein, even according to Einstein himself! (See Investigative Project 2.) On the one hand, he says that he was motivated by electromagnetic problems. Later in his life, however, he said that the Michelson–Morley experiments were important in convincing him of the validity of Maxwell's theory. He also wondered what it would be like to ride alongside a light wave.

†*The Evolution of Physics* (Simon & Schuster, New York, 1938), page 196.

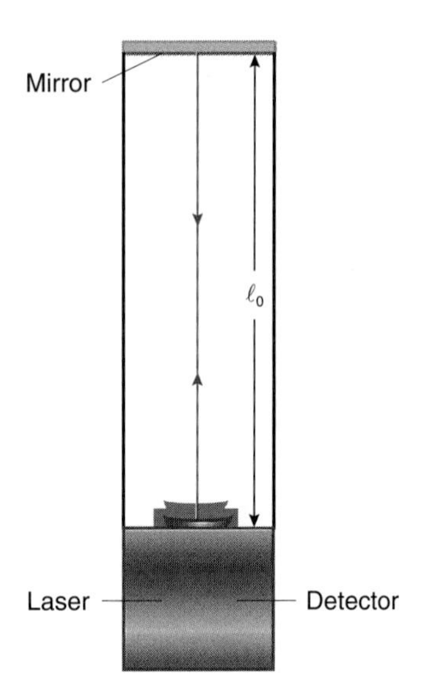

FIGURE 25.9 A special light clock.

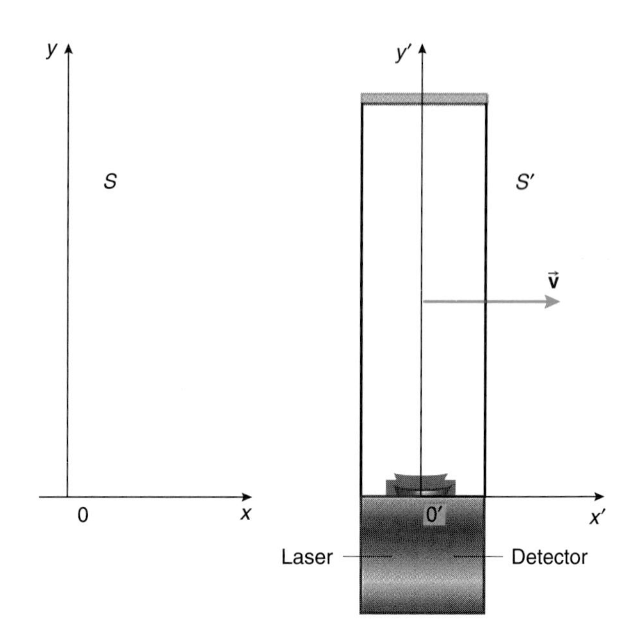

FIGURE 25.10 A clock at rest in S' is moving as seen from S.

The three events are specified by indicating where and when they occur. In the reference frame S', the space and time coordinates of each event are as follows:

1. emission of the pulse from the laser:

$$x_1' = 0 \text{ m}$$
$$y_1' = 0 \text{ m}$$
$$z_1' = 0 \text{ m}$$
$$t_1' = 0 \text{ s}$$

2. reflection of the pulse from the mirror:

$$x_2' = 0 \text{ m}$$
$$y_2' = \ell_0$$
$$z_2' = 0 \text{ m}$$
$$t_2' = \frac{\ell_0}{c}$$

3. detection of the pulse and emission of the next pulse:

$$x_3' = 0 \text{ m}$$
$$y_3' = 0 \text{ m}$$
$$z_3' = 0 \text{ m}$$
$$t_3' = \frac{2\ell_0}{c}$$

Identical clocks are at rest in reference frame S in order to measure time intervals in the S frame. We want to describe the three events associated with the moving clock from the perspective of the S reference frame. In particular, we want to find the time interval τ between Event 3 and Event 1, for this is the tick rate of the *moving* clock as measured by the stationary clocks in S.

We arrange things so that Event 1 (emission of the light pulse of the moving clock) occurs at the instant the S' origin is coincident with the origin in S. We start a stopwatch in S at this instant; see Figure 25.11. The stopwatch keeps time in agreement with the identical light clock at rest in this frame. Event 1 is described in S by

1. emission of the light pulse:

$$x_1 = 0 \text{ m}$$
$$y_1 = 0 \text{ m}$$
$$z_1 = 0 \text{ m}$$
$$t_1 = 0 \text{ s}$$

While the light pulse is propagating from the laser to the mirror, the clock at rest in reference frame S' is moving to the right according to frame S. When the stopwatch reads $\tau/2$ in frame S, the light is reflected from the mirror; see Figure 25.12.

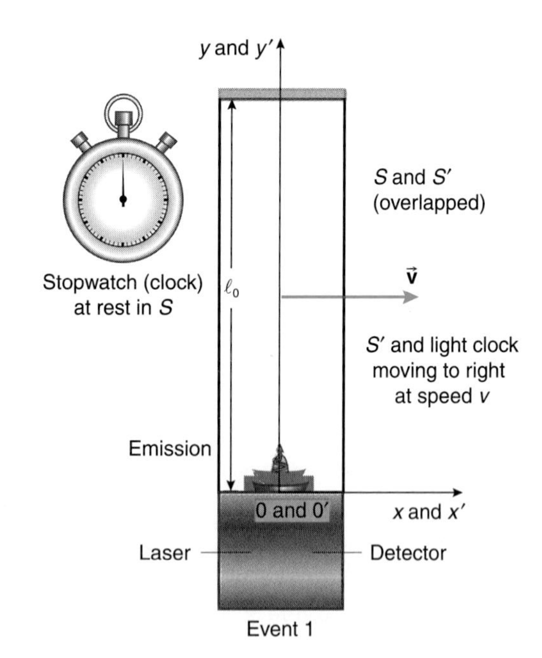

FIGURE 25.11 Emission of the light pulse seen from frame S.

The clock has moved a distance to the right (in S) equal to

$$\frac{v\tau}{2}$$

between Event 1 and Event 2. Event 2 is described in the reference frame S by the following space and time coordinates:

2. reflection of the light by the mirror:

$$x_2 = \frac{v\tau}{2}$$
$$y_2 = \ell_0$$
$$z_2 = 0 \text{ m}$$
$$t_2 = \frac{\tau}{2}$$

Between Events 2 and 3 the clock continues to move to the right in frame S. The light pulse then is detected by the detector, as in Figure 25.13. Event 3 is described in S in the following way:

3. detection of pulse (and emission of the next pulse):

$$x_3 = v\tau$$
$$y_3 = 0 \text{ m}$$
$$z_3 = 0 \text{ m}$$
$$t_3 = \tau$$

Notice in Figure 25.12 that in the S reference frame, the light was emitted at the coordinates $x_1 = 0$ m and $y_1 = 0$ m (Event 1) and reflected at coordinates $x_2 = v\tau/2$ and $y_2 = \ell_0$. From the well-worn Pythagorean theorem, the path following by the light in the reference frame S is of length

$$\sqrt{\ell_0^2 + \left(\frac{v\tau}{2}\right)^2} \tag{25.10}$$

From the first postulate of special relativity, the light travels this distance at the familiar constant speed c. The time for the light to travel any distance is the distance divided by the speed:

$$\frac{\sqrt{\ell_0^2 + \left(\frac{v\tau}{2}\right)^2}}{c}$$

But this is just the time interval between Events 2 and 1 in S, which is $\tau/2$. Thus

$$\frac{\tau}{2} = \frac{\sqrt{\ell_0^2 + \left(\frac{v\tau}{2}\right)^2}}{c} \tag{25.11}$$

Now we solve Equation 25.11 for the time interval τ. Recall that τ represents the time between emission and detection of the light pulse as measured by an observer in reference frame S using clocks at rest in that reference frame (the time between Events 3 and 1). We eliminate the square root by squaring Equation 25.11, obtaining

$$\frac{\tau^2}{4} = \frac{\ell_0^2 + \left(\frac{v\tau}{2}\right)^2}{c^2}$$

Now we turn the proverbial mathematical crank to obtain an expression for τ^2. After some algebra we find

$$\tau^2 = \frac{4\ell_0^2}{c^2 - v^2}$$

For convenience we factor out c^2 from the denominator, obtaining

$$\tau^2 = \frac{\frac{4\ell_0^2}{c^2}}{1 - \frac{v^2}{c^2}}$$

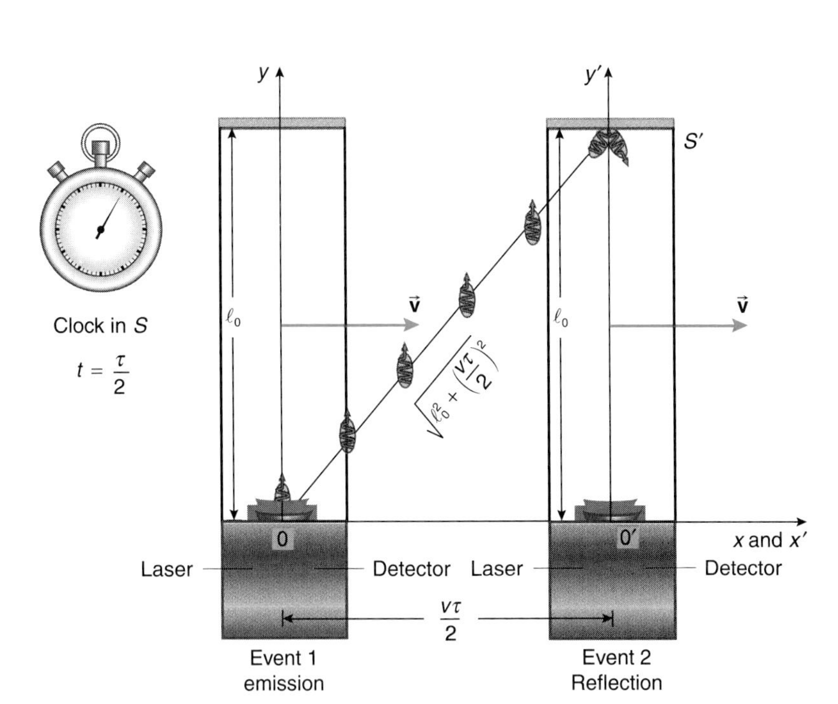

FIGURE 25.12 Light reflected from mirror, as seen in frame S.

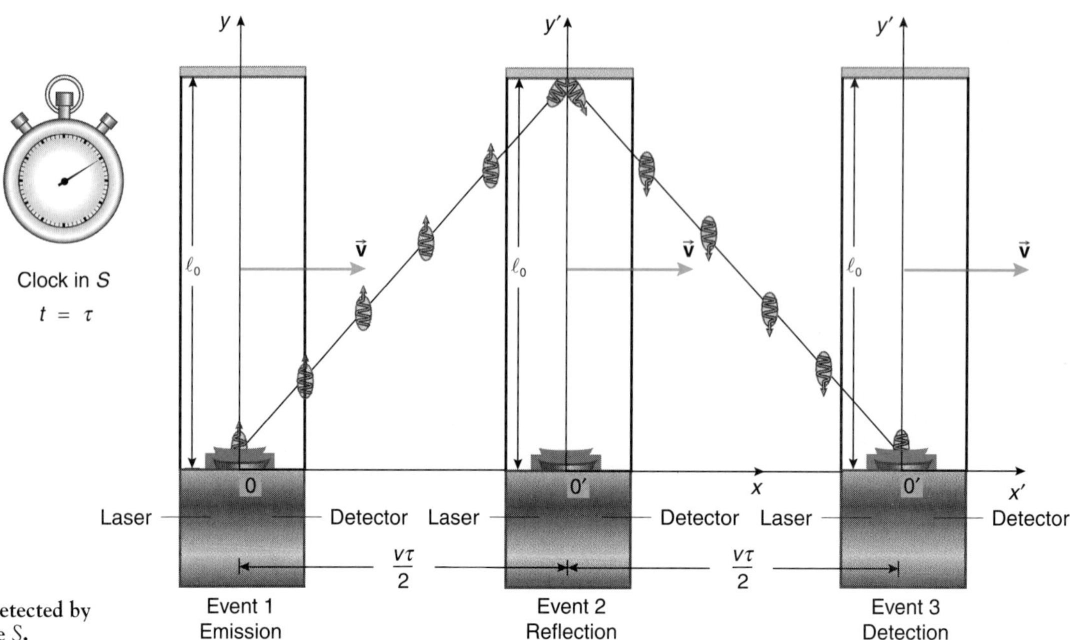

FIGURE 25.13 Light detected by detector, as seen in frame S.

We take the (positive) square root to obtain τ itself:

$$\tau = \frac{\frac{2\ell_0}{c}}{\sqrt{1 - \frac{v^2}{c^2}}}$$

But the quantity $2\ell_0/c$ is the fundamental tick rate τ_0, the proper time of the clock as measured in the reference frame S' in which the clock is at rest (Equation 25.9). The quantity $2\ell_0/c$ also is the fundamental tick rate of the identical clocks at rest in S.

Thus the relationship between the time intervals is

$$\tau = \frac{\tau_0}{\sqrt{1 - \frac{v^2}{c^2}}} \qquad (25.12)$$

The square root is a pure real number less than 1 when $v < c$. Thus the time interval between the emission and detection of the light in the moving clock, as measured by the clock at rest in S, is *longer* than the time interval between the two events as measured in the reference frame S' where the clock stays at rest. The moving clock is seen to tick more slowly than the identical clock at rest. The moving clock has a greater time interval between its ticks than the clock that is at rest; a moving clock runs slow. This effect is called **time dilation**.

Thus time intervals in relativity are not absolute or universal but depend on whether the clock is moving or not: we say time is a relative not an absolute quantity. If the clock is at rest with respect to you, such as the watch on your wrist, it shows no time dilation and nothing is peculiar about its rate of ticking. It is only the tick rate of a clock moving with respect to you that ticks more slowly than your own clock at rest. The slower tick rate is determined and measured by clocks at rest with respect to you—the clocks in your own reference frame.

Time dilation has been confirmed by many experiments, some using radioactive particles as natural clocks. In another experiment, the U.S. Naval Observatory flew two atomic clocks around the world in opposite directions (first class, mind you, but with no champagne or caviar, even for its government chaperons) and compared them with a clock left behind. The experiment involved both the predictions of special and general relativity; the clocks confirmed the predictions of both theories.

At first glance, you might think you can cheat nature and extend your lifetime using time dilation simply by moving fast. Sorry! Your aging process is determined by natural biological clocks that *always are at rest* with respect to you. The relativity of time is real only for clocks moving with respect to you. Thus your own individual aging always proceeds quite normally, no matter what speed you happen to be traveling.

We get another useful piece of information from this analysis. In particular, notice that if $v > c$ in Equation 25.12, the square root becomes a purely *imaginary number*. So, if the moving clock has a speed $v > c$, τ becomes an imaginary time. Frankly, we do not know what to make physically of this imaginary time. Ordinary time and other physical quantities are represented by real numbers, since the measurements we make from meter sticks, clocks, and other instruments always yield real numbers. Nonetheless, the imaginary prediction when $v > c$ has not prevented theorists from speculating about the possible existence of faster-than-light particles, called **tachyons**. Despite feverish theoretical and experimental activity, no experimental evidence exists for such tachyon particles (see Investigative Project 3).

Later, in Section 25.17, we see that it is impossible for a particle initially traveling at a speed less than c to attain a speed equal to c.* Hence in special relativity the speed of light evidently is the ultimate speed limit.

The inverse of the purely numerical square root in Equation 25.12 occurs so frequently in relativity that it is given its own special symbol: γ.

By definition, we take

$$\gamma \equiv \frac{1}{\sqrt{1-\frac{v^2}{c^2}}} \tag{25.13}$$

With the definition of γ, the time dilation Equation 25.12 is succinctly written as

$$\tau = \gamma\tau_0 \tag{25.14}$$

A graph of γ versus the speed ratio v/c is shown in Figure 25.14. Notice that if $v = 0$ m/s, then $\gamma = 1$. As v approaches c, γ increases dramatically in value, diverging to ∞ when $v = c$.

PROBLEM-SOLVING TACTIC

25.1 Be sure to remember that γ is the *inverse* of the square root

$$\sqrt{1-\frac{v^2}{c^2}}$$

The thing to keep in mind about γ is that it is *always* greater than or equal to 1:

$$\gamma \geq 1$$

When calculating γ with your calculator, do not forget to take the inverse of the square root.

*We will see that for a particle initially with speed $v < c$, it takes an *infinite* amount of work to increase its speed to $v = c$. Since such an amount of energy transfer simply is unavailable, any particle with a speed less than c will remain at speeds less than c.

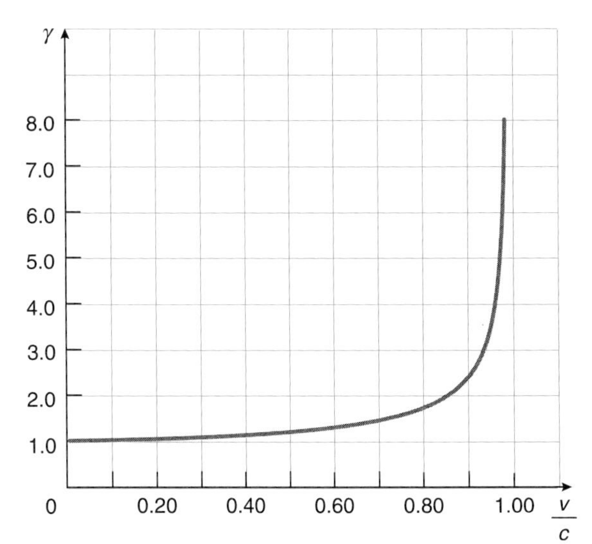

FIGURE 25.14 A graph of γ versus v/c.

QUESTION 3

Moving clocks run slow. Does time dilation depend on the velocity of the clock or simply on its speed?

EXAMPLE 25.1

Clocks in a moving S' reference frame indicate an interval between noon (12 h 0 min 0 s) and 12 h 1 min 0 s. Clocks in reference frame S indicate the same interval as noon till 12 h 2 min 0 s. How fast is frame S' moving with respect to frame S?

Solution

Use Equation 25.14. The one minute on the S' moving clocks is measured to be two minutes according to clocks in S. The proper time τ_0 of the moving clocks is one minute, since this one minute interval is indicated on their dials. In S, clocks indicate the interval τ of the moving clock to be two minutes. Hence Equation 25.14 becomes

$$2.00 \text{ min} = \gamma(1.00 \text{ min})$$

Thus you have

$$\gamma = 2.00$$

Use the definition of γ, Equation 25.13, to find v/c:

$$\gamma = \frac{1}{\sqrt{1-\frac{v^2}{c^2}}}$$

$$2.00 = \frac{1}{\sqrt{1-\frac{v^2}{c^2}}}$$

After squaring, solve for v/c:

$$4.00 = \frac{1}{1-\frac{v^2}{c^2}}$$

$$\frac{v^2}{c^2} = 0.750$$

$$v = 0.866c$$

The S' clocks are moving at 86.6% the speed of light, or 2.60×10^8 m/s.

STRATEGIC EXAMPLE 25.2

A day on the Earth has 24.000 hours. How fast must a spacecraft travel so that the spacecraft clocks tick through 23.000 hours (as indicated by the clock hands on the spacecraft clocks) while the Earth clocks tick through 24.000 hours from the viewpoint of an observer at rest on the Earth?

Solution

Identify the Earth with reference frame S and the spacecraft with reference frame S', as in Figure 25.15. Since the spacecraft (frame S')

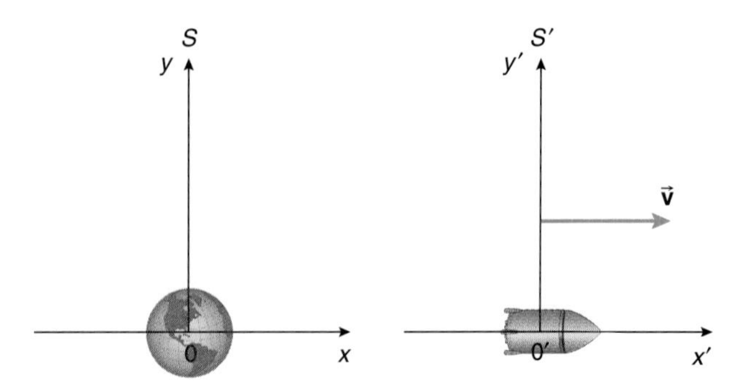

FIGURE 25.15 The standard geometry.

is moving relative to the Earth (frame S), the spacecraft clocks run slow compared with the clocks on the Earth. You can find the value of γ in two ways.

Method 1
The clocks on the spacecraft tick through 23 hours while the Earth clocks tick through 24 hours. Thus the spacecraft hours are dilated or are longer than the Earth hours by the factor

$$\frac{24.000 \text{ h}}{23.000 \text{ h}} = 1.0435$$

In other words, it takes 1.0435 hours according to the Earth clocks for the spacecraft clocks to tick through one hour. In this way, when the Earth clocks have ticked through 24.000 hours, the spacecraft clocks have ticked through only 23.000 hours.

So, summarizing, the moving spacecraft clocks are slow according to an observer at rest with respect to the Earth clocks. Hence the moving spacecraft clock hours are dilated by the factor 1.0435 compared with the Earth clock hours. The dilated time interval τ is related to the proper time interval τ_0 by the time dilation equation (Equation 25.14):

$$\tau = \gamma\tau_0$$

Since $\gamma > 1$, you know $\tau > \tau_0$. Thus you can write

$$1.0435 \text{ h} = \gamma(1.0000 \text{ h})$$

and so

$$\gamma = 1.0435$$

Method 2
Identify the 23.000 h on the spacecraft clock as the proper time interval τ_0, since this clock is at rest at a fixed location on the spacecraft. The Earth observer measures a dilated time interval τ of 24.000 h. The time dilation equation (Equation 25.14) becomes

$$24.000 \text{ h} = \gamma(23.000 \text{ h})$$

and so

$$\gamma = 1.0435$$

Having found γ, you can solve for the speed since

$$\gamma = \frac{1}{\sqrt{1 - \dfrac{v^2}{c^2}}}$$

$$1.0435 = \frac{1}{\sqrt{1 - \dfrac{v^2}{c^2}}}$$

Solving for v/c, you find

$$\frac{v}{c} = 0.2857$$

Hence the speed is

$$\begin{aligned} v &= 0.2857c \\ &= 0.2857 \times 3.00 \times 10^8 \text{ m/s} \\ &= 0.857 \times 10^8 \text{ m/s} \end{aligned}$$

25.5 LENGTHS PERPENDICULAR TO THE DIRECTION OF MOTION

Wait a minute. There is a residual question that needs to be addressed in connection with our derivation of the relativity of time. If something as basic as time is affected by relative motion, who is to say that lengths are not affected as well? In particular, in analyzing the light clock in the previous section, a tacit assumption was made: we assumed the length ℓ_0 of the clock was unaffected by the motion. That is, the length of the moving light clock attached to frame S′ was assumed (implicitly) to be the same as for the clock at rest in frame S.

If the length of the moving clock is affected by the motion in some way, that could easily account for the difference in the tick rates of the two clocks. In particular, if the length of the moving clock was longer as measured in frame S than its length at rest (in frame S′), then the time interval between emission and detection of the light would, of course, be longer as measured in S. So we have to be sure nothing funny is going on associated with a moving length ℓ_0 oriented perpendicular to the direction of motion. We now show that lengths measured *perpendicular* to the direction of the motion are unaffected by the motion.*

First, we have to clarify what we mean by the term "moving length." Take two identical sticks of length ℓ_0, cut from the same stock at the same time. Glue one of the sticks to the y′-axis in reference frame S′ and the other to the y-axis in reference frame S, as in Figure 25.16.

Now let the S′ reference frame go past the S frame at speed v. An observer in reference frame S compares the length of the stick nailed down to the y-axis in the S frame with the length of the stick nailed down to the y′-axis in the S′ frame. The comparison

*In Section 25.6 we shall have something significant to say about lengths *parallel* to (along) the direction of motion.

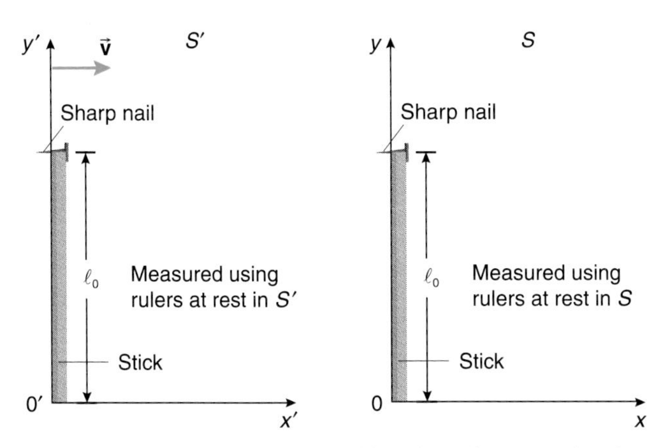

FIGURE 25.16 Two sticks of equal length, each measured in their respective reference frames. Note that S' is here on the left, the origins have not yet coincided, so both t and t' are < 0 s, but we are not concerned with time here, only length.

of the lengths can be made in the following way. A sharp nail is attached to the top of each stick. When the two sticks go by each other, if they are not the same length, the nail on the shorter stick will make a scratch mark on the longer stick.

We prove the sticks are the same length by contradiction. That is, we make the hypothesis that such a length *is* affected by the motion and then show that a contradiction develops, so that the hypothesis must be false.

We make the hypothesis that a moving stick, oriented perpendicular to the direction of motion, is *longer* than an identically placed stick at rest.

An observer in reference frame S then makes the following observations (see Figure 25.17):

1. The stick glued to the y'-axis in S', perpendicular to the direction of motion of the S' frame, is moving.
2. By hypothesis, the moving stick glued to the y'-axis in S' is longer (as measured in S) than the stick at rest in S; this is shown in Figure 25.17.
3. When the moving stick passes by the stick at rest in S, the nail on the stick in S will make a scratch on the stick in S'.
4. The nail on the stick in S' will *miss* the stick in S.

The observer in S therefore concludes

a. the stick in S' has a scratch on it; and
b. the stick in S has no scratch on it.

Now we switch reference frames. An observer stationed in S' sees the S frame moving to the left, as shown in Figure 25.18. The observer in S' makes the following observations:

1. The stick in S, perpendicular to the direction of motion of the S frame, is moving.
2. By hypothesis, the moving length glued to the y-axis in reference frame S is measured in S' to be longer than the stick at rest in S'.
3. When the moving stick passes by the stick in S', the nail on the stick in S' will make a scratch on the stick in the S frame.
4. The nail on the stick in the S frame misses the stick in the S' frame.

Therefore the observer in the S' frame concludes

a. the stick in S has a scratch on it; and
b. the stick in S' has no scratch on it.

The results in each reference frame can be communicated to the other reference frame at their leisure (by radio signals or e-mail), so that a comparison of their conclusions can be made. But notice that the observers in S and S' come to *contradictory* conclusions about which stick has the scratch on it. Thus the hypothesis must be false.

Contradictory conclusions also ensue if we assume that a moving length oriented perpendicular to the direction of its motion is *shorter* than the length at rest; you might want to check the reasoning for this scenario yourself.

Thus the only legitimate conclusion we can draw is the following.

> Lengths measured perpendicular to the direction of motion are *unaffected* by the motion.

The two nails on the tops of the sticks pass by each other at the same height. Thus our assumption about the length of the light clock was legitimate; the length is not affected by the motion

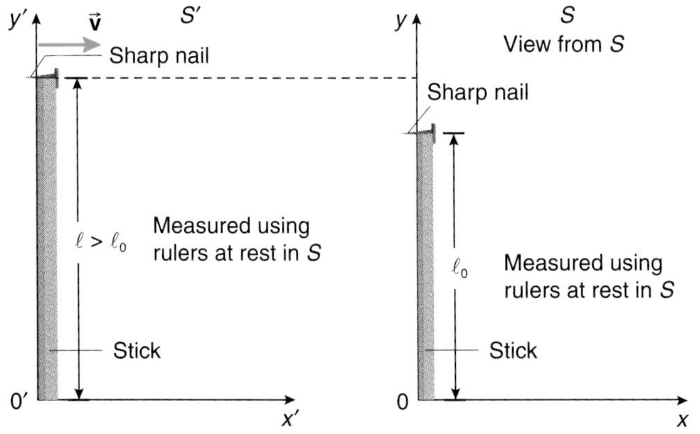

FIGURE 25.17 The view from S: frame S' is moving. Hypothesis: Frame S measures the moving stick (in S') to be longer than the stick at rest in S.

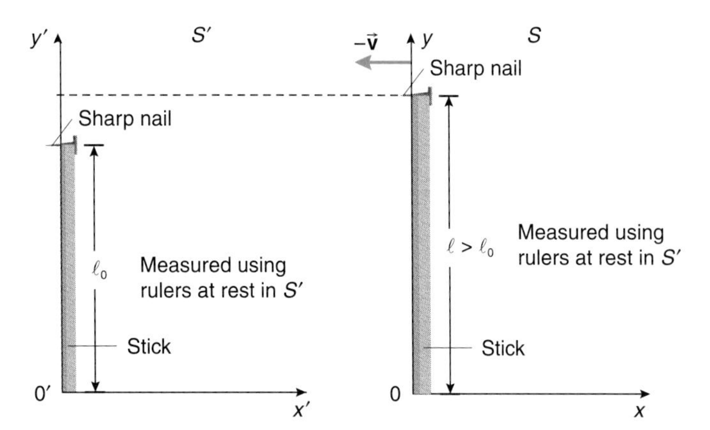

FIGURE 25.18 The view from S': frame S is moving. Hypothesis: Frame S' measures the moving stick (in S) to be longer than the stick at rest in S'.

because the length of the light clocks is oriented perpendicular to the direction of the motion!

25.6 Lengths Oriented Along the Direction of Motion: Length Contraction

In the previous section, we proved that lengths perpendicular to the direction of motion are *unaffected* by the motion. Here we find—surprisingly—that if oriented along the direction of motion, a length *is* affected by the motion!

To show this, we imagine a particle (consider it to be our fairy godmother) that "lives" for only a fixed period of time τ_0 according to her own watch, a clock at rest with respect to her.* The time interval τ_0 thus is a proper time interval or rest time interval.

The fairy godmother particle travels at high speed v across our laboratory during its lifetime and we note that it traverses a distance ℓ_0 (say, from one end of the lab to the other) during its brief existence. The distance ℓ_0 is a **rest length** or **proper length** since the lab is at rest with respect to us when we measure its dimensions with rulers also at rest with respect to us in the lab.

Our clocks in the lab measure the lifetime of the particle to be the dilated time $\tau = \gamma\tau_0$, according to the clocks on the wall of the laboratory, because the particle lifetime (its watch) is a moving clock. We calculate the speed v of the particle by dividing the measured distance ℓ_0 it travels by the time $\tau = \gamma\tau_0$ it exists according to our clocks:

$$v = \frac{\ell_0}{\tau}$$

$$= \frac{\ell_0}{\gamma\tau_0} \qquad (25.15)$$

Notice that to determine v, we used a distance and a time measured using rulers and clocks at rest in our reference frame.

Now take the view of our fairy godmother particle. She sees the length of our lab moving past her at the same relative speed v. Using rulers at rest with respect to her, she measures the length of our lab to be ℓ, whose value is to be determined. According to our fairy godmother (using her watch), the lab passes by during her lifetime τ_0. Therefore she determines the speed of the lab by dividing its measured length ℓ by the time τ_0 it takes to pass by her:

$$v = \frac{\ell}{\tau_0} \qquad (25.16)$$

Notice that to determine v, she used a distance and a time measured using rulers and clocks at rest in her reference frame.

The relative speeds in Equations 25.15 and 25.16 are the same. Equating them yields

$$\frac{\ell}{\tau_0} = \frac{\ell_0}{\gamma\tau_0}$$

*Rather than unobservable fairy godmothers, physicists use radioactive particles with a well-defined half-life for this purpose; we discuss radioactivity in Chapter 26.

The moving length is

$$\ell = \frac{\ell_0}{\gamma} \qquad (25.17)$$

The proper length ℓ_0 is *shorter* when measured in a frame in which it is moving. This is called relativistic **length contraction.**

Length contraction only occurs for those lengths (or components of lengths) oriented along the direction of motion.

Lengths (or components of lengths) perpendicular to the direction of motion are unaffected by it, as we showed in Section 25.5. We measure the contracted moving length using rulers at rest with respect to ourselves.

Example 25.3

A javelin with a proper (rest) length of 2.00 m is moving past you at the incredible speed of 0.95c. What do you measure to be the length of the javelin, using rulers at rest with respect to you?

Solution

Identify the frame in which the javelin is at rest as S', where it has the proper length $\ell_0 = 2.00$ m. Your reference frame you identify as S, in which the javelin is moving at speed 0.95c.

Determine the value of the factor γ from its definition, Equation 25.13:

$$\gamma = \frac{1}{\sqrt{1 - \frac{v^2}{c^2}}}$$

Since $v/c = 0.95$, this becomes

$$\gamma = \frac{1}{\sqrt{1 - (0.95)^2}}$$

$$= 3.2$$

The javelin is moving in your reference frame and so appears contracted by the factor $1/\gamma$. You measure its length ℓ (using rulers at rest with respect to you) to be

$$\ell = \frac{\ell_0}{\gamma}$$

$$= \frac{2.00 \text{ m}}{3.2}$$

$$= 0.63 \text{ m}$$

25.7 The Lorentz Transformation Equations

The old mechanics is valid for small velocities and forms the limiting case of the new one.

Albert Einstein and Leopold Infeld†

In Section 25.3 we saw that the Galilean transformation equations relating the space and time coordinates of two inertial reference

†*The Evolution of Physics* (Simon & Schuster, New York, 1938), page 204.

frames needed to be replaced with another transformation to account for the peculiar constancy of the speed of light, an invariance that violates the Galilean velocity component addition rule. However, since Galilean relativity is quite valid for speeds $v << c$, the new transformation equations must approach the Galilean equations in the limit of small speeds. The new transformation equations are known as the Lorentz transformation equations. In this section we develop their form.

The relative motion of the two reference frames is along the x- and x'-coordinate axes in the standard geometry. The y- and y'-axes, as well as the z- and z'-axes all are perpendicular to the direction of motion. Since we showed in Section 25.5 that lengths measured perpendicular to the direction of motion are unaffected by it, the Lorentz transformation must have

$$y' = y \tag{25.18}$$

and

$$z' = z \tag{25.19}$$

It remains for us to determine how the spatial coordinates x and x' are related, as well as the time instants t and t'.

The Galilean transformation gives us a clue about these new relationships. Notice in Equation 25.3 that the relationship between x' and x also involves t. Thus it is tempting to write

$$x' = Ax + Bt \tag{25.20}$$

where A and B are constants (which may depend on the constant relative speed v) whose values we wish to find.

From the viewpoint of reference frame S, the origin of the S′ reference frame is located using the equation

$$x = vt$$

because the origins coincide when $t = 0$ s in the standard geometry, as indicated in Figure 25.19. We use this fact in Equation 25.20 in the following way. The origin in S′ has coordinate $x' = 0$ m, in which case x has the value vt. Making these substitutions in Equation 25.20, we obtain

$$\begin{aligned} 0 \text{ m} &= Avt + Bt \\ &= t[Av + B] \end{aligned} \tag{25.21}$$

The only way Equation 25.21 can be true for *any* and *every* instant t is if

$$Av + B = 0 \text{ m/s}$$

or

$$B = -Av$$

Thus Equation 25.20 becomes

$$\begin{aligned} x' &= Ax + (-Av)t \\ &= A[x - vt] \end{aligned} \tag{25.22}$$

For values of v that are small compared with c, the value of A must approach 1, and so the transformation equations reduce to the Galilean transformation, Equation 25.3. Likewise, the inverse equation for x in terms of x' and t' must have the form

$$x = A[x' + vt'] \tag{25.23}$$

which then also reduces to the Galilean result (Equation 25.2) for small v as A approaches 1.

Now let us use our newfound knowledge about length contraction to determine the value of A. Consider a proper (rest) length ℓ_0 in reference frame S′ as shown in Figure 25.20. The x'-coordinates of the ends of the proper length are x_1' and x_2' in S′, as indicated in Figure 25.20.

We write Equation 25.22 for each x'-coordinate:

$$x_1' = A[x_1 - vt_1] \tag{25.24}$$

$$x_2' = A[x_2 - vt_2] \tag{25.25}$$

Now we subtract Equation 25.24 from 25.25:

$$x_2' - x_1' = A[x_2 - x_1] - Av[t_2 - t_1] \tag{25.26}$$

But from Figure 25.20, the left-hand side of Equation 25.26 is the proper length ℓ_0. Hence

$$\ell_0 = A[x_2 - x_1] - Av[t_2 - t_1] \tag{25.27}$$

The length ℓ_0 in S′ is moving in frame S. To measure the moving length in S (with rulers at rest in S, of course), it is necessary to determine where the ends of the moving length are *at the same time in* S. It does no good to determine where one end of the moving length is at one instant and where the other end is at another instant; to measure the length of a moving car, we must determine where its bumpers are at the same time. Hence in S, to measure the moving length, we *must* make the measurements simultaneously—that is, when $t_1 = t_2$. Call ℓ the length measured in S (using rulers at rest in S). Then Equation 25.27 becomes, with $x_2 - x_1 = \ell$ and $t_1 = t_2$,

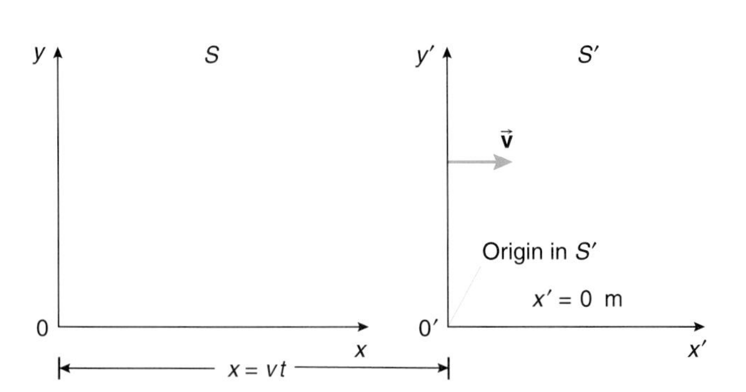

FIGURE 25.19 The view in S of the origin of S′.

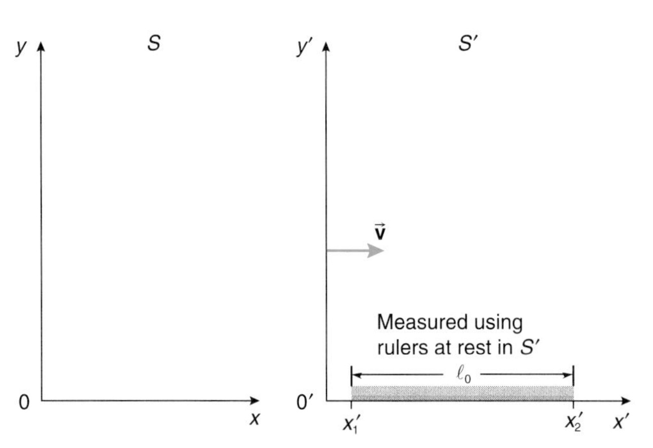

FIGURE 25.20 A proper length in frame S′.

$$\ell_0 = A\ell \quad (25.28)$$

But we already know that the proper (rest) and moving lengths are related by Equation 25.17:

$$\ell = \frac{\ell_0}{\gamma}$$

Comparing Equations 25.28 and 25.17, we find

$$A = \gamma$$

Thus the Lorentz transformation equations (Equations 25.22 and 25.23) become

$$x' = \gamma(x - vt) \quad (25.29)$$

$$x = \gamma(x' + vt') \quad (25.30)$$

To find the relationship between t' and t, we take Equations 25.29 and 25.30 and eliminate x' between them. After some tedious (and somewhat tricky) algebra that should keep you out of trouble for awhile, we eventually find

$$t' = \gamma\left(t - \frac{v}{c^2}x\right) \quad (25.31)$$

Now we solve Equation 25.31 for t in terms of x' and t', after substituting Equation 25.30 for x. With proper algebraic footwork, we find

$$t = \gamma\left(t' + \frac{v}{c^2}x'\right) \quad (25.32)$$

We summarize these results.

The set of Lorentz transformation equations is

$$x' = \gamma(x - vt) \quad (25.33)$$

$$y' = y \quad (25.34)$$

$$z' = z \quad (25.35)$$

$$t' = \gamma\left(t - \frac{v}{c^2}x\right) \quad (25.36)$$

PROBLEM-SOLVING TACTIC

25.2 Equations 25.33–25.36 are useful when the space and time coordinates of an event are known in the reference frame S and you need to know the corresponding coordinates of the same event in S'.

The inverse equations also are useful. Solving Equations 25.33–25.36 for x, y, z, and t in terms of x', y', z', and t', we obtain

$$x = \gamma(x' + vt') \quad (25.37)$$

$$y = y' \quad (25.38)$$

$$z = z' \quad (25.39)$$

$$t = \gamma\left(t' + \frac{v}{c^2}x'\right) \quad (25.40)$$

PROBLEM-SOLVING TACTIC

25.3 Equations 25.37–25.40 are useful if the space and time coordinates of an event are known in the reference frame S' and the corresponding coordinates of the event in S are needed.

In Equations 25.33–25.36 and Equations 25.37–25.40, the relative speed v is positive. The transformation equations for going from S to S' (Equations 25.33–25.36) and those from S' to S (Equations 25.37–25.40) differ only by the sign of terms involving v, other than switching primes for unprimes. This is because in one case S' is moving at velocity $v\hat{\imath}$ with respect to S, while in the other case S is moving with velocity $-v\hat{\imath}'$ with respect to S'. The only difference is the sign of the velocity component.

For reassurance, notice that when $v << c$,

$$\gamma \to 1$$

$$\frac{v}{c^2} \to 0 \text{ s/m}$$

So, for the familiar limit of low speeds, the Lorentz transformation Equations 25.33–25.40 reduce to the familiar Galilean transformation equations (and their inverses):

$$x' = x - vt$$

$$y' = y$$

$$z' = z$$

$$t' = t$$

The Lorentz transformation equations easily give us time dilation and length contraction, which they must since we used these effects to find the equations. For example, we revisit the peculiar relationship between stationary and moving clocks, first addressed in Section 25.4 with a pair of light clocks. We want to show that the results obtained using the light clocks do not depend on our using that special type of clock.

Imagine two events that take place at instants t_1' and t_2' *at the same place* x_0' in reference frame S'. The events might correspond to, say, two successive ticks on any clock in S'; or the events might correspond to the creation of a radioactive particle at rest at instant t_1' and its subsequent decay at t_2', where we identify frame S' as the one with the particle at rest at coordinate x_0'. The time interval τ_0 between the two events in S' is $\tau_0 = t_2' - t_1'$; this is the rest time interval or proper time interval because the two events happen at the same place in S' (the particle is at rest in this frame). We want to calculate the temporal interval between the two events as measured by the clocks in reference frame S.

According to the Lorentz transformation equations (Equation 25.40, in particular), the time instants the events occur in the reference frame S are

$$t_1 = \gamma\left(t_1' + \frac{v}{c^2}x_0'\right)$$

and

$$t_2 = \gamma\left(t_2' + \frac{v}{c^2}x_0'\right)$$

We call τ the temporal separation of the two events as measured by clocks at rest in S; the two events occur at different locations in S because S′ is moving, but we are interested here only in the time interval between the events as determined from the clocks distributed throughout the S reference frame. The time interval of interest is

$$\tau = t_2 - t_1$$

$$= \gamma(t_2' - t_1') + \frac{v}{c^2}(x_0' - x_0')$$

$$= \gamma(t_2' - t_1')$$

But $t_2' - t_1' = \tau_0$, the proper time interval between the two events. Hence we have the same relationship (Equation 25.14) between the time intervals that we first derived with the special light clocks of Section 25.4:

$$\tau = \gamma\tau_0$$

Notice that this argument makes no mention of special clocks, and so it applies to *all* clocks. The tick rate of a moving clock is slower than a clock at rest with respect to you.

Example 25.4

A firecracker goes off in reference frame S at a point with coordinates $x = 20.0$ m, $y = 5.0$ m, and $z = 0.0$ m when clocks (stopwatches) distributed throughout that frame indicate 0 s. Captain Kirk in reference frame S′ is traveling by at a speed $v = 0.900c$. What are the space and time coordinates of the event according to rulers and clocks at rest with respect to Captain Kirk?

Solution

First find the factor γ, using its definition, Equation 25.13:

$$\gamma = \frac{1}{\sqrt{1 - \frac{v^2}{c^2}}}$$

$$= \frac{1}{\sqrt{1 - (0.900)^2}}$$

$$= 2.29$$

Since the space and time coordinates of the event are known in S and you want those in S′, use Problem-Solving Tactic 25.2 and Equations 25.33–25.36. The speed v is $0.900c$ or 2.70×10^8 m/s:

$$x' = \gamma(x - vt)$$

$$= 2.29[20.0 \text{ m} - (2.70 \times 10^8 \text{ m/s})(0 \text{ s})]$$

$$= 45.8 \text{ m}$$

$$y' = y$$

$$= 5.0 \text{ m}$$

$$z' = z$$

$$= 0.0 \text{ m}$$

$$t' = \gamma\left(t - \frac{v}{c^2}x\right)$$

$$= 2.29\left[0 \text{ s} - \frac{(2.70 \times 10^8 \text{ m/s})(20.0 \text{ m})}{(3.00 \times 10^8 \text{ m/s})^2}\right]$$

$$= -1.37 \times 10^{-7} \text{ s}$$

Notice that in reference frame S′, the event occurs before $t' = 0$ s on the clocks in S′. The event occurs at the point with coordinates $x' = 45.8$ m, $y' = 5.0$ m, and $z' = 0.0$ m according to rulers at rest in S′.

25.8 The Relativity of Simultaneity

The importance of doing things simultaneously is quite apparent when playing in an orchestra. Beethoven's famous Ninth Symphony would be mere cacophony if each musician began to play when they felt like it. Every musician must begin when the conductor commands.

Imagine two musicians in the Cleveland Symphony separated by a distance ℓ_0 along the x-axis in frame S as shown in Figure 25.21. The two musicians begin to play simultaneously and harmoniously when $t = 0$ s, in their frame and that of the conductor. We define two events and write their space and time coordinates in S:

Event 1	Event 2
Musician 1 begins to play	Musician 2 begins to play
$x_1 = 0$ m	$x_2 = \ell_0$
$t_1 = 0$ s	$t_2 = 0$ s

You are rushing past the performance at high speed v carrying reference frame S′ along with you; you are the S′ reference frame. We use the Lorentz transformation, Equations 25.33 and 25.36, to find the x'- and t'-coordinates of the two events in S′. We find

Event 1	Event 2
Musician 1 begins to play	Musician 2 begins to play
$x_1' = 0$ m	$x_2' = \gamma\ell_0$
$t_1' = 0$ s	$t_2' = -\gamma v\ell_0/c^2$

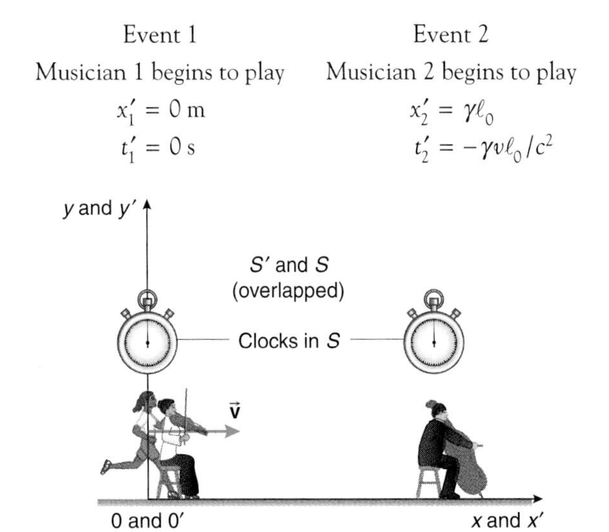

FIGURE 25.21 Two musicians in S, separated by distance ℓ_0 along the x-axis. You are the S′ frame and rush by at speed v. Situation shown when $t = 0 \text{ s} = t'$.

Note that the two events are *not simultaneous* in S', unless $\ell_0 = 0$ m, in which case the two events really are the same musician—that is, the same event! This is a characteristic and surprising feature of special relativity: the notion of simultaneity is relative.

> If two events are simultaneous in one inertial reference frame, they will *not* be simultaneous to any other inertial frame moving at nonzero speed v with respect to the first.

25.9 A Relativistic Centipede

Here we examine a hypothetical problem that ties together much of the kinematics of relativity. The problem is pedagogically useful and playful at the same time. It involves a high-speed centipede (*Scolopendra cingulata*) and a compassionate butcher (*Homo sapiens*).* The centipede is 10.0 cm long, measured with rulers at rest with respect to the centipede; thus, the proper length of the centipede is 10.0 cm. A butcher holds two meat cleavers 9.0 cm apart. The centipede runs at such a high speed v across a chopping block that the butcher measures the (contracted) length of the centipede to be 8.0 cm, using rulers inlaid on the chopping block. According to length contraction, Equation 25.17, we have

$$\ell = \frac{\ell_0}{\gamma}$$

$$8.0 \text{ cm} = \frac{10.0 \text{ cm}}{\gamma}$$

Note that we can use units of convenience here. Solving for γ, we obtain

$$\gamma = \frac{10.0 \text{ cm}}{8.0 \text{ cm}} = 1.25$$

The definition of γ is Equation 25.13. We use this to solve for v/c:

$$\gamma = \frac{1}{\sqrt{1 - \frac{v^2}{c^2}}} = 1.25$$

We find

$$\frac{v}{c} = 0.600$$

Talk about fleet feet! Sports considerations aside (let alone coordinating all the feet), the instant the tail of the centipede is at cleaver A (see Figure 25.22), the butcher immediately swings *both* cleavers instantaneously down on the chopping block simultaneously (kerchunk!) and immediately raises them up again. (Obviously, we need a butcher with quick reflexes to avoid having the centipede collide with cleaver B.) The butcher argues that since the centipede is only 8.0 cm long according to the rulers inlaid on the chopping block in the butcher's reference frame, the centipede neatly fits between the cleavers and no bug juice is shed.

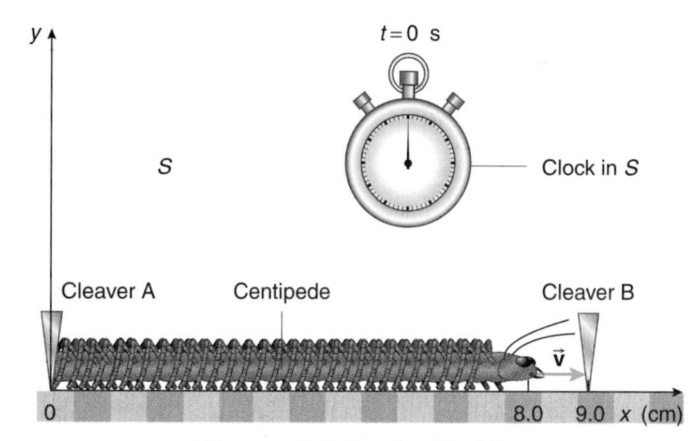

FIGURE 25.22 The view of the butcher, the S frame.

On the other hand, the relativistically literate centipede is quite worried about the situation. From the viewpoint of the centipede, the meat cleavers and butcher are approaching at a high speed v. The centipede sees the separation of the cleavers, not as the 9.0 cm according to the butcher, but as a contracted length

$$\ell = \frac{\ell_0}{\gamma} = \frac{9.0 \text{ cm}}{1.25} = 7.2 \text{ cm}$$

The separation of the cleavers is only 7.2 cm according to the centipede; but the centipede knows she is 10.0 cm long! Uh oh.

Either there *is* bug juice on the chopping block or there is *none*. The butcher says no juice was shed; so how does the centipede manage to wiggle out of this relativistic dual guillotine? We will see in what follows.

We call the frame of the butcher S and that of the centipede S'. We define two events corresponding to the cleavers hitting the chopping block. These events are most easily described in the reference frame S of the butcher. Let the x-coordinates of the cleavers be $x_A = 0$ cm $= 0$ m and $x_B = 9.0$ cm $= 0.090$ m, as in Figure 25.22.

Let the cleavers hit the chopping block when $t = 0$ s according to the butcher. Here we use SI units since we intend to use the Lorentz transformation equations. Then the space–time coordinates of the two events in S are as follows (the y and z coordinates are not relevant so we do not specify them; we can set them both equal to zero):

Event 1. Cleaver A hits the block:

$$x_1 = 0 \text{ m}$$
$$t_1 = 0 \text{ s}$$

Event 2. Cleaver B hits the block:

$$x_2 = 0.090 \text{ m}$$
$$t_2 = 0 \text{ s}$$

*The author believes the idea for this problem originates with Professor George Ruff, Bates College, Lewiston, Maine 04240.

The events are *simultaneous* in the reference frame S of the butcher, so that $t_1 = t_2$. The events occur, however, at two different locations in S.*

Now we transform the events into the reference frame S' of the centipede using the Lorentz transformation, Equations 25.33 and 25.36; remember that γ has the value 1.25.

Event 1. Cleaver A hits the block:

$$x_1' = 1.25[0\text{ m} - v(0\text{ s})]$$
$$= 0\text{ m}$$
$$t_1' = 1.25\left[0\text{ s} - \frac{v}{c^2}(0\text{ m})\right]$$
$$= 0\text{ s}$$

Event 2. Cleaver B hits the block:

$$x_2' = 1.25[0.090\text{ m} - v(0\text{ s})]$$
$$= 0.11\text{ m} = 11\text{ cm}$$

*If two events occur at the same time *and* place, then the two events are identical as far as relativity is concerned.

$$t_2' = 1.25\left[0\text{ s} - \frac{v}{c^2}(0.090\text{ m})\right]$$

$$t_2' = -2.3 \times 10^{-10}\text{ s}$$

The two events are *not simultaneous* to the centipede, since $t_2' \neq t_1'$. Event 2 (cleaver B hits the block) occurs before Event 1 (cleaver A hits the block). Thus the fact that the separation of the cleavers is only 7.2 cm according to the 10.0 cm centipede is no problem, since the cleavers do not descend simultaneously in her frame. Since the centipede is stretched out in S', the event that occurs first according to the centipede (Event 2) takes place away from her head (her head is located at $x' = 10.0$ cm when cleaver B hits at $x_2' = 11$ cm), as in Figure 25.23.

The subsequent event (Event 1) occurs at the end of her tail (located at $x_2' = 0$ cm). The centipede escapes, as in Figure 25.24.

Note that the separation of the meat cleavers according to the centipede (7.2 cm) is *not* $x_2' - x_1' = 11$ cm, because the events associated with the coordinates x_2' and x_1' *do not happen at the same time in* S'. To measure the separation of the (*moving*) cleavers in S', it is necessary for the centipede to determine the locations of the two cleavers *at the same time in her own reference frame*.

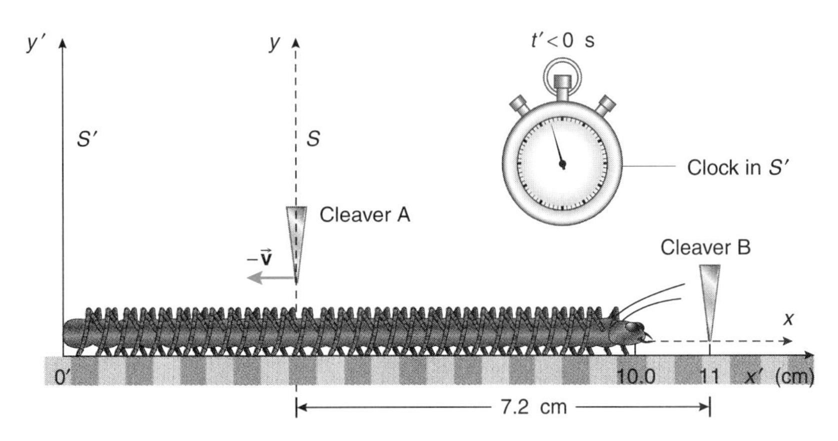

FIGURE 25.23 Event 2 occurs first according to the centipede. All coordinates and lengths are as measured by the centipede in S'.

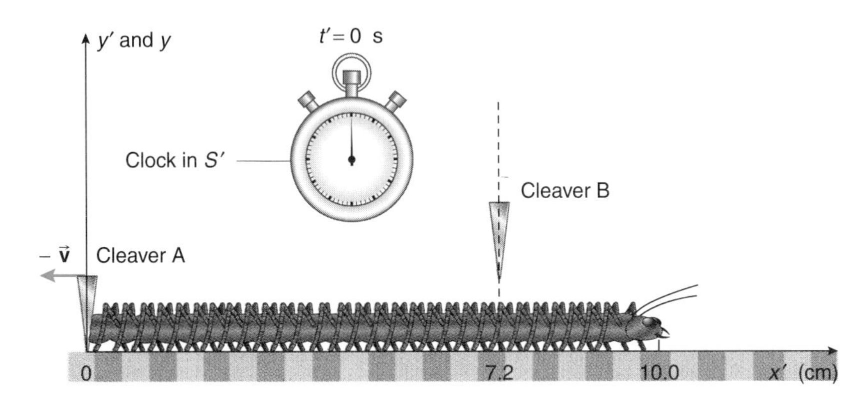

FIGURE 25.24 Event 1 occurs later according to the centipede. Coordinates are according to the centipede in S'.

25.10 A Relativistic Paradox and Its Resolution*

Speeding through space, speeding through heaven and the stars
Walt Whitman (1819–1892)†

Time dilation means that clocks in one inertial reference frame that is moving with respect to another inertial reference frame run more slowly than the clocks at rest in the latter frame. Length contraction means that moving lengths oriented along the direction of motion are shorter than identical lengths at rest. Time dilation and length contraction apply *whenever* clocks or lengths (oriented along the motion) are *moving* with respect to another inertial reference frame.

Now here's the rub.

1. Imagine yourself in reference frame S in the standard geometry, shown in Figure 25.25. Reference frame S′ is moving to the right at speed v with respect to you in S. Thus, according to time dilation, you say the clocks in S′ tick slow compared with your clocks in S. Likewise, if you compare the length of a meter stick nailed down to the x'-axis in S′ with a meter stick nailed down to your x-axis in S, then length contraction means the moving meter stick (in S′) is shorter than your own meter stick at rest in S.
2. But now imagine a friend in the S′ reference frame. From their viewpoint, you and your reference frame S are moving to the left at the speed v, as in Figure 25.26. Thus, according to time dilation, your friend in S′ says *your* clocks in S tick slow compared with the clocks at rest in S′. Likewise, if your friend in S′ compares the length of your meter stick nailed down to the x-axis in S with a meter stick nailed down to the x'-axis in S′, then length contraction to him means *your* moving meter stick (in S) is shorter than the meter stick at rest in S′.

†*Leaves of Grass*, Song of Myself, #33, line 790 (David McKay, Philadelphia, 1900), page 67.

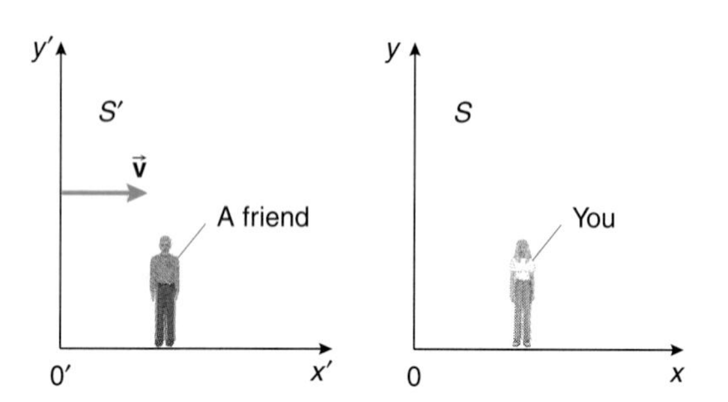

FIGURE 25.25 The standard geometry; the view in the S frame (you).

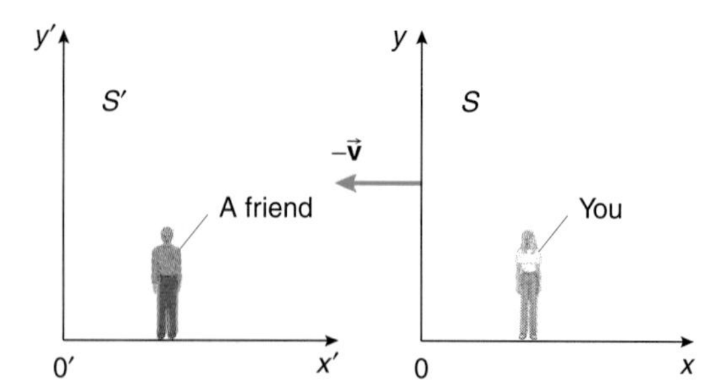

FIGURE 25.26 The standard geometry; the view in the S′ frame of a friend.

> *Each* reference frame says
> - the clocks in the *other* reference frame run slow compared with the clocks at rest; and
> - the lengths in the *other* frame moving along the direction of motion are shorter.
>
> Ridiculous, you might think! How can each of you say the *other* clocks are slow and the *other* lengths are shorter? Yet this apparent paradox is indeed what happens; both views are correct!

To demonstrate there is no contradiction or paradox, we consider a particular example in some detail.

The example involves an interstellar spaceflight, so that it will be convenient to measure time in years (y) and lengths in light-years (LY). A light-year is the distance light travels during an interval of one year. Let t_0 be the number of seconds in one year:

$$t_0 \approx 3.156 \times 10^7 \text{ s}$$

Then one light-year is a distance ct_0:

$$\begin{aligned} 1 \text{ LY} &= ct_0 \\ &= (3.00 \times 10^8 \text{ m/s})(3.156 \times 10^7 \text{ s}) \\ &= 9.47 \times 10^{15} \text{ m} \end{aligned}$$

To convert a distance x in meters to a distance X in light-years, we divide x by ct_0. To convert a time t in seconds to a time T in years, we divide t by t_0. Hence Equation 25.33 of the Lorentz transformation becomes*

$$\frac{x'}{ct_0} = \frac{\gamma(x - vt)}{ct_0}$$

$$X' = \gamma\left[X - \frac{v}{c}T \text{ (LY/y)}\right] \tag{25.41}$$

$$(X \text{ and } X' \text{ in LY, } T \text{ in y})$$

Likewise, Equation 25.36 of the Lorentz transformation becomes†

$$\frac{t'}{t_0} = \frac{\gamma\left(t - \frac{v}{c^2}x\right)}{t_0}$$

$$T' = \gamma\left[T - \frac{v}{c}X \text{ (y/LY)}\right] \tag{25.42}$$

$$(T \text{ and } T' \text{ in y, } X \text{ in LY})$$

*We explicitly include the units LY/y with the T term so the units for X' emerge properly in LY.

†Likewise, we explicitly include the units y/LY in the X term so the units for T' emerge properly in y.

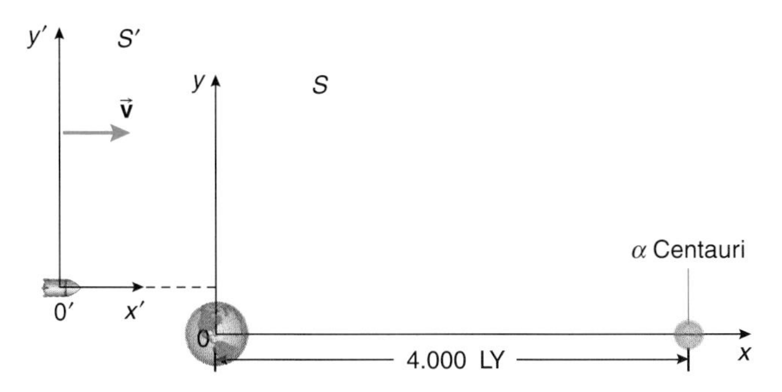

FIGURE 25.27 The standard geometry. We shift the x- and x'-axes slightly here for clarity.

The distance from our solar system to α Centauri (the closest star system to our solar system) is about 4.2 LY, which means light takes 4.2 y to travel this distance, according to clocks at rest on the Earth (or at rest on α Centauri). Just to keep the numbers clean, we consider the distance to be precisely 4.000 LY, so that we can focus on the physics rather than awkward numbers.

Let the Earth and α Centauri define a reference frame S, as in Figure 25.27. The distance between the Earth and α Centauri is 4.000 LY as measured by rulers at rest in the S frame. For convenience, we place the Earth at the origin in S, so that α Centauri is 4.000 LY out along the positive x-axis. Let a spacecraft travel at constant velocity through our region of the galaxy on its way past the Earth and α Centauri. Let S' be a reference frame attached to the spacecraft with it at the S' origin, as in Figure 25.27.

Let the speed of the spacecraft (frame S') be $v = 0.8660c$. After a short calculation, we find that this speed yields $\gamma = 2.000$. We want to resolve the paradox that *each* reference frame says the *other* frame's clocks are slow and the other frame's lengths (oriented parallel to the direction of motion) are contracted.

To compare the results of measurements in each frame, we need to communicate information between the frames. This can be accomplished in the following way: when the spacecraft is flying by the Earth, set both the clocks on the spacecraft and the clocks on the Earth to zero. When the spacecraft reaches α Centauri, a light signal is sent by the spacecraft back to the Earth indicating the time on the spacecraft clock and the results of calculations made in this reference frame, so a comparison can be made with those in the Earth–α Centauri frame S.

We approach the problem by (1) defining several events, (2) writing the space–time coordinates of the events in one of the reference frames (whichever frame in which it is most convenient to do so), and then (3) transforming the space–time coordinates of these events to find the space–time coordinates of the same events in the other reference frame using the customized Lorentz transformation given by Equations 25.41 and 25.42. We then interpret the results.

We define the following events:

Event 1: The spacecraft has the Earth outside its window.
Event 2: The spacecraft has α Centauri outside its window and sends a light signal back to the Earth indicating the time on the spacecraft clocks and the results of its calculations in the S' frame.
Event 3: The Earth receives the light signal sent by the spacecraft when α Centauri was outside its window.

The space–time coordinates of the events are most easily expressed in the reference frame S, the Earth–α Centauri system.

Event 1 is that the spacecraft has the Earth outside its window, shown in Figure 25.28. The space–time coordinates of this event are easy to enumerate:

$$X_1 = 0 \text{ LY}$$
$$T_1 = 0 \text{ y}$$

We use the customized Lorentz transformation, Equations 25.41 and 25.42, to find the space–time coordinates of this event in the S' reference frame (the spacecraft). After substituting for X_1 and T_1, we find

$$X_1' = 0 \text{ LY}$$
$$T_1' = 0 \text{ y}$$

Event 2 is that the spacecraft has α Centauri outside its window and sends a light signal back to Earth, as shown in Figure 25.29. In reference frame S, this event occurs at the location

$$X_2 = 4.000 \text{ LY}$$

The time for the spacecraft to fly to α Centauri, according to Earth clocks, is determined by dividing the 4.000 LY distance that the spacecraft must travel in this frame by the speed of the spacecraft in this frame ($0.8660c$). Because the distance is expressed in LY, the time involved is (be careful with the units):

$$\frac{\text{distance}}{\text{speed}} = \frac{4.000 \text{ LY}}{0.8660 \text{ LY/y}} = 4.619 \text{ y}$$

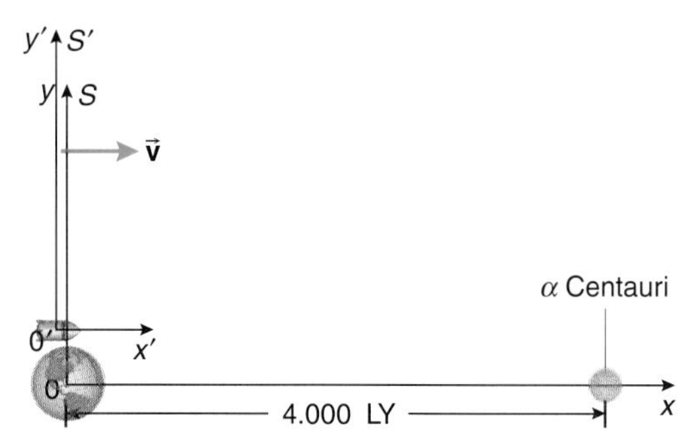

FIGURE 25.28 Event 1 from the viewpoint of the S reference frame

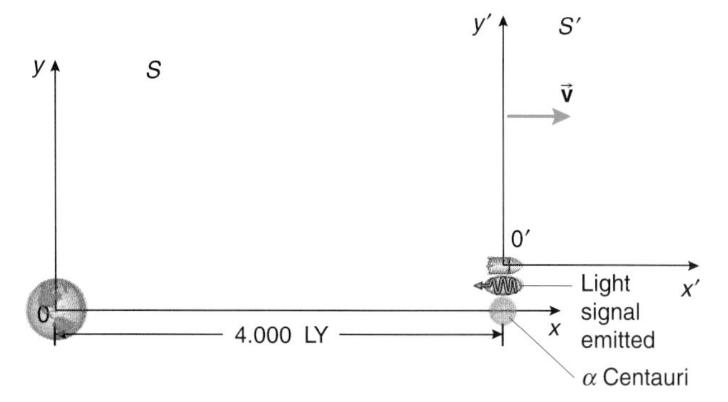

FIGURE 25.29 Event 2 from the viewpoint of the S reference frame.

Thus

$$T_2 = 4.619 \text{ y}$$

We use the customized Lorentz transformation, Equations 25.41 and 25.42, to find the space–time coordinates of this event in the S' frame (the spacecraft):

$$\begin{aligned} X_2' &= 2.000\left[4.000 \text{ LY} - 0.8660\,(4.619 \text{ y}) \text{ LY/y}\right] \\ &= 0 \text{ LY} \end{aligned}$$

and

$$\begin{aligned} T_2' &= 2.000\left[4.619 \text{ y} - 0.8660\,(4.000 \text{ LY}) \text{ y/LY}\right] \\ &= 2.310 \text{ y} \end{aligned}$$

The result $X_2' = 0$ LY is not surprising since the spacecraft is at the origin in its own S' frame! The result $T_2' = 2.310$ y means that the clocks on the spacecraft indicate that the trip took only 2.310 years. The Earth clocks say the trip took 4.620 y. Notice that the Earth sees the spacecraft moving, and so says the spacecraft clocks tick slower than the Earth clocks; they tick slower by the factor $1/\gamma$, here 1/2.000. According to the Earth, the moving clocks (in the spacecraft and its reference frame) indeed are ticking slower than the clocks on the Earth and in its reference frame.

From the perspective of the spacecraft (S'), the trip is described differently. The spacecraft sees the S frame moving toward it at the speed $0.8660c$, as in Figure 25.30. Thus the 4.000 LY proper length measured in S is contracted by the factor $1/\gamma$; the spacecraft measures the moving distance between the Earth and α Centauri (using rulers at rest in S') to be only

$$\ell = \frac{\ell_0}{\gamma} = \frac{4.000 \text{ LY}}{2.000} = 2.000 \text{ LY}$$

When the Earth is outside the spacecraft window, the spacecraft sets its clock to 0 y. The time (according to the spacecraft clocks) for the moving Earth–α Centauri distance (measured to be 2.000 LY in S') to go past the spacecraft thus is*

$$\frac{\text{distance}}{\text{speed}} = \frac{2.000 \text{ LY}}{0.8660 \text{ LY/y}} = 2.309 \text{ y}$$

*The slight discrepancy is due to rounding in the calculations.

Hence the S frame (the Earth–α Centauri system) interprets the reading on the spacecraft clock as due to time dilation, because the Earth sees the spacecraft clock as a moving clock; the spacecraft explains the reading on its clock as due to a length contraction of the moving Earth–α Centauri distance.

Now we consider Event 3, in which the light signal from the spacecraft arrives back at the Earth, as in Figure 25.31. The event occurs at the location of the Earth, and so the spatial coordinate in S is

$$X_3 = 0 \text{ LY}$$

According to the Earth clocks, the spacecraft arrived at α Centauri when the time was 4.619 y and then emitted a light signal to travel back to the Earth. The light signal takes 4.000 y to travel the 4.000 LY distance back to the Earth, according to the Earth clocks. Thus the total elapsed time when the light gets back to the Earth (according to the Earth clocks) is

$$4.619 \text{ y} + 4.000 \text{ y} = 8.619 \text{ y}$$

Thus

$$T_3 = 8.619 \text{ y}$$

We use the customized Lorentz transformation, Equations 25.41 and 25.42, to find the space–time coordinates of this event in the S' frame:

$$\begin{aligned} X_3' &= 2.000\left[0 \text{ LY} - 0.8660\,(8.619 \text{ y}) \text{ LY/y}\right] \\ &= -14.93 \text{ LY} \\ T_3' &= 2.000\left[8.619 \text{ y} - 0.8660\,(0 \text{ LY}) \text{ y/LY}\right] \\ &= 17.24 \text{ y} \end{aligned}$$

The light signal sent by the spacecraft returns to the Earth when the Earth is located in S' at coordinate $X_3' = -14.93$ LY and when the spacecraft (S') clock indicates a time of 17.24 y, as shown in Figure 25.32.

Why did it take so long for the light signal to return to the Earth from the viewpoint of the spacecraft? According to it, the light was sent on its way toward an Earth that was receding from the spacecraft at $0.8660c$. Thus it took quite some time (17.24 y − 2.310 y = 14.93 y) for the light to travel the 14.93 LY distance to the Earth in the S' frame.

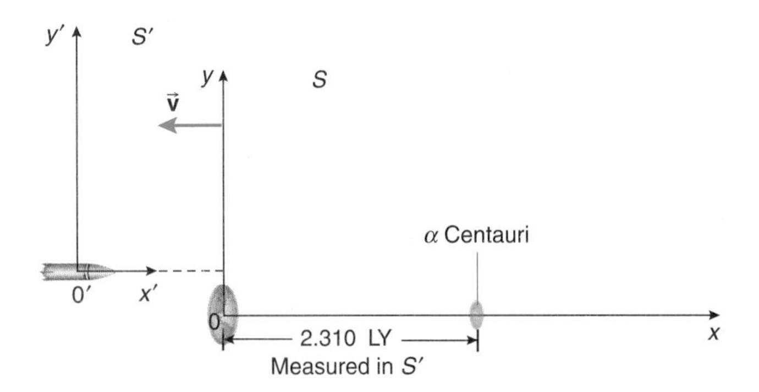

FIGURE 25.30 Perspective of the spacecraft, the S' reference frame.

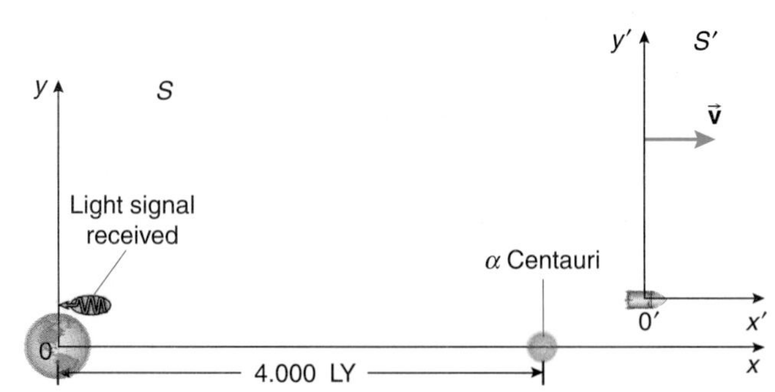

FIGURE 25.31 Event 3: The view from frame S.

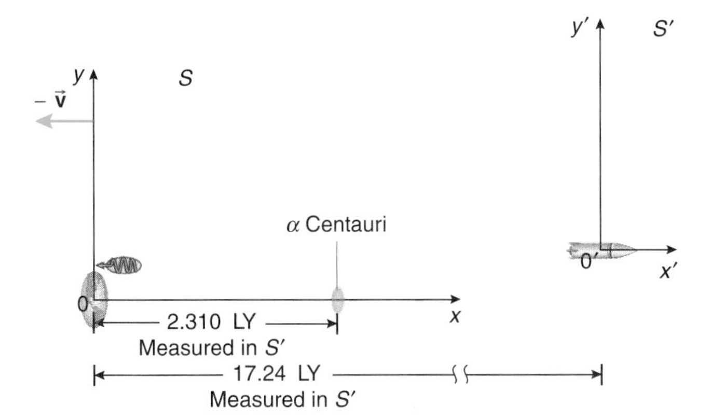

FIGURE 25.32 Event 3: The view from frame S'.

Notice that now the tables are turned. The Earth clock reads 8.619 y when the signal arrives; the spacecraft clock reads 17.24 y. The spacecraft, viewing the moving Earth, says the moving clocks on the Earth tick slower than the spacecraft clocks by the factor $1/\gamma$, here equal to 1/2.000. The Earth clocks thus read half the time of the spacecraft clocks.*

There is only one residual question to resolve. According to the Earth (S frame), how distant is the spacecraft when the light from the spacecraft returns to the Earth? The signal was sent from the spacecraft when it was 4.000 LY distant (at α Centauri); the light took 4.000 y to reach the Earth (according to the Earth clocks). Meanwhile the spacecraft went on its merry way for an additional 4.000 y at the speed of 0.8660c. During those 4.000 y, the spacecraft thus traveled an additional (0.8660 LY/y)(4.000 y) = 3.464 LY. Thus the distance of the spacecraft in the S frame is 4.000 LY + 3.464 LY = 7.464 LY when the light reaches the Earth. The 14.93 LY proper length between the spacecraft and the Earth as measured in S' is contracted to 14.93 LY/γ = 14.93 LY/2.000 = 7.465 LY† according to the Earth, since the Earth sees the 14.93 LY distance as a moving length oriented along the direction of motion.

Thus each frame says that the moving clocks in the other frame run slow compared with clocks at rest. Each frame also says that lengths moving along the direction of motion are contracted. Both are right. There is no paradox.

25.11 Relativistic Velocity Addition

The velocity of light forms the upper limit of velocities for all material bodies. . . . The simple mechanical law of adding and subtracting velocities is no longer valid or, more precisely, is only approximately valid for small velocities, but not for those near the velocity of light. The number expressing the velocity of light appears explicitly in the Lorentz transformation, and plays the role of a limiting case, like the infinite velocity in classical mechanics.

Albert Einstein and Leopold Infeld‡

*There is a slight discrepancy due to rounding in the calculations.

†The discrepancy between 7.460 LY and 7.465 LY is due to rounding in the calculations.

‡*The Evolution of Physics* (Simon & Schuster, New York, 1938), page 202.

The Achilles' heel of Galilean relativity became apparent in the addition of velocity components. The Galilean rule (Equation 25.6) is not correct except at speed $v << c$; we discover the new rule in this section. Just as the Galilean rule stemmed from the Galilean transformation equations, the relativistic rule follows from the Lorentz transformation equations.

Velocity Parallel or Antiparallel to the Direction of Motion of the Two Inertial Reference Frames

We use the standard geometry. Imagine a particle in the frame S' moving along the x'-axis with a velocity component u_x', as in Figure 25.33. We want to determine the velocity component u_x of the particle as measured in the reference frame S. The Galilean result was (Equation 25.6)

$$u_x = u_x' + v$$

but this only yields the correct result if the velocity components v and u_x' are small compared with the speed of light c.

The velocity component of the object in the S' frame is the time rate at which the position of the object changes in the S' frame:

$$u_x' = \frac{dx'}{dt'} \tag{25.43}$$

The velocity component of the object in the S frame is the time rate at which the position of the object changes in the S frame:

$$u_x = \frac{dx}{dt} \tag{25.44}$$

To find u_x we need to find the differentials dx and dt and then take their quotient. The differentials are found from the Lorentz transformation equations (Equations 25.37 and 25.40):

$$x = \gamma(x' + vt') \qquad \text{and} \qquad t = \gamma\left(t' + \frac{v}{c^2}x'\right)$$

We take the differentials; remember that the speed of light c and the relative speed v of the two reference frames are constant:

$$dx = \gamma(dx' + v\,dt') \qquad \text{and} \qquad dt = \gamma\left(dt' + \frac{v}{c^2}dx'\right)$$

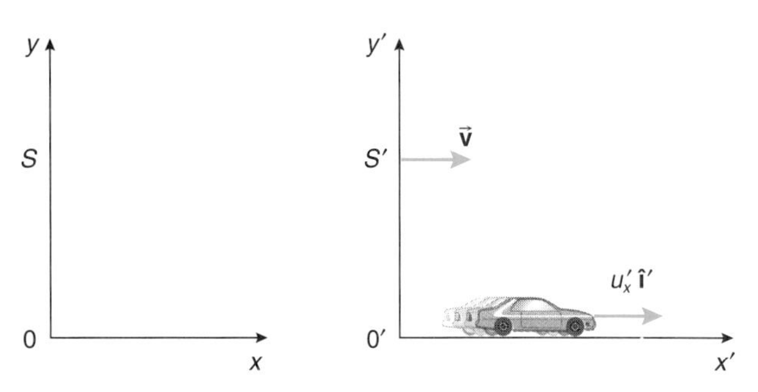

FIGURE 25.33 A particle moving in frame S' with velocity component u_x'.

Taking their quotient, we obtain

$$u_x = \frac{dx}{dt} = \frac{\gamma(dx' + v\,dt')}{\gamma\left(dt' + \frac{v}{c^2}\,dx'\right)}$$

Things begin to fall together if we divide the numerator and denominator of the right-hand side by dt':

$$u_x = \frac{\frac{dx'}{dt'} + v}{1 + \frac{v}{c^2}\frac{dx'}{dt'}}$$

But

$$u'_x = \frac{dx'}{dt'}$$

is the velocity component of the object in the S' frame.

Thus we have for the x-components of the velocities:

$$u_x = \frac{u'_x + v}{1 + \frac{vu'_x}{c^2}} \qquad (25.45)$$

Equation 25.45 is the new velocity component addition rule along the common x- and x'- axes.

Notice happily that if the velocity components v and u'_x are small compared with the speed of light c, the second term in the denominator is negligible and Equation 25.45 reduces to the Galilean result (Equation 25.6), as it must for small speeds.

PROBLEM-SOLVING TACTIC

25.4 If $u'_x > 0$ m/s, then the object is moving to the right in S' toward increasing values of x'. If the object is moving toward decreasing values of x' in S', then substitute a negative velocity component for u'_x in Equation 25.45. A few examples (see the following) will (hopefully!) make the application of Equation 25.45 clear and comfortable for you.

Velocity Perpendicular to the Direction of Motion of the Two Inertial Reference Frames

The Galilean transformation for velocities transverse to the direction of motion v was simple (Equation 25.7):

$$u_y = u'_y$$

In special relativity, even though the y- and y'-coordinates in the two frames are the same ($y = y'$), the velocity components are *not* the same because of the relativity of time ($t \neq t'$).

In particular, if a particle has a velocity component u'_y in S', then

$$u'_y = \frac{dy'}{dt'}$$

The velocity component of this particle along the y-axis in the S frame is

$$u_y = \frac{dy}{dt}$$

To find the relationship between them, we use the Lorentz transformation equations once again:

$$y = y' \qquad \text{and} \qquad t = \gamma\left(t' + \frac{v}{c^2}x'\right)$$

Taking differentials of these equations, we find

$$dy = dy' \qquad \text{and} \qquad dt = \gamma\left(dt' + \frac{v}{c^2}\,dx'\right)$$

From these differentials, we can construct u_y:

$$u_y = \frac{dy}{dt} = \frac{dy'}{\gamma\left(dt' + \frac{v}{c^2}\,dx'\right)}$$

Dividing the numerator and denominator of the right-hand side of this result by dt', we obtain

$$u_y = \frac{\frac{dy'}{dt'}}{\gamma\left(1 + \frac{v}{c^2}\frac{dx'}{dt'}\right)} \qquad (25.46)$$

But

$$\frac{dy'}{dt'} = u'_y$$

and

$$\frac{dx'}{dt'} = u'_x$$

and so Equation 25.46 becomes

$$u_y = \frac{u'_y}{\gamma\left(1 + \frac{vu'_x}{c^2}\right)} \qquad (25.47)$$

Note that this differs from the transformation for u_x (Equation 25.45).

Thus the transverse velocity components are different in the two reference frames. Notice, in particular, that even if $u'_x = 0$ m/s, so that the particle has only a y'-component for its velocity in S', the y and y' velocity components still are not the same! If $u'_x = 0$ m/s, then

$$u_y = \frac{u'_y}{\gamma} \qquad (25.48)$$

Since the z- and z'-axes also are transverse to the motion, an equation similar to Equation 25.47 exists for the velocity components along z and z':

$$u_z = \frac{u_z'}{\gamma\left(1 + \frac{vu_x'}{c^2}\right)} \tag{25.49}$$

For a particle moving with a velocity

$$\vec{u}' = u_x'\,\hat{\imath}' + u_y'\,\hat{\jmath}' + u_z'\,\hat{k}'$$

in frame S', each velocity component must be transformed separately and appropriately to find the velocity of the particle in the frame S:

$$\vec{u} = u_x\hat{\imath} + u_y\hat{\jmath} + u_z\hat{k}$$

The u_x and u_x' components are related via Equation 25.45; u_y and u_y' are related via Equation 25.47, and u_z and u_z' via Equation 25.49.

PROBLEM-SOLVING TACTIC

25.5 The relativistic velocity component addition equations always yield velocity components less than or equal to the speed of light. If you ever find that you have calculated a speed greater than c with these equations, you can be sure a mistake has been made, and you can set about finding it and making repairs.

Example 25.5

A laser is traveling at a high speed v and emits light in the direction of its motion, as indicated in Figure 25.34. Use Equation 25.45 to show that the speed of light in the reference frame in which the laser is moving still is c.

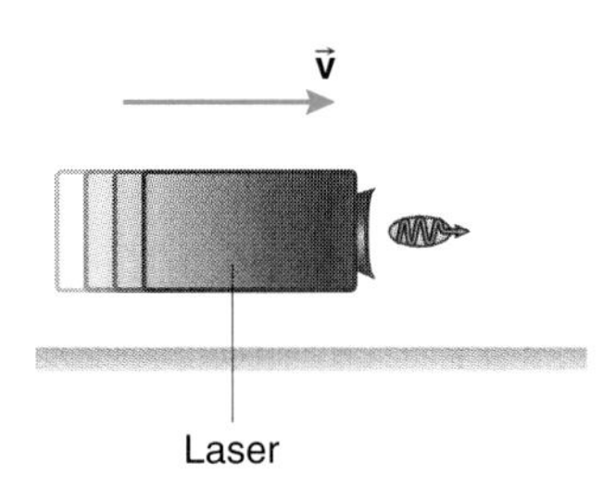

FIGURE 25.34

Solution

The speed of the light with respect to the laser is, of course, c. Let the laser be bolted to frame S', so that you have the standard geometry indicated in Figure 25.35. You can then make the following identification:

$u_x' = c$ (the velocity component of the light moving in S')

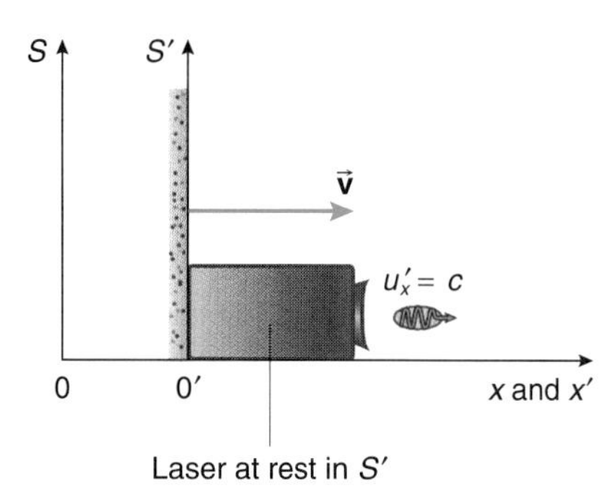

FIGURE 25.35

To find the velocity component of the light with respect to the frame S, use the velocity component addition Equation 25.45:

$$u_x = \frac{u_x' + v}{1 + \frac{vu_x'}{c^2}}$$

Making the substitution $u_x' = c$, you find

$$u_x = \frac{c + v}{1 + \frac{vc}{c^2}}$$

This simplifies to

$$u_x = \frac{c + v}{\frac{c^2 + cv}{c^2}}$$

$$= \frac{c + v}{c(c + v)}c^2$$

$$= c$$

The speed of light is c in frame S and also in frame S', in conformity with the first postulate of special relativity.

Strategic Example 25.6

Two spacecraft are approaching the Earth from opposite directions with speeds 0.80c and 0.50c, as measured by Earth observers, as shown in Figure 25.36. What is the speed of one spacecraft as seen by the other?

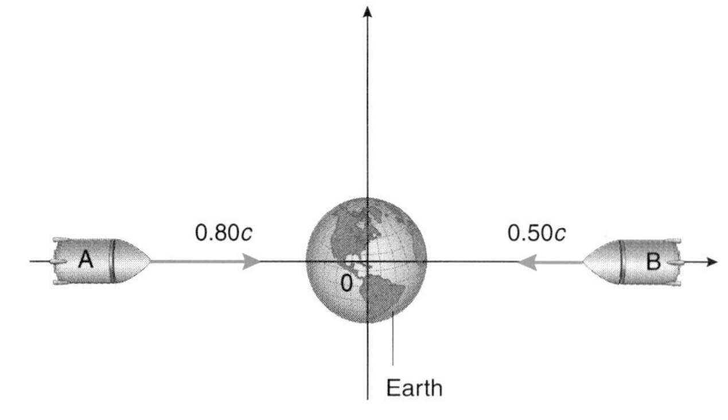

FIGURE 25.36

Solution

Note, incidentally, that the Galilean prediction for the relative speed of approach of the two spacecraft is 1.30c. Their speed of approach must be less than the speed of light according to the special theory of relativity.

Method 1

Assign reference frames and objects so that you have the standard geometry. Imagine yourself inside the spacecraft B on the right in Figure 25.37. Then the Earth and the other spacecraft A are traveling to the right. That is, from the viewpoint of spacecraft B, the situation is as shown in Figure 25.37.

Thus you identify a reference frame attached to spacecraft B as S; the Earth is S′ (and is moving to the right with velocity component $v = 0.50c$); the spacecraft A is moving to the right in S′ with a velocity component $u'_x = 0.80c$. Thus you can apply Equation 25.45 with all the velocity components positive:

$$\begin{aligned} u_x &= \frac{u'_x + v}{1 + \dfrac{vu'_x}{c^2}} \\ &= \frac{0.80c + 0.50c}{1 + \dfrac{(0.50c)(0.80c)}{c^2}} \\ &= \frac{1.30c}{1 + 0.40} \\ &= \frac{1.30}{1.40}c \\ &= 0.929c \end{aligned}$$

Thus spacecraft B sees spacecraft A approaching at a speed of (only!) 0.929c. Spacecraft A sees B approaching at the same speed (but in the opposite direction).

Method 2

For this method you identify the frame S with the Earth and the spacecraft on the left as the frame S′, to have the standard geometry as in Figure 25.38.

Frame S′ thus is moving to the right according to S with a speed of $v = 0.80c$. The spacecraft on the right is moving in S toward decreasing values of x; thus, the velocity component is negative: $u_x = -0.50c$. Here you want to find u'_x, the velocity component of the spacecraft B as seen by spacecraft A. You can anticipate that u'_x also will be negative since spacecraft B is moving in the negative x' direction. Having made these identifications, you can use Equation 25.45:

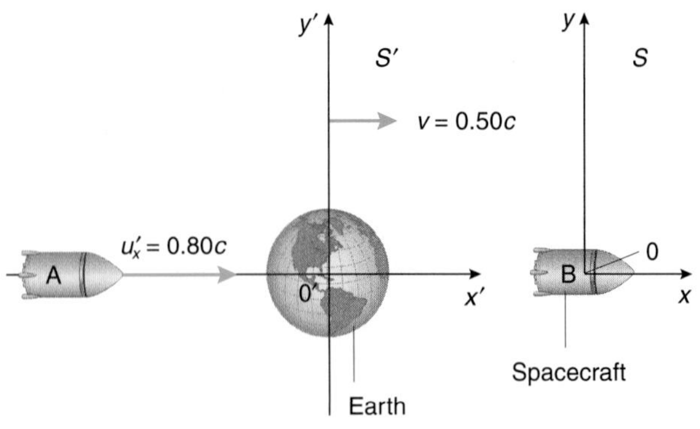

FIGURE 25.37

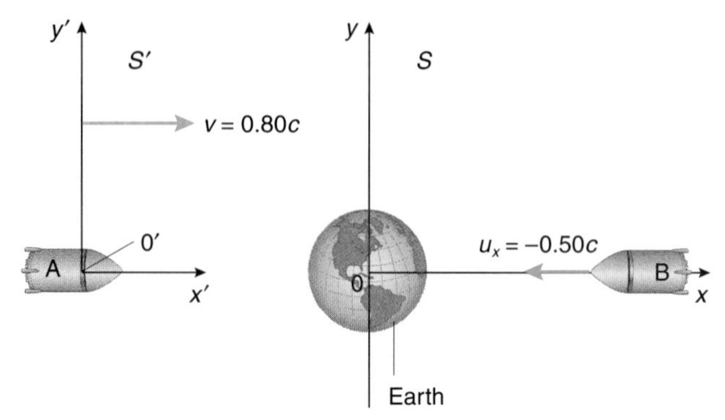

FIGURE 25.38

$$u_x = \frac{u'_x + v}{1 + \dfrac{vu'_x}{c^2}}$$

$$-0.50c = \frac{u'_x + 0.80c}{1 + \dfrac{(0.80c)u'_x}{c^2}}$$

Now solve for u'_x. After a bit of algebraic tedium, you will find that

$$\begin{aligned} u'_x &= -\frac{1.30}{1.40}c \\ &= -0.929c \end{aligned}$$

Thus spacecraft A says spacecraft B is approaching at a speed of 0.929c.

25.12 Cosmic Jets and the Optical Illusion of Superluminal Speeds*

Quasars are thought to be young and active galaxies and are among the most distant objects in the universe; they are located billions of light-years from us, far beyond our own Milky Way Galaxy. Some of these quasars emit jets of ionized gas that appear to be moving at speeds faster than the speed of light,† an observation that contradicts what we have learned about relative velocity addition in special relativity. In fact, the apparent superluminal speeds of these jets is an *illusion*; here we show how and why.‡

Consider a source of light located at point A, a distance d from the Earth located far away at distant point O in Figure 25.39.

†See the article by Roger D. Blandford, Mitchell C. Begelman, and Martin J. Rees, "Cosmic jets," *Scientific American*, 246, #5, pages 124–142 (May 1982).
‡This argument is based upon a paper by Richard M. Helsdon, "Cosmic jets," *Physics Education*, 18, #4, pages 169–170 (July 1983).

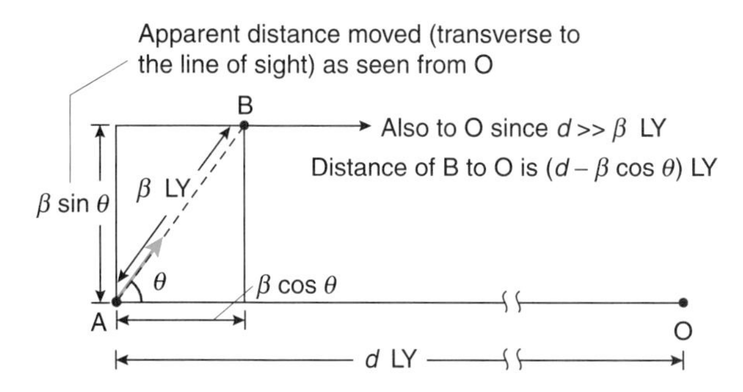

FIGURE 25.39 A very distant light source moves from A to B during one year.

The distance d is very large (billions of light-years). If we express the distance d in light-years, light from the source point A reaches the Earth after a time in years numerically equal to d. Let the source at A move at high speed v, which we express as a fraction of the speed of light: $v/c \equiv \beta$. After *one year*, the source will move a distance in light-years numerically equal to β, to position B. The distance β LY is less than 1 LY because the speed of the source must be less than the speed of light.

Since the distance d is many billions of light-years and the distance β is less than 1 LY, the line from B to the Earth at O is essentially parallel to the line from A to O. Thus, if we examine the geometry of Figure 25.39, the distance (in LY) of point B from the Earth is

$$d - \beta\cos\theta$$

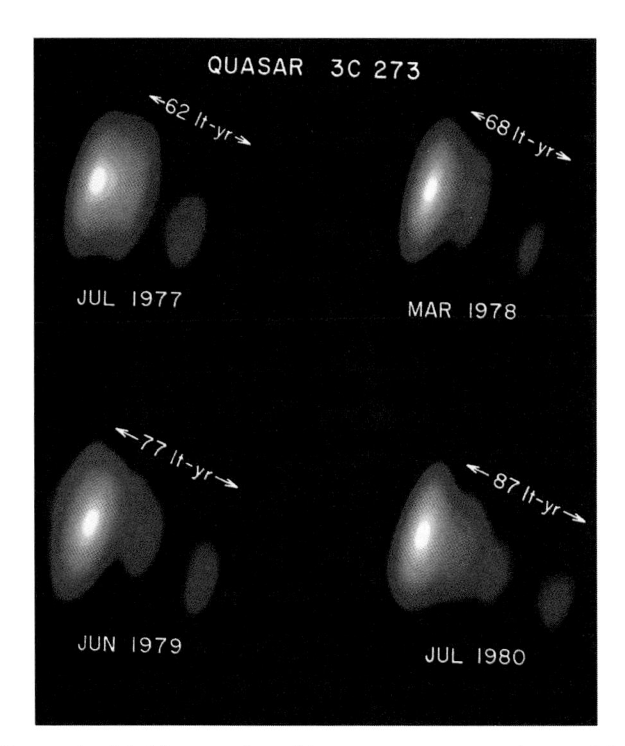

Some celestial objects, such as this quasar, appear to emit material at speeds exceeding the speed of light; the effect is an illusion.

Once light has been emitted at point B, it takes $d - \beta\cos\theta$ years to reach the Earth. Since the source took one year to get from point A to point B, the time of arrival of the light from B at the Earth is $1 + d - \beta\cos\theta$ years after the emission of the light from point A (which takes d years to get to the Earth). Thus the time *interval* ΔT between the arrival at the Earth of the light emitted when the source was at A and the arrival at the Earth of the light emitted when the source was at point B is

$$\Delta T = (1 + d - \beta\cos\theta) - d \qquad \text{(in y)}$$
$$= 1 - \beta\cos\theta \qquad \text{(in y)}$$

During this time, the source is seen by an observer on the Earth to move perpendicular (transverse) to the line of sight a distance ΔX given by

$$\Delta X = \beta\sin\theta \qquad \text{(in LY)}$$

Thus an observer on the Earth says the object moved an apparent distance $(\beta\sin\theta)$ LY during an interval $(1 - \beta\cos\theta)$ y. The *apparent fractional speed* $\beta' = \Delta X/\Delta T$ of the object (expressed as a fraction of the speed of light) is

$$\beta' = \frac{\beta\sin\theta}{1 - \beta\cos\theta} \tag{25.50}$$

Now we fix β and see if the apparent fractional speed β' has a maximum value as a function of θ. We maximize Equation 25.50 as a function of θ by setting its derivative with respect to θ equal to zero:

$$\frac{d\beta'}{d\theta} = 0$$

$$\frac{d\beta'}{d\theta} = \frac{(1 - \beta\cos\theta)(\beta\cos\theta) - (\beta\sin\theta)(\beta\sin\theta)}{(1 - \beta\cos\theta)^2} = 0$$

This yields the following equation:

$$\beta\cos\theta - \beta^2\cos^2\theta = \beta^2\sin^2\theta \tag{25.51}$$

Since

$$\sin^2\theta + \cos^2\theta = 1 \tag{25.52}$$

Equation 25.51 can be reduced to

$$\beta = \cos\theta \tag{25.53}$$

as the condition for maximizing the apparent fractional speed β'.

If $\beta = \cos\theta$, then Equation 25.52 also implies

$$(1 - \beta^2)^{1/2} = \sin\theta \tag{25.54}$$

Using Equation 25.53 and 25.54 in Equation 25.50 for the apparent fractional speed β', we obtain

$$\beta'_{\text{max}} = \frac{\beta}{(1 - \beta^2)^{1/2}} \tag{25.55}$$

Now we can see how the apparent fractional speed β'_{max} depends on the real fractional speed β of the object. In particular $\beta'_{max} = 1$, implying an apparent speed of the object equal to the speed of light, when

$$\beta = \frac{1}{\sqrt{2}} = 0.707$$

For this value of β, the angle θ (found from Equation 25.53) is 45°. For values of β greater than 0.707, then, it is possible (for some angle θ) for the apparent fractional speed β'_{max} to be greater than 1, implying an apparent speed greater than the speed of light. For example, if $\beta = 0.90$, then the angle $\theta = 26°$ (since $\beta = \cos\theta$) and the maximum apparent fractional speed is, using Equation 25.55,

$$\beta'_{max} = 2.1$$

The apparent speed of the object to an Earth observer thus is over *twice* the speed of light! But notice that the *actual* speed of the object moving from A to B is 0.90 times the speed of light. Thus the apparent superluminal speed of the real object is not its real speed.

Superluminal speeds do not need galactic astronomy (or nuclear disintegrations) for their occurrence. All you need is the realization that they represent motion, not of real objects, but of ideas. We conceptualize ideas, like points of intersection, which are free to move as fast as fantasy. A garden hose, for example, can easily produce superluminal motion. Hold it as you stand watering the lawn. Water follows a leisurely (parabolic) arc and hits the lawn at a distant spot. Next, aim the hose downward quickly so that the spot of impact is near your feet. With a little skill you can make the stream strike near you while it is still arriving at the previous point of impact. The point of impact can, with practice, be caused to move toward you at any speed, and certainly faster than c. You might call it a superluminal splash! But it is just the intersection point that moves, and that is merely a geometric concept.

25.13 The Longitudinal Doppler Effect

We studied the acoustic Doppler effect in Chapter 12. There we discovered that because sound waves travel in a material medium, the change in frequency was different if the source was moving rather than the observer. Light waves need no medium and can travel through a vacuum, since it is a well-known observation that we can see the Sun (on clear days!). We will find that it is only the relative motion of the source and observer that is important in the Doppler effect for light. There is no preferred inertial reference frame, just as Einstein said. In this section we see how the frequency of a light source is affected by the relative motion of the source and observer.

Imagine a well-trained and fleet-winged all-American firefly (*Photuris pennsylvanica*) moving at high speed v relative to an inertial frame S, as in Figure 25.40. Let inertial frame S′ be attached to the firefly, with the bright bug at the origin in S′.

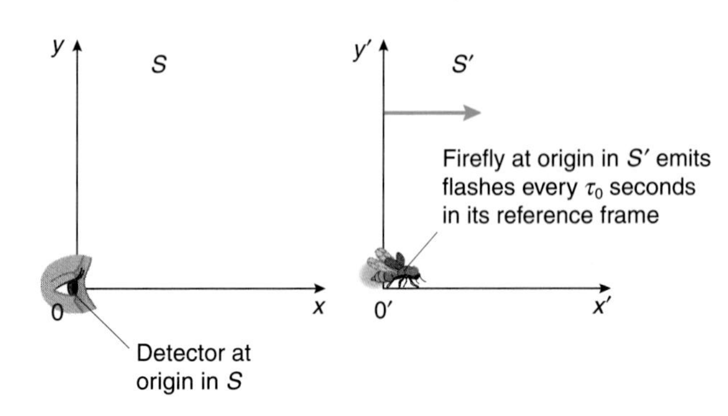

Figure 25.40 The standard geometry.

The firefly emits a flash of light every τ_0 seconds in its reference frame (S′). Thus the proper period of the firefly (in S′) is τ_0 and its proper frequency ν_0 is the reciprocal of τ_0:

$$\nu_0 = \frac{1}{\tau_0} \tag{25.56}$$

We specify a whole series of events in S′ corresponding to successive flashes of the firefly. Since the firefly is at the origin of the S′ frame, x' and t' space–time coordinates of these events are (if we let the initial flash, the zeroth flash, occur when $t' = 0$ s) as follows:

Event 0. The zeroth flash:

$$x'_0 = 0 \text{ m}$$
$$t'_0 = 0 \text{ s}$$

Event 1. The first flash:

$$x'_1 = 0 \text{ m}$$
$$t'_1 = \tau_0$$

Event 2. The second flash:

$$x'_2 = 0 \text{ m}$$
$$t'_2 = 2\tau_0$$

And so on for additional flashes.

Let your eye (a light detector) be at the origin of inertial reference frame S. You see the firefly receding from your eye at the speed v. We want to determine the frequency ν of the flashes *detected by your eye* (or another suitable light detector) at the origin of frame S.

To find this frequency, we transform the events in S′, just enumerated, to frame S using the appropriate Lorentz transformation (Equations 25.37 and 25.40):

$$x = \gamma(x' + vt')$$

and

$$t = \gamma\left(t' + \frac{v}{c^2}x'\right)$$

We substitute the space and time coordinates (x' and t') of each event into these Lorentz transformation equations to determine where and when each event occurs in the frame S. Since the x'-coordinate of every event is zero, this is relatively easy to do (pun intended). The results are as follows.

Event 0. The zeroth flash:

$$x_0 = 0 \text{ m}$$
$$t_0 = 0 \text{ s}$$

Event 1. The first flash:

$$x_1 = \gamma v \tau_0$$
$$t_1 = \gamma \tau_0$$

Event 2. The second flash:

$$x_2 = 2\gamma v \tau_0$$
$$t_2 = 2\gamma \tau_0$$

And so on.

Your eye at the origin in S sees the zeroth flash when $t_0 = 0$ s on your watch, since this flash occurs where your eye is located in S (at $x = 0$ m). But your eye does *not see* the first flash when $t_1 = \gamma\tau_0$. Why not? The first flash occurs at the location $x_1 = \gamma v \tau_0$ when the clocks in the frame S (including your watch) all read time t_1, as shown in Figure 25.41. For your eye at the origin actually to *see* the flash, the light at position x_1 must propagate from x_1 to the origin. The light thus needs to travel the distance x_1 at the speed c, and so takes an additional time $\gamma v \tau_0/c$ to get to the origin in S. Therefore the total time interval τ between the zeroth flash and the first flash as seen by your eye at the origin in S is the sum of t_1 and the propagation time:

$$\tau = \gamma\tau_0 + \frac{\gamma v \tau_0}{c} \qquad (25.57)$$

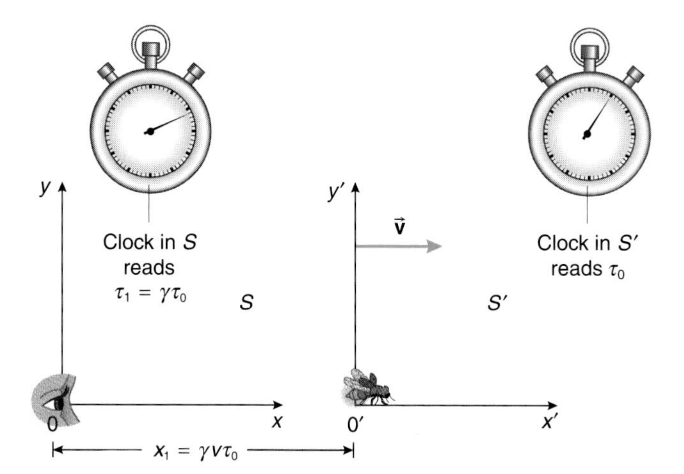

FIGURE 25.41 Event 1 in the S frame.

The second flash (Event 2) occurs in S at the location $x_2 = 2\gamma v\tau_0$ when the clocks in frame S read $t_2 = 2\gamma\tau_0$. This flash also must propagate the distance x_2 to be detected by your eye at the origin in S. Therefore the time on the clocks in S that the second flash arrives at your eye at the origin is

$$2\gamma\tau_0 + 2\frac{\gamma v \tau_0}{c}$$

which is 2τ.

Therefore the time interval between the flashes seen by your eye at the origin in frame S (the period of the flashes observed in S) is τ, given by Equation 25.57. The frequency ν of the flashes seen at the origin is $1/\tau$ or

$$\nu = \frac{1}{\gamma\tau_0 + \dfrac{\gamma v \tau_0}{c}} = \frac{1}{\gamma\tau_0\left(1 + \dfrac{v}{c}\right)}$$

But $1/\tau_0$ is the proper frequency ν_0 of the firefly. Thus the detected frequency from the receding bug is

$$\nu_{\text{recede}} = \frac{\nu_0}{\gamma\left(1 + \dfrac{v}{c}\right)}$$

Now we substitute for γ using its definition, Equation 25.13, and simplify:

$$\nu_{\text{recede}} = \frac{\nu_0\sqrt{1 - \dfrac{v^2}{c^2}}}{1 + \dfrac{v}{c}} = \frac{\nu_0\sqrt{1 - \dfrac{v}{c}}\sqrt{1 + \dfrac{v}{c}}}{1 + \dfrac{v}{c}}$$

Hence the receding source is observed to have a frequency given by

$$\nu_{\text{recede}} = \nu_0\sqrt{\frac{1 - \dfrac{v}{c}}{1 + \dfrac{v}{c}}} \qquad \text{(source receding)} \qquad (25.58)$$

The frequency observed by your eye in S is less than the proper frequency of the receding firefly.

If the firefly is *approaching* your eye, you can replace v in Equation 25.58 with $-v$ to obtain

$$\nu_{\text{appro}} = \nu_0 \sqrt{\frac{1 + \frac{v}{c}}{1 - \frac{v}{c}}} \quad \text{(source approaching)} \qquad (25.59)$$

The observed frequency ν_{appro} is greater than the proper frequency.

What we have analyzed, as you may have surmised (if only from the title of this section), is the relativistic **longitudinal Doppler effect**. The flashes play the role of the peak values of a light wave.

It makes no difference whether the source is moving away from the observer (detector), or the observer is moving away from the source. In other words, if we redid the problem with a firefly at the origin in S and the detector receding at speed v in S', the frequency of the flashes seen by the detector is less than the proper frequency of the source and is given by Equation 25.58. Likewise it makes no difference whether it is the source approaching the observer, or the observer approaching the source; the frequency seen by the observer is higher than the proper frequency and is given by Equation 25.59.

Of course, since the frequency and wavelength are related by

$$c = \nu\lambda$$

if the frequency decreases, the wavelength increases.

Any shift to a longer (larger) wavelength (i.e., smaller frequency) is known as a **red shift**,[†] regardless of the actual color or wavelength of the light. Red shifts occur if the distance between the source and observer is increasing with time (i.e., they are *receding* from each other) regardless of whether the source, observer, or both are moving.

Correspondingly, if the frequency increases, the wavelength decreases. Any shift to a shorter (smaller) wavelength (i.e., increased frequency) is known as a **blue shift**, regardless of the actual color or wavelength of the light. Blue shifts occur when the distance between a source and observer is decreasing with time (i.e., they are *approaching* each other) regardless of whether the source, observer, or both are moving.

The relativistic longitudinal Doppler effect for light in a vacuum is much simpler than the corresponding acoustic Doppler effect for sound (see Chapter 12). In the latter case, distinctly different quantitative effects occur even at low velocities (1) when the source is in motion; (2) when the observer is in motion; and (3) if there is a wind or motion of the medium itself in combination with motion of the source and/or observer. No such complications occur in the relativistic Doppler effect for light in a vacuum: there is no medium, and it is only the relative motion of the source and observer that is important. However, when the source and light travel in a material medium (such as water or glass), the Doppler effect becomes more complicated and takes on several aspects of the acoustical Doppler effect.

[†]The term arose because the red wavelengths are longer than blue wavelengths, defining roughly the extremes of the visible region of the electromagnetic spectrum.

EXAMPLE 25.7

A very distant quasar is receding from us. You find that the frequency of the light from hydrogen atoms in the quasar is 75% that from a hydrogen source at rest in your lab. At what speed is the quasar receding? Express your answer as a fraction of the speed of light. All observed Doppler shifts from distant galaxies are red shifts, indicating these sources are receding from us: the universe is "expanding."

Solution

The light is red-shifted. Since $\nu = 0.75\nu_0$, Equation 25.58 becomes

$$0.75\nu_0 = \nu_0 \sqrt{\frac{1 - \frac{v}{c}}{1 + \frac{v}{c}}}$$

After squaring, you can solve for v/c. You find that

$$v = 0.28c$$

The quasar source is receding at an incredible 28% the speed of light. Such large speeds for quasars are not uncommon.

25.14 THE TRANSVERSE DOPPLER EFFECT*

In the preceding section, the relative motion of the source or observer was along the line of sight. What happens if our relativistic firefly is moving transverse to the line of sight, as when the firefly is at point P in Figure 25.42?

The proper period of the firefly is τ_0 and its proper frequency (the inverse of the period) is ν_0 in its reference frame S', since the firefly is at rest in this frame. We imagine the frequency ν_0 is so great that many flashes are emitted while the firefly is in the immediate vicinity of point P in frame S. In this way, the propagation time of the pulses sent from the vicinity of P to your eye at the origin in S is the same for all the flashes because they all travel the same distance.

In S, you can think of the firefly as a moving clock. Thus, because of time dilation, the period τ between the flashes in S is measured to be longer by the factor γ:

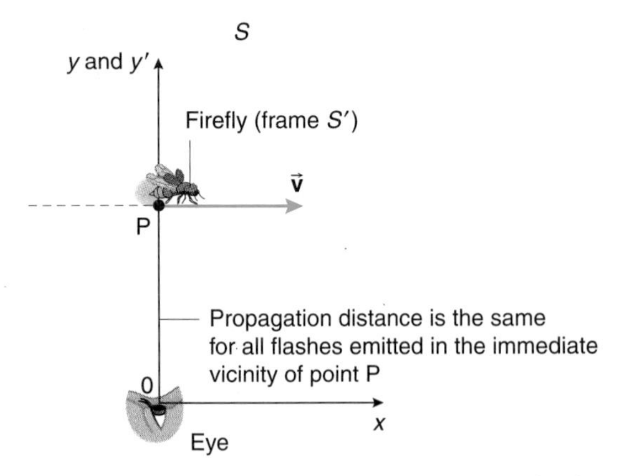

FIGURE 25.42 A source transverse to the line of sight.

$$\tau = \gamma\tau_0$$

The frequency of the firefly is the inverse of the period, and so the frequency ν of the flashes observed at the origin in S when the source is at P is

$$\begin{aligned}\nu &= \frac{1}{\tau}\\ &= \frac{1}{\gamma\tau_0}\\ &= \nu_0\sqrt{1-\frac{v^2}{c^2}} \qquad (25.60)\end{aligned}$$

Thus, when the light source is moving transverse to the line of sight, the frequency is *smaller* than the proper frequency; this effect is called the **transverse Doppler effect**. The effect is a consequence of time dilation and thus has no classical analog.

Classically, if a source is moving transverse to the line of sight there is no change in frequency; however, special relativity indicates there is!

According to Equation 25.60, the transverse Doppler effect *always* is a red shift, since the observed frequency is less than the proper frequency (i.e., the observed wavelength is longer than the proper wavelength).

It is interesting to compare the longitudinal and transverse Doppler effect red shifts. Figure 25.43 is a plot of the fractional frequency ratio ν/ν_0 versus v/c for the two effects.

Note that the transverse Doppler effect red shift always is smaller than the longitudinal Doppler effect red shift for the same v/c (the ratio of the frequencies ν/ν_0 always is closer to 1 for the transverse effect).

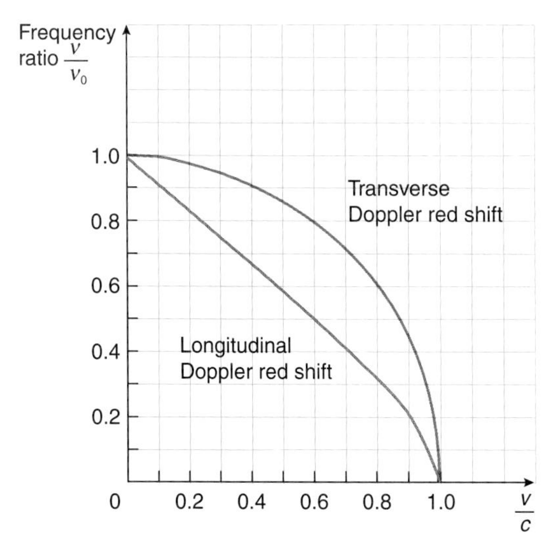

FIGURE 25.43 Longitudinal and transverse Doppler effects.

EXAMPLE 25.8

A light source moves at speed v perpendicular to your line of sight. The frequency of the light is only 75% that from a comparable source at rest with respect to you. What is the speed of the source, expressed as a fraction of the speed of light?

Solution

Since the source is moving perpendicular to the line of sight, the frequency changes because of the transverse Doppler effect. Use Equation 25.60 with $\nu = 0.75\nu_0$:

$$0.75\nu_0 = \nu_0\sqrt{1-\frac{v^2}{c^2}}$$

After squaring, you can solve for v/c. You find

$$\frac{v}{c} = 0.66$$

The source is moving at 66% the speed of light across your line of sight. Compare this with Example 25.7. Notice that for a given percentage frequency change, the source must move much faster for the change to occur via the transverse Doppler effect than via the longitudinal Doppler effect. Viewed another way, for a source with constant speed v, the longitudinal Doppler effect produces a greater frequency change (i.e., smaller frequency ratio) than does the transverse Doppler effect, as shown in Figure 25.43.

25.15 A GENERAL EQUATION FOR THE RELATIVISTIC DOPPLER EFFECT*

The red and blue shifts for the longitudinal Doppler effect and the red shift for the transverse relativistic Doppler effect can be incorporated into a single equation. We let the relative motion of the source be as indicated in Figure 25.44.

The angle θ is the angle between the velocity of the source and the line of sight from the source to the observer. Notice that while the source and observer are approaching each other, the angle θ ranges from 0° to 90° (acute); when the two are receding from each other, the angle θ ranges from 90° to 180° (obtuse).

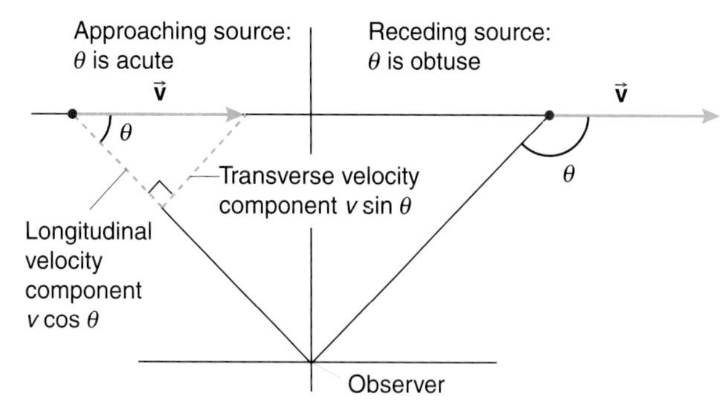

FIGURE 25.44 A source approaching not along the direct line of sight has both transverse and longitudinal velocity components.

With this definition of θ, the general equation for the relativistic Doppler effect is*

$$\nu = \nu_0 \frac{\sqrt{1 - \frac{v^2}{c^2}}}{1 - \frac{v}{c}\cos\theta} \quad (25.61)$$

When a source is approaching an observer the *longitudinal* Doppler effect predicts a *blue shift* in the spectrum, caused by the component of the velocity along the line of sight. On the other hand, the *transverse* Doppler predicts a *red shift*, caused by the component of the velocity transverse to the line of sight. Thus, somewhere among the possible angles of approach, there must exist an angle θ_0 at which the two effects exactly cancel each other and *no* shift in frequency is observed.

To find this special angle, we take the general equation (Equation 25.61) and set ν equal to ν_0. The resulting expression is

$$1 = \frac{\sqrt{1 - \frac{v^2}{c^2}}}{1 - \frac{v}{c}\cos\theta_0} \quad (25.62)$$

Solving this expression for $\cos\theta_0$, we find

$$\cos\theta_0 = \frac{1 - \sqrt{1 - \frac{v^2}{c^2}}}{\frac{v}{c}}$$

The solution of this equation for θ_0 for various values of v/c is shown in Figure 25.45.

For small speeds (small v/c), the transition from an approaching blue shift to a receding red shift occurs essentially at an angle of 90°, since the transverse Doppler effect is small compared with the longitudinal effect. As the speed increases, however, the transition from blue to red shift occurs at decreasing values of θ_0.

In other words, for high-speed sources not moving directly along the line of sight, the transition from blue to red shift occurs *before* the source and observer pass each other.

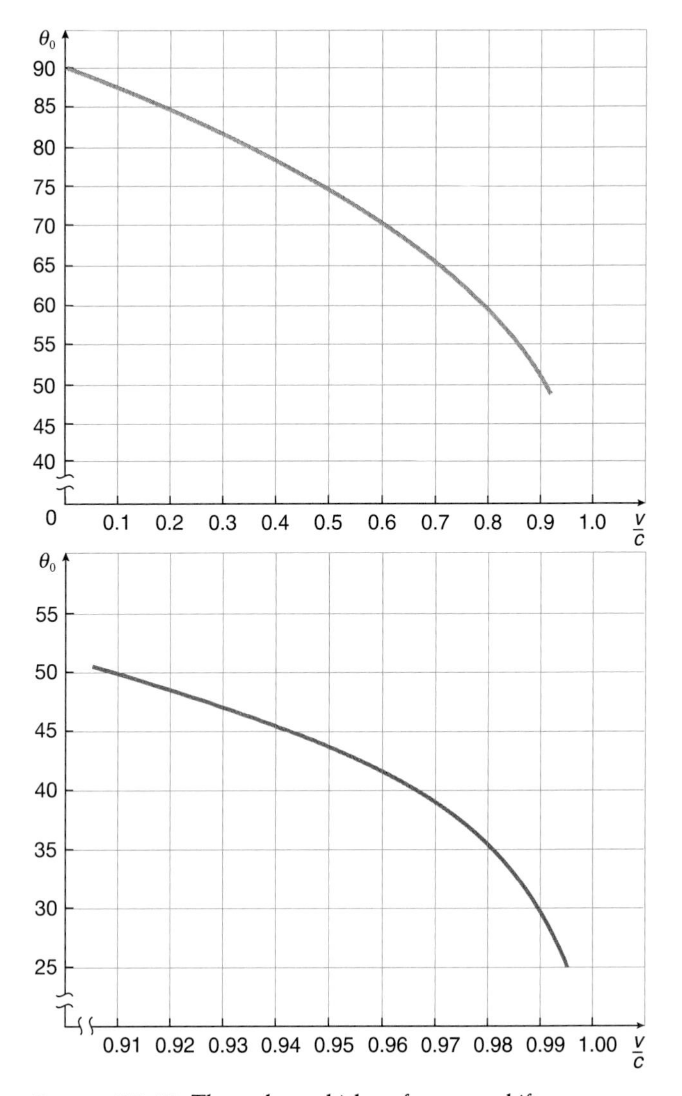

FIGURE 25.45 The angle at which no frequency shift occurs.

25.16 Relativistic Momentum

Our analysis of the special theory of relativity thus far has examined kinematical connections between the descriptions of events in space and time in two different inertial reference frames. We now turn to a consideration of the implications of relativity for the dynamics of a single particle, the effect of forces, the work done by these forces, and the consequent changes in momentum and kinetic energy of the particle.

Recall the classical momentum $\vec{\mathbf{p}}_{\text{class}}$ of a particle with a velocity $\vec{\mathbf{u}}$ is defined to be

$$\vec{\mathbf{p}}_{\text{class}} = m\vec{\mathbf{u}} \quad (25.63)$$

We also know from our study of collisions of two or more particles in Chapter 9 that when the total force on the system of colliding particles is zero, the total momentum of system is conserved.

A easily visualized experiment is sufficient to show that the classical definition of the momentum cannot be valid at high speeds if we accept the postulates of the special theory of relativity (in particular, Postulate 2). Imagine you are in reference frame S and a friend is in reference frame S′, moving past at speed v with respect to you in the standard geometry (see Figure 25.46).

Each of you throws a particle of the same standard mass m (such as a billiard ball) at speed u_0 along your respective y- or y'-axis to hit the other ball. Each of you determines the magnitude of u_0 using rulers and clocks at rest in your respective reference frames. The particles make an elastic collision and return to each of you

*We have not derived this equation, but pluck it from thin air. You can see that when $\theta = 0°$ (source approaching along the line of sight), Equation 25.61 reduces to Equation 25.59. When $\theta = 180°$ (source receding along the line of sight), Equation 25.61 reduces to Equation 25.58. Likewise, when $\theta = 90°$ (source transverse to the line of sight), Equation 25.61 reduces to Equation 25.60. See Problem 34.

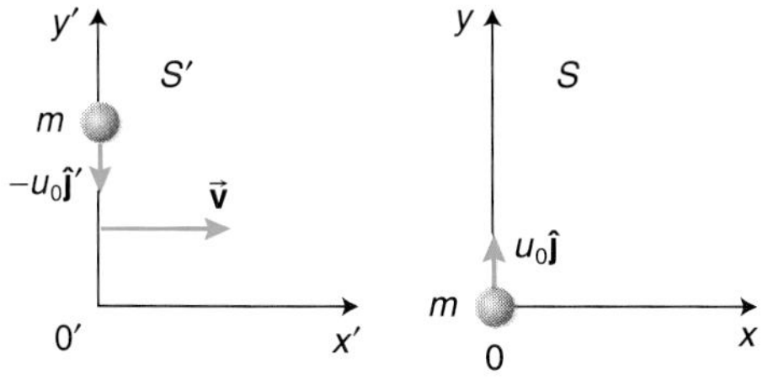

FIGURE 25.46 The standard geometry.

along the y- and y'-axes with their velocities reversed. Because of the symmetry of each of your viewpoints, we would like to think each of you says that the total y-component of momentum of the system of two particles is zero before and after the collision, since each of you says that equal mass particles were thrown with opposite y-velocities in your respective reference frames.

But knowing what we now do about how velocity components transform between the two reference frames (Section 25.11), if you (or your friend) analyze the elastic collision using the classical definition of the momentum, you find that the total momentum is *not* zero before or after the collision in your frame. Why not? The ball you throw has a classical momentum of $mu_0\hat{\mathbf{j}}$ according to rulers and clocks in your reference frame S. (Analogously, your friend says the ball she throws has classical momentum $-mu_0\hat{\mathbf{j}}'$ according to rulers and clocks in her reference frame S′.) But you say your friend's billiard ball has a y-velocity component in your reference frame of

$$-\frac{u_0}{\gamma}\hat{\mathbf{j}}$$

because of the new rule for adding velocity components (Equation 25.48). Hence you say her billiard ball has a classical momentum (when measured using rulers and clocks in your reference frame) of

$$-m\frac{u_0}{\gamma}\hat{\mathbf{j}}$$

and this has a *different* magnitude from that of the momentum of the ball you threw ($mu_0\hat{\mathbf{j}}$) in the opposite direction. As a result, the total classical momentum before the collision is *not* zero, though you both would expect it to be zero because of the planned symmetry of the encounter; nor is the total momentum zero after the collision.

The question then becomes, what expression or definition for the momentum can enable each of you to note that the total momentum before and after the collision is conserved and in fact zero? Measure the velocity of a particle to be $\vec{\mathbf{u}} = u_x\hat{\mathbf{i}} + u_y\hat{\mathbf{j}} + u_z\hat{\mathbf{k}}$ in your reference frame (using, of course, rulers and clocks in your frame).

We define the **relativistic momentum** $\vec{\mathbf{p}}$ of a particle to be

$$\vec{\mathbf{p}} \equiv \gamma m\vec{\mathbf{u}} \qquad (25.64)$$

where

$$\gamma \equiv \frac{1}{\sqrt{1-\dfrac{u^2}{c^2}}}.$$

Notice that γ involves the *speed* of the particle.* The magnitude of its velocity as measured in your reference frame is $u = (u_x^2 + u_y^2 + u_z^2)^{1/2}$. Observe that if $u << c$, in which case $\gamma \to 1$, the relativistic momentum reduces to the familiar classical expression for the momentum: $m\vec{\mathbf{u}}$.

You need to be aware that many texts distinguish between a so-called rest mass m of a particle (what we have called simply the mass, an intrinsic property of a particle) and a so-called relativistic mass $m_{rel} \equiv \gamma m$, so that the relativistic momentum is written as the product of the relativistic mass times the velocity: $\vec{\mathbf{p}} = m_{rel}\vec{\mathbf{u}}$; this makes the equation for the relativistic momentum look identical to the classical form for the momentum. Despite the historical popularity of such an approach, even Einstein suggested simply, "It is better to introduce no other mass concept than the 'rest mass' m."† We use this approach: the mass m (the **rest mass**) is the *only* mass we will use for a particle, with the relativistic momentum defined by Equation 25.64.

We now satisfy your curiosity and show that the definition of the relativistic momentum (Equation 25.64) means that the total y-component of the momentum of the two colliding equal mass particles in our collision experiment is zero in each reference frame. The ball you throw along the y-axis has velocity $u_0\hat{\mathbf{j}}$ and speed u_0 as measured by you. Its relativistic momentum then is

$$\vec{\mathbf{p}}_1 = m\gamma u_0\hat{\mathbf{j}} \qquad (25.65)$$

where

$$\gamma = \frac{1}{\sqrt{1-\dfrac{u_0^2}{c^2}}}$$

since the speed of the particle is u_0 according to you. Since your friend in the S′ frame is moving at speed v, the ball she throws at speed u_0 (as measured by her rulers and clocks) along her y'-axis has *two* velocity components in your reference frame (using rulers and clocks in your frame):

$$u_x = v \qquad (25.66)$$

and, according to the rule for transforming y-velocity components, Equation 25.48,‡

$$u_y = -u_0\sqrt{1-\frac{v^2}{c^2}} \qquad (25.67)$$

With Equation 25.64 for the definition of the relativistic momentum, you say that her particle has a y-component of momentum in your reference frame given by

$$p_2 = m\frac{1}{\left[1-\dfrac{(u_x^2+u_y^2)}{c^2}\right]^{1/2}}\left(-u_0\sqrt{1-\frac{v^2}{c^2}}\right) \qquad (25.68)$$

*This γ involves the speed of the *particle* and must be distinguished from the γ that involves the relative speed v of the reference frames S and S′, unless the particle *is* one of these frames.

†The quotation by Einstein appears in an article by Lev B. Oken, "The concept of mass," *Physics Today*, 42, #6, pages 31–36 (June 1989); the quotation is on page 32.

‡Notice that here only the relative speed v of the two reference frames is involved in the square root.

We use Equation 25.66 for u_x and Equation 25.67 for u_y in Equation 25.68. After making these substitutions and simplifying, we find

$$
\begin{aligned}
p_2 &= m \frac{1}{\left[1 - \dfrac{v^2 + u_0^2 - \dfrac{u_0^2 v^2}{c^2}}{c^2}\right]^{1/2}} \left(-u_0\sqrt{1 - \frac{v^2}{c^2}}\right) \\
&= m \frac{1}{\left[1 - \dfrac{v^2}{c^2} - \dfrac{u_0^2}{c^2} + \dfrac{u_0^2 v^2}{c^4}\right]^{1/2}} \left(-u_0\sqrt{1 - \frac{v^2}{c^2}}\right) \\
&= m \frac{1}{\sqrt{1 - \dfrac{v^2}{c^2}}\sqrt{1 - \dfrac{u_0^2}{c^2}}} \left(-u_0\sqrt{1 - \frac{v^2}{c^2}}\right) \\
&= -m \frac{u_0}{\sqrt{1 - \dfrac{u_0^2}{c^2}}} \qquad (25.69)
\end{aligned}
$$

When you find the total y-component of the momentum before the collision in your reference frame, using Equations 25.65 and 25.69, you now pleasingly discover the total y-component of the momentum before the collision is zero. A similar argument used by your friend in S' shows that the total y'-component of the momentum of the two particles is zero when measured by her using clocks and rulers in her frame.

Hence the new definition of the momentum, Equation 25.64, is consistent with the postulates of special relativity and we can conserve momentum in collisions within any inertial reference frame.

There is a more heuristic way to see how Equation 25.64 comes about for the relativistic momentum. For simplicity, we consider a particle subject to a constant total force $\vec{\mathbf{F}}_{\text{total}}$. In the Galilean and Newtonian theory of motion, a constant total force acting on a particle of constant mass m produces a constant acceleration $\vec{\mathbf{a}}$ in the same direction as the total force; this is Newton's second law of motion:

$$\vec{\mathbf{F}}_{\text{total}} = m\vec{\mathbf{a}}$$

The constant acceleration changes the velocity of the particle according to the familiar kinematic relation

$$\vec{\mathbf{v}} = \vec{\mathbf{v}}_0 + \vec{\mathbf{a}}t$$

Given enough time, such a force (and resulting acceleration) can increase the speed of the particle without limit, according to this classical view.

Later, in Section 25.17, we shall see that the special theory of relativity indicates there is an *upper limit* to the speed a particle can attain: the speed of light. Thus the question arises, how do we reconcile the Newtonian description of motion and Newton's second law with the special theory of relativity? The key is through the more general statement of the Newtonian second law of motion: the total force is equal to the time rate of change of the momentum of the particle:

$$\vec{\mathbf{F}}_{\text{total}} = \frac{d\vec{\mathbf{p}}}{dt}$$

Given enough time, a constant total force will increase the *momentum* of a particle without limit, but *not* its speed. In classical physics the momentum of a particle was defined to be the product of its mass m and its velocity $\vec{\mathbf{u}}$:

$$\vec{\mathbf{p}}_{\text{class}} \equiv m\vec{\mathbf{u}}$$

The mass of a particle is an innate characteristic of the particle (much like electric charge). Although under the action of a constant total force, the speed of a particle can increase up to a limit, we want the momentum of the particle to be able to increase without limit under the action of a constant force to follow Newton's law, $\Delta\vec{\mathbf{p}} = \vec{\mathbf{F}}_{\text{total}}\,\Delta t$. Thus the classical expression for the momentum of a particle needs to be modified or changed. How do we make the change? Certainly, for small speeds, the classical expression for the momentum is legitimate, and so whatever we take for the relativistic momentum must reduce to the classical expression when $u << c$.

We need to multiply the classical expression for the momentum by a dimensionless factor that

- increases without limit as $u \rightarrow c$; and
- approaches 1 when $u << c$ (or equivalently, as $u \rightarrow 0$ m/s).

These constraints are just the characteristics of the ubiquitous factor γ in the special theory of relativity. Hence we try defining the relativistic momentum of a particle with a velocity $\vec{\mathbf{u}}$ to be

$$\vec{\mathbf{p}}_{\text{rel}} = \gamma m\vec{\mathbf{u}} = \frac{m\vec{\mathbf{u}}}{\sqrt{1 - \dfrac{u^2}{c^2}}}$$

where u is the speed of the particle (the magnitude of its velocity); this is Equation 25.64. With this definition, under the action of a constant force, the momentum of a particle of intrinsic (constant) mass m can increase indefinitely as $u \rightarrow c$ whereas its speed u has an upper bound.

The justification for this definition of the relativistic momentum is, of course, the correct results of all experiments to test it. As we saw earlier, experiments involving the collisions of particles traveling at high speeds indicate that the definition of the relativistic momentum in Equation 25.64 is the only way momentum can be defined so that it is conserved in relativistic collisions.

EXAMPLE 25.9

A cosmic ray proton is traveling with a speed of 0.99900*c*. Determine the magnitude of its relativistic momentum.

Solution
First calculate the factor γ associated with its speed:

$$\gamma = \frac{1}{\sqrt{1 - \frac{u^2}{c^2}}}$$
$$= \frac{1}{\sqrt{1 - (0.99900)^2}}$$
$$= 22.4$$

The magnitude of the relativistic momentum is found from Equation 25.64:

$$p = \gamma mu$$
$$= 22.4(1.67 \times 10^{-27}\ \text{kg})(0.99900 \times 3.00 \times 10^8\ \text{m/s})$$
$$= 1.12 \times 10^{-17}\ \text{kg}\cdot\text{m/s}$$

25.17 The CWE Theorem Revisited

The CWE theorem (see Chapter 8) states that if we neglect thermal effects, the work done by all the forces acting on a system is equal to the change in the kinetic energy of the system. What implications does this theorem have in relativity?

We consider a single force acting on a single particle system. Let the force have only a positive x-component, $\vec{\mathbf{F}} = F\hat{\imath}$, so the problem is one-dimensional. The differential work dW done by the force when the position vector of the particle changes by $d\vec{\mathbf{r}} = dx\,\hat{\imath}$ is

$$dW = \vec{\mathbf{F}} \cdot d\vec{\mathbf{r}}$$
$$= F\,dx$$

According to the CWE theorem, the differential work done by the total force on a system is equal to the differential change in the kinetic energy, $d(\text{KE})$, of the system:

$$F\,dx = d(\text{KE}) \qquad (25.70)$$

Keeping Newton's second law, the time rate of change of the relativistic momentum $\vec{\mathbf{p}}$ still is equal to the applied total force:

$$\vec{\mathbf{F}} = \frac{d\vec{\mathbf{p}}}{dt}$$

where the relativistic momentum* is $\vec{\mathbf{p}} = \gamma m\vec{\mathbf{v}}$.

Substituting for F in the CWE theorem, Equation 25.70, and writing the change in the kinetic energy first, we obtain

$$d(\text{KE}) = \frac{dp}{dt}\,dx \qquad (25.71)$$

*We switch to the more conventional notation of $\vec{\mathbf{v}}$ for the velocity of the particle, since we can take the particle to be at rest in S'. Remember that the γ in the relativistic momentum involves the speed v of the particle.

We want to integrate this equation over distance to find an expression for the changes in the relativistic kinetic energy caused by the force. We anticipate that since the relativistic momentum of the particle increases without limit as $v \to c$, the relativistic kinetic energy will do likewise. We see already that the classical expression for the kinetic energy,

$$\text{KE}_{\text{class}} = \frac{mv^2}{2}$$

like the classical momentum, *cannot* be the correct expression for the relativistic kinetic energy of a particle, since the speed of a particle cannot exceed c, a fact we justify shortly. The classical expression for the kinetic energy would imply an upper bound, equal to $mc^2/2$, on the kinetic energy of a particle, even though F could do an infinite amount of work on it over unbounded distance. In reality, such an upper bound on the kinetic energy does not exist. The expression for the relativistic kinetic energy must be different from the classical expression. Nonetheless, for speeds $v \ll c$, the relativistic expression must reduce to the classical expression, since the latter expression serves us well in mechanics, as long as the speeds of the particles are such that $v \ll c$.

The problem at hand is to integrate Equation 25.71 to find an expression for the relativistic kinetic energy. It is not immediately apparent how to proceed with this integration. One approach suggests itself: since the relativistic momentum is a function of the velocity of the particle, it is convenient to convert the integration from one involving the change in the position vector component dx to one involving the change in the velocity component dv. To accomplish this change, we use the chain rule from calculus in the following way.

The right-hand side of Equation 25.71 can be expressed as

$$\frac{dp}{dt}\,dx = \frac{dp}{dv}\frac{dv}{dt}\,dx \qquad (25.72)$$

Likewise, the derivative

$$\frac{dv}{dt}$$

can be written with the chain rule as

$$\frac{dv}{dt} = \frac{dv}{dx}\frac{dx}{dt}$$

Hence Equation 25.72 becomes

$$\frac{dp}{dt}\,dx = \frac{dp}{dv}\frac{dv}{dx}\frac{dx}{dt}\,dx$$

But

$$\frac{dx}{dt} = v$$

and so

$$\frac{dp}{dt}\,dx = \frac{dp}{dv}\frac{dv}{dx}\,v\,dx$$
$$= \frac{dp}{dv}\,v\,dv$$

Thus the CWE theorem (Equation 25.71) becomes

$$d(\text{KE}) = \frac{dp}{dv} v\,dv \qquad (25.73)$$

Now we just have to find the derivative

$$\frac{dp}{dv}$$

The magnitude p of the relativistic momentum component in one dimension is

$$p = \gamma m v \qquad (25.74)$$

We put in the explicit meaning of the factor γ:

$$p = \frac{mv}{\sqrt{1 - \frac{v^2}{c^2}}}$$

Remember that c and m are constants; we take the derivative of p with respect to v, using the usual calculus rules for the derivative of a quotient:

$$\frac{dp}{dv} = \frac{\sqrt{1 - \frac{v^2}{c^2}}\, m - mv\left[\frac{1}{2}\left(1 - \frac{v^2}{c^2}\right)^{-1/2}\left(-2\frac{v}{c^2}\right)\right]}{1 - \frac{v^2}{c^2}}$$

After a bit of algebra (maybe a big bit!), this ungainly expression reduces to

$$\frac{dp}{dv} = \frac{m}{\left(1 - \frac{v^2}{c^2}\right)^{3/2}}$$

Substituting this derivative into Equation 25.73 for the CWE theorem, we obtain

$$d(\text{KE}) = \frac{mv\,dv}{\left(1 - \frac{v^2}{c^2}\right)^{3/2}}$$

This expression can now be integrated. When $v = 0$ m/s the kinetic energy also is zero, and so the definite integrals are

$$\int_{0\text{ J}}^{\text{KE}} d(\text{KE}) = m\int_{0\text{ m/s}}^{v} \frac{v\,dv}{\left(1 - \frac{v^2}{c^2}\right)^{3/2}}$$

The integral on the right-hand side can be put into the standard form:

$$\int u^n\,du = \frac{u^{n+1}}{n+1}$$

where we let

$$u = 1 - \frac{v^2}{c^2}$$

The result is

$$\text{KE} = \frac{m}{\left(-\frac{2}{c^2}\right)}\int_{0\text{ m/s}}^{v} \frac{-\frac{2}{c^2}\, v\,dv}{\left(1 - \frac{v^2}{c^2}\right)^{3/2}}$$

$$= mc^2\left(\frac{1}{\sqrt{1 - \frac{v^2}{c^2}}} - 1\right)$$

The first term in the brackets is γ.

Thus the **relativistic kinetic energy** is

$$\boxed{\text{KE} = (\gamma - 1)mc^2} \qquad (25.75)$$
$$= \gamma mc^2 - mc^2$$

The second term on the expanded right-hand side of Equation 25.75 is *independent of the speed* of the particle and is intrinsic to the particle itself. Hence mc^2 is called the **rest energy** of the particle.

It is from this term that Einstein made the famous conclusion that energy is the mass times the square of the speed of light. Mass is equivalent to energy. There is a subtlety to this equivalence that we explore subsequently in Section 25.18.

Consider a particle with $v < c$. It has a finite amount of relativistic kinetic energy. If we try to increase its kinetic energy so that $v = c$, at which speed the factor γ becomes infinite, the particle then would have an infinite amount of relativistic kinetic energy. This means we need to do an infinite amount of work on the particle to accomplish this feat! Such an amount of energy clearly is unavailable. Hence a particle initially confined to a speed less than c can *never* attain a speed even equal to c in special relativity.

The speed of light thus is an unreachable upper bound on the speed of a particle (with nonzero mass) in special relativity.

This is much like absolute zero is an unreachable lower bound on the temperature of a system.

Transposing the rest energy term in Equation 25.75, we find

$$\gamma mc^2 = \text{KE} + mc^2 \qquad (25.76)$$

The right-hand side of Equation 25.76 contains two terms: the kinetic energy; and the rest energy mc^2, which is independent of the speed of the particle.

The sum of the kinetic energy of the particle and its rest energy is called the **total relativistic energy** E of the particle:

$$E \equiv \text{KE} + mc^2 \tag{25.77}$$

Thus, from Equation 25.76, the total relativistic energy is

$$\boxed{E = \gamma mc^2} \tag{25.78}$$

The relativistic kinetic energy thus is the difference between the total relativistic energy E and the rest energy mc^2:

$$\begin{aligned} \text{KE} &= \gamma mc^2 - mc^2 \\ &= E - mc^2 \end{aligned} \tag{25.79}$$

What is the relationship between the relativistic kinetic energy and the classical expression for the kinetic energy? To elucidate this relationship, we begin with Equation 25.75 and rearrange it as follows:

$$\begin{aligned} \text{KE} &= mc^2(\gamma - 1) \\ &= mc^2\left(\frac{1}{\sqrt{1 - \frac{v^2}{c^2}}} - 1\right) \\ &= mc^2\left[\left(1 - \frac{v^2}{c^2}\right)^{-1/2} - 1\right] \end{aligned} \tag{25.80}$$

We use the binomial theorem to expand the term

$$\left(1 - \frac{v^2}{c^2}\right)^{-1/2}$$

The binomial theorem states that

$$(1 + x)^n = 1 + nx + \frac{n(n-1)}{2!}x^2 + \frac{n(n-1)(n-2)}{3!}x^3 + \cdots$$

where we assign $x = -v^2/c^2$ and $n = -1/2$. The expansion is

$$\begin{aligned} \left(1 - \frac{v^2}{c^2}\right)^{-1/2} &= 1 + \left(-\frac{1}{2}\right)\left(-\frac{v^2}{c^2}\right) + \frac{\left(-\frac{1}{2}\right)\left(-\frac{3}{2}\right)\left(-\frac{v^2}{c^2}\right)^2}{2!} \\ &\quad + \frac{\left(-\frac{1}{2}\right)\left(-\frac{3}{2}\right)\left(-\frac{5}{2}\right)\left(-\frac{v^2}{c^2}\right)^3}{3!} + \cdots \\ &= 1 + \frac{1}{2}\frac{v^2}{c^2} + \frac{3}{8}\frac{v^4}{c^4} + \frac{5}{16}\frac{v^6}{c^6} + \cdots \end{aligned}$$

Substituting the expansion into Equation 25.80 for the kinetic energy, we obtain

$$\text{KE} = mc^2\left(1 + \frac{1}{2}\frac{v^2}{c^2} + \frac{3}{8}\frac{v^4}{c^4} + \frac{5}{16}\frac{v^6}{c^6} + \cdots - 1\right)$$

After removing the parentheses, we find

$$\text{KE} = \frac{1}{2}mv^2 + \frac{3}{8}m\frac{v^4}{c^2} + \frac{5}{16}m\frac{v^6}{c^4} + \cdots \tag{25.81}$$

If the speed v of the particle is small compared with the speed of light c—that is, if $v << c$—then only the first term of Equation 25.81 for the kinetic energy is significant. This term we recognize as the classical expression for the kinetic energy:

$$\text{KE} \approx \frac{1}{2}mv^2 = \text{KE}_{\text{class}} \qquad (v << c)$$

PROBLEM-SOLVING TACTIC

25.6 Remember that the relativistic kinetic energy is

$$\text{KE} = (\gamma - 1)mc^2$$

The relativistic kinetic energy is *not* $(1/2)\,mv^2$.

There are two important and frequently used expressions relating the magnitude of the relativistic momentum $p = \gamma mv$ and the total relativistic energy $E = \gamma mc^2$ of a single particle moving in one dimension. They are exact relations, not just approximations.

1. We solve Equation 25.78 for γ, obtaining

$$\gamma = \frac{E}{mc^2} \tag{25.82}$$

Likewise, using Equation 25.74 and solving for γ, we find

$$\gamma = \frac{p}{mv} \tag{25.83}$$

Equating the expressions for γ in Equations 25.82 and 25.83, we find

$$p = \frac{v}{c^2}E \tag{25.84}$$

2. With Equations 25.78 and 25.74, we form the expression

$$E^2 - c^2p^2 = \gamma^2m^2c^4 - c^2\gamma^2m^2v^2$$

After substitution for γ^2, and a bit (!) of algebra to remove v, we find that the right-hand side reduces to just m^2c^4. Hence we have

$$E^2 - p^2c^2 = m^2c^4$$

or

$$E^2 = p^2c^2 + m^2c^4 \tag{25.85}$$

EXAMPLE 25.10

Calculate the rest energy of an electron. Express your result in both joules and in millions of electron-volts (MeV).

Solution
The rest energy of an electron is

$$\begin{aligned} E_{\text{rest}} &= m_e c^2 \\ &= (9.11 \times 10^{-31}\ \text{kg})(3.00 \times 10^8\ \text{m/s})^2 \\ &= 8.20 \times 10^{-14}\ \text{J} \end{aligned}$$

To convert to electron-volts, use the conversion factor

$$1.602 \times 10^{-19}\ \text{J} = 1.000\ \text{eV}$$

Thus the rest energy in electron-volts is

$$\begin{aligned} E_{\text{rest}} &= \frac{8.20 \times 10^{-14}\ \text{J}}{1.602 \times 10^{-19}\ \text{J/eV}} \\ &= 5.12 \times 10^5\ \text{eV} \\ &= 0.512\ \text{MeV} \end{aligned}$$

STRATEGIC EXAMPLE 25.11

A proton is moving at speed $v = 0.900c$.

a. Find its total relativistic energy E. Express your result in both J and MeV.
b. Find its kinetic energy. Express your result in both J and MeV.
c. Determine the magnitude p of its relativistic momentum.

Solution
First find the factor γ:

$$\begin{aligned} \gamma &= \frac{1}{\sqrt{1 - \dfrac{v^2}{c^2}}} \\ &= \frac{1}{\sqrt{1 - (0.900)^2}} \\ &= 2.29 \end{aligned}$$

a. The total energy is found from Equation 25.78:

$$\begin{aligned} E &= \gamma mc^2 \\ &= 2.29(1.67 \times 10^{-27}\ \text{kg})(3.00 \times 10^8\ \text{m/s})^2 \\ &= 3.44 \times 10^{-10}\ \text{J} \end{aligned}$$

Converting to electron-volts,

$$\begin{aligned} E &= \frac{3.44 \times 10^{-10}\ \text{J}}{1.602 \times 10^{-19}\ \text{J/eV}} \\ &= 2.15 \times 10^9\ \text{eV} \\ &= 2.15 \times 10^3\ \text{MeV} \end{aligned}$$

b. The kinetic energy is found using Equation 25.75:

$$\begin{aligned} \text{KE} &= (\gamma - 1)mc^2 \\ &= (2.29 - 1)(1.67 \times 10^{-27}\ \text{kg})(3.00 \times 10^8\ \text{m/s})^2 \\ &= 1.94 \times 10^{-10}\ \text{J} \\ &= 1.21 \times 10^3\ \text{MeV} \end{aligned}$$

c. The magnitude of the relativistic momentum can be found in two ways.

Method 1

Use the definition of the relativistic momentum, Equation 25.64:

$$\begin{aligned} p &= \gamma mv \\ &= 2.29(1.67 \times 10^{-27}\ \text{kg})(0.900 \times 3.00 \times 10^8\ \text{m/s}) \\ &= 1.03 \times 10^{-18}\ \text{kg}\cdot\text{m/s} \end{aligned}$$

Method 2

Use Equation 25.84:

$$\begin{aligned} p &= \frac{v}{c^2}E \\ &= \frac{v}{c}\frac{E}{c} \\ &= (0.900)\frac{3.44 \times 10^{-10}\ \text{J}}{3.00 \times 10^8\ \text{m/s}} \\ &= 1.03 \times 10^{-18}\ \text{kg}\cdot\text{m/s} \end{aligned}$$

25.18 Implications of the Equivalence Between Mass and Energy

The equivalence between mass and energy alluded to in Section 25.17 warrants further discussion.* Mass, like charge, is a property or attribute of a particle. Mass and charge cannot exist independently of a particle. Particles may or may not have the attribute or property called mass; they also may or may not have charge. Likewise, particles may or may not have various kinds of energy.† For example:

- Some particles have mass (the electron, proton, neutron, nerf ball, *you*, etc.). If a particle has mass, we imagine the particle as being tangible. If a particle has mass, it has rest energy mc^2 and may or may not have other varieties of energy such as kinetic and potential energy.
- Other particles have energy but zero mass; for example, the particle of light, the photon,‡ has zero mass but nonetheless has energy (although not rest energy, since $m = 0$ kg means $mc^2 = 0$ J).

In an isolated system the *total* amount of energy possessed by its constituent particles is conserved; this statement is the generalized first law of thermodynamics. The total energy is the sum of the individual varieties of energy of all the particles of the isolated system. This includes the individual rest energies of particles that have mass, kinetic energies of the particles, various potential energies of the particles, and so on.

*An extensive discussion can be found in Ralph Baierlein, "Teaching $E = mc^2$," *The Physics Teacher*, 29, #3, pages 170–175 (March 1991).
†Waves also can have the attribute of energy.
‡We introduce the photon particle in a formal way in Chapter 26.

It is possible to convert an isolated system of particles with mass to a system of particles with less mass, even zero mass, and—remarkably—vice versa. These processes are what is typically meant by statements such as: "mass is equivalent to energy," "mass is converted into energy," or "energy is converted into mass." What is meant is that *rest energy* is converted into other types of energy, or other types of energy are converted into rest energy.

Examples include nuclear fusion and fission processes.† Consider the fission process of the $^{235}_{92}U$ isotope of uranium. With the addition of a slow neutron, the nucleus of this atom spontaneously splits (fissions) into a host of other particles, some with mass, some without mass. The total energy is conserved in the process. The total energy of the two initial particles, the slow neutron and the $^{235}_{92}U$ atom (including rest energy, kinetic energy, and various potential energies), is equal to the total energy of all the particles resulting from the fission process (once again, including rest energy, kinetic energy, energy associated with photons, and various potential energies).

The conversion of rest energy into other forms of energy is what happens in nuclear explosions.

The point is that the total energy is apportioned among the resulting particles in different ways than it was for the initial particles. One finds that the total *rest energy* of the initial particles is greater than the total rest energy of the resulting particles by some amount $(\Delta m)c^2$. Energy has *not* been created out of mass, since the energy was all there to begin with. The energy merely has been shuffled or distributed among the constituent particles in different ways than it was originally. The idea is no more subtle or profound from a conceptual point of view than the conversion of potential energy into kinetic energy when, say, a rock falls freely under the influence of the gravitational force.

The product particles have less rest energy but a correspondingly greater amount of the other kinds of energy (kinetic energy, energy of photons, etc.). The increase in the other kinds of energy in the product particles is precisely compensated for by the decrease in the rest energies.

The total energy is conserved in the isolated system.

Question 4

When browsing through some artifacts in your parents' attic, you come across the physics book used by your great-grandfather in 1904. In leafing through its pages, you come across a statement that "matter can neither be created nor destroyed." Is this statement still true in view of the special theory of relativity? Comment.

25.19 Space–Time Diagrams*

A **space–time diagram** is a graphical depiction of the space and time coordinates of a reference frame. Since there are three spatial coordinates and one temporal coordinate, a complete space–time diagram is four-dimensional! Since our powers of visualization in four dimensions are limited, we typically suppress one or more spatial dimensions, and plot the time coordinate vertical with one or two spatial coordinates horizontally, as in Figure 25.47.

An event is represented by a point on a space–time diagram. The path followed by a system on a space–time diagram is called a **world-line**. A space–time diagram is dynamic. Time stops for no man (or woman), and so if you simply stand or lie at a particular point in space as a couch potato, a vertical world-line is traced out on the space–time diagram, as shown in Figure 25.48. If you walk at constant speed along the x-axis of the coordinate grid, your world-line is inclined to the vertical, as also shown in Figure 25.48.

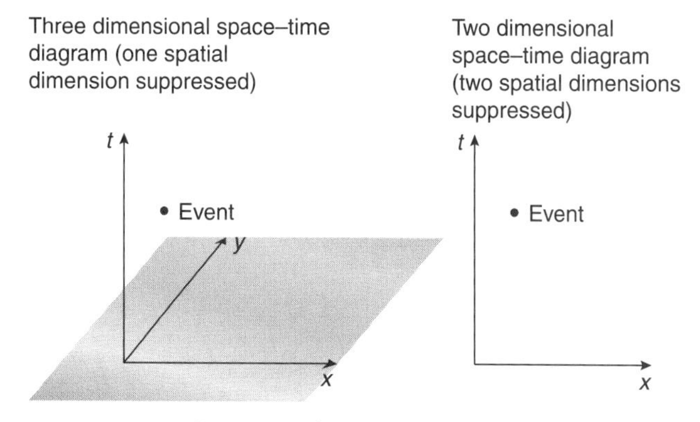

FIGURE 25.47 Space–time diagrams.

†Another example is the annihilation of matter with antimatter, as when an electron meets its antiparticle the positron. Both particles are annihilated and form two (or more) particles of light (photons). Since the total energy is conserved, the total energy associated with the electron–positron pair of particles (kinetic energy + rest energy) is transferred to the photon particles that have no rest energy, since they have zero mass. Also, it is possible for particles with no mass to create particles with mass. A sufficiently energetic photon (with zero mass) can create an electron–positron pair in a process known as pair production; this process is seen frequently in nuclear physics. In this case the energy of the photon is transferred to the total relativistic energy associated with the electron and positron. The photon must have an energy at least equal to the sum of the rest energies of the electron and positron for this process to occur (since the minimum value for the total relativistic energy of an electron and positron is the sum of their rest energies with zero kinetic energy).

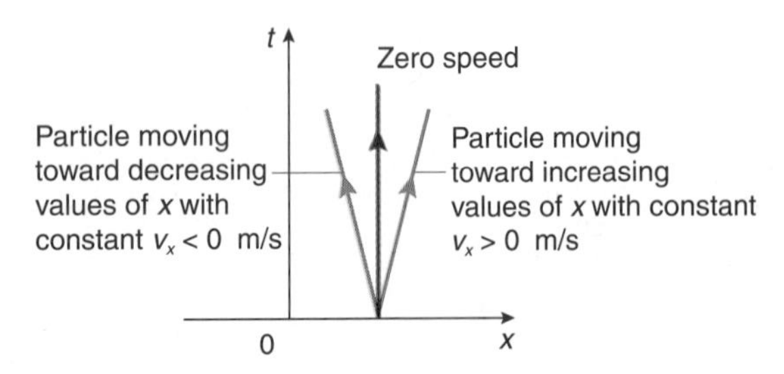

FIGURE 25.48 World-lines of particles moving with zero and nonzero constant speeds.

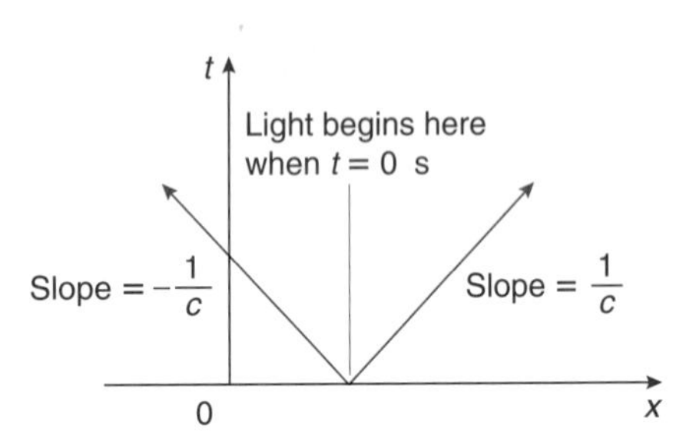

FIGURE 25.49 World-lines of light on a two-dimensional space–time diagram.

The greater the speed, the smaller the slope of the world-line on the space–time diagram. Indeed, the inverse of the slope of the world-line on a space–time diagram is the velocity component of the particle. We saw in Section 25.17 that the ultimate speed is the speed of light. If we let light propagate parallel and antiparallel to $\hat{\imath}$, the path of the light on the space–time diagram is represented by straight world-lines whose slopes are $\pm 1/c$, as shown in Figure 25.49. Nothing can have a world-line on a space–time diagram that has a slope with an absolute magnitude less than that of the world-line for light, because it takes an infinite amount of work to change the kinetic energy of a particle with mass so that its speed is c.

A space–time diagram is a simple way to show that travel into your past (a topic of much science fiction) is impossible according to special relativity.† The point at the origin on a space–time diagram signifies "here and now." The future is $t > 0$ s and the past is $t < 0$ s. To travel to the past involves a world-line on a space–time diagram similar to that shown in Figure 25.50.

As the slope of the world-line decreases, the speed of the particle increases. Eventually, you reach a point P on the world-line of Figure 25.50 where the slope is equal to that of the light-line. Beyond point P, the slope of the world-line is less than that of the light-line, meaning a speed greater than c has been achieved. But such speeds exceeding c are prohibited according to special relativity because infinite amounts of energy are unavailable. Hence such a loop on a space–time diagram is impossible to complete. Thus travel into the past is *verboten* according to special relativity; the past is gone forever.

†There still is considerable theoretical debate about whether time travel into the past is permitted under the *general* theory of relativity, but that is too long a story for us to consider here.

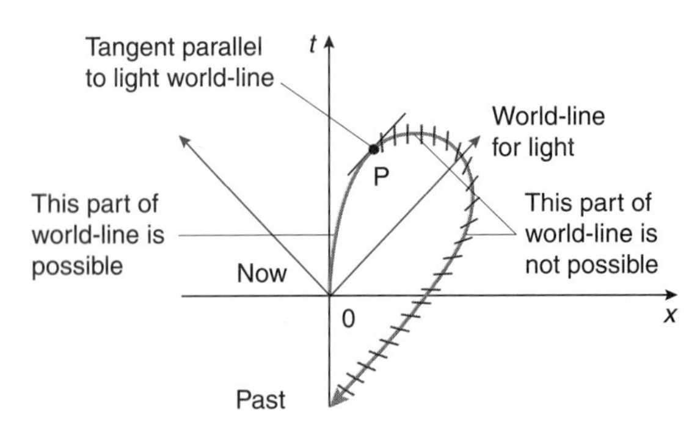

FIGURE 25.50 A hypothetical world-line into the past.

Ironically, nothing in relativity prohibits travel (at speeds $v < c$) into the future region on the space–time diagram. But once there, the future instant becomes the new "now" and travel back to the past is prohibited.

PROBLEM-SOLVING TACTIC

25.7 Do not confuse a space–time diagram, which is drawn for a particular reference frame, with the standard geometry showing the relative motion of two reference frames. The latter diagrams do not show a time axis.

25.20 Electromagnetic Implications of the Special Theory*

Questions about electromagnetism led Einstein to relativity. One of the problems that perplexed him we examine here.

Imagine a positive charge q located at coordinate y' on the y'-axis some distance from an (infinite) array of closely spaced positive charges Q distributed uniformly along the x'-axis of a reference frame S', as in Figure 25.51. There are λ' coulombs of charge along every meter of the x'-axis in frame S'. The charge q experiences an electrical force $\vec{\mathbf{F}}'_{\text{elec}} = q\vec{\mathbf{E}}'$, where the electric field $\vec{\mathbf{E}}'$ that q experiences is caused by the array of positive charges distributed along the x'-axis. The electrical force $\vec{\mathbf{F}}'_{\text{elec}}$ on q is directed away from the array of like charges along the x'-axis as indicated in Figure 25.51.

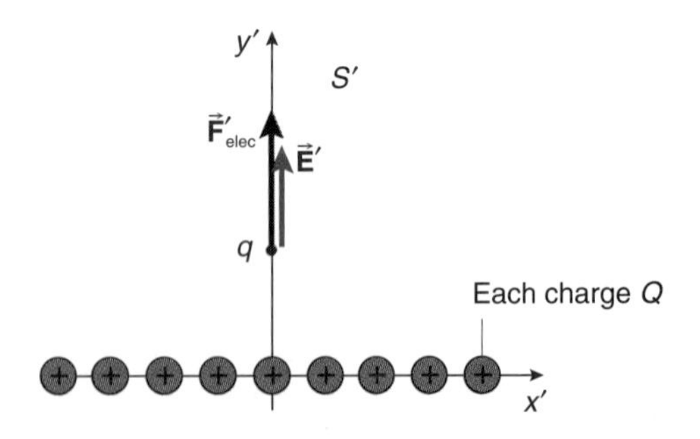

FIGURE 25.51 An array of positive charges in frame S'.

The electric field of an infinite line charge (from Equation 16.19, with $r = y'$) at the position of q has a single component along the y'-axis:

$$E_{y'} = \frac{\lambda'}{2\pi\varepsilon_0 y'}$$

Thus the repulsive force on q, measured in the S' frame, has a y'-component equal to

$$F'_{\text{elec}} = \frac{q\lambda'}{2\pi\varepsilon_0 y'} \tag{25.86}$$

Now we imagine the S' inertial reference frame moving at constant speed v with respect to another inertial reference frame S in the standard geometry of Figure 25.52. From the viewpoint of reference frame S, the array of charges along the x'-axis is moving at speed v; there are *two* forces acting on q in this reference frame.

First, in the frame S, the λ' coulombs of charge on each meter in S' is length contracted into a length $(1\text{ m})/\gamma$ in frame S. Hence the line charge density in S, the number of coulombs of charge in every meter in S, *increases* to $\lambda = \gamma\lambda'$. The change in the line charge density does not violate charge conservation: the total amount of charge along the entire line is the same in the two systems; it is just distributed differently in the two reference frames. In other words, conservation of charge is unaffected by relativity. The array of charges in S produces an electrical force on q. The electrical force is determined from

$$\vec{\mathbf{F}}_{\text{elec}} = q\vec{\mathbf{E}}$$

where $\vec{\mathbf{E}}$ is the electric field produced by the array of moving charges along the x-axis in S. The electric field in S also is that of an infinite line charge and has a single component at the position of q:

$$E_y = \frac{\lambda}{2\pi\varepsilon_0 y}$$

Since $\lambda = \gamma\lambda'$ and $y = y'$ (from the Lorentz transformation equations), the electric field component in S is

$$E_y = \frac{\gamma\lambda'}{2\pi\varepsilon_0 y'}$$

The electrical force on q in frame S thus has only the y-component

$$F_{\text{elec}} = \frac{q\gamma\lambda'}{2\pi\varepsilon_0 y'}$$

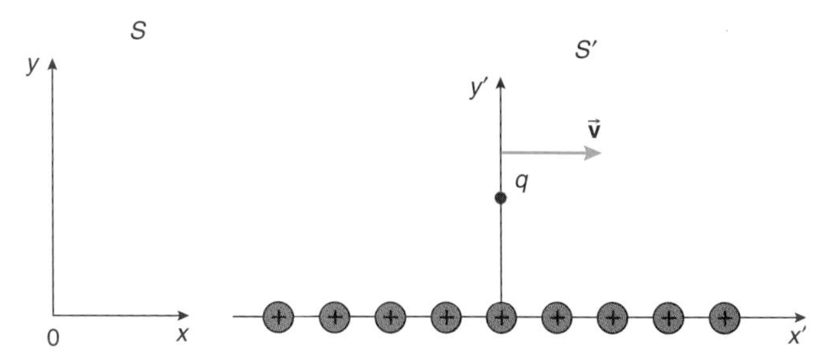

FIGURE 25.52 The standard geometry.

The electrical force in frame S is repulsive since the charges have the same sign.

Now for the second force on q. Since the charges Q are moving in S, they constitute an electrical current $I = \lambda v = \gamma\lambda' v$. This current produces a magnetic field $\vec{\mathbf{B}}$ surrounding the x-axis, directed as shown in Figure 25.53. The magnitude of the magnetic field is that of an infinite straight current (Example 20.11, with $d = y$)

$$B = \frac{\mu_0}{4\pi}\frac{2I}{y}$$

According to frame S, the charge q also is moving with velocity $\vec{\mathbf{v}}$ in this magnetic field. Therefore q experiences a magnetic force given by

$$\vec{\mathbf{F}}_{\text{magnet}} = q\vec{\mathbf{v}} \times \vec{\mathbf{B}}$$

By using the vector product right-hand rule, the direction of this magnetic force on the positive charge q is *toward* the array of charges (see Figure 25.54). The magnitude of the magnetic force on q is (since $y = y'$)

$$\begin{aligned} F_{\text{magnet}} &= qv\frac{\mu_0}{4\pi}\frac{2I}{y} \\ &= qv\frac{\mu_0}{4\pi}\frac{2\gamma\lambda' v}{y'} \end{aligned}$$

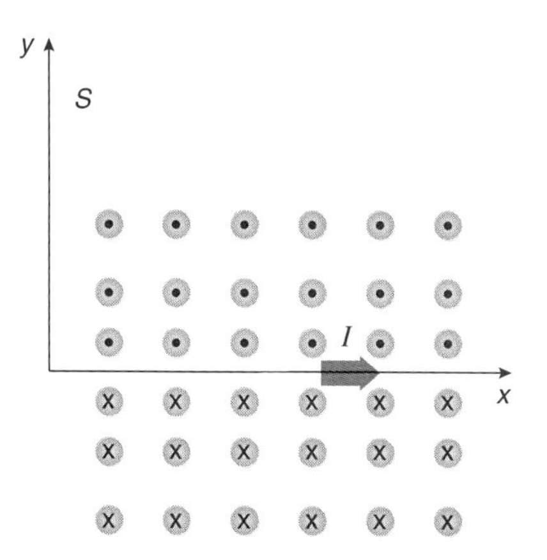

FIGURE 25.53 The magnetic field in S caused by the moving charges.

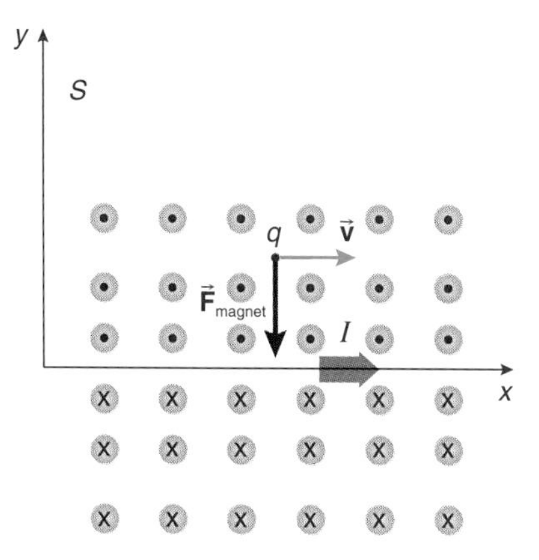

FIGURE 25.54 The magnetic force on q, as viewed in the S frame.

The total force $\vec{\mathbf{F}}$ on q in frame S is the vector sum of the magnetic and electrical forces calculated in frame S:

$$\vec{\mathbf{F}} = \vec{\mathbf{F}}_{\text{elec}} + \vec{\mathbf{F}}_{\text{magnet}}$$
$$= \frac{q\gamma\lambda'}{2\pi\varepsilon_0 y'}\hat{\jmath} - qv\frac{\mu_0}{4\pi}\frac{2\gamma\lambda' v}{y'}\hat{\jmath}$$

The total force has a single component F_y directed away from the array of charges along the x-axis:

$$F_y = \frac{q\gamma\lambda'}{2\pi\varepsilon_0 y'} - qv\frac{\mu_0}{4\pi}\frac{2\gamma\lambda' v}{y'}$$
$$= \frac{q\lambda'}{2\pi\varepsilon_0 y'}\gamma(1 - \varepsilon_0\mu_0 v^2)$$

From Equation 21.26,

$$c^2 = \frac{1}{\mu_0\varepsilon_0}$$

We use this to write the component of the total force in S as

$$F_y = \frac{q\lambda'}{2\pi\varepsilon_0 y'}\gamma\left(1 - \frac{v^2}{c^2}\right)$$
$$= \left(\frac{q\lambda'}{2\pi\varepsilon_0 y'}\right)\frac{1}{\gamma}$$

The term in brackets is the repulsive electrical force component F'_{elec} on q in the S' frame (Equation 25.86). Hence the repulsive force components in the two reference frames are related by

$$F_S = \frac{F_{S'}}{\gamma} \tag{25.87}$$

The point is this. In frame S' the force on q is purely an electrical interaction. On the other hand, in frame S the interaction is electric *and* magnetic. Thus electric fields in one inertial reference frame transform into electric *and* magnetic fields in another inertial reference frame. Likewise, magnetic fields in one inertial frame transform into magnetic *and* electric fields in another inertial frame. Thus magnetic fields can be thought of as manifestations of moving electric fields. There is no reason to take one or the other field as more basic or privileged.

Note that Equation 25.87 implies that as $v \to c$, so $\gamma \to \infty$, the total force on q in S approaches zero. Since the total force in S is the vector sum of the electric and magnetic forces, the two interactions are of equal strength when $v = c$.†

Relativity indicates that the way we calculate the electric fields of static charges (essentially as a manifestation of Coulomb's law) and the way in which we calculate the magnetic field of moving charges (currents) (the Biot–Savart law) are really the same fundamental interaction. Thus, instead of talking about electricity and magnetism as if they are distinct, we speak of electromagnetism to emphasize their unity.

The details of the transformation equations associated with electric and magnetic fields are a wonderful story you will learn in a subsequent course in electromagnetic theory or relativity.

†The limit $v = c$ cannot be reached for a charge q, since charge always is associated with mass and a mass cannot attain $v = c$ because then its total relativistic energy E would be infinite.

25.21 THE GENERAL THEORY OF RELATIVITY*

One of the postulates of the special theory of relativity states that the laws of physics must be the same in all *inertial* reference frames. Einstein eventually was able to broaden this postulate to state that the laws of physics must be the same in *all* reference frames, not only inertial ones. This generalization is called the general theory of relativity. The extension was not easy to accomplish; it took Einstein over a decade to figure it out. Unlike special relativity, the general theory is quite difficult mathematically, and so we consider it only qualitatively and will briefly describe only some of its successful predictions.

The general theory is based on a seemingly common observation about gravity and accelerations. When you drop any mass near the surface of the Earth, the mass accelerates downward with an acceleration of magnitude $g \approx 9.81$ m/s^2. Recall from Chapter 6 that the local acceleration due to gravity at a point in space is the same thing as the gravitational field at that point. Now imagine yourself in a spaceship, well away from any other masses, so that the gravitational field at the position of the spacecraft is zero. If its rockets are ignited and the spacecraft is given an acceleration $\vec{\mathbf{a}} = -\vec{\mathbf{g}}$, there is no way, through mechanics experiments strictly confined to the interior of the spacecraft, that you can distinguish the effects of the acceleration on a physical system (such as a free mass within the spacecraft) and the effects we attribute to gravitation near the surface of the Earth; see Figure 25.55.

The general theory of relativity is based on a generalization of this observation, called the **principle of equivalence**‡: the effects

‡Another (equivalent) way to state the principle of equivalence is to say the gravitational and inertial masses of a system are identical. The principle is restricted to small regions of space over which variations in the magnitude of the gravitational field are negligible.

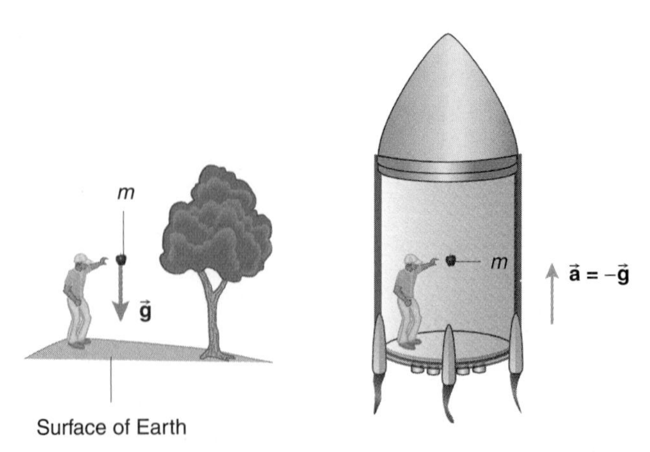

FIGURE 25.55 The effects of gravitation are indistinguishable from those of an acceleration.

of any gravitational field $\vec{g}$ are indistinguishable from the effects of an acceleration $\vec{a} = -\vec{g}$ in the absence of any gravitational field. We explore some of the consequences of the general theory, recognizing that the details involve sophisticated mathematics that we do not confront.

The Gravitational Deflection of Light

One of the predictions of the general theory is that the path of a light ray in a gravitational field will be curved or bent; we call this the **gravitational deflection of light.**

To see why the deflection occurs, imagine yourself in a uniformly *accelerating* spacecraft well away from any other masses, such as the Earth or Sun, as in Figure 25.56. A light pulse traverses the spacecraft in a direction perpendicular to its acceleration. Let the time interval during which the light travels across the spacecraft be divided into a number of equal smaller intervals Δt. Because of its acceleration, the spacecraft moves through increasingly greater distances during each successive interval Δt. If the position of the light pulse is noted after each successive interval Δt, the path of the light pulse across the spacecraft is parabolic, as shown in Figure 25.56.

According to the principle of equivalence, the effects of accelerations are indistinguishable from those of gravitation. Thus Einstein realized that a light beam will be deflected when traveling in a gravitational field (unless it is moving parallel or antiparallel to the field direction). In the gravitational field of the Earth, the deflection of a horizontally directed light ray is very small because the speed of light is so great and the gravitational field is quite weak compared with, say, the gravitational field at the surface of the Sun. Indeed, over a 100 m distance, the vertical deflection of an initially horizontal light beam on the Earth is on the order of the diameter of a typical nucleus of an atom ($\sim 10^{-14}$ m). Thus the deflection on the Earth is negligible and too small to detect.

Einstein suggested that when light from distant stars passes near the Sun (which has a much stronger gravitational field at its surface than the Earth), the light also would be deflected but by an amount that was detectable with the astronomical technology of that time (c. 1916). For a ray grazing the solar surface tangentially, he predicted the deflection in the path of the light would be about 1.7 arc seconds ($\sim 4.7 \times 10^{-4}$ °). The effect of the deflection is to cause the star to appear at a position slightly removed from its true position, as shown in Figure 25.57.

Einstein also suggested that such a deflection could be detected by comparing star positions in the sky before and during a total solar eclipse. During such an eclipse, the light of the Sun is blocked by the Moon and stars can be seen quite near the position of the Sun in the sky. The successful observation of the deflection in 1919, and its quantitative agreement with the prediction by Einstein,* propelled his name to the fame it has worthily enjoyed ever since. The gravitational deflection of light by the very strong gravitational fields of white dwarf stars, neutron stars, and black holes is much greater than that near the Sun. Indeed, the theory predicts that compact massive objects such as black holes, neutron stars, and galactic cores even can act as *gravitational lenses* for light, producing multiple images of more distant objects along the line of sight. Such gravitationally induced multiple images also have now been observed by astronomers.

*One also can predict that light is deflected by the gravity of the Sun using essentially Newtonian physics, but the result obtained is exactly *half* that predicted by Einstein's general theory. The agreement with Einstein's prediction was the cause for celebration.

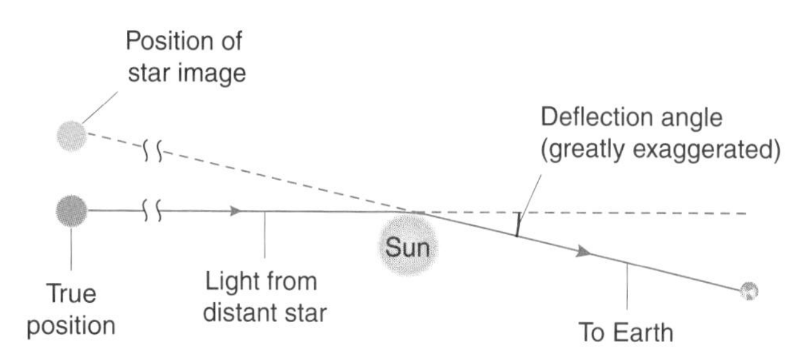

FIGURE 25.57 Gravitational deflection of light grazing the Sun.

FIGURE 25.56 Light follows a parabolic path across the accelerating spacecraft.

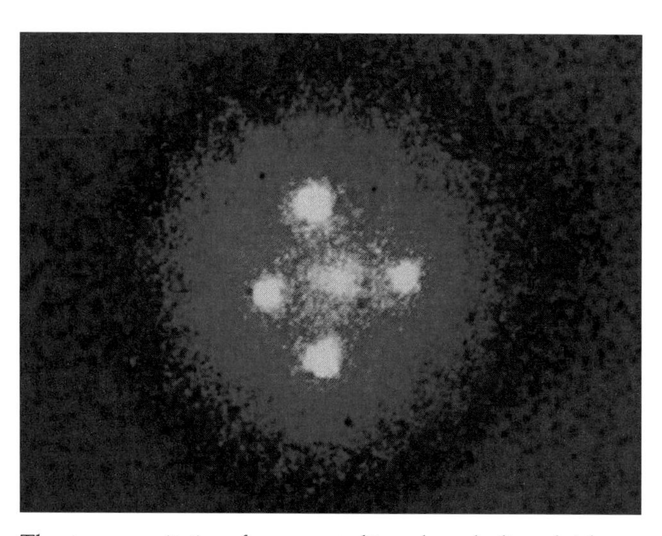

The strong gravitation of a compact object along the line of sight produces multiple images of this very distant quasar.

The Precession of the Perihelion Point of Mercury

The general theory also was able to explain a long-standing discrepancy in the motion of the planet Mercury that was apparent by about 1845. The major axis of the elliptical orbit of Mercury gradually rotates about the Sun (which is at one focus of the ellipse, of course), causing the position of the perihelion and aphelion points of the orbit to change; see Figure 25.58. The angular speed of the rotation is quite slow, about 574 arc seconds per *century*.

The gravitational influences of the other planets on Mercury cause most of this precession of the perihelion point, about 531 arc seconds per century. The excess precession, amounting to 574 arc seconds − 531 arc seconds = 43 arc seconds per century, is completely explained by the general theory of relativity.

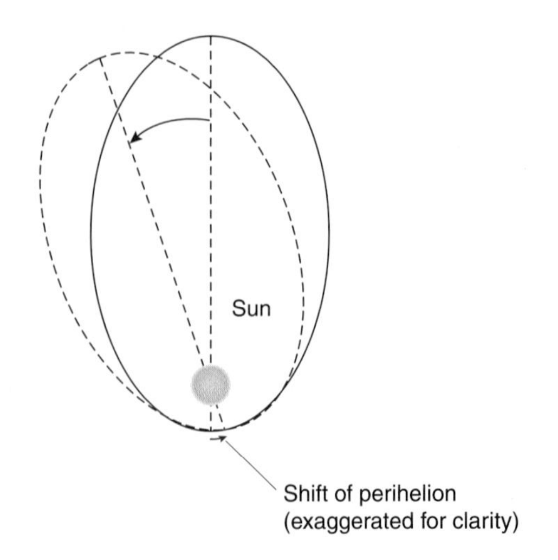

FIGURE 25.58 Long-term motion of the major axis of the orbit of Mercury.

The effect exists for the other planets as well, including the Earth, but is greatest for the planet Mercury because the gravitational field of the Sun that Mercury experiences is greater than for any other planet. Mercury also has the greatest speed of all the planets. The perihelion motion predicted by general relativity also depends on the orbital eccentricity, and Mercury's orbit has an eccentricity second only to Pluto's in magnitude.

Gravitational Red Shifts and Blue Shifts

General relativity also predicts that clocks run slower in stronger gravitational fields than in weaker ones. Clocks on the basement of your building run slightly slower than those on the roof! The consequences of this for light are a **gravitational red shift** in its wavelength (to longer wavelength) for light escaping from a strong gravitational field, such as from a white dwarf star or a neutron star, and a corresponding **gravitational blue shift** (to shorter wavelength) for light falling on such objects.

The red shift of light escaping from strong gravitational fields was discovered in the 1920s, using the white dwarf companion to the bright star Sirius (α Canis Majoris). The gravitational red shift also has been verified experimentally within the confines of a single laboratory on the Earth in an ingenious series of experiments over vertical distances of only several tens of meters.*

General relativity also predicts bizarre behavior of light and particles near black holes, and other objects such as collapsed star cores and compact galactic cores, whose gravitational fields in their vicinity are exceedingly large.

Despite the technical difficulties of performing experimental tests of general relativity, the theory has met every test to which it has been placed.

QUESTION 5

When the position of a star is changed by the gravitational deflection of its light by the gravitational field of the Sun (see Figure 25.57), is the image of the star we see in the sky a real or virtual image?

*These experiments were first made by R. V. Pound and G. A. Rebka Jr., "Apparent weight of photons," *Physical Review Letters*, 4, #7, pages 337–341 (1 April 1960), now called the *Rebka–Pound experiment*.

CHAPTER SUMMARY

A *reference frame* is a coordinate system together with a collection of synchronized clocks, distributed at rest throughout the coordinate system. An *inertial reference frame* is one in which Newton's first law of motion is valid: zero total force on a particle means the particle has zero acceleration.

To specify an *event*, one states where and when it occurs. Once a reference frame is chosen, an event is specified by indicating its three spatial coordinates and its single time coordinate.

The special theory of relativity is based on two postulates:

1. The speed of light in a vacuum has the same numerical value c, equal to exactly 299 792 458 m/s, when measured in any inertial reference frame, independent of the motion of the source and/or the observer.
2. The fundamental laws of physics must be the same in all inertial reference frames.

Time dilation means that moving clocks run slow when compared with clocks at rest. A *proper (or rest) time interval* τ_0 is registered by a clock at rest. If that clock then has a speed v, the interval is found to be a longer time interval τ when measured using clocks at rest in the frame that sees the moving clock, where

$$\tau = \gamma\tau_0 \tag{25.14}$$

The quantity γ is a pure number, always greater than or equal to 1:

$$\gamma \equiv \frac{1}{\sqrt{1 - \frac{v^2}{c^2}}} \quad (25.13)$$

Particles with nonzero mass are restricted to speeds less than the speed of light because it takes an infinite amount of work to increase their speed to the speed of light c, at which speed their kinetic energy would be infinite. Thus γ is real and finite, since v must be less than c.

Moving lengths oriented perpendicular to the direction of their motion are unchanged by the motion.

Length contraction means that the length of a moving object (or the component of its length) oriented along the direction of its motion is measured to be shorter than the identical length at rest. A moving *proper length* ℓ_0 (as determined with rulers at rest with respect to it) is measured to be a length ℓ (as determined by rulers at rest in the frame that sees the moving object), where

$$\ell = \frac{\ell_0}{\gamma} \quad (25.17)$$

We use a *standard geometry* that has an inertial reference frame S' traveling to the right at speed v with respect to another inertial reference frame S, with their x'- and x-axes collinear. Stopwatch clocks in each frame indicate $t = t' = 0$ s the instant the two origins of the two reference frames coincide.

The *Lorentz transformation equations* express the relationship between the space and time coordinates of an event in an inertial frame S and those of the same event in another inertial reference frame S':

$$x' = \gamma(x - vt) \quad (25.33)$$
$$y' = y \quad (25.34)$$
$$z' = z \quad (25.35)$$
$$t' = \gamma\left(t - \frac{v}{c^2}x\right) \quad (25.36)$$

The corresponding inverse equations are

$$x = \gamma(x' + vt') \quad (25.37)$$
$$y = y' \quad (25.38)$$
$$z = z' \quad (25.39)$$
$$t = \gamma\left(t' + \frac{v}{c^2}x'\right) \quad (25.40)$$

The Lorentz transformation equations reduce to the *Galilean transformation equations* for speeds $v << c$, in which case $\gamma \rightarrow 1$:

$$x' = x - vt \quad (25.3)$$
$$y' = y \quad (25.4)$$
$$z' = z \quad (25.5)$$
$$t' = t \quad (25.1)$$

If a particle is moving with velocity $\vec{u}' = u_x' \hat{\imath}' + u_y' \hat{\jmath}' + u_z' \hat{k}'$ in reference frame S', then the corresponding velocity components of the particle in reference frame S are

$$u_x = \frac{u_x' + v}{1 + \frac{vu_x'}{c^2}} \quad (25.45)$$

$$u_y = \frac{u_y'}{\gamma\left(1 + \frac{vu_x'}{c^2}\right)} \quad (25.47)$$

The expression for the velocity component u_z is similar to that for u_y. The quantity

$$\gamma = \frac{1}{\sqrt{1 - \frac{v^2}{c^2}}}$$

involves the relative speed v of the two reference frames.

When $v << c$, these velocity component transformation equations reduce to the Galilean transformation equations:

$$u_x = u_x' + v \quad (25.6)$$
$$u_y = u_y' \quad (25.7)$$
$$u_z = u_z'$$

Imagine a light source at rest that emits wavelength λ_0, corresponding to frequency ν_0. The *longitudinal Doppler effect* states that if the source is receding from an observer, and/or an observer is receding from the source at relative speed v, the light detected by the observer is *red-shifted* to longer wavelengths λ and smaller frequency ν:

$$\nu = \nu_0 \sqrt{\frac{1 - \frac{v}{c}}{1 + \frac{v}{c}}} \quad \text{(source and/or observer receding)} \quad (25.58)$$

If the source is approaching the observer and/or the observer is approaching the source at relative speed v, the light detected by the observer is *blue-shifted* to shorter wavelength λ and greater frequency ν:

$$\nu = \nu_0 \sqrt{\frac{1 + \frac{v}{c}}{1 - \frac{v}{c}}} \quad \text{(source and/or observer approaching)} \quad (25.59)$$

The *transverse Doppler effect* indicates that light from a source moving at speed v perpendicular to the line of sight is red-shifted. The red-shifted frequency ν is

$$\nu = \nu_0 \sqrt{1 - \frac{v^2}{c^2}} \quad (25.60)$$

For a given speed v, the transverse Doppler effect red shift produces a smaller frequency change than a longitudinal Doppler effect red shift.

The relativistic momentum $\vec{p}$ of a particle of mass m moving with velocity $\vec{v}$ is

$$\vec{p} = \gamma m \vec{v} \quad (25.64)$$

where γ involves the speed v of the particle.

The relativistic kinetic energy of such a particle is

$$\text{KE} = (\gamma - 1)mc^2 \quad (25.75)$$

Only if $v << c$ does the relativistic kinetic energy reduce to the familiar classical expression

$$\frac{1}{2}mv^2$$

The *rest energy* of a particle of mass m is mc^2. The *total relativistic energy* E of such a particle is the sum of its kinetic and rest energies:

$$E \equiv \text{KE} + mc^2 \quad (25.77)$$
$$= \gamma mc^2 \quad (25.78)$$

The total relativistic energy of a particle and the magnitude of its momentum are related by

$$p = \frac{v}{c^2} E \tag{25.84}$$

as well as by

$$E^2 = p^2c^2 + m^2c^4 \tag{25.85}$$

A *space–time diagram* is a graph with the time axis vertical and, typically, one or two spatial coordinate axes horizontal. An event is represented by a point on a space–time diagram. The trajectory of a single particle system on a space–time diagram is called a *world-line*.

The general theory of relativity is based on the *principle of equivalence*, which states that the effects of a gravitational field $\vec{g}$ are equivalent to the effects of an acceleration $\vec{a} = -\vec{g}$ in the absence of a gravitational field. General relativity predicts that the path of light is bent in a gravitational field (if the path is not parallel or antiparallel to the field), called the *gravitational deflection of light*. The general theory also predicts that clocks in stronger gravitational fields run slower than clocks in weaker gravitational fields. This effect results in a *gravitational red shift* of light (to longer wavelengths) when it escapes from massive objects (into regions with a weaker gravitational field) and a corresponding *gravitational blue shift* when light falls onto such objects (into regions with a stronger gravitational field).

SUMMARY OF PROBLEM-SOLVING TACTICS

25.1 (page 1159) Be sure to remember that γ is the *inverse* of the square root

$$\sqrt{1 - \frac{v^2}{c^2}}$$

The thing to keep in mind about γ is that it is *always* greater than or equal to 1:

$$\gamma \geq 1$$

25.2 (page 1164) Equations 25.33–25.36 are useful when the space and time coordinates of an event are known in the reference frame S and you need to know the corresponding coordinates of the same event in S′.

25.3 (page 1164) Equations 25.37–25.40 are useful if the space and time coordinates of an event are known in the reference frame S′ and the corresponding coordinates of the event in S are needed.

25.4 (page 1172) If $u_x' > 0$ m/s, then the object is moving to the right in S′ toward increasing values of x'. If the object is moving toward decreasing values of x' in S′, then substitute a negative velocity component for u_x' in Equation 25.45.

25.5 (page 1173) The relativistic velocity component addition equations *always* yield velocity components less than or equal to the speed of light.

25.6 (page 1185) Remember that the relativistic kinetic energy is

$$\text{KE} = (\gamma - 1)mc^2$$

The relativistic kinetic energy is *not* $(1/2)mv^2$.

25.7 (page 1188) Do not confuse a space–time diagram, which is drawn for a particular reference frame, with the standard geometry showing the relative motion of two reference frames.

QUESTIONS

1. (page 1151); 2. (page 1155); 3. (page 1159); 4. (page 1187); 5. (page 1192)
6. Describe several reference frames that are *not* inertial reference frames. Explain why they are not inertial.
7. Using radio telescopes, we have the technological ability to detect extraterrestrial intelligent civilizations located anywhere in the galaxy, should they wish to make their presence known to us (which in itself is a good topic for discussion). Since our physical theories are creations of the human intellect, can you expect such civilizations to express the laws of physics the same way we do? If so, explain your reasons. If not, does this violate the second postulate of the special theory of relativity?
8. Explain why special relativity implies that a perfectly rigid body cannot exist.
9. If you observe a gold brick moving past you at a very high speed, will the density of the moving brick be measured to be greater than, less than, or equal to the density of a gold brick at rest with respect to you?
10. You are in a spaceship moving at a speed 0.900*c* away from the Sun. (a) At what speed does sunlight stream past you? (b) Do you measure your pulse rate to be slower than normal? (c) Will an observer on Earth think your pulse rate is slower than normal?
11. Time dilation means that moving clocks run slow, compared with clocks at rest with respect to you. Do observers at rest with respect to the moving clocks perceive time to run slow? Explain.
12. The speed of light was first deduced to be finite by Ole Römer about 1730 by noting a systematic discrepancy between the times the moons of Jupiter were predicted to be eclipsed and the times they actually were eclipsed. Prior to Römer, the speed of light was thought to be essentially infinite. If the speed of light were infinite, what would happen to time dilation, length contraction, and velocity component addition?
13. The equation for the relativistic kinetic energy of a mass m moving at speed v is $\text{KE} = (\gamma - 1)mc^2$. A sharp student notes that the equation apparently does not have the speed v

explicitly in it. Does this observation mean the relativistic kinetic energy is independent of the speed v? Explain.

14. The photon is the particle of light (an energy bundle). Why can you never be in a reference frame with the photon at rest?

15. When two events occur simultaneously, must they occur at the same place?

16. Discuss whether it is proper to say that "mass is energy and energy is mass."

17. Explain why a particle of mass m cannot travel at speeds equal to or greater than the speed of light c in special relativity.

18. Explain why it is not possible to travel into the past according to special relativity.

19. A broom handle of length ℓ_0 makes an angle θ' with the x'-axis of a spacecraft moving at $v = 0.99c$ with respect to reference frame S in the standard geometry. In reference frame S, is the angle θ of the broom handle with the x-axis greater than, less than, or the same as θ'? Explain your reasoning.

20. The horizontal beam from a lighthouse sweeps around the horizon at an angular speed of $(\pi/2 \text{ rad})/\text{s}$. The beam sweeps along the hull of a distant ship anchored with its hull perpendicular to the beam. At what distance from the lighthouse is the speed of the beam along the hull equal to the speed of light? If the ship were anchored further away, the speed of the beam along the hull would be *greater* than the speed of light. Does this violate relativity?

21. If you approach a plane mirror along the normal line at a speed of $0.800c$, does your image approach you at $1.60c$ since the image is virtual and not real?

22. Can you say correctly that the proper time interval between two events that occur at the same place is the smallest time interval between the two events?

23. When something will never occur, you may have heard someone say that it "will only occur light-years from now." What is wrong with this all-too-common statement?

24. The total mass of an electron and a proton when they are well separated from each other is slightly greater than when they are bound together in a hydrogen atom. Why?

25. Astronomers can *see* the past, but cannot go into the past. Explain this apparent paradox.

26. When light is red- or blue-shifted in the relativistic Doppler effect or in the gravitational red shift or blue shift, what is shifted and with respect to what?

27. Why can you *not* use length contraction to calculate the shortened wavelength of blue-shifted light in the Doppler effect?

28. The Concorde supersonic airplane is speeding past you at 0.50 km/s. Explain an experimental procedure you can employ to measure its length while it is in motion. If you compared your measurement with its length when at rest, would you expect a detectable difference? Explain.

29. Imagine yourself in a spacecraft orbiting the Earth every 120 minutes flying east. If you readjust your calendar watch every time you cross a time zone, you quite soon are into tomorrow and then the next day. How does the International Date Line [at longitude 180° E (or west)] adjust for such apparent time travel into the future?

30. When you fly west across a time zone, you have to adjust your watch back an hour, thus giving you a longer day than normal. Is this because of time dilation?

31. The speed of light in transparent materials is less than its speed c in vacuum. Can a particle with a nonzero mass travel faster than the speed of light (but less than c) in such materials?

32. For interplanetary spaceflight, NASA must create many complex computer programs based on the dynamics of the spacecraft. The speeds of such spacecraft occasionally are quite large, upward of 25 km/s. Can the computer programmers use mv for the magnitude of the spacecraft momentum, or must they used γmv? Explain your answer.

33. Neutrinos are particles created prolifically in nuclear reactions in stellar interiors as well as during the spectacular death throes of massive stars when they undergo supernova explosions and more nucleosynthesis. Neutrinos may have zero mass, in which case they travel at the speed of light, but there are reasons now to suspect they have a small, nonzero mass. If neutrinos do have a small mass, can they travel at the speed of light? Explain why or why not.

34. (See Question 33 as well.) About 30 000 years ago, a massive star underwent a supernova explosion in the irregular satellite galaxy of the Milky Way called the Large Magellanic Cloud. Light from the explosion first arrived at the Earth on 23 February 1987. Neutrinos are extraordinarily difficult to detect because of their weak interaction with matter. However, slightly over a dozen neutrinos from the explosion were detected on the Earth at essentially the same time as the arrival of the light (within a few hours). Discuss what this observation implies about: (a) the speed of the neutrinos; (b) whether they have nonzero mass.

35. If the chronological ordering of two events in inertial frame S indicates event A occurred before event B, does another inertial reference frame S′ exist that has the chronological ordering reversed? Distinguish between the following situations: two events that are *causally* related (event A *caused* event B) and two events that are not causally related.

36. Will a pendulum clock in an accelerating elevator tick slow or fast if the acceleration is upward? Downward? Is this what we mean by time dilation in special relativity? Is this what is meant by the relativity of time in general relativity?

PROBLEMS

Sections 25.1 Reference Frames
25.2 Classical Galilean Relativity

1. You place a collection of new clocks at rest throughout a coordinate system to form a reference frame. Describe an experimental procedure to synchronize the clocks, so that they all indicate the same time.

•2. You are riding your bicycle at speed 5.0 m/s along a straight, level section of the Greenbriar River Trail in West Virginia. You toss a ball at 2.0 m/s horizontally. Determine the speed of the ball with respect to the ground immediately after the ball leaves your hand if you toss the ball: (a) in the forward direction; (b) in the backward direction; (c) to the side, perpendicular to your velocity.

•3. A straight river has a current of speed v_0 that is the same across its width ℓ. You can swim at constant speed v with respect to the water. Assume $v > v_0$. (a) If you swim a distance ℓ downstream, how long an interval does the trip take? (b) How long will it take to swim back to the point you began? (c) What is the round-trip travel time? Assume you reverse direction instantaneously. (d) If you swim in such a direction that your motion with respect to the river bank is directly across the river and back along the same path, how long does the round trip take? Again, assume you can reverse direction instantaneously. (e) Which of the times you calculated in (c) and (d) is the shorter time?

Sections 25.3 The Need for Change and the Postulates of the Special Theory
25.4 Time Dilation
25.5 Lengths Perpendicular to the Direction of Motion
25.6 Lengths Oriented Along the Direction of Motion: Length Contraction

4. Your spaceship is traveling off to another star and moving directly away from the Sun at a speed of 0.90*c*. How fast is the light from the Sun passing you?

5. How fast must a clock be moving with respect to you so that it ticks off a 1.00 s interval while your watch indicates 3.00 s have elapsed?

6. An astronaut traveling at a speed of 0.90*c* is holding a meter stick in her hand along the line of motion. What does she measure to be the length of the meter stick?

7. A meter stick whizzes by you at a speed of 0.80*c* with its length oriented parallel to its velocity. (a) What do you measure to be the length of the moving meter stick using rulers at rest relative to you? (b) How long does it take the moving meter stick to pass by you?

•8. A clock at rest in an inertial reference frame S' is moving directly away from you (in another inertial reference frame S) at speed $v = 0.949c$. (a) What is the factor γ? (b) If the moving clock indicates an interval of 24.0 h has elapsed, what interval has elapsed on your watch? (c) Are your answers the same if the moving clock is coming directly toward you?

•9. Show that as $v/c \to 1$, an approximate expression for the relativistic factor γ is

$$\gamma \approx \frac{1}{\left[2\left(1 - \frac{v}{c}\right)\right]^{1/2}}$$

Evaluate γ using this expression for $v/c = 0.999\,999$.

•10. Close to the San Andreas Fault south of San Francisco lies the Stanford Linear Accelerator (SLAC for short), used to project subatomic particles to great speed. The accelerator is 3.0 km long and actually passes under a major highway. The accelerator is capable of accelerating electrons to speeds that have a relativistic factor $\gamma = 1.00 \times 10^4$. (a) Calculate the ratio of the speed of the electrons to the speed of light; use the result of Problem 9 to find v/c. Can you easily find v/c on your calculator *without* the result of Problem 9? Comment. (b) In a reference frame in which the electrons are at rest, what is the length of the accelerator? Assume (incorrectly) that the speed of the electrons is constant along the length of the accelerator.

•11. A beam of morons is moving at a speed of 2.90×10^8 m/s. They may be dumb, but they sure are fast. At this speed, the half-life of the morons is measured to be 2.50×10^{-6} s. [The half-life is the time during which half the particles change (we say decay) into other particles.] What is the half-life of a collection of morons that is at rest?

Sections 25.7 The Lorentz Transformation Equations
25.8 The Relativity of Simultaneity
25.9 A Relativistic Centipede
25.10 A Relativistic Paradox and Its Resolution*

•12. In an inertial reference frame S, the following observations are made: (i) your professor passes out a horrible test at the origin and starts a stopwatch simultaneously; (ii) 10.0 s later, you have a fit at $x = 9.0 \times 10^8$ m. The Dean of Deans (located at the origin in another inertial reference frame S' in the standard geometry) is cruising by at a speed of 0.98*c*. (a) Calculate the relativistic factor γ. (b) Specify the space and time coordinates of the two events in S. (c) Use the Lorentz transformation equations to find the space and time coordinates of these events in the Dean of Deans' reference frame.

•13. You are rushing off to an 8 A.M. class at a constant speed. Your roommate, at rest, nonchalantly observes that you travel the 100.0 m length of a long corridor in 5.00×10^{-7} s. (a) What is your speed? (b) According to rulers at rest with respect to you, what is the length of the corridor? (c) How long does it take you to cover the distance according to your watch?

•14. A U.S. marshal is riding along a straight trail into town at 0.867*c*. Two gunslingers are glaring at each other with only 50.0 m of dirt between them (oriented parallel to the line along which the marshal is riding), according to witnesses cowering behind hitching posts along the street. The two desperadoes draw and fire simultaneously according to them. (a) Determine the value of the relativistic factor γ. (b) How far apart are the bad guys according to the marshal? (c) Carefully sketch and define two appropriate reference frames. Then define two events corresponding to the shots fired by the two gunslingers. Transform the events into the other reference frame. (d) Which gunslinger fired first according to the reference frame of the marshal?

•**15.** At what speed (expressed as v/c) could an astronaut (at least in principle) travel across the diameter of our Milky Way Galaxy, a distance of about 1.0×10^5 LY, in a mission lifetime of 25 y?

•**16.** Two plane mirrors are located in the S' reference frame. One mirror M_1 is at $x' = 0$ m while the other (mirror M_2) is at $x' = \ell_0$ as indicated in Figure P.16. The S' reference frame with its mirrors is moving at high speed v relative to a reference frame S in the standard geometry. The origins of the two reference frames coincide when $t = t' = 0$ s. A photon (a particle of light) bounces back and forth between the mirrors and is reflected from mirror M_1 when $t = t' = 0$ s. (a) In reference frame S', when does the photon arrive at mirror M_2? (b) Specify the space–time coordinates of two events in S' corresponding to the photon at mirror M_1 and the photon at mirror M_2. (c) In reference frame S, when does the photon arrive at mirror M_2? (d) In reference frame S, where on the x-axis is mirror M_2 when the photon arrives at that mirror? (e) Find the distance Δx in S that the photon travels while moving from mirror M_1 to mirror M_2. (f) Find the interval Δt in S that it takes the photon to travel from mirror M_1 to mirror M_2. (g) Take the ratio of

$$\frac{\Delta x}{\Delta t}$$

using the results of parts (e) and (f). Is this ratio what you expected to obtain?

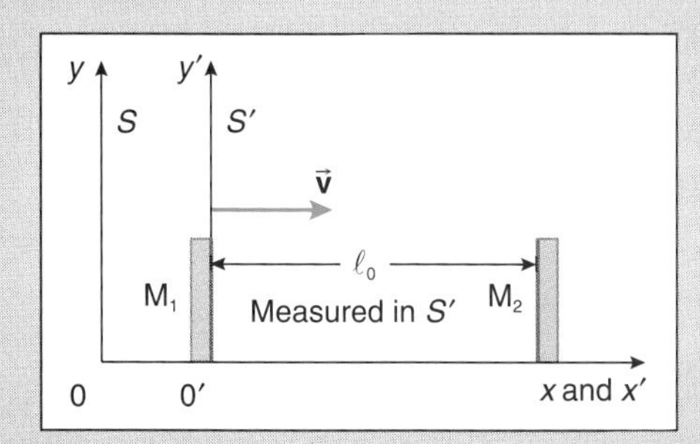

FIGURE P.16

•**17.** You decide to stop the world and get off by heading for α Centauri, the nearest star to the Sun, located 4.2 LY away. (a) At what constant speed must you travel to get to α Centauri in 3.0 y according to a calendar and watch you carry along in your pocket? (b) How long does your journey take according to your friends left behind on the Earth?

•**18.** Two unforgettable events occur 1.00 s apart at positions separated by 2.40×10^8 m along the x-axis in the well-known inertial reference frame S. The equally prestigious inertial reference frame S' is moving with the standard geometry. Both events occur at the same position on the x'-axis. (a) Delineate the space and time coordinates of the two events in the reference frame S. (b) How fast is the S' reference frame moving with respect to S? (c) What is the time interval between the events in S'?

•**19.** Two events occur in the inertial reference frame S and have space and time coordinates (x_1, t_1) and (x_2, t_2). (a) Find the speed of another inertial reference frame S' in which the two events occur simultaneously. (b) Find the spatial separation $\Delta x'$ of the two events in this reference frame. (c) Discuss the feasibility of finding yet another reference frame in which the two events occur at the same place.

•**20.** You are in a spaceship (of length 100 m) moving at speed 0.900c to visit the bright star Arcturus (α Boötis). Back home, your mother notices that you forgot your socks. She sends a light signal to the spaceship. Let S' be the reference frame of the spaceship and S be the reference frame of home-sweet-home, back on the Earth. The signal arrives at the tail of the spaceship when $t = t' = 0$ s. (a) When does the light signal reach the front of the spaceship according to the spaceship clocks? (b) When does the light signal reach the front of the spaceship according to your mother's clocks? (c) Are the answers to (a) and (b) related by the time dilation equation? Why or why not? (d) The light signal is reflected back to the tail by a mirror in the nose of the spaceship. When does the light signal reach the tail according to the spaceship clocks? When does it arrive at the tail according to the Earth-based clocks? (e) Determine the total distance traveled by the light signal as it travels from the tail to the nose and then back to the tail according to: (i) you, inside the spaceship; (ii) your mother, on the Earth.

•**21.** A meteor crashes into your lecture room leaving a gaping hole in the ceiling. Precisely 1.10 s later (according to clocks in the classroom), hawk-eyed astronomers at your college notice that a meteor crashes into the Moon (3.84×10^5 km away), creating a fresh crater. A tabloid newspaper hypothesizes that both events were caused by a strafing run from a passing flying saucer dropping meteors when near the Moon and near the Earth. Assume (unrealistically) the fall-time of the rocks from the saucer is negligible. (a) How long does it take light to propagate from the Moon to the Earth according to your classroom clock? (b) What is the time interval between the impacts as measured by clocks in the reference frame of the Earth? Which event occurred first in the Earth–Moon reference frame? (c) Could the impacts have been caused by such a flying saucer? Explain your answer.

•**22.** When $t = 0$ s, you notice a speeder go by in a new Ferrari at speed 0.200c. Later, when $t = 10.0$ s, a state trooper goes by at a speed of 0.500c with flashing lights atop his Yugo. The lawyers, of course, bring up the rear. (a) At what time (according to your clocks) does the trooper catch up with the speeder? (b) How far from you does the trooper nab the speeder?

•**23.** In reference frame S', a flashbulb goes off at the origin and the light propagates toward positive x' and negative x' to two detectors, each located 24.0×10^8 m from the S' origin as measured in S' (see Figure P.23). (a) Define two events in S'

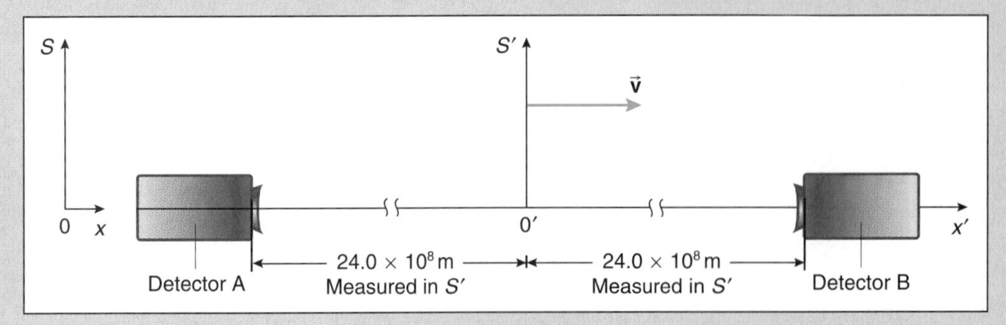

FIGURE P.23

corresponding to the detectors' reception of the light. (b) Reference frame S' is moving at speed $v = 0.995c$ relative to reference frame S in the standard geometry. Find γ. (c) Find the spatial separation Δx between the two events in S. (d) Find the interval Δt between of the two events in S. (e) Which event occurs first in S?

24. If the transformation equations between the inertial reference frames S and S' are *not* linear, peculiar things happen. For example, suppose the transformation equation for x has the quadratic form

$$x^2 = Ax' + Bt'$$

where A and B are two constants that depend on the relative speed v of the two frames. Imagine two one-meter sticks nailed down in reference frame S, one meter stick between coordinates $x = 2$ m and $x = 3$ m and the other between $x = 4$ m and $x = 5$ m, as in Figure P.24. (a) Write the given transformation equation for each given value of x. (b) Since the meter sticks are at rest in reference frame S, they are moving in S'. Hence, to measure their length, you must determine where the ends of the sticks are at the same time in S', say $t_1' = t_2' = t_3' = t_4' = 0$ s. Find the lengths of the two sticks in S' and note that they are not the same. This situation is ridiculous: equal lengths in S have *different* lengths in S' depending on where the objects are located! The only way we can avoid this unpleasant result is to have linear transformation equations between the reference frames.

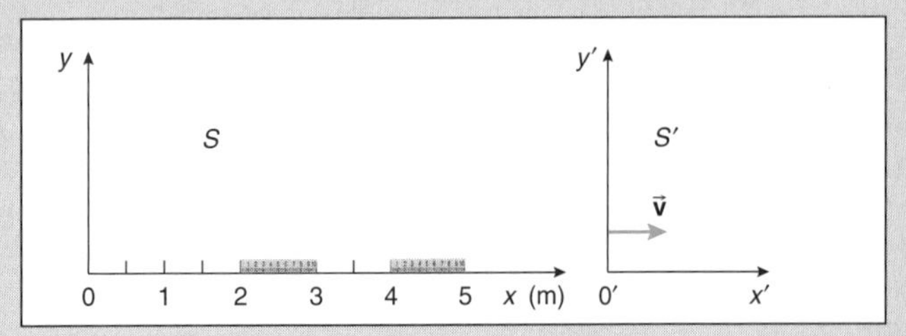

FIGURE P.24

25. (a) Use the binomial expansion to show that if $v/c << 1$, then the amount $\Delta\ell \equiv \ell - \ell_0$ that a proper length ℓ_0 is contracted is approximately

$$\Delta\ell \approx -\frac{1}{2}\frac{v^2}{c^2}\ell_0$$

(b) The Earth has a diameter of 12.7×10^3 km and an orbital speed of about 30 km/s. By about how much is the diameter of the planet length contracted because of its orbital motion, when viewed from the Sun? Is every measure of the diameter so contracted?

Section 25.11 Relativistic Velocity Addition
25.12 Cosmic Jets and the Optical Illusion of Superluminal Speeds*

•26. Two spacecraft are approaching the Earth at equal speeds from opposite directions. Each spacecraft sees that the other spacecraft approaching at a speed $0.90c$. How fast are the spacecraft approaching the Earth?

•27. The chase is on. The Lone Ranger, galloping at $0.950c$, is chasing a desperado, moving at speed $0.900c$, in a direction directly away from Dodge City, Kansas. What is the speed at which the desperado sees the Lone Ranger approaching her?

•28. Two protons in a laboratory particle accelerator are fired toward each other, each with speed $0.990\,000c$ when measured in the lab. What is the speed at which one proton sees the other approaching?

•29. In inertial reference frame S', a particle is fired along the y'-axis at speed $0.995c$ parallel to $\hat{\jmath}'$, measured using rulers and clocks in the S' frame. The S' reference frame is moving at speed $0.990c$ with respect to inertial reference frame S in the standard geometry. (a) Determine the velocity components u_x and u_y of the particle in S. (b) What is the angle the velocity vector $\vec{u}$ makes with the x-axis in S?

•30. A high-frequency lighthouse beacon sweeps around the horizon with a frequency of 1.43 kHz. At what distance is the beam sweeping perpendicular to the line of sight from the lighthouse with a speed of $1.5c$? Does this result violate the first postulate of the special theory of relativity? Explain.

•31. A particle in the S' reference frame has a velocity $\vec{u}' = 0.779c\hat{\imath}' + 0.450c\hat{\jmath}'$. The S' frame is moving in the standard geometry with speed $0.950c$ with respect to reference frame S. (a) Sketch the situation. (b) What is the angle θ' that $\vec{u}'$ makes with the x'-axis in S'? (c) What is the speed u' of the particle in the S' frame when measured with clocks and rulers at rest in S'? (d) Find the velocity components u_x and u_y of the particle in the S reference frame. (e) What is the speed u of the particle in the S reference frame, measured using rulers and clocks in that frame? (f) What angle θ does the velocity vector $\vec{u}$ make with the x-axis in S?

32. A light source at the point P in Figure P.32 is turned on at time $t = 0$ s for a very short interval and then extinguished. (a) At what time does the light arrive at the origin O? (b) At what time does the light arrive at a point A located a distance x from the origin along the $+x$-axis? (c) The spot of light moves from point O to point A in a time given by the difference between the answer to part (b) and the answer to part (a). Show that the *average* speed $\langle v\rangle$ of the light spot in moving along the x-axis from O to A is

$$\langle v\rangle = c\,\frac{a + (a^2 + x^2)^{1/2}}{x}$$

(d) Show that when x is small, $\langle v\rangle \to \infty$ m/s, and when $x \to \infty$ m, $\langle v\rangle$ approaches c. The spot of light, therefore, begins to move out along the x-axis at a speed *greater* than the speed of light and *slows down* to the speed of light as $x \to \infty$ m. Does this result contradict the special theory of relativity? Explain.

This problem was inspired by Gilbert W. Kessler, "Shadows," *The Physics Teacher*, *17*, #5, pages 315–316 (May 1979).

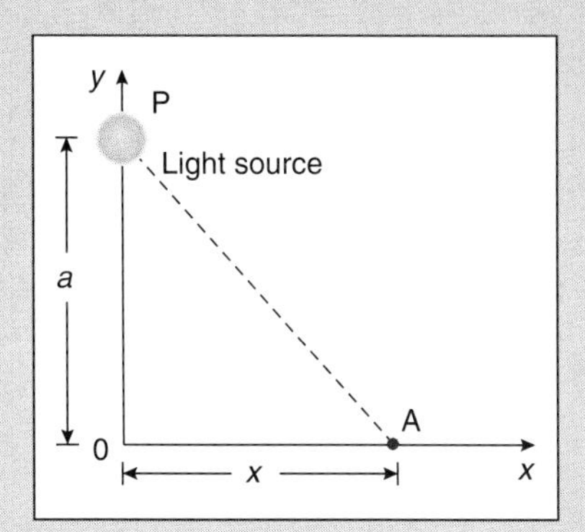

FIGURE P.32

33. In special relativity two reference frames S and S′ are inertial reference frames. A particle with velocity component u_x in inertial reference frame S has an acceleration component

$$a_x = \frac{du_x}{dt}$$

The acceleration component of the same particle in reference S′ is

$$a'_x = \frac{du'_x}{dt'}$$

Corresponding definitions exist for the other components of the acceleration. The relative speed v of the two inertial frames is constant. By differentiating the velocity component transformation equations (Equations 25.45 and 25.47 on page 1172), show that the acceleration components are related by the following ghastly expressions:

$$a_x = \frac{a'_x}{\gamma^3\left(1 + \dfrac{vu'_x}{c^2}\right)^3}$$

$$a_y = \frac{\left(1 + \dfrac{vu'_x}{c^2}\right)a'_y - \dfrac{vu'_y}{c^2}a'_x}{\gamma^3\left(1 + \dfrac{vu'_x}{c^2}\right)^3}$$

where

$$\gamma = \frac{1}{\sqrt{1 - \dfrac{v^2}{c^2}}}$$

Sections 25.13 The Longitudinal Doppler Effect
25.14 The Transverse Doppler Effect*
25.15 A General Equation for the Relativistic Doppler Effect*

•34. (a) Show that when $\theta = 0°$ in Equation 25.61 on page 1180, a purely longitudinal Doppler blue shift results—that is, Equation 25.61 reduces to Equation 25.59. (b) Show that when $\theta = 180°$ in Equation 25.61, a purely longitudinal Doppler red shift results—that is, Equation 25.61 reduces to Equation 25.58. (c) Show that when $\theta = 90°$ in Equation 25.61, a purely transverse Doppler effect (a red shift) results—that is, Equation 25.61 reduces to Equation 25.60.

•35. For a given value of v/c, show that the ratio of the fractional frequency changes ν/ν_0 produced by the red shifts of the transverse Doppler to the longitudinal Doppler effect is

$$1 + \frac{v}{c}$$

•36. On the same graph, plot ν/ν_0 versus v/c for the red shifts produced by the longitudinal Doppler effect and the transverse Doppler effect for v/c between 0.90 and 1.00. Discuss and interpret the graphs.

•37. Using the firefly as in Section 25.13 and the standard geometry, show that a decrease in frequency (a red shift to longer wavelengths) occurs for a detector at the origin in S′, moving away from a stationary firefly at the origin in S. Hint: Define suitable events in S, transform them to S′, and proceed as in Section 25.13.

•38. You are vigorously testing a new sports car. Your speed is so great that a red traffic light ($\lambda \approx 620$ nm) appears green ($\lambda \approx 540$ nm). An officer of the law pulls you over and gives you a ticket for running a red light. In court, you contest the charge, indicating with relativistic calculations that the red light certainly looked green as you approached the light. The ticket should have been written for speeding instead; the charge is dismissed on this legal technicality. Now that is practical relativity! Calculate the speed at which you approached the traffic light.

•39. Quasars are among the most perplexing objects in astronomy. They are thought to be at the fringes of the observable universe and are being carried away from us by the expansion of the universe itself. Their speeds of recession produce enormous red shifts in their spectra. The red shift z is formally defined as

$$z \equiv \frac{\Delta\lambda}{\lambda_0}$$

where $\Delta\lambda \equiv \lambda - \lambda_0$ and λ_0 is the proper wavelength of the light. Some quasars have red shifts with $z = 4.0$. Calculate the speed at which these objects are receding from us.

•**40.** If $v << c$, use the binomial theorem to show that the red shift in wavelength $\Delta\lambda \equiv \lambda - \lambda_0$ caused by the *transverse* Doppler effect is approximately

$$\Delta\lambda \approx \lambda_0 \frac{v^2}{2c^2}$$

where λ_0 is the wavelength of the light measured with the source at rest.

•41. A star emits light from hydrogen with a wavelength $\lambda_0 = 656.282$ nm. How fast must the star move *transverse* to the line of sight for you to measure its wavelength to be 656.382 nm?

•**42.** If $v << c$, use the binomial theorem to show that the red shift in wavelength $\Delta\lambda \equiv \lambda - \lambda_0$ caused by the *longitudinal* Doppler effect is approximately

$$\Delta\lambda \approx \lambda_0 \frac{v}{c}$$

where λ_0 is the wavelength of the light measured with the source at rest.

•**43.** A star emits light from hydrogen with a wavelength $\lambda_0 = 656.282$ nm. How fast must the star move *along* your line sight for you to measure its wavelength to be 656.382 nm? Is the distance between the source and receiver increasing or decreasing with time?

Sections 25.16 Relativistic Momentum
25.17 The CWE Theorem Revisited
25.18 Implications of the Equivalence Between Mass and Energy

44. The speed of a particle is increased by a factor of two from 1.25×10^8 m/s to 2.50×10^8 m/s. (a) By what factor does its momentum increase? (b) By what factor does its kinetic energy increase?

45. Calculate your rest energy and the price it would bring at $0.10 per kilowatt-hour.

46. (a) Calculate the rest energy of a penny with a mass of 5.0 g. (b) If a 1.0 kg mass had a kinetic energy equal to the rest energy of the penny, at what speed would the kilogram be moving?

47. The luminosity of the Sun is 3.83×10^{26} W. By how much does the mass of the Sun decrease each second because of its luminosity?

48. How much mass must be completely converted from rest energy into other forms of energy in order to keep a 100 W light bulb burning for one century?

49. Calculate the kinetic energy, total relativistic energy E, and the magnitude of the momentum p of an electron traveling at a speed of $0.600c$.

50. Show that the momentum of a particle with *zero* mass is

$$p = \frac{E}{c}$$

where E is its relativistic energy.

•**51.** A 1.0 g sugar cube travels past you at a speed of $0.90c$. How sweet it is! You measure its density while it is moving. By what numerical factor is the density of the cube changed because of its speed?

•**52.** (a) To double the relativistic kinetic energy, how is the factor gamma at the higher speed (say γ') related to the gamma at the slower speed (say γ)? (b) Show that in the nonrelativistic limit, when $v << c$, the expression derived in part (a) implies an increase in speed by a factor of $\sqrt{2}$ doubles the nonrelativistic kinetic energy.

•53. (a) At what speed is the total relativistic energy of a proton 10% greater than its rest energy? (b) How much work is necessary to accelerate the proton from rest to this speed?

•**54.** (a) At what speed must an electron travel so that its kinetic energy is equal to its rest energy? (b) At what speed must a proton travel so that its kinetic energy is equal to its rest energy?

•**55.** Show that if E is the total relativistic energy of any particle, then

$$\frac{dE}{dp} = v$$

where p is the magnitude of the relativistic momentum and v is the speed of the particle.

•**56.** One of the most energetic cosmic ray protons detected had a kinetic energy estimated to be about 1.0 J. For such a proton, by what amount is v/c less than exactly 1?

•57. An electron at rest is given a present of 1.60×10^{-15} J of kinetic energy. (a) Find the speed of the electron. (b) Express the kinetic energy of the electron in units of keV. (c) What is the magnitude of the momentum of the electron?

•**58.** An electron moves horizontally and parallel to a wall at speed $0.90c$ through your physics lab (of dimensions 10.0 m × 10.0 m × 3.0 m) on its way to its annihilation with a positron next door. (a) What is the volume of your lab as measured by you? (b) What is the volume of your lab in a reference frame at rest with respect to the electron? (c) Find the kinetic energy of the electron as measured in the lab. (d) Through what potential difference did the electron accelerate from rest to attain the speed of $0.90c$?

•**59.** A proton has a total relativistic energy of 1500 MeV. (a) What is its rest energy in MeV? (b) What is its speed? (c) What is the magnitude of its momentum?

•**60.** (a) How much work must be done to bring an electron from rest to a speed of $0.600c$? Express your result in both joules and keV. (b) How much additional work is needed to increase its speed from $0.600c$ to $0.800c$? Express your result in both joules and keV. (c) What is the ratio of the kinetic energy of the electron at the speed $0.800c$ to its kinetic energy at the speed $0.600c$?

•61. A politician of mass 70.0 kg is ejected from office and sent at a speed of $0.980c$ to a planet located 20.0 LY from the Earth. (a) What is the kinetic energy of the politician according to us on the Earth? (b) What is the kinetic energy of the politician according to the politician? (c) How many years (measured with clocks at rest on the Earth) will the one-way trip take? (d) According to the politician, how far away will the Earth be when the trip is completed (but the politician is still moving at the same speed)? (e) How many years will the trip take according to clocks at rest with respect to the politician?

•**62.** The energy associated with the eruption of the Mt. St. Helens volcano (in southern Washington state) on 18 May 1980 was estimated to be about 10^{17} J. At about what speed would a 10^3 kg spacecraft have a kinetic energy equal to this?

•**63.** (a) Through what potential difference must an electron be accelerated so that its total relativistic energy is 1.0% greater than its rest energy? (b) Will the result calculated in part (a) be different if the particle is a proton? (c) Repeat the calculation for a total relativistic energy that is 10% greater than the rest energy.

•**64.** Two protons approach each other at equal speeds 0.800c in some reference frame. What is the kinetic energy of one proton as seen from a reference frame with its origin on the other proton?

•**65.** (a) Beginning with the left-hand side of the following relationship, show that it is equal to the right-hand side using the definition of the relativistic factor γ:

$$\frac{\gamma^2 v^2}{\gamma + 1} = c^2(\gamma - 1)$$

(b) Use the result in part (a) to show that the total relativistic energy E can be expressed as

$$E = \gamma mc^2 = mc^2 + \frac{m\gamma^2 v^2}{\gamma + 1}$$

(c) Use the result of part (b) to show that the relativistic kinetic energy KE is related to the nonrelativistic kinetic energy

$$\mathrm{KE}_{\text{class}} = \frac{mv^2}{2}$$

in the following way:

$$\mathrm{KE} = \mathrm{KE}_{\text{class}} \frac{2\gamma^2}{\gamma + 1}$$

This problem was inspired by Wendell G. Holladay, "The derivation of relativistic energy from the Lorentz γ," *American Journal of Physics*, 60, #3, page 281 (March 1992).

•**66.** A mass m at rest spontaneously disintegrates into two masses m_1 and m_2 with speeds v_1 and v_2 respectively. Show that conservation of energy implies that $m > m_1 + m_2$.

•**67.** Nuclear weapons convert rest energy into other forms of energy. The nuclear weapon used in the bombing of Hiroshima on 6 August 1945 had an estimated yield equivalent to the detonation of about 10^4 metric tons of TNT,* and so was called a 10 kiloton weapon. Each metric ton of TNT yields about 4.9×10^9 J when detonated. (a) Estimate the amount of mass that vanished when the nuclear weapon exploded. (b) If 10^4 metric tons of TNT exploded, does the same amount of mass vanish?

•**68.** On the same graph, make schematic plots of the total relativistic energy E versus the magnitude p of the momentum for each of the following: (a) A particle with mass m. What is the meaning of the intercept of this graph? For large p, what is the slope of this graph? (b) A particle with zero mass. What is the slope of this graph? (c) A classical particle with zero rest energy. What is the intercept of this curve? What is the shape of this curve?

Section 25.19 Space–Time Diagrams*

69. Draw a space–time diagram with its time axis vertical and one spatial dimension. Schematically illustrate the world-lines that represent you: (a) while sleeping; (b) while walking in a straight line to class; (c) running in a straight line from class; (d) pacing back and forth when nervous about an upcoming exam.

•**70.** Draw a space–time diagram that schematically illustrates the world-line of a ball tossed vertically from the origin near the surface of the Earth. What is the mathematical shape of the world-line?

•**71.** Draw a space–time diagram and schematically sketch the world-line of a particle executing simple harmonic motion. The equilibrium position is at $x = 0$ m. The particle is released at $x = A$ when $t = 0$ s. What is the mathematical shape of the world-line?

•**72.** Draw an appropriate space–time diagram and schematically sketch the world-line of a particle executing uniform circular motion in the x–y plane. Describe the shape of the world-line in a cogent sentence or two.

•**73.** A flash of light occurs at $x = 0$ m when $t = 0$ s and moves toward increasing values of x. Draw a space–time diagram that accurately plots the world-line of the light.

•**74.** The starship *Enterprise* in *Star Trek* occasionally traveled at "warp speeds" of, say, three times the speed of light (which is not permissible in special relativity). On the same space–time diagram, sketch the world-line of light propagating along the x-axis and the world-line of the starship when traveling at such a fictional speed in the same direction. Explain why such speeds are not possible in special relativity.

---

*Trinitrotoluene ($C_7H_5N_3O_6$).

Investigative Projects

A. Expanded Horizons

1. For millennia, the speed of light was thought to be infinite. Ole Römer (1644–1710) was the first to realize that light travels at a finite speed with his investigations of the eclipses of the Galilean satellite Io of the planet Jupiter in 1676. Investigate Römer's discovery and various other historical and contemporary methods for measuring the speed of light.
   Andrzej Wróblewski, "De mora luminis: a spectacle in two acts with a prologue and an epilogue," *American Journal of Physics*, *53*, #7, pages 620–630 (July 1985); this article corrects many historical errors in the physics literature regarding the nature of the discovery by Römer.
   Harry E. Bates, "Resource letter RMSL-1: Recent measurements of the speed of light and the redefinition of the meter," *American Journal of Physics*, *56*, #8, pages 682–687 (August 1988); this contains an extensive bibliography.

2. Investigate the background, details, and results of the Michelson–Morley experiment (c. 1880s) and the impact (if any) of the experiment on Einstein's development of the special theory of relativity.
   Bernard Jaffe, *Michelson and the Speed of Light* (Anchor Books, Garden City, New York, 1960).
   Yoram Kirsh and Meir Meidav, "The Michelson–Morley experiment and the teaching of special relativity," *Physics Education*, *22*, #5, pages 270–273 (September 1987).
   Gerald Holton, "Einstein and the 'crucial' experiment," *American Journal of Physics*, *37*, #10, pages 968–982 (October 1969).
   R. S. Shankland, "Michelson–Morley experiment," *American Journal of Physics*, *32*, #1, pages 16–35 (January 1964).
   Gerald Holton, "On the origins of the special theory of relativity," *American Journal of Physics*, *28*, #7, pages 627–636 (October 1960).
   Isaac Asimov, *How Did We Find Out About the Speed of Light?* (Walker, New York, 1986).
   R. S. Shankland, "Michelson: America's first Nobel Prize winner in science," *The Physics Teacher*, *15*, #1, pages 19–25 (January 1977).
   Arthur I. Miller, *Albert Einstein's Special Theory of Relativity: Emergence (1905) and Early Interpretation (1905–1911)* (Springer-Verlag, New York, 1997).

3. Hypothetical particles that travel *faster* than the speed of light (so-called *tachyons*) have intrigued physicists. The characteristics of such hypothetical particles have been investigated in a number of different ways. One way uses some new math involving *perplex numbers* whose absolute value is -1, somewhat like the *complex* numbers stemming from $\sqrt{-1}$. This method is explored in the following:
   Paul Fjelstad, "Extending special relativity via the perplex numbers," *American Journal of Physics*, *54*, #5, pages 416–422 (May 1986).
   Other ways of investigating such hypothetical particles can be found in the following references:
   Olexa-Myron Bilaniuk and E. C. George Sudarshan, "Particles beyond the light barrier," *Physics Today*, *22*, #5, pages 43–51 (May 1969); *22*, #12, pages 47–52 (December 1969).
   Laurence M. Feldman, "Short bibliography on faster-than-light particles (tachyons)," *American Journal of Physics*, *42*, #3, pages 179–182 (March 1974).
   Edwin F. Taylor, "Why does nothing move faster than light? Because ahead is ahead!" *American Journal of Physics*, *58*, #9, pages 889–890 (September 1990).

4. Investigate the nature of the expansion of the universe. In particular, address the question of whether expansion speeds really are Doppler-type speeds and whether expansion speeds can exceed the speed of light without violating one of the postulates of relativity.
   Edward R. Harrison, *Cosmology* (Cambridge University Press, Cambridge, England, 1981), Chapter 10, pages 206–230.
   H. S. Murdoch, "Recession velocities greater than light," *Quarterly Journal of the Royal Astronomical Society*, *18*, #2, pages 242–247 (June 1977).

5. Investigate in greater detail the ideas involved in the general theory of relativity. In particular, discover the postulates of the theory and the investigate the three classical tests of the theory discussed in Section 25.21 in greater detail: (1) the gravitational deflection of light (and how it was first detected during a solar eclipse in 1919); (2) the motion of the perihelion point of the planet Mercury (and why it is that, of all the planets, it is Mercury that is most affected by the theory); and (3) the gravitational red shift of light (and how it was first detected with light from the white dwarf star known as Sirius B in the constellation Canis Major).
   Wolfgang Rindler, *Essential Relativity* (Van Nostrand Reinhold, New York, 1969), pages 130–151.
   Edward R. Harrison, *Cosmology* (Cambridge University Press, Cambridge, England, 1981), Chapter 8, pages 160–184.
   S. Chandrasekhar, "Einstein and general relativity: historical perspectives," *American Journal of Physics*, *47*, #3, pages 212–217 (March 1979).
   Clifford M. Will, "Testing general relativity: 20 years of progress," *Sky and Telescope*, 66, #4, pages 294–299 (October 1983).
   P. W. Worden Jr. and C. W. F. Everitt, "Resource letter GI-1: Gravity and inertia," *American Journal of Physics*, *50*, #6, pages 494–500 (June 1982); this contains many references to the topic.
   S. Chandrasekhar, "On the derivation of Einstein's field equations," *American Journal of Physics*, *40*, #2, pages 224–234 (February 1972).

6. Discover the association between the general theory of relativity, gravity, and geometry. In particular, investigate why the geometry around massive objects (such as the Sun) is not Euclidean but of positive curvature. Also research contemporary ideas about the possible geometry of the universe itself.
   George Gamow, *Gravity* (Anchor, Garden City, New York, 1962), pages 115–146.
   P. K. MacKeown, "Gravity is geometry," *The Physics Teacher*, *22*, #9, pages 557–564 (December 1984).
   Edward R. Harrison, *Cosmology* (Cambridge University Press, Cambridge, England, 1981), Chapter 7, pages 147–159.
   Richard H. Price, "General relativity primer," *American Journal of Physics*, *50*, #4, pages 300–329 (April 1982); and Edward P. Tryon, "Comment on 'General relativity primer,'" *American Journal of Physics*, *52*, #4, pages 366–367 (April 1984).
   P. W. Worden Jr. and C. W. F. Everitt, "Resource letter GI-1: Gravity and inertia," *American Journal of Physics*, *50*, #6, pages 494–500 (June 1982); this contains many references to the topic.
   Peter G. Bergmann, *The Riddle of Gravitation* (Dover, New York, 1992).

B. Lab and Field Work

7. There are various ways to measure the speed of light within the confines of a laboratory. Consult your instructor about such methods; then design and perform an experiment to measure c.
   Harry E. Bates, "Resource letter RMSL-1: Recent measurements of the speed of light and the redefinition of the meter," *American Journal of*

Physics, 56, #8, pages 682–687 (August 1988); this contains an extensive bibliography.

8. Visit a professor in the astronomy department at your college or university and secure spectra that illustrate the red and blue shifts of light from stars. Use data from the spectra to calculate the speed of recession and approach of the given stars. Also secure the spectrum of a quasar and determine its recessional speed.

C. Communicating Physics

9. Within the genre of science fiction, it sometimes is difficult for the uninitiated to sort scientifically bogus ideas from those that are scientifically plausible on the basis of the science we now know. Choose a favorite work of science fiction, either in print, film, or video form. Analyze the work, sorting the science into fiction and truth; use your professor as a resource for this if necessary. Can you cite examples of ideas in science fiction that were fiction at the time they were written or produced, but subsequently have been found to be scientifically correct or at least plausible?

Chapter 26

An Aperitif
Modern Physics

[I] tried immediately to weld the elementary quantum of action *h* somehow into the framework of the classical theory. But in the face of all such attempts, this constant showed itself to be obdurate. . . . My futile attempts to fit the elementary quantum of action into the classical theory continued for a number of years, and they cost me a great deal of effort. Many of my colleagues saw in this something bordering on a tragedy. But I feel differently about it. For the thorough enlightenment I thus received was all the more valuable. I now knew for a fact that the elementary quantum of action played a far more significant part in physics than I had originally been inclined to suspect.

*Max Planck**

Physics seemed quite tidy in 1890, quaintly formal and typically Victorian. The approach of the discipline to nature was essentially mechanistic and deterministic. With the successes of Newton's dynamical theory in the 17th century, and the development of thermodynamics, electromagnetism, and more sophisticated ways of solving complicated equations during the 19th century, some physicists seemed almost cocky that the end of physics was almost nigh. What lay ahead could not even be imagined.

Between about 1890 and 1925, a number of startling, perplexing, and seemingly unconnected discoveries were made that eventually converged to provide explanations for a host of divergent phenomena. The time was one of unusual ferment that led us into a totally new and unimagined submicroscopic world and worldview of physics. A study of the scientific history of this period reveals much of how science proceeds: haltingly and with hesitation, carefully and with caution, chaotically and with chance playing a major role. Yet the period was filled with bold new hypotheses and theoretical insight, based fundamentally on the keen and wary eye of scientists at first simply intrigued by unexpected observations coming from seemingly mundane experiments. The many accidental discoveries gradually fueled progress on many unexpected fronts; finally, a clearer picture emerged in the late 1920s of a new and almost incomprehensible submicroscopic world.

It is only with the luxury of hindsight that we can spy and superimpose a somewhat more orderly, but imperfect, pedagogical thread through the maze of new knowledge. The path we choose in presenting this material somewhat distorts the historical record and glosses over the confusion experienced by physicists of the time, but should give you a feeling for how physicists confronted and explored the submicroscopic world of the atom and nucleus. You may get the impression that the development was so logical as to be inevitable, but nothing could be further from the truth. All scientists proceed much like sleepwalkers,[†] simultaneously aware and yet unaware of just what they are doing.[‡] The advance of science is made as much by the "what's that?" and "aha!" experiences as by logic.

26.1 The Discovery of the Electron

Electrons are very small particles, too small to see, too small to feel even if one hit your nose at great speed. How do we know they exist? The discovery of this fundamental particle of nature illustrates how significant discoveries can arise from careful observations made while performing experiments whose purpose is completely different. In this case, the unexpected discovery arose from observations made while investigating the discharge of electricity through low-pressure gases.

The glowing discharge of electricity through gases at low pressure is accomplished by connecting a source of large potential difference to two terminals sealed into a glass tube with the gas, as in Figure 26.1. The positive terminal of the tube is called the **anode**, and the negative one the **cathode**.[§]

If we gradually evacuate the gas from inside the tube, thus decreasing the pressure, a dark region is observed near the cathode as the tube is emptied. With further evacuation, the region extends farther along the tube length, eventually reaching the opposite end. There a diffuse spot of light can be seen, whose color depends on the kind of glass used to form the tube. If several wire screens are placed inside the tube, the spot is quite well defined. Such observations in the late 19th century indicated that apparently something invisible was being emitted by the cathode, traveling across the tube, and colliding with the glass, causing the emission of light. Such **cathode rays** intrigued a number of investigators.

Early experiments in 1895 by Jean Perrin (1870–1942) found that a transverse electric field deflected the spot in a direction opposite to that of the field, indicating the cathode rays were negatively charged particles. Since charge was always associated with mass, the particles had to have mass.[#]

J. J. Thomson (1856–1940) was able to determine the ratio of the charge to mass of these particles in the following way. By subjecting the particles to mutually perpendicular electric and magnetic fields, as in the speed selector we examined in Section 20.2, he was able to select particles with a known speed $v = E/B$. Then by measuring the deflection of the particles in another magnetic field of known magnitude and extent,[¶] the ratio of the absolute value of the charge to its mass was found to be about $|q|/m \approx 1.8 \times 10^{11}$ C/kg.

The largest charge to mass ratio of any particle then known was about 9.6×10^{7} C/kg for hydrogen ions (measured using electrolysis experiments). Thomson conjectured that if the charge of the newly discovered negatively charged particles was typical of those on chemical ions, such as the hydrogen ion, which he had already measured, the large charge to mass ratio of cathode ray particles implied the mass of the individual particles was significantly smaller than that of any atom.

*(Chapter Opener) *Scientific Autobiography and Other Papers*, translated by Frank Gaynor (Philosophical Library, New York, 1949), pages 44–45.

[†]This is an analogy wonderfully made by Arthur Koestler in his classic study of the history of astronomy from the ancient Greeks to Galileo. See *The Sleepwalkers* (Penguin, London, 1959).

[‡]A scientist, unknown to the author, once remarked that science is doing what you do when you do not know what you are doing.

[§]The word anode comes from the Greek ἄνοδος (*anados*), meaning "the way up" (or positive). Cathode comes from the Greek κἄθοδος (*kathodos*), meaning "the way down" (or negative).

[#]At the time, no massless particles were known.

[¶]His apparatus was an early mass spectrometer.

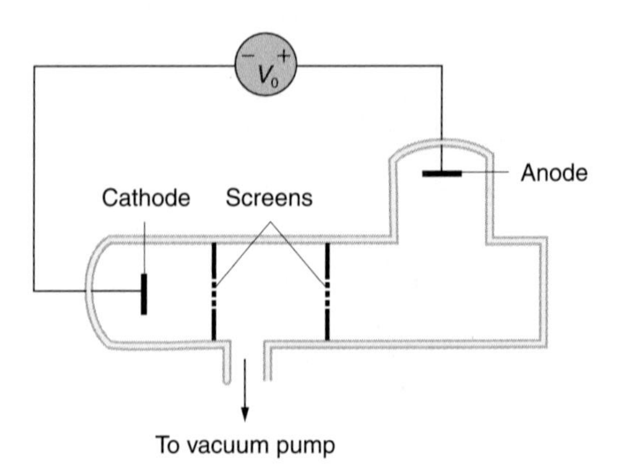

FIGURE 26.1 A gas discharge tube.

He called the particles primordial atoms or corpuscles; we now call them **electrons**.

The charge of electrons finally was measured by Robert Millikan (1868–1953) in 1909 with an elegant series of experiments using charged oil mist droplets floating in air and manipulated with known electric fields. The charge of the electron was found to be a constant quantity, and from the known charge to mass rario their mass also was established. We now know the charge is the negative fundamental unit of charge $q = -e = -1.602 \times 10^{-19}$ C; their mass is 9.11×10^{-31} kg. The first truly fundamental particle of nature had been found.

Since atoms commonly are electrically neutral, the discovery of the electron, presumed to exist within the neutral atom, was the first clear evidence that the heretofore indivisible atoms of nature likely had structure.

The search for an acceptable atomic model thus began in earnest at the end of the 19th century; we discuss several early such models in Sections 26.6 and 26.7.

26.2 The Discovery of X-rays

The importance of x-rays in medicine and dentistry is well known, not only for the diagnosis of fractures and caries, but also for the treatment of malignant and inoperable tumors. Certainly the practice of medicine would be much cruder had x-rays not been discovered over a century ago. Their application to clinical medicine occurred within months of their discovery by Wilhelm Röntgen (1845–1923) in 1895; technology transfer was as rapid then as it is now for significant new tools.

The discovery of x-rays was accidental, as are many great discoveries in science. Indeed, if Röntgen had set out to discover a new means to assist physicians in diagnosing fractures, he almost certainly *never* would have been playing with just the right equipment needed for him to produce and, therefore, to discover x-rays. This shows how pure (basic) research, driven solely by the interest and curiosity of scientists, is in many cases more productive for society in the long run than is directed (applied) research, undertaken with specific uses or goals in mind.

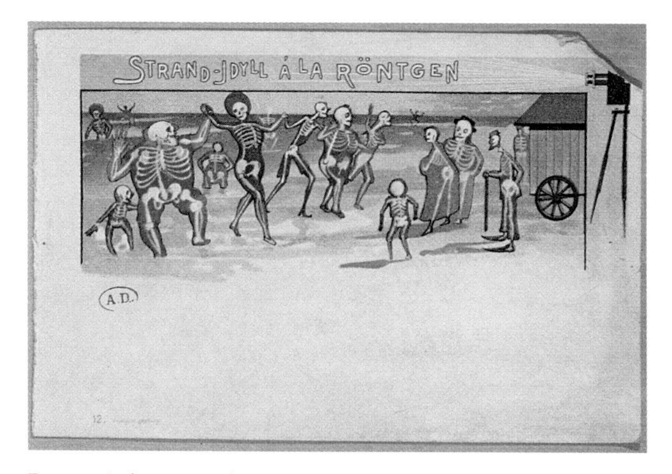

Röntgen's discovery of x-rays precipitated fear that clothing would need to be "x-ray proof" to avoid x-rated images; the fear was groundless. This post card from 1900 depicts "Sun bathing à la Röntgen"!

X-rays were discovered while investigating the discharge of electricity through rarefied gases, the same phenomena and equipment through which cathode rays first were noticed by Perrin and Thomson. Indeed, it is likely that x-rays were generated also in their laboratories and others, but Röntgen was the first person to *notice* them as he was investigating the characteristics of cathode rays. This is yet another example of how a researcher might be deflected by a puzzling observation during an experiment, only to realize later that the puzzle actually is more important than the initial investigation. It is, therefore, important for scientists to be constantly on the lookout for puzzling effects, even if they might seem at first to be an annoyance or nuisance.

Röntgen noticed that a paper screen treated with barium platino cyanide [$BaPt(CN)_4$] lit up quite brilliantly when near the discharge tube in which he had generated cathode rays. Intrigued, he found the effect was visible at distances up to two meters from the tube. He attributed the effect to a new, unknown "x" type of radiation, which he appropriately dubbed **x-rays**; the name stuck. He quickly was able to assess that the radiation appeared to originate from the area where the cathode rays (electrons) struck the glass tube. Through a remarkable series of insightful investigations, an experimental tour de force lasting only several weeks, Röntgen discovered many of salient features of x-rays and reported them in his first publication about their discovery. Appropriately, he won the very first Nobel prize in physics in 1901.

X-rays were shown by Röntgen to have the following characteristics:

1. X-rays are generated whenever high-energy cathode rays (electrons) strike solid materials. Generally, the greater the density of the impacted material, the more x-rays produced.
2. Matter is more or less transparent to x-rays. Wood and flesh are very transparent, bone and metals less so, which makes the use of x-rays in medicine so useful.
3. Photographic film is affected by x-rays, so its use as a detector was assured from the beginning.
4. The rays are undeviated by electric and magnetic fields, and so are uncharged.

Subsequent experiments by Hermann Haga and Cornelius H. Wind in 1899 determined that x-rays could be diffracted by extraordinarily narrow slits, on the order of 10^{-6} m wide. X-rays, therefore, are waves of very small wavelengths, much less than visible light, although the electromagnetic nature of x-rays was only demonstrated later.

The wave nature of x-rays makes them useful tools for the study of the structure of crystals and molecules, where the atoms and molecules act as three-dimensional diffraction gratings. Indeed, it was the diffraction of x-rays by crystals [by Max von Laue (1879–1960) in 1912] that proved crystals were regular arrangements of atoms and molecules. X-ray diffraction techniques, which mushroomed during the early decades of the 20th century, pioneered by Walter Bragg (1862–1942), were used to deduce the complex double helical structure of the famous DNA molecule* at

*Deoxyribonucleic acid.

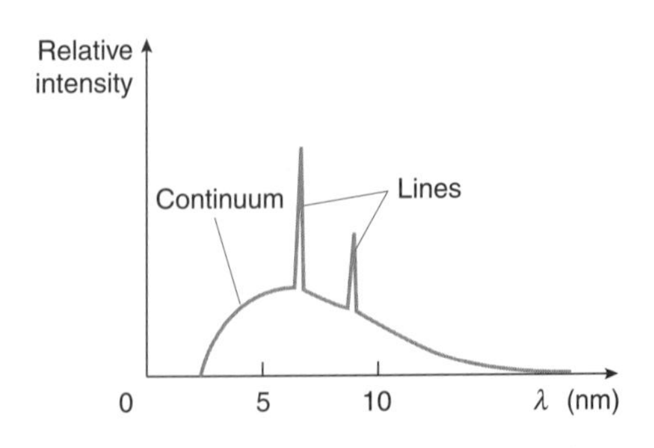

FIGURE 26.2 A typical x-ray spectrum. Note that their wavelengths are about 1/1000 those of visible light.

mid-century. DNA is critical for cellular reproduction, because it carries the molecular blueprints of the genetic code.

The spectrum of x-rays produced when electrons slam into a solid was first determined by Walter Bragg in 1913. The spectrum consists of a small continuous range of wavelengths and several very intense monochromatic (single-wavelength) lines, characteristic of the elements in the solid; an example is shown in Figure 26.2.

The Maxwell theory of electromagnetism predicts that accelerated charges produce electromagnetic radiation. The continuous range of x-ray electromagnetic wavelengths stems from the large and various magnitudes of the accelerations of the electrons as they are slowed by the material. We now know that the monochromatic x-ray wavelengths arise from within and characterize the specific atom itself.

26.3 The Discovery of Radioactivity

The proper disposal of radioactive waste, be it from medical facilities, nuclear power plants, research reactors, or nuclear weapons, is a serious problem of ongoing concern to humanity, if only because of the environmental consequences to us and future generations. However, radioactive materials are *not* solely the result of our inventive, technological prowess; they occur naturally and have been with us on the Earth from the very beginning, albeit in concentrations that typically are not serious environmental hazards.

The discovery of radioactivity, about a century ago, is an interesting tale. In 1896, just a few months after the discovery of x-rays by Röntgen, Henri Becquerel (1852–1908) set out to investigate the way in which the mysterious x-rays caused various substances to emit visible light and expose photographic plates. He took a crystalline sample of a chemical salt of uranium* and placed it near a photographic plate completely wrapped in heavy, opaque black paper, intending to see what happened when he exposed the salts to sunlight. The plates were exposed, as if light had penetrated the heavy wrapping encasing the plates. The effect was the same even when both the wrapped plates and uranium salts were in complete darkness.

Well intrigued by the miraculous exposure of the plates when well shielded from visible light, Becquerel explored the phenomena in some detail. In short order he realized the magnitude of the effect (the exposure) was proportional to the amount of uranium present and that changing the temperature had no effect. The latter observation led him to conclude the effect was an atomic process (originating *within* the atom) rather than a chemical one, since temperature typically has a dramatic effect on the rate of virtually all chemical reactions.

*Potassium uranyl sulfate: $K_2SO_4 \cdot UO_2SO_4 \cdot 2H_2O$.

His publication of the results led others to search for other substances that produced similar effects. In 1898 Marie Curie (1867–1934) noticed that two different uranium ores were significantly more active than pure uranium itself and, as a result, eventually was able to chemically isolate two previously unknown elements, which she called polonium† and radium. She called the phenomenon **radioactivity**. Curie discovered the radiations evidently were very energetic, because a sample of radium remarkably was able to maintain itself in thermal equilibrium several degrees above room temperature! Speculation arose that perhaps radioactivity was the source of the Sun's energy (now known to be false) and a source of energy for the hot interior of the Earth (true).

The substances that produce the effect evidently emit several distinct types of radiations. One is a penetrating radiation, dubbed α, that propagates through several centimeters in air and can even penetrate very thin metal foils. Another less penetrating radiation, dubbed β, is easily stopped by even a sheet of paper. Another type, called γ, was discovered in 1900 and is much more penetrating than even the α radiation. Magnetic fields influenced the trajectories of the α and β radiations, and indicated the α radiation consisted of positively charged particles while the β were negatively charged particles. The γ radiation was unaffected by such fields, and so was electrically uncharged.

A series of experiments with an electromagnetic speed selector showed that the β particles from a given substance emerge with a variety of speeds that are significant fractions of the speed of light (over 90% in many cases). Substances that emit α particles produce them with discrete (specific) speeds, also very large. The extraordinary speeds and energies indicated that the processes causing their emission were not chemical reactions of any known type. Charge to mass ratio experiments proved that the **β particles** were high-speed cathode rays: electrons. The **α particles** were associated with helium atoms (the nucleus had not yet been discovered); helium always seemed to be present in the radioactive materials that emitted α radiation. In fact, α particles are helium nuclei, not helium atoms.

†After Poland, the homeland of Marie Curie. Her maiden name was Marie Sklodowska. She met and married Pierre Curie in Paris.

Becquerel came from a long line of French physicists and occupied the same professorship of physics held by both his father and grandfather!

This symbol indicates the presence of radioactive materials much the way other symbols are used to indicate poisons or biohazards.

By 1908 Ernest Rutherford (1871–1937) was able to count directly the number of α particles emitted per gram of radium* as well as the total charge they represented, concluding correctly that each α possessed twice the magnitude of charge as on a β particle. With a knowledge of the charge and kinetic energy of the α particles emitted by radium, Rutherford had the tools needed to begin probing the structure of the atom using the monoenergetic α particles as high-speed, submicroscopic bullets (see Sections 26.6 and 26.7).

We discuss radioactivity in more detail in Section 26.11 from the viewpoint of the nuclear model of the atom.

26.4 The Appearance of Planck's Constant h

Turn on an electric stove heating element and soon it begins to glow a dark red, eventually becoming bright red as its temperature increases to its maximum. Turn on a light bulb and it glows white hot because its temperature is much hotter than the stove burner. The greater the temperature of a solid, the bluer the light emitted by it. The same is true even of very hot gases: bluish-colored stars, such as Rigel (β Orionis) in the constellation Orion, have higher surface temperatures than reddish-colored stars, such as Betelgeuse (α Orionis) also in Orion.

We imagine a closed oven or cavity with a tiny hole through which we can peek, as in Figure 26.3. As the absolute temperature of the cavity gradually is increased, we begin to notice that light emitted from the hole is dark red in color. As we increase the temperature of the cavity still more, the light becomes bright red, then whitish, then progressively bluish. When the entire cavity is all in thermal equilibrium at one absolute temperature T, we can analyze the spectrum (or the domain of wavelengths or frequencies) of the light from the hole, knowing very surely the source temperature. Experiments have shown that the spectrum of the light inside the cavity for any fixed absolute temperature T is independent of the material composing the cavity. The spectrum has a characteristic shape (see Figure 26.4) called a **blackbody spectrum**, even though the cavity, if its temperature is high enough, will hardly look black when we peek into it.

The peak of the spectral curve shifts to shorter wavelengths (higher frequencies) for higher absolute temperatures, thus accounting for the change in the color of the light emerging from the cavity. Let λ_{max} be the wavelength at which the blackbody emission spectrum has its peak for a given absolute temperature T. Wilhelm Wien (1864–1929) thermodynamically analyzed what happened to the light inside the cavity if it behaved adiabatically (thermally isolated). He discovered that the product of the peak

*When α particles collide with a phosphor screen, some light is emitted, so Rutherford and others actually could count the collisions visually.

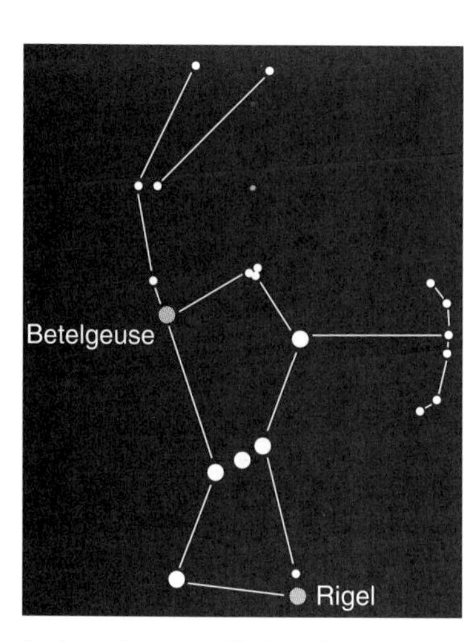

When you next look at the constellation Orion, notice the color difference between bluish Rigel and reddish Betelgeuse, indicative of their different surface temperatures.

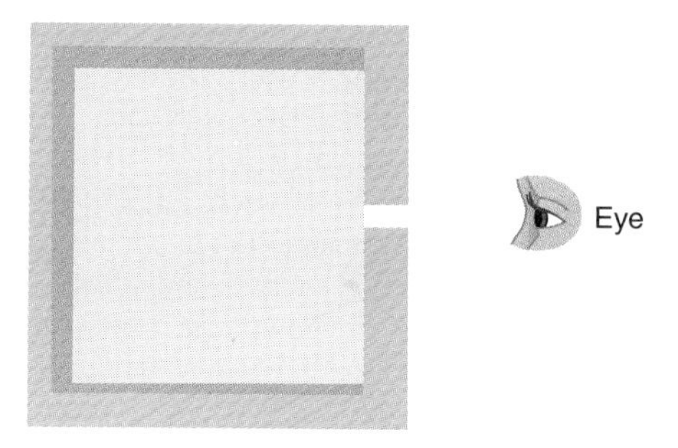

FIGURE 26.3 A cavity.

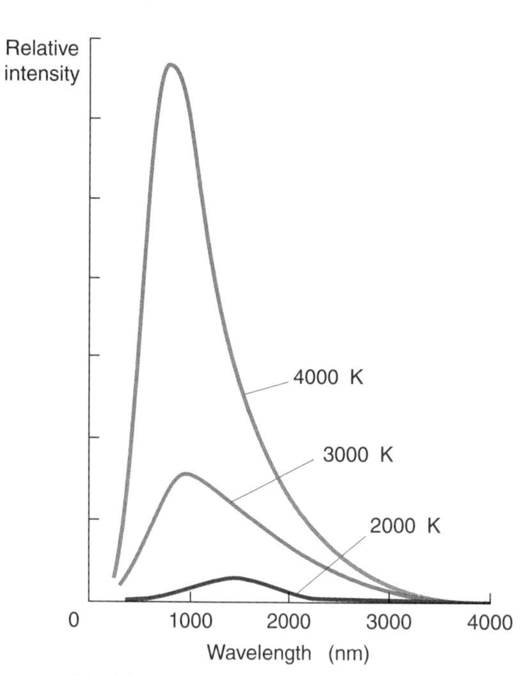

FIGURE 26.4 Blackbody spectra at various absolute temperatures. Visible light has wavelengths from about 400 nm to 700 nm.

wavelength λ_{max} and the absolute temperature T was a constant. By use of experimental spectra, this constant is found to have the approximate value

$$\lambda_{max}T = 0.289\,78 \times 10^{-2}\ \mathrm{m \cdot K} \tag{26.1}$$

Equation 26.1 is called **Wien's displacement law.***

Increase the temperature and the peak wavelength shifts to smaller wavelengths.

Arno Penzias and Robert Wilson serendipitously detected a blackbody spectrum from space in the mid-1960s at the Bell Telephone Laboratories in Holmdel, New Jersey, during experiments tracing the source of peculiar static in a large microwave antenna. The spectrum from space has a characteristic absolute temperature of only about 3 K. The detection of this radiation provided experimental evidence to support the Big Bang theory of the creation of the universe.† The light left over from the Big Bang (some 10 or 20 billion years ago) has cooled significantly as the universe expanded and now has its peak in the microwave region of the electromagnetic spectrum.

In the late 19th century, understanding the characteristic shape of a blackbody spectrum provoked much theoretical interest and controversy. Since the cavity was in thermal equilibrium, it was hoped that an application of the laws of statistical mechanics and thermodynamics might account for the spectrum of the electromagnetic waves inside and emerging from the cavity. Indeed, Wien initially met with some success with the displacement law (Equation 26.1).

However, other results of such an analysis were very troubling. Since the electromagnetic waves inside a cavity were continually being absorbed, reemitted, and reflected by the walls of the cavity, physicists imagined the cavity as filled with standing electromagnetic waves. Recall from our study of waves (Chapter 12) that standing waves in one dimension (such as those on a guitar string) are formed from the superposition of two identical waves of wavelength λ traveling in opposite directions. When confined to a distance ℓ, say one dimension of a cubical cavity, the wavelengths λ_n that form standing waves are

$$\lambda_n = \frac{2\ell}{n}$$

where n is a positive integer ($n = 1, 2, 3, \ldots$). In particular, note that there is, in principle, no limit on how short the wavelengths can be and, thus, on how high the corresponding frequency can be.

In statistical thermodynamics, when we add a particle to a system, the number of degrees of freedom of the system increases; the Maxwell theory of electromagnetism implies that each standing wave in the cavity would be a degree of freedom. When in thermal equilibrium, the equipartition of energy theorem (Section 14.8) assigns an average energy $kT/2$ to each degree of freedom (provided it is not frozen out).

*The term, which arises from the German *Verschiebungsgesetz*, was first used in 1899 by Otto Lummer and Ernst Pringsheim.

†They won the 1978 Nobel prize in physics for their discovery.

Recall from Chapter 7 that the energy of a classical oscillator is proportional to the square of its *amplitude* and is *independent* of its frequency. Hence each electromagnetic standing wave oscillator in the cavity should share in the energy inside the cavity. This leads to two equally unpleasant possible conclusions:

a. The energy available in the cavity is finite, but the cavity has an essentially infinite number of standing waves. Thus each standing wave effectively has zero energy and no electromagnetic waves should emerge from the hole! But electromagnetic waves clearly *are* inside the cavity and are emitted from it.
b. Since there are many more standing waves of high frequency than of low frequency, the light emerging from the hole should be at the extreme high frequency end of the spectrum (with blue and violet light, even x-rays). This interpretation, first realized by Lord Rayleigh (John William Strutt) (1842–1919) and James Jeans (1877–1946) in 1900, came to be known as the **ultraviolet catastrophe**.

Every attempt to explain the blackbody spectrum based on electromagnetic theory and thermodynamics failed to predict the shape of the blackbody spectrum. Clearly, something peculiar was happening that was not amenable to the standard physics of the day.

In 1901 Max Planck (1858–1947) finally was able to account for the shape of the blackbody spectrum using a drastic ad hoc hypothesis made, he said, in desperation.

Quite reluctantly, Planck assumed that the energy E associated with the light inside the cavity was present only in finite packets (bundles) proportional to the frequency ν:

$$E = h\nu \tag{26.2}$$

where h was an unknown constant that he hoped to be able to set equal to zero after taking appropriate mathematical limits.‡

‡He was, in essence, trying to approximate a diverging integration by a summation.

Planck had little to do with rocket science, although it would not appear so from this stamp from the Ivory Coast commemorating his Nobel prize. Planck was an accomplished pianist as well as a physicist; many physicists love the arts, and practice them as well!

With this hypothesis, the energy associated with high-frequency light is very large, and so the degrees of freedom of such standing waves must be frozen out. The finite amount of energy in the cavity then can be apportioned among a smaller, finite number of low-frequency oscillators (standing waves). The hypothesis neatly avoided the ultraviolet catastrophe and correctly predicted the shape of the blackbody spectrum.

By fitting his predictions to the experimental blackbody spectrum, Planck had hoped to show that the value of h was zero. Much to his surprise and consternation, he found the constant could not be set to zero but had an approximate numerical value of

$$h \approx 6.6 \times 10^{-34} \text{ J} \cdot \text{s}$$

Tiny for sure, but *not* zero.

We now know h with more precision to be about

$$h = 6.626 \times 10^{-34} \text{ J} \cdot \text{s} \qquad (26.3)$$

The constant h, now regarded as a fundamental constant of nature, was called by Planck the *quantum of action*, but shortly became simply **Planck's constant.**

The revolutionary and disturbing nature of Planck's hypothesis, Equation 26.2, is apparent in Planck's recollections in the opening quotation of this chapter.

QUESTION 1

Show that the SI units of Planck's constant (J · s) are the same as those of angular momentum (kg · m²/s).

EXAMPLE 26.1

Light from the Sun has a blackbody radiation spectrum with its peak at the visible light wavelength 5.0×10^2 nm. What is the surface temperature of the Sun?

Solution

The Wien displacement law, Equation 26.1, relates the wavelength of the peak of a blackbody spectrum to the absolute temperature:

$$\lambda_{\text{max}} T = 0.289\ 78 \times 10^{-2} \text{ m} \cdot \text{K}$$

Since $\lambda_{\text{max}} = 5.0 \times 10^2$ nm, you have

$$\begin{aligned} T &= \frac{0.289\,78 \times 10^{-2} \text{ m} \cdot \text{K}}{5.0 \times 10^2 \times 10^{-9} \text{ m}} \\ &= 5.8 \times 10^3 \text{ K} \end{aligned}$$

A rather toasty temperature!

26.5 THE PHOTOELECTRIC EFFECT

Planck's strange hypothesis about the way energy is carried in bundles by light soon bore fruit in another area, quite unrelated to the blackbody spectrum: the photoelectric effect. The reappearance of Planck's constant in another context solidified its place as a fundamental constant of physics. Here we investigate this additional thread in the complex tapestry of discoveries at the dawn of the 20th century.

When light of an appropriate frequency (or correspondingly, of an appropriate wavelength) is incident on a metallic surface, electrons are liberated from the surface. This observation is known as the **photoelectric effect.***

The electronic detection of light, even incredibly dim light such as that from stars and distant galaxies, is based on the photoelectric effect for some kinds of sensitive photodetectors. Its discovery was contemporaneous with the work with cathode rays.

In 1888 Wilhelm Hallwachs (1859–1922) noticed that a freshly polished metal (he used zinc), when insulated and connected to an electroscope, lost negative charge and became positive when irradiated with ultraviolet light.† Initially it was thought the charge might be transferred by the gas particles surrounding the metal sample, but the loss of negative charge persisted even when the sample was illuminated while in a good vacuum, thus eliminating that possibility. Perhaps charged atoms of the metal were being expelled by the light? No, because even after extensive illumination, no traces of metallic material could be found elsewhere inside the chamber in which the metal electrode was placed. By 1899 Philipp Lenard (1862–1947)‡ had shown that the emitted particles were negative with the same charge to mass ratio as cathode rays, correctly surmising that the light caused the metal to emit electrons.

Experiments with the photoelectric effect typically are performed with an apparatus schematically illustrated in Figure 26.5. When appropriate light is incident on the metal, the ejected electrons are swept to and collected by the positive terminal

*The effect really should be called the photo-induced liberation of electrons from metals, but this is shortened for convenience to the photoelectric effect.

†The effect evidently was first observed by Heinrich Hertz in the course of his experiments with the production and detection of electromagnetic waves, but he did not follow up on his observations—a missed opportunity!

‡Regrettably, Lenard was an anti-Semite, who made vicious personal attacks on many Jewish scientists, even Einstein. An early admirer of Adolph Hitler, Lenard became the bridge between the Third Reich and the German scientific establishment before and during the Second World War.

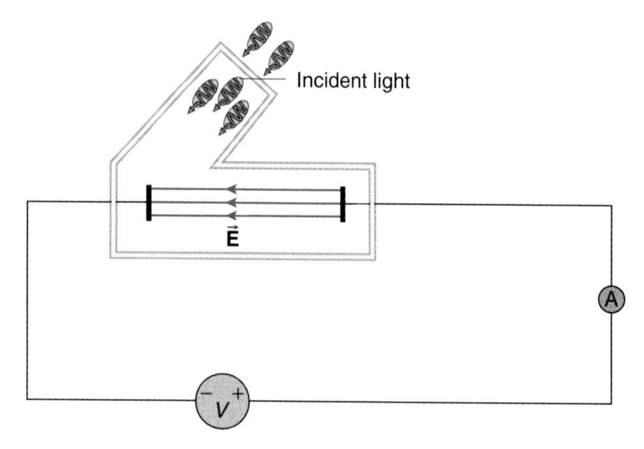

FIGURE 26.5 **Apparatus for investigating the photoelectric effect.**

(the anode). A small current flows in the circuit and is detected with a sensitive ammeter.

Experiments by Lenard and others discovered a number of significant features about the effect:

1. There exists a critical **cutoff frequency** ν_c for the incident light (and corresponding **cutoff wavelength** $\lambda_c = c/\nu_c$). If the frequency of the incident light is less than ν_c, no electrons are liberated from the metal regardless of the intensity of the incident light: there is no current detected by the ammeter.
2. For incident light with frequencies $\nu > \nu_c$, the number of electrons liberated per second is directly proportional to the light intensity. That is, the current in the circuit is directly proportional to the incident light intensity, as shown in Figure 26.6.
3. For a given frequency of light illuminating the metal, if we increase the source potential difference V in the circuit of Figure 26.5, no increase in the current occurs (see Figure 26.7 for $V > 0$ V). The electric field established in the tube by the battery effectively pulls all the liberated electrons to the positive terminal to be collected and registered by the ammeter.

   On the other hand, if we reverse the polarity of the potential difference, the direction of the electric field in Figure 26.5 reverses. As the value of the reversed potential difference increases, the current in the circuit decreases to zero (see Figure 26.7 for $V < 0$ V). The electric field now prevents some of the electrons from reaching the now negative terminal. When this reversed electric field is large enough, no electrons are collected and the current is zero. This observation indicates that the liberated charge is negative and that these electrons have a smooth variety of kinetic energies, ranging from essentially zero up to a maximum value KE_{max}. Indeed, we will show later in this section that the maximum kinetic energy of the electrons is related to the reversed potential difference V_s at which the current of electrons ceases (see Figure 26.7). The quantity V_s is known as the **stopping potential**, since the electric field it creates in the tube stops *all* the liberated electrons from reaching the opposite terminal, even the most energetic ones.
4. Experiments also show that the value of this stopping potential V_s is directly proportional to the frequency of the incident light for frequencies greater than the critical cutoff frequency, as shown in Figure 26.8.

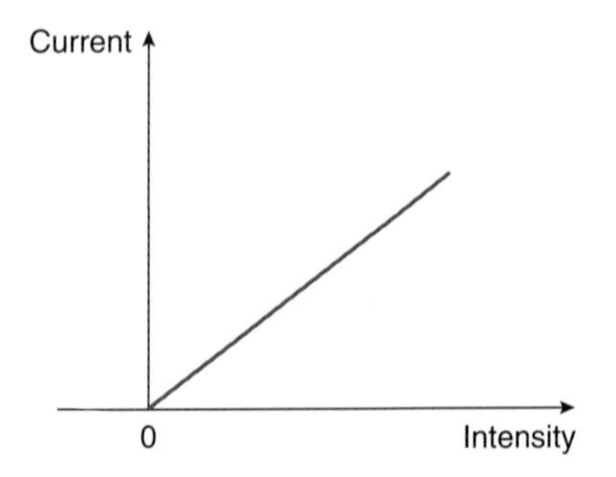

FIGURE 26.6 For $\nu > \nu_c$, the current is proportional to the incident light intensity.

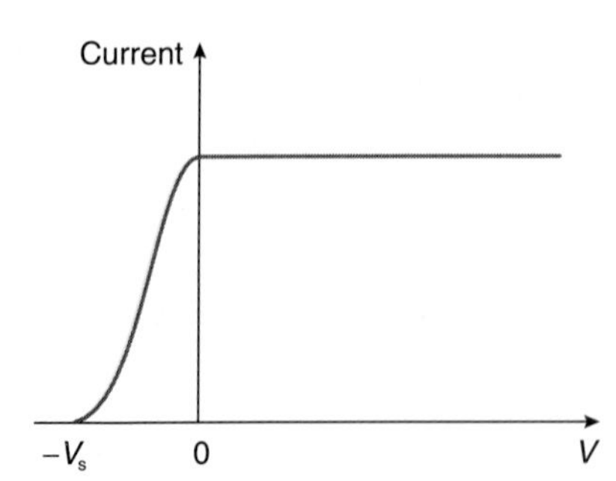

FIGURE 26.7 The current versus the applied potential difference with the arrangement of Figure 26.5 for the photoelectric effect.

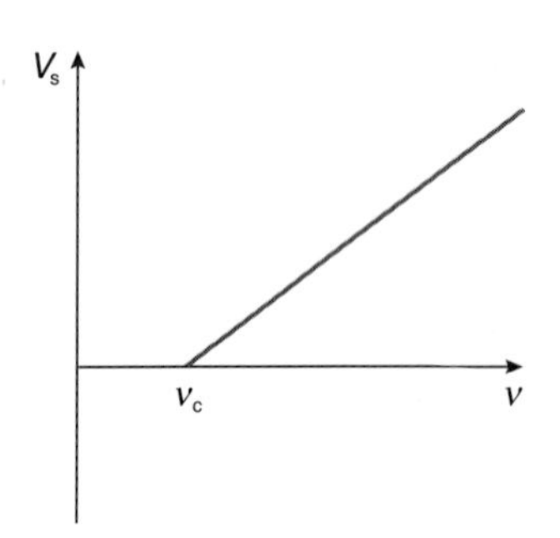

FIGURE 26.8 The stopping potential is proportional to the light frequency for $\nu > \nu_c$.

5. The liberated electrons appear promptly when the metal is illuminated, with no time lag. That is, the electrons ejected from the metal appear essentially instantaneously even when the incident light is of very low intensity.

It is difficult to account for the existence of a cutoff frequency (observation #1) and the prompt emergence of the electrons (observation #5) with a wave model for light for two reasons:

a. As we recalled in Section 26.4, the energy of a classical wave is proportional to the square of its amplitude and frequency. Thus, according to this scheme, electrons should be able to absorb energy from incident light of *any* frequency. The photoelectric effect should be independent of frequency; however, the effect is strongly frequency dependent.
b. For low levels of illumination (light waves of very small amplitude or energy), an electron might have to wait for enough energy to arrive before accumulating enough to become liberated from the metallic surface. In other words, for very low levels of light wave illumination, there likely will be a significant and detectable time delay between the onset of the light and the release of the electrons. However, no such delay is observed with low illumination; the electrons always appear promptly.

The problems with explaining the photoelectric effect experiments with a classical wave model for light forced Einstein to abandon it in 1905.* He invented a bold new model that completely accounted for the observations. It was principally for his

*You may recall that Einstein also published his special theory of relativity in 1905. It was a very productive year for him—and for physics.

explanation of the photoelectric effect that he won the 1921 Nobel prize in physics. Einstein's explanation is all the more remarkable in retrospect, because the charge on the electron was unknown in 1905. Many of Einstein's ideas about the effect were not confirmed by experiment for another decade.

Einstein built on and extended the ad hoc hypothesis about light energy used by Planck to avoid the ultraviolet catastrophe of the blackbody spectrum (Section 26.4). Einstein imagined light to consist of wavelike particles, packets (bundles) of electromagnetic energy, called **photons**. Planck initially thought that light photons only existed when confined to a cavity, and that once light left the cavity it was an electromagnetic wave. Einstein, on the other hand, extended Planck's idea to say that such photons (light quanta) existed in free space, even outside of any cavity. The particles of light each have an energy E_{photon} proportional to the frequency of the light:

$$E_{photon} = h\nu \tag{26.4}$$

where h is Planck's constant. A beam of light thus consists of an incredible flood of photons (see Example 26.2).

When light is absorbed by matter (i.e., by particles with mass), Einstein hypothesized that the matter-particle "swallows" the *entire* photon, thus destroying (annihilating) it completely (no leftovers). A matter-particle cannot nibble just a piece of a passing photon: it is all or nothing, like swallowing a pill. Energy is conserved because the matter-particle now has the energy previously possessed by the photon. Of course, the matter-particle also has other forms of energy, such as rest energy; but the change in the energy of the matter-particle is precisely the energy of the photon.

How do these ideas account so neatly for the experimental observations of the photoelectric effect?

When an electron in the metal totally absorbs an incoming photon and acquires its energy, some of it subsequently is lost in collisions as the electron escapes from the metal. This lost energy (which can be any fraction, since it is not associated with a photon but is the kinetic energy of the electron) is transferred to other atoms in the metal.* We sweep the details of these complex interactions under the rug and say that the electron does a certain amount of work to escape the metal, and so loses some of the energy gained from the photon. The leftover energy, after escape, is the kinetic energy of the liberated electron. In other words, we conserve energy for *each* electron that absorbs a photon by writing

$$\begin{matrix}\text{energy} \\ \text{of the} \\ \text{photon} \\ \text{absorbed}\end{matrix} = \begin{matrix}\text{work needed to} \\ \text{get the electron} \\ \text{liberated from} \\ \text{the metal}\end{matrix} + \begin{matrix}\text{kinetic energy} \\ \text{of the} \\ \text{liberated} \\ \text{electron}\end{matrix} \tag{26.5}$$

Depending on the particular circumstances of an individual electron, it may have to give up part or all of its newly acquired additional energy to get out of the bulk metal. Collectively, the emerging electrons thus appear with a variety of kinetic energies. Also because of energy conservation, the privileged electrons that escape from the metal with the *minimum* amount of energy loss (the first term on the right-hand side of Equation 26.5) must be the ones with the *maximum* amount of kinetic energy after escape, since the energy picked up from the absorbed photon is a fixed amount. For these privileged electrons, Equation 26.5 becomes

$$\begin{matrix}\text{energy} \\ \text{of the} \\ \text{photon} \\ \text{absorbed}\end{matrix} = \begin{matrix}\textit{minimum} \\ \text{work needed to} \\ \text{get the electron} \\ \text{liberated from} \\ \text{the metal}\end{matrix} + \begin{matrix}\textit{maximum} \\ \text{kinetic energy} \\ \text{of the} \\ \text{liberated} \\ \text{electron}\end{matrix} \tag{26.6}$$

The minimum work needed to liberate an electron from a metal is called the **work function** W of the metal. Thus Equation 26.6 is rewritten as

$$\boxed{h\nu = W + KE_{max}} \tag{26.7}$$

If we now decrease the frequency of the incoming light, thus decreasing the energy of the incoming photons (since $E_{photon} = h\nu$), the maximum kinetic energy of the emergent electrons also decreases. Eventually we reach a frequency ν_c such that the electrons barely escape and (each) have zero kinetic energy:

$$\boxed{h\nu_c = W + 0\ \text{J}} \tag{26.8}$$

The frequency ν_c is the cutoff frequency for the photoelectric effect. For frequencies $\nu < \nu_c$, the energy of the incoming photons is not sufficient to give the electron even the minimum energy W needed to escape from the metal. As a result, no electrons emerge and the photoelectric effect disappears.

Thus the work function of a metal is equal to the energy of a photon with the cutoff frequency.

Experimentally, this is how the work functions of various metals are determined. Table 26.1 presents the approximate work functions of various metals.

How is the maximum kinetic energy of the emitted electrons measured? We use the apparatus of Figure 26.5, but with polarity of the voltage source reversed, as in Figure 26.9. The voltage source produces an electric field between the terminals inside the tube. The liberated electrons find themselves in this electric field, and thus are subjected to an electrical force

$$\vec{\mathbf{F}}_{elec} = q\vec{\mathbf{E}} = -e\vec{\mathbf{E}}$$

in the direction opposite to the field, since the charge on the electron is negative. This force drives them back to the plate from which they emerged.

As the source potential difference increases from zero toward the value of the stopping potential, the current in the circuit gradually diminishes (see Figure 26.7, to the left of the origin); only the electrons ejected with sufficiently high kinetic energy can reach the opposite plate. The potential difference V_s that stops all current from the ejected electrons is the stopping potential. The liberated electrons are subjected to only the conservative electrical force in the region between the plates. As they leave the metal, the most energetic liberated electrons have

*Once the photon energy is gained by the electron, the energy is part of a common pool from which any amount can be tapped.

TABLE 26.1 Approximate Photoelectric Work Functions of Various Metals

Metal	*Work function* (eV)
Aluminum (Al)	4.1
Barium (Ba)	2.5
Cesium (Cs)	2.0
Copper (Cu)	4.7
Gold (Au)	4.8
Iron (Fe)	4.7
Lead (Pb)	4.1
Lithium (Li)	2.4
Mercury (Hg)	4.5
Nickel (Ni)	5.0
Platinum (Pt)	6.4
Potassium (K)	2.2
Rubidium (Rb)	2.1
Silver (Ag)	4.7
Sodium (Na)	2.3
Strontium (Sr)	2.7
Tin (Sn)	3.9
Zinc (Zn)	4.3

Source: CRC *Handbook of Chemistry and Physics*, 50th edition (CRC Press, Cleveland, Ohio, 1963), pp. 2655–2660.

an initial kinetic energy KE_{max} and an initial electric potential energy $PE_{elec} = qV = (-e)V_s = -eV_s$. When these electrons (barely) reach the opposite plate, they have zero kinetic energy, and an electrical potential energy $PE_{elec} = qV = -e(0\text{ V}) = 0\text{ J}$. The CWE theorem becomes

$$\begin{aligned} W_{other} &= \Delta KE + \Delta PE \qquad (26.9) \\ 0\text{ J} &= \Delta KE + \Delta PE \\ &= (KE + PE)_f - (KE + PE)_i \\ 0\text{ J} &= (0\text{ J} + 0\text{ J}) - (KE_{max} - eV_s) \end{aligned}$$

and so

$$\boxed{KE_{max} = eV_s} \qquad (26.10)$$

The stopping potential therefore is a direct measure of the kinetic energy of the most energetic liberated electrons.

Equation 26.7 for these electrons then can be rewritten as

$$h\nu = W + eV_s \qquad (26.11)$$

The work function W is related to the cutoff frequency by Equation 26.8:

$$W = h\nu_c$$

Now we send light of frequency $\nu_1 > \nu_c$ and measure the stopping potential V_{s1} associated with this incident light. We change the light to another frequency $\nu_2 > \nu_c$ and measure its associated stopping potential V_{s2}. We do this for a number of different frequencies: ν_3, ν_4, ..., all $> \nu_c$. We plot the measured stopping potentials versus the associated frequency. According to Equation 26.11, we have

$$V_s = -\frac{W}{e} + \frac{h}{e}\nu \qquad (26.12)$$

The graph of V_s versus ν will be a straight line (as in Figure 26.8), with a slope equal to h/e and an intercept $-W/e$; see Figure 26.10.

Such an experiment, and its resulting plot of V_s versus ν, is one way to measure Planck's constant, given the electronic charge e. The frequency corresponding to $V_s = 0$ V is the cutoff frequency.

PROBLEM-SOLVING TACTICS

26.1 In using Equations 26.7, 26.8, 26.10, or 26.11 for the photoelectric effect, you must express all the energies either in joules (J) or, if you choose, electron-volts (eV). As a practical matter, the photon energy $h\nu$ in SI units is expressed in joules. The work functions of metals typically are expressed in electron-volts (eV) (see Table 26.1). For photoelectricity, electron-volts are very convenient, since the energy of visible light photons is on the order of just a few electron-volts (see Example 26.2).

26.2 From Equation 26.10, the stopping potential in volts is *numerically* equal to the maximum kinetic energy of the electrons expressed in electron-volts. This is always the case, and is one reason that the electron-volt is a convenient unit of energy. See Example 26.4.

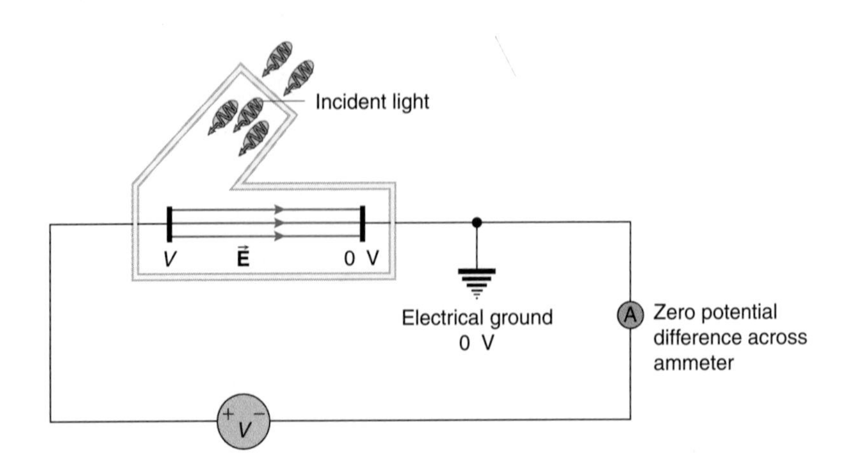

FIGURE 26.9 Apparatus to determine KE_{max}.

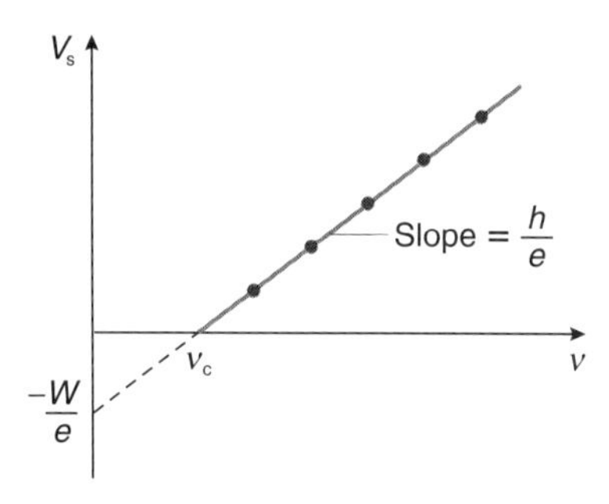

FIGURE 26.10 The stopping potential versus the frequency of the incident light.

QUESTION 2

Which of the metals in Table 26.1 has the longest cutoff wavelength for the photoelectric effect? What is that wavelength? For what wavelengths is the photoelectric effect observed for that metal?

EXAMPLE 26.2

A low-power helium–neon laser has a power output of 1.00 mW of light of wavelength 632.8 nm.

a. Calculate the energy of each photon, expressing your result in both joules and electron-volts.
b. Determine the number of photons emitted by the laser each second.

Solution

a. The energy E of each photon is proportional to its frequency ν, from Equation 26.4:

$$E = h\nu$$

Since the wavelength and frequency are related to the speed of the light waves by

$$c = \nu\lambda$$

Equation 26.4 becomes

$$\begin{aligned} E &= \frac{hc}{\lambda} \\ &= \frac{(6.626 \times 10^{-34}\ \text{J}\cdot\text{s})(3.00 \times 10^{8}\ \text{m/s})}{632.8 \times 10^{-9}\ \text{m}} \\ &= 3.14 \times 10^{-19}\ \text{J} \end{aligned}$$

To convert to electron-volts, divide by 1.602×10^{-19} J/eV; you obtain

$$E = 1.96\ \text{eV}$$

Photons of visible light have energies on the order of several electron-volts.

b. Since the energy of each photon is quite small, we expect the number N of photons emitted by the laser each second to be quite large. The number is equal to the energy emitted by the laser each second (its power output in W = J/s) divided by the energy of each photon:

$$\begin{aligned} N &= \frac{1.00 \times 10^{-3}\ \text{W}}{3.14 \times 10^{-19}\ \text{J/photon}} \\ &= 3.18 \times 10^{15}\ \text{photons/s} \end{aligned}$$

This is quite a large number indeed!

EXAMPLE 26.3

Iron has a work function of 4.7 eV (see Table 26.1). Calculate the cutoff frequency and the corresponding cutoff wavelength for the photoelectric effect for this metal.

Solution

The energy of a photon with a frequency equal to the cutoff frequency ν_c is equal to the work function W of the metal (see Equation 26.8):

$$h\nu_c = W$$

The work function is 4.7 eV; converting this to joules, you obtain

$$(4.7\ \text{eV})(1.602 \times 10^{-19}\ \text{J/eV}) = 7.5 \times 10^{-19}\ \text{J}$$

Therefore Equation 26.8 becomes

$$h\nu_c = 7.5 \times 10^{-19}\ \text{J}$$

Solving for the cutoff frequency, you find

$$\begin{aligned} \nu_c &= \frac{7.5 \times 10^{-19}\ \text{J}}{h} \\ &= \frac{7.5 \times 10^{-19}\ \text{J}}{6.626 \times 10^{-34}\ \text{J}\cdot\text{s}} \\ &= 1.1 \times 10^{15}\ \text{Hz} \end{aligned}$$

The photoelectric effect from iron will occur only with incident light that has a frequency *greater* than 1.1×10^{15} Hz. The wavelength corresponding to this frequency is

$$c = \nu_c\lambda_c$$

or

$$\begin{aligned} \lambda_c &= \frac{c}{\nu_c} \\ &= \frac{3.00 \times 10^{8}\ \text{m/s}}{1.1 \times 10^{15}\ \text{Hz}} \\ &= 2.7 \times 10^{-7}\ \text{m} \\ &= 2.7 \times 10^{2}\ \text{nm} \end{aligned}$$

This is well into the ultraviolet region of the electromagnetic spectrum. The photoelectric effect will occur for iron only if the wavelength of the incident light is *less* than 2.7×10^2 nm; visible light will not produce a photoelectric effect in iron.

STRATEGIC EXAMPLE 26.4

Ultraviolet light of wavelength 200 nm is incident on a freshly polished iron surface. Find:

a. the stopping potential;
b. the maximum kinetic energy of the liberated electrons; and
c. the speed of these fastest electrons.

Solution

The incident photon has an energy of

$$E = h\nu = \frac{hc}{\lambda} = \frac{(6.626 \times 10^{-34}\ \text{J}\cdot\text{s})(3.00 \times 10^{8}\ \text{m/s})}{200 \times 10^{-9}\ \text{m}} = 9.94 \times 10^{-19}\ \text{J} = \frac{9.94 \times 10^{-19}\ \text{J}}{1.602 \times 10^{-19}\ \text{J/eV}} = 6.20\ \text{eV}$$

a. There are two ways to find the stopping potential.

Method 1

Use Equation 26.11 for the photoelectric effect using energy units of joules for each term. The work function W, expressed in joules, is

$$W = (4.7\ \text{eV})(1.602 \times 10^{-19}\ \text{J/eV}) = 7.5 \times 10^{-19}\ \text{J}$$

Thus Equation 26.11 for the photoelectric effect equation becomes

$$h\nu = W + eV_s$$
$$9.94 \times 10^{-19}\ \text{J} = 7.5 \times 10^{-19}\ \text{J} + eV_s$$

Solving for eV_s, you obtain

$$eV_s = 2.4 \times 10^{-19}\ \text{J}$$

The stopping potential V_s then is

$$V_s = \frac{2.4 \times 10^{-19}\ \text{J}}{1.602 \times 10^{-19}\ \text{C}} = 1.5\ \text{V}$$

Method 2

The problem also can be solved using electron-volt units for the energy in Equation 26.11 for the photoelectric effect. The photon energy is 6.20 eV; from Equation 26.4, $E_{photon} = h\nu$:

$$h\nu = W + eV_s$$
$$6.20\ \text{eV} = 4.7\ \text{eV} + eV_s$$

The product eV_s, in electron-volts, is

$$eV_s = 1.5\ \text{eV}$$

This can be converted to joules:

$$eV_s = (1.5\ \text{eV})(1.602 \times 10^{-19}\ \text{J/eV}) = 2.4 \times 10^{-19}\ \text{J}$$

Then solve for the stopping potential:

$$V_s = \frac{2.4 \times 10^{-19}\ \text{J}}{1.602 \times 10^{-19}\ \text{C}} = 1.5\ \text{V}$$

Note that in accordance with Problem-Solving Tactic 26.2, the maximum kinetic energy, expressed in electron-volts, is *numerically* equal to the stopping potential in volts.

b. The maximum kinetic energy of the photoelectrons is eV_s:

$$eV_s = (1.602 \times 10^{-19}\ \text{C})(1.5\ \text{V}) = 2.4 \times 10^{-19}\ \text{J}$$

c. Use the nonrelativistic form of the kinetic energy. Solve for the square of the speed and substitute for the mass of the electron:

$$\frac{mv^2}{2} = 2.4 \times 10^{-19}\ \text{J}$$

So

$$v^2 = \frac{2(2.4 \times 10^{-19})\ \text{J}}{9.11 \times 10^{-31}\ \text{kg}} = 5.3 \times 10^{11}\ \text{m}^2/\text{s}^2$$

Hence the speed is

$$v = 7.3 \times 10^{5}\ \text{m/s}$$

Notice that $v/c = 2.4 \times 10^{-3}$, so that the relativistic factor γ is very close to 1, which justifies using the nonrelativistic expression for the kinetic energy of the liberated electrons.

26.6 THE QUEST FOR AN ATOMIC MODEL: PLUM PUDDING

When diffraction gratings were invented by Henry Rowland (cf. Section 24.9) in the 19th century, they were applied vigorously to the study of light emitted in gas discharges. As we have seen, the study of such discharges led to the discovery of electrons, x-rays, and the photoelectric effect. It was quickly determined that each element emitted a characteristic set of optical wavelengths of light, which could be used to detect the presence of the various elements in light sources such as distant stars.

However, the mechanism for how multiple wavelengths could be emitted simultaneously by a sample of atoms remained totally unknown. Hydrogen, known to be the least massive atom and, therefore, likely the simplest in structure, emits many different characteristic wavelengths. How is light produced in an atom that is supposedly indivisible?* The discovery of the electron and of radioactivity shattered the myth of an indivisible atom, and the search for an acceptable atomic model began in

*Recall that the word atom stems from the Greek word ἄτομος, meaning "indivisible."

earnest.* Certainly one test of any such model would be its ability to predict the spectrum of electromagnetic wavelengths emitted by atoms. The α radiation of radium provided the tool for exploring the structure of atoms.

Since negatively charged particles (electrons) came out of atoms, yet atoms were electrically neutral, the first model of the atom [proposed by J. J. Thomson (1856–1940)] imagined the electrons as discrete entities randomly and uniformly distributed throughout a homogeneous spherical glob of positively charged material (see Figure 26.11). The model came to be known picturesquely as the **plum pudding model**, after the English culinary delicacy.

The experiments of Ernest Rutherford (1871–1937), Hans Geiger (1882–1945), and Ernest Marsden (1889–1970) between 1909 and 1913 concerning the scattering of a beam of α particles by thin gold foils† were the first convincing evidence that the plum pudding model was wrong. The collision of a high-speed, positively charged α particle with a plum pudding model atom was expected to be undramatic. As the α particle plowed through the pudding, it would just as likely be attracted one way as another, since the positive electrical charge within the atom was assumed to be distributed uniformly throughout the pudding and the negative charges were assumed to be at random locations. The deflection or scattering of the α particles was expected to be slight.

The experiments indicated that most α particles indeed *were* scattered through very small angles; a hasty experimentalist would call this proof that the model is correct. However, Rutherford was not hasty with this pudding. He noticed that a very small number of the α particles (estimated to be about one in ten thousand) were scattered through quite a large angle, even in the backward direction. Rather than dismiss these rare events as "bad data" (always a temptation to be resisted), Rutherford took the problem of these few strays very seriously. He described the problem of the backward-scattered α particles quaintly:

> It was quite the most incredible event that has happened to me in my life. It was almost as if you fired a 15-inch [diameter artillery] shell at a piece of tissue paper and it came back and hit you. On consideration I realized that this scattering backwards must be the result of a single collision, and when I made calculations I saw that it was impossible to get anything of that order of magnitude unless you took a system in which the greater part of the mass of the atom was concentrated in a minute nucleus. It was then that I had the idea of an atom with a minute massive center carrying a charge.‡

The annoying data burned the pudding into oblivion and gave us the nuclear model of the atom.

Since the kinetic energy of the α particles was known, Rutherford could calculate the distance of closest approach of the α particle to a relatively massive and motionless nucleus by using our good friend, the CWE theorem. Far from the pointlike nucleus, the α particle has a known kinetic energy, and its electrical potential energy at great distances is zero since the atom is electrically neutral. Hence the initial total mechanical energy of the α particle is $(\text{KE} + \text{PE})_i = \text{KE}_i + 0\ \text{J}$. For a head-on trajectory, the α particle has zero kinetic energy when closest to the nucleus and, at that location, the electrical potential energy of the α particle is

$$\text{PE}_f = qV = \frac{1}{4\pi\varepsilon_0}\frac{Qq}{r}$$

where q is the charge on the α particle and V is the electric potential of the nucleus of charge Q at the distance r of closest approach. From the CWE theorem, then, we have

$$\text{KE}_i = \frac{1}{4\pi\varepsilon_0}\frac{Qq}{r}$$

From this analysis, Rutherford estimated the diameter of the nucleus to be only about 10^{-15} m. The diameter of an atom, however, was known to be ~10^{-10} m, about 100 000 times larger. Thus most of an atom is empty space.

*The nuclear model of the atom is not as obvious as it may seem to you. In fact, the only reason it may seem obvious is that ever since you first encountered the idea back in grade school, you have been *told* that it is the way things are. The nuclear model has been drilled into you since infancy. But how would you convince a nonbeliever?

†Gold foils were used because the metal is very malleable, capable of being rolled into extraordinarily thin foils that are quite transparent to visible light!

‡Transcript of a lecture by Rutherford, "Forty years of physics," in *Background to Modern Science*, edited by Joseph Needham and Walter Pagel (Macmillan, New York, 1938), pages 68–69.

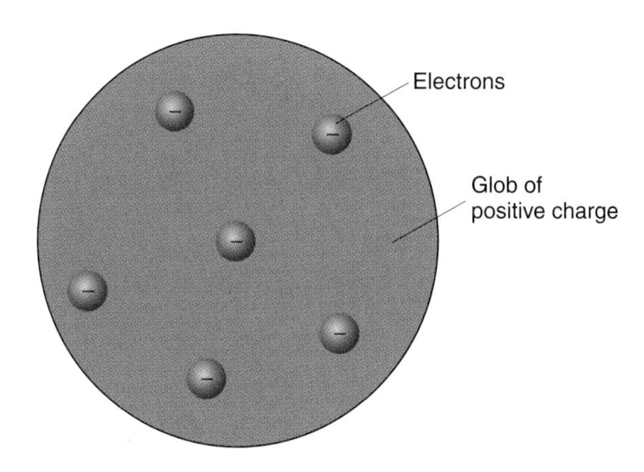

FIGURE 26.11 The plum pudding model of the atom.

The scattering of charged particles by nuclei, now called Rutherford scattering, is celebrated on this Soviet era postage stamp.

The crude pictures typically drawn to represent an atom grossly distort the relative sizes of the nucleus relative to the size of the atom. With a nucleus scaled up and modeled as a sphere a mere 1 mm in diameter, the size of an atom is 100 000 times larger, or 100 m!

QUESTION 3

Explain why it was tempting for Rutherford to dismiss the few wayward, backscattered α particles as bad data. Have *you* ever been tempted to disregard an observation as bad data?

Niels Bohr was smuggled out of Nazi-occupied Denmark during World War II by the Allies. Before leaving, he dissolved his gold Nobel physics prize medal in acid and hid it in an inconspicuous cabinet at his institute in Copenhagen. After the war, the Nobel Institute kindly recast his medal using the very same gold atoms!

26.7 The Bohr Model of a Hydrogenic Atom

There were serious unresolved questions about how atoms emit light. Most light originates from atoms.* The visible spectra of atoms were studied extensively in the late 19th century with the advent of the diffraction gratings of Henry Rowland. Each element produces its own unique spectrum of light, specific wavelengths characteristic of that element or isotope. The least massive element, hydrogen, has a spectrum that is yet quite complex: many different wavelengths are emitted. Indeed, from our modern perspective, how could an atom as simple as hydrogen, consisting of a single electron and a single proton, produce a complex spectrum of specific wavelengths?

Between 1913 and 1915 Niels Bohr developed a quantitative atomic model for the hydrogen atom that could account for its spectrum. The model incorporated the nuclear model of the atom proposed by Rutherford and his collaborators, as well as the concept of light photons developed by Einstein to explain the photoelectric effect.

The **Bohr model** was developed specifically for **hydrogenic atoms**: atoms consisting of a nucleus with positive charge $+Ze$ (where Z is the atomic number, the number of protons in the nucleus) and a *single* electron. Thus more complex electron–electron interactions are nonexistent for hydrogenic atoms.

The model is appropriate for hydrogen, singly ionized helium, doubly ionized lithium, and so forth. We shall see that this model was successful in its ability to predict the gross features of the spectra emitted by such hydrogenic atoms.

However, the Bohr model is not a true picture of even these simple atoms. The true picture is a fully quantum mechanical affair that differs from the Bohr model in a number of fundamental ways, which we touch on in Section 26.10. Since the Bohr model incorporates aspects of some classical and some modern physics, it is now called a semiclassical model.† If the Bohr model is not strictly correct, why, then, do we bother to present it here? For several reasons, other than tradition: (1) The model does account for the gross features of the hydrogen spectrum. (2) The model incorporates in a single problem many of the ideas you have learned from a study of introductory physics. (3) The model illustrates how a theoretical physicist occasionally must quite literally ignore certain problems of an approach in hopes of being able to make some predictions. If the predictions of the theory or model agree with experiment, a theoretician then must somehow hope to explain away or rationalize the problems that were ignored along the way. These considerations make the Bohr model still pedagogically fruitful.

The features of the Bohr model are the following.

The Coulomb Force and Newton's Second Law

First, the Bohr model incorporates and analyzes features of the Rutherford nuclear model. In the Rutherford nuclear model of the atom, the single electron in a hydrogenic atom and the nucleus exert electrical forces (the Coulomb electrostatic force) on each other, and both orbit the center of mass of the electron–nucleus system. However, since the mass of the nucleus is so much greater than the mass of the electron, the center of mass of the system is essentially coincident with the nucleus. Thus, to a first approximation, we consider the electron to be in an orbit about a fixed nucleus. The orbit of the electron is assumed to be circular for simplicity; thus, the acceleration of the electron is the centripetal acceleration and has a magnitude of

$$a_c = \frac{v^2}{r} \tag{26.13}$$

where v is the speed of the electron and r is the radius of its circular orbit. The force producing the acceleration of the electron is the Coulomb force exerted by the nuclear charge on the electron. This force has a magnitude of

$$\begin{aligned} F_{elec} &= \frac{1}{4\pi\varepsilon_0}\frac{|q||Q|}{r^2} \\ &= \frac{1}{4\pi\varepsilon_0}\frac{e(Ze)}{r^2} \end{aligned} \tag{26.14}$$

Now we apply Newton's second law to the motion of the electron:

$$\vec{F} = m\vec{a}$$

*Light also is emitted (a) when charged particles are accelerated (a prediction of the Maxwell theory of electromagnetism), which accounts for the continuous spectrum of x-rays when electrons crash into solids; and (b) when matter and antimatter annihilate, producing γ-rays.

†The model also could be called a semimodern model, but semiclassical has a nicer ring to it.

where m is the mass of the electron. Since the force and acceleration are in the same direction, we take their magnitudes to obtain

$$\frac{1}{4\pi\varepsilon_0}\frac{e(Ze)}{r^2} = m\frac{v^2}{r} \tag{26.15}$$

The Total Energy

In discussing the second ingredient of the Bohr model, we assume that relativistic speeds are not involved, so we can use the classical expressions for the kinetic energy and ignore the constant rest energy of the particles. The total energy E of the electron thus consists of the sum of its kinetic energy and electrical potential energy:

$$E = \frac{mv^2}{2} + \text{PE}_{\text{elec}} \tag{26.16}$$

The electrical potential energy of the electron is found, as usual, in the following way. The pointlike nuclear charge $+Ze$ produces an electric potential V at each point in the surrounding space:

$$V = \frac{1}{4\pi\varepsilon_0}\frac{+Ze}{r}$$

The electron at distance r from the nuclear charge thus has an electrical potential energy given by

$$\begin{aligned}\text{PE}_{\text{elec}} &= qV \\ &= (-e)V \\ &= (-e)\frac{1}{4\pi\varepsilon_0}\frac{Ze}{r} \\ &= -\frac{1}{4\pi\varepsilon_0}\frac{Ze^2}{r}\end{aligned}$$

Thus the total energy E of the electron (Equation 26.16) becomes

$$E = \frac{mv^2}{2} - \frac{1}{4\pi\varepsilon_0}\frac{Ze^2}{r} \tag{26.17}$$

There is nothing fundamentally new in what we have done so far. We used classical electrostatics to determine the force (the Coulomb force), classical dynamics (Newton's second law) to analyze the motion, classical (nonrelativistic) kinetic energy, and once again classical electrostatics to determine the electrical potential energy of the electron.

Here is where Bohr had to turn a blind eye toward one aspect of classical physics and ignore a glaring inconsistency. Classical electrodynamics (the Maxwell theory of electromagnetism) predicts that an accelerated charge will radiate electromagnetic waves and energy as light. Indeed, electromagnetic theory predicts that a charge executing circular motion at frequency ν should radiate light with the same frequency. The electron in the Bohr model is accelerated because of its circular motion; therefore, according to classical electrodynamics, the electron should steadily lose energy by emitting light. From a purely classical viewpoint the electron would have, ages ago, spiraled down into the nucleus and thus obliterating the model.* Bohr was forced to postulate that the electron in such an orbit was in a *stable* state, and simply would not radiate energy in spite of classical electrodynamics. Bohr called this stable state a *stationary state*, but today we call them simply a **state** of the atom.

Yet Bohr used the Coulomb force law that is part and parcel of the classical theory of electromagnetism! He used one part of the theory of electromagnetism and threw out another part; it is tantamount to keeping one's theoretical cake and eating it too.

The Angular Momentum

Now we reach the third key feature of the Bohr model. Part of the genius of Bohr lay in his masterful understanding of the formal, theoretical aspects of mechanics. Bohr had reason to believe† that the magnitude of the orbital angular momentum of the electron was restricted to only certain values; we say the orbital angular momentum of the electron is **quantized**. He therefore took this as a second postulate of the model. In Section 26.15, we will see why his postulate was a reasonable one in retrospect, based on a discovery made almost ten years after he developed his model. For now, we must take the postulate on an ad hoc basis.

Bohr proposed that the magnitude of the orbital angular momentum of the electron L_{orbit} could not have just any value, but instead, only integral multiples of Planck's constant divided by 2π:‡

$$L_{\text{orbit}} = n\frac{h}{2\pi} \tag{26.18}$$

where n is a positive integer known as a **quantum number**. The quantity $h/2\pi$ occurs so frequently in quantum physics that, for convenience, it is given its own designation $\hbar$, pronounced "h-bar":

$$\hbar \equiv \frac{h}{2\pi} \approx 1.055 \times 10^{-34}\ \text{J}\cdot\text{s} \tag{26.19}$$

The Bohr orbital angular momentum postulate then becomes

$$L_{\text{orbit}} = n\hbar \tag{26.20}$$

The orbital angular momentum of a particle is defined to be (from Equation 10.1)

$$\vec{\mathbf{L}}_{\text{orbit}} = \vec{\mathbf{r}} \times \vec{\mathbf{p}}$$

---

*Of course, we want to be able to explain how light is emitted from the atom. Bohr realized that the light emitted from this spiraling into the nucleus has a *continuous* spectrum (analogous to white light), not the observed *discrete* spectrum with a few specific wavelengths. The lifetime of an atom with a circulating and radiating electron is extremely short ($\sim 10^{-11}$ s).

†The reasons are associated with a peculiar integral in the formal structure of mechanics known as the *action integral*. The details are beyond the level of this course.

‡The factor of 2π arises because there are 2π rad in the circle of its orbit, but this connection in mechanics is not one you would be expected to understand at this point in your studies.

In circular motion, the position vector $\vec{\mathbf{r}}$ of a particle is perpendicular to its momentum $\vec{\mathbf{p}}$, and so the magnitude of the orbital angular momentum is

$$\begin{aligned} L_{\text{orbit}} &= rp \sin 90° \\ &= rp \end{aligned}$$

The Bohr angular momentum postulate, Equation 26.20, then becomes

$$rp = n\hbar \qquad (26.21)$$

Thus the Bohr model has three ingredients:

1. an application of Newton's second law (Equation 26.15);
2. the expression for the total energy (Equation 26.17); and
3. the angular momentum postulate (Equation 26.21).

If we mix these ingredients appropriately, two results emerge.

First, we take Equation 26.21 for the magnitude of the angular momentum and substitute for the magnitude of the momentum $p = mv$; we obtain

$$rmv = n\hbar \qquad (26.22)$$

Solving for v, we get

$$v = \frac{n\hbar}{mr} \qquad (26.23)$$

We use Equation 26.23 to substitute for v in Equation 26.15 (Newton's second law); this yields

$$\frac{1}{4\pi\varepsilon_0}\frac{e(Ze)}{r^2} = \frac{m}{r}\left(\frac{n\hbar}{mr}\right)^2$$

Now we solve for the radius r:

$$r = \frac{n^2\hbar^2\,(4\pi\varepsilon_0)}{mZe^2} \qquad (26.24)$$

Since n is an integer, Equation 26.24 implies that the electron correspondingly orbits the nucleus only at specific radii (so-called **allowed orbits**).

Now for the second result. Equation 26.15,

$$\frac{1}{4\pi\varepsilon_0}\frac{e(Ze)}{r^2} = m\frac{v^2}{r}$$

can be simplified slightly by multiplying by the radius r; this yields

$$\frac{1}{4\pi\varepsilon_0}\frac{e(Ze)}{r} = mv^2 \qquad (26.25)$$

Using this expression for mv^2 in the energy equation (Equation 26.17), we find

$$\begin{aligned} E &= \frac{mv^2}{2} - \frac{1}{4\pi\varepsilon_0}\frac{Ze^2}{r} \\ &= \frac{1}{2}\frac{1}{4\pi\varepsilon_0}\frac{e(Ze)}{r} - \frac{1}{4\pi\varepsilon_0}\frac{Ze^2}{r} \\ &= -\frac{1}{2}\frac{1}{4\pi\varepsilon_0}\frac{Ze^2}{r} \end{aligned} \qquad (26.26)$$

Using Equation 26.24 to substitute for the radius of the orbit, we obtain

$$\begin{aligned} E &= -\frac{1}{2}\frac{1}{4\pi\varepsilon_0}\frac{Ze^2}{\dfrac{n^2\hbar^2\,(4\pi\varepsilon_0)}{mZe^2}} \\ &= -\frac{1}{2}\frac{mZ^2e^4}{(4\pi\varepsilon_0)^2 n^2\hbar^2} \end{aligned} \qquad (26.27)$$

Beyond all the several constants, we see that since n is an integer, the energy also is quantized.

Only certain values of the energy are permitted.

The rather ungainly expression for the energy (Equation 26.27) can be simplified if we substitute numerical values for the fundamental constants m (the mass of the electron), e (the fundamental unit of charge), ε_0 (the permittivity of free space), and $\hbar$ (Planck's constant divided by 2π). The result of these substitutions is

$$E = -(21.8 \times 10^{-19}\ \text{J})\frac{Z^2}{n^2} \qquad (26.28)$$

This can be converted to electron-volts by dividing by 1.602×10^{-19} J/eV, yielding

$$\boxed{E = -(13.6\ \text{eV})\frac{Z^2}{n^2}} \qquad (26.29)$$

For hydrogen, $Z = 1$, and we have

$$E_{\text{hydrog}} = -\frac{13.6}{n^2}\ \text{eV} \qquad (26.30)$$

Bohr's permitted values of the energy typically are represented graphically on an **energy level diagram**, shown in Figure 26.12. The vertical axis of the graph is an energy scale; there is no horizontal axis, but horizontal lines are drawn to show the appropriate levels (values) of the energy.

The lowest energy state (corresponding to $n = 1$ in Equations 26.29 and 26.30) is the **ground state**. Notice that as the quantum number n increases, the energy of the electron increases and approaches zero as a limit for $n = \infty$. As $n \to \infty$, the radius of the orbit also increases indefinitely according to Equation 26.24. Thus, when $E \geq 0$ J, the electron is no longer associated with the nucleus and is free; the atom has been **ionized**.

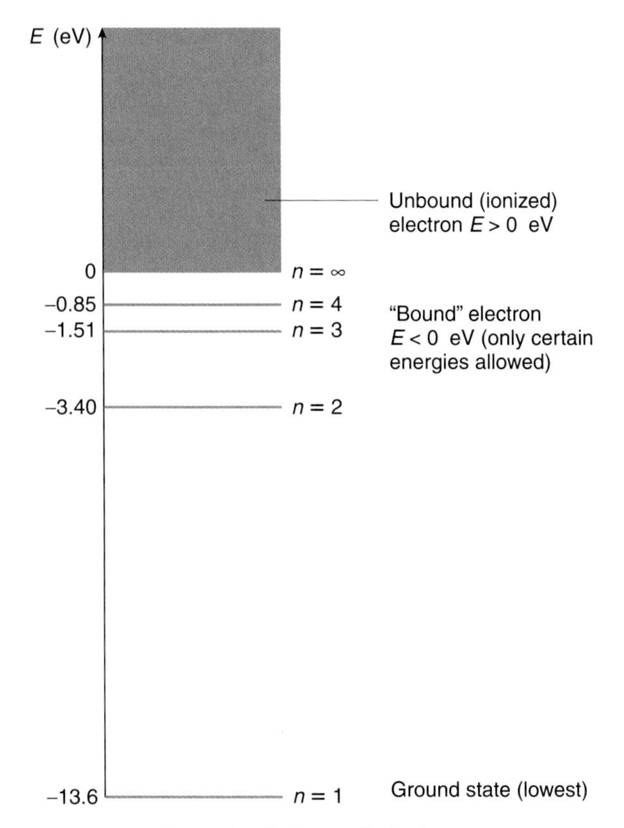

FIGURE 26.12 **Energy level diagram for hydrogen.**

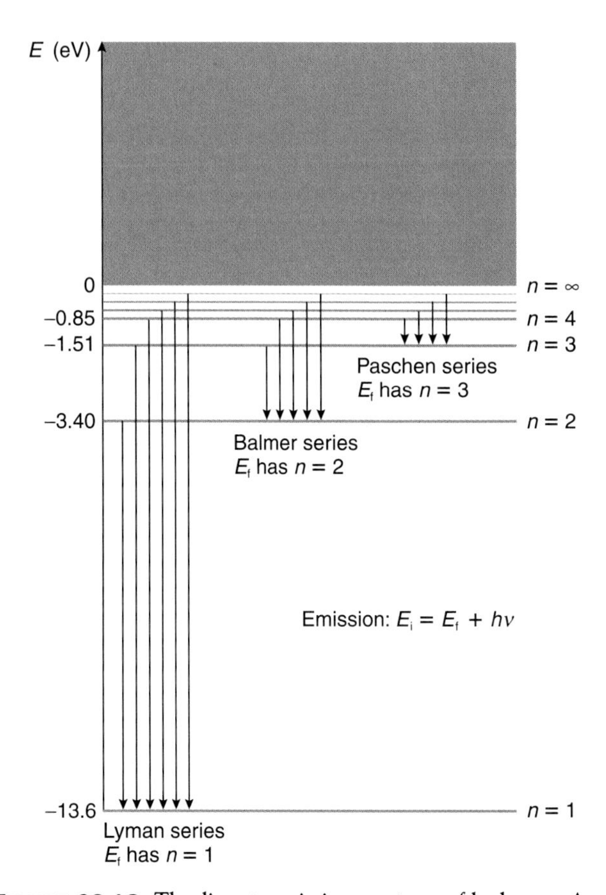

FIGURE 26.13 **The discrete emission spectrum of hydrogen. A photon is emitted when the electron makes a transition to an energy state with a lower value of the quantum number *n*.**

The neutral atom has energies $E < 0$ J, with the electron bound to the nucleus.

Recall that Bohr's objective was to explain the discrete spectrum of hydrogen, the fact that only specific wavelengths of light are emitted by a collection of hydrogen atoms. Since Bohr *postulated* that the electron would not radiate electromagnetic waves despite its acceleration, he had to invent a new mechanism for light emission. The energy of the electron in a particular orbit in the atom depends on the quantum number n; in particular, the higher n, the higher the energy.

Thus an electron with an $n > 1$ can lower its energy by changing to an orbit with a lower value of n, an orbit closer to the nucleus according to Equation 26.24. Such a jump is called a **transition.*** In this process, the electron lowers the amount of its energy (loses energy). Bohr postulated that the energy lost by the electron appears as the energy of a photon emitted by the atom.

In this way total energy is conserved. That is, if an electron is in an orbit with $n > 1$ with energy E_i, in making a transition to a lower energy E_f the atom conserves overall energy according to

$$E_i = E_f + E_{\text{emitted photon}} \quad \text{(emission)} \tag{26.31}$$

*This is the origin of the pedestrian phrase "quantum jumps."

The discrete emission spectrum of hydrogen therefore corresponds to electrons cascading down to lower energy levels in various atoms of hydrogen (see Figure 26.13).

Each transition of the electron to a lower energy level produces a photon with a specific energy or, equivalently, with a definite frequency or wavelength. All transitions to the deep $n = 1$ state emit light that is in the ultraviolet region of the spectrum; this series of transitions is called the **Lyman series**. The visible light spectrum of hydrogen involves transitions to the $n = 2$ state from $n > 2$ higher states, called the **Balmer series**. There are other series too, those ending on $n = 3$, $n = 4$, and so forth, but it is pointless to name them all since there are an infinite number of them.†

Correspondingly, if a photon of exactly the right energy is available, the electron in a lower state can totally absorb the photon (destroying it, but taking its energy) and make a transition to a higher energy level.

Energy again is conserved by writing:

$$E_i + E_{\text{photon absorbed}} = E_f \quad \text{(absorption)} \tag{26.32}$$

†Those transitions ending on $n = 3$ form the Paschen series; those on $n = 4$ are the Brackett series; those on $n = 5$ are the Pfund series, after which we gave up naming them.

This neatly explains the experimental observation that the wavelengths (or frequencies) that hydrogen emits are the same as the wavelengths that hydrogen absorbs. That is, if a white light spectrum (a continuous spectrum) is incident on cool hydrogen gas, the hydrogen will absorb just those photon wavelengths that (hot) hydrogen emits. So the transmitted light is missing certain wavelengths; the wavelengths missing are precisely the same ones that hydrogen emits.

The host of energy levels and the multitude of possible transitions thus explain the complexity of the discrete hydrogen spectrum. The crowning achievement of the Bohr model was its ability to predict the wavelengths of the spectrum of single-electron atoms (hydrogenic atoms).

Refinements subsequently were made to the Bohr model, allowing for the possibility of elliptical electron orbits, much like the elliptical orbits of the planets of the Sun. Corrections made for the finite mass of the nucleus, by treating both the electron and nucleus in orbit about their center of mass, led to small corrections in the predicted spectrum and eventually to the discovery of **deuterium**, an isotope of hydrogen with twice the nuclear mass of ordinary hydrogen.

QUESTION 4

Why is the total energy in the Bohr model of a hydrogenic atom negative?

EXAMPLE 26.5

a. Calculate the orbital radius of the electron orbit closest to the nucleus of the hydrogen atom. This distance is called the *Bohr radius* a_0 of the hydrogen atom.

b. Express the other radii of a hydrogenic atom in terms of the Bohr radius. Use this expression to find the radius of the orbit closest to the nucleus of singly ionized helium.

Solution

a. The orbital radius is given by Equation 26.24:

$$r = \frac{n^2\hbar^2\,(4\pi\varepsilon_0)}{mZe^2}$$

For hydrogen, $Z = 1$. The orbit closest to the nucleus has the smallest n-value: $n = 1$. Hence Equation 26.24 becomes, for the Bohr radius a_0,

$$a_0 = \frac{\hbar^2(4\pi\varepsilon_0)}{me^2} \qquad (1)$$

The mass m is that of the electron. Substituting numerical values for the various quantities, you get

$$\begin{aligned} a_0 &= \frac{(1.055 \times 10^{-34}\ \text{J}\cdot\text{s})^2}{(9.11 \times 10^{-31}\ \text{kg})(1.602 \times 10^{-19}\ \text{C})^2(9.00 \times 10^9\ \text{N}\cdot\text{m}^2/\text{C}^2)} \\ &= 5.29 \times 10^{-11}\ \text{m} \end{aligned}$$

b. To express the other radii of hydrogenic atoms in terms of the Bohr radius a_0, use equation (1) in Equation 26.24:

$$\begin{aligned} r &= \frac{n^2}{Z}\,\frac{\hbar^2 4\pi\varepsilon_0}{me^2} \\ &= \frac{n^2}{Z}\,a_0 \qquad (2) \end{aligned}$$

Singly ionized helium is a hydrogenic atom (a one-electron atom). Helium has $Z = 2$ since there are two protons in its nucleus. The orbit closest to the nucleus has $n = 1$. Hence for helium you have

$$\begin{aligned} r &= \frac{a_0}{2} \\ &= 2.65 \times 10^{-11}\ \text{m} \end{aligned}$$

which is half the Bohr radius of the hydrogen atom.

EXAMPLE 26.6

An electron in the hydrogen atom makes a transition from the $n = 5$ orbit to the $n = 2$ orbit. Calculate the wavelength of the photon emitted by the atom.

Solution

It is easier to use Equation 26.30 for the energy of each state of the hydrogen atom, rather than the more ungainly Equation 26.27. The energy of the $n = 5$ state thus is

$$\begin{aligned} E_5 &= -\frac{13.6}{(5)^2}\ \text{eV} \\ &= -0.544\ \text{eV} \end{aligned}$$

while that of the $n = 2$ state is

$$\begin{aligned} E_2 &= -\frac{13.6}{(2)^2}\ \text{eV} \\ &= -3.40\ \text{eV} \end{aligned}$$

Since the total energy is conserved, you have

$$\begin{aligned} E_{\text{before}} &= E_{\text{after}} \\ E_5 &= E_2 + E_{\text{photon}} \\ -0.544\ \text{eV} &= -3.40\ \text{eV} + h\nu \\ h\nu &= 2.86\ \text{eV} \\ &= (2.86\ \text{eV})(1.602 \times 10^{-19}\ \text{J/eV}) \\ &= 4.58 \times 10^{-19}\ \text{J} \end{aligned}$$

The wavelength λ is found recalling that $c = \nu\lambda$. Hence you have

$$\frac{hc}{\lambda} = 4.58 \times 10^{-19}\ \text{J}$$

Solving for the wavelength, you find

$$\begin{aligned} \lambda &= \frac{hc}{4.58 \times 10^{-19}\ \text{J}} \\ &= \frac{(6.626 \times 10^{-34}\ \text{J}\cdot\text{s})(3.00 \times 10^8\ \text{m/s})}{4.58 \times 10^{-19}\ \text{J}} \\ &= 4.34 \times 10^{-7}\ \text{m} \\ &= 434\ \text{nm} \end{aligned}$$

This is a visible light wavelength; you can see the light as a violet-colored line in the spectrum.

26.8 The Bohr Correspondence Principle

Recall that Bohr swept a crucial prediction of classical electromagnetic theory under the rug, namely, that a charge undergoing circular motion (accelerated motion) should radiate light with a frequency equal to that of its orbital motion. Bohr devoted much thought and effort to the problem of how his new ideas about light emission might be compatible with or correspond to the classical predictions in at least some appropriate limiting situation.

To see how he made the correspondence, it is first necessary to find the orbital frequency of the electron in a hydrogenic atom. The electron travels the circumference $2\pi r$ at the speed v, and so the period T of its motion is

$$T = \frac{2\pi r}{v}$$

The frequency of its orbital motion ν_{orbit} is the reciprocal of its period, so that we have

$$\nu_{\text{orbit}} = \frac{v}{2\pi r} \tag{26.33}$$

The angular momentum postulate,

$$L = pr = mvr = n\hbar$$

indicates that

$$v = \frac{n\hbar}{mr}$$

Making this substitution for v into Equation 26.33 for ν_{orbit}, we find

$$\nu_{\text{orbit}} = \frac{1}{2\pi r}\frac{n\hbar}{mr} = \frac{n\hbar}{2\pi m r^2}$$

If we now substitute for the radius using Equation 26.24, we obtain

$$\nu_{\text{orbit}} = \frac{n\hbar}{2\pi m \dfrac{n^4\hbar^4(4\pi\varepsilon_0)^2}{m^2Z^2e^4}} = \frac{1}{2\pi}\frac{mZ^2e^4}{(4\pi\varepsilon_0)^2}\frac{1}{\hbar^3 n^3} \tag{26.34}$$

For small quantum numbers n this orbital frequency is *not* the frequency of any of the wavelengths emitted by such a hydrogenic atom. But Bohr wondered what happened to the frequency of light emitted by the hydrogen atom in making a transition from the n to the $n - 1$ state, at large values of the quantum number n. For such a transition from n to $n - 1$ (see Figure 26.14) a photon is emitted with an energy that satisfies Equation 26.31:

$$E_i = E_f + E_{\text{emitted photon}}$$

We use the expression for the energy levels of such an atom (Equation 26.27) in Equation 26.31 to find

$$-\frac{1}{2}\frac{mZ^2e^4}{(4\pi\varepsilon_0)^2\hbar^2n^2} = -\frac{1}{2}\frac{mZ^2e^4}{(4\pi\varepsilon_0)^2\hbar^2(n-1)^2} + E_{\text{emitted photon}}$$

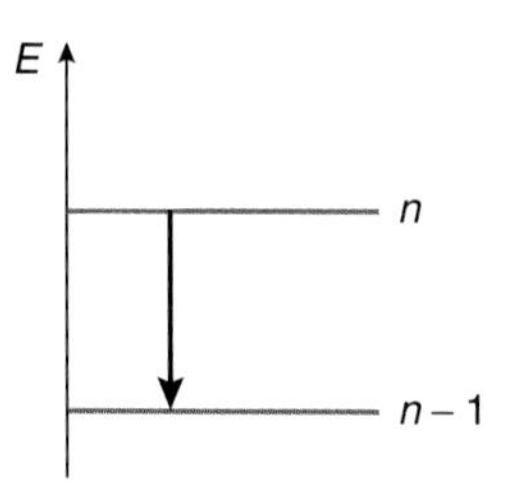

FIGURE 26.14 The transition from a state with quantum number n to one with quantum number $n - 1$.

Rearranging, we have

$$E_{\text{emitted photon}} = \frac{1}{2}\frac{mZ^2e^4}{(4\pi\varepsilon_0)^2\hbar^2}\left[\frac{1}{(n-1)^2} - \frac{1}{n^2}\right] \tag{26.35}$$

The expression in the brackets on the right-hand side of Equation 26.35 can be combined into a single term:

$$\frac{1}{(n-1)^2} - \frac{1}{n^2} = \frac{n^2 - (n-1)^2}{n^2(n-1)^2} = \frac{2n-1}{n^4 - 2n^3 + n^2} \tag{26.36}$$

We take the limit of this expression as n becomes very large. Then the number 1 can be neglected in the numerator of the right-hand side of Equation 26.36, and all but the first term in the denominator also can be neglected,* so that

$$\frac{1}{(n-1)^2} - \frac{1}{n^2} \approx \frac{2n}{n^4} \approx \frac{2}{n^3}$$

Thus, for large n, the energy of the light emitted in this transition is, from Equation 26.35,

$$E_{\text{emitted photon}} = \frac{1}{2}\frac{mZ^2e^4}{(4\pi\varepsilon_0)^2\hbar^2}\frac{2}{n^3}$$

The frequency of this photon is found from Equation 26.4:

$$E_{\text{photon}} = h\nu$$

Hence the emitted photon has a frequency of

$$\nu_{\text{photon}} = \frac{1}{h}\frac{1}{2}\frac{mZ^2e^4}{(4\pi\varepsilon_0)^2\hbar^2}\frac{2}{n^3}$$

*If $n = 1000$, then the difference between

$$\frac{2n-1}{n^4 - 2n^3 + n^2}$$

and

$$\frac{2n}{n^4}$$

is less than 0.15% (which can be verified by direct substitution into the two expressions).

Converting the h to an $\hbar$ by multiplying and dividing by 2π, we obtain

$$\nu_{\text{photon}} = \frac{1}{2\pi}\frac{mZ^2e^4}{(4\pi\varepsilon_0)^2}\frac{1}{\hbar^3 n^3} \tag{26.37}$$

This frequency is identical to the frequency of the orbital motion of the electron given by Equation 26.34.

Thus Bohr discovered that for large quantum numbers n, the light emitted when the electron jumps to the next lower orbit has a frequency that is the same as the classical prediction of electromagnetic theory. The correspondence between classical physics and what happens with large quantum numbers is called **Bohr's correspondence principle.**

The principle is analogous to the situation encountered in relativity: in the limit of speeds small compared with the speed of light, the Lorentz transformation equations reduce to the classical Galilean relativity equations.

26.9 A Bohr Model of the Solar System?*

The motion of the electron around the nucleus of the hydrogen atom in the Bohr model is imagined to be similar to the motion of a planet around the Sun: a planetary atom. What happens if we apply the Bohr model to the two-body gravitational interaction between a planet and the Sun? The only real change is the substitution of the gravitational force for the electrical force. Let's formulate a Bohr model for the Earth in its orbit around the Sun and ask two questions:

1. What is the quantum number for the orbit of the Earth?
2. What is the distance between adjacent permitted orbits for the Earth? Is the next allowed orbit beyond the orbit of the Earth where Mars is located?

We formulate the problem just as we did for the Bohr model of a hydrogenic atom. First, we apply Newton's second law to the motion of the Earth, assuming (to a first approximation) that the orbit of the Earth is a circle of radius r:

$$\vec{\mathbf{F}} = m\vec{\mathbf{a}}$$

The total force is the gravitational force on the Earth by the Sun. Since the force is in the same direction as the centripetal acceleration, we use the magnitudes of the vectors:

$$\frac{GMm}{r^2} = m\frac{v^2}{r} \tag{26.38}$$

where M is the mass of the Sun and m is the mass of the Earth.

The second step is to apply Bohr's angular momentum postulate. This postulate states that the magnitude of the angular momentum is quantized in integral multiples of $\hbar$:

$$L = pr = mvr = n\hbar \tag{26.39}$$

where n is the quantum number of the orbit of the Earth. We solve Equation 26.39 for v:

$$v = \frac{n\hbar}{mr}$$

Substituting for v in the force law expression, Equation 26.38, we find

$$\frac{GMm}{r^2} = \frac{m}{r}\left(\frac{n\hbar}{mr}\right)^2$$

Now we solve for r, obtaining

$$r = \frac{n^2\hbar^2}{GMm^2} \tag{26.40}$$

To find the quantum number n for the orbit of the Earth, we solve Equation 26.40 for n:

$$n^2 = \frac{GMm^2 r}{\hbar^2}$$

We make the following numerical substitutions:

$$\begin{aligned} G &= 6.67 \times 10^{-11}\ \text{N}\cdot\text{m}^2/\text{kg}^2 \\ M_{\text{Sun}} &= 1.99 \times 10^{30}\ \text{kg} \\ m_{\text{Earth}} &= 5.98 \times 10^{24}\ \text{kg} \\ r_{\text{Earth orbit}} &= 1.496 \times 10^{11}\ \text{m} \\ \hbar &= 1.055 \times 10^{-34}\ \text{J}\cdot\text{s} \end{aligned}$$

We finally find that

$$n = 2.52 \times 10^{74} \tag{26.41}$$

The quantum number of the Earth is *quite* large!

Let Δr be the distance between the present orbit of the Earth, with quantum number n, and the next orbit out from the Sun, corresponding to quantum number $n + 1$. From Equation 26.40, we find

$$\begin{aligned} \Delta r &= r_{n+1} - r_n \\ &= \frac{\hbar^2}{GMm^2}\left[(n+1)^2 - n^2\right] \end{aligned} \tag{26.42}$$

Since n is so large, we make the following approximation:

$$\begin{aligned} (n+1)^2 - n^2 &= n^2 + 2n + 1 - n^2 \\ &= 2n + 1 \\ &\approx 2n \end{aligned}$$

Equation 26.42 for Δr then becomes

$$\Delta r \approx \frac{\hbar^2}{GMm^2}2n \tag{26.43}$$

Now make the appropriate substitutions and use $n = 2.52 \times 10^{74}$. From Equation 26.43 we find that

$$\Delta r \approx 1.2 \times 10^{-63}\ \text{m}$$

This is ~48 *orders of magnitude* smaller than the diameter of a typical nucleus (~ 10^{-15} m)! The distance between the quantized orbits of the Earth is *so* tiny as to be essentially zero; the planet has a continuum of orbits in which to orbit.

The Bohr model thus is irrelevant for the celestial mechanics of the solar system, and the spacing of the planets in the solar system has nothing to do with the quantization of angular momentum.

26.10 Problems with the Bohr Model

Alas, the Bohr model with its planet-like electron must be discarded. While the model correctly predicts the gross features of the spectrum of hydrogenic atoms, in particular, the frequencies of the light emitted or selectively absorbed, the model was just a way station to a completely new mechanics, quantum mechanics, and an improved but stranger model of the atom. The limitations of the Bohr model include the following.

The Bohr model is applicable only to hydrogenic (single-electron) atoms. It cannot be easily extended, even to mere two-electron atoms such as helium.

Unlike the situation in the solar system, where planet–planet gravitational forces are very small compared with the gravitational force of the Sun on each planet (because the mass of the Sun is so much greater than the mass of any of the planets), the electron–electron electrical force interaction is comparable in magnitude to the electron–nucleus electrical force, because the charges and distances are of the same order of magnitude.

While the Bohr model correctly predicts the frequencies of the light emitted by hydrogenic atoms, the model says nothing about the relative intensities of the frequencies in the spectrum.

Some of the visible wavelengths have weak intensity, others strong. The question is, of course, why? The experimental fact that the relative intensities of the various frequencies are not the same actually is a relief, since there are infinitely many of them, but it indicates that some transitions are more favored than others. On what basis is this assessment made by the electron? The Bohr model is unable to account for the intensity variations.

The angular momentum postulate of the Bohr model, Equation 26.20, $L = n\hbar$, predicts that the magnitude of the orbital angular momentum of the ground state, corresponding to $n = 1$, is $L_{\text{ground state}} = \hbar$.

In retrospect, after the advent of a quantum mechanical model of the atom in the late 1920s, it was realized that the ground state has an orbital angular momentum equal to *zero*. This called into question the whole conceptual picture of an orbiting planet-like electron and placed the fundamental premise of the Bohr model in jeopardy.

These problems indicated that the model is not a correct rendition of even the simplest atom, hydrogen. Thus, while the Bohr model has a picturesque simplicity and is a model that can be readily visualized, the model has fundamental flaws and presents an incorrect picture of the atom.

26.11 Radioactivity Revisited

The nuclear model of the atom led to an increased understanding of radioactivity, discovered in 1896 by Becquerel, as we saw in Section 26.3. The understanding of radioactivity that emerged was the realization of the age-old quest of ancient and medieval alchemists: the **transmutation** of the elements (the transformation of one chemical element into another). Such transmutations are impossible via chemical techniques, which leave the nucleus quite unaffected.

Early experiments involving radioactive materials indicated that the particle emission rate (the number of particles per unit time) emitted from a radioactive substance

a. is independent of temperature and pressure;
b. is not affected by the presence of electric and/or magnetic fields; and
c. is independent of how the element is chemically bound.

The emission rate depends *solely* on the *concentration* of the radioactive isotope, that is, on the number of atoms of the isotope present. These surprising observations, together with the extraordinary energy of the emitted particles (compared with the energies typically associated with chemical reactions) led to the conclusion that the particles emitted from a radioactive material emerge from the nuclei of the atoms involved.

Atoms are radioactive if their nuclei are unstable and spontaneously (and randomly) emit various particles, the α, β, and/or γ radiations. When naturally occurring nuclei are unstable, we call the phenomena **natural radioactivity**. Other nuclei can be transformed into radioactive nuclei by various means, typically involving irradiation by neutrons; this is called **artificial radioactivity**.

We saw in Section 26.3 that α particles were found to be helium nuclei (two protons and two neutrons), since helium always was present in materials that emitted α particles. Magnetic and electric field experiments with β particles led to the conclusion that they were electrons (called β^- particles). During the 1930s, a particle with the same mass as an electron but with a *positive* charge, $+e$, was discovered in cosmic rays and later in the emission of certain radioactive isotopes. These **positrons** are called β^+ particles. A positron is called the **antiparticle** of the electron.

Gamma particles (γ-rays), unaffected by both electric and magnetic fields, finally were shown to be high-energy photons. Gamma rays have frequencies larger than x-rays, and correspondingly shorter wavelengths; the distinction between γ-rays and x-rays is somewhat arbitrary. X-rays typically result from certain transitions of the electrons surrounding the nucleus in the atom; γ-rays generally are emitted from the nucleus itself.

A radioactive nucleus is called a **parent** nucleus; the nucleus resulting from its decay by particle emission is called the **daughter** nucleus. Daughter nuclei also might be radioactive, producing granddaughter nuclei, and so on. There are no son or grandson nuclei.*

For any nucleus, the atomic number Z is the number of protons in the nucleus, typically written as a numerical subscript together with the chemical symbol for the element. The mass number A is the sum of the number of protons and neutrons in

*While the term parent nucleus conforms to contemporary, politically correct trends for nonsexist language, the term daughter nuclei does not. The term "progeny" nuclei just doesn't hack it.

the nucleus, typically written as a numerical superscript along with the chemical symbol for the element. Thus, for example, we write $^{235}_{92}U$ for the isotope of uranium (atomic number = 92) of mass number 235.

α Decay

The emission of an α particle (a helium nucleus, $^{4}_{2}He$), a process known as **α decay**, means that the mass number of the parent nucleus decreases by 4 and the number of protons by 2. The resulting daughter nucleus thus has a mass number of A − 4 and an atomic number of Z − 2. An example of this is the decay of radium into radon:

$$^{226}_{88}Ra \rightarrow \; ^{222}_{86}Rn \; + \; ^{4}_{2}He$$

parent nucleus → daughter nucleus + α particle

The α particles from a given decay are monoenergetic, that is, all emerge with the same energy.

β Decay

The emission of an electron (β^- particle) in radioactive decay is not the emission of an orbiting electron; the β^- particle is ejected from the nucleus. To conserve electrical charge, we imagine a neutron turning into a proton (which remains in the nucleus) with the ejection of an electron. Thus a parent nucleus with atomic number Z and mass number A decays by β^- emission into a daughter with atomic number Z + 1 and the same mass number A.

Experiments indicated that the energies of the emitted electrons from a given nuclear species have a continuous spectrum of energies. To conserve energy and momentum, it was necessary to hypothesize that another particle was emitted with each β^- decay: an electron antineutrino, designated $\bar{\nu}$. The hypothetical particle was not easily detected, but its existence finally was confirmed in the 1950s (the 1995 Nobel prize in physics was awarded to Frederick Reines for the discovery). Such antineutrinos interact extremely weakly with matter and are extraordinarily difficult to detect. They have no charge, and little (if any) mass; the question of their mass still is hotly debated and the subject of much experimental and theoretical inquiry. The consensus now is that the mass is nonzero, but very small. An example of β^- decay, antineutrino and all, is the decay of carbon 14 into nitrogen:

$$^{14}_{6}C \rightarrow \; ^{14}_{7}N + \beta^- + \bar{\nu}$$

Positron (β^+) emission from a nucleus decreases the atomic number Z by 1 while keeping the same mass number A. The process can be imagined as the conversion of a proton to a neutron within the nucleus, with the positron carrying off the positive charge. Positron emission always is accompanied by the emission of an electron neutrino ν that interacts extremely weakly with matter and has properties similar to the antineutrino of β^- decay (the difference between an electron neutrino and an electron antineutrino has to do with another fundamental property of particles known as their spin). An example of positron emission is the decay of another isotope of carbon into boron:

$$^{11}_{6}C \rightarrow \; ^{11}_{5}B + \beta^+ + \nu$$

Both β^- and β^+ emission are known as **β decay**.

γ Decay

The emission of a γ-ray photon from a nucleus, called **γ decay**, does not transmute the element. Nuclei that emit γ-ray photons typically are the daughter nuclei of another radioactive decay process. The daughter nuclei created in this other decay are in an excited state (analogous to the excited states of the orbiting electrons, except that the energy levels associated with the nucleus have much larger energy differences than those involved with the atomic electrons). The subsequent transition of the nucleus to a lower energy state results in the emission of a γ-ray photon whose energy is the energy lost by the nucleus.

An example of γ decay is that associated with cobalt 60, a radioisotope of cobalt widely used in medicine. The cobalt 60 decays into nickel by β^- decay† but the nickel nucleus is formed in an excited state [designated with a star (*) superscript] that subsequently emits two γ-ray photons:

$$^{60}_{27}Co \rightarrow (^{60}_{28}Ni)^* + \beta^- + \bar{\nu}$$
$$\hookrightarrow \; ^{60}_{28}Ni + 2\gamma$$

Radioactive Decay Law

Since all the radioactive nuclei of a given radioactive material do not decay instantaneously, but do so over finite, even extended periods of time, the mechanism governing the process has to be fundamentally different from classical physics.

If you drop 1000 marbles all at once, they all fall immediately. Indeed, it would be quite surprising if they did not! With 1000 radioactive nuclei, they do *not* all decay immediately. The mechanism governing radioactivity is *statistical* in nature rather than deterministic. It is impossible to say which specific nuclei will decay during a given time interval; it is only possible to measure how many out of the large population of nuclei do so. The process is much like life. Everyone will die, since life is fatal (we all are unstable particles). But we cannot predict which people will live for how long, although we can deduce statistical averages on the basis of observations of a large number of people.‡

Experimentally, the small number of nuclei dN that decay during a short time interval dt is found to be proportional to the number N of nuclei present (that is, to the concentration of the nuclei), as well as to the interval dt. That is, we have

$$dN \propto N\, dt$$

A proportionality constant is introduced, called the **disintegration constant** λ,§ so that we write

$$dN = -\lambda N\, dt \qquad (26.44)$$

†This decay can proceed by two different paths resulting in the emission of β^- particles with two different energies and γ-rays (from the decay of the nickel nucleus) of two different energies. Each individual cobalt nucleus proceeds along one or the other of these two paths.

‡The accuracy and dependability of such demographic statistics are what make issuing life insurance a viable economic activity.

§Do not confuse the disintegration constant λ with the wavelength of light.

The minus sign indicates that the number of nuclei present (of a particular kind) is decreasing with time. The time rate at which the nuclei decay is

$$\frac{dN}{dt}$$

and is called the **activity** of the sample; it is proportional to the number of nuclei present:

$$\boxed{\frac{dN}{dt} = -\lambda N} \tag{26.45}$$

The SI unit of activity is the **becquerel** (Bq). One becquerel is one disintegration per second. A more common and convenient unit of activity is the **curie** (Cu), defined to be exactly 36 billion disintegrations per second:

$$1 \text{ Cu} \equiv 36 \times 10^9 \text{ Bq} \tag{26.46}$$

Equation 26.45 can be separated and integrated; we rearrange the equation slightly to find

$$\frac{dN}{N} = -\lambda\, dt$$

If we begin with N_0 nuclei at time $t = 0$ s and have N left at time t, then

$$\int_{N_0}^{N} \frac{dN}{N} = -\lambda \int_{0\text{ s}}^{t} dt$$

After integration, we find

$$\ln N - \ln N_0 = -\lambda t$$

or

$$\ln \frac{N}{N_0} = -\lambda t$$

Taking the antilogarithms,* we obtain

$$\frac{N}{N_0} = e^{-\lambda t}$$

or

$$\boxed{N = N_0 e^{-\lambda t}} \tag{26.47}$$

Equation 26.47 is known as the **radioactive decay law**.

The number of parent radioactive nuclei decreases exponentially with time, as shown in Figure 26.15.

The time $\tau_{1/2}$ at which only half the parent nuclei remain is known as the **half-life** of the parent.

We use Equation 26.47 and substitute $N = N_0/2$ when $t = \tau_{1/2}$; this yields

$$\frac{N_0}{2} = N_0 e^{-\lambda \tau_{1/2}}$$

*Do not confuse the base of natural logarithms e, a pure number approximately equal to 2.718 2818. . ., with the fundamental unit of charge $e \approx 1.602 \times 10^{-19}$ C.

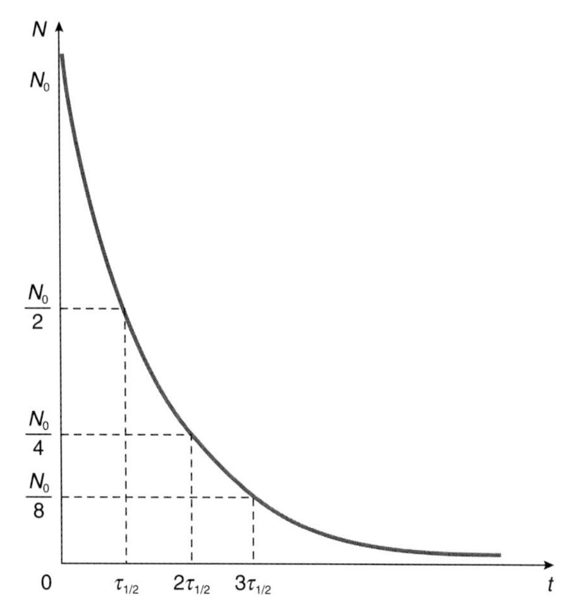

FIGURE 26.15 The number of radioactive nuclei decays exponentially with time.

or

$$\frac{1}{2} = e^{-\lambda \tau_{1/2}}$$

Taking the natural logarithm of this equation, we find

$$\ln 1 - \ln 2 = -\lambda \tau_{1/2}$$

Since $\ln 1 = 0$, we have

$$\boxed{\begin{aligned}\tau_{1/2} &= \frac{\ln 2}{\lambda} \\ &= \frac{0.6931}{\lambda}\end{aligned}} \tag{26.48}$$

which relates the disintegration constant λ to the half-life. After a time equal to two half-lives, only 1/4 of the original nuclei is present. After three half-lives, only $(1/2)^3 = 1/8$ is present; and so on. After a time equal to ten half-lives, only $(1/2)^{10} = 1/1024$ is present, not quite 0.1%.

Experimentally, it is easier to measure the activity, rather than the number of atoms present. Let the original activity of the sample at $t = 0$ s be

$$\left(\frac{dN}{dt}\right)_0$$

According to Equation 26.45, the original activity is proportional to the original number N_0 of radioactive nuclei present:

$$\left(\frac{dN}{dt}\right)_0 = -\lambda N_0 \tag{26.49}$$

Later, at time t, the activity is

$$\frac{dN}{dt}$$

This activity is proportional to the number of number N present at time t:

$$\frac{dN}{dt} = -\lambda N \qquad (26.50)$$

According to Equation 26.47, $N = N_0 e^{-\lambda t}$. Making this substitution for N in Equation 26.50, we find

$$\frac{dN}{dt} = -\lambda N_0 e^{-\lambda t} \qquad (26.51)$$

But $-\lambda N_0$ is the original activity

$$\left(\frac{dN}{dt}\right)_0$$

according to Equation 26.49. Hence Equation 26.51 becomes

$$\boxed{\frac{dN}{dt} = \left(\frac{dN}{dt}\right)_0 e^{-\lambda t}} \qquad (26.52)$$

In other words, the activity of the sample also decays exponentially with time with the same half-life as the number of nuclei.

The half-life of a sample of radioactive nuclei is the time for half of the sample to disintegrate. The average or **mean life** of the sample is different. Some of the atoms in the sample exist much longer than others before decaying. To determine the mean life, consider an analogy. Imagine a collection of 10 people with the following death statistics: 4 die at age 65 y, 3 at age 75 y, 2 at 85 y, and 1 at a venerable 95 y. The average age is found by multiplying the number dying at each age, summing the results, and dividing the sum by the total sample size:

$$\text{average age} = \frac{4(65\ \text{y}) + 3(75\ \text{y}) + 2(85\ \text{y}) + 1(95\ \text{y})}{10}$$
$$= 75\ \text{y}$$

The average lifetime of an initial sample of N_0 radioactive atoms is found in a similar way. Let N be the number of atoms that still exit at time t. Between t and $t + dt$, we lose a few of these hearty atoms: dN of them decay. Thus the number of atoms that live a time t is dN. The sum in the numerator of the average is really an integration of the quantity $t\ dN$ between N_0 and 0 particles; the denominator is the sum of the particles, or the integration of dN over all the particles:

$$\text{mean or average life} = \frac{\int_{N_0}^{0} t\ dN}{\int_{N_0}^{0} dN} \qquad (26.53)$$

For the integration in the numerator, we use Equation 26.45 to find dN:

$$\frac{dN}{dt} = -\lambda N$$
$$dN = -\lambda N\ dt$$

We make this substitution for dN in the numerator of Equation 26.53. For the average life, we then have

$$\text{mean or average life} = \frac{-\int_{0\ \text{s}}^{\infty\ \text{s}} t\lambda N\ dt}{-N_0} \qquad (26.54)$$

But $N = N_0 e^{-\lambda t}$ from Equation 26.47. Thus Equation 26.54 becomes

$$\text{mean life} = \frac{\int_{0\ \text{s}}^{\infty\ \text{s}} t\lambda N_0 e^{-\lambda t}\ dt}{N_0} \qquad (26.55)$$

This integration is done by parts. The result is

$$\text{mean life} = \frac{1}{\lambda} \qquad (26.56)$$

The mean life is the reciprocal of the disintegration constant.

Notice that the mean life is longer than the half-life; from Equation 26.47, the mean life $t = 1/\lambda$ is the time when there are $N_0 e^{-\lambda(1/\lambda)} = N_0 e^{-1} = N_0/e$ nuclei remaining, as shown in Figure 26.16.

PROBLEM-SOLVING TACTIC

26.3 In the radioactive decay law for the number of nuclei remaining or for the activity, the units you use for $\tau_{1/2}$ and $1/\lambda$ must be the same. The dimensions for the half-life are time while those for the disintegration constant λ are inverse time. The units need not be seconds. Example 26.7 at the end of the next section illustrates this point.

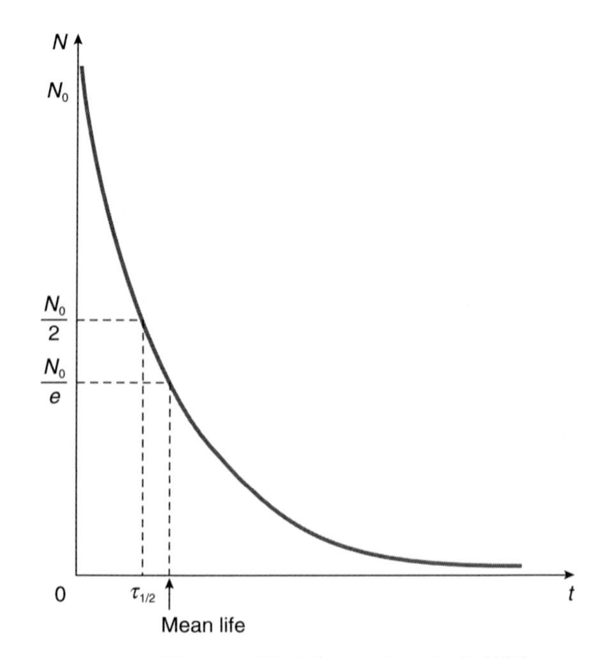

FIGURE 26.16 The mean life is longer than the half-life.

26.12 Carbon Dating

One of the most common methods for dating organic artifacts and remains makes use of a radioactive isotope of carbon, ${}^{14}_{6}C$, commonly called carbon 14. This radioactive isotope is created in the upper atmosphere continuously via cosmic ray* bombardment from space. Interactions of cosmic ray protons (designated p) with nuclei in the upper atmosphere produce, among many other particles, neutrons. Some of these uncharged neutrons (designated n) are captured easily by plentiful nitrogen nuclei in the atmosphere to form carbon 14 via the nuclear reaction

$$ {}^{14}_{7}N + n \rightarrow {}^{14}_{6}C + p $$

The radioactive carbon 14 decays by β^- emission:

$$ {}^{14}_{6}C \rightarrow {}^{14}_{7}N + \beta^- + \bar{\nu} $$

Since carbon 14 is created by cosmic rays and simultaneously destroyed by radioactive decay, an equilibrium concentration of carbon 14 exists in the atmosphere. The equilibrium concentration is about 1.5×10^{-12} kg of carbon 14 per kilogram of carbon.† Atmospheric gases such as carbon dioxide thus contain a small amount of the radioactive isotope. Plant life uses carbon dioxide for photosynthesis; in turn, plants are eaten by other forms of life and the radioactive carbon works its way into and up the food chain. As a result, all living things contain small amounts of carbon 14. The carbon 14 decays, of course, but is replenished continuously with fresh carbon 14 as the plant or organism breathes and eats.‡ Thus living things have a small equilibrium concentration of carbon 14. Most of the carbon in living things is the much more plentiful and stable isotope of carbon, ${}^{12}_{6}C$.

It is a sad but well-known fact that, among other things, the simple pleasures of eating and breathing cease when death occurs. At this juncture, the organism no longer can replenish its carbon 14, but the decay of the existing ingested radioactive isotope continues unabated. Thus the concentration of carbon 14 decreases exponentially with time upon the onset of death. Hence old bones have less carbon 14 than new bones; old wood has less carbon 14 than new wood; and so on.

The decay of carbon 14 has a half-life of 5.73×10^3 y. Once the carbon content of the sample is known, we can compare the relative concentration of carbon 14 in the sample (or the activity of the sample) with that of a fresh sample of organically derived carbon. The concentrations (or activities) are related via the radioactive decay law of Equation 26.47 (or Equation 26.52 for the activities). Since little of a radioactive material is left after a time span of ten half-lives (only about 0.1%; see Section 26.11), carbon 14 dating techniques only can be used on samples with ages no greater than about $10 \times 5.73 \times 10^3$ y or about 6×10^4 y B.P. (Before the Present).§

*Cosmic rays are high-energy charged particles (principally protons—hydrogen nuclei) from space whose origins have mystified astronomers for decades.

†*Experimental Nuclear Physics*, edited by Emilio Segrè (John Wiley, New York, 1960), Volume I, page 615.

‡It is well known that we cannot exist entirely on purely inorganic materials such as rocks and water.

§For some purposes, the time before the present is more important, and using B.P. is preferable to the B.C. and A.D. epochs.

The carbon 14 dating technique has many practical difficulties.

1. The experiments to measure the carbon concentrations and activities involve destroying the sample; in other words, to determine the concentration of carbon present, chemical techniques that destroy the integrity of the sample are necessary. If the sample has artistic, cultural, or intrinsic value of some kind, only a small piece of the object is used. Of course, the smaller the piece used, the smaller the amount of carbon 14 present and the more difficult the measurements become. More recent techniques using particle accelerators count the carbon 14 atoms directly and are capable of determining ages up to about 100 000 y B.P. These techniques also have the advantage of being able to use very small mass samples (measured in milligrams).
2. Another significant problem that must be addressed is assessing if the sample is indeed pure and not contaminated with fresh carbon via leaching from surrounding materials.
3. The technique assumes that the carbon 14 content of the atmosphere is independent of time. Since carbon 14 is created by cosmic rays, the assumption is, then, that the cosmic ray flux on the atmosphere never changes. This is a bold assumption and, unfortunately, is only a first and crude approximation. In fact, the cosmic ray flux does change with time and the ages derived from carbon 14 analysis are only approximate. With the use of samples of carbon from tree rings, however, the carbon 14 dates for the most recent epochs are calibrated against ages determined from tree ring counts. That is, a correction can be applied to the carbon 14 derived dates to determine the age of the sample more accurately. The tree ring corrections, however, only can be used for several millennia B.P. Work continues (with the discovery and analysis of old wood samples) to extend these corrections further back in time.
4. In more recent epochs, the extensive use of fossil fuels has increased the stable carbon isotope content of the atmosphere relative to the carbon 14 content. Such fossil fuels are derived from carbon deposits in which all the carbon 14 effectively has decayed. On the other hand, atmospheric testing of nuclear weapons during the decades immediately after World War II significantly increased the present carbon 14 content of the atmosphere. These changes have affected the background value of the ${}^{14}C/{}^{12}C$ ratio.

Question 5

Why can nuclei easily capture neutrons but not protons?

Example 26.7

The earthly remains of a long-expired dean have an activity of 13 disintegrations per minute per gram of carbon. A freshly sacrificed dean has an activity of 16 disintegrations per minute per gram of carbon. Determine the approximate date of the administrator's demise.

Solution

Since the activities are known, Equation 26.52 is used:

$$ \frac{dN}{dt} = \left(\frac{dN}{dt}\right)_0 e^{-\lambda t} $$

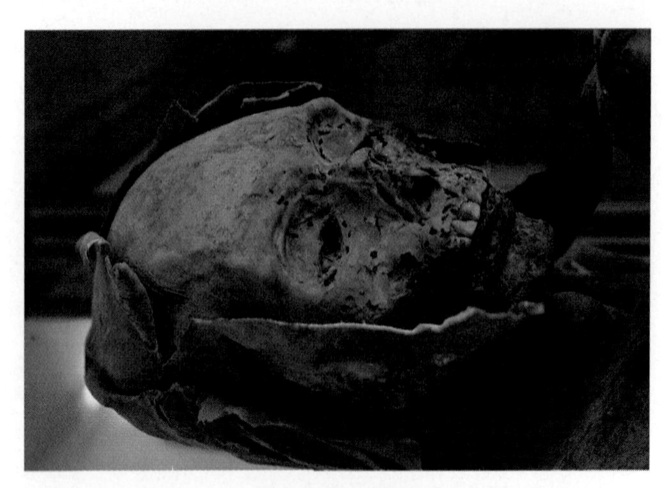

The decay of carbon 14 helps scientists date organic remains, such as this rather well-preserved Incan mummy.

$$13 \text{ disintegrations/min} = (16 \text{ disintegrations/min})e^{-\lambda t}$$

$$0.81 = e^{-\lambda t}$$

Take the natural logarithm of this equation:

$$\ln 0.81 = -\lambda t$$
$$-0.21 = -\lambda t$$

Solving for t, you get

$$t = \frac{0.21}{\lambda} \qquad (1)$$

The disintegration constant λ is related to the half-life $\tau_{1/2}$ via Equation 26.48:

$$\tau_{1/2} = \frac{0.6931}{\lambda}$$

For carbon 14 the half-life is 5.73×10^3 y, so that

$$5.73 \times 10^3 \text{ y} = \frac{0.6931}{\lambda}$$

The disintegration constant thus is

$$\lambda = \frac{0.6931}{5.73 \times 10^3 \text{ y}} = 1.21 \times 10^{-4} \text{ y}^{-1}$$

Making this substitution into equation (1) for the age t, you find

$$t = \frac{0.21}{1.21 \times 10^{-4} \text{ y}^{-1}} = 1.7 \times 10^3 \text{ y}$$

Notice that even though the activities were given in disintegrations per minute per gram, we can use other units (here years) in the exponentials for both t and $1/\lambda$, in accordance with Problem-Solving Tactic 26.3.

26.13 Radiation Units, Dose, and Exposure*

The activity of a sample is measured in becquerels (Bq) [the number of disintegrations per second] or curies (Cu) [one curie is exactly 36×10^9 Bq]. While the curie is a useful unit for activity, it does not give an indication of the amount of energy deposited when a substance absorbs the radiation. For this purpose, a new SI unit, the **gray** (Gy), was introduced about 1975. One gray is equal to one joule of energy absorbed per kilogram of material (see Table 26.2).† The unit was named for the English medical physicist Harold Gray who, in 1955, discovered the *oxygen effect*: cells exposed to radiation in the presence of oxygen are more easily killed than the same cells exposed in the absence of oxygen.‡

The gray is a physical unit that quantifies the **dose** D of radiation. Unfortunately, the biological effects of radiation are quite difficult to measure and assess. For example, γ radiation deposits energy over relatively long path lengths through material; α particles deposit their energy in quite short path lengths of material. Thus the likelihood of a given sample of biological tissue suffering damage is far greater from a dose of 1 Gy of α particles than from a dose of 1 Gy of γ radiation. To account for these differences, a **quality factor** (QF) is used to measure the amount of energy deposited per unit path length of material.§ The quality factor depends on (a) the type of radiation, (b) the energy of the radiation, and (c) to some extent the nature of the target material (soft tissue versus bone, etc.). Radiations that deposit relatively little energy per unit path length (β and γ) have quality factors of about 1, while α particles have quality factors of about 20. (See Table 26.3.)

The effect of various radiations in biological tissues depends on the product of the dose D (in grays) and the quality factor (QF), forming the **dose-equivalent** (DE):

$$\text{DE} = D\,\text{QF} \qquad (26.57)$$

For near total confusion, the dose-equivalent is measured in **sieverts** (Sv),#¶ named for the 20th-century Swedish radiologist,

† An older unit used for energy absorption is the rad (for *r*adiation *a*bsorbed *d*ose); 1 Gy is equivalent to 100 rad. Do not confuse the radiation unit (rad) with the unit abbreviation for radian, also rad.

‡ The oxygen effect has implications in the treatment of tumors with radiation. Cells at the center of large tumors are oxygen deprived, since the blood flow to them is hogged by tumor cells near the periphery of the tumor. Thus radiation is effective in killing tumor cells on the periphery but not tumor cells at its center, leaving them free to reproduce and make the nasty tumor reoccur.

§ An older way of measuring the biological effectiveness of various radiations is with a unit called the RBE (*r*elative *b*iological *e*ffectiveness); the RBE attempts to assess the effects of the radiation in the tissue compared with x-rays that produce the same effects. The RBE is quite difficult to measure and so the quality factor is coming into style.

Alas, another way of measuring the dose-equivalent is to express the dose in rads and use the RBE instead of the quality factor. In this scenario, the dose-equivalent is

$$\text{DE} = D\,\text{RBE}$$

with the result expressed in a unit known as the rem (for *r*öntgen *e*quivalent *m*an). The conversion between sieverts and rems is

$$1 \text{ Sv} = 100 \text{ rem}$$

¶ One sievert is the γ-ray dose received during one hour at a distance of 1 cm from a point source of 1 mg of radium enclosed in a shell of platinum 0.5 mm thick.

TABLE 26.2 Common Radiation Units and Conversions

	Traditional unit	*SI unit*
Activity	curie (Cu) = 36 × 10^9 disintegrations/s	becquerel (Bq) = 1 disintegration/s
Absorbed dose *D*	rad	gray (Gy)
	100 rad = 1 Gy	
Dose-equivalent DE	rem	sievert (Sv)
	100 rem = 1 Sv	

TABLE 26.3 Quality Factors for Various Absorbed Radiation

Radiation	*Quality factor*
X-rays, β, γ	1
Low-energy protons and neutrons (~keV energies)	2–5
High-energy protons and neutrons (~MeV energies)	5–10
α particles	20

TABLE 26.4 Dose-equivalents of Common Exposures to Radiation

Exposure	*Dose-equivalent**
Natural background from cosmic rays and naturally occurring radioactive materials in the environment (soils, concrete, etc.)	0.1–0.2 rem/y 1–2 × 10^{-3} Sv/y
Absorption by bone marrow (very sensitive tissue)	
diagnostic chest x-ray	0.05 rem 5 × 10^{-4} Sv
dental x-ray	0.002 rem 2 × 10^{-5} Sv

*The International Commission on Radiation Protection recommends a whole-body dose-equivalent of no more than 0.5 rem/y = 5 × 10^{-3} Sv/ y for the general public and 5 rem/y = 5 × 10^{-2} Sv/y for people who work with radiation routinely.

Data from Kenneth S. Krane, *Introductory Nuclear Physics* (John Wiley, New York, 1988), page 187.

Rolf Sievert (1896–1966), long active in the problem of quantifying radiation doses in biological materials. Table 26.4 lists the dose-equivalents of some common forms of radiation exposure.

The long-term effects of radiation on biological tissues is a matter of some concern and much research. The problem of assessing safe exposures is quite difficult, if only because the same dose-equivalent can have different effects in different individuals. This is quite apparent in medical treatment; the same radiation dose-equivalent given to two individuals can result in widely different responses and side effects.

26.14 The Momentum of a Photon

There is no wind in the vacuum of space but, nonetheless, it is possible to sail to the stars (if you have lots of time on your hands). Huh? When light reflects from a surface, a small amount of momentum is transferred to the surface. This idea is the principle behind a concept for effortlessly propelling a spacecraft between the planets or off to the stars by using a large reflecting surface (perhaps several square kilometers in area) to gain the momentum change associated with the reflection of sunlight from the sail.

In this section we see how the idea of momentum associated with a photon of light arose from seemingly totally unrelated experiments concerning the interaction of x-rays with matter.

While investigating the interaction (scattering[†]) of x-rays with graphite in 1923, Arthur H. Compton (1892–1962) noticed that if monochromatic x-rays were incident, the spectrum of the scattered x-rays consisted of the incident wavelength as well as one shifted slightly to a longer wavelength, as shown in Figure 26.17. This is called the **Compton effect**. The reason for the new wavelength intrigued Compton.

[†] Collision experiments involving fundamental particles such as photons and electrons are examples of *scattering experiments* in physics.

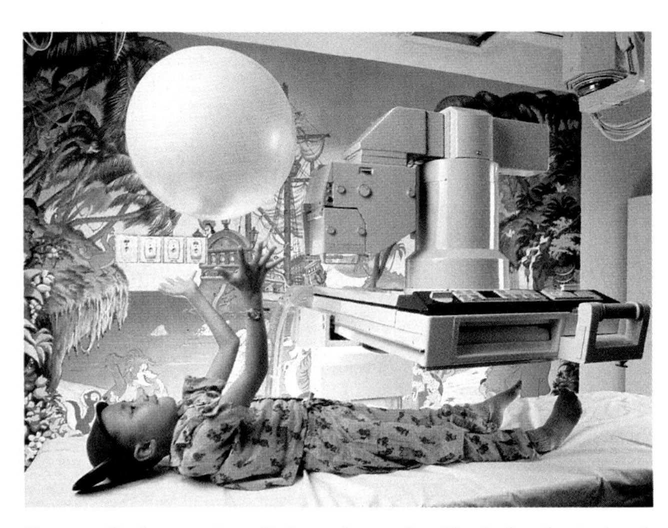

For a medical career in radiology, the study of both the physical and biological effects of radiation is necessary. Here a child is having a diagnostic x-ray.

Reflecting photons from sails can propel spacecraft such as in this artist's conception of a solar sail freighter.

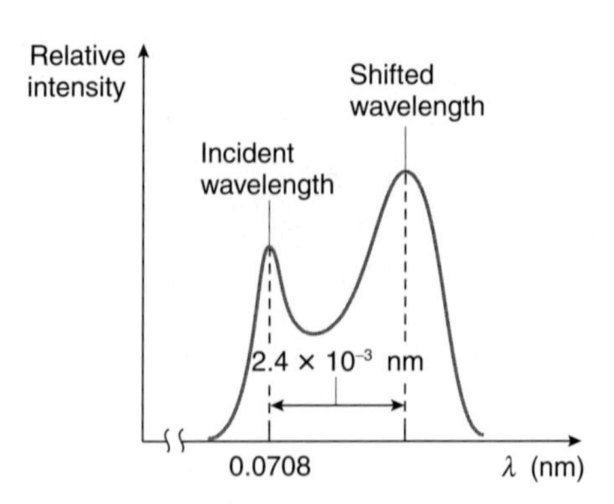

FIGURE 26.17 The scattered x-ray intensity versus wavelength in Compton's original experiment.

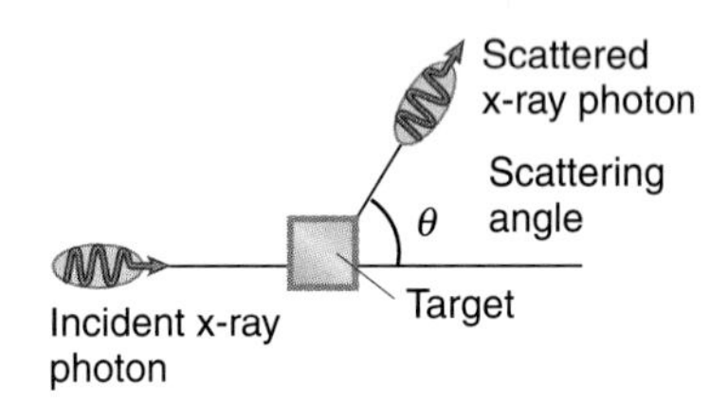

FIGURE 26.18 Geometry for the Compton effect.

The amount of the shift in wavelength depends on the angle θ at which the scattered x-rays were observed, relative to the incident direction, as in Figure 26.18. No shift is observed if the scattering angle θ is 0°, while the maximum shift occurs if scattered x-rays are detected in the backward direction where the scattering angle θ is 180°.

In order to account for these observations, Compton boldly applied the photon concept invented earlier by Einstein for the photoelectric effect, and used by Bohr with great success in his model of the hydrogen atom. Compton's analysis provided further confirmation that the photon concept was real and legitimate.

According to Einstein's treatment of the photoelectric effect, a photon is a particle of light with an energy $E = h\nu$ (Equation 26.4). There is no particle more relativistic (high speed) than a photon, of course, since it travels at the speed of light. We discovered in Chapter 25 that the total relativistic energy E of a particle is related to the magnitude p of its momentum and its mass m via (Equation 25.85)

$$E^2 = p^2c^2 + m^2c^4$$

A photon has zero mass, and so for it this equation yields

$$E = pc \tag{26.58}$$

Thus the magnitude of the momentum of a photon (or any other particle with zero mass) is its energy divided by the speed of light:

$$p = \frac{E}{c} \tag{26.59}$$

For a photon, $E = h\nu$, and so the magnitude of the photon momentum is

$$p_{\text{photon}} = \frac{h\nu}{c} \tag{26.60}$$

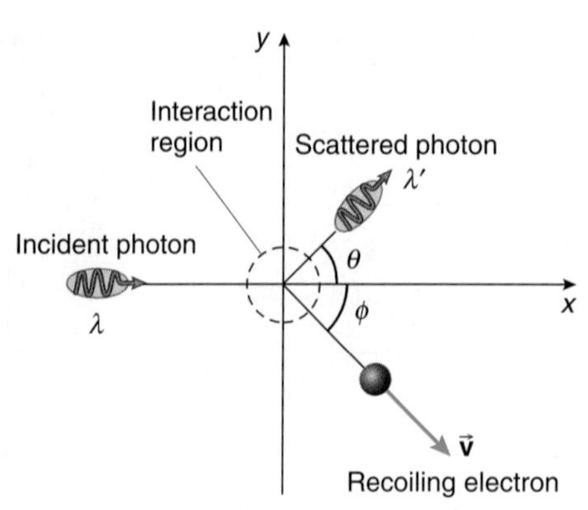

FIGURE 26.19 Geometry for the analysis of the Compton effect.

Since $c = \nu\lambda$, this also can be written in the equivalent form

$$p_{\text{photon}} = \frac{h}{\lambda} \tag{26.61}$$

Compton assumed that an incoming x-ray photon collided elastically with an electron, conserving both energy and momentum. A recoiling electron meant the scattered x-ray photon would have less energy and, therefore, a longer wavelength than the incident x-ray.

What actually happens in the collision is an absorption of the incident photon by the electron and a subsequent (almost instantaneous) reemission of a different photon, conserving energy and momentum in the process; Compton shrewdly avoided the details of the photon absorption and reemission and treated the interaction simply as a collision. The electron binding energy (holding it to the nucleus) is trivial in comparison with the energy of the x-ray photon (see Example 26.8 below). With a collision viewpoint, a photon of definite energy and momentum is sent into a "black box" containing an electron assumed to be essentially at rest,* as in Figure 26.19. After the interaction, a photon with a different energy and momentum is observed leaving the interaction region, as well as a recoiling electron that is not detected in the experiment.†

In an elastic collision between two classical particles we could ignore the details of the interaction between the particles and simply conserve momentum and energy (see Chapter 9). We do likewise here, treating the interaction between the photon and electron particles as an elastic collision to see if what emerges from the calculations agrees with experiment or not. In fact, as we will see, the analysis is confirmed by the observations. To analyze the collision, we set up the coordinate system indicated in Figure 26.19.

*This condition is satisfied if the incident photon energy is much greater than any initial kinetic energy of the electron. This is the case for x-ray photons (see Example 26.8). Since the electrons in atoms are moving, however, the scattered x-rays are not strictly monochromatic but have a spread in their wavelengths at a given scattering angle.

†It was the observation of a different frequency emerging from the interaction that led Compton to hypothesize about the momentum of a photon and to treat the interaction as an elastic collision.

First we look at conservation of momentum. The incident photon of wavelength λ has momentum

$$\frac{h}{\lambda}\,\hat{\imath}$$

Since the electron is assumed to be initially at rest, it has zero momentum. Thus the total momentum of the system of two particles before the collision is

$$\vec{p}_i = \frac{h}{\lambda}\,\hat{\imath} \qquad (26.62)$$

The photon of wavelength λ' emerging from the experiment at an angle θ to the x-axis has momentum

$$\left(\frac{h}{\lambda'}\cos\theta\right)\hat{\imath} + \left(\frac{h}{\lambda'}\sin\theta\right)\hat{\jmath}$$

The electron, recoiling from the collision at speed v, and making an angle ϕ with the x-axis, has *relativistic* momentum

$$(\gamma m v\cos\phi)\hat{\imath} - (\gamma m v\sin\phi)\hat{\jmath}$$

Thus the total momentum after the collision is

$$\vec{p}_f = \left(\frac{h}{\lambda'}\cos\theta + \gamma m v\cos\phi\right)\hat{\imath} + \left(\frac{h}{\lambda'}\sin\theta - \gamma m v\sin\phi\right)\hat{\jmath} \qquad (26.63)$$

Momentum conservation means that the change in the total momentum is zero:

$$\vec{p}_f - \vec{p}_i = 0\ \text{kg}\cdot\text{m/s}$$
$$\vec{p}_i = \vec{p}_f \qquad (26.64)$$

Since two vectors are equal if and only if their respective components are equal, Equation 26.64 means the following:

1. The x-components of the total momentum before and after the collision are equal:

$$\frac{h}{\lambda} = \frac{h}{\lambda'}\cos\theta + \gamma m v\cos\phi \qquad (26.65)$$

2. The y-components of the total momentum before and after the collision are equal:

$$0\ \text{kg}\cdot\text{m/s} = \frac{h}{\lambda'}\sin\theta - \gamma m v\sin\phi \qquad (26.66)$$

Now we examine (relativistic) energy conservation. The incident photon has an energy $h\nu$ or, in terms of its wavelength,

$$\frac{hc}{\lambda}$$

The electron, assumed to be initially at rest, has a total (relativistic) energy equal to its rest energy mc^2, since it has no kinetic energy (actually, a negligible amount when compared with its rest energy). Thus the total initial energy is

$$E_i = \frac{hc}{\lambda} + mc^2 \qquad (26.67)$$

After the collision, the emerging photon has energy

$$\frac{hc}{\lambda'}$$

and the emerging electron has total relativistic energy

$$\gamma mc^2$$

Thus the final total energy is

$$E_f = \frac{hc}{\lambda'} + \gamma mc^2 \qquad (26.68)$$

Energy conservation means that the change in the total energy is zero:

$$E_f - E_i = 0\ \text{J}$$

or

$$E_i = E_f \qquad (26.69)$$

Making the appropriate substitutions, we find

$$\frac{hc}{\lambda} + mc^2 = \frac{hc}{\lambda'} + \gamma mc^2 \qquad (26.70)$$

Equations 26.65 and 26.66, from momentum conservation, and Equation 26.70, from energy conservation, summarize the physics. Since it is the emerging x-ray photon that is detected, we want to eliminate the speed v and the angle ϕ associated with the recoiling electron in order to find how the emerging photon wavelength λ' is related to the incident photon wavelength λ and the scattering angle θ. The problem is a bit of an algebraic quagmire, but the solution is outlined here in case you like an algebraic challenge and have your mathematical boots on. The result is Equation 26.78 if you want to skip the details.

Here is the solution for the emerging photon wavelength; check it out.

1. To eliminate the angle ϕ associated with the recoiling electron:
   a. Solve Equation 26.65 for $\gamma m v\cos\phi$ and square the result. You should get

$$\gamma^2 m^2 v^2\cos^2\phi = \left(\frac{h}{\lambda} - \frac{h}{\lambda'}\cos\theta\right)^2 \qquad (26.71)$$

   b. Solve Equation 26.66 for $\gamma m v\sin\phi$ and square the result. You should get

$$\gamma^2 m^2 v^2\sin^2\phi = \left(\frac{h}{\lambda'}\sin\theta\right)^2 \qquad (26.72)$$

   c. Add Equations 26.71 and 26.72 to eliminate ϕ. After a bit of algebraic simplification, you should be able to put this into the form

$$\gamma^2 m^2 v^2 = \frac{h^2}{\lambda'^2} + \frac{h^2}{\lambda^2} - \frac{2h^2\cos\theta}{\lambda\lambda'} \qquad (26.73)$$

2. To eliminate the speed of the recoiling electron v:
   a. Take the energy equation (Equation 26.70) and rearrange

it into the following form (some transposing and canceling is involved):

$$\left(\frac{h}{\lambda} - \frac{h}{\lambda'}\right) + mc = \gamma mc \qquad (26.74)$$

b. Square Equation 26.74, obtaining

$$\left(\frac{h}{\lambda} - \frac{h}{\lambda'}\right)^2 + m^2c^2 + 2mc\left(\frac{h}{\lambda} - \frac{h}{\lambda'}\right) = \gamma^2m^2c^2 \qquad (26.75)$$

c. Now subtract Equation 26.73 from Equation 26.75. After some simplification, you should obtain

$$-\frac{2h^2}{\lambda\lambda'} + m^2c^2 + 2mc\left(\frac{h}{\lambda} - \frac{h}{\lambda'}\right) + \frac{2h^2 \cos\theta}{\lambda\lambda'}$$
$$= \gamma^2m^2c^2 - \gamma^2m^2v^2 \qquad (26.76)$$

d. Substitute for γ^2 in the right-hand side of Equation 26.76. The right-hand side of the equation then simplifies to just m^2c^2:

$$-\frac{2h^2}{\lambda\lambda'} + m^2c^2 + 2mc\left(\frac{h}{\lambda} - \frac{h}{\lambda'}\right) + \frac{2h^2 \cos\theta}{\lambda\lambda'} = m^2c^2 \qquad (26.77)$$

e. Now simplify and rearrange Equation 26.77. You should eventually obtain

$$\lambda' - \lambda = \frac{h}{mc}(1 - \cos\theta) \qquad (26.78)$$

The wavelength of the x-ray photon changes by an amount given by Equation 26.78.

In Equation 26.78, the quantity

$$\frac{h}{mc}$$

where m is the mass of the electron, has the dimensions of a length and is known picturesquely as the **Compton wavelength** of the electron. It is *not* a wavelength of light; the quantity simply has the same dimensions as a wavelength.

The mass of the electron m appears in the denominator of Equation 26.78. This is the reason the change in wavelength is more apparent for scattering from electrons rather than from, say, protons or the nucleus. The proton and nuclear masses are much larger than the electron mass; so any shift in the frequency of the photon in its scattering from protons or nuclei is much less than from electrons and accounts for the unshifted wavelength in Figure 26.17. Another way of saying this is that the Compton wavelength of a proton or a nucleus is much less than the Compton wavelength of an electron because of the drastic difference in the masses of the two particles.

The experiment is not done with a single photon but with a huge number of photons. A beam of photons is incident on a whole collection of electrons. Each photon scatters from an individual electron and scatters to some definite angle θ. The photons emerging from the interaction region at the angle θ have the wavelength λ' given by Equation 26.78. Photons emerging at different angles (from encounters with different electrons) emerge with different wavelengths.

In agreement with experiment, the change in the wavelength is a maximum when $\theta = 180°$, in which case the emerging photon moves antiparallel to the incident photon. In this case the change in wavelength is equal to twice the Compton wavelength of the electron (substitute $\theta = 180°$ into Equation 26.78).

The result of Compton's bold theoretical analysis of the problem in terms of photon energy and momentum, Equation 26.78, beautifully confirmed his experimental results. He received the 1927 Nobel prize in physics for the discovery.

Another way to see that photons of any wavelength have momentum is to observe the force they impart to a reflector. A floodlight can produce measurable movement of a delicate mirror suspension in high vacuum. We reflect a beam of photons off a mirror surface of mass m, as in Figure 26.20. The reflection is essentially a collision between each incoming photon and the mirror. The Compton wavelength of the mirror is extremely small because of its huge macroscopic mass. Thus the reflected photon is effectively unchanged in wavelength. The incident photon momentum is $(h/\lambda)\hat{\imath}$ with the coordinates of Figure 26.20. The exiting photon momentum is $-(h/\lambda)\hat{\imath}$. We call the momentum of the recoiling mirror $\vec{p}_{mirror}$. Conserving momentum in the collision means that

$$\vec{p}_f - \vec{p}_i = 0 \text{ kg·m/s}$$
$$-\frac{h}{\lambda}\hat{\imath} + \vec{p}_{mirror} - \frac{h}{\lambda}\hat{\imath} = 0 \text{ kg·m/s}$$

Solving for $\vec{p}_{mirror}$, we find

$$\vec{p}_{mirror} = +2\frac{h}{\lambda}\hat{\imath} \qquad (26.79)$$

The mirror picks up some momentum as a kick from the reflected photon. If instead of a single photon, we reflect a flood

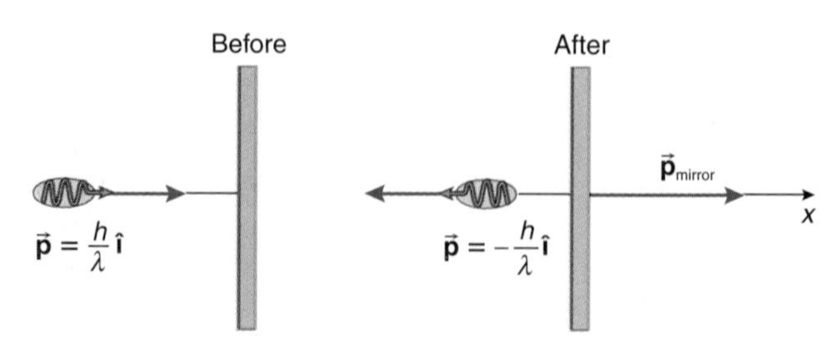

FIGURE 26.20 Reflection of photons from a mirror.

of photons each second from the mirror, the momentum gained by the mirror will increase correspondingly. If we reflect N photons each second from the mirror, the mirror gains momentum each second equal to N times that of Equation 26.79. Since the photons are changing the momentum of the mirror by a certain amount each second, they exert a force on the mirror equal to the time rate of change of the momentum of the mirror (Newton's second law).

The magnitude of the force per unit area of the light on the mirror is known as the **radiation pressure** exerted by light. It is the principle behind sailing between the planets and off to the stars. Radiation pressure also is apparent in some comet tails. When sunlight reflects from small particles, such as those in the tails of comets, it drives the particles in directions away from the Sun. Hence some dusty comet tails are directed away from the Sun, even as the comet recedes from the Sun.*

Example 26.8

Compton used photons with an incident wavelength of 0.0708 nm.

a. Determine the energy of each incident photon in joules and in electron-volts. Compare this energy with the sum of the kinetic and potential energies of an electron in the hydrogen atom.

b. Determine the magnitude of the momentum of each incident photon.

Solution

a. The photon energy is

$$\begin{aligned} E &= h\nu \\ &= \frac{hc}{\lambda} \\ &= \frac{(6.626 \times 10^{-34}\ \text{J}\cdot\text{s})(3.00 \times 10^{8}\ \text{m/s})}{7.08 \times 10^{-2} \times 10^{-9}\ \text{m}} \\ &= 2.81 \times 10^{-15}\ \text{J} \\ &= \frac{2.81 \times 10^{-15}\ \text{J}}{1.602 \times 10^{-19}\ \text{J/eV}} \\ &= 17.5\ \text{keV} \end{aligned}$$

This is much greater than the sum of the kinetic and potential energies of the electron in the hydrogen atom (typically on the order of only a few electron-volts, and negative; see the Bohr model of the hydrogen atom in Section 26.7). Similar arguments apply to electrons in other atoms. This comparison justifies Compton's neglect of the any initial kinetic energy associated with the electron.

b. The magnitude p of the momentum of the photon is found from Equation 26.59:

$$\begin{aligned} p &= \frac{E}{c} \\ &= \frac{2.81 \times 10^{-15}\ \text{J}}{3.00 \times 10^{8}\ \text{m/s}} \\ &= 9.37 \times 10^{-24}\ \text{kg}\cdot\text{m/s} \end{aligned}$$

*Some comets have multiple tails, each caused by different physical effects.

Example 26.9

Note that in Figure 26.17 the wavelength shift of the x-rays in Compton's original experiment was 2.4×10^{-3} nm. Use Equation 26.78 to determine the approximate angle θ at which Compton measured the scattered x-rays.

Solution

The shift in wavelength of the scattered x-rays is given by Equation 26.78:

$$\lambda' - \lambda = \frac{h}{mc}(1 - \cos\theta)$$

Hence you have

$$\begin{aligned} 1 - \cos\theta &= \frac{mc}{h}(\lambda' - \lambda) \\ &= \frac{(9.11 \times 10^{-31}\ \text{kg})(3.00 \times 10^{8}\ \text{m/s})(2.4 \times 10^{-12}\ \text{m})}{6.626 \times 10^{-34}\ \text{J}\cdot\text{s}} \\ &= 0.99 \end{aligned}$$

Solving for $\cos\theta$, you find

$$\cos\theta = 0.01$$

or

$$\theta \approx 90^\circ$$

Indeed, from Compton's paper in 1923,† the scattering angle used was 90°.

26.15 The de Broglie Hypothesis

Now a purely corpuscular theory [of light] contains nothing that enables us to define a frequency; for this reason alone, therefore, we are compelled in the case of Light, to introduce the idea of a corpuscle and that of a periodicity simultaneously.

On the other hand, determination of the stable motion of electrons in the atom introduces integers; up to this point the only phenomena involving integers in Physics were those of interference and of normal modes of vibration. This fact suggested to me the idea that electrons too could not be regarded simply as corpuscles, but that periodicity must be assigned to them also.

Louis de Broglie (1892–1987)‡

You easily can picture a wave if you think of a surface water wave. You also easily can picture a particle, be it a baseball, peanut, or a sand grain, since such particles are quite visible to us. It is vastly more difficult for us to picture a wave (such as light) as a particle (a photon). To do so, we have to extrapolate into a regime that is not directly accessible to our sense experiences. Perhaps then, we should be cautious about extending our common ideas about particles into a realm in which particles are not amenable to our direct senses, the domain of the atom,

† Arthur H. Compton, "The spectrum of scattered x-rays," *Physical Review* (2nd series), *22*, #5, pages 409–413 (November 1923); Figure 26.17 is from page 411.

‡ Louis de Broglie, *Matter and Light*, translated by W. H. Johnston (Norton, New York, 1939), pages 168–169.

with its constituent electrons, protons, neutrons, and other submicroscopic entities.

The true nature of light is difficult to assess. The double slit experiments by Young in 1801 showed that light exhibited the macroscopic wavelike properties of diffraction and interference, although he was unaware of the nature of the waves themselves. Much later in the century, Maxwell discovered that light was a wave of oscillating electric and magnetic fields: an electromagnetic wave. On the other hand, we have seen in this chapter that the photoelectric and Compton effects indicate that light has the microscopic aspects of a particle, the photon, with both energy and momentum. Thus light exhibits a **wave–particle duality**. This duality profoundly taxes our imagination. Both the photoelectric and Compton effects are not amenable to our direct senses: we cannot literally *see* the processes take place.

The wave–particle duality was extended to particles in the early 1920s with the brilliantly original work of a young French physicist, Louis de Broglie (1892–1987).* His theoretical insight into the nature of particles and waves was the last critical thread in the tapestry that led to the invention of a totally new mechanics of particles: quantum mechanics, which we touch on in Chapter 27.

In a sudden inspiration de Broglie reasoned that if what we think of as waves have particle properties, perhaps what we think of as particles (such as electrons) have wave properties. This symmetry argument is called the **de Broglie hypothesis**.

De Broglie was then able to extend this concept into the analysis of particle motion. Using relativistic arguments, we saw in Section 26.14 with the Compton effect that the particle of light, the photon, has a momentum whose magnitude is given by Equation 26.61:

$$p_{\text{photon}} = \frac{h}{\lambda}$$

Thus the wavelength of light is related to the magnitude of its momentum via

$$\boxed{\lambda = \frac{h}{p_{\text{photon}}}} \qquad (26.80)$$

Taking this fact as a cue, de Broglie proposed that a wavelength is associated with *any* particle and is related to the magnitude of its momentum by an equation of the same form:

$$\boxed{\lambda = \frac{h}{p_{\text{particle}}}} \qquad (26.81)$$

The wavelength associated with a particle is called its **de Broglie wavelength**; it is *not* a wavelength of light, nor is it the so-called Compton wavelength of a particle (of the previous section). He also assumed that the energy E of the particle is proportional to the frequency ν of the associated wave:

$$E = h\nu \qquad (26.82)$$

just as for a photon.

*His last name is pronounced like "broy" (brô y), a close rhyme to "oily." He won the 1929 Nobel prize in physics.

Louis de Broglie began to study history at the Sorbonne in Paris at the age of 17, but he became enthralled with physics. He won the Nobel prize in physics in 1929 for his theoretical ideas about particle-waves.

The true nature of the wave associated with a particle remained a mystery, yet to be determined (much like when Young discovered the wave nature of light). The de Broglie hypothesis implies that the wave–particle duality has a universal and symmetrical character: waves have particle properties, particles have wave properties.

The boldness of de Broglie's hypothesis is even more astounding when you realize that Equation 26.81 implies that the particle speed v (which determines its momentum) is *not* the same as the speed v' of the de Broglie waves associated with the particle! The de Broglie wave speed v' is related to the frequency and wavelength of the de Broglie wave by $v' = \nu\lambda$, so $\lambda = v'/\nu$. From Equation 26.81, we have

$$\frac{v'}{\nu} = \frac{h}{p}$$

or

$$v' = \frac{h\nu}{p}$$

From Equation 26.82, this becomes

$$v' = \frac{E}{p} \qquad (26.83)$$

If we write the energy of the particle as $E = \gamma mc^2$ and the magnitude of its momentum as $p = \gamma mv$, then Equation 26.83 yields

$$v' = \frac{\gamma mc^2}{\gamma mv}$$

or

$$vv' = c^2 \qquad (26.84)$$

Equation 26.84 indicates the wave speed v' thus is *inversely proportional* to the particle speed (through its momentum). Since the particle speed v is restricted to be less than the speed of light c (from relativity), the de Broglie waves associated with the particle apparently travel at superluminal speeds ($> c$)!

These superluminal speeds may seem peculiar until you recall from Chapter 12 that the superposition of a collection of waves with slightly different wavelengths (or frequencies) forms a *wave group*; the speed of the wave group typically is different from that of the waves that form the group.* An individual de Broglie wave thus does not remain long with the particle. We associate the particle speed with the *group speed* of the de Broglie waves, and this always is less than the speed of light.

You might be tempted to think that things really are getting a little far-fetched. What is the wavelength associated with common and easily seen particles such as a baseball, a bowling ball, or even a particle as exotic as a physics student? To get a feeling for the size of the de Broglie wavelength of various particles, look at Examples 26.10 and 26.11. You will see that the de Broglie wavelength of macroscopic particles is completely negligible, but the wavelength associated with electrons and other small particles (that are too small for us to see directly) can be significant on the atomic or nuclear scales of length.

The de Broglie hypothesis casts the Bohr angular momentum hypothesis in a new light, as de Broglie himself realized. Recall that the Bohr angular momentum postulate (Equation 26.20) stated that the magnitude of the angular momentum L of the orbiting electron in a hydrogenic atom was an integral multiple of $\hbar$:

$$L = n\hbar$$

How do the de Broglie waves reconcile with the Bohr model? For an electron in a circular orbit of radius r, the magnitude of its orbital angular momentum $\vec{\mathbf{L}} = \vec{\mathbf{r}} \times \vec{\mathbf{p}}$ is mvr, since the two vectors $\vec{\mathbf{r}}$ and $\vec{\mathbf{p}}$ are mutually perpendicular; recall that the electron in the Bohr model is nonrelativistic, so the momentum is just mv. Hence the Bohr postulate becomes

$$mvr = n\frac{h}{2\pi} \tag{26.85}$$

Since the magnitude of the momentum is $p = mv$, Equation 26.85 can be rewritten as

$$2\pi r = n\frac{h}{p} \tag{26.86}$$

The de Broglie wavelength λ of the electron is given by Equation 26.81, so that Equation 26.86 can be rewritten as

$$2\pi r = n\lambda \tag{26.87}$$

In other words, the allowed orbits of the electron are precisely those for which an integral number of electron de Broglie wavelengths fit along the circumference. This realization gave credence to the de Broglie hypothesis and to the quantization of angular momentum. Nonetheless, a definitive experiment was needed to confirm the existence of the waves associated with particles such as electrons. De Broglie suggested an experiment with crystals acting as diffraction gratings for the particle-waves, but his idea lay fallow for several years.

Wave phenomena such as diffraction and interference become apparent when the waves interact with (i.e., pass through or around) obstacles or openings on the order of the wavelength in size. Clearly it is impossible for the student in Example 26.10 to pass through or

*This is commonly seen with water waves, where the speeds of the waves that form a group are greater than that of the group itself. The waves thus arise at the trailing edge of the group disturbance, propagate through it, and disappear at the leading edge of the group. When at the beach, you might note that every so often, a very large amplitude wave appears; this is the maximum amplitude of the group of waves formed by the superposition of the waves that create the group and travel faster than the group itself. When you are canoeing on a lake and watch waves from a bag distant boat coming toward you, the component waves seem to move faster than the big swell itself.

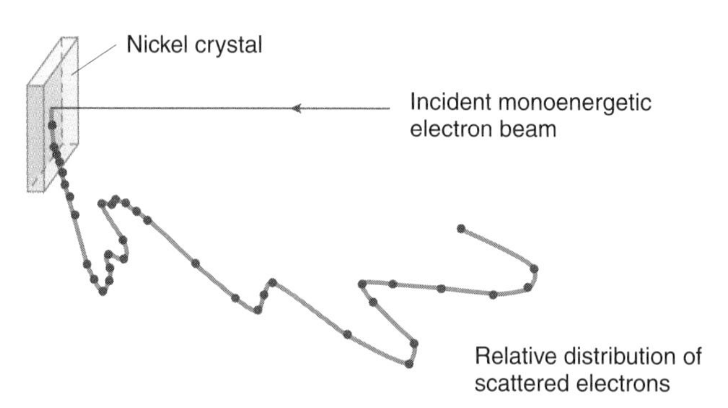

FIGURE 26.21 Distribution (relative number) of electrons scattered at various angles in the Davisson and Germer experiment.

around objects as small as 10^{-36} m (even if objects so small existed!), and so the wavelike aspect of a physics student particle never is manifest. The de Broglie wavelength of any macroscopic particle is *so* small that you always can neglect the wave aspects of these large particles; the wavelike aspects never become apparent.

On the other hand, Example 26.11 indicates that the de Broglie wavelength of electrons can be quite large compared with the sizes of nuclei (10^{-15} m) and on the order of the size of atoms (10^{-10} m = 0.1 nm). The wavelike behavior of electrons thus is very important in atomic physics.

The experimental verification of the wavelike aspects of electrons first was made accidentally by Clinton Davisson and Lester Germer in 1927 in studies of the reflection of electrons from a nickel target, some three years after de Broglie published his conjectures.† After some initial experiments, the accidental explosion of a bottle of liquid air oxidized the nickel target. To purge the nickel, the target was subjected to prolonged heating. It then crystallized upon cooling, and the electrons then scattered by the nickel target had an angular distribution totally unlike that detected before the accident. Much to their initial consternation, Davisson and Germer detected maximum and minimum peaks and valleys in the angular distribution of electrons scattered from the crystal array (see Figure 26.21), which they finally and successfully interpreted by treating the crystalline array of nickel atoms as a three-dimensional diffraction grating for the electron de Broglie waves.

Working independently, George Thomson (1892–1975)‡ also detected the diffraction and interference of electron waves through a thin film of crystallized metal. These discoveries confirmed the de Broglie hypothesis. As we have seen on many occasions, from such innocuous squiggles come Nobel prizes.§

An important technological application of the short wavelengths associated with electrons became apparent to Ernst Ruska in 1931: an electron microscope.# We saw in Chapter 24 that the limiting resolution angle of a microscope

†Clinton Davisson and Lester H. Germer, "Diffraction of electrons by a crystal of nickel," *Physical Review*, 30, #6, pages 705–740 (December 1927).

‡Thomson was the son of J. J. Thomson, the inventor of the plum pudding model of the atom (see Section 26.6).

§George Thomson shared the Nobel prize with Clinton Davisson in 1937 for the discovery of the electron de Broglie waves. One wonders why Germer did not share in the prize.

#He eventually recieved the Nobel prize in physics (in 1986) for the invention along with Gerd Binning and Heinrich Rohrer, who (in 1981) invented an improved and sophisticated version of the instrument known as a scanning tunneling electron microscope.

Electron microscopes exploit the wave nature of electrons to improve resolution.

of circular aperture a depends on the wavelength according to Equation 24.5:

$$\theta = \frac{1.220\,\lambda}{a} \qquad \text{(circular aperture)}$$

Since electron wavelengths can be made so much smaller than visible light wavelengths, the limiting resolution angle is correspondingly that much smaller for electron-waves than for visible light. Electron microscopes thus have the potential for probing smaller details than optical microscopes. The electrons can be manipulated (refracted) much like light waves except that the electromagnetic "lenses" used consist of electric and magnetic fields. The details are involved, as one might imagine, but the essence of the ability of electron microscopes to fathom greater detail lies in the much smaller size of the electron wavelengths used. You might wonder why x-ray or γ-ray microscopes have not been invented or developed to increase the resolution afforded by electromagnetic radiation, since their wavelengths are much smaller than those of visible light. But it is difficult to manipulate or focus such short electromagnetic wavelengths, whereas it is quite easy to manipulate electrons by electric and magnetic fields.*

The de Broglie hypothesis and the wave–particle duality of nature raise several questions:

1. How can a particle be diffracted? What does it mean when we say a *particle* undergoes diffraction because of its wavelike properties?
2. What is it that is jiggling or oscillating in these particle waves? What does the particle-wave represent?
3. Do the particle-waves satisfy the classical wave equation (Equation 12.7) or some other equation?

These are interesting and profound questions about the nature of the wave–particle duality. We shall have more to say about these questions shortly (Chapter 27). We raise them here because you may be wondering about them; you are not alone. Physicists *still* find these questions quite profound and not easily answered.

*On the other hand, ingenious x-ray and γ-ray telescopes have been invented over the last 30 years to collect and study these radiations coming from distant sources in the galaxy and beyond.

Example 26.10

You are late and rushing off to class at a speed of 3.0 m/s; assume your mass is 60 kg. Determine the de Broglie wavelength of the you-particle.

Solution

The magnitude of your momentum is

$$\begin{aligned} p &= mv \\ &= (60\ \text{kg})(3.0\ \text{m/s}) \\ &= 1.8 \times 10^{2}\ \text{kg}\cdot\text{m/s} \end{aligned}$$

Substituting this into Equation 26.81 for the de Broglie wavelength, you find

$$\begin{aligned} \lambda &= \frac{h}{p} \\ &= \frac{6.626 \times 10^{-34}\ \text{J}\cdot\text{s}}{1.8 \times 10^{2}\ \text{kg}\cdot\text{m/s}} \\ &= 3.7 \times 10^{-36}\ \text{m} \end{aligned}$$

This is 21 orders of magnitude smaller than the diameter of a typical nucleus ($\sim 10^{-15}$ m)! Your wavelike properties are negligible beyond belief.

Example 26.11

Electrons initially at rest are accelerated through a potential difference of 100 V, as in Figure 26.22. Find the de Broglie wavelength of the electrons when they reach their greatest speed.

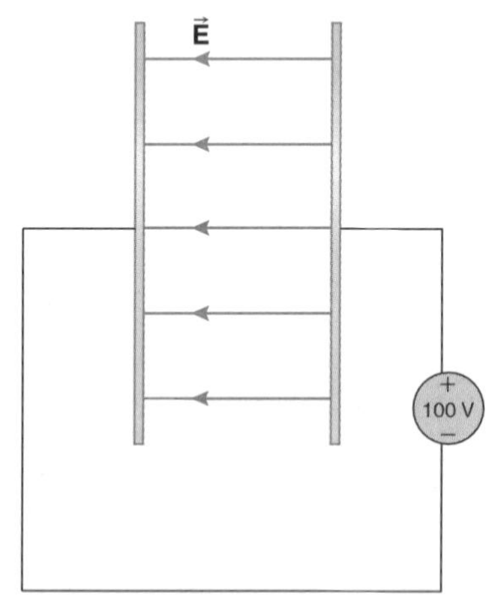

FIGURE 26.22

Solution

To find the de Broglie wavelength of the electrons at their greatest speed, you need to find their momentum; this, in turn, means you have to calculate their maximum speed. The easy way to find the speed is to realize that the kinetic energy of the emerging electrons is changed by 100 eV, because they were accelerated through a potential difference of 100 V; see Equation 17.25. Since their initial kinetic energy is zero, their final kinetic energy is 100 eV. This corresponds to a kinetic energy in joules of

$$(100\ \text{eV})(1.602 \times 10^{-19}\ \text{J/eV}) = 1.60 \times 10^{-17}\ \text{J}$$

Assume the electrons are traveling at a nonrelativistic speed, so for each electron you can write

$$\frac{mv^2}{2} = 1.60 \times 10^{-17}\ \text{J}$$

Substituting for the mass of an electron ($m = 9.11 \times 10^{-31}$ kg) and solving for the speed v, you find

$$v = 5.93 \times 10^{6}\ \text{m/s}$$

Notice that $v/c = 0.0198$, implying the relativistic factor γ has the value $\gamma = 1.000$. The low speed justifies the use of the nonrelativistic expressions for the kinetic energy and momentum.

The electrons each have a momentum of magnitude

$$\begin{aligned} p &= mv \\ &= (9.11 \times 10^{-31}\ \text{kg})(5.93 \times 10^{6}\ \text{m/s}) \\ &= 5.40 \times 10^{-24}\ \text{kg}\cdot\text{m/s} \end{aligned}$$

The de Broglie wavelength of the electrons then is

$$\begin{aligned} \lambda &= \frac{h}{p} \\ &= \frac{6.626 \times 10^{-34}\ \text{J}\cdot\text{s}}{5.40 \times 10^{-24}\ \text{kg}\cdot\text{m/s}} \\ &= 1.23 \times 10^{-10}\ \text{m} \\ &= 0.123\ \text{nm} \end{aligned}$$

Although the wavelength of the electrons is quite small, it is comparable in size to that of an atom and much larger than the de Broglie wavelength of macroscopic masses (see Example 26.10).

CHAPTER SUMMARY

The discoveries of the *electron* and *radioactivity* were the first clues that atoms had structure.

Planck's constant $h \approx 6.626 \times 10^{-34}$ J·s made its first appearance in physics in the successful attempt by Max Planck to avoid the *ultraviolet catastrophe* associated with blackbody radiation.

When light with a frequency greater than a *cutoff frequency* ν_c illuminates a metal, electrons are liberated; this is the *photoelectric effect*. The cutoff frequency depends on the specific metal used in the effect. Einstein explained the features of the effect by assuming light consists of a collection of massless particles, called *photons*, each with an energy proportional to the frequency of the light:

$$E_{\text{photon}} = h\nu \qquad (26.4)$$

where h is Planck's constant. The emerging electrons, called photoelectrons, have a variety of speeds. For the electrons with the greatest speed, energy conservation implies that

$$h\nu = W + \text{KE}_{\text{max}} \qquad (26.7)$$

where W is the *work function* of the metal and KE_{max} is the kinetic energy of the fastest emergent electrons. The work function is related to the cutoff frequency ν_c for the photoelectric effect, and represents the minimum amount of work needed to liberate an electron from the metal:

$$h\nu_c = W \qquad (26.8)$$

For incident light of frequency ν_c, the (barely) liberated electrons have 0 J of kinetic energy. The maximum kinetic energy of the emergent photoelectrons is related to the stopping potential V_s:

$$\text{KE}_{\text{max}} = eV_s \qquad (26.10)$$

Experiments by Rutherford using α particles scattered from thin gold foils led to the development of the *nuclear model* of the atom.

Bohr developed a nuclear model for *hydrogenic atoms* (single-electron atoms), called the *Bohr model*, that correctly predicted the wavelengths of light emitted and absorbed by such atoms. To do this, Bohr quantized the magnitude of the orbital angular momentum of the electron:

$$L_{\text{orbit}} = n\hbar \qquad (26.20)$$

where n is a positive, nonzero integer called a quantum number, and $\hbar$ is Planck's constant divided by 2π:

$$\hbar \equiv \frac{h}{2\pi} = 1.055 \times 10^{-34}\ \text{J}\cdot\text{s} \qquad (26.19)$$

As a result of the quantization of angular momentum, the electron orbits the nucleus at only specific radii given by

$$r = \frac{n^2\hbar^2(4\pi\varepsilon_0)}{mZe^2} \qquad (26.24)$$

The total energy also is quantized:

$$E_n = -\frac{1}{2}\frac{mZ^2e^4}{(4\pi\varepsilon_0)^2 n^2\hbar^2} \qquad (26.27)$$

$$= -(13.6\ \text{eV})\frac{Z^2}{n^2} \qquad (26.29)$$

The *ground state* has the lowest energy and corresponds to the quantum number $n = 1$. A photon of light is emitted if the electron lowers its energy by changing to an orbit with a lower value of n:

$$E_i = E_f + E_{\text{emitted photon}} \quad \text{(emission)} \qquad (26.31)$$

An atom absorbs light of the same frequency the atom emits, in which case the electron changes to an orbit with a higher value for n:

$$E_i + E_{\text{absorbed photon}} = E_f \quad \text{(absorption)} \qquad (26.32)$$

The *Bohr correspondence principle* states that as a quantum number such as n becomes very large, the results of quantum theory approach those secured from classical physics.

The Bohr model is unable to account for the relative intensities of the wavelengths emitted by hydrogenic atoms, and so is an incomplete model of the atom.

Radioactive materials emit three types of radiation:

- α particles (helium nuclei ${}^4_2\text{He}$);
- β particles (electrons β^-; or positrons β^+, which are positively charged electrons); and
- γ rays (high-energy photons).

The emission of α or β particles by a nucleus transmutes the element into a different element, realizing the dream of ancient and medieval alchemy.

The *activity*

$$\frac{dN}{dt}$$

of a radioactive material measures the number of nuclei decaying per unit time. The activity is measured in SI units of *becquerels* (Bq), equal to one disintegration per second. Another common and convenient unit of activity is the curie (Cu), equal to 36×10^9 Bq.

The activity is proportional to the number of nuclei of the radioactive isotope present:

$$\frac{dN}{dt} = -\lambda N \qquad (26.45)$$

where λ is the *disintegration constant.* The number of radioactive nuclei N decreases exponentially with time:

$$N = N_0 e^{-\lambda t} \qquad (26.47)$$

where N_0 is the number of nuclei present when $t = 0$ s. The *half-life* $\tau_{1/2}$ is the time at which only half the original number of nuclei remain:

$$\tau_{1/2} = \frac{\ln 2}{\lambda} \qquad (26.48)$$

The mean lifetime is the average age of the entire sample of nuclei after their decay:

$$\text{mean lifetime} = \frac{1}{\lambda} \qquad (26.56)$$

The mean life is longer than the half-life.

The activity of a sample decays exponentially with time with the same half-life as the number of nuclei:

$$\frac{dN}{dt} = \left(\frac{dN}{dt}\right)_0 e^{-\lambda t} \qquad (26.52)$$

The biological effects of radiation are difficult to assess, predict, and measure. The absorbed dose is measured in *grays* (Gy) while the dose-equivalent is measured in *sieverts* (Sv).

A photon has a momentum of magnitude p, related to its energy E:

$$p = \frac{E}{c} \qquad (26.59)$$

Since the photon energy is $E = h\nu = hc/\lambda$, the momentum and wavelength of a photon are related by

$$p = \frac{h}{\lambda} \qquad (26.61)$$

The *Compton effect* is observed during the scattering of high-energy photons (x-rays) from electrons. The emerging x-ray wavelength λ' is greater than the incident x-ray wavelength λ. The change in the wavelength of the incident photon is related to the scattering angle θ of the emerging x-ray and the *Compton wavelength* h/mc of the electron:

$$\lambda' - \lambda = \frac{h}{mc}(1 - \cos\theta) \qquad (26.78)$$

Compton derived Equation 26.78 by assuming the x-ray photon undergoes an elastic collision with a stationary electron. The Compton wavelength of a particle is *not* a wavelength of light, nor is it the de Broglie wavelength of the particle. The Compton wavelength merely is a way to say that the quantity h/mc has the dimensions of a length. It has no other physical significance.

The *de Broglie hypothesis* associates wavelike properties with particles. The *de Broglie wavelength* λ of a particle with momentum p is

$$\lambda = \frac{h}{p} \qquad (26.81)$$

The de Broglie wavelength is not a wavelength of light nor is it the Compton wavelength of a particle. The Bohr angular momentum postulate implies there are an integral number of de Broglie wavelengths of the electron around the circumference of an allowed orbit.

The de Broglie wavelength of macroscopically massive particles is completely negligible, since it is vanishingly small compared with even the size of a nucleus. On the other hand, the wave nature of particles of small mass, such as electrons, is not negligible on the atomic or nuclear length scales. The wave aspect of material particles was first confirmed in the electron scattering experiments of Davisson and Germer, and of Thomson and led to the development of electron microscopes.

Summary of Problem-Solving Tactics

26.1 (page 1214) In using Equations 26.7, 26.8, 26.10, or 26.11 for the photoelectric effect, you must express all the energies either in joules (J) or, if you choose, electron-volts (eV).

26.2 (page 1214) From Equation 26.10, the stopping potential in volts is *numerically* equal to the maximum kinetic energy of the electrons expressed in electron-volts.

26.3 (page 1228) In the radioactive decay law for the number of nuclei remaining or for the activity, the units you use for $\tau_{1/2}$ and $1/\lambda$ must be the same.

Questions

1. (page 1211); 2. (page 1215); 3. (page 1218); 4. (page 1222); 5. (page 1229)
6. Television picture tubes, oscilloscopes, and computer monitors all are called CRTs. Guess or determine what the meaning of CRT is.
7. Why is the term radioactivity somewhat of a misnomer? Invent a new term to describe the phenomenon.
8. All objects emit a blackbody spectrum characteristic of their absolute temperature. Why then can we not see most things in the dark?

9. The stars Antares (α Scorpii in the constellation Scorpius) and Aldebaran (α Tauri in the constellation Taurus) are slightly reddish in color while Vega (α Lyrae in the constellation Lyra) and Arcturus (α Boötis in the constellation Boötes) are slightly bluish in color. What does this observation imply about their relative surface temperatures?
10. What is the smallest unit of currency in your country? Is money quantized? Explain how the smallest unit of money plays the role of Planck's constant in measuring economic activity. What is the role of a quantum number in this case?
11. Is there a smallest unit of a living organism? Is life quantized?
12. Which has greater energy: a microwave photon or a radio wave photon?
13. Ultraviolet light causes sunburn. Visible light does not. Explain this observation in terms of photons.
14. Can you get a sunburn while sitting in your car with the windows up? What does this tell you about the optical properties of glass?
15. Explain why plastic bags do not deteriorate when exposed to incandescent lights inside your home but will degrade when exposed to sunlight for long periods. What photons likely are responsible for the degradation of the plastic?
16. Is the dark night sky empty of all photons?
17. If a light source is approaching you, a frequency higher than the true frequency of the light is observed. This means the photon you observe has a higher energy than the one in a reference frame in which the light source is at rest. Where did the additional energy come from? Does this observation violate energy conservation?
18. You perform the photoelectric effect experiment with light of frequency $\nu > \nu_c$ for a metal with a work function W_0. You now use light of the same frequency on another metal with work function W. In order to produce photoelectrons with the replacement metal using light of the same frequency as before, what must be the relationship between W and W_0?
19. What is the difference (if any) between an electron and a photoelectron?
20. Why are many spacecraft wrapped in gold foil rather than much cheaper aluminum or tin foils?
21. Does the stopping potential V_s depend on the frequency of light used in the photoelectric effect? Does V_s depend on the intensity of the light? Explain your answers.
22. Does the current detected by the ammeter in Figure 26.5 depend on the intensity of the light? Does the current depend on the frequency of the light as long as $\nu > \nu_c$? Explain your answers.
23. Would you expect the work function of a metal to be a function of the temperature of the metal? If so, in what way?
24. When zinc is illuminated by the light from a mercury vapor lamp, the photoelectric effect is noticed. When a thick piece of glass is placed between the lamp and the zinc, the photoelectric effect stops even though light still reaches the zinc. Develop a hypothesis to explain this effect.
25. The nuclear model of the atom is not as obvious as it may seem. How would you convince a nonbeliever of its reality?
26. A collection of hydrogen atoms is illuminated with high-energy photons with energies greater than 13.6 eV. Can such photons be absorbed by the electrons? What happens if they are?
27. An antihydrogen atom consists of a positron (a positively charged electron) orbiting an antiproton (a negatively charged proton). Will the spectrum of antihydrogen be different from the spectrum of ordinary hydrogen? Can we be certain that the hydrogen in distant stars is ordinary hydrogen and not antihydrogen? One of the great mysteries of the universe is why it is principally composed of ordinary matter and not antimatter. We think we know why; see Edward R. Harrison, *Cosmology* (Cambridge University Press, Cambridge, England, 1981), pages 354–357.
28. The electromagnetic and gravitational forces are long-range forces. Masses exert gravitational forces and charges exert electrical forces on each other whatever their separation. Explain why the existence of multiple protons in a nucleus is evidence for the existence of a very strong nuclear force with a very short effective range.
29. What evidence suggests that radioactivity is not an atomic process but a nuclear process?
30. In the typical carbon dating technique of organic materials, explain why it is typically necessary to destroy (at least part of) the sample in order to determine its age.
31. In β^- decay, what experimental evidence indicates the electron originated from the nucleus rather than from the electrons surrounding the nucleus?
32. Some radioactive materials, like $^{238}_{92}U$, have very long half-lives (for $^{238}_{92}U$ it is 4.47 *billion* years). Obviously, we did not actually wait billions of years to measure such a long half-life, since we have only known about radioactivity for about a century. Describe how it is possible to measure such very long half-lives.
33. What is the difference between the photoelectric effect and the Compton effect?
34. Why is the Compton effect not observed with visible light?
35. If the recoiling electron in the Compton effect has the maximum kinetic energy, in what direction is it moving relative to the direction of the incident x-rays?
36. A proton and electron have the same speed. Which has the longer de Broglie wavelength and by how much?
37. The electron and proton have the same magnitude of electric charge. What factors make the development of a proton microscope less appealing than an electron microscope?
38. A photon in vacuum has a certain momentum. If the same photon is in water, is its momentum changed? This is not an easy question to answer! Why? You might discuss what it means to say a photon is in water.
39. The wavelength of a photon is given by

$$\frac{hc}{\text{total energy}}$$

but the de Broglie wavelength of a neutron is *not* given by

$$\frac{hc}{\text{total energy}}$$

Why not? What property of the neutron makes for the distinction?

PROBLEMS

Sections 26.1 The Discovery of the Electron
26.2 The Discovery of X-rays
26.3 The Discovery of Radioactivity
26.4 The Appearance of Planck's Constant h

1. If you double the temperature of a blackbody radiator, by what factor is the peak wavelength emitted changed?

2. Your normal body temperature is about 37 °C. According to Wien's law, what is the wavelength of the peak of the blackbody spectrum that you emit? In what region of the electromagnetic spectrum is this wavelength?

3. What is the peak wavelength of the blackbody radiation left over from the creation of the universe, whose temperature is 3 K? Is this radio wave, microwave, infrared, visible, ultraviolet, x-ray, or γ-ray electromagnetic radiation?

4. The stars with the highest surface temperatures are called O-type stars and appear bluish in color. The peak of their blackbody spectrum occurs at a wavelength of about 1.0×10^2 nm. What is the approximate surface temperature of such stars?

5. A red giant star such as Antares (α Scorpii in the constellation Scorpius) has the peak of its spectrum at about 6.5×10^2 nm. What is the surface temperature of the star?

6. The tungsten filament of a common light bulb has a temperature of 2.0×10^3 K. What is the wavelength of the peak of its blackbody spectrum? What does this result imply about the relative efficiency of the bulb in producing visible light compared with infrared light?

7. At what wavelength is the peak of the electromagnetic spectrum of a blackbody radiator with a temperature of 20 °C?

Section 26.5 The Photoelectric Effect

8. How many radiofrequency photons per second are emitted by a radio station broadcasting at 89.1 MHz with a power output of 50 kW?

9. The visible light spectrum ranges from about 400 nm to 700 nm. What is the corresponding energy range of visible light photons in electron-volts?

•10. When 280 nm light is incident on a particular metal, the stopping potential is found to be 0.90 V. (a) What is the maximum kinetic energy of the photoelectrons in eV? (b) What is the maximum kinetic energy of the photoelectrons in J? (c) What is the maximum speed of the photoelectrons? (d) What is the work function of the material in eV?

•11. The cutoff wavelength for photoelectric emission from a certain metal is found to be 656 nm. (a) What is the work function of the material in eV? (b) When the material is illuminated with light with a wavelength of 430 nm, what is the maximum kinetic energy of the photoelectrons in eV? (c) What is the corresponding stopping potential? (d) If the number of 430 nm photons illuminating the material is doubled, what happens to the stopping potential?

•12. Photons with an unknown wavelength are used in the photoelectric effect from a sodium surface. Sodium has a work function of 2.3 eV. It is noted that a stopping potential of 5.0 V causes the photocurrent to decrease to zero. (a) Determine the wavelength of the incident photons in nm. (b) Determine the maximum speed of the emitted photoelectrons.

•13. Light with a wavelength of 200 nm falls on an aluminum metallic smile. The work function of aluminum is 4.1 eV. (a) What is the maximum kinetic energy of the photoelectrons in eV? (b) What is the stopping potential? (c) What is the cutoff wavelength in nm?

•14. Photons with an energy of 3.1 eV are incident on a metal surface. Zap. The maximum kinetic energy of the emitted photoelectrons is found to be 1.0 eV. Find the cutoff wavelength for the photoelectric effect on this metal.

•15. After many hours of diligent research, you obtain the following data on the photoelectric effect for a certain material:

Wavelength of light (nm)	*Stopping potential* (V)
360	1.40
300	2.00
240	3.10

(a) Plot the stopping potential (vertical axis) versus the *frequency* of the light. (b) What is the cutoff frequency for the photoelectric effect in this material? (c) What is the cutoff wavelength? (d) What is the work function (in eV) for the metal used in this experiment? (e) Given the value of the fundamental unit of charge $e = 1.602 \times 10^{-19}$ C, what value for Planck's constant is implied by this data?

•16. You doesn't mess around with low-class metals. When you do the photoelectric effect, you does it with class: on platinum. The work function for platinum is 6.4 eV. (a) What is the cutoff wavelength for the photoelectric effect on platinum? (b) If light of wavelength 488 nm is incident on the platinum surface, what happens?

•17. Light with a frequency of 5.0×10^{14} Hz strikes a metal surface that has a work function of 2.0 eV. (a) Will photoelectrons be ejected or not? Show your reasoning. (b) Sketch a graph of the greatest kinetic energy (in eV) of the photoelectrons versus the frequency of light used in the photoelectric effect on this metal. (c) If the light intensity is doubled, how does this affect the graph of part (b)?

•18. Professor I. M. Shirley Wright has given you the task of designing a photodetector that is sensitive to blue light (~450 nm) but not to red light (~600 nm). (a) In order to choose a suitable photocathode metal, determine the range of possible work functions that would be acceptable for this detector. (b) Sketch a complete simple circuit to detect this light using an ideal voltage source (battery), an ammeter (to detect the photocurrent), wire, and metal parts within the photocell. Show where the incident light hits and which way the standard current flows through the ammeter. (c) Without using selective light-transmitting filters, is it possible to design a photodetector that is sensitive to the red light but not the blue light? Explain your answer.

•19. An aluminum tool is adrift in the vacuum of space, left basking in the unshielded sunlight by a careless butterfingered astronaut. (a) The tool gradually acquires an electrical charge. By what mechanism does it acquire a charge and what is the sign of the charge? (b) Will the amount of charge accumulate indefinitely? Explain. (c) Calculate the range of solar wavelengths that are involved in the process. (d) Solar radiation is most intense at about 550 nm but some radiation exists down to about 300 nm and below. To prevent charge accumulation on a spacecraft, you are asked by your supervisor at NASA to consider coating the craft with a thin layer of gold. The work function for gold is 4.8 eV. What range of wavelengths will *not* produce any photoelectric charging of the gold surface? (e) A gold salesperson insists that a gold layer 1000 times thicker would be 1000 times more effective in preventing charge accumulation by sunlight. Assess this option from the standpoint of the physics involved and make a recommendation to your supervisors about the wisdom of the sales proposal.

Sections 26.6 The Quest for an Atomic Model: Plum Pudding
26.7 The Bohr Model of a Hydrogenic Atom
26.8 The Bohr Correspondence Principle
26.9 A Bohr Model of the Solar System?*
26.10 Problems with the Bohr Model

•**20.** An incident α particle with an initial kinetic energy of 5.0 MeV is in a head-on trajectory with a gold nucleus ($Z = 79$). Use the CWE theorem to calculate the distance of closest approach of the α particle to the gold nucleus.

•**21.** An electron is in the $n = 2$ state of the Bohr model for hydrogen atom. Determine: (a) the orbital radius of the electron; (b) the orbital speed of the electron; (c) the magnitude of the orbital angular momentum of the electron; (d) the magnitude of the centripetal acceleration of the electron; (e) the ratio of the magnitude of the centripetal acceleration to that of the local acceleration of gravity g near the surface of the Earth; (f) the kinetic energy of the electron (in eV); (g) the potential energy of the electron (in eV).

•**22.** The various wavelengths emitted by hydrogen in its spectrum are categorized in the following way. The wavelengths emitted when the final state of electron is the $n = 1$ state are part of the Lyman series; those wavelengths emitted when the final state of the electron is the $n = 2$ state are part of the Balmer series; the wavelengths emitted when the final state of the electron is the $n = 3$ state are part of another series. Strictly speaking, each series of wavelengths contains an infinite number of wavelengths. (a) Explain why the various wavelengths in a given series converge to a short wavelength limit. (b) Calculate the short wavelength limit of the Lyman and Balmer series.

•**23.** In the Bohr model of the hydrogen atom, use Newton's second law and the expressions for the centripetal acceleration and the (nonrelativistic) kinetic energy to find (algebraically) the pure number ratio of the kinetic to the potential energy of the circulating electron.

•**24.** An electron is quite Bohr(ed) tooling around in the lowly $n = 2$ state of the hydrogen atom and decides to make life more exciting by absorbing an appropriate passing photon and making a transition to the $n = 5$ state. Determine the wavelength (in nm) of the photon absorbed.

•**25.** (a) What is the wavelength of a photon that barely can ionize a hydrogen atom with its electron initially in the $n = 2$ state? (b) Will longer wavelength photons ionize the atom if the electron is initially in the $n = 2$ state?

•**26.** A hydrogen atom becomes excited by a passing photon, absorbs it, and makes a transition from the $n = 1$ state (of Virginia) to the $n = 5$ state (also in Virginia). (a) What energy must be absorbed for the electron to make this transition? (b) What was the wavelength of the photon absorbed?

•**27.** Cool atomic hydrogen just gobbles up (absorbs) light with a wavelength of about 122 nm yet turns its nose up at nearby wavelengths, which pass unimpeded through the gas. Explain why this is the case with the assistance of an appropriate calculation.

•**28.** In the Bohr model of the hydrogen atom: (a) Show that the ratio of the nth orbital radius to the *longest* wavelength light that can be emitted from that state is

$$(5.80 \times 10^{-4})\left[\frac{n^2}{(n-1)^2} - 1\right]$$

(b) Show that the ratio of the nth orbital radius to the *shortest* wavelength light that can be emitted from that state is

$$5.80 \times 10^{-4}(n^2 - 1)$$

•**29.** Photons with wavelengths less than about 10 nm typically are classified as x-rays. Prove that none of the light emissions from the transitions of the orbital electron in the hydrogen atom produce x-rays.

•**30.** Hydrogen atoms with very large quantum numbers n are called *Rydberg atoms*. If $n = 1.00 \times 10^3$, find the radial distance of the electron from the nucleus.

31. An exotic hydrogenic atom consists of an electron and positron (a particle with the same mass as an electron but with a positive charge $+e$); the atom is known as *positronium*. The electron–positron pair orbit their center of mass (located halfway between them (see Figure P.31) at the same speed v. We can develop a Bohr model for positronium, assuming the mutual orbit is circular. Let $2r$ be the distance between the

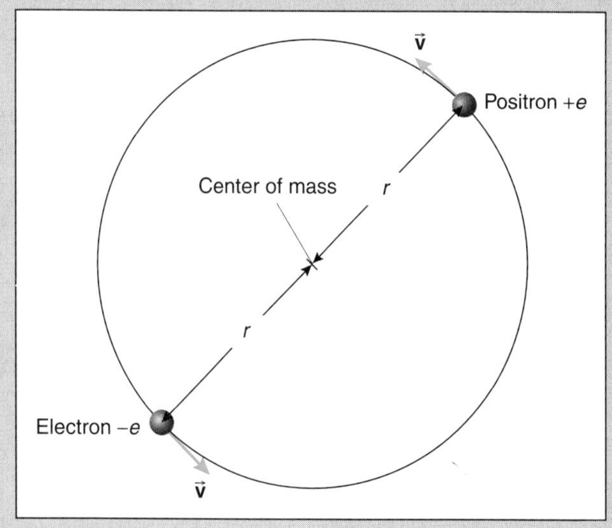

FIGURE P.31

electron and positron, so r is the radius of their mutual circular orbit as indicated in Figure P.31. (a) The force on each particle is the Coulomb force of electrical attraction. Apply Newton's second law to the electron (or the positron) to show that

$$\frac{1}{4\pi\varepsilon_0}\frac{e^2}{4r} = mv^2$$

where v is the common speed of the electron and positron. (b) The total energy E of the system consists of the kinetic energy of the electron and that of the positron (assume the speeds are nonrelativistic), plus the mutual electric potential energy of the pair of charges. Show that the total energy E of the system can be written as

$$E = -\frac{1}{4}\frac{1}{4\pi\varepsilon_0}\frac{e^2}{r}$$

(c) Quantize the magnitude of the total orbital angular momentum of the electron and positron by setting it equal to a positive integer n times $\hbar$. Show that this quantization leads to the expression

$$v = \frac{n\hbar}{2mr}$$

(d) Show that the total energy E then can be written in the form

$$E = -\frac{1}{2}\left[\frac{1}{2}\frac{1}{(4\pi\varepsilon_0)^2}\frac{me^4}{n^2\hbar^2}\right]$$

The expression in the bracket is the total energy for the electron in the normal hydrogen atom. Thus the energy levels for the positronium system are precisely *one-half* those of the normal hydrogen atom. (e) Show that the energy levels of positronium can be expressed in electron-volts as

$$E = -\frac{6.8 \text{ eV}}{n^2}$$

(f) Calculate the wavelength in nm of a photon emitted by positronium in making a transition from the $n = 2$ state to the $n = 1$ state. Is this photon in the visible region of the electromagnetic spectrum? The existence of positronium was experimentally confirmed in 1951 by Martin Deutsch (1917–). After its creation, the positronium atom is quite short-lived before the electron and positron annihilate each other to produce γ-rays.

32. A particle called a *muon* has a mass 207 times that of the electron but also with a single negative fundamental unit of charge ($-e$). The muon can be captured by a proton to form an exotic hydrogenic atom known as *muonium*. (a) Modify the Bohr expression for the energy levels of a hydrogenic atom to find the energy levels of muonium in electron-volts; for simplicity, assume the nucleus is fixed. Explain why this assumption is a crude first approximation. (b) What energy is needed to ionize muonium if the muon is in the ground state? (c) If the muon makes a transition from the $n = 2$ state to the $n = 1$ state, what is the wavelength of the photon emitted?

Sections 26.11 Radioactivity Revisited
26.12 Carbon Dating
26.13 Radiation Units, Dose, and Exposure*

•**33.** Radon $^{222}_{86}$Rn has a half-life of 3.82 d. It is produced naturally from the decay of uranium, present in small amounts in stone, bricks, concrete, and other building materials. Long-term exposure to even low doses of radon has been implicated as a possible cause of lung cancer. Since radon is a clear, odorless gas, it cannot be detected by its odor in unventilated spaces such as basements. (a) What is the disintegration constant λ for the decay of radon? (b) What fraction of a sample of radon remains after a month (30 days)?

•**34.** Radium $^{226}_{88}$Ra was first isolated by Marie and Pierre Curie in 1898. Its half-life is 1.60×10^3 y. (a) What is its disintegration constant λ in y^{-1}? (b) Approximately what percentage of their original sample has not yet decayed?

•**35.** Radium $^{226}_{88}$Ra decays by α particle emission with a half-life of 1.60×10^3 y. (a) What is its half-life in s? (b) Find the disintegration constant λ in s^{-1}. (c) How many atoms of the substance will provide an activity of 1.00 Cu? (d) What is the mass of the collection of atoms found in part (c)? Your answer is an indication that radium in large quantities is an extraordinarily dangerous radioactive material.

•**36.** For a radioactive material with a half-life $\tau_{1/2}$, how long will it take until the activity of a given sample is only 10% of its initial value? Express your answer in terms of the half-life of the substance.

•**37.** Provide the details of the integration in Equation 26.55 to show that the mean life is equal to the reciprocal of the disintegration constant.

•**38.** Show that after a time t equal to the mean life, the number of radioactive particles remaining in the sample is N_0/e, where e is the base of natural logarithms.

•**39.** Show that the difference between the mean life and the half-life is approximately

$$\frac{0.3069}{\lambda}$$

•**40.** The element $^{226}_{88}$Ra decays by a emission with a half-life of 1.60×10^3 y. (a) What is its daughter? (b) How long will it take until only 1.0% of an original sample has not decayed?

•**41.** The isotope $^{238}_{92}$U has a half-life of 5.0×10^9 y. What fraction of the $^{238}_{92}$U present on the Earth 2.5×10^9 years ago has not yet decayed?

•**42.** Indicate the type of particle emitted in each of the following radioactive decay equations by completing the reaction: (a) $^{56}_{27}\text{Co} \rightarrow ^{56}_{26}\text{Fe} +$ ______; (b) $^{212}_{84}\text{Po} \rightarrow ^{208}_{82}\text{Pb} +$ ______; (c) $^{232}_{90}\text{Th} \rightarrow ^{228}_{88}\text{Ra} +$ ______.

•**43.** The radioactive isotopes of sodium, $^{22}_{11}$Na and $^{24}_{11}$Na, both decay by b⁻ emission. Determine the daughter products of the decays of each isotope.

•**44.** Radioactive materials are used widely in cancer therapy and for biological research. For these purposes, it is useful to know the *biological half-life* of the radioactive substance, defined as the time during which the organism expels half the ingested material. Assume the biological expulsion is similar to the natural physical decay. When ingested, some of the material decays and some is expelled. Assume the activity remaining is proportional to the number of nuclei remaining, via

$$\frac{dN}{dt} = -(\lambda + \lambda_b)N$$

where λ_b is the biological disintegration constant and λ is the physical disintegration constant. Show that the effective half-life τ_{eff} of the sample within the organism is related to the biological half-life by $\tau_b = (\ln 2)/\lambda_b$, and the natural half-life τ by

$$\frac{1}{\tau_{eff}} = \frac{1}{\tau} + \frac{1}{\tau_b}$$

•45. An animal is given a small dose of radioactive iodine $^{131}_{53}I$, which has a physical half-life of 8.1 d. Measurements are made of the activity of the animal and it is determined that the effective half-life of the material in the organism is 6.2 d. What is the biological half-life of the material within the organism? (Refer to Problem 44.)

•**46.** A 10 mCu pointlike source of $^{60}_{27}Co$ is implanted into a tumor. A detector with a collecting area of 25 cm^2 is placed 15 cm from the source, external to the patient. Assume the γ-rays emitted by the source are equally likely to be emitted in any direction. What is the approximate activity registered by the detector assuming only a negligible (but biologically significant) fraction of the radiation is absorbed by the tumor and surrounding tissue?

•**47.** An archeological dig in the basement of a physics laboratory uncovers the bones of a college student who expired some time ago. A careful analysis of the remains indicates a carbon 14 activity of 0.15 Bq per gram of carbon. A freshly sacrificed student has a carbon 14 activity of 0.27 Bq per gram of carbon. The half-life of carbon 14 is 5.73×10^3 y. (a) What is the disintegration constant λ in y^{-1}? (b) What is the age of the old student bones? (c) What will be the activity per gram of carbon of the old student bones when the class of A.D. 3268 graduates?

•**48.** Because of burnout, a sample of deans decreases its activity from 8000 memos per day to 1000 memos per day in 10 days. Assume their memo output is governed by exponential decay. What is the half-life of the sample?

•49. You uncover a cache of yellowed lecture notes on papyri written by the ancient and apocryphal natural philosopher Claudius Mediocratus. By careful analysis, the notes are discovered to have an activity (all their own) of 750 disintegrations per lecture hour per gram of carbon. Fresh notes on papyrus have an activity of 960 disintegrations per lecture hour per gram of carbon. The half-life of carbon 14 is 5.73×10^3 y. (a) Find the disintegration constant λ. (b) Determine the age of the lecture notes.

•**50.** An unusual sample population of Methuselahs decays exponentially with half-life of 50 y. (a) How long will it take till only 10% of the original population is left? (b) If the initial collection consisted of 1000 persons, what is the death rate (deaths/year) of the sample when only 10% remain?

•**51.** Uranium ores on the Earth at the present time typically have a composition consisting of 99.3% of the isotope $^{238}_{92}U$ and 0.7% of the isotope $^{235}_{92}U$. The half-lives of these isotopes are 4.47×10^9 y and 7.04×10^8 y respectively. If these isotopes were equally abundant when the Earth was formed, estimate the age of the Earth.

•**52.** The radioactive gas radon $^{222}_{86}Rn$ decays by α particle emission with a half-life of 3.82 days. (a) What is the daughter? (b) Compute the activity in curies of 1.00 mg of radon. This activity indicates why radon, even in small concentrations, may be a health hazard.

•53. The half-life of a sample of carbon 14 atoms is 5.73×10^3 y. If you had a collection of only *five* carbon 14 atoms, how many will still be present after 5.73×10^3 y? Explain the meaning of your answer.

Section 26.14 The Momentum of a Photon*

54. Calculate the magnitude of the momentum of a photon with a vacuum wavelength of 500 nm.

55. Photon A has twice the energy of photon B. What is the ratio of their momenta?

56. Massive atoms are struck by x-ray photons each with an energy of 40 keV and electrons are easily knocked away. What is the change in the wavelength of an x-ray photon scattered through an angle of 120° to the direction of the incident beam?

57. An x-ray photon with a wavelength of 0.1000 nm makes an elastic collision with an electron and is scattered through an angle of 90.0°. What is the wavelength of the scattered photon?

•**58.** Light of wavelength 632.8 nm is incident normally on a black mass and is totally absorbed. The number of photons per second that strike the mass is 1.50×10^{18}. (a) What is the magnitude of the force exerted on the black mass by the photons? (b) What is the power supplied to the black mass?

•**59.** Light from the Sun has an effective wavelength of 5.5×10^2 nm. The luminosity of the Sun is 3.83×10^{26} W and the distance between the Earth and the Sun is 149.6×10^6 km. (a) About how many joules per second are incident on each square meter of a sphere centered on the Sun with a radius equal to the distance between the Earth and Sun? (b) About how many photons per second are incident on each such square meter? (c) If the photons reflect from a sail oriented perpendicular to the direction to the Sun at the distance of the Earth, how large a square sail area is needed for it to experience a force of magnitude 1.00 N?

•**60.** A helium–neon laser emitting light at a wavelength of 632.8 nm has a power output of 500 mW. (a) How many photons per second are being emitted by the laser? (b) The laser has a mass of 5.00 kg and is so far away from any large masses that gravitational effects can be neglected. The laser is initially at rest and then is turned on for a year. Calculate the final speed of the laser.

•61. Show that the momentum of a photon can be interpreted as the product of Planck's constant h times the number of wavelengths in a meter. The number of wavelengths per meter is called the *wavenumber*.

•**62.** If the Earth (of radius $R = 6730$ km) absorbed all photons incident on it from the Sun, what magnitude of force would the solar radiation exert on the Earth? Assume the effective wavelength of solar radiation is 550 nm and the solar luminosity is 3.83×10^{26} W.

63. Interplanetary dust in the solar system is simultaneously subjected to at least two forces: the gravitational attraction of the Sun on the particles and a repulsive force caused by the absorption of solar radiation and the consequent transfer of photon momentum to the particle. Here we investigate the conditions under which one or the other of these forces dominates. To do this, first consider the force that solar photons exert on a dust particle. For simplicity, consider a spherical particle of radius R located a distance r from the Sun. Let the solar luminosity be L (equal to the power output of the Sun: 3.83×10^{26} W). Let ν be the average or effective frequency of solar radiation (the frequency of light with a wavelength of about 550 nm). (a) Show that the number of photons per second intercepted by the particle of radius R located a distance r from the Sun is

$$\frac{L}{4\pi r^2}\frac{\pi R^2}{h\nu}$$

(b) Each photon has a momentum $h\nu/c$. If all the intercepted photons are absorbed by the particle, show that the magnitude of the force on the particle (the magnitude of the time rate of change of its momentum) due to the solar radiation is

$$F_{\text{radiation}} = \frac{L}{4\pi r^2 c}\pi R^2$$

Note that the magnitude of this force of repulsion is proportional to the cross-sectional area of the particle and so increases with the square of its radius. (c) Let ρ be the density of the spherical dust particle. Show that the magnitude of the gravitational force of the Sun (of mass $M = 1.99 \times 10^{30}$ kg) on the particle is

$$F_{\text{grav}} = \frac{GM}{r^2}\frac{4}{3}\pi R^3\rho$$

Note that the magnitude of this attractive force is proportional to the volume of the particle and so increases with the cube of its radius. (d) Show that the *ratio* of the magnitude of the repulsive force due to solar radiation to that of the attractive force due to gravitation is

$$\frac{F_{\text{radiation}}}{F_{\text{grav}}} = \frac{3L}{16\pi GMcR\rho}$$

Note that this ratio is *independent of the distance r of the particle from the Sun*. (e) Let the density of each particle be $\rho = 5.0 \times 10^3$ kg/m^3, which is on the order of the density of many minerals. Show that the repulsive force of radiation is equal in magnitude to the attractive force of gravitation if the radius R of the particle is on the order of 1.2×10^{-7} m. Particles smaller than this are repelled from the Sun since the solar radiation force dominates; particles larger than this are attracted to the Sun, since the gravitational force then dominates. Radiation forces on small particles were very important factors in the early history of the solar system.

Section 26.15 The De Broglie Hypothesis

64. For a local fund-raiser, you volunteer to be the target at a pie-throwing booth. Calculate the de Broglie wavelength of a 0.50 kg banana cream pie thrown at you with a speed of 10 m/s.

65. For an electron in a Bohr model energy state with quantum number n, determine how many de Broglie wavelengths of the electron fit around the circumference of the orbit.

•**66.** An electron is accelerated from rest through a potential difference of V to a speed that is nonrelativistic. In terms of V, h, e, and the mass of the electron m_e, find expressions for: (a) the kinetic energy of the electron; (b) the magnitude of the momentum of the electron; and (c) the de Broglie wavelength of the electron.

•**67.** An electron beam in an electron microscope has electrons with individual kinetic energies of 50 keV. Determine the de Broglie wavelength of such electrons.

•**68.** A photon and an electron have the same wavelength λ. Assume the electron has a nonrelativistic speed. In terms of λ and fundamental constants: (a) determine the speeds of the photon and the electron; (b) determine the magnitudes of the momenta of the photon and the electron; and (c) determine the energy of the photon and the kinetic energy of the electron.

•**69.** (a) What is the wavelength of a photon that has the same magnitude of momentum as an electron moving with a speed of 2.0×10^6 m/s? (b) What is the de Broglie wavelength of the electron?

•**70.** An electron and a proton are accelerated so that they have equal kinetic energies. Find the ratio of their de Broglie wavelengths. Assume that the speeds of the particles are nonrelativistic.

•**71.** (a) At what speed must a neutron move to have a de Broglie wavelength of 1.5×10^{-11} m? (b) What is the magnitude of the momentum of the neutron of part (a)? (c) At what speed must a photon move to have a wavelength of 1.5×10^{-11} m? (d) What is the magnitude of the momentum of the photon of part (c)?

•**72.** What is the ratio of the photon wavelength emitted in the $n = 2$ to $n = 1$ transition in the Bohr model of hydrogen to the de Broglie wavelength of the electron in the $n = 1$ state?

Investigative Projects

A. Expanded Horizons

1. Investigate the use of radioactive isotopes in determining the age of the Earth.
Ludwik Kowalski, "Radioactivity and nuclear clocks," *The Physics Teacher, 14*, #7, pages 409–416 (October 1976).
Robert J. Packhurst, "Radiometric dating in geology," *Physics Education, 15*, #7, pages 340–343 (November 1980).
Hans E. Suess, "Radiocarbon geophysics," *Endeavor, 4*, #3, pages 113–117 (1980).

2. Investigate the so-called *natural units* known as the *Planck length*, the *Planck mass*, and the *Planck time*.
Edward R. Harrison, *Cosmology* (Cambridge University Press, Cambridge, England, 1981), page 33.
Robert L. Wadlinger and Geoffrey Hunter, "Max Planck's natural units," *The Physics Teacher, 26*, #8, pages 528–529 (November 1988).

3. Write a précis about the following interesting article about the origins of particle waves.
Bruce R. Wheaton, "Louis de Broglie and the origins of wave mechanics," *The Physics Teacher, 22*, #5, pages 297–301 (May 1984).

4. Investigate the use of particle accelerators in improving the carbon 14 dating technique.
Harry E. Gove, "A new accelerator-based mass spectrometry," *The Physics Teacher, 21*, #4, pages 237–245 (April 1983).

5. Tree ring counts are used to fine-tune carbon 14 dating techniques. Investigate how this done.
Hans E. Suess, "Radiocarbon geophysics," *Endeavor, 4*, #3, pages 113–117 (1980).

6. Investigate the use of modern physics in the analysis of art work to decipher painting techniques as well as uncover forgeries.
Maurice J. Cotter and Kathleen Taylor, "Neutron activation analysis of paintings," *The Physics Teacher, 16*, #5, pages 263–271 (May 1978).

7. Investigate the exposure to radiation from background, diagnostic x-rays, and other sources.
Stewart C. Bushong, "Radiation exposure in our daily lives," *The Physics Teacher, 15*, #3, pages 135–144 (March 1977).

8. An early model of the universe (from the 1950s) imagined a dual universe, half of which was composed of matter, the neighboring half of antimatter. Investigate this model and report on the evidence that indicates this model is not likely to be correct.
Hannes Alfvén, *Worlds–Antiworlds: Antimatter Is Cosmology* (W. H. Freeman, San Francisco, 1966).

9. Why is the universe composed of matter and not antimatter?
Edward R. Harrison, *Cosmology* (Cambridge University Press, Cambridge, England, 1981), pages 356–357, 365.

10. Discover why the curie unit of activity is defined as 36×10^9 Bq rather than some more convenient number of disintegrations per second.

B. Lab and Field Work

11. Design and perform an experiment to measure the photoelectric effect from a metal with a small work function, so that visible light can be used.

12. Estimate the number of photons per second received by your eye from a bright star. Organize your calculation carefully and state what assumptions you make to secure an answer.

13. With an appropriate photometer, measure the average power output of an American firefly (*Photuris pennsylvanica*). Measure the spectrum of its light and determine an approximately effective wavelength of its light. Then estimate the approximate number of photons per second emitted in each flash.

14. Design and perform an experiment to measure the temperature dependence of the work function of a metal in the photoelectric effect.

15. Design and build a scale model of the Bohr model for hydrogen and place it appropriately on your campus. You may need to seek permission from an appropriate campus official!

16. Fairly short-lived radioactive nuclides are readily available for experimental use; consult your professor about them. Design and perform an experiment to measure the half-life of a radioactive material with a reasonably short half-life. Now do the same for a material with a long half-life.

17. Design and perform an experiment to measure the relative penetrating power of α, β, and γ particles.

18. Discover how commercial radon detectors operate. Measure the radon level at various places in the basement of your physics building and perhaps other structures on campus. Report your findings to the director of buildings and grounds on your campus.

19. Small particles can be levitated with the pressure provided by light. Investigate the particle sizes and powers of light sources necessary. Then demonstrate the effect for your class.
Arthur Ashkin, "The pressure of laser light," *Scientific American, 226*, #2, pages 62–71 (February 1972); page 118 has additional references.

20. By consulting with a local radiologist and/or oncologist, investigate the activities of the sources and dosages associated with radiation therapy for various cancer treatments. How are the dosages measured?
T. S. Curry, J. E. Dowdey, and R. C. Murry, *Christensen's Introduction to the Physics of Diagnostic Radiology* (Lea and Febiger, Philadelphia, 1984).
Harold E. Johns and John R. Cunningham, *The Physics of Radiology* (4th edition, C. C. Thomas, Springfield, Illinois, 1983).
J. W. Boag, "Forty years of development in radiation dosimetry," *Physics in Medicine and Biology, 29*, #2, pages 127–130 (February 1984).
W. J. Meredith, "Forty years of development in radiotherapy," *Physics in Medicine and Biology, 29*, #2, pages 115–120 (February 1984).

21. The electron microscope is an invaluable tool for research. Investigate the characteristics of an electron microscope likely to be at your university. How are samples prepared? What wavelength electrons are used? What is the resolution of the instrument?

C. Communicating Physics

22. The life of Marie Curie is a case study not only of the trials and tribulations, but also of the successes and triumphs of women in physics. After reading about Curie, interview one or more women physicists at your university (or another university) to see if the climate for women in the field has changed during the last century. On the basis of your interview, write a feature for your campus newspaper.

Dictionary of Scientific Biography (Charles Scribner, New York, 1971), Volume III, pages 497–503; see page 503 for additional references.

23. Poll your colleagues to determine common meanings of the phrase "quantum jump." Is the common usage of the term anything like its use in physics?

24. Soon after the discovery of x-rays, René-Prosper Blondlot (1849–1930) reported the discovery of yet another "unknown" type of radiation, which he dubbed N-rays, after Nancy, France, the city and university where he worked. Many other scientists also claimed to observe the mysterious N-rays. The story of their discovery and the subsequent realization that they simply do not exist is a case study of the danger of scientists seeing what they wanted to see, rather than what was actually there. Investigate the case of the "discovery" of N-rays and the convincing way that American physicist Robert Williams Wood (1868–1955) of The Johns Hopkins University "undiscovered" them by deft sleight of hand. Prepare an interesting oral report about the N-ray fallacy for your departmental seminar.

Robert T. Lagemann, "New light on old rays: N rays," *American Journal of Physics*, 45, #3, pages 281–284 (March 1977).

Spencer Weart, "A little more light on N rays," *American Journal of Physics*, 46, #3, page 306 (March 1978).

Robert Williams Wood, "The n-rays," *Nature*, 70, #1822, pages 530–531 (29 September 1904).

RADIOACTIVE DATING OF DAUGHTER PRODUCTS

Chapter 27

An Introduction to Quantum Mechanics

. . . I think I can safely say that nobody understands quantum mechanics.

Richard Feynman (1918–1988)*

The amusing yet sobering statement by Richard Feynman (see previous page), one of the foremost minds in 20th-century physics (and a physics Nobel prize laureate as well), reminds us of Sommerfeld's comments about thermodynamics (cf. the introduction to Chapter 13). We approach quantum mechanics with more than a little awe and certainly respect for the mysteriousness of nature.

In many respects, the world of physics mirrors its surrounding cultural milieu and, to some extent, helps to shape it. The classical and formal art, music, literature, and mathematics created during the Renaissance and Enlightenment periods, until the dawn of the 20th century, were complemented by the classical physics of Newton's dynamics, Maxwell's electromagnetism, and Boltzmann's thermodynamics. Art was realistic, music was harmonious, literature (poetry in particular) followed strict rules of form, meter, and rhyme; mathematics was calculus and number theory. Physics was precise, predictable, and deterministic.

But realistic art gave way to the abstractions of Renoir, Monet, and Picasso. The harmonic and structural formality of Bach, Mozart, and Beethoven evolved into new harmonies and tone poems by Chopin, Liszt, and Mahler as well as the wonderful spontaneity and improvisation of ragtime and jazz. The measured meters of Shakespeare, Shelley, Byron, and Tennyson changed to the free verse of Whitman, Eliot, Frost, and e. e. cummings. Mathematicians developed abstract algebra. And physicists discovered a new abstract formulation of the physical world as Bohr, de Broglie, Schrödinger, Heisenberg, and Dirac elucidated the features of a totally new and unexpected (almost counterintuitive) type of mechanics: quantum mechanics. The new art, music, literature, and mathematics was not always easy to understand or appreciate; it did not make common sense or resonate pleasingly to everyone. The physics of relativity (Chapter 25) and quantum mechanics (which we examine briefly in this chapter) defied common sense as well. Just as creators in the humanities were doing, physicists uncovered and invented a new way of thinking and a new sense of beauty. Strict determinism was replaced by probability, uncertainty, and an unfamiliar new world of nature.

In this chapter we venture a first look into this peculiar natural world. Some of this is quite heady material that will require much reflection, even years to understand; but, as in contemporary art or music, with guidance you certainly are capable of gleaning an appreciation for it. Keep your mind focused on the conceptual forest rather than the mathematical trees.

27.1 THE HEISENBERG UNCERTAINTY PRINCIPLES

I have known what the Greeks did not: uncertainty.

Jorge Luis Borges (1899–1986)†

Light exhibits both wave and particle characteristics. The diffraction and interference experiments of Young and the Maxwell theory of electromagnetism clearly indicate that light has wave properties. Yet Einstein's theory of the photoelectric effect and the Compton effect indicate that light, at least on a submicroscopic level, exhibits the attributes of a particle as well (a photon), having both energy and momentum. Conversely, the de Broglie hypothesis, the Bohr angular momentum postulate, and the experiments of Davisson and Germer, and Thomson indicate that what we traditionally think of as particles on a submicroscopic level (electrons, protons, neutrons, etc.) also exhibit behavior typically associated with waves (diffraction and interference). Thus nature is bilateral: particles are waves and waves are particles. The particle aspect carries with it the traditional concepts of position and momentum; the wave aspect carries with it the concepts of wavelength and frequency.

Here we shall see that the wave-particle aspects are inextricably interwoven in a way that fundamentally limits our ability to know or measure several of these parameters simultaneously with perfect precision. Nature places natural limits on the precision of our measurements; some knowledge and information forever is shrouded from our prying eyes.

The Position–Momentum Uncertainty Principle

Consider the diffraction of a monochromatic wave through a slitlike aperture of width a. A single slit diffraction pattern is observed on a distant screen, as shown in Figure 27.1. From our analysis of single slit diffraction in Section 24.4 we know that the smaller the aperture of the slit, the greater the angular width of the central diffraction peak. Such diffraction is characteristic of all wave phenomena.

Let's look at the diffraction from a slightly different perspective. To the left of the aperture, the wave has a precise wavelength λ and, according to the de Broglie relation, we have

$$\lambda = \frac{h}{|p_x|} \tag{27.1}$$

The absolute value of the momentum component $|p_x|$ in the incident direction of the wave also is well defined or precisely known. Since the wave is traveling along the x-axis, the wave has zero momentum component along the transverse direction, y. On the other hand, the position of the wave along the x-axis to the left of the aperture can be described by saying that the wave is everywhere. It is not possible to say that the wave is at any particular value of x. The wave simultaneously is at *every* value of x to the left of the aperture. Thus the incident wave has a well-defined wavelength or, equivalently, x-component of momentum, but an ill-defined position.

The passage of the wave through the slitlike aperture results in a single slit diffraction pattern on a distant screen. Most of the wave intensity is confined within the central diffraction peak (though

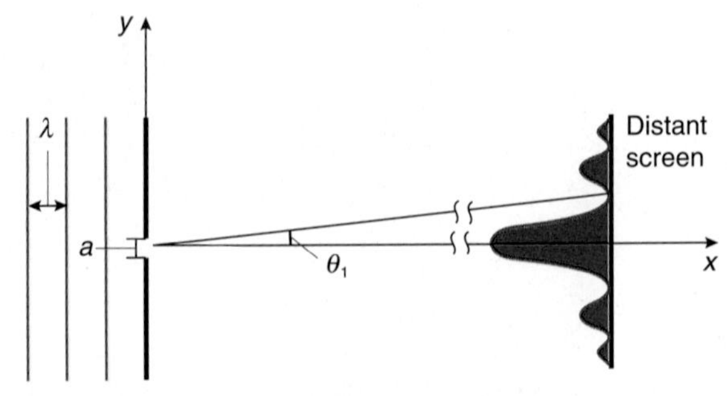

FIGURE 27.1 A single slit diffraction pattern.

*(Chapter Opener) *The Character of Physical Law* (M.I.T. Press, Cambridge, Massachusetts, 1967), page 129.

† *Ficciones*, edited by Anthony Kerrigan (Grove Press, New York, 1962), "The Babylonian Lottery," page 65.

several secondary maxima also exist). The location of the first minimum of the diffraction pattern is found from Equation 24.3:

$$a \sin \theta_1 = \lambda$$

If the angle θ_1 is small, the sine is approximately equal to the angle itself (in radians), which is the small angle approximation. Thus Equation 24.3 becomes

$$a\theta_1 = \lambda \qquad (27.2)$$

Using the de Broglie relation (Equation 27.1) to substitute for λ in Equation 27.2, we find

$$a\theta_1 = \frac{h}{|p_x|}$$

Rearranging this slightly, we obtain

$$a\theta_1 |p_x| = h \qquad (27.3)$$

At the slit, the slit width a is a measure of the extent to which the position of the wave is confined along the y-axis; certainly as the wave goes through the slit, you can say that the position of the wave was known to within $\Delta y = a$ along the y-axis or transverse direction. Correspondingly, the quantity $\theta_1 |p_x|$ is a measure of the degree to which we know the y-component of the momentum of the wave. The *smaller* the aperture, the *larger* θ_1 becomes, which means that the width of the central diffraction peak increases. Hence as θ_1 increases, the greater is the extent to which the wave spreads in the transverse (y) direction and has a nonzero y-component of the momentum.

We call $\theta_1 |p_x| = \Delta p_y$ and $a = \Delta y$.* Then Equation 27.3 becomes

$$\Delta y \, \Delta p_y = h \qquad (27.4)$$

At the slit, the more we try to confine the wave along the y-axis (by using a smaller and smaller slit opening $a = \Delta y$), the better we know the position at the slit of the wave along the y-axis. But the result of decreasing the slit width means that the width of the diffraction peak increases, which means we know less about the value of the y-component of the momentum of the wave. If we account for the fact that *secondary* diffraction maxima also exist, the uncertainty in the y-component of the momentum is even greater: Equation 27.4 really is a best-case scenario. To account for the secondary maxima, we really should write

$$\Delta y \, \Delta p_y \geq h \qquad (27.5)$$

Although what we have just discussed certainly is not a rigorous derivation, the treatment is an illustration of a fundamental aspect of the wave–particle duality.

> The **Heisenberg uncertainty principle** states that there exists a fundamental limit to the extent to which we can simultaneously determine the position and corresponding momentum component of a wave-particle in any given direction.

The principle is named for Werner Heisenberg (1901–1976), who first elucidated the principle in the late 1920s.

*We consider Δy and Δp_y to be intrinsically positive. We really should write $|\Delta y|$ and $|\Delta p_y|$ but prefer to avoid the clutter of the absolute value signs.

There is nothing sacred about the y-axis in Equation 27.5; the principle applies to the position and corresponding momentum components in any direction:

$$\boxed{\Delta x \, \Delta p_x \geq h} \qquad (27.6)$$

Corresponding quantities such as y and p_y, or x and p_x, are known as **complementary variables**.

The Energy–Time Uncertainty Principle

Recall (Equation 12.16) that the mathematical form for a monochromatic wave with wavelength λ and frequency ν is

$$\Psi(x, t) = A \cos(kx - \omega t)$$

where $k = 2\pi/\lambda$ and $\omega = 2\pi\nu$. Writing out these substitutions in Equation 12.16, we get

$$\Psi(x, t) = A \cos\left(\frac{2\pi}{\lambda} x - 2\pi\nu t\right) \qquad (27.7)$$

Now we use the de Broglie relation for particle-waves, Equation 27.1,

$$\lambda = \frac{h}{|p_x|}$$

and substitute for λ in Equation 27.7; we obtain

$$\Psi(x, t) = A \cos\left(\frac{2\pi}{h} |p_x| x - 2\pi\nu t\right) \qquad (27.8)$$

Recall that Einstein hypothesized in the photoelectric effect that the relationship between the energy E of a photon particle and its frequency ν is

$$E = h\nu \qquad (27.9)$$

We take this equation as the relationship between the energy of *any* particle-wave and its associated frequency, as de Broglie did.

Heisenberg received his Ph.D. at the venerable age of 22 and also was a gifted pianist. He was the most prominent active physicist to remain in Nazi Germany during World War II. His role as head of the Nazi efforts to exploit nuclear energy was quite controversial both during and after the war.

Using Equation 27.9 to substitute for the frequency in Equation 27.8, we find

$$\Psi(x, t) = A \cos\left(\frac{2\pi}{h}\left|p_x\right| x - \frac{2\pi}{h} Et\right) \tag{27.10}$$

Notice that the roles of E and t in Equation 27.10 are *mathematically identical* to the roles of x and $|p_x|$. Since there is a Heisenberg uncertainty principle for the complementary pair of variables x and p_x, the mathematical similarity of the quantities E and t in the expression for the wave strongly suggests that an uncertainty principle also exists for the energy and time.

We can find this additional uncertainty principle in the following simplified way. For a free particle with zero potential energy moving at speed v along the x-axis, its (nonrelativistic) energy is

$$E = \frac{1}{2} mv^2$$

$$= \frac{p^2}{2m} \tag{27.11}$$

We take the derivative

$$\frac{dE}{dp}$$

and approximate the differential dE with ΔE and dp with Δp. Then the uncertainty in energy ΔE is related to the uncertainty Δp in the momentum by

$$\frac{\Delta E}{\Delta p} = \frac{2p}{2m}$$

$$\Delta E = \frac{p}{m} \Delta p \tag{27.12}$$

Likewise, the speed v is approximately

$$v = \frac{\Delta x}{\Delta t} \tag{27.13}$$

We use Equation 27.12 for Δp and Equation 27.13 for Δx and substitute into the position–momentum uncertainty principle, Equation 27.6. This yields

$$\Delta x \, \Delta p \geq h$$

$$(v \, \Delta t)\left(\frac{m}{p} \Delta E\right) \geq h$$

Hence we find that the energy–time uncertainty principle is

$$\boxed{\Delta E \, \Delta t \geq h} \tag{27.14}$$

The energy and time have their own uncertainty principle.

In fact Equation 27.14 is a very general equation, not restricted to free particles, although we have not shown this here.

Both the position–momentum and the energy–time uncertainty principles are fundamental manifestations of the underlying mathematics of waves and harmonic functions. We broached this connection in Chapter 12 with a brief introduction to Fourier analysis (Section 12.21). For example, consider a sinusoidal wave with a precisely known wavelength (or momentum since $\lambda = h/p$), as in Figure 27.2. This wave is defined for *all* values of x, and so information about the position of the wave is very uncertain: it is everywhere. On the other hand, if the wave is nonzero over only a very small region of space, and resembles a pulse or wave packet as shown in Figure 27.3, then the wave has a wavelength that is not precisely known (what is called an ill-defined wavelength).

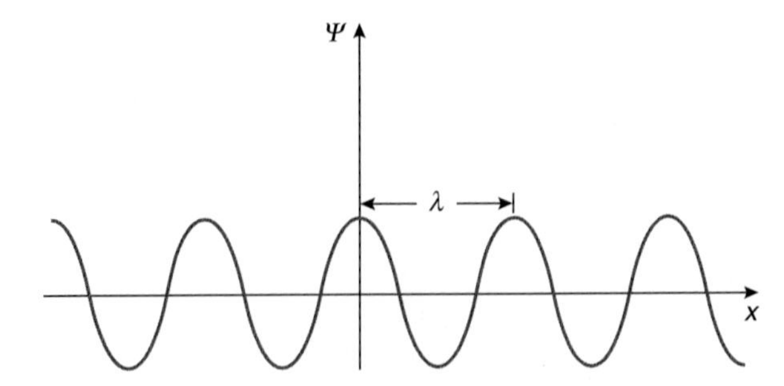

FIGURE 27.2 A sinusoidal waveform.

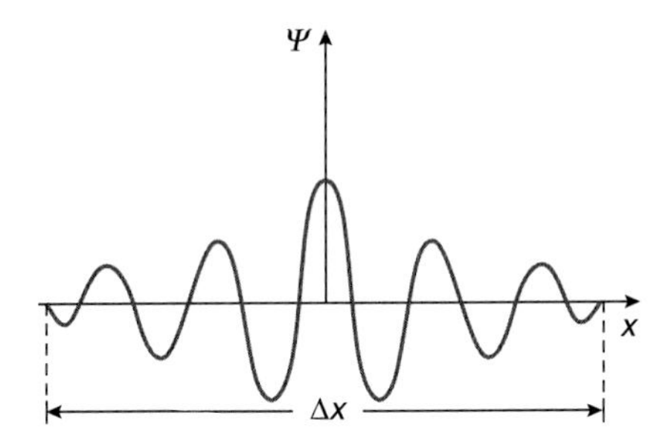

FIGURE 27.3 A wave packet.

Why is that? Many slightly different sinusoidal wavelengths must be superimposed to create the short waveform. The various wavelengths interfere constructively in the general region where the wave is nonzero and interfere destructively everywhere else. Since many different wavelengths are needed to perform the superposition, the wave pulse has a wavelength that is not precise or, equivalently, a momentum (again since $\lambda = h/p$) that is not precisely known. There is no escape. The Heisenberg uncertainty principles are inextricably tied to the mathematical formalism associated with the description of waves. Equally, they describe the inevitable disturbances caused by all experimental measurements of physical quantities like energy and time, or position and momentum.

27.2 Implications of the Position–Momentum Uncertainty Principle

In classical mechanics it was possible to say that a particle has a well-defined position x and momentum p_x (or equivalently, velocity component v_x) at a time t. In principle, the position and momentum component can be measured with arbitrary precision in classical physics. In quantum mechanics, however, this is not strictly true, because of the wave nature of particles. The position and momentum uncertainty principle was given by Equation 27.6:

$$\Delta x \, \Delta p_x \geq h$$

The more precisely we know the position, the less precisely we can know the corresponding momentum component and vice versa.

The position–momentum Heisenberg uncertainty principle enables us to understand one reason that atoms do not collapse with their electrons spiraling into the nucleus.* Recall from Chapter 26 that the Bohr model of the atom simply *postulated* the stability of the nuclear model of the atom.

Within the context of the Heisenberg uncertainty principle, we say that the electron is confined to a region on the order of size r and abandon the idea of an electron orbit. The position of the electron thus is known with a precision of about $\Delta x \approx r$. The Heisenberg uncertainty principle implies that any time a particle is confined or localized, there is an uncertainty associated with its complementary momentum component. Thus, even if we *imagine* the electron to be at rest with zero momentum, the uncertainty principle states that the momentum *cannot* be exactly zero but must be on the order of

$$p \approx \Delta p \approx \frac{h}{\Delta x} \approx \frac{h}{r}$$

The smaller the value of r, the greater the uncertainty in the momentum. If p increases, so does the kinetic energy of the electron. So what does the electron then do? The electron in the hydrogen atom has an electrical potential energy

$$\begin{aligned} \text{PE} &= qV \\ &= (-e)V \\ &= (-e)\frac{1}{4\pi\varepsilon_0}\frac{e}{r} \end{aligned} \quad (27.15)$$

where V is the electric potential of the nuclear charge ($+e$ for hydrogen) at the position of the electron. The electron has a kinetic energy

$$\text{KE} = \frac{1}{2}mv^2$$

which, since $p = mv$, can be rewritten as

$$\begin{aligned} \text{KE} &= \frac{1}{2}m\left(\frac{p}{m}\right)^2 \\ &= \frac{p^2}{2m} \end{aligned} \quad (27.16)$$

Now we consider the total energy E of the electron,

$$\begin{aligned} E &= \text{KE} + \text{PE} \\ &= \frac{p^2}{2m} - \frac{1}{4\pi\varepsilon_0}\frac{e^2}{r} \\ &= \frac{h^2}{2mr^2} - \frac{1}{4\pi\varepsilon_0}\frac{e^2}{r} \end{aligned}$$

and find what value of r minimizes the total energy. To do this, we take the derivative of E with respect to r and set it equal to zero:

$$\frac{dE}{dr} = \frac{h^2}{2m}\left(-\frac{2}{r^3}\right) - \frac{1}{4\pi\varepsilon_0}e^2\left(-\frac{1}{r^2}\right) = 0 \text{ J/m}$$

Solving for r, we find

$$r = 4\pi\varepsilon_0 \frac{h^2}{me^2}$$

This is the same order of magnitude as the fictitious orbits of the Bohr model (Equation 26.24) for low values of the quantum number n. Indeed the first Bohr orbital radius a_0 (with $n = 1$) was found in Example 26.5 to be

$$a_0 = 4\pi\varepsilon_0 \frac{\hbar^2}{me^2}$$

Thus one reason that atoms do not collapse is the uncertainty principle.

If you try to squeeze an atom by forcing its electrons closer to the nucleus, the smaller Δx causes a larger Δp_x due to the uncertainty principle; the larger momentum uncertainty prevails against the confinement.

Example 27.1

You measure the diameter of a shotgun pellet to be 1.00 mm ± 0.01 mm with a micrometer. What is the minimum uncertainty of its momentum if the magnitude of its momentum is measured simultaneously?

Solution
According to the Heisenberg uncertainty principle, $\Delta x \Delta p_x \geq h$. Here the minimum uncertainty associated with its position is $\Delta x = 0.01$ mm $= 1 \times 10^{-5}$ m. The minimum uncertainty in its momentum thus is

$$\begin{aligned} \Delta p_x &= \frac{h}{\Delta x} \\ &= \frac{6.626 \times 10^{-34} \text{ J}\cdot\text{s}}{1 \times 10^{-5} \text{ m}} \\ &= 7 \times 10^{-29} \text{ kg}\cdot\text{m/s} \end{aligned}$$

*There is another reason that atoms do not collapse: the *Pauli exclusion principle*, which also arises from quantum mechanics, though we will not show how. The principle prohibits electrons (and other particles belonging to a family of particles known as fermions) from "all doing the same thing," or having identical sets of quantum numbers in an atom. The principle is named for Wolfgang Pauli (1900–1958), who first enunciated it in the late 1920s. It is the Pauli exclusion principle that prevents the electrons in atoms from all ending up in the ground state. The Pauli exclusion principle creates a pressure that resists atomic collapse. This special pressure (that is independent of temperature and so is very different from a thermal pressure) colloquially is called a packing pressure but is more formally known as the *Fermi pressure* (after the great theoretical and experimental physicist Enrico Fermi [1901–1954]). The Fermi pressure is an important factor in the later stages of the evolution of stars.

This is exceedingly small. As a result, for such a macroscopic particle, the limitations placed on the simultaneous determinations of its position and momentum by the Heisenberg uncertainty principle are irrelevant.

STRATEGIC EXAMPLE 27.2

A proton is known to be in the nucleus of an atom. The size of the nucleus is about 1.0×10^{-14} m.

a. What is the minimum magnitude of the momentum the proton must have?
b. If its momentum is nonrelativistic, what is the minimum speed of the proton?
c. What fraction of the speed of light is the speed calculated in part (b)? Does this result justify the use of the nonrelativistic expression for its momentum?
d. What minimum kinetic energy must the proton have because of its confinement? Express your result in MeV.

Solution

a. The minimum magnitude of momentum is found using the Heisenberg uncertainty principle, Equation 27.6 (with an equality). Since the proton is confined to within 1.0×10^{-14} m, you use this value for Δx, and find Δp accordingly:

$$\begin{aligned}\Delta p &= \frac{h}{\Delta x}\\ &= \frac{6.626 \times 10^{-34}\ \text{J}\cdot\text{s}}{1.0 \times 10^{-14}\ \text{m}}\\ &= 6.6 \times 10^{-20}\ \text{kg}\cdot\text{m/s}\end{aligned}$$

b. The magnitude of the nonrelativistic momentum is $p = mv$. Since the minimum uncertainty in the momentum is Δp, this is the minimum possible value for p itself. Hence the minimum proton speed is

$$\begin{aligned}v &= \frac{p}{m}\\ &= \frac{6.6 \times 10^{-20}\ \text{kg}\cdot\text{m/s}}{1.67 \times 10^{-27}\ \text{kg}}\\ &= 4.0 \times 10^{7}\ \text{m/s}\end{aligned}$$

c. The speed is the following fraction of the speed of light:

$$\begin{aligned}\frac{v}{c} &= \frac{4.0 \times 10^{7}\ \text{m/s}}{3.00 \times 10^{8}\ \text{m/s}}\\ &= 0.13\end{aligned}$$

With this ratio, you can calculate the relativistic factor γ:

$$\begin{aligned}\gamma &= \frac{1}{\sqrt{1 - \dfrac{v^2}{c^2}}}\\ &= \frac{1}{\sqrt{1 - (0.13)^2}}\\ &\approx 1.0\end{aligned}$$

Since $\gamma \approx 1$, the use of the nonrelativistic momentum is justified.

d. In view of part (c), you can use the nonrelativistic expression for the kinetic energy:

$$\begin{aligned}\text{KE} &= \frac{1}{2} mv^2\\ &= \frac{1}{2}(1.67 \times 10^{-27}\ \text{kg})(4.0 \times 10^{7}\ \text{m/s})^2\\ &= 1.3 \times 10^{-12}\ \text{J}\\ &= \frac{1.3 \times 10^{-12}\ \text{J}}{1.602 \times 10^{-19}\ \text{J/eV}}\\ &= 8.1 \times 10^{6}\ \text{eV}\\ &= 8.1\ \text{MeV}\end{aligned}$$

This is a very substantial amount of energy and indicates that particles in the nucleus have energies much greater than the energies typically associated with the electrons surrounding the nucleus (which are on the order of a few to a few tens of electron-volts). These large energies of nuclear particles also indicate why the comparably high-energy particles emitted in radioactive decay originate from the nucleus.

27.3 Implications of the Energy–Time Uncertainty Principle

In classical mechanics, it is possible to say that a particle has a definite energy E at time t. The energy of the particle can be precisely determined at a given instant to arbitrary precision. In quantum mechanics, this is not the case. The uncertainty principle for energy and time is (Equation 27.14)

$$\Delta E\ \Delta t \geq h$$

In order to increase the precision with which the energy is determined, it is necessary to increase the time over which the measurement is made. Correspondingly, if we decrease the time over which a measurement of the energy is made, we must put up with less precision in the value of the energy so determined.

There are many implications to the energy–time uncertainty principle; we consider only two here.

The Mass of Fundamental Particles

One implication is the precision with which we can determine the mass of fundamental particles. In relativity we discovered that there is a rest energy associated with a particle with mass:

$$E_{\text{rest}} = mc^2 \tag{27.17}$$

The uncertainty principle implies that there is a nonzero uncertainty associated with any measurement of an energy, here of the rest energy ΔE_{rest}. The uncertainty Δm associated with the mass of the particle is found by taking differentials of Equation 27.17 and approximating the differential changes with ΔE and Δm:

$$\Delta E_{\text{rest}} = c^2\ \Delta m \tag{27.18}$$

The energy–time uncertainty principle (Equation 27.14) becomes

$$c^2\ \Delta m\ \Delta t \geq h \tag{27.19}$$

For example, the mean lifetime of a free neutron* is 888 s. We use this for Δt in Equation 27.19. Thus the mass of a neutron can be determined to a precision no better than Δm, where

$$\Delta m = \frac{h}{c^2\ \Delta t}$$

Evaluating this expression yields

$$\Delta m = \frac{6.626 \times 10^{-34}\ \text{J}\cdot\text{s}}{(3.00 \times 10^8\ \text{m/s})^2(888\ \text{s})} = 8.29 \times 10^{-54}\ \text{kg}$$

This minimum uncertainty associated with the measurement of the mass of the free neutron is 27 orders of magnitude smaller than the mass of the neutron itself (1.675×10^{-27} kg) and so is of no experimental import, since no experiments currently known can determine the mass of the neutron to such precision. The smallness of the fundamental uncertainty arises because the neutron is quite long-lived as a free particle. What this means from a practical standpoint is that if the masses of a large number of neutrons are determined, all the masses are essentially identical, as shown schematically in Figure 27.4.

On the other hand, there are many exotic, unstable, fundamental particles that have very short lifetimes, with the result that the uncertainty in determining their mass can be a significant fraction of it.

In other words, if the masses of a large number of these very short-lived particles are measured, the values obtained have an intrinsic spread because of the Heisenberg uncertainty principle, as shown schematically in Figure 27.5. The shorter the lifetime of a particle, the greater the uncertainty associated with determining its mass.

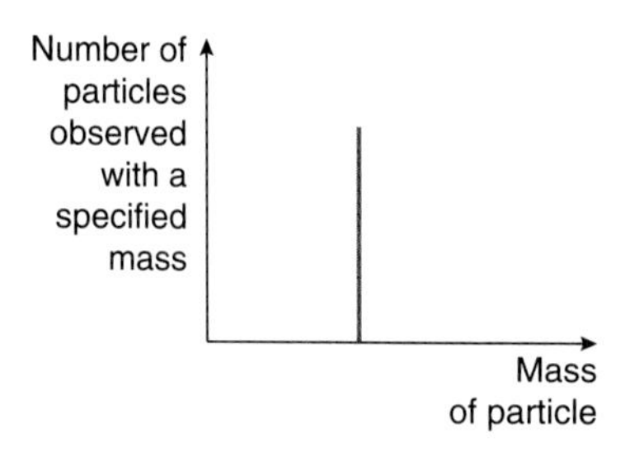

FIGURE 27.4 The mass of the neutron can be determined quite precisely because it is a relatively long-lived particle.

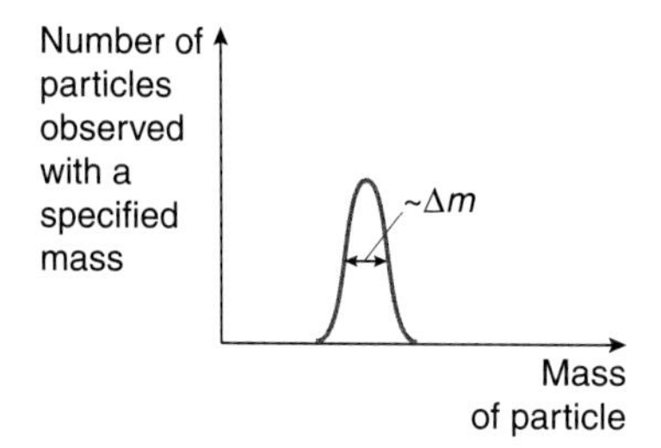

FIGURE 27.5 Short-lived particles have noticeable uncertainty in their masses.

The Nature of a Vacuum

Another strange implication of the energy–time uncertainty principle concerns the nature of a vacuum. The idea of a vacuum in quantum physics is quite different from the classical idea of a vacuum as simply a volume with nothing in it.

The quantum mechanical vacuum is a seething sea of particle–antiparticle pairs, called **virtual particles**, since their existence is ephemeral. The pairs of virtual particles well up out of *nothing*, live for a very short time, and then disappear (annihilate each other).

To create a virtual particle–antiparticle pair, each with mass m, we need to borrow or create an amount of energy ΔE at least equal to the sum of the rest energies of the two particles: $2mc^2$. The virtual particles can live only for a time Δt no longer than the minimum time obtained from the Heisenberg energy–time uncertainty principle (Equation 27.14):

$$\Delta E\ \Delta t = h$$

Solving for Δt, we find

$$\Delta t = \frac{h}{\Delta E} = \frac{h}{2mc^2} \qquad (27.20)$$

The larger the mass of the virtual particle–antiparticle pair, the shorter the time interval during which the virtual particles exist as elements of the vacuum. For example, a virtual electron–positron pair, the least massive particle–antiparticle pair,† can exist only for a time Δt that is at most

$$\Delta t = \frac{6.626 \times 10^{-34}\ \text{J}\cdot\text{s}}{2(9.11 \times 10^{-31}\ \text{kg})(3.00 \times 10^8\ \text{m/s})^2} = 4.04 \times 10^{-21}\ \text{s}$$

This is a very short time indeed.

The vacuum process of particle–antiparticle creation and annihilation does not violate the principle of energy conservation. The process is analogous to kiting a check. You can borrow or create money from your empty checking account (a vacuum) by writing a check. As long as you get the money into the account before the check clears (annihilation) there is no problem (although kiting is technically illegal and banks certainly do not like you to do it!). In the quantum vacuum, you can borrow or create an amount of energy ΔE to create a virtual particle–antiparticle pair as long as the energy loan is repaid (by particle annihilation) before a time Δt elapses given by the minimum of the Heisenberg energy–time uncertainty principle: $\Delta E\ \Delta t = h$.

*Neutrons not bound inside a nucleus are called free neutrons. Free neutrons are radioactive and decay into a proton, an electron, and an electron antineutrino.

†Neutrinos now are thought to have a small nonzero mass, much smaller than that of the electron or positron.

The more massive the virtual particle–antiparticle pair, the shorter the duration of the energy loan. Thus virtual electron–positron pairs can last longer than virtual proton–antiproton pairs.

You might legitimately ask how in heaven such exotic aspects of a vacuum could be verified if the processes occur behind the well-shrouded veil of the uncertainty principle. Suffice it to say that there are experiments that only can be explained via an appeal to such an exotic interpretation of the vacuum. Among these experiments is the **Lamb shift** in the spectrum of hydrogen (for which Willis Lamb Jr. [1913–] won the 1955 Nobel prize in physics.

Near the positively charged nucleus, virtual electron and positron pairs generated from the vacuum ceaselessly are appearing momentarily and disappearing.* Thanks to the positive nucleus, the negative virtual electrons spend more time near the nucleus than the positively charged virtual positrons. This results in a slight or partial shielding (neutralization) of the nuclear charge by a cloak of virtual electrons. Thus the real electron in the ground state feels the electrostatic effect of a nuclear charge that is slightly *less* than $+Ze$. The shielding causes a slight shift in the energy levels and the resulting spectrum of light emitted or absorbed by hydrogen. The shift in the ground state energy is the Lamb shift; it is quite real. The Lamb shift is said to be attributed to a **vacuum polarization**.

EXAMPLE 27.3

For an electron–positron pair created from the vacuum, what is the maximum distance the pair can travel before they annihilate each other?

Solution

You saw in this section that an electron–positron pair can exist for at most 4.04×10^{-21} s. Since the upper limit on their speed is the speed of light, they can travel at most a distance

$$\begin{aligned} c\,\Delta t &\approx (3.00 \times 10^{8}\ \text{m/s})(4.04 \times 10^{-21}\ \text{s}) \\ &\approx 1.21 \times 10^{-12}\ \text{m} \end{aligned}$$

This distance is about a factor of 10^2 to 10^3 greater than the size of the nucleus, which is what permits vacuum polarization.

EXAMPLE 27.4

The uncertainty in the rest energy of some of the most short-lived elementary particles is estimated to be about 3×10^2 MeV. What is the approximate mean lifetime of such particles?

Solution

The energy–time uncertainty principle, Equation 27.14, states that

$$\Delta E\,\Delta t \geq h$$

The uncertainty in the rest energy is 3×10^2 MeV. Hence the mean lifetime is about

$$\begin{aligned} \Delta t &\approx \frac{h}{\Delta E} \\ &\approx \frac{6.626 \times 10^{-34}\ \text{J}\cdot\text{s}}{(3 \times 10^{8}\ \text{eV})(1.602 \times 10^{-19}\ \text{J/eV})} \\ &\approx 1 \times 10^{-23}\ \text{s} \end{aligned}$$

These are among the shortest-lived particles known.

EXAMPLE 27.5

Light of wavelength 632.8 nm is incident on an extremely fast shutter that chops the beam into pulses. The shutter stays open for only 1.5×10^{-9} s. What is the approximate minimum range of wavelengths $\Delta\lambda$ in the light pulses that pass the shutter?

Solution

The energy–time uncertainty principle states that

$$\Delta E\,\Delta t \geq h$$

For the minimum uncertainty, you have

$$\Delta E\,\Delta t \approx h$$

Since $\Delta t \approx 1.5 \times 10^{-9}$ s, the uncertainty in the energy is

$$\begin{aligned} \Delta E &\approx \frac{h}{\Delta t} \\ &\approx \frac{6.626 \times 10^{-34}\ \text{J}\cdot\text{s}}{1.5 \times 10^{-9}\ \text{s}} \\ &\approx 4.4 \times 10^{-25}\ \text{J} \end{aligned}$$

The energy of light is proportional to its frequency:

$$\begin{aligned} E &= h\nu \\ &= \frac{hc}{\lambda} \end{aligned}$$

To see how E varies with a change in wavelength, take differentials of this expression, approximating the differentials with small changes ΔE and $\Delta\lambda$. You can ignore the minus sign resulting from the differentiation, since only the absolute values are of interest:

$$\Delta E = \frac{hc}{\lambda^2}\,\Delta\lambda$$

You know ΔE, h, c, and λ, and so you can find $\Delta\lambda$:

$$\begin{aligned} \Delta\lambda &\approx \frac{\Delta E\,\lambda^2}{hc} \\ &\approx \frac{(4.4 \times 10^{-25}\ \text{J})(632.8 \times 10^{-9}\ \text{m})^2}{(6.626 \times 10^{-34}\ \text{J}\cdot\text{s})(3.00 \times 10^{8}\ \text{m/s})} \\ &\approx 8.9 \times 10^{-13}\ \text{m} \end{aligned}$$

*Since virtual electron–positron pairs can last longer than virtual proton–antiproton pairs, the effect of the former is more prevalent.

The fractional spread in wavelengths, $\Delta\lambda/\lambda$, is quite small but is nonetheless very measurable:

$$\frac{\Delta\lambda}{\lambda} = \frac{8.9 \times 10^{-13}\ \text{m}}{632.8 \times 10^{-9}\ \text{m}}$$
$$= 1.4 \times 10^{-6}$$

27.4 Observation and Measurement

Physics ultimately must "preserve the phenomena," as the ancient Greek natural philosophers attempted to do with their imaginative models of the night sky and solar system. Physics is based on experiments, which, of course, involve observation and measurement.

In the classical domain, such observations and measurements are conceptually easy, at least in principle. For example, to measure the speed of a moving object we simply look at the object and record its changing position as a function of time. The act of looking does not disturb the object under examination; or, even if the measurement did disturb the system, we could account for the disturbance with classical physical laws and techniques to find out what was happening before our measurement.

In other words, in classical physics, the act of measurement either has a negligible effect on the physical system under observation, or can be accounted for, at least in principle, in predictable ways.

We already have encountered several instances where we had to be careful about the act of measurement in classical physics. Specifically:

1. When we studied thermodynamics and the measurement of temperature with thermometers, we mentioned that caution is needed to be sure that the mass of a thermometer is insignificant with respect to the mass of the system whose temperature is to be determined. If the mass of a thermometer is too large compared with the mass of the system, the act of measurement affects the temperature of the system to be measured because of the energy exchange between the system and the thermometer. (Of course, if the system is in contact with a reservoir, this condition is not necessary.) Classical physics can, at least in principle, account for these effects.
2. When analyzing electrical forces, it is necessary to be careful that the charge q placed in an electrical field is not so large that its presence affects the magnitude or direction of the field in which the charge is placed. Nonetheless, classical physics can, in principle, account for the effect of q on the other charges creating the field in which q is placed.
3. In electrical circuits, when measuring potential differences, it is necessary to use voltmeters with a high internal resistance so as not to affect the potential differences in the circuit under study. The same is true with ammeters: an ammeter with low effective resistance is needed in order to minimize the effects of the meter on the current to be measured. Again, it is possible in classical physics to account completely for the effect of a measurement or a disturbance on a physical system.

In the quantum physics of very small systems, the act of measurement or observation inescapably plays a significant role. Even simply looking is no longer innocent or trivial. The mere act of inspecting something means that light must be scattered or reflected from it. A reflected or scattered photon is no big deal to a baseball or to a tasty M&M. But to an electron, an atom, or other small particle, the interaction with even a single photon is tantamount to a sledgehammer. Any measurement on a physical system always involves an interaction of some kind between the system and an apparatus.

Furthermore, the act of measurement in the quantum domain inevitably changes the system in ways that cannot be accounted for in a deterministic sense, *even in principle*.

The system changes in a random and unpredictable way. For example, in Section 27.1, we tried to localize the particle-waves by passing them through a slit aperture. But the greater the extent to which we try to localize the particles with a smaller and smaller aperture, the greater the width of the resulting diffraction pattern and the uncertainty associated with its transverse momentum. Thus the uncertainty principle limits our ability to make measurements of physical observables with infinite precision. The role of an observer in quantum physics is vastly more complex and bumbling than in classical physics. It differs fundamentally in that it is impossible, in principle, to account for the changes in a system induced by the act of measurement.

27.5 Particle-Waves and the Wavefunction

All Nature is but Art unknown to thee;
All chance, direction, which thou canst not see;
All discord, harmony not understood;
All partial evil, universal good;
And spite of Pride, in erring Reason's spite,
One truth is clear, Whatever is, is right.

Alexander Pope (1688–1744)*

We saw in Chapter 26 that particles have aspects of waves (the electron diffraction experiments of Davisson and Germer, and Thomson) and waves have aspects of particles (the Compton effect and Einstein's explanation of the photoelectric effect). When we say that photons, electrons, protons, neutrons, and so forth behave like *particles*, we mean that they have physical properties (which may be zero), such as mass, charge, and momentum, all reminiscent of what we imagine when we think of the very word particle. On the other hand, the behavior of the particles is determined by a wavelike function whose nature we have yet to delineate. That time has come.

**An Essay on Man*, Epistle I, lines 289–294, *The Complete Poetical Works of Alexander Pope*, edited by Henry W. Boynton (Houghton Mifflin, Boston, 1903), page 141.

Here we try to address some of the questions we raised at the conclusion of the last chapter. Do the particle-waves represent anything in a physical context? The superluminal speeds of the particle-waves (see Equation 26.84) give us pause for thought. How are the particle-waves to be represented in a mathematical context by a wavefunction $\Psi(x, t)$? What equation governs the propagation of the particle-waves? Is it the classical wave equation? This last aspect is addressed in the next section.

The wave aspects of matter are shown by diffraction and interference experiments. In order for these effects to stand out, the particles must interact with objects whose size is comparable to the de Broglie wavelength of the particles. What happens in such experiments? What is observed in particle diffraction and interference experiments?

Imagine a horde of identical particles, each with the same momentum (that is, with the same de Broglie wavelength), incident on an appropriately sized single slit aperture, as in Figure 27.1. The shape of the pattern that develops on the screen or detector is identical to the single slit diffraction pattern of a classical wave. The pattern indicates the number of particles detected at any given location along the screen. In other words, even if the particles are sent through the slit *one at a time*, each particle goes somewhere on the screen, and their distribution (the diffraction pattern) builds up over time.

The diffraction pattern is a direct measure of the number of particles arriving at various locations on the screen in a given time; it is a number distribution of particles. The pattern, therefore, is an indication of the likelihood or the *probability* that a particle will arrive at each location on the screen. We *cannot predict* where an individual particle will go when it traverses the slit; we can only assess the odds or the probability that it will arrive at a particular location. We *cannot* say that the wave used to predict the pattern on the screen indeed *is* the particle; the wave model determines the shape of the pattern of particles, or equivalently, the pattern displays the relative probability that a single particle will arrive at any particular location on the screen.

The situation becomes even more perplexing if a double slit is used. We might naively anticipate that since an individual particle goes through either one slit or the other, the pattern observed will be simply two slightly separated single slit diffraction patterns as in Figure 27.6. This is not the case.

The experimental results indicate that, just as in the classical optical double slit experiments discussed in Chapter 24, a double slit *interference* pattern results on the distant screen (or detector), as in Figure 27.7. Even if the particles are sent individually (one at a time) through the double slit arrangement, the double slit pattern emerges gradually as the particles accumulate at various positions across the screen, as in Figure 27.8. The same effect is seen with either electrons or photons.

This defies logic and common sense, because it taxes our conceptual notions about the word particle. We get one pattern on a distant screen if the particles go through a single slit, but a completely different pattern if the particles go through two or more closely spaced slits. Thus we cannot think of the particles as going through either one slit or the other in the double slit arrangement. Why not? If each particle went through one slit or the other, we would logically expect two single slit patterns as if from two incoherent sources, not an interference pattern. However, each particle individually somehow senses *both* slits.

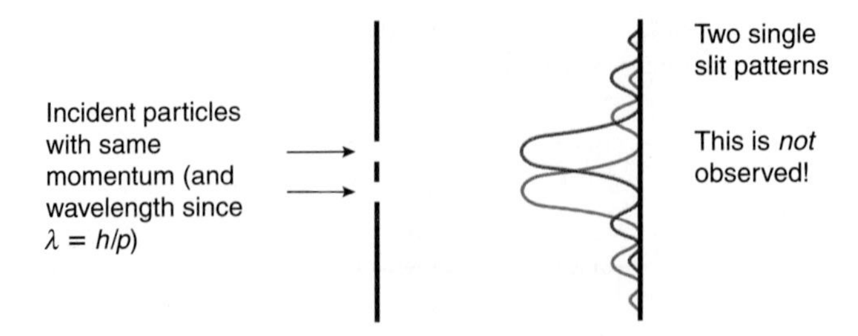

FIGURE 27.6 Two single slit patterns of particles are *not* observed.

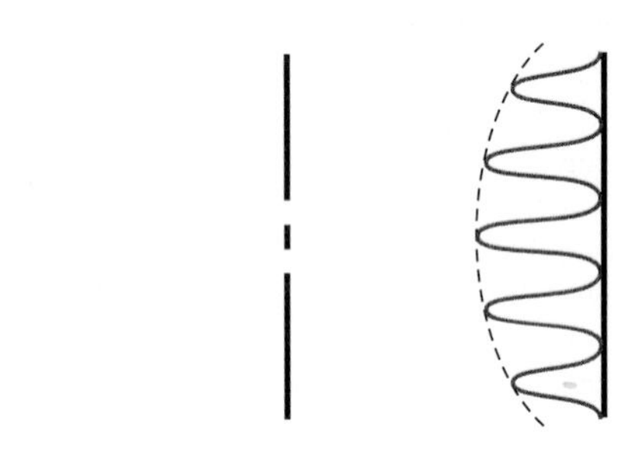

FIGURE 27.7 A double slit pattern of particles *is* observed.

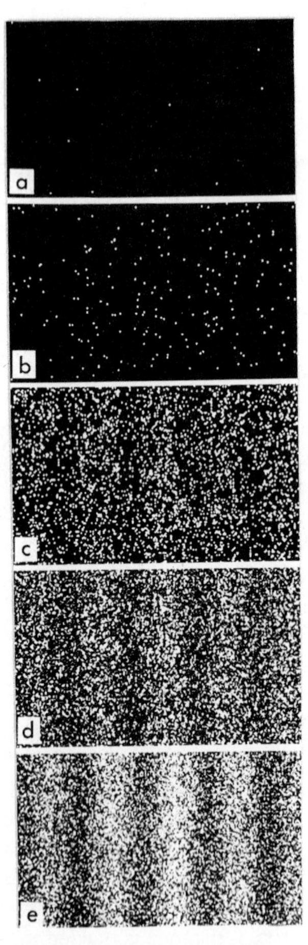

FIGURE 27.8 A double slit pattern gradually emerges as more and more particles are sent through the double slit.

Thus the idea of a particle as a discrete billiard ball entity must be abandoned in favor of some abstract particle-wave which can diffract and interfere with itself.

What characteristics does a wavefunction Ψ representing the particle-wave have? Several come to mind:

1. In order to represent the particle-wave, the amplitude of the wavefunction must be relatively large in regions where the particle is likely to be found and, conversely, of small amplitude in regions where the particle is not likely to be located.
2. The wavefunction must be able to interfere with itself, since a single or individual particle-wave somehow is able to sense the presence of either a single or double slit and know what to do in order to reproduce statistically the single slit diffraction pattern or the double slit interference pattern, as the case may warrant.
3. Since a single particle-wave is capable of sensing the presence of both slits in a double slit experiment, the wavefunction must represent the behavior of single particle-waves and not the statistical distribution of an ensemble of identical particle-waves.
4. For the wavefunction to truly represent all features of the particle-wave, we must be able (somehow) to obtain from the wavefunction Ψ *all* physical characteristics of the particle-wave such as its momentum, energy, and other physical parameters that are observable. In other words, by doing something to the wavefunction, by mathematically operating on it in some way, we ought to be able to extract the ordinary physical parameters or observables that we associate with the particle-wave.

The wavefunction used to predict the pattern on the screen must have something to do with the probability that a particle arrives at any location. The wavefunction itself *cannot* be the probability since, by their very nature, probabilities are intrinsically greater than or equal to zero; while waves are oscillatory in nature and have both positive and negative values.

We consider a special case. What kind of wavefunction can represent an individual particle in free space, traveling in one dimension, say, toward increasing values of x, with a well-defined (i.e., precise) momentum (or equivalently, a well-defined wavelength)?* We ignore any descriptions in the y and z directions. Intuitively, we might think that the monochromatic wavefunction we have used on many occasions in the past could do this:

$$\Psi(x, t) = A\cos(kx - \omega t) \qquad (27.21)$$

Since this function can take on both positive and negative values, perhaps the square of Ψ is the probability of finding a particle at any given place x at time t? Alas, this cannot be the case. Why not? Consider the wavefunction of Equation 27.21 when $t = 0$ s. Then the square of Ψ is

$$\Psi^2 = A^2\cos^2(kx)$$

which is indeed everywhere positive (see Figure 27.9) but has other unacceptable problems, such as unreal periodic bunchings along the line of motion.

*The Heisenberg uncertainty principle then implies that the wave is everywhere, since $\Delta x = \infty$ m.

For such a particle unconstrained in x, it would be equally likely to find it at *any* position x, since the wave is defined for all values of x. In other words, the probability of finding this particle at *any* position x ought to be independent of x, as in Figure 27.10. But the square of Equation 27.21, depicted in Figure 27.9, certainly is not independent of x. How can we avoid this problem?

Recall that a sinusoidal oscillation also can be represented by a *complex* exponential expression via the Euler identity†:

$$\begin{aligned}\Psi(x, t) &= Ae^{i(kx - \omega t)} \\ &= A[\cos(kx - \omega t) + i\sin(kx - \omega t)]\end{aligned} \qquad (27.22)$$

The squared magnitude of Ψ, found by multiplying Ψ by its complex conjugate Ψ^*, then would give us just what we want: a probability that is independent of x and t, as in Figure 27.10:

$$\begin{aligned}\Psi^*\Psi &= A^*e^{-i(kx - \omega t)}Ae^{i(kx - \omega t)} \\ &= A^*A\end{aligned}$$

Although this argument involves a very specific example (a free particle with a well-defined wavelength), there are several important points to note here about the nature of *any* quantum mechanical wavefunction Ψ.

> The wavefunction $\Psi(x, t)$, whatever mathematical form it takes in a particular context, typically is a complex-valued function.

†We did something similar when considering ac circuits in Chapter 22, where the real part of the complex exponential represented the real potential difference or current and the imaginary part came along for a mathematical ride. The complex potential differences and currents were introduced simply to make calculations easier, since exponentials are easier to manipulate mathematically than trigonometric functions.

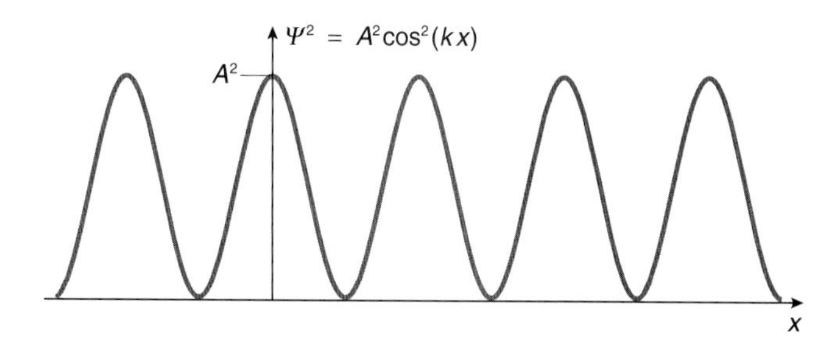

FIGURE 27.9 Graph of the function $A^2\cos^2(kx)$ versus x.

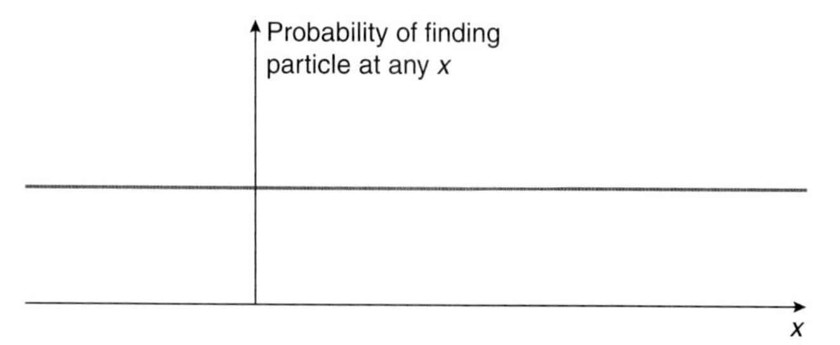

FIGURE 27.10 A free particle is equally likely to be found anywhere.

Since the wavefunction $\Psi(x, t)$ is a complex-valued function, $\Psi(x, t)$ itself is *not* a physical quantity that can be observed directly in any experiment; measurements in experiments always yield *real* numbers.

The quantity $\Psi^*(x, t)\Psi(x, t)\, dx$ is usefully defined as the probability of finding the particle in a region of space between x and $x + dx$ at a particular time t. For this reason, the wavefunction Ψ itself also is known as a **probability amplitude**.

The complex wavefunction for a particle with a well-defined wavelength, Equation 27.22,

$$\Psi(x, t) = Ae^{i(kx - \omega t)}$$

provides clues about the nature of the equation that governs the propagation of the particle-waves. If the particle is traveling in a region where its potential energy is constant, for convenience we can choose the constant potential energy to be zero, as we have many times in the past, because the place where the potential energy is zero is an arbitrary choice. With this choice, the total, nonrelativistic energy E of the particle is just its kinetic energy:

$$E = \frac{1}{2}mv^2 = \frac{p^2}{2m} \qquad (27.23)$$

The particle energy E is related to its frequency by

$$E = h\nu = \hbar\omega \qquad (27.24)$$

The momentum is related to the wavelength via the de Broglie relation

$$p = \frac{h}{\lambda}$$

or in terms of k, since $k = 2\pi/\lambda$,

$$p = \hbar k \qquad (27.25)$$

Thus Equation 27.23 for the energy of the free particle can be put in the form

$$\hbar\omega = \frac{1}{2m}\hbar^2k^2 \qquad (27.26)$$

To obtain this same relationship by suitably differentiating the wavefunction given by Equation 27.22, taking the *first time derivative* of Ψ will yield a term in ω (like the left-hand side of Equation 27.26), and the *second spatial derivative* of Ψ will yield a term in k^2 (like the right-hand side of Equation 27.26). Thus it is likely that the differential equation governing the propagation of the particle-waves relates the *first time derivative* to the *second spatial derivative* of the wavefunction. The implications of this are profound.

A new wave equation is needed to describe the propagation of the wavefunction representing the particle-wave.

Our old familiar classical wave equation relates the *second time derivative* to the second spatial derivative of the wavefunction:

$$\frac{\partial^2\Psi}{\partial x^2} - \frac{1}{v^2}\frac{\partial^2\Psi}{\partial t^2} = 0 \qquad \text{(classical wave equation)} \qquad (27.27)$$

The new wave equation for the particle-wave function is the called the **Schrödinger equation**, after Erwin Schrödinger (1887–1961), who first deduced the equation in the mid-1920s while on a romantic holiday in the Alps with a paramour. In Section 27.7 we develop the specific form of the Schrödinger equation.

Notice that in the quantum domain we lose the ability to predict with certainty what happens to individual particles such as electrons. We encountered this peculiar aspect of quantum mechanical phenomena previously in our study of radioactive decay: the statistical behavior of a large number of radioactive particles can be ascertained, but we cannot predict when an individual particle will decay.

The determinism of classical physics must be abandoned in the quantum world. Does this imply that everything we learned in classical physics was for naught? Fortunately, this is surely not the case.* The situation is analogous to what we encountered in relativity: relativity is a theory of space and time that, in the limit of speeds small compared with the speed of light, reduces to classical Galilean relativity. No one uses relativity to examine the flight of a football; we use the Newtonian representation because it gives the correct results (and so would a relativistic calculation, but that is much harder!).

In quantum mechanics, large masses usually mean small de Broglie wavelengths (see Example 26.10); as long as the wavelengths of the particles are small compared with other lengths involved in a particular situation, our classical physics is typically quite appropriate, and can be used with confidence and correctness. The de Broglie wavelength of a macroscopic object is many

*I can imagine the reaction if you were told, after nearly a full year of hard work, that everything you learned was "wrong"! Arg!

Schrödinger was an Austrian physicist who loved Vienna. His friends at Oxford University helped him to escape the Nazis after the Anschluss. He spent 18 years in Dublin, Ireland, before returning to Vienna in 1956 to a welcoming country and government.

orders of magnitude smaller than even the size of a typical nucleus of an atom; the wavelike behavior of such a particle never is manifest because its wavelength is so small compared with anything interacting with it. No one (in his or her right mind, at any rate) would use quantum mechanics to calculate the motion of a planet in orbit around the Sun.

On the other hand, for an electron in an atom or other atomic, nuclear, or submicroscopic phenomena, the de Broglie wavelength of the particle under consideration typically is comparable to the size of the system itself; then the wavelike aspects are really central to an analysis of the situation, and quantum mechanics is unavoidable for making predictions that agree with experiments.

27.6 OPERATORS*

Perhaps without realizing it, you have been using mathematical operators for some time. A mathematical operator[†] is a symbolic way of representing a mathematical operation. For example, various mathematical operations can be performed on the function $\Psi(x, t)$. These might include the following:

1. Multiplying the function by a scalar. For example, we could multiply the function $\Psi(x, t)$ by x to get

$$x\Psi(x, t)$$

We symbolically represent the multiplication simply by placing x next to $\Psi(x, t)$. No formal symbol is used.[‡]

2. Taking derivatives of the function. For example, when we write

$$\frac{\partial \Psi}{\partial x}$$

what we mean is to take the partial derivative[§] of $\Psi(x, t)$ with respect to x. The operation is symbolized by

$$\frac{\partial}{\partial x}$$

and so this is called a differential operator. Partial differentiation with respect to time is represented by the differential operator

$$\frac{\partial}{\partial t}$$

There are many other types of mathematical operators as well, but we need not be concerned with them in this brief introduction to quantum mechanics.

It is important to realize that the order in which mathematical operators are written and are performed can be important. For example, the results obtained in the following examples of sequential operations are quite different:

a. First multiply Ψ by x and then take the derivative with respect to x. This is symbolized in the following way:

$$\frac{\partial}{\partial x}(x\Psi) = \Psi + x\frac{\partial \Psi}{\partial x} \tag{27.28}$$

b. Reversing the order, first take the derivative of Ψ and then multiply by x. The result is symbolized by

$$x\frac{\partial \Psi}{\partial x}$$

which is quite different from Equation 27.28.

We say, then, that the operators "multiply by x" and "take the derivative with respect to x" *do not commute* since the order in which they are executed affects the result.

On the other hand, certain operators do commute: the order in which they are performed is immaterial. For example, the differential operators

$$\frac{\partial}{\partial x} \quad \text{and} \quad \frac{\partial}{\partial t}$$

commute since

$$\frac{\partial^2 \Psi}{\partial x\, \partial t} = \frac{\partial^2 \Psi}{\partial t\, \partial x}$$

Let's play with the wavefunction that represents a monochromatic (fixed wavelength) free particle to see what operators on the wavefunction yield the momentum and energy of the particle. The free particle wavefunction is given by Equation 27.22:

$$\Psi(x, t) = Ae^{i(kx - \omega t)}$$

The Momentum Operator

We operate on Ψ with the differential operator

$$\frac{\partial}{\partial x}$$

We find

$$\begin{aligned} \frac{\partial}{\partial x}\Psi(x, t) &= \frac{\partial}{\partial x}[Ae^{i(kx-\omega t)}] \\ &= ikAe^{i(kx-\omega t)} \\ &= ik\Psi \end{aligned} \tag{27.29}$$

Then using the definition of k,

$$k = \frac{2\pi}{\lambda}$$

and the de Broglie relation for the momentum, we get

$$\begin{aligned} k &= \frac{2\pi}{h}p \\ &= \frac{p}{\hbar} \end{aligned}$$

[†] We do not mean your math or physics professors.

[‡] We did use formal symbols to represent particular types of multiplication associated with vectors: the familiar scalar product by • and the vector product by ×.

[§] Recall that to differentiate Ψ partially with respect to x, we treat t as a constant and use the familiar differentiation rules. Or to differentiate Ψ partially with respect to t, we treat x as a constant and apply the usual differentiation rules.

Substituting this into Equation 27.29, we obtain

$$\frac{\partial \Psi}{\partial x} = i\frac{p}{\hbar}\Psi$$

Rearranging things slightly, we find

$$\frac{\hbar}{i}\frac{\partial \Psi}{\partial x} = p\Psi$$

We rationalize the complex number on the left-hand side of the equation by multiplying the numerator and denominator by i. This yields

$$-i\hbar\frac{\partial \Psi}{\partial x} = p\Psi \qquad (27.30)$$

Hence the differential operator

$$-i\hbar\frac{\partial}{\partial x}$$

is what extracts the x-component of the momentum from the wavefunction. In quantum mechanics, then, we call the operator

$$-i\hbar\frac{\partial}{\partial x} \qquad (27.31)$$

the (one-dimensional) **momentum operator.**

Although we used a very specific wavefunction to obtain Equation 27.31, the (one-dimensional) momentum operator in quantum mechanics always has this form.

The Energy Operator

We use the differential operator

$$\frac{\partial}{\partial t}$$

on the free particle wavefunction

$$\Psi(x, t) = Ae^{i(kx - \omega t)}$$

We obtain

$$\begin{aligned}\frac{\partial \Psi}{\partial t} &= \frac{\partial}{\partial t}[Ae^{i(kx-\omega t)}] \\ &= -i\omega Ae^{i(kx-\omega t)} \\ &= -i\omega\Psi \qquad (27.32)\end{aligned}$$

Since the particle has an energy $E = h\nu = h(\omega/2\pi) = \hbar\omega$, Equation 27.32 can be rewritten by substituting for ω:

$$\frac{\partial \Psi}{\partial t} = -i\frac{E}{\hbar}\Psi$$

Rearranging this slightly, we obtain

$$-\frac{\hbar}{i}\frac{\partial \Psi}{\partial t} = E\Psi$$

We rationalize the left-hand side of this equation, finding

$$i\hbar\frac{\partial \Psi}{\partial t} = E\Psi \qquad (27.33)$$

Hence the operator

$$i\hbar\frac{\partial}{\partial t} \qquad (27.34)$$

is what extracts the total energy of the particle from the wavefunction, and so is called the **energy operator** in quantum mechanics.

Notice from Equation 27.30 that when the momentum operator acts on the wavefunction, the result is the momentum component times the wavefunction itself. Likewise, when the energy operator acts on the wavefunction, the result is the energy times the wavefunction itself (Equation 27.33). When a mathematical operator acts on a function and simply produces a scalar multiple of the function itself, the relationship is called an **eigenvalue equation.*** An eigenvalue equation is represented symbolically in the following way:

$$\text{operator} \mid \text{function}\rangle = \text{scalar} \mid \text{function}\rangle$$

The fancy brackets $\mid\ \rangle$ around the function serve to set the function apart from the operator and the scalar eigenvalue, and form the basis of a useful and ubiquitous shorthand notation invented for quantum mechanics by Paul Dirac (1902–1984); it is called **Dirac notation.**

*The word *eigen* (pronounced eye-gen) is German and has several meanings: *special, characteristic,* or *specific.*

Dirac's contributions to quantum mechanics were honored with the 1933 Nobel prize in physics, which he shared with Erwin Schrödinger. Dirac was the first person to foretell the existence of antimatter particles such as the positron and antiproton. Dirac was honored with a plaque in 1995, set next to the tomb of Newton in Westminster Abbey.

The scalar is called an **eigenvalue** and the function is called an **eigenfunction** of the mathematical operator. Eigenvalue equations indicate the way we extract physically meaningful information from the complex-valued wavefunction Ψ of the particle-wave.

Various mathematical operators correspond to physically observable properties, such as momentum and energy. The operators act on the wavefunction, and the resulting eigenvalues are the appropriate values of that physical observable.[†] For the eigenvalues to represent physically observable properties, the eigenvalues of the operators in quantum mechanics must be real numbers.

From a purely mathematical viewpoint, the real nature of the eigenvalues places restrictions on the type of mathematical operators that are associated with observable quantities in quantum mechanics.[‡ §]

With the operator formalism, we say that the wavefunction contains *all* the information about the physical system. This explains why such a premium is placed in quantum mechanics on discovering the form for the wavefunction in a particular physical context; the wavefunction Ψ says it all. Discovering what the wavefunction *is*, is the problem we face in quantum mechanics; for many systems, it is not an easy undertaking to find Ψ.

27.7 The Schrödinger Equation*

If you came this way,
Taking any route, starting from anywhere,
At any time or at any season,
It would always be the same: you would have to put off
Sense and notion.

T. S. Eliot (1888–1965)[#]

When de Broglie first surmised that particles might have wavelike properties, the search for the physical nature of the waves as well as an equation that all such particle-waves satisfied began in earnest. The procedure for finding such an equation was by no means obvious then, nor is it really clear even in hindsight.

The best that can be done is to surmise an equation for the particle-waves based on what we know of them (which originally was quite little; see Section 27.5) and then test the equation and its predictions in situations that can be compared with experiment. Just as there is little value in pulling Newton's second law out of thin air, there is little value in doing the same with the Schrödinger equation. Perforce, our development of the Schrödinger equation is not rigorous, but it should serve to make it somewhat plausible to you.

Two of the pillars on which quantum mechanics is built are the de Broglie relation between the wavelength and momentum of a particle-wave,

$$\lambda = \frac{h}{p} \tag{27.35}$$

and the relationship between the energy of the particle-wave and its frequency (generalized from Einstein's treatment of the photon particle-wave of light),

$$E = h\nu = \hbar\omega \tag{27.36}$$

Both of these relationships have roots in relativity.

Classical mechanics is dichotomous: there is relativistic mechanics and nonrelativistic mechanics. The latter is a special case of the former, appropriate when the speeds of the particles are much less than the speed of light. Quantum mechanics has a similar dichotomy depending on whether the speeds of the particles are nearly equal to or small compared with the speed of light. The fundamental equation of nonrelativistic quantum mechanics is the Schrödinger equation; the corresponding equation in relativistic quantum mechanics is called the Klein–Gordon equation.[¶] For reasons of simplicity and time, we will examine only the nonrelativistic scenario (the Schrödinger equation).

The distinction between the relativistic and nonrelativistic regimes is needed because the expressions for the total energy and momentum of a particle are quite different in the two domains. For the nonrelativistic limit, the rest energy of the particle is ignored** and the classical nonrelativistic expression for the kinetic energy of the particle is used:

$$\text{KE} = \frac{1}{2}mv^2 = \frac{p^2}{2m} \tag{27.37}$$

Then total energy of a particle is the sum of the kinetic energy and any potential energies appropriate to the problem. Until now, for clarity we have always used PE as the symbolic representation for the potential energy. Alas, this notation now must be abandoned to conform with standard notation in quantum mechanics where V commonly is used for the potential energy of

[†]We have only demonstrated this feature with a particular wavefunction, but it is a general feature of quantum mechanics.

[‡]Operators that have real eigenvalues are known as *Hermitian operators* in mathematics.

[§]In an allegorical and sarcastic sense, one of the most useful "operators" known is the *clearly* operator: it has one eigenvalue—the *result*. Thus

$$\text{clearly} \mid \text{anything}\rangle = \text{result} \mid \text{anything}\rangle$$

You have seen this "operator" in action many times, in many texts (likely including this one). It typically appears in the following way. At the beginning or in the middle of some complicated calculation, an author will invoke the *clearly* operator, and somehow the *result* magically appears on the very next line of text. Other variations of the *clearly* operator are known as well: "The reader can readily show that ..." or "Obviously, ..." Clearly, it is best *not* to take the contents of this footnote too seriously!

[#]*Four Quartets*, "Little Gidding," lines 39–43 (Harcourt Brace Jovanovich, San Diego, California, 1971) page 50.

[¶]For so-called spin 1/2 particles, the relativistic equation is known as the Dirac equation.

**In the classical limit, the rest energy can be considered constant. It is only changes in energy that are physically significant, so that we can legitimately ignore the rest energy.

a particle.*† Any potential energies involved in a problem are still, of course, related to the presence of conservative forces acting on the particle.

For the nonrelativistic situation, the total energy E of a particle is

$$E = \frac{p^2}{2m} + V(x, t) \tag{27.38}$$

For simplicity we consider only one-dimensional systems, so the potential energies V and wavefunctions Ψ are functions of only one spatial coordinate (x) and the time (t).

If we take the classical wave equation (Equation 27.27) as a paradigm, we expect the Schrödinger equation to involve the partial derivatives of the wavefunction $\Psi(x, t)$ with respect to x and t. We conjectured in the last section that the equation will involve the first partial time derivative and the second spatial derivative of Ψ.

We assume the Schrödinger equation must be a linear equation. The assumption of linearity ensures that if $\Psi_1(x, t)$ and $\Psi_2(x, t)$ are both solutions to the equation, then the sum $\Psi_1(x, t) + \Psi_2(x, t)$ also is a solution. Thus the principle of superposition for waves is retained and follows from the assumption of linearity. The principle of superposition is necessary to account for the experimentally observed interference of particle-waves, such as in the electron diffraction experiments of Davisson and Germer. To be linear, the Schrödinger equation must involve only the first powers of Ψ, and its partial derivatives. That is, terms such as Ψ^2 or squares of the derivatives of Ψ do not appear in the equation.

The Schrödinger equation for particle-waves can be found from Equation 27.38 for the total energy, here rewritten as

$$\frac{p^2}{2m} + V(x, t) = E \tag{27.39}$$

We multiply this expression by the wavefunction $\Psi(x, t)$:

$$\left[\frac{p^2}{2m} + V(x, t)\right]\Psi(x, t) = E\Psi(x, t)$$

*Do not confuse the potential energy V with the electrical potential, which also is represented by V, as you know. Once again, the problem of suitable notation rears its ugly head. To eliminate the potential (pun!) for this confusion early in your study of physics is the reason we have, until now, used PE for potential energy rather than V.

†A famous professor (said to be Julian Schwinger), having covered the blackboard with ambiguous letters meaning different things, once responded: "Oh, we could use the same symbol for everything, provided we know what it means!" Here, we are trying to be honest with our notational ambiguities!

Now we replace the momentum p and energy E with their differential operators, which we found in Section 27.6 (Equations 27.31 and 27.34):

$$p \to -i\hbar\frac{\partial}{\partial x} \quad \text{and} \quad E \to i\hbar\frac{\partial}{\partial t}$$

We obtain

$$\left[\frac{1}{2m}\left(-i\hbar\frac{\partial}{\partial x}\right)\left(-i\hbar\frac{\partial}{\partial x}\right) + V(x, t)\right]\Psi(x, t) = i\hbar\frac{\partial\Psi(x, t)}{\partial t}$$

This can be simplified to the following:

$$-\frac{\hbar^2}{2m}\frac{\partial^2\Psi(x, t)}{\partial x^2} + V(x, t)\Psi(x, t) = i\hbar\frac{\partial\Psi(x, t)}{\partial t} \tag{27.40}$$

Equation 27.40 is the one-dimensional Schrödinger equation for the wavefunction Ψ.

The mathematical operator formed from the expression for the total energy,

$$\frac{p^2}{2m} + V \quad \text{which is} \quad -\frac{\hbar^2}{2m}\frac{\partial^2}{\partial x^2} + V$$

is called the **Hamiltonian** H of the system.‡ Using the Hamiltonian operator, the (one-dimensional) Schrödinger equation is written in a less intimidating and more succinct form as

$$\boxed{H\Psi = i\hbar\frac{\partial\Psi}{\partial t}} \tag{27.41}$$

Just as Newton's second law encompasses classical nonrelativistic mechanics, the Schrödinger equation is the touchstone of nonrelativistic quantum mechanics. In classical mechanics, a knowledge of the forces acting on a system permitted us to find interesting information about the acceleration, velocity, position, energy, and momentum of the system. The key was our ability to identify and describe the forces acting on the system. In quantum mechanics, the wavefunction is the corresponding key to information about the physical system.

‡The term is named for William Hamilton (1805–1865), who during the 19th century developed a formal theory of classical dynamics that showed how to obtain the equations of motion of a system from an expression for its total energy.

CHAPTER SUMMARY

The wavelength λ of a particle-wave is found from the de Broglie relationship:

$$\lambda = \frac{h}{|p_x|} \tag{27.1}$$

The energy E of a particle-wave, like that of a photon, is proportional to its frequency ν, or its angular frequency ω:

$$E = h\nu = \hbar\omega \tag{27.9}$$

Particles have wavelike properties and waves have particle-like properties. The *wave–particle duality* naturally leads to two *Heisenberg uncertainty principles*. The uncertainty in the position x and its complementary momentum component p_x satisfy the following inequality:

$$\Delta x\, \Delta p_x \geq h \tag{27.6}$$

If the wave-particle becomes more localized, Δx decreases and there is an increase in the uncertainty associated with the complementary momentum component. The uncertainty principle fundamentally limits our ability to determine simultaneously the position and corresponding momentum component of a particle-wave. Likewise, if the energy of the particle-wave is determined during a time interval whose uncertainty is Δt, the energy has a corresponding uncertainty ΔE, where

$$\Delta E\, \Delta t \geq h \tag{27.14}$$

Diffraction and interference of particle-waves are statistical processes. If the particles are sent through a single slit or multiple slits, it is impossible, even in principle, to predict where an individual particle will go. However, the single slit or multiple slit pattern gradually appears as more and more particles are sent through the arrangement. The diffraction or interference pattern that results therefore represents the *relative probability* that a particle is detected at a given position.

Observation and measurement of a system in quantum mechanics is fundamentally different from that in classical physics. In classical physics it is possible, at least in principle, to account for the effects of a measurement on a physical system. However, in quantum mechanics, observation and measurement fundamentally alter the system in ways that cannot be predicted with certainty.

A particle-wave is characterized by a wavefunction $\Psi(x, t)$, known as a *probability amplitude*. Since the wavefunction is complex valued, it is not directly observable.

Observable quantities such as momentum and energy are extracted from the wavefunction by operators in quantum mechanics. The (one-dimensional) *momentum operator* is

$$-i\hbar \frac{\partial}{\partial x} \tag{27.31}$$

The *energy operator* is

$$i\hbar \frac{\partial}{\partial t} \tag{27.34}$$

The wavefunction $\Psi(x, t)$ in nonrelativistic quantum mechanics obeys a new wave equation known as the *Schrödinger equation*:

$$-\frac{\hbar^2}{2m}\frac{\partial^2\Psi(x,t)}{\partial x^2} + V(x,t)\Psi(x,t) = i\hbar\frac{\partial\Psi(x,t)}{\partial t} \tag{27.40}$$

The Schrödinger equation also can be written as

$$H\Psi = i\hbar\frac{\partial\Psi}{\partial t} \tag{27.41}$$

where H is an operator called the *Hamiltonian* of the system.

QUESTIONS

1. Particles behave like waves and waves like particles. We have described this situation by calling them both particle-waves. Use your imagination to invent a new, shorter word to associate with the hybrid.
2. An electron, proton, neutron, and α particle all have the same energy. List the particles in the order of increasing de Broglie wavelength.
3. An electron, proton, neutron, and α particle all have the same speed. List the particles in the order of increasing de Broglie wavelength.
4. An electron, proton, neutron, and α particle all have momentum of the same magnitude. What can you say about the de Broglie wavelengths of the particles?
5. Explain why the Bohr model of the atom, with its well-defined, planet-like orbits for the electron, is incompatible with the Heisenberg position–momentum uncertainty principle.
6. Explain why the Heisenberg uncertainty principles are irrelevant for the physics of a macroscopic particle such as a jumping bean.
7. Explain why a radioactive material with a very short half-life corresponds to a large energy associated with the disintegration.
8. The age of the universe is about 15 billion years. What does the uncertainty principle say about the possibility of their being an exact instant when its age was zero?
9. If a single electron is sent through a single slit, explain in what sense the electron forms a single slit diffraction pattern.
10. If a single particle is sent through a double slit, in what sense is a double slit interference pattern formed?
11. If particle-waves are sent one at a time through a double slit arrangement, the double slit interference pattern gradually emerges. Explain why the double slit experiment is absolutely confounding to the notion we have of a particle, which is

what makes quantum mechanics somewhat baffling to our notions of common sense.

12. Discuss the similarities and differences between particle-waves and electromagnetic waves.

13. Is the relationship between geometric optics and physical (wave) optics similar to the relationship between a particle and a particle-wave? Discuss this analogy.

14. Particle-waves have a wavelength. If the wavelength is in the range of visible light wavelengths, do the particle-waves have color? Explain.

15. For a confined particle, its energy never can be zero. What does this imply about the attainability of the absolute zero of temperature?

PROBLEMS

Sections 27.1 The Heisenberg Uncertainty Principles
27.2 Implications of the Position–Momentum Uncertainty Principle
27.3 Implications of the Energy–Time Uncertainty Principle
27.4 Observation and Measurement

1. What are the values of Planck's constant h and $\hbar = h/2\pi$ when expressed in units of eV · s rather than the SI units of J · s?

2. Show that the SI units of the product $\Delta x\, \Delta p_x$ and the product $\Delta E\, \Delta t$ are those of angular momentum (kg · m^2/s) and that these units also are the same as the SI units of Planck's constant (J · s).

3. If the uncertainty Δx in the position of a particle is equal to its de Broglie wavelength, show that the uncertainty in the magnitude of its momentum Δp_x is greater than or equal to the magnitude of its momentum.

•4. Define the *angular wavenumber* k of a de Broglie wave to be

$$k \equiv \frac{2\pi}{\lambda}$$

(a) Show that the magnitude of the momentum of a particle-wave can be expressed as

$$p = \hbar k$$

(b) Show that the position–momentum uncertainty relation can be written as

$$\Delta x\, \Delta k \geq 2\pi$$

(c) Since the energy E of a particle-wave is

$$E = h\nu = \hbar\omega$$

show that the energy–time uncertainty relation can be expressed as

$$\Delta\omega\, \Delta t \geq 2\pi$$

These forms for the uncertainty relationships were discussed in Chapter 12 on waves (Section 12.21).

•5. A particle known unromantically as the $\Delta(1232)$ has a rest energy of 1232 MeV ± 120 MeV. What is the average lifetime of such a particle? These are among the shortest-lived particles known.

•6. An atom represents a region about 2×10^{-10} m wide in which an electron is confined. Use the Heisenberg uncertainty principle to estimate the minimum kinetic energy of the electron, expressing your result in electron-volts (eV).

•7. A passing photon is found to have a length in space (its wavetrain) of about 10 cm. The wavelength of the photon is 633 nm. About how well (to what fractional error $\Delta E/E$) is its energy or frequency determinable?

•8. A 60 kg motorist in a 1940 kg car is accused of speeding at 72 km/h while crossing an intersection 13.2 m wide where the legal speed limit is 36 km/h. A defense lawyer from the firm Chase, Cheatam, and Dunnum contends that because of the Heisenberg uncertainty principle, the speed cannot be exactly determined even with the best of equipment; therefore, the case should be dismissed against her client. Discuss the relevancy of the argument and include a quantitative estimate of the ultimate uncertainty in the speed in this situation.

•9. The lifetime of a certain atom while it exists in an excited state before making a transition to a lower-energy state is about 1.0 ns. The wavelength emitted in the transition is 200 nm. (a) What is the frequency of the light? (b) About how well can the frequency of the light be known? (c) About how well can the wave packet (photon) be localized in space?

•10. (a) The lifetimes of electrons in excited states in atoms are typically on the order of 1.0×10^{-8} s. What is the minimum uncertainty in the energy of states with such lifetimes? (b) What is the corresponding uncertainty in the frequency of the photons emitted in the transition from such a state to the ground state?

•11. Certain atoms have special excited states (called metastable states) in which an electron can exist for relatively long times, on the order of 1.0×10^{-4} s. What is the minimum uncertainty associated with the frequency from such metastable excited states? Such relatively long-lived states make lasers possible; lasers emit light with a very precise frequency.

•12. If the entire universe once were confined to a volume on the order of 1×10^{-90} m across in its early history, what is the corresponding energy uncertainty? Assume this early universe was filled entirely by high-energy photons and no mass.

⁝13. (a) About what magnitude of momentum might an electron have if it were confined inside a box about 1.0×10^{-14} m on a side? (This is about the size of the nucleus of an atom.) (b) Estimate the approximate speed of such an electron. (Hint: Use the relativistic expression $p = \gamma mv$ for the magnitude of the momentum.) (c) Estimate the approximate total (relativistic) energy of such an electron; express your result in MeV. Compare this energy with the rest energy of an electron (0.511 MeV).

Sections 27.5 Particle-Waves and the Wavefunction
27.6 Operators*
27.7 The Schrödinger Equation*

14. Probabilities are dimensionless. What are the dimensions and SI units of the product $\Psi^*\Psi$ of the one-dimensional wavefunction and its complex conjugate? Use this result to determine the dimensions and SI units of the one-dimensional wavefunction Ψ itself.

•15. If $\Psi_1(x, t)$ and $\Psi_2(x, t)$ are both solutions to the Schrödinger equation for a given potential energy function $V(x)$, show that

$$c_1\Psi_1(x, t) + c_2\Psi_2(x, t)$$

also is a solution to the Schrödinger equation, where c_1 and c_2 are constants that may be real or complex numbers. This principle of linear superposition indicates that particle-waves can interfere with each other.

INVESTIGATIVE PROJECTS

A. Expanded Horizons

1. Investigate the statistical behavior of dice with parallelopiped shapes.
Patricia F. Bronson and Robert L. Bronson, "Dice with parallelopiped shapes," *The Physics Teacher*, 28, #5, pages 286–290 (May 1990).

2. Investigate the interesting quantum mechanical problem known as the paradox of *Schrödinger's cat*.
John Gribben, *In Search of Schrödinger's Cat* (Bantam Books, New York, 1984), pages 203–208.
Christopher N. Villars, "The paradox of Schrödinger's cat," *Physics Education*, 21, #4, pages 232–237 (July 1986).

3. Investigate the so-called *cosmic number hypothesis* and clues to several peculiar numerological coincidences among the constants of physics and cosmology.
Edward R. Harrison, *Cosmology* (Cambridge University Press, Cambridge, England, 1981), Chapter 17, pages 329–345.
Herman Bondi, *Cosmology* (Cambridge University Press, Cambridge, England, 1960), Chapter 7.
Edward R. Harrison, "The cosmic numbers," *Physics Today*, 25, #12, pages 30–34 (December 1972).

4. Investigate how the characteristic sizes of many things in nature are related to the magnitudes of fundamental physical constants.
Victor F. Weisskopf, "Of atoms, mountain, and stars: a study in qualitative physics," *Science*, 187, #4177, pages 605–612 (21 February 1975).

5. Investigate speculations about the fundamental constants of nature *varying with time*.
Paul S. Wesson, "Does gravity change with time?" *Physics Today*, 33, #7, pages 32–37 (July 1980).
Thomas C. Van Flandern, "Is gravity getting weaker?" *Scientific American*, 234, #2, pages 44–52, 140 (February 1976).
R. A. Alpher, "Large numbers, cosmology, and Gamow," *American Scientist*, 61, #1, pages 52–58 (January–February 1973).
See also the references under Investigative Project 3.

6. Investigate the effects of gravity on antimatter.
Terry Goldman, Richard J. Hughes, and Michael Martin Nieto, "Gravity and antimatter," *Scientific American*, 258, #3, pages 48–56 (March 1988).

B. Communicating Physics

7. Uncertainty is a perilous concept in life, religion, business, politics, and the legal profession. In an essay, compare and contrast the notion of uncertainty in these contexts with its meaning in physics.

EPILOGUE

. . . to make an end is to make a beginning.
The end is where we start from.

T. S. Eliot (1888–1965)*

To paraphrase Matthew 25:21, *well done thou good and faithful student*. We come full circle, like the orobouros, the serpent swallowing its tail.

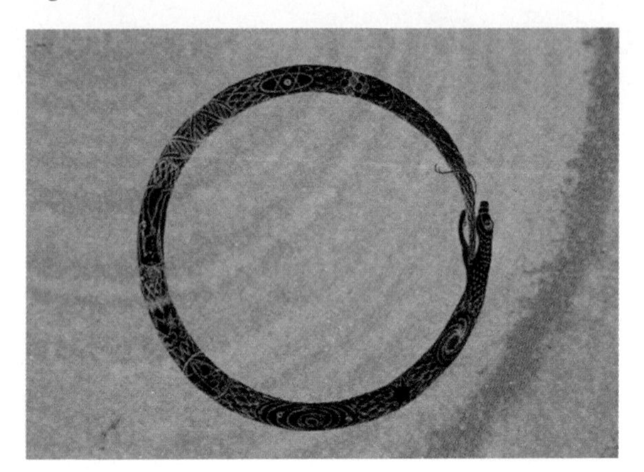

The five great theories of physics with which you now have more than passing acquaintance—classical mechanics, thermodynamics, electromagnetism, relativity, and quantum mechanics—are the beginning, not the end. There really is no end to physics! You may see this either as a horrifying realization or, more optimistically, as a window of opportunity through which to transmit your own creativity and imagination to the field. Whichever, (though I hope it is the latter!), I wish you Godspeed and welcome your correspondence. *Shalom aleichem*.—R.L.R.

We shall not cease from exploration
And the end of all our exploring
Will be to arrive where we started
And know the place for the first time.

T.S. Eliot†

**Four Quartets*, "Little Gidding," lines 215–216 (Harcourt Brace Jovanovich, New York, 1971).

†*Four Quartets*, "Little Gidding," lines 239–242.

APPENDIX A
PROOFS OF THE GRAVITATIONAL SHELL THEOREMS

A.1 A MASS WITHIN A UNIFORM SPHERICAL SHELL

We seek to prove the following theorem:

> If a mass m is located anywhere within the volume enclosed by a thin uniform spherical shell of mass M, the total gravitational force of the shell on m is equal to zero.

Consider a pointlike mass m positioned anywhere *inside* a thin uniform spherical shell of radius R and mass M, as shown in Figure A.1. Draw two narrow, equal apex-angled, differential cones centered on the mass m, as indicated. These small conical constructions are like two narrow ice cream cones. The cones intercept masses dM and dM' on the spherical shell. The differential gravitational force $d\vec{\mathbf{F}}$ of dM on m attracts m toward dM. The differential gravitational force $d\vec{\mathbf{F}}'$ of dM' on m attracts m toward dM'.

We need to calculate the ratio of the magnitudes of these two forces on m. The vector addition of the two forces then will be quite easy, since they are in opposite directions because of how we constructed the differential cones.

If the apex angle of the cones is sufficiently small, the masses dM and dM' are essentially pointlike masses. Hence we use the point-mass gravitational force law deduced by Newton, Equation 6.3, to calculate the gravitational effects of dM and dM' on the mass m inside the shell. Thus the magnitude of the differential force of dM on m is

$$dF = \frac{Gm\,dM}{r^2}$$

where r is the distance of dM from the mass m inside the shell as shown in Figure A.1. Likewise, the magnitude of the differential force of dM' on m is

$$dF' = \frac{Gm\,dM'}{r'^2}$$

where r' is the distance between m and dM', also indicated in Figure A.1.

We take the ratio of the magnitudes of the two forces:

$$\frac{dF}{dF'} = \frac{\dfrac{Gm\,dM}{r^2}}{\dfrac{Gm\,dM'}{r'^2}} = \frac{dM}{dM'}\frac{r'^2}{r^2} \qquad \text{(A.1)}$$

Since the shell is a *uniform* shell, the ratio of the masses intercepted by the differential cones is equal to the ratio of the areas intercepted:

$$\frac{dM}{dM'} = \frac{dS}{dS'}$$

We draw radii R from the center of the sphere to the two areas dS and dS', as shown in Figure A.2. The radii intersect the axis of the double cone, forming a large isosceles triangle with its apex angle at the center of the shell. The angle ϕ that the radii make with the common axis of the cones therefore is the same at area dS and dS'. Let dA be the *projected* area of dS that is perpendicular to the axis of the cone; let dA' be the corresponding projected area at dS'.

The angle between dS and dA is ϕ; likewise, the angle between dS' and dA' is ϕ. The projected areas are related to the areas on the shell itself by

$$dA = dS\cos\phi$$

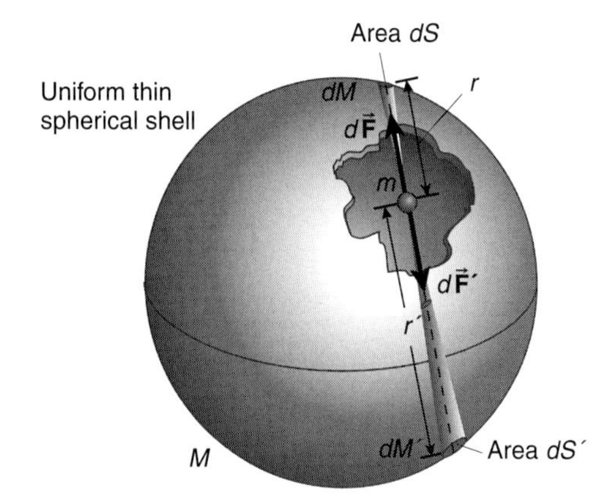

FIGURE A.1 A pointlike mass m inside a thin uniform spherical shell.

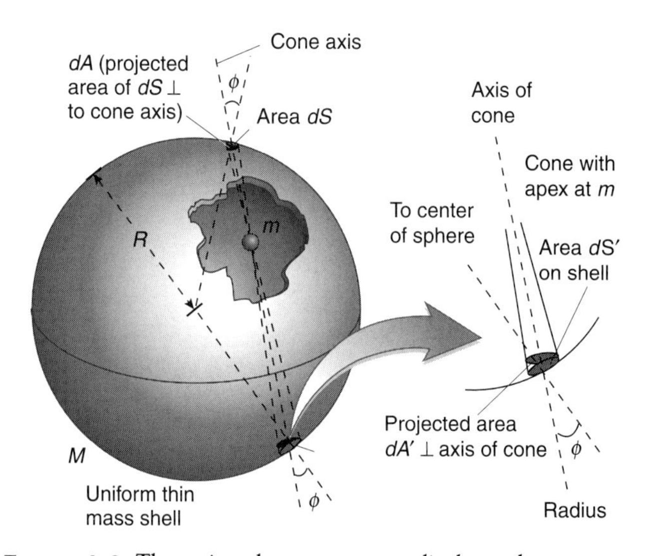

FIGURE A.2 The projected areas are perpendicular to the common axis of the differential cones.

and

$$dA' = dS' \cos \phi$$

So the ratio of the areas dS/dS' is the same as the ratio of the areas dA/dA'. Since the apex angles of the cones are equal, the sizes of the projected areas are proportional to the square of the distance they are from the apex of the cones; that is,

$$\frac{dA}{dA'} = \frac{r^2}{r'^2}$$

Hence the ratio of the masses intercepted by the cones is the same as

$$\frac{dM}{dM'} = \frac{dA}{dA'} = \frac{r^2}{r'^2}$$

The ratio of the magnitudes of the two differential forces is then (from Equation A.1)

$$\frac{dF}{dF'} = \frac{dM}{dM'}\frac{r'^2}{r^2} = \frac{r^2}{r'^2}\frac{r'^2}{r^2} = 1 \tag{A.2}$$

Thus the two differential forces on m have exactly the same magnitude, but point in opposite directions. The two differential forces are *not* a Newton's third law pair since they act on the *same* system, the mass m. Hence the vector sum of the two differential forces on m is *zero*. Although m is closer to dA than to dA', the more distant area dA' has more mass in just the right amount to compensate for its greater distance from m.

Now we sweep the differential cones around in all directions to take into account the effect of the entire spherical shell on m. *Every* pair of little differential masses at opposite ends of the differential cones produce zero total gravitational force on the mass m. Thus we reach the startling conclusion that the total gravitational force of the shell on mass m is zero, as indicated in Figure A.3.

Notice that this remarkable result depended *crucially* on the inverse square law nature of the gravitational force law; if the gravitational force had any other power of the distance in the denominator, the distances in Equation A.2 would not cancel and there would be a nonzero total gravitational force on m unless it were at the very center of the shell. Nature was kind in giving us an inverse square law gravitational force.

A.2 A Mass Outside a Uniform Spherical Shell

We seek to prove the following theorem:

> If a pointlike mass m is located anywhere outside a thin uniform spherical shell of mass M, the total gravitational force of the shell on m is the same as that of a point mass, with a mass equal to the mass M of the shell, located at the center of the shell.

To prove this theorem, we have to calculate the gravitational force of a uniform spherical shell of mass M on a pointlike mass m lying a distance r from the center of the shell, where $r > R$, the radius of the shell (see Figure A.4).

It is necessary to do a vector integration here; we managed to avoid such an integration when the point mass was inside the sphere, but here it is unavoidable. Vector integrations can be difficult because we have to account both for the magnitudes *and* directions of the vectors. So we proceed with some caution; it will be beneficial for you to study this technique with some care, since vector integrations are encountered again later in our course of study. The typical procedure for performing a vector integration involves converting the vector integration into an ordinary scalar integration by subtly looking for ways to exploit or account for the directional effects of the vectors. This only comes with practice and patience.

The amount of mass on any square meter of the shell is known as a *surface mass density* σ. Since the shell is uniform, the surface mass density σ is constant and equal to the total mass of the shell divided by the surface area of the spherical shell:

$$\sigma = \frac{M}{4\pi R^2}$$

We begin by examining a thin differential ring of the shell, as shown in Figure A.5. Every point on the differential ring is the

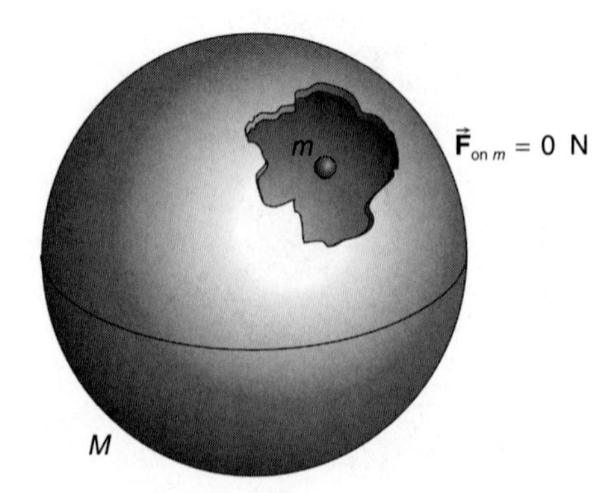

FIGURE A.3 The total gravitational force of the shell on mass m inside the shell is zero.

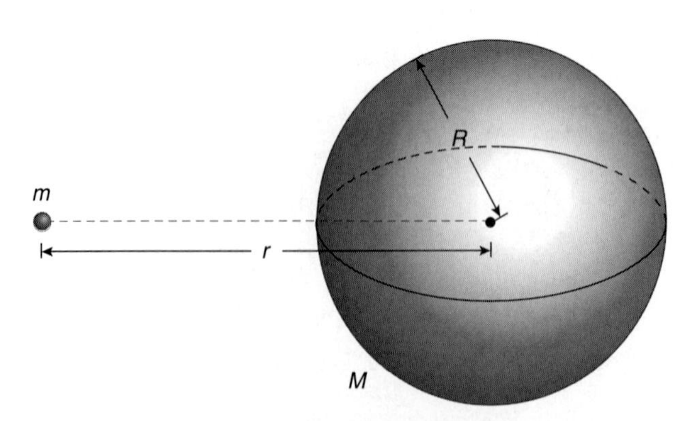

FIGURE A.4 A pointlike mass m outside a thin uniform spherical shell.

same distance s from the point mass m. The radius of the differential ring is $R \sin \theta$. The area of the differential ring is its circumference $(2\pi R \sin \theta)$ times its width $(R\, d\theta)$, or

$$\begin{aligned} dA &= (2\pi R \sin \theta)(R\, d\theta) \\ &= 2\pi R^2 \sin \theta\, d\theta \end{aligned}$$

The mass dM on this ring of material is the product of the area dA and the surface mass density σ:

$$\begin{aligned} dM &= \sigma\, dA \\ &= \sigma\,(2\pi R^2 \sin \theta\, d\theta) \end{aligned}$$

Each little segment of the ring produces a gravitational force on mass m that is directed toward that little piece of the ring, as shown in Figure A.6. Across the diameter of the ring is a twin segment of the ring. Each of the two twins produces a differential force on m. Note that the components of the two force vectors that are perpendicular to the line joining m and the center of the shell will vector sum to zero. The components of the force vectors of the twins directed *along* the line between m and the center of the shell will add, since they point in the same direction.

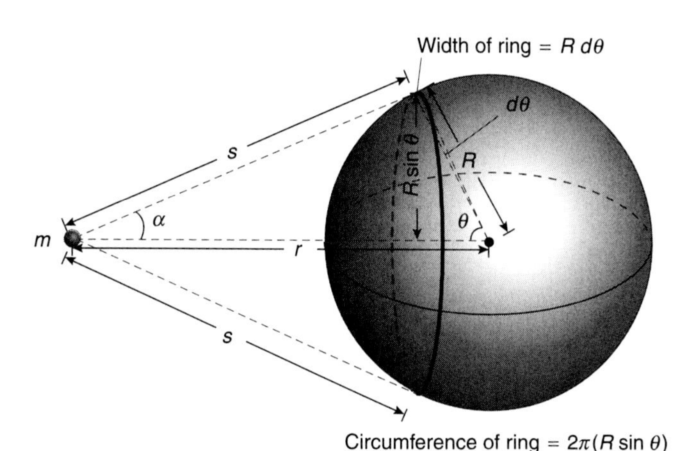

FIGURE A.5 A thin differential ring of the shell.

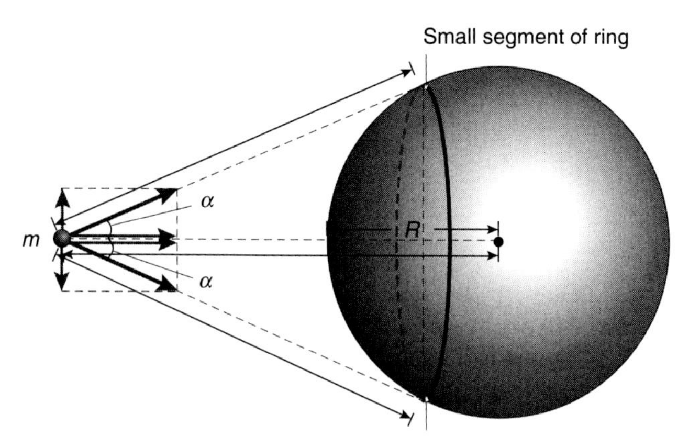

FIGURE A.6 Twin segments produce forces of equal magnitude on m. The force components perpendicular to the line joining m and the center of the shell vector sum to zero; the components along the line add.

Thus, as we vector sum (integrate) around the ring, the components of the differential forces perpendicular to the line will pairwise vector sum to zero, while those along the line will simply add. This is certainly cause for celebration!

The *surviving* component of the differential force on m by the segment of material dM is

$$dF = \frac{Gm\, dM}{s^2} \cos \alpha$$

The $\cos \alpha$ term picks off the surviving part of the differential force as we vector integrate around the ring. Thus we have accounted for the directional effects of the vectors with the $\cos \alpha$ term. The resulting integration is now just an ordinary scalar integration for the *magnitude* of the total gravitational force of the shell on m.

If we substitute for dM, the expression for dF begins to look like a nightmare:

$$dF = Gm \frac{\sigma\,(2\pi R^2) \sin \theta\, d\theta}{s^2} \cos \alpha \qquad \text{(A.3)}$$

Now we integrate this expression over the shell. For all little rings anywhere on the spherical shell, the differential forces given by Equation A.3 simply add together because they all point in the same direction: toward the center of the shell.

It is convenient to express everything in terms of the distance s. It is not obvious that this is the thing to do, but using the variable s enables us to perform the integration easily and that is why we do it.

The law of cosines from trigonometry will help. In the triangle whose sides are r, s, and R in Figure A.5, the law of cosines implies that

$$s^2 = r^2 + R^2 - 2rR \cos \theta$$

As we integrate over the sphere, s and θ change but r and R are fixed constants. Taking differentials of the law of cosines equation, we find

$$2s\, ds = -2rR(-\sin \theta\, d\theta)$$

or

$$s\, ds = rR \sin \theta\, d\theta$$

Hence

$$\sin \theta\, d\theta = \frac{s\, ds}{rR}$$

We apply the law of cosines in a different way in the same triangle; this enables us to find an expression for $\cos \alpha$. Since

$$R^2 = s^2 + r^2 - 2sr \cos \alpha$$

we can solve for $\cos \alpha$:

$$\cos \alpha = \frac{s^2 + r^2 - R^2}{2sr}$$

Substituting the expressions for $\sin\theta\, d\theta$ and $\cos\alpha$ into the expression for dF in Equation A.3, we really get quite a mouthful:

$$dF = \frac{Gm\sigma\,(2\pi R^2)}{s^2}\,\frac{s\,ds}{rR}\,\frac{s^2 + r^2 - R^2}{2sr}$$

Simplifying this slightly, we obtain

$$dF = \frac{Gm\sigma\pi R}{r^2}\,\frac{s^2 + r^2 - R^2}{s^2}\,ds$$

Now we integrate to find the magnitude of the total force on m due to the entire shell. The limits on s are from $r - R$ to $r + R$ (see Figure A.6). The integral becomes

$$F = \frac{Gm\sigma\pi R}{r^2}\int_{r-R}^{r+R}\left[1 + \frac{(r^2 - R^2)}{s^2}\right]ds$$

The two integrals are not too bad to do, since both r and R are constants:

$$F = \frac{Gm\sigma\pi R}{r^2}\left[\int_{r-R}^{r+R} ds + (r^2 - R^2)\int_{r-R}^{r+R}\frac{ds}{s^2}\right]$$

Performing the integrations, we get

$$F = \frac{Gm\sigma\pi R}{r^2}\left[s - \frac{(r^2 - R^2)}{s}\right]\Bigg|_{r-R}^{r+R}$$

Inserting the limits of integration and simplifying, we find

$$F = \frac{Gm\sigma\pi R}{r^2}\,4R$$

We rearrange this a bit, getting

$$F = \frac{Gm}{r^2}\,4\pi R^2\sigma$$

But the quantity $4\pi R^2\sigma$ is the product of the surface mass density σ and the surface area of the shell. This is the total mass M of the shell. Hence the magnitude of the total force on m is

$$F = \frac{GmM}{r^2}$$

Voilà! Notice that this is the same expression as Equation 6.3 for the magnitude of the gravitational force between two pointlike masses separated by a distance r.

Thus the magnitude of the gravitational force of the shell on m is found by imagining the mass M of the shell to be located at its center, since the distance r is the distance between m and the center of the shell. Effectively, we can think of the mass of the shell as if it were at the center of the shell, as shown in Figure A.7. What a relief this result must have been to Newton as he wrestled with how to account for the gravitational effects of the shell on a mass m located outside the shell!

To satisfy Newton's third law of motion, the mass m exerts a gravitational force of the same magnitude but opposite direction *on the shell*, so that we have the third law force diagram shown in Figure A.8.

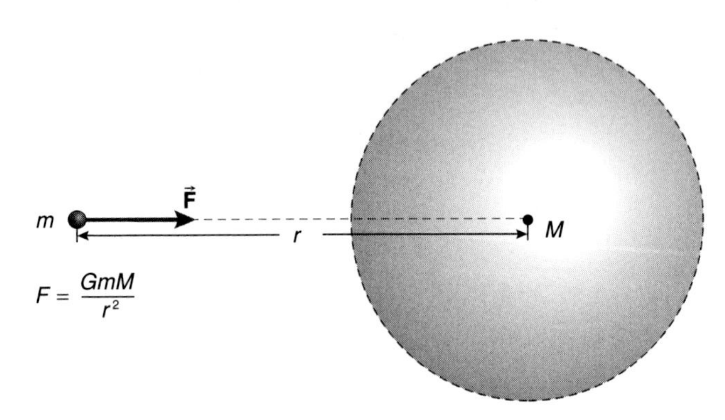

FIGURE A.7 The force of the shell on m can be found by imagining the entire mass of the shell to be at its center.

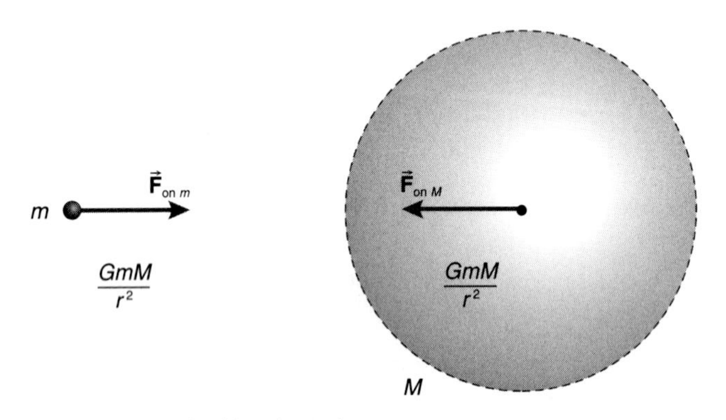

FIGURE A.8 Third law force diagram.

ANSWERS TO PROBLEMS

CHAPTER 1

1. 4.6403×10^{-5} m^3

5. peekaboos, nannygoats, military, millionaires, megaphones, microphones, megabucks, centipedes, microfilm, terrapin, megalopolis, deck of cards, microfiche, microwaves, decimate, microeconomics, to kill a mockingbird, terrible, terra firma, megalomaniacs, millinery, microscopes, megabytes

9. (a) 1.83×10^3 (b) 3.33×10^5 (c) 2×10^{11} (d) 3×10^{11} Parts (c) and (d) are the same order of magnitude.

13. (a) ~10^6 m (b) ~10^5 m

17. (a) A factor of 3 (b) A factor of 9

21. (a) A factor of $\sqrt{3} \approx 1.73$ (b) A factor of $3\sqrt{3} \approx 5.20$

25. 0.447 h·m/(mile·s)

29. 3.4×10^{22} molecules

33. 3.63 m^3

37. (a) 7.77×10^{-4} m^3 (b) 1.6×10^{-4} m^3

41. (a) 6.5×10^8 m^3 (b) 8.7×10^2 m on an edge (c) 3.6×10^8 m^2 (d) 19 km

45. 1.80×10^{12} furlongs/fortnight

49. (a) 9.74×10^{-4} kg (b) 8.0×10^{-5} m (c) 8.2×10^{-7} m^3 (d) 9.74×10^5 kg (e) 8.0×10^4 m (f) 1.0×10^7 m^2 (g) 9.4 m

53. (a) 5×10^{-28} kg/m^3 (b) 3×10^9 m

57. (a) ~12 minutes (b) ~3 years (c) ~10 minutes

61. 1.19×10^{57} proton masses

65. ~3×10^7 ties, 1×10^6 m^3

69. ~7×10^8 km^3 assuming an average depth of 2 km

73. ~several percent

77.

Angle	*Sine*	*Cosine*	*Tangent*
78.0°	0.978	0.208	4.70
78.02°	0.9782	0.2076	4.713
78.024°	0.97823	0.20750	4.7143

CHAPTER 2

1.

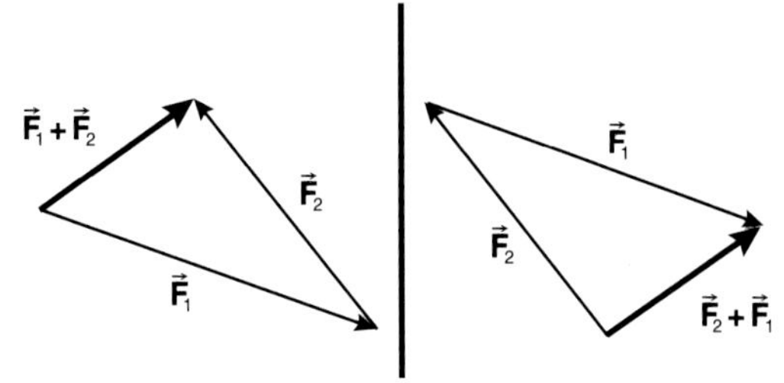

5. 100 m east, 141 m northwest

9. $(\vec{R} + \vec{r}) \cdot (\vec{R} - \vec{r}) = 0$ so $\vec{R} + \vec{r}$ is perpendicular to $\vec{R} - \vec{r}$. The vector $\vec{R} + \vec{r}$ is also perpendicular to $\vec{r} - \vec{R}$.

13. Answers are in Table 2.1.

17. $\dfrac{\hat{\imath} + \hat{\jmath} + \hat{k}}{\sqrt{3}}$

21.

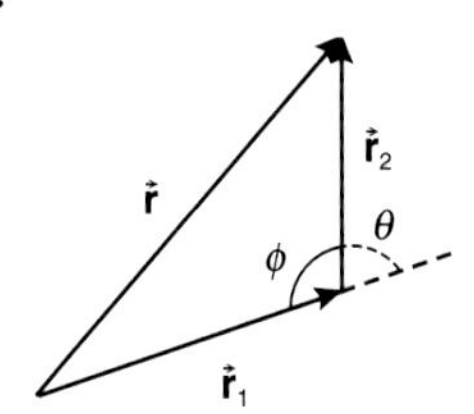

$r^2 = r_1^2 + r_2^2 - 2r_1r_2 \cos \phi$

25. (a) 64.9° (b) 115.1°

29. Answers are given in the statement of the problem.

33. (a) 90° (b) 53° between the vector sum and the vector of magnitude 12. 37° between the vector sum and the vector of magnitude 16

37. $\alpha = -1.00, \beta = -10.00, \gamma = 9.00$

41. $\hat{k}$ or $\alpha\hat{k}$ where α is any scalar; the solution is not unique.

45. (a) 19.1 (b) $-8.2\,\hat{k}$

49. (a) 0 (b) $6\,\hat{\jmath} - 3\,\hat{k}$; The solution is not unique. (c) 0

53. A proof

57. (a) $-3.0\,\hat{\imath} + 3.0\,\hat{\jmath} - 4.0\,\hat{k}$ (b) 4.0 (c) 37°

61. Answers are given in the statement of the problem.

65. $\dfrac{5}{2\sqrt{t}}\,\hat{\imath} - 9\sqrt{t}\,\hat{\jmath}$

69. A proof

CHAPTER 3

1. (a) 0 m (b) 50 m (c) 4.2 m/s (d) 0 m/s

5. 4.0 weeks

9. 6.7 m/s, 2×10^{-2} s

13. Let $\hat{\imath}$ be in the direction the tanker is moving: (a) $\vec{a} = (-9.26 \times 10^{-3}$ m/s$^2)\,\hat{\imath}$ (b) 3.75 km

17. (a) 2.47 s (b) 24.2 m/s (c) 19.8 m/s (d) 25.1 m

21. (a) $v_x \approx -6.0$ m/s (b) Yes, when $t \approx 1.5$ s. (c) No. $x \approx 5.2$ m.

25. 3×10^{24} m/s^2

29. (a) 10 s (b) No. 10 m

33. (a) $\vec{a}_{ave} = (3.3 \times 10^{-2}$ m/s$^2)\hat{\imath}$, $\vec{a} = (3.3 \times 10^{-2}$ m/s$^2)\hat{\imath}$ (b) $\vec{a}_{ave} = (-6.7 \times 10^{-2}$ m/s$^2)\hat{\imath}$, $\vec{a} = (-6.7 \times 10^{-2}$ m/s$^2)\hat{\imath}$ (c) $\vec{a}_{ave} = (3.3 \times 10^{-2}$ m/s$^2)\hat{\imath}$, $\vec{a} = (3.3 \times 10^{-2}$ m/s$^2)\hat{\imath}$ (d) The barge begins at rest and accelerates to a speed 2.0 m/s when $t = 60$ s. The barge then begins to slow down, stops instantaneously when $t = 90$ s, reverses the direction it is moving and reaches a speed of 2.0 m/s in the opposite

direction when t = 120 s. Then the barge again begins to slow down and is brought to a stop when t = 180 s.

(e)

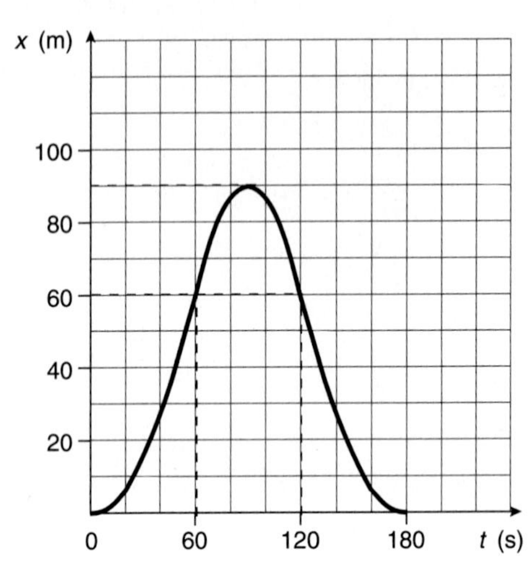

(f)

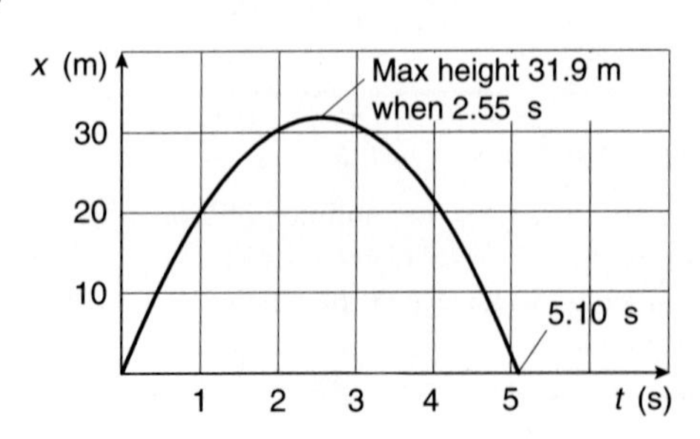

57. 6.5 m

61. 1.7 m

65. 42 m

69. (a) Up (b) Up (c) ≈3.0 s (d) ≈25 m (e) ≈25 m

37. (a) When t = 0 s and 6.0 s (b) During the interval between t = 1.0 s and 4.0 s (c) $\vec{\mathbf{a}} = (-2.0\ \text{m/s}^2)\ \hat{\mathbf{i}}$ (d) 18 m

41. (a) 6.4×10^2 m/s^2 (b) 48 m, 6.2 s

45. (b) 12.0 s and 20.0 s (c) When t = 12.0 s, they first catch up to the back of the train. If they kept running they would pass the back of the train, but when t = 20.0 s, the back end of the accelerating train would pass them for the last time.

49. (a) 1.0 ms (b) 6.0×10^9 m/s^2

53. (a) Let $\hat{\mathbf{i}}$ be upward with the origin at the point of release. (b) 31.9 m (c) 5.10 s
(d)

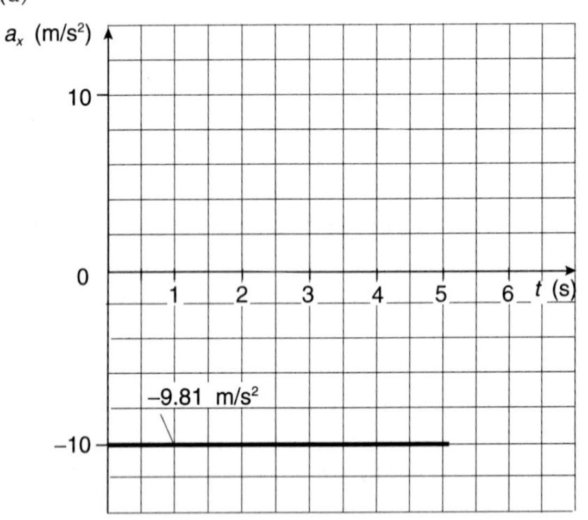

(e)

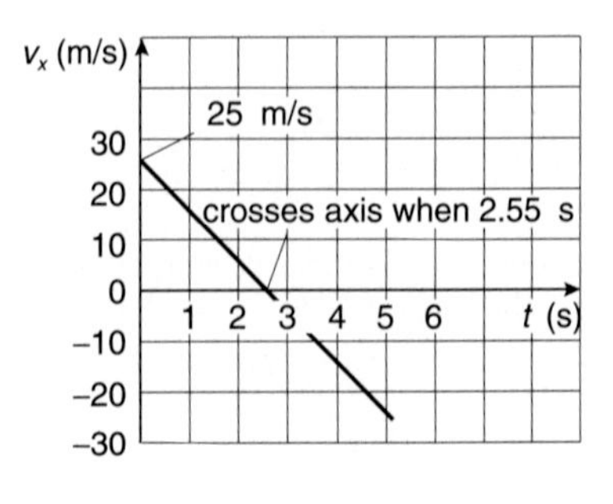

Chapter 4

1. (a) $\vec{\mathbf{r}}_i = (-0.15\ \text{m})\hat{\mathbf{i}}$, $\vec{\mathbf{r}}_f = (0.26\ \text{m})\hat{\mathbf{i}} + (0.15\ \text{m})\hat{\mathbf{j}}$
(b) average speed = 7.5×10^{-3} m/s, $\vec{\mathbf{v}}_{ave} = (6.8 \times 10^{-3}\ \text{m/s})\hat{\mathbf{i}} + (2.5 \times 10^{-3}\ \text{m/s})\hat{\mathbf{j}}$. $|\vec{\mathbf{v}}_{ave}| = 7.2 \times 10^{-3}$ m/s.

5. Answers are given in the statement of the problem.

9. (a)

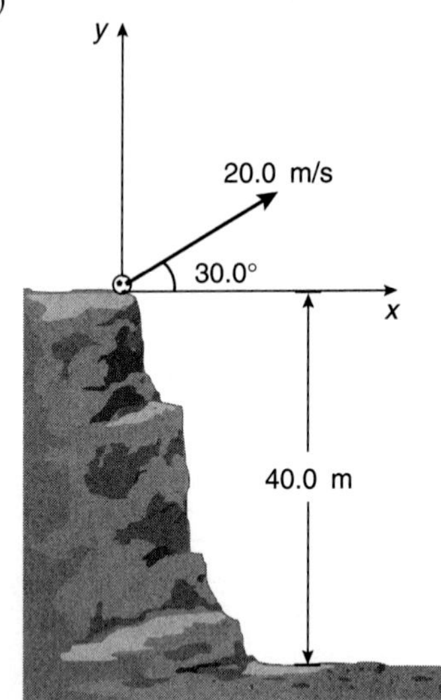

(b) $v_y(t) = 20.0\ \text{m/s}\ (\sin 30.0°) - gt$, $y(t) = 0\ \text{m} + 20.0\ \text{m/s} \times (\sin 30.0°)t - (1/2)gt^2$, $v_x(t) = 20.0\ \text{m/s}\ (\cos 30.0°)$, $x(t) = 20.0\ \text{m/s}\ (\cos 30.0°)\ t$ (c) 4.05 s (d) Using the coordinate system shown in part (a), the coordinates of the impact point are x = 70.1 m and y = −40.0 m.

13. 0.56 s and 2.09 s

17. (a)

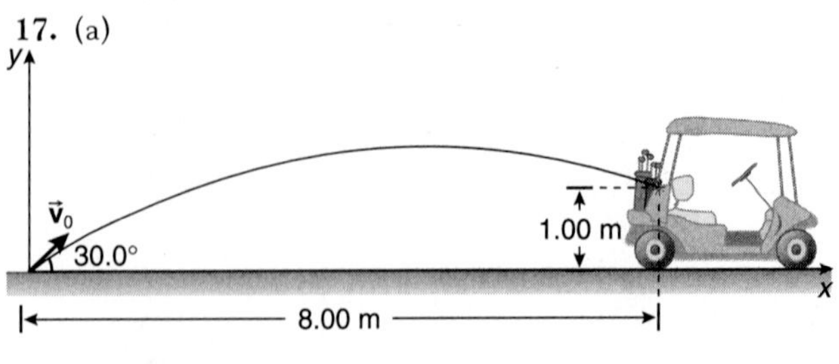

(b) 10.8 m/s (c) 0.856 s

21. (b) 35 m/s (c) 23 m/s^2

25. Answer is given in the statement of the problem.

29. (a) $(2gh)^{1/2}$ (b) $2h$ (c) $h/2$

33. A proof

37. (a) 36.9° to the current (b) 120 s

41. (a) Choose $\hat{\mathbf{i}}$ east and $\hat{\mathbf{j}}$ north, $\vec{\mathbf{v}} = (-1.50 \text{ m/s})\,\hat{\mathbf{i}} + (1.00 \text{ m/s})\,\hat{\mathbf{j}}$ (b) No

45. (a) Let $\hat{\mathbf{i}}$ be east and $\hat{\mathbf{j}}$ be north, $(-1095 \text{ km/h})\,\hat{\mathbf{i}} + (495 \text{ km/h})\,\hat{\mathbf{j}}$ (b) 1.20×10^3 km/h

49. 314.2 rad/s

53. 10 m

57. 4.4×10^8 m/s^2

61. 18 km. If the radius decreases, the centripetal acceleration increases, so the curve must have a radius no smaller than 18 km.

65. (a) Second hand: 0.105 rad/s; minute hand: 1.75×10^{-3} rad/s; hour hand: 1.45×10^{-4} rad/s (b) No (c) Yes, perpendicular to the clock face directed into the clock (d) 60 (e) 12

69. (a) 131 rad/s^2 (b) 75.1 rev

73. (a) 25.0 rad/s^2 (b) 3.52×10^5 rev (c) 8.82×10^6 m/s^2, $a/g = 8.99 \times 10^5$ (d) 43.8 rad/s^2

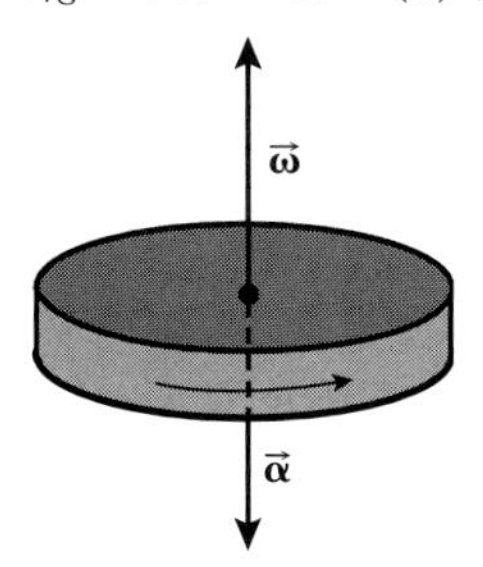

77. (a) A proof (b) $\pi/4$ rad

81. (a) Let $\hat{\mathbf{k}}$ be parallel to $\vec{\boldsymbol{\omega}}$ (b) 6.28 s (c) $\alpha_z = 0.159$ rad/s^2 (d)

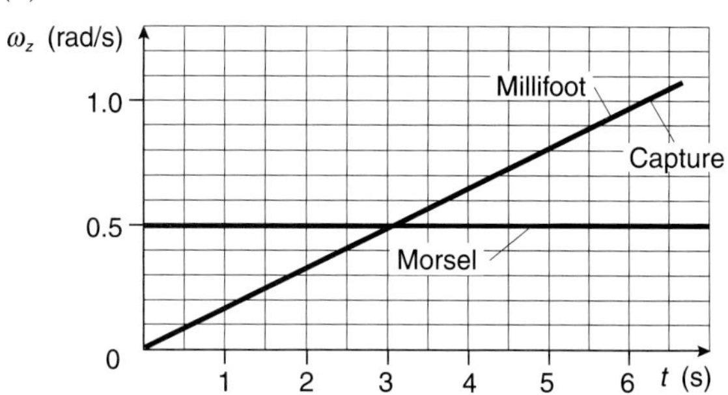

At capture, the areas under each curve are the same size, since both then have traveled through the same angle.

85. Answers are given in the statement of the problem.

Chapter 5

1. 188 N

5. (a) 8.4 N, 73° clockwise from the $+x$-axis (b) 2.1 m/s^2

9. (a)

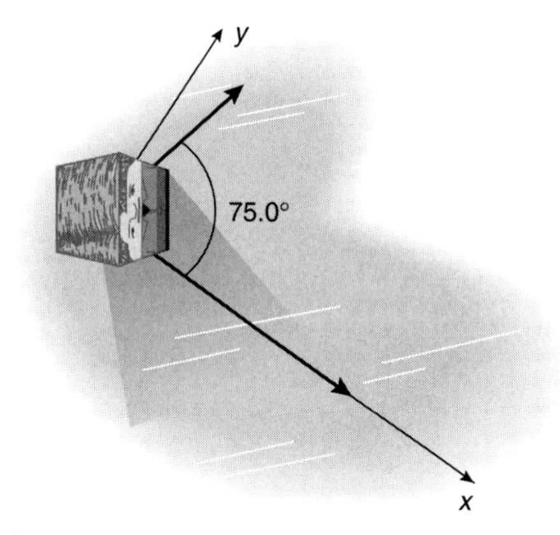

(b) $\vec{\mathbf{F}}_{\text{total}} = (565 \text{ N})\,\hat{\mathbf{i}} + (241 \text{ N})\,\hat{\mathbf{j}}$, $F_{\text{total}} = 614$ N (c) 7.7 m/s^2 (d) $0.920\,\hat{\mathbf{i}} + 0.393\,\hat{\mathbf{j}}$, yes

13. (a)

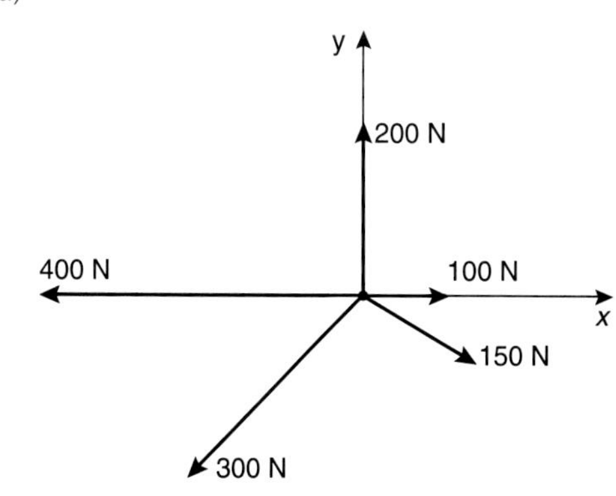

(b) $(-4.2 \text{ m/s}^2)\,\hat{\mathbf{i}} - (1.0 \text{ m/s}^2)\,\hat{\mathbf{j}}$ (c) 4.3 m/s^2

17. 0.102 kg

21. (a)

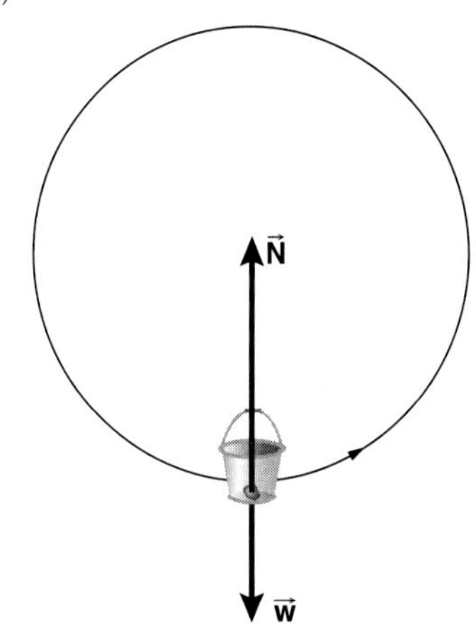

(b) 20.0 N toward the center of the circle (c) 34.7 N vertically up (d) It remains in contact with the bottom of the bucket since the normal force of the bucket bottom on the rock is not zero at this location. Its magnitude is 5.3 N, directed vertically down.

25. (a) 5.85° (b) Yes (c) 1.00 N

29. 10.2 kg

33. (a) 100 N vertically up

(b)

(c) Force of you on the rope, force of you on the surface, gravitational force of you on the Earth. (d) 687 N (e) 587 N (f) the magnitude of the force you exert on the surface; 59.8 kg.

37. Answers are given in the statement of the problem.

41. (a)

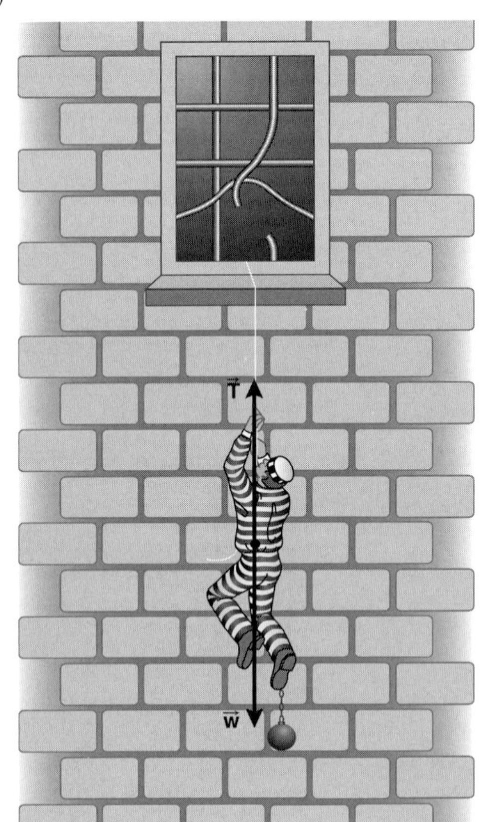

(b) 1.48 m/s^2, minimum value (c) 4.87 m/s

45. (a) 12.4 kN (b) 1.13 kN. Since the oaf is accelerating upward, the normal force of the floor on the oaf is greater than the oaf's weight.

49. (a)

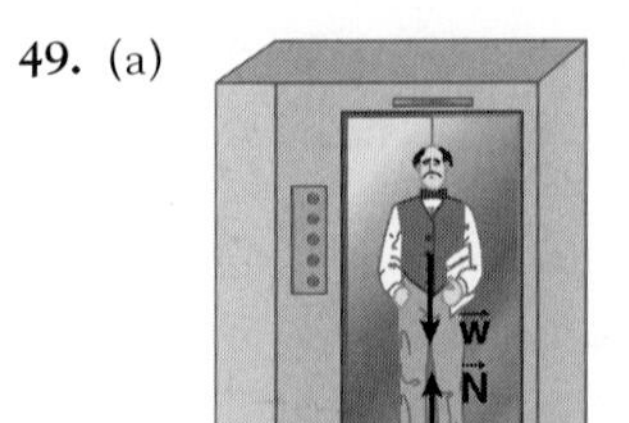

(b) 785 N (c) 785 N

(d)

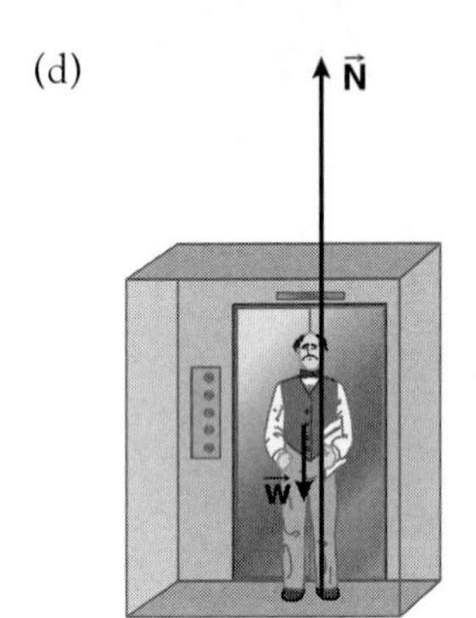

(e) 785 N (f) 865 N
(g) 865 N. The scale measures the magnitude of the normal force, not the magnitude of the weight.

(h)

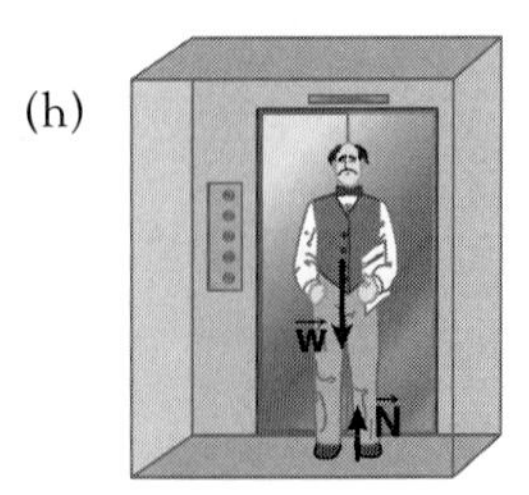

(i) 785 N (j) 705 N
(k) 705 N

53. $\mu_s = 0.71$

57. $\mu_k = 0.09$

61. (a)

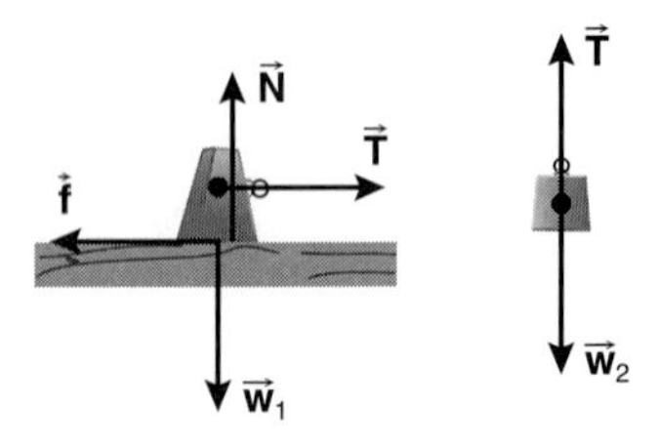

(b) 10 kg (c) 49 N static friction

65. (a)

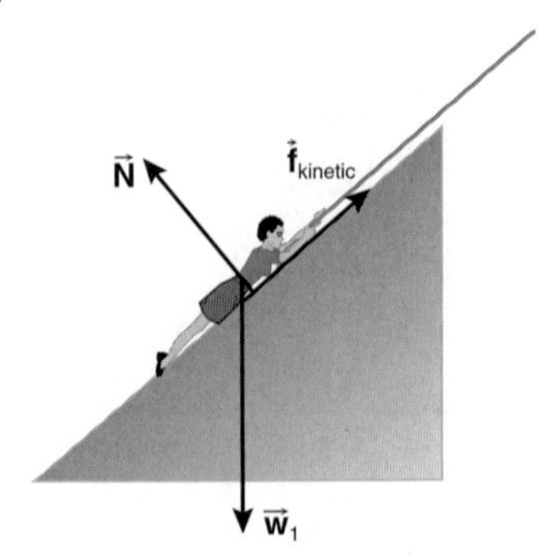

You will slip because the component of your weight down the plane is greater than f_{max}.

(b)

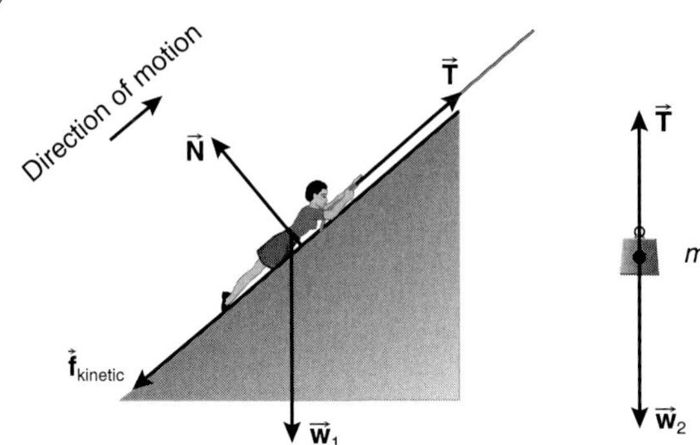

88 kg (c) 7.3×10^2 N

69. 3.4 m/s^2

73. (a) The total force is directed toward the center of the circle since you have a centripetal acceleration in that direction. (b) 2.0 m/s (c) 7.8 m (d) Ground observer: the ball follows a parabolic trajectory. The horizontal component of the velocity of the ball is 2.0 m/s and the vertical component is equal to that given to the ball by the person on the merry-go-round; merry-go-round observer: the ball goes up and hooks to the right as the merry-go-round is spinning counterclockwise when seen from above.

77. (a) and (b)

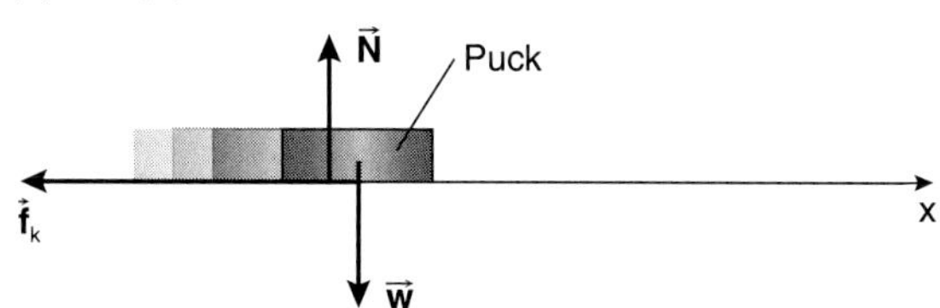

(c) $v_x(t) = v_0 - \mu_k g t$

(d) $t = \dfrac{v_0}{\mu_k g}$

(e) $x(t) = v_0 t - \dfrac{\mu_k g t^2}{2}$

(f) $\dfrac{v_0^2}{2\mu_k g}$

(g) 6.4×10^2 m

81. (a)

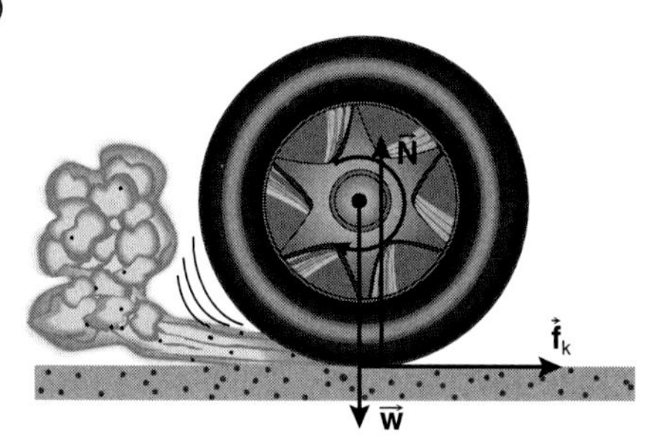

(b) The kinetic frictional force. (c) Answer is given in the statement of the problem.

85. (a) and (b) Answer is given in the statement of the problem.

(c)

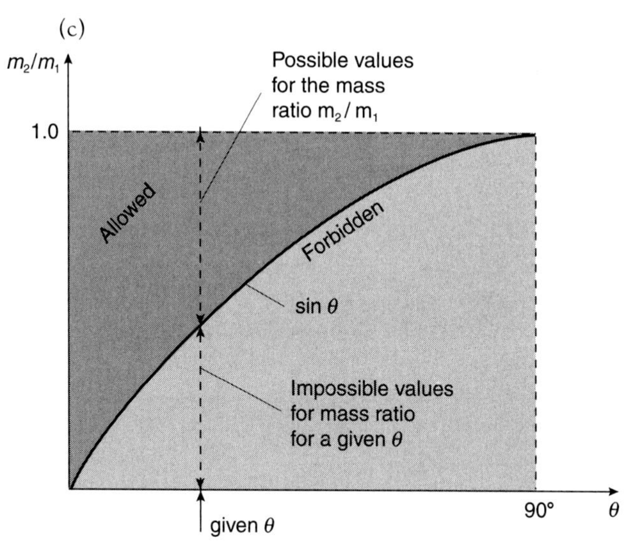

89. 0.0931 — The effects of pseudoforces are more apparent on Jupiter than on the Earth, since this ratio is almost 10%.

Chapter 6

1. 1.85×10^{-10} N

5. (a) $v = (GM_{Sun}/r)^{1/2} = 29.8$ km/s (b) Nothing

9. (a) 1.05×10^{-8} N (b) 1.05×10^{-8} N (c) For the 0.400 kg sphere: 2.63×10^{-8} m/s^2; For the 4.00 kg sphere: 2.63×10^{-9} m/s^2

13. (a) Answer is given in the statement of the problem. (b) 0.16 m/s^2 (c) 1.2×10^2 m, 78 s

17. (a) $\rho = \left(\dfrac{9F}{4\pi^2 G R^4}\right)^{1/2}$

(b) No element is dense enough. (c) 1.61 m, 3.93×10^5 kg

21. 6.43×10^{23} kg, $M_{Mars} = 0.108\, M_{Earth}$

25. (a) 5.00×10^6 km (b) Outside the Sun (c) Perihelion: 1.471×10^8 km, aphelion: 1.521×10^8 km (d) 0.334 cm

29. 1.12

33. (a) Answer is given in the statement of the problem. (b) 0.140, 0.417

37. 1.034

41. 4.7 y

45. (a) 17.9 AU (b) 0.968 (c) 35.2 AU (d) ≈62

49. (a) 4.7×10^3 km from the center (b) 1.7×10^3 km below the surface

53. (a) $\dfrac{GM^2}{64r^2}$

(b) $\dfrac{GM^2}{4r^2}$

(c) $\dfrac{17}{64}\dfrac{GM^2}{r^2}$

(d) $2\pi r/T$

(e) $T^2 = \frac{64\pi^2}{17GM} r^3$

57. 1.86×10^{32} kg, 93.5 solar masses

61. (a) 2.1×10^8 y (b) About 24 times (c) 1.9×10^{41} kg, 9.5×10^{10} solar masses

65. (a) 0.900 d, measured from the center of the Earth (b) No

69. 34 km

73. -6.17×10^{13} m^3/s^2, -5.01×10^{15} m^3/s^2

CHAPTER 7

1. 50 N/m

5. 2.5×10^2 N/m

9. Answers are given in the statement of the problem.

13. (a) 40.0 rad/s (b) 6.37 Hz

(c) $x(t) = (0.050 \text{ m}) \cos\left[(40.0 \text{ rad/s})\, t + \frac{\pi}{2} \text{ rad}\right]$

(d) −0.0363 m

17. (a) $\omega = 11.5$ rad/s, $A = 0.201$ m, $\phi = -1.05$ rad (b) 0.546 s (c) 2.31 m/s, 26.6 m/s^2

(d)

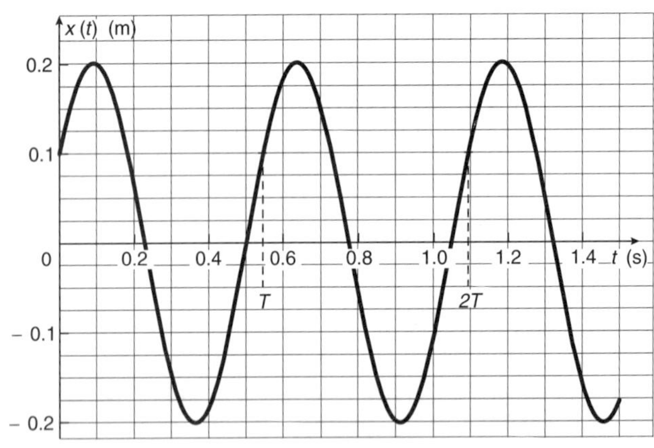

21. $v_x(t) = (-0.500 \text{ m/s}) \sin\left[(7.07 \text{ rad/s})\, t - \frac{\pi}{2} \text{ rad}\right]$

$a_x(t) = (-3.54 \text{ m/s}^2) \cos\left[(7.07 \text{ rad/s})\, t - \frac{\pi}{2} \text{ rad}\right]$

graphs follow:

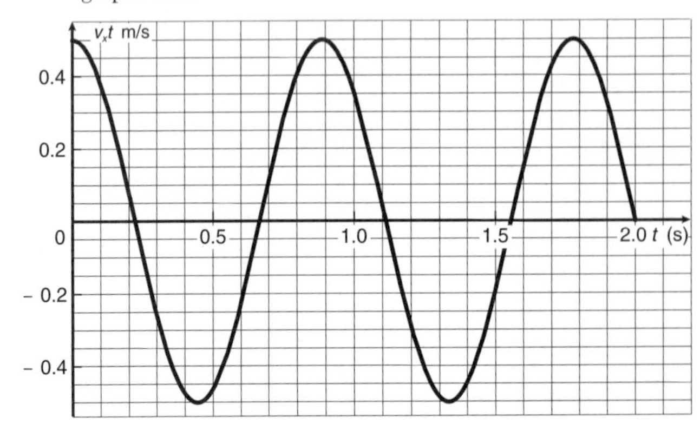

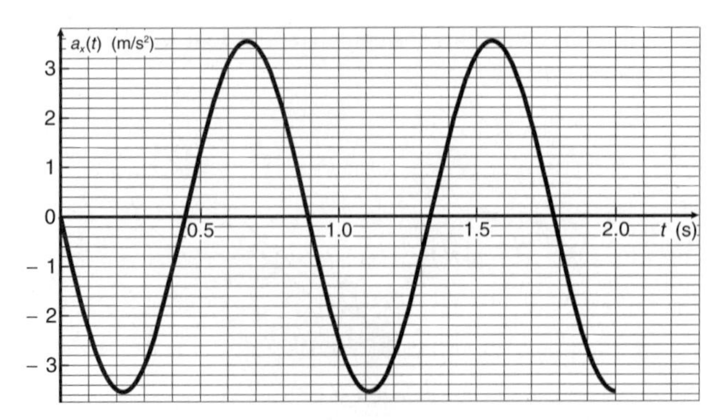

25. (a) 25.1 s, 18.0 s (b) Approximately 10 s between the instants

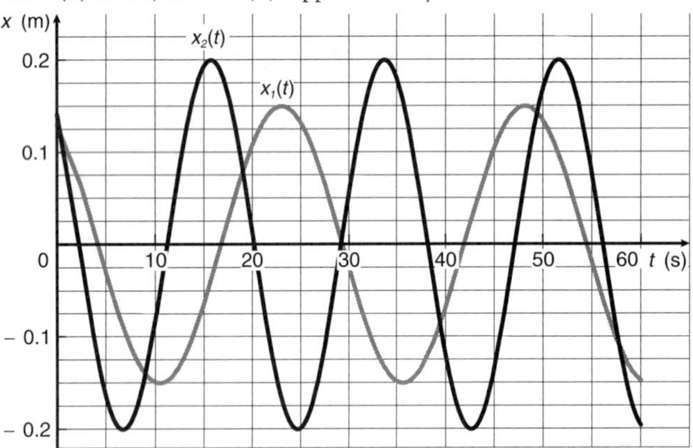

(c) The times change as well as the intervals between them.

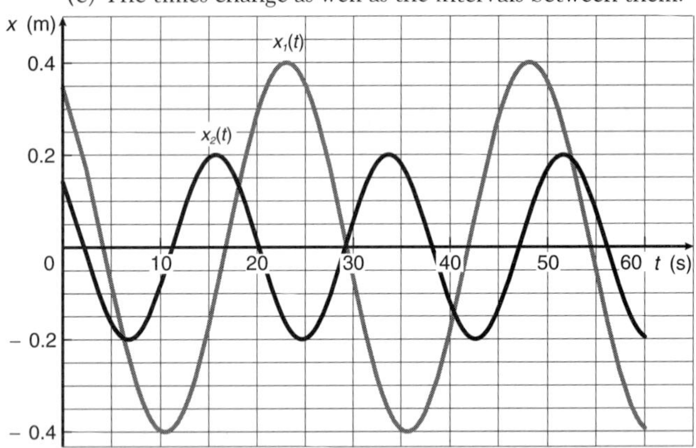

29. 0.496 m from the fixed end

33. (a) 1.30 Hz (b) At the high point of the oscillation (c) At the low point of the oscillation (d) 0.147 m

37. (a) 1.5 m/s (b) 23 m/s^2

41. $R \cos\left(\omega t - \frac{\pi}{6} \text{ rad}\right)$

45. (a) 10 s (b) 0.10 Hz (c) no

49. The equation of motion is not like Equation 7.15.

53. 16.5 s

57. $\frac{T_1 T_2}{T_1 - T_2}$

61. 9.780 m/s^2

65. 526 km

69. An experiment

73. 14 Hz

CHAPTER 8

1. (a) 75.0 J (b) 0 J (c) 65.0 J (d) −53.1 J

5. (a) The graph should be a downward-sloping straight line with a slope of −6.00 N/m, passing through the origin. (b) 0 J

9. 2.4×10^2 J

13. (a) 360 J (b) 0 J (c) −62.5 J (d) 62.5 J

17. Answer is given in the statement of the problem.

21. Answer is given in the statement of the problem.

25. Answer is given in the statement of the problem.

29. 7.91 km/s

33. (a) 7.91 km/s

(b) $\frac{1}{\sqrt{2}} = 0.707$

37. 5.8 km

41. (a) $-\frac{GMm}{r}$

(b) $\left(\frac{GM}{r}\right)^{1/2}$

(c) $\frac{1}{2}\frac{GMm}{r}$

(d) $-\frac{1}{2}$

45. (a) $(2gh)^{1/2}$ (b) $(gR)^{1/2}$ (c) 5/2 (d) $(2gh)^{1/2}$ (e) mgh (f) Only (e)

49. Answer is given in the statement of the problem.

53. (a) $-\frac{GMm}{r_1}$ $-\frac{GMm}{r_2}$

(b) mv_1r_1, mv_2r_2 (c) Answer is given in the statement of the problem.

57. (a) Answer is given in the statement of the problem. (b) 3.33×10^{-3} m/s

61. Answer is given in the statement of the problem.

65. (a) 70.2 N/m (b) 0.585 m

69. (a) 434 W (b) 868 W

73. (a)

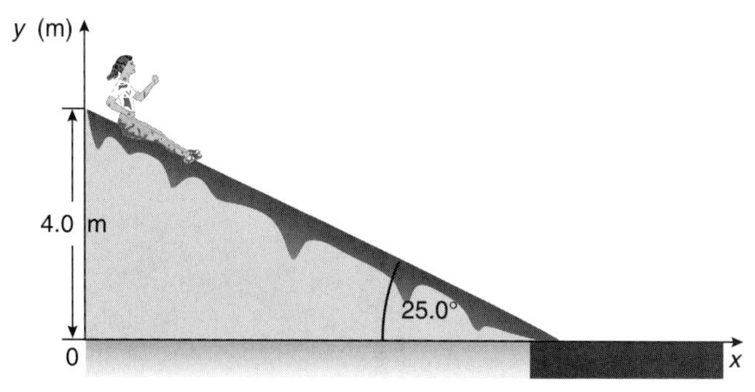

(b) 3.1×10^3 J (c) 8.9 m/s (d) 20 m (e) 0 J along both paths (f) -3.1×10^3 J (g) -1.4×10^3 W; No, since the velocity is changing.

77. 109 kW

81. (a) J/m (b) $-C\,\hat{\imath}$ for $x > 0$ m, $C\,\hat{\imath}$ for $x < 0$ m
(c)

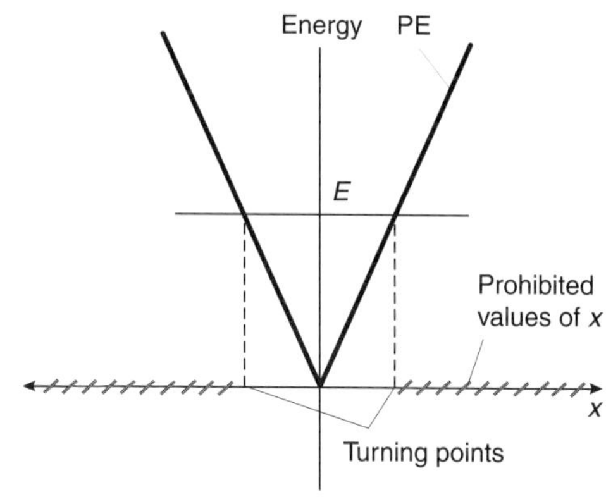

(d) Two turning points (e) The motion is oscillatory but is not simple harmonic oscillation. This force has a constant magnitude whereas the force in simple harmonic oscillation is a function of position.

CHAPTER 9

1. 3.3×10^2 km/h

5. 4.63 m/s = 16.7 km/h

9. 4.96 kg·m/s

13. (a)

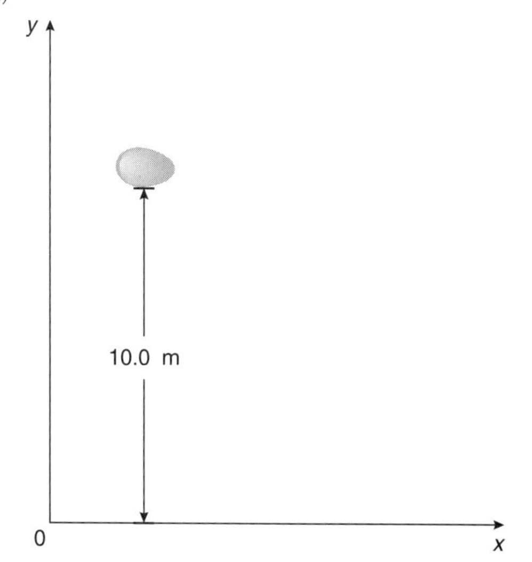

(b) 98.1 J (c) Yes, since the total force is constant. (d) $(-14.0\text{ N}\cdot\text{s})\,\hat{\jmath}$

17. (a) $(15.0\text{ N}\cdot\text{s})\,\hat{\imath}$ (b) $(7.50\text{ N})\,\hat{\imath}$ (c) $(15.0\text{ N}\cdot\text{s})\,\hat{\imath}$ (d) $(3.75\text{ N})\,\hat{\imath}$

21. 87 m/s²

25. 1.718

29. Answer is given in the statement of the problem.

33. (a) $(8.00\text{ m/s})\,\hat{\imath}$ (b) It is an inelastic collision since the kinetic energy is not conserved. $\Delta KE = -2.40 \times 10^4 \text{ J} \neq 0$ J (c) No. The momentum of each particle in a collision is not conserved; only the total momentum of all the particles is conserved in a collision.

37. (a) A stupid one. Completely inelastic.
(b) $(1.67 \times 10^4\ \text{kg}\cdot\text{m/s})\ \hat{\imath} + (1.11 \times 10^4\ \text{kg}\cdot\text{m/s})\ \hat{\jmath}$
(c) The same as (b) (d) 10.0 m/s (e) -1.01×10^5 J

41. (a) Take $\hat{\imath}$ east and $\hat{\jmath}$ north, $(19\ \text{m/s})\ \hat{\imath} - (4\ \text{m/s})\ \hat{\jmath}$
(b) -5×10^4 J

45. (a) No. The speed of the car was 29.0 m/s. (b) 0.930 m/s

49. 17

53. A proof

57. $(-0.129\ \text{m})\ \hat{\imath} + (0.257\ \text{m})\ \hat{\jmath}$

61.

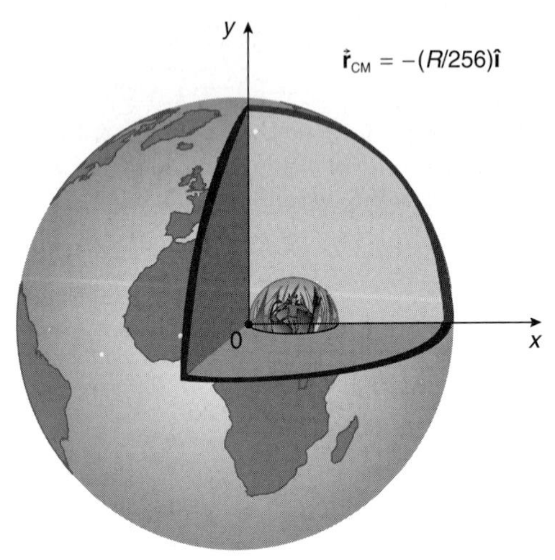

65. 0.71 m/s in a direction opposite to that of his daughter

69. (a) 1.8 m/s (b) $\Delta KE = -3 \times 10^4$ J (c) 1.8 m/s

Chapter 10

1. 0 kg m²/s, orbital

5. Answers are given in the statement of the problem.

9. 4.78×10^4 km

13. $7mR^2/4$

17. Answer is given in the statement of the problem.

21. (a) 1.36×10^6 J (b) 0.047

25. (a) Answer is given in the statement of the problem.
(b) 2.83×10^{42} J (c) Answer is given in the statement of the problem. (d) -7.17×10^{31} W, 1.87×10^5 times the luminosity of the Sun

29. (a) $7mR^2/5$ (b) $3mR^2/2$ (c) $5mR^2/3$

33. (a) 7.293×10^{-5} rad/s (b) $7.08 \times 10^{33}\ \text{kg}\cdot\text{m}^2/\text{s}$ (c) The effect is to lengthen the day (the moment of inertia increases and decreases ω_{spin} because of conservation of angular momentum). (d) 1.99×10^{-7} rad/s (e) $2.64 \times 10^{40}\ \text{kg}\cdot\text{m}^2/\text{s}$
(f) 3.73×10^6

37. $\dfrac{13}{10}\,\omega_0$

41. (a) Answer is given in the statement of the problem.

(b)

	KE_{rot}	KE_{CM}
highest	cylindrical shell	sphere
↓	spherical shell	disk
	disk	spherical shell
lowest	sphere	cylindrical shell

(c) First to last: sphere, disk, spherical shell, cylindrical shell.

45. (a) $1.96 \times 10^3\ \text{N}\cdot\text{m}$ (b) 7.34 rad/s² (c) No. The torque of the weight of the beam about the hinge varies with the orientation of the beam to the vertical. (d) 3.84 rad/s

49. $\left(\dfrac{1-\varepsilon}{1+\varepsilon}\right)^2$

53. (a) 4.67×10^6 m from the center of the Earth along the Earth–Moon line (b) $2.81 \times 10^{34}\ \text{kg}\cdot\text{m}^2/\text{s}$
(c) $3.47 \times 10^{32}\ \text{kg}\cdot\text{m}^2/\text{s}$ (d) $2.37 \times 10^{29}\ \text{kg}\cdot\text{m}^2/\text{s}$
(e) $7.08 \times 10^{33}\ \text{kg}\cdot\text{m}^2/\text{s}$

57. (a) The spherical shell (b) The solid sphere reaches the bottom of the incline before the spherical shell if both are released at rest from the same location.

61. $\dfrac{v_0/R}{1+\dfrac{M}{2m}}$
The lazy-susan rotates in a direction opposite to that in which the mouse is going.

65. (a) $1.0 \times 10^{42}\ \text{kg}\cdot\text{m}^2/\text{s}$ (b) $5 \times 10^{43}\ \text{kg}\cdot\text{m}^2/\text{s}$ (c) 0.7 d

69. Answer is given in the statement of the problem.

73. 59°

77. 0.71 m

Chapter 11

1. 8.0 kg

5. 1.2×10^7 N/m²

9. Answer is given in the statement of the problem.

13. 1.5 m

17. (a) 7.92×10^3 m (b) 0.93 km

21. (a) 1 (b) 10

25. 10.3 m

29. 0.833×10^3 kg/m³

33. 2×10^{-3} kg

37. (a) It floats. (b) 3.00×10^4 kg

41. (a)

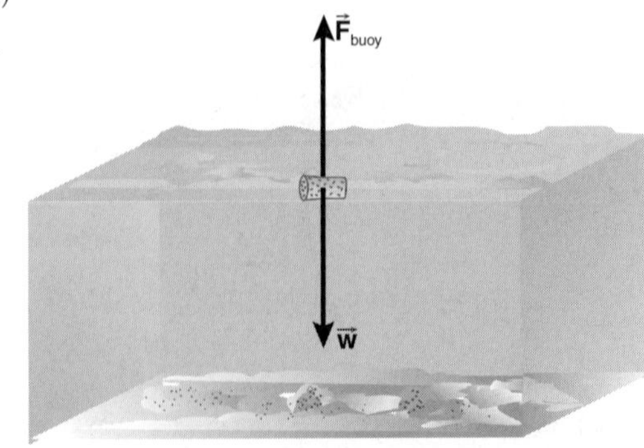

(b) 0.120 (c) 4.2×10^{-2} kg (d) 0.26 N

45. 2.7×10^3 kg/m^3

49. 5.3 cm

53. For 2 mm diameter straw and $h = 1.5$ cm, $\gamma = 0.074$ N/m.

57. 1.2×10^2 m/s

61. 9.0×10^{-3} m^3/s

65. (a) 19.8 m/s (b) 156 liters/s (c) 20.0 m

69. 26 m/s

Chapter 12

1. 1.7×10^2 m

5. (a) 1.8×10^3 km (b) Only the distance to the focus was found. (c) Use three seismometers at different locations.

9. (a)

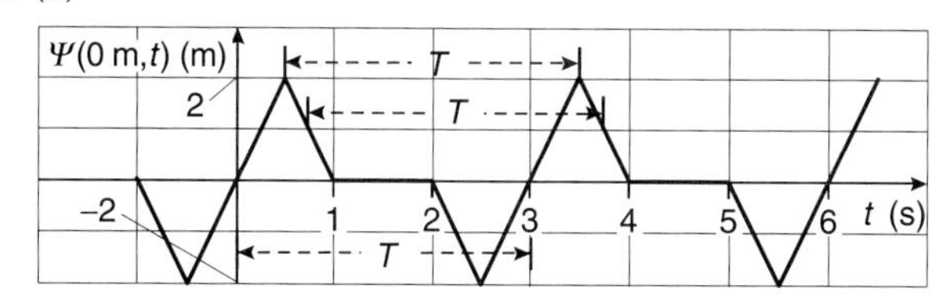

(b) 0.33 Hz

(c)

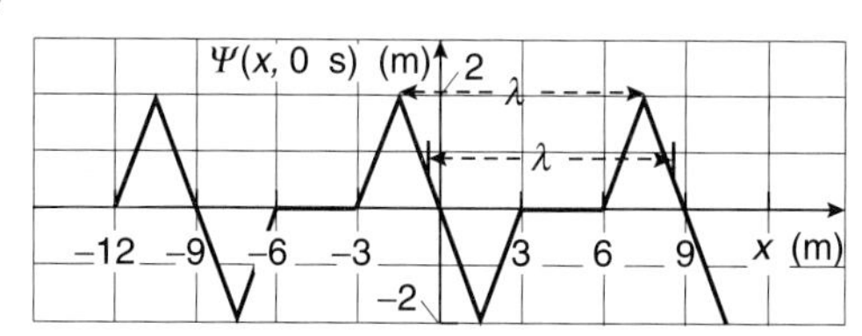

13. (a) and (b)

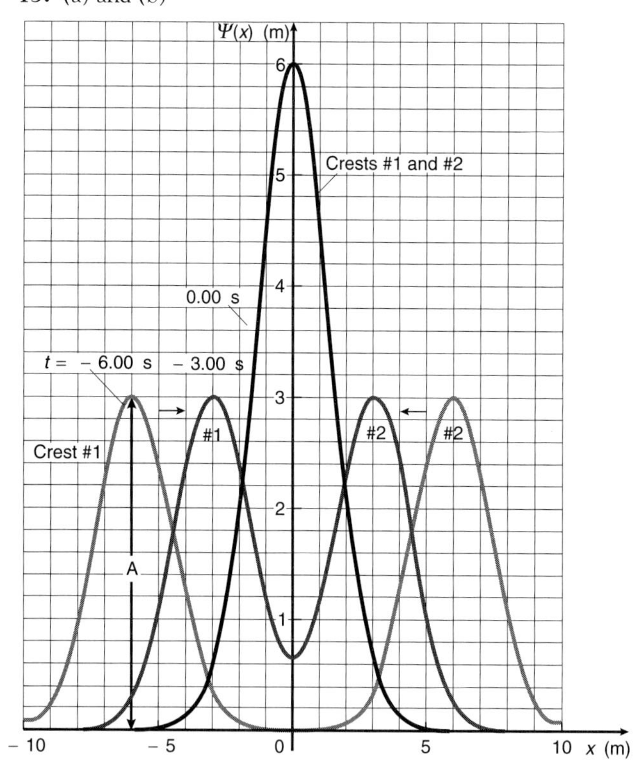

(c)

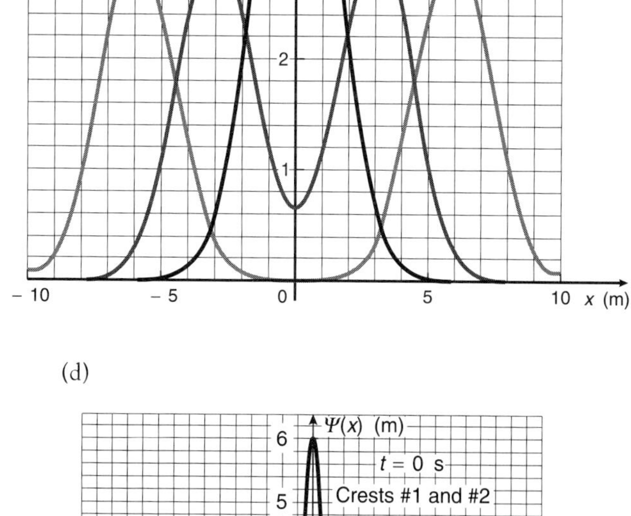

(d)

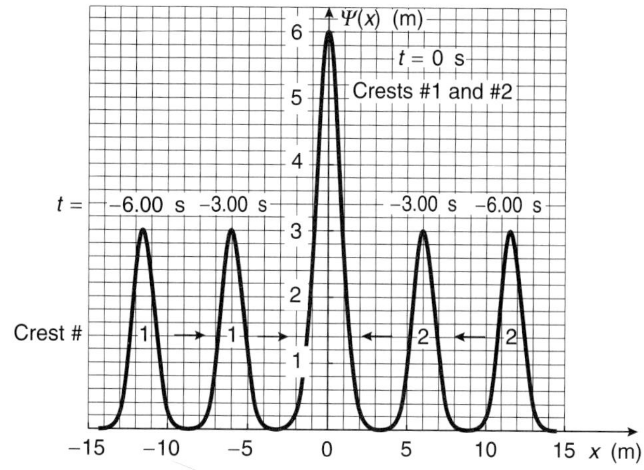

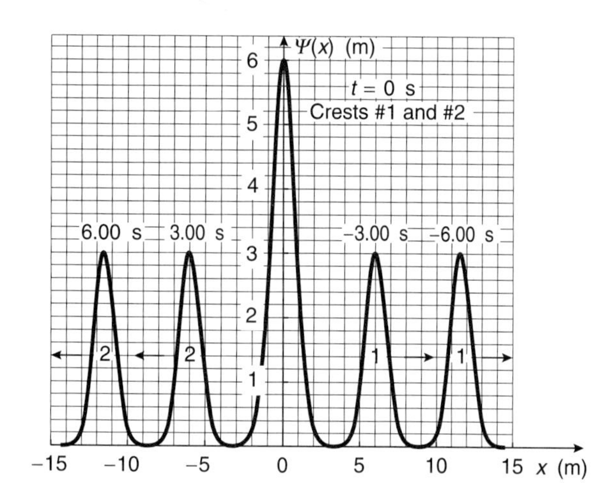

17. (a) 6.00×10^{-3} Pa (b) 8.06 rad/m (c) 8.06 wavelengths (d) 0.780 m (e) 2.76×10^{3} rad/s (f) 439 Hz (g) 2.28×10^{-3} s (h) 342 m/s
(i)

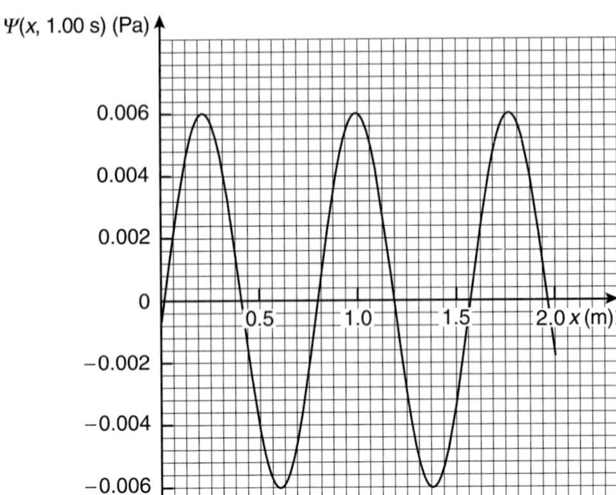

(j)

(k) -1.23×10^{-3} Pa. The negative result implies a rarefaction with an absolute pressure below the average value in the medium in which the sound is propagating.

21. (a) 3.00 Hz (b) $\Psi(x, t) = (0.50 \times 10^{-2}\text{ m}) \cos[(31.4\text{ rad/m}) \times x - (18.8\text{ rad/s})\,t]$

25. 83 N

29. 3.95×10^{-2} s

33. 86 Hz

37. $I = \dfrac{P}{2\pi r \ell}$

41. (a) 3.1×10^{2} W (b) 1.6×10^{2} m

45. 106 dB

49. (a) 2.180×10^{3} Hz (b) 3 m/s

53. (a) 1200 Hz (b) 1.21×10^{3} Hz (c) 1.19×10^{3} Hz

57. 30.2 kHz

61. (a) Mach number = 2.1 (b) 29°

65. Answer is given in the statement of the problem.

69. (a) The first (second) wavefunction represents a wave moving toward increasing (decreasing) values of x. (b) The two will form a standing wave; $\lambda/2 = 0.349$ m (c) 0.175 m, 0.524 m, 0.873 m

73. $\dfrac{v}{\alpha}$

77. (a)

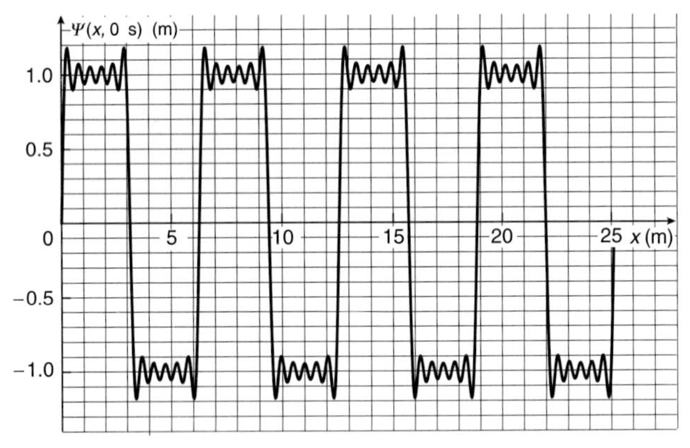

(b)

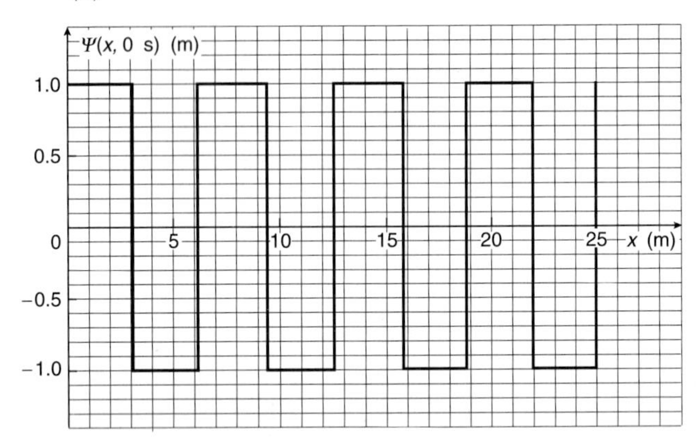

which is called a square wave.

(c) ≈6.3 m

Chapter 13

1. 37.0 °C

5. −13 °F

9. 50.00 °C

13. 0.012%

17. (a) 7.6×10^{-4} m (b) $\Delta T = 63$ K

21. $\Delta T = 4 \times 10^{2}$ K

25. A proof

29. 1.5×10^{4} N

33. A proof

37. 1.4×10^{-23} Pa

41. 9.3 m below the surface

45. (a) It all melts (b) 8.40 °C

49. (a) 9.4×10^7 J (b) 5.2 h

53. 3.9×10^3 W

57. (a) 1 (b) 14 °C

61. 7.4

65. (a) 3.04×10^3 J (b) 1.52×10^3 J (c) 2.28×10^3 J (d) -3.04×10^3 J (e) -1.52×10^3 J (f) -2.28×10^3 J

69. (a) -2.86×10^3 J (b) 2.86×10^3 J

73. (a) A proof (b) 0.11 K (c) 0.213 K

77. 0.398 K

Chapter 14

1. 1.908×10^{16} y

5. 3.1×10^{-16} J

9. 1.93×10^3 m/s; The rms speed of the hydrogen molecules is four times faster than that of the oxygen molecules.

13. 5.14×10^{-23} N·s

17. (a) 2.46×10^{-2} m^3 (b) ~3.4×10^{-9} m

21. 0.86 m/s

25. 2.01×10^4 K

29. 1.93 km/s; Although this rms speed is considerably less than the escape speed, some hydrogen molecules will have speeds exceeding the escape speed and can escape, if moving in the right direction, if no collisions occur. Heat transfer from the Earth continually replenishes the high-speed molecules, so they gradually escape. Such a process also explains the phenomenon of evaporation.

33. 1×10^{-23} Pa

37. (a) The helium atom loses KE while the hydrogen atom gains KE. (b) The KE of each is unchanged.

41. 2

45. 2.67×10^{-2} mol

49. Answer is given in the statement of the problem

53. (a) 1.93×10^3 m/s (b) 1.25×10^6 Pa (c) No. $\Delta U = 0$ J. (d) 6.25×10^5 Pa

57. (a)

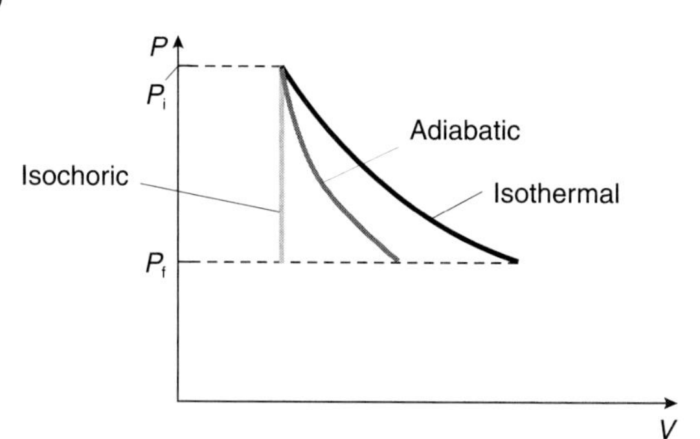

(b) The isochoric process (c) The isothermal process (d) Answer is given in the statement of the problem. (e) The isochoric process

61. A proof

Chapter 15

1. 0.558

5. 0.055

9. 400 K and 500 K

13. (a) 186 MW (b) 1.5×10^4 kg/s, 15 m^3/s

17. The coefficient of performance increases more by warming the colder reservoir than by cooling the warmer reservoir.

21. (a) 0.400 (b) $|Q_H| = 3.20 \times 10^3$ J, $|Q_C| = 2.40 \times 10^3$ J

25. (a) 3.9 J; No (b) 71 J

29. (a) 2.8×10^2 J (b) 7.8×10^2 J

33. -6.05×10^3 J/K, No

37. (a) 141 J/K (b) -136 J/K (c) 5 J/K

41. (a)

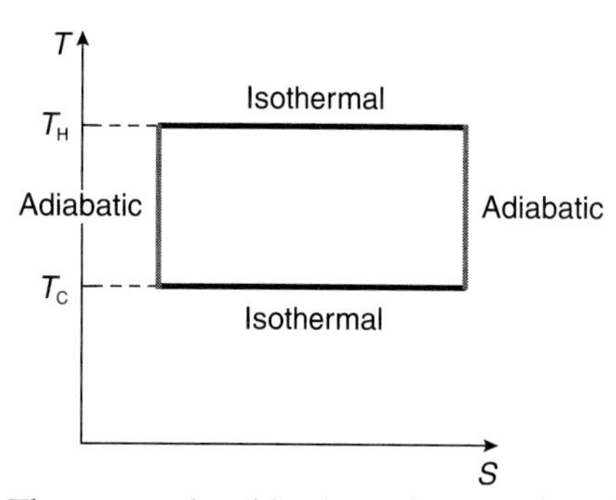

(b) The area enclosed by the cycle is equal to the total heat transfer to the gas system in one cycle.

(c) Take the ratio of $T_H - T_C$ to T_H:

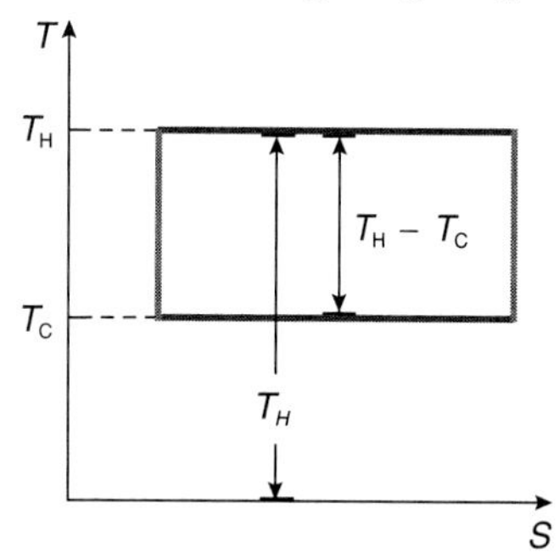

45. (a) 4.00 kW (b) 3.50 W/K

49. 1.6×10^3 J/K

53. (a) $Q = 0$ J (b) $\Delta U = -2000$ J (c) $\Delta S = 0$ J/K (d) The temperature fell. (e) $\Delta T = -48.1$ K

57. $\Delta S = nc_P \ln\left[\frac{(T_1 + T_2)^2}{4T_1T_2}\right]$

61. Shake the box many times and assess the experimental probability of securing each macrostate after a shake. If after many shakes, you *never* see a result with 0 heads, you can begin to become suspicious.

65. (a)

Number on left side	*Number on right side*	*Number of microstates*
6	0	1
5	1	6
4	2	15
3	3	20
2	4	15
1	5	6
0	6	1

(b) 0 J/K (c) The macrostate with three particles in each half: 4.137×10^{-23} J/K

69. (a)

Macrostate *Number of heads*	*Number of microstates* Ω	$\frac{S}{k}$
0	1	0
1	20	3.00
2	190	5.25
3	1140	7.04
4	4845	8.49
5	15 504	9.65
6	38 760	10.57
7	77 520	11.26
8	125 970	11.74
9	167 960	12.03
10	184 756	12.13
11	167 960	12.03
12	125 970	11.74
13	77 520	11.26
14	38 760	10.57
15	15 504	9.65
16	4845	8.49
17	1140	7.04
18	190	5.25
19	20	3.00
20	1	0

(b) The graph of the number of microstates versus the macrostate is shown below:

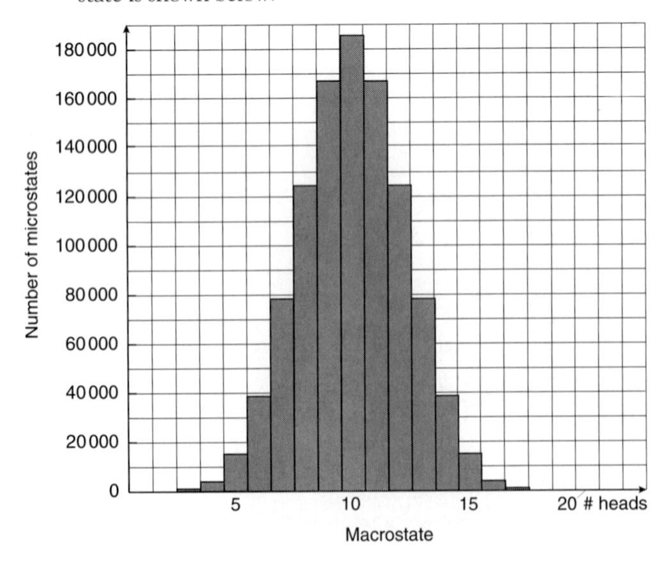

(c) The graph of $\frac{S}{k}$ versus the macrostate is shown below:

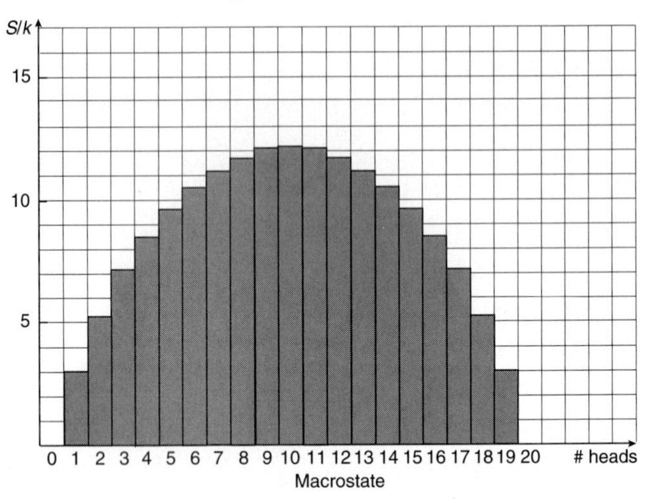

Chapter 16

1. 9.6486×10^{4} C

5. 9.6486×10^{4} C

9. (a) 365 N (b) 365 N

13. 2.95×10^{17} C

17. (a)

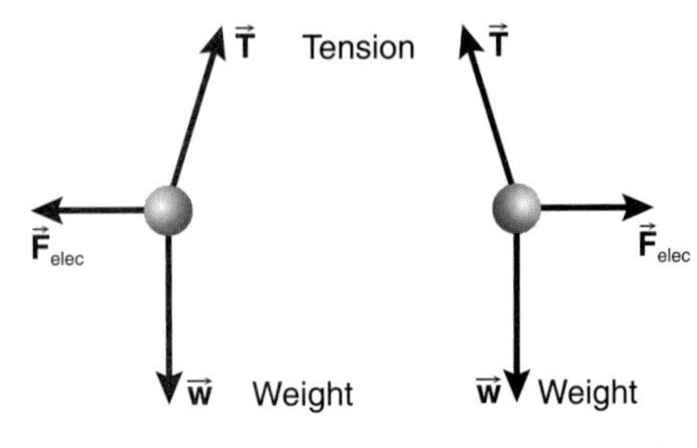

(b) Answer is given in the statement of the problem.
(c) 4.4×10^{-7} C (d) 2.7×10^{12}

21. (a) $\vec{\mathbf{r}}_{\text{center of charge}} = \frac{\sum_i q_i \vec{\mathbf{r}}_i}{Q_{\text{total}}}$

(b) If the total charge is zero, then $\vec{\mathbf{r}}_{\text{center of charge}}$ is undefined.

25. -1.2×10^{-2} C

29. (a) Negative

(b)

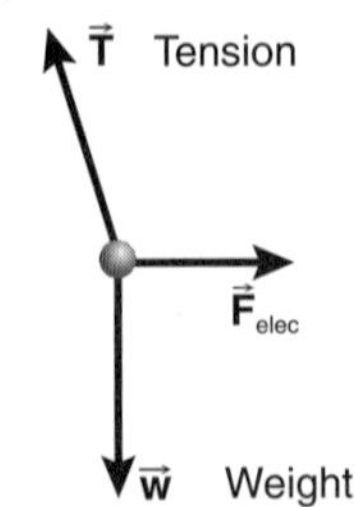

(c) -3.5×10^{-4} C

33. Going left to right: $+2q$, $-q$, $-q$

37. The information implies that the three atoms of the molecule are collinear so the two dipole moment of the two carbon–oxygen bonds point in opposite directions and vector sum to zero.

41. (a) Since the distance from Q to the dipole is much greater than the separation of the charges in the dipole, each charge of the dipole feels essentially the same electric field. Hence the magnitude of the force on the dipole is approximately 0 N. (b) 0 N·m

45. Answer is provided in statement of problem.

49. 1.16×10^{-7} C/m^2

53. 1.81×10^3 N/C

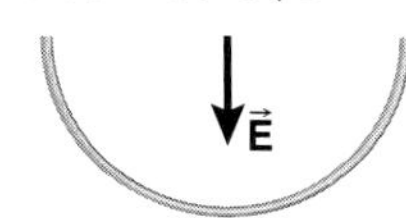

57. (a) Q and σ must be like charges. (b) Answer is given in the statement of the problem.

61. (a) $\vec{F}$ must be antiparallel to $\vec{v}$, so $\vec{E}$ must be parallel to $\vec{v}$.
(b)

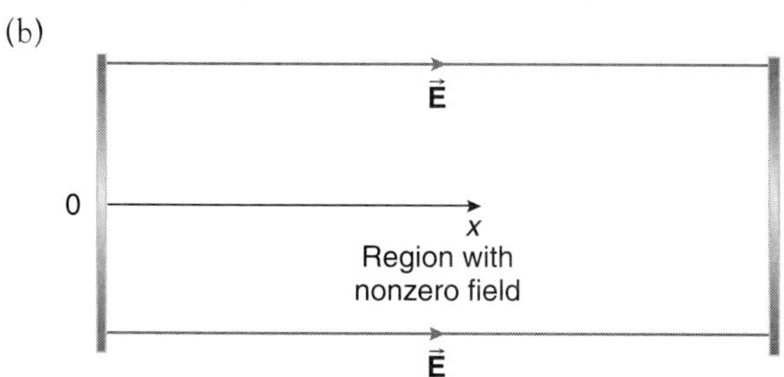

(c) $a_x = -1.76 \times 10^{13}$ m/s^2 (d) 2.84×10^{-7} s (e) 0.71 m (f) The electron will not remain at rest. It will retrace its path and emerge from the region of the field with a speed equal to the initial speed of the electron.

65. 0.142 m

69. 0 N·m^2/C

73. 2.3 N·m^2/C

77. (a)

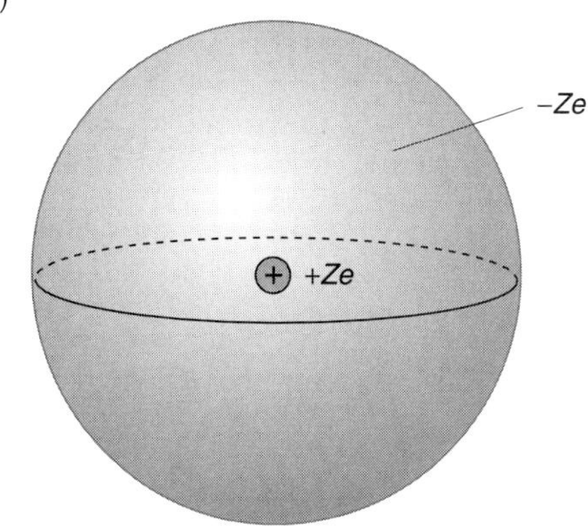

(b) 0 N/C

(c) $E = \dfrac{1}{4\pi\varepsilon_0}\dfrac{Ze}{r^2}\left(1 - \dfrac{r^3}{R^3}\right)$

CHAPTER 17

1. 7.50×10^{-4} J

5. (a) 3.0×10^9 V (b) 54 C

9. (a) The gravitational potential energy per kilogram (b) 49.1 J/kg

13. (a) -8.84×10^{-8} C/m^2 (b) 8.84×10^{-8} C/m^2 (c) -1.00×10^3 V (d) 9.81×10^{-9} C

17. Take $\hat{\jmath}$ to the vertically upward. Then the gravitational potential is gy.

21. 8.3×10^{-3} m If the radius were smaller, the magnitude of the field would exceed the maximum magnitude.

25. (a) 0 V

(b) $\dfrac{Ze}{4\pi\varepsilon_0}\left[\dfrac{1}{r} + \dfrac{r^2}{2R^3} - \dfrac{3}{2R}\right]$

29. (a) 1.08×10^3 V (b) 1.62×10^3 V

33. (a) 2.00 cm sphere: 1.35 kV; 4.00 cm sphere: 675 V (b) 900 V for each, 1.00 nC was exchanged

37. (a) $\dfrac{\lambda}{2\pi\varepsilon_0}\ln\left(\dfrac{a}{r}\right)$

(b) $r < a$ (c) $r > a$ (d) The distribution of charge extends to infinity; likewise, when $r = 0$ m the natural logarithm of ∞ is undefined. (e) A proof.

41. (a) 1.00×10^3 eV (b) 1.88×10^7 m/s

45. (a) The lower plate (b) 200 eV = 3.20×10^{-17} J (c) 8.38×10^6 m/s

49. (a) and (b)

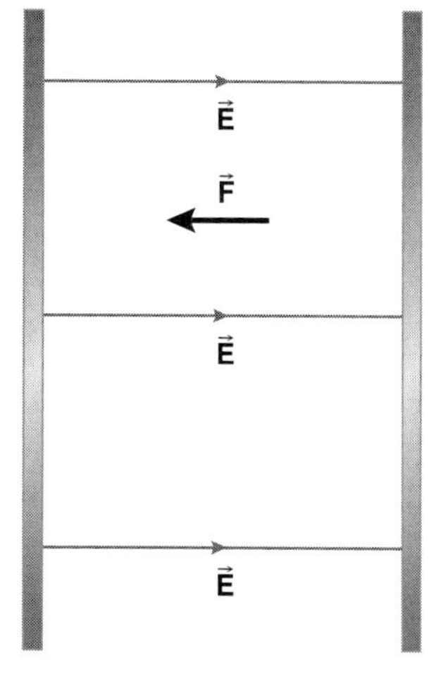

(c) 0.9×10^{-15} J (d) 0.9×10^{-15} J (e) 4×10^7 m/s

53. (a) −40 eV (b) +60 eV

57. (a) $\dfrac{1}{4\pi\varepsilon_0}\dfrac{Q}{R}$

(b) $-\dfrac{1}{4\pi\varepsilon_0}\dfrac{|q|Q}{R}$

(c) $\left(\dfrac{2|q|Q}{4\pi\varepsilon_0 Rm}\right)^{1/2}$

(d) The gravitational potential energy is proportional to the mass, and so appears in both the KE and PE in the CWE theorem and divides out. The electrical potential energy is independent of the mass, and so the mass appears in the KE term, but not the PE term and so does not divide out. (e) 3×10^{-19} C ≈ $2e$ (f) In this case $Q \approx 3 \times 10^{-15}$ C ≈ $2 \times 10^4 e$; Nuclei with such high charge quantum numbers do not exist in nature.

61. (a) $\vec{E} = -(0.01 \times 10^7 \text{ N/C})\,\hat{\imath} - (1.04 \times 10^7 \text{ N/C})\,\hat{\jmath}$
(b) -4.02×10^5 V (c) 6.2×10^{-23} J (d) $(6 \times 10^{-23}\text{ N}\cdot\text{m})\,\hat{k}$

65. 0 J

69. Answer is given in the statement of the problem.

Chapter 18

1. 3 nodes

5. A and D are in parallel; B and E are in parallel.

9. 6 in series, all connected (+) to (–)

13. 23×10^{-12} C

17. C increases by a factor of 2

21. Answer is given in the statement of the problem.

25. 0.92 μF

29. 33 μF

33. (a) 1.20 mC (b) In parallel (c) 8.3×10^2

37.

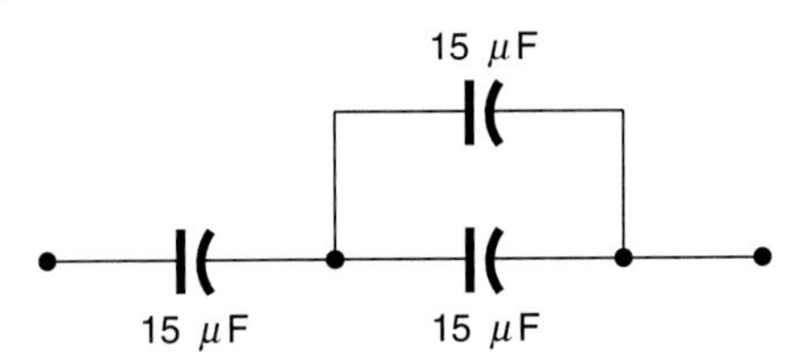

41. 1.4×10^2 μF

45.

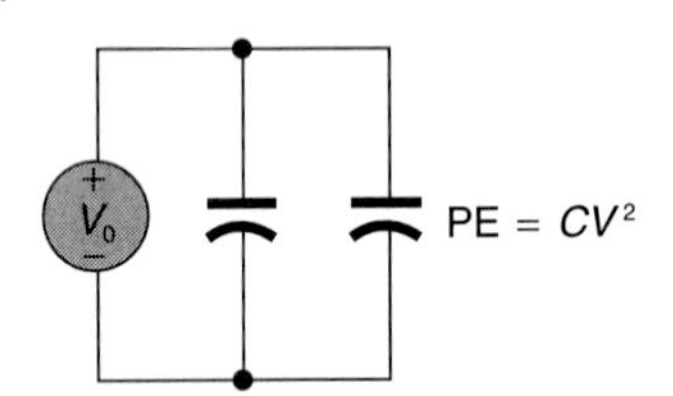

49. κ

53. Answer is given in the statement of the problem.

57. 0.1 μF, 2×10^{12} J

Chapter 19

1. About 0.16 Pa

5. (a) 4.84×10^{-5} m/s (b) 5.74 hours

9. 0.98 Ω

13. 6.0 Ω

17. The smaller-diameter wire has a drift speed four times that of the larger-diameter wire.

21. $R = \dfrac{(\text{resistivity})\,(\text{density})\,\ell^2}{m}$

25. (a) 3.0 kΩ (b) 1.3 Ω (c) $(13/8)R$ (d) 1.3 Ω

29. The filament of the 100 W bulb has the greater resistance.

33. (a) 2.3×10^{-3} m (b) Gauge 10

37. The 100 watt bulb absorbs 15 W and the 60 watt bulb absorbs 25 W. The 60 watt bulb is brighter.

41. (a) and (b)

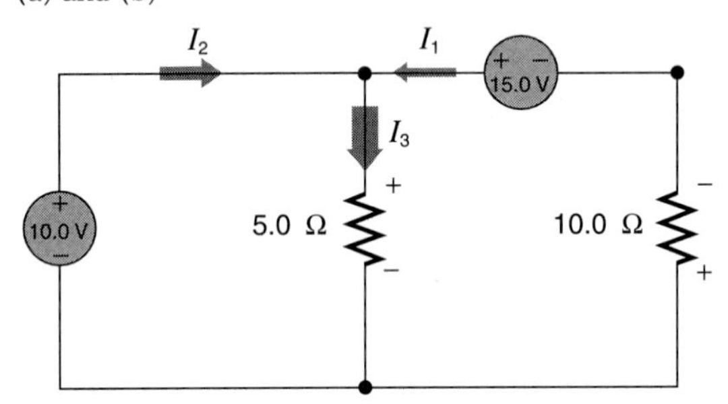

(c) $I_1 = 1.5$ A, $I_2 = 0.50$ A, $I_3 = 2.0$ A
(d) $P_{5\Omega} = 20$ W, $P_{10\Omega} = 2.5$ W, $P_{10\text{V}} = -15$ W, $P_{15\text{V}} = -7.5$ W

45. (a) 7.00 Ω (b) Resistors 5 and 6 (c) 0.29 A
(d)

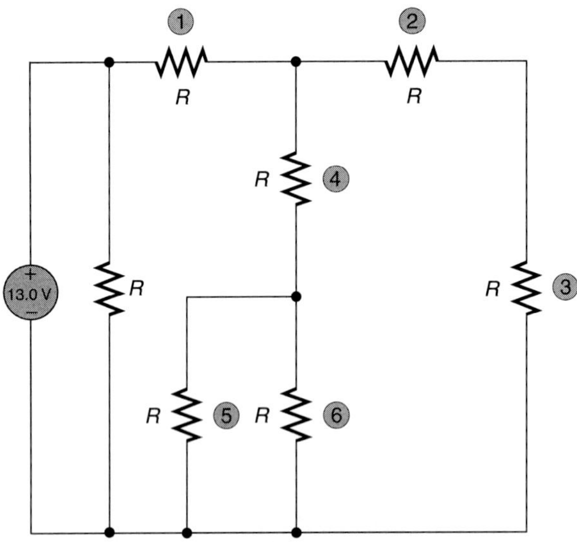

49. (a) 0 A (b) 12.0 V (c) The potential difference between A and B is entirely provided by the 12.0 V independent voltage source since there is no current in the 4.0 Ω resistor. (d) $P_{8\text{V}} = 8.0$ W, $P_{12\text{V}} = 0$ W, $P_{16\text{V}} = -16$ W (e) $P_{\text{all resistors}} = 8.0$ W

53. (a)

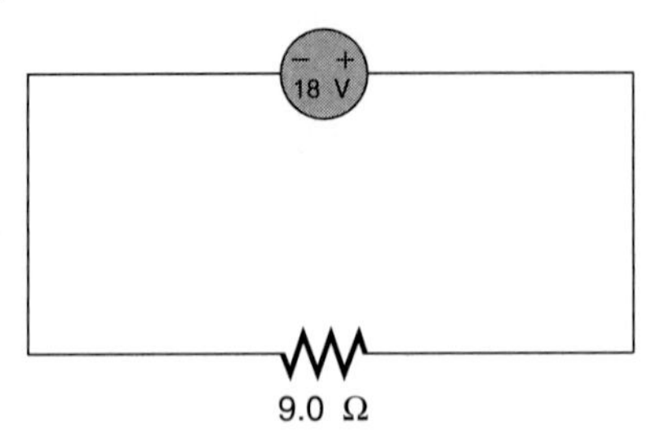

(b) 2.0 A (c) 8.0 W (d) –36 W

57. (a) $\dfrac{R_0}{R_0 + r} V_0$

$-\dfrac{V_0^2}{R_0 + r}$

(b) 0 V, $\dfrac{V_0}{r}$

61. (a)

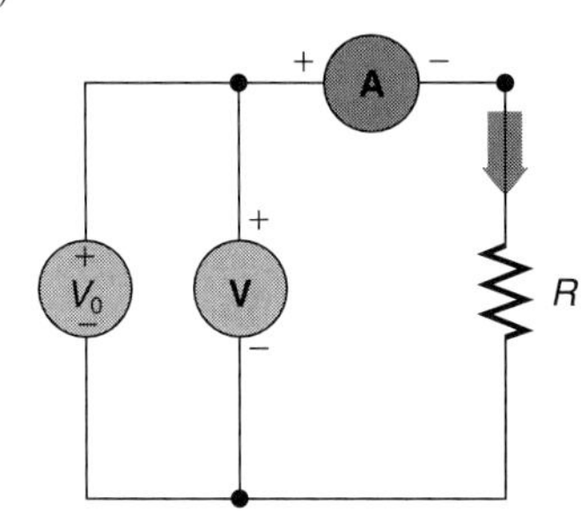

(b) 20.0 Ω (c) 1.25 W (d) 5.00 V

65. (a) 10.0 V, A_1 and A_2 both indicate 0 A. (b) $r = 1.6\ \Omega$, $R_1 = 1.6\ \Omega$, $R_2 = 12\ \Omega$, $R_3 = 8.0\ \Omega$

69. $\tau \ln 2$ ($\approx 0.693\ \tau$); Yes

73. Across C_1: $\dfrac{R_1}{R_1 + R_2} V_0$

Across C_2: $\dfrac{R_2}{R_1 + R_2} V_0$

Chapter 20

1.

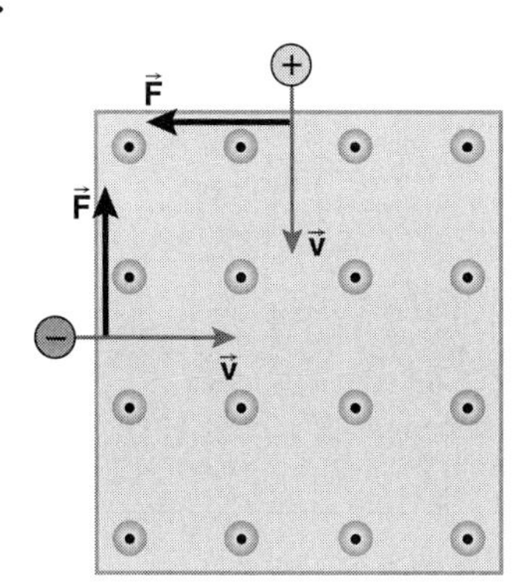

5.

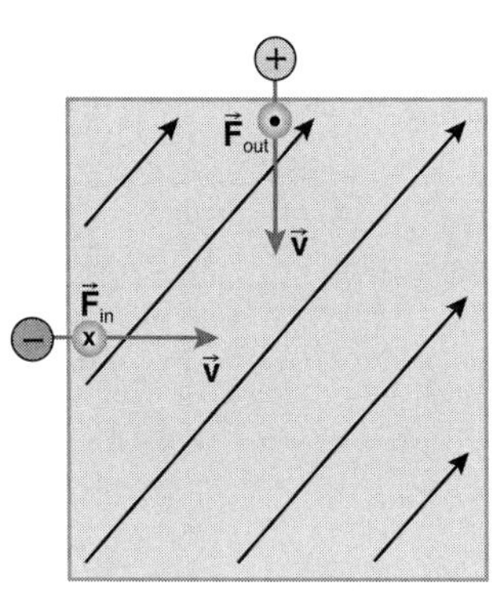

9. $(-3.0 \times 10^{-4}\ \text{T})\ \hat{\mathbf{j}} + (4.0 \times 10^{-4}\ \text{T})\ \hat{\mathbf{k}}$

13. (a)

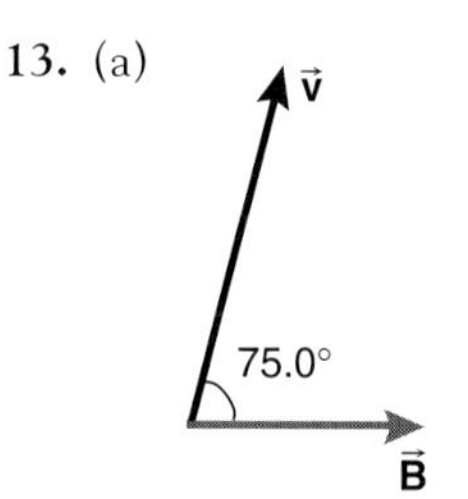

(b) 2.90×10^{-4} N directed perpendicular to and out of the page (c) 0.145 m/s²

17. (a) and (b)

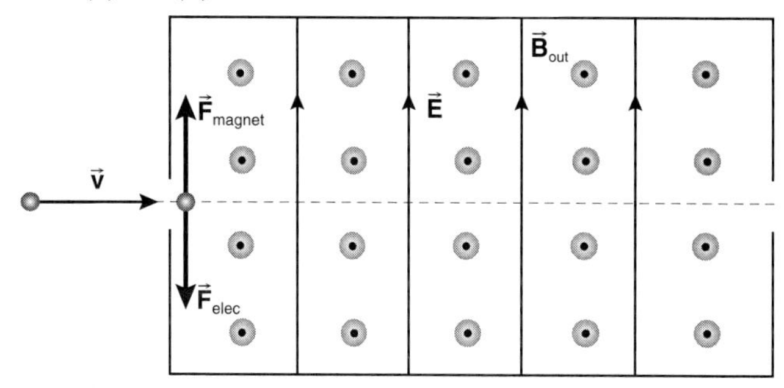

(c) 1.3×10^5 m/s

21. Answers are given in the statement of the problem.

25. (a) $1.2 \times 10^{-2}\ \text{A} \cdot \text{m}^2$ (b) $7.2 \times 10^{-3}\ \text{N} \cdot \text{m}$

(c)

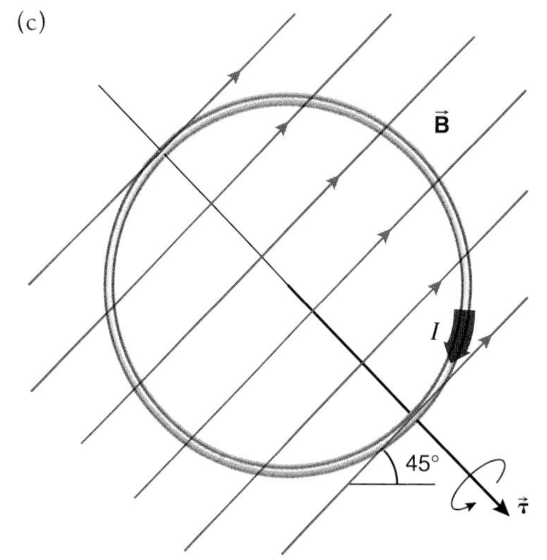

(d) 0 J

29. (a) $I\ell B$

(b)

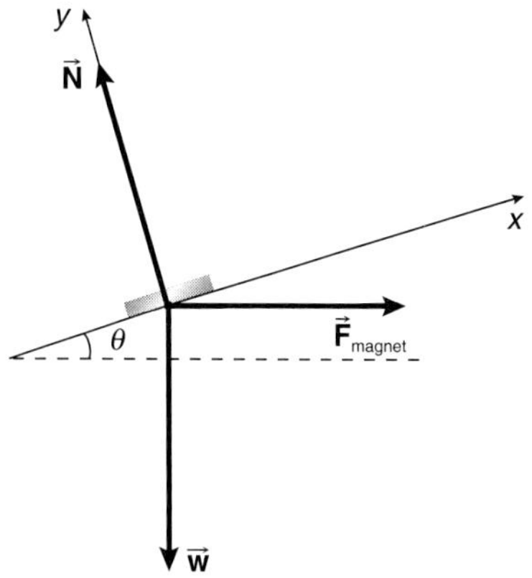

(c) $\left(\dfrac{I\ell B}{m}\cos\theta - g\sin\theta\right)t\hat{\imath}$

(d) $\dfrac{mg\tan\theta}{I\ell}$

33. The area of the coil must be 4.00×10^{-2} m^2.

37. Answer is given in the statement of the problem.

41. 0 T

45. (a) Answer is given in the statement of the problem.

(b)

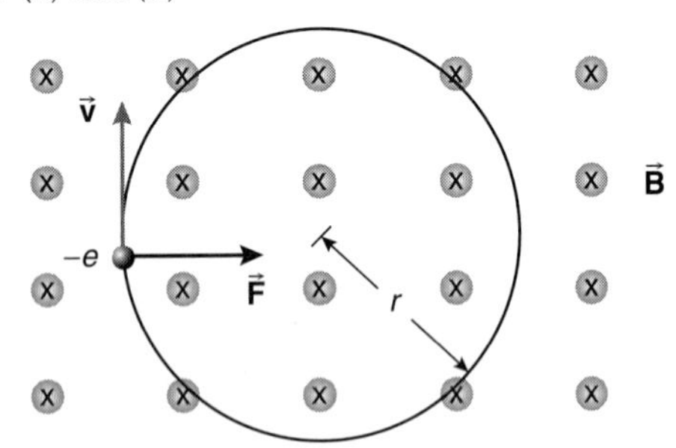

49. (a) 3.0×10^{-4} T (b) 2.4×10^{-15} N, 30° counterclockwise from $\hat{\imath}$ in sketch (c) 2.4×10^{-15} N, 30° counterclockwise from $-\hat{\imath}$ in sketch (d) 5.0×10^{7} m/s

53. (a) 7.5×10^{-6} T perpendicularly out of page (b) 1.9×10^{-5} N toward the infinite wire (c) 15×10^{-6} T perpendicularly out of page (d) 3.8×10^{-5} N away from the infinite wire (e) 0 N (f) 1.9×10^{-5} N away from the infinite wire (g) 1.9×10^{-5} N away from the rectangular loop

57. (a) 1.2×10^{-3} T (b) 0 T

61. Answers are given in the statement of the problem.

65. (a) and (b)

(c) $r = \dfrac{m_e v}{eB}$

(d) 0 J (e) It remains at rest.

69. (a) 90°, 6.00×10^{-14} N (b) 14° (c) 160 eV (d) The same: 160 eV

73. A proof

CHAPTER 21

1. Answer is given in statement of problem.

5. (a) 311 V (b) 50.0 Hz

9. (a) The charge separation creates an electric field that produces an electric force on the charges equal in magnitude but opposite in direction to the magnetic force on the charges. These forces are *not* a Newton's third law force pair.

(b)

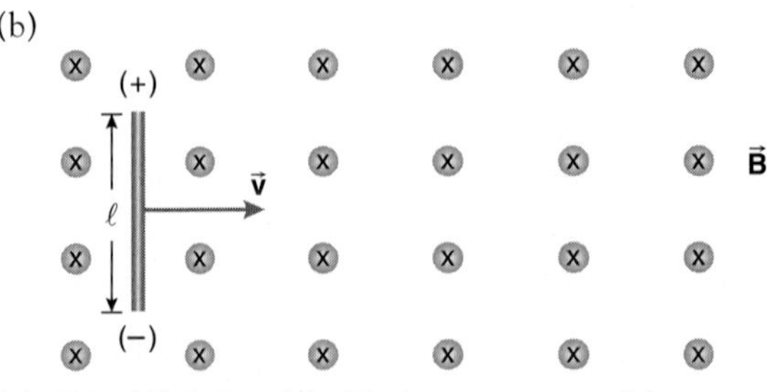

(c) $vB\ell$ (d) $\ell v\,\Delta t$ (e) $vB\ell$, the same as part (c)

13. (a) $\Phi = \pi r^2 (\cos\theta) B_0 e^{-\alpha t}$
(b) Induced emf $= \alpha \pi r^2 (\cos\theta) B_0 e^{-\alpha t}$

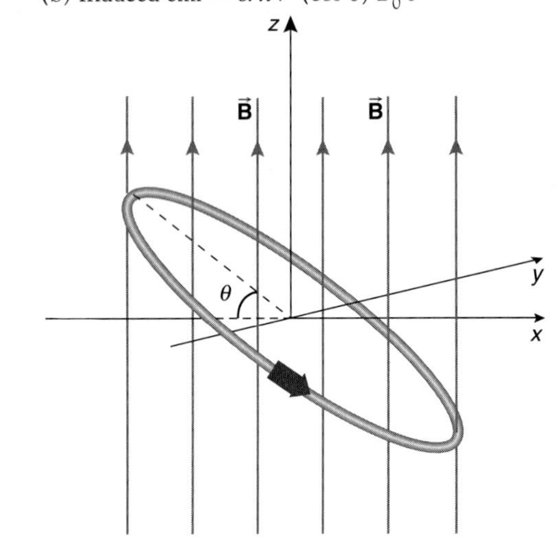

17. (a) -3×10^{-3} V (b) For a hoizontally moving train, only the vertical component contributes to the flux.

21. (a) 314 rad/s (b) 6.4×10^{-3} T

25. The answer depends on the frequency of the station. For 89.1 MHz, $\lambda = 3.37$ m.

29. (a) 9.0 V (b) −7.5 V (c) 6.0 V

33. (a) $\Phi = \dfrac{\mu_0}{4\pi} 2I\ell \ln\left(\dfrac{R_2}{R_1}\right)$
(b) Answer is given in the statement of the problem.
(c) 2.8×10^{-7} H/m

37. (a)

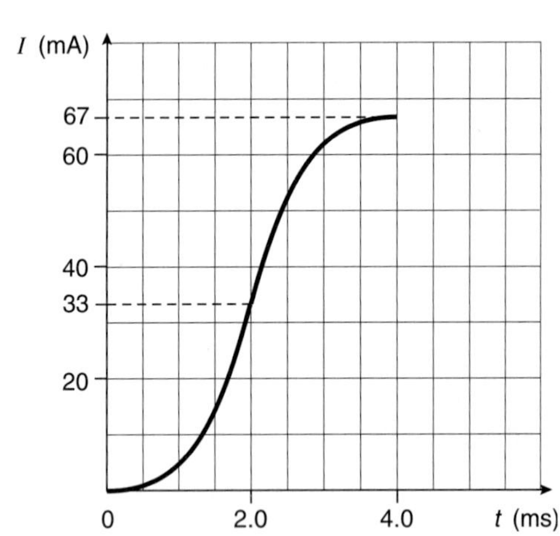

(b) 33 mA, 67 mA

(c) The current is $1/L$ times the area under the $V(t)$ versus t graph.

41. 5.2 time constants

45. (a) 1.67×10^{-3} s (b) 80.0×10^{-3} A (c) $0.69\,\tau = 1.2 \times 10^{-3}$ s

49. (a) For a resistor, V is proportional to I; this is not the case here. (b) For a capacitor, I is proportional to

$$\frac{dV}{dt}$$

this is not the case here. (c) For an inductor V is proportional to

$$\frac{dI}{dt}$$

this *is* the case here. (d) 0.10 H

(e)

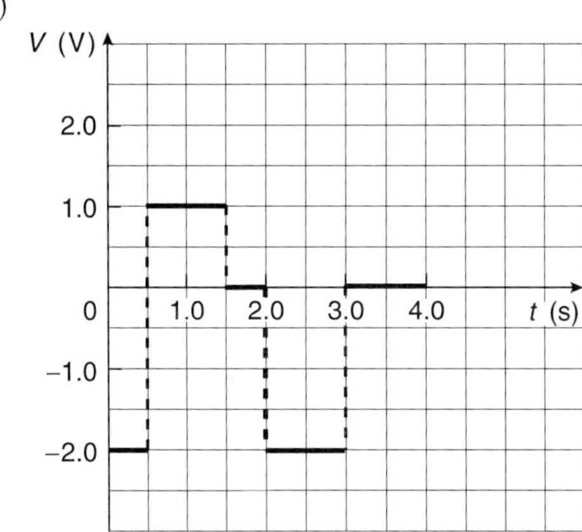

53. 30 pF to 3.1×10^{2} pF

57. 64

61.

Sketch 1	Sketch 2
Primary: emf: 2, V: 1	Primary: emf: 2, V: 1
Secondary: emf: 3, V: 4	Secondary: emf: 3, V: 4

Chapter 22

1. z_1, z_2: (a) $5, \sqrt{29} \approx 5.4$ (b) $5\angle 0.927$ rad, $\sqrt{29}\angle(-1.19 \text{ rad}) \approx 5.4\angle(-1.19 \text{ rad})$ (c) $5\,e^{i(0.927 \text{ rad})}, \sqrt{29}\,e^{-i(1.19 \text{ rad})} \approx 5.4\,e^{-i(1.19 \text{ rad})}$

(d)

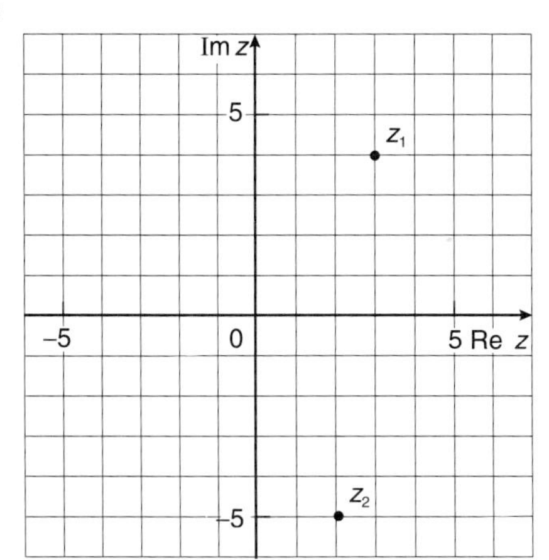

5. (a) −4 (b) $-i(2)$

(c) $\dfrac{i}{2}$

9. A proof

13. A proof

17.

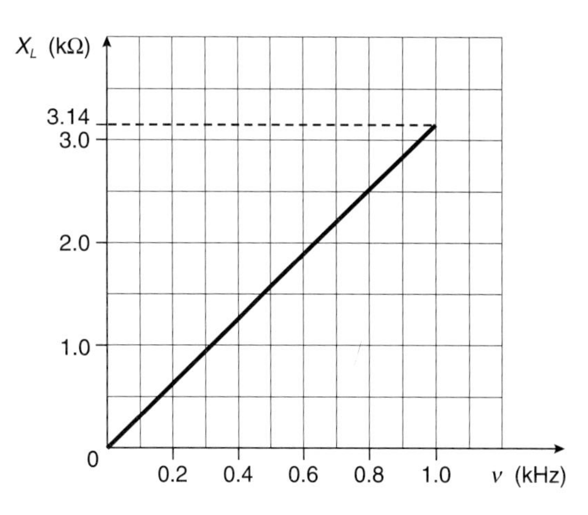

21. $>3.2 \times 10^{2}$ Hz

25. (a) $\nu = \dfrac{1}{2\pi(LC)^{1/2}}$ (b) no

29. (a)

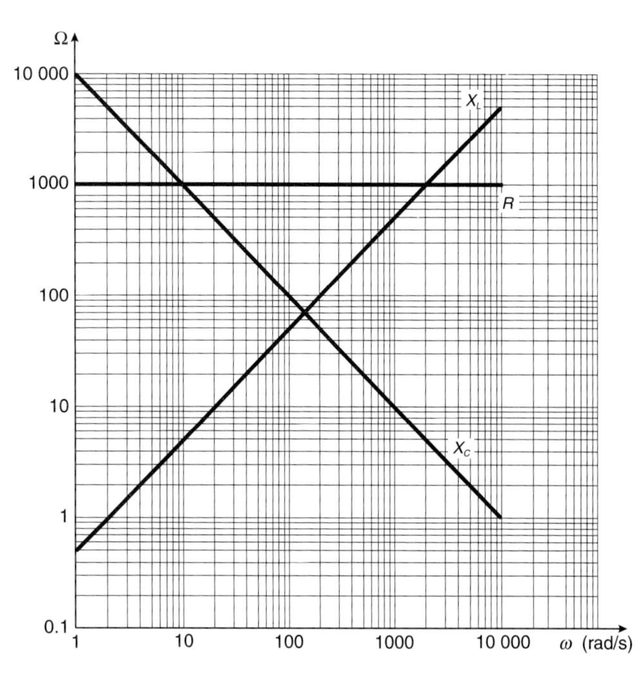

(b) Since $X_L = \omega L$, log X_L will proportional to log ω; Since $X_C = 1/(\omega C) = (\omega C)^{-1}$, log X_C will proportional to −log ω; Since R is a constant, the graph of R versus ω will be a horizontal line. (c) 2.00×10^{3} rad/s, 318 Hz (d) 10.0 rad/s, 1.59 Hz (e) 141 rad/s, 22.4 kHz

33. $\dfrac{I_0}{\sqrt{2}}$

37. (a) 60.0 Hz (b) $i(56.6\ \Omega)$ (c) $(170 \text{ V})\,e^{i[(377 \text{ rad/s})\,t]}$, $170 \text{ V}\angle[(377 \text{ rad/s})\,t]$

(d)

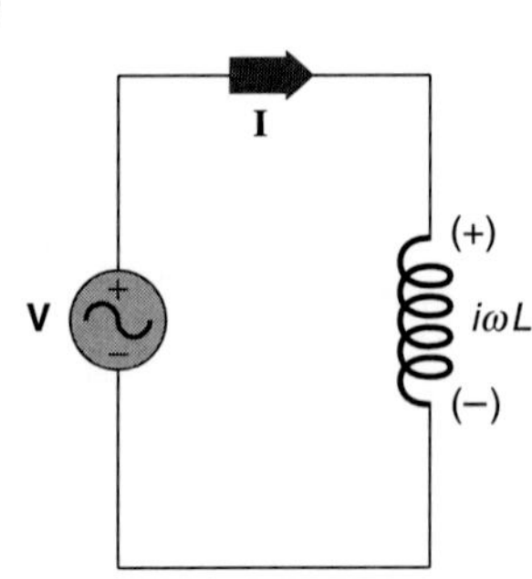

(e) $(3.00\text{ A})\, e^{i\,[(377\text{ rad/s})\,t\,-\,\pi/2\text{ rad}]}$

$(3.00\text{ A}) \angle \left[(377\text{ rad/s})\,t - \frac{\pi}{2}\text{ rad}\right]$

(f) $(170\text{ V})\, e^{i\,[(377\text{ rad/s})\,t]}$, $170\text{ V} \angle [(377\text{ rad/s})\,t]$

(g) $I(t) = (3.00\text{ A}) \cos\left[(377\text{ rad/s})\,t - \frac{\pi}{2}\text{ rad}\right]$

(h) $V(t) = (120\text{ V}) \cos[(377\text{ rad/s})\,t]$

(i) $\frac{\pi}{2}$ rad

(j) 3.00 A (k) 2.12 A (l) 170 V (m) 120 V (n) 0 (o) 0 W

41. $\omega = \frac{\sqrt{3}}{RC}$

voltage gain $= \frac{1}{2}$

45. (a)

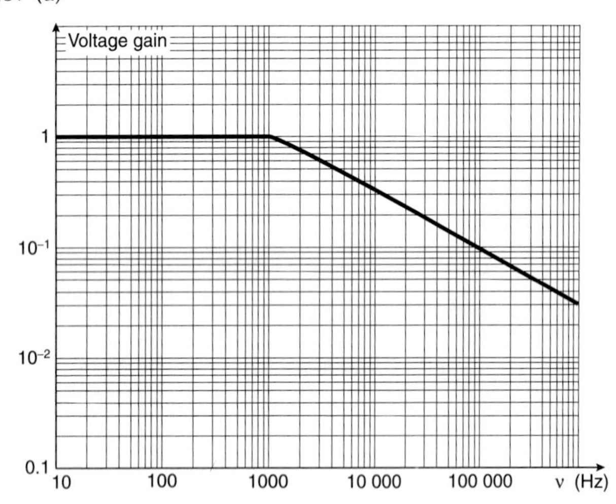

(b) $\nu \approx 5$ kHz; A calculation yields 5.5 kHz.

49. (a)

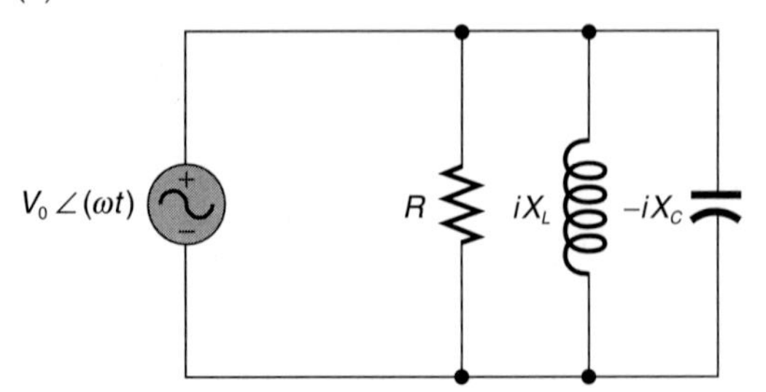

(b) Answers are given in the statement of the problem.
(c) Answer is given in the statement of the problem.
(d) Answer is given in the statement of the problem.

(e) $\omega = \frac{1}{(LC)^{1/2}}$

This is the same as the resonant angular frequency of a series *RLC* circuit.

CHAPTER 23

1. 14.9 W/m^2

5. 4.54 kJ

9. Answer is given in the statement of the problem.

13. Answer is given in the statement of the problem.

17. $d\left(\frac{n^2 - \sin^2\theta}{n^2\cos^2\theta}\right)^{1/2}$

21. 25°

25. Answer is given in the statement of the problem.

29. Answer is given in the statement of the problem.

33. (a) $s' = 4.5$ cm (b) $m = 0.090$ (c) Virtual (e) Upright

37. (a) Virtual (b) $R = \infty$ m

41. $2v\cos\theta$

45. Answers are given in the statement of the problem.

49.

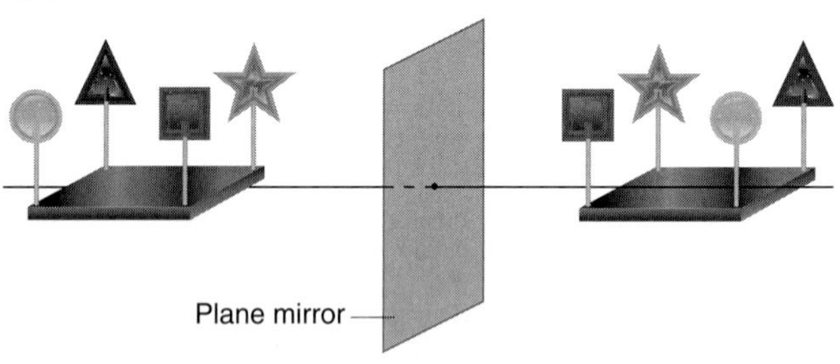

53. (a) Depth of 3.8 cm (b) $m = 1.0$ (c) Virtual, upright

57. (a)

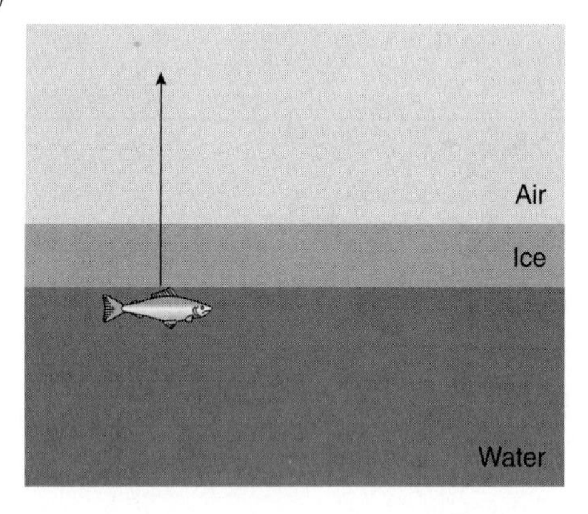

(b)

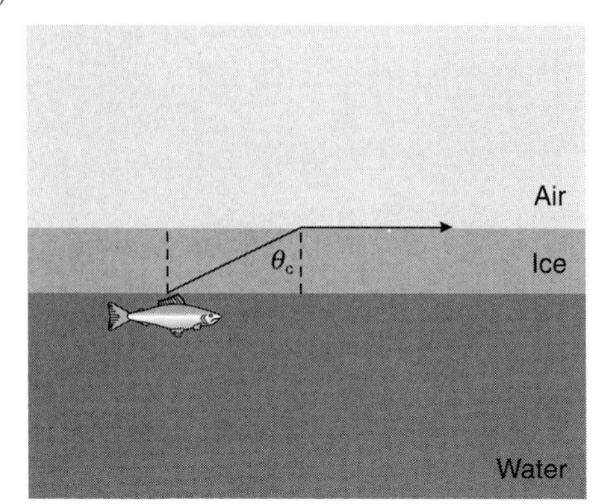

(c) 0.77 m (d) 1.02 m

61. A proof

65. (a) 1×10^2 cm to the left of the diverging lens (b) $m_{total} = 2$ (c) Virtual, upright

69. $f_{eff} = \dfrac{f_{lens} f_{mirror}}{f_{lens} - f_{mirror}}$

73. (a) $m = -44.1$ (b) Inverted. Invert the slide. (c) 10.2 cm (d) 450 cm

77.

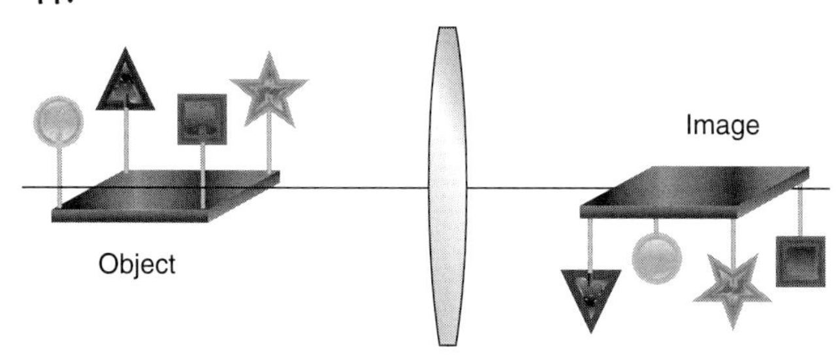

81. The image is 60 cm to the right of the lens system.

85. −6.7 dp

89. 10 mm

93. (a) ≈0.1 rad ≈ 6° (b) 9.06×10^{-3} rad = 0.519° (c) −12.00 m (d) 9.06×10^{-3} rad = 0.519° (e) Real (f) 0.109 m (g) Circular

97. (a) $m = -0.10$ (b) Place the object near the secondary focal point of the objective lens; the system then acts as a microscope, with $|m| \approx 14$.

Chapter 24

1. 3.5×10^{-3} rad, 0.20°

5. (a) 3.2 cm (b) The separation of the fringes is halved to 1.6 cm.

9. (a) 1.31 m (b) 41°

13. (a) 0.73 mm (b) Increase the diameter.

17. A proof

21. (a) The telescope because of its larger aperture (b) 1.0 m (c) Magnification is doubled; resolution is unaffected.

25. 14° for the first order, 29° for the second order

29. 20.8°

33. (a) no (b) 437 nm

37. (a) 422 nm (b) 2.00×10^8 m/s

41. (a) and (c)

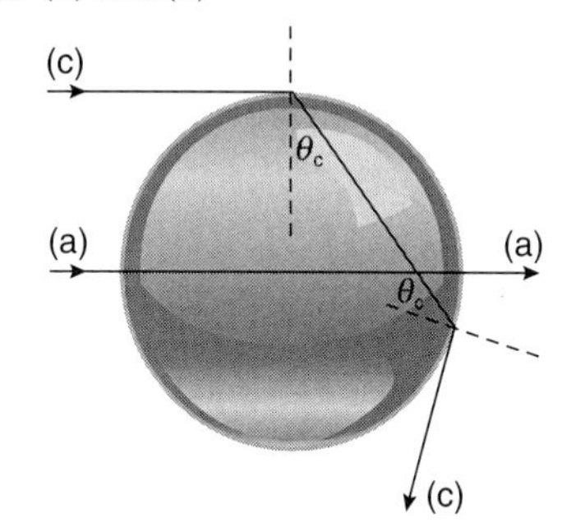

(b) 1.00×10^{-10} s (d) 7.45×10^{-11} s

45. (a) Flat #1 (b) Slightly spherical

49. Answer is given in the statement of the problem.

53.

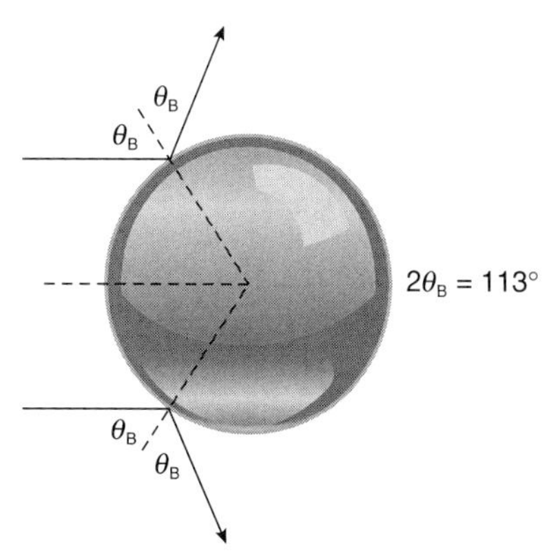

57. (a) and (b) Answers are given in the statement of the problem. (c) 12.5°; This is less than the critical angle of 12.8°.

61. 36.9°

Chapter 25

1. Account for the light travel time between a central clock and the others.

5. $0.943c = 2.83 \times 10^8$ m/s

9. 7×10^2

13. (a) 2.00×10^8 m/s (b) 74.6 m (c) 3.73×10^{-7} s

17. (a) $v = 0.82c = 2.5 \times 10^8$ m/s (b) 5.1 y

21. (a) 1.28 s (b) 0.18 s: The Moon is hit first. (c) No, a single saucer it would have to travel with $v > c$ to cause them both.

25. (a) A proof (b) 0.064 m; No, only the diameter measured in the direction of the motion or the projection of a diameter along the direction of the orbital motion is length contracted.

29. $u_x = 0.990c$, $u_y = 0.14c$ (b) 8.0°

33. Answers are given in the statement of the problem.

37. A proof

41. $0.0175c \approx 5.25 \times 10^3$ km/s

45. Assume a mass of 60 kg: \$$1.5 \times 10^{11}$

49. $KE = 2.05 \times 10^{-14}$ J, $E = 1.02 \times 10^{-13}$ J,
$p = 2.05 \times 10^{-22}$ kg · m/s

53. (a) $0.417c$ (b) 1.5×10^{-11} J

57. (a) $0.19470c$ (b) 9.99 keV (c) 5.42×10^{-23} kg · m/s

61. (a) 2.5×10^{19} J (b) 0 J (c) 20.4 y (d) 4.0 LY (e) 4.1 y

65. Answers are given in the statement of the problem.

69.

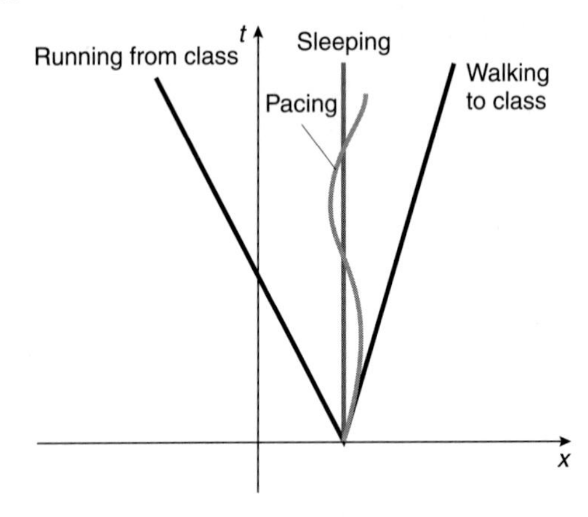

73.

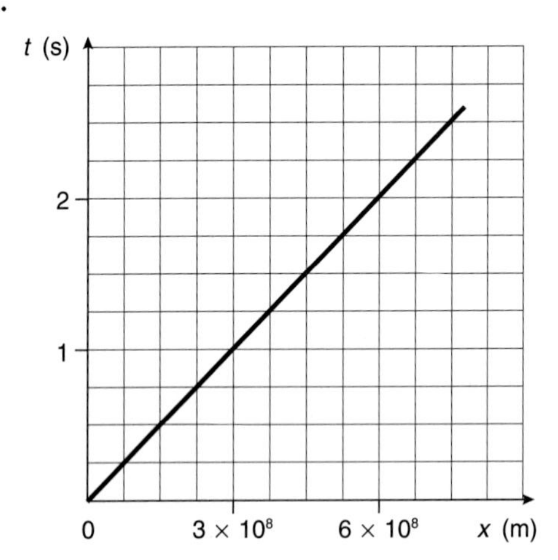

CHAPTER 26

1. λ_{max} is halved.

5. About 4.5×10^3 K

9. For 400 nm: 3.10 eV; For 700 nm: 1.77 eV

13. (a) 2.1 eV (b) 2.1 V (c) 3.0×10^2 nm

17. (a) Yes, since $\nu > \nu_c = 4.8 \times 10^{14}$ Hz.

(b)

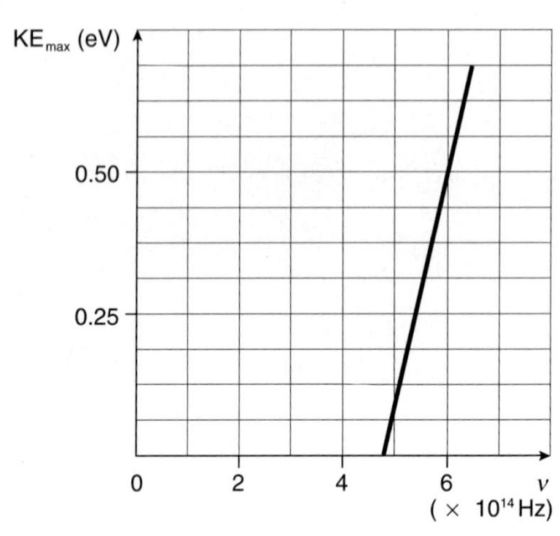

(c) The graph is unaffected.

21. (a) 2.12×10^{-10} m (b) 1.09×10^6 m/s
(c) 2.110×10^{-34} kg · m²/s (d) 5.60×10^{21} m/s²
(e) 5.71×10^{20} (f) 3.38 eV (g) −6.80 eV

25. (a) 365 nm (b) No

29. The shortest wavelength is the series limit of the Lyman series: 91.2 nm. This is too long to be classified as an x-ray wavelength.

33. (a) 0.181 d^{-1} (b) 4.5×10^{-3}

37. Answer is given in the statement of the problem.

41. 0.70

45. 26 d

49. (a) 1.21×10^{-4} y^{-1} (b) 2.04×10^3 y

53. One cannot say with certainty, since the process is statistical. It is possible that all five may have decayed, or none. The most likely number remaining is two or three.

57. 0.1024 nm

61. Answer is given in the statement of the problem.

65. $n = 2\pi r/\lambda$

69. (a) 0.37 nm (b) 0.37 nm

CHAPTER 27

1. $h = 4.136 \times 10^{-15}$ eV · s, $\hbar = 6.583 \times 10^{-16}$ eV · s

5. 3.45×10^{-23} s

9. (a) 1.50×10^{15} Hz (b) $\Delta\nu \geq 1.0 \times 10^9$ Hz (c) 0.30 m

13. (a) 7×10^{-20} kg · m/s (b) $\lesssim c$ (c) 1×10^2 MeV; This energy is approximately 2×10^2 times the rest energy.

QUOTATIONS INDEX

REFERENCE INDEX

Abernethy, Bruce, 379
Able, Kenneth P., State University of New York - Albany, 950
Able, Mary A., State University of New York - Albany, 950
Adair, Thomas W., III, Texas A&M University, 637
Adams, Bryant E., Washington and Lee University, 216
Akinrimisi, Jibayo, University of Lagos, 764
Alerstam, Thomas, 950
Alfvén, Hannes, 1247
Alpher, R.A., 1267
Alvarez, Don, Princeton University, 1094
Amato, Joseph C., Colgate University, 1146
Amann, George A., F.D. Roosevelt High School, 423
Angrist, Stanley W., 703
Archibald, E. Douglas, 1135
Arfken, George, Miami University (Oxford, Ohio), 71
Arons, Arnold, University of Washington, 28, 103, 104, 105, 106, 229, 398, 637, 755, 885
Ashkin, Arthur, Bell Telephone Laboratory, 1247
Asimov, Isaac, 1202
Askenfelt, Anders, 585
Ausubel, Frederick A., 804
Baierlein, Ralph, Wesleyan University, 703, 1187
Baker, James S., 423
Ball, June E., North Staffordshire Polytechnic, 1146
Ballard, S.S., 1056
Barbour, J., 279
Barger, Vernon, University of Wisconsin, Madison, 486
Bartlett, Albert A., University of Colorado, 486, 636
Bartlett, D.F., University of Colorado, 32
Bastien, Joseph, University of Oklahoma, 765
Bate, Roger R., 232
Bates, Harry E., Towson State University, 1202
Baylis, W.E., University of Windsor, 71
Beams, Jesse W., University of Virginia, 280
Becker, Robert O., 1040
Begelman, Mitchell C., Cambridge University, 1174
Behringer, Robert P., Duke University, 529
Behroozi, Feredoon, University of Northern Iowa, 1146
Bell, W.R., 766, 804
Benade, Arthur H., 584
Bender, Paul A., Washington State University, 280, 584
Benzinger, T.H., National Institute of Standards and Technology, 636
Berg, Richard E., University of Maryland, 804
Bergmann, Peter G., Syracuse University, 1202
Berkhout, A.J., 584
Bernard, Josef, 833
Bernstein, Alan D., 1002
Bettis, Clifford, University of Nebraska - Lincoln, 229
Bilaniuk, Olexa-Myron, Swarthmore College, 1202
Black, Eugene B., 893
Blandford, Roger D., California Institute of Technology, 1174
Block, Richard J., 529
Bloxham, Jeremy, Harvard University, 950
Boag, J.W., 1247
Boas, Mary L., DePaul University, 68, 71, 317
Boccippio, Dennis J., 833
Bondi, Herman, 763
Bonomo, R.P., Universita di Catania, 664
Born, Max, Universities of Göttingen and Edinburgh, 664
Borra, Ermanno F., 1094
Borshe, Peter, 33
Boudreau, William, Mankato West High School, 486
Boyer, Rodney F., 804
Boyes, Edward, University of Liverpool, 893
Boyle, W.S., 1099
Boys, Charles V., 1146
Brancazio, Peter J., Brooklyn College of C.U.N.Y., 229
Brickhill, Paul, 478
Brill, Dieter R., University of Maryland, 1100
Bronowski, Jacob, 34
Bronson, Patricia F., 1267
Bronson, Robert L., 1267
Brophy, James J., 893
Brown, Colin, Peers School, 893
Brown, Sanborn C., 637
Brown, Thomas J., Memorial Sloan-Kettering Cancer Center, 636
Bruner, L.J., University of California, Riverside, 528
Buchwald, Jed Z., 1002
Buckmaster, H.A., University of Calgary, 168
Bürger, Wolfgang, 486
Bushong, Stewart C., Baylor College of Medicine, 1247
Butterfield, Herbert, 229
Cade, Tom J., 765
Cady, Walter Guyton, 804
Cameron, John R., University of Wisconsin, 529, 585, 893, 1040
Camps, P., Universite de Montpellier, 932
Cardon, Bartley L., University of Arizona, 1099
Carr, Edward F., University of Maine, 1147
Carrington, Fred, Grant High School, 584
Carruth, John A.S., 1146
Cartwright, Colin M., Dundee Institute of Technology, 1092, 1100
Cartwright, Paul, Chilworth Technology Limited, 804
Case, William B., Grinnell College, 486
Cassiday, George L., University of Utah, 317
Chaisson, Eric, Tufts University, 280
Chambers, Llewelyn G., 71
Chandrasekhar, S., University of Chicago, 1202
Chang, George Y., Washington and Lee University, 32, 568
Cheney, Margaret, 950
Childers, Richard L., University of South Carolina, 665
Chong, Eric W. T., Iolani School, 161
Christensen, Sabinus H., Lincoln University, 528
Christianson, Gale E., 318
Churchill, Ruel V., University of Michigan, 71
Clark, Harold E., 765
Clark, John M., Jr., 804
Clark, Kenneth C., University of Washington, 28
Clotfelter, B.E., Grinnell College, 280
Coe, R.S., University of California, Santa Cruz, 932
Cohen, I. Bernard, 954
Cohen, Richard J., Massachusetts Institute of Technology, 487
Cotter, Maurice J., Queens College, 1247
Crandall, R.E., Reed College, 764
Crane, H. Richard, University of Michigan, 486, 528. 765
Craun, Edwin D., Washington and Lee University, 33
Crawford, Frank S., University of California, Berkeley, 664
Crawford, Gregory P., Kent State University, 1146
Cromer, Alan, Northeastern University, 315

Crowe, Michael J., 71
Crowell, A.D., University of Vermont, 317
Crowley, Ronald J., California State University - Fullerton, 793
Cunningham, John R., 1247
Curry, Stephen M., University of Texas at Dallas, 486
Curry, T.S., 1247
Dashek, William V., 804
Davenport, William H., Harvey Mudd College, 34
Davies, Paul, University of Newcastle upon Tyne, 703
Davis, Michael, North Georgia College, 529
Davis, Michael, Clemson University, 936
Day, Michael A., Lebanon Valley College, 168
De Santillana, Giorgio, 229
Dessauer, John H., 765
Desurvire, Emmanuel, Columbia University, 1099
Devy, Gilbert B., National Science Foundation, 584
Dewey, Mark, 893
Dickman, Steven, 764
Diggins, John Patrick, 229
Dilsaver, John S., Ozark High School, 276
Dircks, Henry, 703
Ditchburn, R.W., 1046
Ditteon, Richard, Rose-Hulman Institute of Technology, 1055
Dittman, Richard H., University of Wisconsin, Milwaukee, 664
Doane, J. William, Kent State University, 1146
Dowdey, J.E., 1146
Draeger, Donn F., 229
Dragovan, Mark, Princeton University, 1094
Drake, Stillman, University of Toronto, 280
Dupuy, David L., Virginia Military Institute, 568
Dyrenforth, W.P., Carpco Research and Engineering, 765
Dyson, Freeman J., Institute for Advanced Study, 637
Edge, R.D., University of South Carolina, 423, 529
Einstein, Albert, 229, 279, 528
Eisenkraft, Arthur, 480
Ellenstein, Marshall, Maine West High School and Ridgewood High School, 878, 1099
Elmore, W.C., Swarthmore College, 32
Erlichson, Herman, The College of Staten Island, 475, 951
Erramilli, Shyamsunder, Princeton University, 804
Evans, Howard E., II, Wright-Patterson Air Force Base, 115
Evans, James, University of Puget Sound, 1100
Everett, Steven M., Washington and Lee University, 33
Everitt, C.W.F., Stanford University, 279, 1202
Fairbank, William M., Jr., Colorado State University, 785
Falk, David S., University of Maryland, 1100
Faller, J.E., Wesleyan University, 764
Farrant, Pat, 584
Feld, M.S., Massachusetts Institute of Technology, 423
Feldman, Laurence M., University of Massachusetts, 1202
Feynman, Richard P., California Institute of Technology, 34, 703
Field, Michael S., 229
Fjelstad, Paul, St. Olaf College, 1039, 1202
Fletcher, Neville H., Australian National University, 584
Folger, Tim, 423
Forbes, Eric G., University Of Edinburgh, 33
Ford, Kenneth W., University of Massachusetts, Boston, 945
Fowles, Grant R., University of Utah, 317, 1147
Franklin, Allan, University of Colorado, 229, 279, 765
Fraser, Alistair B., Pennsylvania State University, 1100
Freeman, Ira M., Rutgers University, 275
French, A.P., Massachusetts Institute of Technology, 279, 486
Fricke, Lynn B., 423
Friedman, Alan, University of California, 34
Frisch, David, Massachusetts Institute of Technology, 280
Frisch, Henry J., University of Chicago, 950
Frohlich, Cliff, University of Texas, 168, 229
Full, Robert J., Stanford University, 168
Fye, W. Bruce, 893
Gamow, George, 68, 665, 703, 1202
Gannon, Robert, 1099
Garrett, Anthony, 71
Gee, J. Ken, 1100
Geisler, C. Daniel., 584
Ghatak, A.K., Indian Institute of Technology, 1100
Giancoli, Douglas C., 791
Gillies, George T., University of Virginia, 279, 280
Glass, Alastair M., AT&T Bell Laboratories, 1099
Gold, T., Cornell University, 703
Golde, Rudolph Heinrich, 765
Goldenbaum, George C., University of Maryland, 528
Golding, William, 1085
Goldhaber, Alfred S., State University of New York - Stony Brook, 950
Goldman, Terry, Los Alamos National Laboratory, 1267
Gordon, Malcolm S., University of California, Los Angeles, 636
Gove, Harry E., University of Rochester, 1247
Gray, Rachel P., 528
Greenler, Robert, University of Wisconsin, Milwaukee, 1099
Greenslade, Thomas B., Jr., Kenyon College, 317, 486
Gregg, William R., Belle Chasse High School, 1100
Gregory, R.A., 954
Gribben, John, 1267
Grundfest, Harry, 765
Gubbins, David, University of Leeds, 950
Hafemeister, David, California Polytechnic State University, 765, 1002
Hahn, G.M., 636
Hahn, H.T., Lockheed Palo Alto Research Laboratory, 765
Hall, A. Rupert, 318
Hall, Stephen S., 950
Hammond, John Winthrop, 1040
Hanson, Michael, Doon Academy, 486
Harnwell, Gaylord P., 317
Harris, Forest K., 317
Harrison, Edward R., University of Massachusetts, 31, 32, 1086, 1202, 1241, 1247, 1267
Harwit, Martin, Cornell University, 664
Haugland, Ole Anton, Tromso Institute, 168
Hauser, Walter, Northeastern University, 71
Hawking, Stephen, Cambridge University, 703
Hawkins, David, 34
Hayn, Carl H., University of Santa Clara, 1147
Hazelton, George L., Chowan College, 229
Heald, Mark A., Swarthmore College, 765
Heering, Peter, Carl von Ossietzky University, 764
Heilbron, J.L., University of California, 32
Heiser, John B., 765
Helm, Hugh, Colgate University, 1146
Helsdon, Richard M., Dorset Institute of Higher Education, 1174
Hendricks, Charles D., 765
Herrin, Michael, Washington and Lee University, 33
Herring, Thomas A., Massachusetts Institute of Technology, 1046

Herschman, Arthur, American Institute of Physics, 168
Hestenes, David, Arizona State University, 71
Hicks, Clifford B., 703
Hill, H.A., Wesleyan University, 764
Hitchins, H.L., 950
Hobbie, Russell K., University of Minnesota, 584, 804
Hodges, Laurent, Iowa State University, 278
Holladay, Wendall G., Vanderbilt University, 1201
Holt, Floyd T., F.D. Roosevelt High School, 423
Holton, Gerald, Harvard University, 1202
Holvenstot, Clyde E., 116
Honeycutt, Richard A., Davidson County Community College, 1040
Hoop, Bernard, Harvard Medical School, 115
Hubbard, William B., University of Arizona, 486
Hughes, Richard J., Los Alamos National Laboratory, 1267
Hunter, Geoffrey, York University, 1247
Huschilt, J., University of Windsor, 71
Hyland, William, 833
Iannotta, Ben, 1094
Infeld, Leopold, 279
Inouye, Carey S., Iolani School, 161
Iona, Mario, University of Denver, 995
Itano, Wayne M., National Institute of Standards and Technology, 32
Jackson, John David, University of California, Berkeley, 764
Jackson, Paul, Natal University, 528
Jacobs, J.A., University of Wales, 950
Jacobs, Kenneth C., Hollins University, 379
Jaeger, Heinrich M., University of Chicago, 529
Jaffe, Bernard, 1202
Jasselette, P., Universite de Liege, 317
Jefimenko, Oleg D., West Virginia University, 755, 766, 804
Jenkins, Francis A., University of California, 1099, 1146, 1147
Johns, Harold E., 1247
Johnston, Ben, 950
Johnson, Frank H., Princeton University, 637
Johnson, Larry D., Northeast Louisiana University, 161
Jones, Arthur Taber, Smith College, 486
Jones, David E. H., 486
Jones, Edwin R., Jr., University of South Carolina, 665
Jones, Ray C., Southwestern Oklahoma State University, 765
Jordinson, R., University of Edinburgh, 167
Judson, Lindsay, 229
Kammer, D.W., Albion College, 584, 804
Känzig, Werner, Swiss Federal Institute of Technology, 804
Katzir, Abraham, Tel Aviv University, 1099
Kaufmann, William J., III, 379
Kennard, Earle H., Cornell University, 664
Kessler, Gilbert W., Canarsie High School, 1199
Khular, E., Indian Institute of Technology, 1100
Kincanon, Eric, Gonzaga University, 168
King, Allen L., Dartmouth College, 486, 703
Kingsbury, Robert F., Bates College, 376
Kippers, Vaughan, 379
Kirkpatrick, Larry D., Montana State University, 480
Kirsh, Yoram, Everyman's University, 1202
Kirshner, Daniel, University of California, Santa Cruz, 486
Kittel, Charles, University of California, Berkeley, 71, 486, 804
Kleban, Peter, University of Maine, 379
Klein, Martin J., Yale University, 703
Kline, Ronald R., 1040
Knight, Walter D., University of California, 486
Kondo, Herbert, 1002
Koppel, Tom, 833
Kordesh, Karl V., Technical University of Graz, 893
Kowal, Charles T., Space Telescope Science Institute, 280
Kowalski, Ludwik, Montclair State College, 831, 1247
Kristopson, Louis J., Kent State University, 584
Kruglak, Haym, Western Michigan University, 664, 1147
Kuhn, Larry, IBM, 765
Ladbury, Ray, 950
Lagemann, Robert T., Vanderbilt University, 1248
Land, Edwin H., 1128
Landau, Lev D., Soviet Academy of Sciences, 379
Lawren, Bill, 950
Lawver, James E., University of Minnesota, 765
Layzer, David, Harvard University, 703
Lebowitz, Joel L., Rutgers University, 703
Leff, Harvey S., California State Polytechnic University, 703
Lehmberg, George, 486
Leigh, James R., 636
Leighton, Robert B., California Institute of Technology, 71
Levin, Eugene, York College of C.U.N.Y., 229
Leyton, Leonard, 515
Liboff, Richard L., Cornell University, 665
Lifshitz, Evgenii M., Soviet Academy of Sciences, 379
Lin, Herbert H., University of Washington, 168, 379
Lipkin, Richard, 528
Lissner, Herbert R., 229
Littler, T.S., 584
Livingston, William, 1099
Lloyd, David A., 765
Lord, A.E., Jr., Drexel University, 528, 584
Lovett, David, 1146
Lowell, J., University of Manchester, 486
Lowry, Thomas M., 1147
Lynch, David K., 1099
Lytle, Vincent, 423
Lyttleton, R.A., 763
Macdonald, C.J., St. George's College, 892
Mach, William H., Pennsylvania State University, 1100
MacKeown, P.K., University of Hong Kong, 1202
Mackinnon, Laurel Traeger, 379
Maddox, John, 280
Madsen, Ernest L., Pembroke State University, 486
Magnusson, Bengt, University of California, 168
Mahajan, Sukhbir, California State University, 584
Majernik, V., 1039
Mak, S.Y., Chinese University of Hong Kong, 529
Malvino, Albert Paul, 893
Mamola, K.C., Appalachian State University, 1100
Mapes, Glynn, 158
Marchand, E.W., 1100
Marino, Andrew A., 1040
Marion, Jerry B., University of Maryland, 379
Markley, F. Landis, Williams College, 664
Marlow, W. C., Perkin-Elmer Corp., 116, 317
Marschall, Emma Beth, 1088
Marschall, Laurence A., Gettysburg College, 1088
Marston, Ray, 833
Mason, Charles M., University of Alaska, 33
Mason, Shirley L., 280
Mather, Kirtley F., 280
Mathes, Dave, California State University, 584
Mathieson, D.L., Tulsa Junior College, 486
Matte, James Allan, 893
May, W.E., 950
McClymer, James P., University of Maine, 1147
McFarland, William N., 765
McGee, Thomas D., Iowa State University, 636
McKell, H.D., University of Manchester, 486
McKenzie, Alan L., 1146
McLaughlin, William I., Jet Propulsion Laboratory, 115
McMillan, Steve, Drexel University, 280
McNair, Ronald E., Johnson Space Center, 229, 423

Shea, Michael, California State University, 584
Shonle, John I., University of Colorado at Denver, 584
Siegal, Peter, California State Polytechnic University, 636
Sieradzan, Andrzej, Central Michigan University, 1093
Siler, Joseph R., Pittsburgh State University, 276
Silvidi, Anthony A., Kent State University, 584
Simanek, Don, Loch Haven University, 489
Simon, A.W., California Polytechnic State University, 804
Sinnott, Roger W., Sky Publishing Corporation, 1099
Skofronick, James G., Florida State University, 529, 585, 893, 1040
Smith, Brian A., University of Sheffield, 317
Smith, Elske v. P., University of Maryland, 379
Smith, Richard C., University of West Florida, 486, 487
Smith, Robert W., 229
Snow, C.P., 34
Snyder, Ralph, University of Connecticut, 486
Sobel, Dava, 33
Spitzer, Lyman, 664
Spolek, Graig A., Portland State University, 528
St. John, Mark, University of California, 114
Stephenson, F. Richard, 33
Stinner, Arthur, University of Toronto, 478
Stix, Gary, 1099
Stoller, Robert, University of Colorado, Boulder, 664
Stone, Judith, 229
Stork, David G., Clark University, 1100
Strangway, David W., 950
Stroink, G., Dalhousie University, 420
Strutt, John William, 1134
Sturges, Leroy D., Iowa State University, 229
Stutzman, W.L., Virginia Polytechnic Institute and State University, 764
Sudarshan, E.C. George, Syracuse University, 1202
Suess, Hans E., 1247
Sündermann, Jurgen, 33
Swanson, Mark A., Grinnell College, 486
Swinson, Derek B., University of New Mexico, 1099
Switzer, Robert L., 804
Symon, Keith R., University of Wisconsin, 317
Tan, A., Alabama Agricultural and Mechanical University, 1091
Tao, P.K., 423
Tape, Walter, University of Alaska - Fairbanks, 1099, 1100
Tarjan, Peter P., 1002
Taylor, Edwin F., Massachusetts Institute of Technology, 1202
Taylor, John R., University of Colorado, 34, 116
Taylor, Kathleen, Brookhaven National Laboratory, 1247
Taylor, Norman Burke, University of Western Ontario, 636
Thom, Alexander, 33
Thomas, Jacques, 486
Thomas, John Meurig, 1002
Thornton, Stephen T., 379
Thyagarajan, K., Indian Institute of Technology, 1100
Tiemann, Bruce, Jet Propulsion Laboratory, 168
Tomantschger, Klaus, Technical University of Graz, 893
Trefil, James, University of Virginia, 1146
Trotter, Donald M., Jr., 833
Trower, W. Peter, Virginia Polytechnic Institute and State University, 950
Tryon, Edward P., City University of New York, 1202
Tsai, W.H., Tatung Institute of Technology, 317
Tuttle, Merlin D., 584
Tyagi, Somdev, Drexel University, 528
Uman, Martin A., 765, 833
Usher, Tim, California State University - San Bernardino, 765
Van Flandern, Thomas C., U.S. Naval Observatory, 1267
Van Heuvelen, Alan, New Mexico State University, 528
Vandermeulen, J., Universite de Liege, 317
Varrey, R.N., Lockheed Palo Alto Research Laboratory, 765
Verance, Percy, 703
Viemeister, Peter E., 765
Viens, Robert E., Rhode Island College, 887
Villars, Christopher N., 1267
Vogelaar, Bruce, Princeton University, 804
Vos, Henk, Ternte University of Technology, 486
Wadlinger, Robert L., York University, 1247
Walker, David K., Waynesburg College, 765
Walker, Jearl D., Cleveland State University, 229, 423
Walker, Martin H., 28th Division Artillery, 168
Walters, Charles, Washington and Lee University, 110
Waring, Gene, Southeast Community College, 703
Weale, Robert A., 1146
Weart, Spencer, AIP Center for the History of Physics, 1248
Webb, John le P., University of Sussex, 664
Webb, Steven, 584
Weber, Robert L., 25
Wehr, M. Russell, Drexel University, 637
Wei, Jiansu, University of Windsor, 71
Weinstock, Harold, Illinois Institute of Technology, 637
Weisskopf, Victor F., Massachusetts Institute of Technology, 1267
Wells, Peter N. T., Avon Area Health Authority, 584
Weltner, Klaus, University of Frankfurt, 528
Wesson, Paul S., University of British Columbia, 1267
West, John B., 636
West, Vita R., 115
Westfall, Richard S., 318
Wheaton, Bruce R., University of California, Berkeley, 1247
Wheeler, David L., 893
Whitaker, Robert J., Southwest Missouri State University, 229
White, Harry J., 765
White, Harvey E., University of California, 1099, 1146, 1147
White, Jerry E., 232
Wiley, P.H., Virginia Polytechnic Institute and State University, 764
Wilk, S.R., University of Rochester, 229, 423
Will, Clifford M., Washington University, 280, 1202
Williams, E.R., Williams College, 764
Williams, Earle, 833
Williams, Edgar M., 765
Williams, J.A., Albion College, 584, 804
Williams, Marian, 229
Williams, Roger E., Colgate University, 1146
Wills, Albert P., 71
Wilson, Alpha E., West Virginia Institute of Technology, 522
Wilson, Don E., 584
Wittaker, Victor Percy, 765
Wittkower, Andrew B., High Voltage Engineering, 804
Wolf, Emil, University of Rochester, 1099
Wong, K.Y., Chinese University of Hong Kong, 529
Wood, Robert Williams, Johns Hopkins University, 1248
Worden, P.W., Jr., Stanford University, 279, 1202
Wróblewski, Andrzej, University of Washington, 1202
Wu, Chau H., Northwestern University, 765
Wu, Mu-Shiang, Tatung Institute of Technology, 317
Yersel, Metin, Lyndon State College, 529
Zayas, Joseph M., University of West Florida, 229
Zebrowski, Ernest, Jr., Pennsylvania College of Technology, 229
Zeidan, Henry M., 804
Zemansky, Mark W., C.U.N.Y., 664, 703
Zuckerman, Edward, 486

General Index

Note: f's, t's, and n's following page numbers refer to figures, tables, and footnotes, respectively.

CREDITS

Chapter 1 Richard Megna/Electronic Publishing Services Inc., NYC
2 Purdue University News Service
4 left ESVA/AIP
4 right © Mary Evans Picture Library/Photo Researchers, Inc.
5 top Scott Camazine/Photo Researchers, Inc.
5 bottom Paul Hanny/Gamma Liaison
6 Ander Tsiaras /Photo Researchers, Inc.
7 NASA
8 top Richard Megna/Fundamental Photographs
8 bottom National Institute of Standards and Technology
9 right Tom Pantages
9 left Michael Newman/Photo Edit
12 Ronald Lane Reese
14 left Earl Roberge/Photo Researchers, Inc.
14 right Tom Pantages
15 M.C. Escher's "Ascending and Descending" © 1998 Cordon Art B. V. - Baarn- Holland. All rights reserved.
16 Jean Loup Charmet/Photo Researchers, Inc.
20 Kathy Ferguson/Photo Edit
23 Linda Harms/Electronic Publishing Services Inc., NYC
26 NASA/Photo Researchers, Inc.
27 top Steve Allen /The Image Bank
27 bottom Ronald Lane Reese
28 top Michael P. Gadomski /Photo Researchers, Inc.
28 bottom Tom McHugh/Photo Researchers, Inc.
30 George Haling/Photo Researchers, Inc.

Chapter 2 Richard Megna/Electronic Publishing Services Inc., NYC. Inset Image: Ronald Lane Reese, courtesy of Barry Knox, Brown Exterminating, Roanoke, Virginia
36 Vanystadt/Photo Researchers, Inc.
64 Photo Take
67 Vic Bider/Photo Edit

Chapter 3 Richard Megna/Electronic Publishing Services Inc., NYC. Inset Image: The Image Bank
74 The Collectors Club
74 The Collectors Club
76 Courtesy of The Fancy Dress Steering Committee of Washington Lee University
107 left NASA
107 right John Kelly/The Image Bank
108 left Romilly Lockyer/The Image Bank
108 right NASA
110 Stephen J. Krasemann/Photo Researchers, Inc.
112 Robert Isaacs/Photo Researchers, Inc.

Chapter 4 Richard Megna/Electronic Publishing Services Inc., NYC. Inset Image: The Image Bank
118 Richard Hutchings/Photo Researchers, Inc.
120 Richard Megna/Fundamental Photos
129 Jim W. Grace/Photo Researchers , Inc.

Chapter 5 Richard Megna/Electronic Publishing Services Inc., NYC. Inset Image: The Granger Collection
170 Michelangelo Fazio
177 Marc Romanelli/The Image Bank
183 top Guy Sauware & Agence Vandystadt/Photo Researchers, Inc.
183 bottom Ronald Lane Reese
202 Courtesy of CBS
206 Sylvie Chappaz/Photo Researchers, Inc.
216 Patrick Behar/Angence/Photo Researchers, Inc.
217 left Agence France Presse/Corbis-Bettmann/Gerard Julien
217 right Michael King/Photo Researchers, Inc.
219 W.B. Spunbarg/Photo Edit
220 W.B. Spunbarg/Photo Edit
224 © Joel Gordon

Chapter 6 Richard Megna/Electronic Publishing Services Inc., NYC. Inset Image: Ronald Lane Reese
232 The Collectors Club
236 Ronald Lane Reese
241 NASA
242 ESVA/AIP
243 Scott Camazine/Photo Researchers, Inc.
245 Ronald Lane Reese
248 M I Morris
255 NASA
255 NASA
260 NASA
267 The Granger Collection
269 Ward's Scientific/Photo Researchers, Inc.

Chapter 7 Richard Megna/Electronic Publishing Services Inc., NYC. Inset Image: The Granger Collection.
282 left Corbis
282 right NASA
298 Bruce Roberts/Photo Researchers, Inc.
304 Ned Haines/Photo Researchers, Inc.
305 Michael Newman/Photo Edit
315 Tom Pantages

Chapter 8 Richard Megna/Electronic Publishing Services Inc., NYC. Inset Image: Photo Disc
322 Photo Researchers, Inc.
343 WARP SCI/Photo Researchers, Inc.
347 NASA
348 NASA
357 Tom Bean/Tony Stone Images

Chapter 9 Richard Megna/Electronic Publishing Services Inc., NYC. Inset Image: NASA.
382 Hubble Space Telescope Team & NASA
383 Bernard Asset/Photo Researchers, Inc.
388 NASA
400 Index Stock Photography
411 Zefa/Index Stock Photography
413 right John Bova/Photo Researchers, Inc.
413 top Richard Hutchings /Photo Edit
413 bottom Ed Braverman/FPG International

Chapter 10 Richard Megna/Electronic Publishing Services Inc., NYC. Inset Image: Photo Disc
444 NASA
461 Peter Skinner/Photo Researchers, Inc.
468 Index Stock Photography
469 top Jonathan Nourok/Photo Edit
469 bottom Mitch Kezar/Photo Take
470 left Ronald Lane Reese
470 right John Neubauer/Photo Edit

Chapter 11 Richard Megna/Electronic Publishing Services Inc., NYC. Inset Image: Photo Disc
488 Corbis
488 Joel Gordon
488 Myrleen Ferguson/Photo Edit
489 Mary Kate Denny/Photo Edit
489 Photo Take

489 Mary Kate Denny/The Image Bank
489 F. Gohier/Photo Researchers, Inc.
493 Jeff Hunter/The Image Bank
494 Index Stock Photography
496 Photo Researchers, Inc.
507 Richard Menga/Fundamental Photographs
508 Kristen Brochmann/Fundamental Photographs
508 Robert Ginn/Photo Edit
516 Paul Silverman/Fundamental Photographs
516 Holt Confer/Photo Take
517 Courtesy of The United States Navy
519 Corbis/Christel Gerstenberg.
525 Eunice Harris/Photo Researchers, Inc.

Chapter 12 Richard Megna/Electronic Publishing Services Inc., NYC.
Inset Image: Photo Disc
532 J.R. Berintenstein Angence Vandystadt/Photo Researchers, Inc.
535 Jonathan Nourok/Photo Edit
536 David Young Wolff /Photo Edit
536 Martin Dohrn/Photo Researchers, Inc.
544 Ronald Lane Reese
547 Steve Satushek/The Image Bank
552 The Collectors Club
557 Photo Researchers, Inc.
557 Photo Edit
558 FP/Fundamental Photographs
560 Mark T. Graham, 1997.
561 Francis Hogan/Electronic Publishing Services Inc., NYC
576 Martin Bough/Fundamental Photographs
581 Palma Match/Photo Researchers, Inc.

Chapter 13 Stephen and Dona O'mera/Photo Researchers, Inc.
Inset Image: Tony Stone Images
589 Francis Hogan/Electronic Publishing Services Inc., NYC
591 Joe Munroe/Photo Researchers, Inc.
593 F. St. Clair Renard/Photo Researchers, Inc.
595 left Tom Pantages
595 top Charles D. Winters/Photo Researchers Inc.
595 bottom Francis Hogan/Electronic Publishing Services Inc., NYC
596 Allen Green/Photo Researchers, Inc.
599 A.S.P./Photo Researchers, Inc.
612 Bachmann/Photo Edit
627 Ronald Lane Reese
629 Michael Melford/The Image Bank

Chapter 14 Richard Megna/Electronic Publishing Services Inc., NYC.
Inset Image: ©1997 Mark Reinstein/Uniphoto
632 Billy E. Barnes/Photo Edit
641 Photo Researchers, Inc.
655 top Jeff Greenberg/Photo Edit
655 bottom George Ranalli/Photo Researchers, Inc.

Chapter 15 Richard Megna/Electronic Publishing Services Inc., NYC.
672 Photo Researchers, Inc.
690 Ronald Lane Reese
694 Tony Freeman/Photo Edit
698 Photo CD Gallery/Expert Software
700 Ted Russell/Image Bank

Chapter 16 Chris Cheale/Tony Stone Images.
Inset Image: From Colliers Encyclopedia, Volume 10, page 329. Copyright © 1997 by Atlas Editions, Inc. Used by Permission. All rights reserved.
706 top Michelangelo Fazio
706 bottom Paul Silverman/Fundamental Photographs
707 The Granger Collection
708 Ronald Lane Reese
Courtesy of C. Vaughan Stanley, Special Collections, Leyburn Library Washington and Lee University Lexington, Virginia 24450
713 Michelangelo Fazio

Chapter 17 Richard Megna/Electronic Publishing Services Inc., NYC.
Inset Image: Hulton Getty/Tony Stone Images
770 left Michelangelo Fazio
770 right Kent Wood/Photo Researchers, Inc.
781 Tom Pantages
792 Tom Pantages

Chapter 18 Photo Edit
806 NASA
807 Charles D. Winters/Photo Researchers, Inc.
809 © Clasos Press/Gamma Liaison
811 Tom Pantages
811 Tony Malanowski

Chapter 19 Provided by Eileen Mitchell/Electronic Publishing Services Inc., NYC.
Inset Image: Photo Edit
813 Charles D. Winters/Photo Researchers, Inc.
815 Paul Silverman/Fundamental Photographs
815 Photo Disc
825 Jerry Howard/Photo Researchers, Inc.
825 Ronald Lane Reese
827 Tom Pantages
836 Jerry Howard/Photo Take
839 Michelangelo Fazio
842 Photo Researchers, Inc.
844 Ronald Lane Reese
846 Photo Disc
854 Michelangelo Fazio
864 Vladimir Lange, M. D./The Image Bank
865 Richard R. Hansen/Photo Researchers, Inc.
869 Paul Silverman/Fundamental Photographs
879 Think Quest
880 Ronald Lane Reese
881 Richard Megna/Fundamental Photographs
883 Mark C. Burnett/Photo Researchers Inc.

Chapter 20 Richard Menga/Fundamental Photographs.
Inset Image: S. Jones/FPG International
896 Nowitz: Richard/Corbis
896 Courtesy of Brio
898 Tony Stone Images
898 Michelangelo Fazio

Chapter 21 Richard Megna/Electronic Publishing Services Inc., NYC.
953 Michelangelo Fazio
962 Jim Steinberg/Photo Researchers, Inc.
963 Michelangelo Fazio
968 Photo Researchers, Inc.
973 Ronald Lane Reese

Chapter 22 Richard Megna/Electronic Publishing Services Inc., NYC.
Inset Image: Bob Schatz/Gamma Liaison
1006 bottom
Used with permission from the Biltmore Estate, Asheville, North Carolina.
1006 top Corbis

Chapter 23 Richard Megna/Electronic Publishing Services Inc., NYC.
Inset Image: Jeff Greenberg/Photo Edit

1039 Ronald Lane Reese
1039 The Collectors Club
1040 Photo courtesy of Coherent Laser Group, Inc., Santa Clara, California.
1040 Andrea Pistolesi/The Image Bank
1044 NASA
1044 JPL
1045 Tony Freeman/Photo Edit
1050 Spencer Grant/Photo Researchers, Inc.
1050 Photo Researchers, Inc.
1053 Keenan Ward © 1998
1071 Ah Rider/Photo Researchers, Inc.
1078 Photo CD Catolog/Expert Software
1081 Francis Hogan/Electronic Publishing Services, Inc. NYC
1089 Courtesy of Mirage® manufactured by Optigone Associates, 402 West Ojai Avenue, Ojai, CA 93023, 805-640-9595, http://www.optigone.com

Chapter 24 Courtesy of Hansen Planetarium/Photograph from the Palomar 48 inch Schmidt telescope.Special processing by D.F. Malin, Anglo Australian Observatory. Copyright© 1983 California Institute of Technology.
1107 Ronald Lane Reese
1113 Paul Silverman/Fundamental Photographs
1115 Arthur Beck/Photo Researchers, Inc.
1116 Ken Kay/Fundamental Photographs
1116 Richard Megna/Fundamental Photographs
1124 Jan Hinsch/Photo Researcher, Inc.
1125 Ronald Lane Reese
1132 Georges Seurat, French, 1859-1891, A Sunday on La Grande Jatte – 1884, oil on canvas, 1884-86, 207.6 x 308 cm, Helen Birch Bartlett Memorial Collection, 1926.224 Photograph © 1998 Art Institute Chicago.

Chapter 25 Photo Disc.
Inset Image: UNIPHOTO
1150 The Collectors Club
1176 Dr. S. C. Unwin/Photo Researchers, Inc.
1188 Photo Disc
1192 NASA

Chapter 26 ©1983 California Institute of Technology/Fundamental Photographs
1207 Michelangelo Fazio
1208 The Collectors Club
1210 Michelangelo Fazio
1217 The Collector's Club
1218 The Granger Collection
1230 Daniele Pellegrini/Photo Researchers, Inc.
1231 left Alexander Tsiaras/Photo Researchers, Inc.
1231 right Julian Baum/Photo Researchers, Inc.
1236 The Granger Collection
1238 Siqui Sanchez/The Image Bank

Chapter 27 Per Eriksson/The Image Bank
1251 ESVA/AIP
1258 Courtesy of Hitachi Ltd. Advanced Research Laboratory/Akira Tonomura, J. Endo, T. Matsuda, T. Kawasaki, H. Ezawa
1260 The Collector's Club
1262 ESVA/AIP
Courtesy of John K. Heyl/Fran Heyl and Associates

Common Unit Abbreviations

Unit	Abbreviation	Unit	Abbreviation	Unit	Abbreviation
ampere	A	gram	g	ohm	Ω
atomic mass unit	u	henry	H	pascal	Pa
atmosphere	atm	hertz	Hz	radian	rad
coulomb	C	hour	h	revolution	rev
day	d	joule	J	second	s
degree (angle)	°	kelvin	K	tesla	T
degree celsius	°C	kilogram	kg	volt	V
degree fahrenheit	°F	meter	m	watt	W
electron-volt	eV	minute	min	year	y
farad	F	newton	N		

Greek Alphabet

Letter	Lower case	Upper case	Letter	Lower case	Upper case	Letter	Lower case	Upper case
alpha	α	A	iota	ι	I	rho	ρ	P
beta	β	B	kappa	κ	K	sigma	σ	Σ
gamma	γ	Γ	lambda	λ	Λ	tau	τ	T
delta	δ	Δ	mu	μ	M	upsilon	υ	Υ
epsilon	ϵ	E	nu	ν	N	phi	ϕ	Φ
zeta	ζ	Z	xi	ξ	Ξ	chi	χ	X
eta	η	H	omicron	o	O	psi	ψ	Ψ
theta	θ	Θ	pi	π	Π	omega	ω	Ω

Planetary, Lunar, and Solar Data

	Mass	Average radius	Spin period	Semimajor axis of orbit		Eccentricity of orbit	Orbital period
Sun	1.99×10^{30} kg	6.96×10^{8} m	≈25 d	—	—	—	—
Mercury	3.34×10^{23} kg	2.45×10^{6} m	58.6 d	5.79×10^{7} km	0.387 AU	0.2056	0.241 y
Venus	4.87×10^{24} kg	6.05×10^{6} m	243 d	1.08×10^{8} km	0.723 AU	0.0068	0.615 y
Earth	5.98×10^{24} kg	6.37×10^{6} m	23.9 h	1.496×10^{8} km	1 AU	0.0167	1.000 y
Moon	7.36×10^{22} kg	1.74×10^{6} m	27.3 d	3.84×10^{5} km	—	0.055	27.3 d
Mars	6.43×10^{23} kg	3.37×10^{6} m	24.6 h	2.28×10^{8} km	1.52 AU	0.0934	1.88 y
Jupiter	1.90×10^{27} kg	6.97×10^{7} m	9.84 h	7.78×10^{8} km	5.20 AU	0.0484	11.9 y
Saturn	5.69×10^{26} kg	5.82×10^{7} m	10.2 h	1.43×10^{9} km	9.54 AU	0.0560	29.5 y
Uranus	8.69×10^{25} kg	2.59×10^{7} m	17.2 h	2.87×10^{9} km	19.2 AU	0.0461	84.0 y
Neptune	1.03×10^{26} kg	2.45×10^{7} m	16.1 h	4.50×10^{9} km	30.1 AU	0.0100	165.8 y
Pluto	1.32×10^{22} kg	1.16×10^{6} m	6.39 d	5.90×10^{9} km	39.4 AU	0.2484	247.7 y